Sea Ice

Sea Ice

Third Edition

EDITED BY

David N. Thomas

School of Ocean Sciences, Bangor University, UK
and
Marine Research Center, Finnish Environment Institute (SYKE), Helsinki, Finland

This edition first published 2017 © 2003, 2010, 2017 by John Wiley & Sons, Ltd

Registered office: John Wiley & Sons, Ltd, The Atrium, Southern Gate, Chichester, West Sussex, PO19 8SQ, UK
Editorial offices: 9600 Garsington Road, Oxford, OX4 2DQ, UK
                The Atrium, Southern Gate, Chichester, West Sussex, PO19 8SQ, UK
                111 River Street, Hoboken, NJ 07030-5774, USA

For details of our global editorial offices, for customer services and for information about how to apply for permission to reuse the copyright material in this book please see our website at www.wiley.com/wiley-blackwell.

The right of the author to be identified as the author of this work has been asserted in accordance with the UK Copyright, Designs and Patents Act 1988.

All rights reserved. No part of this publication may be reproduced, stored in a retrieval system, or transmitted, in any form or by any means, electronic, mechanical, photocopying, recording or otherwise, except as permitted by the UK Copyright, Designs and Patents Act 1988, without the prior permission of the publisher.

Designations used by companies to distinguish their products are often claimed as trademarks. All brand names and product names used in this book are trade names, service marks, trademarks or registered trademarks of their respective owners. The publisher is not associated with any product or vendor mentioned in this book.

Limit of Liability/Disclaimer of Warranty: While the publisher and author(s) have used their best efforts in preparing this book, they make no representations or warranties with respect to the accuracy or completeness of the contents of this book and specifically disclaim any implied warranties of merchantability or fitness for a particular purpose. It is sold on the understanding that the publisher is not engaged in rendering professional services and neither the publisher nor the author shall be liable for damages arising herefrom. If professional advice or other expert assistance is required, the services of a competent professional should be sought.

Library of Congress Cataloging-in-Publication Data

Names: Thomas, David N. (David Neville), 1962- editor.
Title: Sea ice / edited by David N. Thomas.
Description: Third edition | Chichester, UK ; Hoboken, NJ : John Wiley &
  Sons, 2017. | Includes index.
Identifiers: LCCN 2016031586| ISBN 9781118778388 (cloth) | ISBN 9781118778357
  (epub)
Subjects: LCSH: Sea ice.
Classification: LCC GB2403.2 .S43 2017 | DDC 551.34/3 – dc23 LC record available at
  https://lccn.loc.gov/2016031586

A catalogue record for this book is available from the British Library.

Wiley also publishes its books in a variety of electronic formats. Some content that appears in print may not be available in electronic books.

Cover image: Courtesy of David N. Thomas

Set in 8.5/12pt, MeridienLTStd by SPi Global, Chennai, India.

Printed in the UK

Contents

List of contributors, vii

Preface, xi

1 Overview of sea ice growth and properties, 1
Chris Petrich & Hajo Eicken

2 Sea ice thickness distribution, 42
Christian Haas

3 Snow in the sea ice system: friend or foe?, 65
Matthew Sturm & Robert A. Massom

4 Sea ice and sunlight, 110
Donald K. Perovich

5 The sea ice–ocean boundary layer, 138
Miles G. McPhee

6 The atmosphere over sea ice, 160
Ola Persson & Timo Vihma

7 Sea ice and Arctic Ocean oceanography, 197
Finlo Cottier, Michael Steele & Frank Nilsen

8 Oceanography and sea ice in the Southern Ocean, 216
Michael P. Meredith & Mark A. Brandon

9 Methods of satellite remote sensing of sea ice, 239
Gunnar Spreen & Stefan Kern

10 Gaining (and losing) Antarctic sea ice: variability, trends and mechanisms, 261
Sharon Stammerjohn & Ted Maksym

11 Losing Arctic sea ice: observations of the recent decline and the long-term context, 290
Walter N. Meier

12 Sea ice in Earth system models, 304
Dirk Notz & Cecilia M. Bitz

13 Sea ice as a habitat for Bacteria, Archaea and viruses, 326
Jody W. Deming & R. Eric Collins

14 Sea ice as a habitat for primary producers, 352
Kevin R. Arrigo

15 Sea ice as a habitat for micrograzers, 370
David A. Caron, Rebecca J. Gast & Marie-Ève Garneau

16 Sea ice as a habitat for macrograzers, 394
Bodil A. Bluhm, Kerrie M. Swadling & Rolf Gradinger

17 Dynamics of nutrients, dissolved organic matter and exopolymers in sea ice, 415
    Klaus M. Meiners & Christine Michel

18 Gases in sea ice, 433
    Jean-Louis Tison, Bruno Delille & Stathys Papadimitriou

19 Transport and transformation of contaminants in sea ice, 472
    Feiyue Wang, Monika Pućko & Gary Stern

20 Numerical models of sea ice biogeochemistry, 492
    Martin Vancoppenolle & Letizia Tedesco

21 Arctic marine mammals and sea ice, 516
    Kristin L. Laidre & Eric V. Regehr

22 Antarctic marine mammals and sea ice, 534
    Marthán N. Bester, Horst Bornemann & Trevor McIntyre

23 A feathered perspective: the influence of sea ice on Arctic marine birds, 556
    Nina J. Karnovsky & Maria V. Gavrilo

24 Birds and Antarctic sea ice, 570
    David Ainley, Eric J. Woehler & Amelie Lescroël

25 Sea ice is our beautiful garden: indigenous perspectives on sea ice in the Arctic, 583
    Henry P. Huntington, Shari Gearheard, Lene Kielsen Holm, George Noongwook, Margaret Opie & Joelie Sanguya

26 Advances in palaeo sea ice estimation, 600
    Leanne Armand, Alexander Ferry & Amy Leventer

27 Ice in subarctic seas, 630
    Hermanni Kaartokallio, Mats A. Granskog, Harri Kuosa & Jouni Vainio

    Index, 645

List of contributors

David Ainley
H.T. Harvey & Associates,
Los Gatos,
CA, USA

Leanne Armand
Department of Biological Sciences and
MQ Marine Research Centre,
Macquarie University,
North Ryde,
NSW, Australia

Kevin R. Arrigo
Department of Earth System Science,
Stanford University,
Stanford,
CA, USA

Marthán N. Bester
Mammal Research Institute,
Department of Zoology and Entomology,
University of Pretoria,
Hatfield, South Africa

Cecilia M. Bitz
Atmospheric Sciences Department,
University of Washington,
Seattle,
WA, USA

Bodil A. Bluhm
Institute of Arctic and Marine Biology,
UiT – The Arctic University of Norway,
Tromsø, Norway

Horst Bornemann
Alfred-Wegener-Institut,
Helmholtz-Zentrum für Polar- und Meeresforschung,
Bremerhaven, Germany

Mark A. Brandon
Department of Earth and Environmental Sciences,
The Open University,
Walton Hall,
Milton Keynes, UK

David A. Caron
Department of Biological Sciences,
University of Southern California,
Los Angeles,
CA, USA

R. Eric Collins
School of Fisheries and Ocean Sciences,
University of Alaska Fairbanks,
AK, USA

Finlo Cottier
SAMS, Scottish Marine Institute,
Oban,
Argyll, UK

Bruno Delille
Astrophysics, Geophysics and Oceanography Department,
Université de Liège,
Liège, Belgium

Jody W. Deming
School of Oceanography,
University of Washington,
Seattle,
WA, USA

Hajo Eicken
University of Alaska Fairbanks,
Fairbanks,
AK, USA

Alexander Ferry
Department of Biological Sciences and
MQ Marine Research Centre,
Macquarie University,
North Ryde,
NSW, Australia

Marie-Ève Garneau
Department of Biological Sciences,
University of Southern California,
Los Angeles,
CA, USA

List of contributors

Rebecca J. Gast
Department of Biology,
Woods Hole Oceanographic Institution,
Woods Hole,
MA, USA

Maria V. Gavrilo
National Park Russian Arctic,
Archangelsk, Russia

Shari Gearheard
National Snow and Ice Data Center,
University of Colorado Boulder,
Boulder,
CO, USA

Rolf Gradinger
Institute of Arctic and Marine Biology,
UiT – The Arctic University of Norway,
Tromsø, Norway

Mats A. Granskog
Norwegian Polar Institute,
Fram Centre,
Tromsø, Norway

Christian Haas
Department of Earth and Space Science and Engineering,
York University,
Toronto,
ON, Canada

Lene Kielsen Holm
Lene Kielsen Holm, Greenland Climate Research Centre,
Greenland Institute of Natural Resources,
Nuuk, Greenland

Henry P. Huntington
Eagle River,
AK, USA

Hermanni Kaartokallio
Finnish Environment Institute (SYKE),
Helsinki, Finland

Nina J. Karnovsky
Pomona College, Department of Biology,
Claremont,
CA, USA

Stefan Kern
University of Hamburg,
Integrated Climate Data Center – ICDC,
Hamburg, Germany

Harri Kuosa
Finnish Environment Institute (SYKE),
Helsinki, Finland

Kristin L. Laidre
Polar Science Center,
Applied Physics Laboratory and School of
Aquatic and Fishery Sciences,
University of Washington,
Seattle,
WA, USA

Amelie Lescroël
Centre d'Écologie Fonctionnelle et Évolutive,
Centre National de la Recherche Scientifique,
Montpellier, France

Amy Leventer
Department of Geology,
Colgate University,
Hamilton,
NY, USA

Ted Maksym
Department of Applied Ocean Physics and Engineering,
Woods Hole Oceanographic Institution,
Woods Hole,
MA, USA

Robert A. Massom
Australian Antarctic Division,
Kingston, Australia; and
Antarctic Climate & Ecosystems Cooperative Research Centre,
Hobart, Australia

Trevor McIntyre
Mammal Research Institute,
Department of Zoology and Entomology,
University of Pretoria,
Hatfield, South Africa

Miles G. McPhee
McPhee Research Company,
Naches,
WA, USA

List of contributors

Walter N. Meier
Cryospheric Sciences Laboratory,
NASA Goddard Space Flight Center,
Greenbelt, MD, USA

Klaus M. Meiners
Australian Antarctic Division and Antarctic
Climate and Ecosystems Cooperative
Research Centre, University of Tasmania,
Hobart, Australia

Michael P. Meredith
British Antarctic Survey,
High Cross, Madingley Road,
Cambridge, UK

Christine Michel
Freshwater Institute,
Fisheries and Oceans Canada,
Winnipeg,
Manitoba, Canada

Frank Nilson
Department of Arctic Geophysics,
The University Centre in Svalbard,
Longyearbyen, Norway

George Noongwook
Native Village of Savoonga,
Savoonga,
AK, USA

Dirk Notz
Max Planck Institute for Meteorology,
Hamburg, Germany

Margaret Opie
Barrow,
AK, USA

Stathys Papadimitriou
School of Ocean Sciences, Bangor University,
Menai Bridge,
Anglesey, UK

Donald K. Perovich
Thayer School of Engineering,
Dartmouth College; and
ERDC – Cold Regions Research and Engineering Laboratory,
Hanover,
NH, USA

Ola Persson
Cooperative Institute for Research in Environmental
Sciences/NOAA/ESRL,
University of Colorado,
Boulder,
CO, USA

Chris Petrich
Northern Research Institute Narvik,
Narvik, Norway

Monika Pućko
Centre for Earth Observation Science and
Department of Environment and Geography,
University of Manitoba,
Winnipeg, Canada

Eric V. Regehr
Marine Mammals Management,
US Fish and Wildlife Service,
Anchorage,
AK, USA

Joelie Sanguya
Clyde River,
NU, Canada

Gunnar Spreen
Norwegian Polar Institute,
Tromsø,
Norway (now at: Institute of Environmental Physics,
University of Bremen,
Germany)

Sharon Stammerjohn
Institute of Arctic and Alpine Research,
University of Colorado,
Boulder,
CO, USA

Mike Steele
Polar Science Center,
Applied Physics Laboratory,
University of Washington,
Seattle,
WA, USA

Gary Stern
Centre for Earth Observation Science,
University of Manitoba,
Winnipeg, Canada

List of contributors

Matthew Sturm
Geophysical Institute,
University of Alaska Fairbanks, Fairbanks,
AK, USA

Kerrie M. Swadling
Antarctic Climate & Ecosystems Cooperative Research Centre
and Institute for Marine and Antarctic Studies,
Hobart, Australia

Letizia Tedesco
Finnish Environment Institute (SYKE),
Helsinki, Finland

Jean-Louis Tison
Laboratoire de Glaciologie,
Université Libre de Bruxelles,
Bruxelles, Belgium

Jouni Vainio
Finnish Meterological Institute,
Helsinki, Finland

Martin Vancoppenolle
LOCEAN-IPSL,
Sorbonne Universités (UPMC Paris 6),
CNRS/IRD/MNHN,
Paris, France

Timo Vihma
Finnish Meterological Institute,
Helsinki, Finland

Feiyue Wang
Centre for Earth Observation Science
and Department of Environment and Geography,
University of Manitoba,
Winnipeg, Canada

Eric J. Woehler
Institute for Marine and Antarctic Studies,
University of Tasmania, Hobart,
Tasmania, Australia

Preface

The concept for the first edition of *Sea Ice* was developed in 1999, and the idea was to provide a resource where the key aspects of the physics, geophysics, chemistry, biology and geology of research into sea ice were presented in such a way that non-experts in the field could find the basic principles in an easy-to-understand text. This first edition was a compilation of 11 chapters written by 15 authors, who succinctly summarized much of the international effort into sea ice research that had been undertaken in the 1980s through to 2000.

Almost as soon as *Sea Ice* was published in 2003, it was clear that a second edition would be required to capture even a sense of the explosion of research activity into sea ice, fuelled by a rapidly changing appreciation of how seasonal sea ice dynamics were changing in the Arctic Ocean, and the implications of sea ice to large-scale ecosystem and biogeochemical processes. To best reflect this change in research activity the second edition contained 15 chapters, authored by 32 specialists.

The international effort into the study of sea ice continues unabated and the efforts have become increasingly multinational and multidisciplinary, requiring considerable ambition and resources from stretched funding agencies. With so much going on, it became evident that there was, more than ever, a need for a revised resource to which non-specialists can turn to for an introduction to the many specialized facets that contribute to our understanding of *sea ice science*. In this third edition we have 27 chapters written by 63 authors.

That the book has required a revision every 7 years is of course an exciting reflection of the great deal of effort undertaken by an army of researchers around the globe who share a fascination for ice-covered oceans and seas. However, scientific curiosity is not enough to justify the allocation of the massive research and infrastructure budgets that underpin our endeavours. Rather, over the past decades it is the huge impact that sea ice has on the planet, and the consequences for the whole Earth System that drive the increasingly ambitious agendas to study sea ice. The impacts of the polar regions on society are now widely reported, and the consequences of not investing in the study of these frozen waters are well understood outside of the scientific community.

It is an exciting time to be involved in sea ice research and there are many international initiatives that reflect the large international effort. Several of the chapters are collaborations of authors engaged in the Scientific Committee for Ocean Research (SCOR) Working Group, Biogeochemical Exchange Processes at the Sea–Ice Interfaces (BEPSII). Groups like this will be central to setting the agenda for the next phases of sea ice research, and, dare I say it, the need for a fourth edition – as I write this, it is difficult to contemplate, but I said that the last time! From the numbers presented above, and at a similar rate in the accumulation of sea ice knowledge, it looks as though there would need to be between 40 and 50 chapters written by around 120 authors.

Editing a book like this brings the editor into close contact with the trials and tribulations of the experts giving up their time to write. It is not easy to find the time and space for such projects, especially when juggling complicated work–life balance around teaching schedules, searching for funding, managing laboratories and departments, as well as long field seasons on research ships or remote field camps. Added to which, book chapters are not always given their deserved credit in these days of output-driven metrics and evaluations. I have been astounded by the dedication of the authors, and the book is a testament to the underlying passion that the team has put into the project for the benefit of the much wider community.

As the ideas for the book evolved in 2013, the international sea ice community lost three eminent colleagues: Katharine Giles (1978–2013), Tim Boyd (1958–2013) and Seymour Laxon (1963–2013). Tim had been one of the first to sign up to be an author on Chapter 7. As this book goes to the printers, I have just heard of the loss of a dear friend, a sea ice copepod expert, Sigi Schiel (1946–2016). It seems fitting that this third edition of *Sea Ice* is dedicated to the memory of the four of them.

David N. Thomas
School of Ocean Sciences
Bangor University
Menai Bridge
UK

Marine Research Center
Finnish Environment Institute (SYKE)
Helsinki
Finland

CHAPTER 1

Overview of sea ice growth and properties

Chris Petrich[1] and Hajo Eicken[2]

[1]*Northern Research Institute Narvik, Narvik, Norway*
[2]*University of Alaska Fairbanks, Fairbanks, AK, USA*

1.1 Introduction

A recent, substantial reduction in summer Arctic sea ice extent and its potential ecological and geopolitical impacts generated a lot of attention in the media and among the general public. The satellite remote-sensing data documenting such recent changes in ice coverage are collected at coarse spatial scales (Chapter 9) and typically cannot resolve details finer than about 10 km in lateral extent. However, many of the processes that make sea ice such an important aspect of the polar oceans occur at much smaller scales, ranging from the sub-millimetre to the metre scale. An understanding of how large-scale behaviour of sea ice monitored by satellite relates to and depends on the processes driving ice growth and decay requires an understanding of the evolution of ice structure and properties at these finer scales and this is the subject of this chapter.

The macroscopic properties of sea ice are of interest in many practical applications discussed in this book. They are derived from microscopic properties as continuum properties averaged over a specific volume (representative elementary volume) or mass of sea ice. This is not unlike macroscopic temperature and can be derived from microscopic molecular movement. The macroscopic properties of sea ice are determined by the microscopic structure of the ice, i.e. the distribution, size and morphology of ice crystals and inclusions. The challenge is to see both the forest (i.e. the role of sea ice in the environment) and the trees (i.e. the way in which the constituents of sea ice control key properties and processes). In order to understand and project how the forest will respond to changes in its environment, we have to understand the life cycle of its constituents, the trees. Here, we will adopt a bottom-up approach, starting with the trees, characterizing microscopic properties and processes and how they determine macroscopic properties, to lay the groundwork for understanding the forest. In using this approach, we will build up from the sub-millimetre scale and conclude with the larger scales shown in Figure 1.1.

Sea ice would not be sea ice without salt. In fact, take away the salt and we are left with lake ice, differing in almost all aspects that we discuss in this chapter. The microscopic and macroscopic redistribution of ions opens the path to understanding all other macroscopic properties of sea ice. We will therefore start in Section 1.2 by looking at the influence of ions on ice growth at the scale of individual ice crystals, in sea ice growing under both rough and quiescent conditions. We will continue in Section 1.3 by looking at the dynamic feedback system between fluid dynamics and pore volume, both microscopically and at the continuum scale. We will point out that our knowledge is far from exhaustive in this fundamental aspect. However, armed with a basic understanding of crystal structure, phase equilibria and pore structure, we can shed light on ice optical, dielectric and thermal properties and macroscopic ice strength in Section 1.4. One of the most discussed aspects of sea ice is its presence or absence. We will look at the growth and energy budget of sea ice and touch on deformation and decay processes in Section 1.5.

1.1.1 Lake ice versus sea ice

Ice in a small lake tends to form before coastal sea ice at a similar location. This is largely explained by the fact that,

---

Sea Ice, Third Edition. Edited by David N. Thomas.
© 2017 John Wiley & Sons, Ltd. Published 2017 by John Wiley & Sons, Ltd.

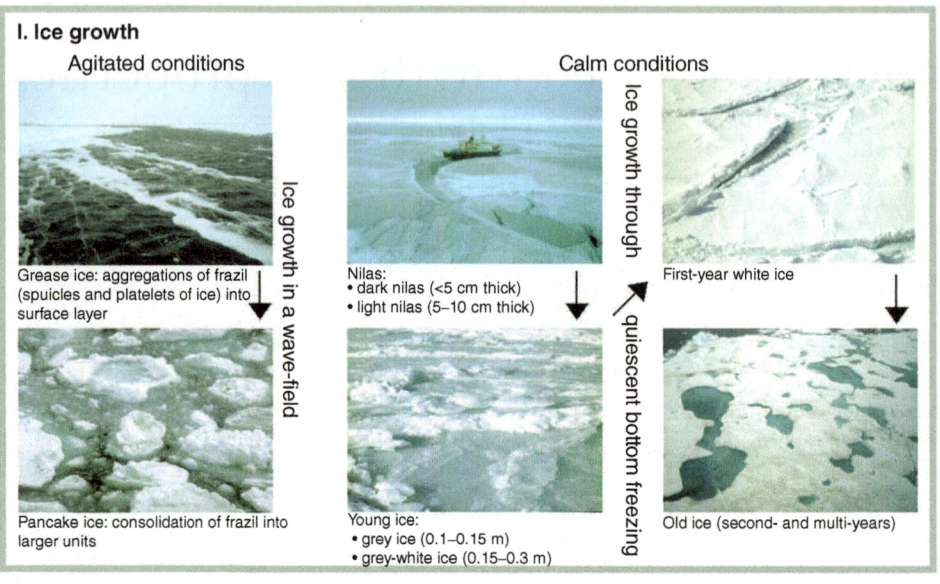
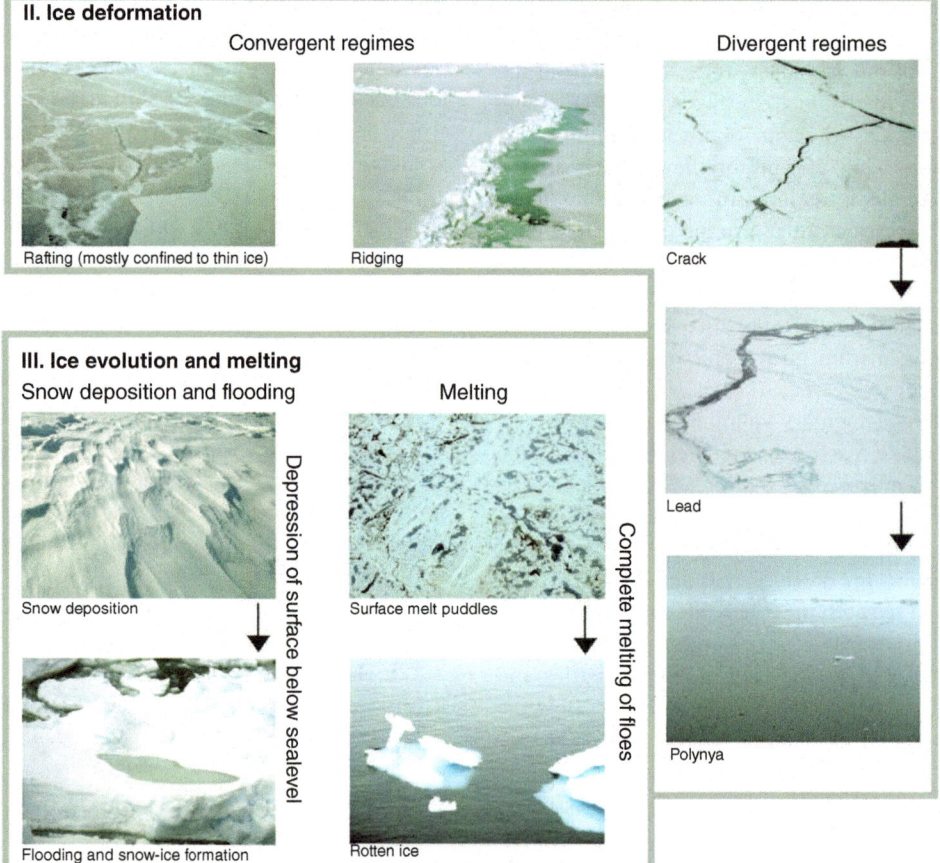

Figure 1.1 Ice types, pack ice features and growth, melt and deformation processes.

in contrast to freshwater, the temperature of maximum density of seawater is not above the freezing point. If a freshwater body is cooled from above then the water body undergoes convective overturning until the temperature reaches +4°C, after which the coldest water stays at the surface where it is cooled rapidly. Hence, ice formation starts relatively early in the season but progresses slowly as the underlying water mass is still above freezing. The situation is different if strong winds continuously overturn the water (e.g. in big lakes), or if ice grows from seawater. In these cases, the entire mixed layer has to be cooled to the freezing point before ice formation sets in. Once this happens, however, thickening progresses relatively quickly.

Salt further impacts ice microstructure. The photographs in Figure 1.2 show the surface of snow-free lake ice and sea ice in spring near Barrow, Alaska. Despite comparable thickness and growth conditions, lake ice, transparent, appears much darker than sea ice, which scatters light. This is also expressed in a large difference in albedo (the fraction of the incident short-wave radiation reflected from a surface; Section 1.4), such that more than three-quarters of the incoming short-wave irradiative flux penetrates the lake ice surface into the underlying water, compared with less than half for a sea ice cover. This has substantial consequences for the heat budget of the ice cover and the water beneath. The fact that sea ice albedo is typically higher than open water albedo by a factor of up to 10 gives rise to the so-called ice–albedo feedback: a perturbation in the surface energy balance resulting in a decreased sea ice extent due to warming may amplify, as the ice cover reduction increases the amount of solar energy absorbed by the system (Chapter 4; Curry et al., 1995; Perovich et al., 2007). For low-albedo lake ice, this effect is less pronounced. What causes these

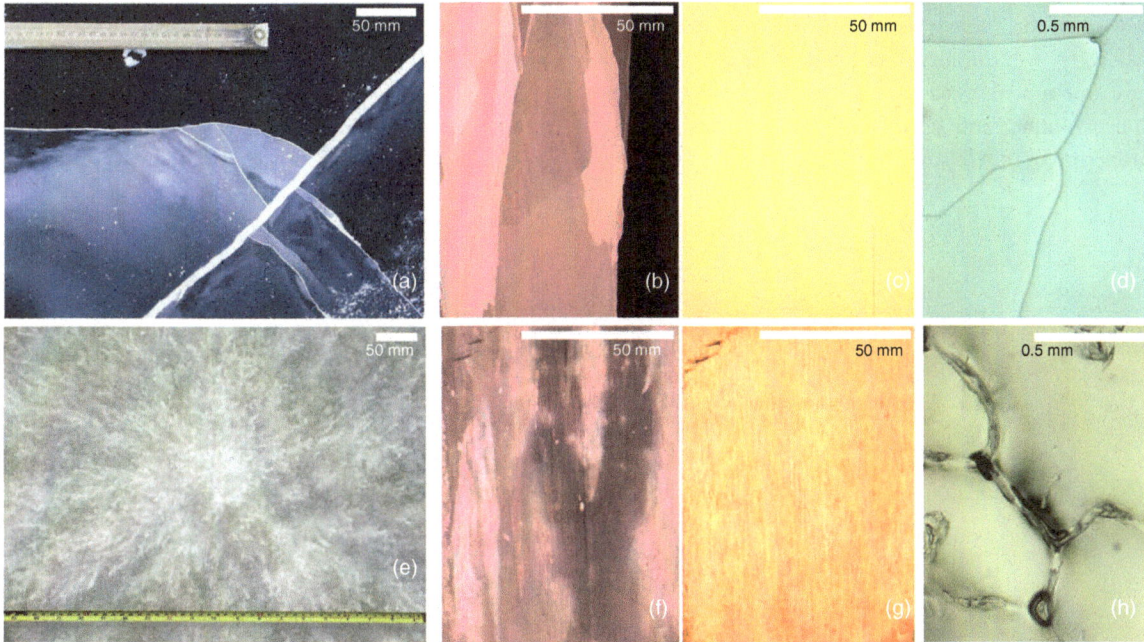

Figure 1.2 Surface appearance and microstructure of winter lake ice (Imikpuk Lake, top, panels a–d) and sea ice (Chukchi Sea landfast ice, bottom, panels e–h) near Barrow, Alaska. The bright features apparent in the lake ice are cracks that penetrate all the way to the bottom of the ice cover (close to 1 m thick), while the clear, uncracked ice appears completely black (a, top). (e) The sea ice surface photograph shows a network of brine channels that join into a few feeder channels. (b, c, f, g) Photographs of vertical thin sections from the two ice covers, with (b) and (f) recorded between crossed polarizers, highlighting different ice crystals in different colours. Panels (c) and (g) show the same section as (b) and (f) in plain transmitted light, demonstrating the effect of brine inclusions on transparency of the ice. (d, h) Photomicrographs showing the typical pore structure at a temperature of −5°C (lake ice) and −15°C (sea ice), with few thin inclusions along grain boundaries in lake ice (d) and a network of thicker brine inclusions in sea ice (h).

contrasts? As the thin-section photographs in Figure 1.2 demonstrate, lake ice is nearly devoid of millimetre and sub-millimetre liquid inclusions, whereas sea ice can contain more than 10 mm^{-3}. The inclusions scatter light due to a contrast in refractive index (Section 1.4). This explains both the high albedo and lack of transparency of thicker sea ice samples.

The crystal microstructure differs between lake ice and sea ice. Lake ice grows with a planar liquid–solid interface rather than a lamellar interface, as is the case of sea ice. In sea ice, brine is trapped between the lamellae at the bottom of the ice, allowing for retention of between 10% and 40% of the ions between the ice crystals. While the differences in bulk ice properties, such as albedo and optical extinction coefficient, are immediately obvious from these images, the physical features and processes responsible for these differences only reveal themselves in the microscopic approach, as exemplified by the thin-section images depicting individual inclusions (Figure 1.2). In the sections that follow, we will consider in more detail how microstructure and microphysics are linked to sea ice growth and evolution, and how both in turn determine the properties of the ice cover as a whole.

1.2 Ions in the water: sea ice microstructure and phase diagram

1.2.1 Crystal structure of ice Ih

The characteristic properties of sea ice and its role in the environment are governed by the crystal lattice structure of ice Ih, in particular its resistance to the incorporation of sea salt ions. Depending on pressure and temperature, water ice can appear in more than 15 different modifications. At the Earth's surface, freezing of water under equilibrium conditions results in the formation of the modification ice Ih, with the 'h' indicating crystal symmetry in the hexagonal system. Throughout this chapter, the term 'ice' refers to ice Ih.

Water molecules (H_2O) in ice are arranged tetrahedrally around each other, with a six-fold rotational symmetry apparent in the so-called basal plane (Figure 1.3). This is why snowflakes have six-fold symmetry. The principal crystallographic axis [referred to either as the corresponding unit vector (0001) or simply as the c-axis] is normal to the basal plane and corresponds to the axis of maximum rotational symmetry (Figure 1.3). The interface of the basal plane is smooth at the molecular level. The basal plane is spanned by the crystal a-axes, and the crystal faces

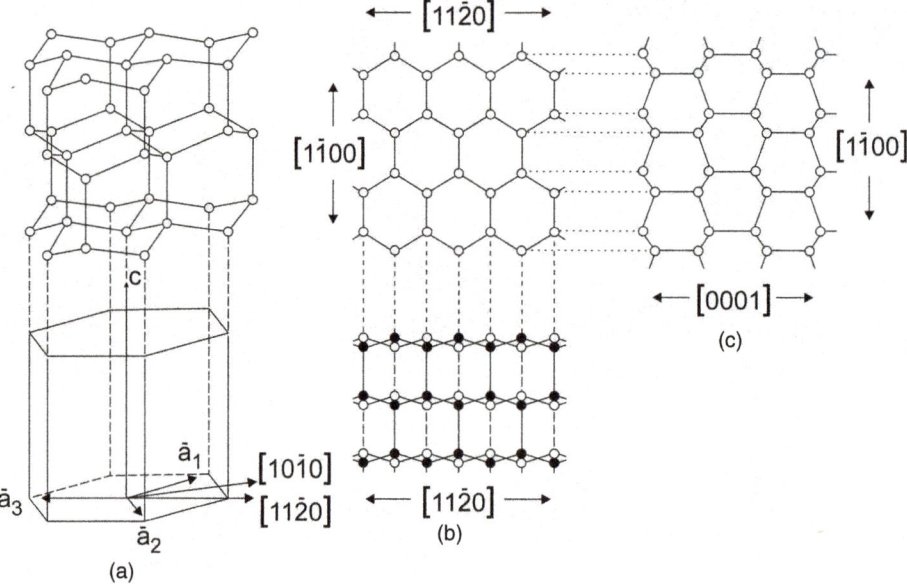

Figure 1.3 Crystal structure of ice Ih (from Weeks & Ackley, 1986). The c-axis is indicated at left and right, and the centre panels correspond to a view along (top) and normal (bottom) to the c-axis.

perpendicular to this plane are rough at the molecular level. The different interface morphologies result in different interface kinetics and are responsible for a pronounced anisotropy in growth rates. For example, the higher growth rates in the basal plane lead to the development of individual frazil ice crystals with thickness-to-width ratios on the order of 1:10 to 1:100 (Hobbs, 1974). Another key aspect of the ice crystal structure is the fact that the packing density of water molecules in ice, and hence its material density, is lower than in the liquid. In the liquid state, water molecules are arranged as hydrate shells surrounding impurities (e.g. sea salt ions) owing to the strong polarity of the water molecule. However, accommodation of sea salt ions is greatly restricted in the ice crystal lattice. Only very few species of ions and molecules are incorporated in the ice crystal lattice in appreciable quantities (either replacing water molecules or filling voids) owing to constraints on size and electric charge. Among them are fluorine and ammonium ions and some gases. However, the major ions present in seawater (Na^+, K^+, Ca^{2+}, Mg^{2+}, Cl^-, SO_4^{2-}, CO_3^{2-}) are not incorporated into the ice crystal lattice, are rejected from the crystal and accumulate at the interface during crystal growth. This has important consequences for the microstructure and properties of sea ice, as part of the salt is retained in liquid inclusions between the ice lamellae, while a larger fraction eventually enters the underlying water column. Both of these processes and their implications will be discussed in the following subsections and in Sections 1.3 and 1.4.

1.2.2 Columnar ice microstructure and texture

As ice grows and the ice–water interface advances downwards into the melt, ions are rejected from the ice. The solute concentration of ions builds up ahead of the advancing interface, increasing the salinity of a thin layer of a few millimetres in thickness. The resulting gradient in salt concentration leads to diffusion of salt away from the interface towards the less saline ocean. Thermodynamic equilibrium dictates that the microscopic ice–water interface itself is at the respective melting/freezing point. As the freezing point decreases with increasing salinity, an increase in salt concentration goes along with a drop in temperature. This leads to a heat flux from the ocean towards the now colder interface.

Heat transport through this boundary layer from the warmer ocean to the colder interface is faster than ion diffusion away from the enriched interface. As a result, a thin layer is established ahead of the interface that is cooled below the freezing point of the ocean but only slightly enriched in salinity above the ocean level. This layer is said to be constitutionally super-cooled as its temperature is below the freezing point of the brine (Figure 1.4).

It is this constitutional super-cooling that distinguishes the growth of lake ice from that of sea ice and helps to explain their respective crystallographic properties. Any small (sub-millimetre) perturbation of a planar ice–water interface that protrudes into the constitutionally super-cooled zone finds itself at a growth advantage, as not only is heat conducted upwards and away from the ice–water interface, but the super-cooled water layer also provides a heat sink. Considering that ice grows fastest in the basal plane, crystals with horizontal c-axes quickly outgrow crystals with c-axes off the horizontal in a process termed geometric selection. By the time ice is thicker than 0.2 m, the c-axes of the remaining crystals are almost exclusively horizontal (Weeks & Wettlaufer, 1996). The solute rejected by a protrusion contributes to a freezing point reduction of the brine along the protrusion boundaries. Consequently, such perturbations can grow into ordered patterns of lamellar bulges at the ice–water interface (for a quantitative analysis of constitutional super-cooling, see Weeks, 2010). The morphology of the interface is mostly reported to be lamellar or cellular (Figure 1.5). In the case of brackish ice grown from

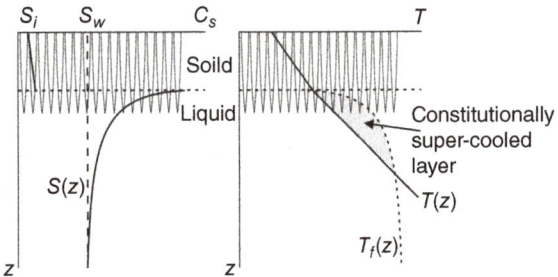

Figure 1.4 Schematic depiction of the lamellar ice–water interface (skeletal layer) and the corresponding salinity (left) and temperature (right) gradients. The freezing temperature profile is shown as a dashed line at right, with a constitutionally super-cooled layer bounded by the actual temperature gradient and the salinity-dependent freezing-point curve.

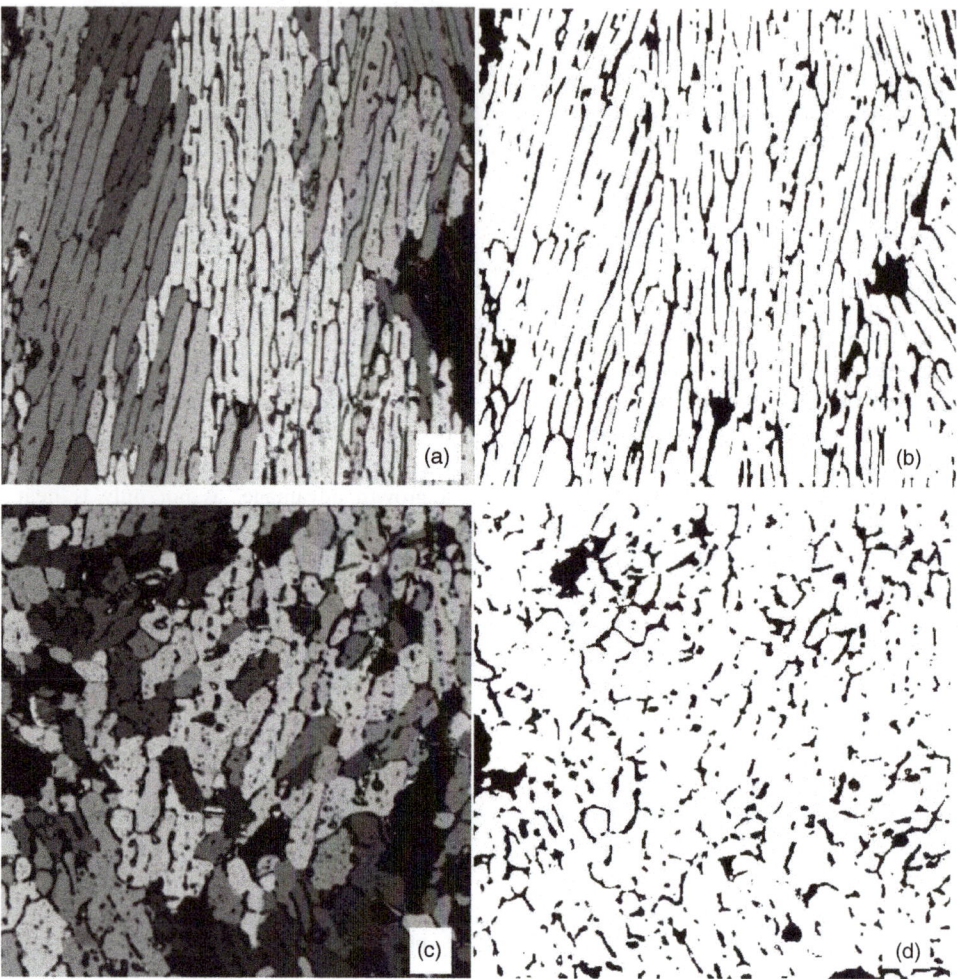

Figure 1.5 (a–d) Thin-section photographs of columnar sea ice grown in a large ice tank (Hamburg Environmental Test Basin, INTERICE experiments) in the absence of an under-ice current (a, b; porosity 0.154, mean pore area 0.096 mm^2) and with a current speed of 0.16 m s^{-1} (c, d; porosity 0.138, mean pore area 0.077 mm^2). Images (a) and (c) have been recorded between crossed polarizers (section is 20 mm wide), with grain boundaries apparent as transitions in grey shades due to different interference colours. Images (b) and (d) show the same section with pores indicated in black based on processing of images recorded in incident light.

water with very low salinities, a planar interface may remain stable throughout the growth process (Weeks & Wettlaufer, 1996).

The growth of individual ice platelets into super-cooled water is most readily observed in the vicinity of Antarctic ice shelves (Jeffries et al., 1993; Leonard et al., 2006) and under Arctic sea ice that is separated from the ocean by a meltwater lens (Notz et al., 2003). Characteristic of the resulting crystal fabric are comparatively large platelets whose c-axes deviate from the horizontal seemingly at random. The process of their formation is poorly understood; one hypothesis is that they are seeded by frazil crystals that formed in the super-cooled water.

When fully developed, as in the case of ordinary columnar sea ice (Figure 1.6), the lamellar interface consists of sub-millimetre-thick blades of ice, separated by narrow films of brine, so-called brine layers. The skeletal layer forms the bottom-most centimetres of sea ice where these brine layers separate individual ice lamellae. It has no appreciable mechanical strength and a porosity of about 30% in its upper reaches. Significant

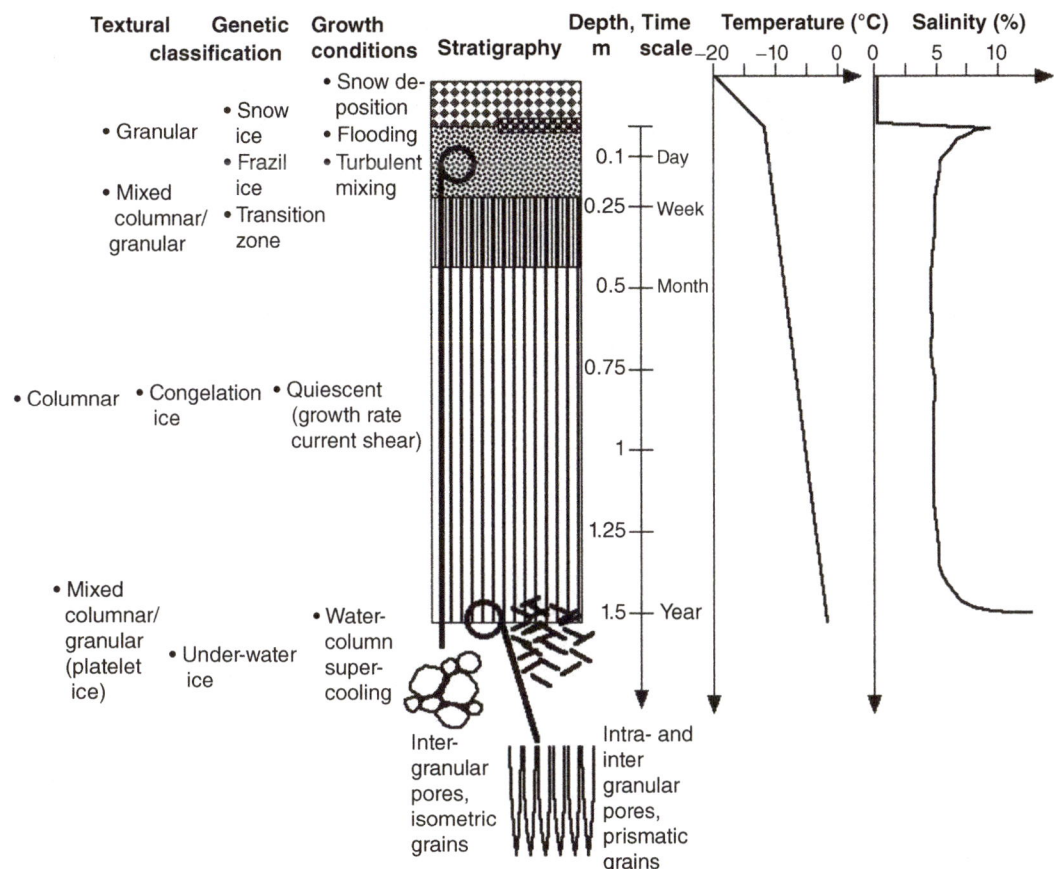

Figure 1.6 Schematic summarizing the main ice textures, growth conditions and timescales and typical winter temperature and salinity profiles for first-year sea ice.

advective brine exchange with the ocean occurs in the skeletal layer and above (Section 1.3).

Consolidation of the skeletal layer follows a trajectory in the phase diagram (Figure 1.7 – although for a smaller bulk salinity than shown in the figure). As the thickness of the ice cover increases, isotherms move downward and the temperature at a given vertical level decreases, ice forms by thickening the lamellae and the fraction of liquid decreases. Eventually, the ice lamellae connect and consolidate into a porous sea ice matrix of strength. During this consolidation process, brine is lost from the ice, as described in more detail in Section 1.3. Although not rigorously accurate, the principal process is the reverse of the warming sequence depicted in Figure 1.8.

The basic crystal pattern laid down in the skeletal layer is retained during ice growth in the form of grain and pore microstructure, i.e. the size and orientation of crystals and the layer spacing of pores. This is illustrated in Figure 1.5, which shows horizontal thin sections of two different varieties of columnar ice grown under the same conditions, except for a difference in the under-ice current. Ice grown in the absence of externally imposed currents exhibits the typical lamellar substructure, with parallel brine layers within individual crystals (also called 'grains'). Along grain boundaries, the size and shape of pores are more heterogeneous, with brine tubes and channels of several millimetres in diameter apparent in the lower right of Figure 1.5(b). The arrangement, shape and size of crystals define the texture of the ice (Figure 1.6).

Ice grown in a current also exhibits a grain substructure delineated by pores, but the degree of parallel alignment of pores and the aspect ratios of individual

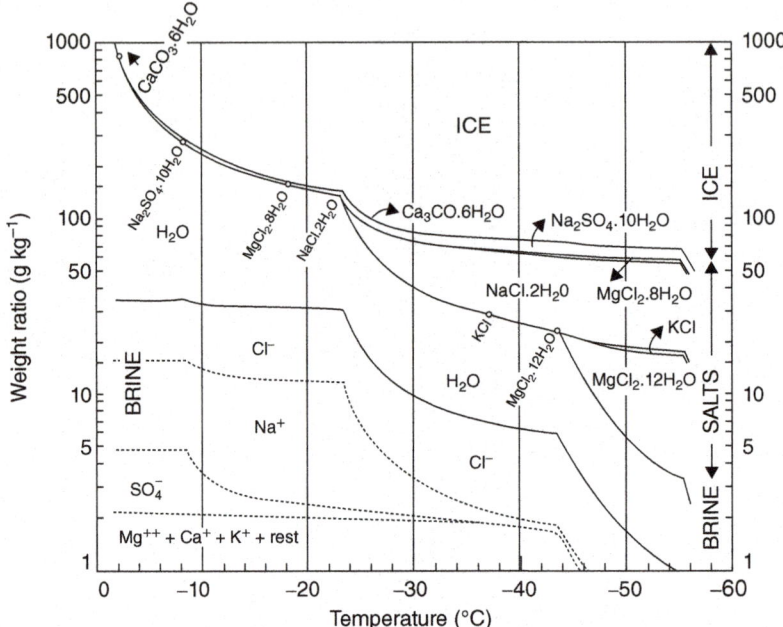

Figure 1.7 Phase diagram of sea ice from Assur (1960). The different curves indicate the mass fraction of solid ice (top), salts (middle) and liquid brine (bottom) present in a closed volume of standard seawater at different temperatures.

inclusions are very different, as are the grain sizes (Figure 1.5c,d). This difference arises from differences in the thickness and degree of super-cooling at the interface, which depend on ice growth rate, seawater salinity and, as illustrated here, the magnitude of currents transporting solute away from the ice. Currents also affect the horizontal orientation of the crystal c-axes.

c-axes tend to point parallel to the direction of a unidirectional under-ice current. It appears that salt transport away from the interface is enhanced for crystals with lamellae oriented perpendicular (c-axes parallel) to the current, providing them with a growth advantage that eventually results in the dominance of c-axes parallel to the current (Langhorne & Robinson, 1986). Hence, under-ice currents enhance the anisotropy of the columnar sea ice.

A further aspect of the lamellar substructure is that the spacing of ice lamellae depends on growth rate. Nakawo & Sinha (1984) demonstrated this for Arctic sea ice, where the down-core reduction in growth rate (see Section 1.5) closely corresponded to an increase in the brine layer spacing a_0. Typically, a_0 is on the order of a few tenths of a millimetre. However, in what is probably the oldest sea ice sampled to date, grown at rates of a few centimetres per year, Zotikov et al. (1980) found brine layer spacings of several millimetres.

The skeletal layer harbours one of the greatest concentrations of phytoplankton in the world's oceans by providing a habitat for diatoms and other microorganisms and, in turn, grazers (Smith et al., 1990; see also Chapters 13–16). As algae depend not only on sunlight but also on nutrients for photosynthetic activity, in many areas the most active layer of ice organisms is found within the bottom few centimetres of the ice cover, where high porosities and permeabilities and the proximity of the ocean reservoir provide a sufficient influx of inorganic nutrients and gas exchange (Chapters 14, 17 and 18). At the same time, this layer offers some protection from the largest grazers (Chapter 16) and presents photosynthetic organisms with a foothold at the top of the water column where irradiative fluxes are highest (Chapter 14; Eicken, 1992a).

1.2.3 Granular ice microstructure and texture

Water at temperatures below the freezing point is called super-cooled. Typically, seawater cannot be

Overview of sea ice growth and properties 9

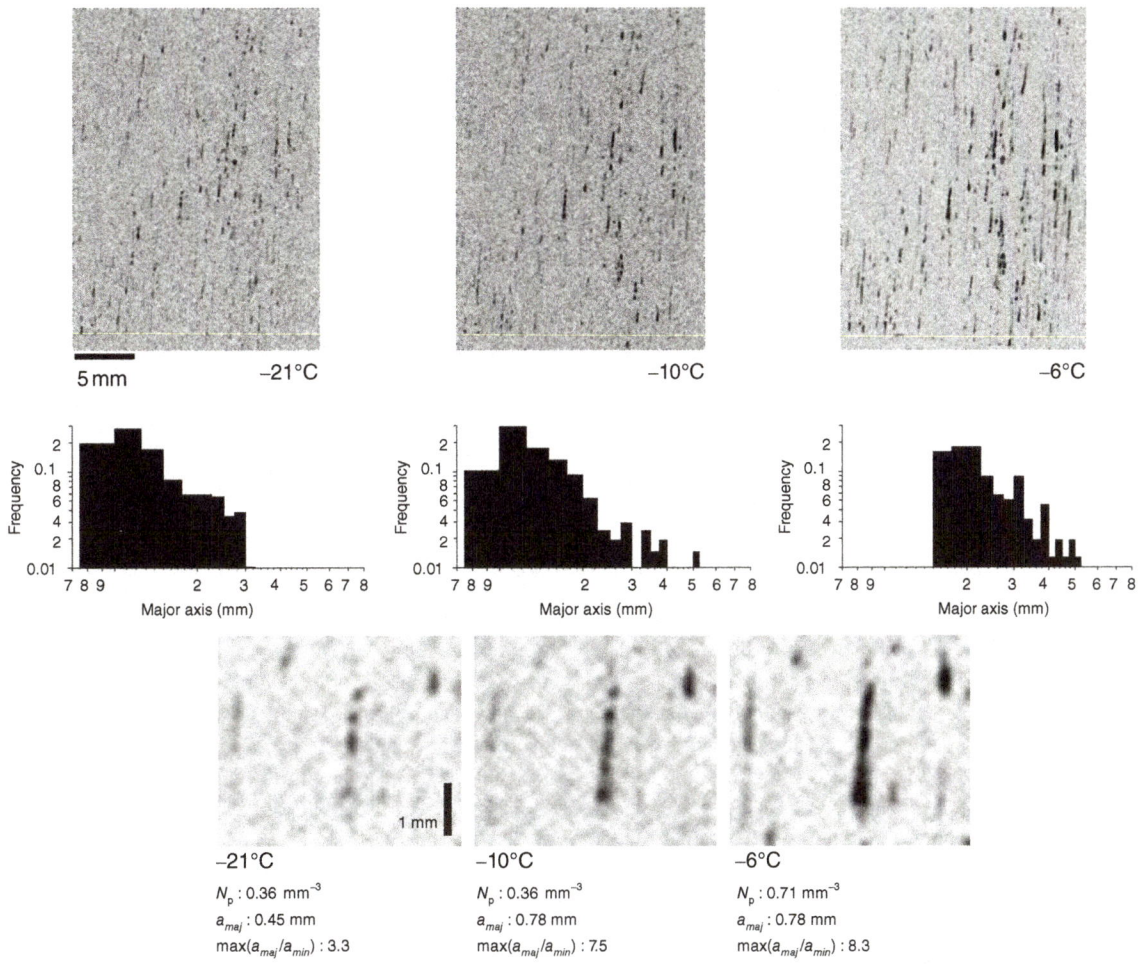

Figure 1.8 Thermal evolution of fluid inclusions in first-year sea ice obtained near Barrow, Alaska (0.13–0.16 m depth, sample obtained in March 1999 and maintained at *in situ* temperatures after sampling up until experiment; for details see Eicken et al., 2000) as studied with magnetic resonance imaging techniques. The upper three panels show a vertical cross-section through the sample as it is warmed, with pores appearing dark. The middle panels show the size distribution of the major pore axes a_{maj} (upper 10th percentile), indicating enlarging and merging of pores in the vertical. The change in pore size, morphology [as indicated by the maximum ratio between major and minor pore axis length, max(a_{maj}/a_{min})] and number density N_p is apparent in the lower panels, which show a smaller subset of pores at 0.15 m depth.

super-cooled by more than 0.1 K because abundant impurities act as nucleation sites for ice crystals (Fletcher, 1970). As a result, ice forming in water subjected to overturning by winds will develop ice crystals that are kept in suspension until a surface layer of ice slush builds up that reduces mixing agitated by wind. These ice crystals take the shape of needles, spicules or platelets, often intertwined into aggregates, and are known as frazil ice (Figure 1.6). Individual crystals are typically a few to a few tens of millimetres in diameter and less than a millimetre in thickness (Weeks, 2010). The surface slush starts to consolidate by freezing of the interstitial brine from the top downwards. Shielded from winds, the ice grows below the slush in a radically different, quiescent environment.

The stratigraphy of a 'typical' ice cover is revealed through analysis of vertical core sections. It consists of a sequence of granular ice (a few centimetres to tens of centimetres at most in the Arctic, but substantially more in other, more dynamic environments such as the Antarctic), with randomly oriented, isomeric or prismatic crystals [see detailed descriptions in Weeks (2010) and Tyshko et al. (1997)], followed by a transitional layer that is underlain by columnar ice (congelation ice), composed of vertically elongated prismatic crystals that can grow to several centimetres in diameter and tens of centimetres in length (Figure 1.6; see above for details).

1.2.4 Frazil ice

Congelation growth of sea ice with columnar texture typically dominates in the Arctic. However, frazil ice growth resulting in granular textures is also common and even more so in the Southern Ocean. Growth of individual platelets and needles of frazil in a super-cooled water column differs from growth of congelation ice insofar as both heat and salt have to be transported away from the interface into the surrounding ocean water. Consequently, beyond a certain size, individual frazil crystals develop rough, dendritic surfaces as a result of solute build-up. Frazil growing in the turbulent uppermost metres of the ocean has the tendency to aggregate into clusters of crystals. The clusters are capable of sweeping particulates and biota from the water column, and carrying them to the surface as a layer of frazil or grease ice accumulates (Reimnitz et al., 1990; Smedsrud, 2001; Chapter 7). Despite its abundance, some aspects of frazil growth are not that well understood: its inherent 'stickiness', enhancing concentrations of biota in Antarctic sea ice, or the conditions governing the growth of larger ice platelets at greater depths (Bombosch, 1998). Frazil ice growth in the presence of melting Antarctic ice shelves is capable of generating large volumes of crystals that contribute to the mass balance of both ice shelves and coastal sea ice (e.g. Leonard et al., 2006).

Another aspect of frazil growth that is currently not well understood is the actual consolidation of loose masses of frazil crystals, with ice volume fractions of between about 10% and 30% in solid granular sea ice. Evidence from oxygen stable-isotope and salinity measurements of individual crystals and layers of granular ice suggests that the consolidation process is a combination of downward freezing of voids among the mesh of frazil crystals and transformations in the size distribution and morphology of the crystals themselves. This is similar to what has been observed to occur in water-saturated snow slush (Eicken, 1998). Recent work (Maus & De la Rosa, 2012; Naumann et al., 2012) has indicated that consolidation of frazil slush can be understood in terms of removal of salt through convective overturning and progressive freezing of remaining voids. Further work to explore these processes is needed, as frazil ice growth is a key process in the interaction of ocean and atmosphere, which is of increasing importance in an ice-diminished Arctic (see Chapter 6).

1.2.5 Formation of sea ice

In the Antarctic, higher wind speeds, the effects of ocean swell penetrating from higher latitudes and the larger number of openings in the pack greatly favour the formation of frazil ice. As a result, fazil ice can constitute as much as 60–80% of the total ice thickness in some regions (Lange et al., 1989; Jeffries et al., 1994). The ice edge advances northwards from the Antarctic continent by as much as 2500 km from austral autumn through to spring (Chapter 10). This dynamic ice-growth environment favours growth of frazil ice and leads to the predominance of the so-called pancake ice (Figure 1.1). Pancake ice forms through accretion of frazil crystals into centimetre-sized floes of ice that in turn accrete into decimetre-sized pans of ice. Under the action of wind and ocean swell penetrating deep into the sea ice zone, these pans bump and grind against one another, resulting in a semi-consolidated ice cover composed of ice discs with raised edges that are from a few centimetres to more than 10 cm thick. These pancakes eventually congeal into larger units (Wadhams et al., 1987). Once the ice cover has consolidated into a continuous, solid sheet or larger floes with snow accumulated on top, only characteristic surface roughness features ('stony fields' or 'rubble ice') betray its dynamic origins. However, stratigraphic analysis of ice cores clearly demonstrates that the ice cover is largely composed of individual pancakes, often tilted or stacked in multiplets on top of one another. The interstices between the individual pancakes eventually consolidate through a combination of frazil growth and freezing of congelation ice (Lange et al., 1989). Typically, these processes account for ice thicknesses of up to 0.5 m (Wadhams et al., 1987; Worby et al., 1998).

In the Arctic, recent reductions in perennial ice extent (Chapter 11) may now increasingly favour the formation of frazil ice. The limited data available to date do suggest an increase in the proportion of granular ice compared with previous studies (Perovich et al., 2008), but more observations are needed to confirm these early indications. In the past, most of the ice cover was composed of congelation ice (Weeks, 2010).

1.2.6 Phase relations in sea ice

Unlike zinc and copper in brass alloys, sea salts and ice do not form a solid solution in which the constituents intermingle in different proportions. Hence, the question arises as to what exactly the fate is of ions in freezing seawater. In order to fully address this problem, one needs to consider the physicochemical phase relations of an idealized or somewhat simplified seawater system.

Sodium and chloride ions (Na$^+$, Cl$^-$) account for roughly 85%, sulphate ions (SO$_4^{2-}$) for 8%, and magnesium, calcium and potassium for another 6% of the mass of salts dissolved in seawater. Owing to the predominance of sodium and chloride ions in seawater, many aspects of sea ice properties and structure can already be observed in a simple sodium chloride solution. More sophisticated representations of seawater typically take into account Na$^+$, K$^+$, Ca^{2+}, Mg^{2+}, Cl$^-$, SO$_4^{2-}$ and CO$_3^{2-}$. In his classical study of the phase relations in sea ice, Assur (1960) assumed a constant 'standard' composition for sea ice. While such an approach is inadequate for geochemical studies (Marion & Grant, 1997; Chapter 17) and does present problems with ice that is strongly desalinated or grown in isolated basins, it is sufficient to predict the most important characteristics of sea ice behaviour upon cooling or warming.

Figure 1.7, taken from Assur's work, serves to illustrate the key aspects of the phase relations in sea ice. In a closed system (i.e. the mass fraction of all components is constant) one would observe that for seawater of salinity 34, cooled below the freezing point at −1.86°C, the ice fraction steadily increases as the temperature is lowered, assuming that the individual phases are in thermodynamic equilibrium. As the ions dissolved in seawater are not incorporated into the ice crystal lattice, their concentration in the remaining brine increases steadily. At the same time, the freezing point of the brine decreases, co-evolving with the increasing salinity of the liquid phase. At a temperature of −5°C, the ice mass fraction in the system amounts to 65% and the salinity of the brine in equilibrium with the ice has risen to 87. At −8.2°C, the concentration of salts has increased to the point where the solution is supersaturated with respect to sodium sulphate, a major component of seawater, resulting in the onset of mirabilite precipitation (Na$_2$SO$_4 \cdot $10H$_2$O; Figure 1.7). If one were to continue lowering the temperature of the system, mirabilite would continue to precipitate in the amounts specified in Figure 1.7. Other salts precipitating during the freezing of seawater include ikaite (CaCO$_3 \cdot $6H$_2$O), the distribution and mineralogy of which we have only learned more about in very recent times (Dieckmann et al., 2010; Papadimitriou et al. 2013, 2014; Hu et al., 2014), as well as hydrohalite (NaCl·2H$_2$O). The latter is predicted to start precipitating at −22.9°C, with roughly 90% of the precipitable sodium chloride present as hydrohalite at −30°C (Figure 1.7). While the mass fraction of brine drops below 8% at −30°C, even at the lowest temperatures typically encountered in sea ice (around −40°C), a small but non-negligible liquid fraction remains. The presence of unfrozen water even at these low temperatures has important consequences, in particular for the survival of microorganisms overwintering in sea ice (Chapters 13–16).

Salinity measurements have undergone changes throughout history (Millero et al., 2008). Originally, the saltiness of ocean water was defined as the ratio of the mass of dissolved material to the mass of the solution. However, the mass of dissolved material is difficult to measure by evaporation as many crystalline salts are bound with water and volatile components may evaporate during heating. A simplified approach has been followed since the beginning of last century, exploiting the fact that ocean water around the world is of almost uniform composition. For most of the last century, the concentration of Cl$^-$ ions has been measured by titration and scaled linearly to salinity (sometimes quoted as ppt or ‰). This relationship is sensitive to the composition of the seawater used for calibration and was corrected slightly in the 1960s. With the advent of conductivity meters, an accurate and even more convenient way opened up for salinity measurements.

Today, the ocean salinity is measured as the ratio of the electrical conductivity of a solution to the conductivity of a reference solution and converted to a practical salinity using an equation provided by UNESCO (1978). As such, it is independent of chlorinity and mass of dissolved material. Practical salinity is defined as a dimensionless quantity and should not be quoted as having a practical salinity unit (psu). However, this is a widespread habit in the literature. IOC et al. (2010) recommend that salinity be stored in archives according to how it has been measured, i.e. in most cases today as practical salinity.

A reference-composition salinity (reference salinity for short) has been introduced that is to be used in the most recent thermodynamic equation of state of seawater and derived properties (IOC et al., 2010). The reference salinity, S_R, is linearly related to the practical salinity u_{PS} through $S_R \approx (35.16504/35)u_{PS}$, is supposed to indicate the actual solute mass dissolved in standard seawater and is quoted in units of g kg^{-1}. IOC et al. (2010) explain that the difference between the numerical values of reference and practical salinities can be traced back to the original practice of determining salinity by evaporation of water from seawater and weighing the remaining solid material. This process also evaporated some volatile components, and most of the 0.16504 g kg^{-1} salinity difference is due to this effect.

The UNESCO salinity definitions apply to the composition of standard seawater and cover the range from 2 to 42 above −2°C. Measurements outside this range may require different equations or procedures (see IOC et al., 2010 for examples). Equations of IOC et al. (2010) are not used in this chapter.

1.3 Desalination and pore microstructure

1.3.1 Salinity profiles of growing and melting sea ice

In a pioneering study, Malmgren (1927)[1] studied the salinity evolution of Arctic first-year sea ice during the course of winter and into the summer melt season. He summarized his observations in a seminal figure that is still commonly shown 90 years after its initial publication (Figure 1.9). In this section, we will briefly consider the processes responsible for the characteristic C-shape of the salinity profile of young and first-year ice as well as the reduction in surface salinities during the first melt season in an ice floe's evolution. The importance of understanding the evolution of an ice cover's salinity profile is rooted in the central role played by temperature and salinity with respect to ice porosity and pore microstructure. Large-scale sea ice and climate

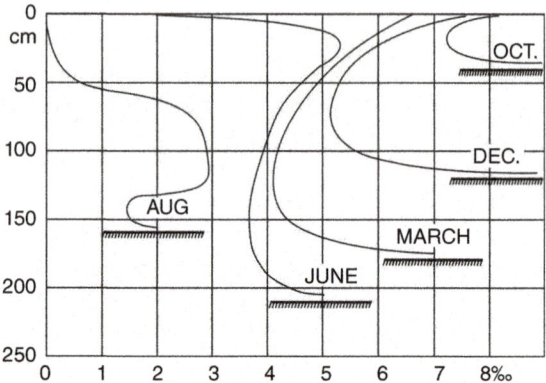

Figure 1.9 Evolution of sea ice salinity profiles (from Malmgren, 1927). Note the characteristic C-shape of the young and first-year ice salinity profile and the reduction in surface salinity due to meltwater flushing with the onset of summer melt.

models are only now beginning to move beyond the assumption of constant ice salinity (originally motivated by the prominence of multi-year sea ice in the Arctic Ocean), creating the need to better understand and simulate salinity and property evolution at the small scale in support of large-scale model efforts. For example, work by Vancoppenolle et al. (2009) found that a change of bulk salinity from 0 to 5 is equivalent to a change in sea ice albedo by 10%.

The importance of the desalination processes is illustrated by comparing the first-year winter sea ice salinity profile in Figure 1.9 with that of summer or multi-year sea ice. As dictated by the phase relationships (Figure 1.7), the transition from winter to summer sea ice generally corresponds to a change in the direction of the conductive heat flux through an ice floe from being directed upwards to being directed downwards (cf. Figure 1.10).

1.3.2 Origin of brine movement in growing sea ice

The microscopic exclusion of ions from a growing ice crystal leads to a local increase in brine salinity but does not change the local bulk salinity. However, observations clearly show that the bulk salinity of growing first-year sea ice (typically 4–6) is only a fraction of the bulk salinity of the ocean water (around 33) (Figure 1.9). Several processes have been considered to explain the reduction of the bulk salinity, i.e. the removal of ions from sea ice. The migration rate of individual brine pockets

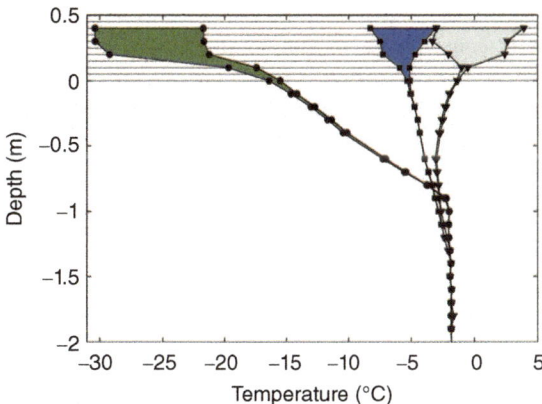

Figure 1.10 Vertical temperature profiles measured with a thermistor probe frozen into the ice in Barrow, 2008. Positive depths are snow and air, while negative depths are sea ice and ocean. Profiles show the temperature range encountered over a 24-hour period in mid-February (day 45, circles), mid-May (day 135, squares) and at the end of May (day 150, triangles).

in a temperature gradient due to solute diffusion inside the pockets was found to be too small (cf. Weeks, 2010). The significance of this process is limited to the microscopic level where it could affect the interconnectivity of the pore space. However, the impacts on that scale are still not well understood. Brine expulsion is the movement of brine in response to volume expansion during ice formation. Notz and Worster (2009) show analytically that the flow rate of brine due to volume expansion is always less than the vertical growth rate of sea ice. Hence, brine expulsion contributes only to solute redistribution in the ice but not to net desalination. This assessment applies during quasi-steady state ice growth, i.e. while the growth rate is essentially constant. A segregation process at the ice–ocean interface (often termed 'initial segregation') that had been assumed in earlier studies to parameterize sea ice salinity development (e.g. Cox & Weeks, 1988) has been found to lack empirical and theoretical basis (Notz & Worster, 2009). Hence, the only significant process contributing to the net desalination prior to the onset of melt is gravity drainage (Notz & Worster, 2009).

Hunke et al. (2011) reviewed key observations and the state of sea ice salinity modelling and found that, in spite of significant recent advances in our understanding of the processes, a simple description remained elusive. Further progress has been made since then.

Considering steady growth conditions, Petrich et al. (2011) and Rees Jones and Worster (2013) used a simple dynamical model to describe the desalination process. Reassuringly, both groups obtained the same expression for the brine flux at the ice–ocean interface. Petrich et al. (2011) were also able to derive an analytical expression for steady-state bulk salinity as a function of ice growth rate that compared well with two-dimensional (2D) computational fluid dynamics (CFD) simulations. However, based on a numerical, 1D desalination model, Griewank and Notz (2013) suggested that quantitative results of these simple quasi-steady models fall short in cases where growth conditions vary rapidly with time (e.g. insolation or diurnal temperature variations seen in spring). One could argue that the first-order description of the bulk salinity profile has been identified for quasi-steady sea ice growth (the zeroth order being depth-independent bulk salinity) and that future attention should be paid to the wide range of higher-order aspects that are relevant to sea ice microbiology, surface chemistry, radar backscatter and pollutant transport (e.g. non-quasi-steady state growth, the role of freeboard, movement of brine towards the surface, banding of inclusions). In this context it should be remembered that, in addition to processes within the ice, processes at the upper surface (e.g. meltwater pooling and vertical flushing through the ice) and beneath the ice (e.g. under-ice shear flow; Feltham et al., 2002) affect bulk salinity profile and evolution.

Let us step back and take a look at the process of natural convection. Why does brine move through sea ice to begin with? For the sake of simplicity, we treat sea ice according to the effective medium approach; we assume that the brine layers and pores in sea ice are interconnected and that no dominant channels exist; fish-tank bubbler stones have such a microstructure. Recall that brine density is primarily a function of brine salinity in sea ice (the higher the salinity, the denser the brine) and that the freezing point of brine depends on salinity (the higher the salinity, the lower the freezing point). In growing sea ice we find a temperature profile with the warmest ice near the ocean and the coldest ice at the upper surface (Figure 1.10). Thus brine salinity and brine density will increase toward the upper surface if brine and ice are in thermodynamic equilibrium. This configuration is hydrostatically unstable, i.e. a small perturbation in the density field will tend to drive flow in a direction that further increases the magnitude of the perturbation (driving natural convection). The flow rate is retarded by friction due to the dynamic fluid viscosity, μ, of the brine and the permeability, Π, of the ice, the latter of which is a single-parameter description of the pore microstructure. While this can slow down the motion, it can, theoretically, not stop the motion. However, as a result of the dependence of the freezing point on salinity, perturbations are not in thermodynamic equilibrium with the surrounding ice. Hence, phase change will take place until thermodynamic equilibrium is obtained again and this process happens to change the brine salinity to reduce the magnitude of the density perturbation. Now, if brine movement is slow (e.g. because friction is high), perturbations can be annihilated completely due to the phase change. The rate at which phase change can take place (and perturbations are reduced) depends on the rate at which heat can be provided to or removed from the microscopic ice–brine interface. This rate is related to the thermal diffusivity of the ice α_{si} and a characteristic distance Δh for heat transport. For the sake of convenience in analytical calculations of fluid movement, a porous medium Rayleigh number can be defined for a homogeneous porous medium (Worster & Wettlaufer, 1997),

$$Ra = \frac{g \Delta \rho \, \Pi \, \Delta h}{\mu \alpha_{si}} \quad (1.1)$$

where $\Delta \rho$ is the density difference, Δh is a characteristic length and g is the acceleration due to gravity. The more the driving forces for fluid movement exceed the retarding forces, the higher the value of Ra. In fact, for natural convection to take place in a porous medium at all, the Rayleigh number has to exceed a critical Rayleigh number Ra_c (for values, see Nield & Bejan, 1998). Wettlaufer et al. (1997) demonstrated in idealized laboratory experiments that brine release from growing saltwater ice set in only after the thickness of the ice (this would be Δh in Eqn (1.1)) exceeded a threshold. It is unclear whether this effect would be observable in thin sea ice growing in its natural environment, given the ubiquity of perturbations in the ice microstructure and at its interfaces. However, heuristically extending equation (1.1) to inhomogenous sea ice (Notz & Worster, 2009) and invoking the concept of a critical Rayleigh number have led to success in 1D numerical modelling of desalination (e.g. Griewank & Notz, 2013).

1.3.3 Brine movement and bulk salinity

Mass conservation dictates that the volume of brine leaving the sea ice due to advection is balanced by brine entering the sea ice from the ocean (we neglect effects due to the density difference between ice and water here). This leads to a turbulent sea ice—ocean interface flux, i.e. an advective flux with no net direction, and there has been one attempt at determining its magnitude experimentally (Wakatsuchi & Ono, 1983). While the experiments showed that the ice–ocean interface flux increased with growth rate, the magnitude of the estimate depended on an assumed mixing factor for the experiment that the authors set by educated guesswork (they chose 0.5). Apart from its relation to the brine flux, the volume flux from the ocean into the ice determines the flux of nutrients, which is of particular relevance for biological processes (Hunke et al., 2011; Chapter 17).

Within sea ice, a downward flow increases the salinity locally, leads to local dissolution of the ice matrix and thereby creates brine channels. Locally, brine is replaced by more saline brine from above, and as the more saline brine is super-heated (i.e. above the freezing point) with respect to the temperature of the surrounding ice, it will dissolve the surrounding ice partially to attain thermodynamic equilibrium. Brine leaving through channels is detectable as distinct brine plumes (streamers) in the underlying water (Wakatsuchi, 1983; Dirakev et al., 2004). Since brine channels develop as a consequence of convection inside the porous medium, brine moving upward through the porous medium is part of the development process of channels (Worster & Wettlaufer, 1997). An upward-directed flow of brine leads to a local reduction of brine salinity, bulk salinity and porosity. Niedrauer & Martin (1979) found in laboratory experiments that downward flow follows a cusp-shape pattern that terminates in brine channels. Consistent with this general pattern of salinity distribution, Cottier et al. (1999) found in high-resolution salinity measurements that the bulk salinity is highest in the presence of brine channels and somewhat lower in between. The salinity difference between the brine leaving the sea ice and the brine entering the sea ice is the cause for the net desalination of sea ice that gives rise to the characteristic bulk salinity profiles (Figure 1.9). Hence, locally varying sea ice bulk salinity is a systematic feature of the growth process. It manifests itself as scattered data when several bulk salinity measurements are performed close to each other on otherwise homogeneous sea ice. Gough et al. (2012) concluded that bulk salinity measurements have to differ by more than 29% to be regarded as different with 90% confidence. This means bulk salinity scatter is to be expected in a range of ±0.5 to ±1.

In growing ice, the rate of bulk desalination becomes insignificant with increasing distance from the sea ice–ocean interface. This gives rise to a quasi-steady state salinity, also termed stable salinity (Nakawo & Sinha, 1981). Two processes have been put forward to explain when bulk flow and desalination cease during ice growth. On the one hand, based on continuum fluid dynamics considerations, the permeability of sea ice may reach values too low to sustain natural convection, similar to arguments leading to equation (1.1). One the other hand, based on percolation considerations, sea ice pores may shrink and finally disconnect from regular fluid motion, retaining their solute content (or precipitates) until the melt season (see later). There are too few experimental or theoretical investigations to make a final call on the dominant process. Numerical 2D CFD simulations have reproduced observed salinity profiles either with or without assuming that the pore space disconnects (Petrich et al., 2007, 2011). The notion of a potentially disconnecting pore space in sea ice has been popularized by Golden et al. (1998). This concept enjoys popularity in the sea ice community because it gives an intuitive explanation for the observation that the sea ice bulk salinity does not change with time where porosities are below 0.05–0.07 (Cox & Weeks, 1988; Arrigo et al., 1993). The sea ice geometry is also consistent with those percolation thresholds (Petrich et al., 2006). However, as the 1D and 2D fluid dynamics simulations mentioned earlier show, the 0.05–0.07 threshold may simply result from a combination of a porosity-dependent permeability in conjunction with basic fluid and thermodynamics.

In general, fluid motion and desalination are confined to a few centimetres at the bottom of growing sea ice (Figure 1.4), possibly <5 cm in winter, but reaching 10–20 cm in spring (Petrich et al., 2013). Textbook bulk salinity profiles such as the ones in Figure 1.9 show systematic characteristics (Eicken, 1992b), such as the typical C-shape during growth, and seem to be predictable from the environmental conditions during growth (see later; Cox & Weeks, 1988). However, there is a considerable amount of variability between cores taken in close proximity (Weeks & Lee, 1962). This variability has been attributed to pores and channels that are large compared with the sample size (Bennington, 1967; Cottier et al., 1999).

1.3.4 Development of the pore microstructure during growth and melt

In spite of its importance for optical properties, there is still no quantitative and complete description of the evolution of the pore microstructure during freezing and melt. However, we can link observations and hypotheses to understand the general process in a piecemeal approach. Here, we focus on sea ice with a columnar rather than granular texture.

The ice microstructure is lamellar close to the ocean and individual ice lamellae are interspersed with liquid brine layers or films. The separation of the lamellae is typically around 0.3–0.5 mm. As the sea ice cools or desalinates, lamellae grow thicker at the expense of brine layers and interconnect by forming bridges, presumably at porosities between 0.1 and 0.3 (see also discussion of Figure 1.11). Horizontal sections of this ice reveal inclusions ranging over several orders of magnitude in size (Perovich & Gow, 1996). What appears as large, high-aspect ratio inclusions in these images is the brine film narrowed by the ice bridges between lamellae. The bridges themselves are interspersed with smaller inclusions at all scales. All inclusions get smaller upon further cooling or desalination and the narrowed brine layers separate into even narrower films and terminate vertically. Some residual connectivity is likely throughout this pore space, at least while volume expansion during freezing creates enough pressure to push brine through the matrix. In addition to the pores between the lamellae, brine channels form that mark the preferred pathways of downward-moving brine during desalination. They are usually vertical (Figure 1.12). Their diameter can exceed the lamellae spacing and evidence is inconclusive as to what extent the lamellar structure affects the development of brine channels. Brine channels are supplied with brine through the pore network that formed between the lamellae (Lake & Lewis, 1970; Niedrauer & Martin, 1979). The direction of flow in brine channels can reverse and oscillation between inflowing and outflowing brine has been observed (Lake & Lewis, 1970; Eide & Martin, 1975). Because brine flowing upward into channels is less saline, it supports the disintegration into pores and closure at the open end, called necking (Eide

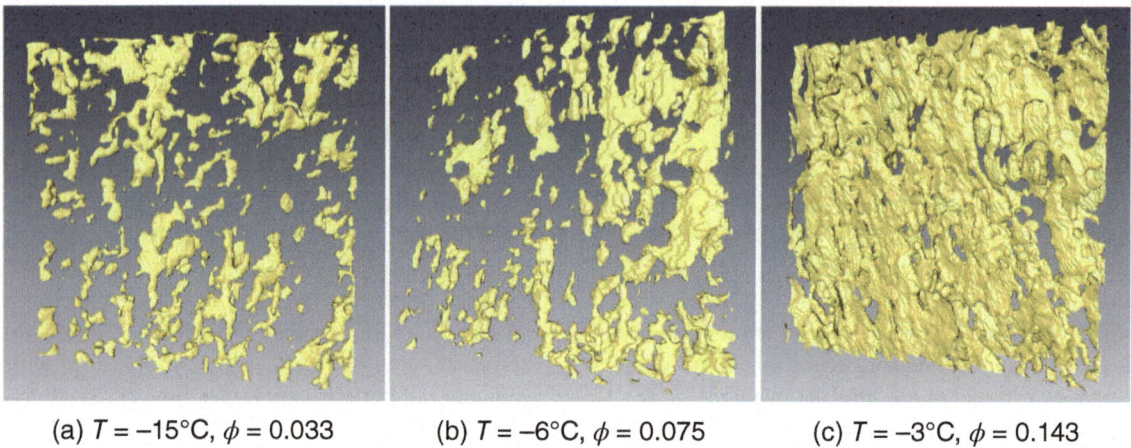

(a) $T = -15°C$, $\phi = 0.033$ (b) $T = -6°C$, $\phi = 0.075$ (c) $T = -3°C$, $\phi = 0.143$

Figure 1.11 X-ray microtomography images of brine layers in sea ice single crystals as a function of temperature. Note how the sample porosity (ϕ) and connectivity increase as it is warmed. Source: Golden et al. 2007. Reproduced with permission of John Wiley & Sons.

Figure 1.12 Photograph of a slab of sea ice obtained from sea ice of approximately 1.4 m thickness near Barrow, Alaska (photo courtesy of D. M. Cole). Note the distinct horizontal layering as well as the parallel rows of vertical brine channels (shown in horizontal cross-section in Figure 1.2e).

& Martin, 1975). Inclusions resembling disintegrated feeder channels are sometimes found leading towards brine channels (Lake & Lewis, 1970). They tend to be inclined around 45° and appear mostly in the upper part of sea ice. Star-like patterns can be observed where they are visible at the ice surface (Figure 1.2e). Overall, small brine inclusions in cold sea ice show a characteristic distribution of length-to-width ratios (Light et al., 2003).

Brine inclusions enlarge upon warming during the melt season and form pathways for brine and meltwater. The slow redistribution of solute in conjunction with a drive towards thermodynamic equilibrium leads to the widening of brine channels that form preferred pathways thereafter (Polashenski et al., 2012). The pore structure of this secondary pore space differs and often contains longer and wider channels than the primary pore space of growing sea ice.

1.3.5 Brine movement during melt

The most prominent, albeit not the only, form of brine movement during the melt season is flushing of surface meltwater (Untersteiner, 1968). As the density of ice is lower than that of water, the surface of free-floating snow-free sea ice of thickness H protrudes above the water level by a distance h_f called freeboard,

$$h_f = H \left(\frac{\rho_{sw} - \rho_{si}}{\rho_{sw}} \right) \quad (1.2)$$

If sea ice is sufficiently permeable or cracks or other flaws exist, then the hydrostatic pressure of brine and meltwater (melting snow and surface ice melt) above the ocean level can result in significant transport through the ice both vertically and laterally. Studies involving different tracers have demonstrated that the vertical and lateral transport of meltwater varies with season as a function of ice permeability. As much as a quarter of the meltwater produced annually at the surface of Arctic sea ice can be retained in pores within the ice cover (Eicken et al., 2002). The bulk salinity reduction of sea ice during melt is due to at least two processes. Infiltration of surface meltwater into the upper 0.1–0.3 m ('flushing') leads to a characteristic linear bulk salinity profile that develops in the very early stages of melt (Figure 1.13), with the lowest salinity at the top (Polashenski et al., 2012). There is modelling evidence that desalination of the lower parts of sea ice is triggered by an increase in sea ice permeability due to warming, resulting in convective overturning (Eicken et al., 2002; Griewank & Notz, 2013). The latter process is not dependent on meltwater from the surface. Repeat desalination in summer leads to extremely low bulk salinities of Arctic multi-year sea ice.

1.3.6 Permeability of sea ice

Fluid moving though sea ice experiences a resistance that is due to both microscopic obstacles in the flow path (which lead to its tortuosity) and viscous drag along the pore and channel walls. The reciprocal of resistivity is the permeability Π, measured in m². Its magnitude is similar to that of the cross-sectional area of an individual duct in a 'porous' medium made up of a close-packed bundle of ducts (i.e. a porous medium consisting entirely of pathways 10^{-4} m in size would have a permeability of around $\Pi = 10^{-8}$ m², not accounting for geometric correction factors). The volume flux q through a homogeneous porous medium can be described by Darcy's law:

$$q = -\frac{\Pi}{\mu}\frac{\partial p}{\partial x} \quad (1.3)$$

where μ is the dynamic viscosity and $\partial p/\partial x$ is the pressure gradient in the direction of flow. The permeability tensor is generally anisotropic in sea ice, i.e. its components depend on the direction of flow; this is due to the anisotropic crystal fabric and pore structure which provide the highest permeability in the vertical direction. Differences of about one order of magnitude seem to be typical (Freitag & Eicken, 2003). While the permeability of the open ocean is infinite, we can expect that the vertical permeability at the bottom of the skeletal sea ice layer remains finite even as the porosity approaches 1: the regularly spaced ice lamellae, no matter how thin, will exert drag forces on the brine moving past.

In fact, the pore space may get clogged to some extent and flow patterns are altered by gelatinous extracellular polymeric substances (EPS) secreted by ice-dwelling organisms (Chapters 13 and 17; Krembs et al., 2011). The influence of EPS on the development of sea ice microstructure and bulk salinity has received little attention from researchers so far.

A possible permeability–porosity relationship for growing sea ice that is consistent with field measurements (Freitag, 1999; Eicken et al., 2004) and effective medium theory (Golden et al., 2007) is

$$\Pi = \Pi_0 \left(\frac{V_b}{V}\right)^3 \quad (1.4)$$

where Π_0 is of the order of $\Pi_0 = 1...3 \times 10^{-8}$ m² and V_b/V is the brine volume (porosity) of the ice. The value of Π_0 depends on whether we consider one component of the permeability tensor in unidirectional flow or an equivalent isotropic permeability for multidimensional flow. The equivalent isotropic permeability is usually taken to be the geometric mean of the individual component. Due to the development of a secondary

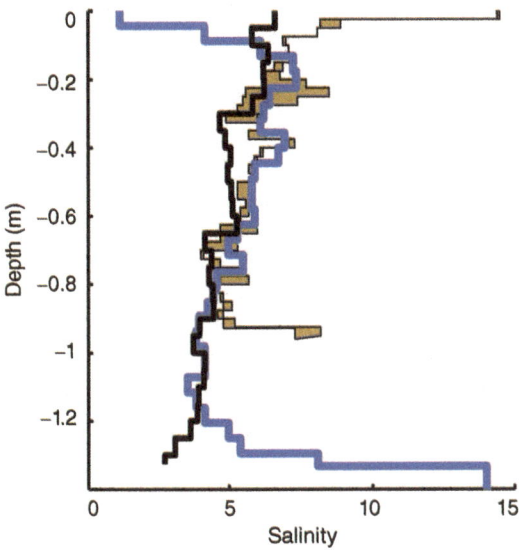

Figure 1.13 Bulk salinity profiles measured in Barrow, Alaska, at the end of May (thick line) and two profiles measured within 2 m in early February (thin lines, difference between profiles is shaded). The solid black line is the expectation based on equation (5) with growth rate from Figure 1.22.

pore space during melt (e.g. large brine channels), the permeability–porosity relationship will probably depend on the history of the ice. In fact, fluid flow through very wide brine channels cannot be described as porous medium flow, as momentum conservation of draining water starts to become apparent (cf. Polashenski et al., 2012).

1.3.7 Sea ice salinity parameterizations

Sea ice bulk salinity is a core state variable of sea ice, and most models of sea ice properties depend on it in some form. As it depends on many aspects of the growth environment, early parameterizations were empirical, statistical correlations based on field measurements or laboratory experiments. One of the first complete, numerical models of sea ice bulk salinity development was motivated by ice strength calculations (Cox & Weeks, 1988). One decade later, salinity modelling started to see a renaissance, probably motivated in part by the advent of general circulation models, climate change and associated scientific interest in sea ice.

Bulk salinity parameterizations take advantage of a steady-state bulk salinity ('stable salinity') that is attained after a few days in growing ice. Most parameterizations are either statistical correlations (Kovacs, 1996; Petrich et al., 2006) or are based on curve-fitted equations derived from theoretical considerations (Nakawo & Sinha, 1981 – following an expression after Burton et al., 1953). Both the equation after Burton et al. (1953) and statistical parameterization typically relate steady-state bulk salinity to growth rate.

Recently, a new type of parameterization was suggested that is based on the analytical solution of the mass conservation equation of gravity-driven fluid motion in sea ice (Petrich et al., 2011). To facilitate this, constant growth rate and a highly simplified permeability profile of sea ice were assumed (i.e. finite and zero permeability above and below a specified porosity threshold, respectively). The results compared favourably with 2D CFD simulations and are reported here for their novel character. The steady-state bulk salinity of sea ice, S_{ice}, growing from seawater of salinity, S_0, is:

$$\frac{S_{ice}}{S_0} = \frac{\rho_0}{\rho_i} \phi_c \left(1 + \frac{h}{z_x}\right) \quad (1.5)$$

where ρ_0 and ρ_i are density of seawater and freshwater ice, respectively, ϕ_c is a critical porosity below which sea ice does not desalinate, and h/z_x is the dimensionless thickness of the desalinating layer,

$$\frac{h}{z_x} = \frac{\phi_c}{2} \frac{v}{\gamma_s w_0} \left[-1 + \sqrt{1 + \frac{2(1-\phi_c)}{\phi_c^2} \frac{\gamma_s w_0}{v}}\right] \quad (1.6)$$

where v is the vertical growth rate of the ice, w_0 is a velocity scale defined as

$$w_0 = \frac{\Pi}{\mu} g \frac{\partial \rho}{\partial C} C_0 \quad (1.7)$$

where C_0 is the solute concentration of seawater, g is the acceleration due to gravity, Π the permeability in the layer of porosity larger than ϕ_c, and γ_s an undetermined parameter that has to be fitted. They set $\phi_c = 0.05$ and determined $\gamma_s w_0 = 4.5 \times 10^{-8}$ m s^{-1} from comparison with 2D CFD model simulations ($C_0 = 34$ kg m^{-3}). Figure 1.13 shows correspondence of growth rate and steady-state bulk salinity from equation (1.5) with data obtained at Barrow, Alaska, in 2008. A feature common to statistical correlations, the relationship after Burton et al. (1953) and equation (1.5) is that the salinity–growth rate dependence approximates a power law, where bulk salinity increases with growth rate. While parameterizations can give a first estimate of bulk salinity, they do not allow for the wide range of variable growth conditions that 1D or 2D numerical simulations can cover explicitly.

1.3.8 Phase fractions and microstructural evolution

A phase diagram, such as that shown in Figure 1.7, in itself contains no direct information on the (microscopic) configuration of individual phases within the system, i.e. its microstructure. In the case of growing, natural sea ice, the latter depends on two principal factors, the growth environment and the boundary conditions at the advancing ice–water interface at the time of growth (Figure 1.4), and the *in situ* temperature and chemical composition of the ice horizon under consideration. This second controlling factor leads to a temperature-dependent porosity and is of prime importance for a wide range of ice properties considered in Section 1.4. Owing to the stark contrast in physical properties between ice, brine, salts and gas inclusions, knowledge of their relative volume fraction can provide us with an estimate of the bulk (macroscopic) properties of sea ice. For example, the mechanical strength of sea ice depends strongly on the relative brine and gas volume fraction, because these two phases effectively have no strength.

A specific example of the microstructural evolution of sea ice as a function of temperature is given in Figure 1.8, which shows the relative volume fraction and microstructure of elongated pores in a core sample of columnar sea ice. The data for these images were obtained using magnetic resonance imaging (MRI) of samples at the indicated temperature. This permitted the microstructural evolution of brine-filled pores to be followed without disturbing or destroying the sample. One particular challenge common to most studies of ice microstructure and properties is the strong dependence of the relative liquid pore volume on temperature. Commonly, samples are cooled to temperatures below –20°C immediately after sampling to avoid loss of brine from the sample, which would also strongly affect properties and microstructure. As evident in Figures 1.8 and 1.11, such temperature changes strongly affect the pore microstructure. Hence it requires either special sampling and sample preparation techniques or non-destructive methods, such as MRI or X-ray tomography, to actually obtain insights into the pore microstructure at the *in situ* temperature (Eicken et al., 2000; Pringle et al., 2009).

As the sample is warmed from its *in situ* temperature, at which it was maintained for the entire period after sampling, the brine volume fraction increases as prescribed by the phase relations. At the same time, however, the size, morphology and connectivity of pores evolve, with pores visibly linking up at higher temperatures (–6°C; Figure 1.8). This process of interconnecting pores of reasonable size as the temperature increases above a critical threshold (for a given salinity) is an important aspect of microstructural evolution and of great importance for sea ice transport properties, such as permeability. In a recent study that combines different modelling approaches and high-resolution X-ray microtomography, Golden et al. (2007) have further examined these linkages between pore connectivity and permeability. The X-ray microtomography data clearly show how sheets of brine segregate into disjunct, isolated pores at low temperatures and how they start to link up at higher temperatures (Figure 1.11). It is important to note that these microtomography images and more detailed data on pore morphology and connectivity derived from them (Pringle et al., 2009), illustrate that the classic pore-space model developed by Assur (see the discussion in Section 1.4) is not sufficient for modelling of ice transport properties.

1.3.9 Temperature and salinity as state variables

From the phase relations (as summarized in Figure 1.7), it follows that the relative volume fraction of brine depends solely on the ice temperature, T, and its bulk salinity, S_{si}, provided the volume is isothermal and in thermodynamic equilibrium and given the pressure (usually atmospheric) and composition of the brine (usually 'standard' composition). The salinity of the brine, S_b, contained within such an ice volume would similarly be prescribed by the phase relations. Hence, temperature and salinity of the ice are the prime controlling or state variables governing not only the phase fractions but, as outlined below, a whole host of other physical properties. Here, use of the term 'state variable' occurs in a very loose sense, as formulation of a true equation of state for sea ice has been elusive to date. For deeper insight into the problem and recent progress, see IOC et al. (2010).

The thermodynamic coupling between these different variables is a key aspect of sea ice as a geophysical material and a habitat. Any temperature change directly affects the porosity and pore microstructure of the ice, as well as the salinity and chemical composition of the brine. Direct measurement of these properties in the field is difficult. Commonly, the *in situ* brine volume fraction and other properties are derived from the bulk salinity of an ice sample and its *in situ* temperature. The latter can be measured by inserting temperature probes into freshly drilled holes in a core or with sensors frozen into the ice. The bulk salinity is typically obtained by dividing an ice core into sections, melting these in the laboratory and then deriving the salinity from electrolytical conductivity measurements. Ideally, one would also measure the density of a sea ice sample, ρ_{si}, in order to determine the air volume fraction V_a/V. In fresh ice, this is typically much smaller than the brine volume fraction, but it can be substantial in multi-year or deteriorated ice, in particular above the water line (Timco & Weeks, 2010).

From the data compiled by Assur (1960) for the phase relations[2] and based on the continuity equations for a multi-phase sea ice mixture, Cox and Weeks (1983) derived a rather useful set of equations describing the brine volume fraction as a function of ice temperature and salinity. Thus, the brine volume fraction is derived as:

$$\frac{V_b}{V} = \left(1 - \frac{V_a}{V}\right) \frac{(\rho_i/1000)S_{si}}{F_1(T) - (\rho_i/1000)S_{si}F_2(T)} \quad (1.8)$$

The density of pure ice is given as:

$$\rho_i = 917 - 0.1403 T \quad (1.9)$$

with ρ_i in kg m^{-3} and T in °C. $F_1(T)$ and $F_2(T)$ are empirical polynomial functions $F_i(T) = a_i + b_i T + c_i T^2 + d_i T^3$, based on the phase relations. The coefficients for different temperature intervals are listed in Table 1.1. The brine salinity and density can be approximated for temperatures above −23°C as:

$$S_b = F_3(T) \quad (1.10)$$

$F_3(T)$ is an empirical polynomial function $F_3(T) = a_3 + b_3 T + c_3 T^2 + d_3 T^3$, based on the phase relations.

The relationships in equations (1.8) and (1.10) give rise to the 'rule of fives' (Golden et al., 1998), which states that in first-year sea ice that formed from ocean water (rather than less saline brackish water) with bulk salinity around $S_{si} = 5$, the porosity is 5% when the ice temperature is near $T = -5°C$. The phase relationship (equation 1.8) is illustrated in Figure 1.14 in the absence of air. Note that the porosity of cold ice is generally small. The brine density ρ_b (in kg m^{-3}) depends on brine salinity S_b according to:

$$\rho_b = 1000 + 0.8 S_b \quad (1.11)$$

Equations (1.8)–(1.11) thus provide us with a simple tool to derive key quantities of importance for a wide

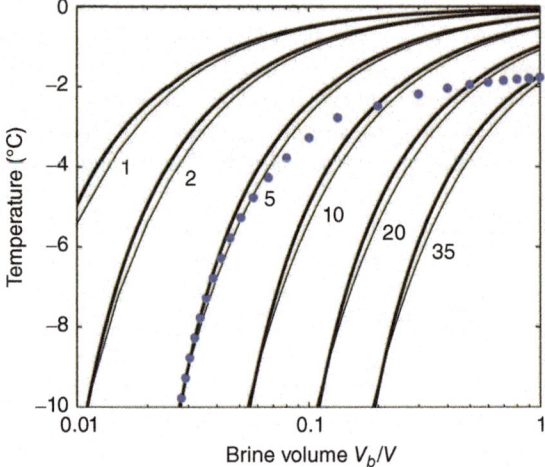

Figure 1.14 Relationship between brine volume fraction and temperature T for various values of bulk sea ice salinity S_{si}. The thick line is after Cox and Weeks (1983), and the thin line follows the linearized relationship $V_b/V = m S_{si}/T$, with $m = -0.054$ K^{-1}. The dots are an artist's impression of a possible desalination pathway in growing sea ice.

Table 1.1 Coefficients for functions $F_1(T)$, $F_2(T)$ and $F_3(T)$ for different temperature intervals. Source: From Cox & Weeks (1983), Leppäranta & Manninen (1988) and Cox & Weeks (1986).

T (°C)	a_1	b_1	c_1	d_1
$0 \geq T > -2$	−0.041 221	−18.407	0.584 02	0.214 54
$-2 \geq T \geq -22.9$	−4.732	−22.45	−0.6397	−0.010 74
$-22.9 > T \geq -30$	9899	1309	55.27	0.7160

T (°C)	a_2	b_2	c_2	d_2
$0 \geq T > -2$	0.090 312	−0.016 111	1.2291×10^{-4}	1.3603×10^{-4}
$-2 \geq T \geq -22.9$	0.089 03	−0.017 63	-5.330×10^{-4}	-8.801×10^{-6}
$-22.9 > T \geq -30$	8.547	1.089	0.045 18	5.819×10^{-4}

T (°C)	a_3	b_3	c_3	d_3
$0 \geq T > -2$	−0.031 6891	−18.3801	0.327 828	0.213 819
$-2 \geq T \geq -22.9$	−3.9921	−22.700	−1.0015	−0.019 956
$-22.9 > T \geq -30$	206.24	−1.8907	−0.060 868	−0.001 0247

range of physical, biological and chemical studies of sea ice.

1.3.10 Caveats and limitations

In arguing the case for the overriding importance of temperature and salinity in determining ice properties, microstructural studies and geochemical analysis indicate that chemical fractionation cannot always be ignored. Consequently, microscopically the fluid composition in sea ice may deviate significantly from its bulk composition, impacting phase evolution, pore microstructure and dependent properties. Such deviations from the evolution of a microscopically homogeneous system are potentially also of significance for biological and chemical processes within the ice cover. Improvements in analytical techniques, with development of *in situ* earth magnetic field nuclear magnetic resonance (NMR) spectroscopy as a particularly interesting example (Callaghan et al., 1999) and advances in geochemical modelling of aqueous solutions at low temperatures (Marion & Grant, 1997; Marion & Kargel, 2008), hold considerable promise in resolving some of these issues. Geochemical modelling is particularly useful, as it allows us to easily treat systems of different chemical composition, whereas the empirical approach outlined earlier is only valid for a single system of a given, standard composition.

1.4 Physical properties

1.4.1 Thermal conductivity

The thermal properties control the magnitude of heat flow through the ice cover and determine its rate of response to variations in surface or bottom forcing. The conductive heat flux through an ice cover, F_c, is given by the product of sea ice thermal conductivity, λ_{si} and the temperature gradient, dT/dz

$$F_c(z) = -\lambda_{si}\left(\frac{dT}{dz}\right)_z \quad (1.12)$$

The thermal conductivity of pure ice as a function of temperature is given by Yen et al. (1991) as:

$$\lambda_i = 1.16 \text{ W m}^{-1}\text{K}^{-1} \times (1.91 - 8.66 \times 10^{-3}T + 2.97 \times 10^{-5}T^2) \quad (1.13)$$

with T in °C. At 0°C λ_i is approximately 2.0 W m^{-1} K^{-1}. The thermal conductivity of brine, on the other hand, is lower by a factor of roughly 4 and can be approximated by:

$$\lambda_b = 0.4184 \text{ W m}^{-1}\text{K}^{-1}(1.25 + 0.030T + 0.00014T^2) \quad (1.14)$$

with T in °C (Yen et al., 1991). The differences in thermal conductivity of ice and brine (and air, if present) make sea ice bulk thermal conductivity sensitive to porosity. Schwerdtfeger (1963) and Ono (1968) developed a model for the thermal conductivity of sea ice based on the thermal properties of the pure end-members, ice and brine, and assumptions about the ice microstructure. For vertically oriented, parallel lamellar brine inclusions, one arrives at the following dependence of λ_{si} on brine volume (no gas inclusions) in a direction parallel to the lamellae:

$$\lambda_{si} = \lambda_i - (\lambda_i - \lambda_b)\frac{S_i \rho_{si}}{m\rho_b T} \quad (1.15)$$

with sea ice salinity S_i, bulk sea ice and brine density, ρ_{si} and ρ_b, respectively, temperature T (in °C) and $m < 0$ the slope of the phase boundary in the phase diagram (Figure 1.7). Note that the fraction in equation (1.15) is a function of the porosity of the ice (cf. Section 1.3). Hence, in this physical configuration, the magnitude of λ_{si} does not depend on the actual size or distribution of the inclusions but only on their mass fraction. Ono (1968) and Schwerdtfeger (1963) also considered the impact of spherical gas inclusions on λ_{si}. These are typically of minor importance in first-year ice, as gas volume fractions are mostly an order of magnitude lower than those of brine.

Untersteiner (1961) introduced a simple parameterization as a function of ice temperature, T, and salinity, S_{si}, that has been employed in some ice-growth models (Maykut, 1986):

$$\lambda_{si} = \lambda_i + 0.13\frac{S_{si}}{T} \quad (1.16)$$

Both equations (1.14) and (1.16) are functionally identical to the continuum approach in Section 1.3 which postulates for porosity V_b/V in thermodynamic equilibrium:

$$\lambda_{si} = \lambda_i + (\lambda_b - \lambda_i)\frac{V_b}{V} \quad (1.17)$$

Recently, accurate measurements of the thermal conductivity of sea ice have been performed by Pringle et al.

(2007). They give the thermal conductivity as:

$$\lambda_{si} = \frac{\rho_{si}}{\rho_i}\left(2.11 - 0.011T + 0.09\frac{S}{T} - \frac{\rho_{si} - \rho_i}{1000}\right) \text{ W m}^{-1}\text{ K}^{-1} \quad (1.18)$$

with T in °C and ρ_{si} and ρ_i (in kg m^{-3}) the densities of sea ice and pure ice, respectively.

1.4.2 Specific heat capacity and latent heat of fusion and of sea ice

The specific heat capacity describes the relationship between an amount of energy, ΔQ, added to a system of mass M and the temperature change, dT, it experiences:

$$\Delta Q = M c\, dT \quad (1.19)$$

As a temperature change of sea ice is always accompanied by a phase transition at the pore scale, the apparent heat capacity of sea ice, c_{si}, is larger than the heat capacity of pure ice, c_i. Given an expression for the apparent heat capacity of sea ice, temperature changes of sea ice can be calculated conveniently from equation (19). Malmgren (1927) introduced an expression for the sea ice heat capacity as a function of heat capacities of ice, brine and water and latent heat released due to phase transition under the assumption of thermodynamic equilibrium and constant bulk salinity. Ono (1967) simplified the expression of Malmgren by assuming a linear relationship between brine salinity and temperature for $T > -8.2$ °C (the onset of mirabilite precipitation), yet allowed for a temperature dependence of the heat capacity of ice and water. The most significant terms in his result are:

$$c_{si} = c_i + \beta T - m_m L \frac{S_{si}}{T^2} \quad (1.20)$$

where he used $c_i = 2.11$ kJ kg^{-1} K^{-1} for the specific heat capacity of ice at 0°C, $\beta = 7.5$ J kg^{-1} K^{-2} as temperature coefficient of the heat capacity of ice; $L = 333.4$ kJ kg^{-1} for the latent heat of fusion of freshwater; and $m_m = -0.05411$ K for the slope of the liquidus curve. S_{si} is the bulk salinity of sea ice and T the temperature in °C. Previously, Untersteiner (1961) proposed an empirical relationship for the heat capacity of sea ice:

$$c_{si} = 2.11 \text{ kJ kg}^{-1}\text{K}^{-1} + 17.2 \text{ kJ K kg}^{-1}\frac{S_{si}}{T^2} \quad (1.21)$$

which is in line with Ono's (1967) result.

It is instructive to trace the origin of this relationship. The specific heat capacity at constant pressure can be expressed in terms of enthalpy (internal energy) H as:

$$c_{si} = \frac{1}{M}\left(\frac{dH}{dT}\right)_P \quad (1.22)$$

where M is the mass containing H. Including sensible heat of liquid and ice and latent heat of fusion, L:

$$H = M[c_w T f_m + c_i T(1 - f_m) - (1 - f_m)L] \quad (1.23)$$

where c_w and c_i are the specific heat capacity of liquid and ice, respectively (assumed to be temperature-independent) and f_m is the liquid mass fraction in thermodynamic equilibrium:

$$f_m = m_m \frac{S_{si}}{T} \quad (1.24)$$

We keep the bulk sea ice salinity S_{si} fixed, insert equation (1.24) into 1.23 and differentiate following equation (1.22) to obtain:

$$c_{si} = c_i - m_m L \frac{S_{si}}{T^2} \quad (1.25)$$

Equation (1.25) is consistent with the result of both Ono (1967) and Untersteiner (1961); it is curious to see that the heat capacity of sea ice does not explicitly depend on the heat capacity of the liquid. This is due to the assumptions of a linear phase relationship (equation 1.24) and constant bulk salinity, which render the sensible heat term of the liquid (the first term in equation 1.23) independent of temperature, i.e. the heat contained in the (variable) liquid mass does not change during temperature change. We should emphasize that the equations given for the specific heat capacity apply to the case of constant bulk salinity only, i.e. not during initial ice formation near the ice–ocean interface. For the sake of simplicity, assuming the bulk salinity decreased linearly as the ice temperature decreases from -2 to -5 °C, we estimate that the effective heat capacity during initial desalination of sea ice could be twice as high as stated in equations (1.20) and (1.21). Given that the heat capacity depends on bulk salinity and that bulk salinity changes with time, it may be easier to perform an energy balance calculation of growing ice based on the effective latent heat of fusion of sea ice, as discussed later.

As is evident from equation (1.21) and from Figure 1.15 showing measurements of c_{si} as a function

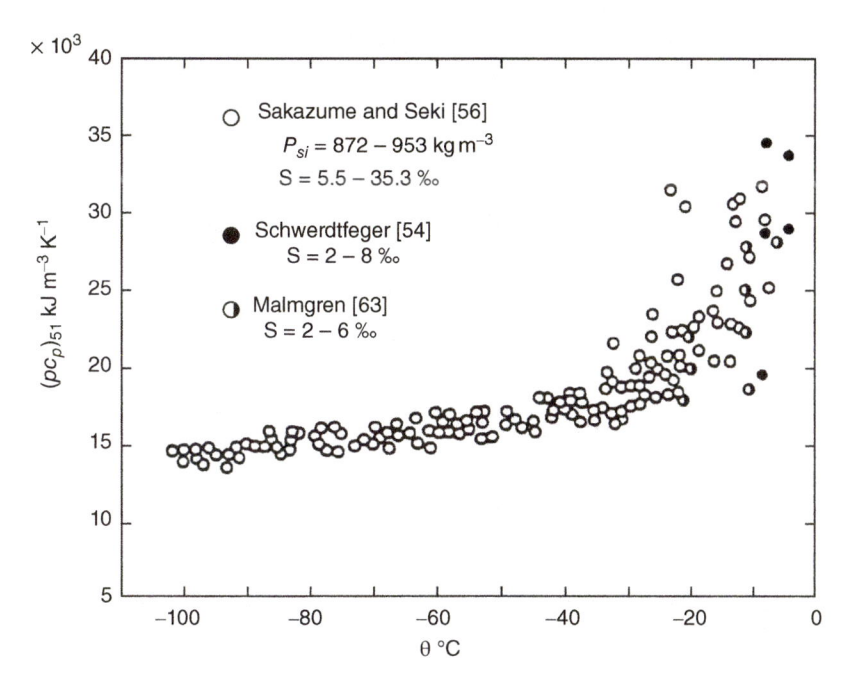

Figure 1.15 Effective specific heat capacity of sea ice c_p (shown as product of c_p with ice density ρ) as a function of temperature θ. Source: Yen et al., 1991.

of temperature, the specific heat capacity increases substantially above about −5°C. This increase implies that it takes substantially more energy to warm an ice cover by 1 K at a temperature close to the freezing point of seawater than at, say, −10°C. The effect of latent heat on temperature change has been found to be crucial for the calculation of the sea ice mass balance in large-scale sea ice or climate models (cf. Vancoppenolle et al., 2009). It also affects the pore space, permeability (and, by extension, albedo), radar signal and optical scattering properties of the ice.

The diffusive propagation of a temperature signal in sea ice can be determined from the heat transfer equation, describing the change in temperature with time dT/dt in terms of c_i, λ_i and ρ_i. In one dimension:

$$\rho_i c_{si} \frac{dT}{dt} = \frac{\partial}{\partial z}\left[\lambda_{si}\frac{\partial T}{\partial z}\right] \quad (1.26)$$

Note that equation (1.26) is energy-conserving, but does not account for density differences between liquid and solid and associated fluid movement. For a medium of homogeneous thermal properties:

$$\frac{dT}{dt} = \frac{\lambda_{si}}{\rho_{si}c_{si}}\frac{\partial^2 T}{\partial z^2} \quad (1.27)$$

The term $(\lambda_{si}/c_{si}\rho_i)$ is commonly referred to as the sea ice thermal diffusivity and describes the rate at which a temperature fluctuation propagates through the ice. It is over an order of magnitude larger in cold ice than it is near the freezing/melting point of sea ice.

The specific heat capacity of sea ice is a useful concept to calculate the temperature change within the ice without the need to consider porosity changes and latent heat explicitly. Similarly, a latent heat of fusion of sea ice can be derived to facilitate the calculation of growth and ablation rates without the need to consider porosity and sensible heat explicitly (Bitz & Lipscomb, 1999; Section 1.5). In this sense, the latent heat of fusion of sea ice can be defined as the temperature integral of the specific heat capacity. This is the difference between the enthalpy of sea ice at a given temperature and salinity and the enthalpy when all the ice is melted at temperature $T_F = m_m S_{si}$. Thus:

$$-L_{si}M = H(T, S_{si}) - H(T_F, S_{si}) \quad (1.28)$$

and using equations (1.23) and (1.24) we find:

$$L_{si} = L - c_i T + c_i m_m S_{si} - m_m L \frac{S_{si}}{T} \quad (1.29)$$

Equation (1.29) contains the dominant terms of the expression derived by Ono (1967). His result is derived from integrating the specific heat capacity of sea ice from temperature T to the melting point of sea ice, keeping bulk salinity S_{si} constant. The most significant terms are:

$$L_{si} = \left(333.4 - 2.11T - 0.114 S_{si} + 18.1 \frac{S_{si}}{T}\right) \text{ kJ kg}^{-1} \quad (1.30)$$

The latent heat of fusion of sea ice takes into account both the reduction of the latent heat due to finite porosity and the transfer of sensible heat between brine and ice. For example, the combined amount of sensible and latent heat required to completely melt sea ice amounts to 246 kJ kg^{-1} at a temperature of $T = -2°$C and a salinity of $S_{si} = 10$.

While equations (1.29) and (1.30) apply only to melting sea ice, the effective latent heat for freezing ice can be obtained similarly if we assume $T_F = m_m S_0$ of the initial liquid of salinity S_0. The result is:

$$L_{si} = L - c_i T + c_i m_m S_{si} - m_m L \frac{S_{si}}{T} + c_w m_m (S_0 - S_{si}) \quad (1.31)$$

Comparing equations (1.29) and (1.31), we find that the effective latent heat of fusion of sea ice is about 2% smaller during ice growth than it is during melt; this is because ocean water of $-2°$C is closer to the temperature of cold ice than the temperature at which all ice (now of bulk salinity below seawater salinity) will be melted. Applications for both sea ice latent heat and heat capacity are given in Section 1.5.

With a liquid phase present in sea ice down to temperatures as low as $-55°$C (Figure 1.7), sea ice is freezing and melting at a microscopic level whenever temperatures change. However, at a macroscopic level, sea ice melt is typically associated with rotting and disintegration of ice floes, largely confined to spring (Petrich et al., 2012a).

1.4.3 Optical properties of sea ice

The pore structure of sea ice has a profound influence on its optical properties, and the physics of absorption and scattering and their relationship with apparent optical properties are described in detail in Chapter 4. Briefly, light is scattered at optical inhomogeneities, in particular air and brine inclusions and precipitated salts (Chapter 4). Light scattered back to the atmosphere gives rise to sea ice albedo (Figure 1.16), while transmitted light is of paramount importance to sea ice microbiota occupying the lower centimetres of Arctic sea ice in spring, sea ice decay and the under-ice light environment. The thin section photographs in Figure 1.2 and the different ice types shown in Figure 1.1 provide some indication of the complexity of the topic of radiative transfer between atmosphere, ice and ocean. Backscatter, absorption and transmission depend on the relative contribution of microscopic absorption and scattering along the path of individual photons (Chapter 4; Perovich, 1998).

While the light field at the underside of sea ice is an important environmental factor for ice and ocean biology, the interpretation of field measurements is challenged by both the anisotropy of the scattering coefficient and the general inhomogeneity of the ice cover. The scattering coefficient of sea ice is highly anisotropic, meaning the probability of scattering is

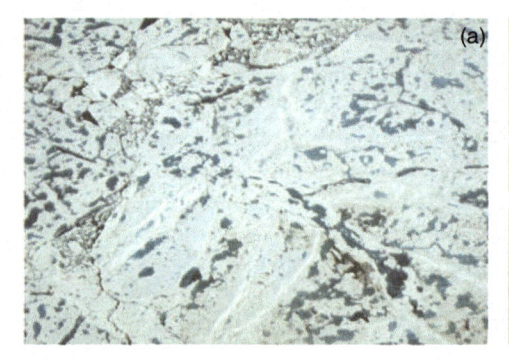

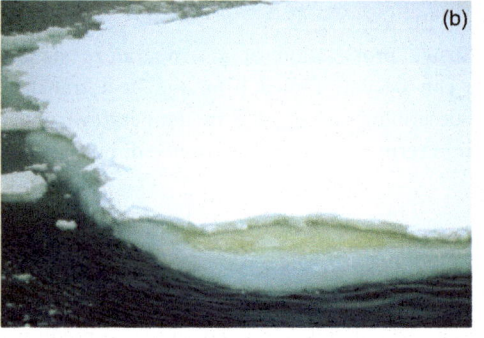

Figure 1.16 (a) Surface of summer Arctic sea ice, aerial photograph, several hundred metres wide. (b) Summer Antarctic sea ice, ship-based photograph, approximately 20 m wide. Note the high albedo of snow-free, well-drained surface ice and the low albedo of the melt ponds in (a). The brown discoloration of the ice visible at freeboard level in (b) is due to high concentrations of sea ice algae in an infiltration layer (Chapter 14).

lower for light travelling vertically through sea ice than for light travelling horizontally (Buckley & Trodahl, 1987; Katlein et al., 2014). This affects the angular distribution of light transmitted through sea ice and has ramifications for the quantitative interpretation of under-ice light measurements (Katlein et al., 2014). At the same time, sea ice and surface optical properties are inhomogeneous, most apparent in the form of snow drifts and melt ponds (Chapter 4).

However, rules on light propagation can be derived that give guidance as to where or how transmission measurements should be performed. For example, under overcast conditions and in a spectral range of low absorption (i.e. blue and green), 10% and 90% of the flux detected under ice is incident on the surface within a radius of less than 0.3 and two times the ice thickness, respectively (Petrich et al., 2012b). Hence, if surface and ice conditions of 1 m thick ice are homogeneous over a radius of 2 m, ice and surface can be treated as homogeneous for the purpose of light transmission measurements. Conversely, a strong surface perturbation of 0.3 m radius directly above the point of measurement will alter the measurements by 10% at most. Transmission measurements near the edge of melt ponds have been found to be consistent with Monte Carlo simulations (Ehn et al., 2011; Petrich et al. 2012b).

1.4.4 Dielectric properties of sea ice

The dielectric properties of sea ice govern the propagation and attenuation of electromagnetic waves which in turn determine both the optical properties of sea ice and sea ice signatures in remote-sensing data sets (Chapter 9). For a detailed treatment of electromagnetic wave propagation in lossy dielectric media, see Schanda (1986) and Hallikainen & Winebrenner (1992).

For an electromagnetic wave propagating in z-direction at the time t, the electric field, E_x, in the x-direction is given by:

$$E_x = E_0 \cos(\omega t - kz) \quad (1.32)$$

where E_0 is the field at $z = 0$, k is the wave number and ω the angular frequency. Maxwell's equations yield the speed of wave propagation (v) in a medium with the relative electric permittivity, ε, and relative magnetic permeability, μ, as:

$$v = \frac{c}{\sqrt{\varepsilon \mu}} \quad (1.33)$$

with the speed of light in terms of the permittivity and permeability of free space:

$$c = \frac{1}{\sqrt{\varepsilon_0 \mu_0}} \quad (1.34)$$

The relative (dimensionless) permittivity ε, also referred to as the dielectric constant, is a complex variable for a medium in which electromagnetic waves are absorbed (a so-called lossy medium), such that

$$\varepsilon = \varepsilon' - i\varepsilon'' = \varepsilon'(1 - i \tan \delta), \quad (1.35)$$

with i denoting the complex part of the permittivity and $i^2 = -1$; ε' describes the contrast with respect to free space ($\varepsilon'_{air} = 1$) and $\tan \delta$ is the so-called loss tangent.

In a non-polar medium, ε' and ε'' are constants, whereas in substances in which the molecules exhibit a permanent dipole moment (e.g. water), ε' and ε'' are frequency-dependent because resonance affects wave decay. Ionic impurities present in seawater and brine also affect the propagation of electromagnetic waves, such that ε'' depends on the conductivity σ:

$$\varepsilon'' = \frac{\sigma}{\varepsilon_0 \omega} \quad (1.36)$$

(Figure 1.17). Ignoring scattering and assuming a homogeneous medium, one can assess the penetration depth, δ_p, of electromagnetic radiation of a given frequency as

$$\delta_p = \frac{1}{\kappa_a} = \frac{\sqrt{\varepsilon'}}{k_0 \varepsilon''} \quad \text{for } \varepsilon' \ll \varepsilon'' \quad (1.37)$$

The attenuation of electromagnetic waves can be described in simplified terms by an extinction coefficient κ, with separate contributions from absorption, κ_a, and scattering, κ_s, similar to the sea ice optical properties. In particular in the single scattering approximation, $\kappa = \kappa_a + \kappa_s$. Despite its gross simplification of radiative transfer, the concept of a penetration depth can help in the interpretation of remote-sensing data, such as radar or passive microwave imagery (Chapter 9). Thus, it is of particular relevance for remote sensing of the oceans and sea ice that the dielectric loss factor ε'' of water varies by more than one order of magnitude in the 1–100 GHz frequency range. Furthermore, at frequencies < 10 GHz, ε'' differs by as much as one order of magnitude for freshwater and seawater or brine, due to the impact of ionic impurities on wave decay. Hence, penetration depths at typical radar or

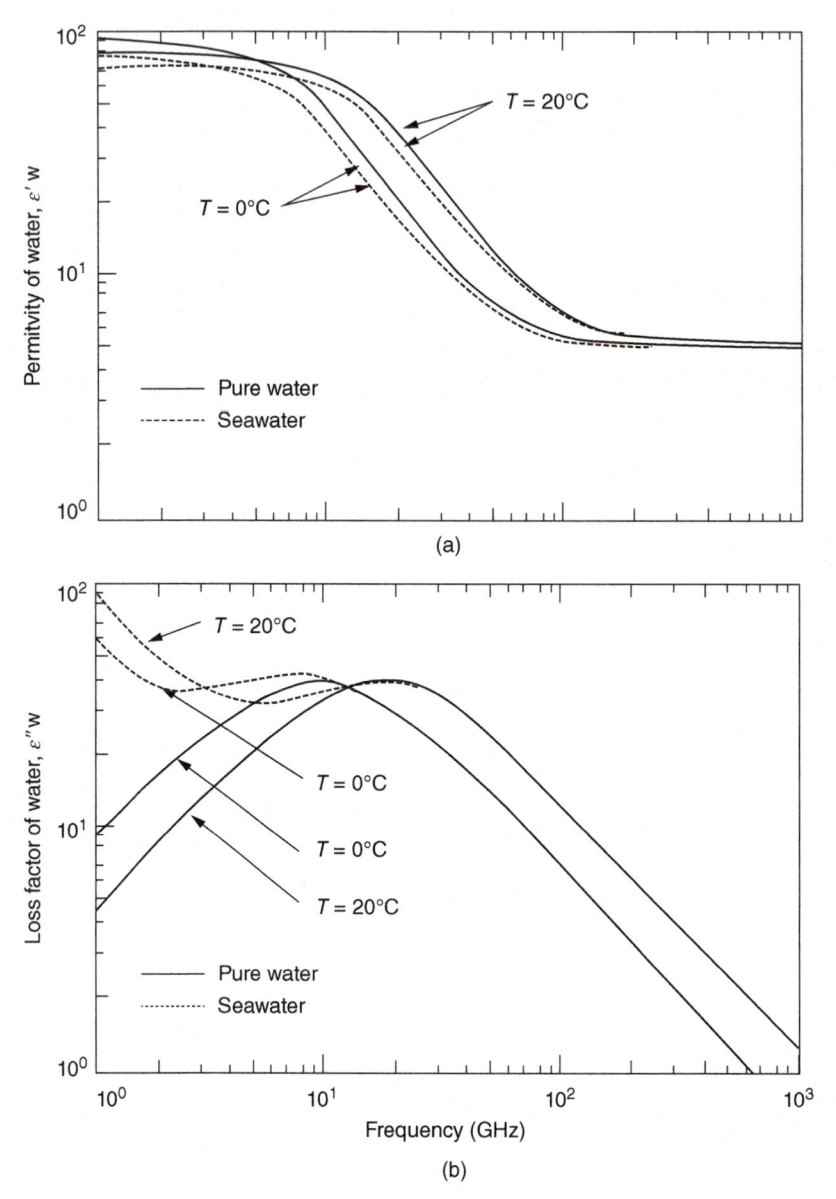

Figure 1.17 Permittivity (a) and loss factor (b) for pure water and seawater at microwave frequencies. Source: Hallikainen & Winebrenner, 1992. Reproduced with permission of John Wiley & Sons.

passive-microwave frequencies are in the order of millimetres to centimetres in seawater. For pure, cold ice ε' is constant at 3.17 for frequencies between 10 MHz and 1000 GHz, whereas ε'' is up to several orders of magnitude smaller than that of water. Consequently, one can derive the dielectric properties of sea ice from the properties of its constituent phases based on an underlying microstructural model (Stogryn, 1987), specifying the shape, size and distribution of brine inclusions as well as their volume fraction. Rather than describing ice microstructure in highly idealized terms such as the Stogryn model, Lin et al. (1988) have derived the emissivity of sea ice at microwave frequencies based on a stochastic representation of sea ice microstructure. In this approach, the sub-parallel arrangement of brine layers within individual ice crystals (Figures 1.4 and 1.5) is rendered in terms of the autocorrelation length of the two-dimensional autocorrelation function.

In simpler, empirical approaches, ε has been parameterized as a function of the fractional brine volume (Hallikainen & Winebrenner, 1992), which in turn is determined by the temperature and salinity of the ice (Section 1.3). The extent to which temperature and ice salinity dominate the penetration depth is illustrated in Figure 1.18. At temperatures of around −5°C, radar waves can thus penetrate several tens of centimetres to more than a metre into low-salinity multi-year ice, whereas penetration depths in saline first-year ice are only on the order of a few centimetres at most. Under such conditions, the radar backscatter signatures of multi-year are dominated by volume scattering from gas bubbles and brine inclusions. In first-year and young sea ice, ice surface scattering predominates and radar backscatter coefficients are mostly determined by the surface roughness of the ice. This allows for discrimination of generally smooth first-year ice and multi-year sea ice in radar satellite imagery. Owing to the polarization and frequency dependence of the dielectric properties (Figure 1.18), more sophisticated multi-frequency and multi-polarization instruments can be of considerable use in distinguishing between different ice types. The same principles apply to thermal emission at microwave frequencies from different ice types and form the basis for commonly employed algorithms in discriminating between first- and multi-year sea ice in passive-microwave data (Chapter 9).

1.4.5 Macroscopic ice strength

Ice strength – at least on the macroscopic level as manifest, for example, in a uniaxial compressive strength test commonly employed in studying the mechanical properties of sea ice (Richter-Menge, 1992) – is typically defined as the peak stress, σ_{max}, sustained by a sample when a load is applied. In simplified terms, σ corresponds to the magnitude of the force, F, applied per unit area, A:

$$\sigma = \frac{F}{A} \quad (1.38)$$

For an ideal elastic response, the strain ε resulting from a given stress σ is proportional to σ, with Young's modulus E as the proportionality constant:

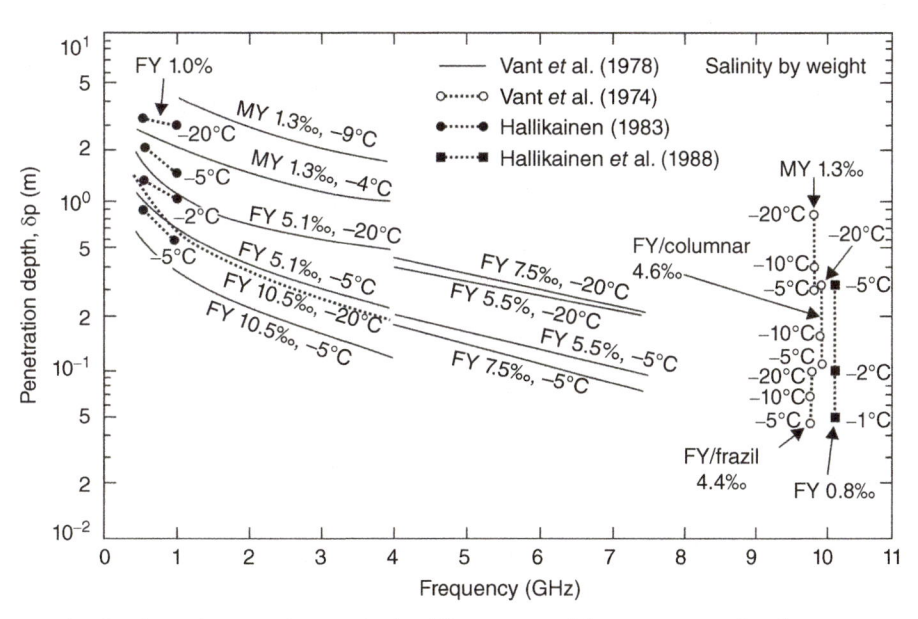

$$\sigma = E\varepsilon \quad (1.39)$$

Figure 1.18 Penetration depth at microwave frequencies for different types of first-year (FY) and multi-year (MY) sea ice. Source: Hallikainen & Winebrenner, 1992. Reproduced with permission of John Wiley & Sons.

Here, strain $\varepsilon = (l - l_0)/l_0$ is defined as the relative change in linear dimension that a body experiences as it is stretched or compressed from l_0 to l. For elastic deformation, the body fully recovers from a given strain in the way an ideal spring would (Figure 1.19a). Upon reaching the yield stress σ_{max}, the sample fails catastrophically, by releasing the stored elastic energy mostly in the form of kinetic energy and sound. This process is referred to as brittle failure. The same mode of failure is experienced by a glass that shatters as it is dropped to the floor. Unlike glass, however, sea ice typically responds to a given stress not only through elastic but also through viscous (non-recoverable) deformation. For a Newtonian liquid, the latter is characterized by a linear relation between stress, σ, and strain rate, $d\varepsilon/dt$, with the viscosity of the medium η as the proportionality constant:

$$\sigma = \eta \frac{d\varepsilon}{dt} \tag{1.40}$$

Hence, deformation of natural sea ice is often best described by a material model that takes into account both elastic and viscous strain components (Mellor, 1986). Equation (1.40) implies a connection between the magnitude of the stress and the strain rate. In fact, as evident from Figure 1.19(a)–(d), the deformation of natural sea ice samples at different strain rates conforms with such a visco-elastic model, with the peak stress σ_{max} reduced by a factor of 3 as the strain rate is dropped from 10^{-2} to 10^{-5} s^{-1}. Also note that at smaller strain rates (Figure 1.19b–d), the ice exhibits ductile failure and is still capable of supporting a load after attaining σ_{max}, in contrast with the brittle failure shown in Figure 1.19(a).

It requires a highly sophisticated approach to translate these results from laboratory experiments into the real world, in which the load could be applied by a person stepping on fairly thin, young ice, with the strain rate dictated by the speed at which the foot is set down on the ice surface. Nevertheless, based on this knowledge one could devise a strategy of how to best cross a very thin, young sea ice cover if one absolutely had to. First, one would not want to apply the full load fast enough to induce catastrophic brittle failure, as this would result in the person breaking through the ice without much of a warning (i.e. prior deflection of the ice sheet; see Figure 1.19a). If strain rates were small enough to induce deformation in a ductile mode, then one would experience the ice giving way underneath after initially appearing quite strong (at least for strain rates corresponding to the transitional brittle–ductile regime characteristic of Figure 1.19b) and, as long as the load would be reduced fast enough, the ice sheet would not fail (i.e. break).

Contrived though this example may appear, polar bears venturing out on nilas (≤10 cm thick) have been observed to make intuitive use of the physics underlying such different sea ice failure modes. Thus, in ambling across the thin ice, the bear sets down each foot just long enough for the peak stress to be surpassed (at the associated strain rates in the order of a few seconds at most) and then lifts it again to place it ahead on yet unstrained ice. Indigenous hunters, such as the Iñupiat of northern Alaska, employ a similar strategy in crossing stretches of ice that would never support a person's weight if stationary for more than a few seconds at most. As the thin, warm ice flexes and is depressed below sea level, brine and seawater forced upwards through the ice onto the surface provide a visible indicator of the amount of accumulated strain.

The deformation process itself and the accommodation of strain on a microscopic level can be quite complicated. Thus, the solid ice matrix experiences deformation of atomic bonds, slip along microcracks, movement of grain boundaries and other processes (Schulson, 1999). It is the combination of all of these individual, microscopic processes that determines the macroscopic response of the ice cover. Hence, the microstructure of the ice plays an important role in determining both ice strength and mode of failure (Schulson, 1999). The contrasting properties of lake and sea ice have already been considered in Section 1.1 and similar contrasts hold for ice deformation. For typical stresses and strain rates, the response of lake ice to an imposed loading is almost exclusively elastic, with failure occurring catastrophically in a brittle mode. This gives no advance warning to anybody who is about to break through the ice, as the yield stress coincides with the complete failure of the material, similar to breaking glass. On the other hand, the lack of brine inclusions, which do not contribute to the overall strength of the material, is responsible for a somewhat higher strength of thin lake ice as compared with sea ice, which in the case of nilas can contain as much as 20% liquid-filled pores.

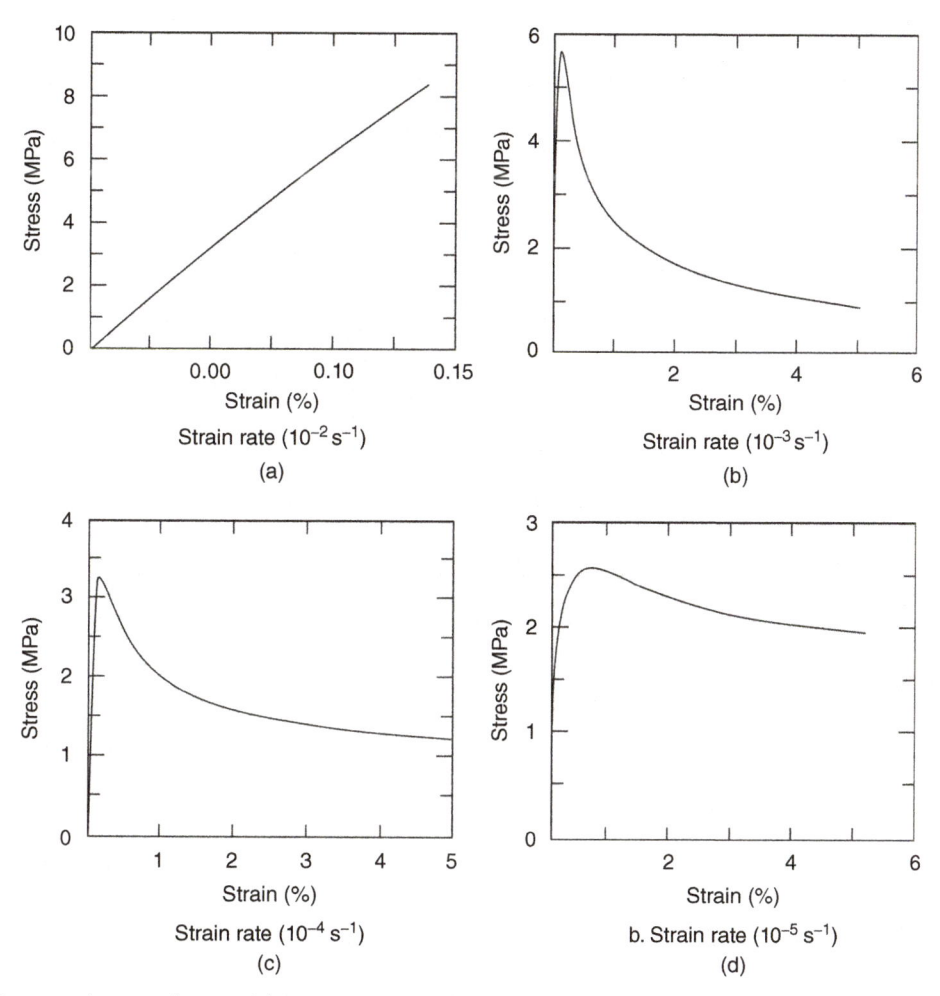

Figure 1.19 Stress–strain curves for uniaxial compression tests of granular sea ice carried out at −10°C at different strain rates. Source: Richter-Menge, 1992; see text for details.

The actual macroscopic strength (i.e. the yield stress σ_{max}; see Figure 1.19) of a sample or volume of sea ice is to a large extent controlled by the volume fraction of gas, V_a/V, and brine, V_b/V, both of which do not contribute to the mechanical strength of the material. As shown in an analysis by Assur (1960) (see summary in Weeks & Ackley, 1986) it is thus the ice cross-sectional area $(1-\psi)$ (with ψ referred to as the 'plane porosity') that determines the magnitude of the *in situ* stress and hence the macroscopic yield stress, σ_{max}, such that:

$$\sigma_f = (1-\psi)\sigma_0 \qquad (1.41)$$

where σ_0 would correspond to the (temperature-dependent) strength of ice with all characteristics of sea ice but zero porosity. Based on Assur's (1960) pore model (Figure 1.20), one can now derive the cross-sectional area of brine inclusions (assuming $V_a/V = 0$) in the vertical plane along which an ice sheet will typically fail under natural conditions. As shown in detail by Weeks and Ackley (1986), assuming the brine cell geometry outlined in Figure 1.20, an expression can be derived for ψ in terms of the platelet spacing, a_0, the minor and major radii of an ellipsoidal inclusion, r_a and r_b, spaced b_0 apart in the direction of the ice lamellae and a vertical separation of inclusions of length g by g_0

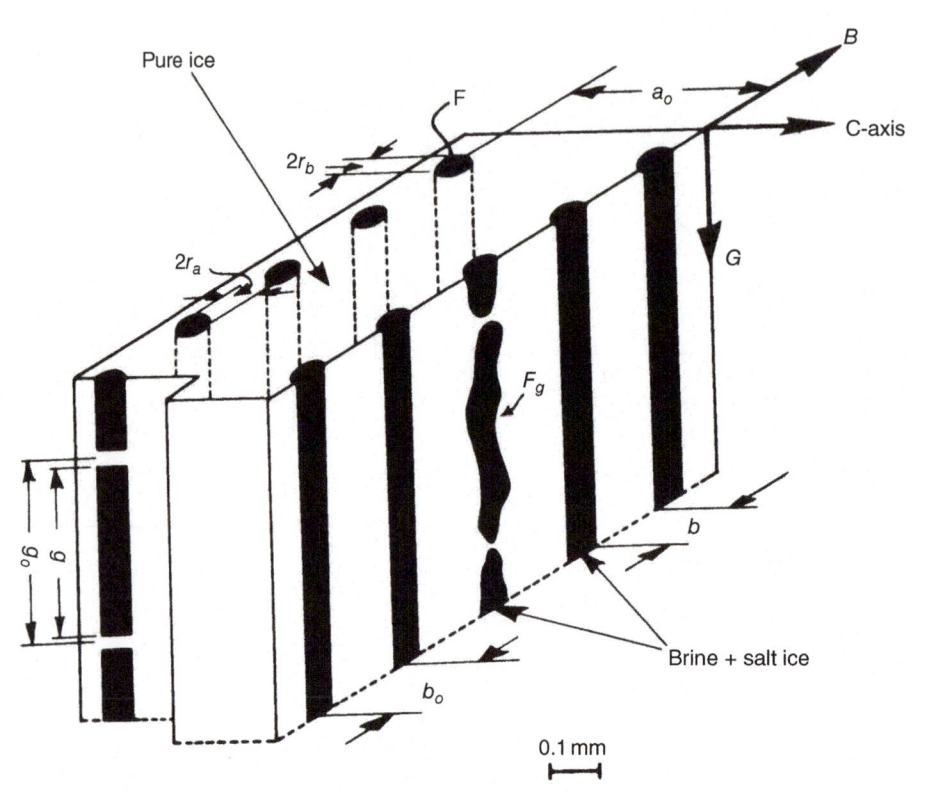

Figure 1.20 Pore microstructural model by Assur (1960).

such that equation (1.41) can be evaluated as:

$$\sigma_f = \left(1 - 2\sqrt{\frac{r_b}{r_a}\frac{g}{g_0}\frac{a_0}{b_0}\frac{V_b}{V}}\right)\sigma_0 \quad (1.42)$$

A model of this form is difficult to evaluate in practice, as both brine volume and pore geometry depend on temperature and history of ice (in particular, melt season vs growth season). In fact, the development of the pore space is far away from trivial (Figure 1.11; Section 1.3). In addition, strength is also highly dependent on the nucleation and propagation of cracks that ultimately lead to failure. The latter depend strongly on other microstructural parameters, such as grain size or preferred alignment of crystals (Schulson, 1999). Empirical relationships are used in practice due to the significant influence of a large number of usually unknown ice properties. For example, Timco and O'Brian (1994) obtained the following equation for flexural strength of cold, growing first-year ice

$$\sigma_f = 1.76 \text{ MPa} \, e^{-5.88\sqrt{V_b/V}} \quad (1.43)$$

On the scale of an individual ice floe, the bending of an ice sheet results in a vertical stress/strain distribution such that maximum tensile stress is found at the top of a flexing ice sheet, with σ decreasing to zero in the interior of the ice sheet and maximum compressive stress found at a point opposite the crest in the ice. As sea ice is stronger in compression than tension by a factor of 2–4, the ice typically fails by cracks developing at the outer surface that is in tension with the crack propagating into the interior of the ice sheet. At the same time, it is the colder, less porous and hence stronger layers that determine the overall strength of an ice sheet (Figures 1.6 and 1.10). Ice strength is a property of sea ice that, probably unlike few others, is highly dependent on the scale at which it is being considered. Thus, at the scale of the ice pack, ice strength is mostly only in the order of tens to hundreds of kPa, whereas it is typically above 1 MPa in laboratory-scale experiments. The state of knowledge on the mechanical properties of sea ice has been reviewed by Timco and Weeks (2010).

1.5 Sea ice growth models, deformation and melt

1.5.1 The role of the snow cover in sea ice growth

Although it typically accounts for less than 10% of the total mass of ice in the polar seas, the snow cover plays a major role in the heat budget of sea ice. First, snow albedos are consistently higher than those of bare sea ice (Chapter 4). More importantly, snow is a good insulator, as the thermal conductivity of snow is generally lower than that of sea ice by roughly one order of magnitude (Chapter 3). The low thermal conductivity is determined by both snow texture and density. The air content is large and results in a snow density around 300 kg m^{-3}. However, most of the variability is related to the metamorphic state of the snow layer (Sturm et al., 2002). The bulk thermal conductivity of snow typically ranges between 0.1 and 0.4 W m^{-1} K^{-1}, as compared with roughly 2 W m^{-1} K^{-1} for sea ice (Massom et al., 2001; Sturm et al., 2002). Even a few centimetres of deposition on 0.5 m ice can impede ice growth by reducing the conductive heat flux, F_c (equation 1.12), by as much as 50%. As a result, thinner ice is often observed beneath snow drifts, at least on Arctic landfast ice (cf. Petrich et al., 2012c). At the same time, the sensitivity of the ice cover to changes in the oceanic heat flux increases with snow depth (equation 1.44). Another consequence of snow deposition on sea ice is a reduction in the amount of short-wave radiation entering the ice and underlying water and a temperature increase in the upper sea ice layers. Both have important ramifications for overall ice properties and the role of sea ice as a habitat for microorganisms (Chapters 13–16).

In most cases, snow-covered ice grows to be thinner than snow-free ice. However, once the load of a thicker snow cover is sufficient to depress the ice surface below the sea surface, seawater and brine may percolate either vertically or laterally through the ice cover (Maksym & Jeffries, 2000). Such seawater flooding shifts the locale of ice growth from the bottom of the ice cover to its top surface and the bottom of the snow pack. The ensuing increase in the conductive heat flux, F_c (equation 1.12), allows more ice to grow per unit area and time than would be possible through freezing at the ice bottom.

In the Arctic, snow is rarely deep enough to allow for surface flooding and snow ice formation. In the Antarctic, however, an overall thinner ice cover and higher snow accumulation rates result in widespread occurrence of this phenomenon (Maksym & Markus, 2008), with more than 50% of the surface flooded in some areas (Eicken et al., 1994; Worby et al., 1998). This has important consequences for sea ice remote sensing as it substantially changes the dielectric properties of sea ice and hence its signature in active and passive microwave remote-sensing data sets (Chapter 9). Similarly, sea ice ecology is also strongly affected by this process (Chapter 14; see infiltration layer community in Figure 1.16b).

As the flooded snow refreezes, the microstructural traces of its meteoric origin are often obliterated and it becomes exceedingly difficult to distinguish such snow ice from similarly fine-grained consolidated frazil ice. Here, the stark contrast in the stable-isotope signatures of snow, which is greatly depleted in the heavy isotopes of oxygen (^{18}O) and hydrogen (D), and ice grown from seawater with its stable, undepleted composition can help to determine the total contribution of precipitation (often referred to as meteoric ice) to the total ice thickness (Eicken et al., 1994). Numerous studies in the Southern Ocean have established that snow ice (the frozen mixture of snow and seawater and/or brine) is very common, accounting for between a few per cent to more than 50% of the total ice thickness (Maksym & Markus, 2008). The actual meteoric ice fraction is generally less than 20%. Both small- and large-scale model simulations (Fichefet & Morales Maqueda, 1999; Maksym & Jeffries, 2000; Powell et al., 2005) indicate that snow ice formation is also important on the global scale.

Snow ice formation may also help to prevent the complete removal of sea ice from areas with high oceanic heat fluxes, F_w. For example, in the eastern Weddell Sea, bottom topography and local hydrography result in winter values of F_w in excess of 100 W m^{-2} for extended periods of time (McPhee et al., 1996). In the mid-1970s this region was the site of a vast polynya that persisted for several years (Chapter 8). Considering that such high heat fluxes would halt ice growth and induce bottom melt even in midwinter, one wonders how an ice cover can survive at all in this region. Observations during the ANZFLUX study (McPhee et al., 1996) demonstrated that level ice of several tens of centimetre thickness does in fact melt at rates of several centimetres per day in the area during intervals of high oceanic heat flux. At the surface of the ice, buffered against the oceanic heat by the ice layers below, snow ice growth seems to be able to compensate for the intermittent substantial bottom ice losses.

1.5.2 Energy budget of sea ice during growth

The fact that in the Southern Ocean, level, undeformed ice typically grows to less than 0.7 m thickness within a single year, compared with as much as 1.8 m in the Arctic, leads us to the question what exactly controls the thickness an ice floe can attain through the freezing of seawater to its underside. We will limit ourselves to the discussion of sea ice during the winter growth season in this section, a time when solar short-wave radiation is negligible.

In the case of congelation ice growth, the growth rate is determined by the energy balance at the lower boundary, i.e. the ice bottom. Here, the conductive heat flux out of the interface into the ice, F_c and the oceanic heat flux, F_w, from the underlying water into the interface are balanced by the release or uptake of latent heat, L_{si}, during freezing or melting, i.e. thickness change dH/dt ($dH/dt > 0$ during freezing):

$$-F_c + F_w + \rho_i L_{si} \frac{dH}{dt} = 0 \quad (1.44)$$

where ρ_i is the density of ice.

The sign of a flux is a matter of convention. When considering the energy balance at one interface only, usually all fluxes are defined as either positive or negative if they are directed away from the interface; however, we consider fluxes between two interfaces and within the bulk in this section. For the sake of consistency we define fluxes as positive if heat flows upwards toward the atmosphere.

In the absence of an oceanic heat flux, the ice cover would thicken as long as heat is removed through the ice to the atmosphere. Without radiative transfer of energy into the ice, this is the case as long as the surface temperature of the ice is less than the freezing point of seawater at the lower interface. However, the ocean underlying the ice typically contains a reservoir of heat that is either remnant from solar heating of the mixed layer in summer (Maykut & McPhee, 1995) or due to transfer of heat from deeper water layers. In particular, near ice shelves, the heat flux from the ocean may be negative, giving rise to the growth of platelets at the bottom of the ice. In the Arctic, where the amount of heat transported into the polar basin and entrained into the surface mixed layer from below the halocline is comparatively small, F_w amounts to a few W m^{-2} in most regions (Steele & Flato, 2000). In the North American Arctic, where advection of heat is minimal, the seasonal cycle of F_w is almost exclusively controlled by the absorption of solar short-wave radiation in the upper ocean, which is transferred to the ice bottom later in the season (Maykut & McPhee, 1995).

In the Southern Ocean, ocean heat flow can be in the order of several tens of W m^{-2} (Martinson & Iannuzzi, 1998). As a result, even an ice cover that is cooled substantially from the atmosphere in winter, may only grow to a maximum thickness that is determined by the balance of ocean and conductive heat flow, $F_c = F_w$. Measurements of ice thickness and surface hydrography indicate that this maximum (winter equilibrium) thickness is on the order of 0.5–0.7 m (Wadhams et al., 1987; Martinson & Iannuzzi, 1998). As shown below, such estimates can also be obtained from simple analytical modelling.

In areas where oceanic heat fluxes can episodically increase to several hundred W m^{-2} due to convective exchange with a deeper ocean well above freezing (McPhee et al., 1996), an ice cover can thin significantly, or vanish entirely, by melting from below. Such extended areas where the ocean is ice-free even in midwinter are referred to as polynyas. The vast Weddell Sea polynya of the 1970s is the most prominent example of open water maintained through active melting and heating of the surface ocean (sensible heat polynya). Alternatively, polynyas can also form dynamically, with strong, steady winds pushing ice away from a coastline or stretches of landfast ice (latent heat polynya).

Oceanic heat and heat released from freezing is conducted to the upper surface of the ice cover and ultimately released to the atmosphere. The rate at which heat can be extracted is determined by the energy balance at the upper surface of the ice floe and snow and the ice thermal properties. For a surface at steady temperature and no solar irradiance, conservation of energy requires that the heat fluxes out of and into the surface be balanced:

$$F_L\!\Downarrow + F_L\!\Uparrow + F_s + F_e - F_c + F_m = 0 \quad (1.45)$$

Here, the individual heat flux terms are the incoming long-wave flux, $F_L\!\Downarrow$; the outgoing long-wave flux,

F_L ⇑; the turbulent atmospheric sensible and latent heat fluxes F_s and F_e, respectively; the heat flux due to melting of ice at the surface, F_m (in the Arctic this is typically only relevant during the summer after the ice surface starts to melt); and the conductive heat flux from the interior of the snow or ice, F_c. Equation (1.45) is formulated with positive fluxes transporting heat upwards for consistency within this section. However, often fluxes directed away from the surface are defined as negative by convention; details on the surface energy balance can be found in Maykut (1986) and Steele and Flato (2000).

A detailed discussion of the magnitude of all these terms is given elsewhere (Maykut, 1986; Persson et al., 2002). Over Arctic multi-year ice, the net radiation balance typically does not drop below −50 W m⁻² during winter and has its maximum in July at just over 100 W m⁻². The other fluxes range between a few to a few tens of W m⁻².

The surface and bottom heat balances are coupled through conductive heat exchange F_c with the snow and ice in between. The energy balance in sea ice is for conductive heat transfer and absorption of shortwave radiation:

$$\frac{\partial H}{\partial t} = \rho_i c_{si} \frac{\partial T}{\partial t} = \frac{\partial}{\partial z}\left(\lambda_{si} \frac{\partial T}{\partial z}\right) - \frac{\partial F_{sw}}{\partial z} \quad (1.46)$$

where F_{sw} is the radiative short-wave flux (for the definition of enthalpy H see equation 1.23; the energy balance for snow is similar). In steady state ($dT/dt = 0$) and in the absence of radiative heat input ($dF_{sw}/dz = 0$), the conductive heat flux is independent of depth within sea ice and F_c is equal at both interfaces. Simplified models that couple both interfaces are discussed next.

1.5.3 Simple models of sea ice growth

A rigorous mathematical treatment of the problem of ice growth requires numerical techniques, because the individual terms in the surface energy balance (equation 1.46) depend either directly or indirectly on surface and air temperatures in a non-linear fashion and the thermal properties in equation (1.46) are temperature-dependent. Nevertheless, as demonstrated more than a century ago by Stefan (Leppäranta, 1993), through some simplifications it is actually possible to arrive at fairly accurate predictions of ice growth. Such simple, so-called degree-day models can also help us in understanding key aspects of the heat budget of sea ice and are therefore discussed in more detail below.

The principal aim of ice-growth modelling is to evaluate the growth rate, dH/dt, of sea ice as a function of time. The temperature at the ice–ocean interface is at the freezing point T_w and we assume the oceanic heat flux F_w to be known (Figure 1.21). The conductive heat flux at the bottom of the ice is largely determined by the heat flux to the atmosphere. At the interface with the atmosphere, one finds that net long-wave radiative fluxes can be described as a function of surface and air temperature during overcast. Likewise, the turbulent heat fluxes F_s and F_e depend on air and surface temperature (for a discussion of the latent heat flux F_e, see Andreas et al., 2002). As a first approximation, the net atmospheric flux can be linearized with respect to the temperature difference between air and surface, i.e. the magnitude of the net atmospheric flux F_a increases with temperature difference between surface and air:

$$F_a = -k(T_a - T_s) + F_a^0(T_a, T_s, \ldots), \quad (1.47)$$

where k is an effective heat transfer coefficient between surface and atmosphere and the residual flux F_a^0 is usually set to 0. Owing to the large ratio between latent and sensible heat stored in sea ice, sea ice grows so slowly that the ice temperature profile adjusts to the increasing ice thickness quasi-instantaneously (Carslaw & Jaeger, 1986). Hence, the temperature profile is linear in the absence of solar heating (e.g. in winter) and rapid temperature fluctuations, provided the thermal conductivity λ_{si} is homogeneous in the ice. Temperature profiles in sea ice are indeed linear during much of the growth season. This is illustrated in Figure 1.10. Considering the two-layer system of snow and ice with

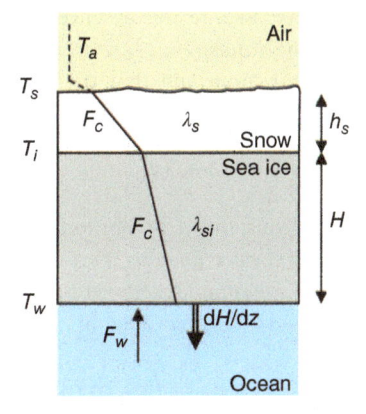

Figure 1.21 Illustration of the two-layer model of sea ice growth.

continuity in temperature and heat flux throughout snow and sea ice (Figure 1.21), Equations (1.12) and (1.47) can be written as:

$$F_a \frac{1}{k} = -(T_a - T_{as}) \quad (1.48)$$

$$F_c \frac{h_s}{\lambda_s} = -(T_{as} - T_{si}) \quad (1.49)$$

$$F_c \frac{H}{\lambda_{si}} = -(T_{si} - T_w) \quad (1.50)$$

where T_{as} and T_{si} are the interface temperatures of air and snow, and snow and ice, respectively, h_s and H are thicknesses of snow cover and sea ice, respectively, and λ_s and λ_{si} are the thermal conductivities of snow and sea ice, respectively. Realizing that $F_a = F_c$ and summing the equations, we find that the net conductive heat flux is:

$$F_a = F_c = -\frac{T_a - T_w}{\frac{1}{k} + \frac{H}{\lambda_{si}} + \frac{h_s}{\lambda_s}} \quad (1.51)$$

since $\lambda_s \approx 0.1\lambda_{si}$ we see that even a layer of snow that is thin with respect to sea ice thickness H can have a profound influence on the heat flux.

Equating (1.51) with the energy balance at the ice bottom (equation 1.44) we find:

$$\frac{dH}{dt}\rho_i L_{si} = -\frac{T_a - T_w}{\frac{1}{k} + \frac{H}{\lambda_{si}} + \frac{h_s}{\lambda_s}} - F_w \quad (1.52)$$

This equation is suitable for numerical modelling of ice thickness if time series of air temperature, snow depth and oceanic heat flux are known. However, further approximations are commonly made to allow ice thickness estimates based solely on the air temperature history: We assume the absence of an oceanic heat flux, that the atmospheric heat transfer coefficient k is constant with time and that the snow depth is proportional to the ice thickness following:

$$h_s = r_s H \quad (1.53)$$

where r_s is the constant of proportionality (note that we would expect surface flooding if r_s exceeded approx. 0.3). Integrating equation (1.52) over time we find that the ice thickness evolves according to:

$$H^2 + \frac{2\lambda_{si}}{k}\left(1 + \frac{\lambda_{si}}{\lambda_s}r_s\right)^{-1} H = \frac{2\lambda_{si}}{\rho_i L_{si}}\left(1 + \frac{\lambda_{si}}{\lambda_s}r_s\right)^{-1} \int -(T_a - T_w) dt \quad (1.54)$$

For the special case of a known and constant surface temperature, i.e. $k \Rightarrow \infty$, constant T_a and absence of snow, i.e. $r_s = 0$, equation (1.54) reduces to the solution of the Stefan problem:

$$H^2 = \frac{2\lambda_{si}(T_w - T_a)}{\rho_i L_{si}}t \quad (1.55)$$

This equation represents the most fundamental growth model of sea ice that states that the thickness of ice increases with the square root of time.

Commonly, the term:

$$\theta = \int_0^{t_e}(T_w - T_a)dt \quad (1.56)$$

in equation (1.54) is computed for discrete time steps $\Delta t = 1$ day; θ is then referred to as freezing degree-days, a variable easily derived from standard meteorological observations. Note that ice thickens approximately linearly with freezing degree-days if the effective heat transfer coefficient k is small and the ice is thin. Reasonable agreement with observations can often be achieved by deriving the factors in equation (1.54) empirically, typically based on ice thickness and air temperature observations at a given location. For example, Anderson (1961) found that:

$$H^2 + 5.1H = 6.7\theta \quad (1.57)$$

where H and θ are in cm and °C days, respectively.

As indicated by the quadratic nature of equation (1.54) (and equation 1.52), the thicker the ice, the lower the growth rate. As a result, differences in thickness between ice floes that started to form at different times during the winter tend to reduce with time. In the absence of ocean (or solar) heat fluxes into the ice and without any summer melt, there would be no limiting thickness for sea ice growth. However, since the oceanic heat flux, F_w, is usually >0, ice thickness has a limiting, maximum value at which conductive heat flux equals oceanic heat flux (equation 1.44) and all ice growth stops. From equations (1.52) and (1.54), this limiting or equilibrium thickness, H_{eq}, is given by:

$$H_{eq} = -\lambda_{si}\frac{T_a - T_w}{F_w}\left(1 + \frac{\lambda_{si}}{\lambda_s}r_s\right)^{-1} \quad (1.58)$$

For typical Antarctic conditions, with $T_a - T_w = -20$°C, $\lambda_{si} = 2.0$ W m^{-1} K^{-1} and $F_w = 20$ W m^{-2}, $\lambda_{si}/\lambda_i = 10$ and $r_s = 0.3$ (i.e. flooding) the limiting thickness is 0.5 m, suggesting that the ocean does in fact limit growth of level ice. In the Arctic F_w is typically smaller by an order of magnitude and the limiting thickness greatly exceeds the amount of ice that can be grown during a single winter.

Figure 1.22 shows a comparison between the Anderson model and measurements. As evident from Figure 1.22, the impact of a snow cover on the ice thickness evolution is substantial. A simple degree-day model following equation (1.54), $H^2 + 25H = 7.3\theta$ ($r_s = 0.08$, $\lambda_{si}/\lambda_i = 6.7$, $k = 10$ W m^{-2} K^{-1}), was used to describe the development of sea ice thickness from the middle of December to June. Varying the parameter r_s, estimates were obtained for the thickness evolution without snow cover ($r_s = 0$) and with a snow cover deep enough to cause flooding ($r_s = 0.3$). Freezing degree-day models have an established place when describing ice growth and break-up but often require location-specific parameters (cf. Petrich et al., 2012a, 2014, and references therein).

Freezing degree-day models imply that ice starting to form earlier in the season will reach greater thickness by the end of the growth season than ice starting to form later in the season. However, this may not be the case if ice formation starts early in the autumn, a time typically chracterized by both high air temperatures and high precipitation (Figure 1.23a). Using a numerical model to simulate ice growth under varing temperature and snow depth conditions, Figure 1.23(b) shows that ice forming at the end of November is expected to grow the thickest at Barrow, Alaska, by 20 May (typical end of the ice growth season). Ice starting to form in October ends up thinner. Using equation (1.5) to predict the steady-state salinity profile, Figure 1.23(b) shows that the thickness-averaged bulk salinity at the end of the growth season is lowest for the ice forming early in the season (i.e. October). As equation (1.5) is not a dynamic model we realize immediately that this finding is not the result of 'older ice having more time to desalinate'. Instead, it is due to slower growth on (a weighted) average.

1.5.4 The importance of ice deformation

Deformation of the ice cover by means of rafting and ridging (Figure 1.1) accounts for the thickest ice observed in the polar oceans and provides a mechanism for thickening even when thermodynamic growth stalls. As a result, deformation processes are particularly important in a warming climate as they may be capable of compensating for some of the thinning due to lesser ice growth and increased ice melt.

First-year pressure ridges can exhibit keels exceeding 10–20 m in depth (Strub-Klein & Sudom, 2012), presenting a particular challenge or even hazard to operations in sea ice-covered waters. When embedded in landfast sea ice and grounded, they stabilize the ice cover into the break-up season (Petrich et al., 2012a). Formation and refreezing of brash ice (ice rubble) as a result of repeat traverses of icebreaker channels in landfast ice enhances ice thickening and generally leads to the undesirable situation of ice being thicker inside the channel than next to it (Sandkvist, 1981). This practical little insight gave rise to the saying: if you break ice, you make ice.

A range of properties and processes of deformed ice differ from level ice: among these are melt-pond development, mechanical strength, ocean and atmospheric drag, and characteristics as a biological habitat. The complex interplay of different deformation processes and their overall impact on the ice thickness distribution are the topic of Chapter 2. Here it suffices to emphasize that both the thermodynamic component of ice growth and melt (potentially including brash ice) and the dynamic component of thickening through rafting or ridging contribute to the characteristics of the sea ice cover.

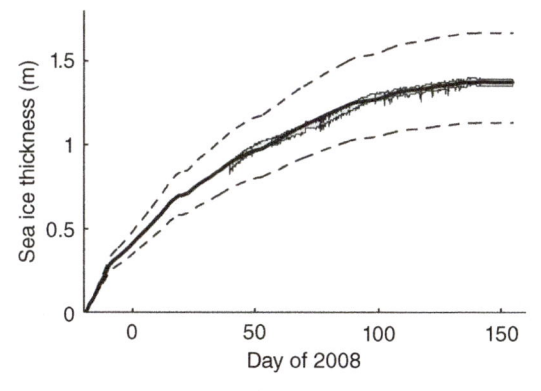

Figure 1.22 Sea ice thickness in Barrow as a function of time based on a degree-day model (thick line) and estimates of thickness development without snow and with heavy snow load (dashed lines; see text for details). Thickness measurements from an acoustic sounder are indicated by a stripe of width 4 cm from day 45 onwards. Air temperatures are based on measurements at 2 m above ground. Source: courtesy Atmospheric Radiation Measurement (ARM) Program.

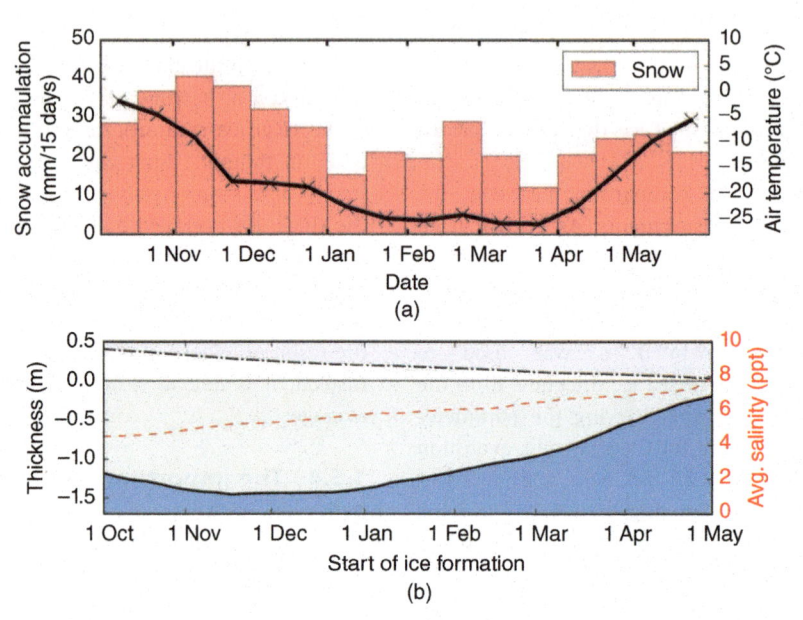

Figure 1.23 (a) Typical development of avaerage air temperature (black line) and precipitation (bars) in recent years at Barrow, Alaska. (b) Dependence of ice characteristics at the end of the growth season (20 May) on the start date of ice formation: modelled ice thickness (solid line), snow depth (dash-dotted line) and depth-averaged bulk salinity (dashed line, red). Ice forming at the end of November will reach the largest thickness at the end of the season and the later the ice starts to form, the higher the average salinity at the end of the season. Note the different scales on the *x*-axes. Plot (b) from Petrich et al. (2011).

1.5.5 Sea ice melt

The melt season is characterized by increasing air temperatures and irradiation. Repeat studies of the melt process of undeformed landfast first-year sea ice at Barrow, Alaska, have been performed following earlier work on the processes leading to pond evolution (Eicken et al., 2002, 2004). A recent series of studies let to detailed descriptions of the processes surrounding melt-pond evolution and sea ice decay (Petrich et al., 2012a,c; Polashenski et al., 2012). Albedo observations of these studies have been summarized and compared with multi-year ice by Perovich & Polashenski (2012). These studies highlighted that a key process (i.e. melt-pond evolution) is, to a significant extent, governed by localized positive feedbacks and spatial inhomogeneity.

Preconditioning of undeformed landfast ice for melt-pond development begins with the development of the snow cover in winter. The locations of snow drifts, once established, remain stationary in spite of episodes of winds and snowfall. On landfast ice at Barrow, the morphology of snow drifts as early as February could be directly related to melt-pond morphology and development in early June. Prior to melt, the snow drifts were, on average, 0.1 m deeper than the average snow cover and separated by some 20 to 30 m. Following a rise in air temperature in May, the thinner patches of snow melted first, starting a positive albedo feedback loop: thin, wet snow and exposed ice are darker and absorb more light than the surrounding thicker snow patches, the ice begins to melt locally and collect melt-water. Meltwater percolating through deeper snow is occasionally observed to form superimposed ice at the ice–snow interface, fostering run-off into adjacent, ponded snow-free areas. Meltwater standing on the surface further increases the heat flux to the surface, promoting surface ablation. By the time the remaining snow has melted, a surface topography has developed, leaving the areas of originally deeper snow at a slightly higher surface elevation than the surrounding ponds. Hence, the decaying ice in these areas is interspersed with air rather than brine inclusions and easily mistaken for snow. The surface meltwater starts to drain preferentially through discrete flaws about 1 week after it first started to appear. As the water is warm, the locations of drainage widen, establishing drainage holes that can reach 10–50 cm in diameter. Surface waters flowing towards these drainage locations contribute to enhanced local surface ablation and establish the outline of a persistent melt-pond pattern. Drainage continues for a few days until most of the surface water has drained to freeboard level. In what follows, the body of ice may either rise or fall with respect to the water level, depending on whether surface ablation or bottom ablation dominates. Ponded ice will rot considerably where particulate matter is present in the ice, locally weakening the integrity of the ice cover until it is easily broken apart by winds. Note that this description applies to undeformed sea ice with a surface that appears flat to visual inspection (except for snow drifts). In the presence of ice surface topography, the surface topography replaces snow drifts in importance (Landy et al., 2014).

A localized weakening of ice beneath melt ponds contrasts with structural weakening of an entire piece of sea ice. The latter is due to a general warming of the ice in spring, even before melt-pond formation. Petrich et al. (2012a) found that identifying the dominant mechanism is crucial to predicting local break-up; disintegration due to melt ponds is related to solar irradiance, while weakening and mechanical break-up due to swell or storms may well be predictable with freezing (or melting) degree-day models.

1.6 Conclusion

This chapter started out with an introduction to the importance of the scale of observation in the study of sea ice. Methodological innovations and advances in our understanding of sea ice as a material and a large-scale phenomenon have now opened the door to studies that are cognizant of both the vast forest and the particulars of each tree. For example, computational limitations have prevented physically realistic representations of sea ice processes in climate models. This is changing now and there is increasing interest and ability by modellers to incorporate the finer details of ice growth and ice properties into these models. Similarly, the increasing sophistication of field research methods and the realization of the importance of key sea ice processes in a global context have prompted a number of investigations that are seeking to bridge the gaps between the microscopic, macroscopic and the regional scales. Much has changed in the world of sea ice in the decade or so that passed between the publishing of the first and third editions of this book. It looks like this trend of increasing interest in the polar sea ice covers is likely to continue.

1.7 Acknowledgements

Chris Petrich acknowledges support of the Norwegian Research Council, project number 195153 (ColdTech). Hajo Eicken is grateful for support from the National Science Foundation, grant OPP-0934683.

Endnotes

1 Malmgren's work and fate may also serve to illustrate how scientific investigation, exploration and the incalculable were intertwined in the early 20th century (and arguably continue to be so to some extent). Finn Malmgren's thorough and ground-breaking study of sea ice properties was prompted by Harald Sverdrup, who had suggested this work as a project to at least gain something out of the unplanned freezing in of Roald Amundsen's vessel *Maud* during the Northeast Passage expedition of 1922–25. Malmgren's experience and scientific prowess predestined him to be one of the key participants in Nobile's ill-fated airship expedition to the North Pole, where Malmgren perished in the ice pack on a trek to reach the Svalbard archipelago.

2 The data assembled by Assur originate mostly from the first half of the 20th century but have been verified to some extent by nuclear magnetic resonance studies of the liquid volume fraction (Richardson, 1976). Nevertheless, some cautions apply.

References

Anderson, D.L. (1961) Growth rate of sea ice. *Journal of Glaciology*, **3**, 1170–1172.

Andreas, E.L., Guest, P.S., Persson, P.O.G. et al. (2002), Near-surface water vapor over polar sea ice is always near ice saturation, *Journal of Geophysical Research*, **107**(C10), 8033, doi:10.1029/2000JC000411.

Arrigo, K.R., Kremer J.N. & Sullivan, C.W. (1993) A simulated Antarctic fast ice ecosystem, *Journal of Geophysical Research*, **98**(C4), 6929–6946.

Assur, A. (1960) Composition of sea ice and its tensile strength. *SIPRE Res. Rep.*, **44**.

Bennington K.O. (1967) Desalination features in natural sea ice. *Journal of Glaciology*, **49**, 845–857.

Bitz, C.M. & Lipscomb, W.H. (1999) An energy-conserving thermodynamic model of sea ice. *Journal of Geophysical Research*, **104**(C7), 15,667–15,677.

Bombosch, A. (1998) Interactions between floating ice platelets and ocean water in the southern Weddell Sea. In: *Ocean, Ice & Atmosphere: Interactions at the Antarctic continental margin* (Ed. S.S. Jacobs) *Antarctic Research Series*, **75**, 257–266.

Buckley, R.G. & Trodahl, H.J. (1987) Scattering and absorption of visible light by sea ice. *Nature*, **326**, 867–869.

Burton, J.A., R.C. Prim, and W.P. Slichter (1953). The distribution of solute in crystals grown from the melt. Pt.I. *Theoretical. Journal of Chemical Physics*, **21**, 1987–1991.

Callaghan, P.T., Dykstra R., Eccles C.D., Haskell T.G. & Seymour, J.D. (1999) A nuclear magnetic resonance study of Antarctic sea ice brine diffusivity. *Cold Regions Science and Technology*, **29**, 153–171.

Carslaw, H.S. & Jaeger, J.C. (1986) *Conduction of Heat in Solids*. Oxford University Press, Oxford.

Cottier, F., Eicken H. & Wadhams, P. (1999) Linkages between salinity and brine channel distribution in young sea ice. *Journal of Geophysical Research*, **104**(C7), 15859–15871.

Cox, G.F.N. & Weeks, W.F. (1983) Equations for determining the gas and brine volumes in sea-ice samples. *Journal of Glaciology*, **29**, 306–316.

Cox, G.F. N. & Weeks, W.F. (1986) Changes in the salinity and porosity of sea-ice samples during shipping and storage, *Journal of Glaciology*, **32**, 371–375.

Cox, G.F.N. & Weeks, W.F. (1988) Numerical simulations of the profile properties of undeformed first-year sea ice during the growth season. *Journal of Geophysical Research*, **93**, 12449–12460.

Curry, J.A., Schramm J.L. & Ebert, E.E. (1995) Sea ice-albedo climate feedback mechanism. *Journal of Climate*, **8**, 240–247.

Dieckmann, G.S., Nehrke, G., Uhlig, C. et al. (2010) Brief Communication: Ikaite ($CaCO_3 \cdot 6H_2O$) discovered in Arctic sea ice. *The Cryosphere*, **4**, 227–230.

Dirakev, S.N., Poyarkov S.G. & Chuvil'chikov, S.I. (2004) Laboratory modeling of small-scale convection under a growing ice cover in winter Arctic leads. *Oceanology*, **44**(1), 62–70.

Drinkwater, M.R. & Xiang, L. (2000) Seasonal to interannual variability in Antarctic sea-ice surface melt. IEEE Transact. *Geoscience and Remote Sensing*, **38**, 1827–1842.

Ehn, J.K., Mundy, C.J., Barber, D.G., Hop, H., Rossnagel, A. & Stewart, J. (2011) Impact of horizontal spreading on light propagation in melt pond covered seasonal sea ice in the Canadian Arctic. *Journal of Geophysical Research*, **116**, C00G02. doi:10.1029/2010JC006908.

Eicken, H. (1998) Factors determining microstructure, salinity and stable-isotope composition of Antarctic sea ice: deriving modes and rates of ice growth in the Weddell Sea. (Antarctic Sea Ice Physical Processes, Interactions and Variability, edited by M.O. Jeffries), *Antarctic Research Series*, **74**, 89–122.

Eicken, H. (1992a) The role of sea ice in structuring Antarctic ecosystems. *Polar Biology*, **12**, 3–13.

Eicken, H. (1992b) Salinity profiles of Antarctic sea ice: field data and model results. *Journal of Geophysical Research*, **97**, 15545–15557.

Eicken, H., Bock C., Wittig R., Miller H. & Poertner, H.-O. (2000) Nuclear magnetic resonance imaging of sea ice pore fluids: Methods and thermal evolution of pore microstructure. *Cold Regions Science and Technology*, **31**, 207–225.

Eicken, H., Grenfell T.C., Perovich D.K., Richter-Menge J.A. & Frey, K. (2004) Hydraulic controls of summer Arctic pack ice albedo. *Journal of Geophysical Research*, **109**, C08007, doi:101029/2003JC001989.

Eicken, H., Krouse H.R., Kadko D. & Perovich, D.K. (2002) Tracer studies of pathways and rates of meltwater transport through Arctic summer sea ice. *Journal of Geophysical Research*, **107**, 10.1029/2000JC000583.

Eicken, H., Lange M.A., Hubberten H.-W. & Wadhams, P. (1994) Characteristics and distribution patterns of snow and meteoric ice in the Weddell Sea and their contribution to the mass balance of sea ice. *Annals of Geophysics*, **12**, 80–93.

Eide, L.I. & Martin, S. (1975) The formation of brine drainage features in young sea ice. *Journal of Glaciology*, **14**(7), 137–154.

Feltham, D.L., Worster M.G. & Wettlaufer, J.S. (2002) The influence of ocean flow on newly forming sea ice. *Journal of Geophysical Research*, **107**(C2), 3009, doi:10.1029/2000JC000559.

Fichefet, T. & Morales Maqueda, M.A. (1999) Modelling the influence of snow accumulation and snow-ice formation on the seasonal cycle of the Antarctic sea-ice cover. *Climate Dynamics*, **15**, 251–268.

Fletcher, N.H. (1970) *The Chemical Physics of Ice*. Cambridge University Press, Cambridge.

Flocco, D., Schroeder D., Feltham D.L. & Hunke, E.C. (2012) Impact of melt ponds on Arctic sea ice simulations from 1990 to 2007. *Journal of Geophysical Research*, **117**, C09032, doi:10.1029/2012JC008195.

Freitag, J. (1999) The hydraulic properties of Arctic sea ice – implications for the small-scale particle transport (in German). *Berichte zur Polarforschung*, **325**, 1–150.

Freitag, J. & Eicken, H. (2003) Meltwater circulation and permeability of Arctic summer sea ice derived from hydrological field experiments. *Journal of Glaciology*, **49**, 349–358.

Golden, K.M., Ackley, S.F. & Lytle, V.I. (1998) The percolation phase transition in sea ice. *Science*, **282**, 2238–2241.

Golden, K.M., Eicken, H., Heaton, A.L., Miner, J., Pringle, D.J. & Zhu, J. (2007) Thermal evolution of permeability and microstructure in sea ice. *Geophysical Research Letters*, **34**, L16501, doi:10.1029/2007GL030447.

Gough, A.J., Mahoney, A.R., Langhorne, P.J., Williams, M.J. M. & Haskell, T.G. (2012) Sea ice salinity and structure: A winter time series of salinity and its distribution. *Journal of Geophysical Research*, **117**, C03008, doi:10.1029/2011JC007527.

Griewank, P.J. & Notz, D. (2013) Insights into brine dynamics and sea ice desalination from a 1-D model study of gravity drainage. *Journal of Geophysical Research Oceans*, **118**, 3370–3386.

Hallikainen, M. & Winebrenner, D.P. (1992) The physical basis for sea ice remote sensing. In: *Microwave Remote Sensing of Sea Ice* (Ed. by F.D. Carsey), pp. 29–46. Geophysical Monograph 68. American Geophysical Union, Washington.

Hobbs, P.V. (1974) *Ice Physics*. Clarendon Press, Oxford.

Hu, Y., Wolf-Gladrow D.A., Dieckmann G.S., Völker C. & Nehrke, G. (2014) A laboratory study of ikaite ($CaCO_3 \cdot 6H_2O$) precipitation as a function of pH, salinity, temperature and phosphate concentration. *Marine Chemistry*, **162**, 10–18.

Hunke, E.C., D. Notz, A.K. Turner, and M. Vancoppenolle (2011). The multiphase physics of sea ice: a review for model developers. *The Cryosphere*, **5**, 989–1009, doi:10.5194/tc-5-989-2011

IOC, SCOR and IAPSO (2010) The international thermodynamic equation of seawater–2010 In: *Calculation and Use of Thermodynamic Properties*. Intergovernmental Oceanographic Commission, Manuals and Guides No. 56. UNESCO (English).

Jeffries, M.O., Shaw R.A., Morris K., Veazey A.L. & Krouse, H.R. (1994) Crystal structure, stable isotopes (d18O) and development of sea ice in the Ross, Amundsen & Bellingshausen seas, Antarctica. *Journal of Geophysical Research*, **99**, 985–995.

Jeffries, M.O., Weeks W.F., Shaw R. & Morris, K. (1993) Structural characteristics of congelation and platelet ice and their role in the development of Antarctic land-fast sea ice, *Journal of Glaciology*, **39**, 223–238.

Katlein, C., Nicolaus, M. & Petrich, C. (2014) The anisotropic scattering coefficient of sea ice. *Journal of Geophysical Research Oceans*, **119**, doi:10.1002/2013JC009502.

Kovacs, A. (1996) Sea ice: Part I. Bulk salinity versus ice floe thickness. *CRREL Report 96-7*.

Krembs, C., Eicken, H. & Deming, J.W. (2011) Exopolymer alteration of physical properties of sea ice and implications for ice habitability and biogeochemistry in a warmer Arctic. *Proceedings of the National Academy of Sciences USA*, **108**, 3653–3658. doi: 10.1073/pnas.1100701108.

Lake, R.A. & Lewis, E.L. (1970) Salt rejection by sea ice during growth. *Journal of Geophysical Research*, **75**, 583–597.

Landy, J., Ehn J., Shields M. & Barber, D. (2014) Surface and melt pond evolution on landfast first-year sea ice in the Canadian Arctic Archipelago. *Journal of Geophysical Research Oceans*, **119**(5), 3054–3075.

Lange, M.A., Ackley S.F., Wadhams P., Dieckmann G.S. & Eicken, H. (1989) Development of sea ice in the Weddell Sea, Antarctica. *Annals of Glaciology*, **12**, 92–96.

Langhorne, P.J. & Robinson, W.H. (1986) Alignment of crystals in sea ice due to fluid motion. *Cold Regions Science and Technology*, **12**(2), 197–215.

Leonard, G.H., Purdie C.R., Langhorne P.J., Haskell T.G., Williams M.J.M. & Frew, R.D. (2006) Observations of platelet ice growth and oceanographic conditions during the winter of 2003 in McMurdo Sound, Antarctica. *Journal of Geophysical Research*, **111**, C04012. doi:10.1029/2005JC002952

Leppäranta, M. (1993) A review of analytical models of sea-ice growth. *Atmosphere-Ocean*, **31**, 123–138.

Leppäranta, M. & Manninen, T. (1988) The brine and gas content of sea ice with attention to low salinities and high temperatures. *Finnish Institute for Marine Research Internal Report*, 88–82, Helsinki.

Light B., Maykut G.A. & Grenfell, T.C. (2003) Effects of temperature on the microstructure of first-year Arctic sea ice, *Journal of Geophysical Research*, **108**(C2), 3051, doi:10.1029/2001JC000887.

Lin, F.C., Kong J.A., Shin R.T., Gow A.J. & Arcone S.A. (1988) Correlation function study for sea ice. *Journal of Geophysical Research*, **93**, 14055–14063.

Maksym, T. & Markus, T. (2008) Antarctic sea ice thickness and snow-to-ice conversion from atmospheric reanalysis and passive microwave snow depth. *Journal of Geophysical Research*, **113**, C02S12, doi:10.1029/2006JC004085.

Maksym, T. & Jeffries, M.O. (2000) A one-dimensional percolation model of flooding and snow ice formation on Antarctic sea ice. *Journal of Geophysical Research*, **105**, 26313–26331.

Malmgren, F. (1927) On the properties of sea-ice. *Norwegian North Polar Expedition "Maud" 1918-1925*, **1**, 1–67.

Marion, G.M. & Kargel, J.S. (2008) *Cold Aqueous Planetary Geochemistry with FREZCHEM: From Modeling to the Search for Life at the Limits*. Springer, Berlin.

Marion, G.M. & Grant, S.A. (1997) Physical chemistry of geochemical solutions at subzero temperatures. *CRREL Special Report*, **97–10**, 349–356.

Martinson, D.G. & Iannuzzi, R.A. (1998) Antarctic ocean-ice interaction: Implications from ocean bulk property distributions in the Weddell Gyre. In: *Antarctic Sea Ice: Physical Processes, Interactions and Variability* (Ed. by M.O. Jeffries). American Geophysical Union, Washington. Antarctic Research Series, **74**, 243–271.

Massom, R.A. et al. (2001) Snow on Antarctic sea ice. *Review of Geophysics*, **39**, 413–445.

Maus, S. & De la Rosa, S. (2012) Salinity and solid fraction of frazil and grease ice. *Journal of Glaciology*, **58**, doi:10.3189/2012JoG11J110.

Maykut, G.A. (1986) The surface heat and mass balance. In: *The Geophysics of Sea Ice*. (Ed. by N. Untersteiner), pp. 395–463. Martinus Nijhoff Publ., Dordrecht (NATO ASI B146).

Maykut, G.A. & McPhee, M.G. (1995) Solar heating of the Arctic mixed layer. *Journal of Geophysical Research*, **100**, 24691–24703.

McPhee, M.G., Ackley, S.F., Guest, P. et al. (1996) The Antarctic Zone Flux Experiment. *Bulletin of the American Meteorological Society*, **77**, 1221–1232.

McPhee, M.G., Stanton, T.P., Morison, J.H. & Martinson, D.G. (1998) Freshening of the upper ocean in the central Arctic:

Is perennial sea ice disappearing? *Geophysical Research Letters*, **25**, 1729–1732.

Mellor, M. (1986) Mechanical behavior of sea ice. In: *The Geophysics of Sea Ice*. (Ed. by N. Untersteiner), pp. 165–281. Martinus Nijhoff Publ., Dordrecht (NATO ASI B146).

Millero, F.J., Feistel, R., Wright, D.G. & McDougall, T.J. (2008) The composition of standard seawater and the definition of the reference-composition salinity scale. *Deep-Sea Research Part* **1**, 55, 50–72.

Nakawo, M. & Sinha, N.K. (1981) Growth rate and salinity profile of first-year sea ice in the high Arctic. *Journal of Glaciology*, **27**, 315–330.

Nakawo, M. & Sinha, N.K. (1984) A note on brine layer spacing of first-year sea ice. *Atmosphere-Ocean*, **22**, 193–206.

Naumann, A.K., Notz, D., Håvik, L. & Sirevaag, A. (2012) Laboratory study of initial sea-ice growth: properties of grease ice and nilas. *The Cryosphere*, **6**, 729–741.

Niedrauer, T.M. & Martin, S. (1979) An experimental study of brine drainage and convection in young sea ice. *Journal of Geophysical Research*, **84**, 1176–1186.

Nield, D.A. & Bejan, A. (1998) *Convection in Porous Media*. Springer-Verlag, NY, USA.

Notz, D., McPhee, M.G., Worster, M.G., Maykut, G.A., Schlünzen, K.H. & Eicken, H. (2003) Impact of underwater-ice evolution on Arctic summer sea ice. *Journal of Geophysical Research*, **108**(C7), 3223, doi:10.1029/2001JC00173.

Notz, D. & Worster, M.G. (2009) Desalination processes of sea ice revisited. *Journal of Geophysical Research*, **114**, C05006, 1–10. doi:10.1029/2008JC004885.

Ono, N. (1967) Specific heat and heat of fusion of sea ice. In: *Physics of Snow and Ice*. (Ed.H. Onura), vol. **1**, 599–610.

Ono, N. (1968) Thermal properties of sea ice: IV, thermal constants of sea ice (in Japanese). *Low Temperature Science Series A*, **26**, 329–349.

Papadimitriou, S., Kennedy, H., Kennedy, D.P. & Thomas, D.N. (2013) Ikaite solubility in seawater-derived brines at 1 atm and sub-zero temperatures to 265 K. *Geochimica et Cosmochimica Acta*, **109**, 241–253

Papadimitriou, S., Kennedy, H., Kennedy, D.P. & Thomas, D.N. (2014) Kinetics of ikaite precipitation and dissolution in seawater-derived brines at sub-zero temperatures to 265 K. *Geochimica et Cosmochimica Acta*, **140**, 199–211.

Perovich, D.K. (1998) Optical properties of sea ice. In: *Physics of ice-covered seas*, vol. **1**. (Ed. by M. Leppäranta), pp. 195–230, University of Helsinki, Helsinki.

Perovich D.K., Grenfell T.C., Richter-Menge J.A., Light B., Tucker III, W.B. & Eicken H. (2003) Thin and thinner: Sea ice mass balance measurements during SHEBA, *Journal of Geophysical Research*, **108**(C3), 8050, doi:10.1029/2001JC001079.

Perovich, D.K. & Gow, A.J. (1996) A quantitative description of sea ice inclusions. *Journal of Geophysical Research*, **101**(C8), 18,327–18,343.

Perovich, D.K., Light, B., Eicken, H., Jones, K.F., Runciman, K. & Nghiem, S.V. (2007) Increasing solar heating of the Arctic Ocean and adjacent seas, 1979–2005: Attribution and role in the ice-albedo feedback, *Geophysical Research Letters*, **34**, L19505, doi:19510.11029/12007/GL031480.

Persson, P.O. G., Fairall C.W., Andreas E.L., Guest P.S. & Perovich, D.K. (2002) Measurements near the Atmospheric surface flux group tower at SHEBA: Near-surface conditions and surface energy budget. *Journal of Geophysical Research*, **107**, doi:10.1029/2000JC000705.

Petrich, C., Langhorne P.J. & Sun, Z.F. (2006) Modelling the interrelationships between permeability, effective porosity and total porosity in sea ice. *Cold Regions Science and Technology*, **44**, 131–144.

Petrich, C., Langhorne P.J. & Haskell, T.G. (2007) Formation and structure of refrozen cracks in land-fast first-year sea ice, *Journal of Geophysical Research*, **112**, C04006, doi:10.1029/2006JC003466.

Petrich, C., Langhorne, P.J. & Eicken, H. (2011) Modelled bulk salinity of growing first-year sea ice and implications for ice properties in spring. In: *Proceedings of the 21st International Conference on Port and Ocean Engineering under Arctic Conditions (POAC), Montreal, Quebec, Canada, 10–14 July 2011*. Paper POAC11-187, 10 pp.

Petrich, C., Eicken, H., Zhang, J., Krieger, J.R., Fukamachi, Y. & Ohshima, K.I. (2012a) Coastal sea ice melt and break-up in northern Alaska: processes and possibility to forecast, *Journal of Geophysical Research*, **117**, C02003, 1–19

Petrich, C., Nicolaus, M. & Gradinger, R. (2012b) Sensitivity of the light field under sea ice to spatially inhomogeneous optical properties and incident light assessed with three-dimensional Monte Carlo radiative transfer simulations. *Cold Regions Science and Technology*, **73**, 1–11.

Petrich, C., Eicken, H., Polashenski, C.M. et al. (2012c) Snow dunes: a controlling factor of melt pond distribution on Arctic sea ice. *Journal of Geophysical Research*, **117**, C09029, 1–10, doi:10.1029/2012JC008192.

Petrich, C., Karlsson, J. & Eicken, H. (2013) Porosity of growing sea ice and potential for oil entrainment. *Cold Regions Science and Technology*, **87**, 27–32.

Petrich, C., Tivy, A.C. & Ward, D.H. (2014) Reconstruction of historic sea ice conditions in a sub-Arctic lagoon. *Cold Regions Science and Technology*, **98**, 55–62.

Polashenski, C., Perovich, D. & Courville, Z. (2012) The mechanisms of sea ice melt pond formation and evolution. *Journal of Geophysical Research*, **117**, C01001, 1–23. doi:10.1029/2011JC007231.

Powell, D.C., Markus, & T., Stössel, A. (2005) Effects of snow depth forcing on Southern Ocean sea ice simulations, *Journal of Geophysical Research*, **110**, C06001, doi:10.1029/2003JC002212.

Pringle, D.J., Eicken H., Trodahl H.J. & Backstrom, L.G.E. (2007) Thermal conductivity of landfast Antarctic and Arctic sea ice. *Journal of Geophysical Research*, **112**, C04017, doi:10.1029/2006JC003641.

Pringle, D.J., Miner J.E., Eicken H. & Golden, K.M. (2009) Pore-space percolation in sea ice single cystals. *Journal of Geophysical Research*, **114**, C12, doi:10.1029/2008JC005145

Rees Jones, D.W. & Worster, M.G. (2013) A simple dynamical model for gravity drainage from growing sea ice. *Geophysical Research Letters*, **40**, 307–311. doi:10.1029/2012GL054301.

Reimnitz, E., Kempema E.W., Weber W.S., Clayton J.R. & Payne, J.R. (1990) Suspended-matter scavenging by rising frazil ice. *CRREL Monographs*, **90-1**, 97–100.

Richter-Menge, J.A. (1992) Compressive strength of frazil sea ice. *Proceedings IAHR Symposium on Ice, Banff, Alberta*, **2**, 1065–1074.

Rothrock, D.A., Yu Y. & Maykut, G.A. (1999) Thinning of the Arctic sea-ice cover. *Geophysical Research Letters*, **26**, 3469–3472.

Sandkvist, J. (1981) Conditions in brash ice covered channels with repeated passages. In: *Proceedings of the 6th International Conference on Port and Ocean Engineering under Arctic Conditions (POAC), Québec City, Canada, 27–31 July 1981*, pp. 244–252.

Schanda, E. (1986) *Physical Fundamentals of Remote Sensing*. Springer-Verlag, Berlin.

Schulson E.M. (1999) The structure and mechanical behavior of ice. *JOM*, **51**, 21–27.

Schwerdtfeger, P. (1963) The thermal properties of sea ice. *Journal of Glaciology*, **4**, 789–807.

Smedsrud, L.H. (2001) Frazil-ice entrainment of sediment: large-tank laboratory experiments. *Journal of Glaciology*, **47**, 461–471.

Smith, R.E.H., Harrison W.G., Harris L.R. & Herman A.W. (1990) Vertical fine structure of particulate matter and nutrients in sea ice of the high Arctic. *Canadian Journal of Fisheries and Aquatic Science*, **47**, 1348–1355.

Steele, M. & Flato, G.M. (2000) Sea ice growth, melt and modeling: a survey. In: *The Freshwater Budget of the Arctic Ocean* (Ed. by E.L. Lewis), pp. 549–587. Kluwer Academic Publishers, Dordrecht.

Stogryn, A. (1987) An analysis of the tensor dielectric constant of sea ice at microwave frequencies. IEEE Transact. *Geoscience and Remote Sensing*, **GE-25**, 147–158.

Strub-Klein, L. & Sudom, D. (2012) A comprehensive analysis of the morphology of first-year sea ice ridges. *Cold Regions Science and Technology*, **82**, 94–109.

Sturm, M., Perovich, D.K. & Holmgren J. (2002) Thermal conductivity and heat transfer through the snow on the ice of the Beaufort Sea. *Journal of Geophysical Research*, **107**, doi:10.1029/2000JC000409.

Timco, G.W. & O'Brian, S. (1994) Flexural strength equation for sea ice. *Cold Regions Science and Technology*, **22**(3), 285–298.

Timco, G.W. & Weeks, W.F. (2010) A review of the engineering properties of sea ice. *Cold Regions Science and Technology*, **60**, 107–129.

Tyshko, K.P., Fedotov V.I. & Cherepanov, I.V. (1997) Spatio-temporal variability of the Arctic seas' ice cover structure. In: *Sea ice* (Eds. I.E. Frolov & V.P. Gavrilo), pp. 191–213. Gidrometeoizdat, St. Petersburg.

UNESCO (1978) *Eighth Report of the Joint Panel On Oceanographic Tables and Standards Unesco Technical Papers in Marine Science*, **28**. Unesco, Paris.

Untersteiner, N. (1961) On the mass and heat budget of Arctic sea ice. *Archives for Meteorology, Geophysics and Bioclimatology Series A*, **12**, 151–182.

Untersteiner, N. (1968) Natural desalination and equilibrium salinity profile of perennial sea ice. *Journal of Geophysical Research*, **73**, 1251–1257.

Vancoppenolle, M., Fichefet, T. & Goosse, H. (2009) Simulating the mass balance and salinity of Arctic and Antarctic sea ice. 2. Importance of sea ice salinity variations. *Ocean Modelling*, **27**, 54–69.

Wakatsuchi, M. (1983) Brine exclusion process from growing sea ice. *Low Temperature Science*, **A33**, 29–65.

Wakatsuchi, M. & Ono, N. (1983) Measurements of salinity and volume of brine excluded from growing sea ice. *Journal of Geophysical Research*, **88**(C5), 2943–2951.

Wadhams, P., Lange, M.A. & Ackley, S.F. (1987) The ice thickness distribution across the Atlantic sector of the Antarctic Ocean in midwinter. *Journal of Geophysical Research*, **92**, 14535–14552.

Weeks, W.F. (2010) *On Sea Ice*. University of Alaska Press, Fairbanks, AK, USA.

Weeks, W.F. & Ackley, S.F. (1986) The growth, structure and properties of sea ice. In: *The Geophysics of Sea Ice*. (Ed. N. Untersteiner), pp. 9–164. Plenum Press, New York (NATO ASI B146).

Weeks, W.F. & Lee, O.S. (1962) The salinity distribution in young sea-ice. *Arctic*, **15**, 92–108.

Weeks, W.F. & Wettlaufer, J.S. (1996) Crystal orientations in floating ice sheets. In: *The Johannes Weertman Symposium*. (Eds. R.J. Arsenault et al.), pp. 337–350. The Minerals, Metals and Materials Society, Warrendale, PA.

Wettlaufer, J.S., Worster, M.G. & Huppert, H.E. (1997) Natural convection during solidification of an alloy from above with application to the evolution of sea ice. *Journal of Fluid Mechanics*, **344**, 291–316.

Worby, A.P., Massom, R.A., Allison, I., Lytle, V.I. & Heil, P. (1998) East Antarctic sea ice: a review of its structure, properties and drift. In: *Antarctic Sea Ice Physical Processes, Interactions and Variability* (Ed. M.O. Jeffries). American Geophysical Union, Washington. Antarctic Research Series, **74**, 41–67.

Worster, M.G. & Wettlaufer, J.S. (1997) Natural convection, solute trapping & channel formation during solidification of saltwater. *Journal of Physical Chemistry B*, **101**, 6132–6136.

Yen, Y.C., Cheng, K.C. & Fukusako, S. (1991) Review of intrinsic thermophysical properties of snow, ice, sea ice & frost. In: *Proceedings 3rd International Symposium on Cold Regions Heat Transfer, Fairbanks, AK, June 11–14, 1991*. (Eds. J.P. Zarling & S.L. Faussett), pp. 187–218. University of Alaska, Fairbanks.

Zotikov, I.A., Zagorodnov, V.S. & Raikovsky, J.V. (1980) Core drilling through the Ross Ice Shelf (Antarctica) confirmed basal freezing. *Science*, **207**, 1463–1465.

CHAPTER 2
Sea ice thickness distribution

Christian Haas
Department of Earth and Space Science and Engineering, York University, Toronto, ON, Canada

2.1 Introduction

Discussions of hemispheric trends and variability of the sea ice cover and their relation to global climate change often ignore the fact that sea ice shows strong regional variability, and that the regions where the ice retreats or advances in a given year can be quite variable. Similarly, the long-term, mean sea ice distribution is characterized by strong zonal (i.e. East-to-West) contrast in sea ice coverage, most prominently in the Arctic. There, sea ice expands southwards along the coasts of east Greenland, eastern Canada, and eastern Siberia and China as far south as 45°N or 40°N or so, while a little further to the east, on the other side of the Atlantic and Pacific Oceans, the Barents Sea, Baffin Bay, or the Bering Sea remain ice-free up to latitudes of almost 80°N (Chapters 7–9, 11, 27). This global sea ice distribution illustrates that the presence or absence of sea ice cannot simply be explained by latitudinal radiation or air temperature gradients. Instead, the actual sea ice distribution is strongly governed by the direction and intensity of prevailing winds and ocean currents which advect the ice, and by atmospheric and oceanic heat fluxes which show strong regional and temporal variability.

The importance of winds and ocean currents extends well into the interior pack ice zones and results in sea ice motion or drift, which in turn lead to sea ice deformation and 'dynamic' changes of the sea ice thickness distribution. These dynamic processes act together with thermodynamic processes discussed in Chapter 1, and govern the overall ice mass balance and ice thickness which initially determine where and how much ice can survive a summer's melt in a given year and region.

Dynamic thickening or thinning is illustrated in Figure 2.1. Like winds or currents, ice drift is not uniform but is subject to divergence or convergence. Due to its 'thin'-ness of only a few metres at best, under motion the ice cover easily breaks up into floes interspersed by cracks or leads. In divergent conditions, leads open and create regions of zero ice thickness, where thin new ice may form and thicken subsequently. In convergent drift, ice floes are pushed together and collide with each other. If the resulting forces in the ice become too large, it will finally break. The resulting ice fragments and blocks will be pushed onto, and below, the edges of the floes, forming so-called pressure ridges (Figure 2.1). Obviously, such dynamically formed ridges are much thicker than the adjacent, thermodynamically grown, undeformed level ice.

The thickness profile in Figure 2.2 shows the result of the various processes shaping the ice thickness distribution. The example is from the Canadian Beaufort Sea, and was obtained during a westward flight from the northwest corner of Banks Island. The profile begins over a recurring coastal polynya with new ice < 0.5 m thick, then continues over thicker first-year ice 1.2 m thick previously formed in and exported from this polynya, and ends over > 2 m thick, second- and multi-year ice incurring from the high Arctic Ocean and Canada Basin. The latter is characterized by a large degree of deformation and numerous pressure ridges > 6 m thick. It can also be seen that the thick ice cover is interspersed with refrozen leads of various thicknesses and widths, resulting from widespread divergent ice motion in this region.

Sea Ice, Third Edition. Edited by David N. Thomas.
© 2017 John Wiley & Sons, Ltd. Published 2017 by John Wiley & Sons, Ltd.

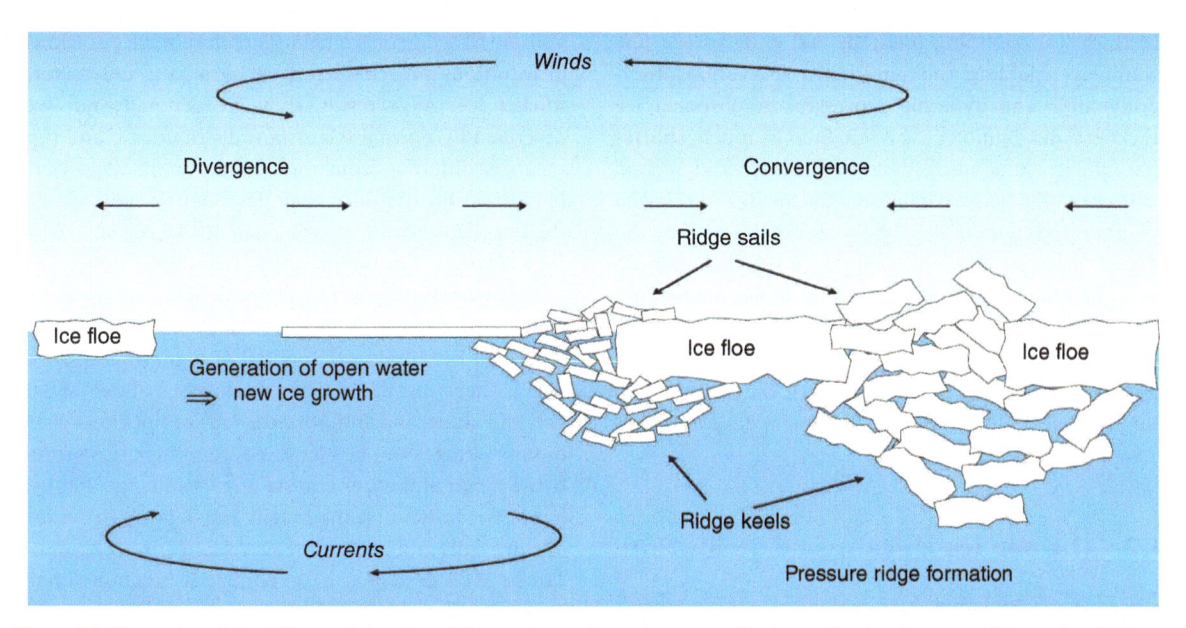

Figure 2.1 Illustration of the different dynamic and thermodynamic processes contributing to the development of the sea ice thickness distribution. Source: from Meier & Haas, 2011.

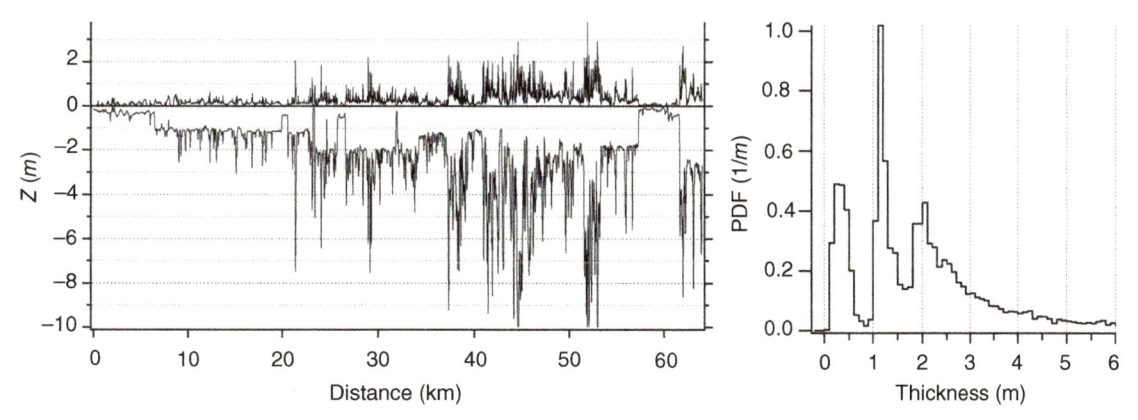

Figure 2.2 Ice thickness profile and thickness distribution [probability density function (PDF), 0.1 m bin width) from the Canadian Beaufort Sea obtained by airborne electromagnetic (EM) profiling. The 65-km-long profile was surveyed on 19 March 2014, and extends from the northwest corner of Banks Island (left) to approximately 73°33′N, 127°29′W (right). Positive Z shows the surface height profile obtained from laser altimetry, while negative Z shows EM ice thickness (see Section 2.3).

Due to the complexity of the various ice growth and deformation processes, the resulting ice thickness distributions can be quite involved (Figure 2.2). However, the thickness of level ice, and the amount and thickness of deformed ice bear information about the integrative formation processes having contributed to the thickness distribution at any given time and location. It is therefore important to be able to observe different ice thickness classes (from thin to thick ice), and to take their importance for sea ice processes into account. It is also important to note that mean ice thickness may insufficiently describe the ice thickness distribution, as the same mean value can be attained by many different arrangements of thin and thick ice.

This chapter summarizes our present understanding of dynamic and thermodynamic changes of the ice thickness distribution and their mathematical and physical representation. It then discusses various sea ice thickness observing methods and ends with a few examples regarding the importance of various thermodynamic and dynamic processes on observed ice thickness distributions. This chapter is much shorter than the corresponding ones in the first and second edition of this book. Therefore, the reader is referred to other chapters in this book covering some of the related material, namely Chapters 9–12. In addition, some excellent complementary text books with more extensive material related to this chapter have been published in the meantime, among them *Field Techniques for Sea Ice Research* (Eicken et al., 2009), *The Drift of Sea Ice* (Leppäranta, 2011), and *Drift, Deformation, and Fracture of Sea Ice* (Weiss, 2013).

2.2 The sea ice thickness distribution

2.2.1 Statistical description

As in the example in Figure 2.2, the thickness distribution in a region is most commonly described by a histogram of the frequency of occurrence of a certain ice thickness class. More formally, it is defined as a probability density function (PDF) $g(h)$ of the areal fraction of ice with a certain ice thickness h in a certain region R (Thorndike et al., 1975). The PDF of ice thickness $g(h)$ is given by:

$$g(h)\,dh = dA(h, h + dh)/R \qquad (2.1)$$

where $dA(h, h + dh)$ is the areal fraction of a region R covered with ice of thickness between h and $(h + dh)$. In practice, the thickness distribution is mostly obtained along linear profiles, and dA and R are one-dimensional, with R being the total length of the profile. $g(h)$ is derived by dividing a frequency histogram of ice thickness data by the bin width (dh). Thus, its dimension is m^{-1}. The advantage of using a PDF instead of a normal frequency distribution is that the numerical value of the height of each thickness bin is independent of the bin width used in calculating the histogram. This may be required if numerical values of thickness histograms are to be compared with other distributions or are used to parameterize the thickness distribution in numerical equations for computer models. For most practical applications, it is sufficient to calculate the frequency distribution and to give results in fractions or as percentages.

Depending on the degree of deformation, pressure ridges can contribute as much as 30–80% to the total ice volume of a floe or ice field. Therefore sea ice thickness distributions are characterized by a long tail towards thicker ice. As a result, they are generally not well described by normal (Gaussian) distributions, and their mean is different from their mode or median. While lognormal distributions may describe the most general thickness distributions (e.g. Tian-Kunze et al., 2014), they clearly fail in the presence of multiple modes.

However, statistical properties of morphological features like pressure ridge sails and keels can be better described by simple exponential or lognormal distributions. These are defined for thicknesses above certain cut-off values, and comprise the tails of thickness distributions. Ridges can be identified from thickness profiles by Rayleigh and other criteria (e.g. Wadhams, 2000). It has been shown that the height and depth distributions of ridge sails and keels above a certain cut-off height h_c can be well described by exponential functions of the form:

$$g(h) = A\,\exp[-B(h - h_c)] \qquad (2.2)$$

while ridge spacing s often has a lognormal distribution of the form:

$$g(s) = \exp(-(\ln(s - s_0) - \mu_s)^2/2\sigma^2)/\sqrt{2\pi}\sigma(s - s_0) \qquad (2.3)$$

where s_0 is a lower bound for the considered ridge spacing s and $\ln(s - s_0)$ is normally distributed (e.g. Wadhams, 2000; Rabenstein et al., 2010; Tan et al., 2012). A and B are variables related to the mean μ and standard deviation σ of ridge height or spacing.

As ridges strongly contribute to the roughness of the ice surface and underside, their height and spacing distribution play a major role in the ice's atmospheric and ocean drag (see equation 2.5). Simple equations relating ridging and surface drag were developed by Arya (1973, 1995) and Hanssen-Bauer and Gjessing (1988). A recent study by Castellani et al. (2014) has combined these concepts with extensive laser altimeter data from the Arctic Ocean to show how ridging and drag vary across the Arctic. However, they also demonstrate that one has to consider not only ridges, but also smaller morphological features and the complete thickness distribution when evaluating surface drag.

2.2.2 Modelling changes of the ice thickness distribution

As mentioned earlier, understanding and predicting the ice thickness distribution requires consideration of both thermodynamic and dynamic processes (Figure 2.1). The temporal development of the ice thickness distribution $\partial g/\partial t$ can be written as (Thorndike et al., 1975):

$$\partial g/\partial t = -\partial(fg)/\partial h - \mathrm{div}(vg) + \Phi \qquad (2.4)$$

Three terms contribute to this equation (Figure 2.3): $f(h, x, t) = \mathrm{d}h/\mathrm{d}t$ is the thermodynamic growth or melt rate of ice of thickness h at a location x and time t; v is the ice drift velocity vector; and Φ is the so-called redistribution function.

Thermodynamics

The thermodynamic growth term of equation (2.4) has been described in detail in Chapter 1. It is important to note that it is dependent on ice thickness itself, i.e. that thin ice grows faster than thick ice due to steeper temperature gradients. On the other hand, thermodynamic growth implies that thick ice exceeding a certain equilibrium thickness (e.g. pressure ridges) will melt, even in winter, if the oceanic heat flux exceeds the (generally low) conductive heat flux through thick, insulating ice. This explains the negative contribution to the ice thickness distribution in Figure 2.3.

Importantly, thermodynamic growth is also strongly modified by the presence of snow, which has a heat conductivity between 0.11 and 0.35 W m^{-1} K^{-1} depending on its density and grain structure (see Chapter 3). This is only one-seventh or less of the heat conductivity of sea ice. Therefore, the presence of snow significantly reduces ice growth and the equilibrium ice thickness. This may be important for future scenarios of ice formation in the Arctic. As freeze-up occurs later and

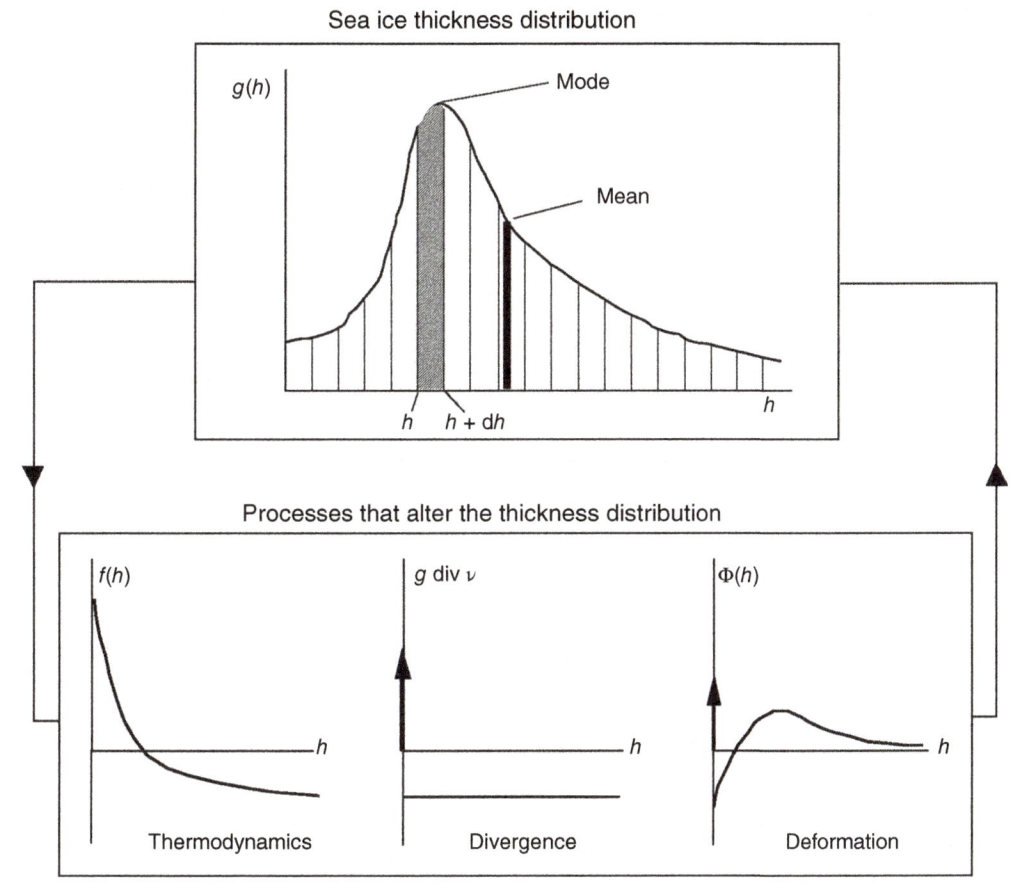

Figure 2.3 Illustration of the contribution of the different terms and processes in equation (2.4) to the ice thickness distribution. Source: adapted from Thorndike et al., 1992.

later in the season, less snow may accumulate on the forming first-year ice, while it can form a thick layer on adjacent, older ice. The first-year ice with less snow cover can therefore have a growth advantage over older ice, and can reach similar thicknesses in the end of the growth season as old ice (e.g. Notz, 2009).

Divergence and advection

The second term in equation (2.4) represents ice divergence and advection due to ice motion. As mentioned earlier, the ice is subject to external forces, mainly due to wind and current action (see equation 2.5). These forces cause the ice to drift. Away from the coast or other obstacles, the ice will drift freely, and drift direction and speed are closely related to the geostrophic wind and Ekman transport. While the Coriolis force turns the ice to the right, surface friction acts to the left of the geostrophic wind vector. By comparing the motion of drifting buoys deployed on ice floes with geostrophic wind fields, it has been shown that ice in the Arctic drifts at 1% of the mean wind speed, and with an angle of 18° to the right (Colony & Thorndike, 1984). Vihma et al. (2012) have found evidence that these numbers have changed somewhat in recent years, possibly due to thinner ice which is less affected by the Coriolis force. For the Weddell Sea, these numbers are 1.6% and 10–15° to the left, respectively (Kottmeier et al., 1992). However, note that ice drift is not free at most times and in most regions, and therefore the complete momentum balance (equation 2.5) has to be taken into account for its complete description (Figure 2.4; Leppäranta, 2011; Weiss, 2013).

Divergence within the ice generates cracks, leads or polynyas with open water where new ice will form. Thus, for a certain region, divergence removes ice of finite thickness and causes a delta pulse at zero thickness in the thickness distribution (Figure 2.3).

Deformation/convergence

The last term in equation (2.4) is the redistribution function describing how thin ice is deformed and transformed into thicker ice classes in the case of ice convergence (negative divergence) and deformation (Figure 2.3). It is the most critical term to realistically model the temporal development of the thickness distribution (e.g. Lipscomb et al., 2007). It is also the most unknown term, as it depends very much on fracture mechanics, and depends critically on factors

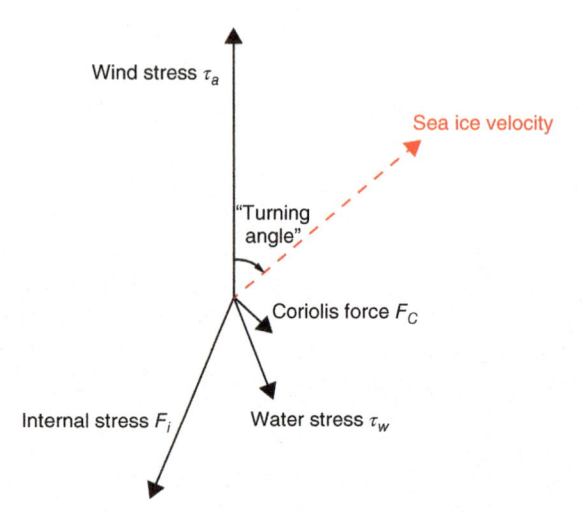

Figure 2.4 Typical force balance for sea ice (cf. equation 2.5) and resulting ice velocity (modified from Colony & Thorndike, 1984; Weiss, 2013). Note that F_t is not shown because it is usually very small.

like small-scale ice properties, friction between ice blocks among each other as well as at the snow and ice interfaces, and deformation energy and scales. A very promising approach to ridge formation modelling has been presented by Hopkins (1994) using a dynamic ridge growth model, where the fate of single ice blocks is computed as a function of external forces. However, thin ice will generally deform more easily than thick ice.

On a regional scale, the large-scale spatial thickness distribution is obtained by solving a force balance equation considering the main forces acting on a unit area of the sea ice cover (Figure 2.4):

$$M\,a = \tau_a + \tau_w + F_C + F_i + F_t, \qquad (2.5)$$

where the a sea ice element of mass M accelerates with a according to the sum of the air and water drags τ_a and τ_w, the Coriolis force F_C, internal ice forces F_i, and of the force due to sea surface tilt F_t. Usually, the first two terms are most dominant by more than an order of magnitude. For every model grid cell, mean ice thickness is derived by solving equation (2.5) for ice motion, and distributing the ice volume drifted into a cell equally over the cell area assuming mass conservation (Chapter 12). Clearly, as with the redistribution term in equation (2.4), ice strength and rheology are of great importance here. The first models involving plastic or viscous-plastic rheologies were developed by

Hibler (1979) and Coon (1980). The rheology describes a viscous flow of an ice field, with plastic deformation once ice concentration and internal ice forces exceed a certain threshold. Explicit ridging schemes have been introduced by, for example, Flato and Hibler (1995) and are reviewed by Lipscomb et al. (2007), Leppäranta (2011) and Weiss (2013). While these first models prescribed the atmospheric and oceanic forces acting on the ice, today complex coupled atmosphere–ice–ocean models exist and are more thoroughly discussed in Chapter 12.

It is noteworthy that the mean drift speed of sea ice in the Arctic has increased in recent years, while there were no corresponding proportional increases in wind speed, the driving force for the drifting ice (Rampal et al., 2009; Spreen et al., 2011; Kwok et al., 2013). These changes are probably related to the concurrent thinning of the ice, as thinner ice is weaker, leading to smaller internal ice forces F_i (equation 2.5). Unfortunately there is little reliable information on changes in Antarctic sea ice drift, which could potentially be linked to the increasing trends of Antarctic sea ice extent (Schwegmann et al., 2011).

Melting

Melting reduces the mean ice thickness and generally moves the thickness distribution shown in Figure 2.3 to the left, i.e. towards thinner ice. This ties the ice thickness distribution closely to summer ice extent, because ice with zero thickness becomes open water. If the amount of summer melt is greater than or equal to the modal thickness, large areas will become ice-free during the summer. Thus there may be accelerated decreases of summer ice extent and larger interannual variations with a thinner ice cover in a future warmer climate (Notz, 2009).

On regional scales, melting patterns correspond to large-scale meteorological conditions and to ocean heat flux regimes. However, even more than with freezing (see earlier), on small scales melt rates depend critically on the ice thickness distribution itself, and are different for different thickness classes and ice types (e.g. Perovich et al., 2003). The heat flux through pressure ridges is lower than through level ice because of their greater thickness. Consequently they would melt faster. As their keels protrude far down into the water, they might even reach into warmer water. More importantly, ridge keels contribute to the roughness of the ice underside, thereby increasing upward turbulent fluxes of heat (Amundrud et al., 2006; see also Chapter 5). The flanks of ridge sails are exposed more normally to the incident solar radiation than are the ridge crests, as solar elevation is low in the polar regions. Therefore, melting may be higher on the flanks. Although the variations of melt rates might seem to be rather small, they can contribute to significantly different thickness changes in the course of the ablation season.

Much stronger differences in melt rates exist on small thickness classes, i.e. on level ice (e.g. Perovich et al., 2003; Eicken et al., 2004). Snow and ice meltwater primarily accumulates at topographic low points to form melt ponds. Even small amounts of snow wetting, and the formation of melt ponds, significantly reduce surface albedo. Typical surface albedos are 0.8 for snow, 0.6 for bare ice and 0.15–0.3 for melt ponds. Thus, once formed, melt ponds absorb more energy than the neighbouring snow or bare ice, thereby increasing local melt rates. Throughout the summer, the surface of melt ponds falls down to sea level, and vertical pond walls form, reaching deep into the floe. This positive feedback causes significant changes to the ice thickness distribution of level ice, and it contributes to an increase in surface roughness (e.g. Polashenski et al. 2012; Landy et al., 2014).

The discussion in this section shows that many factors are responsible for shaping the ice thickness distribution. Thinning in a certain region, for instance, can result from melting, but also from advection of thinner ice. Therefore, any interpretation or forecast of changes of the ice thickness distribution in terms of climate change has to take into account both thermodynamic and dynamic processes.

2.2.3 Global sea ice thickness distributions

The processes described above and the interplay of dynamic and thermodynamic growth shown in Figure 2.3 act at small and large scales, and also shape the mean, Arctic- and Antarctic-wide ice thickness distributions. Mean wintertime fields of ice thickness and drift between 1980 and 2005 are shown in the maps in Figure 2.5. Data were taken from the Community Climate System Model, version 4 (CCSM4; Jahn et al., 2012; Landrum et al., 2012), one of the most advanced and commonly used climate models (e.g. Stroeve et al., 2014). It includes a subgrid-scale ice thickness distribution, elastic–viscous–plastic rheology, and an improved ridging scheme (Lipscomb et al., 2007).

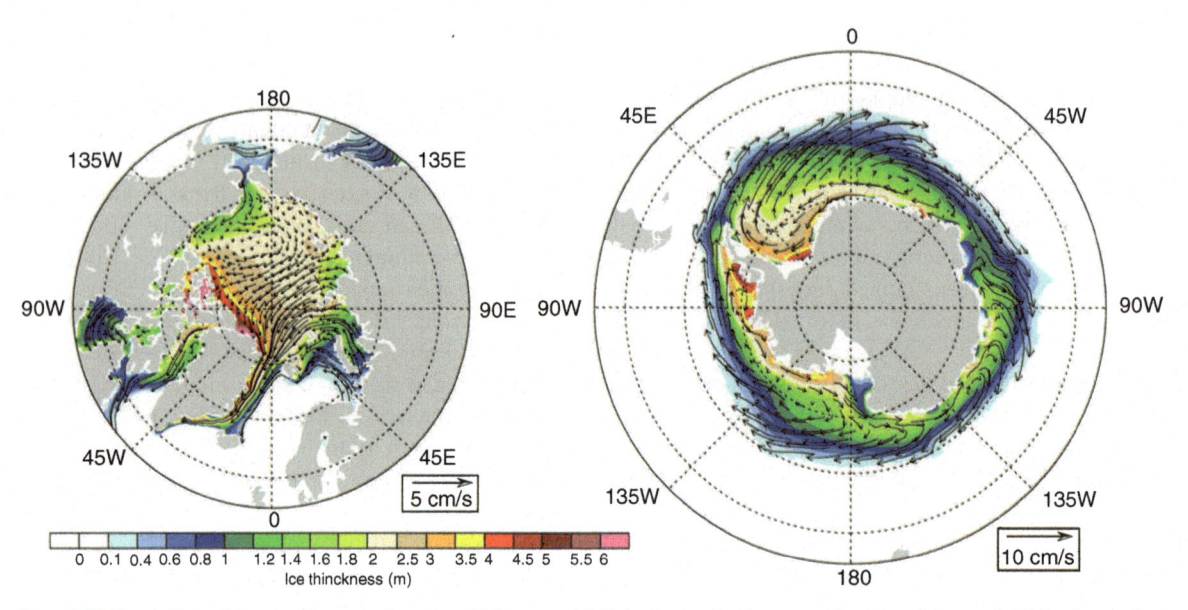

Figure 2.5 Illustration of simulated mean winter ice thickness and drift in the Arctic (January–March) and Antarctic (July–September) for the years 1980–2005 using the Community Climate System Model, version 4 (CCSM4). Data were taken from Jahn et al. (2012; Arctic, ensemble 012) and Landrum et al. (2012; Antarctic, ensemble 007). (Figure compiled by C.C. Bajish, York University.)

It becomes immediately obvious that Arctic sea ice is generally thicker than its counterpart around Antarctica. In the model simulations, most Arctic ice is thicker than 2 m. In contrast, little ice grows as thick in the Southern Ocean. These hemispheric contrasts are due to at least five main differences in the thermodynamic and dynamic boundary conditions of ice growth in the Arctic and Southern Oceans (see also Chapter 10):

- **Ocean heat flux.** One fundamental difference between the Arctic and Southern Oceans is the occurrence of a fresh mixed layer in the Arctic overlying a strong pycnocline. This layer is fed by the inflow of fresh water from large rivers, mainly from the Siberian continent. The Arctic Ocean receives approximately 10% of the world river run-off. The fresh mixed layer is very stable and prohibits any significant heat fluxes from the much warmer Atlantic water underneath. A typical value for the ocean heat flux in the Arctic Ocean is 4 W m^{-2}. The 'Atlantic layer' at a depth of 200–300 m is 1–2°C warm. This heat would be sufficient to melt all ice during the summer (Barry et al., 1993). In the Southern Ocean, no rivers enter the seas and icebergs melt further to the north. Therefore, the mixed layer is much saltier and not well stratified. Mean ocean heat fluxes amount to about 40 W m^{-2}.

- **Snow thickness.** With a thermal conductivity between 0.11 and 0.35 W m^{-1} K^{-1} (Massom et al., 2001) snow is a strong thermal insulator. Therefore, ice with a thick snow cover grows more slowly than if the snow were thin. The Arctic Ocean is surrounded by large continents and mean snow thickness reaches only about 0.3 m in spring (Warren et al., 1999). In contrast, Antarctic sea ice is usually covered by thick snow. On perennial ice in the Pacific sector or in the western Weddell Sea, mean snow thickness can be larger than 0.5 m (Massom et al., 2001; Nicolaus et al., 2009; Kwok & Maksym, 2014). This is due to the fact that the sea ice areas are completely surrounded by oceans, which provide a permanent moisture source. In the south, sea ice may collect the snow blown off the continental ice shelves. As a consequence of the thick, insulating snow and high ocean heat fluxes, Antarctic ice may melt at its underside even during winter, because the temperature gradients through the ice are only small (e.g. Lytle & Ackley, 1996; Maksym et al., 2012).

- **Ice age.** Although the area of old ice in the Arctic has decreased strongly (e.g. Maslanik et al., 2011), most

Arctic ice still drifts for 2 to 5 years (Rampal et al., 2009) until it leaves the Basin mostly through Fram Strait where it melts further south. The older an ice field becomes, the more deformation events it will experience, where it thickens by the accumulation of pressure ridges (Figure 2.1). This dynamic thickening is accompanied by passing through several winters where the ice can also thicken by thermodynamic growth until it reaches an equilibrium thickness. In contrast, most Antarctic ice melts during summer. Thus, it rarely becomes older than 1 year, and only few regions with perennial ice exist in the western Weddell Sea and southern Bellingshausen, Amundsen and Ross Seas.

- *Divergence versus convergence.* As mentioned earlier, the Arctic Ocean is surrounded by continents, and thus ice motion is confined by coasts where the ice converges and thickens by deformation, particularly north of Canada and Greenland. In contrast, ice drift around Antarctica is mostly divergent (Kottmeier et al., 1992) with a net northerly drift component towards the surrounding open oceans. Divergence causes the opening of polynyas and leads, and the addition of thin new ice to the thickness distribution.
- *Latitude.* Most of the ice in the Arctic is at latitudes north of 70°N, whereas in the southern hemisphere most ice extends into much lower latitudes, as far north as 55°S. Thus, air temperatures, total incoming solar radiation and the length of the summer season are generally lower in the Arctic than in the Southern Ocean. However, the Antarctic ice sheet is a giant cold reservoir, and the sea ice region is well isolated from lower latitudes by the atmospheric and oceanic flow regimes of the Circumantarctic Current, so that warm and moist air advection are not as important as they are for the Arctic. Due to these, strong surface melting rarely occurs on sea ice in the Southern Ocean (Nicolaus et al., 2006). This is in stark contrast to conditions in the Arctic, where strong surface melting occurs in summer, even at the North Pole, at much higher latitudes than in the Antarctic.

The order of these points is arbitrary and does not imply any ranking. The final ice thickness depends on the magnitude of, and interrelation between, these different aspects. Clearly, both dynamic and thermodynamic factors are responsible for the hemispheric differences.

The maps in Figure 2.5 also show large regional thickness variations within each hemisphere itself. These are primarily a result of ice motion and deformation. As 30–80% of the volume of an ice field is contained within pressure ridges, the mean thickness of a region is more dependent on the number and thickness of ridges than on the thermodynamic thickness of level ice. In other words, for the overall ice volume within a certain region, dynamics is more important than thermodynamics. Therefore, on a regional scale, the average ice thickness distribution is determined by the prevailing atmospheric circulation regimes, which are responsible for mean ice motion and the dominant drift directions. Where the ice drifts against, or shears along, a coast, there will be strong ice pressure, and the ice will become heavily deformed. As a result, the mean thickness in regions with mean drift convergence (e.g. north of Canada) is larger than in regions with mean divergence (e.g. north of Siberia), where thin new ice is permanently generated and exported.

The arrows in Figure 2.5 show the dominant drift patterns which develop as the result of the prevailing atmospheric circulation. Mean annual ice motion is very similar to the winter circulation shown in the figure. In the Arctic, mainly two drift systems exist. The Beaufort Gyre is an anticyclonic gyre in the Canada Basin north of Canada and Alaska. It is caused by quasi-permanent high atmospheric pressure over the Beaufort Sea. The Beaufort Gyre can transport ice floes for several years before they are exported into the Transpolar Drift. This is the other prominent drift system, which transports ice from the source regions on the Siberian Shelves within about 1–3 years across the North Pole into Fram Strait and the East Greenland Current, where it finally melts. The Transpolar Drift is mainly driven by low-pressure systems passing from the North Atlantic into the Barents and Kara Seas. On average, these drift patterns push the ice against the coasts of northern Canada and Greenland. Consequently, as a result of strong convergence and deformation, the thickest ice is found in these regions. Mean maximum thicknesses are > 4 m, mainly resulting from the large spatial density of thick ridges. The youngest and thinnest ice is found along the Siberian Shelf, where prominent polynyas occur and from where ice is permanently exported into the Transpolar Drift (e.g. Krumpen et al., 2011).

The ice thickness distributions shown in Figure 2.5 are in good qualitative agreement with submarine and airborne observations (e.g. Lindsay & Schweiger, 2015; see Figure 2.10 later), and with satellite-derived ice thickness maps (see Chapter 9).

Figure 2.5 shows that the thickest ice in Antarctica occurs close to the continent, in accordance with the greatest latitude, and in the vicinity of the coast where it is sporadically compressed. The most prominent feature, however, is the thickest ice in the southern and western Weddell Sea. On the one hand, this is one of a few regions possessing perennial ice. On the other hand, it is a region where ice drift is directed towards the coast, and subsequently much deformation occurs. The so-called Weddell Gyre is caused by low average sea level pressure over the central Weddell Sea. It should be noted that both the Beaufort Gyre and the Weddell Gyre rotate clockwise. However, due to the Coriolis force, this results in ice convergence within the gyre centre on the northern hemisphere, whereas clockwise circulation results in net divergence inside the gyre in the south. The great thickness in the western Weddell Sea is therefore caused by ice motion away from the divergent gyre centre, with the Antarctic Peninsula acting as an obstacle for the drifting ice.

In contrast to the Arctic, regional ice thickness distributions in the Antarctic are less well known, because only a few systematic measurements have been performed (Chapter 10). The use of military nuclear submarines is prohibited by the Antarctic Treaty. However, thickness maps derived from visual observations (Worby et al., 2008), airborne measurements (e.g. Kwok & Maksym, 2014), and satellite altimetry (e.g. Giles et al., 2008; Kurtz & Markus, 2012) show similar patterns to those in Figure 2.5.

It should be noted that the drift systems and thickness distributions shown in Figure 2.5 represent long-term average conditions. There is large seasonal, interannual and decadal variability superimposed on these mean patterns, as described in the following and in Chapter 11.

2.3 Measurement techniques

While sea ice extent and concentration can be measured with sufficient accuracy by satellites from space (Chapters 9 and 11), determining its thickness is much more difficult, even from aircraft or while standing on the ice. This is due to its relative thinness, which is a challenge for any geophysical measurement technique. Therefore, most methods are indirect measurements, which observe ice properties closely related to ice thickness. This chapter briefly presents the most important submarine, on-ice, and airborne ice thickness measurement methods (Figure 2.6). Space-based ice thickness measurement methods are discussed in Chapter 9.

One important difference between space-based and other ice thickness observations discussed in this chapter is that the latter can usually provide information about the complete thickness distribution, while space-based methods obtain large-scale mean thickness information at best. Despite the overwhelming spatial and temporal coverage of satellite observations, the methods discussed here are important for revealing the underlying processes of observed thickness variability via the mechanisms affecting the ice thickness distribution discussed above, and for validating satellite data.

Drill-hole measurement provides the most direct and most accurate ice thickness information (e.g. Haas & Druckenmiller, 2009). It is the only method allowing the simultaneous observation of all important components of ice thickness, namely the thickness of both ice and snow, as well as ice draft and freeboard (the depth of the ice underside below the water level and the height of the ice surface above the water level, respectively), with one measurement (Figure 2.6). However, drill-hole measurement is slow, tedious and requires access to the ice, such that the spatial extent and statistical representativeness can be quite limited, and usually does not include information about very thin or thick ice. However, it can provide important information on key aspects of the ice thickness distribution, e.g. modal thickness representing the thickness of the most frequently occurring, thermodynamically grown level ice (e.g. Eicken et al., 1995; Worby et al., 1996). It is also good practice to support measurements made by other means with a few drill-hole measurements for calibration and/or validation.

Ice thickness can also be observed visually by recording ice floes broken by an icebreaker. The broken ice fragments are often moved side-up against the ship's hull, revealing their cross-profile. Coordinated, systematic visual ice observations are encouraged by the international Antarctic Sea Ice Processes and Climate (ASPeCt) programme and have been compiled by

Sea ice thickness distribution

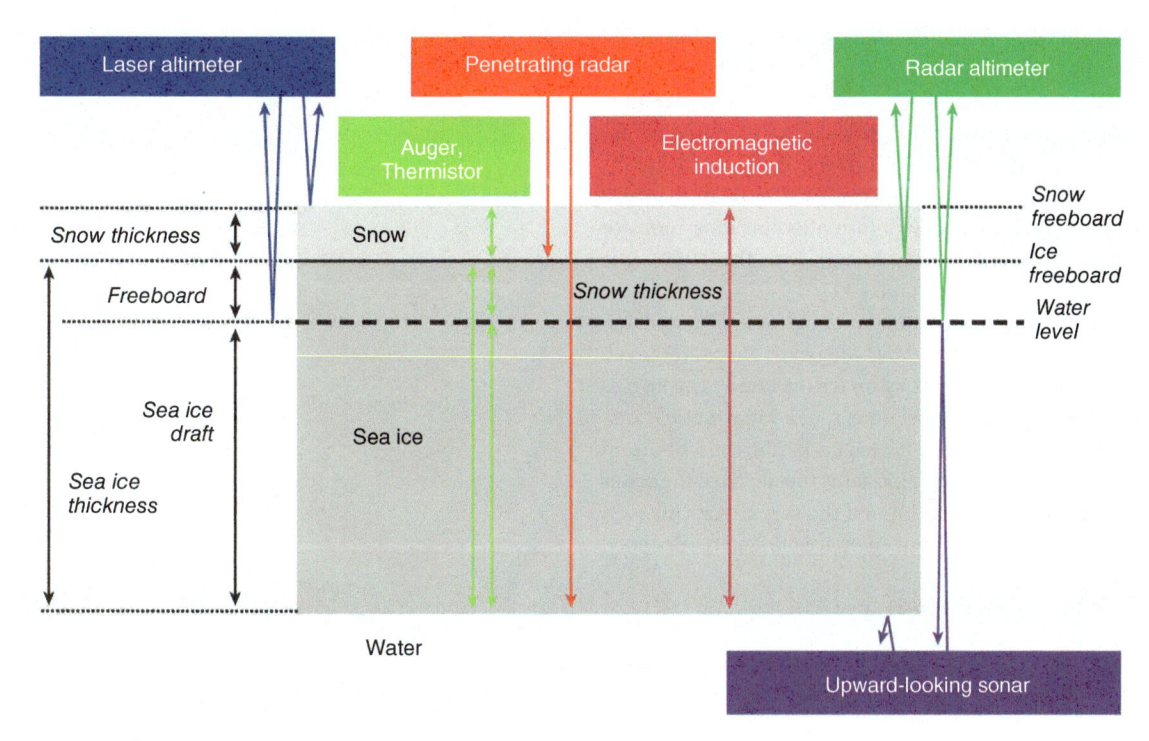

Figure 2.6 Ice floe cross-section showing definitions of ice and snow thickness terms and parameters, and primary measurements of individual ice thickness observing methods. Source: adapted from Meier & Haas, 2011.

Worby et al. (2008) to provide a unique snapshot of circum-Antarctic sea ice thicknesses.

With ground-penetrating radar (GPR), the travel time of radar waves between the radar instrument and reflections from the snow surface, snow–ice interface or ice–water interface is measured. GPR could therefore potentially yield coincident measurements of snow and ice thickness (Galley et al., 2009; Haas & Druckenmiller, 2009). While radar frequencies of a few 100 MHz have to be used to penetrate through the sea ice, higher frequencies (≥ 1 GHz) have to be used to resolve the much thinner snow. Although the references above include a few examples of successful point measurements, widespread application of the method suffers from highly variable dielectric ice properties due to variations in brine volume, brine distribution and other internal interfaces. These lead to highly variable radar propagation speeds and radar absorption, such that acquired radargrams can be difficult to interpret. Recent new technology developments have led to the design of broadband, frequency-modulated, continuous-wave (FMCW) radars for snow and ice thickness measurements, which may improve some of the issues related to absorption and resolution (Kanagaratnam et al., 2007; Holt et al., 2008). In fact, an airborne, ultra-wideband, 2–8 GHz FMCW radar (Panzer et al., 2013) is now routinely flown for snow thickness surveys over Arctic and Antarctic sea ice (Kurtz et al., 2013). This is discussed further below and in Chapter 3.

The following sections provide short overviews of the methods most commonly applied today, which have resulted in the most abundant thickness data so far and have been used in extensive model and satellite data validation studies (e.g. Schweiger et al., 2011; Johnson et al., 2012; Stroeve et al., 2014; Lindsay & Schweiger, 2015). More detailed descriptions of these and other techniques, including data examples, are given by Haas and Druckenmiller (2009).

2.3.1 Submarine and moored upward looking sonars

So far, most ice thickness data have been obtained by means of upward-looking sonars (ULSs), also called ice profiling sonars (IPS). These have been mounted on either military nuclear submarines (e.g. Bourke & Garret 1987; Rothrock et al. 2008) or oceanographic moorings (e.g. Strass & Fahrbach, 1998; Vinje et al., 1998; Melling et al., 2005; Hansen et al., 2013; Krishfield et al., 2014), and have provided some of the most extensive and invaluable information on long-term ice thickness variability and change since the 1950s (see Figure 2.11 later and Chapter 11).

Using this method, estimates of draft are obtained, i.e. of the depth of the ice underside below the water level, which is a reasonable proxy for ice thickness. The instruments measure the travel time, t, of a sonar pulse transmitted by the ULS and reflected back from the ice bottom (Figure 2.6 and 2.7). Additionally, the depth of the sonar beneath the water level, z, and the sound velocity, v, in the water must be known. Then, ice draft, d, is calculated according to:

$$d = z - v^* \, t/2 \qquad (2.6)$$

The depth of the ULS is derived from pressure sensors, whose measurements are also dependent on atmospheric sea level pressure. The sound velocity profile is either assumed constant with a certain sound speed or taken from conductivity–temperature–depth measurements or a mixed-layer model (Strass & Fahrbach, 1998). Measurements are also corrected for sonar tilt which can be caused by currents (Figure 2.7). Data processing can become quite complicated in the case of strong water stratification or when the measurements are performed close to ocean frontal zones. A plausibility test for the depth measurement or the sound velocity profile can be performed when profiling the water level in open leads. Then, the measured sonar distance must equal the ULS depth (Figure 2.7).

Ice thickness, h, is calculated from draft, d, by assuming isostatic equilibrium, a certain snow depth, z_s, and water, ice and snow densities ρ_w, ρ_i and ρ_s:

$$h = (\rho_w d - \rho_s z_s)/\rho_i \qquad (2.7)$$

The values for ice and snow density, as well as snow depth, are reasonably well known so that only small errors arise for h.

Overall, uncertainties of sound velocity profiles, biases due to sonar beam divergence, and other technical issues range between 0.10 m (Melling & Riedel, 1995) and 0.25 m (Rothrock & Wensnahan, 2007). Accuracies have improved with the use of high pulse

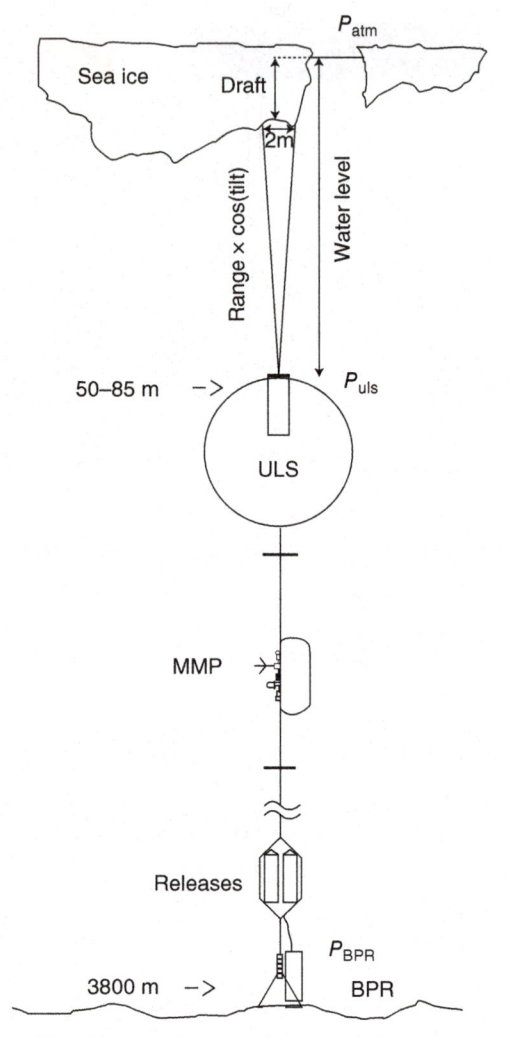

Figure 2.7 Schematic of a moored upward-looking sonar (ULS) with additional sensors required for accurate ice draft measurements (Krishfield et al., 2014): a vertically profiling current meter (McLane Moored Profiler, MMP) and bottom pressure recorder (BPR). Estimating ice draft from ULS requires measurements of pressure at the ULS (P_{ULS}), ULS tilt and range; atmospheric pressure (P_{atm}); and bottom pressure (P_{BPR}). Source: From Krishfield et al., 2014. Reproduced with permission of John Wiley & Sons.

repetition rates, which allow better detection of leads between ice floes, serving as zero thickness reference measurements.

Submarines allow for long-range, basin-scale transects for determining the ice thickness profile. However, so far, submarine surveys have only been performed in conjunction with military cruises. This implies that transects are not designed with scientific priorities and in a systematic manner. As a consequence, measurements often have to be corrected for seasonal variability before they can be compared with each other (Rothrock et al., 2008). The Scientific Ice Expeditions SCICEX programme of the US navy facilitates timely release and scientific analysis of submarine data. Meanwhile, autonomous underwater vehicles are being developed and may provide an alternative to the use of submarines in the near future (Dowdeswell et al., 2008; Williams et al., 2015). They can also be used in Antarctica, where the operation of military, nuclear submarines is prohibited by the Antarctic Treaty.

Upward-looking sonars mounted on oceanographic moorings provide long time-series of ice thickness in a single location. These allow studying the temporal development of the ice thickness distribution, e.g. in the course of the growing season (Melling & Riedel, 2004). Transects can be achieved if several moorings are simultaneously operated across a certain region, such as in the Fram Strait (Hansen et al., 2013) or the Beaufort Sea (Krishfield et al., 2014). Combined with ice drift velocity data retrieved from satellite imagery or buoys, mooring data allow for the calculation of ice volume or freshwater fluxes.

While moored ULSs can provide very valuable continuous data, the operation of the instruments at water depths of 50–150 m for periods of 1 year or more is still a technological challenge. Similarly, the recovery of the instruments is often difficult, or instruments may be lost, e.g. as a result of commercial trawl fishing. Moorings cannot be deployed in shallow waters, where they might be destroyed by scouring from ridge keels or icebergs.

2.3.2 Electromagnetic induction sounding

In contrast to high-frequency radar echo sounding, frequency-domain electromagnetic (EM) induction sounding uses low-frequency EM signals between 1 and 100 kHz and has become a widely applied method. The technique is sensitive to the distribution of electrical conductivity in the underground and is traditionally employed in mineral exploration or groundwater mapping on land. The transmitter coil of an EM instrument generates a 'primary' EM dipole field which penetrates almost unaffectedly through the resistive ice, whose electrical conductivity typically ranges between 0 and 50 milli-Siemens m^{-1} (mS m^{-1}), depending on ice type and season (Haas et al., 1997). However, in the highly conductive seawater, with conductivities ranging between 2400 mS m^{-1} in the western Arctic and 2700 mS m^{-1} in the Southern Ocean, eddy currents are induced. These, in turn, induce a 'secondary' EM field which propagates back through the ice and whose amplitude and phase are measured by a receiver coil, relative to the primary field (Figure 2.8). The amplitude of the secondary field decreases exponentially with increasing distance, d_{EM}, between the instrument and the surface of the water, which coincides with the bottom of the ice. Relations between the (complex) secondary field amplitude and d_{EM} can be derived empirically by means of coincident measurements over drill holes or open water (Haas et al., 1997, 2009), or by forward modelling (Anderson et al., 1979; Pfaffling et al., 2007). In addition, the EM response depends on the signal frequency and coil orientation and spacing. EM sensors cannot distinguish between ice and snow, because both have very low electrical conductivities. Therefore, d_{EM} corresponds to

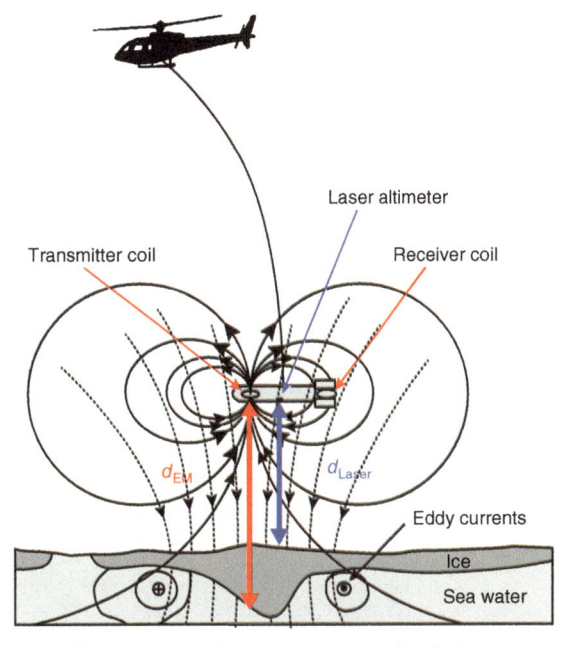

Figure 2.8 Principle of electromagnetic (EM) ice thickness measurements, showing primary (solid) and secondary (stippled) EM fields connected through induced eddy currents in the water, and the derived distance to the ice–water interface, d_{EM}. Total (snow + ice) thickness can be obtained by subtracting the laser-measured height of the instrument above the snow surface, d_{Laser}, from d_{EM}. Source: adapted from Rossiter & Holladay, 1994.

the total, i.e. ice + snow thickness, when the instrument is located on the snow surface. If the sensor is located above the snow surface, its height has to be measured independently, and ice + snow thickness can then be computed from the difference between d_{EM} and the instrument height above the snow (d_{Laser} in Figure 2.8).

Electromagnetic sounding can be performed with an EM instrument located either on the ice or above the ice. On-ice measurements are now routinely performed with various portable, commercially available instruments. The amplitude of the imaginary ('quadrature') component of the measured secondary field is usually converted into an 'apparent conductivity' (McNeill, 1980), and simple, logarithmic equations can be applied to convert apparent conductivity into d_{EM}, which is equal to the ice thickness if the instrument lies on the ice or snow (e.g. Kovacs & Morey, 1991; Haas et al., 1997). If the instrument is carried or mounted on a sledge and towed behind a snowmobile, for example (see Figure 2.14, later), the height above the snow has to be subtracted from the measurement of d_{EM}.

Electromagnetic instruments can be mounted on other platforms and be operated over larger distances or longer times. Their height above the snow is then typically measured with laser or sonic altimeters (Figure 2.8). For example, EM ice thickness measurements can be performed with common instruments from bridges or lighthouses (Haas & Jochmann, 2003), or ships (Haas, 1998; Reid et al., 2003; Uto et al., 2006). However, as the EM field strength decreases exponentially with increasing height above the water, more sensitive instruments have to be used with increasing operating height. Specialized instruments are used for airborne EM thickness surveys. To minimize disturbance from interference with conducting metallic parts of the aircraft and to facilitate operation close to the ice surface, these sensors are either operated as towed 'birds' (e.g. Kovacs & Holladay, 1990; Haas et al., 2009) or flown at very low altitude (Multala et al., 1996; Prinsenberg et al. 2002). Most of these surveys were performed by helicopter, with limited range and weather tolerance. Most recently, an EM bird and winch system was installed on a Basler BT67 airplane, now facilitating long-range, highly accurate ice thickness surveys less affected by adverse weather conditions (Figure 2.9; Haas et al., 2010).

The ice thickness profile shown in Figure 2.2 was obtained with the surveying system shown in Figure 2.9. Figure 2.10 presents another example of an extensive, high-resolution ice thickness survey covering key regions of old ice in the Arctic Ocean, recognizable by its high radar backscatter (Haas et al., 2010). The thickness distributions reveal distinct differences between the thinner, divergent ice in the Fram Strait and the Beaufort and Chukchi Seas, and the convergent ice conditions leading to the thickest ice north of Canada and Greenland. Results also provide validation for

Figure 2.9 Ice thickness survey over the CryoVEx 2014 ice camp north of Greenland with an electromagnetic 'bird' towed by a Basler BT67 aeroplane.

Sea ice thickness distribution 55

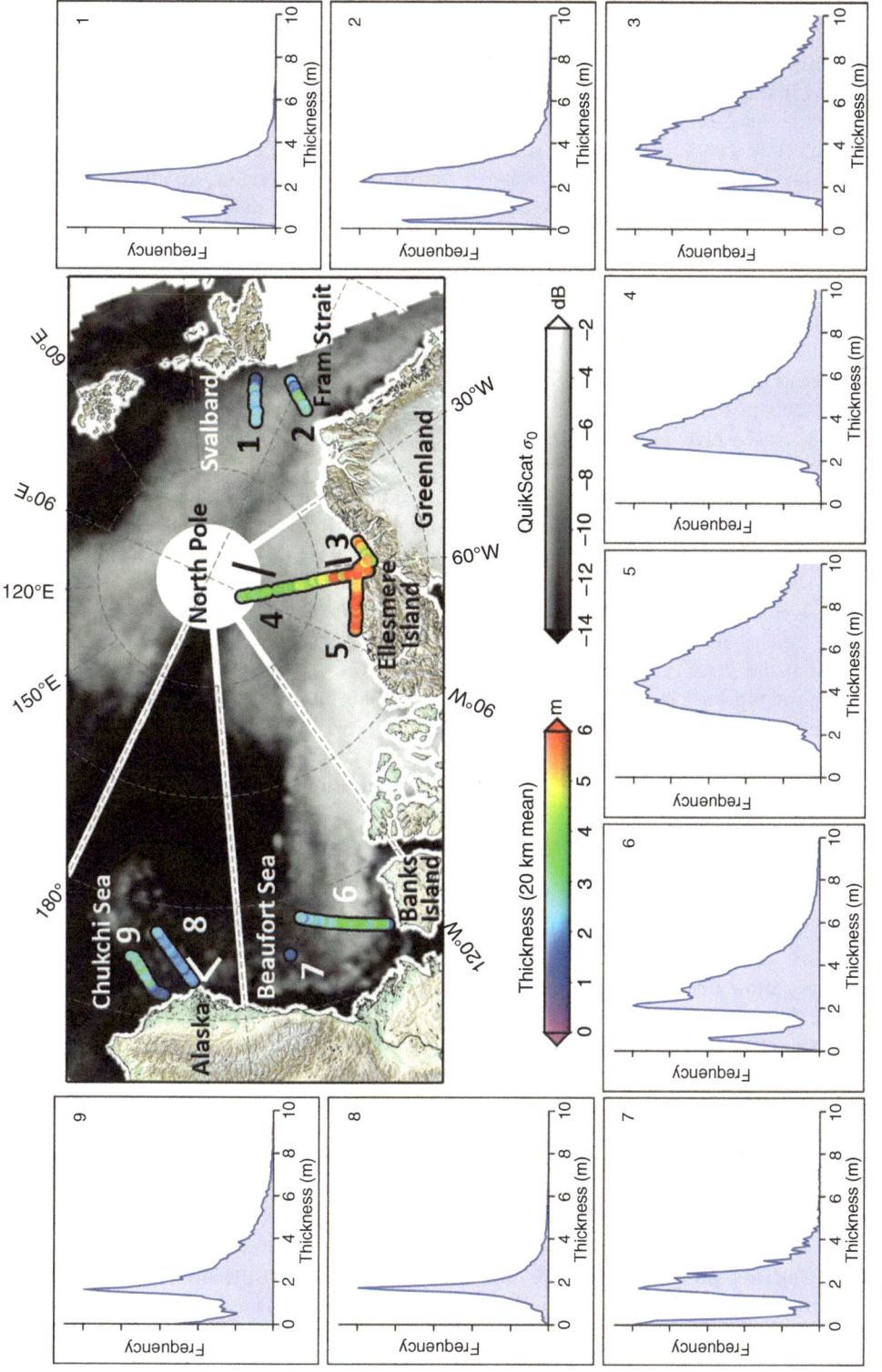

Figure 2.10 Ice thickness distributions in different regions of the Arctic during April 2009. Numbers correspond to flight lines shown on map, superimposed on QuikScat radar backscatter. Source: Haas et al., 2010.

model results (Figure 2.5; see also Chapter 12) and satellite products (Chapter 9).

Comparisons with drill-hole data and theoretical considerations show that the accuracy of EM thickness measurements is better than ±0.1 m over level ice (Haas et al., 1997; Pfaffling et al., 2007). However, due to the spatial extent of eddy currents induced in the water (Figure 2.8), the footprint of the method is quite large and can range between 3.7 and 10 times the distance between the EM instrument and the water surface, depending on the instrument configuration (Kovacs et al., 1995; Reid et al., 2006). Consequently, EM measurements underestimate the maximum thickness of deformed ice such as ridge keels, by as much as 50% (Haas & Jochmann, 2003).

As EM sounding is sensitive to the presence of various layers with contrasting conductivity, there is also the potential to survey more complicated ice conditions undetectable by other methods. For example, multi-channel sensors can be used to detect freshwater layers under the ice derived from river run-off or melting (Prinsenberg et al., 2008), or the thickness and properties of thick ice shelves and platelet ice formed under fast ice in Antarctica (Rack et al., 2013; Hunkeler et al., 2015). These and other sea ice properties will be the focus of future improvements in the EM method and will expand its application far beyond just thickness measurements.

2.4 Airborne laser altimetry

Like draft, sea ice freeboard is closely related to ice thickness via isostasy, and therefore there are significant efforts to survey snow freeboard (Figure 2.6) by means of airborne laser altimetry, most recently as part of NASA's Operation IceBridge mission carried out from 2009 to 2019 (e.g. Hvidegaard & Forsberg, 2002; Kurtz et al., 2013; Richter-Menge & Farrell, 2013). The principles of the method and conversion of freeboard to ice thickness are identical to satellite laser altimetry (Chapter 9). However, due to the lower flying altitudes of aircraft, the spatial resolution and footprint of individual measurements is significantly improved, and the use of scanning lasers provides two-dimensional freeboard and surface roughness information. Therefore the full ice thickness distribution can be resolved, and the data can be well analyzed for surface roughness

and pressure ridge distributions (e.g. Tan et al., 2012; Castellani et al., 2014).

In contrast to ULS measurements of sea ice draft (Section 3.1), the uncertainty in altimetric ice thickness retrievals is much larger due to the small freeboard-to-thickness ratio. For a given freeboard error the corresponding ice thickness uncertainty is approximately ten-times larger (Equations 2 and 3 in Chapter 9.4.2).

In addition, accurate knowledge of snow thickness is required for reliable thickness retrievals. In contrast to satellite measurements, snow thickness has successfully been measured from aircraft with FMCW radars (Panzer et al., 2013; Kurtz et al., 2013; Kwok and Maksym, 2014; Newman et al., 2015; see also Chapter 3), thus significantly improving ice thickness estimates.

Altimetric freeboard measurements also require accurate knowledge of the water level ('Sea surface height') against which freeboard is referenced (Figure 2.6). Due to variable aircraft flying heights, unresolved geoid undulations, and dynamic sea surface topography, the height of the water level is generally poorly known and needs to be obtained from the altimetric measurements themselves. For high-resolution airborne measurements this information can be obtained from the laser echo strength and from auxiliary photography of the location of open leads (Kurtz et al., 2013). Overall, the uncertainty of laser altimetric mean ice thickness retrievals is estimated to be between 0.22 and 0.7 m (Kurtz et al., 2013).

2.5 Ice thickness measurement intercomparison and long-term trends

This chapter has demonstrated how important ice thickness information is for interpretations of sea ice change, and that it provides complementary information about the underlying processes of volume changes. However, due to the different approaches of the methods presented in this chapter, and due to variable spatial and temporal coverage resulting from logistical and funding constraints, it is actually difficult to construct a unified record of long-term sea ice thickness change (Lindsay & Schweiger, 2015). Figure 2.11 shows a compilation of various Arctic data sets published so far, from ULS, satellite laser altimetry (ICESat; see Chapter 9), and EM sounding (Meier et al., 2014). It shows how mean

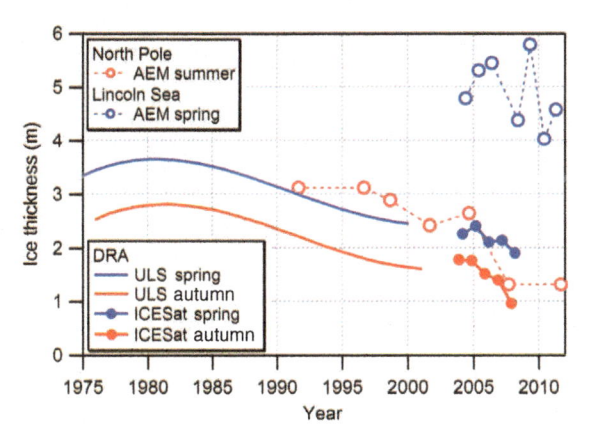

Figure 2.11 Time series of Arctic sea ice thickness changes in spring (February–March) and autumn (October–November) from 1975 to 2012, from submarine upward-looking sonar (ULS), ICESat laser altimeter and airborne electromagnetic (AEM) observations (Meier et al., 2014). The submarine and ICESat laser estimates were averaged over much of the central Arctic basin, the 'data release area' (DRA; Kwok & Rothrock, 2009) while the AEM data are from near the North Pole (Haas et al., 2008) and the Lincoln Sea north of Canada and Greenland (Haas et al., 2010).

summer ice thickness in the central Arctic Ocean has generally decreased from > 2.5 m to < 1.5 m between 1980 and 2012. It can also be seen that seasonal variations of ≤ 1.2 m have to be taken into account when comparing data from different seasons, and that in this example the altimetry data show a smaller amplitude of seasonal change than ULS (Kwok & Rothrock, 2009). The EM measurements (Haas et al., 2008) show the same long-term change in summer; however, they agree more with ULS winter data in the first period and with altimetry summer data in the latter. In addition, EM data are also available from the region of the thickest ice north of Canada and Greenland, and show that there the ice is still ~4 m thick. The observed changes are also clearly related to the retreat of old ice from the North Pole and towards the coast of Canada (Haas et al., 2008; Kwok et al., 2009; Maslanik et al., 2011).

Different methods have rarely been used along the same profiles for direct comparison of their uncertainties. Direct comparison is also very difficult given the different footprint sizes of the methods, and the large small-scale thickness variability of most sea ice sites. However, indirect, large-scale comparisons of the different methods have been performed with regard to the validation of satellite retrieved and modelled ice thicknesses. The latter provide a common reference for the different validation measurements which have been performed in different regions and at different times. Figure 2.12 shows the results of one such effort using CryoSat data (Laxon et al., 2013). It can be seen that correlations between different methods with CryoSat range between 0.6 and 0.9, and that mean differences are between 0.05 and 0.08 m. Note that these good results are affected by averaging over large regions and numbers of samples.

In similar ways, comparisons with model output were performed to identify systematic differences between different data sets and model results (Johnson et al., 2011; Schweiger et al., 2011; Stroeve et al., 2014). These studies found that systematic biases of between 0.4 and 0.7 m exist between models and observations, and between observations themselves. A direct Arctic-wide regression analysis between mean monthly ULS, EM and altimetry data sets showed good agreement (within 0.15) between most measurements, while systematic differences of up to 0.5 m were found for a few (Lindsay & Schweiger, 2015). There is clearly a need for more method intercomparison projects and a better understanding of the uncertainties involved in each method, such that comparability can be improved.

2.6 Dynamic and thermodynamic thickness changes

This chapter has illustrated how dynamic thickening and thinning due to deformation can strongly contribute to the ice mass balance. Indeed, there is evidence that much of the observed early changes in Arctic ice thickness (Kwok & Rothrock, 2009) have been caused by changes in atmospheric circulation, e.g. related to the Arctic Oscillation, and that these have led to a redistribution of thick and thin ice on Arctic-wide scales (e.g. Zhang et al., 2000; Holloway and Sou, 2002). In recent years, the impact of the Arctic Oscillation is less clear, and additional important circulation patterns have been identified, e.g. the Dipole Anomaly (Wu et al., 2006) and Central Arctic Index (Vihma et al., 2012). While these recent studies have shown that the nature of the dominant circulation has changed, they do not imply that the mechanisms of dynamic thickening have changed.

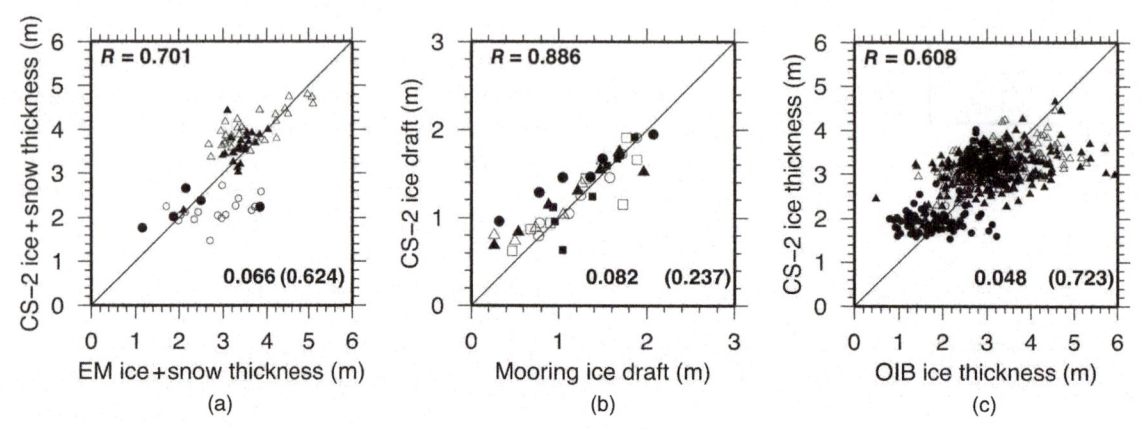

Figure 2.12 Validation of CryoSat (CS-2) sea ice thickness retrievals with: (a) coincident airborne electromagnetic (EM) data; (b) moored upward-looking sonar (ULS) data; and (c) Operation IceBridge (OIB) aircraft laser data (Laxon et al., 2013). Different symbols denote different time periods or ice types. Numbers show correlation coefficients (R) and mean differences and their standard deviations (in brackets).

This is illustrated by the example in Figure 2.13, showing how a positive ice thickness anomaly in the multi-year ice zone (MYZ) north of Canada in 2013/14 is closely related to anomalous ice drift, leading to strong convergence against the coast (Kwok & Cunningham, 2015). The MYZ roughly coincides with the region below the red line nearly parallel to the coast of Canada in Figure 2.13(a) and has been delineated from satellite radar data. Figure 2.13(a) shows that anomalous sea level pressure patterns in the summer of 2013 led to strong ice drift towards the coast of Canada. The net ice area advected across the red flux gate was small for the summers (June–August) of 2011 and 2012, but more than 0.45×10^6 km^2 in 2013, with the largest contribution in June (Figure 2.13b). Consequently the area between the flux gate and the coast decreased by 23.5 %, while there was no change in 2012 and a small expansion of 17.8% in 2011, i.e. net divergence (Kwok & Cunningham, 2015). Figure 2.13(c) shows how these different deformation regimes affected the ice thickness distribution as observed by CryoSat. In the fall of 2011 (October–November), mean multi-year ice thickness was smallest, at 1.59 ± 0.48 m. It was 1.94 ± 0.48 m in 2012 and increased to 2.31 ± 0.78 m in 2013, in good qualitative agreement with the different divergence. However, more importantly, thickness distributions in 2013/14 were characterized by longer tails resulting from larger mean thickness due to increased deformation in the convergent conditions during the 2013 summer. Strong dynamic thickening also coincided with lower air temperatures in the summer of 2013, expressed in melt degree-days in Figure 2.13(d). Combined, these dynamic and thermodynamic conditions have probably contributed to the partial recovery of Arctic sea ice extent after the record minimum year in 2012.

Most consideration of thermodynamic controls of the sea ice mass balance is given to the ice surface energy balance and melt-pond-related albedo feedback mechanisms during summer (see Chapter 4). However, as mentioned earlier, oceanic heat flux plays a very important role in growth and melt of the ice as well, and can vary strongly locally and between regions (Chapter 1). Melt has occasionally been observed to occur at the underside of Antarctic and Arctic pack ice during winter (e.g. Lytle & Ackley, 1996; Jackson et al., 2012). However, Arctic fast ice has been considered to be relatively uniform and its mass balance is mostly controlled by snow (Flato & Brown, 1996).

During extensive snowmobile journeys with a sled-based EM sensor over fast ice in Jones Sound in the Canadian Arctic Archipelago (CAA), large fast ice thickness variations over ice of similar age were observed (Figure 2.14; Melling et al., 2015). In some fjords the ice was only 0.5 m thick, where a thickness of almost 2 m would have been expected from CAA climatology. While differences in snow thickness could explain some of the variability within individual fjords, they were too small to cause the regional ice thickness differences. The regions of thin ice coincide with the occurrence of polynyas during spring melt, with

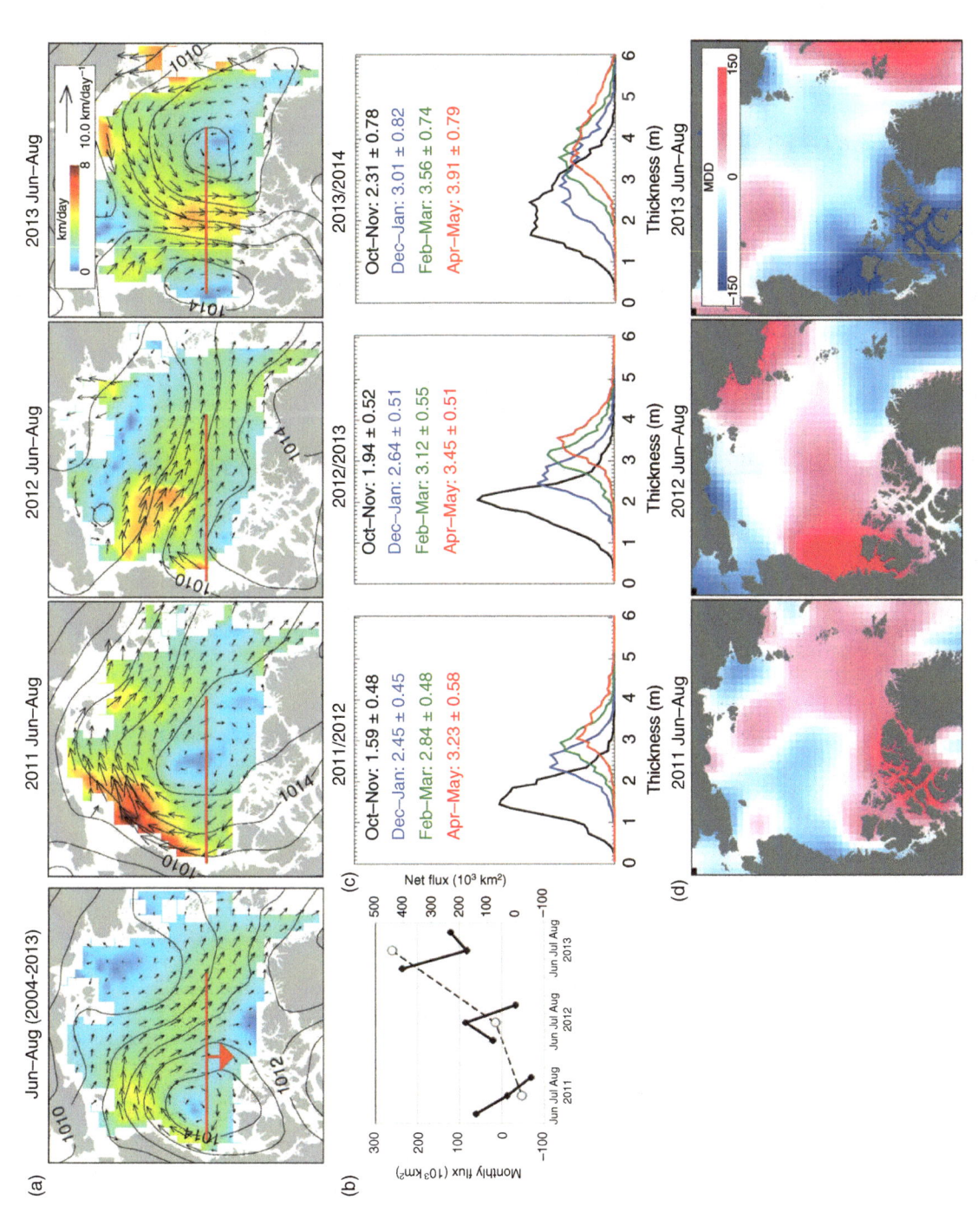

Sea ice thickness distribution 59

Figure 2.13 Changes of Arctic ice drift and ice thickness in the multi-year ice zone (MYZ) between 2011 and 2013 (from Kwok & Cunningham, 2015). (a) Mean (Jun–Aug) ice motion fields and sea level pressure (black lines and labels): decadal average (2004–2013), 2011, 2012 and 2013. (b) Monthly and net ice area flux through red gate shown in (a). (c) CryoSat-derived MYZ ice thickness distributions in the months after summer. (d) Anomalies in melting degree-days for 2011, 2012 and 2013 (June–August).

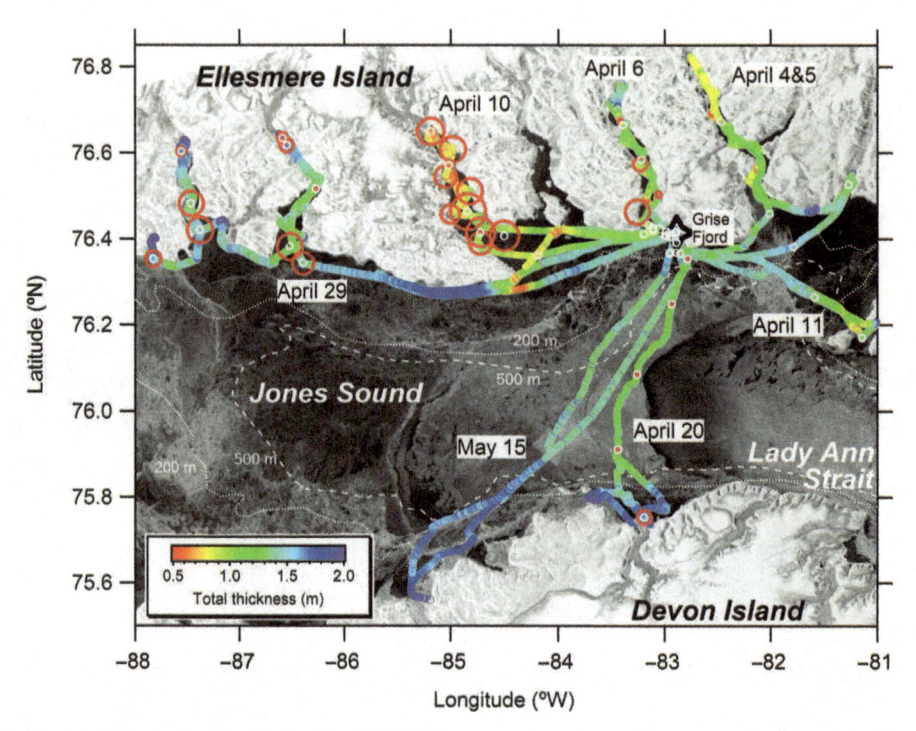

Figure 2.14 Ice + snow thickness profiles obtained in 2013 during extensive snowmobile surveys in Jones Sound in the Canadian Arctic (from Melling et al., 2015). The size of red circles represents manually measured snow thickness (4–39 cm) and white circles show locations of measurements of under-ice water temperature and salinity. Background synthetic aperture radar (SAR) image from RADARSAT-2 acquired on 4 February 2013. (Data and Products © MacDonald Dettwiler and Associates Ltd. (2013) – All Rights Reserved. RADARSAT is an official mark of the Canadian Space Agency.)

important implications for local weather, migration of wildlife and hunting practices.

Measurements of under-ice water temperature and salinity showed that the thin ice regions coincided with regions where water temperatures were elevated as much as 0.1 K above freezing (Melling et al., 2015). This increased oceanic heat flux is related to upwelling caused by tidally forced internal hydraulic flow over sills at the entrances of those fjords. The study by Melling et al. (2015) demonstrated the close relationships between ice thickness and ocean heat flux, and pointed to the potential of ice thickness measurements to identify such upwelling hot spots.

References

Amundrud, T., Melling, H., Ingram, R.G. & Allen, S.E. (2006) The effect of structural porosity on the ablation of sea ice ridges. *Journal of Geophysical Research*, **111**, C06004, doi:10.1029/2005JC002895.

Anderson, W.L., 1979: Computer Program. Numerical integration of related Hankel transforms of orders 0 and 1 by adaptive digital filtering. *Geophysics*, **44**, 1287–1305.

Arya, S.P. S. (1973) Contribution of form drag on pressure ridges to the air stress on Arctic ice. *Journal of Geophysical Research*, **78**(30), 7092–7099.

Arya, S.P. S. (1975) A drag partitioning theory for determining the large-scale roughness parameter and wind stress on the Arctic pack ice. *Journal of. Geophysical Research*, **80**, 3447–3454.

Barry, R.G., Serreze, M.C., Maslanik, J.A. & Preller, R.H. (1993) The Arctic sea ice climate system: observations and modeling. *Reviews of Geophysics*, **31**, 397–422.

Bourke, R.H. & Garrett, R.P. (1987) Sea ice thickness distribution in the Arctic Ocean. *Cold RegionsScience and Technology*, **13**, 259–280.

Castellani, G., Lüpkes, C., Hendricks, S. & Gerdes, R. (2014) Variability of Arctic sea-ice topography and its impact on the atmospheric surface drag. *Journal of Geophysical Research, Oceans*, **119**, 6743–6762.

Colony, R. and Thorndike, A.S. (1984) An estimate of the mean field of Arctic sea ice motion. *Journal of Geophysical Research*, **89**, 10623–10629.

Coon, M.D. (1980) A review of AIDJEX modelling. In: *Sea Ice Processes and Models*, (Ed. R.S. Pritchard), pp. 12–27. University of Washington Press, Seattle, WA.

Dowdeswell, J.A., Evans, J., Mugford, R. et al. (2008) Autonomous underwater vehicles (AUVs) and investigations of the ice-ocean interface: Deploying the Autosub AUV in Antarctic and Arctic waters. *Journal of Glaciology*, **54**, 661–672.

Eicken, H., Gradinger, R., Salganek, M., Shirasawa, K., Perovich, D. & Lepparanta, M., (Eds., 2009) *Field techniques for Sea-ice Research*. University of Alaska Press, Fairbanks.

Eicken, H., Grenfell, T.C., Perovich, D.K., Richter-Menge, J.A. & Frey, K. (2004) Hydraulic controls of summer Arctic pack ice albedo. *Journal of Geophysical Research*, **109**, C08007, doi:10.1029/2003JC001989.

Eicken, H., Lensu, M., Lepparanta, M., Tucker, W.B., Gow, A.J. & Salmela, O. (1995) Thickness, structure and properties of level summer multi-year ice in the Eurasian sector of the Arctic Ocean. *Journal of Geophysical Research*, **100**(C11), 22697–22710.

Flato, G.M. & Brown, R.D. (1996) Variability and climate sensitivity of landfast Arctic sea ice. *Journal of Geophysical Research*, **101**, 25767–25778.

Flato, G.M. & Hibler, W.D., III (1995) Ridging and strength in modeling the thickness distribution of Arctic sea ice. *Journal of Geophysical Research*, **100**, 18,611–18,626.

Galley, R.J., Trachtenberg, M., Langlois, A. et al. (2009) Observations of geophysical and dielectric properties and ground penetrating radar signatures for discrimination of snow, sea ice and freshwater ice thickness. *Cold Regions Science and Technology*, **57**: 29–38.

Giles, K.A., Laxon, S.W. & Worby, A.P. (2008) Antarctic sea ice elevation from satellite radar altimetry. *Geophysical Research Letter*, **35**, L03503, doi:10.1029/2007GL031572.

Haas, C., Hendricks, S., Eicken, H. & Herber, A. (2010) Synoptic airborne thickness surveys reveal state of Arctic sea ice cover. *Geophysical Research Letters*, **37**, L09501, doi:10.1029/2010GL042652 .

Haas, C. & Druckenmiller, M. (2009) Ice thickness and roughness measurements. In: *Field Techniques for Sea-ice Research* (Eds. Hajo Eicken et al.), pp. 49–116. University of Alaska Press, Fairbanks.

Haas, C., Lobach, J., Hendricks, S., Rabenstein, L. & Pfaffling, A. (2009) Helicopter-borne measurements of sea ice thickness, using a small and lightweight, digital EM system. *Journal of Applied Geophysics*, **67**, 234–241.

Haas, C., Pfaffling, A., Hendricks, S., Rabenstein, L., Etienne, J.L. & Rigor, I., (2008) Reduced ice thickness in Arctic Transpolar Drift favors rapid ice retreat. *Geophysical Research Letters*, **35**, L17501, doi:10.1029/2008GL034457.

Haas, C. & Jochmann, P. (2003) Continuous EM and ULS thickness profiling in support of ice force measurements. In: *Proceedings of the 17th International Conference on Port and Ocean Engineering under Arctic Conditions*, POAC '03, Trondheim, Norway, vol. 2 (Eds. S. Loeset, B. Bonnemaire, & Bjerkas, M.), pp. 849–856. Department of Civil and Transport Engineering, Norwegian University of Science and Technology NTNU, Trondheim, Norway.

Haas, C. (1998) Evaluation of ship-based electromagnetic-inductive thickness measurements of summer sea-ice in the Bellingshausen and Amundsen Sea. *Cold Regions Science and Technology*, **27**, 1–16.

Haas, C., Gerland, S., Eicken, H. & Miller, H. (1997) Comparison of sea-ice thickness measurements under summer and winter conditions in the Arctic using a small electromagnetic induction device. *Geophysics*, **62**, 749–757.

Hansen, E., Gerland, S., Granskog, M.A. et al. (2013) Thinning of Arctic sea ice observed in Fram Strait: 1990–2011. *Journal of Geophysical Research, Oceans*, **118**, 5202–5221.

Hanssen-Bauer, I. & Gjessing, Y.T. (1988) Observations and model calculations of aerodynamic drag on sea ice in the Fram Strait. *Tellus, Ser. A*, **40**, 151–161.

Hibler, W.D., III (1979) A dynamic thermodynamic sea ice model. *Journal of Physical Oceanography*, **7**, 987–1015.

Holloway, G. & Sou, T. (2002) Has Arctic Sea Ice Rapidly Thinned? *Journal of Climate* **15**, 1691–1701.

Holt, B., Kanagaratnam, P., Gogineni, S.P., Ramasami, V.C., Mahoney, A. & Lytle, V. (2008) Sea ice thickness measurements by ultrawideband penetrating radar: First results. *Cold Regions Science and Technology*, doi:10.1016/j.coldregions.2008.04.007.

Hopkins, M.A. (1994) On the ridging of intact lead ice. *Journal of Geophysical Research*, **99**, 16351–16360.

Hunkeler, P.A., Hendricks, S., Hoppmann, M., Paul, S. & Gerdes, R., (2015) Towards an estimation of sub-sea-ice platelet-layer volume with multi-frequency electromagnetic induction sounding. *Annals of Glaciology* **56**(69), 137–146.

Hvidegaard, S.M. & Forsberg, R. (2002) Sea-ice thickness from airborne laser altimetry over the Arctic Ocean north of Greenland. *Geophysical Research Letters*, **29**(20), 1952, doi:10.1029/2001GL014474.

Jackson, J.M., Williams, W.J. & Carmack, E.C. (2012) Winter sea-ice melt in the Canada Basin, Arctic Ocean. *Geophysical Research Letters*, **39**, L03603, doi:10.1029/2011GL050219.

Jahn, A., Sterling, K., Holland, M.M., Kay, J.E., Maslanik, J.A., Bitz, C.M., Bailey, D.A., Stroeve, J., Hunke, E.C., Lipscomb, W.H. & Pollak, D.A., (2012) Late-twentieth-century simulation of Arctic sea ice and ocean properties in the CCSM4. *Journal of Climate*, **25**, 1431–1452. doi:10.1175/JCLI-D-11-00201.1.

Johnson, M., Proshutinsky, A., Aksenov, Y. et al. (2012). Evaluation of Arctic sea ice thickness simulated by Arctic Ocean Model Intercomparison Project models. *Journal of Geophysical Research*, **117**, C00D13, doi:10.1029/2011JC007257.

Kanagaratnam, P., Markus, T., Lytle, V. et al.(2007) Ultra-wideband radar measurements of thickness of snow over sea ice. *Transactions on Geoscience and Remote Sensing*, **45**, 2715–2724.

Kottmeier, C., Olf, J., Frieden, W. & Roth, R. (1992) Wind forcing and ice motion in the Weddell Sea region. *Journal of Geophysical Research*, **97**, 20373–20382.

Kovacs, A., Holladay, J.S. & Bergeron, C.J.J. (1995) The footprint/altitude ratio for helicopter electromagnetic sounding of sea ice thickness: comparison of theoretical and field estimates. *Geophysics*, **60**, 374–380.

Kovacs, A. & Holladay, J.S. (1990) Sea-ice thickness measurements using a small airborne electromagnetic sounding system. *Geophysics*, **55**, 1327–1337.

Kovacs, A. & Morey, R.M. (1991) Sounding sea-ice thickness using a portable electromagnetic induction instrument. *Geophysics*, **56**, 1992–1998.

Krishfield, R.A., Proshutinsky, A., Tateyama, K. et al.(2014) Deterioration of perennial sea ice in the Beaufort Gyre from 2003 to 2012 and its impact on the oceanic freshwater cycle. *Journal of Geophysical Research Oceans*, **119**, 1271–1305.

Krumpen, T., Hölemann, J.A., Willmes, S. et al. (2011) Sea ice production and water mass modification in the eastern Laptev Sea. *Journal of Geophysical Research*, **116**, C05014, doi:10.1029/2010JC006545.

Kurtz, N.T., Farrell, S.L., Studinger, M. et al. (2013) Sea ice thickness, freeboard, and snow depth products from Operation IceBridge airborne data. *The Cryosphere*, **7**, 1035–1056.

Kurtz, N.T. & T. Markus (2012) Satellite observations of Antarctic sea ice thickness and volume. *Journal of Geophysical Research*, **117**, C08025, doi:10.1029/2012JC008141.

Kwok, R. & Cunningham, G.F. (2015) Variability of Arctic sea ice thickness and volume from CryoSat-2. *Philosophical Transactions of the Royal Society*, in press.

Kwok, R. & Maksym, T. (2014) Snow depth of the Weddell and Bellingshausen Sea ice covers from IceBridge surveys in 2010 and 2011: an examination. *Journal of Geophysical Research*, **119**, doi: 10.1002/2014JC009943

Kwok, R., Spreen, G. & Pang, S. (2013) Arctic sea ice circulation and drift speed: Decadal trends and ocean currents. *Journal of Geophysical Research Oceans*, **118**, 2408–2425.

Kwok, R. & Rothrock, D.A. (2009) Decline in Arctic sea ice thickness from submarine and ICESat records: 1958–2008. *Geophysical Research Letters*, **36**, L15501, doi:10.1029/2009GL039035.

Kwok, R., Cunningham, G.F., Wensnahan, M., Rigor, I., Zwally, H.J. & Yi, D. (2009) Thinning and volume loss of the Arctic Ocean sea ice cover: 2003–2008. *Journal of Geophysical Research*, **114**, C07005, doi:10.1029/2009JC005312.

Landrum, L. Holland, M.M., Schneider, D.P & Hunke, E., (2012) Antarctic sea ice climatology, variability, and late twentieth-century change in CCSM4. *Journal of Climate*, **25**, 4817–4838.

Landy, J., Ehn, J., Shields, M. & Barber, D. (2014) Surface and melt pond evolution on landfast first-year sea ice in the Canadian Arctic Archipelago. *Journal of Geophysical Research Oceans*, **119**, 3054–3075.

Laxon, S.W., Giles, K.A., Ridout, A.L. et al. (2013) CryoSat-2 estimates of Arctic sea ice thickness and volume. *Geophysical Research Letters*, **40**, doi:10.1002/grl.50193.

Leppäranta, M. (2011) *The Drift of Sea Ice*. Springer-Verlag, Berlin.

Lindsay, R. & Schweiger, A. (2015) Arctic sea ice thickness loss determined using subsurface, aircraft, and satellite observations. *The Cryosphere*, in press.

Lipscomb, W.H., Hunke, E.C., Maslowski, W. & Jakacki, J. (2007): Ridging, strength, and stability in high-resolution sea ice models. *Journal of Geophysical Research*, **112**, C03S91, doi:10.1029/2005JC003355.

Lytle, V.I. & Ackley, S.F. (1996) Heat flux through sea ice in the western Weddell Sea: convective and conductive transfer processes. *Journal of Geophysical Research*, **101**(C4), 8853–8868.

Maksym, T., Stammerjohn, S.E., Ackley, S. & Massom, R., (2012) Antarctic sea ice – A polar opposite? *Oceanography* **25**(3): 140–151.

Maslanik J, Stroeve J, Fowler C, & Emery W. (2011) Distribution and trends in Arctic sea ice age through spring 2011. Geophysical. *Research Letters*, **38**, L13502, doi:10.1029/2011GL047735.

Massom, R.A., Eicken, H., Haas, C. et al. (2001) Snow on Antarctic sea ice. *Reviews of Geophysics*, **39**, 413–445.

McNeill, J.D. (1980), Electromagnetic terrain conductivity measurements at low inductions numbers, *Technical Note TN-6*. Geonics Limited, Mississauga, Ontario.

Meier, W.N., Hovelsrud, G.K., Van Oort, B.E.H. et al. (2014) Arctic sea ice in transformation: a review of recent observed changes and impacts on biology and human activity. *Reviews of Geophysics*, doi: 10.1002/2013RG000431.

Meier, W. & Haas, C. (2011). Changes in the physical state of sea ice. In: *AMAP, 2011. Snow, Water, Ice and Permafrost in the Arctic (SWIPA): Climate Change and the Cryosphere*. Arctic Monitoring and Assessment Programme (AMAP), Oslo, Norway. xii + 538 pp. 9–4 to 9–18.

Melling, H., Haas, C. & Brossier, E. (2015) Invisible 'polynyas': modulation of fast ice thickness by ocean heat flux on the Canadian polar shelf. *Journal of Geophysical Research, Oceans*, **120**, doi:10.1002/2014JC010404.

Melling, H., Riedel, D.A. & Gedalof, Z. (2005) Trends in the draft and extent of seasonal pack ice, Canadian Beaufort Sea. *Geophysical Research Letters*, **32**, L24501, doi:10.1029/2005GL024483.

Melling, H. & Riedel, D.A., (2004) Draft and movement of pack ice in the Beaufort Sea: a time-series presentation April 1990–August 1999. *Canadian Technical Reports Hydrographic and Oceanic Science*, **238**, 24 pp.

Melling, H. & Riedel, D.A. (1995) The underside topography of sea ice over the continental shelf of the Beaufort Sea in the winter of 1990. *Journal of Geophysical Research*, **100**, 13 641–13 653.

Multala, J., Hautaniemi, H., Oksama et al. (1996) An airborne electromagnetic system on a fixed wing aircraft for sea ice thickness mapping. *Cold Regions Science and Technology*, **24**, 355–373.

Newman, T., Farrell, S.L., Richter-Menge, J. et al. (2014) Assessment of radar-derived snow depth over Arctic sea ice. *Journal of Geophysical Research Oceans*, **119**, doi:10.1002/2014JC010284.

Nicolaus, M., Haas, C. & Willmes, S. (2009) Evolution of first-year and second-year snow properties on sea ice in the Weddell Sea during spring-summer transition. *Journal of Geophysical Research*, **114**, D17109, doi:10.1029/2008JD011227.

Nicolaus, M., Haas, C., Bareiss, J. & Willmes, S. (2006) A model study of differences of snow thinning on Arctic and Antarctic first-year sea ice during spring and summer. *Annals of Glaciology*, **44**, 146–153.

Notz, D. (2009) The future of ice sheets and sea ice: between reversible retreat and unstoppable loss. *Proceedings of the National Academy of Sciences, U.S.A.*, **106**, 20 590–20 595.

Panzer, B., Gomez-Garcia, D., Leuschen, C. et al. (2013) An ultra-wideband, microwave radar for measuring snow thickness on sea ice and mapping near-surface internal layers in polar firn. *Journal of Glaciology*, **59**(214), doi: 10.3189/2013JoG12J128.

Perovich, D.K., Grenfell, T.C., Richter-Menge, J.A., Light, B., Tucker III, W.B. & Eicken H. (2003) Thin and thinner: Sea ice mass balance measurements during SHEBA. *Journal of Geophysical Research*, **108**(C3), 8050, doi:10.1029/2001JC001079.

Pfaffling, A., Haas, C. & Reid, J.E. (2007). A direct helicopter EM sea ice thickness inversion, assessed with synthetic and field data. *Geophysics*, **72**, F127-F137.

Polashenski, C.M., Perovich, D.K. & Courville, Z., (2012) The mechanisms of sea ice melt pond formation and evolution. *Journal of Geophysical Research*, **117**, C01001, doi:10.1029/2011JC007231.

Prinsenberg, S.J., Peterson, I.K. & Holladay, J.S. (2008) Measuring the thicknesses of the freshwater-layer plume and sea ice in the land-fast ice region of the Mackenzie Delta using helicopter-borne sensors. *Journal of Marine Systems*, **74**, 783–793.

Prinsenberg, S.J., Holladay, J.S. & Lee, J. (2002) Measuring ice thickness with EISFlowTM, a fixed-mounted helicopter electromagnetic-laser system. In: *Proceedings of the Twelfth (2002) International Offshore and Polar Engineering Conference, May 26–May 31, 2002*, **1**, 737–740, Kitakyushu, Japan.

Rabenstein, L., Hendricks, S., Martin, T., Pfaffhuber, A. & Haas, C. (2010) Thickness and surface-properties of different sea-ice regimes within the Arctic Trans Polar Drift: data from summers 2001, 2004 and 2007. *Journal of Geophysical Research*, **115**, C12059, doi:10.1029/2009JC005846.

Rack, W., Haas, C. & Langhorne, P.J. (2013) Airborne thickness and freeboard measurements over the McMurdo Ice Shelf, Antarctica, and implications for ice density. *Journal of Geophysical Research Oceans*, **118**, doi:10.1002/2013JC009084.

Rampal, P., Weiss, J. & Marsan, D. (2009) Positive trend in the mean speed and deformation rate of Arctic sea ice, 1979–2007. *Journal of Geophysical Research*, **114**, C05013, doi:10.1029/2008JC005066.

Reid, J.E., Pfaffling, A. & Vrbancich, J. (2006) Airborne electromagnetic footprints in 1D earths. *Geophysics*, **71**, G63–G72.

Reid, J.E., Worby, A.P., Vrbancich, J. & Munro, A.I.S (2003) Shipborne electromagnetic measurements of Antarctic sea-ice thickness. *Geophysics*, **68**, 1537–1546.

Richter-Menge, J.A., & Farrell, S.L. (2013) Arctic sea ice conditions in spring 2009–2013 prior to melt. *Geophysical Research Letters*, **40**, 5888–5893.

Rossiter, J.R. & Holladay, J.S. (1994) Ice thickness measurements. In: *Remote Sensing of Sea Ice and Icebergs* (Eds. Haykin, S. et al.), pp. 141–176. Wiley.

Rothrock, D.A., Percival, D.B. & Wensnahan, M. (2008) The decline in arctic sea-ice thickness: Separating the spatial, annual, and interannual variability in a quarter century of submarine data. *Journal of Geophysical. Reserach*, **113**, C05003, doi:10.1029/2007JC004252.

Rothrock, D.A. & Wensnahan, M. (2007) The accuracy of sea-ice drafts measured from U.S. Navy submarines. *Journal of Atmospheric and Oceanic Technology*, doi:10.1175/JTECH2097.1.

Schweiger, A., Lindsay, R., Zhang, J., Steele, M., Stern, H. & Kwok, R. (2011) Uncertainty in modeled Arctic sea ice volume. *Journal of Geophysical Research*, **116**, C00D06, doi:10.1029/2011JC007084.

Schwegmann, S., Haas, C., Fowler, C. & Gerdes, R. (2011). A comparison of satellite-derived sea-ice motion with drifting-buoy data in the Weddell Sea, Antarctica. *Annals of Glaciology*, **52**, 103–110.

Spreen, G., Kwok, R. & Menemenlis, D. (2011) Trends in Arctic sea ice drift and role of wind forcing: 1992–2009. *Geophysical Research Letters*, **38**, L19501, doi:10.1029/2011GL048970.

Strass V.H. & Fahrbach, E. (1998) Temporal and regional variation of sea ice draft and coverage in the Weddell Sea obtained from Upward Looking Sonars. In: *Antarctic Sea Ice: Physical Processes, Interactions and Variability*, (Ed. M.O. Jeffries), American Geophysical Union, Washington, D.C. *Antarctic Research Series*, **74**, 123–140.

Stroeve, J., Barrett, A., Serreze, M. & Schweiger, A. (2014) Using records from submarine, aircraft and satellites to evaluate climate model simulations of Arctic sea ice thickness. *The Cryosphere*, **8**, 1839–1854.

Tan, B., Li, Z., Lu, P., Haas, C. & Nicolaus, M. (2012) Morphology of sea ice pressure ridges in the northwestern Weddell Sea in winter. *Journal of. Geophysical Research*, **117**, C06024, doi:10.1029/2011JC007800.

Thorndike, A.S., Rothrock, D.A, Maykut, G.A & Colony, R. (1975) Thickness distribution of sea ice. *Journal of Geophysical Research*, **80**, 4501–4513.

Thorndike, A.S., Parkinson, C. & Rothrock, D.A. (Eds.) (1992) *Report of the Sea Ice Thickness Workshop, 19–21 November 1991*, New Carrollton, Applied Physics Laboratory, University of Washington, Seattle.

Tian-Kunze, X., Kaleschke, L., Maaß, N. et al. (2014) SMOS-derived thin sea ice thickness: algorithm baseline, product specifications and initial verification. *The Cryosphere*, **8**, 997–1018.

Uto, S., Toyota, T., Shimoda, H., Tateyama, K. & Shirasawa, K. (2006) Ship-borne electromagnetic induction sounding of sea-ice thickness in the southern Sea of Okhotsk. *Annals of Glaciology*, **44**, 253–260.

Vihma, T., Tisler, P. & Uotila, P. (2012) Atmospheric forcing on the drift of Arctic sea ice in 1989–2009. *Geophysical Research Letters*, **39**, L02501, doi:10.1029/2011GL050118.

Vinje, T., Nordlund, N. & Kvambekk, A. (1998) Monitoring ice thickness in Fram Strait. *Journal of Geophysical Research*, **103**, 10437–10450.

Wadhams, P. (2000) *Ice in the Ocean*. Gordon and Breach Science Publishers, Amsterdam.

Warren, S.G., Rigor, I.G., Untersteiner, N. et al. (1999) Snow depth on Arctic sea ice. *Journal of Climate*, **12**, 1814–1829.

Weiss, J. (2013) *Drift, Deformation, and Fracture of Sea Ice*. Springer.

Williams, G., Maksym, T., Wilkinson, J. et al.(2015) Thick and deformed Antarctic sea ice mapped with autonomous underwater vehicles. *Nature Geoscience*, **8**, 61–67.

Worby, A.P., Geiger, C.A., Paget, M.J., Van Woert, M.L., Ackley, S.F. & DeLiberty, T.L. (2008) Thickness distribution of Antarctic sea ice. *Journal of Geophysical Research*, **113**, C05S92, doi:10.1029/2007JC004254.

Worby, A.P., Jeffries, M.O., Weeks, W.F., Morris, K. & Jaña, R. (1996) The thickness distribution of sea ice and snow cover during late winter in the Bellingshausen and Amundsen Seas, Antarctica. *Journal of Geophysical Research*, **101**, 28 441–28 455.

Wu, B., Wang, J. & Walsh, J.E. (2006) Dipole anomaly in the winter Arctic atmosphere and its association with sea ice motion. *Journal of Climate*, **19**, 210–225.

Zhang, J., Rothrock, D. & Steele, M. (2000) Recent changes in Arctic sea ice: the interplay between ice dynamics and thermodynamics. *Journal of Climate*, **13**, 3099–3114.

CHAPTER 3

Snow in the sea ice system: friend or foe?

Matthew Sturm[1] and Robert A. Massom[2,3]

[1] *Geophysical Institute, University of Alaska Fairbanks, Fairbanks, AK, USA*
[2] *Australian Antarctic Division, Kingston, Tasmania, Australia*
[3] *Antarctic Climate & Ecosystems Cooperative Research Centre, Hobart, Tasmania, Australia*

3.1 Introduction

There is an exquisite timing, a dance if you will, between snow and sea ice. In the autumn, and depending on the recent weather, snow can fall on newly formed ice or into the ocean if the ice has yet to form. It can pile up on the ice, depress the ice surface below sea level and become saturated with seawater. The wind can blow the snow about, either after or before an ice deformation event, to create deep drifts or sweep the snow off ice floes and into leads. If present, the snow can protect the ice from the sun's rays in spring and summer, or, if it has melted away, leave the ice exposed and vulnerable to melt itself. When drifts form, they can control if, and where, there will be solar-absorbing melt ponds in the Arctic, accelerating the ice melt long after the snow has gone away. This symbiosis between the snow and ice has profound implications for the fate of sea ice as the Earth's climate continues to warm. In this chapter, we examine the characteristics of that snow cover and the nature of that dance (Figure 3.1).

Most sea ice older than a few days is covered with snow, the only prerequisite being that the ice is thick enough to hold the snow that falls on it. The strong multi-faceted nature of the Arctic snow–sea ice system is shown in Figure 3.1. Once in place, the snow loads the ice and modifies the physical, climatic and biological processes taking place. When the ice is thin, the snow load may dominate this interaction (see numbers 2 and 3 in Figure 3.1), but as both the snow cover and ice thicken, the snow comes to control the thermal balance of the ice (4, 5 and 6). In large measure, this is due to a number of unique properties of snow: it has one of the highest albedos of any natural substance on Earth (see Chapters 1 and 4), it is one of the best thermal insulators per unit thickness, and it has very high specific and latent heat capacities that can buffer the ice temperature and retard or prevent ice melt. In the Arctic, before the ice can melt (7, 8 and 9), the snow has to melt (6), and as it does so, it supplies freshwater (7 and 8) to create melt ponds and reduce the salinity of the upper ocean. In the Antarctic, in particular, surface flooding by seawater is a widespread phenomenon (Massom et al., 2001), due in large part to the weight of the snow overburden on a relatively thin ice cover. Because of its properties and these physical interactions, snow is an integral component of the sea ice system, so much so, in fact, that when most people speak of the sea ice they really mean the ice *and* snow.

The snow itself is also important from a natural and human ecological perspective. The snow provides a habitat for bear and seal dens and lairs in the Arctic (Smith & Stirling, 1975; Amstrup & Gardner, 1994) and affects seal pup survival (Iacozza & Ferguson, 2014), and its presence reduces the amount of light penetrating into and through the ice (Järvinen & Leppäranta, 2011), thereby affecting biota within and under the ice (Arrigo et al., 2003, 2012). Moreover, snow on sea ice plays an important role in the biogeochemical system as a chemical reactor and reservoir of chemical components precipitated from the atmosphere (e.g. Lannuzel et al., 2007; Nomura et al., 2010a) and in the deposition of mercury in the Arctic (Chapter 19; Douglas et al., 2008; Pratt et al., 2013).

Most human travel for hunting and subsistence is done while the ice is covered with snow, which

---

Sea Ice, Third Edition. Edited by David N. Thomas.
© 2017 John Wiley & Sons, Ltd. Published 2017 by John Wiley & Sons, Ltd.

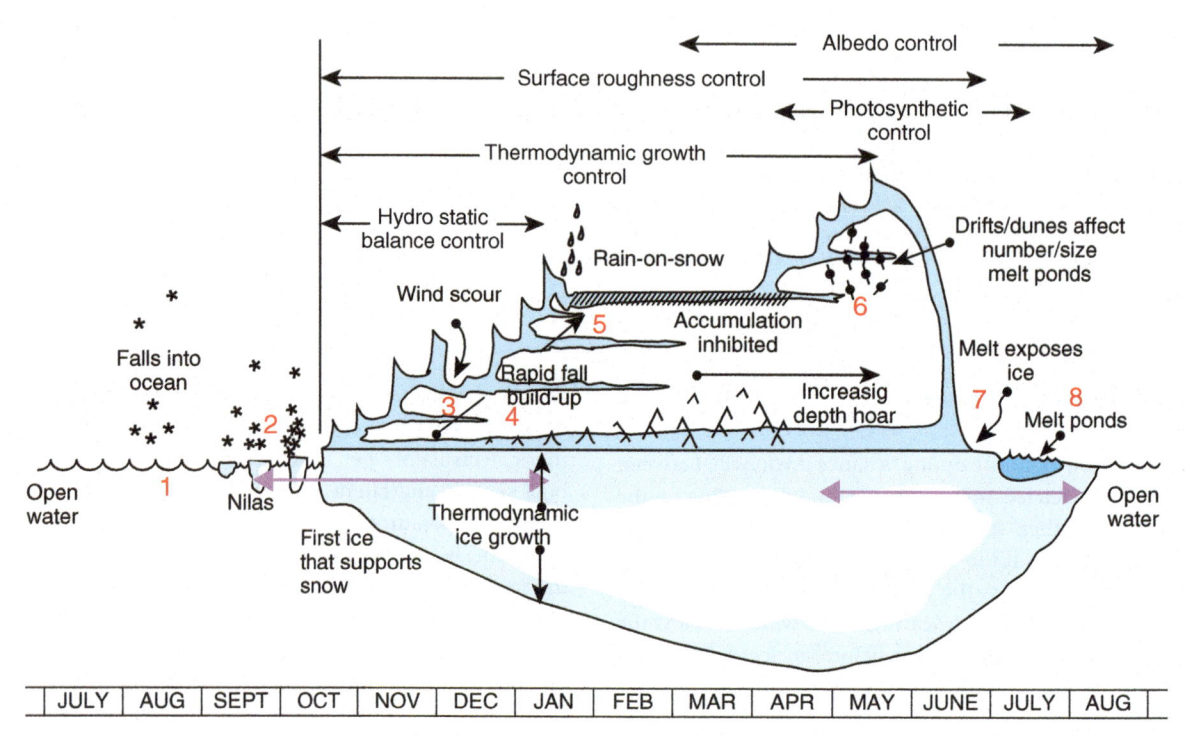

Figure 3.1 The Arctic snow and sea ice system through the winter. Snow–ice interactions are stochastic early in the winter (1 and 2). During the shoulder periods (purple arrows at 3 and 4, 6 and 7) snow–ice interactions are at their most dynamic and exert the greatest degree of control on the ice. By midwinter, congelation ice growth (4, 5 and 6) becomes a steady process controlled by the air temperature and moderated by the depth and metamorphic state of the snow. In spring, snowmelt sets the timing for ice melt and pre-conditions the ice.

can make travel easier (Chapter 25; Gearheard et al., 2006, 2013). By the same token, the snow reduces the efficiency of icebreakers through increased skin friction (Jones et al., 2007), and it alters how an oil spill might interact with the ice to affect clean-up efforts (Buist et al., 2009).

Regarding remote sensing, the snow complicates the interpretation and retrieval of sea ice geophysical parameters from satellite data by visually and electromagnetically cloaking the ice surface [for more in-depth coverage of remote-sensing issues, see Lewis et al. (1994), Barber et al. (1998), Lubin and Massom (2006) and Chapter 9]. This complicating factor is particularly true with respect to retrieval of sea ice thickness from laser and radar altimeter data, whereby independent knowledge of snow thickness and density is required (see Chapter 9).

Snow on ice would be a relatively simple topic were it not for its heterogeneity. For one thing, the snow is constantly redistributed by the wind (Leonard & Maksym, 2011), creating 10-fold variations in depth and snow properties over short distances. Research papers about snow on sea ice tend to report an average depth value, but in fact, it is rare to find a patch of snow more than a few metres across that has a constant depth. Large variations in depth occur with floe age and roughness, and the mechanical, thermal and optical properties of the snow vary in a similar manner due to the tendency of snow to metamorphose over time. Nonetheless, superimposed on this local spatial variability are large-scale patterns in snow cover, and distinct differences between these in the Arctic and Antarctic.

As detailed here, and in Chapters 4 and 9, we may be entering an era when the attributes and characteristics of the sea ice snow cover, as well as its heterogeneity, can be determined remotely. If this comes to pass, it will be a much-needed improvement in our abilities to study and understand critical processes related to changing sea ice conditions. However, such an improvement is also likely to mean that, increasingly, researchers will not experience the ice and snow first hand. Many subtleties concerning the snow cover are lost from a height of > 1000 m above the ice. It is beyond the scope of this chapter to provide detailed information on the techniques used to directly observe and measure snow on sea ice, but the reader is referred to Sturm (2010) for a primer on those techniques, to Fierz et al. (2009) for a general classification of deposited snow, and to Male and Gray (1981) for a wide-ranging treatment of snow properties and processes. The optical properties of snow on sea ice are covered in detail in Chapter 4 of this book. Here, we merely encourage readers to seek this direct, rather than remote, experience when possible.

3.1.1 Roots: a short history of snow on sea ice studies

Intrepid sailors who entered the pack ice more than 2000 years ago were undoubtedly the first observers to note that sea ice is usually covered by snow, though it is uncertain how much attention they paid to this detail. The first recorded account is of Pytheas, a Greek from Massilia, who sailed a bireme northwest of Scotland in 325 BC and encountered pack ice southeast of Iceland (Mowat, 1960). There were undoubtedly other early voyages, but the next recorded run-in with the ice seems to be that of the Celtic priest, Brendan, who sailed a leather currragh not only to Iceland and Greenland but also to Jan Mayen Island, all in 550 AD. In his extensive travels, Brendan must surely have noted that until late in the summer, the Arctic ice floes were covered by snow. By the 1550s, mariners were plying the Northeast Passage up and around Scandinavia and Russia for the purposes of trading, and occasionally they were forced to winter over on the northern coast. During those dull and interminable dark winters, they would have had an excellent first-hand look at the snow, as well as plenty of time to make observations on the snow cover, but if they did, no accounts have survived. The first proven account of snow on sea ice we have found is work of William Scoresby (1820), a whaling captain from England who travelled extensively in the Arctic waters near Greenland (Figure 3.2). Scoresby was one of a legion of navigators who tried to find and sail through the Northwest Passage (Cooke & Holland, 1978).

The search for the Northwest Passage continued until 1906 and is recounted in several engaging books, including *Ordeal by Ice* (Mowat, 1960) and *Arctic Grail* (Berton, 1989). Early voyages venturing south to the Antarctic ice pack started with Captain Cook's amazing journey in 1773 (Beaglehole, 1974), and reached a crescendo with Ross's epic voyages of 1841 and 1842 (Ross, 1847). Here again, explorers encountered sea ice with snow on the ice.

3.1.2 Snow on ice in northern cultures

Western explorations notwithstanding, we must look to northern indigenous cultures for the earliest in-depth knowledge of the sea ice snow. There are about 75 words for snow and snow features in Iñupiaq, the Eskimo language spoken in northern Alaska (Webster and Zibell, 1970; Sturm 2009a,2009b; Krupnik & Müller-Wille, 2010). A similar list exists for Inuktitut, the language of Greenland and eastern Canada, and for Yu'pik (Fienup-Riordan & Rearden, 2010). There is good reason to believe that these languages are thousands of years old, suggesting that these cultures had an appreciation of many of the features and subtleties of snow on ice that would not be learned by western science until the late 20th century (Table 3.1).

3.2 The Relationship Between Snow and Sea Ice

Sea ice requires only below-freezing ocean temperatures to form, but the snow cover needs three conditions: low temperatures to ensure that precipitation falls as snow, sleet or hail; precipitation itself; and an existing ice cover with sufficient bearing capacity to support the fallen snow. Regarding this latter requirement, much depends on the sequence of weather in autumn when the sea ice is beginning to form (Figure 3.1). If the weather is cold but with little snowfall, then rapid thermodynamic sea ice growth will occur because the insulating properties of the snow cover will not fully retard the ice growth (Holtzmark, 1955; Maykut & Untersteiner, 1971). With warmer but snowy autumn weather, first snowfalls are likely to coincide with the initial sea ice formation. The snow accumulating on the thin, newly formed sea ice will significantly slow the thermodynamic growth of the ice (Holtzmark, 1955; Wu et al., 1999; Fichefet et al., 2000) and at the same time depress it below sea level. In this scenario, the base of the snow becomes flooded with seawater and the formation of snow ice takes place – a common occurrence in Antarctica (Maksym & Jeffries, 2000; Jeffries et al., 2001). The two

Figure 3.2 Snowflakes observed by Captain William Scoresby while whaling near Greenland, 1810–1820. From: Scoresby (1820), p. 588, Plate VIII.

trajectories have the potential to produce quite different snow and ice thicknesses under the same snowfall regime. Looking at Figure 3.1, one can see that highly variable, essentially stochastic, weather events (the cycling of storms and cold periods), superimposed on the more predictable and slowly varying seasonal cycle of temperature, determine which pathway is followed.

Two key equations, linked by snow and ice thickness, snow density and thermal conductivity, describe this intimate interplay between snow and ice. The simplicity of these equations belies the many snow and ice trajectories that can be produced. The first equation describes the isostatic balance between floating ice and an overburden of snow, assuming that the ice is in

Table 3.1 Iñupiaq words for snow features (from Sturm, 2009a,b).

Word	Meaning
Akillukkak	Soft snow behind a drift
Apun	Snow cover; snow (generic)
Aluktinniq	Windswept snow-free area on the ice
Avoorik	A whaleback drift
Kamoruk	A long snow drift useful for navigation.
Katagagnatuaq	Deep melting snow or wet slushy area (a snow swamp)
Kiyokluk	An 'anvil head' drift or sastrugi
Katikronuk	Migrating dunes, still soft and eroding by the wind
Isrriqutit	Diamond dust or ice crystals in the air
Masallak	Snow damp enough to stick, as in making snowballs
Natigvik	Low drifting snow, no higher than the knee.
Nataovuoaluka	Saltating and suspended snow in a full-on blizzard
Nutagak	Fresh powder snow
Nutaak	Soft snow
Nutagun	New/fresh snow covering a hole in the ice
Pukak	Basal snow that has metamorphosed into depth hoar
Kayuktak	Ripple marks on the snow surface
Silliq	Hard, icy snow
Silliqruk	Old icy snow, extra hard

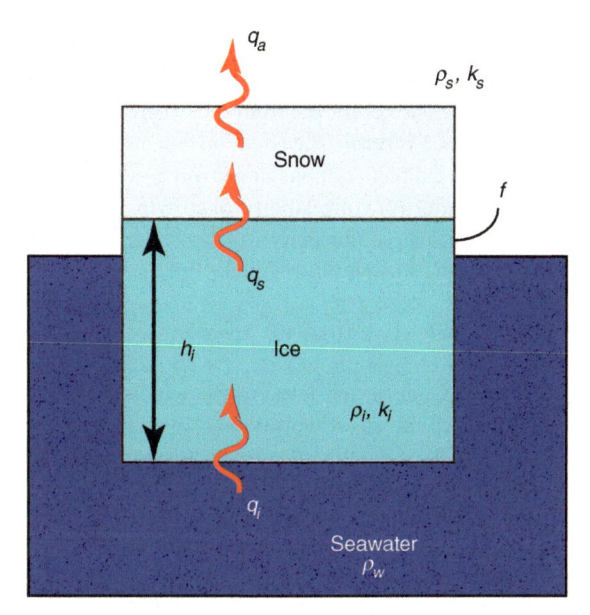

Figure 3.3 A schematic of a sea ice floe covered by a snow cover. The red lines indicate heat flow, which here is assumed to be only in the vertical direction, and hence must be equal at all levels along a vertical pathway, i.e. $q_a = q_s = q_i$. The heat flux from the ocean is assumed to be zero, so that all heat lost upwards is converted into ice growth at the base of the ice. ρ, density; q, heat loss; h, depth; k, thermal conductivity. Subscripts a, s, w and i refer to air, snow, water and ice, respectively.

hydrostatic equilibrium (Figure 3.3):

$$f = [h_i(\rho_w - \rho_i) - h_s\rho_s]/\rho_w \quad (3.1a)$$

where f is the ice freeboard, h is thickness, ρ is density, and the subscripts i, w and s refer to ice, water and snow, respectively. Equation (3.1a) states that ice freeboard is controlled by the difference between the ice buoyancy (first term on the right-hand side) and the weight of the snow (second term). Snow flooding potentially occurs when f is negative, or when:

$$h_s \geq h_i \left\{ \frac{\rho_w - \rho_i}{\rho_s} \right\} \quad (3.1b)$$

Approximate mean values of ρ_w, ρ_i and ρ_s are 1000, 900 and 300 kg m^{-3}, respectively, suggesting that if the snow depth to ice thickness ratio exceeds 1:3, then seawater flooding could occur. Flooding also depends on there being conduits between the ocean and ice surface, which in turn depends on the ice temperature and salinity (Golden et al., 2007), and/or state of deformation (Massom et al., 2001). However, a key point here is that seawater flooding diminishes the insulating capacity of the snow. Water-saturated snow conducts heat far better than dry snow, resulting (under freezing conditions) in a rapid refreezing of the 'snow-slush', converting it into snow ice. Through this thermodynamic shortcut, sea ice formation can, paradoxically, occur in the presence of a thick insulative snow cover which in normal circumstances would inhibit ice growth. A key difference between snow ice formation and 'conventional' thermodynamic (or congelation) ice formation is that the former occurs at the surface of the ice and the latter at the base.

The second equation describes the relationship between sea ice growth and heat loss (q):

$$\frac{dh_i}{dt} = \frac{1}{\rho_i L_i} \left\{ -k_s \frac{\Delta T_s}{h_s} \right\} \quad (3.2a)$$

where t is time, k_s is the bulk thermal conductivity of the snow (W m^{-1} K^{-1}), ΔT_s is the vertical temperature gradient across a snow cover h_s thick, and ρ_i and L_i are the density and the latent heat of fusion of ice, respectively

(Pounder, 1965). This simple equation for ice growth assumes that the heat flow through the ice equals that through the snow, and that there is no significant heat flow into the base of the ice from the ocean below (i.e. q_i in Figure 3.3 is zero). The equation essentially asserts that there is a direct relationship between heat loss and thermodynamic (or congelation) ice growth.

Snow thermal conductivity can be related to snow density (kg m^{-3}) using (Sturm et al., 1997):

$$k_s = 0.138 - 0.00101\rho_s + 0.000003233\rho_s^2 \quad (3.2b)$$

which when substituted into equation (3.2a) links ice growth directly to snow density and inversely to snow depth. For a fixed temperature drop across the snow, the denser the snow or the thinner the snow pack, the faster the rate of ice growth. In practice, as the ice gets thicker, the snow–ice interface temperature will decrease, and the temperature difference across the snow pack (ΔT_s) will be reduced, slowing the rate of ice growth. Snow thermal conductivity is discussed in more detail later.

The interactions described by equations (3.1) and (3.2) have been explored in numerous models. The one-dimensional sea ice model of Maykut and Untersteiner (1971) was the first to explicitly treat snow, prescribing a seasonal accumulation rate that resulted in a time-varying snow depth on the ice. Reducing the snow depth to zero in their model runs resulted in an increase in ice thickness over the base run, but so too did increasing the snow depth, suggesting there was a 'sweet spot', or an optimal snow depth, that produced the thinnest ice. Semtner (1976), building on the Maykut and Untersteiner model, identified this depth to be about 0.6 m. For depths less than this, the insulating property of the snow was reduced, producing more thermodynamic ice growth; for greater depths, the snow shielded the ice from spring and summer melting, to retain its thickness through the summer and produce thicker ice the following winter.

Ledley (1985) identified that these earliest models, in failing to include the isostatic balance between snow and ice, produced unrealistic multi-year scenarios of ice thickness because they allowed for snow depths that in reality would not have been possible. In reality, snow ice formation would in fact convert snow to ice. In her later papers (Ledley, 1991, 1993), Ledley also added the albedo effect of the snow (it is higher than that of the ice – see Chapter 4) and found even more complex snow–ice interactions, with the reflection of short-wave radiation interacting with the insulating properties, as well as snow ice production, to produce two sea ice thickness minima – one due to a lack of snow and another due to just enough snow to reduce congelation growth. Snow albedo was the most dominant aspect of the snow controls in her modelling work.

Using a snow model that included sublimation and snow blowing into leads, Fichefet and Morales Maqueda (1999) found striking differences between the impact of increasing snow precipitation rates in the Arctic versus the Antarctic, with the latter area being more sensitive due to the prevalence of snow ice formation. The insensitivity they found for the Arctic, however, might no longer exist now that the Arctic ice is thinner (Maslanik et al., 2007) and potentially more sensitive to snow loading rates. Powell et al. (2005) confirmed the results of Fichefet and Morales Maqueda for Antarctica, and in fact their work suggested a return to the simpler scenario described by Semtner, wherein decreased snow precipitation produces thicker ice due to reduced insulation by snow, while increased precipitation produces thicker ice due to snow ice formation (see also Wu et al., 1999).

The modelling work to date suggests that the Antarctic snow–ice system may be delicately balanced in a way that minimizes the current ice thickness, where either a decrease or an increase in snowfall rates could lead to an increase in ice thickness. In the Arctic, the state of the snow–ice system, as reflected in models, is less clear. Hezel et al. (2012), using CMIP5 results (Taylor et al., 2012), suggested that snow depths on Arctic sea ice are likely to decrease because a proportion of what used to fall as snow will come as rain in the future, or will fall into an unfrozen sea. They further suggested that the reduction in snow on the ice would promote congelation growth, slowing the climate change-driven thinning of Arctic sea ice. The first part of their prediction seems to be validated by the recent observations of Webster et al. (2014), who compared modern and past snow climatologies and found less snow now. However, recent unpublished results (Derksen et al., 2014) indicate that the snow radar used by Webster et al. has a low bias, calling into question these early snow climatological results. The second part of the Hezel et al. (2012) prediction remains to be seen.

The most recent snow and ice model results (Blazey et al., 2013), based on a fully coupled ice–ocean–climate

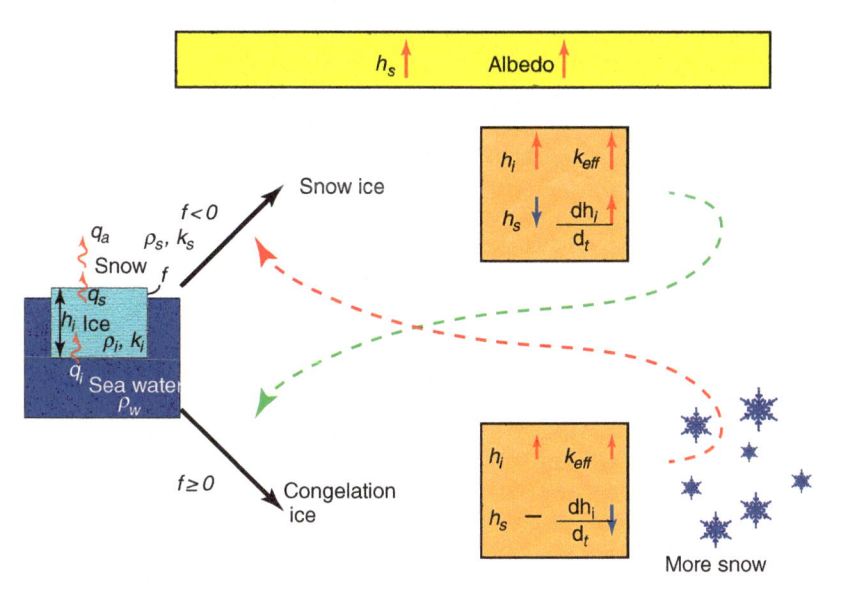

Figure 3.4 The effect of snow on sea ice, based on the models described in the text. More ice is produced through congelation growth when the snow is thin, or when the snow is so thick that snow ice is produced. A minimum amount of ice is produced at some configuration of snow and ice between these two alternatives. The actual shape of the snow ice function is not known. ρ, density; q, heat loss; h, depth; k, thermal conductivity. Subscripts a, s, w, eff and i refer to air, snow, water, effective and ice, respectively.

model (CCSM), continue to suggest that reduced snow cover indeed promotes thicker ice in the Arctic, but in a complex way that is strongly moderated by ice thickness. These most recent studies emphasize that changes in snow depth, or other snow properties, are likely to produce a highly non-linear response in the ice thickness, and caution over-interpreting current model findings. So, despite 45 years of snow and ice modelling, the subtleties of how snow affects ice thickness remain an unsolved problem, with potentially different (though possibly converging) answers in the Arctic and the Antarctic. Figure 3.4 illustrates the general impact of snow on ice growth, and in addition why this linked behaviour is both endlessly fascinating and confoundedly difficult to pin down.

3.3 General characteristics of snow on ice

3.3.1 Snowflakes

The impact of the snow on ice thickness may require more research, but we know a lot about the nature of individual snow grains and snow pack properties. The raw material from which the snow cover is built comes from snowflakes, those falling gems of exquisite symmetry that have fascinated humans for thousands of years (see Figure 3.2). Snowflakes have been the subject of numerous books (e.g. Nakaya, 1954; Libbrecht & Rasmussen, 2003), and several snowflake classification systems have been proposed (e.g. LaChapelle, 1992; Magono & Lee, 1966). Over 120 snowflake types have now been identified (Kikuchi et al., 2013). These can be grouped into two basic classes: those with hexagonal symmetry and euhedral crystalline forms; and those with more amorphous forms. The former class includes sector plates, stellar dendrites (Figure 3.5a: the 'classic' six-armed snowflake of Christmas decorations), columns and needles. The latter class includes graupel, ice pellets and hail.

3.3.2 Snow metamorphism

With respect to the sea ice snow cover, it hardly matters which class or type of snowflake falls because, once deposited, snowflakes quickly lose their original form, through the action of wind, gravitational settling and breakage, thermal effects and wettening. What is important is the amount of snow that falls; the rate of snow accumulation; whether the snow falls with or without wind; and whether any rain or above-freezing

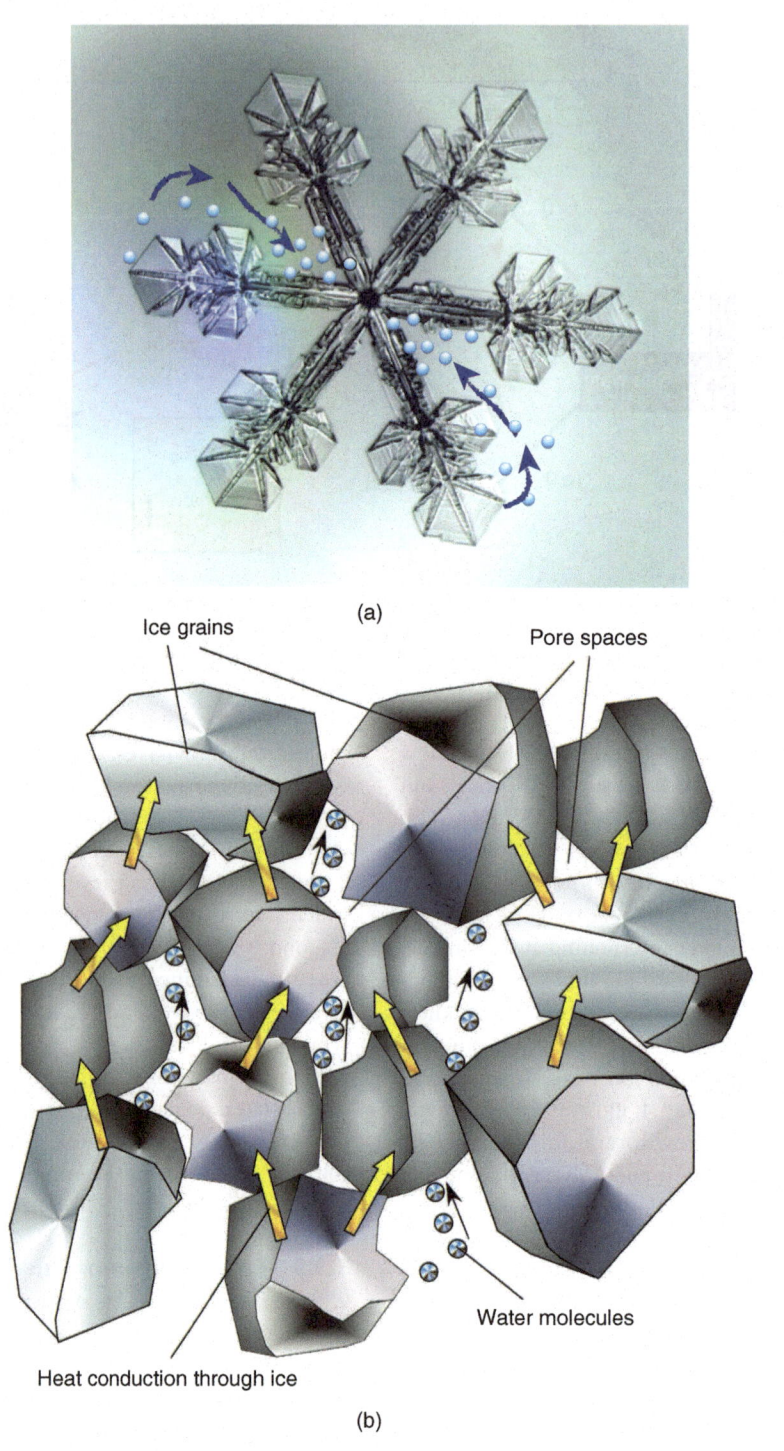

Figure 3.5 (a) A stellar dendrite (branched snowflake), showing schematically how vapour diffusion driven by the Kelvin (curvature) effect leads to a net transfer of water molecules (blue spheres) from branch points to cusps near the centre of the snowflake, rapidly degrading and rounding the crystal, until the crystal looks like the grains in (b). (b) Macroscopic temperature gradients across the snow pack produce a net upward flux of water molecules (small circles), accreting on the lower side of grains and sublimating from the upper sides, thereby changing the size and shape of the grains. The yellow arrows indicate heat flow by conduction through the ice network of grains. The black arrows indicate heat flow resulting from the vapour flux. Drawing by M. Sturm.

temperatures occur during the snowfall event and/or after the snow arrives. The high surface curvature of sharp-pointed snowflakes drives a net movement of water molecules from snowflake branch tips to cusps (negative curvature), rounding the grains (the Kelvin effect: see Figure 3.5a). Gravity-driven settling further breaks off fragile snowflake arms and aids snow compaction. Wind can even more effectively break and fragment snowflakes, creating sharp-point shards that can sinter (bond) rapidly to other grains. Temperature gradients across the snow pack drive a macroscopic diffusive flux of vapour from the base of the snow upwards, creating more rounding or, when the temperature gradient is strong enough, faceted crystalline grains (Figure 3.5b). Above-freezing temperatures can rapidly round grains and create thick bonds between grains. In short, in just a few hours, snowflakes are replaced by new forms by metamorphism. These changes impact the grain size, density and thermal conductivity of the snow, as well as its optical and electromagnetic properties.

Four types of snow metamorphism have been identified, though several interchangeable sets of terms have been used to describe two of these pathways (Table 3.2). One (called destructive metamorphism, equitemperature metamorphism or equilibrium growth) produces rounded grains when temperatures are below freezing but temperature gradients are weak. Another [constructive metamorphism, temperature gradient (TG) metamorphism or kinetic growth] produces ornate faceted grains that grow rapidly when a strong temperature gradient is imposed on the snow cover. Wind metamorphism produces small fragmented grains that sinter (bond) strongly. Lastly, wet snow metamorphism, which occurs when liquid water is introduced into the snow cover by above-freezing temperatures or rain or the presence of brine/seawater on the ice surface, results in rounded grains that can be extremely well bonded. Moreover, and particularly on Antarctic sea ice, a layer of wet snow occurs above the saturated snow-slush layer when flooding occurs (Massom et al., 2001).

3.3.3 Snow grain types

The International Classification for Snow on the Ground (Colbeck et al., 1990; Fierz et al, 2008) provides comprehensive descriptions of snow grains as well as pictures of each type. The sea ice environment differs from that on land, and just six basic types are typically found there (both Arctic and Antarctic: Massom et al., 2001; Sturm et al., 2002a). These are:

- new and recent snow;
- fine-grained snow;
- wind slab;
- faceted grains including depth hoar;
- icy layers;
- damp/wet snow and slush.

Micrographs of four of the most common types on sea ice are shown in Figure 3.6.

New/recent snow

New snow is generally defined as having fallen within the previous 24 hours, and is always at the top of the snow pack. Recent snow can retain features of new snow (i.e. snowflake forms) for days to weeks and is usually

Table 3.2 Types of metamorphic pathways for snow on sea ice with alternate names.

Type 1	Type 2	Type 3	Type 4
Destructive metamorphism[a]	Constructive metamorphism	Wind metamorphism	Wet snow metamorphism
Equitemperature metamorphism (ET)[b]	Temperature gradient metamorphism (TG)[b]		
Equilibrium growth[c]	Kinetic growth[c]	Wind slabs formed by sintering	Corn snow; melt-freeze
Rounded grains	Faceted grains	Small shard-like grains with percussive features	Rounded grains with water films that become ice bonds

[a]Seligman (1936); von Eugster (1950).
[b]Sommerfeld (1970).
[c]Colbeck (1982).

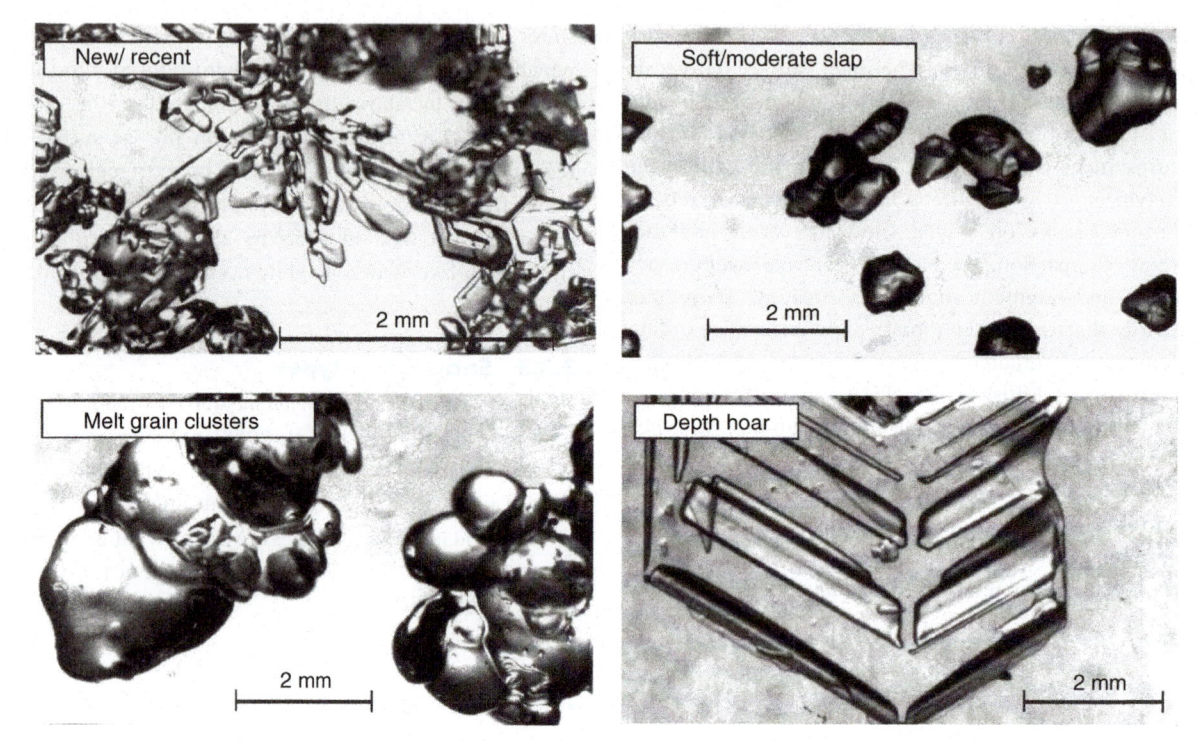

Figure 3.6 Four common types of snow grains found on sea ice. Source: Sturm et al., 1998. Reproduced with permission of John Wiley & Sons.

a little deeper in the pack. In the absence of wind, gravitational settlement is the primary metamorphic force acting on new and recent snow, with the amount of newly fallen snow (i.e. the overburden) determining the settlement rate and the density. Kojima (1966; see also Sturm & Holmgren, 1998) has modelled the settling of new snow as a viscous fluid, where the viscosity is highly dependent on the snow density. As the snow densifies, it becomes more viscous and therefore settles and compacts ever more slowly. This results in a characteristic exponential settling curve that depends on the initial density of the snow. Table 3.3 lists the range of densities of new and recent snow.

Fine-grained snow

Fine-grained snow is frequently a default identification when snow grains are small (<1 mm); when they show neither obvious characteristics of the initial snowflakes nor indications of wind action; and when they have no facets or other obvious characteristics of kinetic growth (Table 3.3). One source of possible confusion in identifying fine-grained snow is that a wind slab, while in

Table 3.3 New and recent snow densities (after McKay & Gray, 1981).

Snow type	Density (kg m^{-3})
Wild snow (still falling)	10–30
New snow	50–65
Recent snow	70–90
New snow with wind	60–80

the process of being formed, will be both fine-grained and soft, but within 24 hours of being deposited, the slab will sinter (see below) into a hard cohesive unit, while the fine-grained snow will not. The density of a fine-grained layer of snow rarely exceeds 350 kg m^{-3}, while wind slabs are likely to be denser.

Wind slab

Strong winds (> 8 m s^{-1}) both during and following snowfall events produce significant mechanical fragmentation and compaction of the snow (Massom et al., 1997), which creates medium- to high-density hard layers known as wind slabs (Seligman 1936; Benson & Sturm, 1993). This snow type is characterized by very small snow grains (0.1–0.5 mm) that are well bonded, has low permeability, and is ideal snow for constructing igloos. The wind transports snow grains by creep (rolling and sliding), saltation and suspension (Kobayashi, 1972; Pomeroy & Gray, 1995; Sturm & Stuefer, 2013). Saltation, which is also a mechanism that transports sand grains in the desert (Bagnold, 1954), occurs because the wind speed between 0.1 to 0.5 m above the snow surface is greater than at the surface itself. Grains ejected upward by grain–grain collisions are accelerated by the faster wind, sustaining the transport process.

When the wind finally ceases, the layer of fragmented snow grains that has been in transport will sinter into a hard mass. Sintering is a process by which a fine-grained material is brought to a temperature that is within a few tens of degrees of its melting point, then allowed to cool (German, 1996). Vapour and volume diffusion during the heating and cooling transfers molecules to the cusps formed where sharp-pointed grains touch each other, bonding the material into a solid mass (Blackford, 2007). Snow sinters readily because it is often within 20°C of its melting temperature. In general, the finer the grains and closer the temperature is to the freezing point, the harder the resulting wind slab. In some cases, these slabs can become so hard they cannot be shovelled, but have to be cut with a saw.

Faceted grains including depth hoar

Temperature gradients produce water vapour density gradients that lead to snow grain growth (Figure 3.5b). When the gradients exceed ~25°C m^{-1} (e.g. a drop of 10°C from the ice surface to the top of a 0.4 m deep snow pack), faceted grains will grow (Akitaya, 1974). Facets and other euhedral and ornate growth structures (Figure 3.7) form because the growth is limited by crystallographic growth mechanisms rather than the vapour supply (Colbeck, 1982). The resulting crystals are beautiful, distinctive and poorly bonded. Both Arctic and Antarctic sea ice snow covers exceed the critical gradient threshold for much of the winter (Massom et al., 2001; Sturm et al., 2001), so faceted grains are common. This is also the case in the sub-polar Sea of Okhotsk (Toyota et al., 2007). Particularly strong temperature gradients exist in autumn when the ice is new, the ice surface temperature is near-freezing,

Figure 3.7 A depth hoar cluster from the base of snow cover that formed over an ice surface. Note the euhedral (geometric) shapes, the striae, and the razor-sharp edges indicating rapid, kinetic growth. The clear crystal in the foreground is about 11 mm across. Source: Photograph by J. Holmgren.

and the snow cover is quite thin (<0.2 m). Sturm et al. (2001) found gradients exceeding 100°C m^{-1}, four times the critical gradient, under such conditions. This is the reason why the basal layers of snow on sea ice are often metamorphosed into large (3–10 mm) faceted grains called depth hoar. Depth hoar formation is enhanced under ice layers due in part to their high thermal conductivity (Sturm et al., 1997) and the fact that these layers impede upward vapour flow (Colbeck, 1991).

There is a continuum of faceted grains. When kinetic (or TG) growth begins, the first grains that form are small (<1 mm) and rounded with a few incipient facets. With increasing time (and higher temperature gradients), fully faceted grains, then hollow skeletal grains and striated pyramidal cups will form (1–8 mm). These can be at the base of the snow (depth hoar), at the surface (surface hoar), or beneath an ice lens. Eventually, complex and ornate grain clusters like those shown in Figure 3.7 will develop. By midwinter, layers composed of grains like those in Figure 3.7 will have undergone total recrystallization 10 or more times over (Sturm & Benson, 1997; Pinzer et al., 2012).

Icy layers

Icy layers and ice lenses are found in the snow during all seasons in both Arctic and Antarctic snow covers, but are more common in the Antarctic. They form in several ways. Above-freezing temperatures (thaws) can produce liquid water at the snow surface, while sleet/rain can introduce water into the pack even if the temperature is slightly below freezing (Massom et al., 1997; Sturm et al., 1998). The water will either penetrate into the snow pack in discreet pipes (this happens if the snow is well below freezing; Wankiewicz, 1978) or move downward as a planar wetting front (if the snow is isothermal at 0°C). In both cases, multiple internal ice lenses and layers can be formed in a single thaw or rain-on-snow event. The telltale sign of this occurrence is that several ice layers will be connected by vertical percolation columns. Two other mechanisms have also been proposed to explain the formation of thin (<0.5 mm) ice layers: wind glazing due to kinetic heating of the snow surface by wind-driven (saltating) drift snow (Goodwin, 1990); and radiation effects (in spring–summer; Colbeck, 1991). Our direct experience has been that virtually all Arctic wind glaze layers could be traced back to episodes of thaw, rain-on-snow, or freezing mist (*minniq* in Iñupiaq). In the Antarctic, we have observed minor topographic perturbations like hoarfrost 'flowers' transformed into icy rippled and 'nubbly' layers consisting of irregular lumps of ice, 0.01–0.02 m high and spaced every few centimetres, but again only when liquid water was present in the air (Sturm et al., 1998). Wind-blasted, upwind faces of sastrugi and dunes are particularly susceptible to icy crust formation (Massom et al., 1997).

Although icy layers overall represent only a small fraction of the total snow cover thickness on sea ice, they play an inordinately important role in the evolution of the snow pack. For example, less than 3% of the snow on the East Antarctic sea ice comprised icy layers in the winter of 1995 (Massom et al., 1998), but these layers effectively 'locked' the snow in place and reduced wind erosion and sublimation losses (Eicken et al., 1994; Massom et al., 1997). Similar results have been observed in the Arctic (Sturm et al., 2002a). Wind crusts and ice layers also intercept meltwater migrating downward in percolation columns (Colbeck, 1991), even in winter (Jeffries et al., 1997a; Sturm et al., 1998). Where intercepted, this meltwater will spread out laterally and can create quite thick (~30 mm) icy layers. These horizontal ice layers form largely impermeable barriers to the upward flux of water vapour, leading to enhanced depth-hoar formation underneath them (Colbeck, 1991; Massom et al., 2001). Such features can also have a significant effect on polarization and emission in microwave remote sensing (Garrity, 1992; Montpetit et al., 2013).

Damp/wet snow and slush

The introduction of liquid water, either fresh or saline, into the snow pack produces a predictable change in grain size and shape. Colbeck et al. (1990) present a general classification of liquid water content in snow: dry when the percentage of liquid water is 0%, moist when it is < 3%, wet at 3–8%, very wet at 8–15%, and saturated (slush) at > 15%. An alternate system recognizes two wetting regimes: a low-water regime in which the pore spaces between snow grains still retain air channels that are connected, with the liquid water a meniscus at grain junctions (pendular regime), and a high-water regime where the water films are continuous and air spaces are isolated bubbles within the water film (funicular regime). Denoth (1982) indicates that the transition between regimes occurs when the water content exceeds 7%. Regardless of which water regime is in effect, the net result is that the snow grains become rounded. The initial rounding should not be mistaken for melting. By definition, water-surrounded grains are isothermal at 0°C, a condition highly conducive to volume diffusion as well as the destruction of sharp-edged features like those in Figure 3.7. Low-water-content metamorphism produces clusters of rounded crystals held together (after refreezing) by large ice-to-ice bonds (Colbeck, 1986; Figure 3.8). High-water-content metamorphism and subsequent freezing lead to rounded melt–freeze particles, whereby individual crystals are frozen into solid polycrystalline grains. Melt-grain clusters up to 10 mm across have been observed in the Antarctic in all seasons but mainly in summer (Haas et al., 2001), and in the Arctic in spring.

The most extreme stage of wet metamorphism is slush. In slush, the ice grains are no longer bonded to one another and may not even be in contact. The grains are oblate ellipsoids. Slush on sea ice can be fresh (as when a melt pond is formed) or saline (when the slush is due to seawater flooding). Capillary 'wicking' can also wet/dampen the snow to a height of 0.1–0.2 m above the saturated layer (see figure 6 in Massom et al., 2001).

Grain size

There has been considerable research in recent years in how to measure, and even how to define, grain size (cf. Montpetit et al., 2012), with those interested in remote sensing preferring metrics like optical grain sizes (Mätzler, 2002) or specific surface area values (see Mätzl & Schneebeli, 2006). These approaches, however, remain an unsettled topic. Here, we present size data based on hand lens and microscope observations for the Antarctic (Table 3.4). They vary over a narrow range (1.5 to ~3.0 mm) and are spatially and temporally similar. With the exception of the Weddell Sea in the winter of 1992 (characterized by both significant melt episodes and a large proportion of depth hoar/faceted snow), summer mean grain sizes tend to be larger than winter grain sizes due to the preponderance of grain clusters formed by strong melting and freeze cycling that occurs predominately in summer (Haas et al., 2001; Nicolaus et al., 2009).

Figure 3.8 A melt-grain cluster showing the rounded grains and thick ice-to-ice bonds typical of low to moderately wet snow (now refrozen). The cluster is ~10 mm across.

Table 3.4 Mean grain sizes for various snow types.

Snow type	Grain size (mm)	No. of samples (snow pits)	Snow density (kg m^{-3})
New/recent snow	0.1 ± 0.6	2	310
Soft-moderate wind slab	1.1 ± 0.4	21	340
Hard wind slab	0.9 ± 0.3	4	410
Depth hoar	2.0 ± 0.9	29	290
Melt clusters	1.3 ± 0.6	8	350

Source: Sturm et al. (1998). Reproduced with permission of John Wiley & Sons.

3.3.4 Metamorphic pathways

There are four basic metamorphic pathways (Figure 3.9). While the 'pure' outcome of each of the individual metamorphic pathways listed in Table 3.2 is distinctive and differentiable snow grains, actual metamorphic trajectories are complex, producing grains with compound characteristics. For example, new snow falling on the sea ice might first undergo kinetic growth (TG metamorphism), becoming faceted or, if large negative temperature gradients prevail longer, forming striated cups. When the temperature rises and reduces the temperature gradient across the snow, equilibrium (TG) metamorphism will be initiated, and the faceted grains will begin to round. With time, the grains may become fully rounded. If the temperature rises further (to above freezing), then melt-grain clusters, or perhaps even slush, will form. If the air temperature drops, to strengthen the temperature gradient across the snow cover, then kinetic growth will start again. Based on likely ambient conditions, the most complex snow metamorphic pathways occur:

- in the marginal ice zone (where snow cover is also affected by wave–ice interaction processes which can remove the snow);
- in the Antarctic ice pack due to the frequent passage of storms;
- in the central Arctic ice pack during the 'shoulder' sea ice seasons (autumn and spring) when ambient conditions are most changeable.

3.3.5 Snow layering, salinity and wetness

Snow grains are the building blocks of snow layers, but layers are the fundamental components of a snow cover. They are easy to recognize, but harder to define. Typically, grain size, shape and multi-grain structures differ from one layer to another, leading to visually distinct units with sharp boundaries (Figure 3.10). The source of these textural differences is the weather. Each layer has a different depositional history and follows a unique post-depositional metamorphic trajectory. In the absence of wind transport, melting or significant ice topography, snow layers tend to be planar and laterally continuous, but on sea ice these conditions are rare. More often, wind-generated features such as ripples, sastrugi and dunes produce undulatory layers. For example, all three of the layers in Figure 3.10(a) are tapered because they were deposited or reworked by the wind into a long whaleback-shaped dune. One reason layers are important is that physical and thermal properties of the snow are assigned by layer, with bulk properties often being computed from individual layer values as a weighted average.

There are no simple rules for connecting snow-layer characteristics to weather events, but the connection is strong. Of the eight layers making up the snow cover on the sea ice of the Chukchi Sea during the Surface Heat Budget of the Arctic (SHEBA) project (Perovich et al., 1999), shown in Figure 3.11, six formed during snowfall with wind, while two formed from snowfall without wind (see Figure 3.12). Significant snow transport takes place when 10 m height wind speeds are in excess of 8 m s^{-1} (Andreas & Claffey, 1995) – a threshold often exceeded over Antarctic sea ice in particular (Massom et al., 2001) – but precipitation seems to be the one necessary ingredient for layer formation. Our experience has been that wind transport events alone cannot produce layers (although they can and do erode existing layers of snow). This is borne out by the example shown in Figure 3.12 of an episode in late March where there were winds in excess of 12 m s^{-1} but no snow layer formed.

Nonetheless, wind is still crucial in the formation of snow layers on the ice. A whole continuum of drift forms (Figure 3.13) produced by the wind have been identified (Doumani, 1966), ranging from ripple marks a few centimetres high to dunes that can be > 1 m high and ~20 m long. The wind produces both depositional and erosional features. Depositional features include dunes (of various shapes and orientations), barchans (crescent-shaped dunes with the horns pointing downwind), and drift aprons (Figure 3.13 bottom). The latter form in the lee of ice blocks and pressure ridges

Snow in the sea ice system: friend or foe? 79

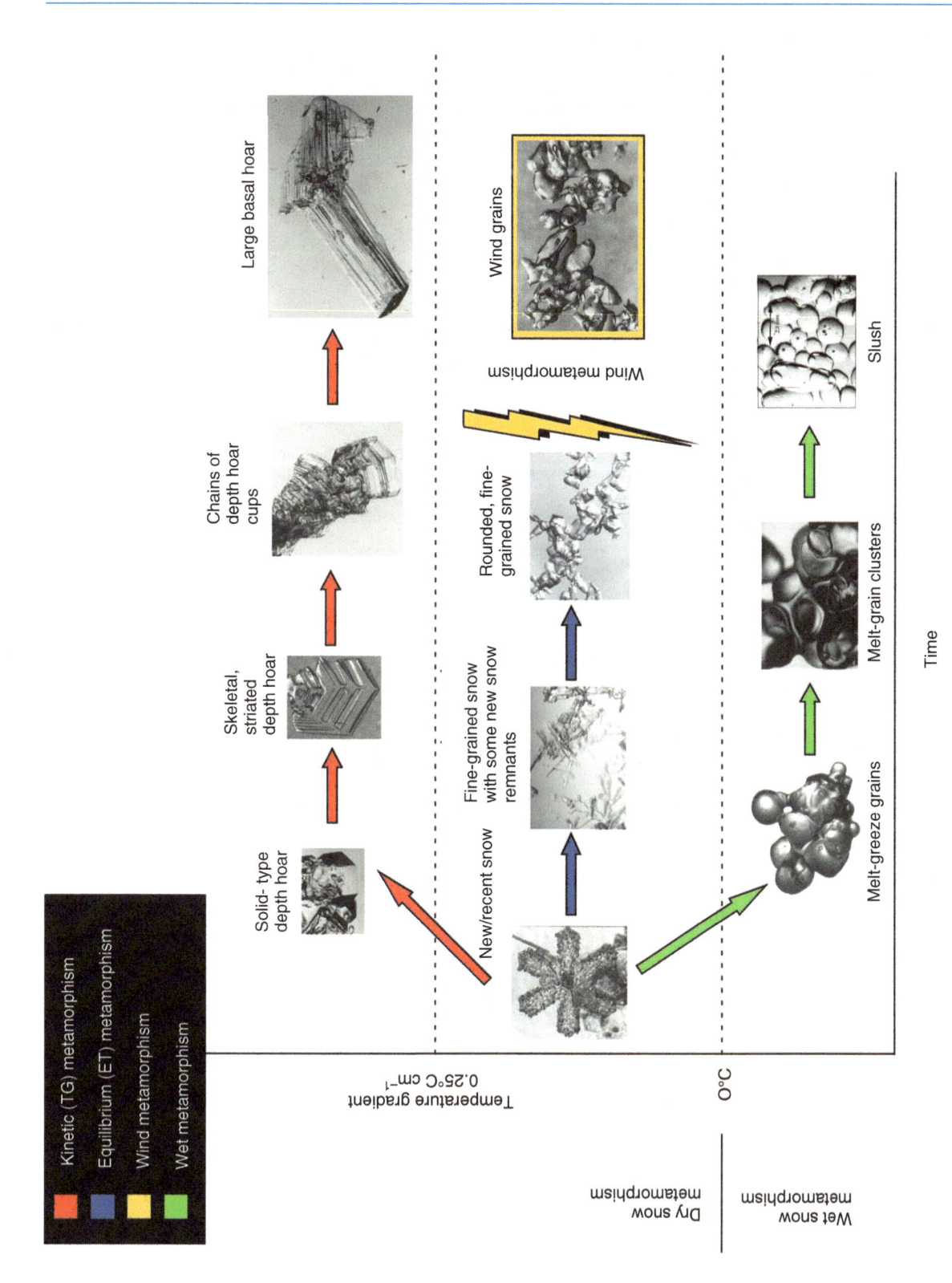

Figure 3.9 Metamorphic pathways for snow grains under different environmental conditions. Cross-connections are not shown to avoid clutter, but virtually all nodes in the figure could be interconnected.

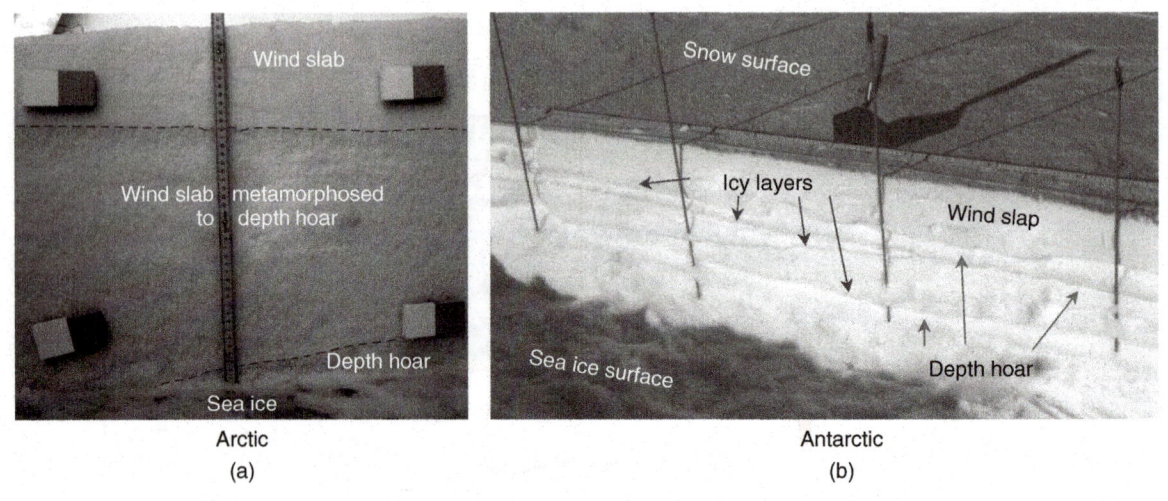

Figure 3.10 (a) The snow cover on sea ice in the Chukchi Sea off the coast of Barrow, Alaska, photographed using near-infrared photography (Mätzl & Schneebeli, 2006). The grey and white markers are a material of known reflectance. The snow is 0.43 m deep. (Photograph by Matthew Sturm.) (b) The snow cover on the ice of the Amundsen Sea, Antarctica, showing distinct layering. The snow is 0.25 m deep. Source: Massom et al., 2001. Reproduced by permission of American Geophysical Union.

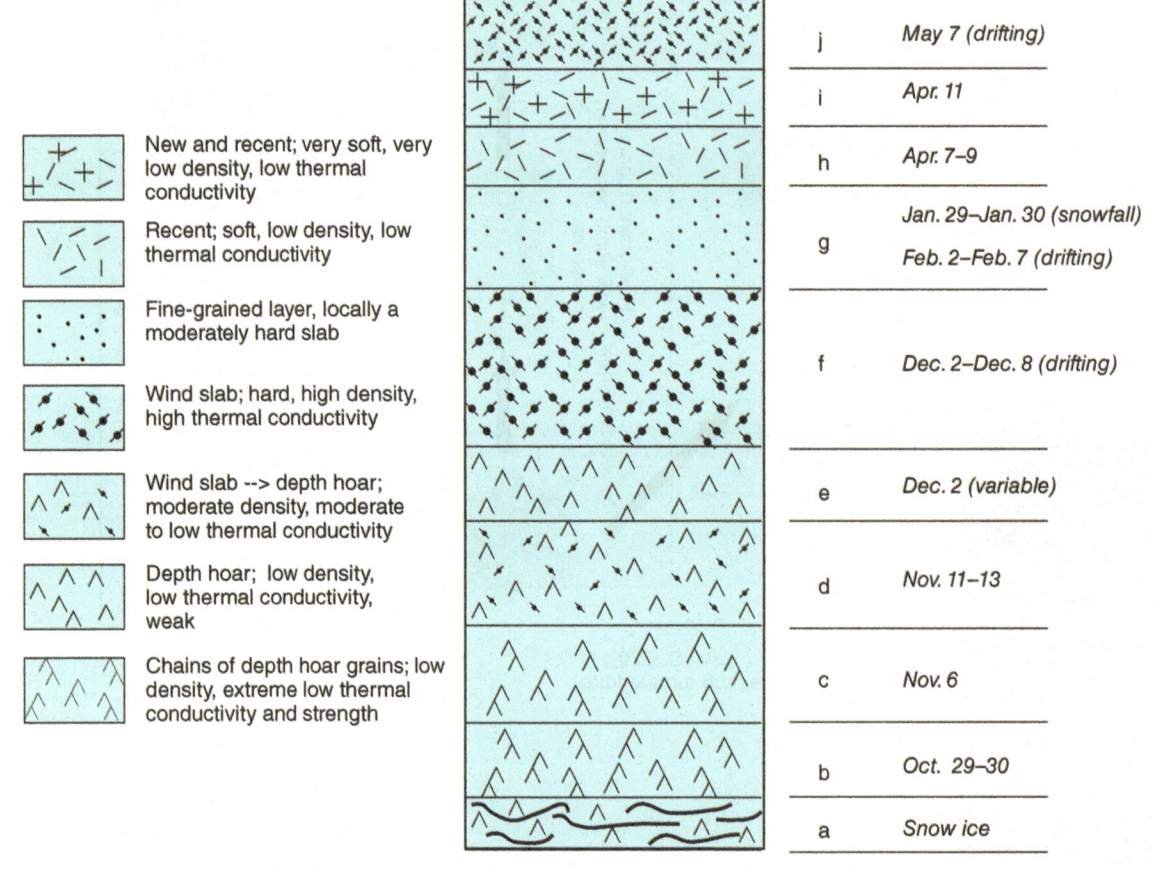

Figure 3.11 Generalized snow stratigraphy column for the Surface Heat Budget of the Arctic (SHEBA) project area, 1997–1998. Symbols follow the Colbeck et al. (1990) snow classification (updated by Fierz et al., 2009). Dates indicate the approximate time of layer deposition. The thin layer of snow ice (a) was in this case formed from snow that had fallen on wet ice prior to the experiment. Source: Sturm et al., 2002a. Reproduced by permission of American Geophysical Union.

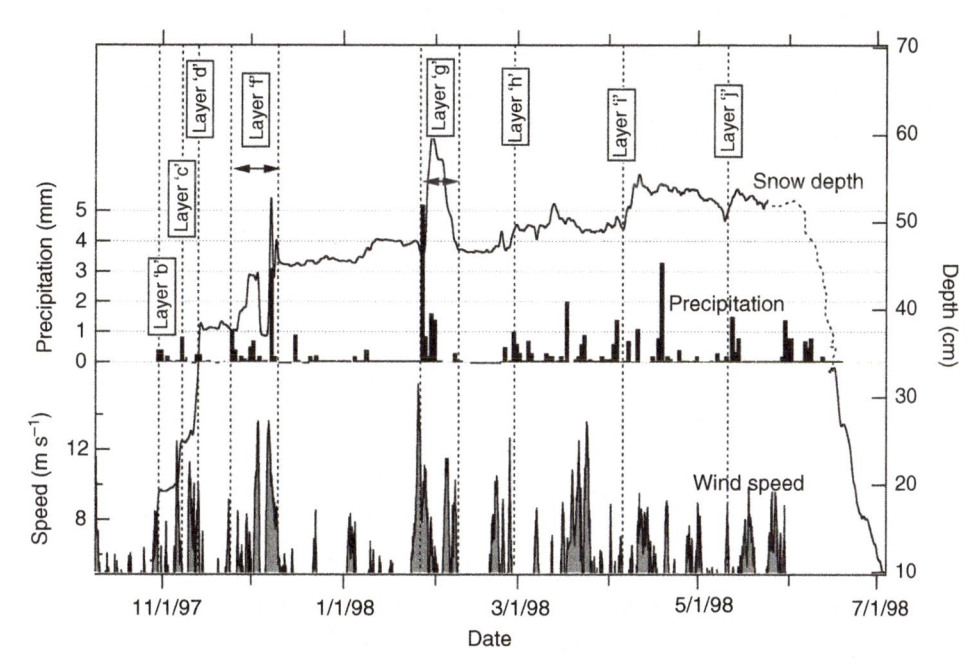

Figure 3.12 Snow layer formation during the winter of 1997–1998 on the sea ice during the Surface Heat Budget of the Arctic (SHEBA) project (Chukchi Sea) as determined from correlation of weather records with the snow depth measured by a sonic sounder (dark line). Layer labels correspond to layers shown in Figure 3.11. In total, the record contained ~30 precipitation events and at least 26 contiguous periods when wind speed was sufficient to transport unconsolidated new snow. Source: Sturm et al., 2002a. Reproduced by permission of American Geophysical Union.

(Adolphs, 1999; Tin & Jeffries, 2001). Because wind directions change and ice floes rotate, these aprons often occur on both sides of pressure ridges. Barchans form in large fields on wide expanses of undeformed flat first-year ice (Massom et al., 2001) or during erosional events. Erosional features include sastrugi (dunes sculpted into fluted forms), pitting and areas of ice swept free of snow. In cross-section in a snow pit, both erosional and depositional drift features appear as layers that pinch and swell at a variety of scales.

Another salient layer feature, particularly in the Antarctic, is the presence of saline and moist basal snow layers, even in the absence of seawater flooding (Massom et al., 1998; Toyota et al., 2011). Brine can move to the surface of the ice due to its upward expulsion during initial stages of ice freezing (Perovich & Richter-Menge, 1994). This brine can be wicked up into the snow by capillary suction. Similarly, seawater at the ice surface due to snow loading and flooding or wave overwashing (Massom et al., 2001) can do the same. High salinities (>10) can occur up to 0.2 m high in the snow cover, but are mainly found in the bottom 0.05 m, as shown in Massom et al. (2001) for Antarctica and Toyota et al. (2007) for the sub-polar Sea of Okhotsk. Brine supply depends on ice temperature and salinity (Golden et al., 1998), which is affected by snow depth. Seawater supply is also determined by freeboard (see Sections 3.2 and 3.6), which can also be controlled by snow depth, and by ice deformation and wave overwashing (Massom et al., 2001). Snow also controls the ice thermal field and thus the intensity and depth of brine convection within the ice (Notz & Worster, 2008).

Mean salinities observed for the bottom 30 mm of Antarctic snow on ice range from 13 to 24, although with considerable lateral variability (Massom et al., 1998). Basal snow salinities ranging from 4.7 to 34.8 have been measured across a smooth, unridged first-year floe. Lower 'background' salinities (of ≤ 1) that are found higher in the snow column are probably the result of blowing snow which has wicked a little salt and/or sea spray from adjacent leads and polynyas during strong wind events. Toyota et al. (2007) found comparable values on ice in the Sea of Okhotsk (salinity of 0–25.5, $n = 64$).

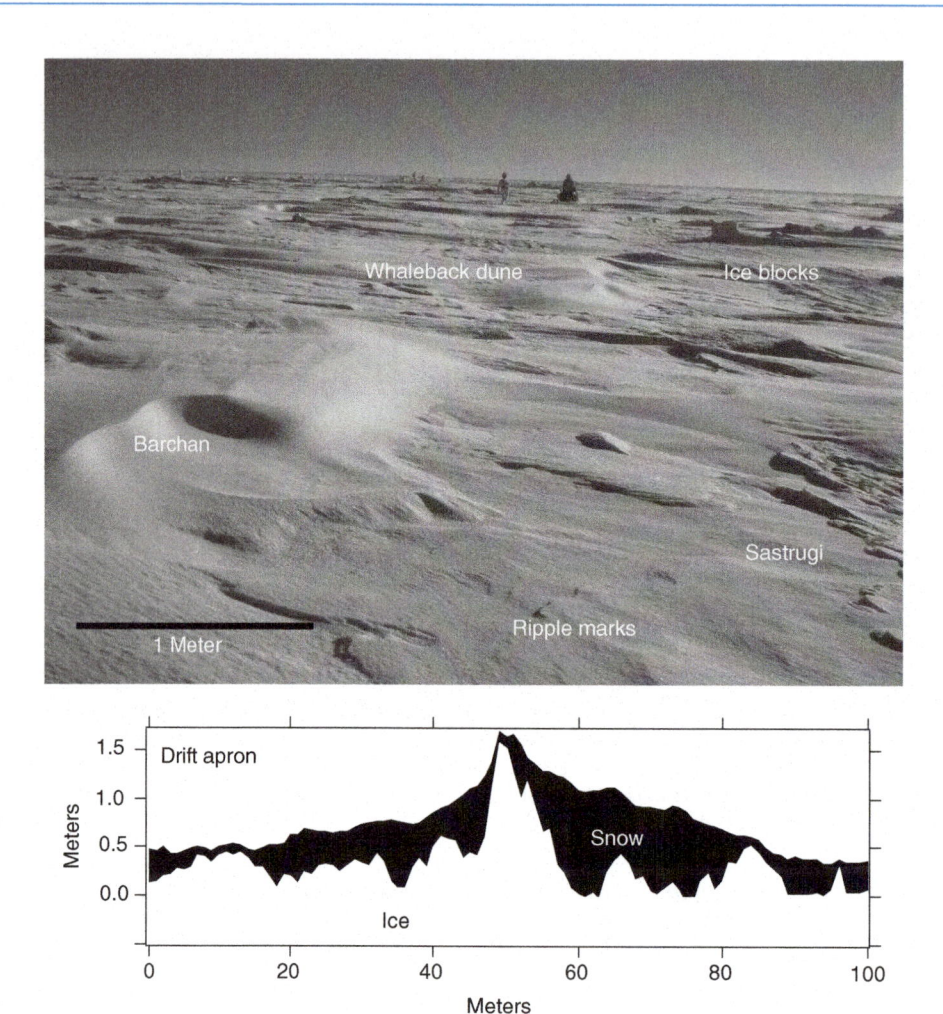

Figure 3.13 Snow drift features common on sea ice. The near-infrared photograph is from the sea ice near Barrow, Alaska, but could easily have been taken in Antarctica. The cross-section at the bottom shows a drift apron in the lee of a 1.5 m ice ridge in the Chukchi Sea (0 m on the *y*-axis is sea level). Comparable drifts occur on Antarctic sea ice. Source: Sturm et al., 1998. Reproduced with permission of Jon Wiley & Sons.

Particularly high snow basal salinities (>34) can occur when frost flowers are incorporated into the base of the snow cover (Perovich & Richter-Menge, 1994; Massom et al., 1997). Frost flowers, which form extensive carpets on nilas in both Arctic and Antarctic sea ice, attain salinities as high as 110 (Drinkwater & Crocker, 1988; Alvarez-Aviles et al., 2008). Enhanced snow salinities depress the snow melting point, leading to a persistent dampness at the base of the snow, even at lower temperatures. This dampness is tangible: even at −10°C, a snowball can be made from saline snow. Mean bulk snow salinities measured in Antarctica are compiled in Table 3.5.

One direct impact of 'wet snow' is a lowering of the snow reflectance/albedo. The actual presence of liquid water at < 10% by volume is minimal at visible and near-infrared wavelengths, as the absorption of water is similar to that of ice in this spectral region (Green et al., 2002). However, liquid water, in the form of either snow meltwater or brine/seawater, has an indirect effect on snow optical properties by changing ice–air interfaces into ice–water interfaces and causing snow grains to cluster. This leads to an increase in effective grain size (Colbeck, 1986), with the net effect of significantly reducing scattering and decreasing reflectance/albedo

Table 3.5 Mean bulk snow salinities from various Antarctic sea ice experiments, as a function of region and season.

Sector	Season (year)	Snow Salinity, psu	Range (psu)	n	Reference
Indian and West Pacific oceans					
75–150°E	Spring (1994)	8.8 ± 13.3	0.1–66.4	41	Worby & Massom (1995)
139–141°E, 144–150°E	Autumn (1993)	8.4 ± 13.5	0.1–47.1	51	Worby & Massom (1995)
138–142°E	Winter (1995)	8.5 ± 10.6	0.1–54.5	202	Massom et al. (1998)
109–118°E	Winter–spring (2003)	4.22	0.1–38.1	197	Toyota et al. (2011)
115–130°E	Winter–spring (2007)	4.36 ± 7.42	0.0–41.3	197	Toyota et al. (2011)
Bellingshausen, Amundsen and Ross Seas					
75–110°W	Autumn–winter (1993)	8.7 ± 10.6	0.2–36.3	84	Jeffries et al. (1997b)
165–180°W	Autumn–winter (1995)	8.7 ± 10.2	0.02–39.8	193	Morris & Jeffries (2001)
175°E–175°W	Autumn–winter (1998)	13.1 ± 18.5	0.01–91.8	179	Morris & Jeffries (2001)
109–171°W	Winter–spring (1994)	2.6 ± 6.5	0.1–40.5	333	Jeffries et al. (1997a); Sturm et al. (1998)
70–130°W	Late–summer (1994)	0.02 ± 0.08	0.0–0.36	37	Haas et al. (1996)
Weddell Sea					
0–55°W	Spring (1989)	4.0	0.0–41.9	87	Eicken et al. (1994)
20–55°W	Late summer (1997)	0.15 ± 0.8	0.0–8.6	130	Haas et al. (1998)
5°E–55°W	Autumn–winter (1992)	2.6	–	–	Lytle & Ackley (1992)
0–50°W	Winter (1992)	8.7 ± 10.7	0.0–48.8	189	Massom et al. (1997)

psu, practical salinity unit.
Source: Adapted from Massom et al. (2001). Reproduced by permission of American Geophysical Union.

at near-infrared wavelengths – an effect that remains when the wet snow refreezes (Dozier et al. 1981; Perovich et al., 2002). Also in regions of floe flooding, waterlogging of the lower snow horizons reduces the effective snow thickness, which has much the same effect as melting snow (Brandt et al. 2005), i.e. it lowers the albedo.

3.4 The temporal evolution of the snow pack

3.4.1 Arctic

In the first edition of this book, we used the results from the SHEBA experiment (based around an icebreaker frozen into the Arctic Ocean; see Perovich et al., 1999; Sturm et al., 2002a) as a model for the temporal evolution of the Arctic sea ice snow cover. These results were thought to be representative of most of the central Arctic Basin – perhaps excluding the Eastern Arctic where greater winter snowfall rates and snow depths occur (Serreze & Maslanik, 1997). Fifteen years after SHEBA, and following the loss of > 20% of the summer ice cover and most of the multi-year ice (MYI) in the Arctic (Stroeve et al., 2007; Maslanik et al., 2011), this assumption is no longer tenable (Figure 3.14). Unfortunately, few new data are available, so we present the old data here with the caveat that there may have been significant changes in the Arctic Ocean snow cover since SHEBA.

During SHEBA, the snow cover initially built up in October and November, attaining near-maximum depth (0.34 m) by mid-December (Figure 3.12). From then until snowmelt onset in late May, the depth and properties changed little. Seasonal snowmelt was rapid, removing the snow in less than 20 days (Figure 3.12). This seasonal pattern is similar to that described for the Arctic Basin by Warren et al. (1999) from Russian drifting ice camp observations and closely matches earlier observations from the Beaufort Sea (Untersteiner, 1961). Today, we might expect that some of the October and November build-up would instead fall into an open ocean, and that the melt would start a few weeks earlier (Webster et al., 2014).

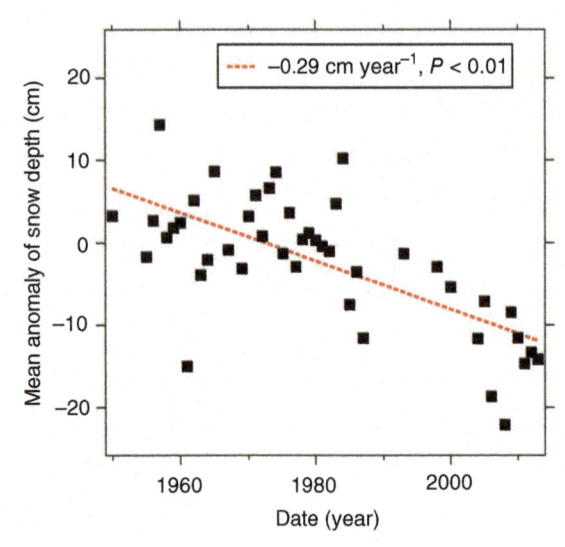

Figure 3.14 The reduction in Arctic spring snow-on-ice depths, based on Webster et al. (2014). This reduction is largely thought to be due to the fact that early season snow, which formerly fell on the ice, now falls into the ocean. Anomalies in snow depth were calculated using data from Soviet drifting ice stations (1950–1987), ice mass balance buoys (1993–2013), and Operation IceBridge snow depth products (2009–2013; see Kurtz & Farrell, 2011). They are shown relative to the 1950–1987 average (see Warren et al., 1999). Used by permission, American Geophysical Institute.

The snow cover during SHEBA comprised three basic snow types: depth hoar (Figures 3.10 and 3.11: layers b, c, d, and e); 2) wind slab (layers f, g, and j); or recent (layers h and i). Strong snow temperature gradients in early winter metamorphosed layers b–d into depth hoar, while diurnal temperature cycling in April and May produced near-surface kinetic growth (Birkeland, 1998) comprising small (0.5 mm) faceted crystals in the upper layers. The snow layers were produced by 10 fairly discrete weather events (defined as continuous periods of precipitation, wind or wind plus precipitation). Most events were short-lived (<24 hours), although multi-day storms were responsible for the formation of the two most prominent layers (wind slabs f and g). The combined time for all the key weather events constituted only 6% of the winter period, highlighting again how sensitive the snow cover properties are to the timing of snowfall and wind events (Figure 3.1).

Cold dry conditions lasted at SHEBA from November until May, during which time the albedo of the snow remained uniformly high (0.8–0.9; Perovich et al., 2002). Moderate warming over the April–May period led to an increase in snow grain size accompanied by a gradual decrease in albedo. Although brief, a rainfall event in late May had an immediate effect, resulting in rapid grain-size coarsening to ~1 mm, and a reduction in albedo from 0.8 to 0.7. Snow melt onset occurred in late May, marking the start of a period of abrupt changes in snow properties. Snow albedo continued to decrease to about 0.5 and exhibited more spatial variability as the melt phase proceeded. Eventually, the ice surface was transformed into a mosaic of bare ice and melt ponds whose location was in some manner tied to the pre-melt snow distribution (Eicken et al., 2004; Polashenski et al., 2012; Petrich et al., 2012). The return to below-freezing temperatures and new snow accumulation at the end of August led to a rapid increase in albedo, with average values returning to their springtime maxima (0.8–0.9) and exhibiting spatial uniformity by the end of September. The reader is referred to Chapters 1 and 4 for more information.

3.4.2 Antarctic

Unfortunately, similar information on snow cover evolution over an annual cycle of Antarctic sea ice is lacking, although seasonal patterns can be pieced together from data collected during various field campaigns, albeit separated in space and time (reviewed by Massom et al. 2001). While there are some similarities to the evolution of the Arctic sea ice snow cover, a number of things differ, most notably a lack of wholesale snow melt, and the absence of surface melt ponds in summer (Andreas & Ackley, 1981), along with the widespread year-round occurrence of icy layers and basal flooding/dampness.

These differences may be largely explained by the weather. Due to the constant passage of storms around the circumpolar Southern Ocean (Simmonds et al., 2003), weather conditions over the Antarctic sea ice zone are highly dynamic and variable, with heavy snowfall being accompanied by strong wind events and generally warmer temperatures than in the Arctic. Changes in air temperature before, during and after wind events are often large and abrupt, including melt episodes in winter (such as the one described below). The more dynamic nature of the Antarctic weather cycles is noticeable in the snow stratigraphy as well (Figure 3.10b), which typically has more layers, and more melt features, than its Arctic counterpart (Figure 3.10a).

As in the Arctic, wind slab and depth hoar form a large proportion of the snow cover in autumn and winter (Table 3.6). Relatively high air temperatures and persistent winds during snowfall account for the relative abundance of soft to moderate slab (Sturm et al., 1998), with the occurrence of hard slabs being substantially lower than in the Arctic where wind conditions are similar but there are significantly lower temperatures (Benson & Sturm, 1993). Depth hoar and faceted snow types account for up to 55% of the Antarctic snow cover in cold seasons. Newly deposited snow quickly undergoes rapid faceting and depth-hoar formation due to the ready availability of moisture at the snow base and the strong temperature gradients (Massom et al., 1998). In fact, the largest percentage of faceted snow and depth hoar has been observed in autumn, possibly linked to larger snow temperature gradients at this time (Massom et al., 2001).

The large proportion of icy snow (ranging from 9–46% in Table 3.7) observed in all seasons reflects both the high incidence of melt–freeze events, the frequency of high winds in the case of crust formation, and the impact of moisture/water (brine/seawater) incursion from below (Sturm et al., 1998; Massom et al., 2001). Sturm et al. (1998) observed an increase in the proportion of icy snow with increasing proximity to the ice edge during winter–spring cruises (consistent with increasing temperature), and an increase in depth hoar with increasing southerly latitude. The other snow types exhibited no such pattern. Results from Toyota et al. (2011) for East Antarctica suggest that snow-type proportions also exhibit large interannual variability.

Flooding and snow ice formation in Antarctica appear to vary seasonally, but with substantial regional differences. Lower percentages of snow ice have been observed early in the winter (Sturm et al., 1998; Massom et al., 2001) when the snow cover has yet to attain maximum thickness. With time (and more snowfall and accumulation), these percentages increase. Come summer, ice surface flooding becomes more extensive (Drinkwater & Lytle, 1997), but now due to large rates of basal ice melt related to high oceanic heat fluxes (Martinson & Iannuzzi, 1998). An exception to this pattern seems to be the West Antarctic Peninsula (WAP) sector, which represents a 'warm end member' in terms of flooding and snow ice formation (Massom et al., 2006; Perovich et al., 2004). According to Perovich et al. (2004), optimal conditions in the WAP region for snow ice formation (sufficient snow, thin ice and cold temperatures for slush freezing) only occur in late autumn–early winter. Although heavy snowfall and flooding continue through mid-to-late winter, slightly warmer air temperatures and increased snow depth (insulation) can prevent the salty slush from freezing.

As the result of the accelerated formation rate of snow ice in the Antarctic, there is a large uncertainty in reported snow depths (Maksym & Markus, 2008).

Table 3.6 Snow cover characteristics observed during selected sea ice field campaigns in Antarctica; numbers represent percentages of the snow comprising each snow type. After Massom et al. (2001). Modified with permission from American Geophysical Union.

	Spring East Antarctica	Spring 1994, Amundsen/ Ross seas	Autumn 1993, East Antarctica	Autumn–winter 1995, Ross Sea	Winter 1995, East Antarctica	Winter–spring 1994, Bellingshausen/ Amundsen seas	Winter–spring 1995 Bellingshausen/ Ross seas
No. of snow pits	10	164	16	73	46	58	45
Icy	10	20	18	26	18	9	46
New and recent	16	14	12	8	23	30	4
Soft to moderate slab	11	23	7	10	8	–	13
Hard slab	7	6	11	0	9	–	0
Faceted/depth hoar	9	31	48	47	39	55	29
Slush	54	6	4	9	3	6	8
Mean depth (m)	0.15	0.28	0.17	0.24	0.13	0.20	0.22
Mean temperature gradient (°C m^{-1})	–	−23	−66	−82	−47	−41	−44

Source: Adapted from Massom et al. (2001). Reproduced by permission of American Geophysical Union.

Table 3.7 Snow densities (mean and standard deviation) measured over Antarctic sea ice. Modified after Massom et al. (2001).

Sector	Season (year)	Snow density (kg m^{-3})	Range (kg m^{-3})	n	Reference
Indian and West Pacific oceans					
139–141°E, 144–150°E	Autumn (1993)	390 ± 170	120–660	55	Worby & Massom (1995)
138–142°E	Winter (1995)	360 ± 110	120–760	170	Massom et al. (1998)
109–118°E	Winter–spring (2003)	306	110–839	215	Toyota et al. (2011)
115–130°E	Winter–spring (2007)	322 ± 38	154–539	200	Toyota et al. (2011)
Bellingshausen, Amundsen and Ross seas					
95–165°W	Autumn (1992)	340 ± 80	99–543	134	Jeffries et al. (1994)
165–180°W	Autumn–winter (1995)	350 ± 80	240–817	73	Sturm et al. (1998)
75–110°W	Winter–spring (1993)	247 ± 210	108–467	210	Jeffries et al. (1997b)
109–171°W	Winter–spring (1994)	360 ± 40	300–440	255	Sturm et al. (1998)
80–110°W, 155–180°W	Winter–spring (1995)	380 ± 80	290–480	45	Sturm et al. (1998)
70–130°W	Late summer (1994)	391 ± 71	212–518	37	Haas et al. (1996)
Weddell Sea					
5°E–55°W	Autumn–winter (1992)	290	200–600	–	Lytle (unpubl. data, 2001)
0–50°W	Winter (1992)	320	180–670	132	Massom et al. (1997)
0–55°W	Spring (1989)	290 ± 70	110–550	206	Eicken et al. (1994)
20–57°W	Late summer (1997)	349 ± 72	130–496	130	Haas et al. (1998)

Source: Adapted from Massom et al. (2001). Reproduced by permission of American Geophysical Union.

Jeffries et al. (2001), however, suggest that perhaps as much as half the snow on ice in the Pacific Ocean sector of Antarctica is eventually converted into snow ice, a point discussed further in Section 3.6.

While the remarkable June transformation of the snow–sea ice system in the central Arctic converts high-albedo snow cover into low-albedo melt ponds in a few short weeks (Perovich et al., 2002), no similar transformation takes place in Antarctica. Satellite-based studies indicate that seasonal surface melt there is relatively short-lived and is neither spatially nor temporally contiguous (Drinkwater & Liu, 2000), with albedo values of surviving ice remaining relatively high throughout the summer. Moreover, Antarctic ice melts largely from the bottom up or laterally from floe edges, due to relatively large ocean heat fluxes (Martinson & Iannuzzi, 1998), with cool and windy surface conditions preventing melt-pond formation (Andreas & Ackley 1981; Nicolaus et al., 2006). As a result, the summer Antarctic snow cover remains intact through February over 2–3 million km^2 of ice in the Amundsen, western Weddell and eastern Ross seas (see Chapter 10). This intact snow cover retards summer surface ablation (Eicken et al. 1995), in sharp contrast to what takes place in the Arctic.

Snow melting does occur on Antarctic sea ice during the austral summer (Haas et al., 2001; Morris & Jeffries, 2001), but not enough to completely remove the snow cover, so melt ponds are rare. This strong summer snow melt is reflected in sharp decreases in snow salinity in summer (Table 3.5), and the widespread occurrence of superimposed ice whereby downward-percolating melt-water refreezes at the ice surface (Jeffries et al., 1994; Kawamura et al., 2004). Haas et al. (2001) attribute the latter to a seasonal reversal in the snow temperature gradient. Superimposed ice is also reported from Arctic regions e.g. the Baltic and Beaufort seas (Eicken et al., 2004; Granskog et al., 2006) and fast ice in Svalbard (Nicolaus et al., 2003).

Also in sharp contrast to the central Arctic, episodic synoptic-scale incursions of warm maritime air occur over the Antarctic ice pack that can cause almost complete snowmelt (ephemeral removal) even in winter. The dramatic impact of one such episode, in the northwestern Weddell Sea in July 1992 (Massom et al., 1997), is documented in Figure 3.15. An increase

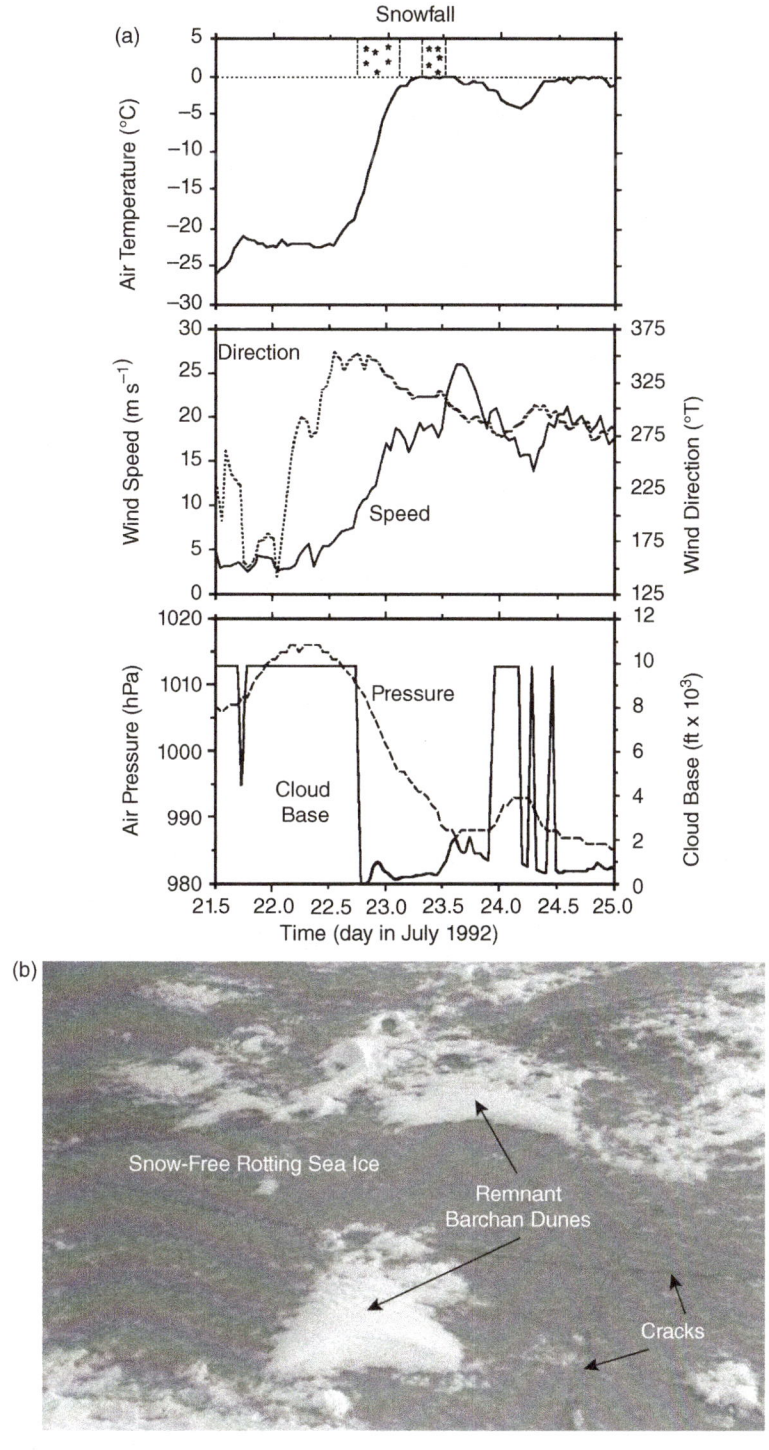

Figure 3.15 (a) Near-surface meteorological variables measured from R/V *Polarstern* while drifting with the pack ice in the northwester Weddell Sea (at ~61°S, 43°W), July 21–25, 1992. (b) Photograph taken on 25 July 1992, showing extensive snowmelt (nearly complete snow removal) associated with synoptic warm air advection. The surviving high-density barchan dune in the foreground is ~5 m across. Source: Massom et al., 2001. Reproduced by permission of American Geophysical Union.

in temperature from –22 to 0°C over a 12-hour period was followed by 3 days of periodic rain and sleet. The end result was a layer of slush on rotting sea ice with a pitted, porous grey surface, but no melt ponds (due to drainage through enlarged brine channels). A return to freezing conditions occurred a few days later, with new snowfall subsequently obscuring the refrozen melt surface. Such events can affect much of the circumpolar Antarctic sea ice zone, although they are more prevalent in the marginal ice zone and sectors where the ice extends to lower latitudes.

3.5 Bulk snow pack properties

3.5.1 Density and snow-water equivalent

Snow density and snow-water equivalent (SWE), which are typically measured by weighing a known volume of snow, provide a measure of the load on the ice, the amount of fresh meltwater that might be released into the ocean in spring, and index values from which thermal conductivity, air permeability and other bulk snow properties might be inferred. Sturm et al. (2010) provide a comprehensive discussion of how snow depth can be converted to SWE using density measurements.

The density of snow layers constantly evolves through snow metamorphism, though in general the trajectory is one of constant or increasing density with time. Layer densities can range from less than 100 kg m^{-3} for new snow (Table 3.4) to 760 kg m^{-3} for icy layers (Sturm et al., 1998). The mean value from a number of Antarctic cruises (Table 3.7) was 355 kg m^{-3} (Massom et al., 2001). Sturm et al. (2002a) report the density and SWE values shown in Figure 3.16. Warren et al. (1999, figure 11) found similar bulk density values for the Arctic Ocean, although lower mean densities have been reported for the sub-polar Sea of Okhotsk, i.e. 225 kg m^{-3} in winter, which has been attributed to the high proportion of depth hoar present (Toyota et al., 2007).

The highest reported Antarctic mean densities (about 390 kg m^{-3}) were measured in late summer, reflecting the predominance of icy snow at this time of year (Haas et al., 2001). Unfortunately, the density of snow columns containing substantial icy snow tends to be underestimated due to difficulties in sampling very hard layers. A typical wintertime frequency distribution from the Indian Ocean sector of the Antarctic pack shows a mode at 300–350 kg m^{-3} with a minor peak at ∼ 700 kg m^{-3} for the icy layers, while another from the Weddell Sea exhibits less spread and a lower mode (250–300 kg m^{-3}) as a result of a greater depth hoar contribution (figure 8 in Massom et al., 2001).

It is possible to roughly estimate the mean density of the snow from the sequence of prevailing weather conditions, if this is known. Low-density snow (200–300 kg m^{-3}) predominates for cold dry conditions; warm, windy conditions produce medium/high density snow (350–500 kg m^{-3}), while melt–refreeze conditions produce very high density (400–700 kg m^{-3}). Of course, there are many exceptions to these rules.

3.5.2 Bulk thermal properties

There are three thermal properties of importance for the snow cover, all interrelated: thermal conductivity (k_s), thermal diffusivity (λ_s), and the volumetric heat capacity ($\rho c)_s$. Thermal conductivity is the proportionality constant that relates the rate at which heat flows through a material to the temperature gradient across the material. For snow, the heat transfer mechanisms include: molecular conduction through the solid network of ice grains (Figure 3.5b); the transfer of heat across pore spaces as water vapour sublimates from a lower grain and condenses on a higher one (Fig 3.5b); conduction across the air spaces; and the convective movement of air through the pore spaces in the snow (driven by natural air buoyancy and/or by surface winds) (de Quervain, 1973; Sturm, 1991; Waddington et al., 1996). Because two of the four transfer mechanisms are not strictly conductive, it is customary to report their combined effect as the effective thermal conductivity of the snow (k_{eff}). k_{eff} varies primarily with snow density and snow type (Figure 3.17), and slightly with temperature, but greatly if the temperature gradient is strong enough to produce natural buoyancy-driven convection.

Thermal diffusivity describes the rate at which temperature waves propagate through a material. For snow, this rate is a balance between the heat flow (as represented by k_{eff}) and the volumetric heat capacity $(\rho c)_s$, a product of the snow density and the specific heat capacity of the snow (c_s):

$$\lambda_s = \frac{k_{eff}}{(\rho c)_s} \tag{3.3}$$

The heat capacity of the snow is given by:

$$(\rho c)_s = (\rho c)_i (1 - \varphi) + (\rho c)_a (\varphi) \tag{3.4}$$

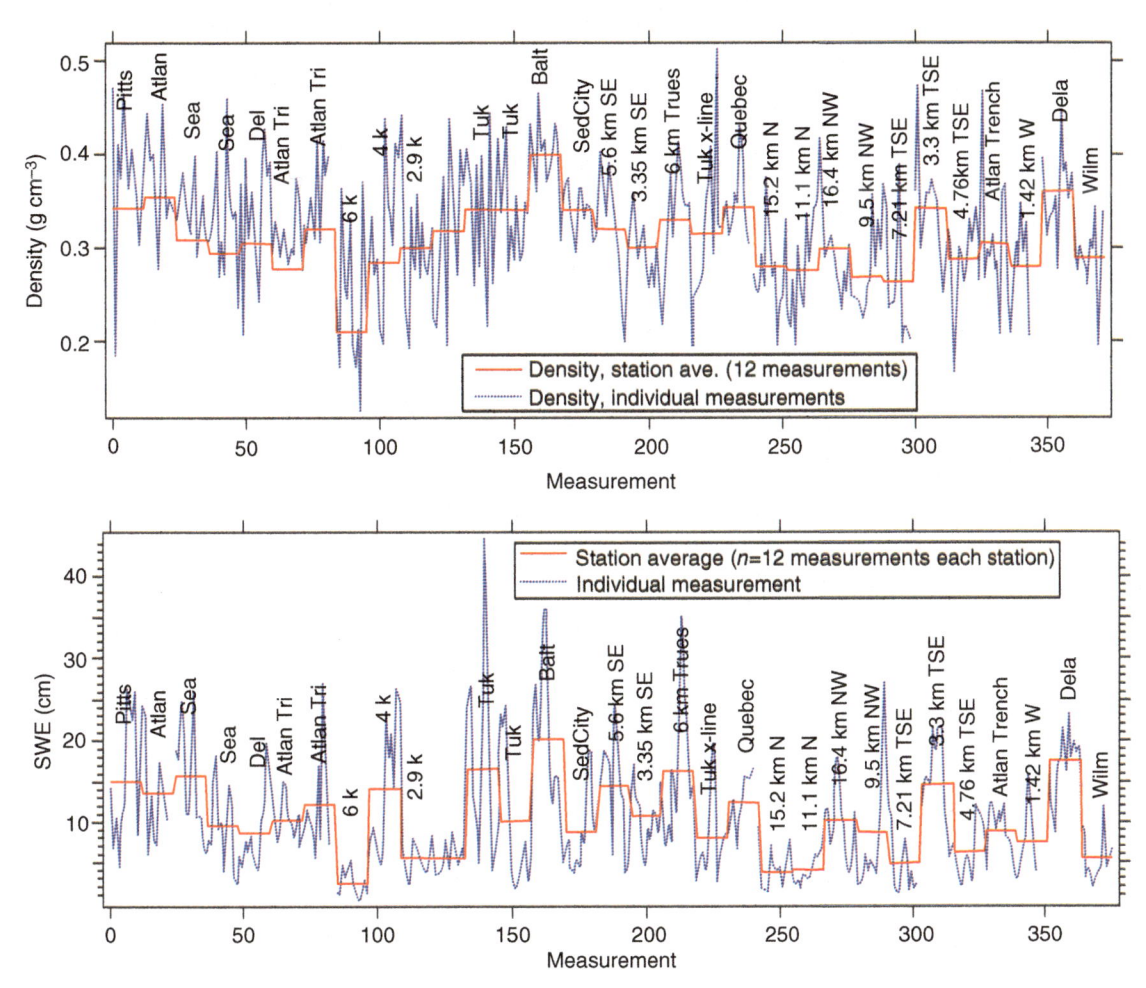

Figure 3.16 (a) Individual and average snow density values from 31 widely spaced stations at the Surface Heat Budget of the Arctic (SHEBA) project. (b) Snow-water equivalent for the same stations. Source: Sturm et al., 2002a. Reproduced by permission of American Geophysical Union.

where ϕ is the snow porosity, and the subscripts i and a indicate the ice and air components of the snow, respectively. The heat capacity of the ice fraction $(\rho c)_i$ is 1.805×10^6 J m^{-3} K^{-1}, or about 1000 times higher than the heat capacity of the air in the snow $(\rho c)_a$, which is 1.507×10^3 J m^{-3} K^{-1} (List, 1951), so the far right-hand term can be dropped from equation (3.4). Moreover, $(1 - \phi)$ can be approximated by the snow density divided by the density of ice (~ 917 kg m^{-3}), allowing equation (3.3) to be rewritten as:

$$\lambda_s = 0.000508 \frac{k_{eff}}{\rho_s} \qquad (3.5)$$

Thermal diffusivities predicted by equation (3.5) range from 2×10^{-7} to 5×10^{-7} m^2 s^{-1}, decreasing with snow density and increasing with thermal conductivity. Thermal diffusivity values are chiefly used in computing the time it takes a change in temperature to propagate through the snow cover.

Clearly, it is essential to have accurate values of k_{eff} in order to compute any of the thermal parameters in sea ice models (Wu et al., 1999; Fichefet et al., 2000; Lecomte et al., 2013). The lowest and highest values of k_{eff} are found in layers of depth hoar (0.08 W m^{-1} K^{-1}) and hard wind slab (0.40 W m^{-1} K^{-1}) respectively (Figure 3.17). These are the two most prevalent types

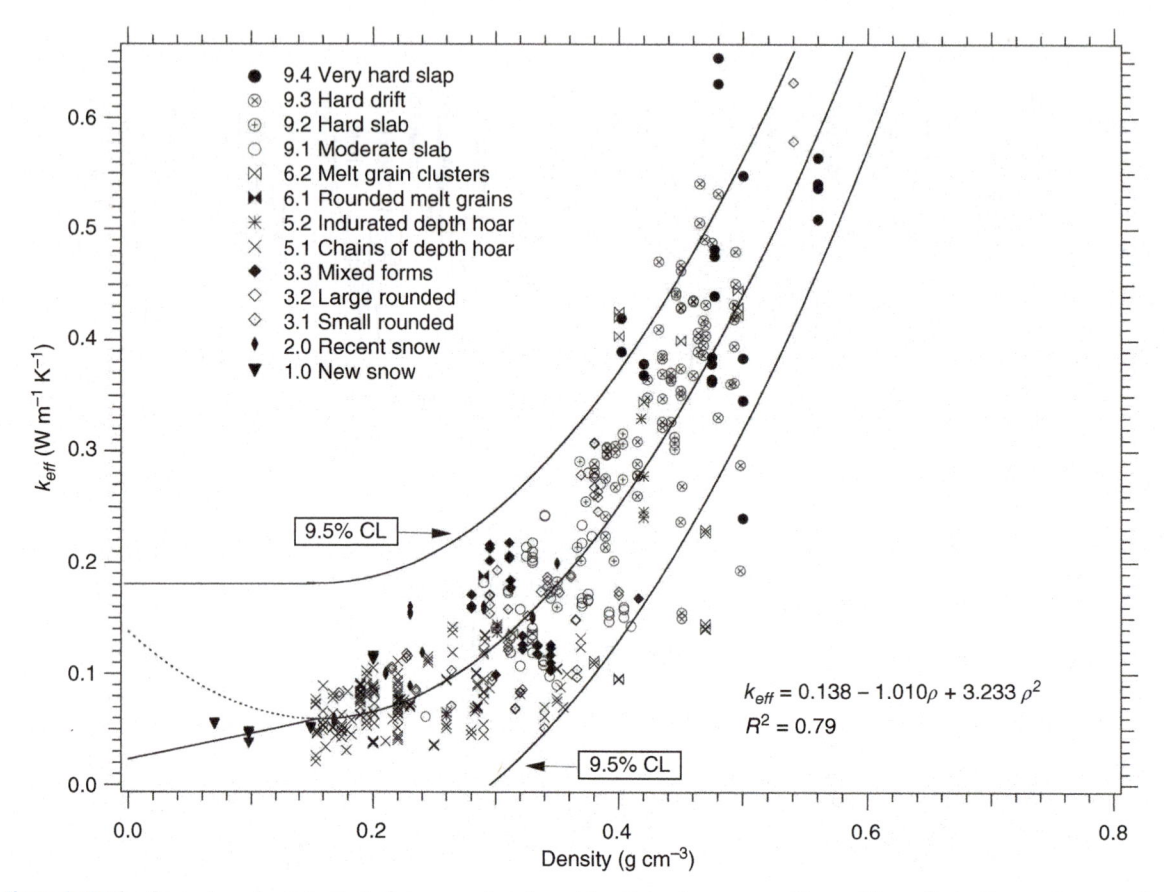

Figure 3.17 The thermal conductivity (k_{eff}) of snow as a function of density and snow type. CL, confidence limit. Source: Sturm et al., 1997. Reproduced with permission of the International Glaciological Society.

of snow found on the Arctic sea ice, with depth hoar layers comprising 36–51% of all layers, and wind slab comprising 35–58% (Sturm et al., 2006). In the Antarctic, depth hoar is also prevalent (up to 55% – see Table 3.6), but wind slab is a more minor component of the snow cover (13–29%) (Massom et al., 2001), while icy layers make up a substantial fraction. Maykut and Untersteiner (1971) used a value of 0.30 W m^{-1} K^{-1} in their modelling and this value is still widely used, but it remains problematic (see later) as to the best value to use.

Thermal conductivity values are used most frequently to compute the vertical heat flow through the snow (q_s) in order to calculate ice growth (Figure 3.3 and equation 3.2a):

$$q_s = -k_{eff}\frac{dT}{dz} \qquad (3.6)$$

where T is temperature and z is the vertical coordinate with zero at the ice surface and positive values up. Here, the negative sign accounts for the fact that a negative temperature gradient, which prevails most of the winter, produces a positive (upward) flux of heat. In practice, $\frac{dT}{dz}$ is often approximated by $\frac{\Delta T}{\Delta z}$. It is assumed that q_s is equal to q_i (Figure 3.3). This then allows equation (3.6) to be used to describe the rate of heat loss from the ice, which can be compared with the thickening of the ice by congelation growth (equation 3.2, with allowance made for oceanic heat flow).

When measured ice thickening is compared with computed thickening using equations (3.2) and (3.6) (Sturm et al., 2001, 2002b) and the most recent measured values of k_{eff} (Sturm et al., 1997), it is often found that k_{eff} needs to be increased in order to achieve agreement. Essentially, the snow seems to be allowing more heat to escape from the ice than would be predicted from the measurements. There are a number of possibilities that might explain this contradiction. The first is that the more recent values (all measured using a needle probe method) are too low (Riche & Schneebeli, 2010), but the older work (Sturm et al., 1997) has been validated by more recent studies (Morin et al., 2010; Domine et al., 2011). Anisotrophic snow texture (Calonne et al., 2011; Holbrook, 2011) might also affect the thermal conductivity, but this seems to be a somewhat limited explanation. Other plausible explanations for the mismatch between computed and measured ice growth rates are air convection in the snow, and wind pumping at the scale of the full thickness of the snow cover. Powers et al. (1985) and Sturm (1991) have shown that natural, buoyancy-driven convection occurs in thin, permeable snow covers subjected to strong temperature gradients, a description that aptly describes snow on sea ice. Similarly, Colbeck (1989), Albert and McGilvary (1992) and Waddington et al. (1996) have all documented that wind blowing over an uneven snow surface will induce wind pumping, again an apt description of the conditions found on the sea ice. Perhaps the problem, however, is a simple scaling issue (Sturm el al., 2002b) whereby the geometry of the snow, thick in some places, thin in others, can explain the discrepancy. Sturm et al. (2002b), using a finite-element heat transfer model, showed that uneven snow depths (basically drifts) could lead to heat flow that was enhanced across an ice floe by a factor of 1.4 or more. To date, the issue of what is the 'best value' remains unresolved, so the 1995 values of k_{eff} shown in Figure 3.17 remain the best available.

3.6 Snow depth distribution

We have made reference several times to the uneven distribution of snow on the ice. In many ways, it is this distribution that matters most when considering snow in the energy balance of the ice, the biological and biogeochemical processes taking place, and many of the issues related to remote sensing. For example, the snow depth distribution plays a major role in limiting the under-ice light and therefore the under-ice ecosystem (Arrigo et al., 2003; Mundy et al., 2005) and it is critical in maintaining the high surface albedo of the ice in the spring (Brandt et al., 2005). As indicated in the previous section, relatively minor changes in snow depth (h_s) can have a disproportionately large impact on the surface heat flux compared with changes in ice thickness (Lytle et al., 2000). In the Antarctic summer, h_s is also thought to play a role in the retention of snow-covered sea ice in certain regions (Massom et al., 2006), and areas of deeper snow are more likely to be the loci for snow ice formation (Maksym & Jeffries, 2000).

Despite this importance, we have only limited understanding of the regional-to-global depth distribution of snow on sea ice and its spatial and temporal evolution. This is because:

- of the vast extent and remoteness of the ice;
- the seasonal, year-to-year and spatial variability of the snow (making it difficult to obtain comparable measurements);
- the difficulty of sampling snow in very thick ice regions;
- remote sensing of h_s remains problematic.

The seasonal increase and decrease in snow depth (Figures 3.1 and 3.12) complicates the collection of snow depth distribution data. Depending on when measurements are made, there can be large apparent year-to-year variations in h_s that may simply reflect the timing of the measurements. Colony et al. (1998), examining seasonal data from Russian Arctic sea ice stations, found that h_s could vary by ±0.1 m (±1σ) from the multi-year mean (0.1–0.3 m) at virtually any time during the winter, a coefficient of variation that ranges up to 100%. Similarly, Warren et al. (1999), analyzing multi-year data from the Russian North Pole Station 22 (NP-22), found that, during a 4-year period, the peak in h_s varied by more than a factor of 2 (from 0.38 to 0.80 m), and that the pattern of snow cover build-up was quite different from one year to the next. Similar findings come from Antarctica, e.g. the Bellingshausen Sea (Perovich et al., 2004) and East Antarctica (Toyota et al., 2011).

Independent of the temporal variability, spatial variations in h_s can be considerable. Figure 3.18 shows the nature of this heterogeneity. Differences in ice floe type correlate with distinct differences in h_s, but the variation in depth is also considerable within floes, a fact that is true for both the Arctic (e.g. Mundy et al., 2005) and Antarctic (e.g. Lewis et al., 2011).

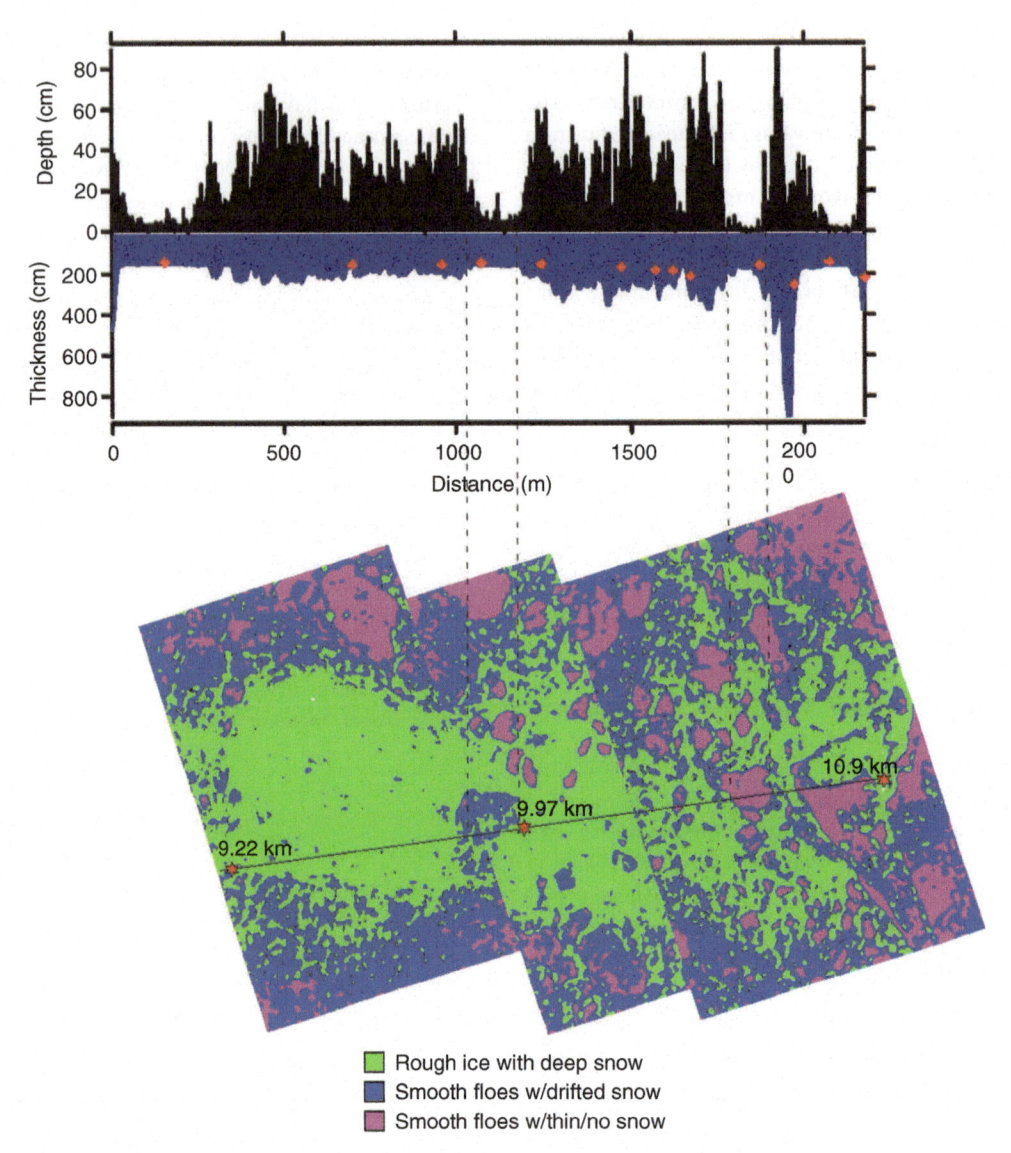

Figure 3.18 The snow depth distribution on the ice of the Beaufort Sea northeast of Barrow, Alaska. (a) The snow depth (black) is shown along with the ice thickness as recorded by EM-31. The red dots are drill hole validation thicknesses. (b) A map of the local ice types. Source: Sturm et al., 2006. Reproduced by permission of TGARSS. © IEEE, 2006.

These results in Figure 3.18 highlight that single depth measurements can be almost meaningless. The measurements also suggest that the roughness characteristics of the underlying sea ice play a key role in determining the h_s distribution even when that roughness is of only moderate amplitude (see also Toyota et al., 2011). As a result, the type of ice, whether it is first- or multi-year, is a reasonable predictor of the snow-depth distribution, at least in the central Arctic (Iacozza & Barber, 1999; Sturm et al., 2002a). Of course, a key factor in the snow-depth distribution is the wind (Leonard & Maksym, 2011), which through snow transport and redistribution produces localized variability that interacts with the ice roughness variability (Massom et al., 1997, 1998; Wadhams et al., 1987; Sturm et al., 1998).

3.6.1 Regional- to hemispheric-scale snow depth distribution

Although we have reasonable knowledge of global sea ice extent dating back to the late 1970s, reliable estimates of even the average h_s on sea ice remain elusive. One approach to obtaining these distributions has been to use satellite passive-microwave (SPMW) remote sensing. These remote sensing products are available on a daily basis but with a grid resolution of 12.5–25 km (Iacozza & Ferguson, 2014; Comiso et al., 2003). Five-day mean products are available from the US National Snow and Ice Data Center for the Antarctic and Arctic (excluding Arctic perennial ice regions) for the period June 2002 to October 2011 (Cavalieri et al., 2014). An example of a SPMW map is shown in Figure 3.19. The SPMW patterns of h_s correspond to similar patterns in ice type and age, e.g. deepest snow on second-/multi-year ice near the Antarctic Peninsula, and deeper snow overall in Antarctica than in the Arctic (where the snow cover appears to be surprisingly shallow and homogeneous in these maps).

Markus et al. (2011) have shown that the large-scale patterns in these satellite-derived maps are in reasonable agreement with ship-borne observations of h_s compiled within the Antarctic Sea Ice Processes and Climate (ASPeCt) dataset (Worby et al., 2008a), except in East Antarctica. However, current SPMW estimates are limited to non-marginal ice zone regions and maximum snow depths of 0.5 m in the Antarctic (Markus & Cavalieri, 1998; Comiso et al., 2003) and 0.6 m in the Arctic (Cavalieri et al., 2011). Moreover, the absolute accuracy of SPMW estimates is questionable. Comparison with detailed *in situ* measurements suggest serious underestimates in areas with a high fraction of rough sea ice (Stroeve et al., 2006) e.g., by a factor of 2–3 in East Antarctica (Worby et al., 2008b). Work is under way to improve the accuracy, possibly by incorporating ice surface roughness information acquired on a similar scale by satellite radar scatterometers. Powell et al. (2006) showed that the accuracy of snow depth retrievals from passive microwave data could also be improved by taking into account the observed variability of snowpack properties (e.g. density and

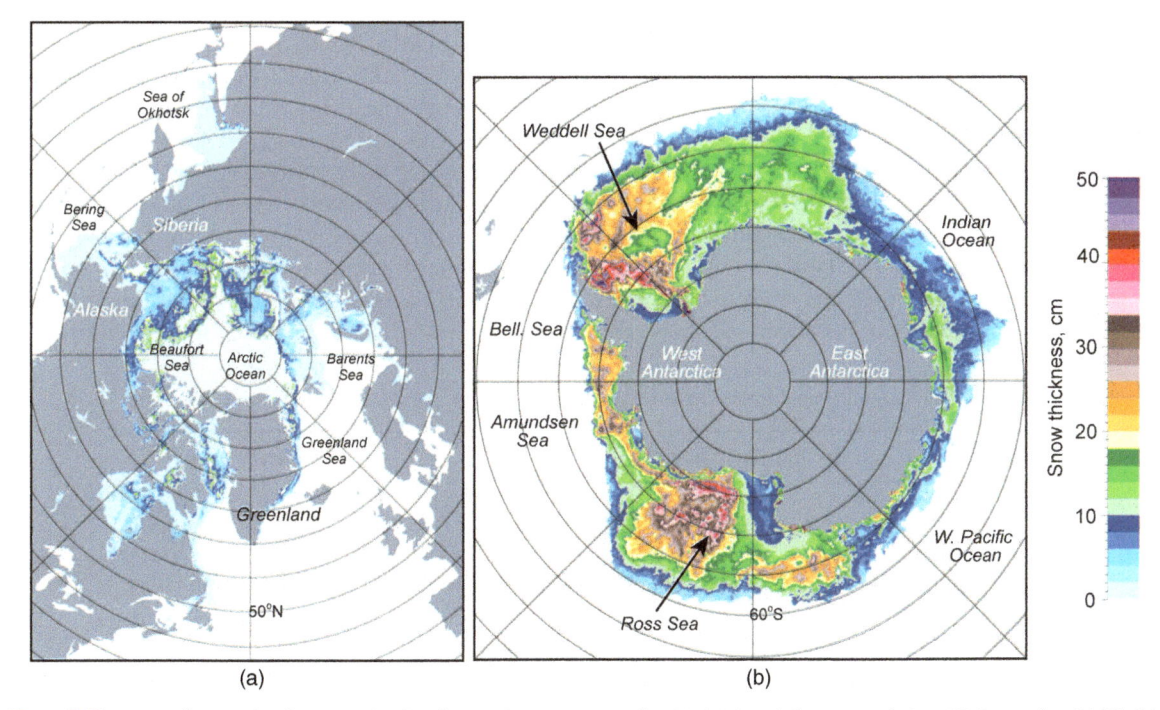

Figure 3.19 Maps of snow depth on sea ice for the Arctic (on 15 March 2007) (a) and the Antarctic (on 17 September 2007) (b), retrieved from Advanced Microwave Scanning Radiometer – Earth (AMSR-E) brightness temperature and derived ice-concentration data. Source: Imagery courtesy Thorsten Markus (NASA Goddard Space Flight Center, USA).

grain size) and sea ice emissivity, although this is not a simple task.

Shipborne observations of h_s compiled within the ASPeCt dataset again provide an alternative means of obtaining large-scale patterns around Antarctica, but are limited both spatially and temporally and to level-ice regions only (Worby et al., 2008a). In situ measurements typically made at 1 m intervals along series of 100 m transects during sea ice field campaigns are more accurate but are even more spatially and temporally limited and also tend to avoid ice regions thicker than ~3 m (Ozsoy-Cicek et al., 2013). Given these limitations, compilations of such data superficially suggest little significant difference in mean h_s between the Arctic and the Antarctic at the hemispheric scale.

Massom et al. (2001, table 1) compiled in situ depth statistics from 42 studies in all sectors of Antarctica and for all seasons ($n = 69\,273$). A wide range of depths was reported, with a mean value of 0.21 m. When only late-winter and spring values were used, the mean value was 0.20 m ($n = 45\,374$). No comparable compilation for Arctic sea ice snow depths exists, but based on a number of studies (Table 3.8) using only March or April (spring) values, the mean snow depth for the Arctic was 0.29 m ($n \approx 38\,000$), a value that is not much different from the Antarctic.

This comparison needs to be used with caution, however, for as much as 50% of the Antarctic snow depth values are likely to be low-biased. This effect is compounded in areas of high ocean heat flux, wherein surface snow ice growth mirrors ice basal melt to create an ice growth 'conveyor belt' such that much of the entire ice column can, in more extreme cases, be composed of snow ice (Lytle & Ackley, 2001), while the observed snow depth will be a small fraction of the real value.

Hemispheric-scale snow depth distributions could potentially be determined from winter precipitation values (Maksym & Markus, 2008), but a poor understanding of the processes that control snow accumulation limits the usefulness of this approach. For example, snow transport, erosion and redistribution by the wind can result in a large loss of snow to leads (Leonard & Maksym, 2011). Indeed, a modelling study by Déry and Tremblay (2004) has shown that up to 100% of all snow mass in transport by the wind may be lost into leads, while a combined observational and modelling study by Leonard and Maksym (2011) estimated that approximately half of the total precipitation falling over the Antarctic sea ice zone may be lost. Clearly, more field measurements of precipitation rate, blowing-snow frequency and magnitude, wind and snow accumulation over sea ice are required to sort this out (Leonard & Maksym, 2011).

Autonomous ice mass balance buoys (IMBs) can also be used to gain some sense of regional variations and the spatiotemporal evolution of snow on ice. Deployments have been limited to date [particularly in the Antarctic; e.g. Perovich et al. (2004)], but two long-lived IMB deployments in the Amundsen and Weddell Seas in 2009 illustrate the delicate balance and relationship between snow accumulation, sea ice and ocean heat (Maksym et al., 2012; Figure 3.20). The IMB record from the Amundsen Sea (Figure 3.20a) represents a relatively warm ice regime with a thick snow cover. Although the ice floods, snow accumulation is too rapid for the slush layer to completely freeze, and ice 'rots' from within as the ocean melts it from below; this

Table 3.8 Snow depths from the Arctic Basin, derived from sources given in the reference column.

Area	Mean depth (m)	n	Reference
Central Arctic Basin	0.34	Thousands	Colony et al. (1998)
Central Arctic Basin	0.34	Thousands	Warren et al. (1999)
Canadian Archipelago	0.23–0.29 (FYI)	2,756	Iacozza & Barber (1999)
Canadian Archipelago	0.24–0.35 (MYI)	2,048	Iacozza & Barber (1999)
Canadian Archipelago	0.23	Dozens	Iacozza & Barber (2001)
Beaufort Sea	0.34	21,169	Sturm et al. (2002a)
Beaufort Sea	0.21	1,906	Sturm et al. (2006)
Average:	0.29	About 38,000	

FYI, first-year ice; MYI, multi-year ice.

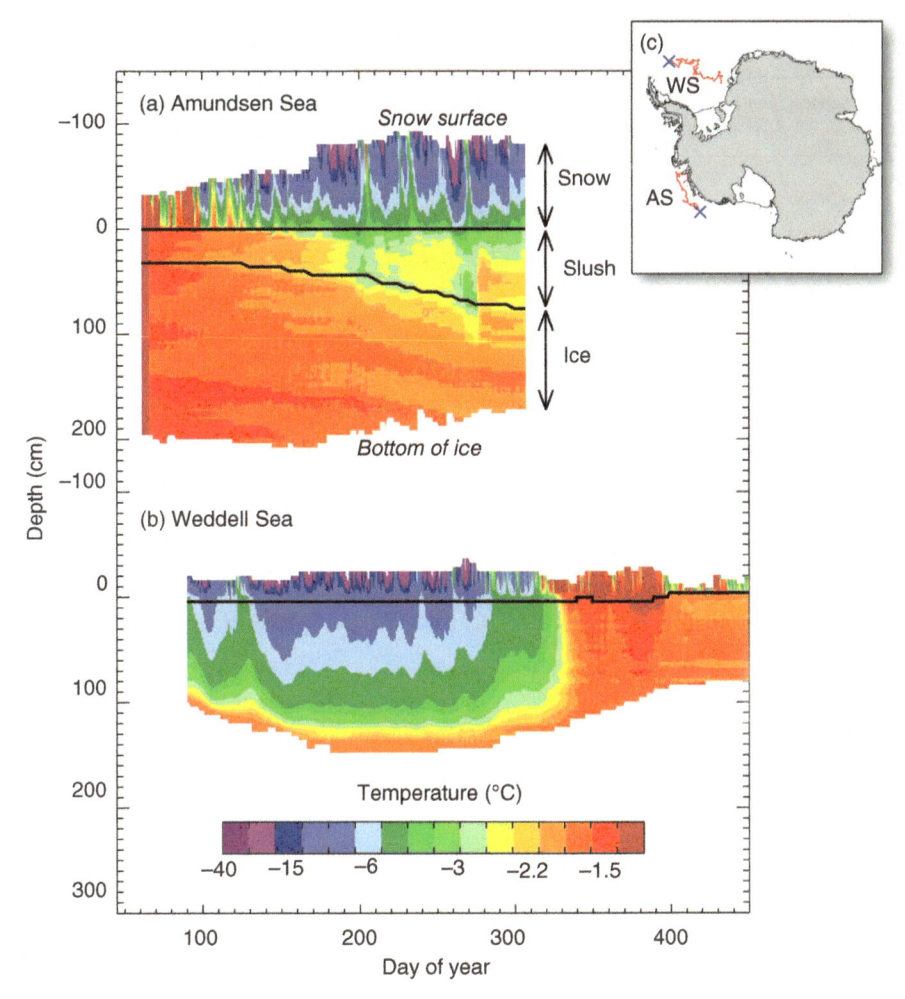

Figure 3.20 Changes in internal sea ice temperature obtained by ice mass balance buoys (IMBs) deployed in sea ice floes in the Amundsen Sea (a) and the Weddell Sea (b) in February 2009. In (b), the increase in height of the black line denotes thickening of the ice from above by refreezing of summer snowmelt on the ice surface to form superimposed ice. The corresponding drift of the IMBs is shown by the red lines in the inset map (c). The blue crosses denote their final positions when the IMB failed. Source: Adapted from Maksym et al., 2012.

overwhelms the ability of the 'conveyor belt' growth mechanism to maintain the ice thickness. In contrast, the IMB record from the Weddell Sea (Figure 3.20b) illustrates a cold ice regime wherein a relatively thin snow cover provides poor insulation, enabling the ice to thicken; this scenario is similar to that observed in the Arctic winter. Minimal surface melt occurred in the subsequent summer, such that the ice actually thickened as snow meltwater refroze at the snow–ice interface to form 'superimposed' ice (e.g. Kawamura et al., 2004).

Any meaningful data on snow depth distribution requires a companion distribution for snow ice formation, but this latter distribution is as hard to come by as the former. Table 3.9 shows the percentage of *in situ* measurements showing negative ice freeboard from various Antarctic cruises. Between 11% and 51% of drill holes in the data set had negative freeboard, or there was slush present. Sturm et al. (1998) suggest that there is a 'self-balancing' mechanism in snow ice formation, whereby negative freeboards are in general likely to be short-lived phenomena, as snow ice will form shortly

Table 3.9 Percentage of drill holes and snow pits with negative freeboard and slush, and the proportion of snow ice measured in sea ice core analyses, from various studies around Antarctica and derived from sources given in the reference column.

Sector	Season (year)	% of pits with slush (n)	% $z_f \leq 0$ m (n)	Mean h_s/h_i (n)	% of snow ice	Reference
Indian and West Pacific oceans						
136–138°E	Early spring (1991)	–	–	0.37 (7)	30	Worby & Massom (1995)
64–106°E	Late spring (1991)	–	–	0.10 (39)	18	Worby & Massom (1995)
62–102°E	Spring (1992)	–	39 (23)	0.07 (67)	23	Worby & Massom (1995)
75–150°E	Spring (1994)	40 (10)	35 (54)	0.18 (106)	–	Worby & Massom (1995)
139–150°E	Autumn (1993)	53 (17)	11 (134)	0.09 (186)	9	Worby & Massom (1995)
138–142°E	Winter (1995)	29 (46)	18 (577)	0.19 (584)	–	Massom et al. (1998)
109–118°E	Winter–spring (2003)	–	15	–	–	Massom (unpubl. data)
115–130°E	Winter–spring (2007)	–	16 (1500)	–	–	Toyota et al. (2011)
Bellingshausen, Amundsen, Ross seas						
66–68°W	Winter (2001)	–	52(488)	0.26 (488)	15	Perovich et al. (2004)
66–68°W	Winter (2002)	–	20(564)	0.20 (564)	–	Perovich et al. (2004)
165–180°W	Autumn–winter (1995)	36 (73)	29 (3671)	0.21/0.31	25	Jeffries & Adolphs (1997); Sturm et al. (1998); Jeffries et al. (1998)
175°E–175°W	Autumn–winter (1998)	0 (43)	17 (3253)	0.18 (3253)	12	Jeffries et al. (2001); Morris & Jeffries (2001)
80–110°W, 155–180°W	Winter–spring (1995)	51(45)	34 (4025)	0.32 (4176)	–	Sturm et al. (1998)
75–110°W						
109–171°W	Winter–spring (1993)	35	18	0.29	24	Jeffries et al. (1997b)
	Winter–spring (1994)	46 (165)	51 (2227)	0.35	38	Sturm et al. (1998); Jeffries et al. (1998)
89–95W						
70–130°W	Winter–spring (2007)	–	–	–	~20	Lewis et al. (2011)
	Late summer (1997)	12 (170)	17 (170)	–	–	Haas et al. (1996)
Weddell Sea						
0–55°W	Winter–spring (1989)	50	40	0.27 (5339)	8	Ackley et al. (1990); Lange et al. (1990)
7°E–5°W	Winter (1986)	–	17	0.18 (4238)	–	Wadhams et al. (1987)
0–50°W	Winter (1982)	–	11.5 (468)	0.14 (386)	–	Massom et al. (1997)
46–54°W	Spring (1988)	–	15	0.23 (1553)	–	Eicken et al. (1994)
45–55°W	Summer (1992)	(~100 holes)	–	–	–	Drinkwater & Lytle (1997)

h_s, snow depth; h_i, ice depth; z_f, freeboard distance.
Source: Adapted from Massom et al. (2001). Reproduced by permission of American Geophysical Union.

after flooding and will re-establish the isostatic balance and produce a zero freeboard. Such relationships may not hold for regions of heavily deformed or multi-year ice, where mean total freeboards ∼ 0.10–0.30 m greater than the snow depth have been reported based on analysis of airborne IceBridge data in October (Kwok & Maksym, 2014).

By comparison, the contribution of snow ice to the overall sea ice mass balance is thought to be smaller in the Arctic, due to the generally thicker ice, but there is little hard evidence on this topic. In the Baltic Sea, however, Granskog et al. (2004) estimated that snow ice accounted for 20–35% of the total thickness of fast ice (see also Saloranta, 2000). A lower value (9%) has been reported from the sub-polar Sea of Okhotsk in winter by Toyota et al. (2007).

3.6.2 Regional snow depth distribution

Regional variations in h_s for the Arctic Ocean for the period 1954–1991 have been presented by Warren et al. (1999). For that period, the deepest snow is found north of Greenland and in the eastern Canadian Archipelago, with depth decreasing towards Wrangell Island and Siberia. Webster et al. (2014) suggest that the pattern of deep and shallow snow has not changed, but that everywhere the snow has thinned, with the maximum thinning north of Alaska and the least amount of thinning north of Greenland and the Canadian Archipelago.

In the Antarctic, *in situ* data collated in Massom et al. (2001) suggest that maximum snow depths occur in the Bellingshausen, Amundsen and Ross Sea sectors. Airborne snow-thickness radar data acquired by the NASA IceBridge project in the Octobers of 2010 and 2011 indicate mean depths of 0.33–0.48 m in the Weddell Sea, with the average snow depth thinning from the northern coast southward to the southeastern Weddell Sea (Kwok & Maksym, 2014). This pattern is consistent with *in situ* observations (e.g. Lange & Eicken, 1991; Massom et al., 1997; Haas et al., 2011). In the Bellingshausen Sea, the thickest snow (ranging from 0.52 to 0.78 m) was found in the nearshore zone where there was deformed older ice (Kwok & Maksym, 2014), an area of high snowfall (Maksym & Markus, 2008).

Of some concern is the fact that the new IceBridge values of h_s are significantly higher than values reported in the literature based on field measurements (e.g. in table 1 of Massom et al., 2001). One explanation for this is that field observations (both *in situ* and ship-based) are biased towards thinner snow due to the difficulty of accessing and sampling regions of heavily deformed and perennial ice where the snow is deeper (Ozsoy-Cicek et al., 2013). The strong dependence of h_s on ice type makes this explanation plausible.

Snow and freeboard distributions are equally limited, but they do illustrate the noted difference between Antarctic and Arctic snow covers. An example is the

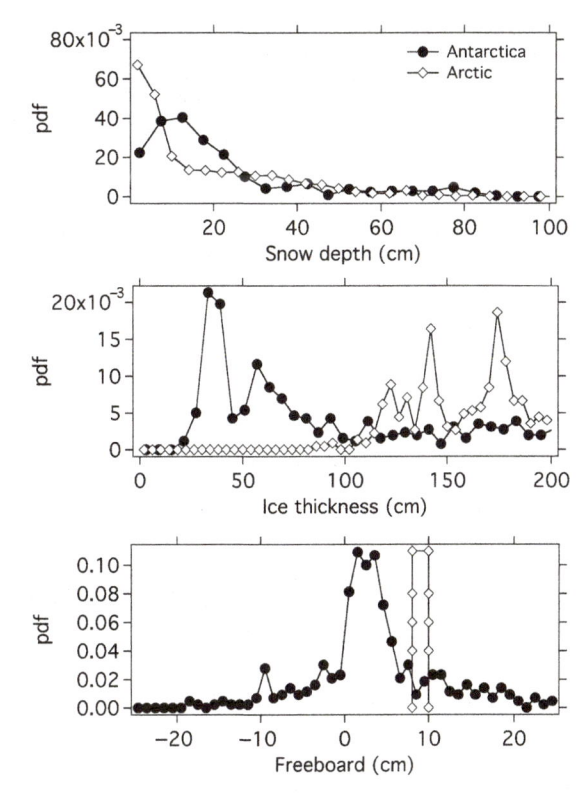

Figure 3.21 Probability distributions (pdf) of snow depth (a), ice thickness (b) and freeboard (c) determined from survey profiles in Marguerite Bay (Antarctica) and in the Beaufort Sea (Arctic Basin). Antarctic data courtesy Don Perovich (CRREL, USA).

comparison of probability distributions from Marguerite Bay in Antarctica (August–September 2002) (Perovich et al., 2004) and a US Navy Ice Camp in the Beaufort Sea in the Arctic Basin (March 2003) (Sturm et al., 2006) shown in Figure 3.21. In this particular case, the Arctic ice, both first- and multi-year, was three to four times thicker than the Antarctic ice, with a consistent positive freeboard of 8–10 cm compared with a mean freeboard of 4.5 cm in Marguerite Bay, where 17% of the freeboard measurements were negative.

Wide spatial and temporal variability in h_s is also the norm for Antarctic fast ice. For example, in Lützow-Holm Bay in East Antarctica, snowfall is heavy (Enomoto et al., 2002) yet there is minimal snow accumulation due to removal by strong and persistent katabatic winds (table 2 in Massom et al., 2001). In this region, h_s tends to increase with distance offshore, rising from 0.29–1.65 m in observations by Kawamura et al.

(1995, 1997) due to the weakening of katabatic-wind strength with increasing distance from the coast. In contrast, near Troll Base at ~5°E, snow depths on the fast ice are generally about 0.40 m but with values up to 2.0 m (Heil et al., 2011). Concerted efforts are under way to coordinate improved collection of snow and ice thickness information from Antarctic bases within the Antarctic Fast Ice Network programme (Heil et al., 2011) to give an improved circum-Antarctic comparison.

3.6.3 Floe-scale snow depth distribution

In many ways, we know more about the local, floe-scale distribution of h_s than we do its regional to hemispheric distribution because these former data are usually derived directly from field measurements. This depth information is often presented as depth cross-sections and distribution curves (Sturm et al., 1998; Iacozza & Barber, 1999; Sturm et al., 2006). Using these, the depth can frequently be discriminated by ice type (Sturm et al., 2002a; see Figure 3.22). Structural length scales in the snow depth fields have been examined using semi-variograms (Sturm et al., 1998; Iacozza & Barber, 1999; Toyota et al., 2011). For the Amundsen, Bellingshausen and Ross Seas, these ranged from 3 to 70 m, compared with ~4–6 m for the underlying ice surface. The longer lengths indicate that the addition of snow increased the floe topographic relief, but tended to 'smooth out' the higher-frequency (i.e. rougher) ice surface features. Sturm et al. (1998) concluded that any ice-surface roughness feature with a relief of > 0.2 m 'nucleated' a significant snow surface feature. Similar results have been reported for East Antarctica by Toyota et al. (2011). For snow on the ice of the Beaufort Sea, structural ranges of 7–120 m were found (Sturm et al., 2007), again with the amplitude increasing with ice roughness (deformation) but with the range seemingly independent of the degree of deformation.

3.7 Remote sensing

Because of the importance of the snow depth distribution in the sea ice system, accurate remote determination of snow depth is clearly a high priority (Iacozza & Ferguson, 2014; Kwok & Maksym, 2014; Kern & Spreen, 2015). Chapter 9 provides further details on how this is being pursued, and why it is difficult to achieve (see also Chapter 5 of Lubin & Massom, 2006). A few key points are worth noting here: the techniques that have been developed for snow are primarily focused on large-scale spatiotemporal patterns (e.g. SPMW estimates shown in Figure 3.19; see also Brucker & Markus, 2013). Another application utilizes the sensitivity of radar backscatter and/or microwave emissivity to liquid-water content in snow on sea ice (preferably in tandem with air temperature information) to map the onset of seasonal snow melt and freeze (e.g. Smith, 1998; Forster et al., 2001; Drinkwater & Liu, 2001; Drobot & Anderson, 2001). Illustrative examples, from the Arctic and for two different atmospheric regimes, are given in Figure 3.23. The same techniques are more problematic when applied to the Antarctica due to less pronounced seasonal changes in the electromagnetic properties of snow on sea ice there compared with the central Arctic (Comiso et al., 1992; Drinkwater, 1995), and due to the difficulty of melt detection when there is widespread basal snow wetness (Massom et al., 2001) or surface flooding (Maksym & Markus, 2008). Having said this, rapid backscatter increases are observed in Antarctic perennial ice regions in November and December due to superimposed ice formation (Haas et al., 2001). Developing improved techniques for monitoring global snow melt and freeze-up patterns is highly desirable, given their sensitivity to climate change and their physical, ecological and biogeochemical importance.

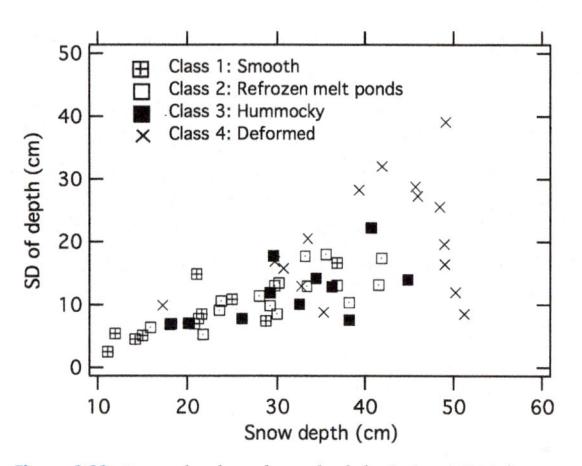

Figure 3.22 Snow depth and standard deviation (SD) of snow depth as a function of ice type for snow on sea ice in the Chukchi Sea at the Surface Heat Budget of the Arctic (SHEBA) project Source: Sturm et al., 2002a. Reproduced by permission of American Geophysical Union.

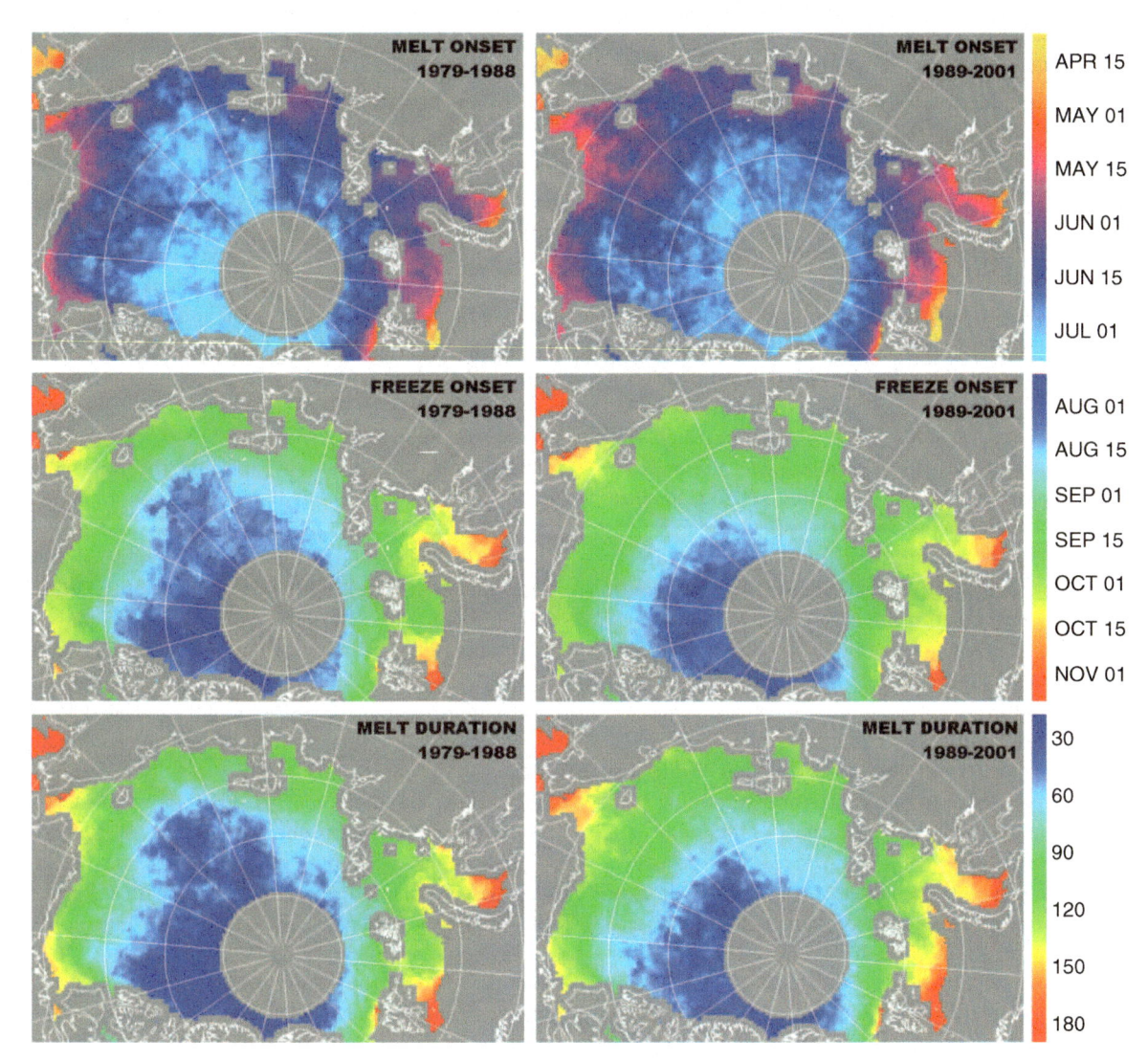

Figure 3.23 Maps of average sea ice melt onset date, freeze onset date and melt season duration in the Arctic Ocean and surrounding seas during the low-index Arctic Oscillation (AO) period (1979–1988) and high-index AO period (1989–2001). These were estimated using coincident satellite passive-microwave brightness temperature data and buoy-derived surface air temperature data (Rigor et al., 2000). Source: From Belchansky et al., 2004. Reproduced by permission of American Meteorological Society. Copyright 2004 American Meteorological Society.

3.8 Ecological and biogeochemical significance of snow

Snow plays a number of important yet contrasting ecological roles in ice-covered oceans, and at all trophic levels and over a wide range of scales (Massom & Stammerjohn, 2011; Iacozza & Ferguson, 2014). On the one hand, it insulates the ice from large and cyclical synoptic-scale fluctuations in air temperature while maintaining a relatively high ice temperature (Massom et al., 1997). This in turn affects the sea ice microstructure, salinity and permeability (Golden et al., 2007), thereby influencing the space available for colonization by microorganisms as well as nutrient transport within the ice (see Chapters 13–16). In addition, extensive seawater flooding due to snow loading favours the formation of productive surface 'infiltration' communities, by seeding the snow–ice interface with macronutrients, algae and other microorganisms (Ackley & Sullivan, 1994). Brine overturning within the sea ice microstructure during flooding and snow ice formation (Lytle & Ackley, 1996) also plays a key role in supplying nutrients to internal algal communities and fuelling ice algal blooms into autumn (Fritsen et al., 1994).

Snow also strongly impacts primary productivity and biological activity both within and under the sea ice by regulating the intensity and spectral composition of downwelling photosynthetically active radiation (PAR) (Arrigo et al., 2003; Mundy et al., 2007; Tremblay & Gagnon, 2009) and by contributing to the pulse of fresh meltwater each spring–summer that stabilizes the ocean mixed layer and fuels ice-edge blooms (see Chapter 14). Even minor variations in snow depth can have a major impact on total transmitted PAR due to the exponential nature of PAR extinction (Iacozza & Barber, 1999). At the same time, snow-covered sea ice dramatically decreases the penetration of potentially harmful ultraviolet radiation into the ice cover and upper ocean, as discussed in Chapter 4.

In the Arctic, snow thickness distribution during winter and spring is an important factor determining habitat selection by predators at the apex of the food chain, such as the ringed seal (*Phoca hispida*) and polar bear (*Ursus maritimus*) (see Chapter 21). Thick snowdrifts on the sea ice, for example, are required by ringed seals for the construction of birthing and resting lairs (Furgal et al., 1996). Moreover, the snow (and ice) must remain sufficiently stable throughout the critical 6-week lactation period, and insufficient snow cover due to low snowfall or melt can also leave seals subject to increased exposure and predation by polar bears (Stirling & Smith, 2004; Chambellant et al., 2012). Snow cover is also an important component of polar bear habitat in that it, too, provides insulation and cover for young bears, with a proportion of the bear population building their maternity dens on multi-year sea ice (Derocher et al., 2004). Changing snow cover and sea ice conditions have potentially dire consequences for these magnificent creatures (Stirling & Derocher, 2012; see Chapter 21).

In addition to its ecological role, snow on sea ice plays a key biogeochemical role in both hemispheres (see Chapter 17). In Antarctica, deposition of atmospheric dust in snow (Sedwick & DiTullio, 1997; Fitzwater et al., 2000; Sedwick et al., 2000; Edwards & Sedwick, 2001) contributes to concentrations of dissolved and particulate iron (Fe) in the sea ice that are one to two orders of magnitude higher than in seawater (e.g. Grotti et al., 2005; Lannuzel et al., 2007, 2010; van der Merwe et al., 2009). When sea ice melts, this Fe is released in the surface waters (Lannuzel et al., 2008; van der Merwe et al., 2011; Kanna et al., 2014) and may contribute to extensive phytoplankton blooms observed in the marginal ice zone and coastal zone in spring (Sedwick & DiTullio, 1997; De Jong et al., 2013; Lannuzel et al., in press).

By regulating and/or stimulating phytoplankton (primary) productivity, snow further plays a role in atmospheric composition by affecting the release by ice algal communities of dimethylsulfoniopropionate (DMSP) into the atmosphere (Curran & Jones, 2000; Trevena & Jones, 2006). This is of considerable climatic significance, in that DMSP is a precursor of dimethyl-sulphide, which is the main source of ocean-derived sulphates and thus affects cloud-condensation nuclei and solar insolation (Liss et al., 2004). It follows that any significant change in sea ice and associated snow cover distribution and conditions could potentially affect climate via this path.

The snow on the ice plays other important roles: it affects the ice–atmosphere exchange of biogases (Semiletov et al., 2004) and bromine release and sea salt aerosol production from sea ice (Yang et al., 2008). In fact, ice layers in the snow can impede sea ice–atmosphere gas exchanges (Tison et al., 2008; Nomura et al., 2010b). Due to its relative proximity to pollutant sources and atmospheric circulation patterns, snow on northern hemisphere sea ice also accumulates soot, nitrogen compounds and ammonium (Warren & Clarke, 1986; Grenfell et al., 2002; Ehn et al., 2004; Zhou et al., 2012; Nomura et al., 2011). Black carbon (a major component of soot) is strongly light-absorbing, and its presence in snow on Arctic sea ice influences the surface albedo (Marks & King, 2013) and is currently an issue of great concern (Goldenson et al., 2012; Bond et al., 2013; Jiao et al., 2014).

3.9 The outlook

The outlook for snow on sea ice is unsettled and uncertain, for three basic reasons:

1. There are large uncertainties in future predictions of precipitation change. In short, while models predict that there will be more precipitation in both polar regions due to the greater water-holding capacity of a warmer atmosphere and changes in storm tracks (Bracegirdle et al., 2008; Rinke & Dethloff, 2008; Walsh, 2008), we really don't know if this is happening over the Arctic and Antarctic sea ice as yet – or if it even will materialize. For one thing, there are almost no data on precipitation rates and snow accumulation and redistribution on sea ice (Leonard & Maksym, 2011). Moreover, it is unclear whether more of the precipitation will fall as rain rather than snow.
2. Even if we knew how the precipitation was changing, we don't know how the timing of autumn precipitation will change with respect to the timing of freeze-up, particularly in the Arctic. Consequently, it could start snowing more, but increasingly later ice freeze-up dates might result in increased snowfall into an unfrozen ocean. Therefore, more snowfall might not mean a deeper snow cover on the ice (Hezel et al., 2012).
3. Models of snow–sea ice interaction remain rudimentary (although some progress is being made using sub-models for snow (e.g. Cheng et al., 2008; Lecomte et al., 2011). While it is widely agreed that there is some optimal snow thickness that produces the least ice, there is no real agreement on or confidence in what that value is, or whether the current Arctic or Antarctic systems are close to this value. What is clear is that snow on sea ice is one of the most important areas where research is needed. It is needed because changing patterns of snow depth and snow properties on sea ice are likely to have very extensive ramifications (cf. Vancoppenolle et al., 2013; Iacozza & Ferguson, 2014). The research is needed because the snow complicates remote sensing, and measurement of its thickness from space is a challenging proposition, making it difficult to use remote sensing as a tool for measuring and monitoring the sea ice mass balance over time (see Chapter 9). It is needed because the snow can so easily be turned into ice, or alternatively shield the ice from cold weather and thereby prevent it from growing thicker by limiting basal accretion. These two deceptively simple yet competing aspects of snow on sea ice remain difficult to model and confound simple measurements. When we add the interplay of the (changing) progression of the season and the time and spatial history of sea ice (which is itself highly dynamic) to the pattern of snow, the interactions of snow on ice are difficult to predict, but endlessly fascinating.

3.10 Acknowledgements

We are indebted to the many colleagues and friends who have collected snow data over the years, and to the captains, crews and helicopter pilots of the vessels involved. Unfortunately, they are far too numerous to list here. For Robert Massom, this work was supported by the Australian Government's Cooperative Research Centres Programme through the Antarctic Climate and Ecosystems Cooperative Research Center (ACE CRC), and contributed to AAS Project 3024. Robert Massom thanks his colleagues in the sea ice group at the Antarctic and ACE CRCs over the years (notably Ian Allison, Petra Heil, Tony Worby, Jan Lieser, Klaus Meiners, Vicky Lytle and Barry Giles). Matthew Sturm thanks his colleagues at CRREL (D. Perovich, J. Richter-Menge, C. Hiemstra, T. Douglas and C. Polashenski) and the University of Alaska (UAF) (A. Mahoney and H. Eicken) for support and thoughtful discussions. His work was supported by UAF, CRREL, and several grants from NASA and the National Science Foundation. We are very grateful to Don Perovich (US CRREL) for supplying his Antarctic data from 2002. We both thank the editor David Thomas for his patience and help. This chapter is dedicated to our wives Betsy and Yuko, and our children Skye, Eli, Adelle and Kaiki.

References

Ackley, S.F., Lange, M. & Wadhams, P. (1990) Snow cover effects on Antarctic sea ice thickness. In: *Sea Ice Properties and Processes, CRREL Monograph 90–1*, (Ed: S.F. Ackley and W.F. Weeks), pp. 16–21. CRREL, New Hampshire.

Ackley, S.F. & Sullivan, C.W. (1994) Physical controls on the development and characteristics of Antarctic sea ice biological communities – a review and synthesis. *Deep-Sea Research Part I*, **41**(10), 1583–1604.

Adolphs, U. (1999) Roughness variability of sea ice and snow cover thickness profiles in the Ross, Amundsen, and Bellingshausen Seas. *Journal of Geophysical Research*, **104**(C6), 13 577–13 591.

Akitaya, E. (1974) Studies on depth hoar. *Contributions from the Institute of Low Temperature Science*, Series A, No. **26**, 1–67.

Albert, M.R. & McGilvary, W.R. (1992) Thermal effects due to air flow and vapour transport in dry snow. *Journal of Glaciology*, **38**, 273–281.

Alvarez-Aviles, L., Simpson, W.R., Douglas, T.A., Sturm, M., Perovich, D.K. & Domine, F (2008) Frost flower chemical composition during growth and its implications for aerosol production and bromine activation. *Geophysical Research Letters*, **113**, D21304, doi:10.1029/2008JD010277.

Amstrup, S.C. & Gardner, C. (1994) Polar bear maternity denning in the Beaufort Sea. *The Journal of Wildlife Management*, **58**, 1–10.

Andreas, E.L., & Ackley, S.F. (1981) On the difference in ablation seasons of Arctic and Antarctic sea-ice. *Journal of Atmospheric Science*, **39**, 440–447.

Andreas, E.L., & Claffey, K.J. (1995) Air-ice drag coefficients in the western Weddell Sea 1. Values deduced from profile measurements. *Journal of Geophysical Research*, **100**(C3), 4821–4831.

Arrigo, K.R. (2003) Physical control of chlorophyll a, POC, and TPN distributions in the pack ice of the Ross Sea, Antarctica. *Journal of Geophysical Research*, **108**(C10), 3316. doi:10.1029/2001JC001138.

Arrigo, K.R., Perovich, D.K., Pickart, R.S., Brown, et al. (2012) Massive phytoplankton blooms under Arctic sea ice. *Science*, **336**, 1408–1408.

Bagnold, R.A. (1954) *The Physics of Blown Sand and Desert Dunes*. Courier Dover Publications.

Barber, D.G., Fung, A.K., Grenfell, T.C. et al. (1998) The role of snow on microwave emission and scattering over first-year sea ice. *IEEE Transactions on Geoscience and Remote Sensing*, **36**, 1750–1763.

Beaglehole, J.C. (1974) *The Life of Captain James Cook*. Stanford University Press, Stanford, USA.

Belchansky, G.I., Douglas, D.C. & Platonov, N.G. (2004) Duration of the Arctic sea ice melt season: Regional and interannual variability, 1979–2001. *Journal of Climate*, **17**, 67–80.

Benson, C.S. & Sturm, M. (1993) Structure and wind transport of seasonal snow on the Arctic slope of Alaska. *Annals of Glaciology*, **18**, 261–267.

Berton, P. (1989) *The Arctic Grail: The Quest for the North West Passage and the North Pole, 1818–1909*. Anchor Canada.

Birkeland, K. (1998) Terminology and predominant processes associated with the formation of weak layers of near-surface faceted crystals in the mountain snowpack. *Arctic and Alpine Research* **30**, 193–199.

Blackford, J.R. (2007). Sintering and microstructure of ice: a review. *Journal of Physics D: Applied Physics*, **40**, R355, doi:10.1088/0022-3727/40/21/R02.

Blazey, B.A., Holland, M.M. & Hunke, E.C. (2013) Arctic Ocean sea ice snow depth evaluation and bias sensitivity in CCSM. *The Cryosphere Discussions*, **7**, 1495–1532.

Bond, T.C., Doherty, S.J., Fahey, D.W., Forster, P.M. et al. (2013). Bounding the role of black carbon in the climate system: a scientific assessment. *Journal of Geophysical Research: Atmospheres*, **118**, 5380–5552.

Bracegirdle, T.J., Connolley, W.M. & Turner, J. (2008) Antarctic climate change over the twenty first century. *Journal of Geophysical Research*, **113**, D03103, doi:10.1029/2007JD008933.

Brandt, R.E., Warren, S.G., Worby, A.P. et al. (2005) Surface albedo of the Antarctic sea-ice zone. *Journal of Climate*, **18**, 3606–3622.

Brucker, L. & Markus, T. (2013) Arctic-scale assessment of satellite passive microwave-derived snow depth on sea ice using Operation IceBridge airborne data, *Journal of Geophysical Research: Oceans*, **118**, 2892–2905.

Buist, I., Belore, R., Dickins, D., Guarino, A., Hackenberg, D., & Wang, Z. (2009) Empirical weathering properties of oil in ice and snow. *Proceedings Arctic and Marine Oilspill Program Technical Seminar*, **32**, 67–107. Environment Canada, Ottawa.

Calonne, N., Flin, F., Morin, S., Lesaffre, B., du Roscoat, S.R., & Geindreau, C. (2011) Numerical and experimental investigations of the effective thermal conductivity of snow. *Geophysical Research Letters*, **38**(, L23501, doi:10.1029/2011GL049234.

Cavalieri, D.J., Markus, T. & Comiso, J.C. (2014) *AMSR-E/Aqua Daily L3 12.5 km Brightness Temperature, Sea Ice Concentration, and Snow Depth Polar Grids*. Version 3. NASA DAAC at the National Snow and Ice Data Center, Boulder, CO USA.

Chambellant, M., Stirling, I., Gough, W.A. & Ferguson, S.H. (2012) Temporal variations in Hudson Bay ringed seal (*Phoca hispida*) life-history parameters in relation to environment. *Journal of Mammalogy*, **93**, 267–281.

Cheng, B., Zhang, Z., Vihma, T., Johansson, M., Bian, L., Li, Z. & Wu, H. (2008) Model experiments on snow and ice thermodynamics in the Arctic Ocean with CHINARE 2003 data. *Journal of Geophysical Research*, **113**(C9), C09020, 10.1029/2007JC004654.

Colbeck, S.C. (1982) An overview of seasonal snow metamorphism. *Reviews of Geophysics and Space Physics*, **20**, 45–61.

Colbeck, S.C. (1986) Statistics of coarsening in water-saturated snow. *Acta Metallurgica*, **34**, 347–352.

Colbeck, S.C. (1989). Air movement in snow due to windpumping. *Journal of Glaciology*, **35**(120), 209–213.

Colbeck, S.C. et al. (1990) The international classification for seasonal snow on the ground. *International Commission on Snow and Ice*. International Association of Scientific Hydrology, Wallingford, Oxon.

Colbeck, S.C. (1991) The layered character of snow covers. *Reviews of Geophysics*, **29**, 81–96.

Colony, R., Radionov, V. & Tanis, F.J. (1998) Measurements of precipitation and snowpack at Russian North Pole drifting stations. *Polar Record*, **34**, 3–14.

Comiso, J.C., Grenfell, T.C., Lange, M. et al. (1992) Microwave remote sensing of the Southern Ocean ice cover. In: *Microwave Remote Sensing of Sea Ice*, (Ed. F.D. Carsey), pp. 243–259. American Geophysical Union, Washington, DC.

Comiso, J.C., Cavalieri, D.J. & Markus, T. (2003) Sea ice concentration, ice temperature, and snow depth, using AMSR-E data. *IEEE Transactions on Geoscience and Remote Sensing*, **41**, 243–252.

Cooke, A. & Holland, C. (1978) *The Exploration of Northern Canada: 500–1920 A Chronology*. The Arctic History Press, Toronto.

Curran, M.A.J. & Jones, G.B. (2000) Dimethylsulphide in the Southern Ocean: seasonality and flux. *Journal of Geophysical Research*, **105**(D16), 20451–20459.

De Jong, J., Schoemann, V., Maricq, N. et al.(2013) Iron in land-fast sea ice of McMurdo Sound derived from sediment resuspension and wind-blown dust attributes to primary productivity in the Ross Sea, Antarctica. *Marine Chemistry*, **157**, 24–40.

Denoth, A. (1982) Effect of grain geometry on electrical properties of snow at frequencies up to 100 MHz. *Journal of Applied Physics*, **53**, 7496–7501.

Derksen C., King, J., Howell, S., Toose, P., Silis, A. & Rutter, N. (2014) *Evaluation of Icebridge snow radar measurements over sea ice in the Canadian Arctic*. Fall AGU Meeting, San Francisco.

Derocher, A.E., Lunn, N.J., & I. Stirling (2004) Polar bears in a warming climate. *Integrative and Comparative Biology*, **44**, 163–176.

Déry, S.J. & Tremblay, L.-B. 2004. Modeling the effects of wind redistribution on the snow mass budget of polar sea ice. *Journal of Physical Oceanography*, **34**, 258–271.

Domine, F., Bock, J., Morin, S., & Giraud, G. (2011). Linking the effective thermal conductivity of snow to its shear strength and density. *Journal of Geophysical Research: Earth Surface*, **116**, F04027, doi:10.1029/2011JF002000.

Doumani, G.A. (1966) Surface structures in snow. In: *International Conference on Low Temperature Science: I. Physics of Snow and Ice*, pp. 1119–1136. Sapporo, Japan.

Dozier, J., S.R. Schneider, S.R. , & McGinnis, D.F. Jr., (1981) Effects of grain size and snowpack water equivalence on visible and near-infrared satellite observations of snow. *Water Resources Research*, **17**, 1213–1221.

Douglas, T.A., Sturm, M., Simpson, W.R. et al. (2008) The influence of snow and ice crystal formation and accumulation on mercury deposition to the Arctic. *Environmental Science and Technology*, **42**, 1542–1551.

Drinkwater, M.R. (1995) Airborne and satellite SAR investigations of sea-ice surface characteristics. In: *Oceanographic Applications of Remote Sensing* (Eds. M. Ikeda & F.W. Dobson), pp. 339–357. CRC Press, Boca Raton, Florida.

Drinkwater, M.R., & Crocker, G.B. (1988) Modeling changes in the dielectric and scattering properties of young snow-covered sea ice at GHz frequencies. *Journal of Glaciology*, **34**(118), 274–282.

Drinkwater, M.R., & Liu, X. (2000) Seasonal to interannual variability in Antarctic sea-ice surface melt. *IEEE Transactions on Geoscience and Remote Sensing*, **38**, 1827–1842.

Drinkwater, M.R., & Lytle, V.I. (1997) ERS-1 radar and field-observed characteristics of autumn freeze-up in the Weddell Sea. *Journal of Geophysical Research*, **102**, 12,593–12,608.

Drobot, S. & M.A. Anderson (2001) An improved method for determining snowmelt onset dates over Arctic sea ice using Scanning Multichannel Microwave Radiometer and Special Sensor Microwave/Imager data. *Journal of Geophysical Research*, **106**, 24033–24049.

Edwards, R., & Sedwick, P.N. (2001) Iron in East Antarctic snow: implications for atmospheric iron deposition and algal production in Antarctic waters. *Geophysical Research Letters*, **28**, 3907–3910.

Ehn, J., Granskog, M.A., Reinart, A. & Erm, A. (2004) Optical properties of melting landfast sea ice and underlying seawater in Santala Bay, Gulf of Finland. *Journal of Geophysical Research*, **109**, C09003. 10.1029/2003JC002042.

Eicken, H., Lange, M.A., Hubberton, H.-W. et al. (1994) Characteristics and distribution patterns of snow and meteoric ice in the Weddell Sea and their contribution to the mass balance of sea ice. *Annales Geophysicae*, **12**, 80–93.

Eicken, H., Fischer, H. & Lemke, P. (1995) Effects of the snow cover on Antarctic sea ice and potential modulation of its response to climate change. *Annals of Glaciology*, **21**, 369–376.

Eicken, H., Grenfell, T.C., Perovich, D.K. et al. (2004) Hydraulic controls of summer Arctic pack ice albedo. *Journal of Geophysical Research*, **109**(C08007), doi: 10.1029/2003JC001989.

Enomoto, H., Nishio, F., Warashina, H. & Ushio, S. (2002) Satellite observation of melting and break-up of fast ice in Lützow-Holm Bay, East Antarctica. *Polar Meteorology and Glaciology*, **16**, 1–14.

von Eugster, H.P. (1950) *On the Form and Metamorphosis of Snow*. Report 113. Institute for Snow and Avalance Research, Davos, Switzerland.

Fichefet, T., & Maqueda, M.M. (1999) Modelling the influence of snow accumulation and snow-ice formation on the seasonal cycle of the Antarctic sea-ice cover. *Climate Dynamics*, **15**, 251–268.

Fichefet, T., Tartinville, B. & Goosse, H. (2000) Sensitivity of the Antarctic sea ice to the thermal conductivity of snow. *Geophysical Research Letters*, **27**, 401–404.

Fienup-Riordan, A. & Rearden, A. (2010) The ice is always changing: Yup'ik understandings of sea ice, past and present. In *SIKU: Knowing Our Ice*, pp. 295–320. Springer, Netherlands.

Fierz, C., Armstrong, R.L., Durand, Y. et al. (2009) *IACS International Classification for Seasonal Snow on the Ground*. Technical Documents in Hydrology. International Association of Cryospheric Sciences/UNESCO.

Fitzwater, S., Johnson, K., Gordon, R., Coale, K.H., & Smith, W. (2000) Trace metal concentrations in the Ross Sea and their relationship with nutrients and phytoplankton growth. *Deep-Sea Research II*, **47**, 3159–3179.

Forster, R.R., Long, D.G., Jezek, K.C., Drobot, S.D., & Anderson, M.R. (2001) The onset of Arctic sea-ice snowmelt as detected with passive- and active-microwave remote sensing. *Annals of Glaciology*, **33**, 85–93.

Fritsen, C.H., Lytle, V.I., Ackley, S.F. & Sullivan, C.W. (1994) Autumn bloom of Antarctic pack-ice algae. *Science*, **266**, 782–784.

Furgal, C.M., Innes, S. & Kovacs, K.M. (1996) Characteristics of ringed seal, Phoca hispida, subnivean structures and breeding habitat and their effects on predation. *Canadian Journal of Zoology*, **74**, 858–874.

Garrity, C. (1992) Characterization of snow on floating ice and case studies of brightness temperature changes during the onset of melt. In: *Microwave Remote Sensing of Sea Ice* (Ed. F.D. Carsey), pp. 313–326. American Geophysical Union, Washington, DC, USA.

Gearheard, S., Matumeak, W., Angutikjuaq, I. et al. (2006) "It's not that simple": a collaborative comparison of sea ice environments, their uses, observed changes, and adaptations in Barrow, Alaska, USA, and Clyde River, Nunavut, Canada. *Ambio*, **35**, 203–211.

Gearheard, S., Kielsen Holm, L., Huntington, H.P. et al. (Eds.) (2013) *The Meaning of Ice: People and Sea Ice in Three Arctic Communities*. International Polar Institute, Hanover, New Hampshire.

German, R.M. (1996) *Sintering Theory and Practice*. John Wiley & Sons, Inc., New York.

Golden, K.M., Ackley, S.F. & Lytle, V.I. (1998) The percolation phase transition in sea ice. *Science*, **282**, 2238–2241.

Golden, K.M., Eicken, H., Heaton, A.L., Miner, J., Pringle, D.J. & Zhu, J. (2007) Thermal evolution of permeability and microstructure in sea ice. *Geophysical Research Letters*, **34**, L16501, doi:10.1029/2007GL030447.

Goldenson, N., Doherty, S.J., Bitz, C.M., Holland, M., Light, B. & Conley, A.J. (2012) Arctic climate response to forcing from light-absorbing particles in snow and sea ice in CESM. *Atmospheric Chemistry and Physics*, **12**, 7903–7920.

Goodwin, I.D. (1990) Snow accumulation and surface topography in the katabatic zone of eastern Wilkes Land, Antarctica. *Antarctic Science*, **2**, 235–242.

Granskog, M.A., Leppäranta, M., Kawamura, T., Ehn, J. & Shirasawa, K. (2004) Seasonal development of the properties and composition of landfast sea ice in the Gulf of Finland, the Baltic Sea. *Journal of Geophysical Research*, **109**, C02020, doi:10.1029/2003JC001874.

Granskog, M.A., Vihma, T., Pirazzini, R. & Cheng, B. (2006) Superimposed ice formation and surface energy fluxes on sea ice during the spring melt-freeze period in the Bering Sea. *Journal of Glaciology*, **52**, 119–127.

Green, R.O., Dozier, J., Roberts, D. et al. (2002) Spectral snow reflectance models for grain-size and liquid-water fraction in melted snow for the solar-reflected spectrum. *Annals of Glaciology*, **34**, 71–73.

Grenfell, T.C., Light, B. & Sturm, M. (2002) Spatial distribution and radiative effects of soot in the snow and sea ice during the SHEBA experiment. *Journal of Geophysical Research*, **107**(C10), 8032, doi 10.1029/2002JC000414.

Grotti, M., Soggia, F., Ianni, C. & Frache, R. (2005). Trace metals distributions in coastal sea ice of Terra Nova Bay, Ross Sea, Antarctica. *Antarctic Science*, **17**, 289–300.

Haas, C., Rebhan, H., Thomas, D.N. & Viehoff, T. (1996) Sea ice. In: *The Expedition ANTARKTIS-XI/3 of RV "Polarstern" in 1994* (Eds. H. Miller & Grobe, H.), pp. 29–43. Report on Polar Research, 188/96.

Haas, C., Thomas, D.N., Steffens, M. et al. (1998) Physical and biological investigations of sea-ice. In: *The Expedition ANTARKTIS-XIV of RV "Polarstern" in 1997, Report of Leg ANT-XIV/3*, (Eds. W. Jokat & H. Oerter), pp. 18–30. Report on Polar Research, 267/98.

Haas, C., Thomas, D.N., & Bareis, J. (2001) Surface properties and processes of perennial Antarctic sea ice in summer. *Journal of Glaciology*, **47**, 623–625.

Haas, C., Nicolaus, M., Friedrich, A., Pfaffling, A., Li, Z. & Toyota, T. (2011) Sea ice measurements during Polarstern cruise ANT-XXIII/7 (Winter Weddell Outflow Study), doi:10.1594/PANGAEA.771247.

Heil, P., Gerland, S. & Granskog, M.A. (2011) An Antarctic monitoring initiative for fast ice and comparison with the Arctic. *The Crysophere Discussions*, **5**, 2437–2463.

Hezel, P.J., Zhang, X., Bitz, C.M., Kelly, B.P. & Massonnet, F. (2012) Projected decline in spring snow depth on Arctic sea ice caused by progressively later autumn open ocean freeze-up this century. *Geophysical Research Letters*, **39**, L17505, doi:10.1029/2012GL052794.

Holbrook, J. (2011) *The Measurement of Anisotropic Thermal Conductivity in Snow with Needle Probes*. Master's thesis, University of Alaska, Fairbanks.

Holtzmark, B.E. (1955) Insulating effect of a snow cover on the growth of young sea ice. *Arctic*, **8**, 60–65.

Iacozza, J. & Barber, D.G. (1999) An examination of the distribution of snow on sea ice. *Atmosphere-Ocean*, **37**, 21–51.

Iacozza, J. and Barber, D.G. (2001) Ablation patterns of snow cover over smooth first-year sea ice in the Canadian Arctic. *Hydrological Processes*, **15**, 3359–3569.

Iacozza, J. & Ferguson, S.H. (2014). Spatio-temporal variability of snow over sea ice in western Hudson bay, with reference to ringed seal pup survival. *Polar Biology*, **37**, 817–832.

Järvinen, O. & Leppäranta, M. (2011) Transmission of solar radiation through the snow cover on floating ice. *Journal of Glaciology*, **57**, 861–870.

Jeffries, M.O. & Adolphs, U. (1997) Early winter snow and ice thickness distribution, ice structure and development of the western Ross Sea pack ice between the ice edge and the Ross Ice Shelf. *Antarctic Science*, **9**, 188–200.

Jeffries, M.O., Veazey, A.L., Morris, K. & Krouse, H.R. (1994) Depositional environment of the snow cover on West Antarctic pack-ice floes. *Annals of Glaciology*, **20**, 33–38.

Jeffries, M.O., Worby, A.P., Morris, K. & Weeks, W.F. (1997a) Seasonal variations in the properties, and structural and isotopic composition of sea ice and snow cover in the Bellingshausen and Amundsen Seas, Antarctica. *Journal of Glaciology*, **43**, 138–151.

Jeffries, M.O., Worby, A.P., Morris, K. et al. (1997b) Late winter snow and ice characteristics of first-year floes in the Bellingshausen and Amundsen Seas, Antarctica. Results of investigations during R.V. Nathaniel B. Palmer cruise NBP 93–5 in

August and September 1993. *Report UAG-R-325*. Geophysical Institute, University of Alaska, USA.

Jeffries, M.O., Hurst-Cushing, B., Krouse, H.R. et al. (1998) The role of snow in the thickening and mass budget of first-year floes in the eastern Pacific sector of the Antarctic pack ice. *Report UAG-R-327*. Geophysical Institute, University of Alaska Fairbanks, USA.

Jeffries, M.O., Krouse, H.R., Hurst-Cushing, B. et al. (2001) Snow-ice accretion and snow-cover depletion on Antarctic first-year sea-ice floes. *Annals of Glaciology*, **33**, 51–60.

Jiao, C. et al. (2014) An AeroCom assessment of black carbon in Arctic snow and sea ice. *Atmospheric Chemistry and Physics*, **14**, 2399–2417.

Jones, A.K. et al. (2007). *Polar Icebreakers in a Changing World: An Assessment of U.S. Needs*. National Academies Press, Washington, DC, USA.

Kanna, N., Toyota, T. & Nishioka, J. (2014) Iron and macro-nutrient concentrations in sea ice and their impact on the nutritional status of surface waters in the southern Okhotsk Sea. *Progress in Oceanography*, **126**, 44–57.

Kawamura, T., Takizawa, T., Ohshima, K.I. & Ushio, S. (1995) Data of sea-ice cores obtained in Lützow-Holm Bay from 1990 to 1992 (JARE-31, -32) in the period of Japanese Antarctic climate research. *JARE Data Report 204 (Glaciology 24)*. National Institute of Polar Research, Tokyo.

Kawamura, T., Ohshima, K., Takizawa, T. & Ushio, S. (1997) Physical, structural and isotopic characteristics and growth processes of fast ice in Lützow-Holm Bay, Antarctica, *Journal of Geophysical Research*, **102**, 334–355.

Kawamura, T., Jeffries, M.O., Tison, J.-L. & Krouse, H.R. (2004) Superimposed-ice formation in summer in Ross Sea pack-ice floes. *Journal of Glaciology*, **39**, 563–568.

Kern, S. & G. Spreen (2015) Uncertainties in Antarctic sea-ice thickness retrieval from ICESat. *Annals of Glaciology*, **56**, doi: 10.3189/2015AoG69A736

Kikuchi, K., Kameda, T., Higuchi, K. & Yamashita, A. (2013) A global classification of snow crystals, ice crystals, and solid precipitation based on observations from middle latitudes to polar regions. *Atmospheric Research*, **132**, 460–472.

Kobayashi, D. (1972) Studies of snow transport in low-level drifting snow. *Contributions from the Institute of Low Temperature Science*, **24**, 1–58.

Kojima, K. (1966) Densification of seasonal snow cover. In: *Physics of Snow and Ice*, pp. 929–951. Institute of Low Temperature Science, Saporro, Japan.

Krupnik, I., & Müller-Wille, L. (2010) Franz Boas and Inuktitut terminology for ice and snow: From the emergence of the field to the "Great Eskimo Vocabulary Hoax". In *SIKU: Knowing our ice*, pp. 377–400. Springer, the Netherlands.

Kurtz, N. & Farrell, S. (2011) Large-scale surveys of snow depth on Arctic sea ice from Operation IceBridge, *Geophysical Research Letters*, **38**, L20505, doi:10.1029/2011GL049216.

Kwok, R., & Maksym, T. (2014) Snow depth of the Weddell and Bellingshausen sea ice covers from IceBridge surveys in 2010 and 2011: An examination. *Journal of Geophysical Research: Oceans*, **119**, 4141–4167.

LaChapelle, E.R. (1992) *Field Guide to Snow Crystals*. International Glaciological Society, Cambridge, UK.

Lange, M.A., & Eicken, H. (1991). The sea ice thickness distribution in the northwestern Weddell Sea. *Journal of Geophysical Research: Oceans*, **96**, 4821–4837.

Lange, M.A., Schlosser, P., Ackley, S.F., Wadhams, P. & Dieckmann, G.S. (1990) ^{18}O concentrations in sea ice of the Weddell Sea, Antarctica. *Journal of Glaciology*, **36**, 315–323

Lannuzel., D., Schoemann, V., De Jong, J. et al. (2007) Distribution and biogeochemical behaviour of iron in the East Antarctic sea ice. *Marine Chemistry*, **106**, 18–32.

Lannuzel, D., Schoemann, V., de Jong, J. et al. (2008) Iron study during a time series in the western Weddell pack ice. *Marine Chemistry*, **108**, 85–95.

Lannuzel, D., Schoemann, V., de Jong, J. et al. (2010). Distribution of dissolved iron in Antarctic sea ice: Spatial, seasonal, and inter-annual variability. *Journal of Geophysical Research*, **115**(G3), G03022. doi:10.1029/2009JG001031

Lannuzel D., Chever, F., van der Merwe P., Janssens, J., Cavagna, A.J., Roukaerts A., Bowie A. & Meiners, K. (in press). Iron biogeochemistry in Antarctic pack ice during SIPEX-2. *Deep-Sea Research II*, doi:10.1016/j.dsr2.2014.12.003.

Lecomte, O., Fichefet, T., Vancoppenolle, M. & Nicolaus, M. (2011) A new snow thermodynamic scheme for large-scale sea-ice models. *Annals of Glaciology*, **52**, 337–346.

Lecomte, O., Fichefet, T., Vancoppenolle, M. et al. (2013) On the formulation of snow thermal conductivity in large-scale sea ice models. *Journal of Advances in Modeling Earth Systems*, **5**, 542–557.

Ledley, T.S. (1985) Sea ice: Multiyear cycles and white ice. *Journal of Geophysical Research: Atmospheres*, **90**, 5676–5686.

Ledley, T.S. (1991) Snow on sea ice: competing effects in shaping climate. *Journal of Geophysical Research*, **96**, 17195–17208.

Ledley, T.S. (1993) Variations in snow on sea ice: A mechanism for producing climate variations. *Journal of Geophysical Research*, **98**, 10 401–10 410.

Leonard, K.C. & Maksym, T. (2011) The importance of wind-blown snow redistribution to snow accumulation on Bellingshausen Sea ice. *Annals of Glaciology*, **52**, 271–278.

Lewis, M.J., Tison, J.-L., Weissling, B. et al. (2011) Sea ice and snow cover characteristics during the winter-spring transition in the Bellingshausen Sea: an overview of SIMBA 2007. *Deep-Sea Research II*, **58**, 1019–1038.

Lewis, E.O., Livingstone, C.E., Garrity, C. et al. (1994) Properties of snow and ice. In: *Remote Sensing of Sea Ice and Icebergs* (Eds. S. Haykin et al.). John Wiley and Sons, New York, pp. 21–96.

Libbrecht, K. & Rasmussen, P. (2003). *The Snowflake: Winter's Secret Beauty*. Voyageur Press.

List, R.J. (1951). *Smithsonian Meteorological Tables*, Smithsonian Institute Publications, Washington, DC.

Lubin, D. & Massom, R.A. (2006) *Polar Remote Sensing. Volume 1: Atmosphere and Polar Oceans*. Praxis/Springer, Chichester & Berlin.

Lytle, V.I. & Ackley, S.F. (1992) Snow properties and surface elevation profiles in the Western Weddell Sea, (NBP92-2). *Antarctic Journal of the US*, **XXVII**, 5, 93–94.

Lytle, V.I. & Ackley, S.F. (1996) Heat flux through sea ice in the western Weddell Sea: Convective and conductive transfer processes. *Journal of Geophysical Research*, **101**, 8853–8868.

Lytle, V.I. & Ackley, S.F. (2001) Snow-ice growth: A fresh-water flux inhibiting deep convection in the Weddell Sea, Antarctica. *Annals of Glaciology*, **33**, 45–50.

Lytle, V.I., Massom, R., Bindoff, N. et al. (2000) Wintertime heat flux to the underside of East Antarctic pack ice. *Journal of Geophysical Research*, **105**, 28 759–28 769.

Magono, C. & Lee, C.W. (1966) Meteorological classification of natural snow crystals. *Journal of the Faculty of Science, Hokkaido University, Series 7, Geophysics*, **2**, 321–335.

Maksym, T. & Jeffries, M.O. (2000) A one-dimensional percolation model of flooding and snow-ice formation on Antarctic sea ice. *Journal of Geophysical Research*, **105**, 26 313–26 331.

Maksym, T. & Markus, T. (2008) Antarctic sea ice thickness and snow-to-ice conversion from atmospheric reanalysis and passive microwave snow depth. *Journal of Geophysical Research*, **113**, C02S12, doi:10.1029/2006JC004085.

Maksym, T., Stammerjohn, S.E., Ackley, S. & Massom, R. (2012) Antarctic sea ice – a polar opposite? *Oceanography*, **25**, 140–151.

Male, D.H. & Gray, D. (1981) *Handbook of Snow: Principles, Processes, Management and Use*. Pergamon Press.

Marks, A.A. & King, M.D. (2013) The effects of additional black carbon on the albedo of Arctic sea ice: variation with sea ice type and snow cover. *The Cryosphere*, **7**, 1193–1204, doi:10.5194/tc-7-1193-2013.

Markus T. & Cavalieri, D. (1998) Snow depth distribution over sea ice in the Southern Ocean from satellite passive microwave data. *Antarctic Research Series*, **74**, 19–39. American Geophysical Union, Washington, DC.

Markus, T., Massom, R., Worby, A., Lytle, V., Kurtz, N., & Maksym, T. (2011) Freeboard, snow depth, and sea ice roughness in East Antarctica from in-situ and multiple satellite data. *Annals of Glaciology*, **52**, 242–248.

Martinson, D.G. & Iannuzzi, R.A. (1998) Antarctic ocean-ice interaction: implications from ocean bulk property distributions in the Weddell Gyre. In: *Antarctic Sea Ice: Physical Processes, Interactions and Variability* (Ed. M. Jeffries), pp. 243–271. American Geophysical Union, Washington, DC, USA.

Maslanik, J.A., Fowler, C., Stroeve, J., et al. (2007) A younger, thinner Arctic ice cover: Increased potential for rapid, extensive sea-ice loss. *Geophysical Research Letters*, **34** (L24501), doi:10.1029/2007GL032043.

Maslanik, J., Stroeve, J., Fowler, C. & Emery, W. (2011) Distribution and trends in Arctic sea ice age through spring 2011. *Geophysical Research Letters*, **38**, L13502, doi:10.1029/2011GL047735.

Massom, R.A. & Stammerjohn, S.E. (2010) Antarctic sea ice change and variability – Physical and ecological implications. *Polar Science*, **4**, 149–186.

Massom, R.A., Drinkwater, M.R. & Haas, C. (1997) Winter snow cover on sea ice in the Weddell Sea. *Journal of Geophysical Research*, **102**, 1101–1117.

Massom, R.A., Lytle, V.I., Worby, A.P. et al. (1998) Winter snow cover variability on East Antarctic sea ice. *Journal of Geophysical Research*, **103**, 24 837–24 855.

Massom, R.A., Eicken, H., Haas, C. et al. (2001) Snow on Antarctic sea ice. *Reviews of Geophysics*, **39**, 413–445.

Massom, R.A., Stammerjohn, S.E., Smith, R.C., et al. (2006) Extreme anomalous atmospheric circulation in the West Antarctic Peninsula region in austral spring and summer 2001/2, and its profound impact on sea ice and biota. *Journal of Climate*, **19**, 3544–3571.

Mätzl, M. & Schneebeli, M. (2006) Measuring specific surface area of snow by near-infrared photography. *Journal of Glaciology*, **52**, 645–648.

Mätzler, C. 2002. Relation between grain-size and correlation length of snow. *Journal of Glaciology*, **48**, 461–466.

Maykut, G.A. & Untersteiner, N. (1971) Some results from a time-dependent thermodynamic model of sea ice. *Journal of Geophysical Research*, **76**, 1550–1575.

McKay, G.A. & Gray, D.M. (1981) The distribution of snowcover. In: *Handbook of Snow: Principles, Processes, Management and Use* (Eds. D.M. Gray and D.H. Male), pp. 153–190. Pergamon, Oxford.

van der Merwe, P., Lannuzel, D., Mancuso Nichols, C.A. et al. (2009) Biogeochemical observations during the winter–spring transition in East Antarctic sea ice: Evidence of iron and exopolysaccharide controls. *Marine Chemistry*, **115**, 163–175.

van der Merwe, P., Lannuzel, D., Bowie, A.R. & Meiners, K.M. (2011). High temporal resolution observations of spring fast ice melt and seawater iron enrichment in East Antarctica. *Journal of Geophysical Research*, **116**, G03017.

Montpetit, B., Royer, A., Langlois, A. et al. (2012) New shortwave infrared albedo measurements for snow specific surface area retrieval. *Journal of Glaciology*, **58**, doi:10.1016/j.coldregions.2010.01.004.

Montpetit, B., Royer, A., Roy, A., Langlois, L. & Derksen, D. (2013) Snow microwave emission modeling of ice lenses within a snowpack using the microwave emission model for layered snowpacks, *IEEE Transactions on Geoscience and Remote Sensing*, **51**, 4705–4717.

Morin, S., Domine, F., Dufour, A. et al. (2013). Measurements and modeling of the vertical profile of specific surface area of an alpine snowpack. *Advances in Water Resources*, **55**, 111–120.

Morris, K. & Jeffries, M.O. (2001) Seasonal contrasts in snow-cover characteristics on Ross Sea ice floes. *Annals of Glaciology*, **33**, 61–68.

Mowat, F. (1960) *Ordeal By Ice: The Search for the Northwest Passage (Volume I of the Top of the World Trilogy)*. McClelland and Stewart, Toronto.

Mundy, C.J., Barber, D.G. & Michel, C. (2005) Variations of thermal, physical and optical properties pertinent to sea ice algae biomass during spring. *Journal of Marine Systems*, **58**, 107–120.

Mundy, C.J., Ehn, J.K., Barber, D.G. et al. (2007) Influence of snow cover and algae on the spectral dependence of transmitted irradiance through Arctic landfast first-year sea ice. *Journal of Geophysical Research*, **112**, C03007, doi:10.1029/2006JC003683.

Nakaya, U. (1954) *Snow Crystals*. Harvard University Press, Boston, USA.

Nicolaus, M., Haas, C. & Bareiss, J. (2003) Observations of superimposed ice formation at melt-onset on fast ice on Kongsfjorden, Svalbard. *Physics and Chemistry of the Earth*, **28**, 1241–1248.

Nicolaus, M., Haas, C., Bareiss, J. & Willmes, S. (2006) A model study of differences of snow thinning on Arctic and Antarctic first-year sea ice during spring and summer. *Annals of Glaciology*, **44**, 147–153.

Nicolaus, M., C. Haas, S. Willmes (2009) Evolution of first-year and second-year snow properties on sea ice in the Weddell Sea during spring-summer transition. *Journal of Geophysical Research*, **114**, D17109. doi.org/10.1029/2008JD011227.

Nomura, D., Nishioka, J., Granskog, M.A. et al. (2010a) Nutrient distributions associated with snow and sediment-laden layers of sea ice of the southern Sea of Okhotsk. *Marine Chemistry*, **119**, 1–8.

Nomura, D., Eicken, H., Gradinger, R. & Shirasawa, K. (2010b) Rapid physically driven inversion of the air-sea ice CO2 flux in the seasonal landfast ice off Barrow, Alaska after onset of surface melt. *Continental Shelf Research*, **30**, 1998–2004.

Nomura, D., McMinn, A., Hattori, H., Aoki, S. & Fukuchi, M. (2011) Incorporation of nitrogen compounds into sea ice from atmospheric deposition. *Marine Chemistry*, **127**, 90–99.

Notz, D. & Worster, M.G. (2008) In situ measurements of the evolution of young sea ice. *Journal of Geophysical Research*, **113**, C03001, doi.org/10.1029/2007JC004333.

Ozsoy-Cicek, B., Ackley, S., Xie, H., Yi, D. & Zwally, J. (2013) Sea ice thickness retrieval algorithms based on in situ surface elevation and thickness values for application to altimetry. *Journal of Geophysical Research: Oceans*, **118**, 3807–3822.

Perovich, D.K. & Richter-Menge, J. (1994) Surface characteristics of lead ice, *Journal of Geophysical Research*, **99**, 16 341–16 350.

Perovich, D.K., Andreas, E.L., Curry, J.A. et al. (1999) Year on ice gives climate insights. *Eos, Transactions of the American Geophysical Union*, **80**, 481–486.

Perovich, D.K., Grenfell, T.C., Light, B. & Hobbs, P.V. (2002) The seasonal evolution of Arctic sea ice albedo. *Journal of Geophysical Research*, **107**, doi:10.1029/2000JC000438.

Perovich, D.K., Elder, B.C., Claffey, K.J. et al. (2004) Winter sea-ice properties in Marguerite Bay, Antarctica. *Deep-Sea Research II*, **51**, 2023–2039.

Petrich, C., Eicken, H., Polashenski, C.M. et al. (2012). Snow dunes: A controlling factor of melt pond distribution on Arctic sea ice. *Journal of Geophysical Research: Oceans*, **117**, C09029, 1–10.

Pinzer, B.R., Schneebeli, M. & Kaempfer, T.U. (2012). Vapor flux and recrystallization during dry snow metamorphism under a steady temperature gradient as observed by time-lapse micro-tomography. *The Cryosphere Discussions*, **6**, 1673–1714.

Polashenski, C., Perovich, D. & Courville, Z. (2012). The mechanisms of sea ice melt pond formation and evolution. *Journal of Geophysical Research: Oceans*, **117**, C01001, doi:10.1029/2011JC007231.

Pomeroy, J.W. & Gray, D.M. (1995) *Snowcover Accumulation, Relocation and Management*. National Hydrology Research Institute, Saskatoon, Canada.

Pounder, E.R. (1965) *Physics of Ice*. Pergamon Press, Oxford.

Powell, D.C., Markus, T. & Stössel, A. (2005) Effects of snow depth forcing on Southern Ocean sea ice simulations. *Journal of Geophysical Research: Oceans*, **110**, C06001, doi:10.1029/2003JC002212.

Powell, D.C., Markus, T., Cavalieri, D.J. et al. (2006) Microwave signatures of snow on sea ice: Modeling. *IEEE Transactions on Geoscience and Remote Sensing*, **44**, 3091–3102.

Powers, D., O'Neill, K., & Colbeck, S.C. (1985). Theory of natural convection in snow. *Journal of Geophysical Research: Atmospheres*, **90**, 10 641–10 649.

Pratt, K.A., Custard, K.D., Shepson, P.B. et al. (2013) Photochemical production of molecular bromine in Arctic surface snowpacks. *Nature Geoscience*, **6**, 351–356.

de Quervain, M.R. (1973). Snow structure, heat and mass flux through snow. *IAHS Publication*, **107**, 203–226.

Riche, F., & Schneebeli, M. (2010) Microstructural change around a needle probe to measure thermal conductivity of snow. *Journal of Glaciology*, **56**, 871–876.

Rigor, I.G., Colony, R.L. & Martin, S. (2000) Variations in surface air temperature observations in the Arctic 1979-97. *Journal of Climate*, **13**, 896–914.

Rinke A. & Dethloff, K. (2008) Simulated circum-Arctic climate changes by the end of the 21st Century. *Global and Planetary Change*, **62**, 173–186.

Ross, J.C. (1847). *A Voyage of Discovery and Research in the Southern and Antarctic Regions, During the Years 1839–43* (Vol. **2**). Murray.

Saloranta, T.M. (2000) Modeling the evolution of snow, snow ice and ice in the Baltic Sea. *Tellus A*, **52**, 93–108.

Scoresby, W. (1820) *An Account of the Arctic Regions With a History and Description of the Northern Whale-Fishery*. Archibald Constable, Edinburgh.

Sedwick, P.N. & DiTullio, G.R. (1997). Regulation of algal blooms by the release of iron from in Antarctic shelf melting sea ice. *Geophysical Research Letters*, **24**, 2515–2518.

Sedwick, P.N., DiTullio, G.R. & Mackey, D.J. (2000). Iron and manganese in the Ross Sea, Antarctica. *Journal of Geophysical Research*, **105**, 11,321–11,336.

Seligman, G. (1936) *Snow Structure and Ski Fields*. International Glaciological Society, Cambridge.

Semtner, A.J. Jr., (1976). A model for the thermodynamic growth of sea ice in numerical investigations of climate. *Journal of Physical Oceanography*, **6**, 379–389.

Serreze. M.C. & Maslanik, J.A. (1997) Arctic precipitation as represented in the NCEP/NCAR reanalysis. *Annals of Glaciology*, **25**, 429–433.

Simmonds, I., Keay, K. & Lim, E.-P. (2003) Synoptic activity in the seas around Antarctica. *Monthly Weather Review*, **131**, 272–288.

Smith, D.M. (1998) Observation of perennial Arctic sea ice melt and freeze-up using passive microwave data. *Journal of Geophysical Research*, **103**, 27 753–27 769.

Smith, T.G. & Stirling, I. (1975) The breeding habitat of the ringed seal (Phoca hispida): The birth lair and associated structures. *Canadian Journal of Zoology*, **53**, 1297–1305.

Sommerfeld, R.A. (1970) The classification of snow metamorphism. *Journal of Glaciology*, **9**, 3–17.

Stirling, I. & Smith, T.G. (2004) Implications of warm temperatures, and an unusual rain event for the survival of ringed seals on the coast of Southeastern Baffin Island. *Arctic*, **57**, 59–67.

Stirling, I. & Derocher, A.E. (2012) Effects of climate warming on polar bears: A review if the evidence. *Global Change Biology*, **18**, 2694–2706.

Stroeve, J.C., Markus, T., Maslanik, J.A., et al. (2006) Impact of surface roughness on AMSR-E sea ice products. *IEEE Transactions on Geoscience and Remote Sensing*, **44**, 3103–3117.

Stroeve, J., Holland, M.M., Meier, W., Scambos, T. & Serreze, M. (2007) Arctic sea ice decline: Faster than forecast. *Geophysical Research Letters*, **34**, L09501, doi:10.1029/2007GL029703.

Sturm, M. (1991) *The Role of Thermal Convection in Heat and Mass Transport in the Subarctic Snow Cover*. U.S. Army CRREL, Hanover, NH.

Sturm, M. (2009a) *Apun: The Arctic Snow*. University of Alaska Press, Fairbanks.

Sturm, M. (2009b) *Composite List of Iñupiaq Snow Words*. Unpublished manuscript.

Sturm, M. (2010) Field techniques for snow observations on sea ice. In: *Field Techniques for Sea Ice Research* (Eds. H. Eicken et al.), pp. 25–48. University of Alaska Press.

Sturm, M., & Benson, C.S. (1997) Vapor transport, grain growth and depth hoar development in the subarctic snow. *Journal of Glaciology*, **43**, 42–59.

Sturm, M. & Holmgren, J. (1998) Differences in compaction behavior of three climate classes of snow. *Annals of Glaciology*, **26**, 125–130.

Sturm, M., & Stuefer, S. (2013). Wind-blown flux rates derived from drifts at Arctic snow fences. *Journal of Glaciology*, **59**, 21–34.

Sturm, M., Holmgren, J., König, M. & Morris, K. (1997) The thermal conductivity of seasonal snow. *Journal of Glaciology*, **43**, 26–41.

Sturm, M., Morris, K. & Massom, R. (1998) The winter snow cover of the West Antarctic pack ice: its spatial and temporal variability. In: *Antarctic Sea Ice Physical Processes, Interactions and Variability* (Ed. M.O. Jeffries), pp. 1–18. American Geophysical Union, Washington, DC.

Sturm, M., Holmgren, J. & Perovich, D.K. (2001) Spatial variations in the winter heat flux at SHEBA: estimates from snow-ice interface temperatures. *Annals of Glaciology*, **33**, 213–220.

Sturm, M., Holmgren, J., & Perovich, D.K. (2002a) Winter snow cover on the sea ice of the Arctic Ocean at the Surface Heat Budget of the Arctic Ocean (SHEBA): temporal evolution and spatial variability. *Journal of Geophysical Research*, **107**, doi:10.1029/2000JC000400.

Sturm, M., Perovich, D.K., & Holmgren, J. (2002b) Thermal conductivity and heat transfer through the snow and ice of the Beaufort Sea. *Journal of Geophysical Research*, **107**, doi:10.1029/2000JC000409, 8043.

Sturm, M., Maslanik, J.A. & Perovich, D.K. et al. (2006) Snow depth and ice thickness measurements from the Beaufort and Chukchi Seas collected during the AMSR-Ice03 Campaign. *IEEE Transactions on Geoscience and Remote Sensing*, **44**, 3009–3020.

Sturm, M., Taras, B., Liston, G.E., Derksen, C., Jonas, T. & Lea, J. (2010) Estimating snow water equivalent using snow depth data and climate classes. *Journal of Hydrometeorology*, **11**, 1380–1394.

Taylor, K.E., Stouffer, R.J. and Meehl, G.A. (2012) An overview of CMIP5 and the experiment design. *Bulletin of the American Meteorological Society*, **93**, 485–498.

Tin, T. & Jeffries, M.O. (2001) Sea ice thickness and roughness in the Ross Sea, Antarctica. *Annals of Glaciology*, **33**, 187–193.

Tison, J.-L. et al. (2008) Temporal evolution of decaying summer first-year sea ice in the Western Weddell Sea, Antarctica. *Deep-Sea Research II*, **55**, 975–987.

Toyota, T., Takatsuji, S., Tateyama, K. et al. (2007) Properties of sea ice and overlying snow in the Southern Sea of Okhotsk. *Journal of Oceanography*, **63**, 393–411.

Toyota, T., Massom, R., Tateyama, K., Tamura, T. & Fraser, A. (2011) Properties of snow overlying the sea ice off East Antarctica in late winter, 2007. *Deep-Sea Research II*, **58**, 1137–1148.

Tremblay, J.-E. & Gagnon, J. (2009) The effect of irradiance and nutrient supply on the productivity of Arctic waters: A perspective on climate change. In: *Influence of Climate Change on the Changing Arctic and Sub-Arctic Conditions* (Eds. J.C.J. Nihoul & A.G. Kostianoy), pp. 73–94. Springer, New York.

Trevena, A.J. & Jones, G.B. (2006) Dimethylsulphide and dimethylsulphoniopropionate in Antarctic sea ice and their release during sea ice melting. *Marine Chemistry*, **98**, 210–222.

Untersteiner, N. (1961) On the mass and heat budget of Arctic sea ice. *Archiv für Geophysiks und Bioklimatologie*, **A12**, 151–182.

Vancoppenolle, M., Meiners, K.M., Michel, C. et al. (2013) Role of sea ice in global biogeochemical cycles: emerging views and challenges. *Quaternary Science Reviews*, **79**, 207–230.

Waddington, E.D., Cunningham, J. & Harder, S.L. (1996) The effects of snow ventilation on chemical concentrations. In: *NATO ASI Series: Chemical Exchange between the Atmosphere &*

Polar Snow (Eds. E.W. Wolff &\ R.C. Bales) pp. 403–451. Springer-Verlag, Berlin.

Wadhams, P., Lange, M. & Ackley, S.F. (1987) The ice thickness distribution across the Atlantic of the Antarctic Ocean in mid-winter. *Journal of Geophysical Research*, **92**, 14 535–14 552.

Walsh, J.E. (2008) Climate of the Arctic marine environment. *Applied Ecology*, **18**, S3–S22.

Wankiewicz, A. (1978) A review of water movement in snow. In: *Modeling of Snow Cover Runoff* (Eds: S.C. Colbeck & M. Ray), pp. 222–237. U.S. Army CRREL, Hanover, NH.

Warren, S.G., & Clarke, A.D. (1986) Soot from Arctic haze: radiative effects on the Arctic snowpack. *Glaciological Data*, **18**, 73–77.

Warren, S.G., Rigor, I.G., Untersteiner, N. et al. (1999) Snow depth on Arctic sea ice. *Journal of Climate*, **12**, 1814–1829.

Webster, D.H., & Zibell, W. (1970) *Iñupiat Eskimo Dictionary*. Summer Institute of Linguistics, Fairbanks, Alaska.

Webster, M.A., Rigor, I.G., Nghiem, S.V. et al. (2014). Interdecadal changes in snow depth on Arctic sea ice. *Journal of Geophysical Research: Oceans*, **119**, 5395–5406.

Worby, A.P. & Massom, R.A. (1995) The structure and properties of sea ice and snow cover in East Antarctic pack ice. *Antarctic CRC Research Report 7*. Antarctic CRC, Hobart, Australia.

Worby, A.P., Geiger, S., Paget, M.J. et al. (2008a) Thickness distribution of Antarctic sea ice. *Journal of Geophysical Research*, **113**, C05S92, doi:10.1029/2007JC004254.

Worby, A.P., Markus, T., Steer, A.D. et al. (2008b) Evaluation of AMSR-E snow depth product over East Antarctic sea ice using in situ measurements and aerial photography. *Journal of Geophysical Research*, **113**, C05S94, doi:10.1029/2007JC004181.

Wu, X., Budd, W.F., Lytle, V.I. & Massom, R.A. 1999. The effect of snow on Antarctic sea ice simulations in a coupled atmosphere–sea ice model. *Climate Dynamics*, **15**, 127–143.

Yang, X., Pyle, J.A. & Cox, R.A. (2008) Sea salt aerosol production and bromine release: Role of snow on sea ice. *Geophysical Research Letters*, **35**, L16815, doi:10.1029/2008GL034536.

Zhou, C., Penner, J.E., Flanner, M.G. et al. (2012) Transport of black carbon to polar regions: Sensitivity and forcing by black carbon. *Geophysical Research Letters*, **39**, L22804, doi:10.1029/2012GL053388.

CHAPTER 4
Sea ice and sunlight

Donald K. Perovich

Thayer School of Engineering, Dartmouth College; and ERDC – Cold Regions Research and Engineering Laboratory, Hanover, NH, USA

4.1 Introduction

Sea ice is a translucent material with an intricate structure consisting primarily of ice, brine inclusions and air bubbles. Biota and particulates may also be present in the ice. The ice structure is temperature-dependent and exhibits considerable spatial and temporal variability. Sea ice structure is discussed in detail in Chapter 1 of this book. The physical structure directly affects the optical properties of sea ice and its interaction with sunlight.

Understanding the reflection, absorption and transmission of sunlight by sea ice is important to a diverse array of scientific problems. This partitioning of solar radiation is a key parameter in the heat budget and mass balance of sea ice (Untersteiner, 1961; Maykut & Untersteiner, 1971; Maykut, 1986; Ebert & Curry, 1993; Perovich et al. 2007, 2008, 2011). Sunlight absorbed in the ice can result in surface and interior melting, while sunlight deposited in the ocean can cause melting on the bottom and lateral edges of the ice (Maykut & Perovich, 1987; Steele, 1992). The sea ice albedo feedback is of particular concern because of its potential impact on climate (Ebert & Curry, 1993; Bitz, 1999; Holland & Bitz 2003; Holland et al., 2006, 2012; Pedersen et al., 2009; Winton et al., 2008; Eisenman & Wettlaufer, 2009; Hudson, 2011). Thus there is a need for information on the large-scale evolution of solar partitioning by a sea ice cover. The optical properties of sea ice are important for the interpretation of visible and near-infrared satellite imagery. Such imagery can provide insight into the large-scale properties of a sea ice cover (Fetterer & Untersteiner, 1998; Kwok, 2014). As discussed in Chapter 14, the amount and spectral composition of short-wave radiation transmitted through sea ice strongly impact primary productivity and biological activity in and under a sea ice cover (Arrigo, 2014; Arrigo et al., 2008; Mundy et al., 2007, 2009; Ehn et al., 2008). Visible light benefits ice biota by contributing to photosynthesis, while ultraviolet light can damage organisms.

Optical refers to the portion of the electromagnetic spectrum that is coincident with the wavelength range of radiation from the sun, from roughly 250 to 2500 nm (Figure 4.1). This portion of the electromagnetic spectrum is also referred to as solar radiation, short-wave radiation or sunlight. The optical portion of the electromagnetic spectral can be divided into three segments: ultraviolet light from 250 to 400 nm, visible light from 400 to 750 nm, and near-infrared light from 750 to 2500 nm. The colours of visible light are shown in Figure 4.1 from violet (400 nm) to blue (450 nm) to green (550 nm) to yellow (600 nm) to red (650 nm). Depending on cloud conditions, approximately 50% to 60% of sunlight that reaches the Earth's surface is visible light (Grenfell & Perovich, 2008). Ultraviolet light can be further divided into three smaller segments; UV-C from 200 to 280 nm, UV-B from 280 to 320 nm, and UV-A from 320 to 400 nm. Because of strong absorption in the atmosphere, essentially no UV-C reaches the Earth's surface. UV-B light levels are substantially enhanced by the depletion of stratospheric ozone (Frederick & Lubin, 1988; Lubin et al., 1989; Tsay & Stamnes, 1992) and may negatively affect living organisms (Smith, 1989; Smith et al., 1992; Boucher et al., 1996; Lazzara et al., 2007).

In terms of the wavelength range, visible light is only a small portion (16%) of the total solar spectrum. However, visible light accounts for about 55% of the incident

---

Sea Ice, Third Edition. Edited by David N. Thomas.
© 2017 John Wiley & Sons, Ltd. Published 2017 by John Wiley & Sons, Ltd.

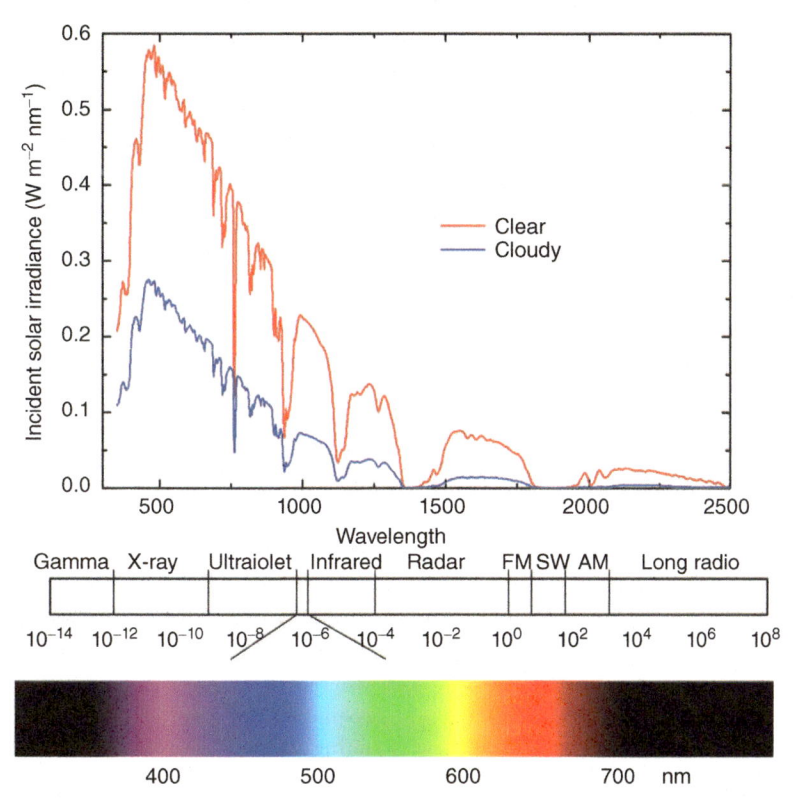

Figure 4.1 The electromagnetic spectrum, the solar spectrum and the visible spectrum. Incident solar irradiance is from Grenfell and Perovich (2008). Clear sky incident is from 16 August at 78.3°N and cloudy sky is from 18 August at 78.2°N.

energy under clear skies and about 60% of the incident energy under cloudy skies. Your eyes provide valuable information on the optical properties of sea ice and on the partitioning of solar radiation.

Properties refer to the parameters that are used to describe solar radiation. These include terms defining incident, reflected, absorbed and transmitted radiation fields plus quantities defining scattering and absorption of light. There is an extensive radiative transfer terminology that can vary somewhat across different fields of study. The physics is the same, but the labels may vary from biology to oceanography to astrophysics. The focus here is on the terms needed for a basic understanding of the optical properties and the terminology follows the conventions of the sea ice literature.

There are inherent optical properties (IOPs) and apparent optical properties (AOPs). IOPs are intrinsic properties of the medium and do not depend on the light field. Key IOPs are the index of refraction, absorption coefficient, the scattering coefficient and the phase function. AOPs depend on both the properties medium and on the directional structure of the illumination field. Key AOPs are albedo and transmittance.

This chapter focuses specifically on the optical properties of sea ice and more generally on the interaction of sunlight with sea ice. The goal is to provide an introductory overview to the topic, not to be a complete compendium of all work in the field. First, optical basic terms are defined, and then the physical principles underlying radiative transfer in sea ice, including scattering and absorption, are discussed. Observational results are presented, with an emphasis on explaining the wide variability in sea ice optical properties in terms of ice physical properties and radiative transfer theory. Snow optical properties are briefly discussed due to the pervasive presence of snow on sea ice and its impact on light reflection and transmission. An overview is presented of selected sea ice radiative transfer models summarizing their basic characteristics, solution schemes, strengths and limitations. The ice albedo feedback and aggregate scale partitioning of solar radiation are discussed. Finally, current and future research areas in sea ice optical properties are discussed.

4.2 Radiative transfer theory

The key to understanding the optical properties of sea ice lies in understanding the physical state of sea ice and radiative transfer. Chapter 1 comprehensively covers sea ice structure. Radiative transfer theory will now be introduced. The schematic in Figure 4.2 illustrates the interaction of sunlight with sea ice. The incident radiation field consists of two components: a direct beam from the sun and diffuse radiation from the sky and clouds. The incident solar radiation is partitioned among reflection from the ice, absorption in the ice and transmission through the ice to the ocean. Some of the reflected sunlight is due to specular reflection directly from the surface. The amount of specular reflection depends on sun angle, cloudiness and surface state. The partitioning of the incident sunlight depends on the physical properties of the ice and on the wavelength of the light.

The transfer of sunlight in sea ice is governed by two processes: absorption and scattering. As a ray of light passes through sea ice, some of the light is absorbed by the ice and some of it is scattered from the beam in different directions. This is expressed more formally as the equation of radiative transfer for a plane parallel medium (Chandrasekhar, 1960):

$$-\mu \frac{dI(\tau,\mu,\varphi,\lambda)}{d\tau} = I(\tau,\mu,\varphi,\lambda) - S(\tau,\mu,\varphi,\lambda) \qquad (4.1)$$

where I is the radiance, λ is the wavelength, μ is the cosine of the zenith angle (θ), and ϕ is the azimuth angle. The spectral radiance $I(\theta,\varphi,\lambda)$ is the radiant flux per nanometer per unit area per unit solid angle in a particular direction and has units of W m^{-2} sr^{-1} nm^{-1}. A plane parallel medium is horizontally homogeneous, but can have variations in the vertical. Scattering is included in the S term, which is referred to as the source function. τ is the non-dimensional optical depth and is defined as:

$$\tau(\lambda) = (k(\lambda) + \sigma(\lambda))z$$

where k is the absorption coefficient, σ is the scattering coefficient, and z is the physical depth. The single scattering albedo:

$$\varpi_o(\lambda) = \frac{\sigma(\lambda)}{k(\lambda) + \sigma(\lambda)}$$

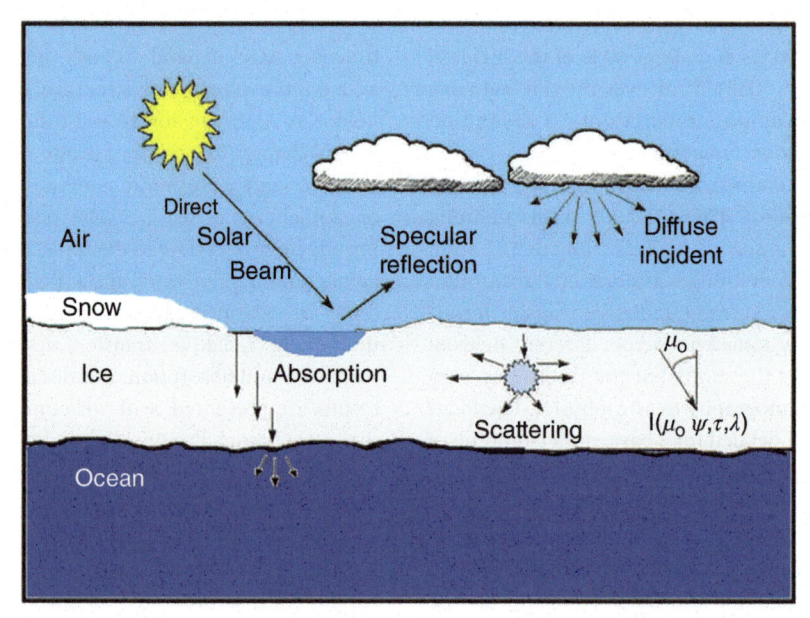

Figure 4.2 Schematic of the interaction of sunlight with sea ice. The three primary sea ice surface types of snow-covered ice, bare ice and ponded ice are shown.

gives the fractional loss due to scattering (Chandrasekhar, 1960; Mobley, 1994). ϖ_o ranges from 0 for a purely absorbing medium to 1 for a purely scattering medium. The absorption coefficient, scattering coefficient and single scattering albedo are all IOPs.

The spectral irradiance, $(F(\lambda))$, is a parameter of interest for energy balance studies. It is simply the radiance projected onto a plane surface integrated over a hemisphere. The radiance is scaled by $\cos(\theta)$ when it is projected onto a plane. The downwelling irradiance, $(F_d(\lambda))$, is the radiance integrated over downward directions (e.g. from the sky), and the upwelling irradiance, $(F_u(\lambda))$, is the radiance integrated over upward directions (e.g. from the surface). This can be expressed as:

$$F_d(\lambda) = \int_{\phi=0}^{2\pi} \int_{\theta=0}^{\pi/2} I(\theta,\varphi,\lambda) \cos\theta \sin\theta d\theta d\varphi$$

$$F_u(\lambda) = \int_{\phi=0}^{2\pi} \int_{\theta=\pi/2}^{\pi} I(\theta,\varphi,\lambda) \cos\theta \sin\theta d\theta d\varphi$$

4.2.1 Absorption

Consider the case of a non-scattering, absorbing medium, illuminated by a normally incident direct beam of light. Since there is no scattering $S = 0$ and $\sigma = 0$. For normal incidence, $\theta = 0$, which means $\mu = 1$. Equation (4.1) then reduces to:

$$\frac{-dI(z,\lambda)}{k(\lambda)dz} = I(z,\lambda)$$

which, when solved, gives the familiar exponential decay law:

$$I(z,\lambda) = I(0,\lambda)e^{-kz} \qquad (4.2)$$

known as Beer's law or the Bouguer–Lambert law. Radiative transfer in a purely absorbing medium is straightforward to describe and model. The radiance decreases exponentially with depth in the medium, with the rate of decrease dependent on the absorption coefficient. For a composite material such as sea ice, the spectral absorption coefficients for its constituent components of ice, brine and air must be known. If present, particulates and biota also contribute to absorption in sea ice.

4.2.2 Scattering

At visible wavelengths, sea ice is a highly scattering medium (ϖ_o close to 1). Scattering adds considerable complexity to the equation of radiative transfer as is evident in the full expansion of the source term S in equation (4.1). For a plane-parallel medium with a direct incident beam the source function is expressed as:

$$S(\tau,\mu,\varphi,\lambda) = \frac{\varpi_o}{4\pi} \int_{-1}^{1} \int_{0}^{2\pi} p(\mu,\mu',\varphi,\varphi')I(\tau,\mu',\varphi',\lambda)d\mu'd\varphi$$
$$- \frac{E_o(\lambda)}{4} p(\mu_o,\mu,\varphi_o,\varphi)e^{-\tau(\lambda)/\mu_o}$$

where $p(\mu,\mu',\varphi,\varphi')$ is the phase function and E_0 is the radiance of the direct beam component of the incident radiation field. Solving the full form of equation (4.1) with scattering is complicated, but we can qualitatively examine the scattering term. The double integral term treats the diffuse radiation field and simply represents scattering of the radiance field $I(\tau,\mu',\phi',\lambda)$ from different directions into the direction of the solution (μ,θ). The scattering coefficient σ is similar to the absorption coefficient and is a measure of the amount of scattering over a unit distance. The phase function $(p(\mu,\mu',\phi,\phi'))$ describes the direction in which the light is scattered. The light is scattered from direction μ,φ to direction μ',φ'. The phase function is normalized so that its integral over angle is equal to 1. The second term provides the contribution of scattered light from the attenuated direct beam $(E_o(\lambda))$ and is only needed if there is direct beam incident irradiance.

Scattering results from differences in the real component of the index of refraction (n). The larger the difference in index of refraction between the inclusion and the ice, the stronger the scattering. As was shown in Chapter 1, sea ice is not a monolithic slab of pure ice. It has an intricate structure consisting of an ice matrix with inclusions of brine, air and perhaps solid salts. As these inclusions have different indices of refraction than the surrounding ice, they scatter light. Sea ice has a multitude of brine pockets and air bubbles and is a highly scattering medium. For sea ice, the relevant real indices of refraction are ice ($n \sim 1.31$), brine ($n \sim 1.33$) and air ($n \sim 1.00$). With a greater difference in index of refraction, air bubbles ($n \sim 1.0$) are more strongly scattering than brine pockets. However, sea ice typically has many more brine pockets than air bubbles. The real part of the index of refraction for brine depends weakly on both temperature and salinity, increasing from 1.34 at $-2°C$ to 1.40 at $-32°C$ (Maykut & Light, 1995). At ice temperatures less than $-23°C$, hydrohalite forms and contributes significantly to scattering (Perovich & Grenfell, 1981; Weeks & Ackley, 1982; Grenfell, 1983; Light et al., 2004).

Scattering in sea ice is assumed to be independent of wavelength (Grenfell, 1983; Light et al., 2004). Two factors support this assumption. The wavelength dependence of the real portion of the index of refraction for ice, brine and air is very weak at optical wavelengths (Maykut & Light, 1995) and typically is assumed to be constant with wavelength in sea ice radiative transfer models (Grenfell, 1983, 1991; Light et al., 2004). Optical wavelengths are on the order of tenths of a micron to microns. The inclusions in sea ice have sizes on the order of tenths of a millimetre for brine pockets to millimetres for air bubbles. As the scatterers are much bigger than the wavelength and the scatterers are far apart, contributions due to diffraction and interference can be ignored (Bohren & Huffman, 1983; Grenfell, 1983; Light et al., 2004). A similar argument is made when analysing scattering in snow (Bohren & Barkstrom, 1974; Wiscombe & Warren, 1980; Warren, 1982). Light (2010) presents a detailed observational and theoretical treatment of scattering in first-year sea ice. A general discussion of light scattering can be found in van de Hulst (1981) and Bohren and Huffman (1983).

Perovich, 1981; Perovich & Govoni, 1991); in the field deep in the ice at the South Pole (Ackermann et al., 2006); and inferred from light transmission profiles in Antarctic snow (Warren et al., 2006). Ice is extremely transparent at the blue end of the spectrum introducing uncertainty and making precise measurements of absorption coefficient difficult. Even very small amounts of impurities or scatterers can affect the measurements. Spectral absorption coefficients compiled from these experiments are shown in Figure 4.3. For comparison, absorption coefficients of clear Arctic water (Tyler & Smith, 1970; Smith & Baker, 1981) are also plotted in Figure 4.3. These seawater absorption coefficients are similar in shape and magnitude to those of pure ice. Brine fractions in sea ice are typically in the 0–0.1 range, so the absorption coefficient of sea ice will be dominated by ice.

Spectral changes in absorption coefficient from 250 to 1400 nm span several orders of magnitude. Differences in transmitted radiance ($I(z, \lambda)$) are exponentially greater than changes in absorption coefficient (equation 4.2). The impact of absorption can be assessed by considering the e-folding length, the amount of ice needed to

4.3 Inherent optical properties

4.3.1 Absorption coefficients

Grenfell (1983) and Light et al. (2004) discussed absorption in sea ice from a modelling perspective. They stated that absorption at any given wavelength (λ) was a volume-weighed sum of absorption of the individual components. Absorption in sea ice (k_{si}) is a combination of absorption in the primary components of ice, brine, gas, and particulates as expressed in equation (4.3). Assuming that absorption in the air bubbles is negligible gives:

$$k_{si} = v_i k_i + v_b k_b + v_s k_s + v_{bio} k_{bio} + v_{sed} k_{sed} \quad (4.3)$$

where k represents the absorption coefficient, v the volume fractions of ice and brine, and the subscripts refer to ice (i), brine (b), solid salts (s), biology (bio) and sediment (sed).

Ice is the primary component of sea ice. For pure, bubble-free ice with no scattering, absorption coefficients can be determined using equation (4.2). Absorption coefficients were determined in the laboratory using pure, bubble-free, fresh ice (Grenfell &

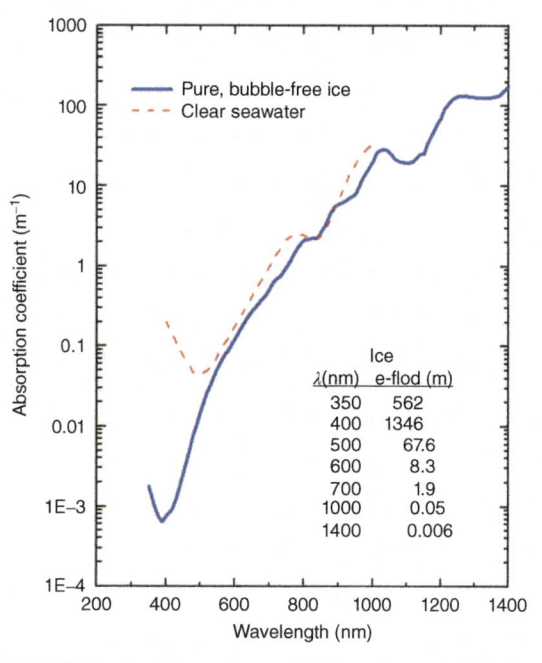

Figure 4.3 Spectral absorption coefficients of pure bubble-free ice and clear water. The insert shows e-folding lengths for selected wavelengths in pure bubble-free ice.

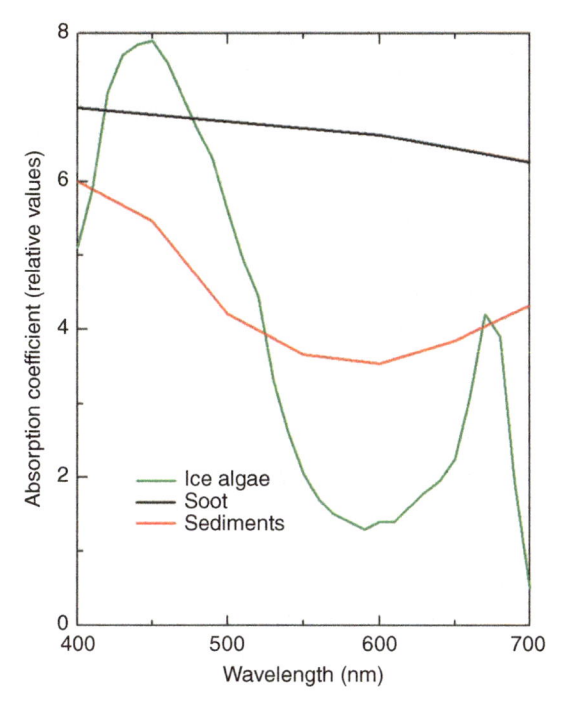

Figure 4.4 Relative values of spectral absorption coefficients for ice algae, soot and sediments. The absolute absorption coefficients depend on the amount of the materials.

reduce the incident light (I(0, λ)) by 1/e (i.e. the transmitted light is 37% of the incident). For pure ice e-folding lengths decrease sharply from 1567 m at 390 nm, to 8 m at 600 nm, to 2 m at 700 nm, to 0.05 m at 1000 nm, to 0.006 m at 1400 nm – a change of six orders of magnitude.

Little is known about absorption by solid salts that may be a factor for cold ice (Light et al., 2004). Particulates can include sediment, dissolved organics, black carbon and biota. If these materials are present in sufficient quantity, their impact on absorption should be included. There is a wide variability in absorption coefficients for these materials, as the examples in Figure 4.4 illustrate. The key points are that particulates absorb more light than ice and water, and their absorption coefficient has a different spectral shape.

4.3.2 Scattering coefficients and phase functions

Even a small sample of sea ice will probably include multiple scatterers making direct measurements of scattering coefficients and phase functions extremely difficult. As a consequence, scattering properties are typically calculated using scattering theory and a description of sea ice microstructure (Grenfell, 1983, 1991; Light et al., 2004) or inferred from bulk measurements of sea ice AOPs (Perovich & Grenfell, 1982; Light et al., 2008; Light et al., 2015). Both of these techniques have difficulties and limitations. Theoretical estimates of scattering coefficient and phase function require not just the volume of brine, air and solid salts, but also how these components are distributed in terms of the number and size distribution of the inclusions. Determining scattering properties from measurements of albedo and transmittance also requires information on vertical profiles of ice structure and physical properties.

Particulates, sediment and ice biota mainly impact absorption and their contribution to scattering is often neglected. Light et al. (2004) developed a detailed structural–optical model of first-year sea ice that included scattering by brine pockets, salt crystals and gas bubbles in brine and in ice. Similar to absorption, scattering in sea ice can be considered as a weighed sum of scattering by the primary components of ice, brine, gas and solid salts. The total scattering coefficient is then represented as:

$$\sigma_{si} = v_i\sigma_i + v_{bp}\sigma_{bp} + v_{bt}\sigma_{bt} + v_{gi}\sigma_{gi} + v_{gb}\sigma_{gb} + v_{sm}\sigma_{sm} + v_{sh}\sigma_{sh}$$

where σ is the scattering coefficient and the new subscripts refer to brine pockets (bp), brine tubes (bt), gas bubbles in ice (gi), gas bubbles in brine (gb), mirabilite crystals (sm, and hydrohalite crystals (sh).

A Henyey–Greenstein (Henyey & Greenstein, 1941) phase function is commonly used in sea ice radiative transfer models. It has the form:

$$p(\theta) = \frac{1}{4\pi} \frac{1-g^2}{(1+g^2 - 2g\cos(\theta))^{3/2}}$$

where θ is the angle between the incident and scattered beam, g is the asymmetry parameter and can vary from −1 for complete backscattering to 1 for complete forward scattering. $g = 0$ gives isotropic scattering.

Direct observations of scattering parameters are limited. Grenfell and Hedrick (1983) used small samples of young ice to directly measure phase functions for sea ice. They found that the phase function is strongly forward-peaked, with forward scattering being more than a factor of 50 greater than side or backward scattering. Even though they used small samples, there was still multiple scattering, and consequently the results represent only an approximation to the true single scattering albedo and phase function. Multiple scattering tends to smooth and reduce the angular dependence of the measured phase function. Grenfell (1983) calculated sea ice phase functions using a Mie scattering model (Bohren & Huffman, 1983) and input parameters of refraction indices for ice and brine and inclusion size distributions. As expected, the calculated phase function is more strongly forward-peaked than the observations, where there was multiple scattering.

Perovich and Grenfell (1982) estimated scattering coefficients of young ice from laboratory observations of albedo and transmittance. They found total scattering coefficients that ranged from 8.9 m^{-1} for melting young ice to 19.6 m^{-1} for cold young ice to 420 m^{-1} for very cold young ice with precipitated solid salts present. Light et al. (2008) determined vertical profiles of scattering coefficients by combining observations of spectral albedo and transmission with a three-layer radiative transfer model. Figure 4.5 shows a time series of σ for melting bare ice and ponded ice assuming a $g = 0.94$. For bare ice there were distinct differences in σ for the three layers. Scattering coefficients were largest, between 350 and 1100 m^{-1} in the surface scattering layer. In the drained ice layer beneath the surface scattering layer, coefficients ranged from 40 to 200 m^{-1}, while in the ice interior, scattering coefficients were only 8–30 m^{-1}. Scattering coefficients in melt ponds were much smaller, 15–30 m^{-1}, and constant with depth.

In a modelling effort, Light et al. (2004) calculated scattering coefficients for the various scatterers present in first-year sea ice as a function of temperature (Figure 4.6a). At temperatures colder than –22.9°C, hydrohalite crystals dominate the scattering coefficient, with mirabilite also making a large contribution. Total values of σ exceed 1000 m^{-1} at these cold temperatures. Scattering contributions from brine tubes exceed brine pockets at all temperatures. Contributions from gas bubbles in the ice increase by a factor of a few hundred as the ice warms. Overall scattering coefficients are quite large ranging from 400 to just over 1000 m^{-1}. Scattering in multi-year ice is probably different, with gas bubbles in the drained upper layer playing a significant role. Henyey–Greenstein g-values for these different scatterers as a function of temperature were also computed (Figure 4.6b). Values were large for all scatterers ranging from 0.82 for gas bubbles to a $g \sim$

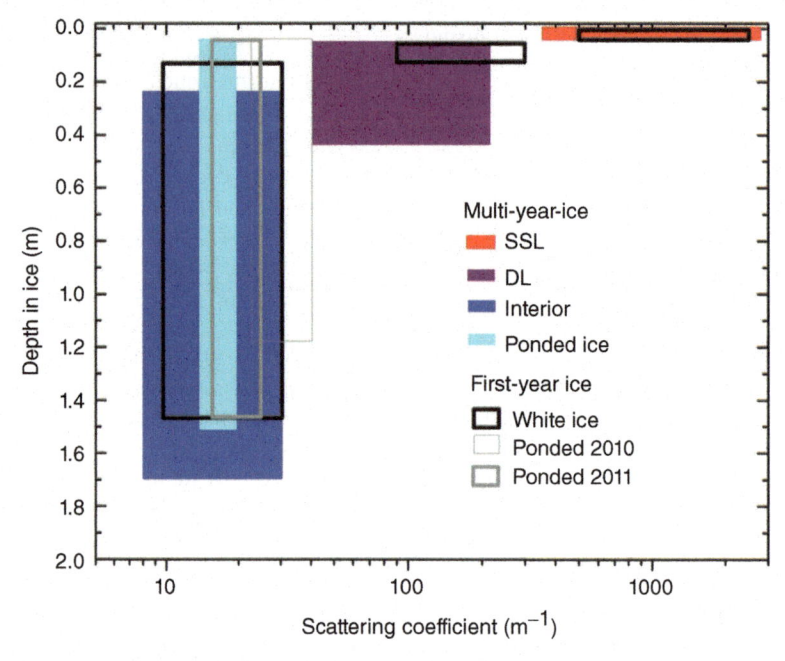

Figure 4.5 Depth dependence of scattering coefficient for: (a) bare ice, (b) ponded ice. SSL, surface scattering layer; DL, drained layer. Source: Light et al., 2008. Reproduced by permission of American Geophysical Union.

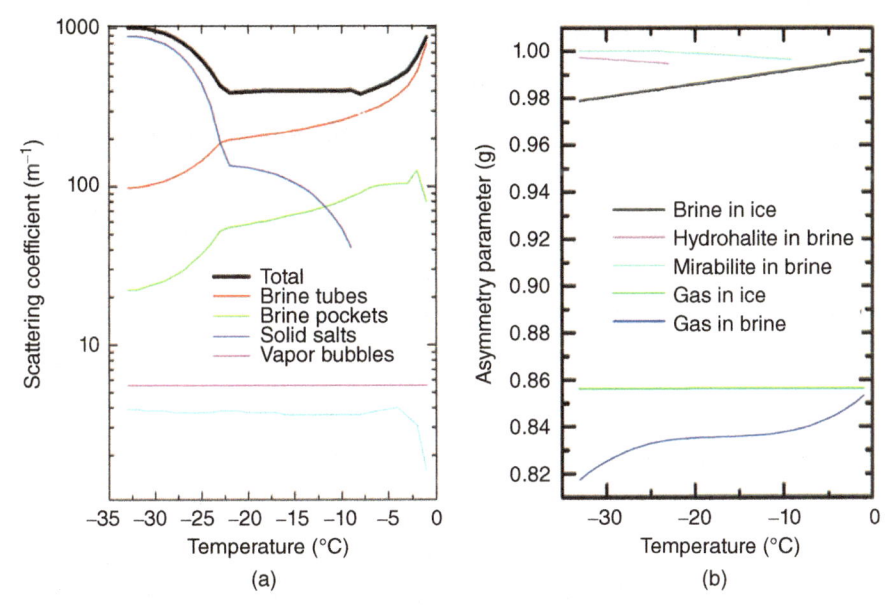

Figure 4.6 Scattering coefficients (a) and asymmetry parameters (b) for scatterers found in sea ice. Source: Light et al., 2004. Reproduced with permission of John Wiley and Sons.

1 for mirabilite and hydrohalite. Sea ice is strongly forward scattering.

Further research on scattering coefficients and phase functions is needed to improve our quantitative understanding of the relationship between scattering and ice physical properties. The important qualitative points with respect to absorption and scattering are as follows:

- Absorption coefficients for ice and brine depend strongly on wavelength.
- Scattering coefficients and phase functions for sea ice are constant with wavelength.
- Increasing the number of inclusions in sea ice increases the amount of scattering.
- Scattering in sea ice is strongly forward-peaked.

4.4 Apparent optical properties

4.4.1 Albedo

The most studied, most used and arguably most important AOP of sea ice is the albedo (α). With a high-quality instrument and the proper technique, the albedo is straightforward to determine. Perovich (2009) provides a detailed description of the proper technique for measuring albedo. The key points are to measure the incident irradiance from the sky and the reflected irradiance from the surface, and then divide the reflected by incident. More formally the spectral albedo is defined as the fraction of the incident irradiance that is reflected:

$$\alpha(\lambda) = \frac{F_u(0, \lambda)}{F_d(0, \lambda)}$$

where the 0 designates a position at the surface. For a perfectly reflecting surface, the albedo is one and for a perfectly absorbing surface the albedo is zero.

In many studies the wavelength-integrated, or total albedo (α_t) is a quantity of interest, as it is a measure of the total solar energy absorbed by the ice and ocean. It can be expressed by integrating the spectral albedo, weighed by incident irradiance, over the full solar spectrum:

$$\alpha_t = \frac{\int \alpha(\lambda) F_d(0, \lambda) d\lambda}{\int F_d(0, \lambda) d\lambda}$$

The total albedo depends on the spectral distribution of the incident irradiance as well as on the spectral albedo of the surface. Thus a change in cloud conditions, and thereby the incident spectral irradiance, can result in changes in the total albedo. For a snow or ice surface, the albedo is 5–10% greater on a cloudy day than on a clear day (Grenfell & Maykut, 1977; Grenfell & Perovich, 1984).

Because of its importance to the surface heat budget, albedo observations have long been a routine part of sea ice field campaigns (Grenfell & Maykut, 1997; Hanesiak et al., 2001; Perovich et al., 2002; Grenfell & Perovich, 2004; Winther et al., 2004; Ehn et al., 2006; Nicolaus et al., 2010). Albedos have also been measured autonomously from the surface (Nicolaus et al., 2010) and estimated from satellites (Laine, 2004; Riihela et al., 2013). There is a wealth of albedo data for different ice types and surface conditions. Albedo is strongly influenced by surface conditions as well as scattering and absorption in the interior of the ice.

Young ice

Albedos are sensitive to thickness during the initial stages of sea ice growth. This thickness dependence has been investigated in the laboratory (Perovich & Grenfell, 1981), in the Arctic (Weller, 1972; Perovich, 1991; Ehn et al., 2004, 2007) and in the Antarctic (Schlosser, 1988; Allison et al., 1993; Brandt et al., 1999; Brandt et al., 2005). Brandt et al. (2005) measured albedos for thin ice, with and without snow. There is a small increase in total albedo during the initial growth phase as leads (α = 0.07) freeze, forming grease ice (α = 0.09), then nilas (α = 0.16), then young ice (α = 0.27) and eventually ice of thickness > 0.7 m (α = 0.54). For snow-covered ice, albedos were as large as 0.87.

In a series of laboratory experiments, Perovich & Grenfell (1981) showed an asymptotic increase in albedo with thickness (Figure 4.7). The asymptotic thickness was inversely related to the absorption coefficient (smaller absorption coefficient = larger albedo and asymptotic thickness). Ehn et al. (2007) found that for young ice up to 0.3 m thick, the reflectance integrated from 350 to 1050 increased exponentially with thickness and had the form:

$$R_0 = 0.031 e^{(0.11 H_i)}$$

where R_0 is the reflectance at nadir and H_i is the ice thickness. The coefficient of determination for this fit is 0.69.

Spectral albedo measurements of a freezing lead (Perovich, 1991) showed an increase in albedo at visible wavelengths as the ice grew thicker. As the ice grows thicker, there are more opportunities for scattering and the albedo increases. The albedo in the near-infrared did not increase until the ice was thick enough that the surface was not saturated with brine. The largest albedo increase occurred when there was a 4 cm snowfall and visible albedos increased from 0.4 to 0.85. A snow layer of about 0.1 m is optically thick, giving wavelength-integrated albedos of about 0.85. Snow is ubiquitous for most of the year both in the Arctic (Sturm et al., 2002) and the Antarctic (see Chapter 3). Because of the pervasive presence of snow, surface conditions of growing bare, thin ice are quite limited both in space and in time. However, snow-free melting ice is quite common in the summer Arctic. The properties and processes of snow on sea ice are covered in Chapter 3.

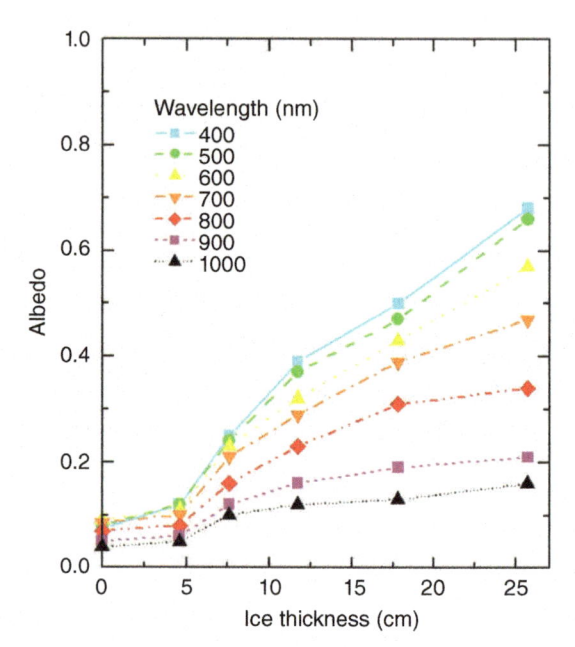

Figure 4.7 Spectral albedo as a function of ice thickness for young ice (Perovich & Grenfell, 1981).

Multi-year Arctic ice

Eight different Arctic sea ice conditions and their spectral albedos are shown in Figure 4.8: cold dry snow, melting snow, bare melting ice, light blue melt pond, blue melt pond, dark melt pond and lead. These cases encompass the dominant surface conditions and ice types for multi-year ice. Most of these measurements were made during the Surface Heat Budget of the Arctic (SHEBA) field experiment (Pegau & Paulson, 2001;

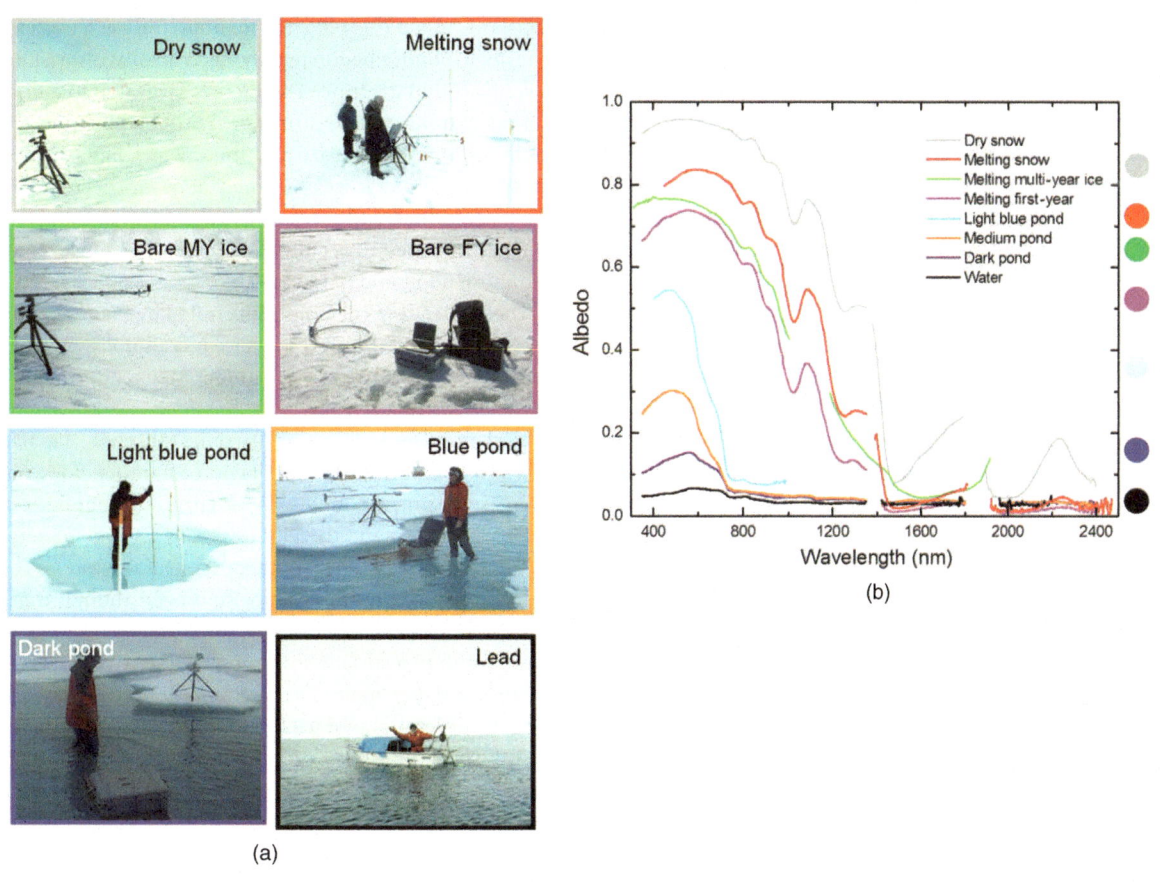

Figure 4.8 (a) Photographs illustrating the seasonal evolution of multi-year sea ice. (b) Spectral albedos of seven multi-year cases and open water. The colour of the curve is coordinated with the border surrounding the photographs in (a). The dots on the right-hand side of the plot show the wavelength-integrated albedo for that surface type.

Perovich et al., 2002; Uttal et al., 2002). These cases encompass expected Arctic multi-year sea ice conditions occurring from early spring to the end of summer melt. The colour-coded dots on the right-hand side of the graph in Figure 4.8 denote the wavelength-integrated albedo. Values extend from 0.07 for leads to 0.85 for snow-covered ice, encompassing 78% of the full range of possible albedos. The clear ocean waters of the Arctic are among the smallest albedos on Earth, while the snow-covered sea ice is one of the largest.

Spectral observations show that the largest albedos occur at visible wavelengths, with values decreasing towards the ultraviolet and the near-infrared. Cold dry snow over multi-year ice has the largest albedos across the spectrum with peak values of about 0.95 at visible wavelengths. When the snow melts, albedos decrease at all wavelengths. Removing the snow cover further reduces the albedo. Spectral albedos of melting multi-year ice are slightly larger than melting first-year ice. Melt-pond albedos are complex. For wavelengths > 800 nm, albedos for all three melt ponds are small and the same as for leads. Peak pond albedos in the visible vary greatly from 0.15 to 0.55. This is not surprising looking at the photographs in Figure 4.8. The light pond is noticeably brighter than the dark pond. Leads have the smallest albedos. For clear ocean water, the albedo is primarily due to specular reflection from the surface. If the water has sediments or biota, visible albedos can be slightly larger.

To explain these differences in albedo, and more generally to explain any sea ice AOP, we must consider the IOPs and the basic radiative transfer processes of absorption and scattering. Sea ice absorption coefficients show a strong spectral dependence. Sea ice scattering coefficients tend to be large, with no dependence on wavelength. Therefore spectral changes in apparent properties are due to absorption and changes in magnitude are due to scattering. A melt pond can be considered as an absorbing, non-scattering surface water layer over an absorbing, scattering ice layer. Pond albedos are smaller than snow or bare ice, as there is no surface scattering layer present.

Because of the significant absorption in water at longer wavelengths, it is unlikely that a photon in the near-infrared will travel downwards for the depth of the pond, be reflected, travel upwards the depth of the pond and be emitted. For example, at 900 nm the absorption coefficient of clear water is 6.55 m^{-1}, and less than 2% of the incident photons would survive a round-trip through a 0.3-m-deep melt pond, even if the pond bottom was a perfect reflector. That is why albedos in the near-infrared for all ponds are essentially the same as lead albedos.

The large differences in pond albedo in the visible are due to scattering, not absorption. The underlying ice in the light melt pond was 2.5 m thick and had more air bubbles, while the dark pond was 0.5 m thick with a large brine volume and few air bubbles. More scattering in the underlying ice results in increased visible albedos. At 500 nm, the e-folding distance in water is 22.5 m. Due to the small values of absorption coefficient, albedos at visible wavelengths are not sensitive to pond depth.

The photographs in Figure 4.9 show the evolution of local surface conditions from early melt to late in the melt season. Simplifying somewhat, the surface is composed primarily of four basic types; snow-covered ice, bare ice, ponded ice and leads. A time series of total albedos for a bare ice and ponded ice evolutionary sequence (Perovich et al., 2002) is shown in Figure 4.9. Lead albedo is plotted for reference. Before melt begins the ice surface is snow-covered and the total albedo is spatially uniform and equal to that of dry snow. There are minor fluctuations in total albedo resulting from changes in sky conditions from clear ($\alpha = 0.80$) to cloudy ($\alpha = 0.85$) and from small changes in snow grain size. Both locations show a drop in albedo at the end of May when the snow begins to melt, though the albedo remains spatially uniform.

Once the snow has melted there is a bifurcation in the albedo time series. The bare ice albedo is relatively constant at 0.65 ± 0.05 throughout the melt season. The melt-pond albedo steadily decreases, with a short respite around 25 June. The increase was due to a brief period with air temperatures below freezing when the surface of the pond froze and there was a light dusting of snow. Temperatures warmed and the decrease in pond albedo continued. In mid-August freeze-up began and snow started to accumulate, increasing albedos and reuniting pond and bare ice values.

There are a few key points illustrated by Figure 4.9. In the spring over 90% of the sea ice has an optically thick snow cover (Sturm et al., 2002). Melt-pond albedos steadily decrease throughout the melt season as the underlying ice melts and becomes thinner. Finally, and most importantly, the albedo of bare ice does not decrease during the melt season. This is evident both in the photographs and the plot in Figure 4.9. The bare ice in the 7 August photograph appears just as white as the bare ice in the 17 June photograph even though there was 0.5 m of surface melt. This large and steady albedo is the result of a surface scattering layer few centimetres deep. The layer looks like metamorphosed snow, but actually is deteriorated ice resulting from solar radiation absorbed in the upper portion of the ice. This surface scattering layer is self-renewing, growing deeper on sunny days with more penetrating solar radiation and thinning on cloudy days. Experiments scraping away the surface scattering layer demonstrated that a new layer would form within hours (Untersteiner, 1961; Perovich et al., 2002). The surface scattering layer is critical to the health of the sea ice cover. The albedo of melting bare ice without a surface scattering layer is only about 0.45.

Profound changes in albedo occur during the seasonal cycle of Arctic sea ice. Figure 4.10 displays a time series from April to October of wavelength-integrated albedo averaged over a 200-m-long line (Perovich et al., 2002). The line included a representative mix of bare ice and ponded ice. These results show the seasonal evolution of albedo observed at only one location in 1 year. Specific values of albedo and the timing of albedo changes will probably vary from place to place and year to year. However, the character of the changes and the sequence of the albedo phases are generally representative of the seasonal evolution of albedo.

The seasonal evolution of albedo is defined by five distinct phases: dry snow, melting snow, pond formation, pond evolution and fall freeze-up. These five

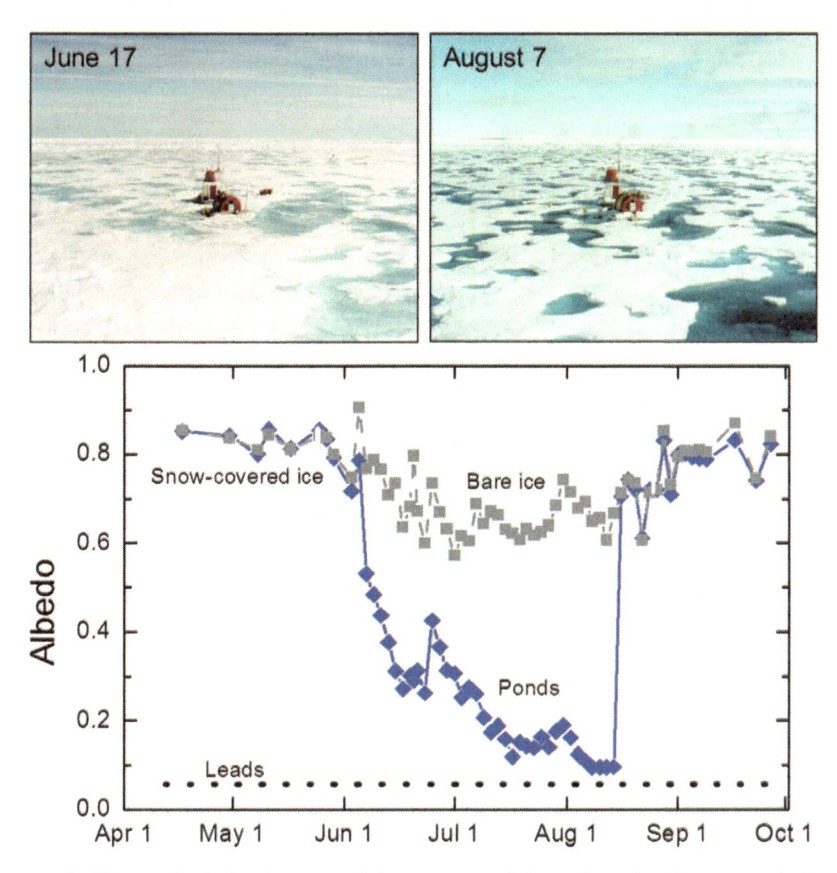

Figure 4.9 The photographs illustrate both the changes and the constants of the surface of multi-year ice during summer melt. The melt ponds grow deeper and darker, while the appearance of the ice surface remains unchanged. The seasonal evolution of multi-year ice for bare and ponded ice is plotted.

phases were defined by our observations of the physical state of the snow and ice. In the first phase (dry snow, April–May), the albedo was large (0.8–0.9) and spatially uniform. There was a slight and gradual decrease in albedo as the snow cover warmed and the snow grain size increased during this period. Rain at the end of May initiated phase 2 melting snow. There was a rapid coarsening of the snow to about 1-mm-diameter grains, an increase in the liquid water content of the snowpack, and a decrease in the average albedo from 0.8 to 0.7. Phase 3, melt-pond formation started in mid-June, when much of the snow cover had melted and ponds began to form. The result was a sharp drop in albedo from 0.7 to 0.5 in only a week. Phase 4 melt-pond evolution came after the initial period of melt-pond formation. This was a long period of a steady albedo decline as the melt ponds grew deeper and larger in areal extent. During phase 4 there were brief periods of cooler temperatures and light snow, resulting in short-lived increases in albedo. By the end of phase 4, the average albedo along the line was only 0.4. The spatial variability of albedo was greatest at this time, ranging from 0.1 for very dark ponds to 0.65 for bare ice. The combination of bare ice and ponds of varying albedo resulted in a standard deviation of 0.25 along the albedo line. The final phase fall freeze-up began in mid-August as air temperatures dropped below freezing. Ice skims formed on the surface of ponds and there were snow flurries, resulting in an increase of 0.2 in average albedo. The albedo continued to increase as the snow cover grew deeper. By the end of September, albedos returned to their springtime maxima of 0.8–0.9 and were spatially uniform (Perovich et al., 2002).

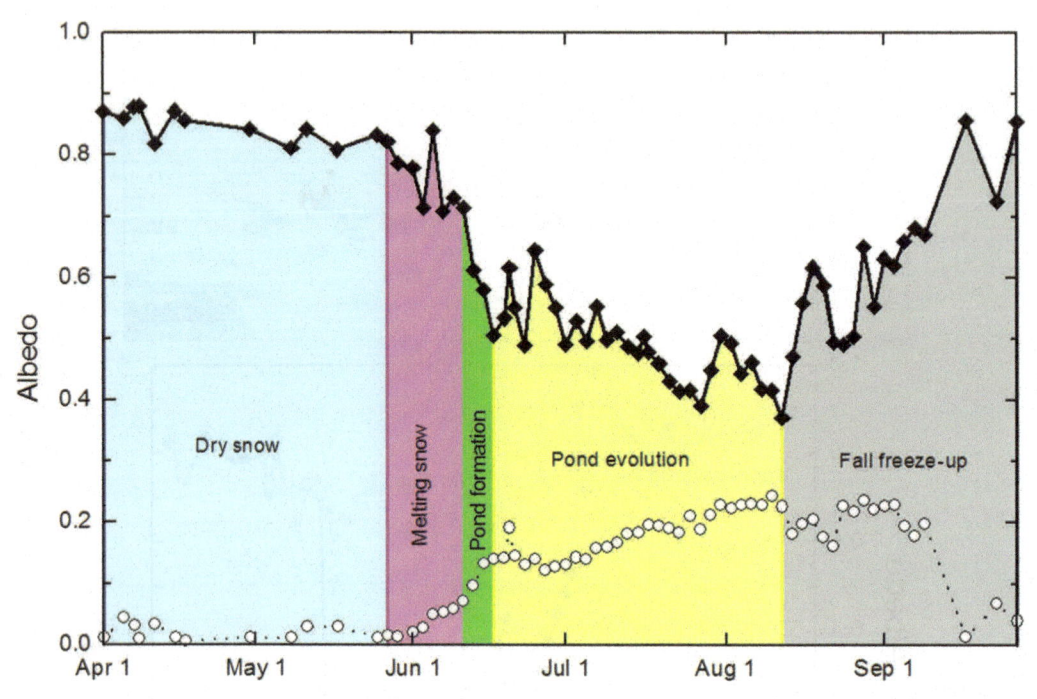

Figure 4.10 The five phases of albedo evolution for multi-year ice: dry snow, melting snow, pond formation, pond evolution and fall freeze-up. The white dots plot the standard deviation of albedo along the 200-m-long line. Source: Perovich et al., 2002. Reproduced by permission of American Geophysical Union.

First-year Arctic sea ice

The Arctic sea ice cover has undergone a fundamental change in character from a predominantly multi-year ice pack to one composed primarily of seasonal ice (Rigor & Wallace, 2004; Maslanik et al., 2007, 2011; Nghiem et al., 2007; Comiso, 2012). These changes have increased the interest in the albedo of first-year ice (Grenfell & Perovich, 2004; Hamre et al., 2004; Ehn et al., 2007; Perovich & Polashenski, 2012).

Figure 4.11 shows a sequence of first-year ice spectral albedos similar to the multi-year ice albedos displayed in Figure 4.8 (Grenfell & Perovich, 2004). There are similarities and differences in albedos due to similarities and differences in the physical state of the ice cover. Cold snow and melting snow albedos are the same for both first-year and multi-year ice. If there is an optically thick snow cover, the albedo is not affected by the substrate. First-year and multi-year bare ice albedos have the same spectral shape. First-year albedos are slightly smaller due to the thinner surface drained layer in first-year ice. First-year ponds have the distinctive spectral shape of multi-year ponds with peak albedo at blue wavelengths. The magnitude of the peak tends to be less than for multi-year ponds due to the thinner ice underlying the melt water.

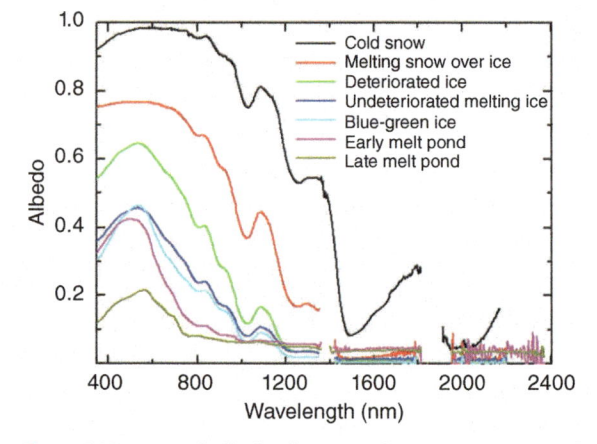

Figure 4.11 Spectral albedos for an evolutionary sequence of first-year ice.

Figure 4.12 shows the seasonal evolution of first-year ice albedo measured along a 200 m long line (Perovich

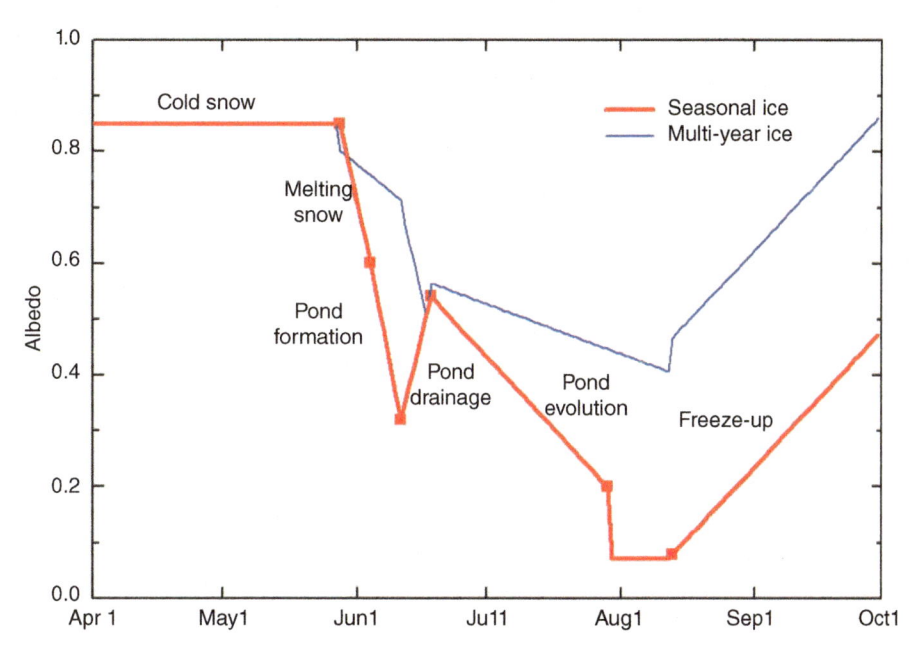

Figure 4.12 The season evolution of first-year ice albedo. This sequence has the first-year ice completely melting during August. Source: Perovich & Polashenski, 2012. Reproduced with permission of John Wiley & Sons.

& Polashenski, 2012).The stages of first-year albedo evolution are similar to multi-year ice. The first stage is cold, optically thick snow with an albedo of ~0.85 followed by melting snow and a decrease in albedo to about 0.7. During the final days of snowmelt, there is a sharp drop in albedo from 0.7 to 0.3, as ponds rapidly spread to cover as much as 80% of the ice surface. The extensive pond coverage is short-lived and there is a rapid rebound of albedo to about 0.6 as ponds drain and reduce their extent. After this drainage event, albedos decrease over time as the melt ponds gradually widen and the ice underlying the ponds thins (Polashenski et al., 2012). As the first-year ice melt progresses into its latest stages, the unponded ice thins to the point where it no longer has sufficient freeboard to maintain a surface scattering layer of drained ice crystals. Once the surface scattering layer is gone, the albedo drops sharply and melt accelerates, resulting in open water. As the unponded ice fraction becomes very thin (< ~0.5 m), the steady decline in albedo shown in Figure 4.12, would greatly accelerate. We represent this rapid change as a step decrease in albedo to 0.07 (open ocean), occurring once albedo drops below 0.2. A seasonal ice albedo of 0.2 denotes thin, highly porous ice that has been observed to completely melt in just a few days. Finally fall freeze-up begins and the albedo increases.

The multi-year ice albedo evolution discussed earlier is plotted for comparison. The seasonal snow cover on first-year ice is not as deep as on multi-year ice (Sturm et al., 2002; Webster et al., 2014), so it melts away faster. As ponds form, albedo drops much more rapidly on first-year ice, because the flat topography encourages ponds to extend horizontally. During pond evolution, the albedo of first-year ponds is typically less than that of multi-year ponds, as the underlying ice is thinner. Also, the areal extent of melt ponds typically remains more extensive on first-year than on multi-year ice during this stage (Grenfell & Perovich, 2004). The fall freeze-up increase in albedo is slower for first-year ice, because it is starting from open water. Eventually, the first-year ice becomes thick enough to support an optically thick snow layer and the albedos of first-year and multi-year ice converge.

Antarctic sea ice

Most of the albedo discussion has focused on results from Arctic sea ice. As discussed in Chapters 1 and 3, there are differences in ice structure and snow cover between Antarctic and Arctic sea ice. Antarctic sea ice has much more frazil ice than does Arctic sea ice, tends to be thinner and is somewhat more saline (Gow & Tucker, 1990; Chapter 1). There are significant difference in surface conditions, with deeper snow (Chapter 3) and very little ponding on Antarctic sea ice (Andreas & Ackley, 1982). Because of the deeper snow and fewer ponds, Antarctic sea ice does not exhibit the large decrease in albedo due to ponding during summer melt.

Figure 4.13 presents spectral albedos from 350 to 1050 nm for Antarctic sea ice cases including several young ice types, thick bare ice, and ice with a thin snow cover (Allison et al., 1993; Brandt et al., 2005). These results show the same spectral behaviour as Arctic sea ice. Even a very thin snow layer (about 10 mm) is enough to increase the visible albedo to about 0.9. There is a steady increase in albedo as the ice grows thicker. As shown earlier, even a thin snow layer greatly increases the albedo. Total albedos under cloudy skies are slightly larger than clear-sky values due to increased infrared absorption in the clouds. Thus, under cloudy skies a larger fraction of the incident irradiance is at visible wavelengths, where albedos are larger than in the infrared (Grenfell & Perovich, 1984; Brandt et al., 2005).

4.4.2 Transmittance

Light transmission through the ice is characterized by the transmittance $T(\lambda)$, which is the fraction of the incident irradiance that is transmitted through the ice. This is similar to the definition of albedo, except that it addresses transmitted light not reflected light. Like albedo, the transmittance is strongly influenced by surface conditions. Unlike albedo it is also strongly influenced by snow depth and ice thickness. For example, increasing the snow depth on 2-m-thick ice from 0.1 to 0.2 m would not change the albedo, but would cause a decrease in visible transmittance by roughly a factor of 5. Transmittance decreases roughly exponentially as the ice thickness increases or snow gets deeper.

Spectral transmittances for a variety of ice types and surface conditions are plotted in Figure 4.14 (Light et al.,

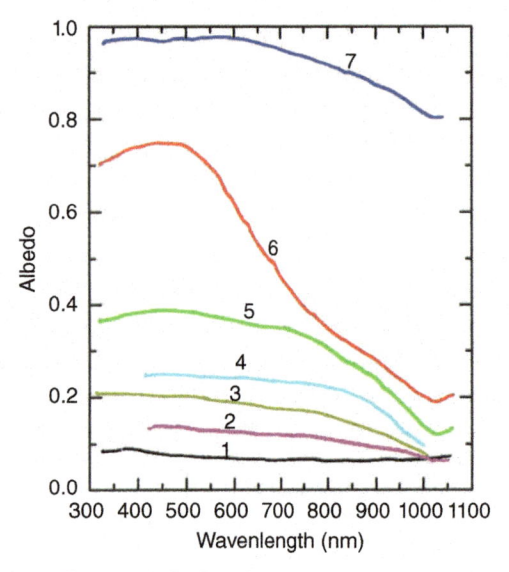

Figure 4.13 Spectral albedos of snow-covered ice, snow free ice and open water for Antarctic sea ice. Results are from Allison et al. (1993) and Brandt et al. (2005). Results are: 1, open water, $\alpha_T = 0.067$; 2, 3 cm nilas, $\alpha_T = 0.12$; 3, 5 cm nilas and slush, $\alpha_T = 0.14$; 4, young grey ice, $\alpha_T = 0.20$; 5, thicker grey ice, $\alpha_T = 0.30$; 6, 140 cm thick ice, $\alpha_T = 0.49$; and 7, deep snow on the Antarctic Plateau, $\alpha_T = 0.83$ (Grenfell et al. 1994).

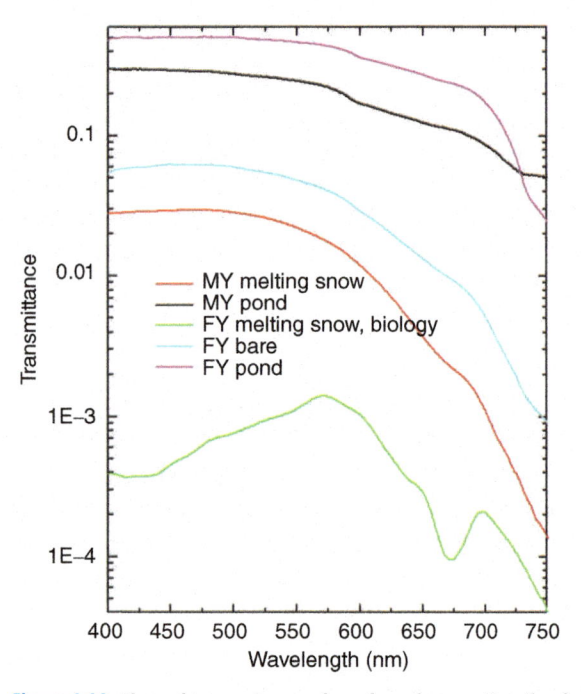

Figure 4.14 Plots of transmittance for selected cases. Details of the cases are as follows: multi-year (MY) melting snow had 10 cm of snow over 186 cm of ice; MY pond had 25 cm of water over 133 cm of ice; first-year (FY) snow has 11 cm of snow over 106 cm of ice with a flourishing ice algae layer; FY bare had 149 cm of ice; and FY pond had 8 cm of water over 112 cm of ice.

2008; Nicolaus et al., 2013). Values range by several orders of magnitude both between cases and across the spectrum. The greatest transmittance of nearly 0.5 was for a first-year melt pond, with transmittances for a multi-year pond roughly half as large. Transmittance through multi-year ice with a melting snow cover was only a few percent. Interestingly the smallest transmittance plotted was for first-year ice with a melting snow cover. The reduced transmittance was due to the presence of a dense ice algae layer, as evidenced by the markedly different spectral shape.

Figure 4.15 demonstrates the impact of surface conditions on transmittance (Frey et al., 2011). The bare ice was 1.27 m thick and the pond was 0.83 m, with 0.15 m surface water depth. The top photograph shows a mix of bare and ponded first-year ice. There are sharp boundaries between the ponds and the bare ice. The second photograph is below the ice looking up. It appears to be a negative of the surface photograph, as the ponds are bright and the bare ice is dark. This illustrates that the ponds are not absorbing more sunlight than the bare ice, but they are transmitting more sunlight to the upper ocean. The boundaries between bare and ponded ice in the under-ice photograph are blurred due to scattering in the ice. The graph in Figure 4.15 shows the spatial variability of transmittance at selected wavelengths over the transition from bare ice to ponded ice (along the red line in the top photograph, Figure 4.15). For visible wavelengths, the pond transmittance (0.45) was almost an order of magnitude larger than the bare ice transmittance (0.05). The transition from bare ice transmittance to ponded ice transmittance occurred over a distance of two to four times the ice thickness. The transmitted light field under ponded ice has large spatial variability (Ehn et al., 2011; Frey et al., 2011).

4.4.3 Extinction coefficients

Transmittance is an intrinsically limited descriptor of ice optical properties because of the dependence on ice thickness and snow depth or pond depth. An irradiance extinction coefficient (k) is often used to characterize the transmission of light through the ice cover. The extinction coefficient includes losses due to both scattering and absorption, has units of m^{-1}, and is defined as:

$$k(z, \lambda) = \frac{-1}{F_d(z, \lambda)} \frac{dF_d(z, \lambda)}{dz}$$

where $F_d(z, \lambda)$ is the downwelling spectral irradiance at depth z in the ice. Extinction coefficients can be directly calculated from a vertical profile of transmission in the ice. They can also be determined from measurements of incident, reflected and transmitted irradiance combined with a radiative transfer model.

Wavelength-integrated extinction coefficients (k_t) were determined for snow and sea ice decades ago. Sea ice values are in the 1.1–1.5 m^{-1} range (Untersteiner, 1961; Chernogovskii, 1963; Thomas, 1963; Weller & Schwerdtfeger, 1967). Snow values are much larger, varying from 4.3 m^{-1} for dense Antarctic snow (Weller & Schwerdtfeger, 1967) to as high as 40 m^{-1} in freshly fallen snow (Thomas, 1963). Wavelength-integrated extinction coefficients are simpler to measure and easier to use than spectral values, but they are severely limited. The wavelength-integrated extinction coefficient combines contributions from different wavelengths and therefore depends on the spectral distribution of incident, reflected and transmitted irradiance. On sunny days the incident irradiance has a larger long-wave component, which is absorbed rapidly in the ice, resulting in higher values of k_t. There is a strong depth dependence of k_t near the surface due to the rapid extinction at longer wavelengths (Grenfell & Maykut, 1977). Most of the light transmitted below the upper 0.1 m is at visible wavelengths, where the spectral dependence of k_λ is weaker, and changes in k_t with depth are small.

As optical instruments improved and requirements increased, the focus shifted to determining spectral extinction coefficients. Figure 4.16 shows spectral extinction coefficients for snow and sea ice from several sources (Grenfell & Maykut, 1977; Perovich & Grenfell, 1981; Light et al., 2008). Snow extinction coefficients are more than an order of magnitude greater than values for bare or ponded, first-year or multi-year ice. Spectral extinction coefficients for a particular snow or ice type vary by as much as a factor of 5. There is also variability in extinction coefficients within an ice type depending on the physical properties of a particular sample.

Like other optical properties, extinction coefficients depend on the internal structure of the ice (Zaneveld, 1966; Grenfell & Maykut, 1977; Perovich & Grenfell, 1981; Gilbert & Buntzen, 1986). Extinction coefficients decrease during warming as the brine volume increases and the number of inclusions decrease. Also, at a given

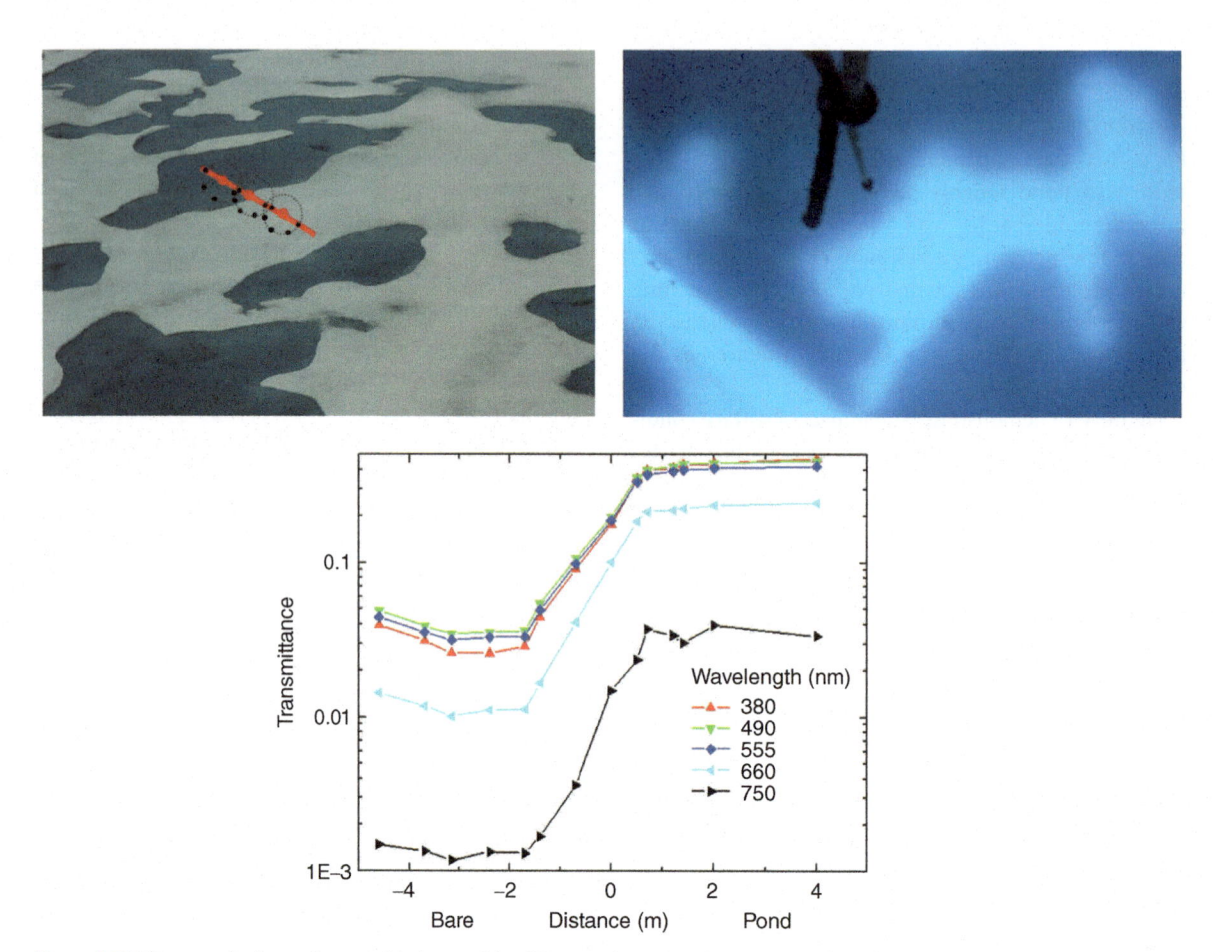

Figure 4.15 Photographs from above and below melting first-year ice. Transmittances at selected wavelengths are plotted. The measurements were made along the red line from bare ice to ponded ice (Frey et al., 2012).

brine volume extinction, coefficients are larger for faster-grown ice, which has more inclusions. As brine drains from above freeboard during melt, ice–brine interfaces become ice–air interfaces; there is enhanced scattering, and extinction coefficients increase in this layer. This has been observed in both Arctic (Grenfell & Maykut, 1977; Light et al., 2008) and Antarctic sea ice (Trodahl et al., 1987, 1990).

4.5 Radiative transfer modelling

Models are essential in interpreting observations and in advancing from a collection of observations to a physically based understanding of radiative transfer in sea ice. A great deal of work has been performed on sea ice radiative transfer, in response to the need for models. Space restrictions preclude a detailed discussion of all the varied models and their results. This section provides a brief overview and guide to that work. Radiative transfer models have been applied to a wide range of problems, including estimating the absorption of solar radiation in sea ice (Maykut & Untersteiner, 1971), studying the relationship between changes in ice physical properties and changes in optical properties (Grenfell, 1983, 1991), investigating bio-optical interactions (Arrigo et al., 1991, 2012), examining the transmission of visible and ultraviolet light through sea ice (Perovich, 1990, 1991, 1993), assessing radiative interactions between the atmosphere, ice and ocean (Jin et al., 1994) and implementing sea ice optics into climate models (Briegleb & Light, 2007; Holland et al., 2012).

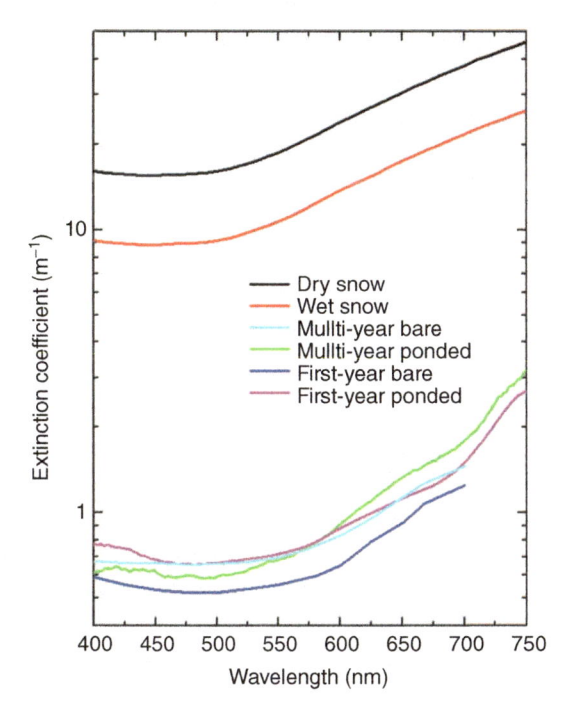

Figure 4.16 Extinction coefficients for various ice types, including dry snow, wet snow, multi-year bare ice, multi-year ponded ice, first-year bare ice and first-year ponded ice.

Sea ice radiative transfer models range in complexity from a simple exponential decay law to numerically detailed solutions of the angular distribution of radiance above, in and below the sea ice cover. While there are many sea ice radiative transfer models, with a variety of solution schemes and different input and output parameters, the same physics underlies all of them. The different models have attributes that favour a model for some applications and restrict them for others.

One of the distinguishing features of radiative transfer models is the number of 'streams' they consider. The number of streams refers to the number of angles where the radiance is calculated. Two-stream models are quite common, where the upwelling and downwelling irradiances are computed. More streams means more angular detail in the calculated radiance field and more detail in the treatment of scattering. There is a computational cost to this additional detail.

The simplest sea ice radiative transfer model is the exponential decay relationship. However, there is an implicit assumption in this relationship that the medium is infinitely thick, and, consequently, the exponential decay law does a mediocre job of representing radiative transfer in thin ice (Grenfell, 1979). Grenfell (1979) developed a two-stream, three-layer model, based on the work of Dunkle and Bevans (1957) that improved the treatment of thin ice with only a modest increase in computational complexity. The two-stream formulation was expanded into an n-layer model (Perovich, 1990) and extended into the ultraviolet (Perovich, 1993). The major deficiency of a two-stream model is the assumption of isotropic scattering.

Models based on the discrete ordinates method (DOM) of Chandrasekhar (1960) have been used to treat scattering in more detail and examine the angular distribution of radiance. In this method, the phase function is approximated by a series of Legendre polynomials (Liou, 1973, 1974; Mobley, 1994) and the discrete ordinates refer to particular angles at which the radiance is computed. This method allows for non-diffuse radiance fields and non-isotropic phase functions.

Using the DOM approach, Perovich and Grenfell (1982) developed a two-layer, four-stream model (radiances at two upward and two downward angles). This model was extended by Grenfell (1983, 1991), who developed a single-layer, 16-stream model and a multi-layer, four-stream model to explore relationships between ice physical properties and ice optical properties. The four-stream (Grenfell, 1991) model included vertically varying ice properties. The single-layer, 16-stream model (Grenfell, 1983) generated a more detailed angular description of radiance, better represented the phase function, and improved the treatment of refraction at the air-ice interface for a homogeneous ice cover.

Light et al. (2004) formulated a temperature-dependent, structural-optical model of first-year sea ice. The model incorporates results from laboratory studies of first-year ice physical and optical properties. Scattering from brine pockets, air bubbles and precipitated salts are all included in the model. This model explicitly treats the optical impacts of the different processes governing the physical state of the ice as the temperature changes. Studies using this model established that scattering in sea ice is strongly forward-peaked, with scattering coefficients reaching maximum values of $3000 \, m^{-1}$ for cold ice and averaging $450 \, m^{-1}$ for moderate and warm ice.

The Monte Carlo method is another approach to radiative transfer modelling. As the name implies,

Monte Carlo models take a probabilistic approach to solving the equation of radiative transfer (equation 4.1). The absorption coefficient, the scattering coefficient and the phase function are transformed into the probability that over a given distance a photon is absorbed or scattered in a particular direction. The cumulative results of a large number of photons 'shot' into the ice define radiative transfer in the ice. Monte Carlo models are simple conceptually, straightforward to program, and widely applicable (Mobley, 1994). This method is particularly well suited for complex geometries or boundary conditions, where other solutions to the equation of radiative transfer are difficult or impossible (Trodahl et al., 1987; Trodahl & Buckley, 1989; Kaempfer et al., 2007). Light et al. (2004) developed a sea ice Monte Carlo model to interpret optical measurements made of sea ice cylinders into scattering properties of the ice. This model can be extended to include a variety of geometries.

Sea ice bio-optical models have also been developed (Arrigo et al., 1991; Arrigo & Sullivan, 1994; Zeebe et al., 1996; Ehn et al., 2008) investigating the interdependence between ice properties, light fields within and under the ice, and biological activity. These models include relationships for light extinction by biota and detritus and for light- and nutrient-driven biological activity. In the model, microalgae growth is determined by temperature, salinity, spectral irradiance and nutrients. Zeebe et al. (1996) coupled with a one-dimensional thermodynamic model and a bio-optical model to investigate the relationship between ice algae on ice temperature, brine volume and melting.

There has been considerable effort and progress in improving the treatment of sunlight and sea ice in climate models. Briegleb and Light (2007) formulated a Delta-Eddington, multiple scattering parameterization that was incorporated into the Community Climate System Model. They use IOPs for sea ice and snow to create a physically based treatment of radiative transfer Using this improved sea ice radiative transfer model, Holland et al. (2012) explored the impact of melt ponds and aerosols on Arctic sea ice. They found that while the direct radiative impact of aerosols and ponds was small, their influence is significant due to the ice albedo feedback.

4.6 Changing ice, changing light

As discussed in other chapters and many articles, the Arctic sea ice cover is undergoing large changes towards less summer ice, thinner ice and younger ice. These changes both impact and are impacted by the interaction of sunlight with sea ice. We will explore a few of these interrelated impacts.

4.6.1 Sea ice albedo feedback

The sea ice albedo feedback is a positive feedback and thus can amplify changes in the ice cover. It is important from a climate perspective and contributes to Arctic amplification of warming (Curry et al., 1995; Holland & Bitz, 2003; Holland et al., 2006; Winton 2006, 2008). The ice albedo feedback results from the large contrast in albedo between snow-covered ice (0.85), melting ice (0.65), melt ponds (~0.4), and open water (0.07). It comprises two components, one due to changes in ice area and the other due to changes in ice albedo. In a warming Arctic there is less sea ice and longer melt seasons. As the sea ice area decreases, highly reflecting ice is replaced by dark open water and more solar energy is absorbed. In addition there are changes in the sea ice albedo during melt as illustrated in Figure 4.17. As melt begins and progresses, the highly reflecting, snow-covered sea ice evolves into a darker mixture of bare ice, melt ponds and open water. The decrease in ice albedo and the increase in the amount of open ocean result in more absorbed solar energy and more melt, leading to a positive feedback loop.

There is evidence of an ongoing ice albedo feedback in the Arctic. The decrease in summer ice extent and area is well documented (Serreze et al., 2007; Stroeve et al., 2007, 2012; Comiso et al., 2012). Melt seasons are starting earlier and ending later (Markus et al., 2009). There is a shift from multi-year ice to first-year ice (Rigor & Wallace 2004; Nghiem et al., 2007; Maslanik et al., 2011). All of these changes result in increased solar energy input to the ice and ocean. Perovich et al. (2011) examined trends from 1980 to 2007 in solar energy input to the ice and to open water for Arctic sea ice (Figure 4.18). Incident solar irradiance was determined from reanalysis products. The solar energy input to the ice was calculated based on an albedo evolution sequence (Perovich et al., 2002, 2011) triggered by satellite-derived onset dates of melt and freeze-up. Differences between first-year and multi-year ice albedo

Figure 4.17 Ice albedo feedback in the interior of the ice cover. The top photograph was taken on 17 April well before the onset of melt. The bottom photograph is from 8 August at the height of the melt season.

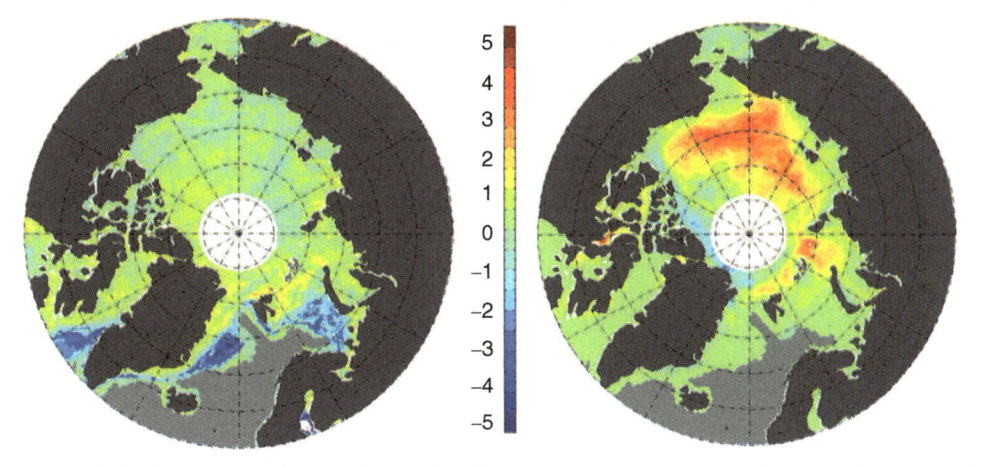

Figure 4.18 Trends of solar heat input to the ice (left panel) and directly to the ocean (right panel) (Perovich et al., 2011).

evolution were not considered; 65% of the area showed a modest increase in the solar energy input to the ice, averaging 0.5% year^{-1}. This increase was due to earlier onset dates of summer melt seasons. The timing of the onset of melt is much more important that the onset of freeze-up (Perovich et al., 2007). There are areas of negative trends (less solar energy input to the ice) in the Chukchi and Beaufort seas, the Fram Strait and near Greenland. These negative trends are mainly due to increased ice loss earlier in the summer; less ice cover means less energy input to the ice.

Over 90% of the study area shows a trend of increasing solar energy input to the ocean, with an average increase of about 1% year^{-1}. The largest increases of 4% year^{-1} were found in the Beaufort, Chukchi and East Siberian seas. This enhanced solar energy input to the upper ocean results in increases in the ocean heat flux and the amount of bottom melting (Perovich et al., 2008, 2011; Steele et al., 2010). Not all of the deposited solar energy is used in melting and the remaining heat could slow freeze-up in the fall.

4.6.2 Shifting from multi-year ice to first-year ice

In recent decades there has been a decrease in multi-year ice in the Arctic. The shift from multi-year to first-year ice impacts the partitioning of solar energy. First-year ice tends to be thinner than multi-year ice and to have a shallower snow cover. As was explained in an earlier section, the albedo of first-year ice tends to be less than that of multi-year ice throughout the summer melt season. In addition, scattering and extinction coefficients tend to be smaller for first-year ice (Light et al., 2008). Taken together, these physical and optical changes combine to result in smaller albedos and larger transmittance for first-year ice compared with multi-year ice.

Differences in solar partitioning between first-year and multi-year ice are explored in Figure 4.19. The wavelength-integrated fraction of sunlight reflected, absorbed in the snow and ice, and transmitted to the upper ocean is presented for the three basic surface conditions – snow-covered ice, bare ice and ponded ice. The partitioning for snow-covered first-year and snow-covered multi-year ice is quite similar. If an optically thick snow cover is present, albedos for first-year and multi-year ice are the same. Slightly more light is transmitted through the snow-covered first-year ice (0.01 vs. 0.002) due to both the snow cover and the ice being thinner.

Bare ice albedos are also similar for first-year and multi-year ice, with multi-year ice being slightly larger due to a thicker drained surface layer. The fraction of light absorbed in the ice is the same, but more is transmitted to the ocean through first-year ice (0.05 vs. 0.01). The largest differences are for melt ponds where ponded first-year ice transmits as much as an order of magnitude more sunlight to the upper ocean than ponded multi-year ice. Melting bare and ponded first-year ice will transmit much more sunlight to the upper ocean than multi-year ice, where the additional heat will probably cause an increase in melting on the underside of the ice.

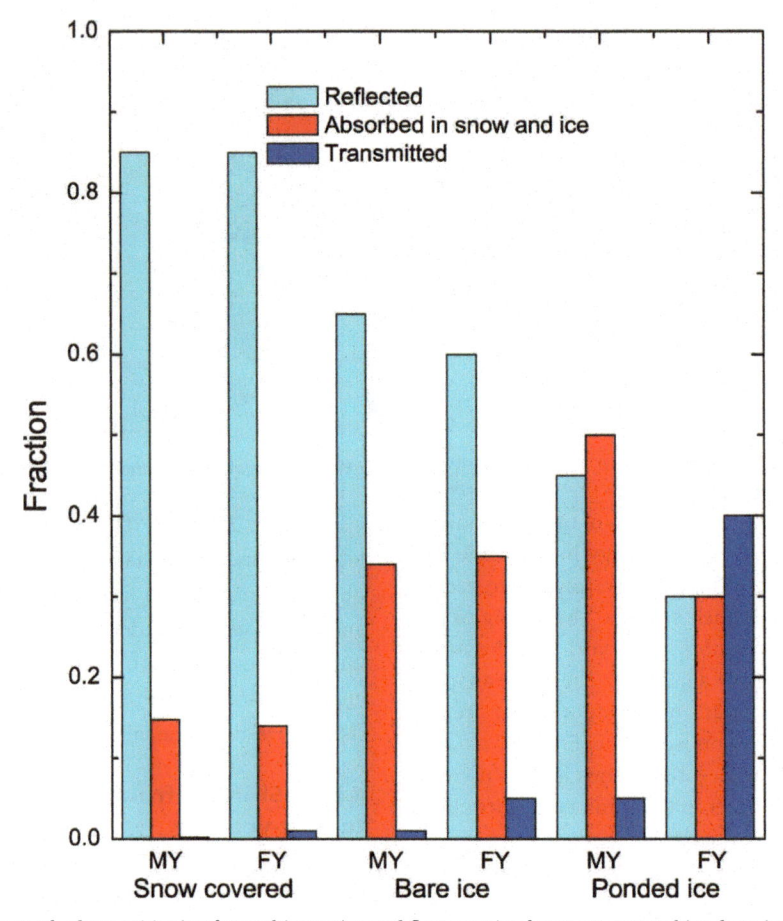

Figure 4.19 Comparison of solar partitioning for multi-year ice and first-year ice for snow-covered ice, bare ice and ponded ice.

The solar energy transmitted through the ice to the upper ocean is substantial and must be considered in heat budget calculations. Nicolaus et al. (2012) compared first-year and multi-year ice transmittances on a basin-wide scale. They found transmittances were nearly three times greater for first-year ice than for multi-year ice, due to the larger melt-pond fraction of first-year ice.

Even though the differences in solar partitioning between snow-covered first-year and multi-year ice are small from a heat budget perspective, they can still be significant for other reasons. The extra amount of light transmitted through snow-covered first-year ice may be enough to support a robust ice algae bloom (Arntsen et al., 2014). The shift from multi-year to first-year ice may have major impacts on ecosystems.

4.6.3 Solar partitioning

The aggregate scale partitioning of solar radiation among reflection to the atmosphere, absorption in the snow and ice, and transmission to the upper ocean is of great interest for sea ice heat and mass balance considerations. Perovich (2005) examined solar partitioning on the aggregate scale considering an area of hundreds of km². The analysis was based primarily on field observations and examined contributions from snow, bare ice, melt ponds and leads. Daily values of reflected, absorbed and transmitted light were determined from 1 April to 5 October and are plotted in Figure 4.20, with the contributions from ice, ponds and leads highlighted.

Snow-covered ice and bare ice are responsible for most of the energy reflected back to the atmosphere (Figure 4.20a). Leads, with their combination of small albedo and area fraction, contribute very little to reflection – about 1% at most. Ponds contribute less than 10%. In April and May the surface is covered by highly reflecting snow and the amount of sunlight absorbed in the ice was between 15% and 20% (Figure 4.20b). Once melt begins and the albedo decreases, the absorbed fraction increases to peak values of 30–40% from mid-June to the end of July. Because of their small albedo, ponds made a substantial contribution to the energy absorbed in the ice cover from mid-June to mid-August. The energy absorbed in the ice decreased in mid-August, as the ponds began to freeze and snow fell and accumulated on the surface.

Before the onset of melt there was negligible transmittance through the snow-covered ice, leads only covered a few per cent of the total ice cover and the total transmission to the ocean was small (Figure 4.20c). There was a slow, but steady, increase in transmission to the ocean in June and July, as the ice thinned and the melt ponds developed. The percentage transmitted continued to increase into mid-August, reaching maximum values of 15–27% of the incident solar energy. At the beginning of August, the transmittance almost doubled as a direct result of an ice divergence event that increased the open water fraction from 0.05 to 0.20. Both ponded and bare ice contributed substantially to transmittance. For example, during July, the transmittance contributions from leads, ponds and bare ice were roughly comparable.

Integrating over time from 1 April to 5 October 1998 the incident solar radiation was partitioned with 68% reflected to the atmosphere, 24% absorbed in the snow and ice, and 8% transmitted to the ocean. Snow and bare ice dominated reflection back to the atmosphere, with ponds contributing only 3% and leads less than 1%. Of the solar energy absorbed in the ice cover, 82% was absorbed in snow-covered or bare ice and 18% in ponds. Most of the solar energy transmitted to the ocean was through leads (61%), though significant portions were also transmitted through bare ice (23%) and ponded ice (16%).

Nicolaus et al. (2012) used field data to generate Arctic-wide maps of light transmission through the ice. Arndt and Nicolaus (2014) extended these results and examined solar energy penetrating through the ice cover to the upper ocean on an Arctic-wide scale from 1979 to 2011 using satellite data, reanalysis products and a light transmission parameterization. They found that May to August were the key months, accounting for 96% of the energy input through the ice into the upper ocean. The amount of solar energy input through the ice into the ocean has increased over the past few decades due to thinning of the ice, changes in ice age and shifts in the timing of the melt season. The average trend was 1.5% per year, with a maximum value of 4% year^{-1}.

4.7 Summary

The optical properties of sea ice are governed by scattering and absorption. In general, spectral variations result

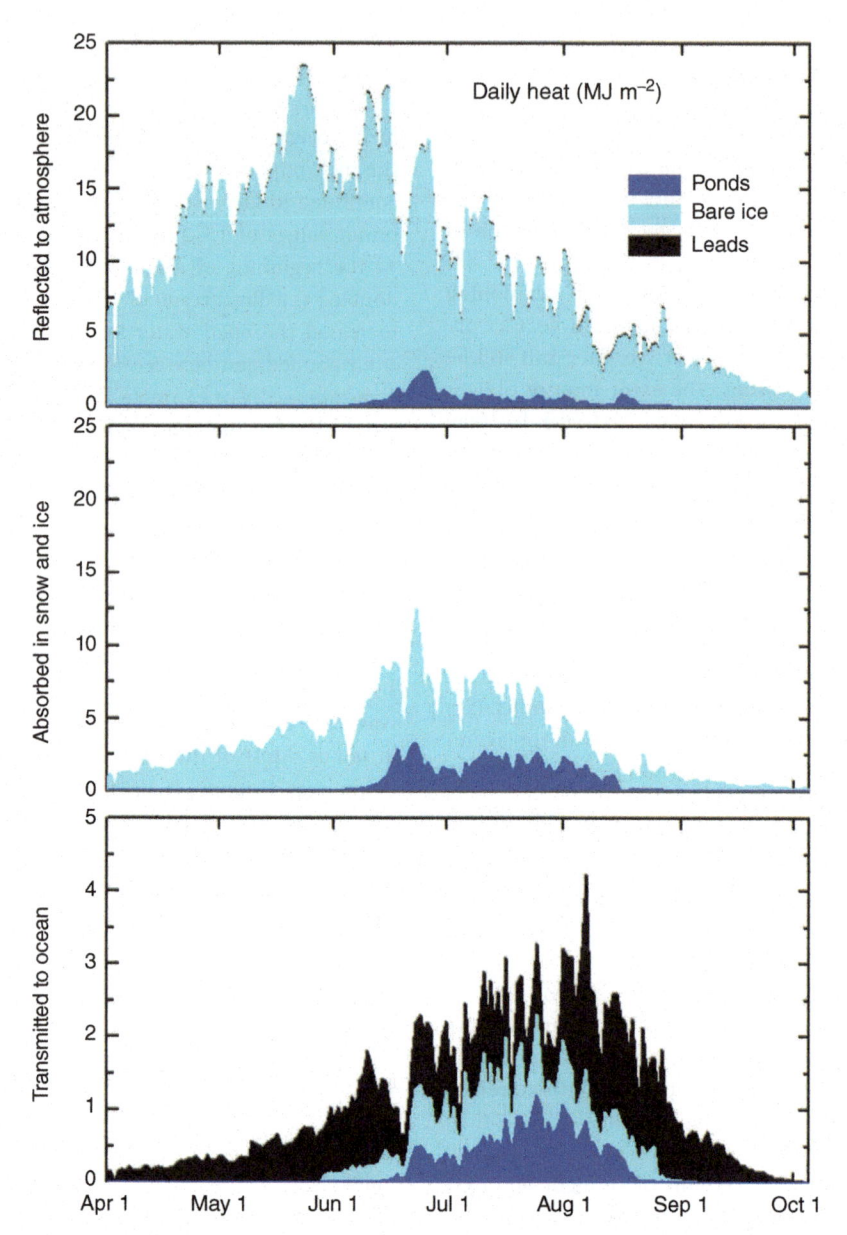

Figure 4.20 Time series of solar partitioning during summer melt in the Beaufort Sea: (a) reflected to the atmosphere, (b) absorbed in the snow and ice, and (c) transmitted to the ocean. Contributions from leads, ponds and bare and snow-covered ice are highlighted. Source: Perovich et al., 2005. Reproduced by permission of American Geophysical Union.

from absorption, while pan-spectral changes in magnitude are due to scattering. The albedo is highly sensitive to the surface state. There are significant differences in albedo among snow (0.85), bare ice (0.65), melt ponds (0.2–0.5) and leads (0.05–0.10).

The partitioning of solar radiation by sea ice is simple for much of the year. From September to late May, the sea ice surface is covered by an optically thick layer of snow. The snow layer reflects approximately 85% of the incident sunlight, absorbing most of the remainder and transmitting very little. Everything changes with the onset of melt, as the surface transforms into a complex, evolving mosaic of bare ice, melt ponds and open water. During the melt season, albedos and transmittances exhibit large spatial and temporal variability, with substantial amounts of solar radiation absorbed in the ice and transmitted to the ocean. The ongoing changes in the physical state of the Arctic sea ice cover result in an increase in solar energy absorbed in the ice and upper ocean and an ice albedo feedback.

There are large uncertainties associated with the ongoing shift in the Arctic from multi-year ice to first-year ice. While there are numerous measurements of the optical properties of first-year ice as well as multi-year ice, there is little known about the atmosphere–ice–ocean interactions in an Arctic sea ice cover that is predominantly composed of first-year ice. What will be the large-scale solar partitioning under these new conditions? How will addition solar energy transmitted to the upper ocean impact ice summer ice melt, autumn freeze-up and primary productivity? Technological improvements will help to answer these and other key questions. Computation advances facilitate more detailed theoretical treatments of radiative transfer in sea ice. In recent years, observational technology has greatly advanced, with improved spectroradiometers and enhanced capabilities for autonomous measurements (Wang et al., 2014). It is now possible to survey spectral albedos from autonomous aerial platforms and spectral transmission from underwater autonomous or tethered vehicles.

4.8 Acknowledgements

The author thanks the Office of Naval Research, the National Science Foundation, and the National Aeronautics and Space Administration for their research support.

Glossary

α = albedo
α_t = wavelength-integrated, or total, albedo
$\tau = (k + \sigma)z$ (non-dimensional optical depth)
μ = cos(zenith angle)
z = physical depth
τ = optical depth
E_o = radiance of the direct beam component of the incident radiation field
F = irradiance
F_d = downwelling irradiance
F_u = upwelling irradiance
I = radiance (flux in a beam, W m^{-2} sr^{-1})
I_r = reflected radiance
k = absorption coefficient – amount of absorption (m^{-1})
k_b = absorption coefficient of brine
k_i = absorption coefficient of ice
k_{si} = absorption coefficient of sea ice
N = real index of refraction
$p(\mu, \mu', \phi, \phi')$ = phase function = angular distribution of scattered light
R = bidirectional reflectance distribution function (BRDF)
R_o = normalized reflected radiance at nadir
S = source function
T = transmittance
ϕ = azimuth angle
ϕ_o = solar azimuth angle
k_t = wavelength-integrated, or total, extinction coefficients
k = extinction coefficient
λ = wavelength
v_i = volume fraction of ice
v_b = volume fraction of brine
θ = zenith angle (0 pointing downward, π pointing upward)
θ_o = solar zenith angle
μ = cosine of the zenith angle, θ
σ = scattering coefficient – amount of scattering (m^{-1})
ϖ_o = single scattering albedo, measure of relative contributions of absorption and scattering

References

Ackermann, M., Ahrens, J., Bai, X. et al. (2006) Optical properties of deep glacial ice at the South Pole. *Journal of Geophysical Research*, **111**, D13203, doi:10.1029/2005JD006687

Allison, I., Brandt, R.E. & Warren, S.G. (1993) East Antarctic sea ice: Albedo, thickness distribution and snow cover, *Journal of Geophysical Research*, **98**(C7), 12 417–12 429.

Andreas, E.L. & Ackley, S.F. (1981) On the differences in ablation seasons of the Arctic and Antarctic sea ice. *Journal of Atmospheric Science*, **39**, 440–447.

Arndt, S. & Nicolaus, M. (2014) Seasonal cycle and long-term trend of solar energy fluxes through Arctic sea ice. *The Cryosphere*, **8**, 2219–2233.

Arntsen, A., Stwertka, C., Polashenski, C. & Perovich, D. (2014) Optical measurements and analysis of sea ice in the Chukchi Sea during the onset of 2014 melt. Proceedings of the Fall Meeting of American Geophysical Union, 2014.

Arrigo, K.R., van Dijken, G.L. & Bushinsky, S. (2008) Primary production in the Southern Ocean 1997–2006. *Journal of Geophysical Research*, **113**, C08004, doi:10.1029/2007JC004551.

Arrigo, K., Perovich, D.K., Pickart, R.S. et al. (2012) Massive phytoplankton blooms under Arctic sea ice. *Science*, **336**,

Arrigo, K.R. & Sullivan, C.W. (1994) A high resolution bio-optical model of microalgal growth: Tests using sea-ice algal community time-series data. *Limnology and Oceanography*, **39**, 609–63.

Arrigo, K.R., Sullivan, C.W. & Kremer, J.N. (1991) A bio-optical model of Antarctic sea ice, *Journal of Geophysical Research*, **96**, 10581–10592.

Arrigo, K.R. (2014) Sea ice ecosystems, *Annual Review of Marine Science*, **6**, 439–467.

Bitz, C.M. & Lipscomb, W.H. (1999) An energy-conserving thermodynamic model of sea ice. *Journal of Geophysical Research*, **104**(C7), 15669–15677.

Bohren, C.F. & Barkstrom, B.R. (1974) Theory of the optical properties of snow. *Journal of Geophysical Research*, **79**, 4527–4535.

Bohren, C.F. & Huffman, D.R. (1983) *Absorption and Scattering of Light by Small Particles*. Wiley, New York.

Boucher, N.P. & Prézelin, B.B. (1996) Spectral Modeling of UV Inhibition of In Situ Antarctic Primary Production Using A Field-Derived Biological Weighting Function. *Photochemistry and Photobiology*, **64**, 407–418.

Brandt, R.E., Roesler, C.S. & Warren, S.G. (1999) Spectral albedo, absorptance, and transmittance of Antarctic sea ice. *5th Conference on Polar Meteorology and Oceanography, Dallas, TX, American Meteorological Society*, 456–459.

Brandt, R.E., Warren, S.G., Worby, A.P. & Grenfell, T.C. (2005) Surface albedo of the Antarctic sea ice zone. *Journal of Climate*, **18**, 3606–3622.

Briegleb, B.P. & Light, B. (2007) A Delta-Eddington Mutiple Scattering Parameterization for Solar Radiation in the Sea Ice Component of the Community Climate System Model. *NCAR Technical Note NCAR/TN-472+STR*.doi: 10.5065/D6B27S71.

Chandrasekhar, S.C. (1960) *Radiative Transfer*. Dover, New York, 393 pp.

Chernigovskii, N.T. (1963) Radiational properties of the central Arctic ice cover.*Trudy Arkticheskogo I Antarkticheskogo Nauchno-Issledovatel'skogo Instituta*, **253**, 249–260.

Comiso, J.C. (2012) Large decadal decline of the Arctic multi-year ice cover. *Journal of Climate*, **25**, 1176–1193.

Curry, J.A., Schramm, J.L.&Ebert, E.E. (1995) Sea ice-albedo climate feedback mechanism. *Journal of Climate*, **8**, 240–247.

Dunkle, R.V. & Bevans, J.T. (1957) An approximate analysis of the solar reflectance and transmittance of a snow cover, *Journal of Meteorology*, **13**, 212–216.

Ebert, E.E., & Curry, J.A. (1993) An intermediate one-dimensional thermodynamic sea ice model for investigating ice-atmosphere interactions. *Journal of Geophysical Research*, **98**, 10 085–10 109.

Ehn, J.K., Hwang, B.J., Galley, R., & Barber, D.G. (2007) Investigations of newly formed sea ice in the Cape Bathurst polynya: 1. Structural, physical, and optical properties. *Journal of Geophysical Research*, **112**, C05002s, doi:10.1029/2006JC003702.

Ehn, J.K., Mundy, C.J. & Barber, D.G. (2008) Bio-optical and structural properties inferred from irradiance measurements within the bottommost layers in an Arctic landfast sea ice cover, *Journal of Geophysical Research*, **113**, C03S03, doi:10.1029/2007JC004194.

Ehn, J.K., Mundy, C.J., Barber, D.G., Hop, H., Rossnagel, A. & Stewart, J. (2011) Impact of horizontal spreading on light propagation in melt pond covered seasonal sea ice in the Canadian Arctic. *Journal of Geophysical Research*, **116**, C00G02, doi:10.1029/2010JC006908.

Ehn, J.K., Granskog, M.A., Papakyriakou, T., Galley, R. & Barber, D.G. (2006) Surface albedo observations of Hudson Bay landfast sea ice during the spring melt. *Annals of Glaciology*, **44**, 23–29.

Ehn, J., Granskog, M.A., Reinart, A. & Erm, A. (2004) Optical properties of melting landfast sea ice and underlying seawater in Santala Bay, Gulf of Finland. *Journal of Geophysical Research*, **109**, C09003, doi:10.1029/2003JC002042.

Eisenman, I. & Wettlaufer, J.S. (2009) Nonlinear threshold behavior during the loss of Arctic sea ice. *Proceedings of the National Academy of Science USA*, **106**, 1, 28–32.

Fetterer, F.&Untersteiner, N. (1998) Observations of melt ponds on Arctic sea ice. *Journal of Geophysical Research*, **103**, 24 821–24 835.

Frederick, J.E. & Lubin, D. (1988) The budget of biologically active ultraviolet radiation in the earth-atmosphere system. *Journal of Geophysical Research*, **93**, 3825–3832.

Frey, K.E., Perovich, D.K. & Light, B. (2011), The spatial distribution of solar radiation under a melting Arctic sea ice cover. *Geophysical Research Letters*, **38**, L22501, doi:10.1029/2011GL049421

Gilbert, G.D. & Buntzen, R.R. (1986) In-situ measurements of the optical properties of Arctic sea ice. *Proceedings of SPIE Ocean Optics VIII*, **637**, 252–263.

Gow, A.J. & Tucker III., W.B. (1990) Sea ice in the polar regions. *In: Polar Oceanography, Part A: Physical Science* (Ed. W.O. Smith), pp. 47–122. Academic Press, San Diego.

Grenfell, T.C. (1979) The effects of ice thickness on the exchange of solar radiation over the polar oceans. *Journal of Glaciology*, **22**, 305–320.

Grenfell, T.C. (1983) A theoretical model of the optical properties of sea ice in the visible and near infrared. *Journal of Geophysical Research*, **88**, 9723–9735.

Grenfell, T.C. (1991) Radiative transfer model for sea ice with vertical structure variations. *Journal of Geophysical Research*, **96**, 16991–17001.

Grenfell, T.C. & Hedrick, D. (1983) Scattering of visible and near infrared radiation by NaCl ice and glacier ice. *Cold Regions Science and Technology*, **8**, 119–127.

Grenfell, T.C. & Maykut, G.A. (1977) The optical properties of ice and snow in the Arctic Basin. *Journal of Glaciology*, **18**, 445–463.

Grenfell, T.C. & Perovich, D.K. (1981) Radiation absorption coefficients of polycrystalline ice from 400–1400 nm. *Journal of Geophysical Research*, **86**, 7447–7450.

Grenfell, T.C. & Perovich, D.K. (2004) The seasonal evolution of albedo in a snow-ice-land-ocean environment. *Journal of Geophysical Research Oceans*, **109**, No. C1, C01001 10.1029/2003JC001866.

Grenfell, T.C. & Perovich, D.K. (2008) Incident spectral irradiance in the Arctic Basin during the summer and fall. *Journal of Geophysical Research*, **113**, D12117, doi:10.1029/2007JD009418.

Grenfell, T.C., Warren, S.G. & Mullen, P.C. (1994) Reflection of solar radiation by the Antarctic snow surface at ultraviolet, visible, and near-infrared wavelengths. *Journal of Geophysical Research*, **99**(D9), 18669–18684.

Hamre, B., Winther, J.-G., Gerland, S., Stamnes, J.J. & Stamnes, K. (2004) Modeled and measured optical transmittance of snow-covered first-year sea ice in Kongsfjorden, Svalbard. *Journal of Geophysical Research Oceans*, **109**(C10). doi:10.1029/2003JC001926

Hanesiak, J.M., Barber, D.G., De Abreu, R.A. & Yackel, J.J. (2001) Local and regional albedo observations of arctic first-year sea ice during melt ponding, *Journal of Geophysical Research*, **106**, 1005–1016.

Henyey, L.G. & Greenstein, J.L. (1941) Diffuse radiation in the galaxy. *The Astrophysical Journal*, **93**, 70–83.

Holland, M. & Bitz, C.M. (2003) Polar amplification of climate change in coupled models. *Climate Dynamics*, **21**, 221–232.

Holland, M.M., Bitz, C.M. & Tremblay, B. (2006). Future abrupt reductions in the summer Arctic sea ice. *Geophysical Research Letters*, **33**, L23503, doi:10.1029/2006GL028024.

Holland, M.M., Bailey, D.A., Briegleb, B.P., Light, B. & Hunke, E. (2012) Improved Sea Ice Shortwave Radiation Physics in CCSM4: The Impact of Melt Ponds and Aerosols on Arctic Sea Ice. *Journal of Climate*, **25**, 1413–1430.

Hudson, S.R. (2011), Estimating the global radiative impact of the sea ice–albedo feedback in the Arctic. *Journal of Geophysical Research*, **116**, D16102, doi:10.1029/2011JD015804.

Jin, Z., Stamnes, K. & Weeks, W.F. (1994) The effect of sea ice on the solar energy budget in the atmosphere-sea ice-ocean system: A model study. *Journal of Geophysical Research*, **99** (C12), 25281–25294.

Kaempfer, T.U., Hopkins, M.A. & Perovich, D.K. (2007), A three-dimensional microstructure-based photon-tracking model of radiative transfer in snow. *Journal of Geophysical Research*, **112**, D24113, doi:10.1029/2006JD008239

Kwok, R. (2014) Declassified high-resolution visible imagery for Arctic sea ice investigations: An overview. *Remote Sensing of Environment*, **142**, 44–56.

Laine, V. (2004) Arctic sea ice regional albedo variability and trends 1982–1998. *Journal of Geophysical Research*, **109**, C06027, doi:10.1029/2003JC001818.

Lazzara, L., Nardello, I., Ermanni, C., Mangoni, O. & Saggiomo, V. (2007) Light environment and seasonal dynamics of microalgae in the annual sea ice at Terra Nova Bay, Ross Sea, Antarctica. *Antarctic Science*, **19**, 83–92.

Light, B. (2010) Theoretical and observational techniques for estimating light scattering in first-year Arctic sea ice. In: *Light Scattering Reviews, vol. 5* (Ed. A. Kokhanovsky), pp. 331–392. Springer-Praxis, Berlin.

Light, B., G.A. Maykut, & T.C. Grenfell (2003), A two-dimensional Monte Carlo model of radiative transfer in sea ice, *Journal of Geophysical Research*, **108**, 3219, doi:10.1029/2002JC001513, C7.

Light, B., Maykut, G.A. & Grenfell, T.C. (2004) A temperature-dependent, structural-optical model of first-year sea ice. *Journal of Geophysical Research*, **109**, C06013, doi:10.1029/2003JC002164.

Light, B., Grenfell, T.C. & Perovich, D.K. (2008) Transmission and absorption of solar radiation by Arctic sea ice during the melt season. *Journal of Geophysical Research*, **113**, C03023, doi:10.1029/2006JC003977.

Liou, K.N. (1973) A numerical experiment on Chandrasekhar's discrete ordinates method for radiative transfer: Applications to cloudy and hazy atmospheres. *Journal of Atmospheric Science*, **30**, 1303–1326.

Liou, K.N. (1974) Analytic two-stream and four-stream solutions for radiative transfer. *Journal of Atmospheric Science*, **31**, 1473–1475.

Lubin, D., Frederick, J.E. & Krueger, A.J. (1989) The ultraviolet radiation environment of Antarctica: McMurdo Station during September–October 1987. *Journal of Geophysical Research*, **94**, 8491–8496.

Markus, T., Stroeve, J.C. & Miller, J. (2009) Recent changes in Arctic sea ice melt onset, freezeup, and melt season length. *Journal of Geophysical Research*, **114**, C12024, doi:10.1029/2009JC005436.

Maslanik, J.A., Fowler, C., Stroeve, J., Drobot, S., Zwally, J., Yi, D. & Emery, W. (2007) A younger, thinner Arctic ice cover: Increased potential for rapid, extensive sea-ice loss. *Geophysical Research Letters*, **34**, L24501, doi:10.1029/2007GL032043.

Maslanik, J., Stroeve, J., Fowler, C. & Emery, W. (2011) Distribution and trends in Arctic sea ice age through spring 2011. *Geophysical Research Letters*, **38**, L13502, doi:10.1029/2011GL047735.

Maykut, G.A. (1986) The surface heat and mass balance. In: *The Geophysics of Sea Ice* (Ed. N. Untersteiner), pp. 395–464. Plenum Press, New York.

Maykut, G.A., & Perovich, D.K. (1987) The role of shortwave radiation in the summer decay of a sea ice cover. *Journal of Geophysical Research*, **92**, 7032–7044.

Maykut, G.A. & Light, B. (1995) Refractive-index measurements in freezing sea-ice and sodium chloride brines. *Applied Optics*, **34**, 950–961.

Maykut, G.A. & Untersteiner, N. (1971) Some results from a time dependent, thermodynamic model of sea ice. *Journal of Geophysical Research*, **76**, 1550–1575.

Mobley, C.D. (1994) *Light and Water, Radiative Transfer in Natural Waters*. Academic Press, San Diego.

Mundy, C.J., Ehn, J.K., Barber, D.G. & Michel, C. (2007) Influence of snow cover and algae on the spectral dependence of transmitted irradiance through Arctic landfast first-year sea ice. *Journal of Geophysical Research*, **112**, C03007, doi:10.1029/2006JC003683.

Mundy, C.J., Gosselin, M., Ehn, J. et al. (2009) Contribution of under-ice primary production to an ice-edge upwelling phytoplankton bloom in the Canadian Beaufort Sea. *Geophysical Research Letters*, **36**, L17601, doi:10.1029/2009gl038837.

Nghiem, S.V., Rigor, I.G., Perovich, D.K., Clemente-Colón, P., Weatherly, J.W. & Neumann, G. (2007), Rapid reduction of Arctic perennial sea ice. *Geophysical Research Letters*, **34**, L19504, doi:10.1029/2007GL031138.

Nicolaus, M., Petrich, C., Hudson, S.R. & Granskog, M. (2013) Variability of light transmission through Arctic land-fast sea ice during spring. *The Cryosphere*, **7**, 977–986.

Nicolaus, M., Katlein, C., Maslanik, J. & Hendricks, S. (2012) Changes in Arctic sea ice result in increasing light transmittance and absorption. *Geophysical Research Letters*, **39**, L24501, doi:10.1029/2012GL053738.

Nicolaus, M., Hudson, S.R., Gerland, S. & Munderloh, K. (2010) A modern concept for autonomous and continuous measurements of spectral albedo and transmittance of sea ice. *Cold Regions Science and Techology*, **62**, 14–28, 2010.

Nicolaus, M., Gerland, S., Hudson, S.R., Hanson, S., Haapala, J. & Perovich, D.K. (2010)Seasonality of spectral albedo and transmittance as observed in the Arctic Transpolar Drift in 2007. *Journal of Geophysical Research: Oceans*, **115**(C11011), doi:10.1029/2009JC006074.

Pedersen, C.A., Roeckner, E., Lüthje, M. and Winther, J.-G. (2009) A new sea ice albedo scheme including melt ponds for ECHAM5 general circulation model. *Journal of Geophysical Research*, **114**, D08101, doi:10.1029/2008JD010440.

Pegau, W.S. & Paulson CA. (2001) The albedo of Arctic leads in summer. *Annals of Glaciology*, **33**, 221–224.

Perovich, D.K. (1990) Theoretical estimates of light reflection and transmission by spatially complex and temporally varying sea ice covers. *Journal of Geophysical Research*, **95**, 9557–9567.

Perovich, D.K. (1991) Seasonal changes in sea ice optical properties during fall freeze-up. *Cold Regions Science and Technology*, **19**, 261–273.

Perovich, D.K. (1993) A theoretical model of ultraviolet light transmission through Antarctic sea ice, *Journal of Geophysical Research*, **98**, 22 579–22 587.

Perovich, D.K. (2005) On the aggregate-scale partitioning of solar radiation in Arctic Sea Ice during the SHEBA field experiment. *Journal of Geophysical Research*, **110**, C03002, 10.1029/2004JC002512.

Perovich, D.K. (2009) Sea ice optics measurements. In: *Field Techniques in Sea Ice Research*, pp. 215–230. University of Alaska Press, Fairbanks..

Perovich, D.K. & Govoni, J.W. (1991) Absorption coefficients of ice from 250 to 400 nm. *Geophysical Research Letters*, **18**, 1233–1235.

Perovich, D.K. & Grenfell, T.C. (1981) Laboratory studies of the optical properties of young sea ice. *Journal of Glaciology*, **27**, 331–346, 1981.

Perovich, D.K. & Grenfell, T.C. (1982) A theoretical model of radiative transfer in young sea ice. *Journal of Glaciology*, **28**, 341–357.

Perovich, D.K.&Polashenski, C. (2012) Albedo evolution of seasonal Arctic sea ice. *Geophysical Research Letters*, **39**, L08501, doi:10.1029/2012GL051432.

Perovich, D.K., Grenfell, T.C., Light B. & Hobbs, P.V. (2002) Seasonal evolution of the albedo of multiyear Arctic sea ice *Journal of Geophysical Research*, 107, doi:10.1029/2000JC000438.

Perovich, D.K., Light, B., Eicken, H., Jones, K.F., Runciman, K. & Nghiem, S.V. (2007) Increasing solar heating of the Arctic Ocean and adjacent seas, 1979–2005. Attribution and role in the ice-albedo feedback. *Geophysical Research Letters*, **34**, (L19505), doi:10.1029/2007GL031480.

Perovich, D.K., Richter-Menge, J.A., Jones, K.F. & Light, B. (2008) Sunlight, water, and ice: Extreme Arctic sea ice melt during the summer of 2007.*Geophysical Research Letters*, **35**, L11501, doi:10.1029/2008GL034007.

Perovich, D.K., Richter-Menge, J.A., Jones, K.F., Light, B., Elder, B.C., Polashenski, C.M., LaRoche, D., Markus, T. &Lindsay, R. (2011) Arctic sea ice melt in 2008 and the role of solar heating. *Annals of Glaciology*, **52**, 355–359.

Polashenski, C., Perovich, D. and Z. Courville (2012), The mechanisms of sea ice melt pond formation and evolution. *Journal of Geophysical Research*, **117**, C01001, doi:10.1029/2011JC007231.

Rigor, I.G. &Wallace, J.M. (2004) Variations in the age of Arctic sea ice and summer sea-ice extent. *Geophysical Research Letters*, **31**, (L09401): doi:10.1029/2004GL019492.

Riihela, R., Manninen, T. & Laine, V. (2013) Observed changes in the albedo of the Arctic sea ice zone for the periodn 1982 – 2009. *Nature Climate Change*, **3**, 895–898.

Schlosser, E. (1988) Optical studies of Antarctic sea ice.*Cold Regions Science and Technology*, **15**, 289–293, 1988.

Serreze M.C., Holland, M.M. &Stroeve, J. (2007) Perspectives on the Arctic's shrinking sea-ice cover. *Science*, **315**, 1533–1536.

Smith, R.C. (1989) Ozone, middle ultraviolet radiation and the aquatic environment. *Photochemistry and Photobiology*, **50**, 459–468.

Smith, R.C. & Baker, K.S. (1981)Optical properties of the clearest natural waters (200–800 nm). *Applied Optics*, **20**, 177–184.

Smith, R.C., Prezelin, B.B., Baker, K.S. et al. (1992) Ozone depletion: Ultraviolet radiation and phytoplankton biology in Antarctic waters. *Science*, **255**, 952–959.

Steele, M. (1992) Sea ice melting and floe geometry in a simple ice-ocean model. *Journal of Geophysical Research*, **97**, 17729–17738.

Steele, M., Zhang, J., & Ermold, W. (2010) Mechanisms of summertime upper Arctic Ocean warming and the effect on sea ice melt. *Journal of Geophysical Research*, **115**, C11004, doi:10.1029/2009JC005849.

Stroeve, J., Holland, M.M., Meier, W., Scambos T. & Serreze, M. (2007) Arctic sea ice decline: Faster than forecast. *Geophysical Research Letters*, **34**, (L09501), doi:10.1029/2007GL029703.

Stroeve, J.C., Serreze, M.C., Holland, M.M., Kay, J.E., Maslanik, J. & Barrett, A.P. (2012) The Arctic's rapidly shrinking sea ice cover: a research synthesis. *Climatic Change*, **110**, 1005–1027.

Sturm, M., Holmgren, J. & Perovich, D. (2002) The winter snow cover on the sea ice of the Arctic Ocean at SHEBA: temporal evolution and spatial variability. *Journal of Geophysical Research*, **107** (C10), doi 10.1029/2000JC000400.

Thomas, C.W. (1963) On the transfer of visible radiation through sea ice and snow. *Journal of Glaciology*, **4**, 481–484.

Trodahl, H.J. & Buckley, R.G. (1989) Ultraviolet levels under sea ice during the Antarctic spring. *Science*, **245**, 194–195.

Trodahl, H.J. & Buckley, R.G. (1990) Enhanced ultraviolet transmission of Antarctic sea ice during the austral spring. *Geophysical Research Letters*, **17**, 2177–2179.

Trodahl, H.J. & Buckley, R.G. & Brown, S. (1987) Diffusive transport of light in sea ice. *Applied Optics*, **26**, 3005–3011.

Tsay, S. & Stamnes, K. (1992) Ultraviolet radiation in the Arctic: The impact of potential ozone depletions and cloud effects. *Journal of Geophysical Research*, **97**, 7829–7840.

Tyler, J.E.&Smith, R.C. (1970) *Measurements of Spectral Radiance Underwater*. Gordon and Breach, New York.

Untersteiner, N. (1961) On the mass and het budget of Arctic sea ice. *Archives for Meteorology, Geophysics and Bioclimatology Series A*, **12**, 151–182.

Uttal, T. et al. (2002) Surface heat budget of the Arctic Ocean. *Bulletin of the American Meteorological Society*, **83**, 255–275.

van de Hulst, H.C. (1981) *Light Scattering by Small Particles*. Dover, New York.

Wang, C., Granskog, M.A., Gerland, S. et al. (2014)Autonomous observations of solar energy partitioning in first-year sea ice in the Arctic Basin. *Journal of Geophysical Research: Oceans*, **119**: 2066–2080.

Wang, C., Granskog, M.A., Gerland, S. et al. (2014) Autonomous observations of solar energy partitioning in first-year sea ice in the Arctic Basin. *Journal of Geophysical Research: Oceans*, **119**, 2066–2080.

Warren, S.G. (1982) Optical properties of snow. *Review of Geophysics and Space Physics*, **20**, 67–89.

Warren, S.G., Brandt, R.E. & Grenfell, T.C. (2006) Visible and near-ultraviolet absorption spectrum of ice from transmission of solar radiation into snow. *Applied Optics*, **45**, 5320–5334.

Webster, M.A., Rigor, I.G., Nghiem, S.V. et al. (2014) Interdecadal changes in snow depth on Arctic sea ice. *Journal of Geophysical Research*, **119**, 5395–5406.

Weeks, W.F., & Ackley, S.F. (1982). The growth, structure, and photosynthesis of sea ice. *Cold Regions Research & Engineering Laboratory Monographs*, **82**(1), 1–131.

Weller, G. & Schwerdtfeger, P. (1967) Radiation penetration in antarctic plateau and sea ice. In: *Polar Meteorology, World Meteorological Organization Technical Note*, **87**, 120–141.

Weller, G., (1972) Radiation flux investigations. *AIDJEX Bulletin*, **14**, 28–30.

Winther, J.-G., Edvardsen, K., Gerland, S. & Hamre, B. (2004) Surface reflectance of sea ice and under-ice irradiance in Kongsfjorden, Svalbard. *Polar Research*, **23**, 115–118.

Winton, M. (2006) Amplified Arctic climate change: What does surface albedo feedback have to do with it? *Geophysical Research Letters*, **33**, L03701, doi:10.1029/2005GL025244.

Winton, M. (2008) Sea Ice–Albedo Feedback and Nonlinear Arctic Climate Change. In *Arctic Sea Ice Decline: Observations, Projections, Mechanisms, and Implications* (Eds. E.T. DeWeaver, C.M. Bitz & L.-B. Tremblay), American Geophysical Union, Washington, DC.

Wiscombe, W.J. & Warren, S.G. (1980) A model for the spectral albedo of snow, 1, Pure snow, *Journal of Atmospheric Science*, **37**, 2712–2733.

Zaneveld, J.R.V. (1964) The Transparency of Sea Ice in the Visible Region. MSc Thesis, Massachusetts Institute of Technology.

Zeebe, R.E., Eicken, H., Robinson, D.H., Wolf-Gladrow, D. & Dieckmann, G.S. (1996) Modeling the heating and melting of sea ice through light absorption by microalgae. *Journal of Geophysical Research*, **101**(C1), 1163–1181.

CHAPTER 5

The sea ice–ocean boundary layer

Miles G. McPhee

McPhee Research Company, Naches, WA, USA

5.1 Introduction

Sea ice exerts profound influence on air–sea interaction by effectively insulating the upper ocean from the cold polar atmosphere, thus reducing both outgoing long-wave radiation and sensible heat exchange, and by drastically changing the surface albedo, inhibiting the oceanic absorption of shortwave radiation. There is often shear (relative motion) between the ice and underlying ocean, either in response to wind forcing in the extensive ice packs of polar oceans, or from tidally induced currents or other circulation under shorefast ice. The shear induces a boundary layer beneath the ice in which transfers of momentum, heat and salt are controlled by turbulent exchange processes. This ice–ocean boundary layer (IOBL) shares many characteristics with open ocean boundary layers, except that the upper boundary conditions are imposed by a solid–liquid interface instead of a gas–liquid interface, and as a consequence surface gravity waves play a minor role in IOBL dynamics, except in marginal ice zones or with low ice concentrations.

In the open ocean, vertical displacements and orbital velocities from surface waves present a formidable obstacle to direct measurement of turbulent fluxes. In contrast, drifting pack ice presents a stable measurement platform, often moving with the maximum velocity in the water column, from which it is relatively easy to deploy clusters of instruments through the entire extent of a rotating planetary boundary layer. Since the 1970s, the 'laboratory' aspects of drifting sea ice have been exploited to measure the small deviatory changes in velocity and scalar properties necessary for covariance estimates of turbulent (Reynolds) fluxes. Coupled with modern techniques for measuring profiles of temperature, salinity and velocity (by acoustic profiler) on a more or less continuous basis for weeks or months at a time, this has provided us an unprecedented view of how the IOBL works, and much insight into planetary boundary layers in general.

Ice-covered polar oceans are cold, often quite near their salinity-determined freezing temperatures, with small temperature contrast between the upper ocean and the base of the ice cover. Compared with instantaneous radiative and even turbulent convective exchanges at the upper surface of the ice, instantaneous ocean (basal) heat flux magnitudes are generally modest, yet Maykut and Untersteiner (1971) found in their thermodynamic sea ice model that about 2 W m^{-2} heat flux from the ocean was required to maintain a steady-state ice thickness. Nevertheless, during the initial planning stages for the year-long Surface Heat Budget of the Arctic (SHEBA) project, some participants questioned the necessity of measuring ice–ocean exchanges at all. As it turned out, our measurements over the SHEBA year demonstrated much variability in basal heat flux, with the annual average of about 7.2 W m^{-2} (McPhee, 2008b; Shaw et al., 2009), coinciding closely with the heat absorbed by a net ice loss of about 75 cm at the SHEBA station reported by Perovich et al. (2003). Clearly, it is important to understand exchanges between the ice and ocean, particularly in a time when the state of the ice pack appears to be rapidly changing.

There are two main themes in this chapter. Section 5.2 investigates the structure of the IOBL, mainly from the standpoint of scales that govern turbulent transfer with emphasis on the drag relationship required for estimating interface transfers of heat and salt as well

---

Sea Ice, Third Edition. Edited by David N. Thomas.
© 2017 John Wiley & Sons, Ltd. Published 2017 by John Wiley & Sons, Ltd.

as momentum. Section 5.3 deals with the immediate ice–ocean interface from the perspective of the ocean. It emphasizes the role played by salt in controlling transfer rates through an infinitesimal control volume following the interface, and formulates boundary conditions for underlying turbulent boundary layer.

5.2 Turbulent exchange in the IOBL

Sea ice (as well as floating glacial ice) interacts with the ocean through a turbulent boundary layer influenced directly by rotation. During the 1893–1896 *Fram* expedition, F. Nansen noted that the vessel consistently drifted in a direction 20–40° clockwise from the surface wind, which he surmised was an obvious manifestation of the Earth's rotation. A young Swedish researcher, V.W. Ekman (1905), working with Nansen and V. Bjerkness, solved the fluid equations with friction in a rotating reference frame, in the process discovering an elegantly simple spiral structure for velocity in the upper ocean forced by stress at the surface. In his truly remarkable paper, Ekman not only deduced the velocity spiral that now bears his name, but also foretold the presence of inertial oscillations, and laid the foundation for understanding coastal upwelling and storm surges.

Despite ample indirect evidence for Coriolis deflection of winds in the atmospheric boundary layer gathered during the first half of the 20th century, the first unequivocal example of an Ekman spiral in nature was published by Hunkins (1966) from a composite of current profiles measured under pack ice over a 2-month period at Arctic Drift Station Alpha during the International Geophysical Year in 1958. Subsequently, numerous measurements made from drifting sea ice have demonstrated spiral-like IOBL structures, in both velocity (e.g. McPhee & Smith, 1976) and turbulent Reynolds stress (McPhee & Martinson, 1994). In spite of the significant logistical barriers to operating in polar oceans, the unique capability for measuring from a stable platform at the surface, without the complicating impact of surface gravity waves (with significant orbital velocities and platform motion), has provided much insight into how turbulent exchange in rotating boundary layers occurs. In this respect, the IOBL studies carry on a tradition founded by Nansen and Ekman.

5.2.1 Basic equations

Equations most pertinent to ice–ocean interaction are derived from basic conservation equations for momentum, heat and salt, and are well presented in standard oceanographic texts (Gill, 1982; Pedlosky, 1987). Derivations specifically for the IOBL are found in McPhee (2008b; Chapter 2). For fully developed turbulence in natural planetary boundary layers (i.e. in a rotating reference frame attached to Earth), the Bousinesq form of the momentum equation (ignoring molecular diffusion but not friction) may be written

$$\frac{\partial \boldsymbol{u}}{\partial t} + \boldsymbol{u} \cdot \nabla \boldsymbol{u} + f \boldsymbol{k} \times \boldsymbol{u} = -\nabla p/\rho - g\frac{\rho'}{\rho}\boldsymbol{k} + \nabla \cdot \underset{\sim}{\boldsymbol{\tau}} \quad (5.1)$$

where $\boldsymbol{u}$ is vector velocity of the *mean* flow (i.e. slowly varying in comparison with turbulent fluctuations); $f = 2\omega \sin \phi$ is the Coriolis parameter (ϕ is latitude, $\omega = 7.292 \times 10^{-5}$ s^{-1} is Earth's angular rotation speed); ∇p is the large-scale pressure gradient; ρ' is a small density perturbation from the background density; and g is the acceleration due to gravity. The last term in equation (5.1) is not viscous stress divergence (which would normally be expressed as $\nu \nabla^2 \boldsymbol{u}$), but rather the divergence of the Reynolds stress tensor, which follows from applying the Reynolds decomposition ($\boldsymbol{U} = \boldsymbol{u} + \boldsymbol{u}'$ where $\boldsymbol{U}$ is total velocity) to the advective term $\boldsymbol{U} \cdot \nabla \boldsymbol{U}$. The components of Reynolds stress tensor are:

$$\tau_{ij} = -\langle u_i' u_j' \rangle$$

Summing the trace of $\underset{\sim}{\tau}$, ($\langle u_i' u_i' \rangle$ repeated indices imply summation) yields twice the turbulent kinetic energy per unit mass. Usually, the main interest for turbulent exchange is the vertical derivative of a horizontal traction vector:

$$\begin{aligned}\boldsymbol{\tau} &= -\langle u'w' \rangle \boldsymbol{e}_x - \langle v'w' \rangle \boldsymbol{e}_y \\ &= -\langle u'w' \rangle - i\langle v'w' \rangle\end{aligned} \quad (5.2)$$

where the second representation expresses the two-dimensional vector as a complex number with the x-component along the real axis and the y-component along the imaginary axis.

The pressure gradient term in equation (5.1) provides a definition of geostrophic current

$$f\boldsymbol{k} \times \boldsymbol{v}_g = -g\nabla \eta \quad (5.3)$$

where η is sea-surface elevation. The same pressure gradient term acts on drifting sea ice, so in a state with no other forcing (e.g. no wind stress or ice stress gradient), the upper ocean and ice will drift with u_g, and there will be no shear to generate turbulence. It is thus often convenient to express the boundary layer equations in terms of horizontal velocity relative to geostrophic flow, eliminating the large-scale pressure gradient term:

$$\frac{\partial u}{\partial t} + u \cdot \nabla u + f k \times u = \nabla \cdot \underset{\sim}{\tau} \quad (5.4)$$

Similar conservation equations may be derived for scalar IOBL properties. For heat (temperature):

$$\frac{\partial T}{\partial t} + u \cdot \nabla T = Q^H/(\rho c_p) \quad (5.5)$$

where Q^H is a heat source within the IOBL (e.g. solar heating), and c_p is the specific heat of seawater. For temperatures near freezing, c_p is close to 4 kJ kg^{-1} K^{-1} (see Gill, 1982, table A3.1). A similar conservation equation for salinity is:

$$\frac{\partial S}{\partial t} + u \cdot \nabla S = Q^S/\rho \quad (5.6)$$

where for generality a source term for salinity is included (a possible source might be from nucleation of frazil crystals).

5.2.2 Impacts of rotation

If IOBL flow is steady and horizontally homogeneous, equation (5.4) is simply:

$$if u = \frac{\partial \tau}{\partial z} \quad (5.7)$$

where $u = u + iv$ in complex notation. At some level near the far extent of the boundary layer, turbulent stress approaches zero, so integrating equation (5.7) from that level to the surface:

$$if \int_{Z_{bl}}^{0} u \, dz = if M = \tau_0 \quad (5.8)$$

where M is the vector volume transport in the IOBL and τ_0 is kinematic stress at the boundary. Thus in a steady state, average transport is at right angles to the stress (clockwise in the northern hemisphere), regardless of the details of turbulence in the IOBL. A shallow layer requires higher mean velocity than a deep layer to effect the same transport, which places an important constraint on IOBL scales.

By vector manipulation of equation (5.8) with the addition of the continuity equation, it follows that:

$$\nabla \cdot M = w_{EP} \approx \frac{1}{f} \nabla \times \tau_0 \quad (5.9)$$

where w_{EP} is vertical (Ekman pumping) velocity at the base of the mixed layer relative to the surface. Over the past few decades, late summer (August–October) wind stress curl over the Canada Basin (western Arctic) has become increasingly negative, resulting in convergence of near surface water (McPhee, 2013). Freshwater from more intense ice melt and continental run-off is herded by the stress curl mechanism toward the center of the Beaufort Gyre, resulting in stronger stratification (shallower mixed layers) and increased sea-surface elevation gradients. Disappearance of perennial pack ice in the Canada Basin (Maslanik et al., 2011; Nghiem et al., 2007) appears to have been strongly influenced by more intense currents in the Beaufort Gyre, which in turn are linked directly to the stress curl via equation (5.9) (McPhee, 2013).

Besides relating stress and volume transport, equation (5.8) reveals little about the structure of currents and momentum flux in the IOBL. Ekman (1905) addressed equation (5.7) by positing that stress and velocity were related by an 'eddy viscosity' that behaved like kinematic viscosity (which depends on the molecular velocity times mean free path) except at much larger scales. His elegant solution revealed a spiral in current structure, in which surface velocity veered 45° to the right of surface stress, seemingly explaining Nansen's drift observations. It is sometimes overlooked that Ekman considered eddy viscosity to be constant only with respect to the vertical dimension. Based partly on the setup of coastal currents during storms, he postulated that it should vary as the square of wind speed (i.e. the surface stress) and suggested quantitative values not far different from what we have subsequently measured with modern instrumentation.

Boundary layer currents measured under drifting pack ice often exhibit what may be interpreted as Ekman spirals. An example from a storm during the Arctic Ice Dynamics Joint Experiment (AIDJEX) Pilot Study in 1972 (Figure 5.1a) shows current vectors at several levels drawn with respect to the current measured at 32 m, where the absolute current was small (McPhee & Smith, 1976). The reference frame was chosen so that the stress acting on the IOBL was in the negative x direction. Profiles in Figure 5.1(b) show that although the surface velocity was largely dominated by the shear between the interface and 2 m level, the integrated transport in the direction of

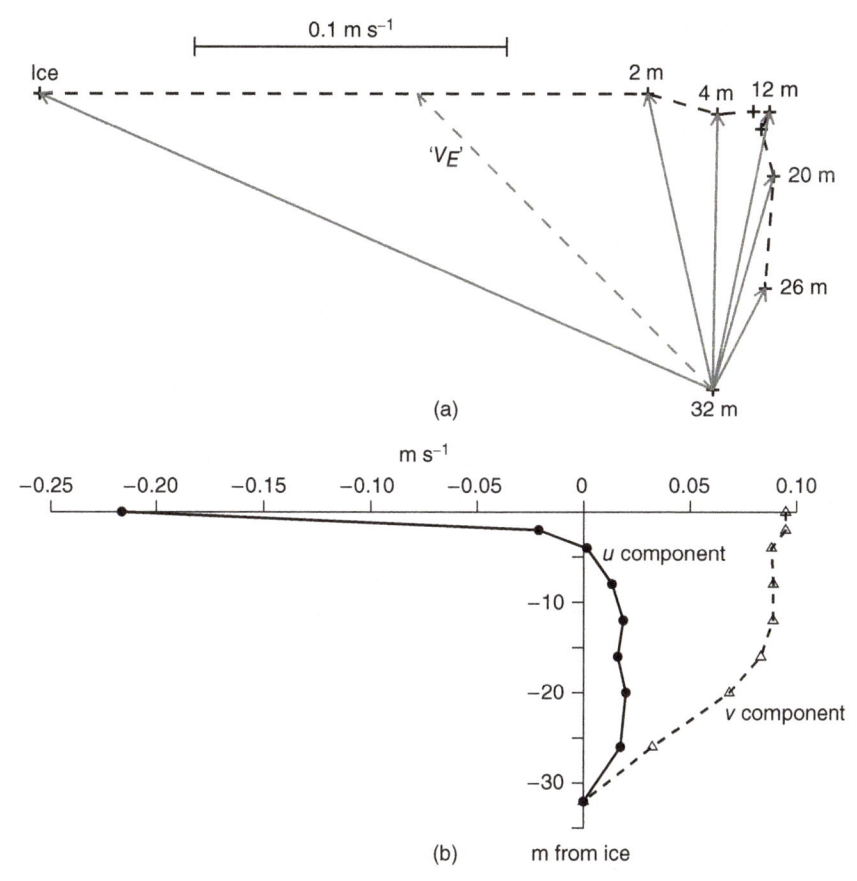

Figure 5.1 (a) Plan view of currents measured for 5 hours during a storm at the 1972 Arctic Ice Dynamics Joint Experiment (AIDJEX) Pilot Study site. All currents are shown relative to the current measured at 32 m. The dashed vector is drawn where the velocity is 45° clockwise from the surface stress. (b) Current profiles with the x-axis aligned antiparallel to the surface stress.

stress was near zero. The dashed vector labelled 'V_E' lies 45° clockwise from surface stress and falls within the upper 2 m. Turbulent stress at that level is nearly the same as at the interface, so this can be taken as the upper limit of the Ekman spiral. Basic aspects of Ekman turning are nearly always present to some degree in our measurements from drifting ice. A further example, this one from the southern hemisphere (Figure 5.2), was constructed from hundreds of hours of acoustic Doppler current profiler current data obtained during the Ice Station Polarstern (ISPOL) project in the western Weddell Sea (McPhee, 2008a). The vectors shown were formed by non-dimensionalizing (by complex division) 3 hour samples of currents at 2 m intervals from 10 to 28 m by the current measured at 30 m, then averaging all of the non-dimensional profiles.

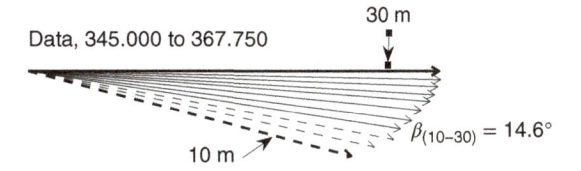

Figure 5.2 Average non-dimensional current hodograph (plan view) of complex currents measured relative to drifting ice, sampled every 2 m from 10 to 28 m, after dividing by the current at 30 m. Source: Adapted from McPhee, 2008a. Reproduced with permission of Elsevier.

5.2.3 Inertial oscillation

Ekman (1905, with credit to Fredholm) also presented a solution to the time-dependent problem, showing circular currents oscillating with a period of a 'half-pendulum day' ($2\pi/f$) about the mean currents of the steady solution. A much simplified illustration of these circular currents follows from the time-dependent version of the horizontally homogeneous volume transport equation:

$$\frac{\partial M}{\partial t} + ifM = \tau_0 \qquad (5.10)$$

If the upper ocean initially at rest is subjected to an impulsive stress in the imaginary (y) direction, $i\tau_0$, at time $t = 0$, the complex solution to equation (5.10) is:

$$M = \frac{\tau_0}{f}(1 - e^{-ift})$$

tracing a circle in the complex plane in one inertial period (half-pendulum day) about the steady-state balance $M_{ss} = \tau_0/f$. Because there is no friction in the system, it continues to oscillate about M_{ss}, but never reaches the steady-state value. While this might seem patently unrealistic, it is useful to consider some numbers. Kinematic surface stress from a fast-moving atmospheric system near the North Pole might have a magnitude $\tau_0 = 2 \times 10^{-4}$ m^{2}s^{-2} with maximum volume transport (at $t = 6, 18, 30$ h, etc.) of about 2.75 m^2 s^{-1}. If the summer mixed layer was 25 m thick, the depth-averaged velocity would be around 11 cm s^{-1}.

Satellite tracking of ice motion often shows 'scallops' in drift trajectories, indicating cycloidal motion, particularly in summer when mixed layers tend to be shallow and internal ice forces small. An example from an unmanned buoy, initially deployed near the North Pole (Figure 5.3), shows that for about four inertial periods beginning at time 270.0 (year day 2002), the buoy trajectory can be reasonably reproduced by integrating the simple expression for velocity, comprising a mean drift plus an inertial oscillation as listed in the figure. The mean ice velocity is dominated by shear between the ice cover and upper mixed layer in the direction of surface stress, so the actual velocity in the mixed layer was probably not much different from the highly idealized example above.

Given a time series of drift positions, complex demodulation provides a useful means of separating inertial (or tidal) components of ice drift from the underlying mean motion. The technique, described in detail by McPhee (1988, 2008b), involves fitting the observed positions to basis functions comprising preferred inertial and/or tidal frequencies, for example:

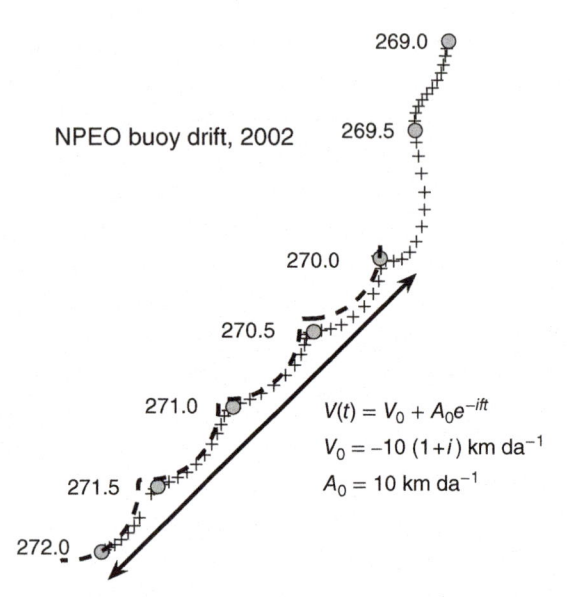

$$X(t) = X_0 + V_0 t + (i/f)[S_{cw}(e^{-ift} - 1) + S_{ccw}(1 - e^{ift})] + \ldots$$
$$(i/\omega)[D_{cw}(e^{-i\omega t} - 1) + D_{ccw}(1 - e^{i\omega t})] \qquad (5.11)$$

Figure 5.3 Track of an unmanned buoy from 26 to 29 September 2002. + symbols indicate satellite navigation positions every half hour. The dashed curve is the integral of the simple velocity expression from the initial position at time 270.0.

where ω is the diurnal tidal frequency. The coefficients are determined by least-squares error minimization from position fixes over a time comparable to the longest period. An example, calculated by differentiating equation (5.11) evaluated from SHEBA GPS positions every 3 h, is shown in Figure 5.4. Note that V_0 is an estimate of the 'mean' velocity with inertial and tidal components removed. As the upper ocean typically oscillates in phase with the ice, V_0 is often a better indicator of shear, hence turbulence, in the upper part of the boundary layer than is the total ice velocity. With frequent, accurate navigation fixes, other methods provide good velocity estimates, but the main advantage of complex demodulation is that it recognizes the inherent inertia of the rotating ice/IOBL system, and provides a means of rationally separating inertial and tidal motions from the velocity due to shear in the IOBL.

5.2.4 Turbulence scales and eddy viscosity

The simplest approach to describing and modelling turbulent transport perpendicular to mean flow in boundary layer shear flows is by relating flux of a quantity directly to its mean gradient using a proportionality factor with units length squared over time, suggesting that

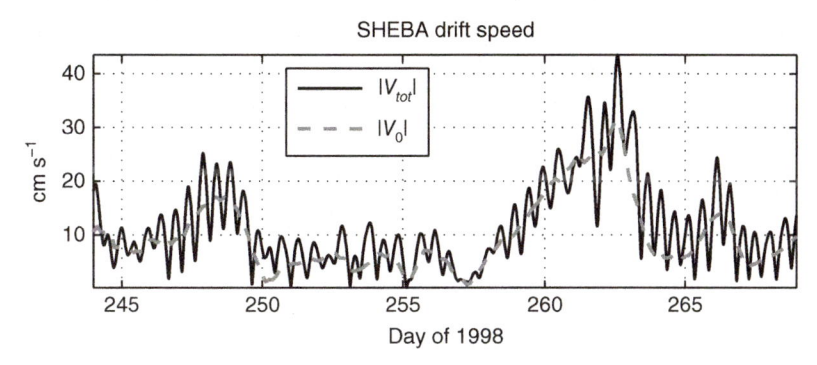

Figure 5.4 Surface Heat Budget of the Arctic (SHEBA) project station drift speed in September 1998. The dashed curve is the 'mean' velocity after removal of inertial and diurnal tidal components via complex demodulation.

the factor (eddy viscosity if the property is momentum flux, and eddy diffusivity for scalar fluxes) is the product of turbulent velocity and length scales, e.g. kinematic stress:

$$\langle w'u' \rangle = -K u_z = -u_\tau \lambda u_z \quad (5.12)$$

where $u_z = \partial u / \partial z$ is shear, u_τ is a turbulent velocity scale, and λ is a length scale expressing the vertical displacement over which the turbulent eddies are effective at exchanging momentum (not necessarily the actual vertical excursion of fluid in the eddies). A major thrust of IOBL turbulence measurements over the past three decades has been aimed at elucidating what controls these scales.

5.2.5 Neutral stratification

The IOBL that forms when there is shear between the ice and underlying ocean is often relatively unaffected by changes in density, when melting or freezing is slow, and when the upper ocean is well mixed in both temperature and salinity to reasonably large depths. These so-called neutral boundary layers (meaning that gravitational forces are negligible compared with friction arising from inertial forces in the fluid) provide a good starting point for considering the general problem of ice–ocean exchange.

It is well established that in the lower tens of metres of the neutral boundary layer of the atmosphere, (i) turbulent stress is relatively constant, equal to boundary stress; (ii) stress and velocity are nearly aligned; and (iii) the velocity profile follows the 'law of the wall':

$$u(z) = \frac{u_{*_0}}{\kappa} \log \frac{z}{z_0} \quad (5.13)$$

where u_{*_0} is friction velocity, a vector defined in terms of the vector kinematic boundary stress, $u_{*_0} u_{*_0} = \tau_0$; z_0 is the hydraulic roughness of the boundary, typically about 1/30th the size of roughness elements on the surface; and κ is von Kàrmàn's constant, ~0.4. An obvious choice of turbulent velocity scale is u_{*_0}, so applying these stipulations to equation (5.12) with $u_\tau = u_{*_0}/(\kappa z)$, it follows from equation (5.13) that $\lambda = \kappa z$ in the region of the flow where (i)–(iii) hold, by convention called the surface layer. In terms of dimensional analysis, this implies that shear is a function of only two independent variables, u_{*_0} and z, with independent dimensions, hence by the Pi theorem (e.g. Barenblatt, 1996), a dimensionless group formed from u_z, z, and u_{*_0} is a constant (equal to $1/k$), and:

$$\phi_m = \frac{\kappa z u_z}{u_{*_0}} = 1 \quad (5.14)$$

In atmospheric terminology, ϕ_m is the dimensionless wind shear in the surface layer.

The surface layer approximation assumes rotation (Coriolis force) is unimportant in formulating the relation between shear and stress. However, from the perspective of the Ekman approach, i.e. that eddy viscosity is invariant with depth, the outer part of the boundary layer behaves quite differently from the surface layer, a view corroborated by measurements like those shown in Figures 5.1 and 5.2 where rotational effects are obvious. If we assume that away from the surface layer, shear depends on f as well as u_{*0} and z, then dimensional analysis leads to

$$\frac{z u_z}{u_{*_0}} = \Phi\left(\frac{fz}{u_{*_0}}\right) = \Phi(\xi) \quad (5.15)$$

introducing the planetary scale, u_{*_0}/f. For the IOBL, with moderate stress ($u_{*_0} = 0.01$ m s^{-1}), the planetary scale is ~70 m at high latitudes. When sea ice moves in a state of free drift (internal stress gradients negligible in the force balance), IOBL boundary stress is usually comparable to wind stress. The ratio of friction velocities then equals the inverse square root of the density ratio, and hence the planetary scale in the atmosphere is roughly 30 times that of the IOBL. This large disparity between the respective boundary-layer scales demonstrates why IOBL measurements are almost always made either beyond the surface layer or at its outer fringes (i.e. IOBL measurements at 2 m would correspond to atmospheric measurements atop a 60 m tower).

A simple way of reconciling the surface and outer layers is to assume that when $\xi = z/(u_{*_0}/f)$ is small (i.e. in the surface layer), λ is governed by the geometric scale, kz, but that in the outer layer, eddies no longer sense their distance from the boundary, and λ is governed instead by some small fraction of the planetary scale, $\Lambda_* u_{*_0}/f$. An extensive series of measurements in the neutrally stratified IOBL has reinforced this theme (McPhee, 1994, 2008b; McPhee & Smith, 1976; McPhee & Martinson, 1994). An approximate value emerging from these studies for the similarity parameter Λ_* is 0.028, which for a typical planetary scale implies that mixing length in the outer layer is about 2 m.

A practical result from comparisons between directly measured stress in the IOBL and variance (energy) wavenumber spectra of the vertical velocity component has been the emergence of a close inverse relationship between λ and wavenumber, k_{max}, at the peak in the area-preserving w spectrum (McPhee & Smith, 1976; McPhee & Martinson, 1994): $\lambda = c_\lambda/k_{max}$, with $c_\lambda \approx 0.85$. With this relationship, it is then possible to estimate the local eddy viscosity:

$$K_{local} = c_\lambda u_* / k_{max}$$

where u_* is obtained either from direct covariance estimates of Reynolds stress or from the w spectrum itself (e.g.McPhee, 2004). With the latter, measuring just vertical velocity spectra at particular levels provides estimates of eddy viscosity. McPhee and Martinson (1994) showed that discrete-level estimates made by this method agreed reasonably well with a bulk estimate of eddy viscosity from an exponential fit of measured Reynolds stress, and with an estimate of thermal eddy diffusivity obtained by dividing average upward heat flux by the negative temperature gradient.

Another application of using the 'measured' mixing length to infer important characteristics of IOBL scales comes from the ISPOL experiment in the western Weddell Sea in 2004–2005 (McPhee, 2008a). Turbulence was measured at several levels, providing covariance estimates of $u_{*(cov)} = |\langle u'w' \rangle + i\langle v'w' \rangle|^{1/2}$ and $\lambda = 0.85/k_{max}$ from averaged w spectra versus $u_{*(cov)}$ (Figure 5.5); we can gauge to what degree the turbulence scales are controlled dynamically by the planetary ($\Lambda_* u_{*_0}/f$) scaling as opposed to the geometric limit of distance from the boundary. Despite typically large scatter in the turbulence data, a least-squares fit through the origin indicates that λ is often controlled by rotation, and is on average relatively predictable.

If λ scales with u_{*_0}/f through most of the IOBL beyond a thin surface layer, the average eddy viscosity is $\overline{K} = \overline{u}_* \lambda_{max} \approx \Lambda_* u_{*_0}^2/(\sqrt{2}f)$, as stress varies roughly linearly from its surface value to zero near the base of the IOBL. Thus the non-dimensional mean eddy viscosity is:

$$K_* = f \, \overline{K}/u_{*_0}^2 \approx 0.02$$

consistent with various estimates from IOBL data (e.g. McPhee & Martinson, 1994). Note that it is also consistent with Ekman's (1905) assertion that eddy viscosity in the ocean varied with the square of the wind speed.

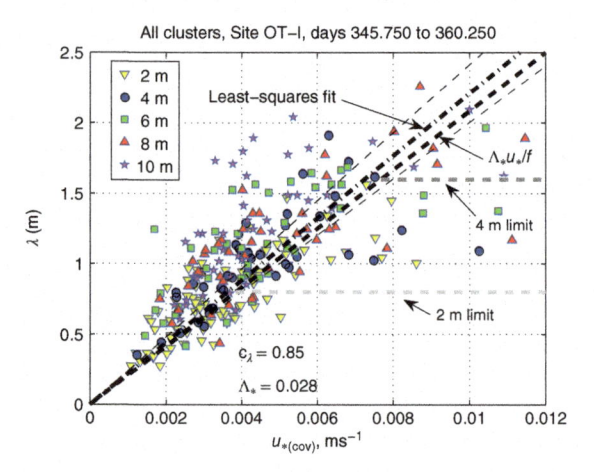

Figure 5.5 Mixing length as inferred from k_{max} (from vertical velocity spectra) versus u_* from 3 hour average turbulence statistics and spectra. The dot-dashed line is a least-squares regression through the origin with 95% confidence interval indicated by light dashed lines. The heavy dashed line indicates the dynamic (planetary) maximum mixing length. Grey dashed horizontals indicate geometric (kz) limits at 2 and 4 m.

5.2.6 Buoyancy impact on turbulence scales

Ice melting or freezing modifies IOBL turbulence in much the same way that nocturnal surface cooling or diurnal heating either quells or enhances turbulence in the atmospheric boundary layer. When ice is melting, turbulent eddies must work against gravity to transport momentum and scalar properties downward, and consequently, for the same stress, the turbulence scales are suppressed and exchange is limited to depths smaller than would be expected under neutral conditions. By the same token, freezing enhances turbulence scales by converting potential energy of the water column to turbulent energy. These tendencies are quantified by considering a third length scale, introduced by Obukhov (1971, English translation):

$$L_0 = \frac{u_{*_0}^3}{\kappa \langle w'b' \rangle_0} \quad (5.16)$$

L_0 is positive when ice is melting, as salinity flux is upward, and negative for freezing. Atmospheric boundary-layer studies have established that non-dimensional shear and analogous scalar gradients across the surface layer follow Monin–Obukhov similarity, i.e.:

$$\frac{z u_z}{u_{*_0}} = \phi_m(\zeta); \quad \zeta = z/L_0 \quad \text{[surface layer]} \quad (5.17)$$

In the surface layer, $\lambda_{sl} = \kappa z/\phi_m$. The form of the similarity parameter ϕ_m has been extensively studied for the atmospheric surface layer, resulting in a variety of empirical functions (Businger et al., 1971; Lettau, 1979). The impact of small positive ζ (stabilizing) is to decrease λ_{sl} and increase shear, while the opposite holds for ζ small and negative (destabilizing).

By comparison with the atmospheric surface layer, there is a paucity of data on the impact of boundary buoyancy flux on the IOBL. However, measurements of turbulence during rapid melting (McPhee et al., 1987; Sirevaag, 2009) and at the edges of freezing leads (McPhee, 1994; McPhee & Stanton, 1996) have provided an important insight into how buoyancy affects turbulence, including in the region beyond the surface layer. For the IOBL stably stratified by surface buoyancy flux (melting), a similarity theory incorporating the three scales discussed so far adequately describes how positive buoyancy flux influences IOBL scales (McPhee, 1981, 1994, 2008b). In that approach the scales for the outer layer are:

Length : $\eta_* u_{*_0}/f$
Velocity : u_{*_0}/η_*
Eddy viscosity : $(u_{*_0} \eta_*)^2/f$
Kinematic stress : $u_{*0} u_{*_0}$

$$\eta_* = \left(1 + \frac{\Lambda_* \mu_*}{\kappa R_c}\right)^{-1/2} \quad (5.18)$$

where $\mu_* = u_{*_0}/(fL_0)$ is the ratio of the planetary scale to the Obukhov length, R_c is the critical flux Richardson number (~0.2), and u_{*_0} is vector (complex) friction velocity. The stability parameter η_* represents a harmonic mean for the maximum mixing length in the outer layer that asymptotically approaches the following limits:

$$\lambda_{\max} \to \Lambda_* u_{*_0}/f \quad \text{for } L \to \infty$$
$$\lambda_{\max} \to R_c \kappa L \quad \text{for } L \to 0^+ \quad (5.19)$$

so that:

$$\lambda_{\max} = \eta_*^2 \Lambda_* u_{*_0}/f \quad (5.20)$$

Stabilizing buoyancy flux ($\eta_* < 1$), is a sink for turbulent kinetic energy, and thus shear is required for turbulence. However, destabilizing buoyancy flux (e.g. from rapid freezing) is a turbulent kinetic energy source, and it is possible to have significant turbulence with little or no ice–ocean shear. In that case, dimensional analysis suggests that the scale velocity in pure free convection ought to be proportional to the cube root of a scale velocity multiplied by the buoyancy flux magnitude. A more common condition is mixed forcing where both density driven convection and shear contribute to turbulence. Again, from dimensional analysis and considering asymptotic behaviour, a workable expression for the scale velocity under these conditions is:

$$w_* = (u_*^3 - \lambda_{\max} \langle w'b' \rangle)^{1/3}; \quad \langle w'b' \rangle < 0 \quad (5.21)$$

When L is negative, $\eta_* > 1$, and according to equation (5.20), $\lambda_{\max}$ increases rapidly. A practical limit on the vertical extent of turbulent eddies is the depth of the well-mixed layer (WML), so we posit that $\lambda_{\max}$ expands until it reaches a limiting value that is some fraction ($0.2 \leq c_{ml} \leq 0.4$) of the depth of the WML, i.e. $\lambda_{\max} \leq c_{ml}|z_{pyc}|$. This allows mixing length to smoothly transition between neutral and unstable scaling.

5.2.7 Rossby similarity

A problem that often arises in considering ice–ocean interaction is how stress at the interface is related to ice velocity (considered with respect to undisturbed geostrophic current). While this is often described by a drag coefficient (e.g. $\tau_0 = c_D V_0^2$), it is convenient in the context of ice–ocean interaction to consider instead a non-dimensional fluid velocity formed by dividing complex surface velocity by vector u_{*_0}. The non-dimensional velocity is in effect an inverse square root of the drag coefficient:

$$U_0 = V_0/u_{*_0} = 1/\sqrt{c_D} \quad (5.22)$$

As an example, if V_0 were the velocity of the ice relative to a level, $z = -2$ m, within the surface layer governed by the law of the wall (equation 5.13), the non-dimensional velocity would be:

$$U_0 = \frac{1}{\kappa} \log \frac{2}{z_0} = \frac{1}{\sqrt{c_h}}$$

In this case U_0 is real, as V_0 and u_{*_0} are assumed to be aligned (in reality, at the far extent of the surface layer, they generally are not). A typical undersurface roughness for multi-year pack ice is about 0.04 m, for which $U_0 = 9.8$ and $c_2 = 0.0105$.

Although it is common practice to express ocean drag on ice as proportional to the square of velocity relative to the current at the far extent of the IOBL (e.g. the geostrophic current) with a constant turning angle (Hibler, 1979), there are several drawbacks to a quadratic drag approach, particularly in light of recent rapid change in the character of the Arctic ice pack (McPhee, 2012). Starting from the premise that surface velocity depends on variables f, u_{*_0}, and z_0, two with independent dimensions, elementary dimensional analysis suggests that non-dimensional surface velocity should be a function of the ratio of the planetary scale to surface roughness:

$$\frac{V_0}{u_{*_0}} = U_0\left(\frac{u_{*_0}}{fz_0}\right) \quad (5.23)$$

Note that equation (5.23) is a quadratic drag expression only if its right-hand side is constant. The interplay between hydraulic roughness and the planetary scale is illustrated in Figure 5.6. Figure 5.6(a) shows measurements from the 1972 AIDJEX IOBL currents (see Figure 5.1). The vector labelled V_E corresponds to the

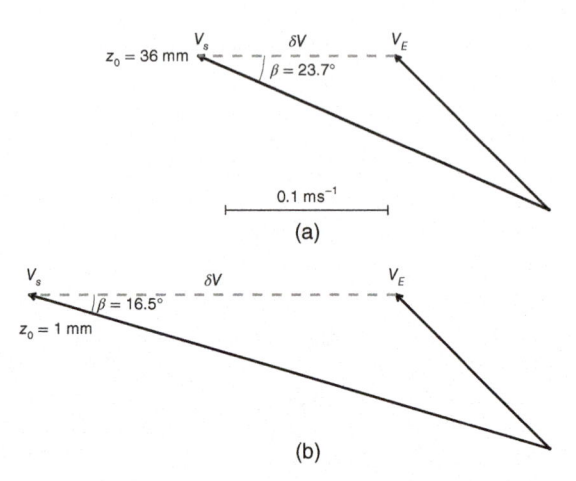

Figure 5.6 (a) Estimating undersurface hydraulic roughness from the Arctic Ice Dynamics Joint Experiment (AIDJEX) measurements (Figure 5.1), assuming the shear between the V_E and V_s follows the law of the wall. (b) For the same Ekman velocity and planetary scale, a smoother surface results in higher surface speed and reduced Coriolis turning.

velocity (relative to the base of the IOBL) at the upper limit of the Ekman layer. We estimated that friction speed during this time was $u_{*_0} = 0.01$ m s^{-1}. The near surface velocity field was distorted by local under-ice morphology (McPhee & Smith, 1976), but as the ice moved as a unit over the entire area observable from the ice station, total shear represented integration over a much larger area than in the immediate vicinity of the instrument mast. If the vector velocity difference $\delta V = V_s - V_E$ is the shear across the logarithmic surface layer, the inferred larger-scale roughness is about 0.04 m. In Figure 5.6(b), we consider an Ekman layer with the same planetary scale, u_{*_0}/f, but much smaller roughness length, $z_0 = 1$ mm, typical of first-year ice in the Weddell Sea (McPhee et al., 1999). The shear across the surface layer is increased, and for the same stress, ice moves about 40% faster, 7° closer to the surface stress direction.

The non-dimensional ice velocity has a direct analogue in the relationship between atmospheric geostrophic wind (proportional to the gradient of the surface pressure field) and wind stress at the surface. From the same principles as discussed qualitatively above, asymptotically matching surface and outer layers (Rossby & Montgomery, 1935; Blackadar & Tennekes, 1968) leads to a functional form for U_0 (representing either the non-dimensional geostrophic wind or ice velocity relative to the undisturbed ocean). In complex notation, Rossby similarity drag is:

$$U_0 = \frac{1}{\kappa}(\log Ro_* - A \mp iB) \quad (5.24)$$

where $Ro_* = u_{*_0}/(fz_0)$ is the surface friction Rossby number. The sign of the imaginary part depends on the hemisphere (northern negative for the IOBL). The similarity constants, A and B, may be obtained empirically from measurements or derived analytically from similarity scaling principles (McPhee, 1990; 2008b). Typical estimated values from IOBL studies are $A = 2.3$ and $B = 2.1$, but these are subject to a fair amount of variability because other factors, including the impact of boundary buoyancy flux, surface inhomogeneity or relatively shallow mixed layers, are often present and unaccounted for in the simple similarity approach.

The right-hand side of equation (5.24) depends on the planetary scale, so that for a given surface roughness, the drag coefficient varies with surface stress. In similarity terms, the non-dimensional surface roughness, fz_0/u_{*_0}, decreases with increasing stress (i.e. the surface appears smoother), so that the drag coefficient and turning angle also decrease (Figure 5.7). At higher ice speeds, this can represent a significant reduction in drag over what would be expected if the relationship between stress and velocity were quadratic (constant drag coefficient).

When ice melts rapidly, as it often does when blown across temperature fronts existing in the marginal ice zone, the surface buoyancy flux reduces turbulence scales and, in effect, reduces the frictional coupling between the IOBL and the underlying ocean. From dimensional analysis, this adds a second dimensionless parameter, the ratio of planetary to Obukhov lengths, to equation (5.23):

$$U_0 = U_0\left(\frac{u_{*_0}}{fz_0}, \frac{u_{*_0}}{fL_0}\right) = U_0\ (Ro_*, \mu_*)$$

Consistent application of the similarity scaling equation leads to:

$$U_0 \simeq -i\delta(1 + \delta\xi_{sl}) + \frac{\eta_*}{\kappa}\left[\log\frac{\xi_{sl}}{\xi_0} - (a - \delta)\xi_{sl} - \frac{a\delta}{2}\xi_{sl}^2\right] \quad (5.25)$$

where $\delta = (\pm i/\Lambda_*)$, $a = \frac{\kappa(1-\eta_*)}{\eta_*\Lambda_*}$, and $\xi_{sl} = -\Lambda_*/\kappa$ (McPhee, 1981, 2008b). Manipulation of equation (5.25) then results in a modified Rossby similarity expression

$$U_0 = \frac{1}{\kappa}[\log Ro_* - A(\mu_*) \mp iB(\mu_*)] \quad (5.26)$$

Drag reduction from rapid melting may account for the separation of distinct ice edge bands that form when wind pushes ice across the marginal ice zone into relatively warm water (McPhee, 1981, 2012; Mellor et al., 1986). Bands separate from the pack because water in the WML is rapidly cooled by passage of the melting ice and hence presents less 'grease' for ice following the initial vanguard.

5.2.8 Discussion

The interaction of sea ice with the IOBL requires keeping track of fluxes into and out of the IOBL as well as the ice column, and for useful applications requires coupled numerical modelling. Several approaches exist for modelling the IOBL. 'Slab' models consider properties uniform in a (literally) mixed layer, and exchanges at its base depend on bulk properties (including velocity). Examples for the open ocean include Pollard et al. (1973), Niiler and Kraus (1977), Price et al. (1986) and, for the IOBL, Toole et al. (2010). Second moment turbulence closure models (Mellor & Yamada, 1982; Burchard & Baumert, 1995) solve conservation equations on a vertical grid, including separate conservation equations for turbulence kinetic energy and master turbulent length. At a higher level of complexity, large-eddy simulation (LES) models parameterize only sub-grid-scale processes in fully three-dimensional simulations of the equations of motion (Skyllingstad et al., 2003; Harcourt, 2005). LES models require comparatively fine grids and are computationally expensive.

First-order closure models, which provide a computational compromise between the slab models and more complex closure schemes, express turbulent fluxes of momentum and scalar properties directly in terms of the vertical property gradients. In an approach based on IOBL measurements, called local turbulence closure (McPhee, 2008b, chapters 6–8), eddy viscosity in the bulk of the IOBL (where property gradients are small) is the product of the local friction speed (square root of the Reynolds stress) and a mixing length, which in turn depends on surface flux conditions, f, and, within the surface layer, distance from the boundary. Its formulation depends on limiting asymptotic scales identified in equation (5.19). In the fully turbulent flow of the WML (where scalar gradients are small, but not generally zero), we often assume scalar eddy diffusivity is the same as eddy viscosity (Reynolds analogy). In the pycnocline (usually defined as beginning at the depth

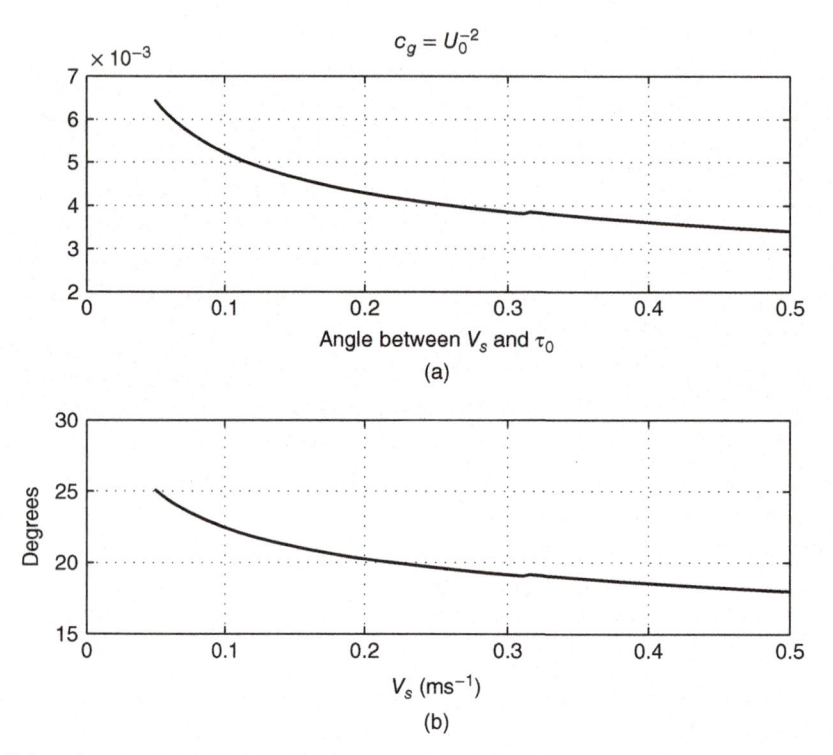

Figure 5.7 Drag coefficients from Rossby similarity with $A = 2.3$, $B = 2.1$, and $z_0 = 0.03$ m. (a) Effective drag coefficient. (b) Amplitude of the turning angle.

where buoyancy frequency exceeds a given threshold), the algorithm for λ is like the WML formulation, but is based on fluxes computed at the top of the pycnocline. When the density gradient is stabilizing, eddy diffusivity is reduced from eddy viscosity following a formula based on local Richardson number, derived from laboratory measurements reported by Turner (1973). The basic mixing length parameterization is illustrated if Figure 5.8. Note the gradients in density drawn in the WML for stable (Figure 5.8b) and statically unstable (Figure 5.8c) cases. When we measure scalar fluxes in the IOBL, we invariably see finite, albeit small, mean scalar gradients.

A key element for the IOBL is z_0, the hydraulic roughness of the ice undersurface. Estimates vary widely depending on ice age, location and the amount of local deformation (e.g. Shirasawa, 1986, table 3; McPhee, 1990, table 6.1). For undeformed, first-year year ice grown in place (Svalbard fjords, Canadian Arctic Archipelago), the undersurface is often found to be hydraulically smooth (Langleben, 1982; McPhee et al., 2008; Crawford et al., 1999). Surface roughness for first-year ice in the eastern Weddell Sea was estimated to be about 1 mm (McPhee et al., 1999). For the 11-month SHEBA deployment (1997–1998), a local value of z_0 for undeformed, multi-year ice was found to be 6 mm (McPhee, 2002). The method specifically excluded the effects of pressure ridge keels and refrozen leads.

In multi-year pack ice, IOBL measurements are typically made in relatively smooth locations away from obvious obstacles and thus may tend to systematically underestimate the overall roughness of a large floe or regional area representative of a model grid cell. We estimated a regional value for z_0 in multi-year ice in the western Weddell Sea of about 4 cm, using a method that compares modelled angular shear in the outer part of the IOBL with observed results over relatively long averaging periods (McPhee, 2008a). The method applied to the SHEBA data set yielded a similar value, 4.9 cm (McPhee, 2008b), and a value of about 9 cm for rough ice surrounding a buoy deployed near the North Pole (Shaw

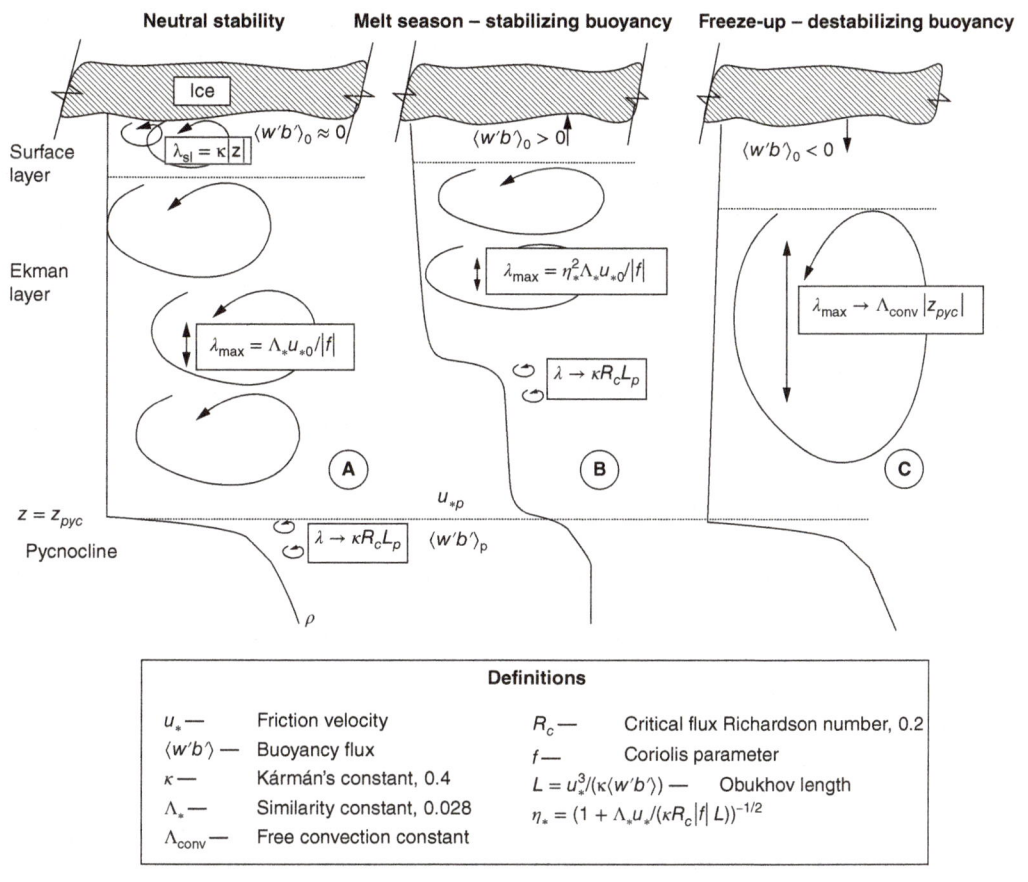

Figure 5.8 Mixing length representation under different conditions of surface buoyancy flux: (a) neutral, (b) stable, and (c) statically unstable.

et al., 2008). The latter is comparable to estimates from the marginal ice zone (Johannessen, 1970; Pease et al., 1983). Lu et al. (2011) described an alternate method for parameterizing sea ice drag by partitioning it among bottom skin friction and form drag on pressure ridge keels and floe edges.

It is useful to recall that it is $\log z_0$ that appears in the equations, not the actual length. Even so, the range is quite large. While the scatter in z_0 estimates makes categorization risky, in lieu of better information a rough guide might be to assign the following values: for undeformed fast ice, $z_{0s} = (\nu/u_{*0})e^{-2}$ (hydraulically smooth); for first-year sea ice, ~ 1 mm; for typical multi-year pack ice, 5 cm; for highly deformed pack ice, ~ 10 cm (McPhee, 2012).

5.3 The ice–ocean interface

5.3.1 Enthalpy and salt balance at the interface

In an idealized control volume following the interface (Figure 5.9), three terms dominate the heat equation, which may be written in 'kinematic' form (i.e. after dividing by ρc_p, the product density and specific heat of seawater) as

$$-\dot{q} + \langle w'T' \rangle_0 = w_0 Q_L \qquad (5.27)$$

where $\dot{q} = -K_{\text{ice}} T_z/(\rho c_p)$ represents heat conduction in the ice column with thermal conductivity K_{ice} and temperature gradient T_z; $\langle w'T' \rangle_0$ is kinematic heat flux from the ocean; $Q_L = L_{\text{ice}}/c_p$ is the latent heat of sea ice divided by specific heat (with temperature units); and w_0 is vertical velocity of the interface due to melting or freezing

Thermal Balance at the Ice/Ocean Interface

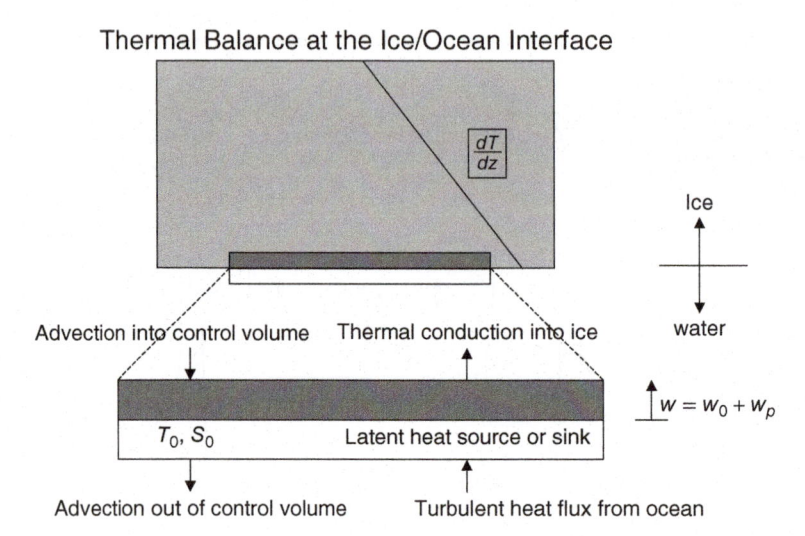

Figure 5.9 Simplified schematic of the heat balance at the ice–ocean interface, with control volume following the boundary.

(i.e. the isostatically balanced basal melt rate). Advective flux associated with w_0 is neglected, assuming that temperatures of fluid entering and leaving the control volume are nearly the same. Thermal properties of sea ice are functions of brine volume and temperature (Notz & Worster, 2009). Here we use simplified expressions for thermal conductivity from Untersteiner (1961):

$$K_{ice} \approx K_{fresh} + \beta S_{ice}/T_{ice}$$

where K_{fresh} is 2.04 J m^{-1} K^{-1} s^{-1} and $\beta = 0.117$ J K^{-1} s^{-1} psu^{-1} °C, and for latent heat from Maykut (1985):

$$Q_L = \frac{L_{fresh}}{c_p}(1 - 0.03 S_{ice})$$

When sea ice melts, the interface is typically very smooth and hard, and turbulent flux of salt ($\langle w'S' \rangle_0$) into the control volume is balanced mainly by an advective flux given by w times the difference between the salinity of water at the interface and ice salinity:

$$\langle w'S' \rangle_0 + w(S_{ice} - S_0) = 0 \quad (5.28)$$

For generality, the vertical interface velocity, $w = w_0 + w_p$, includes a second term, w_p, intended to incorporate the idea that a percolation flow of fresh or brackish water driven by a pressure head in the ice column could induce upward interface motion (and negative salinity flux) without direct melting. In most applications and it what follows, w_p is ignored.

When ice freezes, there is substantial evidence that the simple balance expressed by equation (5.28) does not adequately account for the salt transfer in an infinitesimal control volume following the interface. The leading edge of growing sea ice appears to be a combination of crystalline ice and entrapped liquid brine in a so-called mushy layer (Wettlaufer et al., 1997; Feltham et al., 2006), and that near the growing interface, S_{ice} approaches the salinity of the melt (Notz & Worster, 2009); hence the advective flux is small. What this implies for the IOBL boundary condition is discussed below.

5.3.2 Measurements

Instrument clusters, including highly accurate ocean thermometers mounted near three-dimensional arrays of current meters, were deployed in the IOBL during the Marginal Ice Zone Experiment (MIZEX) in 1984, so that in addition to turbulent stress, accurate estimates of turbulent heat flux were available for the first time (McPhee et al., 1987). Near the end of MIZEX, northerly winds pushed our floe south into water where the mixed layer temperature was well above freezing. Measured heat flux and basal melt rate increased substantially, but were still considerably smaller than we had anticipated from previous modelling studies, and when the floe on which we were drifting did not melt away, it was clear that we had to re-examine our treatment.

The sea ice–ocean boundary layer

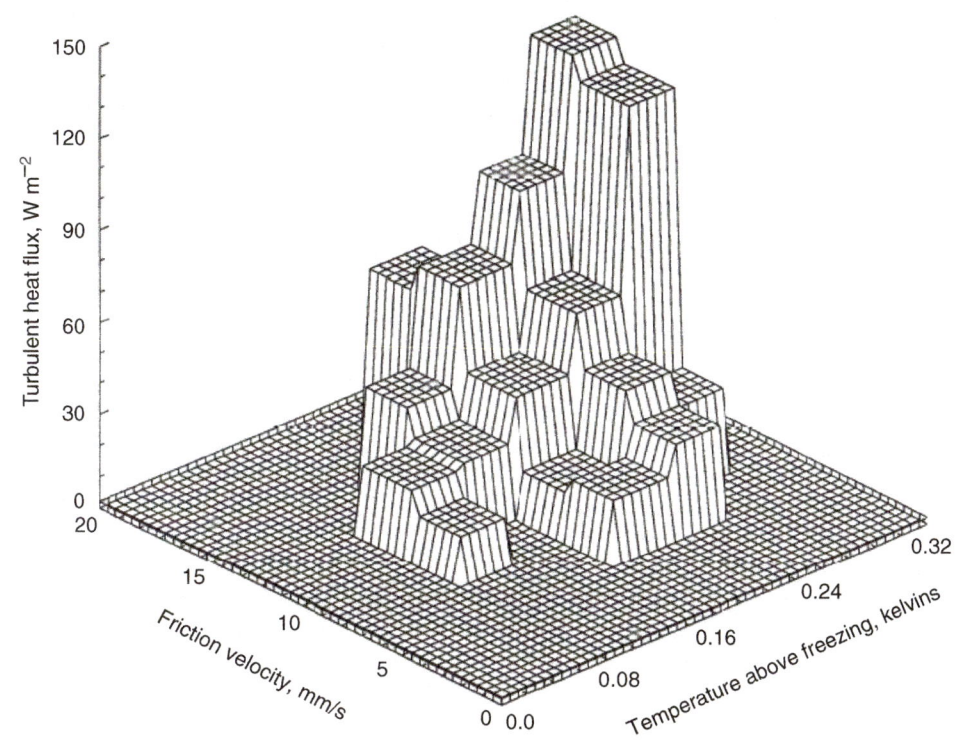

Figure 5.10 Directly measured heat flux averaged in bins of ΔT and u_* for all data from Marginal Ice Zone Experiment (MIZEX) in 1984. Source: Adapted from McPhee, 1992. Reproduced with permission of American Geophysical Union.

Results from the combination of heat flux and Reynolds stress measurements during MIZEX are summarized in Figure 5.10 (McPhee, 1992), intended to illustrate that ocean heat flux depends on both temperature elevation from freezing and friction speed, u_*. MIZEX and subsequent studies suggested a simple empirical parameterization of kinematic heat flux of the form:

$$\frac{H_f}{\rho c_p} = \langle w'T' \rangle_0 = St_* u_{*_0} \Delta T \qquad (5.29)$$

where $\Delta T = T_{ml} - T_f(S_{ml})$ is the elevation of mixed layer temperature above its salinity determined freezing temperature, and St_* is a turbulent Stanton number. IOBL measurements under widely varying conditions of both u_* and ΔT have shown St_* to be relatively invariant. Over the year-long SHEBA experiment its mean value was estimated to be $St_* = 0.0057 \pm 0.0004$ (McPhee, 2008b).

5.3.3 Interface approximations

The first serious efforts at thermodynamic modelling of sea ice indicated that heat flux from the ocean was important for maintaining a realistic equilibrium ice thickness. Maykut and Untersteiner (1971) found that without a steady basal flux of about 2 W m^{-2}, ice continued to grow to unrealistically large thickness in their thermodynamic ice model. Using a similar model for Southern Ocean sea ice, Parkinson and Washington (1979) required a constant ocean heat flux about an order of magnitude larger to limit ice growth to observed thickness.

Measurements over several decades have shown that the IOBL, even in the central Arctic with perennially high ice concentrations, spends a sizeable fraction of the year at temperatures above its salinity-determined freezing temperature (Figure 5.11). Thus early in the summer (at maximum insolation) a significant fraction of the solar radiation making its way into the ocean is heating the upper water column, rather than melting ice. For example, during June 1998 at the SHEBA station, we estimated that energy expended in heating the mixed layer was roughly equivalent to the total basal heat transfer. Obviously, the albedo feedback effect (heating creates more open water, absorbing

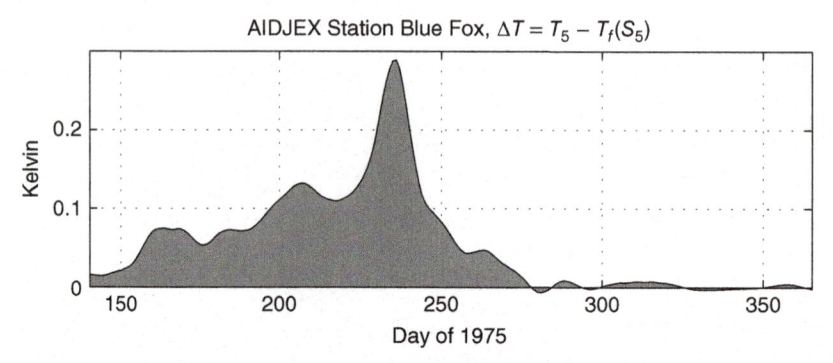

Figure 5.11 Elevation of mixed layer temperature above freezing at the Arctic Ice Dynamics Joint Experiment (AIDJEX) station Blue Fox during the summer of 1975 in the Canadian Basin of the western Arctic.

more solar radiation, opening more water, and so on) is most effective at high sun angles, and the capacity of the upper ocean to absorb heat then and release it to the ice later provides a presumably significant brake on the feedback process.

Critical links in understanding the balances at the interface are the turbulent heat and salinity fluxes. A combination of dimensional analysis and measurements showing little or no Reynolds number dependence for St_* suggests that the following dimensionless expression for the kinematic turbulent heat flux at the interface in terms of a thermal exchange coefficient, α_h, is a function of the Prandtl number:

$$\frac{\langle w'T'\rangle_0}{u_{*_0}\delta T} = \alpha_h\left(\frac{\nu}{\nu_T}\right) = \alpha_h(\mathrm{Pr}) \quad (5.30)$$

where $\delta T = T_{ml} - T_0$ is the difference in temperature between the far-field fluid and interface (the thermal driving); ν is molecular viscosity; and ν_T is molecular thermal diffusivity. Similarly, we can write a functional expression for dimensionless salinity flux using molecular haline diffusivity, ν_S:

$$\frac{\langle w'S'\rangle_0}{u_{*_0}\delta S} = \alpha_S\left(\frac{\nu}{\nu_S}\right) = \alpha_S(\mathrm{Sc}) \quad (5.31)$$

where $\delta S = S_{ml} - S_0$ is the difference between far-field and interface salinities, and Sc is the Schmidt number.

5.3.4 The 'two-equation' parameterization

Assuming that water at the interface is at its freezing temperature, $\alpha_h = St_*$ only if salinity there is the same as S_{ml}. For slow melt rates, this is often an adequate approximation, and by calculating $\langle w'T'\rangle_0$ from equation (5.30)

with $\alpha_h = St_*$, an expression for the isostatically adjusted melt rate is:

$$w_0 = \frac{-\dot{q} + St_* u_{*_0}\Delta T}{Q_L} \quad (5.32)$$

from which salinity flux follows:

$$\langle w'S'\rangle_0 = w_0(S_{ml} - S_{ice}) \quad (5.33)$$

These may be combined to estimate surface buoyancy flux (an important parameter for the IOBL turbulence):

$$\langle w'b'\rangle_0 = \frac{g}{\rho}\langle w'\rho'\rangle_0 = g(\beta_S\langle w'S'\rangle_0 - \beta_T\langle w'T'\rangle_0) \quad (5.34)$$

where g is the acceleration of gravity, and β_S and β_T are the haline contraction and thermal expansion factors for cold seawater, respectively. For water near freezing β_T is so small that buoyancy flux is controlled almost exclusively by salinity flux, which further simplifies equation 5.34.

5.3.5 The 'three-equation' parameterization

An obvious paradox accompanies the two-equation approach. If there is enough heat flux to initiate melting, salinity of water in the control volume must be less than in the far field, and $\delta T < \Delta T$ so $\alpha_h/St_* = \Delta T/\delta T$, i.e. the thermal exchange factor is larger than St_*. If the Prandtl number is important in the heat transfer process, then it is reasonable to assume that the Schmidt number plays a role in salt transfer. At cold temperatures, the molecular thermal diffusivity is roughly 200 times greater than salt diffusivity, and thus we suspect that if heat and salt exchange depend on Pr and Sc, $\alpha_h \neq \alpha_S$ and the process is inherently 'double-diffusive.'

If water in the control volume is at freezing, the dependence of freezing temperature on salinity, $T_0 = T_f(S_0) \doteq -mS_0$ (where m is the local slope of the freezing curve), may be combined with equations (5.28), (5.30) and (5.31) to obtain a quadratic in, for example, S_0:

$$mS_0^2 + (T_H + T_L - mS_{ice})S_0 - T_H S_{ice} - T_L S_w = 0 \quad (5.35)$$

where

$$T_H = T_{ml} - \frac{\dot{q}}{\alpha_h u_{*_0}}; \qquad T_L = \frac{\alpha_S Q_L}{\alpha_h} \quad (5.36)$$

T_H and T_L are temperature-scale parameters expressing sensible and latent heat forcing, respectively (assuming w_p is negligible as earlier). If α_h and α_S are known, w_0 follows from equation (5.33), providing the melt rate and buoyancy flux.

If heat conduction in the ice column is small, the temperature and salinity at the interface depend on the ratio $R = \alpha_h/\alpha_S = (\nu_T/\nu_S)^n$, where the exponent is thought to range from 2/3 to 0.8 based on a combination of laminar theory and empirical results, in turn suggesting a range $35 \leq R \leq 70$ (Notz et al., 2003). R is a measure of the strength of double diffusion, i.e. as R increases, heat transfer increases relative to salt transfer. For fresh ice with no temperature gradient ($\dot{q} = S_{ice} = 0$), equation (5.35) reduces to

$$S_0 = S_{ml}/\left(1 + {R\delta T}/{Q_L}\right) \quad (5.37)$$

Q_L is the latent heat of ice divided by specific heat of seawater, about 84 K for fresh ice. In this case if $R = 1$, $Q_L \gg \delta T$, so $S_0 \to S_{ml}$. On the other hand, in water a degree or two above freezing, if R is in the range given above, then $Q_L/R \sim \delta T$ and S_0 may be considerably less than S_{ml}.

5.3.6 Double-diffusion during melting

Most measurements of turbulent heat flux have been made at relatively small values of ΔT: generally from 0.1 to 0.3 K. In areas with low ice concentrations (e.g. in the marginal ice zone) or where ice encounters inflowing warm water (e.g. Whaler's Bay north of Svalbard), ΔT may reach values an order of magnitude or more greater, with much higher melt rates (decimetres day^{-1}) and presumably increased double-diffusive tendency.

When melting depresses salinity at the interface, $\alpha_h/St_* > 1$. For commonly encountered ranges of stress and thermal forcing ($5 \leq u_{*_0} \leq 15$ mm s^{-1}; $0.05 \leq \Delta T \leq 0.5$ K) observed during the field programmes, α_h differs from St_* by less than 2% when $R = 1$ (no double diffusion), while for $R = 70$, it is more than twice as large (McPhee, 2008a, figure 6.6).

The rate-limiting impact of salt diffusion can be examined (Figure 5.12) by considering solutions to equation (5.35) for moderate interface stress ($u_{*_0} = 0.015$ m s^{-1}) with bulk thermal driving, ΔT, ranging from 0 (freezing) to 3 K (as might be encountered when ice drifts across a front in the marginal ice zone), and limiting values of R. In the first case with no double diffusion, we specify $R = 1$, $\alpha_h = 0.0058$. For the second, strong double diffusion: $R = 70$, $\alpha_h = 0.0137$. In both cases, the heat exchange coefficient is chosen so that the 'three-equation' solution matches the bulk expression in equation (5.29) (with $St_* = 0.0057$) for small values of ΔT. We emphasize that without direct knowledge of either R or α_h, the main observational constraint is the measured bulk Stanton number. For $R = 1$, the interface thermal driving, δT, (Figure 5.12a, dashed curve) is almost indistinguishable from the bulk driving, i.e. T_0 remains very close to the freezing temperature of the far field. For $R = 70$ (solid curve), δT is much reduced and exhibits positive curvature with increasing ΔT. As might be expected, the saline driving, δS, (Figure 5.12b) is highly dependent on the double-diffusive strength. For $R = 70$, δS approaches 18 psu for large ΔT. For $R = 1$, it remains within about 1 psu of the far field, even with strong thermal forcing. Basal heat flux (Figure 5.12c) follows directly from δT. At high values of ΔT, heat flux for $R = 70$ is half as large again as for $R = 1$. In the latter case, the bulk relation (overlain grey curve in Figure 5.12c) is essentially the same as the three-equation solution.

Because of large temporal and spatial gradients in marginal ice zones, it is notoriously difficult to measure both $\langle w'T' \rangle$ and $\langle w'S' \rangle$ accurately in a rapidly melting environment; however, by careful analysis of turbulence measurements in Whaler's Bay north of Svalbard, Sirevaag (2009) obtained independent estimates of the exchange coefficients: $\alpha_h = 0.0131$ and $\alpha_S = 4 \times 10^{-4}$ with $R \approx 33$ when the average heat flux was 268 W m^{-2}. He also reported a direct bulk Stanton number: $St_* = 0.0084$, which is reasonably close to $(\delta T/\Delta T)\alpha_h = 0.0088$ obtained from reported values. His results thus confirm that heat flux increases with thermal driving in a superlinear fashion, with

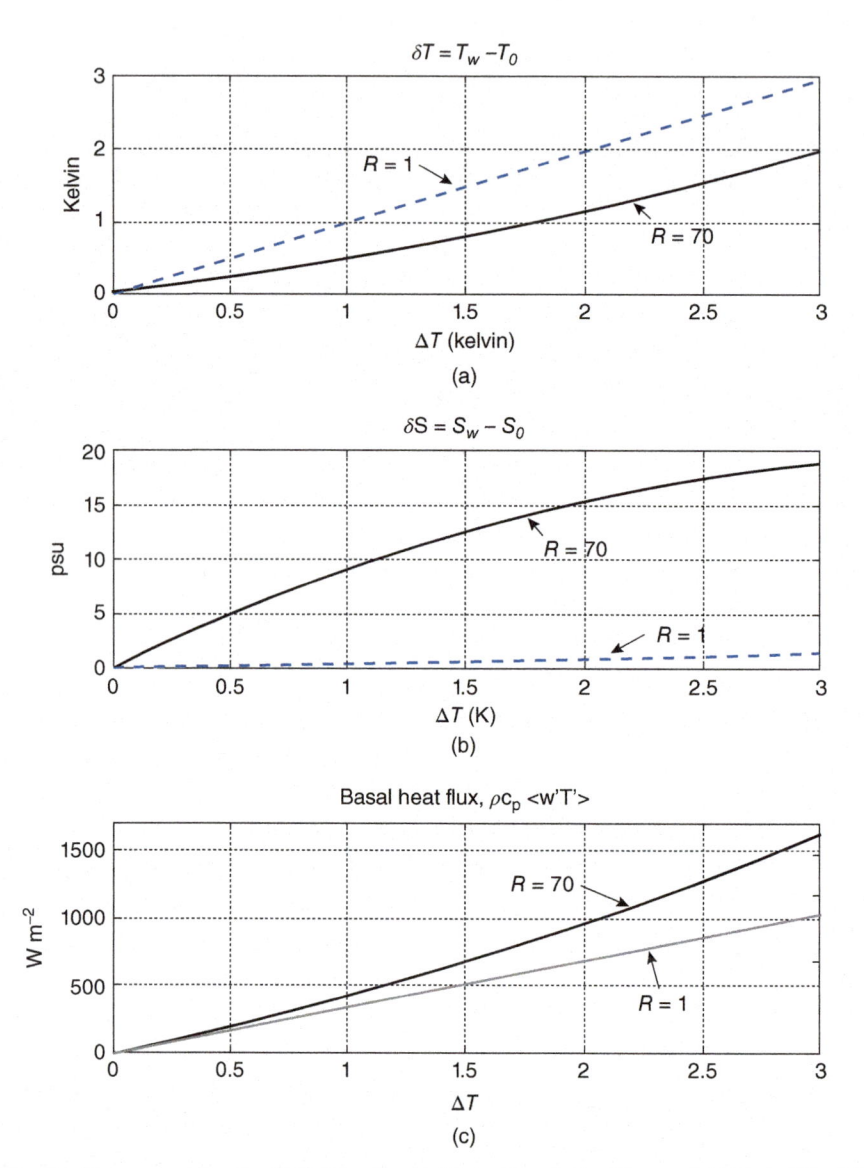

Figure 5.12 (a) $\delta T = T_w - T_0$ versus $\Delta T = T_w - T_f(S_w)$ for $R = 1$, (dashed) and $R = 70$, $\alpha_h = 0.0137$ (solid). Other parameters: $u_{*_0} = 0.015$ m s^{-1}; $S_w = 33$ psu; $S_{ice} = 4$ psu. (b) δS vs ΔT. (c) Basal heat flux versus ΔT. The curve labelled $R = 1$ is overlain in grey by the linear relation: $H_f = \rho c_p St_* u_{*_0} \Delta T$ with $St_* = 0.0057$.

potential importance for estimating ice melt and IOBL stratification in low ice concentration conditions.

5.3.7 False bottoms

Double diffusion also appears to be the controlling factor in the persistence and migration of false bottoms. These occur when concavities in the ice underside fill with fresh meltwater, forming a thin ice layer at the interface between the fresh water and underlying, colder seawater. Observations of the bottom elevation changes at AIDJEX station Big Bear during the summer of 1975 by A. Hanson showed that gauges deployed at the start of summer in relatively thick ice (280–300 cm) indicated relatively continuous bottom melt over the course of the summer, totalling 30–40 cm. In contrast, gauges deployed in thinner ice (160–175 cm) showed little net bottom melt and even, in some cases, net growth. Hanson attributed this to the formation of so-called false bottoms that occur when concavities in the ice underside fill with fresh meltwater, which then forms a thin layer of ice at the interface with colder seawater. During one 10-day period in August 1975, four of the shallow thickness gauges showed similar changes in bottom elevation, averaging about 15 cm, while the three thick-ice gauges each showed bottom ablation of about 5 cm.

Notz et al. (2003) investigated evolution of false bottoms in both laboratory and natural settings, including a simulation of the AIDJEX event just described. They found that the observed upward migration of the false bottoms could be well simulated with $R = 50$, but that with $R = 1$ (no double diffusion) and observed stress, the initial false bottoms disappeared within a short time. Our experience, particularly at SHEBA, has instead been that they form repeatedly and migrate upwards during the summer, making space for new false bottoms. This produces interleaved layers of ice and fresh water by the time freeze-up begins; numerous examples were observed during the setup for SHEBA in October 1997.

False bottoms may have an important impact on the ice–albedo feedback. First, they shield the thinnest pack ice from direct contact with above-freezing seawater, greatly reducing bottom ablation and reducing the area of low-albedo open water when sun angles are highest. Second, depending on how extensively they cover the ice undersurface, because of the large positive temperature gradient across the fresh ice layer, they act as a source of heat for the upper ocean (Figure 5.13a). The effective bulk Stanton number for ice with area fraction A_{fb} of the undersurface covered by false bottoms or meltwater ponds is:

$$St_{*_{\text{eff}}} = H_{total}/(\rho c_p u_{*_0} \Delta T) = \frac{(1 - A_{fb})\langle w'T'\rangle_0 + A_{fb}\langle w'T'\rangle_{fb}}{u_{*_0}\Delta T}$$

As illustrated in Figure 13(b) using results from the double-diffusive false bottom model of Notz et al. (2003), relatively small A_{fb} may significantly decrease the effective heat transfer from the ocean, again lessening the amount of melting when sun angles are high.

5.3.8 Heat and salt exchange during freezing

If the ice–water interface retained its double-diffusive character when ice was forming rather than melting, heat would be extracted from the water column faster than salt was added, and a mixed layer that was initially at freezing would become super-cooled, presumably forming frazil ice. With exchange coefficients appropriate for melting and moderate growth rates, the three-equation formulation indicated significant frazil formation at the same time, slowing direct congelation growth (Mellor et al., 1986; Steele et al., 1989). Holland et al. (1997) considered double-diffusive effects during freezing with a coupled sea ice–ocean numerical model with several ice thickness categories. Frazil accreted equally on all thickness categories with a net increase in heat loss and equilibrium thickness because of more conductive flux through thinner categories.

Observations of multi-year pack ice in the Arctic reveal that neither super-cooling nor frazil production is extensive (Weeks & Ackley, 1986), suggesting that the freezing process was not much affected by double diffusion, at least not to the degree suggested by the models. An experiment designed to test this was conducted in Van Mijenfjorden, Svalbard, in tidal flow under growing fast ice. As reported by McPhee et al. (2008), turbulence data showed an average heat flux of about 1.5 W m^{-2} when upward heat conduction in the ice was estimated to be about 21 W m^{-2}, producing a (downward) salinity flux of about -2×10^{-6} psu m s^{-1}. These data ruled out anything but weak double diffusion, and a more extensive modelling study, using data from the main project and the subsequent student exercises, confirmed that the three-equation interface solution could adequately describe the measured heat flux 1 m below the ice only for $R = 1$.

The IOBL response thus corroborates studies showing that salinity increases rapidly near the base of growing ice (Notz & Worster, 2009) and that the simple advective approach for describing the salt balance at the interface (equation 5.28) is not adequate for growing ice. If equation (5.35) is retained as the governing description, then when ice is freezing, the VMF data and modelling suggest that $R = 1$, which implies that $\langle w'S'\rangle_0 = -\langle w'T'\rangle_0/m$.

An interesting twist on heat transfer at the interface of growing ice was provided by turbulence measurements 1 m beneath fast ice near Erebus Glacier Tongue,

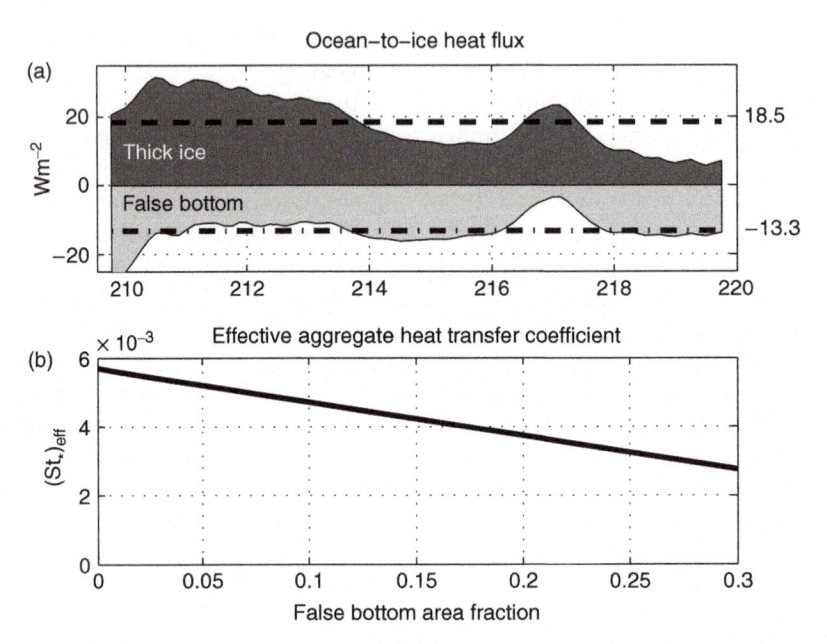

[Figure 5.13: Two-panel plot. (a) Ocean-to-ice heat flux (Wm⁻²) vs time (days 210–220), showing thick ice (dark shading, mean 18.5) and false bottom (light shading, mean −13.3). (b) Effective aggregate heat transfer coefficient $(St_*)_{\mathrm{eff}}$ (×10⁻³) vs false bottom area fraction (0 to 0.3), decreasing linearly from ~6 to ~3.]

Figure 5.13 (a) Upward heat flux under thick ice (dark shading) and under false bottoms (light shading) from the Arctic Ice Dynamics Joint Experiment (AIDJEX) simulation. Mean values are listed on the right. (b) Corresponding effective bulk Stanton number $(St_*)_{\mathrm{eff}}$ as a function of undersurface false bottom (or melt pond) area coverage.

Antarctica, in October 2010. The upper 35 m of the water column was isothermal such that the top 15 m was super-cooled, i.e. below its pressure-dependent freezing temperature. Super-cooling is often observed near ice shelves debouching into McMurdo Sound as water that has cooled as it circulated under thick ice advects from under the shelf and rises. Although limited in duration by ice accretion on the instrumentation, our results indicated tidally modulated, but consistently downward heat flux (Figure 5.14). Our interpretation is that upward conduction of heat away from the interface in the ice column (~2.3 m thick with 10 cm snow) was insufficient to balance heat released as the super-cooled water finds suitable nucleation sites on the ice undersurface (Robinson et al., 2010). Measured heat flux is shown in Figure 5.14(c) (pentagram symbols with standard deviation error bars), whereas the shaded curve indicates the bulk heat flux, $H_{\mathrm{bulk}} = c_H u_* \Delta T$, where $c_H = 0.01$ is the heat transfer coefficient, and ΔT is the temperature measured 1 m below the ice minus the freezing temperature at 3 dbar. While the record is too short to draw definitive conclusions, the transfer coefficient is similar, if somewhat larger, than $St_{*\mathrm{eff}}$ found for melting. As freezing is adding some negative buoyancy, an increase in the transfer coefficient is not unexpected. Tidal modulation of the heat flux illustrates the role of shear stress in removing heat from the interface, and suggests that platelet growth on the ice underside may be rate-limited by heat flux into the ocean when conductive flux in the ice column slows.

5.3.9 Discussion

Understanding of thermodynamics and salt transfer at the growing or melting ice–ocean boundary is far from complete. Even the assumption that fluid at the interface is at the liquidus temperature is open to question: Worster (personal communication, 2006) has pointed out that during ice growth, if heat influx outpaces salt loss in fluid draining through brine channels traversing a negative vertical temperature gradient, then it may enter the upper ocean at a temperature slightly above its salinity-determined freezing point.

While recognizing these limitations, McPhee et al. (2008) suggested that a parameterization of boundary fluxes for sea ice in direct contact with the IOBL is as follows: with T_w, S_w, and $\mathbf{u}_{*_0}$ specified either from measurements or modelling/parameterization, along with pertinent ice parameters (S_{ice}, .q), solve equation (5.35) with $\alpha_h = 0.0093$ and $R = 35$ (to satisfy the St_* constraint). If the resulting heat balance grows

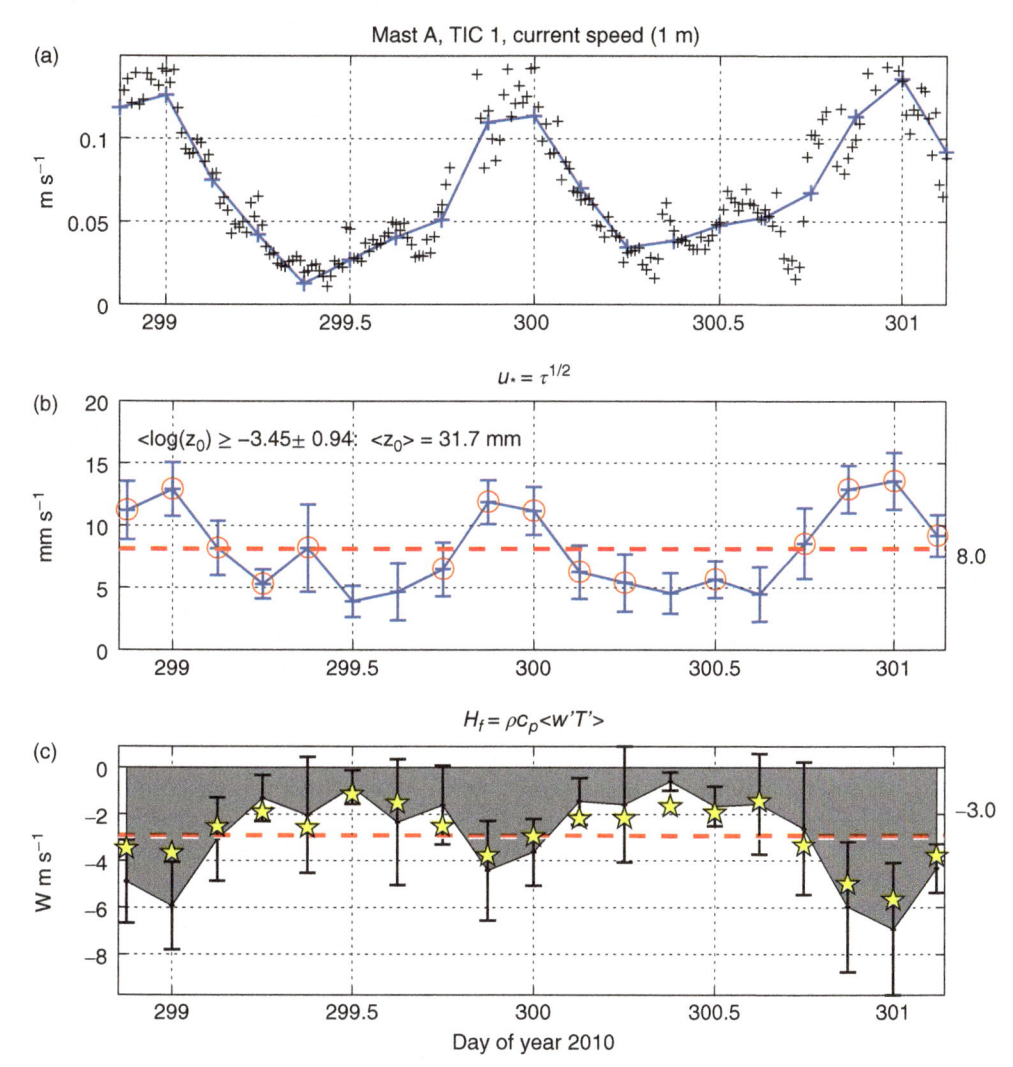

Figure 5.14 Current speed (a), friction speed (b) and turbulent heat flux (c) 1 m below fast ice near Erebus Glacier Tongue, Antarctica. Plus symbols (+) in (a) are current speeds for each 15-minute average used for covariance realizations. For direct flux measurements [circles in (b) and pentagrams in (c)], error bars indicate the standard deviation of 15-minute covariance realizations in each 2-hour average. The shaded curve in (c) is the bulk heat flux relation discussed in the text.

ice, then redo the calculation with $R = 1$ to eliminate double-diffusive effects.

References

Barenblatt, G.I. (1996) *Scaling, Self-similarity and Intermediate Asymptotics*. Cambridge University Press, Cambridge.

Blackadar, A.K. & Tennekes, H. (1968) Asymptotic Similarity in Neutral Barotropic Planetary Boundary Layers. *Journal of the Atmospheric Sciences*, **25**, 1015–1020.

Burchard, H. & Baumert, H. (1995) On the performance of a mixed-layer model based on the k-ε turbulence closure. *Journal of Geophysical Research: Oceans*, **100**(C5), 8523–8540.

Businger, J.A., Wyngaard, J.C., Izumi, Y. & Bradley, E.F. (1971) Flux-profile relationships in the atmospheric surface layer. *Journal of the Atmospheric Sciences*, **28**, 181–189.

Crawford, G., Padman, L. & McPhee, M.G. (1999) Turbulent mixing in Barrow Strait. *Continental Shelf Research*, **19**, 205–245.

Ekman, V.W. (1905) On the influence of the earth's rotation on ocean currents. *Arkiv för Matematik, Astronomi och Fysik*, **2**, 1–52.

Feltham, D.L., Untersteiner, N., Wettlaufer, J.S. & Worster, M.G. (2006) Sea ice is a mushy layer. *Geophysical Research Letters*, **33**, L14501.

Gill, A.E. (1982) *Atmosphere-Ocean Dynamics, XV*. Academic Press, New York.

Harcourt, R.R. (2005) Thermobaric cabbeling over Maud Rise: theory and large eddy simulation. *Progress in Oceanography*, **67**, 186–244.

Hibler, W.D. (1979) A dynamic thermodynamic sea ice model. *Journal of Physical Oceanography*, **9**, 815–846.

Holland, M.M., Curry, J.A. & Schramm J.L. (1997) Modeling the thermodynamics of a sea ice thickness distribution: 2. Sea ice/ocean interactions. *Journal of Geophysical Research: Oceans*, **102**, 23093–23107.

Hunkins, K. (1966) Ekman drift currents in the Arctic Ocean. *Deep Sea Research and Oceanographic Abstracts*, **13**, 607–620.

Johannessen, O.M. (1970) Note on some vertical profiles below ice floes in the Gulf of St. Lawrence and near the North Pole. *Journal of Geophysical Research*, **75**, 2857–2861.

Langleben, M.P. (1982) Water drag coefficient of first-year sea ice. *Journal of Geophysical Research: Oceans*, **87**(C1), 573–578.

Lettau, H. (1979) Wind and temperature profile prediction for diabatic surface layers including strong inversion cases. *Boundary-Layer Meteorology*, **17**, 443–464.

Lu, P., Li, Z., Cheng, B. & Leppäranta M. (2011) A parameterization of the ice-ocean drag coefficient. *Journal of Geophysical Research: Oceans*, **116**, C07019.

Maslanik, J., Stroeve, J., Fowler, C. & Emery, W. (2011) Distribution and trends in Arctic sea ice age through spring 2011. *Geophysical Research Letters*, **38**, L13502.

Maykut, G.A. (1985) *An Introduction to Ice in Polar Oceans. Reports of the Applied Physics Laboratory*. University of Washington, Seattle, WA.

Maykut, G.A., & Untersteiner, N. (1971) Some results from a time-dependent thermodynamic model of sea ice. *Journal of Geophysical Research*, **76**, 1550–1575.

McPhee, M.G. (1981) An analytic similarity theory for the planetary boundary layer stabilized by surface buoyancy. *Boundary-Layer Meteorology*, **21**, 325–339.

McPhee, M.G. (1988) Analysis and Prediction of Short-Term Ice Drift. *Journal of Offshore Mechanics and Arctic Engineering*, **110**, 7.

McPhee, M.G. (1990) Small Scale Processes. In: *Polar Oceanography, Part A Physical Science* (Ed. W.O. Smith), pp. 287–334. Academic Press, San Diego, CA.

McPhee, M.G. (1992) Turbulent heat flux in the upper ocean under sea ice. *Journal of Geophysical Research: Oceans*, **97**, 5365–5379.

McPhee, M.G. (1994) On the turbulent mixing length in the oceanic boundary layer. *Journal of Physical Oceanography*, **24**, 2014–2031.

McPhee, M.G. (2002) Turbulent stress at the ice/ocean interface and bottom surface hydraulic roughness during the SHEBA drift. *Journal of Geophysical Research: Oceans*, **107**(C10), 8037.

McPhee, M.G. (2004) A spectral technique for estimating turbulent stress, scalar flux magnitude, and eddy viscosity in the ocean boundary layer under pack ice. *Journal of Physical Oceanography*, **34**, 2180–2188.

McPhee, M.G. (2008a) Physics of early summer ice/ocean exchanges in the western Weddell Sea during ISPOL. *Deep Sea Research Part II*, **55**, 1075–1097.

McPhee, M.G. (2008b) *Air-Ice-Ocean Interaction Turbulent Ocean Boundary Layer Exchange Processes*, IX.Springer, New York.

McPhee, M.G. (2012) Advances in understanding ice–ocean stress during and since AIDJEX. *Cold Regions Science and Technology*, **76–77**, 24–36.

McPhee, M.G. (2013) Intensification of geostrophic currents in the Canada Basin, Arctic Ocean. *Journal of Climate*, **26**, 3130–3138.

McPhee, M.G. & Smith, J.D. (1976) Measurements of the turbulent boundary layer under pack ice. *Journal of Physical Oceanography*, **6**, 696–711.

McPhee, M.G. & Martinson, D.G. (1994) Turbulent mixing under drifting pack ice in the Weddell Sea. *Science*, **263**, 218–221.

McPhee, M.G. & Stanton, T.P. (1996) Turbulence in the statically unstable oceanic boundary layer under Arctic leads. *Journal of Geophysical Research: Oceans*, **101**, 6409–6428.

McPhee, M.G., Maykut, G.A. & Morison, J.H. (1987) Dynamics and thermodynamics of the ice/upper ocean system in the marginal ice zone of the Greenland Sea. *Journal of Geophysical Research: Oceans*, **92**, 7017–7031.

McPhee, M.G., Kottmeier, C. & Morison, J.H. (1999) Ocean heat flux in the central Weddell Sea during Winter. *Journal of Physical Oceanography*, **29**, 1166–1179.

McPhee, M.G., Morison, J.H. & Nilsen, F. (2008) Revisiting heat and salt exchange at the ice-ocean interface: Ocean flux and modeling considerations. *Journal of Geophysical Research: Oceans*, **113**, C06014.

Mellor, G.L. & Yamada, T. (1982) Development of a turbulence closure model for geophysical fluid problems. *Reviews of Geophysics*, **20**, 851–875.

Mellor, G.L., McPhee, M.G. & Steele, M. (1986) Ice-Seawater Turbulent Boundary Layer Interaction with Melting or Freezing. *Journal of Physical Oceanography*, **16**, 1829–1846.

Nghiem, S.V., Rigor, I.G., Perovich, D.K., Clemente-Colón, P., Weatherly, J.W.& Neumann, G. (2007) Rapid reduction of Arctic perennial sea ice. *Geophysical Research Letters*, **34**, L19504.

Niiler, P.P. & Kraus, E.B. (1977) One-dimensional models of the Upper Ocean. In: *Modelling and Prediction of the Upper Layers of the Ocean* (Ed. E.B. Kraus). Pergamon Press, Oxford.

Notz, D. & Worster, M.G. (2009) Desalination processes of sea ice revisited. *Journal of Geophysical Research: Oceans*, **114**, C05006.

Notz, D., McPhee, M.G., Worster, M.G. et al. (2003) Impact of underwater-ice evolution on Arctic summer sea ice, *Journal of Geophysical Research: Oceans*, **108**, 3223.

Obukhov, A.M. (1971) Turbulence in an atmosphere with a non-uniform temperature, *Boundary-Layer Meteorology*, **2**, 7–29.

Parkinson, C.L. & Washington, W.M. (1979) A large-scale numerical model of sea ice. *Journal of Geophysical Research: Oceans*, **84**, 311–337.

Pease, C.H., Salo, S.A. & Overland, J.E. (1983) Drag measurements for first-year sea ice over a shallow sea. *Journal of Geophysical Research: Oceans*, **88**, 2853–2862.

Pedlosky, J. (1987) *Geophysical fluid dynamics*, 2nd edn., **xiv**, Springer-Verlag, New York.

Perovich, D.K., Grenfell, T.C., Richter-Menge, J.A., Light, B., Tucker, W.B. & Eicken, H. (2003) Thin and thinner: Sea ice mass balance measurements during SHEBA. *Journal of Geophysical Research: Oceans*, **108**, 8050.

Pollard, R.T., Rhines, P.B. & Thompson, R.O. R.Y. (1973) The deepening of the wind mixed layer, *Geophysical Fluid Dynamics*, **3**, 381–404.

Price, J.F., Weller, R.A. & Pinkel, R. (1986) Diurnal cycling: Observations and models of the upper ocean response to diurnal heating, cooling, and wind mixing, *Journal of Geophysical Research: Oceans*, **91**, 8411–8427.

Robinson, N.J., Williams, M.J. M., Barrett, P.J. & Pyne, A.R. (2010) Observations of flow and ice-ocean interaction beneath the McMurdo Ice Shelf, Antarctica. *Journal of Geophysical Research: Oceans*, **115**, C03025.

Rossby, C.-G. & Montgomery, R.B. (1935) The layer of frictional influence in wind and water current. *Papers in Physical Oceanography and Meteorology, Massachusetts Instituteof Technology and Woods Hole Oceanographic Institute*, **3**, 1–100.

Shaw, W.J., Stanton, T.P., McPhee, M.G. & Kikuchi, T. (2008), Estimates of surface roughness length in heterogeneous under-ice boundary layers, *Journal of Geophysical Research: Oceans*, **113**, C08030.

Shaw, W.J., Stanton, T.P., McPhee, M.G., Morison, J.H. & Martinson, D.G. (2009) Role of the upper ocean in the energy budget of Arctic sea ice during SHEBA. *Journal of Geophysical Research: Oceans*, **114**, C06012.

Shirasawa, K. (1986) Water stress and ocean current measurements under first-year sea ice in the Canadian Arctic. *Journal of Geophysical Research: Oceans*, **91**, 14305–14316.

Sirevaag, A. (2009) Turbulent exchange coefficients for the ice/ocean interface in case of rapid melting.*Geophysical Research Letters*, **36**, L04606.

Skyllingstad, E.D., Paulson, C.A., Pegau, W.S., McPhee, M.G. & Stanton, T. (2003) Effects of keels on ice bottom turbulence exchange, *Journal of Geophysical Research: Oceans*, **108**, 3372.

Steele, M., Mellor, G.L. & McPhee, M.G. (1989) Role of the molecular sublayer in the melting or freezing of sea ice. *Journal of Physical Oceanography*, **19**, 139–147.

Toole, J.M., Timmermans, M.L., Perovich, D.K., Krishfield, R.A., Proshutinsky, A. & Richter-Menge, J.A. (2010), Influences of the ocean surface mixed layer and thermohaline stratification on Arctic Sea ice in the central Canada Basin. *Journal of Geophysical Research: Oceans*, **115**, C10018.

Turner, J.S. (1973) *Buoyancy Effects in Fluids*, Cambridge University Press, Cambridge.

Untersteiner, N. (1961) On the mass and heat budget of Arctic sea ice. *Archives for Meteorology, Geophysics and Bioclimatology Series A*, **12**, 151–182.

Weeks, W.F & Ackley, S.F. (1986) The growth, structure, and properties of sea ice, in *The Geophysics of Sea Ice* (Ed. by N. Untersteiner), pp. 9–164, Plenum Press, New York.

Wettlaufer, J.S., Worster, M.G. &Huppert, H.E. (1997) The phase evolution of young sea ice. *Geophysical Research Letters*, **24**, 1251–1254.

CHAPTER 6

The atmosphere over sea ice

Ola Persson[1] and Timo Vihma[2]

[1] *Cooperative Institute for Research in Environmental Sciences/NOAA/ESRL, University of Colorado, Boulder, CO, USA*
[2] *Finnish Meteorological Institute, Helsinki, Finland*

6.1 Introduction

For interactions with atmospheric dynamics and thermodynamics, sea ice has unique characteristics when compared with other types of Earth surfaces. Sea ice cover is seldom uniform but typically consists of floes of varying size and thickness, and is broken by smaller and larger areas of open water: cracks, leads and polynyas. As a result of interactive dynamic and thermodynamic processes, the sea ice surface may change rapidly. The dynamic processes include ice convergence and divergence as well as rafting and ridging of ice floes, which may change ice thickness faster than any thermodynamic process (Hunke & Zhang, 1999; Babko et al., 2002). Although ice growth and melt are slower processes, the surface temperature may change quickly, except during the melt season when the latent heat of melting/freezing strongly reduces surface temperature variations (Persson, 2012). Rapid variations in other seasons (Persson et al., 2002a) are related to the low heat conductivity of sea ice and, in particular, snow. As snow is a good insulator, it has a strong effect on sea ice growth and melt.

For most of the year, the atmospheric boundary layer (ABL) over sea ice has a stable or near-neutral stratification. Over leads and polynyas, however, the open ocean releases much heat to the atmosphere, and localized convection takes place. In winter, from the point of view of temperature, ice-covered seas are the most heterogeneous surface type with a broad global coverage (Overland et al., 2000). Open water areas have a surface temperature near the freezing point of sea water (approximately −1.8°C), while the surface temperature of thick, snow-covered ice floes in the central Arctic can be as low as −44°C; transitions between the two occur as step changes. Furthermore, the surface albedo varies significantly: from less than 0.1 (open water) up to 0.85 (sea ice covered by dry snow). The surface roughness of ice-covered seas is relatively invariant compared with many land areas, although some regional variations in surface roughness occur from ice ridging, summer melt ponds and marginal ice zones (Andreas et al., 2010a).

In the climate system, sea ice is both an important component and a highly sensitive indicator of climate variability and change. The ABL plays an important role as a buffer between sea ice and the bulk of the atmosphere; understanding the physics of the ABL is a prerequisite for reliable modelling of the present and future climate in both polar regions. For short-term weather prediction, detailed information on sea ice conditions is necessary to correctly model the air–mass modification over the sea, the formation of fog and clouds, and the effects on the wind field. For operational modelling of sea ice conditions, a realistic description of the ABL is essential, as it affects the wind forcing of sea ice dynamics, as well as the radiative and turbulent forcing of ice thermodynamics.

Our knowledge of the ABL over sea ice is based on a variety of sources, including meteorological observations from ice camps, ships, drifting buoys and coastal stations. Important data have originated from manned stations based on ice floes. In the periods 1937–1991 and 2003–2015, Russian scientists organized a total of 41 drifting ice stations in the Arctic Ocean, and data from these stations have been widely utilized (Jordan et al., 1999; Makshtas et al., 1999; Vihma et al., 2008). Multidisciplinary observations were made at a 1-year ice station during the 'Surface Heat Budget of the Arctic

---

Sea Ice, Third Edition. Edited by David N. Thomas.
© 2017 John Wiley & Sons, Ltd. Published 2017 by John Wiley & Sons, Ltd.

Ocean (SHEBA; Uttal et al., 2002) project, yielding major advances in the understanding of the atmosphere over sea ice (Persson et al., 2002a, 2012; Andreas et al., 2002; Intrieri et al., 2002; Shupe et al., 2005, 2006; Grachev et al., 2005, 2007, 2013). Other major experimental activities include the Arctic Ocean expeditions by the Swedish icebreaker Oden in 1991, 1996, 2001 and 2008 (Leck et al., 1996, 2001; Tjernström et al., 2004, 2012), with some resulting studies focusing on the late-summer ABL (Nilsson, 1996; Tjernström, 2005, Shupe et al., 2013), and the 15-month drift of the schooner *Tara* (Gascard et al., 2008; Vihma et al., 2008, Jakobson et al., 2013). Airborne measurement campaigns of the ABL over sea ice have provided spatially averaged turbulent and radiative fluxes over the heterogeneous ice cover (Overland 1985; Walter & Overland 1991; Brümmer et al., 2002a).

The thermal stratification of the ABL over sea ice varies in space and time, being affected by snow cover, ice thickness and presence of open water; the diurnal cycle of solar radiation; synoptic-scale changes in weather conditions; and seasonal cycles of solar radiation, free-tropospheric temperatures and various surface characteristics. Hence, cryospheric and oceanographic variability impacts the ABL from below, while free tropospheric variability affects it from above. In winter, the presence of sea ice generates a strong climatological signal in the spatial distributions of near-surface air temperature, as illustrated in Figure 6.1. The air over sea ice is climatologically clearly colder than over the open ocean but warmer than the coldest surrounding archipelago and continental regions. The latter is due to the heat flux from the ocean, which reaches the marine atmosphere partly via leads and polynyas and partly via heat conduction through snow and ice.

6.2 Surface energy budget

6.2.1 Current knowledge

Energy fluxes affecting Arctic sea ice come from either the atmosphere or the ocean. Atmospheric energy fluxes, F_{atm}, are either radiative fluxes or turbulent energy fluxes and can be expressed as:

$$F_{atm} = SW_d - SW_u + LW_d - LW_u - H_s - H_l \quad (6.1)$$

where SW_d/SW_u and LW_d/LW_u are the downwelling/upwelling, short-wave/long-wave radiative fluxes, respectively; and H_s/H_l are the turbulent sensible/latent heat fluxes. The downwelling radiative fluxes depend on solar zenith angle (i.e. season, time of day) and characteristics of the atmosphere, including cloud macro- and microphysical structure, atmospheric temperature and humidity, and aerosol content. The upwelling surface radiative fluxes depend on the surface characteristics, such as surface type (snow, sea ice, open water) and specific properties of these surfaces [e.g. albedo (α_s), emissivity (ε_s), temperature (T_s)]. The turbulent energy fluxes depend on both surface characteristics [e.g. surface type (T_s), surface roughness (z_0)] and atmospheric characteristics [e.g. atmospheric temperature (T_a), atmospheric stability, atmospheric specific humidity (Q_a), boundary-layer winds (U)]. By reformulating equation (6.1), these dependencies are more easily seen in:

$$F_{atm} = SW_d(1 - \alpha_s) + (1 - \varepsilon_s)LW_d - \varepsilon_s \sigma T_s^4 \\ - \rho_a c_p C_H U(T_s - T_a) - \rho_a L_v C_v U(Q_s - Q_a) \quad (6.2)$$

where σ is the Stefan–Boltzmann constant, ρ_a is the atmospheric density, c_p is the atmospheric heat capacity, C_H is the heat transfer coefficient, L_v is the latent heat of vaporization, C_v is the moisture transfer coefficient, and Q_s is the surface specific humidity (generally determined from T_s and the assumption of moisture saturation at the Arctic Ocean surface). C_H and C_v both depend on z_0 and atmospheric stability. The last two terms are bulk parameterizations of H_s and H_l, respectively, and are used here only to illustrate some of the major dependencies of these terms. Both the atmospheric and surface characteristics vary in time and space over significant ranges of temporal and spatial scales, and this variability depends on a variety of atmospheric and surface processes. Understanding this variability and how the processes combine to produce the net energy flux is key to understanding atmosphere–surface interactions over the Arctic Ocean. While the common two-band (broadband) model is used here for simplicity, some discussions will require consideration of specific narrow spectral bands or the full electromagnetic spectrum. The variability in many of the key terms and parameters in equation (6.2) occurs in narrow bands of the electromagnetic spectrum, such as for the cloud effects on SW_d and surface albedo (Grenfell & Perovich 1984; Perovich et al., 2002).

The surface of the Arctic Ocean consists of either snow-covered sea ice, sea ice with or without surface

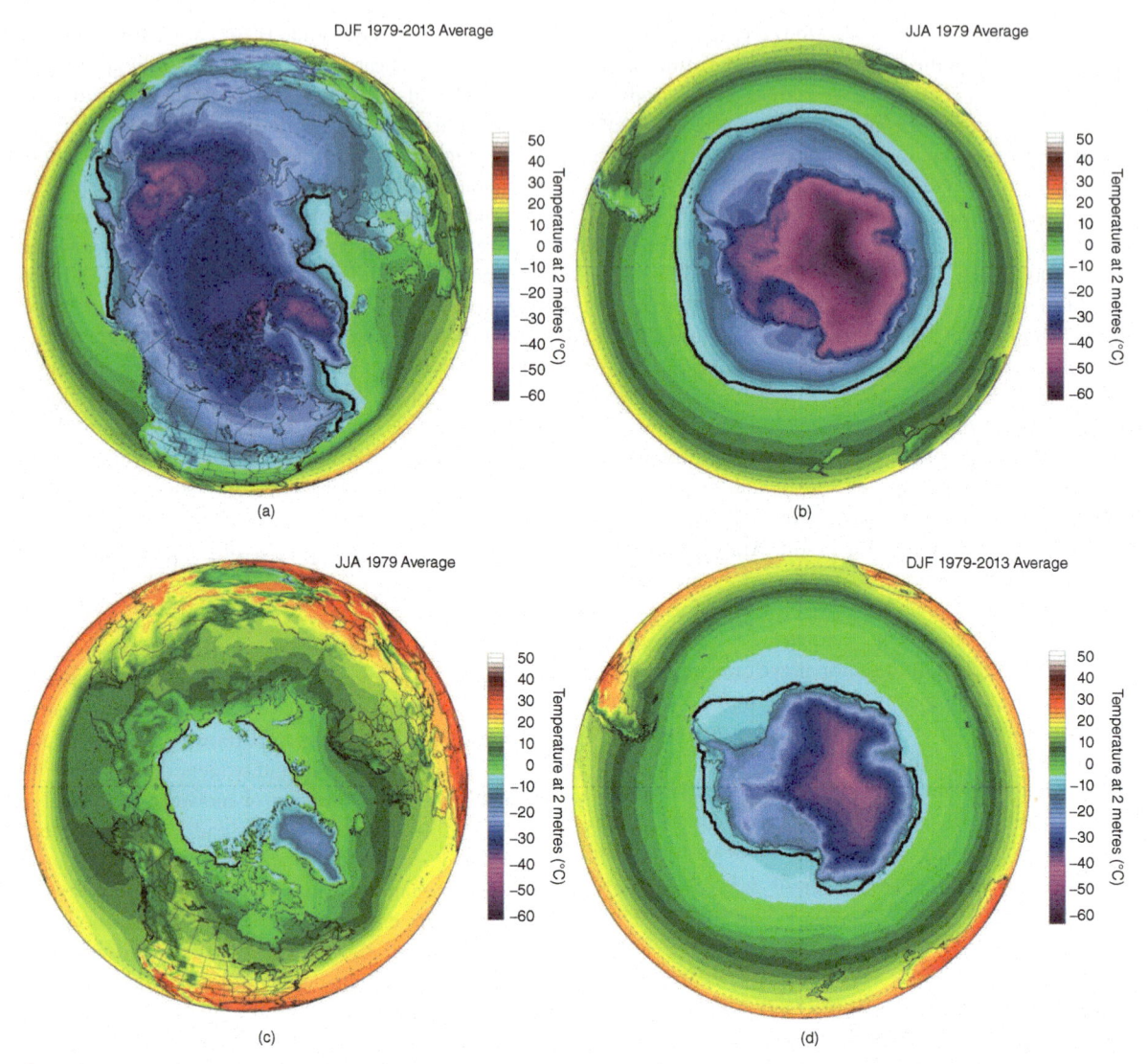

Figure 6.1 Seasonal mean 2 m air temperature in Arctic winter (December–February) (a), Arctic summer (June–August) (b), Antarctic winter (June–August) (c) and Antarctic summer (December–February) (d) from National Center for Atmospheric Research-Climate Forecast System Reanalysis (NCEP-CFSR) reanalysis averaged over 1979–2013. The heavy black line shows the approximate extent of sea ice.

melt ponds, or open water. The open water may be either large extensive areas persisting for weeks to months or more, or smaller more transitory areas associated with leads lasting hours to days. Polynyas are intermediate-sized, open water areas that can persist for weeks to months, and are often forced by persistent terrain-relative winds in coastal areas. For areas covered by sea ice, the conductive heat flux at the atmosphere–surface interface (F_{c0}, positive upward) is theoretically in balance with the atmospheric heat flux after considering short-wave radiative fluxes transmitted into the sea ice and ignoring phase changes, though this balance is rarely seen in observations, i.e.:

$$F_{net} = F_{c0} + F_{atm} - SW_t \approx 0 \quad (6.3a)$$

where the short-wave transmission through the ice, SW_t, can be estimated from

$$SW_t = SW_d(1 - \alpha_s)\, f(D_s, D_i)$$
$$= SW_d(1 - \alpha_s)\, e^{-k_{xs} D_s} I_0 e^{-k_{xi} D_i} \quad (6.3b)$$

The parameters represent snow depth (D_s), ice thickness (D_i), solar extinction coefficient in snow (kx_s) and in ice (kx_i), and ice surface transmission parameter (I_0) (Light et al., 2008). This solar transmission was found to be larger than anticipated during the SHEBA campaign on multi-year ice (~5 W m^{-2}), especially during the late summer when the ice no longer had snow cover (Perovich et al., 2003; Light et al., 2008; Sirevaag et al., 2011). It is also believed to be even more significant for first-year ice (FYI), which has recently replaced large areas of multi-year ice (MYI), as FYI is thinner and may not contain as many occlusions that scatter light. The solar transmission is also believed to be a major contributor to thermal absorption and biological activity in the upper ocean under the sea ice (e.g. Jackson et al., 2010, 2012).

Persson (2012) estimated the conductive heat flux, F_{c0}, directly from the measured surface and snow–ice interface temperatures (T_s and T_{ice}) and the measured snow depth and ice thickness by the relations:

$$F_{c0} = -k_s[(T_s - T_{ice})/D_s], \quad D_s > 0.06 \text{ m} \quad (6.4a)$$

$$F_{c0} = -[r_s k_s (T_s - T_{ice})/D_s \quad (6.4b)$$
$$+ (1 - r_s) k_i (T_s - T_w)/D_i], \quad 0.005 < D_s \leq 0.06$$

$$F_{c0} = -k_i[(T_s - T_w)/D_i], \quad D_s \leq 0.005 \text{ m} \quad (6.4c)$$

where k_s (= 0.3 W m^{-1} K^{-1}; Sturm et al., 2002) and k_i (= 2.0 W m^{-1} K^{-1}) are the thermal conductivities of snow and ice, respectively; T_w ($\approx$ −1.8°C) is the temperature of the ocean water below the ice, and r_s is the snow depth-to-threshold ratio [= min(D_s/0.06, 1)]. Equations (6.4a–c) illustrate the inverse dependence of F_{c0} on snow depth and ice thickness.

Figure 6.2 shows direct measurements of surface energy budget terms and parameters during the year-long deployment at SHEBA. The net long-wave radiative fluxes (LW$_{net}$ = LW$_d$ − LW$_u$) are generally negative throughout the year, and net short-wave radiative fluxes (SW$_{net}$ = SW$_d$ − SW$_u$) provide the primary positive energy flux during 6 months of the year. The conductive and turbulent sensible heat fluxes provide the only positive energy flux to the surface during the winter, and are less important in summer. The latent heat flux has a small cooling effect in summer, while the transmission of solar radiation through the ice is only significant in late summer when the SHEBA MYI was snow-free. In winter, there is essentially a pseudo-balance between positive contributions from the conductive heat flux F_{c0} and the turbulent sensible heat flux and the LW$_{net}$ radiative loss, with only a small negative net flux F_{net}, likely representing snow-layer heat loss. In summer, the negative LW$_{net}$ only partially compensates for the strongly positive SW$_{net}$, while the turbulent and conductive heat fluxes are restricted by the nearly invariant surface temperature (Persson, 2012), resulting in a large positive F_{atm} and F_{net}. In spring and autumn, all terms are at times significant. The summer positive F_{net} reaches greater magnitudes than the winter negative values, though it is significantly positive for only 2–3 months. The freezing of ice in winter and melting of ice in summer are determined by these net energy imbalances, and the latter occurs much more rapidly than the former because of the larger summer F_{atm} magnitudes. On an annual basis, LW$_{net}$ and SW$_{net}$ are about an order of magnitude greater than the other terms (Table 6.1), although the magnitude of the net radiative energy flux (Q_{net} = LW$_{net}$ + SW$_{net}$) is very similar to those of the other fluxes.

The short-term variability of the individual terms is quite large. The data in Figure 6.2 have been smoothed with a 3-day running mean, which allows the impact of atmospheric synoptic variability (3–5 days) to be clearly evident. Radiative energy fluxes vary by 20–40 W m^{-2} over this timescale, and conductive and turbulent heat fluxes vary by 5–20 W m^{-2}. The variability on even shorter timescales can be more than twice these values, indicating the importance of transient atmospheric events on the net surface energy balance. The large, sudden increases in T_s (Figure 6.2c) occur because of long-wave radiative impacts from super-cooled liquid water clouds (Sverdrup, 1933; Persson et al., 1999, 2016; Stramler et al., 2011). The wintertime minimum T_s over sea ice is believed to be ~−49 to −44°C, which is significantly higher than over Arctic land areas because of the conductive heat flux from the relatively warm ocean only a few metres below (Sverdrup, 1933) and because of heat fluxes from leads. This difference illustrates the importance of F_{c0}. Summertime albedo changes, both gradual and rapid, are due to melting/accumulation of surface snow and formation/freezing of surface melt ponds.

Figure 6.3 summarizes estimates of regional energy fluxes for an ice-covered Arctic Ocean system. The values of these energy fluxes are obtained from a variety of sources, and many of them have significant

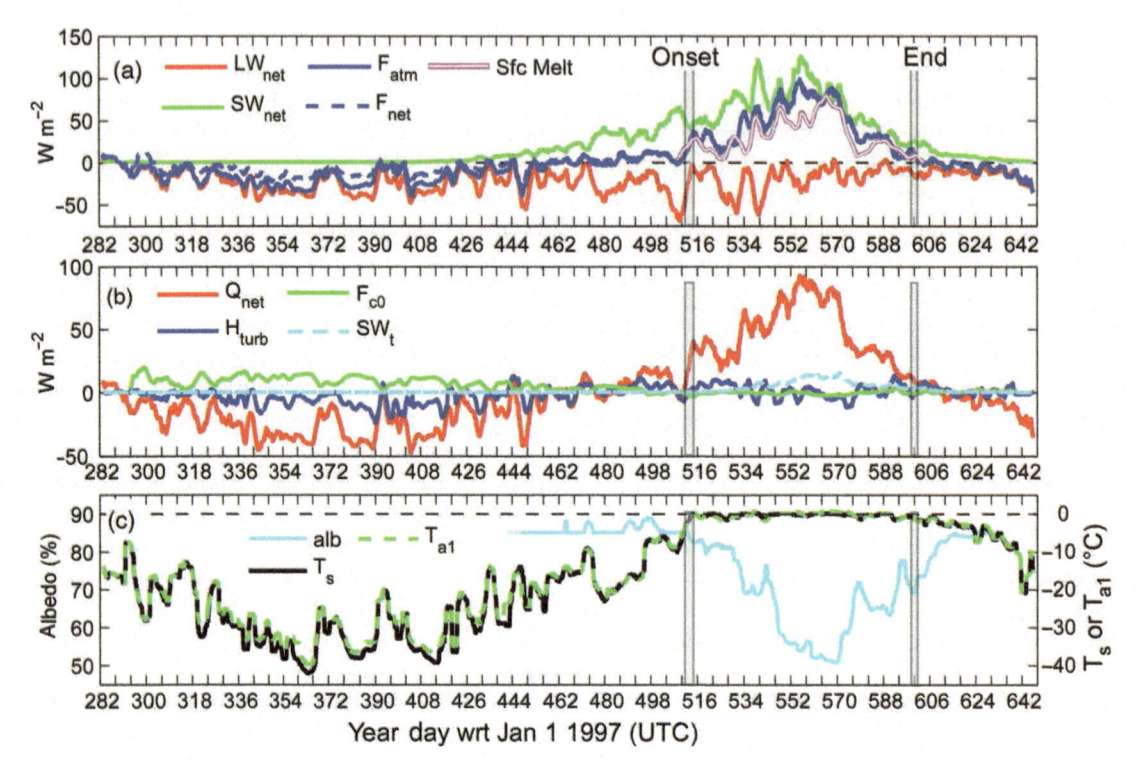

Figure 6.2 Annual cycle of 3-day running means of key Surface Heat Budget of the Arctic Ocean (SHEBA) project surface energy budget terms: (a) net long-wave radiative flux (LW_{net}), net short-wave radiative flux (SW_{net}), atmospheric energy flux (F_{atm}), net atmospheric energy flux (F_{net}); and (b) net radiative energy flux (Q_{net}), turbulent heat flux (H_{turb}), conductive heat flux at the atmosphere–surface interface (F_{c0}), short-wave transmission through the ice (SW_t). (c) Three-day running medians of albedo (□), surface temperature (T_s) and near-surface atmospheric temperature (T_{a1}) are shown. The outlined solid magenta line in the summer in (a) shows the energy flux equivalent of the measured surface snow and ice melt. The labelled vertical shaded areas show the times of the onset and end of the summer melt. Note that dashed lines in all panels, denoting F_{net}, SW_t and T_{a1}, respectively, are at times obscured by the respective solid lines denoting F_{atm}, F_{c0}, and T_s due to having similar values. Source: From Persson, 2012. Reproduced with permission of Springer.

uncertainties. For instance, the annual surface downwelling solar radiation varies from 92 to 101 W m^{-2} in the sources listed in Table 6.1, resulting in an annual surface net solar energy flux of 21.8–29.5 W m^{-2} after the variations in surface albedo are considered. However, the values illustrate the complexity of the Arctic system, the relative importance of the various flux processes, and the interdependence of processes in the system. Furthermore, atmospheric transports of sensible and latent (moisture) energy from lower latitudes are significant energy contributors to the Arctic system (e.g. Woods et al., 2013), and are generally believed to represent transport into the Arctic Basin by transient cyclones. These transport values are quite similar to the net Arctic energy loss to space.

This discussion will focus on surface energy fluxes, as these are key for impacts on sea ice. Some surface energy fluxes are quite large while others are up to two orders of magnitude smaller (e.g. LW radiative fluxes and ocean heat fluxes, respectively), and others are intermediate (e.g. solar fluxes). Figure 6.3 also does not explicitly include ocean heat fluxes due to local Arctic sources, such as solar heating of the upper ocean below thinner, snow-free sea ice, in leads, and in open water regions of the emerging marginal ice zone (MIZ) (e.g. Steele et al., 2010).

In a steady-state system, the annual net energy flux to the sea ice from both the atmosphere and the ocean should be 0 W m^{-2}, and an error or imbalance of 10 W m^{-2} is approximately equivalent to an annual growth or melt of 1 m of sea ice. In fact, Kwok and

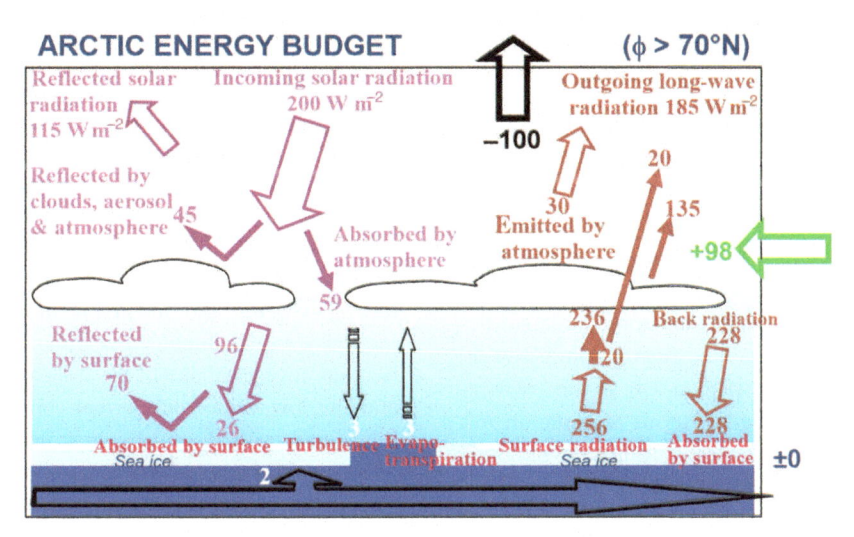

Figure 6.3 Estimates of regional annual mean Arctic energy fluxes over a mostly ice-covered Arctic Ocean assuming a cylinder over the Arctic at approximately 70°N.

Table 6.1 Annual surface energy fluxes over multi-year ice from Surface Heat Budget of the Arctic Ocean (SHEBA) project and pre-SHEBA sources.[a]

Parameter	SHEBA	SHEBA ($\alpha_{July} = 0.64$)	B66	MU71	M82 3m	M82 0.8 – ∞ m	EC93	L98
SW_d	91.88	91.88	98.1	100.04	99.85	99.85	101.3	96.83
SW_{net}	23.36	21.77	23.5	24.16	23.74	23.74	29.49	23.23
LW_d	230.80	230.80	211.5	220.22	220.35	220.35	215.34	219.31
LW_{net}	−21.32	−21.32	−24.4	−24.42	−20.66	−22.06	−28.41	−22.74
H_s	−2.20	−2.20	−1.3	−3.58	−3.59	−3.59	−1.84	−3.01
H_l	1.06	1.06	0.9	4.25	4.11	3.49	1.55	2.32
F_{atm}	3.18 [4.17]	1.59 [2.29]	−0.5	−0.92	2.56	1.79	1.33	1.18
F_{c0}	2.45 (5.04)	2.45 (5.04)	N/A	7.96	6.31	5.82	8.12	5.68
F_{net} [F_{net}]	5.63 (8.22) [6.57 (9.12)]	4.04 (6.63) [4.69 (7.24)]	N/A	7.04	8.87	7.60	9.45	6.86

B66, Badgley (1966); MU71, Maykut and Untersteiner (1971); M82, Maykut (1982); EC93, Ebert and Curry (1993); L98, Lindsay (1998).
[a]Values are given as annual average fluxes (W m⁻²). They were computed from monthly mean components, with SHEBA values for the month of October interpolated from the September and November values. The third column is identical to the second, except that the SW_u for July was determined assuming an albedo of 0.64 (Perovich et al., 2002). F_{c0} and F_{net} in columns 2 and 3 use $k_s = 0.14$ W m⁻¹ K⁻¹ (0.3 W m⁻¹ K⁻¹). The square brackets show annually averaged hourly atmospheric flux and F_{net} values (requiring all components to be measured that hour), rather than the summation of monthly mean budget components. Two ice categories from M82 are shown. $F_{c0} + F_{net}$ are not available for B66.
Source: Persson et al. (2002a). Reproduced with permission of American Geophysical Union.

Untersteiner (2011) estimated that an excess of 1 W m⁻² in the net annual surface energy flux over the past 30 years can account for the observed reduction in sea ice extent and mass over that time. Many of the terms, including the relatively small long-distance ocean heat flux and atmospheric turbulent heat fluxes depicted in Figure 6.3, have uncertainties that are greater than this 1 W m⁻², as illustrated by the range of net surface solar radiation estimates discussed earlier.

The complexity of the Arctic system is due not only to the numerous processes in the system but also

166 Chapter 6

to the interdependence between these processes, as implied by Figure 6.3. We are only now starting to recognize some of these interdependencies; there are likely many others we have not yet recognized, and the interdependencies are probably changing with the climate. An interdependency example is the wintertime relationships between the downwelling long-wave radiative flux (forced by clouds), and the processes driven by the surface temperature response to that forcing, which are the outgoing long-wave radiative flux, the turbulent sensible heat flux, and the conductive flux through the ice and snow (Figure 6.4). In the dry winter atmosphere, the first and last terms of equation (6.2) are zero and very small, respectively, leaving only three flux terms and the conductive flux given by equation (6.4). LW_d represents the external forcing by clouds and atmospheric thermal structure, and the system responds by distributing the forcing energy between the other three surface terms and a change in T_s in accordance with the properties of the system. These properties include C_H, atmospheric surface layer stability and wind speed, for H_s; ε_s for LW_u; and D_s, D_i, k_s and k_i for F_{c0}. LW_{net} ($= LW_d - LW_u$) can be viewed as the portion of LW_d that is available to a response in the two non-radiative flux processes.

Figure 6.4 shows that the system response to LW_d forcing from wintertime mixed-phase clouds is distributed to each of these two fluxes, with the magnitude of each response dependent on the associated properties. For the SHEBA environment shown, F_{c0} decreases by about 9 W m^{-2} and H_{turb} ($= H_s + H_l$) increases by about 15 W m^{-2} to compensate for the 40 W m^{-2} increase in LW_{net} as clouds with significant liquid water pass over the site. The response in T_s is then dependent not only on the forcing LW_d, but also on the magnitudes of the system responses to the forcing. For instance, because of the typically strong stability of the atmospheric surface layer and the low values of thermal conductivity of snow, the responses in H_s and F_{c0} may not be as large as in other environments, thereby producing much larger responses in T_s for this Arctic environment. In addition, some properties, such as stability, wind speed, D_s and D_i, may have spatial, temporal, or interannual variability. These interdependencies become even more complex when SW_d and H_l become important, and significant surface heterogeneity is introduced for α, D_s, D_i, T_s and other characteristics.

Furthermore, limitations on the surface energy system produce different responses in the system. For example, the surface temperature of sea ice cannot exceed 0°C. Hence, during the melt season, H_{turb} and

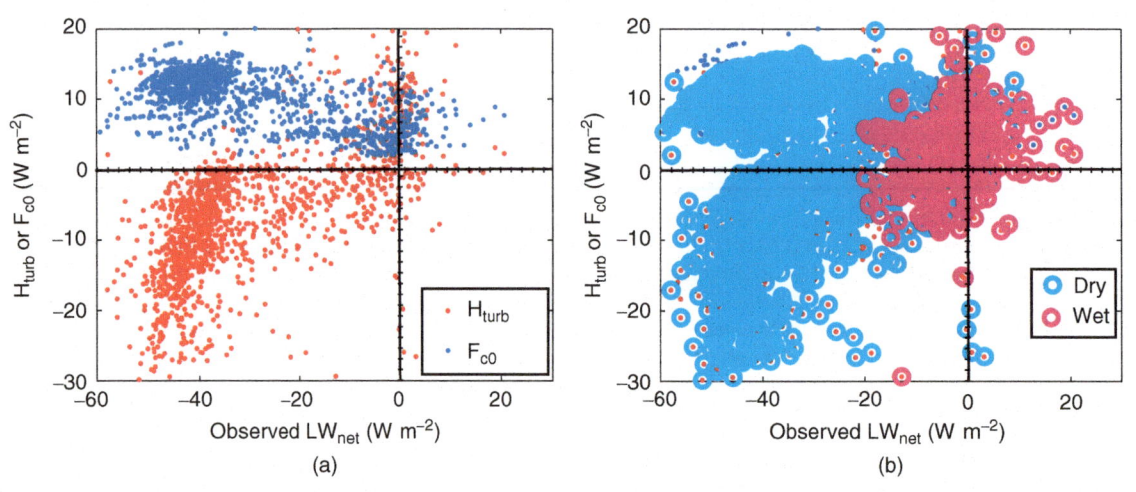

Figure 6.4 Observed hourly values of turbulent heat flux ($H_{turb} = H_s + H_l$; red) and conductive heat flux, F_{c0} (blue), as a function of observed net long-wave radiative flux ($LW_{net} = LW_d - LW_u$) (a) and the liquid water path (LWP) forcing (coloured circles; dry, LWP < 5 g m^{-2}; wet, LWP > 10 g m^{-2}) (b) for each point. The LWP threshold produces a bimodal LW_d (LW_{net}) distribution. The data are from November 1997 to February 1998 at the Surface Heat Budget of the Arctic Ocean (SHEBA) project (from Persson et al., 2016).

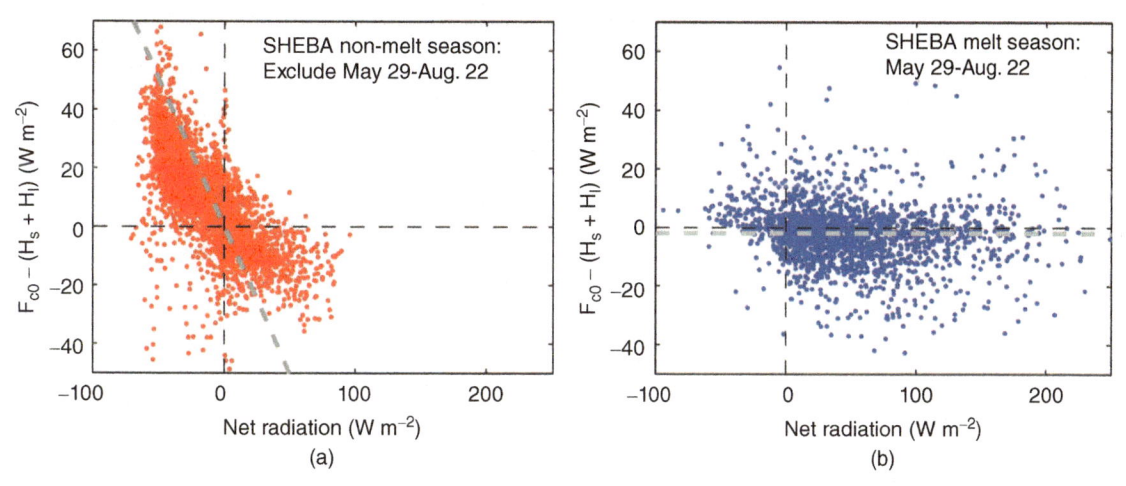

Figure 6.5 Hourly values of conductive – turbulent heat flux (F_{c0} - H_{turb}) at the Surface Heat Budget of the Arctic Ocean (SHEBA) project as a function of net radiative energy flux, Q_{net} [= net short-wave (SW_{net}) + net long-wave radiative flux (LW_{net})] during: (a) the non-melt season (all days excluding May 29 – Aug. 22); and (b) the melt season (May 29 – Aug. 22). Source: From Persson, 2012. Reproduced with permission of Springer.

F_{c0} cannot respond to and compensate for surface radiation changes (Figure 6.5), so increases in net surface radiation fluxes can all be used to melt ice and episodes of large net radiation can be enhanced by positive contributions from H_{turb} and F_{c0}. Therefore, the length of the melt season (when $T_s \approx 0°C$) is extremely important for determining the amount of melt, and only short increases can provide significant additional melt as the compensating flux responses in the system available during other times of the year are not available during the melt season. Changes in melt-season transition times and understanding transition processes are therefore crucial.

6.2.2 Changes in a new Arctic

The energy fluxes and relationships discussed in the previous section are estimates from a system that include primarily a MYI component. The uncertainty of these values is even larger for the developing 'New Arctic', which has much greater amounts of FYI and for which various processes may have different significance. The surface albedo value and its annual cycle may be significantly different, the solar transmission is probably more important, the turbulent atmospheric energy fluxes are possibly quite different, and cloud characteristics may be different, altering both the long-wave and short-wave radiative fluxes. The interdependencies in the atmosphere–surface system have probably changed, as properties such as stability, wind speed, D_s and D_i may have changed as thinner, smoother FYI has become more dominant. Surface heterogeneity in the New Arctic certainly includes greater amounts of FYI and may involve greater variability of energetically important surface types (greater amounts of thin ice and open water). This new heterogeneity probably impacts the radiative and turbulent atmospheric heat fluxes in addition to the energy fluxes in the upper ocean.

The greater areas of seasonal ice in the New Arctic have made ocean heat fluxes due to local Arctic sources, such as solar heating of the upper ocean below thinner, snow-free sea ice, in leads, and in open water regions of the emerging MIZ (e.g. Steele et al., 2010), much more important. Heat storage and altered stratification in the upper ocean due to changes in radiative heat fluxes may impact the flux of energy from deeper in the ocean and may impact the formation or melt of sea ice throughout the year (Jackson et al., 2012). Figure 6.3 illustrates the complexity of the Arctic energy system and the intricate coupling of the various processes that lead to an approximate balance for the surface energy fluxes. Clearly, the changes that have occurred over the past two to three decades have probably altered the significance of various processes and either produced a new balance for the New Arctic or a system that is still in imbalance as it continues to evolve.

6.3 Vertical structure of the atmosphere (from surface layer to inversion top)

6.3.1 Effects of stability on surface-layer profiles

The turbulent sensible and latent heat fluxes (H_s and H_l) in the surface energy budget equation (6.1) and the related turbulent momentum flux or stress τ are key parameters for describing the surface–atmosphere interactions in the atmospheric surface layer. The surface layer is classically considered to be the lowest 10% of the boundary layer and is also known as the constant flux layer. These parameters are defined by:

$$H_s = \rho c_p \overline{wt} \equiv -\rho c_p u_* t_* \quad (6.5)$$

$$H_L = \rho L_v \overline{wq} \equiv -\rho L_v u_* q_* \quad (6.6)$$

$$\tau = -\rho \overline{uw} \equiv \rho u_*^2 \quad (6.7)$$

where ρ is the density of air; c_p is the specific heat of air at constant pressure; L_v is the latent heat of vaporization or sublimation; and u, w, τ, and q are turbulent fluctuations in longitudinal velocity, vertical velocity, temperature, and specific humidity, respectively. The overbar indicates a time-averaged covariance measurement for, nominally, 10–60 minutes. Once equation (6.7) yields u_*, the friction velocity, equations (6.5) and (6.6) provide the related scalar flux scales t_* and q_*.

Estimations of τ, H_s, and H_L from readily available non-turbulence parameters are frequently desired and are usually based on the bulk aerodynamic method, in which:

$$\tau = \rho C_{Dr} S^2 \quad (6.8)$$

$$H_s = \rho c_p C_{Hr} S(\Theta_s - \Theta_r) \quad (6.9)$$

$$H_L = \rho L_v C_{Er} S(Q_s - Q_r) \quad (6.10)$$

Here Θ_s is the time-averaged surface potential temperature; Θ_r is the average potential temperature at reference height r; Q_s is the average surface specific humidity; Q_r is the average specific humidity at height r; and S is a velocity scale that represents the wind velocity and the gustiness. The heart of the bulk aerodynamic method is finding the transfer coefficients (C's) in equations (6.8)–(6.10). C_{Dr} is the drag coefficient appropriate at height r, and C_{Hr} and C_{Er} are the corresponding sensible and latent heat transfer coefficients for height r. These coefficients depend on the reference height r and on the stability, which is quantified by the stability parameter $\zeta = r/L$, where L is the Obukhov length (Monin & Obhukov 1954). That is (e.g. Andreas & Murphy 1986; Launiainen & Vihma 1990):

$$C_{Dr} = c_{Dr}^2 = \left[\frac{k}{\ln(r/z_0) - \psi_m(r/L)}\right]^2 \quad (6.11)$$

$$C_{Hr} = c_{Dr} c_{Hr} = \left[\frac{k}{\ln(r/z_0) - \psi_m(r/L)}\right]\left[\frac{k}{\ln(r/z_T) - \psi_h(r/L)}\right] \quad (6.12)$$

$$C_{Er} = c_{Dr} c_{Er} = \left[\frac{k}{\ln(r/z_0) - \psi_m(r/L)}\right]\left[\frac{k}{\ln(r/z_Q) - \psi_h(r/L)}\right] \quad (6.13)$$

where k (= 0.4) is the von Kármán constant, z_0 is the momentum roughness length, z_T and z_Q are the scalar roughness lengths, and ψ_m and ψ_h are integrals of $(1 - \varphi_m)$ and $(1 - \varphi_h)$, respectively, and φm and φh are empirical momentum and scalar stability correction functions, respectively. L is computed by:

$$L = -\frac{u_*^3 \, \theta_v}{\kappa g <w'\theta_v'>} \quad (6.14)$$

where θ_v' is the fluctuation of virtual potential temperature and g is the gravitational constant. L is negative for unstable conditions and positive for stable conditions, with L decreasing in magnitude for increasing stability.

Equations (6.11)–(6.13) show that the turbulent transfer in the atmospheric surface layer is a logarithmic function of height for neutral conditions (ψ_m and $\psi_h = 0$). The turbulent transfer coefficients are dependent on the roughness lengths and stability, with increasing roughness producing enhanced turbulent transfer. The functions ϕ_m and ϕ_h are positive for stable conditions, and increase with increasing stability. The corresponding changes in ψ_m and ψ_h initially increase the transfer coefficients with increasing stability for weakly stable conditions but then decrease them with increasing stability for greater stability (Andreas & Murphy 1986; Andreas 1998; Grachev et al., 2005). Physically, this decrease in transfer coefficients for very stable conditions represents a decrease in turbulence (Grachev et al., 2005).

The stable boundary layer (SBL) has elicited increased attention in the last few decades, initially to understand nocturnal mid-latitude conditions and later to understand polar ABLs. Based on Monin–Obukhov similarity theory (Monin & Obhukov, 1954) and mid-latitude observations and conceptual models (Mahrt, 1999), Grachev et al. (2005) utilized the year-long SHEBA turbulence data to identify four vertical mixing regimes in the stable ABL based on stability and height (Figure 6.6). For weakly stable conditions, the traditional constant flux layer exists in the lowest few tens of metres above a blending layer that exists between heights approximately equal to the roughness elements and those in the surface layer. However, as stability increases, this constant flux layer is replaced by a transition regime where similarity theory still holds but is based on local similarity and fluxes at height z. Fluxes are no longer constant with height within this regime and the stability correction functions are more accurately expressed with local fluxes. However, turbulence remains nearly continuous in this transition regime. As stability increases, turbulence weakens but is still present, the fluxes vary with height, and the wind is strongly influenced by the Coriolis force. Hence, this is referred to as the turbulent Ekman regime. At even stronger stability or higher heights, turbulence is absent or only intermittent, and the wind is governed by Coriolis turning and the Ekman spiral, even down to the surface. Surface stress becomes negligible, although intermittent and sporadic turbulence does occur. This is the super-critical stable regime or very stable regime as described by Mahrt (1999) for mid-latitude nocturnal conditions. These basic turbulence regimes are sometimes perturbed by gravity waves, detached elevated turbulence and inertial oscillations.

Observations over sea ice show stability conditions up to $\zeta \approx O(100)$ (Grachev et al., 2005, 2007). Unfortunately, in the SBL, empirical stability correction functions vary widely even for much smaller values of ζ (Figure 6.7). Some of these functions are from early classical studies (e.g. Dyer, 1974) and others have been specifically determined later for SBLs (e.g. Beljaars & Holtslag, 1991; Cheng & Brutsaert, 2005). Two have even been determined from polar

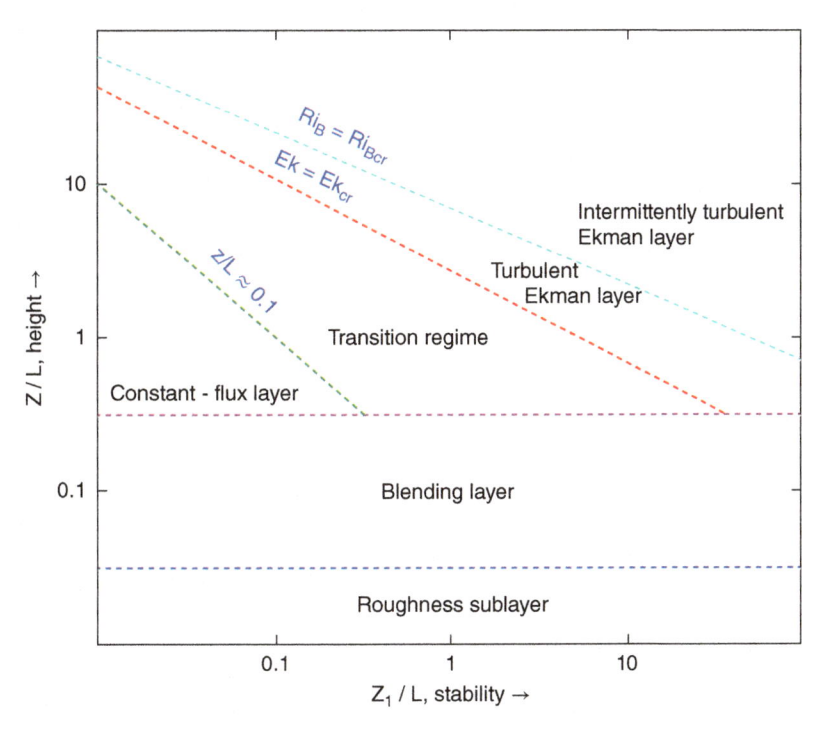

Figure 6.6 Schematic diagram of the stable boundary layer scaling regimes as functions of the stability (z_1/L) and height (z). Here $z_1 \approx 2$ m Dividing lines between scaling regimes are drawn based on the Ekman number (Ek) and the bulk Richardson number (Ri_B), with critical values of each Ek_{cr} ~1 and Ri_{Bcr} ~0.2. Source: From Grachev et al., 2005. Reproduced with permission of Springer.

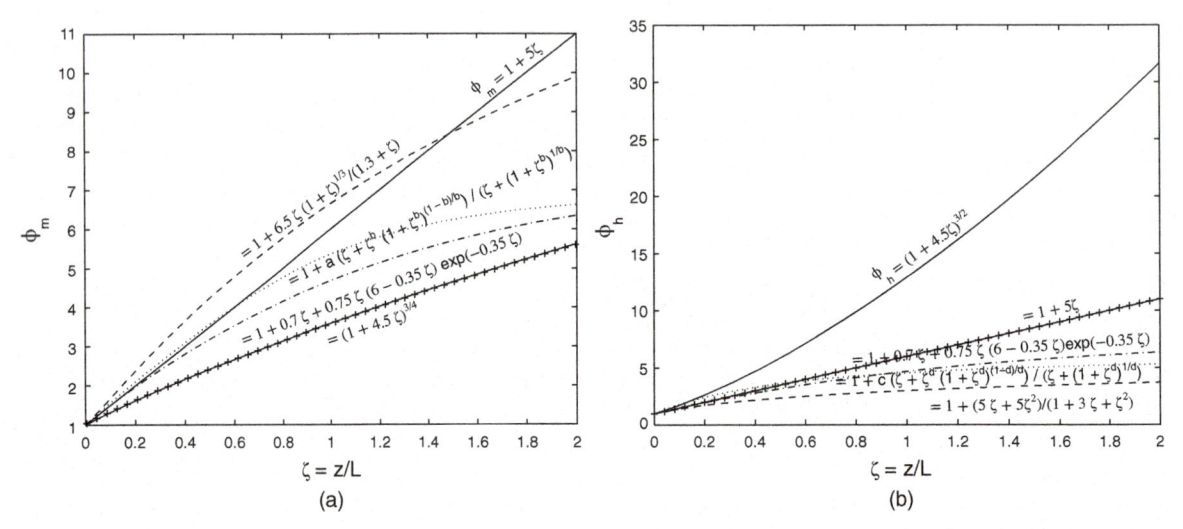

Figure 6.7 Dependence of several stability correction functions on the stability parameter (z/L) for momentum (φ_m) (a) and scalars (φ_h) (b) as a function of stability (adapted from Andreas, 1998). The curves from Chang and Brutsaert (2005) (dotted) and Grachev et al. (2007) (dashed) have been added. Coefficients for the Chang and Brutsaert curves are as follows: a = 6.1, b = 2.5, c = 5.3, and d = 1.1.

observations (Lettau 1979; Grachev et al., 2007). Most of the functions are valid for $\zeta \sim 1–10$, although those of Grachev et al. use data where $\zeta \approx O(100)$. Clearly, because of the large variability, validation of these stability correction functions is necessary for the SBLs encountered over sea ice.

6.3.2 Temperature and humidity profiles above the surface layer

The structure of the lower troposphere over the perennial Arctic pack ice is unique in the world because of the year-round persistence and large spatial extent of the lower ice boundary. Hence, it requires some special definitions. Temperature profiles in the lowest 2 km over the pack ice show a consistent general inversion with a top between 500 and 1500 m altitude (Figure 6.8). This inversion layer is formed because of the relatively cold pack-ice surface, exists in all seasons, and large parts of it can be locally inactive (Persson et al., 1992; Serreze et al., 1992; Andreas et al., 2000; Tjernström & Graversen, 2009). We will refer to this layer as the Arctic inversion (AI) (Figures 6.8 and 6.9). Although the AI layer has, at times, also been referred to as the Arctic boundary layer, we argue here that a true ABL that represents the local surface mixing processes is also frequently observed and is much shallower than the AI layer, and is hence distinct from the AI. This has been previously recognized (e.g. Mahrt, 1981; Andreas et al., 2000). The base of the AI can be very near the surface at times, especially during clear, calm, wintertime conditions when even mechanical mixing has collapsed except in the lowest few metres (e.g. Figure 6.9a) (Grachev et al., 2005; Tjernström & Graversen, 2009). However, the AI base is often elevated because the lower portion of the AI becomes modified by local near-surface buoyant or mechanical mixing and/or from cloud processes. This near-surface mixing layer will be referred to as the ABL, although others (Andreas et al., 2000) have called this the surface layer to show its direct connection to the surface. In our discussion, we will reserve the term 'surface layer' for its classical meaning, i.e. for the lowest part of the ABL in which surface stress is approximately constant and which is, at most, a few tens of metres deep over the pack ice and frequently even shallower (Persson et al., 2002a; Grachev et al., 2005). Profile relationships in the surface layer were described in the previous section. A cloud-induced mixed layer resulting from 'upside-down convection' (Smedman, 1988) is observed above the ABL extending from cloud top to a few hundred metres below cloud base (Harrington et al., 1999; Solomon et al., 2011). This cloud mixed layer (CML) is often distinct from the ABL below, but sometimes couples with it to form a deeper, well-mixed coupled ABL and CML (e.g. Wang et al., 2001; Shupe

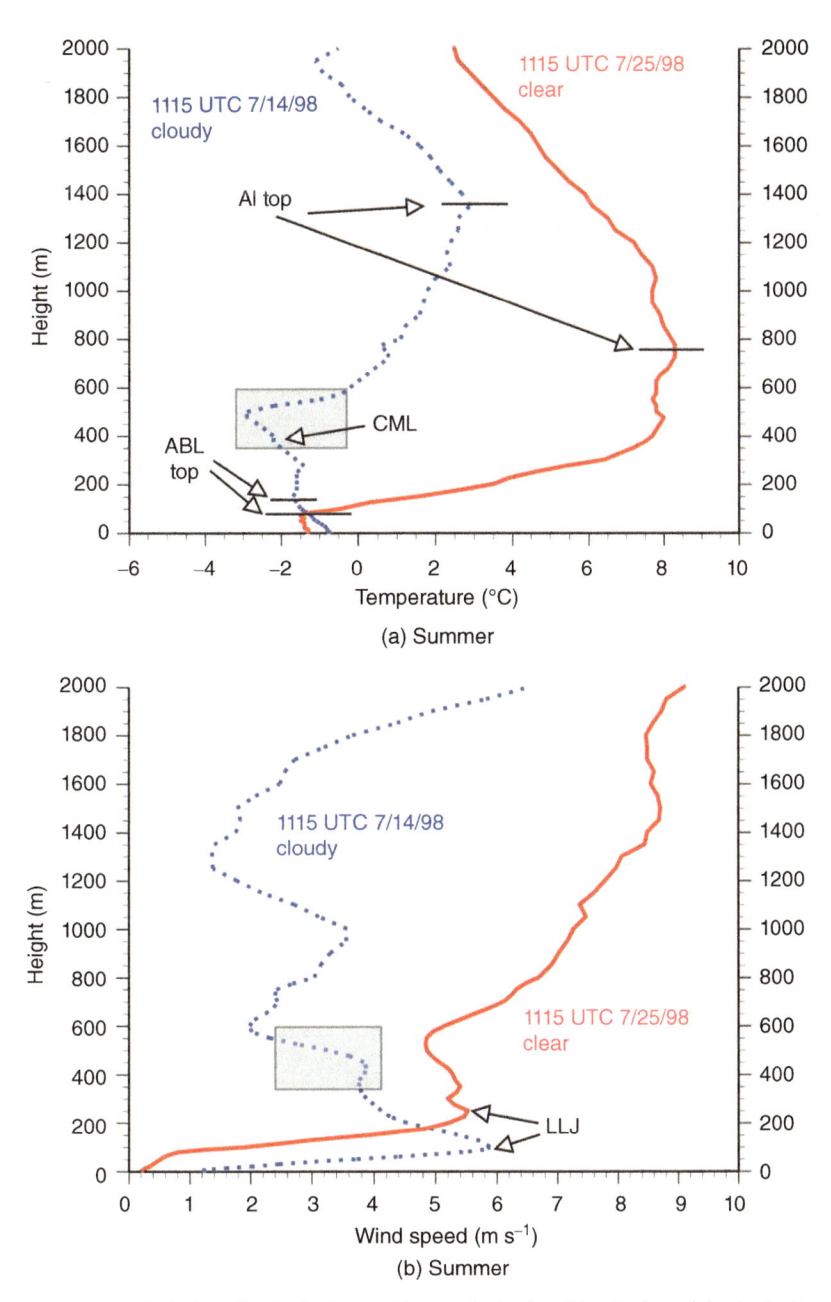

Figure 6.8 Sample temperature and wind profiles in the lowest 2 km at the Surface Heat Budget of the Arctic Ocean (SHEBA) project site for summer (a, b) and winter (c, d) for clear (solid) and cloudy (dotted) conditions. The shaded grey region represents the observed clouds for the respective soundings. The Arctic inversion tops (AI), the atmospheric boundary layer tops (ABL), the cloud mixed layer (CML) and the low-level jets (LLJ) are given for each profile.

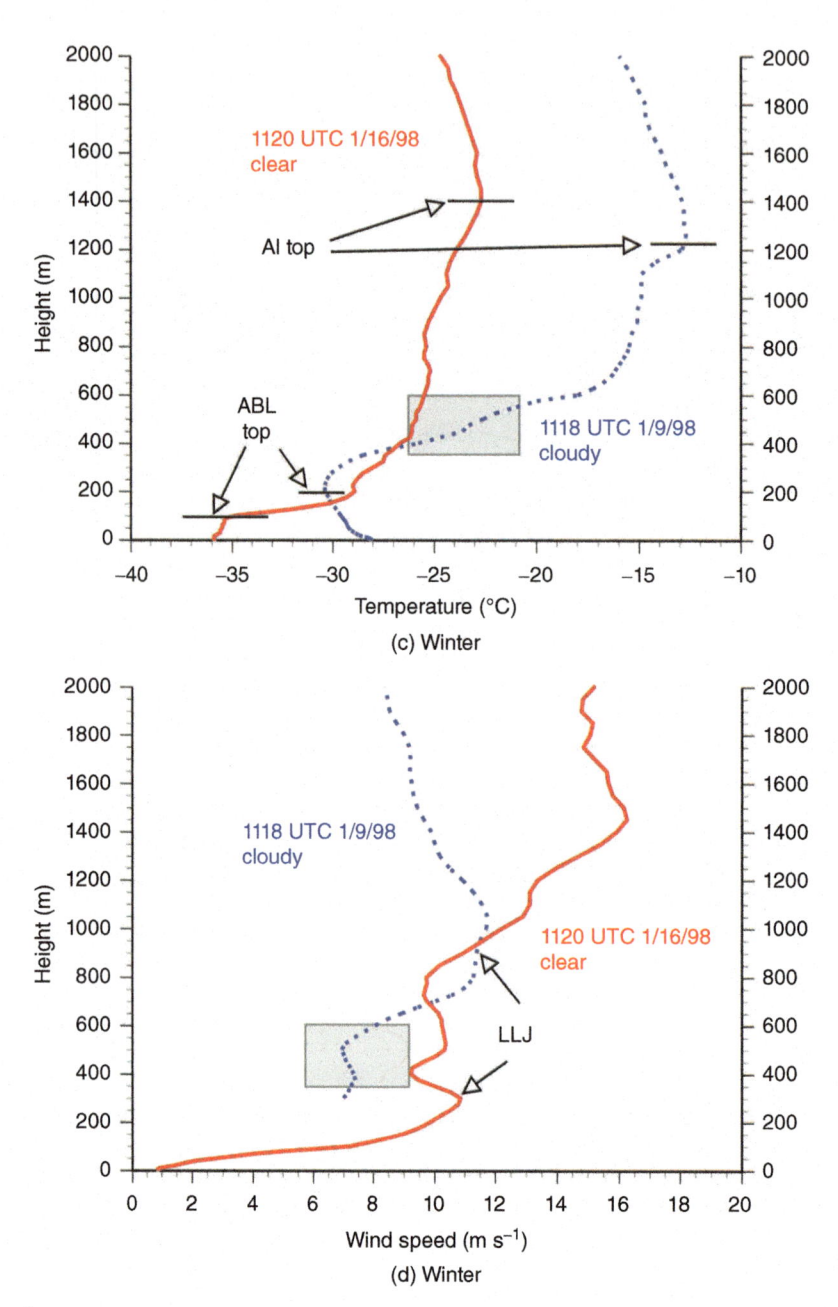

Figure 6.8 (*Continued*)

et al., 2013). Hence, both the surface-based ABL and the elevated CML reside within the deeper AI and modify its thermal structure through local surface and cloud processes affecting the lower tropospheric stability. At times, the cloud is unable to produce a CML and an inversion exists through the cloud layer, as in the winter example in Figure 6.8(c) (Sotiropoulu et al., 2014). The base of the AI generally coincides with the top of the ABL for clear skies and the top of the CML when clouds are present. The free troposphere is above the AI.

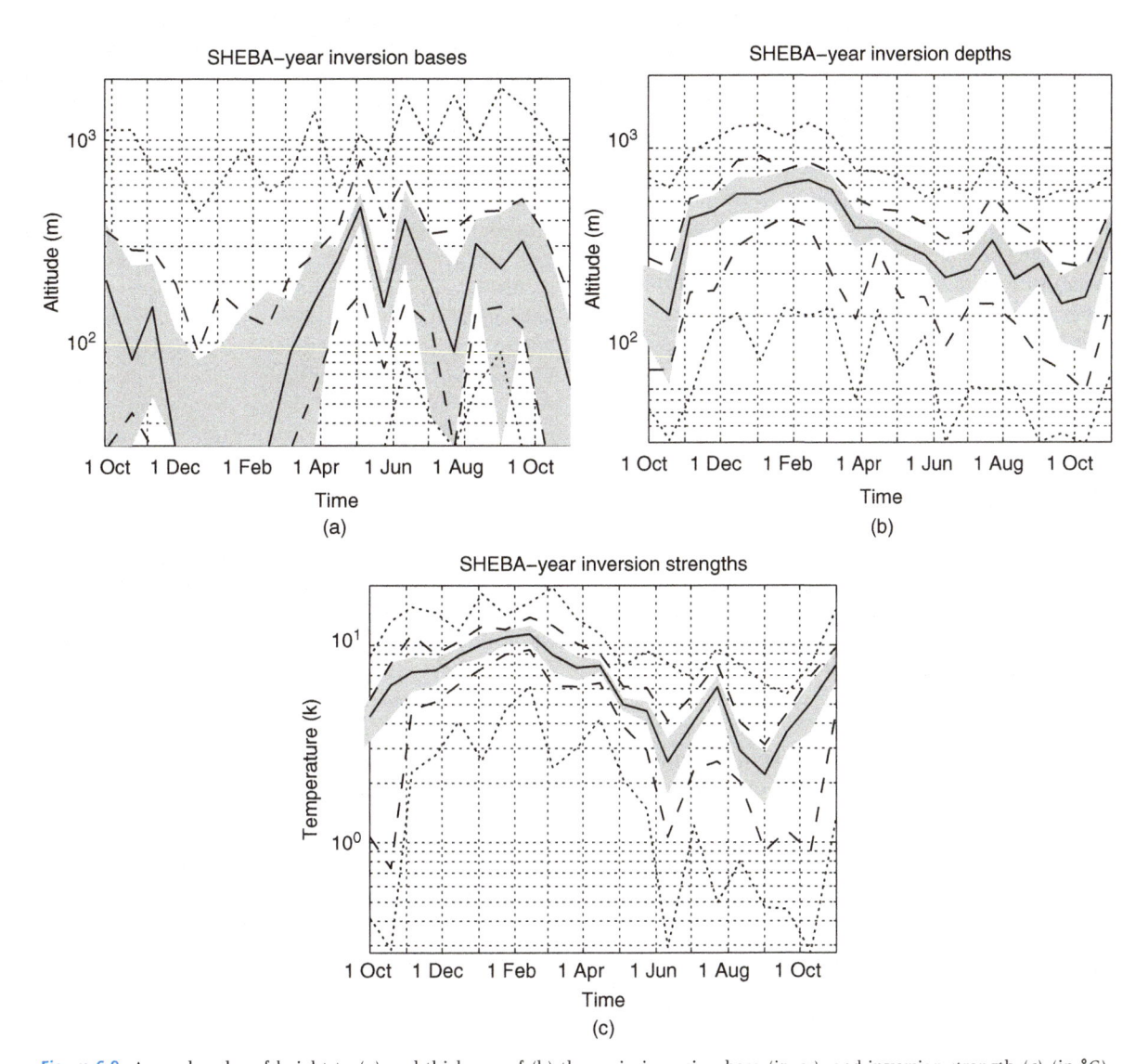

Figure 6.9 Annual cycles of height to (a) and thickness of (b) the main inversion base (in m), and inversion strength (c) (in °C). The plots show the median (solid black), 25th and 75th percentiles (dashed) and the 5th and 95th percentiles (dotted). The grey area outlines the 95% confidence interval, using a double-sided Student's *t*-test. Source: From Tjernström & Graversen, 2009. Reproduced with permission from John Wiley and Sons.

The depth of the combined ABL and CML (i.e. height to the inversion base) is typically between 0 and 800 m, and has a clear seasonal variability (Figure 6.9a; Tjernström & Graversen, 2009). In winter or at night, under clear skies and weak winds with nearly non-existent turbulence, the ABL may be only ~5–10 m deep or less (Grachev et al., 2005) and the AI base is often considered to be at the surface. Clear sky conditions with surface mechanical mixing (possibly intermittent) from stronger winds can produce a 40–150 m deep, near-isothermal or weakly stable ABL even in winter (Overland 1985; Walter & Overland, 1991; Persson et al., 2002b; also see Figure 6.8c), while solar radiation can produce a shallow but well-mixed ABL (Figure 6.8a). During some spring or summertime conditions, the ABL can be well mixed or slightly unstable and several hundred metres deep. Long-wave radiative forcing from clouds can produce a shallow well-mixed ABL even in mid-winter (Persson et al., 2002a) (Figure 6.8c). With significant buoyant mixing, as occurs in late spring (Ruffieux et al., 1995; Persson et al., 2002a), the ABL reaches its maximum depth (Figure 6.9a). In the mean, the ABL consists of an inversion during November through April, a stable isothermal layer in July, and a shallow well-mixed layer during the other months (Persson et al., 2002a). Enhanced turbulence is a characteristic of the ABL, either throughout or in sublayers (Overland 1985; Walter & Overland, 1991). The well-mixed CML produced by cloud-top radiative cooling of shallow stratocumulus clouds can either form an elevated mixed layer (Figure 6.8a) or couple with the ABL (Vihma et al., 2005). The coupling processes have been studied for late summer conditions (Shupe et al., 2013), although the modulation of the coupling for any time of the year is not understood. The CML seems to occur for stratocumulus clouds in all seasons.

Some spatial and temporal characterization of the AI structure variability has been done using soundings from Soviet North Pole drifting stations, the year-long SHEBA observations, and coastal sites. Over the perennial pack ice, AIs occur 100% of the time during the winter half of the year, reaching a minimum of 85% of the time in August (Serreze et al., 1992; Tjernström & Graversen, 2009). A well-mixed ABL occurs 85–90% of the time during the summer and 20–30% of the time during winter. Serreze et al. (1992) finds that the inversions are thicker over the interior pack ice compared with coastal sites. With considerations for some interannual and regional variability, many of these findings are consistent with studies with more limited temporal and spatial data (e.g. Kahl 1990; Persson et al., 1992; Nilsson, 1996; Persson et al., 2002a), including observations over the Weddell Sea in the Antarctic (Andreas et al., 2000). Serreze et al. (1992) and others conclude that the spatial and temporal distribution of the AI characteristics vary with synoptic activity, cloud cover, and proximity to open water, consistent with the above discussion. They state that there is a need to relate the inversion characteristics to synoptic activity more physically and provide more detail on the temporal and regional variations of AI characteristics in order to assess the potential for significant impact on boundary layer structure by various climate change processes.

6.3.3 Wind profiles above the surface layer

A low-level jet (LLJ) is a common feature between the top of the ABL and the top of the AI, most frequently observed just above the ABL. Figures 6.8(b) and (d) illustrate many of the characteristics discussed in the following. On the basis of the ERA-Interim reanalysis, the Arctic sea ice margin is among the zones of highest occurrence of LLJs in the mid- and high latitudes of the northern hemisphere (Tuononen et al., 2015). The observed frequency of occurrence of LLJs over sea ice away from terrain or the ice edge in the Arctic and Antarctic has ranged from about 25–50% of the time (Tjernström et al., 2004; Jakobson et al., 2013) to 60–90% of the time (Nilsson 1996; Andreas et al., 2000), although some of this variability is due to different criteria when defining a LLJ. The height of LLJs has typically been within an inversion layer above the ABL, often in the 200–500 m height range, although a few have been reported near the top of the AI or slightly above (Persson et al., 1992; Andreas et al., 2000). LLJs over sea ice have been observed below 100 m and at heights of ~1000 m, and have been observed in all seasons, including spring/summer (Jakobson et al., 2013), late summer/autumn (Nilsson 1996; Andreas et al., 2000; Tjenström et al., 2004) and winter (Persson et al., 2002b; see also Figure 6.8). Many of the observed LLJs have an average velocity only 2–3 m s^{-1} faster than the free-stream wind above, and so are fairly subtle. A few, however, have a much more distinct maximum. Subcritical Richardson numbers are frequently observed near the LLJs, especially below them. Nilsson (1996) reported subcritical Richardson numbers occurring in 40–60% of soundings below the LLJ maximum, where shear increases, while aircraft observations of LLJs within the inversion just above the ABL showed enhanced turbulence just above or below the LLJs (Walter & Overland, 1991).

Over the level pack ice away from terrain and ice-edge effects, LLJs are believed to be caused by inertial oscillations in a layer decoupled from the surface by stable stratification (e.g. Andreas et al., 2000). The initiation of an LLJ implies a transition from a fairly deep well-mixed ABL to a much shallower ABL. When the decoupling occurs, an inertial oscillation begins, causing air in the upper layer to accelerate to supergeostrophic speeds. An LLJ then develops at heights formerly in the upper portion of the ABL but which are now decoupled from the surface. The inertial oscillation causes the wind in this upper layer to oscillate in speed and direction around those of the geostrophic wind. Once an oscillation is initiated, it can continue for a long time, with the period of variation equal to the inertial period, which is about 12–13 hours in polar regions. When the LLJ is near its peak speed, shear causing turbulence below the LLJ may occur which could impact the longevity of the LLJ. Changes in the synoptic conditions would also influence the LLJ through changes in geostrophic wind. To initiate an LLJ by this mechanism requires the stabilization of a previously well-mixed ABL, which could occur in the Arctic through diurnal cycles during some times of the year and through changes in radiative forcing attendant to the dissipation of low-level clouds during any time of the year. The occurrence of LLJs in association with transitions from a deep coupled ABL–CML to a decoupled CML, as was observed in August in the Central Arctic (Figure 6.10), would be consistent with this process. Baroclinicity near transient cyclones was the primary forcing of LLJs observed during the Tara drift in the Central Arctic (Jakobson et al., 2013).

Such baroclinicity, ABL and CML height variability, and coupling transitions appear to be common characteristics of the Arctic lower troposphere, but may be especially common near the MIZ. Studies suggest that the strength and characteristics of LLJs near the MIZ will vary with baroclinicity and the ice-edge relative wind direction (Langland et al., 1989). Brümmer and Thiemann (2002) showed that the formation of an ABL structure 300 m deep occurred as an internal boundary layer when air flows from the open ocean to 300 km over the pack ice within a warm sector of a storm system. An LLJ forms within the stable layer within 50 km of the ice edge, and other AI substructure is produced by the effects of clouds, variations in the ice thickness and the presence of leads.

In summary, the sporadic elevated mixed layers or enhanced-stability layers observed within the AI are produced by episodic radiatively driven processes within boundary-layer clouds (e.g. Pinto, 1998; Harrington et al., 1999; Wang et al., 2001; Shupe et al., 2013), by possible intermittent mixing processes associated with an LLJ or by clear-air radiative processes (André & Mahrt 1982). Many researchers (e.g. Arya 1981; Mahrt 1981; Andreas et al., 2000) have defined the height of the SBL as the height to the LLJ, typically found above the ABL but well below the top of the AI (e.g. Nilsson 1996; Andreas et al., 2000; see also Figure 6.8), because intermittent turbulence exists above the ABL to this level. Hence, the AI appears to consist of the lower turbulent ABL, a layer between the ABL and the LLJ with intermittent turbulence (in either clear or cloudy conditions), a CML layer during cloudy conditions producing modifications to this thermodynamic profile either above (Figure 6.8a,b) or below (Figure 6.8c,d) the LLJ, and a quasi-inactive upper layer above the LLJ dominated by radiative cooling.

6.3.4 Leads, polynyas and thin ice

The thermodynamic effects of leads, polynyas and areas of thin ice on the ABL and AI strongly depend on the season, large-scale flow conditions and the size of the areas of anomalous surface temperature. The maximum effects are felt in cold winter conditions over large polynyas/leads. In such conditions, the turbulent surface fluxes of sensible and latent heat may reach several hundreds of W m^{-2}, which results in localized convection. Although the surface fluxes are large, the convection has to work against stable stratification in the inversion layer, and remains shallow in most cases (Lüpkes et al., 2012). The convection may penetrate through the Arctic wintertime inversion only in rare cases (Schnell et al., 1989) and its vertical penetration and downwind reach are dependent on the ambient upwind ABL stability and AI depth (e.g. Burk et al., 1997). The air temperature increase over an individual lead is typically modest. A lead 0.5 km wide may cause an increase of 0.5–2 K in the 2 m air temperature (Ruffieux et al., 1995; Pinto et al., 2003; Vihma & Pirazzini, 2005). A large number of leads under an air-mass trajectory may, however, generate much larger effects. Via column-model experiments, Lüpkes et al. (2008) showed that during the polar night under clear skies, a 1% decrease in sea ice concentration results in up to a 3.5 K increase of the near-surface air temperature, if the air-mass flows for a sufficiently long time over the broken sea ice field (~48 h). Three-dimensional model experiments have yielded results of the same order of magnitude (Valkonen et al., 2008; Bromwich et al., 2009).

Detailed understanding of localized convection over leads and polynyas requires accurate observation and models that can resolve individual convective eddies (Esau, 2007) or at least the heat plume originating from the warm surface (Lüpkes et al., 2008, 2012). The issue is further complicated by the fact that leads

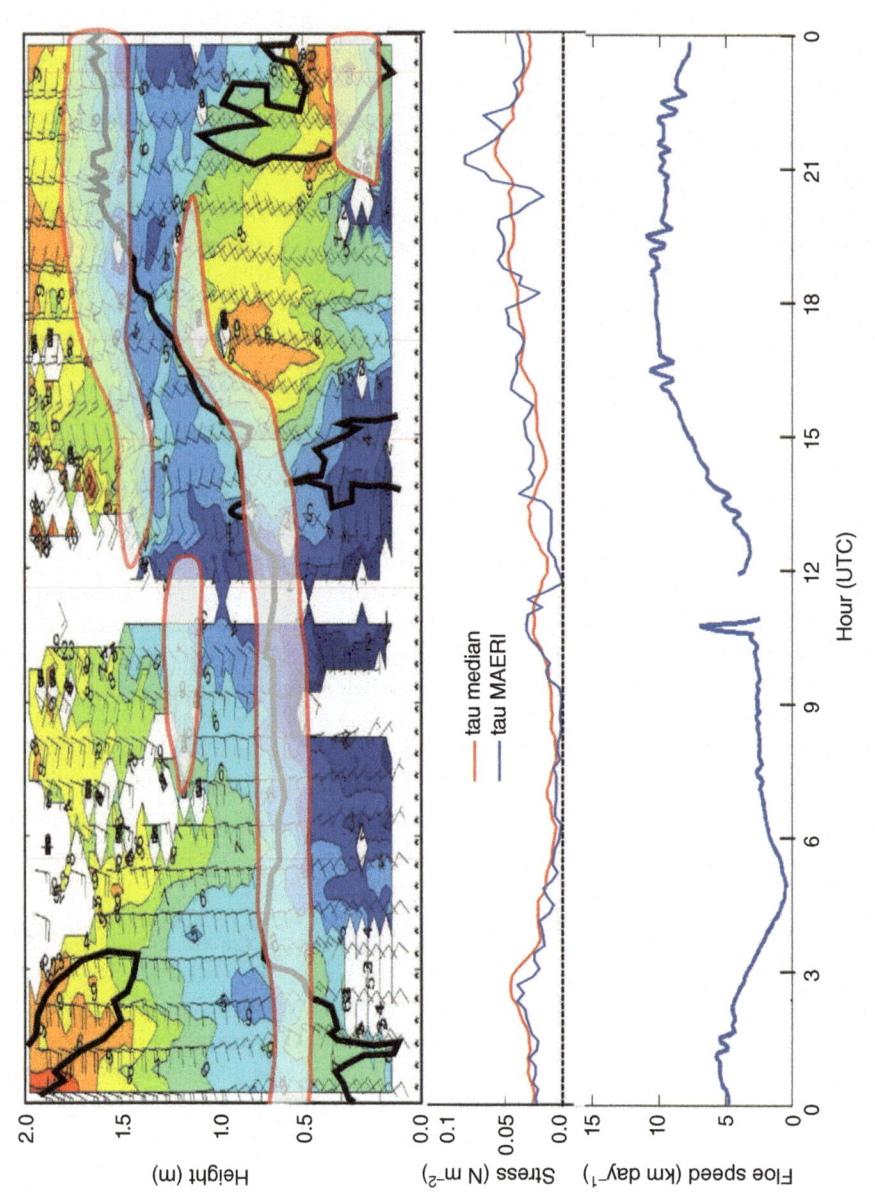

Figure 6.10 (a) Isotach analysis (isopleth interval 1 m s^{-1}) and wind barbs from the 449 MHz wind profiler on the R/V *Oden* on 25 August 2008, during the Arctic Summer Clouds-Ocean Experiment (ASCOS) at 87.5°N. Thirty-minute wind profiles are used, although the wind barbs are subsampled in time and height for clarity. The cloud boundaries from the cloud radar (heavy black line) and the temperature inversions from the scanning radiometer and soundings (grey areas) are also shown. (b) Surface stress measurements from on-ice tower measurements (red) and bulk estimates from ship and Marine Atmospheric Emitted Radiance Interferometer (MAERI) observations (blue). (c) Ice-floe movement (km day^{-1}).

and polynyas are seldom entirely open, but are at least partly covered by thin ice, which has a surface temperature between those of thick ice and open water. Hence, the surface fluxes of sensible and latent heat also have intermediate magnitudes. This complicates the simulation of the effects on the atmosphere in numerical weather prediction (Burk & Thompson, 1995; Tisler et al., 2008) and climate models.

A possibly unique feature of the lower atmosphere over sea ice is the presence of 'microfronts'. These are frontal features with changes in temperature, humidity and cloud characteristics in the lowest 500 m or so, and they have only been documented in the high Arctic during late summer (Tjernström & Mauritsen, 2009). They are suggested to be produced by thermal and moisture gradients near leads, but their formation mechanism and frequency are unknown.

6.3.5 Marginal ice zone

In the MIZ, the ABL structure and processes strongly depend on the flow direction with respect to the ice margin. During cold air outbreaks from the sea ice zone to the open ocean, the main process is the development of a convective boundary layer (CBL) with increasing distance from the ice edge. To successfully simulate the CBL development, it is essential to account for the counter-gradient transport (Lüpkes & Schlünzen, 1996), cloud physics and heat fluxes from the open sea and from leads upwind of the large-scale ice edge (Vihma & Brümmer, 2002). Also, the ice edge is often diffuse during cold air outbreaks, as the wind generates ice drift at about 30° to the right of the wind. Seasonal and interannual variability of cold-air outbreaks is affected by the variability of the 700 hPa air temperature and sea surface temperature, the former being more important (Kolstad et al., 2009). Serreze et al. (2011) found that, due to the Arctic warming, recent cold air outbreaks are not as cold as before. The model results of Vavrus et al. (2006) suggest, however, regional increases in the occurrence of cold air outbreaks during the 21st century, such as over the Atlantic Ocean.

During off-ice flows, the upwind snow/ice surface is usually close to a thermal equilibrium with the ABL, and in the downwind region, due to the large heat capacity of the sea and the vertical mixing of the ocean mixed layer, the surface temperature of the open sea is only slowly affected by heat loss from the advected cold air. In contrast, during advection of a warm marine air mass over the sea ice, the snow/ice surface temperature is strongly affected, while an ice surface of a limited fetch does not always have large thermodynamic effects on the ABL (Vihma & Brümmer, 2002; Persson 2012; Tjernström et al., 2015). During warm air advection from the open ocean to over sea ice, a shallow, SBL develops with downward turbulent flux of sensible heat. Because of cooling of the base of this warm and typically moist air, clouds frequently form, producing both radiative effects on the surface and CMLs. With sufficient radiative effects, a surface mixed layer may form, and the CML may or may not couple with this surface mixed layer. Essential issues in modelling include a good vertical resolution in the ABL, accurate information on sea ice roughness (Vihma et al., 2003), and a realistic description of the thermal interaction between the snow surface and the ABL with a good model resolution in snow and ice (Cheng & Vihma, 2002). A cloud-microphysics scheme that allows appropriate interactions with the boundary-layer scheme and can represent mixed-phase processes is usually necessary (e.g. Solomon et al., 2009). Cases of on-ice air flow are typically associated with a distinct ice margin; the ice pack is usually rather compact downwind of the ice edge.

Both observational and modelling studies have suggested that the baroclinicity in the MIZ can produce mesoscale atmospheric features unique to the MIZ, such as ice-edge parallel LLJs (e.g. Langland et al., 1989; Drüe & Heinemann 2001). LLJs may, in turn, enhance wave fracturing, ice drift and surface turbulent heat fluxes. The surface curl of strong low-level winds may enhance upwelling of warm subsurface water near the ice edge (Liu et al., 1993; Steele et al., 2011), which could enhance lateral and bottom melt of the ice along the ice edge, thereby accelerating the seasonal northward retreat of the ice edge and the MIZ or retard the autumn southward advance. The primary mechanisms for MIZ-generated LLJs are the decrease (increase) of the easterly (westerly) flow with height due to thermal wind effects of colder air to the north over sea ice, combined with an easterly flow generated by either the Coriolis turning of the near-surface branch of a direct (off-ice) thermal circulation or the larger-scale pressure field associated with synoptic storms (e.g. Økland 1998; Grønås & Skeie, 1999). Wintertime cases with strong baroclinicity ($\sim \Delta T\ 20 - 30°C$) have produced observed surface wind speeds over 30 m s^{-1} and sensible heat fluxes over 1200 W m^{-2} even in the absence of a storm (Grønås & Skeie, 1999). Significant effects on the surface stress could also be expected. Even though thermal gradients in the Marginal Ice Zone Experiment (MIZEX), conducted during July 1983, were only ~4°C modest LLJs ($\Delta u \sim 3-6\,\mathrm{m\,s^{-1}}$) were noted near the ice edge and over the sea ice from rawinsonde and aircraft data (Kellner et al., 1987; Fairall & Markson 1987).

6.3.6 Geographical differences

Compared with the Arctic, the major differences in the lower boundary conditions for the atmosphere over Antarctic sea ice are the following. As the sea ice zone surrounds the continent, the sea ice cover is, on average, thinner, the ice concentration is lower, the sea ice zone occurs at lower latitudes than in the Arctic, and katabatic winds advect cold continental air masses over the sea ice. These factors probably reduce the thermal stratification in the ABL, but the temporal and spatial distribution of stratification is not well known.

Except for ship-based observations in summer and year-round buoy observations, there are very few meteorological observations from the Antarctic sea ice zone. The most extensive data set was gathered from the western Weddell Sea during the US–Russian drifting station in February – June 1992 (Andreas et al., 2000, 2004, 2005). There has also been a shorter drifting station in December 2004–January 2005 (Vihma et al., 2009). In addition, good data have been obtained from recent aircraft observations and the Weddell Sea winter cruise in June–August 2013, which included short ice stations with unmanned aerial vehicle observations. Turbulent fluxes have also been estimated on the basis of (a) ship observations mostly in summer (Andreas & Makshtas, 1985; Wendler et al., 2004); (b) wind and temperature observations from drifting buoys (Kottmeier & Engelbart, 1992; Launiainen & Vihma, 1994; Vihma et al., 2002); and (c) kite measurements applying a budget method for fluxes over leads (Guest, 2007). Results from these studies suggest that the lower troposphere over the Antarctic sea ice zone resembles that in the Arctic. Andreas et al. (2000) obtained statistics for Antarctic temperature inversions that are not far from those in the Arctic (Kahl, 1990; Serreze et al., 1992). Vihma et al. (2009) compared early summer observations from the western Weddell Sea against SHEBA data from the same season, and concluded that in the Antarctic the air was colder, the surface albedo was higher, and a larger portion of the absorbed solar radiation was returned to the atmosphere via turbulent heat fluxes. The limited melt allowed larger diurnal cycles. We have to remember, however, that the best data sets from the Antarctic sea ice zone are from the western Weddell Sea, where sea ice concentration is high, ice is thick, and the climate is colder than in most other parts of the Antarctic sea ice zone. We do not yet have enough observations to understand how much the lower latitudes, lower sea ice concentrations and thinner ice in the Antarctic than in the Arctic affect the structure of and processes in the ABL. Section 6.6 describes model results for the atmosphere over Antarctic sea ice.

Baltic Sea

The Baltic Sea is located between 54 and 66°N and is mostly isolated from the North Atlantic Ocean by the Scandinavian Peninsula. The sea ice extent has a large interannual variability, but is most persistent in the northern and eastern sections, especially the Gulf of Bothnia and the Gulf of Finland. Large-scale weather conditions over the Baltic Sea are strongly controlled by the North Atlantic Oscillation (NAO). In winters with a prevailing positive NAO, the sea ice extent remains small and the ABL conditions are often characterized by advection of warm, moist air masses over the small areas of sea ice. In winters with a prevailing negative NAO, an SBL under clear skies and weak winds is more common, and cold air outbreaks from the sea ice zone to the open sea are common. Due to synoptic-scale variations, both types of ABL naturally occur each winter. Large fluxes of sensible and latent heat from the open water modify the ABL, and occasionally generate convective snow bands with hazardous effects (Gustafsson et al., 1998; Savijärvi, 2012; Mazon et al., 2014).

Except for landfast ice in coastal regions, the ice cover in the Baltic Sea is usually fractured. This reduces the representativeness of local point measurements, and aircraft observations are essential to measure spatially averaged fluxes in the ABL. The results of Brümmer et al. (2002b) revealed significant spatial and temporal variations in the surface fluxes; the fluxes depended, above all, on the large-scale weather conditions and on the state of the surface, i.e. either open water, compact landfast sea ice or broken sea ice. Brümmer et al. (2005) summarized the structure of the lower troposphere over Baltic Sea ice on the basis of aircraft observations. A temperature inversion is almost always present, being surface-based for two-thirds of the time and extending to the height of 165 m on average. The inversion strength is on average 2.1 K, i.e. much smaller than in the Arctic. A LLJ occurs about 86% of the time with an average height of 245 m and speed of 13.3 m s^{-1} (Brümmer et al., 2005). Aerodynamic roughness lengths over Baltic Sea ice have been measured by Launiainen et al. (2001).

Sea of Okhotsk

The Sea of Okhotsk is located east of Siberia and west of the Kamchatka Peninsula between 50 and 60°N, and is the southernmost seasonal ice zone in the northern hemisphere. The extent of sea ice shows significant interannual variability. As a seasonal ice zone similar to the Baltic Sea, the ABL over the Sea of Okhotsk is dominated by heat and moisture fluxes from the MIZ area and the numerous leads over the ice-covered part. The atmospheric structure over the ice is determined by the thermal and moisture structure of the air being advected over the ice; that is, it is determined by whether the air flow is from Siberia to the north and west or from the open ocean to the south and east. Studies (Inoue et al., 2005) show the importance of leads and open water for the warming and moistening of the cold, dry air masses advecting southeastwards from Siberia. For midwinter cases of cold air advection, a well-mixed ABL of a few hundred metres' depth is found over the ice to the northwestern side of the sea. The surface heat fluxes can increase this depth to approximately 1000 m on the downwind (southeastern) edge of the sea ice, which may be several hundreds of kilometres distant. The heat flux from the leads within the ice-covered area can be sufficient to produce roll vortices and cloud streets for the right wind conditions. However, the importance of seasonal sea ice in the Sea of Okhotsk, once it forms, is that it reduces the atmospheric warming and moistening of cold air outbreaks and makes the air mass transformation more gradual and occur further to the east. While atmospheric conditions over the Sea of Okhotsk during on-ice (southeasterly) flow have not been studied, one would expect the lower troposphere to have similar structures to those observed elsewhere for on-ice flow near a MIZ. That is, one would expect the generation of strong inversions and an ice-generated internal boundary layer nearest the surface.

6.3.7 New Arctic

Changes in the character and extent of the Arctic sea ice during the last few decades have been significant enough to warrant a brief discussion of the likely impact of these changes on the ABL in this 'New Arctic'. In brief, the ice extent currently has much greater seasonal variability, the speed of seasonal meridional movement of the ice edge and MIZ has increased, the region of seasonal sea ice has greatly increased, and that of multi-year sea ice has greatly decreased (e.g. Stroeve et al., 2012). This means that the conductive heat flux through the ice, which is inversely dependent on ice thickness, is generally greater, with energy balances producing warmer surface temperatures. This could lead to more frequent and deeper well-mixed ABLs, especially during winter, and potentially to more frequent coupling between the ABL and the CML. The greater extent of open water means that MIZ processes are active during a greater part of the year than was previously the case, so one might expect to encounter more frequent LLJs. There are some suggestions that increased advection of warm, moist air from lower latitudes to over the sea ice and the associated turbulent and radiative processes discussed above have caused some of these sea ice changes (e.g. Graversen et al., 2011), including the earlier onset of the summer melt (Markus et al., 2009; Maksimovich & Vihma, 2012; Persson 2012). However, there is still significant debate on this point. Likewise, the Arctic sea ice decline has affected the weather and climate in mid-latitudes, but there is large uncertainty with regard to the magnitude, timing and spatial extent of these effects, as well as their physical mechanisms (Francis & Vavrus 2012; Cohen et al., 2014; Vihma, 2014; Perlwitz et al., 2015).

6.4 Cloud–radiation–turbulence–wave interactions

The stability and depth of the AI would appear to present a significant impediment to the downward transport of heat, moisture, momentum and aerosols from the free troposphere to levels where these quantities can impact clouds or the surface, although evidence suggests that such transport does occur. This transport appears to involve cloud and radiation processes as well as interactions between waves and turbulence, with some of these events probably being transitory.

Evidence for transitory mixing events in the lower troposphere has been found during several field programmes. During IAOE-91, sudden changes in near-surface concentrations of cloud condensation nuclei of a factor of two or more in 1 hour or less were frequent, despite the long distances from significant sources (Bigg et al., 1996). These changes often appeared in groups separated by 1 hour or more. Bigg et al. (1996) hypothesize that many of these changes are the result of a two-step process. First, large vertical gradients of aerosol are produced in stable layers in the AI. These gradients may be produced by interactions between clouds, evaporating drizzle, and aerosol, and by differences in times or source regions for trajectories arriving at different heights in the AI. Secondly, transient mixing processes in the AI tap these large aerosol gradients, producing large changes of aerosol concentrations measured near the surface. These mixing processes will occur on timescales of an hour or less, and may only mix part of the AI. The transient mixing processes proposed include Kelvin–Helmholtz (K–H) waves, gravity waves, lee waves, a collapse or intermittent breakdown of the boundary layer, and horizontal roll vortices. The various mixing processes would produce different temporal variations in aerosol concentrations, such as a single decrease and then recovery, and sequences of maxima and minima with different periods.

A key mixing process not considered by Bigg et al. (1996) is the generation of a CML by cloud-top radiative cooling (e.g. Pinto 1998; Wang et al., 2001). This cloud-induced mixed layer resulting from 'upside-down convection' (Smedman, 1988) extends from cloud top to a few hundred metres below cloud base, which, when coupled with the ABL, can transport aerosols and other constituents to the surface (Shupe et al., 2013). Clouds are very common in the Arctic and over sea ice (Intrieri et al., 2002; Shupe et al., 2011), especially low-level stratocumulus clouds, and these clouds frequently contain CML. However, this mechanism only works when no cloud layers exist above the low-level cloud, producing maximum long-wave radiative cooling, so information on cloud-layering is crucial. CMLs have been successfully simulated in models (e.g. Harrington et al., 1999; Solomon et al., 2011) and shown to depend on the clouds being liquid or mixed-phase. While recent studies have improved our understanding of the CML, coupling between the CML and ABL is not yet understood. We also do not understand whether and how interactions between the CML and LLJ modify the AI thermal and kinematic structure.

The portions of the AI above the CML or near the LLJs are especially enigmatic in their structure, intermittent mixing events and development. Analyses of soundings can only suggest possible mechanisms, but have not provided the temporal resolution to confirm which mechanisms produce the mixing events. K–H waves will develop at boundaries with different densities and wind shear, such as exist near the top of the well-mixed ABL and below the LLJ in the AI. K–H waves have periods of the order of 1 minute, and may grow in amplitude and break, producing turbulence and vertical mixing. The shear associated with the LLJ can also produce gravity waves with longer periods, which may radiate down to the surface and mix air from higher altitudes. Such turbulence would be sporadic and would have lower frequencies than turbulence described by standard Monin–Obukhov similarity theory. Examination of turbulence spectra collected during SHEBA show the presence of low-frequency turbulence possibly resulting from such waves (Grachev et al., 2005). Interactions between gravity waves and the mean shear within the stable AI or in the ABL can produce intermittent turbulence as well (Finnigan et al., 1984). Sodar data with high temporal resolution have shown the existence of K–H and gravity waves in the Arctic, and revealed their vertical extent and evolution in some cases (Cheung, 1991; Ruffieux et al., 1994). It is unclear whether the occurrence of gravity waves is frequent enough and their vertical extent deep enough to be the main transport process from the free atmosphere to within the AI and/or within the AI above the CML. Their vertical extent and frequent occurrence may be limited to regions below the LLJ.

Evidence has frequently been found for horizontal roll vortices oriented at a slight angle to the mean geostrophic wind during Arctic field programmes (e.g. Fett et al., 1994; Nilsson 1996). The transverse wavelength and depth of the roll vortices are dependent on the depth of the ABL and the wind speed. If horizontal roll vortices are common, they could have a profound effect on aerosol concentrations because of the increased path length of the aerosols and the repeated condensation and evaporation within the air parcels in the rolls (Bigg et al., 1996) as they repeatedly move from the top to the bottom of the ABL and back.

Mixing events or sudden large changes in surface aerosol concentrations have at times coincided with frontal passages (e.g. Bigg et al., 1996; Tjernström et al.,

2004), suggesting that some of these mixing processes are frontally generated. Gravity waves are known to be generated at fronts, while changes in stability, LLJs and mesoscale circulations are also features associated with atmospheric fronts. Finally, scavenging by precipitation, frequently generated at fronts, is a rapid mechanism for downward transport of aerosol from above the AI to the ABL.

The physics of the SBL is poorly understood, and our knowledge is mostly based on observations of the mid-latitude nocturnal SBL. Stable stratification in the Arctic winter ABL is often long-lived, in contrast to the nocturnal stable ABL at lower latitudes; the latter is separated from the free atmosphere by the residual layer. This makes the Arctic ABL more liable to the effects of propagating internal gravity waves (Zilitinkevich, 2002). Zilitinkevich and Esau (2005) presented a parameterization to better take this into account. The Obukhov length (L) was replaced by a stability parameter that is a combination of L, the Coriolis parameter f, and the free-atmosphere Brunt–Väisälä frequency (N), which is essential for the occurrence of gravity waves in the free atmosphere. Another attempt to account for the role of gravity waves is the quasi-normal scale elimination theory (Sukoriansky et al., 2005, 2006), which is a spectral theory that does not differentiate between waves and turbulence but treats them as one entity.

6.5 Wind and ice motion

6.5.1 Effects of wind speed, roughness and stratification

The drift of sea ice is affected by the air-ice and ice-ocean momentum fluxes, the Coriolis force, sea surface tilt, and the internal resistance of the ice field. The air–ice momentum flux (τ) is usually the primary forcing factor and depends on the wind speed at height z in the surface layer (V_z) and the drag coefficient C_D (at height z):

$$\tau = \rho C_D V_Z^2 \quad (6.15)$$

where C_D increases with increasing aerodynamic roughness of sea ice (z_0) and decreases with increasing stratification in the surface layer, characterized by the Ψ_M function (see Section 3.1):

$$C_D = \frac{k^2}{\left[\log\left(z/z_0\right) - \Psi_M\left(z/L\right)\right]^2} \quad (6.16)$$

If the sea ice field includes numerous ridges, floe edges and sastrugi, z_0 is larger than for an undeformed ice or snow field. Due to more floe edges, z_0 is generally found to be greater with an ice concentration of ~50% (Andreas et al., 2010a). Aerodynamic roughness of sea ice has been studied extensively since the 1970s (e.g. Arya, 1975), but few studies have addressed the effects of stratification on the wind stress over Arctic sea ice. Equations (6.15) and (6.16) include the direct effect of stratification on the momentum flux, i.e. the effect originating from the stratification in the layer between the ice surface and the wind observation height z. In addition, the stratification higher up in the atmosphere affects the vertical transport of momentum and, accordingly, the near-surface wind speed V_z. Enhanced air–ice momentum flux could therefore be expected with strong low-level winds (e.g. strong atmospheric pressure gradients, LLJs), enhanced boundary-layer mixing and enhanced surface heterogeneity. Figure 6.10 shows an example of increasing momentum flux enhancing ice movement in association with an LLJ and changing boundary-layer stability.

Roughness is usually larger and stratification more stable over sea ice than over open ocean; these differences tend to compensate for each other in their effect on surface momentum flux. Some observations (Brümmer & Thiemann, 2002) and model results (Tisler et al., 2008; Kilpeläinen et al., 2011) show larger momentum fluxes over the open ocean, suggesting that the effect of stratification dominates. The stability dependence of the turbulent momentum transfer is rather well understood for unstable conditions, but there is a lot of uncertainty for very stable conditions (Section 6.3.1). The main challenge is to better understand how fast the momentum flux decreases with increasing stratification. It is also unknown whether intermittent turbulence during generally very stable conditions, as can occur during winter over sea ice (Grachev et al., 2005), is important for enhancing the momentum flux and sea ice movement. Data to quantify the relative importance of and interactions between the wind forcing mechanism, the stability and the surface heterogeneity for modulating the surface stress and sea ice movement are generally not available, but are needed. For recent recommendations for parameterization of C_D, see Andreas et al. (2010a,b), Lüpkes et al. (2012) and Lüpkes and Gryanik (2015).

6.5.2 Observed relationships

Except in coastal regions, ice drift on synoptic timescales is usually strongly related to the near-surface wind and momentum flux, with correlation coefficients frequently near 0.80–0.90 (e.g. Thorndike & Colony, 1982; Vihma et al., 1996; Kwok et al., 2013). On timescales of weeks and longer, however, correlation is often much weaker, particularly in regions where the ocean current has a rather constant direction (Vihma & Launiainen, 1993). If the ocean current and the internal resistance of the ice field are weak, a typical ratio between ice drift speed and near-surface wind speed is of the order on 0.02–0.04 in both the Arctic (Nansen, 1902) and the Antarctic (Martinson & Wamser, 1990). In the Arctic (Antarctic) the ice drift deviates approximately 30° to the right (left) of the wind vector. This deviation angle, caused by the Coriolis force, happens to be approximately the same, but of the opposite sign, as the deviation angle between the near-surface wind and geostrophic wind, caused by surface friction. Hence, the ice drift approximately follows sea-level pressure isobars in both the Arctic (Zubov, 1943) and the Antarctic (Vihma & Launiainen, 1993; Kottmeier & Sellmann, 1996).

The greatest surface stresses over open waters of the North Atlantic Ocean occur in association with strong winds in warm sectors of storms, despite the enhanced stability in these regions. Secondary peaks in surface stress occur in unstable cold sectors with slightly weaker winds but stronger vertical mixing (Persson et al., 2005). It is unknown to what extent such stability and storm-relative considerations modulate the surface momentum fluxes over sea ice. Any asymmetry between the surface stress on the two sides of a storm may also impact the net advective movement of the sea ice. Furthermore, mesoscale (~10–100 km) deformational features of the low-level wind field, including those produced by atmospheric fronts (e.g. Wakimoto & Murphey, 2008) and divergence due to surface roughness gradients, may lead to divergence in the sea ice drift (Brümmer et al., 2008). Clearly, transient cyclones have a significant impact on the sea ice evolution (Kriegsmann & Brümmer, 2014).

On a larger scale, the interannual variations in Arctic sea ice drift speeds are closely related to wind forcing. The atmospheric variable best explaining the variations is the pressure gradient across the Arctic Ocean, along longitudes 90° and 270°E (Vihma et al., 2012).

6.5.3 New Arctic

The drift of Nansen's Fram across the central Arctic in 1893–1896 took almost three times longer than the drift of the French schooner Tara in 2006–2007 along almost the same path (Gascard et al., 2008). Buoy data clearly demonstrate that sea ice drift speeds in the Arctic have increased since the 1950s (Häkkinen et al., 2008). This is due to combined effects of increasing wind speeds and the sea ice becoming thinner. The effect of increased wind speeds seems to be limited to earlier decades, as the increasing trend in drift speeds since 1989 cannot be explained by atmospheric forcing (Vihma et al., 2012) and even less so for 2001–2009 (Kwok et al., 2013). According to Rampal et al. (2009), Spreen et al. (2011) and Gimbert et al. (2012), the increase in drift speeds since 1979 are mostly due to thinner ice with a reduced mechanical strength. Kwok et al. (2013) observed an increase in the ratio of the ice drift speed and geostrophic wind speed, which is consistent with the thinner and weaker ice cover. They further found that part of the ice motion variance, which is explained by the geostrophic wind, has increased during 1982–2009. With the further reduction of Arctic sea ice thickness and concentration, we may expect ice drift in the New Arctic that is even more strongly controlled by wind forcing, with higher drift–wind speed ratios (due to reduced internal resistance of the ice field) and smaller deviation angles between drift and wind vectors (due to reduced ice thickness).

6.6 Modelling of the atmosphere over sea ice

6.6.1 Main challenges

Representing the atmosphere over sea ice in numerical models has been challenging because of difficulties representing clouds, especially mixed-phase clouds, correctly modelling the stable ABL, and correctly incorporating interactions with the surface and its various forms (e.g. sea ice, snow-covered sea ice, melt ponds, leads). Distinct negative biases in downwelling short-wave

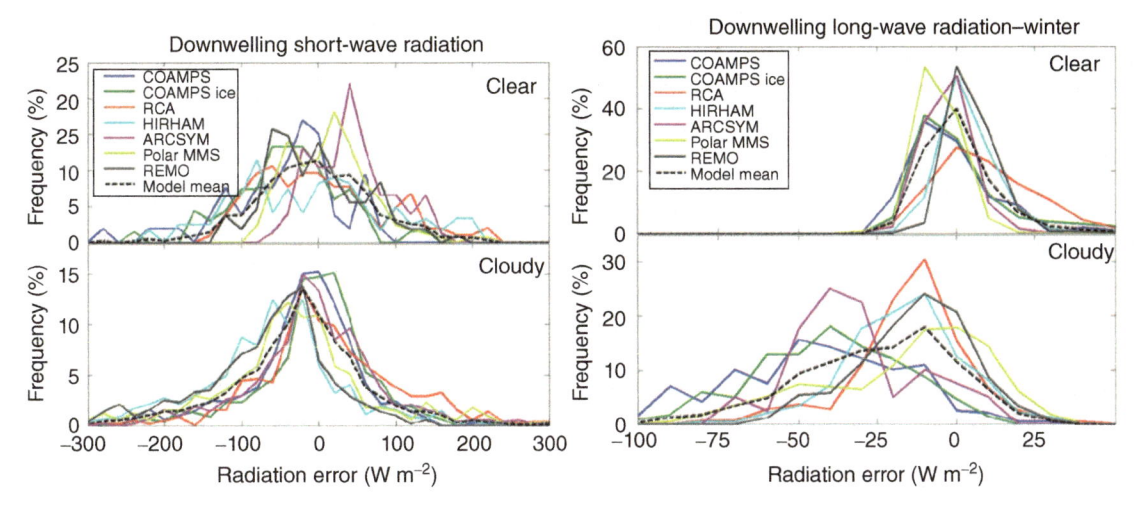

Figure 6.11 Probability distribution functions of downwelling short-wave (a) and long-wave (b) radiation for clear (top) and cloudy (bottom) skies for a number of regional climate models simulating the surface radiation for the Surface Heat Budget of the Arctic Ocean (SHEBA) project year. Source: From Tjernström & Graversen, 2009. Reproduced with permission from John Wiley and Sons.

and long-wave radiation for cloudy conditions are seen for many models (Figure 6.11) (Tjernström et al., 2005, 2008), while broad error distributions and large inter-model differences are also found for cloudy conditions and even for clear-sky short-wave radiation. The radiation errors are often caused by difficulties in producing and maintaining the mixed-phase nature of the ubiquitous low-level Arctic stratocumulus clouds, and in simulating cloud occurrence, phase and altitude (e.g. Karlsson & Svensson 2013; deBoer et al., 2013a). Studies have found that some improvement in the representation of cloud phase in super-cooled conditions may be obtained by incorporating double-moment representation of the microphysical parameters (e.g. Morrison et al., 2005; Solomon et al., 2009) or through improving the representation of aerosols in the droplet and ice nucleation processes (e.g. Birch et al., 2012; deBoer et al., 2013b). Obtaining a correct net surface radiation is further complicated by the heterogeneity of the sea ice surface. As mentioned previously, surface albedo can vary greatly, especially during the summer with the development of melt ponds and leads, while wintertime surface temperatures can vary greatly because of the large variability in ice thickness and snow cover, thereby producing great variability in outgoing long-wave radiation. Some models do not include melt pond parameterizations, so the summer surface albedo is too large, producing too little surface heating or melt. Melt-pond parameterizations have recently been developed and used in both regional and global climate models (Pedersen et al., 2009; Flocco et al., 2010; Holland et al., 2012). For instance, Holland et al. (2012) reported regional Arctic summertime surface energy flux differences of $10\,W\,m^{-2}$ and significantly thinner ice using a melt-pond parameterization. Furthermore, some models do not correctly represent snow on the ice surface, producing too low albedos (deBoer et al., 2013a), or parameterize surface albedo with inaccurate parameters (e.g. Birch et al., 2009). Timing of the seasonal transitions between snow-covered sea ice, melt ponds and bare sea ice are also often poorly captured in models, and so resulting albedo errors can change signs, impacting the simulation of the important length of the summer melt season.

The modelling of this unique environment is further complicated by the complex responses of the turbulent ABL and the conductive heat flux to the radiative forcing. As Figure 6.4(a) shows, turbulent heat flux during winter responds to changes in the net long-wave radiative forcing by the clouds, with liquid-containing clouds producing a near-zero net long-wave environment for which the turbulent heat flux will become near zero or slightly positive. For clear-sky conditions and strongly negative net long-wave radiation, the turbulent heat flux can be either strongly negative or near zero, depending on the stability of the surface layer. However, many models are not capable of reducing the turbulence for very stable conditions, producing too strong H_s for stable conditions. A few stability correction functions are able to account for this decrease in turbulence (Beljaars & Holtslag 1991; Grachev et al., 2007) and this issue may improve as they become incorporated in models. Other models do not represent the snow-covered sea ice correctly. For instance, the European Center for Medium Range Weather Forecasting operational model does not include snow on sea ice, so the conductive heat flux is too large and hence warms the surface too much. Again, as more sophisticated sea ice models that do include snow cover (e.g. CICE; Hunke & Lipscomb 2010) become coupled to atmospheric models, improvements in atmosphere–ice interactions should be seen. Due to a combination of surface temperature impacts of errors in other energy fluxes, such as radiation, and due to inadequate turbulent flux parameterizations, comparisons have shown that modeled turbulent heat fluxes have significant errors and are poorly correlated to observed values (Tjernström et al., 2005).

In the Antarctic in winter, Valkonen et al. (2008) applied the Polar MM5 mesoscale model [a predecessor of Polar Weather Research and Forecasting (WRF) model] and found that accurate sea ice concentration is very important for controlling modelled 2 m air temperature (T_{2m}) over sea ice during cold air advection from the continent. Large differences in sea ice concentration for available data sets produced sensitivity of T_{2m} occasionally reaching 13 K. For wind from the open ocean, T_{2m} is not sensitive to sea ice concentration, as differences between air and water temperatures are small. Tastula et al. (2012) carried out experiments for the Antarctic winter applying Polar WRF with comparisons against observations at the Ice Station Weddell (ISW) in February–June 1992. They identified several modelling challenges. As for Arctic conditions, latent and sensible heat fluxes had poor correlation with observations, and the model produced much fewer LLJs than observed, and they seldom (about 30% of time) matched the observed jets in time and altitude. The simulations were more successful with respect to the occurrence and base height of temperature inversions, but the inversion depth was a challenging variable. Polar WRF was also applied to the Antarctic sea ice zone in summer (Valkonen et al., 2013). The model performed well for synoptic-scale variations, but there were major challenges with clouds, ABL processes as well as sea ice and snow layer thermodynamics.

As a summary, although models perform reasonably well for synoptic-scale systems in both the Arctic and the Antarctic, significant challenges exist in simulation of ABL quantities and processes. Further development of the sea ice schemes applied in atmospheric models may help in some of the problems (Valkonen et al., 2008, 2013), but a major advance would require better understanding and parameterization of key processes, especially for unique polar conditions such as at low temperatures, low specific humidities or low aerosol concentrations.

Modelling to understand some of these processes usually include very high-resolution limited area models, using either idealized or real-case conditions. Process modelling of the effects of leads (Glendening & Burk 1992; Burk et al., 1997) have found that the thermal internal boundary layer and the associated cloud plumes from leads are primarily sensitive to the relative humidity and lapse rate of the environmental air. For leads of a significant width (e.g. 1 km) and for very stable conditions (i.e. surface-based inversions) with high relative humidity, the lead-induced internal boundary layer is only able to rise a few tens of metres, is destroyed within a few tens of kilometres downwind of the lead, and cloud plumes are readily damped. However, if the lapse rate is nearly adiabatic near the surface, the internal thermal plumes can rise to several hundred metres, and can advect significantly further downwind, with cloud plumes that consist initially of liquid water but later of primarily ice. That is, the extent and height of lead plumes seen in satellite images can provide information on the atmospheric environmental conditions near the lead. Process modelling of Arctic clouds has focused on the generation and maintenance of low-level, mixed-phase, stratocumulus clouds. Morrison et al. (2005) and Morrison & Pinto (2006) found that representing the microphysical constituents with both a mass and number concentration produced greater amounts of super-cooled liquid water in these clouds, while Solomon et al. (2009) showed that this enhanced degree of freedom of the model produced a large reduction in the number of ice and snow particles, thereby reducing the efficiency of the removal of small liquid drops. Solomon et al. (2011) furthermore showed that different scales of turbulence allowed the creation of the CML and the entrainment of moisture at the cloud top from the moisture inversion generally present (e.g. Sedlar & Tjernström, 2009), while Solomon et al. (2014) showed that the mixed-phase stratocumulus clouds are long-lived and resilient because they can exist through entrainment of moisture either from the surface or from the moisture inversion above.

6.6.2 Process validations

In order to understand or model the response of the sea ice surface system to a forcing, numerous processes and their key parameters must be understood, well represented and measured. A measure of how well the processes are represented in a model can be found by examining how well the model can produce the observed process relationships (e.g. Figures 6.4 and 6.5). Recent studies recognize the self-regulatory nature of these interdisciplinary process relationships for Arctic systems (Morrison et al., 2012), continue trying to understand their dependencies (Sterk et al., 2013), and have begun identifying model deficiencies through comparisons of process relationships in observations and models (deBoer et al., 2012, 2013a; Rinke et al., 2012; Pithan et al., 2014).

6.6.3 Reanalyses

Reanalyses combine model output with available observations to obtain the best estimate of the state of the atmosphere. However, in data-sparse areas, such as the Earth's polar regions, reanalyses primarily represent model output. Atmospheric reanalyses are applied in research of the Arctic and Antarctic sea ice zones to analyse the trends, variability and processes in the atmosphere over sea ice, to analyse relationships among ocean, sea ice and atmospheric variables, and to provide forcing for sea ice and ocean models. The accuracy of reanalyses is, however, not well known; it depends on the variable, reanalysis, region, decade, season and altitude.

Considering the Arctic sea ice zone, on the basis of Francis (2002), Bromwich and Wang (2005) and Bromwich et al. (2007), we may conclude that after the start of assimilation of satellite data in 1979 reanalyses perform reasonably well for synoptic-scale variability of atmospheric pressure, winds and temperature in the troposphere. The studies of Walsh and Chapman (1998) and Curry et al. (2002) indicated, however, that there are much larger errors in clouds, precipitation, short-wave and long-wave radiative fluxes, as well as near-surface air temperature and humidity. These studies examined the older reanalyses by the National Center for Environmental Prediction and National Center for Atmospheric Research (NCEP-NCAR) and the European Centre for Medium-Range Weather Forecasts (ECMWF; their first two reanalyses ERA-15 and ERA-40). Since then a lot of work has been done on developing new, more accurate reanalyses. These include the ECMWF ERA-Interim reanalysis, the Japanese Meteorological Agency Climate Data Assimilation System (JCDAS/JRA-25), the NCEP / Department of Energy reanalysis (NCEP-DOE, also known as NCEP2), the NCEP Climate Forecast System Reanalysis (NCEP-CFSR), and the NASA Modern Era Retrospective-Analysis for Research and Applications (MERRA).

On the basis of radiosonde soundings from R/V *Polarstern* over the Arctic sea ice over three summers, Lüpkes et al. (2010) detected large warm and moist biases in the ABL in ERA-Interim, even though the radiosonde observations were available for assimilation by ERA-Interim. Jakobson et al. (2012) applied independent tethersonde sounding data from the Tara expedition in April–August 2007, and found almost similar results for ERA-Interim, including a warm bias (up to 2°C in the lowermost 400 m layer) and a moist bias (0.3–0.5 g kg^{-1} throughout the lowermost 900 m), but other reanalyses included even larger errors. Jakobson et al. (2012) noted that when all the observed variables (air temperature, specific humidity, relative humidity and wind speed) were considered throughout the lowermost 900 m layer, ERA-Interim was the best reanalysis, outperforming JCDAS, NCEP-DOE, NCEP-CFSR and MERRA. Focusing on the near-surface variables, which are essential for forcing of sea ice models, NCEP-CFSR was, however, the best reanalysis. A 3-week late-summer comparison of ERA-Interim, NCEP and NCEP-DOE reanalyses with surface energy fluxes measured during the late-summer Arctic Summer Clouds-Ocean Experiment showed significant errors and low correlations in all surface energy flux terms, although the biases for ERA-Interim were clearly smaller (deBoer et al., 2013a). Another reanalysis evaluation based on independent observations was that by Wesslén et al. (2014). It was also the first evaluation of the new Arctic System Reanalysis (ASR; Bromwich et al., 2010) over sea ice. The results indicated that both ERA-Interim and ASR had difficulties with air moisture and clouds. Although ASR has a more sophisticated cloud microphysics scheme, it did not outperform ERA-Interim in simulating the mixed-phase stratocumulus in summer.

In Curry et al. (2002) and Makshtas et al. (2007) the evaluation of reanalyses was complemented by experiments applying a thermodynamic sea ice model. This demonstrated the impact of errors in reanalyses on sea ice growth and melt. Makshtas et al. (2007) concluded that the equilibrium ice thickness based on forcing from NCEP-NCAR reanalyses was only half of that based on forcing from ice station observations.

Considering the Antarctic sea ice zone, Vihma et al. (2002) and Vancoppenolle et al. (2011) have evaluated the old NCEP-NCAR reanalysis. Vihma et al. (2002) detected an annual cold bias of 3°C in the surface and 2 m air temperature, and Vancoppenolle et al. (2011) detected a too low cloud fraction, resulting in too much downward solar radiation and too little downward long-wave radiation. Tastula et al. (2013) evaluated the most widely used reanalyses (ERA-40, ERA-Interim, JRA-25, NCEP-DOE, NCEP-CFSR and MERRA) against the Ice Station Weddell observations in February–May 1992, which represent the most extensive set of meteorological observations available from the Antarctic sea ice zone. Considering all the observed quantities (near-surface air temperature, humidity and wind speed, the radiative and turbulent surface fluxes, and the cloud fraction) as a whole, NCEP-CFSR yielded the best results. Considering the diurnal cycle in these quantities, ERA-Interim yielded the best results.

It is noteworthy that reanalyses are not in hydrological balance over the Arctic and Antarctic sea ice zones, i.e. the seasonal/annual means of net precipitation and atmospheric moisture flux convergence do not match. The imbalance is severe (of the order of 50%) in NCEP-NCAR, NCEP-DOE, ERA-15 (Cullather et al., 2000; Serreze & Barry, 2005) and MERRA (Cullather & Bosilovich, 2011) but much smaller in ERA-40 (Tietäväinen & Vihma, 2008; Jakobson & Vihma, 2010). The imbalance originates from the fact that net precipitation in a reanalysis is based on short-term forecasts of precipitation and evaporation, whereas moisture flux convergence is an analysis variable directly affected by data assimilation. It is difficult to evaluate whether the reanalysis net precipitation or moisture flux convergence is a more reliable estimate for the true net precipitation, to be used, for example, as forcing for sea ice and ocean models. Cullather et al. (2000) and Serreze et al. (2006) consider the moisture flux convergence more reliable than the net precipitation. Jakobson and Vihma (2010) agree on this over the open ocean, but point out that the situation may be different over sea ice, as the absolute error in evaporation forecasts is smaller than over the open ocean and reanalyses do not benefit as much from assimilation of remote sensing data.

The ERA-Interim and NCEP-CFSR, which are the reanalyses evaluated as best over sea ice (Jakobson et al., 2012; deBoer et al., 2013; Tastula et al., 2013), agree on the correlation between sea ice concentration and 2 m air temperature but not on the correlation between sea ice concentration and 10 m wind speed. The situation is qualitatively similar for both the Arctic (Figure 6.12) and the Antarctic (Figure 6.13). During winter, a reduction in sea ice concentration results in an increase in the heat flux from the ocean to the atmosphere, seen as an increase in the near-surface air temperature. This should also result in less stable or more unstable ABL stratification, which would enhance the vertical mixing of momentum, resulting in stronger near-surface wind speeds. This seems to take place over large areas in ERA-Interim, but not in NCEP-CFSR. The lack of negative correlation may be related to the role of wind in modifying sea ice concentration; in contrast to ERA-Interim and other reanalyses, CFSR is based on a dynamic sea ice model coupled with the atmospheric model.

6.7 Summary

The atmosphere over sea ice has special characteristics because of its interactions with this unique surface and the high latitudes of sea ice occurrence. The lower boundary conditions of the atmosphere are variable, as the sea ice cover and snow pack on top of it often change. The fastest changes are due to ice advection and deformation, causing opening and closing of leads and larger-scale ice-cover changes; slower ones occur from snow accumulation, snow and ice melt, and ice growth. Another unique feature of broken sea ice cover is the extreme heterogeneity of surface temperature during cold seasons and of surface albedo during spring and summer. Ice and snow properties most essential for atmospheric thermodynamics are those affecting

The atmosphere over sea ice **187**

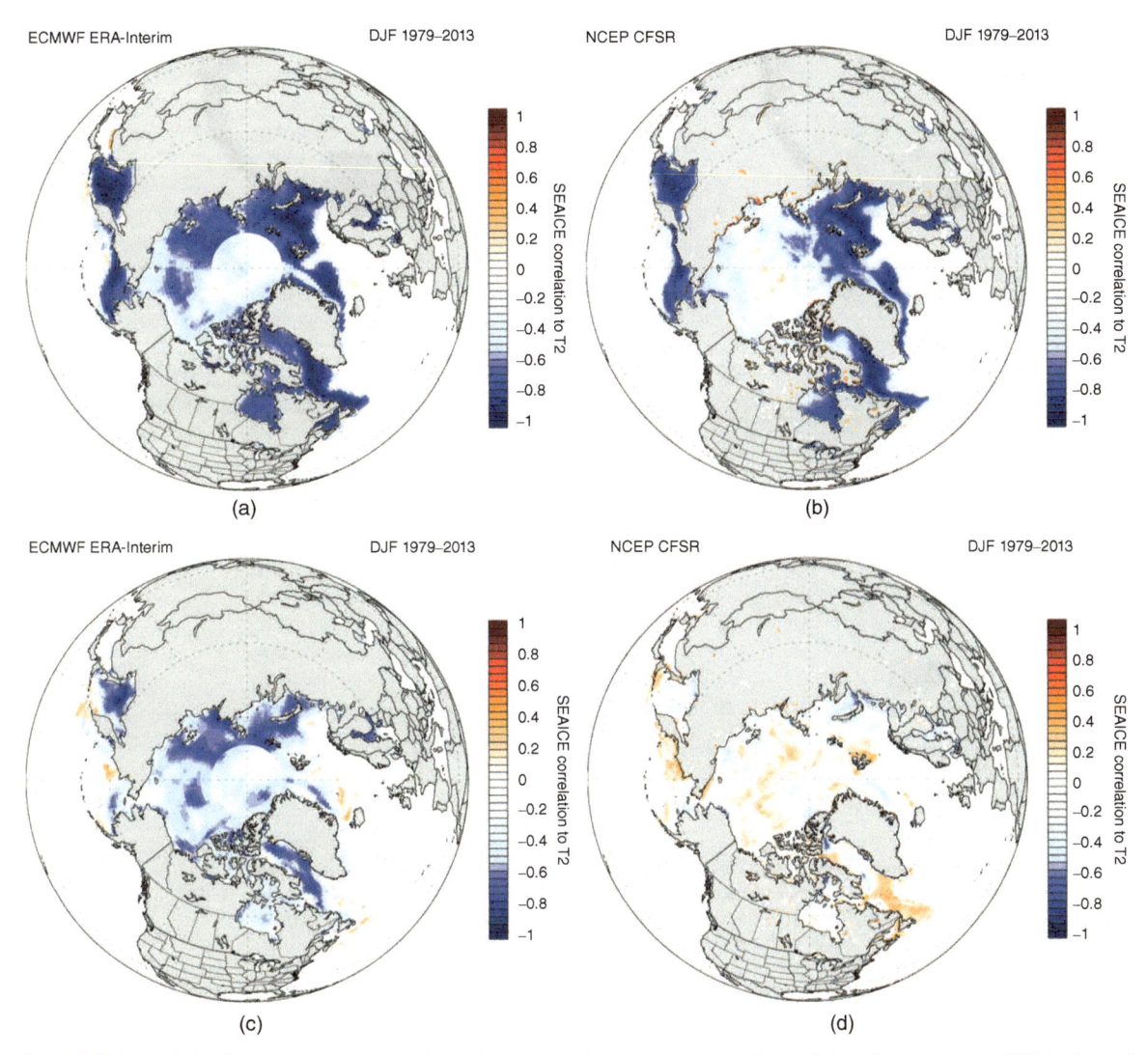

Figure 6.12 Correlation between Arctic December–February (DJF) sea ice concentration and 2 m air temperature (T2) (a, b) and between December–February (DJF) sea ice concentration and 10 m wind speed (WS10) (c, d) according to ERA-Interim (a, c) and National Center for Atmospheric Research-Climate Forecast System Reanalysis (NCEP-CFSR) (b, d).

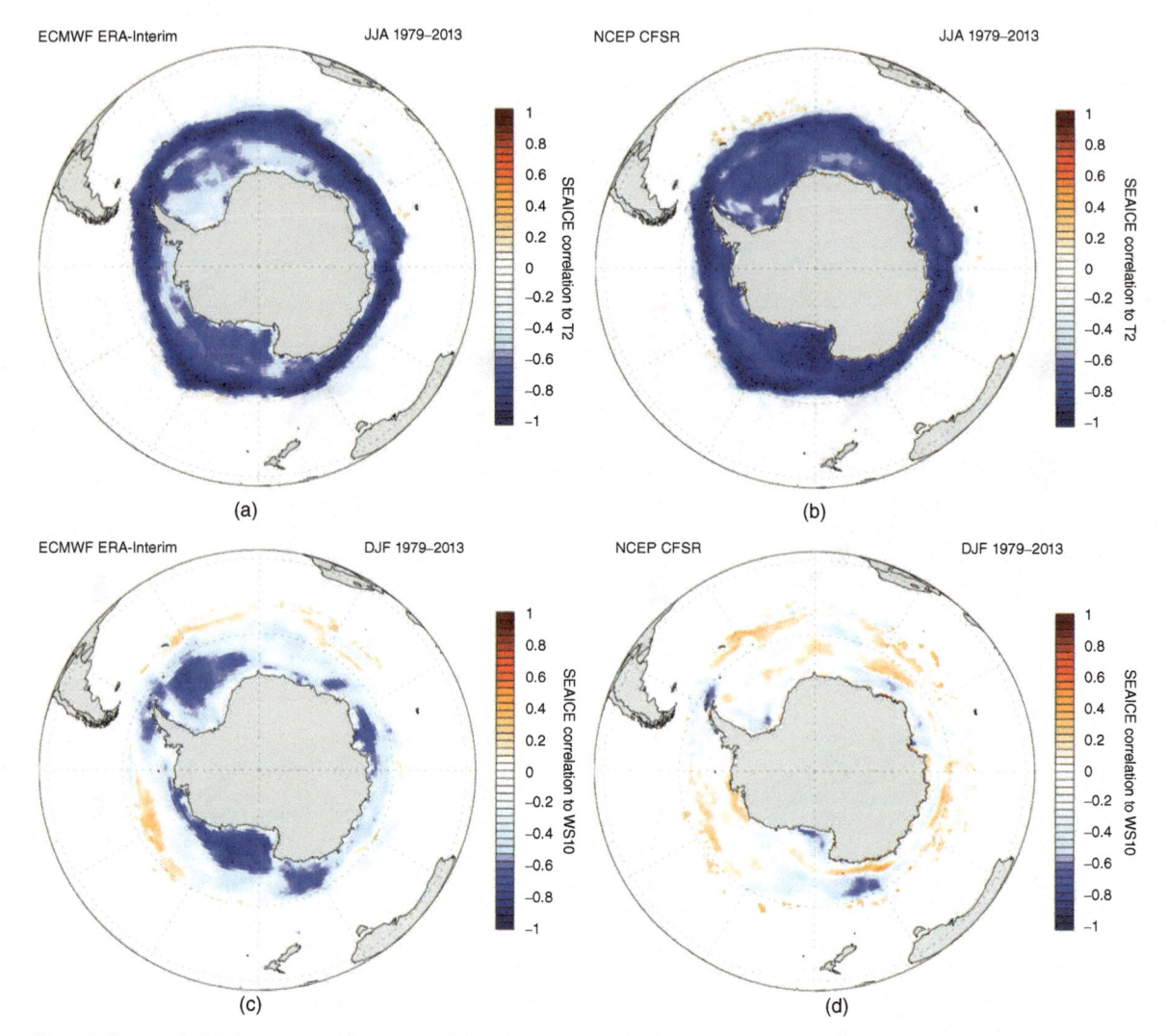

Figure 6.13 Correlation between Antarctic December–February (DJF) sea ice concentration and 2 m air temperature (T2) (a, b) and between December–February (DJF) sea ice concentration and 10 m wind speed (WS10) (c, d) according to ERA-Interim (a, c) and National Center for Atmospheric Research-Climate Forecast System Reanalysis (NCEP-CFSR) (b, d).

the surface temperature: the ice concentration and snow and ice thickness. From the point of view of dynamics of the ABL, the essential surface properties affecting the aerodynamic surface roughness are the ice concentration, freeboard, occurrence and height of pressure ridges and sastrugi, as well as floe, lead and melt-pond size distributions.

In addition to surface forcing, the structure of and processes in the ABL are governed by the large-scale pressure gradient and advection of heat, air moisture, clouds and aerosols. Complex interactions occur among radiative transfer, cloud microphysics, turbulence and waves. Turbulence is generated by the wind shear, convection over leads and polynyas, cloud-top radiative cooling and absorption of solar radiation in clouds. In winter, the state of the lower atmosphere over sea ice is governed by the air-mass advection from lower latitudes, cloud cover (which may be of advective or local origin) and the wind speed. Air temperature and specific humidity typically increase with height through the lowest kilometre. The warmest near-surface temperatures develop under overcast skies with strong winds, and the lowest ones under clear skies with weak winds. Over leads and polynyas, large upward fluxes of sensible and latent heat occur. During summer melt season, the thermal differences between the atmosphere, sea ice and open water are small, and the surface and near-surface temperatures remain close to the melting point. Air temperature and specific humidity also increase with height in summer.

Properties of and processes in the ABL play a major role in sea ice dynamics and thermodynamics. Wind speed and direction, ABL stratification and surface roughness control the air–ice momentum flux, which is a primary factor controlling sea ice drift. In winter, the net radiation on snow surface is typically negative, balanced by a sensible heat flux from air to snow and heat conduction from the ocean through ice and snow. In spring or early summer, the melt onset is frequently triggered by a large downward long-wave radiation from clouds associated with cases of warm air-mass advection from lower latitudes. Due to a decrease in surface albedo and solar zenith angle, the cloud radiative forcing becomes negative for at least part of the summer melt season. Surface melt is caused by the combined effects of downward solar and long-wave radiation and, occasionally, the sensible heat flux.

Due to the lack of accurate observations and complexity of highly interactive physical processes related to mixed-phase clouds, radiative transfer, turbulence, waves and air–snow–ice coupling, serious challenges remain in modelling the atmosphere over sea ice. Numerical weather prediction models, atmospheric reanalyses and climate models all have large errors in the surface and near-surface variables and, especially, fluxes. To improve the situation, more year-round *in situ* observations from the Arctic and Antarctic sea ice zone are needed, as well as a close collaboration between observational (including remote sensing) and modelling communities.

References

André, J.C. & Mahrt, L. (1982) The nocturnal surface inversion and influence of clear-air radiative cooling. *Journal of the Atmospheric Sciences*, **39**, 864–878.

Andreas, E.L (1998) The atmospheric boundary layer over polar marine surfaces. In: *Physics of Ice-Covered Seas*, Vol. 2. (Ed. M. Leppäranta), pp. 715–773. University of Helsinki Press, Helsinki.

Andreas, E.L, Claffey, K.J. & Makshtas, A.P. (2000) Low-level atmospheric jets and inversions over the western Weddell Sea. *Boundary Layer Meteorology*, **97**, 459–486.

Andreas, E.L., Guest, P.S., Persson, P.O.G. et al. (2002) Near-surface water vapor over polar sea ice is always near ice-saturation, *Journal of Geophyical. Research*, **107** (C8), doi:10.1029/2000JC000411.

Andreas, E.L, Horst, T.W., Grachev, A.A. et al. (2010a) Parameterising turbulent exchange over summer sea ice and the marginal ice zone. *Quarterly Journal of the Royal Meteorology Society*, **136B**, 927–943.

Andreas, E.L, Jordan, R.E., & Makshtas, A.P. (2004) Simulations of snow, ice, and near-surface atmospheric processes on Ice Station Weddell. *Journal of Hydrometeorology*, **5**, 611–624.

Andreas, E.L., Jordan, R.E. & Makshtas, A.P. (2005) Parameterizing turbulent exchange over sea ice: the Ice Station Weddell results. *Boundary-Layer Meteorology*, **114**, 439–460.

Andreas, E.L. & Makshtas, A.P. (1985) Energy exchange over Antarctic sea ice in the spring. *Journal of Geophysical Research*, **90**, 7199–7212.

Andreas, E.L, & Murphy, B. (1986) Bulk transfer coefficients for heat and momentum over leads and polynyas. *Journal of Physical Oceanography*, **16**, 1875–1883.

Andreas, E.L., Persson, P.O. G., Jordan, R.E., Horst, T.W., Guest, P.S., Grachev, A.A. & Fairall, C.W. (2010b), Parameterizing turbulent exchange over sea ice in winter. *Journal of Hydrometeorology* **11**, 87–104.

Arya, S.P.S. (1975) Drag partition theory for determining the large-scale roughness parameter and wind stress on Arctic pack ice. *Journal of Geophysical Research*, **80** 3447–3454.

Arya, S.P.S. (1981) Parameterizing the height of the stable atmospheric boundary layer. *Journal of Applied Meteorology*, **20**, 1192–1202.

Babko O., Rothrock, D.A. & Maykut, G.A. (2002) Role of rafting in the mechanical redistribution of sea ice thickness *Journal of Geophysical Research*, **107** (C8), 3113, doi: 10.1029/1999JC000190

Badgley, F.I. (1966) Heat budget at the surface of the Arctic Ocean. In: *Proceedings of the Symposium on the Arctic Heat Budget and Atmospheric Circulation* (Ed. J.O. Fletcher), pp. 267–277. The RAND Corporation, Memorandum RM-5233-NSF.

Beljaars, A.C.M. & Holtslag, A.A.M. (1991) Flux parameterization over land surfaces for atmospheric models. *Journal of Applied Meteorology*, **30**, 327–341.

Bigg, E.K., Leck, C. & Nilsson, E.D. (1996) Sudden changes in arctic atmospheric aerosol concentration during summer and autumn. *Tellus B*, **48**, 254–271.

Birch, C.E., Brooks, I.M., Tjernström, M. et al. (2009) The performance of a global and mesoscale model over the central Arctic Ocean during late summer. *Journal of Geophysical Research-Atmospheres*, **114**, D13104, doi:10.1029/2008JD010790

Birch, C.E., Brooks, I.M., Tjernström, M. et al. (2012) Modelling atmospheric structure, cloud and their response to CCN in the

Central Arctic: ASCOS case studies. *Atmospheric Chemistry and Physics*, **12**, 3419–3435.

Bromwich, D.H. & Wang, S.-H. (2005) Evaluation of the NCEP-NCAR and ECMWF 15- and 40-yr reanalyses using rawinsonde data from two independent Arctic field experiments. *Monthly Weather Review*, **133**, 3562–3578.

Bromwich, D.H., Fogt, R.L., Hodges, K.I. & Walsh, J.E. (2007) A tropospheric assessment of the ERA-40, NCEP, and JrA-25 global reanalyses in the polar regions. *Journal of Geophysical Research*, **112** (D10), doi: 10.1029/2006JD007859

Bromwich, D.H., Hines, K.M. & Bai, L-S. (2009) Development and testing of Polar Weather Research and forecasting model: 2. Arctic Ocean. *Journal of Geophysical Research*, **114** (D8), doi: 10.1029/2008JD010300

Bromwich D.H., Kuo, Y.-H., Serreze, M.C. et al. (2010) Arctic system reanalysis: call for community involvement, *EOS*, **91**, 13–20.

Brümmer, B., Schröder, D., Müller, G., Spreen, G., Jahnke-Bornemann, A. & Launiainen, J. (2009) Impact of a Fram Strait cyclone on ice edge, drift, divergence, and concentration: Possibilities and limits of an observational analysis. *Journal of Geophysical Research*, **113**, C12003, doi:10.1029/2007JC004149.

Brümmer, B. & Thiemann, S. (2002) The atmospheric boundary layer in an Arctic wintertime on-ice flow. *Boundary-Layer Meteorology*, **104**, 53–72.

Brümmer, B., Kirchgässner, A., Müller, G., Schröder, D., Launiainen, J. & Vihma, T. (2002a) The BALTIMOS field campaigns over the Baltic Sea during all four seasons. *Boreal Environmental Research*, **7**, 371–378.

Brümmer, B., Schröder, D., Launiainen, J., Vihma, T., Smedman, A.-S. & Magnusson, M. (2002b) Temporal and spatial variability of surface fluxes over the ice edge zone in the northern Baltic Sea. *Journal of Geophysical Research*, **107**(C8), 3096, doi:10.1029/2001JC000884.

Brümmer, B., Kirchgässner, A. & Müller, G. (2005) The atmospheric boundary layer over Baltic Sea ice. *Boundary-Layer Meteorology*, **117**, 91–109.

Burk, S.D., Fett, R.W. & Englebretson, R.E. (1997) Numerical simulation of cloud plumes emanating from Arctic leads. *Journal of Geophysical Research*, **102**(D14), doi:10.1029/97JD00339

Burk, S.D., & Thompson, W.T. (1995) Passage of a shallow front across a Beaufort Sea polynya. *Journal of Geophysical Research*, **100** (C3), 4461–4472.

Cheng, Y. & Brutsaert, W. (2005) Flux-profile relationships for wind speed and temperature in the stable atmospheric boundary layer. *Boundary-Layer Meteorology*, **114**, 519–538.

Cheng, B. & Vihma, T. (2002) Idealized study of a 2-D coupled sea-ice/atmosphere model during warm-air advection. *Journal of Glaciology*, **48**, 425–438.

Cheung, T.K. (1991) Sodar observations of the stable lower atmospheric boundary layer at Barrow, Alaska. *Boundary-Layer Meteorology*, **57**, 251–274.

Cohen, J., Screen, J.A., Furtado, J.C. et al. (2014) Recent Arctic amplification and extreme mid-latitude weather. *Nature Geoscience*, **7**, 627–637.

Cullather, R.I. & Bosilovich, M.G. (2011) The moisture budget of the polar atmosphere in MERRA. *Journal of Climate*, **24**, 2861–2879.

Cullather, R.I., Bromwich, D.H. & Serreze, M.C. (2000) The atmospheric hydrologic cycle over the Arctic basin from reanalyses. Part I: comparison with observations and previous studies. *Journal of Climate*, **13**, 923–937.

Curry, J.A., Schramm, J.L., Alam, A., Reeder, R., Arbetter, T.E. & Guest, P. (2002) Evaluation of data sets used to force sea ice models in the Arctic Ocean. *Journal of Geophysical Research*, **107**(C10), 8027, doi:10.1029/2000JC000466.

deBoer, G., Hashino, T., Tripoli, G.J. & Eloranta, E.W. (2013b) A numerical study of aerosol influence on mixed-phase stratiform clouds through modulation of the liquid phase. *Atmospheric Chemistry and Physics*, **13**, 1733–1749.

de Boer, G., Chapman, W., Kay, J.E. et al. (2012) A characterization of the present-day Arctic atmosphere in CCSM4. *Journal of Climate*, **25**, 2676–2695.

de Boer, G., Shupe, M.D., Caldwell, P.M. et al. (2013a) Near-surface meteorology during the Arctic Summer Cloud ocean Study (ASCOS): evaluation of reanalyses and global climate models. *Atmospheric Chemistry and Physics*, **14**, 427–445.

Dyer, A.J. (1974) A review of flux-profile relationships. *Boundary-Layer Meteorology*, **7**, 363–372.

Drüe, C. & Heineman, G. (2001) Airborne investigation of Arctic boundary-layer fronts over the marginal ice zone of the Davis Strait. *Boundary-Layer Meteorology*, **101**, 261–292.

Ebert, E.E., & Curry, J.A. (1993) An intermediate one-dimensional thermodynamic sea-ice model for investigating ice-atmosphere interactions. *Journal of Geophysical Research*, **98**, 10 085–10 109.

Esau, I.N. (2007) Amplification of turbulent exchange over wide arctic leads: Large-eddy simulation study. *Journal of Geophysical Research Atmospheres*, **112**, doi: 10.1029/2006JD007225

Fairall, C.W. & Markson, R. (1987) Mesoscale variations in surface stress, heat fluxes, and drag coefficients in the marginal ice zone during the 1983 Marginal Ice Zone Experiment. *Journal of Geophysical Research*, **92**, C7, 6921–6932.

Fett, R.W., Burk, S.D., Thompson, W.T. & Kozo, T.L. (1994) Environmental phenomena of the Beaufort Sea observed during the Leads Experiment. *Bulletin of the American Meteorological Society*, **75**, 2131–2145.

Finnigan, J.J., Einaudi, F. & Fua, D. (1984) The interaction between an internal gravity wave and turbulence in the stably stratified nocturnal boundary layer. *Journal of the Atmospheric Sciences*, **41**, 2409–2436.

Flocco, D., Feltham, D.L. & Turner, A.K. (2010) Incorporation of a physically based melt pond scheme into the sea ice component of a climate model. *Journal of Geophysical Research*, **115**, C08012, doi:10.1029/2009JC005568.

Francis, J.A. (2002) Validation of reanalysis upper-level winds in the Arctic with independent rawinsonde data. *Geophysical Research Letters*, **29**, 1315, doi:10.1029/2001GL014578.

Francis, J.A. & Vavrus, S.J. (2012) Evidence linking Arctic amplification to extreme weather in mid-latitudes. *Geophysical Research Letters*, **39**, L06801, doi:10.1029/2012GL051000.

Gascard, J.C., Festy, J., le Goff, H. et al. (2008) Exploring Arctic transpolar drift during dramatic sea-ice retreat. *EOS Transactions of the American Geophysical Union*, **89**, 21, doi: 10.1029/2008EO030001.

Gimbert, F., Jourdain, N.C., Marsan, D., Weiss, J. & Barnier, B. (2012). Recent mechanical weakening of the arctic sea ice cover as revealed from larger inertial oscillations. *Journal of Geophysical Research Oceans*, 117, doi: 10.1029/2011JC007633

Glendening, J.W. & Burk, S.D. (1992) Turbulent transport from an Arctic lead: a large-eddy simulation. *Boundary-Layer Meteorology*, **59**, 315–339.

Grachev, A.A., Fairall, C.W., Persson, P.O.G., Andreas, E.L. & Guest, P.S. (2005) Stable boundary-layer scaling regimes: The SHEBA data. *Boundary-Layer Meteorology*, **116**, 201–235.

Grachev, A.A., Andreas, E.L, Fairall, C.W., Guest, P.S. & Persson, P.O.G. (2007) SHEBA flux-profile relationships in the stable atmospheric boundary layer. *Boundary-Layer Meteorology*, , **124**, 315 - 333.

Grachev, A.A., Andreas, E.L, Fairall, C.W., Guest, P.S. & Persson, P.O.G. (2013) The critical Richardson number and limits of applicability of the local similarity theory in the stable boundary layer. *Boundary-Layer Meteorology*, **147**, 51–82,

Graversen, R.G., Mauritsen, T., Drijfhout, S., Tjernstrom, M. & Martensson, S. (2011). Warm winds from the pacific caused extensive arctic sea-ice melt in summer 2007. *Climate Dynamics*, **36**, 2103–2112.

Grenfell, T.C., & Perovich, D.K. (1984) Spectral albedos of sea ice and incident solar irradiance in the southern Beaufort Sea. *Journal of Geophysical Research*, **89**, 3573–3580.

Grønås, S.R., & Skeie, P. (1999) A case study of strong winds at an arctic front. *Tellus.Series A*, **51**, 865–879.

Guest, P.S. (2007) Measuring turbulent heat fluxes over leads using kites, *Journal of Geophysical Research*, **112**, C5, C05021, doi:10.1029/2006JC003689.

Gustafsson, N., Nyberg, L. & Omstedt, A. (1998) Coupling of a high-resolution atmospheric model and an ocean model for the Baltic Sea. *Monthly Weather Review*, **126**, 2822–2846.

Hakkinen, S., Proshutinsky, A., & Ashik, I. (2008) Sea ice drift in the Arctic since the 1950s. *Geophysical Research Letters*, **35**, doi: 10.1029/2008GL034791

Harrington, J.Y., Reisin, T., Cotton, W.R. & Kreidenweis, S.M. (1999) Cloud resolving simulations of Arctic stratus Part II: Transition-season clouds. *Atmospheric Research*, **51**, 45–75.

Holland, M.M., Bailey, D.A., Briegleb, B.P., Light, B. & Hunke, E. (2012) Improved sea ice shortwave radiation physics in CCSM4: The impact of melt ponds and aerosols on Arctic sea ice. *Journal of Climate*, **25**, 1413–1430.

Hunke, E.C. & Lipscomb, W.H. (2010) CICE: The Los Alamos sea ice model, documentation and software user's manual, Version 4.1. *Technical Report LACC-06–012*, T-3 Fluid Dynamics group, Los Alamos National Laboratory, Los Alamos, NM. (Available at http://climate.lanl.gov/models/cice)

Hunke, E.C. & Zhang, Y. (1999) A comparison of sea ice dynamics models at high resolution. *Monthly Weather Review*, **127**, 396–408.

Inoue, J., Kawashima, M., Fujiyoshi, Y. & Wakatsuchi, M. (2005) Aircraft observations of air-mass modification over the Sea of Okhotsk during sea-ice growth. *Boundary-Layer Meteorology*, **117**, 111–129.

Intrieri, J.M., Shupe, M.D., Uttal, T. & McCarty, B.J. (2002) Annual cycle of Arctic cloud geometry and phase from radar and lidar at SHEBA. *Journal of Geophysical Research*, **107**(C10), doi: 10.1029/2000JC000423.

Jackson, J.M., Carmack, E.C., McLaughlin, F.A., Allen, S E. & Ingram, R.G. (2010) Identification, characterization, and change of the near-surface temperature maximum in the Canada Basin, 1993–2008, *Journal of Geophysical Research*, **115**, C05021, doi:10.1029/2009JC005265

Jackson, J.M., Williams, W.J. & Carmack, E.C. (2012) Winter sea-ice melt in the Canada Basin, Arctic Ocean. *Geophysical Research Letters*, **39**, L03603, doi:10.1029/2011GL050219

Jakobson, E., Vihma, T., Palo, T., Jakobson, L., Keernik, H. & Jaagus, J. (2012) Validation of atmospheric reanalyses over the central Arctic Ocean, *Geophysical Research Letters*, **39**, L10802, doi:10.1029/2012GL051591.

Jakobson, E. & Vihma, T. (2010) Atmospheric moisture budget in the arctic based on the ERA-40 reanalysis. *International Journal of Climatology*, **30**, 2175–2194.

Jakobson, L., Vihma, T., Jakobson, E., Palo, T., Männik, A. & Jaagus, J. (2013) Low-level jet characteristics over the Arctic Ocean in spring and summer. *Atmospheric Chemistry and Physics*, **13**, 11089–11099.

Jordan, R.E., Andreas, E.L. & Makshtas, A.P. (1999), Heat budget of snow-covered sea ice at North Pole 4, *Journal of Geophysical Research*, **104**(C4), 7785–7806.

Kahl, J.D. (1990) Characteristics of the low-level temperature inversion along the Alaskan Arctic coast. *International Journal of Climatology*, **10**, 537–548.

Karlsson, J., & Svensson, G. (2013) Consequences of poor representation of Arctic sea-ice albedo and cloud-radiation interactions in the CMIP5 model ensemble. *Geophysical Research Letters*, **40**, 4374–4379.

Kellner, G., Wamser, C. & Brown, R.A. (1987) An observation of the planetary boundary layer in the marginal ice zone. *Journal of Geophysical Research*, **92**(C7), 6955–6965.

Kilpeläinen, T., Vihma, T. & Olafsson, H. (2011) Modelling of spatial variability and topographic effects over Arctic fjords in Svalbard, *Tellus A*, **63**, 223–237.

Kolstad, E.W., Bracegirdle, T.J. & Seierstad, I.A. (2009) Marine cold-air outbreaks in the north atlantic: temporal distribution and associations with large-scale atmospheric circulation. *Climate Dynamics*, **33**, 187–197.

Kottmeier, C. & Engelbart, D. (1992) Generation and atmospheric heat exchange of coastal polynyas in the weddell sea. *Boundary-Layer Meteorology*, **60**, 207–234.

Kottmeier, C. & Sellmann, L. (1996) Atmospheric and oceanic forcing of Weddell Sea ice motion. *Journal of Geophysical Research*, **101**, 20809–20824.

Kriegsmann, A. & Brümmer, B. (2014) Cyclone impact on sea ice in the central Arctic Ocean: a 2988 statistical study. *The Cryosphere*, **8**, 303–317.

Kwok, R., Spreen, G. & Pang, S. (2013), Arctic sea ice circulation and drift speed: decadal trends and ocean currents, *Journal of Geophysical Research-Oceans*, **118**, 2408–2425.

Kwok, R. & Untersteiner, N. (2011) The thinning of Arctic sea ice. *Physics Today*, **64**, doi:10.1063/1.3580491

Langland, R.H., Tag, P.M. & Fett, R.W. (1989) An ice breeze mechanism for boundary-layer jets. *Boundary-Layer Meteorology* **48**, 177–195.

Launiainen, J. & Vihma, T. (1990) Derivation of turbulent surface fluxes – an iterative flux-profile method allowing arbitrary observing heights. *Environmental Software* **5**, 113–124.

Launiainen, J. & Vihma, T. (1994) On the surface heat fluxes in the Weddell Sea, In: *The Polar Oceans and Their Role in Shaping the Global Environment*, Nansen Centennial Volume (Eds. O.M. Johannessen, R. Muench & J.E. Overland). American Geophysical Union Geophysical, Monograph Series, **85**, 399–419.

Launiainen, J., Cheng, B., Uotila, J. & Vihma, T. (2001) Turbulent surface fluxes and air--ice coupling in the Baltic AirSea-Ice Study (BASIS). *Annals of Glaciology*, **33**, 237–242.

Leck, C., Bigg, E.K., Covert, D.S. et al. (1996) Overview of the atmospheric research program during the International Ocean Expedition of 1991 (IAOE-1991) and its scientific results. *Tellus B*, **48**, 136 – 155.

Leck, C., Nilsson, E.D., Bigg E.K. & Bäcklin, L. (2001) The atmospheric program of the Arctic Ocean Expedition 1996 (AOE-1996) - an overview of scientific objectives, experimental approaches and instruments. *Journal of Geophysical Research*, **106**, 32 051–32 067.

Lettau, H.H. (1979) Wind and temperature profile prediction for diabatic surface layers including strong inversion cases. *Boundary-Layer Meteorology*, **17**(4), 443–464.

Light B., Grenfell, T.C. & Perovich D.K. (2008) Transmission and absorbtion of solar radiation by Arctic sea ice during the melt season. *Journal of Geophysical Research*, **113**, C03023. doi:1029/2006JC003977

Lindsay, R.W. (1998) Temporal variability of the energy balance of thick Arctic pack ice. *Journal of Climate*, **11**, 313-333.

Liu, A.K., Häkinnen, S. & Peng, C.Y. (1993) Wave effects on ocean-ice interaction in the marginal ice zone. *Journal of Geophysical Research*, **98**(C6), 10 025–10 036.

Lüpkes, C. & Schlünzen, K.H. (1996) Modelling the arctic convective boundary-layer with different turbulence parameterisations. *Boundary-Layer Meteorology*, **79**, 107–130.

Lüpkes, C., Vihma, T., Jakobson, E., König-Langlo, G. & Tetzlaff, A. (2010) Meteorological observations from ship cruises during summer to the central Arctic: a comparison with reanalysis data. *Geophysical Research Letters*, **37**, L09810, doi:10.1029/2010GL042724

Lüpkes, C., Vihma, T., Birnbaum, G. et al. (2012) Mesoscale modelling of the Arctic atmospheric boundary layer and its interaction with sea ice, In: *Arctic Climate Change - The ACSYS Decade and Beyond* (Eds. Lemke, P. & Jacobi, H.-W), Atmospheric and Oceanographic Sciences Library, **43**, doi: 10.1007/978–94–007–2027–5

Lüpkes, C., Vihma, T., Birnbaum, G. & Wacker, U. (2008) Influence of leads in sea ice on the temperature of the atmospheric boundary layer during polar night, *Geophysical Research Letters*, **35**, L03805, doi:10.1029/2007GL032461.

Lüpkes, C., Gryanik, V.M., Hartmann, J. & Andreas, E.L. (2012) A parametrization, based on sea ice morphology, of the neutral atmospheric drag coefficients for weather prediction and climate models. *Journal of Geophysical Research* **117**, D13112, doi:10.1029/2012JD017630.

Lüpkes, C. & Gryanik, V.M. (2015) A stability-dependent parametrization of transfer coefficients for momentum and heat over polar sea ice to be used in climate models. *Journal of Geophysical Research*, **120**, 1–30.

Mahrt, L. (1981) Modelling the depth of the stable boundary layer. *Boundary-Layer Meteorology*, **21**, 3–19.

Mahrt, L. (1999) Stratified atmospheric boundary layers. *Boundary-Layer Meteorology*, **90**, 375 - 396.

Makshtas, A.P., Andreas, E.L, Svyashchennikov, P.N., & Timachev, V.F. (1999) Accounting for clouds in sea ice models. *Atmospheric Research*, **52**, 77–113.

Makshtas, A., Atkinson, D., Kulakov, M., Shutilin, S., Krishfield, R. & Proshutinsky, A. (2007) Atmospheric forcing validation for modeling the central Arctic. *Geophysical Research Letters*, **34**, doi: 10.1029/2007GL031378

Maksimovich, E., & Vihma, T. (2012) The effect of surface heat fluxes on interannual variability in the spring onset of snow melt in the central Arctic Ocean. *Journal of Geophysical Research*, **117**, C07012, doi:10.1029/2011JC007220.

Markus, T., Stroeve, J.C., & Miller, J. (2009) Recent changes in Arctic sea ice melt onset, freezeup, and melt season length. *Journal of Geophysical Research*, **114** C12024. doi: 10.1029/2009JC005436

Martinson, D.G., & Wamser, C. (1990). Ice drift and momentum exchange in winter Antarctic pack ice. *Journal of Geophysical Research*, **95**, 1741–1755.

Maykut, G.A. (1982) Large-scale heat exchange and ice production in the Central Arctic. *Journal of Geophysical Research*, **87** (C10), 7971–7984.

Maykut, G.A., & Untersteiner, N. (1971) Some results from a time-dependent thermodynamic model of sea ice. *Journal of Geophysical Research*, **76**, 1550–1575.

Mazon, J., Niemelä, S., Pino, D., Savijärvi, H. & Vihma, T. (2015) Snow bands over the Gulf of Finland in wintertime. *Tellus A*, **67**, 25102, doi: 10.3402/tellusa.v67.25102

Monin, A.S. & Obukhov, A.M. (1954) Basic laws of turbulent mixing in the surface layer of the atmosphere, *Trudy Geofiz. Inst. Acad. Nauk SSSR.* **24**, 163 – 187.

Morrison, H., Curry, J.A. & Khvorostyanov, V.I. (2005) A new double-moment microphysics scheme for application in cloud and climate models. Part I: Description. *Journal of the Atmospheric Sciences*, **62**, 1665–1677.

Morrison, H., de Boer, G., Feingold, G., Harrington, J.Y., Shupe M.D. & Sulia, K. (2012) Resilience of persistent Arctic mixed-phase clouds, *Nature Geosciences.*, **5**, 11–17.

Morrison, H., & Pinto, J.O. (2006) Intercomparison of bulk cloud microphysics schemes in mesoscale simulations of springtime arctic mixed-phase stratiform clouds. *Monthly Weather Review*, **134**, 1880–1900.

Nansen, F. (1902) *The Norwegian North Polar Expedition 1893–1896, Scientific Results, Vol.* **3**, *The Oceanography of the North Polar Basin*. Longman Green & Co., Kristiania, Norway.

Nilsson, E.D. (1996) Planetary boundary layer structure and air mass transport during the International Arctic Ocean Expedition 1991. *Tellus B*, **48**, 178–196.

Økland, H. (1998) Modification of frontal circulations by surface heat flux. *Tellus A*, **50**, 211–218.

Overland, J.E. (1985) Atmospheric boundary layer structure and drag coefficients over sea ice. *Journal of Geophysical Research*, **90**(C5), 9029–9049.

Overland, J.E., McNutt, S., Groves, J., Salo, S., Andreas, E.L. & Persson, P.O.G. (2000) Regional sensible and radiative heat flux estimates for the winter Arctic during SHEBA. *Journal of Geophysical Research*, **105** (C6), 14093–14102.

Pedersen, C.A., Roeckner, E., Lüthje, M. & Winther, J.-G. (2009) A new sea ice albedo scheme including melt ponds for ECHAM5 general circulation model. *Journal of Geophysical Research*, **114**, D08101, doi:10.1029/2008JD010440.

Perlwitz, J., Hoerling, M. & Dole, R. (2015) Arctic tropospheric warming: Causes and linkages to lower latitudes. *Journal of Climate*, **28**, 2154–2167.

Perovich, D.K., Grenfell, T.C., Light, B. & Hobbs, P.V. (2002) Seasonal evolution of the albedo of multiyear Arctic sea ice. *Journal of Geophysical Research*, **107** (C10), 8044. doi:10.1029/2000JC000438

Perovich, D.K., Grenfell, T.C., Richter-Menge, J.A., Light, B., Tucker III., W.B. & Eicken, H. (2003) Thin and thinner: sea ice mass balance measurements during SHEBA. *Journal of Geophysical Research* **108** (C3) 8050, doi:10.1029/2001JC001079,

Persson, P.O.G. (2012) Onset and end of the summer melt season over sea ice: Thermal structure and surface energy perspective from SHEBA. *Climate Dynamics*, **39**, 1349–1371.

Persson, P.O.G., Bao, J.-W. & Michelson, S. (2002b) Mesoscale modeling of the wintertime boundary layer structure over the Arctic pack ice. *Preprints, 15th Symposium on Boundary Layers and Turbulence*, 15–19 July, Wageningen, the Netherlands, pp. 335–338.

Persson, P.O.G., Fairall, C.W., Andreas, E.L., Guest, P.S. & Perovich, D.K. (2002a) Measurements near the Atmospheric Surface Flux Group tower at SHEBA: near-surface conditions and surface energy budget. *Journal of Geophysical Research*, **107**, 8045, doi:10.1029/2000JC000705.

Persson, P.O.G, Hare, J.E., Fairall, C.W. & Otto, W. (2005) Air-sea interaction processes in warm and cold sectors of extratropical cyclonic storms observed during FASTEX. *Quarterly Journal of the Royal Meteorological Society*, **131**, 877–912.

Persson, P.O.G., Ruffieux, D. & Davidson, K. (1992) Characteristics of the lower troposphere during LeadEx 92. *Preprints, Third Conference Polar Meteorology and Oceanography*, 29 Sept–2 Oct, Portland, OR, 50–53

Persson, P.O.G., Shupe, M., Perovich, D. & Solomon, A. (2016) Linking atmospheric synoptic transport, cloud phase, surface energy fluxes, and sea-ice growth: Observations of midwinter SHEBA conditions. *Climate Dynamics*, DOI: 10.1007/s00382-016-3383-1, Accepted.

Persson, P.O G., Uttal, T., Intrieri, J.M., Fairall, C.W., Andreas, E.L. & Guest, P.S. (1999) Observations of large thermal transitions during the Arctic night from a suite of sensors at SHEBA. *Preprints, 3rd Symp. on Integrated Observing Systems.*, Jan. 10–15, Dallas, TX, 171–174.

Pinto, J.O. (1998) Autumnal mixed-phase cloudy boundary layers in the Arctic. *Journal of the Atmospheric Sciences*, **55**, 2016–2038.

Pinto, J.O., Alam, A., Maslanik, J.A., Curry, J.A. & Stone, R.S. (2003) Surface characteristics and atmospheric footprint of springtime arctic leads at SHEBA. *Journal of Geophysical Research Oceans*, **108**, doi:. 10.1029/2000JC000473

Pithan, F., Medeiros, B. & Mauritsen, T. (2014) Mixed-phase clouds cause climate model biases in arctic wintertime temperature inversions. *Climate Dynamics*, **43**(1–2), 289–303.

Rampal, P., Weiss, J. & Marsan, D. (2009) Positive trend in the mean speed and deformation rate of Arctic sea ice: 1979–2007, *Journal of Geophysical Research*, **114**, C05013.

Rinke, A., Xin, Y., Bian, L., Ma, Y., Dethloff, K., Persson, P.O.G., Lüpkes, C. & Xiao, C. (2012) Evaluation of atmospheric boundary layer-surface process relationships in a regional model along an East Antarctic traverse. *Journal of Geophysical Research*, **117**, D09121, doi:10.1029/2011JD016441.

Ruffieux, D.R., Persson, P.O.G., Fairall, C.W. & Wolfe, D.E. (1995) Ice pack and lead surface energy budgets during LEADEX 92. *Journal of Geophysical Research*, **100**, C3, 4593–4612.

Savijärvi, H.I. (2012) Cold air outbreaks over high-latitude sea gulfs. *Tellus A*. **64**, 12244.

Schnell, R.C., Barry, R.G., Miles, M.W. e#. (1989) Lidar detection of leads in Arctic sea ice. *Nature*, **339**, 530–532.

Sedlar, J. & Tjernström, M. (2009) Stratiform cloud-inversion characterization during the Arctic melt season. *Boundary-Layer Meteorology*, **132** (3), 455–474.

Serreze, M.C., Kahl, J.D. & Schnell, R.C. (1992) Low-level temperature inversions of the Eurasian Arctic and comparisons with Soviet drifting station data. *Journal of Climate*, **5**, 615–629.

Serreze, M.C. & Barry, R.G. (2005) *The Arctic Climate System.* Cambridge University Press, Cambridge.

Serreze, M.C., Barrett, A.P., Slater, A.G. et al. (2006) The large-scale freshwater cycle of the Arctic, *Journal of Geophysical Research-Oceans,* **111,** C11010, doi:10.1029/2005JC003424.

Serreze, M.C., Barrett, A.P., & Cassano, J.J. (2011) Circulation and surface controls on the lower tropospheric air temperature field of the Arctic *J. Geophysical Research-Atmospheres,* **116,** doi: 10.1029/2010JD015127 DOI:10.1029/2010JD015127 #_NEW

Shupe, M.D., Uttal, T. & Matrosov, S.Y. (2005) Arctic cloud microphysics retrievals from surface-based remote sensors at SHEBA. *Journal of Applied Meteorology,* **44,** 1544–1562.

Shupe, M.D., Matrosov, S.Y. & Uttal, T. (2006) Arctic mixed-phase cloud properties derived from surface-based sensors at SHEBA. *Journal of the Atmospheric Sciences,* **63,** 697–711.

Shupe, M.D., Persson, P.O.G., Brooks, I.M. et al. (2013) Cloud and boundary layer interactions over the Arctic sea-ice in late summer. *Atmospheric Chemistry and Physics,* **13,** 9379–9399

Shupe, M.D., Walden, V.P., Eloranta, E. et al. (2011) Clouds at Arctic Atmospheric Observatories, Part I: Occurrence and macrophysical properties. *Journal of Applied Meteorology and Climatology,* **50,** 626–644.

Sirevaag, A., de la Rosa, S., Fer, I., Nicolaus, M., Tjernström, M. & McPhee, M.G. (2011) Mixing, heat fluxes, and heat content evolution of the Arctic Ocean mixed layer. *Ocean Science,* **7,** 335–349.

Smedman, A.-S. (1988) Observations of a multi-level turbulence structure in a very stable atmospheric boundary layer, *Boundary-Layer Meteorology,* **44,** 231–253.

Solomon, A., Morrison, H., Persson, O., Shupe, M. & Bao, J.-W. (2009) Investigation of microphysical parameterizations of snow and ice in Arctic clouds during M-PACE through model-observation comparisons. *Monthly Weather Review,* **137,** 3110–3128

Solomon, A., Shupe, M.D., Persson, P.O.G. & Morrison, H. (2011) Moisture and dynamical interactions maintaining decoupled Arctic mixed-phase stratocumulus in the presence of a humidity inversion. *Atmospheric Chemistry and Physics,* **11,** 10127–10148.

Solomon, A.S., Shupe, M.D., Persson, P.O.G. et al. (2014) The sensitivity of springtime Arctic mixed-phase stratocumulus clouds to surface layer and cloud-top inversion layer moisture sources. *Journal of the Atmospheric Sciences,* **71,** 574–595.

Sotiropoulu, G., Sedlar, J., Tjernström, M., Shupe, M.D., Brooks, I.M. & Persson, P.O.G. (2014) The thermodynamic structure of summer Arctic stratocumulus and the dynamic coupling to the surface. *Atmospheric Chemistry and Physics,* **14,** 12573–12592.

Spreen, G., Kwok, R., & Menemenlis, D. (2011). Trends in arctic sea ice drift and role of wind forcing: 1992–2009. *Geophysical Research Letters,* **38,** doi: /10.1029/2011GL048970

Steele, M., Zhang, J. & Ermold, W. (2010) Mechanisms of summertime upper arctic ocean warming and the effect on sea ice melt. *Journal of Geophysical Research,* **115,** C11004, doi: /10.1029/2009JC005849

Steele, M., Ermold, W. & Zhang, J. (2011) Modeling the formation and fate of the near-surface temperature maximum in the canadian basin of the arctic ocean. *Journal of Geophysical Research,* **116,** C11015, doi.org/10.1029/2010JC006803

Sterk, H.A.M., Steeneveld, G.J. & Holtslag, A A.M. (2013) The role of snow-surface coupling, radiation, and turbulent mixing in modeling a stable boundary layer over Arctic sea ice. *Journal of Geophysical Research-Atmospheres,* **118,** 1199–1217.

Stramler, K., del Genio, A.D. & Rossow, W.B. (2011) Synoptically driven Arctic winter states. *Journal of Climate,* **24,** 1747–1762.

Stroeve, J.C., Serreze, M.C., Holland, M.M., Kay, J.E., Maslanik, J. & Barrett, A.P. (2012) The Arctic's rapidly shrinking sea ice cover: A research synthesis. *Climatic Change,* **110,** 1005–1027.

Sturm, M., Perovich, D.K., & Holmgren, J. (2002) Thermal conductivity and heat transfer through the snow on the ice of the Beaufort Sea. *Journal of Geophysical Research,* **107** (C21), 8043, doi:10.1029/2000JC000409,

Sukoriansky, S., Galperin, B., & Perov, V. (2005) Application of a new spectral theory of stably stratified turbulence to the atmospheric boundary layer over sea ice. *Boundary-Layer Meteorology,* **117**(2), 231–257.

Sukoriansky, S., Galperin, B. & Perov, V. (2006) A quasi-normal scale elimination model of turbulence and its application to stably stratified flows. *Nonlinear Processes in Geophysics,* **13,** 9–22.

Sverdrup, H.U. (1933) *The Norwegian North Polar Expedition with the "Maud" 1918–1925, Scientific Results,* Vol. **II,** Meteorology, Part I: Discussion. Geofysisk Institut, Bergen.

Tastula, E.-M, Vihma, T. & Andreas, E. L (2012) Modeling of the atmospheric boundary layer over Antarctic sea ice in autumn and winter. *Monthly Weather Review,* **140,** 3919–3935.

Tastula, E.-M., Vihma, T., Andreas, E.L. & Galperin, B. (2013) Validation of the diurnal cycles in atmospheric reanalyses over Antarctic sea ice. *Journal of Geophysical Research-Atmospheres,* **118,** 4194–4204.

Thorndike, A.S. & Colony, R. (1982) Sea ice motion in response to geostrophic winds. *Journal of Geophysical Research,* **87**(C8), 5845–5852

Tietäväinen, H. & Vihma, T. (2008) Atmospheric moisture budget over Antarctica and Southern Ocean on the basis of ERA-40 reanalysis. *International Journal of Climatology,* **28,** 1977–1995.

Tisler, P., Vihma, T., Müller, G. & Brümmer, B. (2008) Modelling of warm-air advection over Arctic sea ice. *Tellus A,* **60,** 775–788.

Tjernström, M. (2005) The summer Arctic boundary layer during the Arctic Ocean Experiment 2001 (AOE-2001). *Boundary-Layer Meteorology,* **117,** 5–36.

Tjernström, M., Birch, C.E. , Brooks, I.M. et al. (2012) Meteorological conditions in the central Arctic summer during the

Arctic Summer Cloud Ocean Study (ASCOS). *Atmospheric Chemistry and Physics*, **12**, 6863–6889.

Tjernström, M., Leck, C., Persson, P.O.G., Jensen, M.L., Oncley, S.P. & Targino, A. (2004) The summertime Arctic atmosphere: Meteorological measurements during the Arctic Ocean Experiment 2001 (AOE-2001). *Bulletin of the American Meteorological Society*, **85**, 1305–1321.

Tjernström, M. & Graversen, R.G. (2009) The vertical structure of the lower arctic troposphere analysed from observations and the ERA-40 reanalysis. *Quarterly Journal of the Royal Meteorological Society*, **135**(639), 431–443.

Tjernström, M. & Mauritsen, T. (2009) Mesoscale variability in the summer Arctic boundary layer. *Boundary-Layer Meteorology*, **130**:383–406.

Tjernström, M., Sedlar, J. & Shupe, M.D. (2008) How well do regional climate models reproduce radiation and clouds in the Arctic? An evaluation of ARCMIP simulations. *Journal of Applied Meteorology and Climatology*, **47**, 2405–2422.

Tjernström, M., Shupe, M.D., Brooks, I.M. et al. (2015) Warm-air advection, air mass transformation and fog causes rapid ice melt. *Geophysical Research Letters*, **42**, 5594–5602.

Tjernström, M., Zagar, M., Svensson, G. et al. (2005) Modelling the Arctic boundary layer: An evaluation of six ARCMIP regional-scale models with data from the SHEBA project. *Boundary-Layer Meteorology*, **117**, 337–381.

Tuononen, M., Sinclair, V.A. & Vihma, T. (2015) A climatology of low-level jets in the mid-latitudes and polar regions of the Northern Hemisphere. *Atmospheric Science Letters*, Published online, doi: 10.1002/asl.587.

Uttal, T., Curry, J.A., McPhee, M.G. et al. (2002) The surface heat budget of the Arctic, *Bulletin of the American Meteorological Society*, **83**, 255–275.

Valkonen, T., Vihma, T., Johansson, M. & Launiainen, J. (2013) Atmosphere - sea ice interaction in early summer in the Antarctic: evaluation and challenges of a regional atmospheric model. *Quarterly Journal of the Royal Meteorological Society*, Published online, doi: 10.1002/qj.2237.

Valkonen, T., Vihma, T., & Doble, M. (2008) Mesoscale modelling of the atmospheric boundary layer over the Antarctic sea ice: a late autumn case study. *Monthly Weather Review*, **136**, 1457–1474.

Vancoppenolle, M., Timmermann, R., Ackley, S.F. et al. (2011) Assessment of radiation forcing data sets for large-scale sea ice models in the southern ocean. *Deep Sea Research Part II*, **58**, 1237–1249.

Vavrus, S., Walsh, J.E., Chapman, W.L. & Portis, D. (2006). The behavior of extreme cold air outbreaks under greenhouse warming. *International Journal of Climatology*, **26**, 1133–1147.

Vihma, T. (2014) Effects of Arctic sea ice decline on weather and climate: a review. *Survey of Geophysics* **35**, 1175–1214.

Vihma, T. & Launiainen, J. (1993) Ice drift in the Weddell Sea in 1990–1991 as tracked by a satellite buoy. *Journal of Geophysical Research*, **98**, 14471–14485.

Vihma, T., Launiainen, J. & Uotila, J. (1996) Weddell Sea ice drift: kinematics and wind forcing, *Journal of Geophysical Research*, **101**, 18279–18296.

Vihma, T. & Brümmer, B. (2002) Observations and modelling of on-ice and off-ice flows in the northern Baltic Sea. *Boundary-Layer Meteorology*, **103**, 1–27.

Vihma, T., Lüpkes, C., Hartmann, J. & Savijärvi, H. (2005) Observations and modelling of cold-air advection over Arctic sea ice in winter, *Boundary-Layer Meteorology*, **117**, 275–300.

Vihma, T., Uotila, J., Cheng, B. & Launiainen, J. (2002) Surface heat budget over the Weddell Sea: buoy results and comparisons with large-scale models. *Journal of Geophysical Research*, **107** (C2), 3013, doi: 10.1029/2000JC00037.

Vihma, T., Hartmann, J. & Lüpkes, C. (2003) A case study of an on-ice air flow over the Arctic marginal sea ice zone, *Boundary-Layer Meteorology*, **107**, 189–217, 2003.

Vihma, T. & Pirazzini, R. (2005) On the factors controlling the snow surface and 2-m air temperatures over the Arctic sea ice in winter. *Boundary-Layer Meteorology*, **117**, 73–90.

Vihma, T., Jaagus, J., Jakobson, E. & Palo, T. (2008) Meteorological conditions in the Arctic Ocean in spring and summer 2007 as recorded on the drifting ice station Tara. *Geophysical Research Letters*, **35**, L18706, doi: 10.1029/2008GL034681.

Vihma, T., Johansson, M.M. & Launiainen, J. (2009) Radiative and turbulent surface heat fluxes over sea ice in the western Weddell Sea in early summer. *Journal of Geophysical Research*, **114**, C04019, doi:10.1029/2008JC004995.

Vihma, T., Tisler, P. & Uotila, P. (2012) Atmospheric forcing on the drift of Arctic sea ice in 1989–2009, *Geophysical Research Letters*, **39**, L02501, doi:10.1029/2011GL050118.

Wakimoto, R.M. & Murphey, H.V. (2008) Airborne Doppler radar and sounding analysis of an oceanic cold front. *Monthly Weather Review*, **136**, 1475–1491.

Walter, B. & Overland, J.E. (1991) Aircraft observations of the mean and turbulent structure of the atmospheric boundary layer during spring in the central Arctic. *Journal of Geophysical Research*, **96** (C3), 4663–4673.

Walsh, J.E. & Chapman, W.L. (1998) Arctic cloud–radiation–temperature associations in observational data and atmospheric reanalyses. *Journal of Climate*, **11**, 3030–3045.

Wang, S., Wang, Q., Jordan, R.E. & Persson, P.O.G. (2001) Interactions among longwave radiation of clouds, turbulence and snow surface temperature in the Arctic: A model sensitivity study. *Journal of Geophysical Research*, **106**, 15323–15333.

Wendler, G., Moore, B., Hartmann, B., Stuefer, M. & Flint, R. (2004) Effects of multiple reflection and albedo on the net radiation in the pack ice zones of Antarctica. *Journal of Geophysical Research.D.Atmospheres*, **109**, doi:10.1029/2003JD003927

Wesslén, C., Tjernström, M., Bromwich, D.H. et al. (2014) The Arctic summer atmosphere: an evaluation of reanalyses using ASCOS data. *Atmospheric Chemistry and Physics*, **14**, 2605–2624.

Woods, C., Caballero, R. & Svensson, G. (2013) Large-scale circulation associated with moisture intrusions into the Arctic during winter. *Geophysical Research Letters*, **40**, 4717–4721.

Zilitinkevich, S.S. (2002) Third-order transport due to internal waves and non-local turbulence in the stably stratified surface layer. *Quarterly Journal of the Royal Meteorological Society*, **128**, 913–925.

Zilitinkevich, S.S. & Esau, I.N. (2005) Resistance and heat-transfer laws for stable and neutral planetary boundary layers: Old theory advanced and re-evaluated. *Quarterly Journal of the Royal Meteorological Society*, **131**, 1863–1892.

Zubov, N.N., *L'dy Arktiki* (*Arctic Ice*, in Russian). Glavsevmorputi, Moscow, 1943. (English translation 1945 by U.S. Naval Oceanographic Office, Suitland, Md. Available as NTIS AD 426972 from Natl. Tech. Inf. Serv., Springfield, Va.)

CHAPTER 7
Sea ice and Arctic Ocean oceanography

Finlo Cottier,[1] Michael Steele[2] and Frank Nilsen[3]

[1] *SAMS, Scottish Marine Institute, Oban, Argyll, UK*
[2] *Polar Science Center, Applied Physics Laboratory, University of Washington, Seattle, WA, 98105, USA*
[3] *Department of Arctic Geophysics, The University Centre in Svalbard, Longyearbyen, Norway*

7.1 Introduction

The nature of Arctic oceanography and the processes that shape it are fundamentally linked to the presence, formation, growth and melt of sea ice. Indeed to treat them as distinct entities would grossly underplay the extent of the interactions occurring between them. Some researchers might consider one (the ocean or ice) to be merely a boundary condition of the other. Here, we look at them in a more coupled manner and describe some of the key interactions between them.

The formation and decay of sea ice are at the heart of this chapter. The impact of sea ice on Arctic oceanography is essentially through three principal mechanisms: (i) the release of dense, salt-enriched brine during ice formation causing the upper water column to become unstable, leading to convective mixing; (ii) the export or release of freshwater during melt, which re-stratifies and stabilizes the water column, inhibiting vertical mixing; and (iii) changes in the extent of ice cover modifying the insulating effect of the ice between the atmosphere and the ocean, controlling the input of heat into the ocean surface. Other impacts include enhanced stirring and mixing of the water when the ice and ocean move relative to each other and the variation in momentum input through wind acting on open or ice-covered water. In this closely coupled system with multiple feedbacks, the structure and properties of the ocean also impact on the formation and retention of sea ice.

Our contemporary motivation is the reported rapid changes in the extent, volume and seasonal persistence of Arctic sea ice. The forecasts of its continued demise reinforce the need to consider the coupling between ice and ocean (Overland & Wang, 2013). The extent of summer sea ice is shrinking at an average rate of around 12% per decade (Stroeve et al., 2012) with compounding reductions in sea ice thickness (Chapter 2), volume (Laxon et al., 2013) and age (Maslanik et al., 2007). The change is so extensive that some commentators have predicted a nearly ice-free summer Arctic Ocean within a few decades (Wang & Overland, 2012).

The ocean is a critical element in the resulting feedbacks associated with the reduction of sea ice (Carmack et al., 2015). Nevertheless, the Arctic and subarctic seas can appear to be rather remote and isolated elements of the global ocean system, so we must emphasize that formation of sea ice and the resulting local water mass transformations can make a significant contribution to the global ocean circulation (Schmitz, 1995) and to weather events in the northern hemisphere (Greene & Monger, 2012; Liu et al., 2012).

There have been many excellent works published on polar oceanography (e.g. Smith Jr. 1990; Smith Jr & Grebmeier, 1995), and this chapter will focus on the coupled nature of sea ice and oceanography, updating and developing the work of Brandon et al. (2010). To accommodate what is now a vast field of research, we focus on just a few selected aspects to illustrate how sea ice and oceanography are each able to modify and influence the properties and processes of the other. The chapter is structured as follows: Section 7.2 provides a short historical introduction to Arctic oceanography with a simplified description and background of the key aspects given in Section 7.3. We look in more detail at the relationship between the Arctic Ocean and sea ice, particularly with respect to heat content and fluxes in Section 7.4, before discussing two key oceanographic

Sea Ice, Third Edition. Edited by David N. Thomas.
© 2017 John Wiley & Sons, Ltd. Published 2017 by John Wiley & Sons, Ltd.

features of the Arctic oceans: the halocline in Section 7.5, and polynyas in Section 7.6. For further information on the role of polynyas in the Southern Ocean see Chapter 8.

7.2 Historical perspective

Knowledge of the Arctic Ocean prior to and during most of the 19th century was sparse, even as to the extent of the ocean covered by sea ice. Observations and mapping of the region were limited to its margins, in the straits and passages of the Northwest and Northeast passages, and the hunting and fishing for whales and walruses were typically confined to the peripheral seas (for an excellent historical review see recent works by Rudels, 2012, 2013). Indeed, in the early part of the 19th century it was thought that the pack ice was limited to a relatively narrow band adjacent to the coast, formed from the discharge of fresh river water. Towards the North Pole the Arctic Ocean was considered to be ice-free, with open water being maintained through the heat transported by warm ocean currents into the Arctic through the Barents Sea and the Bering Strait (Figure 7.1).

Arctic oceanography and the role that sea ice plays in setting the structure and transforming the water has its formal origins in the 19th century. It was the Norwegian explorer and scientist Fridtjof Nansen, during the drift of the *Fram* (1893–1896), who described the characteristic layers of water in the Arctic Ocean. He discussed the different physical properties, the possible interactions between the ocean and the ice leading to water mass transformation and the response of ice drift to wind through the action of Earth's rotation (Nansen, 1897). Nansen observed the presence of an Atlantic

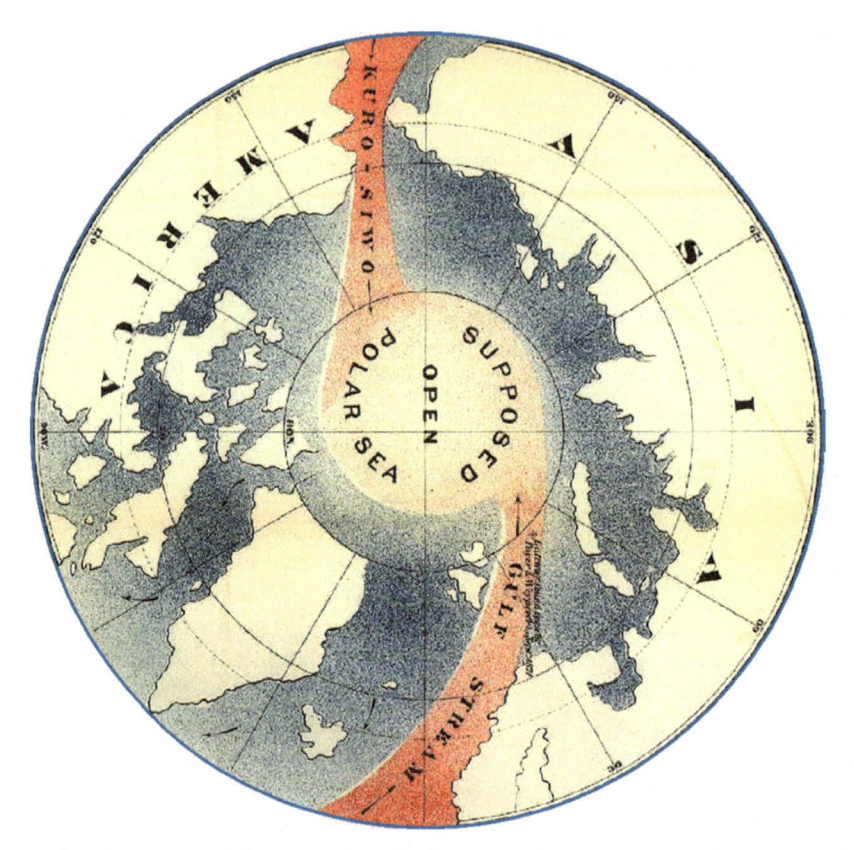

Figure 7.1 Historic map by Silas Bent, 1872, illustrating the path of warm ocean currents entering the Arctic via the Barents Sea and the Bering Strait and their presumed impact on the Arctic ice cover. From Overland et al. (2011). (Image libr0568, Treasures of the National Oceanic and Atmospheric Administration Library Collection)

Water (AW) layer across the arctic, with a temperature maximum between 150 and 250 m depth. The AW layer is isolated from the ice by a layer of cold, low-salinity water now known as Polar Surface Water. Identifying this arrangement of stably stratified waters explained the lack of a 'supposed open sea' presumed in earlier times (Figure 7.1).

Arguably, Nansen laid the foundation for Arctic oceanography and the techniques he employed proved to be enduring. The biggest difficulties in ship-based observations are those caused by the drift and convergence of sea ice. Using the ice itself as the ideal platform for trans-Arctic sampling was pioneered during the drift of the *Fram*. This was followed in the late 1930s by the first of the manned Russian ice drift stations. Since then, there have been around 40 such drift stations established, each with a programme of sampling for oceanographic and meteorological parameters, the most recent ending in June 2013. Drift stations have also been set up to investigate particular research challenges, an example being the Surface Heat Budget of the Arctic Ice Station experiment (SHEBA), which aimed to measure the rate at which heat is exchanged between the ocean, the sea ice and the atmosphere throughout the polar year. The year-long study was supported by the Canadian Coast Guard vessel *Des Groseilliers*, which drifted over 2000 km from 2 October 1997 until 11 October 1998 (Macdonald et al., 2008).

7.3 General oceanographic description

The essential geography of the Arctic Ocean is one of a landlocked sea (Figure 7.2). There are four topographic gaps connecting the Arctic to the global oceans, and thus there are rather limited routes for exchange and communication. By far the deepest and most significant gap is Fram Strait between Greenland and the archipelago of Svalbard, where water depths are over 3000 m. The three smaller but still important gaps are the Barents Sea Opening between the Norwegian mainland and Svalbard, the relatively narrow (~85 km) and shallow (~50 m deep) Bering Strait between North America and Asia, and the numerous channels between the Arctic Ocean and Baffin Bay through the Canadian archipelago.

There are also four distinct deep basins (the Canada Basin, the Makarov Basin, the Amundsen Basin and the Nansen Basin) separated by mid-ocean ridges (Sohn et al., 2008). A dominant feature of the Arctic Ocean is the extensive shallow continental shelf that is essentially circumpolar except for the deep connection at Fram Strait. The shelf areas account for approximately 50% of the Arctic sea floor (Jakobsson, 2002) and dominate the eastern hemisphere where there are large shallow regions less than 500 m deep (and <100 m in most areas) north of the Russian coastline. For an overview of Arctic oceanography and the adjacent shelf areas see (Carmack et al., 2006).

The distribution and growth–melt cycle of Arctic sea ice are well described and reported (Perovich, 2011). In addition to the perennial pack, a seasonal sea ice cover forms each winter, reaching its maximum extent in March before diminishing through melt and export to reach a minimum extent in September. The winter maximum extent is typically 15 million km^2, whilst the summer minimum is around 5 million km^2 with a currently observed record minimum in 2012 of 3.4 million km^2 (Parkinson & Comiso, 2013). However, the details of ice formation, melt, break-up and export can be highly regional, giving rise to variation in the dominating ice–ocean coupling processes.

The general circulation patterns of the Arctic Ocean are dominated by two wind-driven surface currents: the anti-cyclonic Beaufort Gyre over the Canada Basin and the Transpolar Drift which flows from the Siberian coast across the north pole, exiting through the western Fram Strait (Figure 7.2). The major inflow to the Arctic is the warm and saline AW entering via the Barents Sea Opening and eastern Fram Strait. This forms a warm boundary current that encircles the Arctic Basin (Carmack, 1990). The Atlantic inflow is the principal source of heat and salt into the Arctic Ocean (Beszczynska-Möller et al., 2011) forming the warm AW layer of the Arctic, typically found between 150 and 900 m deep with variable temperatures (Polyakov et al., 2010; see also Figure 7.3). There is also a relatively fresh inflow to the Arctic Ocean from the Pacific through the Bering Strait that brings a third of the freshwater into the Arctic and is a significant source of inorganic nutrients (Beszczynska-Möller et al., 2011). As a consequence of its freshness, Pacific Water (PW) tends to occupy the upper water column, contributing to the stable stratification of the Arctic. The Pacific inflow also brings heat into the Arctic during summer and has seen a period of warming in recent years that may act to

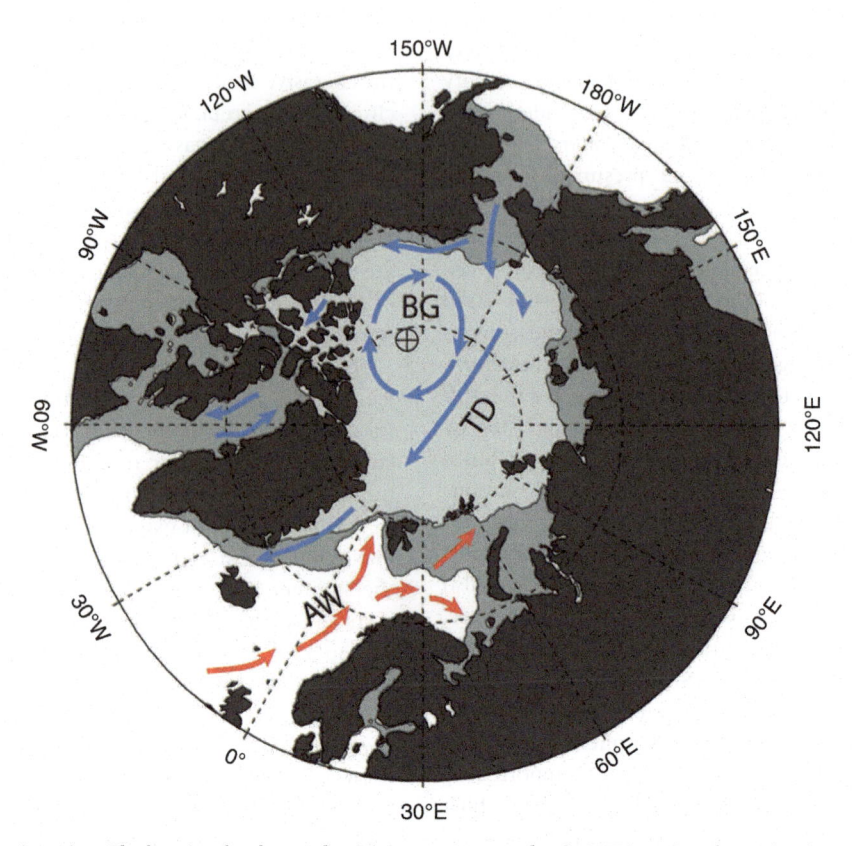

Figure 7.2 The Arctic region. The location for the conductivity–temperature–depth (CTD) station shown in Figure 7.8 is marked by the target symbol in the Beaufort Gyre. Blue arrows represent cold currents and red arrows the inflowing Atlantic Water. BG, the Beaufort Gyre; TD, the Transpolar Drift; AW, Atlantic Water. First published by Brandon et al. (2010)

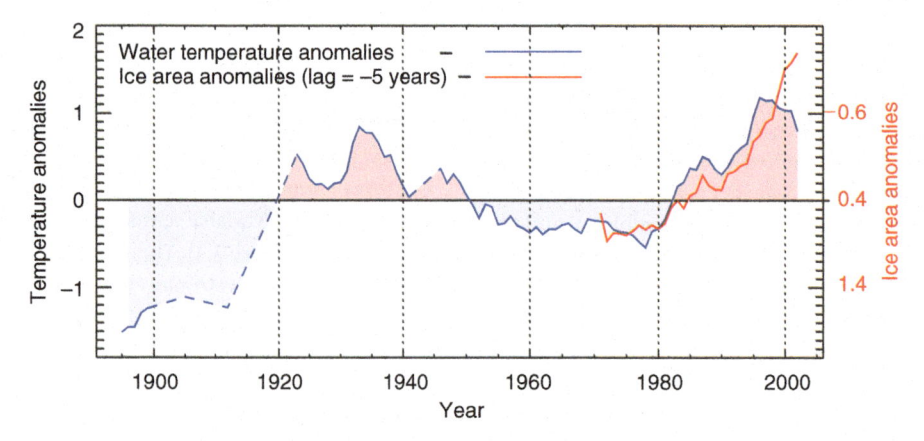

Figure 7.3 Seven-year running mean normalized Atlantic Water temperature anomalies adapted from Polyakov et al. (2004) (dashed segments represent gaps in the record) and the September Arctic sea ice area anomalies (10^6 km^2; reverse vertical axis is used, lagged by 5 years). Source: From Polyakov et al., 2010. ©American Meteorological Society. Used with permission.

modify the ice cover (Woodgate et al., 2010; Bourgain & Gascard, 2012; Timmermans et al., 2014).

The role of the Arctic within the global freshwater cycle is also of growing significance. Freshwater input to the Arctic Ocean is increasing via river discharge (Peterson et al., 2006) and coupled model simulations project Arctic-wide freshening (Holland et al., 2007) over the coming century. Recent work indicates the beginnings of a freshening trend, although the magnitude is still a matter of debate (Haine et al., 2015). Whilst clear evidence exists of a build-up in freshwater in the Beaufort Gyre of the Canadian Basin (Giles et al., 2012), the magnitude of partial compensation (i.e. salinification) in other areas such as the shelves and the Eurasian Basin is less clear (Morison et al., 2012; Rabe et al., 2014). Further, partial release of Beaufort Gyre freshwater has occurred in the last few years (Timmermans et al., 2011; Krishfield et al., 2014).

7.4 Ocean heat and sea ice

The role of ocean heat in recent sea ice loss occurs through either upper ocean solar heating and consequent bottom melting (a manifestation of the ice–albedo feedback mechanism) or increased influx of warm ocean waters with thermodynamic coupling between the ice and the oceanic interior (Polyakov et al., 2012; Carmack et al., 2015). However, as we will see in Section 7.5, across much of the Arctic these warmer waters reside within or below a strong pycnocline that suppresses upward heat flux; consequently, the precise coupling between the ocean and the sea ice remains to be fully understood. In this section we consider some of the important forcings and feedbacks that act on Arctic sea ice.

7.4.1 Surface warming

The cycle of feedbacks between the ocean, sea ice and the atmosphere by which summer sea ice extent is reducing is illustrated in Figure 7.4 from the work by Stroeve et al. (2012). A fundamental parameter in this framework is the thickness of the spring ice cover, influenced by growth-season air temperatures which are found to be rising. The higher temperatures in autumn are linked to there being more open water at the end of the summer and a transfer of heat from the ocean mixed layer to the atmosphere. The impact of this

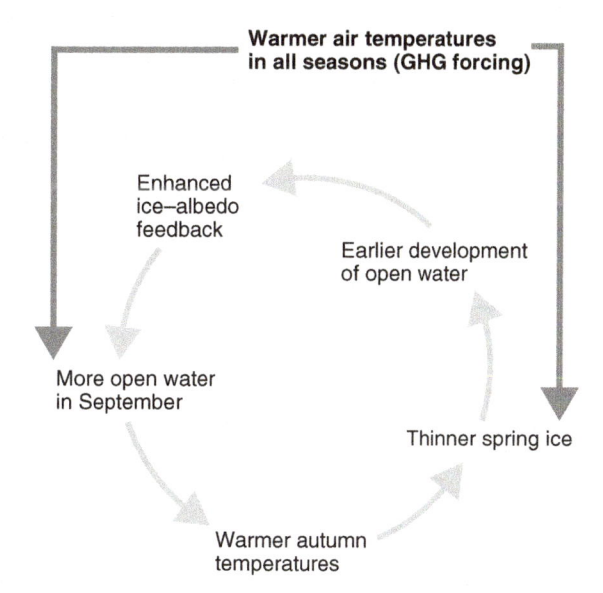

Figure 7.4 The feedback processes in the ocean–ice–atmosphere system that control the cycle of sea ice growth and decay. GHG, greenhouse gas. Source: From Stroeve et al., 2012, figure 2, p. 1007. Reproduced with permission from Springer Science+Business Media.

is a delay in the onset of freeze-up (Stroeve et al., 2014) which in turn leads to a thinner spring sea ice cover. Transition of the Arctic to having a greater proportion of thin first-year ice (Maslanik et al., 2007) further accelerates its decay in summer. Beginning in spring, the melt season has been shown to be lengthening by ~ 1–2 weeks per decade (Stroeve et al., 2014). The impact is a greater area of open water during the summer melt season leading to an increase in heat stored in the upper ocean (Stroeve et al., 2014) and an enhancement of the ice–albedo feedback cycle (Perovich et al., 2007). In the summer months, over the last 30 years, there has been an increase in the total absorbed solar radiation of around 20–30 W m^{-2} per decade. The cumulative effect of this increase in solar energy over a 5-year period from 2007 to 2011 is equivalent to melting ~1 m of ice during the summer (Stroeve et al., 2014), thus promoting the occurrence of more open water in September.

Whilst significant sea ice retreat allows for unprecedented summertime warming of the upper ocean through increased input of solar energy, it is necessary to understand the relative roles of advection and solar heating to evaluate their impact on sea ice melt. The heating of the surface by incident radiation is primarily

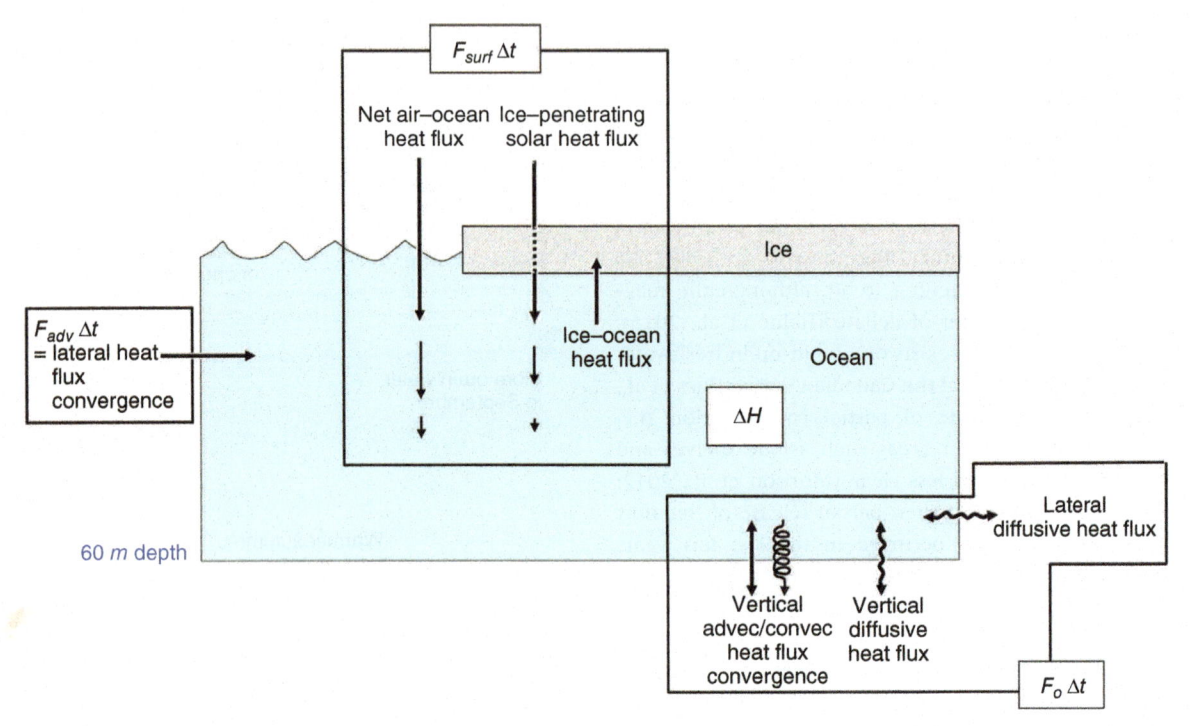

Figure 7.5 Schematic illustration of the key components contributing to the upper ocean heat budget which are relevant to ice-ocean interactions. Source: From Steele et al., 2010, figure 3. Reproduced with permission of John Wiley & Sons.

confined to the upper 50–60 m, which absorbs 99.8% of the incoming solar energy (Steele et al., 2010) and the change in heat content of this layer (ΔH) can be expressed by:

$$\Delta H = \Delta time \text{ (net surface heat flux + net advected heat flux + sum of ocean heat fluxes).}$$

With reference to Figure 7.5, the net surface heat flux is comprised of the air–sea flux (turbulent + radiative) with the subsurface heating occurring through penetrating solar radiation in open water and a smaller but significant flux from the solar radiation penetrating through sea ice (Nicolaus et al., 2012) and the ice–ocean heat flux. The net ocean advective term is the lateral heat flux convergence within the upper 60 m of ocean arising from lateral gradients in heat content, ocean velocity or a combination of both. The advective term can be large where there are significant warm currents. The other ocean heat fluxes include those from vertical heat flux convergence, from lateral and vertical diffusion, and from vertical convection. It also includes the heat flux into the upper layer from warmer deeper layers (Steele et al., 2010). The bulk of summertime heating occurs after the ice has been lost when the surface heat flux over open water is generally large. The balance of terms will vary regionally and it has been shown for the Pacific Sector that 80% of ocean warming during the summer comes from ocean surface heat flux, and the other 20% comes from lateral heat flux convergence (Steele et al., 2010).

One relatively recent oceanographic manifestation of increased summer surface warming is a warm water mass identified at depths typically 25–35 m and referred to as the near-surface temperature maximum (NSTM) (Jackson et al., 2010). It forms seasonally, typically in the Pacific Sector of the Arctic, when solar radiation penetrates the ocean surface and warms the ocean below a cold mixed layer in contact with sea ice. The absorption of radiation below the surface is critical to the formation of NSTM, otherwise it is susceptible to forced convection and wind stirring during autumn and winter (Steele et al., 2011). The greater heat content in the surface enhances ice melt to form a near-surface halocline during summer. The increased surface stratification caps the heat gained from solar radiation during the summer which is stored through the following winter.

Recent work has demonstrated that far from being an oceanographic curiosity, the NSTM can have a profound effect on sea ice conditions. Initial observations of the NSTM showed that it did not survive beyond autumn (Maykut & McPhee, 1995), but more recently the NSTM has warmed by up to 1.5°C and expanded northwards and the stored heat now resides year-round. In some locations Ekman downwelling acts to deepen the NSTM such that it is not eroded by cooling and mixing during autumn. The greater longevity of the NSTM, just under the winter mixed layer, has the capacity to produce elevated heat fluxes into the surface layer through entrainment during intense winter mixing events and thus affect winter ice growth, further reducing the thickness of spring ice cover (Jackson et al., 2012; Timmermans, 2015).

The impact of the feedback cycle presented in Figure 7.4 is an ice cover that is younger, thinner and potentially heavily decayed. This mechanically weaker first-year ice is then susceptible to mechanical break-up by winds (Barber et al., 2009). As areas of open water develop during the summer, a further feedback cycle comes into play where the open water affords a greater fetch, resulting in more effective break-up of ice cover (Asplin et al., 2012; Asplin et al., 2014). Increasing fetch distances support an effective increase in the long-wave energy that is then able to propagate deep into the sea ice. The resulting flexural swell promotes fracturing and failure of the ice depending on the original floe size and its physical properties, such that the floe size distribution is shifted to smaller sizes. Smaller, more mobile floes are more susceptible to lateral melting by radiative forcing, as the newly formed open water around the now fractured pack is heated through solar radiation (Steele, 1992; Perovich et al., 2008).

Further oceanic impacts on the ice result from the fractured ice cover, which leads to changes in the ability of the wind to transfer momentum into the ocean (Martin et al., 2014) as well as a greater potential for storm-driven upwelling of ocean heat (Pickart et al., 2009) and mixing of the ocean surface mixed layer (Rainville et al., 2011). One particular event in 2012 was the 'Great Arctic Cyclone' which was implicated in the break-up and additional sea ice decay through enhanced bottom melt (Zhang et al., 2013). Further, greater mobility of smaller ice floes may result in them being advected southward into areas with greater melt rates (Rampal et al., 2011). The resulting increase in the sea ice kinematics (velocity and deformation rates) through wave break-up is a property of sea ice which needs to be appropriately captured in climate models so that ice properties are appropriately estimated (Rampal et al., 2011).

7.4.2 Oceanic heat

In certain regions of the Arctic, inflowing warm water masses provide the dominating interaction between the ocean and the sea ice (Carmack et al., 2015). For example, the Barents Sea is categorized as an inflow shelf (Carmack et al., 2006) where one branch of warm AW enters the Arctic. This heat transport sets the boundary of the ice-free Atlantic domain and hence the ice extent (Årthun et al., 2012) as illustrated in Figure 7.2. Helland-Hansen and Nansen (1909) argued that the sea ice extent in spring is highly dependent on the quantity of heat contained in the water masses of the Barents Sea during winter, and less dependent on variations in air temperature. During the period 1998–2008, there was a pronounced reduction in Barents Sea ice area of up to 50% which was reflected in the variability of the Atlantic heat transport in terms of both volume transport and temperature (Årthun et al., 2012).

Another branch of AW enters the Arctic west of Spitsbergen. Here it encounters and melts sea ice drifting south, such that the upper part of the AW is transformed into less saline and colder surface water (Rudels et al., 1996). However, only a fraction of the heat lost by AW goes into ice melt, with the rest transferred directly to the atmosphere (Rudels, 2012). Two buoyancy inputs are thus generated – one is positive due to melt and the other is negative due to cooling. The AW heat loss in this region is sufficient to maintain the mean annual ice boundary at about 80°N (Untersteiner, 1988) and we will see later on that this is an example of a sensible-heat polynya.

Surface buoyancy fluxes and wind-stirring are agents for eroding the base of the mixed layer, whereas tides over rough topography (Rippeth et al., 2015) with corresponding diapycnal mixing and unstable shelf waves associated with slope currents (Nilsen et al., 2006; Teigen et al., 2011), with corresponding lateral heat and salt exchange by isopycnal diffusion, play an important role in preconditioning the AW for excessive winter heat loss

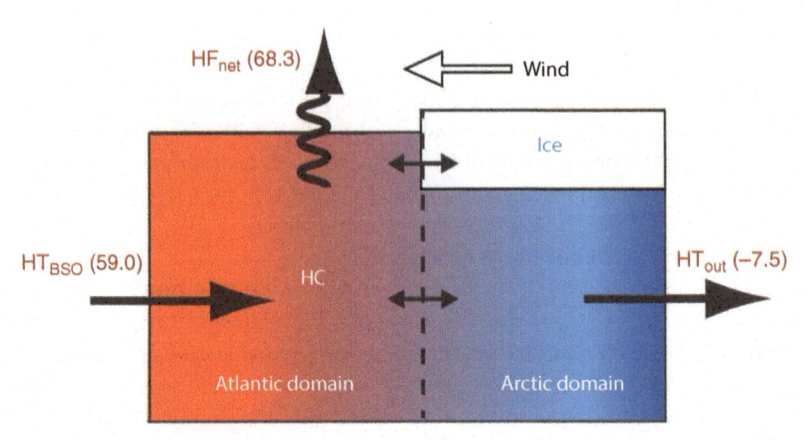

Figure 7.6 Schematic illustration of the Barents Sea long-term mean heat flux, heat transport and factors related to the integrated heat content, HC, and the variable sea ice area; HT is the mean heat transport (TW) of the Atlantic inflow (denoted BSO) and the balancing outflow, respectively; also, HF$_{net}$ is the net heat flux to the atmosphere (TW). The sea ice extent defines the boundary between the Atlantic and the Arctic domain. Source: From Årthun et al., 2012, figure 1, p. 4737. © American Meteorological Society. Used with permission.

to the atmosphere by reducing the vertical stability. This may enhance convection and is an important mechanism for cooling the main body of the AW entering the Arctic Ocean at depths of between 40 to 200 m.

Recent observations of the ice cover to the north of Svalbard further demonstrate the intimate link between the heat of the AW and the distribution of sea ice. Recent work has shown that the sea ice area north of Svalbard has been decreasing for all months since 1979, with the largest ice reduction occurring during the winter months at a rate of 10% per decade (Onarheim et al., 2014). This observed reduction is concurrent with a gradual warming of 0.3°C per decade of the AW entering the Arctic Ocean; hence, there is more heat available for melting. As seen in Figure 7.6, the northward transport of heat meets sea ice advected out of the Arctic Ocean driven by wind. In addition to melting the ice from below, the additional oceanic heat further inhibits ice growth during winter as the ocean needs more cooling to reach the freezing point. This is also noted for the Barents Sea where the winter ice edge has retreated (Årthun et al., 2012).

The AW layer within the Arctic Ocean is reported to show a general warming during recent decades (Polyakov et al., 2010), although this is not without contention (Bourgain & Gascard, 2012). The warming is associated with a shoaling of the upper boundary in the central Arctic Ocean by up to 75–90 m and a weakening in the Eurasian basin stratification. The warming, shoaling and weakening suggest the potential for greater upward transfer of AW to the ocean surface layer and there is observational evidence suggesting that the pycnocline is susceptible to enhanced heat flux in winter (Polyakov et al., 2013). Three-dimensional modelling approaches (Polyakov et al., 2010) indicate a potential loss of 28–35 cm of ice thickness over 50 years in response to an increase in ocean heat flux of 0.5 W m^{-2}, which is comparable to losses from atmospheric warming.

In contrast, observations from ice-tethered platforms (Timmermans et al., 2008; Lique et al., 2014) indicate that the vertical heat flux from the AW layer in the central Canada basin is unlikely to have a significant impact on the surface heat budget, and that mixing rates have not changed over the past few decades (Guthrie et al., 2013). However, vertical transport of heat may be enhanced at margins and continental shelves via upwelling (Aagaard et al., 1981; Carmack & Chapman, 2003; Cottier et al., 2007; Falk-Petersen et al., 2015).

How does PW influence sea ice? As for AW, summer PW meets the ice edge with enough heat to cause substantial melting (Brugler et al., 2014). It also has the potential to flux heat into the surface layer far north of the ice edge depending in part on its circulation pattern within the Arctic Ocean (Morison & Smith, 1981; Steele, 2004). In recent years with generally reduced ice cover, the coupling between winds, ice, and ocean has changed (Figure 7.7), allowing more summer PW to directly enter the Canada Basin and melt ice, as opposed to moving eastward along the

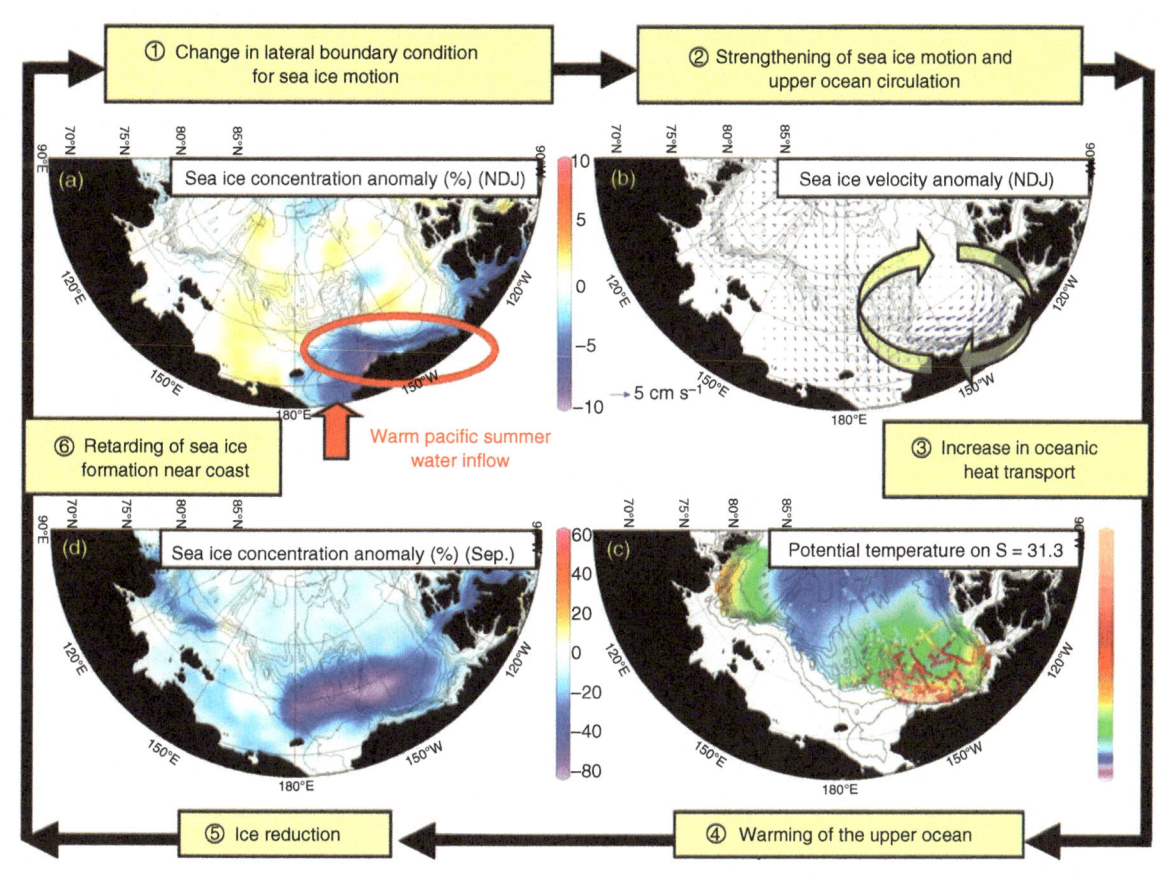

Figure 7.7 The feedback system in the Canada Basin as proposed by Shimada et al (2006). (a) Sea ice concentration anomaly for November to January (this is the mean concentration from the period 1997–2003 minus the mean concentration from the period 1979–1997). The concentration is greatly reduced in the shelf regions. (b) Sea ice velocity anomaly derived from the same time periods as (a) showing that the anticyclonic circulation has increased into the central Arctic Basin. (c) The potential temperature on the isohaline layer S = 31.3. The background colour is the climatology from the Environmental Working Group Arctic Ocean Atlas and dotted circles are from measurements for the period 1998–2004. Clearly the dots on the shelf are up to 2°C warmer than the atlas data. (d) Sea ice concentration anomaly for September derived from concentration (1998–2003) minus the concentration (1979–1997) showing that, in summer, the ice concentration is greatly reduced, particularly over the shelf region. Source: From Shimada et al., 2006, figure 5, page 3. Reproduced with permission from John Wiley & Sons.

North American shelf break (Shimada et al., 2006). Further, an increase in the temperature of summer PW (Bourgain & Gascard, 2012), as well as its thickness and core depth (Timmermans et al., 2014) may have enhanced this effect, both at the ice edge and under the perennial ice pack.

7.5 The Arctic halocline

The growth and melt of Arctic sea ice are intimately linked to the oceanographic structures beneath it. One of the most important of these structures is the halocline, the boundary between cold and fresh Polar Surface Waters and the warm and saline AW layer. A vertical profile of salinity from the northern Canada Basin shows the strong halocline located at depths between 50 and 200 m (Figure 7.8). Whilst there is a rapid increase in salinity with depth within the halocline, in some locations (the eastern Nansen Basin, the Amundsen Basin and, at times, part of the Makarov Basin) the temperature remains almost isothermal and usually very close to the freezing point. This gives rise to its widely accepted name, the cold halocline layer

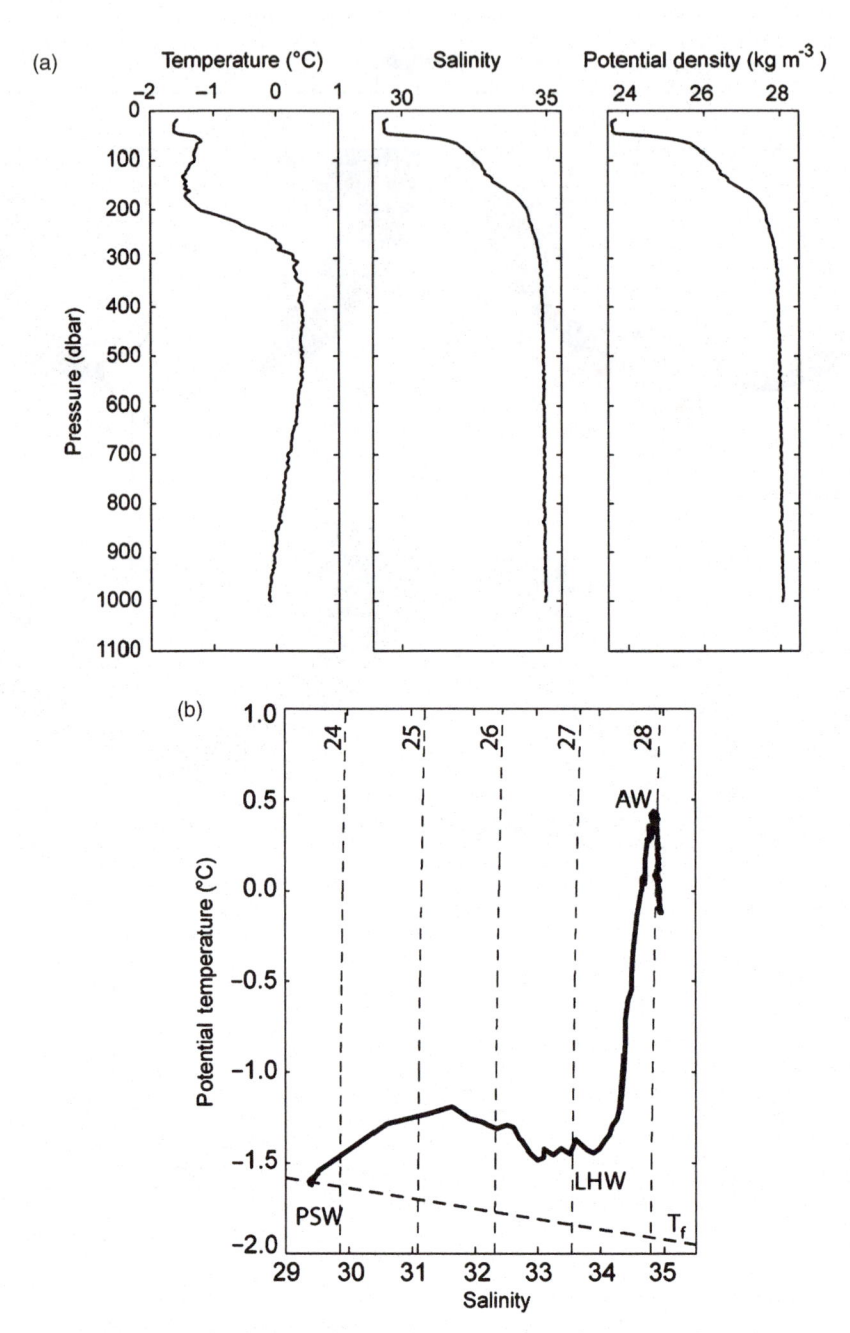

Figure 7.8 A conductivity–temperature–depth station showing the upper 1000 m of the water column from the central Arctic Ocean (81°10′N, 138°52′W marked on Figure 7.2) recorded by the submarine USS *Pogy* in September 1996). (a) Profiles of the temperature, salinity and potential density against pressure. (b) Potential temperature–salinity plot of the same data in (a). The water masses of Polar Surface Water (PSW) and Atlantic Water (AW) are indicated and the position of the Lower Halocline Water (LHW) is shown. The surface freezing line is shown as a dashed line labelled T_f. The near-vertical dashed lines are lines of potential density referenced to the surface in units of kg m^{-3}. Source: From Brandon et al., 2010.

(CHL; Aagaard et al., 1981). In areas influenced by PW (e.g. Figure 7.8), the halocline contains some heat and so might be termed a 'cool halocline'.

At low temperatures, the density of seawater is determined primarily by salinity – as seen by the almost vertical orientation of the potential density lines in the potential temperature salinity (TS) plot of Figure 7.8. Hence, the Arctic Ocean halocline gives rise to a steep gradient in water density (or pycnocline), acting as a barrier between the surface water and the AW. Consequently, one of the critical roles that the halocline plays in Arctic oceanography is to inhibit the upward mixing of warm AW into the surface water, which promotes winter sea ice growth and ultimately helps maintain the sea ice cover. Above the halocline, the surface waters are quite fresh (S < 31) owing to river discharge, the relatively fresh Bering Sea inflow, excess precipitation over evaporation, and (in summer) fresh water addition through sea ice melt. As we shall see in Section 7.3, in those regions of the Arctic where a halocline is not well developed, e.g. around the Svalbard margin, enhanced ocean heat flux in winter to the surface can be sufficient to prevent the formation of sea ice.

It is obvious to ask what is the origin of the Arctic Ocean halocline? Figure 7.8(b) shows a TS plot of the Arctic waters where the surface water and AW are clearly identifiable, one as a cold, fresh end-point (in this case with potential temperature ~ −1.5°C, salinity ~29.5) and one as a relatively warm, salty end-point (potential temperature ~0.4°C, salinity ~34.9). But between these two water masses are several kinks in the TS curve. If the halocline was simply a product of mixing between the surface water and AW then there would be a straight mixing line between the two water masses (Kikuchi et al., 2004). Clearly, as Figure 7.8(b) shows, this is not the case and we must look for alternative explanations. The first kink above the AW layer in the TS curve defines the lower halocline water (LHW), which is the most saline water within the halocline at or close to the surface freezing point. The next shallower layer is winter PW, also cold but fresher than LHW, while the layer above that is the relatively warm (although much cooler than AW) and fresher still summer PW. Finally, cold (in winter or in summer under the perennial sea ice pack) and fresh surface water is found on the left end of the TS curve. In the next section, we focus on explanations for LHW formation. We review two proposed mechanisms and expand on the role that sea ice plays in forming this key oceanographic structure.

7.5.1 Advection mechanism

One of the earliest mechanisms proposed identified the marginal shelf seas surrounding the Amundsen and Nansen basins (collectively referred to as the 'Eurasian Basin') as the source of halocline waters (Aagaard et al., 1981). Winter ice formation on the shelves, particularly in areas of sea ice divergence, would lead to high rates of brine production, increasing the shelf water salinity. These cold, saline waters would then advect laterally from the shelf break, entering the Arctic basin below the surface water to form LHW (Figure 7.9). By implication, any removal of ice from the shelves through wind stress would promote new ice formation and thus enhance the rates of brine release.

A number of studies that have tried to interpret the contribution of brine-enriched shelf waters to LHW (Rudels et al., 1996; Steele & Boyd, 1998; Winsor & Björk, 2000). One problem identified with this mechanism is the differing salinity characteristics of the shelf waters compared with those of the LHW in the Eurasian Basin such that LHW could only form in the Barents Sea (Steele & Boyd, 1998). However, recent modelling work investigating the impact of landfast ice on brine production and halocline stability has shown that moving the region of brine release offshore results in an increased brine export from the shelf that can contribute to LHW formation (Itkin et al., 2015).

7.5.2 Convection mechanism

Another mechanism for the formation of LHW invokes winter convection processes (Rudels et al., 1996). Figure 7.10 illustrates the steps in this mechanism and the evolution of the resulting TS distribution. The first stage is the 'upstream condition' and involves the inflow of warm saline AW into the Nansen Basin through Fram Strait and the Barents Sea. When the northward-flowing AW encounters sea ice at Fram Strait and in the Barents Sea, melting occurs and a surface 'wedge' of fresher water is created that is well mixed and at near-freezing temperatures. In Figure 7.10(a), the surface layer is termed pre-existing surface water (pSW) below which is the remaining warm salty core of AW. At this point in TS space the curve simply shows a straight line between these two end-members.

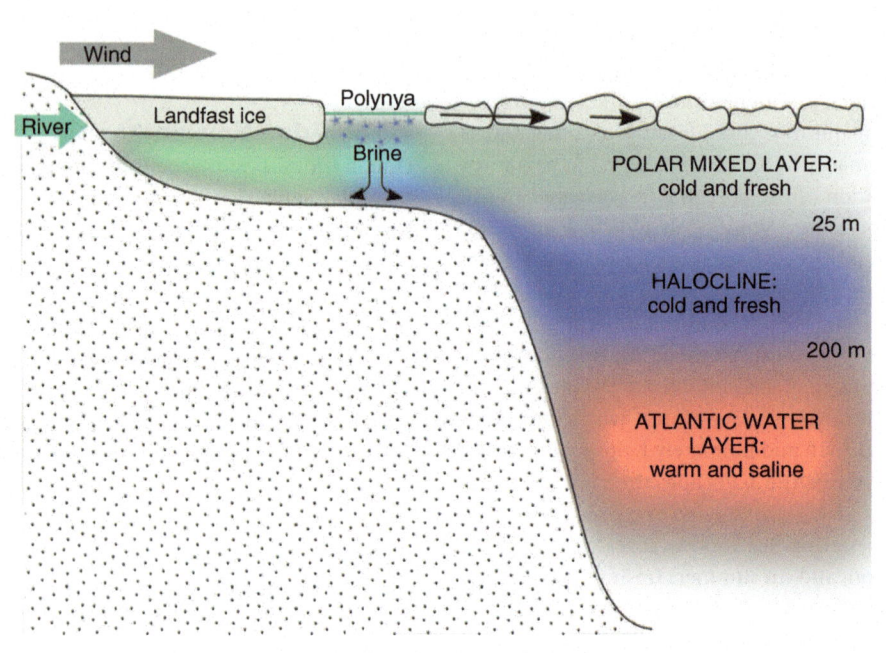

Figure 7.9 Schematic to illustrate the advection mechanism for the formation of cold, salty lower halocline waters. Source: From Itkin et al., 2015, figure 2, page 2623. Reproduced with permission from John Wiley & Sons.

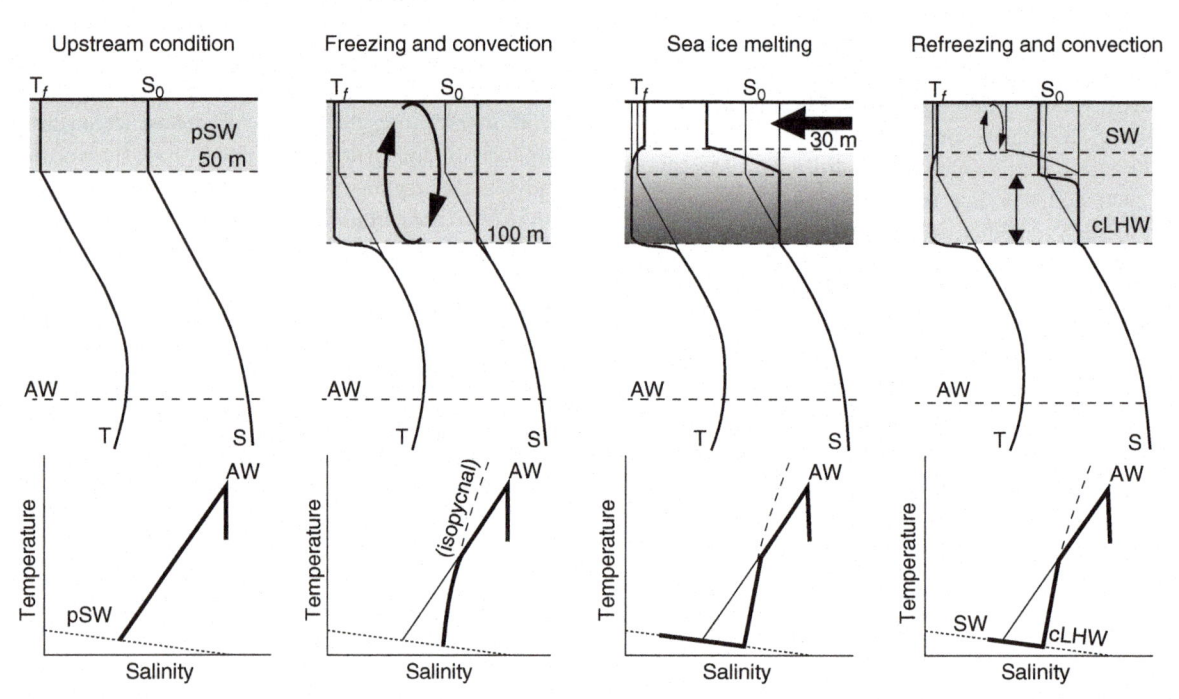

Figure 7.10 Schematic view of the formation process of the cold halocline through convection processes. The upper panel shows typical temperature and salinity profiles from before ice generation processes to the end product, where T_f is the surface freezing temperature and S_0 is the initial surface water salinity. The lower panel shows the θ–S corresponding to the upper panels with the dotted lines indicating the surface freezing point, T_f. pSW, pre-existing Surface Water; SW, Surface Water; AW, Atlantic Water; cLHW, convectively formed Lower Halocline Water. Source: From Kikuchi et al., 2004, figure 4, page 5. Reproduced with permission from John Wiley & Sons.

Further downstream north of the Barents Sea, the second stage involves winter ice formation. Brine release gives rise to an unstable convective situation to form an approximately 100 m thick mixed layer with increased salinity and with temperatures at the surface freezing point. The increase in salinity and departure of the mixed layer from the pSW–AW mixing line is shown in the TS curve (Figure 7.10b). Sea ice melt through the following spring and summer decreases the salinity at the surface, forming a shallow seasonal halocline (Steele & Boyd, 1998). Figure 7.10(c) shows that this is represented in TS space as an extension of the less dense surface water towards lower salinity along the surface freezing point line. There is further brine release during ice formation in the following winter to produce instabilities and mixing down to the permanent pycnocline, as shown in Figure 7.10(d). This cycle of melt and freezing can occur several times; the resulting TS curve is shown in the final panel of Figure 7.10(d). The observed range of Eurasian shelf salinities (salty in the west, fresher in the east) leads to a regional variation in LHW formation and its influence on the halocline (Nguyen et al., 2012; Bauch et al., 2014; Itkin et al., 2015).

Some indication of the variation in the formation process of LHW can be found by examining conductivity–temperature–depth (CTD) data from sections across the Arctic shelves and into the central basins. Data from such a section in the Nansen Basin are shown in the TS plot in Figure 7.11 (Woodgate

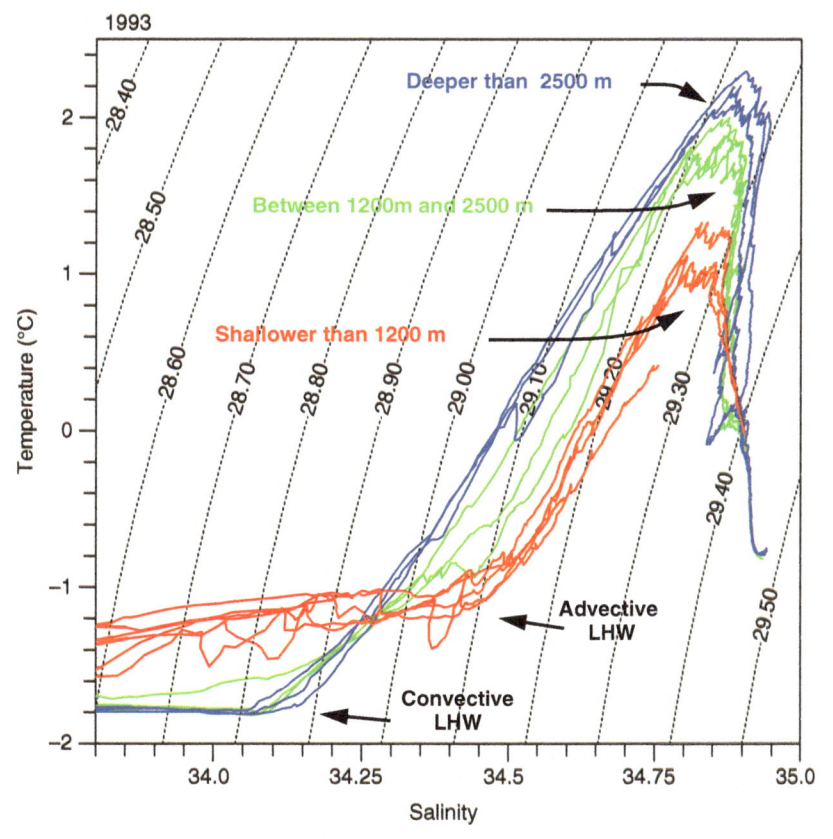

Figure 7.11 Curves of *in situ* temperature versus salinity for a CTD section running north from the Siberian shelf into the Eurasian Basin at 118°E recorded in 1993. Red profiles are the six stations closest to the continental shelf (water depth < 1200m), blue curves are for deepest stations (water depth was > 2500m). Green profiles are for three stations at intermediate depths showing characteristics of both deep and shallow stations. Dotted lines are isopycnals referenced to 290 db. LHW, Lower Halocline Water. Source: From Woodgate et al., 2001, figure 9, page 1772. Reproduced with permission from Elsevier.

et al., 2001). In this example, the origin of the LHW is determined from the thermal signature. In the convective process, the LHW is near to its freezing point as the convection is driven by the brine released through freezing. In advection, the temperature of the halocline need not necessarily be at the surface freezing point. Offshore, the TS data show LHW close to surface freezing point, which is indicative of formation through convection. Inshore, the TS data show that the LHW is warmer and saltier, which is indicative of a shelf origin and thus an advective source. In this example, shelf processes feed the LHW within the boundary current, whilst in deeper water, convection is predominant.

As already noted, the Arctic Ocean halocline is important in isolating the underlying warm AW from the surface waters and sea ice, and in fact one definition for the presence of a CHL is that the winter oceanic heat flux is small (Steele & Boyd, 1998). Clearly any weakening of the CHL or appropriate mechanisms for mixing across the pycnocline will increase the potential for enhanced mixing of warm AW into the surface layer. The result will be increased ocean heat flux and a significant impact on the sea ice cover: both the thickness of winter growth and the total period for growth will be reduced. Steele and Boyd (1998) used a simple one-dimensional ocean model to quantify the potential impact on winter ice formation under conditions of a weakened CHL by using temperature and salinity profiles derived from field data and a surface mixing term. With a weak CHL structure, the wintertime ocean heat flux to the underside of the sea ice increased from a presumed winter value of zero to $2\,W\,m^{-2}$. To put this increase in context, the average heat flux generated during the short Arctic summer months can be up to $5\,W\,m^{-2}$. Therefore an enhanced winter heat flux could account for up to an additional 40% of the annual average ocean heat flux. This addition to the annual ocean heat flux term could switch the sea ice growth–melt condition from one of net freezing to one of net melting in certain regions (Polyakov et al., 2013).

The Arctic halocline represents a strong oceanographic control on sea ice formation and decay operating over seasonal and interannual time scales. In contrast, polynyas are a mechanism where sea ice formation has one of the greatest impacts on ocean salinity and operate at time scales of weeks to months.

7.6 Polynyas

The word polynya originates from a Russian term for 'ice hole' and are areas of enhanced air–sea fluxes in winter relative to the neighbouring ice-covered regions. This is caused by direct contact between the cold atmosphere and a relatively warm ocean, resulting in heat exchange two orders of magnitude greater than the exchange through the ice. Polynyas can be regions of recurring open water and/or a reduced ice concentration in the seasonal sea ice zone in locations where a more consolidated and thicker ice cover would be climatologically expected (Barber & Massom, 2007; Smith Jr. & Barber, 2007). In area they range in size from tens to tens of thousands of square kilometres and tend to form at particular geographical locations (Maqueda et al., 2004). They are important locations for deep-water formation (Shcherbina et al., 2003; Skogseth et al., 2005b), there is enhanced gas ventilation and they respond sensitively to thermodynamic and dynamic forcing by both the ocean and atmosphere. Polynyas are also ecologically important polar oases that enable birds and mammals to overwinter in high latitudes and encourage enhanced primary production in spring (Smith Jr. & Barber, 2007).

Traditionally polynyas have fallen into two classes: sensible-heat polynyas which are thermally driven, and latent-heat polynyas which are mechanically driven through ice divergence. Sensible-heat polynyas are typically formed in regions of upwelling, vigorous vertical mixing or where there is a strong interaction between ocean currents and topographic features. They are called sensible because the heat that is transferred from the ocean to the atmosphere causes the water temperature to decrease and thus causes a reduction of the sensible heat content of the surface waters. The size of a sensible-heat polynya is limited to the dimensions of the flux of warm water to the surface and is dependent on the rate of melting being greater than the advection rate of any ice into it. Locations where warm water comes to the surface in winter can be regions of low ice concentration or have thinner ice relative to the surrounding ice (Melling et al., 2015). An example of a large sensible-heat polynya is Whaler's Bay north of Svalbard where the warm inflowing AW dominates the area and has a direct influence on the sea ice which retreats above the pathway of the AW (Onarheim et al., 2014).

Latent-heat polynyas require a divergent sea ice cover. Therefore they tend to form adjacent to coastlines swept by offshore winds (called coastal or shore polynyas) or downwind or down-current of landfast ice, glacier tongues or grounded icebergs (called flaw polynyas) (Dmitrenko et al., 2015; Itkin et al., 2015). They are called latent-heat polynyas because the energy that is lost from the ocean stems from the removal of latent heat from the water so that it can freeze. New ice that forms in the open water is carried away by the forcing mechanism so an area of open water is maintained. The temperature of the water within a latent heat polynya stays at the surface freezing point and there is no change in its sensible heat content. It is only the ice divergence through wind stress or ocean currents that maintains the open water. Therefore the polynya size is governed by the balance of ice formation and export: accumulation of frazil ice tends to make the polynya narrower whereas offshore wind-forcing of the pack ice tends to make it wider.

With surface waters already at freezing point, strong cold winds promote high rates of sea ice formation and continually advect the new ice away as it forms (Dmitrenko et al., 2015; Ito et al., 2015). This results in substantially higher net ice production rates than would be measured in open regions in the ice cover away from such winds. Indeed, latent-heat polynyas have often been described as the 'ice factories'. Consequently, there is a high rate of brine release into the water column under a mechanically forced latent-heat polynya and highly efficient water mass transformation takes place. Along the 2×10^5 km coastal margin of the Arctic there are many coastal (mainly flaw) polynyas on the shallow continental shelves (Barber & Massom, 2007; Tamura & Ohshima, 2011), e.g. the Canadian flaw lead (Barber et al., 2010) and the Laptev Sea polynya (Dmitrenko et al., 2010). Brine rejection associated with ice formation results in brine-enriched shelf waters which will tend to spill over the shelf edge and descend, or cascade, down the shelf slope (Shapiro et al., 2003; Ivanov et al., 2004). As noted earlier, these waters make an important contributors to the halocline of the Arctic Ocean (Winsor & Björk, 2000).

Regional case studies are an important tool for understanding and documenting the ice production and brine formation in polynyas. Such a case study is in Storfjorden, Svalbard, which is one of the few areas in the easily accessible Barents Sea where there is a recurrent wind-driven polynya that forms brine-enriched shelf waters (Skogseth et al., 2008). Once formed in the polynya, this dense water can, if the density excess permits, cascade down the continental slope into the deep Norwegian Sea and northward along the eastern slope of Fram Strait towards the Arctic Ocean (Fer et al., 2003; Akimova et al., 2011). Although relatively small, it is estimated that this polynya alone supplies between 5% and 20% of the all the newly formed dense water that enters the Arctic Ocean (Skogseth et al., 2005a).

Measurements near the edge of fast ice in Storfjorden (McPhee et al., 2013) and the Laptev Sea polynya (Dmitrenko et al., 2012) reveal mixing processes associated with tidal advection of a sharp front in salinity, including possible super-cooling induced by double diffusion in a fully turbulent water column. The measurements in Storfjorden indicate a novel ice production process along the edge of tidally induced latent heat polynyas where salinity fronts are generated (Skogseth et al., 2013). The persistence of super-cooling and frazil ice crystals is probably a general process that occurs in many Arctic and Antarctic polynyas, given that some of the brine-enriched shelf water is advected below thicker ice. Although ice growth and super-cooling are several orders of magnitude smaller than those in the open polynya, a small and new volume of frazil ice may also grow from the suggested double-diffusion process (McPhee et al., 2013).

7.7 Conclusion

As summer sea ice continues to retreat, the area of open water and the duration for which there is open water will increase. This will change the way in which ice can impact the Arctic Ocean. The loss of summer sea ice removes the insulating properties of the ice such that there will be increasing heat content of the surface waters by autumn. The effect is a delay in the onset of freezing as ocean heat is removed to the atmosphere. Once freezing starts, there is now a greater area of open water for ice formation and thus the generation of brine-enriched waters is occurring over a much more extensive area, both over the shelves and in the deep ocean. The delay in onset of freezing leads to a shorter freezing period and thus the ice at the end of the following winter is thinner and more susceptible to breaking up, exposing the ocean surface to solar heating. Therefore the fate of sea ice is inextricably linked to its relationship with the oceanography of the Arctic Ocean.

Dedication

This chapter is dedicated to our friend and colleague, Tim Boyd – an enthusiastic and dedicated polar oceanographer who gave generously of his time and ideas. It had been intended that Tim would be a co-author of this chapter and indeed some of the changes contained in this third edition were those proposed by Tim shortly before he died in January 2013. Tim was a great educator and we hope that those who read this chapter and the rest of the book are inspired to pursue their love of the polar oceans.

References

Aagaard, K., Coachman, L. & Carmack, E. (1981) On the halocline of the Arctic Ocean. *Deep-Sea Research, Part A*, **28**, 529–545.

Akimova, A., Schauer, U., Danilov, S. & Núñez-Riboni, I. (2011) The role of the deep mixing in the Storfjorden shelf water plume. *Deep Sea Research Part I*, **58**, 403–414.

Årthun, M., Eldevik, T., Smedsrud, L.H., Skagseth, Ø. & Ingvaldsen, R. (2012) Quantifying the influence of atlantic heat on barents sea ice variability and retreat. *Journal of Climate*, **25**, 4736–4743.

Asplin, M.G., Galley, R., Barber, D.G. & Prinsenberg, S. (2012) Fracture of summer perennial sea ice by ocean swell as a result of Arctic storms. *Journal of Geophysical Research: Oceans*, **117**, C06025.

Asplin, M.G., Scharien, R., Else, B. et al. (2014) Implications of fractured Arctic perennial ice cover on thermodynamic and dynamic sea ice processes. *Journal of Geophysical Research: Oceans*, **119**, 2327–2343.

Barber, D. & Massom, R. (2007) The role of sea ice in Arctic and Antarctic polynyas. *Elsevier Oceanography Series*, **74**, 1–54.

Barber, D., Galley, R., Asplin, M.G. et al. (2009) Perennial pack ice in the southern Beaufort Sea was not as it appeared in the summer of 2009. *Geophysical Research Letters*, **36**, L24501.

Barber, D., Asplin, M., Gratton, Y. et al. (2010) The International Polar Year (IPY) Circumpolar Flaw Lead (CFL) system study: overview and the physical system. *Atmosphere-Ocean*, **48**, 225–243.

Bauch, D., Torres-Valdes, S., Polyakov, I. et al. (2014) Halocline water modification and along-slope advection at the Laptev Sea continental margin. *Ocean Science*, **10**, 141–154.

Beszczynska-Möller, A., Woodgate, R.A., Lee, C., Melling, H. & Karcher, M. (2011) A synthesis of exchanges through the main oceanic gateways to the Arctic Ocean. *Oceanography*, **24**, 82–99.

Bourgain, P. & Gascard, J.C. (2012) The Atlantic and summer Pacific waters variability in the Arctic Ocean from 1997 to 2008. *Geophysical Research Letters*, **39**, L05603.

Brandon, M.A., Cottier, F.R. & Nilsen, F. (2010) Sea ice and oceanography. In: *Sea Ice* (Eds. Thomas, D.N. & Dieckmann, G.S.), pp. 79–111. Wiley-Blackwell.

Brugler, E.T., Pickart, R.S., Moore, G., Roberts, S., Weingartner, T.J. & Statscewich, H. (2014) Seasonal to interannual variability of the Pacific water boundary current in the Beaufort Sea. *Progress in Oceanography*, **127**, 1–20.

Carmack, E. (1990) Large-scale physical oceanography of polar oceans. In: *Polar Oceanography, Part A: Physical Science* (Ed. Smith Jr, W.O.), pp. 171–222, Academic Press, San Diego.

Carmack, E. & Chapman, D. (2003) Wind-driven shelf/basin exchange on an Arctic shelf: The joint roles fo ice cover extent and shelf-break bathymetry. *Geophysical Research Letters*, **30**, 1778, doi:1710.1029/2003GL017526.

Carmack, E., Barber, D., Christensen, J., Macdonald, R., Rudels, B. & Sakshaug, E. (2006) Climate variability and physical forcing of the food webs and the carbon budget on panarctic shelves. *Progress in Oceanography*, **71**, 145–181.

Carmack, E., Polyakov, I., Padman, L. et al. (2015) Towards quantifying the increasing role of oceanic heat in sea ice loss in the new Arctic. *Bulletin of the American Meteorological Society*, doi: 10.1175/BAMS-D-13-00177.1.

Cottier, F.R., Nilsen, F., Inall, M.E., Gerland, S., Tverberg, V. & Svendsen, H. (2007) Wintertime warming of an Arctic shelf in response to large-scale atmospheric circulation. *Geophysical Research Letters*, **34**, L10607.

Dmitrenko, I.A., Wegner, C., Kassens, H. et al. (2010) Observations of supercooling and frazil ice formation in the Laptev Sea coastal polynya. *Journal of Geophysical Research: Oceans*, **115**, C05015.

Dmitrenko, I.A., Kirillov, S.A., Bloshkina, E. & Lenn, Y.-D. (2012) Tide-induced vertical mixing in the Laptev Sea coastal polynya. *Journal of Geophysical Research: Oceans*, **117**, C00G14.

Dmitrenko, I.A., Kirillov, S.A., Rysgaard, S. et al. (2015) Polynya impacts on water properties in a Northeast Greenland fjord. *Estuarine, Coastal and Shelf Science*, **153**, 10–17.

Falk-Petersen, S., Pavlov, V., Berge, J., Cottier, F., Kovacs, K.M. & Lydersen, C. (2015) At the rainbow's end: high productivity fueled by winter upwelling along an Arctic shelf. *Polar Biology*, **38**, 5–11.

Fer, I., Skogseth, R., Haugan, P.M. & Jaccard, P. (2003) Observations of the Storfjorden overflow. *Deep-Sea Research I*, **50**, 1283–1303.

Giles, K.A., Laxon, S.W., Ridout, A.L., Wingham, D.J. & Bacon, S. (2012) Western Arctic Ocean freshwater storage increased by wind-driven spin-up of the Beaufort Gyre. *Nature Geoscience*, **5**, 194–197.

Greene, C.H. & Monger, B.C. (2012) An Arctic wild card in the weather. *Oceanography*, **25**, 7–9.

Guthrie, J.D., Morison, J.H. & Fer, I. (2013) Revisiting internal waves and mixing in the Arctic Ocean. *Journal of Geophysical Research Oceans*, **118**, 3966–3977.

Haine, T.W., Curry, B., Gerdes, R. et al. (2015) Arctic freshwater export: Status, mechanisms, and prospects. *Global and Planetary Change*, **125**, 13–35.

Helland-Hansen, B. & Nansen, F. (1909) The Norwegian Sea. Its physical oceanography based upon the Norwegian researches 1900–1904. *Report on Norwegian Fishery and Marine Investigations*, **11**, 1–390.

Holland, M.M., Finnis, J., Barrett, A.P. & Serreze, M.C. (2007) Projected changes in Arctic Ocean freshwater budgets. *Journal of Geophysical Research – Biogeosciences*, **112**, G04S55.

Itkin, P., Losch, M. & Gerdes, R. (2015) Landfast ice affects the stability of the Arctic halocline: Evidence from a numerical model. *Journal of Geophysical Research: Oceans*, **120**, 2622–2635.

Ito, M., Ohshima, K.I., Fukamachi, Y. et al. (2015) Observations of supercooled water and frazil ice formation in an Arctic coastal polynya from moorings and satellite imagery. *Annals of Glaciology*, **56**, 307–314.

Ivanov, V.V., Shapiro, G.I., Huthnance, J.M., Aleynik, D. & Golovin, P. (2004) Cascades of dense water around the world ocean. *Progress in Oceanography*, **60**, 47–98.

Jackson, J., Carmack, E., McLaughlin, F., Allen, S.E. & Ingram, R. (2010) Identification, characterization, and change of the near-surface temperature maximum in the Canada Basin, 1993–2008. *Journal of Geophysical Research: Oceans*, **115**, C05021.

Jackson, J., Williams, W. & Carmack, E. (2012) Winter sea-ice melt in the Canada Basin, Arctic Ocean. *Geophysical Research Letters*, **39**, L03603.

Jakobsson, M. (2002) Hypsometry and volume of the Arctic Ocean and its constituent seas. *Geochemistry, Geophysics, Geosystems*, **3**, 1–18.

Kikuchi, T., Hatakeyama, K. & Morison, J.H. (2004) Distribution of convective Lower Halocline Water in the eastern Arctic Ocean. *Journal of Geophysical Research: Oceans*, **109**, C12030.

Krishfield, R.A., Proshutinsky, A., Tateyama, K. et al. (2014) Deterioration of perennial sea ice in the Beaufort Gyre from 2003 to 2012 and its impact on the oceanic freshwater cycle. *Journal of Geophysical Research: Oceans*, **119**, 1271–1305.

Laxon, S.W., Giles, K.A., Ridout, A.L. et al. (2013) CryoSat-2 estimates of Arctic sea ice thickness and volume. *Geophysical Research Letters*, **40**, 732–737.

Lique, C., Guthrie, J.D., Steele, M., Proshutinsky, A., Morison, J.H. & Krishfield, R. (2014) Diffusive vertical heat flux in the Canada Basin of the Arctic Ocean inferred from moored instruments. *Journal of Geophysical Research: Oceans*, **119**, 496–508.

Liu, J., Curry, J.A., Wang, H., Song, M. & Horton, R.M. (2012) Impact of declining Arctic sea ice on winter snowfall. *Proceedings of the National Academy of Sciences*, **109**, 4074–4079.

Maqueda, M., Willmott, A. & Biggs, N. (2004) Polynya dynamics: a review of observations and modeling. *Reviews of Geophysics*, **42**, RG1004.

Martin, T., Steele, M. & Zhang, J. (2014) Seasonality and long-term trend of Arctic Ocean surface stress in a model. *Journal of Geophysical Research: Oceans*, **119**, 1723–1738.

Maslanik, J., Fowler, C., Stroeve, J. et al. (2007) A younger, thinner Arctic ice cover: Increased potential for rapid, extensive sea-ice loss. *Geophysical Research Letters*, **34**, L24501.

Maykut, G. & McPhee, M.G. (1995) Solar heating of the Arctic mixed layer. *Journal of Geophysical Research: Oceans*, **100**, 24691–24703.

McPhee, M.G., Skogseth, R., Nilsen, F. & Smedsrud, L.H. (2013) Creation and tidal advection of a cold salinity front in Storfjorden: 2. Supercooling induced by turbulent mixing of coldwater. *Journal of Geophysical Research: Oceans*, **118**, 3737–3751.

Melling, H., Haas, C. & Brossier, E. (2015) Invisible polynyas: Modulation of fast ice thickness by ocean heat flux on the Canadian polar shelf. *Journal of Geophysical Research: Oceans*, **120**, 777–795.

Morison, J. & Smith, J.D. (1981) Seasonal variations in the upper Arctic Ocean as observed at T-3. *Geophysical Research Letters*, **8**, 753–756.

Morison, J., Kwok, R., Peralta-Ferriz, C. et al. (2012) Changing Arctic Ocean freshwater pathways. *Nature*, **481**, 66–70.

Nansen, D.F. (1897), *Farthest North*, Archibald, Constable and Company, Westminster.

Nguyen, A.T., Kwok, R. & Menemenlis, D. (2012) Source and pathway of the Western Arctic upper halocline in a data-constrained coupled ocean and sea ice model. *Journal of Physical Oceanography*, **42**, 802–823.

Nicolaus, M., Katlein, C., Maslanik, J. & Hendricks, S. (2012) Changes in Arctic sea ice result in increasing light transmittance and absorption. *Geophysical Research Letters*, **39**, L24501.

Nilsen, F., Gjevik, B. & Schauer, U. (2006) Cooling of the West Spitsbergen Current: isopycnal diffusion by topographic vorticity waves. *Journal of Geophysical Research: Oceans*, **111**, C08012.

Onarheim, I.H., Smedsrud, L.H., Ingvaldsen, R.B. & Nilsen, F. (2014) Loss of sea ice during winter north of Svalbard. *Tellus A*, **66**, 23933.

Overland, J.E., Wood, K.R. & Wang, M. (2011) Warm Arctic-cold continents: climate impacts of the newly open Arctic Sea. *Polar Research*, **30**, 15787.

Overland, J.E. & Wang, M. (2013) When will the summer Arctic be nearly sea ice free? *Geophysical Research Letters*, **40**, 2097–2101.

Parkinson, C.L. & Comiso, J.C. (2013) On the 2012 record low Arctic sea ice cover: Combined impact of preconditioning and an August storm. *Geophysical Research Letters*, **40**, 1356–1361.

Perovich, D.K. (2011) The changing Arctic sea ice cover. *Oceanography*, **24**, 162–173.

Perovich, D.K., Light, B., Eicken, H., Jones, K.F., Runciman, K. & Nghiem, S.V. (2007) Increasing solar heating of the Arctic

Ocean and adjacent seas, 1979–2005: attribution and role in the ice-albedo feedback. *Geophysical Research Letters*, **34**, L19505.

Perovich, D.K., Richter-Menge, J.A., Jones, K.F. & Light, B. (2008) Sunlight, water, and ice: Extreme Arctic sea ice melt during the summer of 2007. *Geophysical Research Letters*, **35**, L11501.

Peterson, B.J., McClelland, J., Curry, R., Holmes, R.M., Walsh, J.E. & Aagaard, K. (2006) Trajectory shifts in the Arctic and subarctic freshwater cycle. *Science*, **313**, 1061–1066.

Pickart, R., Moore, G., Torres, D.J., Fratantoni, P.S., Goldsmith, R.A. & Yang, J. (2009) Upwelling on the continental slope of the Alaskan Beaufort Sea: Storms, ice, and oceanographic response. *Journal of Geophysical Research: Oceans*, **114**, C00A13.

Polyakov, I., Alekseev, G., Timokhov, L. et al. (2004) Variability of the intermediate Atlantic water of the Arctic Ocean over the last 100 years. *Journal of Climate*, **17**, 4485–4497.

Polyakov, I., Timokhov, L.A., Alexeev, V.A. et al. (2010) Arctic Ocean warming contributes to reduced polar ice cap. *Journal of Physical Oceanography*, **40**, 2743–2756.

Polyakov, I., Walsh, J.E. & Kwok, R. (2012) Recent changes of Arctic multiyear sea ice coverage and the likely causes. *Bulletin of the American Meteorological Society*, **93**, 145–151.

Polyakov, I., Pnyushkov, A.V., Rember, R., Padman, L., Carmack, E.C. & Jackson, J.M. (2013) Winter convection transports Atlantic Water heat to the surface layer in the eastern Arctic Ocean. *Journal of Physical Oceanography*, **43**, 2142–2155.

Rabe, B., Karcher, M., Kauker, F. et al (2014) Arctic Ocean basin liquid freshwater storage trend 1992–2012. *Geophysical Research Letters*, **41**, 961–968.

Rainville, L., Lee, C.M. & Woodgate, R.A. (2011) Impact of wind-driven mixing in the Arctic Ocean. *Oceanography*, **24**, 136–145.

Rampal, P., Weiss, J., Dubois, C. & Campin, J.-M. (2011) IPCC climate models do not capture Arctic sea ice drift acceleration: consequences in terms of projected sea ice thinning and decline. *Journal of Geophysical Research: Oceans*, **116**, C00D07.

Rippeth, T.P., Lincoln, B.J., Lenn, Y., Green, J.A.M., Sundfjord, A. & Bacon, S. (2015) Tide-mediated warming of Arctic halocline by Atlantic heat fluxes over rough topography. *Nature Geoscience*, **8**, 191–194.

Rudels, B. &erson, L.G. & Jones, E.P. (1996) Formation and evolution of the surface mixed layer and halocline of the Arctic Ocean. *Journal of Geophysical Research: Oceans*, **101**, 8807–8821.

Rudels, B. (2012) Arctic Ocean circulation and variability-advection and external forcing encounter constraints and local processes. *Ocean Science*, **8**, 261–286.

Rudels, B. (2013) Arctic Ocean circulation, processes and water masses: A description of observations and ideas with focus on the period prior to the International Polar Year 2007–2009. *Progress in Oceanography*, **132**, 22–67.

Schmitz, W.J. (1995) On the interbasin-scale thermohaline circulation. *Reviews of Geophysics*, **33**, 151–173.

Shapiro, G.I., Huthnance, J.M. & Ivanov, V.V. (2003) Dense water cascading off the continental shelf. *Journal of Geophysical Research: Oceans*, **108**, 3390.

Shcherbina, A.Y., Talley, L.D. & Rudnick, D.L. (2003) Direct observations of North Pacific ventilation: Brine rejection in the Okhotsk Sea. *Science*, **302**, 1952–1955.

Shimada, K., Kamoshida, T., Itoh, M. et al. (2006) Pacific Ocean inflow: Influence on catastrophic reduction of sea ice cover in the Arctic Ocean. *Geophysical Research Letters*, **33**, L08605.

Skogseth, R., Fer, I. & Haugan, P.M. (2005a) Dense-water production and overflow from an Arctic coastal polynya in Storfjorden. In: *The Nordic Seas: An Integrated Perspective* (Eds. Drange, H. et al.), pp. 73–88. American Geophysical Union, Washington, DC.

Skogseth, R., Haugan, P. & Jakobsson, M. (2005b) Watermass transformations in Storfjorden. *Continental Shelf Research*, **25**, 667–695.

Skogseth, R., Smedsrud, L.H., Nilsen, F. & Fer, I. (2008) Observations of hydrography and downflow of brine-enriched shelf water in the Storfjorden polynya, Svalbard. *Journal of Geophysical Research: Oceans*, **113**, C08049.

Skogseth, R., McPhee, M.G., Nilsen, F. & Smedsrud, L.H. (2013) Creation and tidal advection of a cold salinity front in Storfjorden: 1. Polynya dynamics. *Journal of Geophysical Research: Oceans*, **118**, 3278–3291.

Smith Jr.,, W.O. (1990), *Polar Oceanography, Part A: Physical Science*. Academic Press, London.

Smith Jr.,, W.O. & Grebmeier, J.M. (1995), *Arctic oceanography: marginal ice zones and continental shelves*, American Geophysical Union, Washington, DC.

Smith Jr.,, W.O. & Barber, D. (2007), *Polynyas: Windows to the World: Windows to the World*. Elsevier, Amsterdam.

Sohn, R.A., Willis, C., Humphris, S. et al. (2008) Explosive volcanism on the ultraslow-spreading Gakkel ridge, Arctic Ocean. *Nature*, **453**, 1236–1238.

Steele, M. (1992) Sea ice melting and floe geometry in a simple ice-ocean model. *Journal of Geophysical Research: Oceans*, **97**, 17729–17738.

Steele, M. & Boyd, T.J. (1998) Retreat of the cold halocline layer in the Arctic Ocean. *Journal of Geophysical Research: Oceans*, **103**, 10419–10435.

Steele, M. (2004) Circulation of summer Pacific halocline water in the Arctic Ocean. *Journal of Geophysical Research: Oceans*, **109**, C02027.

Steele, M., Zhang, J. & Ermold, W. (2010) Mechanisms of summertime upper Arctic Ocean warming and the effect on sea ice melt. *Journal of Geophysical Research: Oceans*, **115**, C11004.

Steele, M., Ermold, W. & Zhang, J. (2011) Modeling the formation and fate of the near-surface temperature maximum in the Canadian Basin of the Arctic Ocean. *Journal of Geophysical Research: Oceans*, **116**, C11015.

Stroeve, J., Serreze, M.C., Holland, M.M., Kay, J.E., Malanik, J. & Barrett, A.P. (2012) The Arctic's rapidly shrinking sea ice cover: a research synthesis. *Climatic Change*, **110**, 1005–1027.

Stroeve, J., Markus, T., Boisvert, L., Miller, J. & Barrett, A. (2014) Changes in Arctic melt season and implications for sea ice loss. *Geophysical Research Letters*, **41**, 1216–1225.

Tamura, T. & Ohshima, K.I. (2011) Mapping of sea ice production in the Arctic coastal polynyas. *Journal of Geophysical Research: Oceans*, **116**, C07030.

Teigen, S.H., Nilsen, F., Skogseth, R., Gjevik, B. & Beszczynska-Möller, A. (2011) Baroclinic instability in the West Spitsbergen Current. *Journal of Geophysical Research: Oceans*, **116**, C07012.

Timmermans, M.-L., Toole, J., Krishfield, R. & Winsor, P. (2008) Ice-tethered profiler observations of the double-diffusive staircase in the Canada Basin thermocline. *Journal of Geophysical Research: Oceans*, **113**, C00A02.

Timmermans, M.-L., Proshutinsky, A., Krishfield, R.A. et al. (2011) Surface freshening in the Arctic Ocean's Eurasian Basin: An apparent consequence of recent change in the wind-driven circulation. *Journal of Geophysical Research: Oceans*, **116**, C00D03.

Timmermans, M.-L., Proshutinsky, A., Golubeva, E. et al. (2014) Mechanisms of Pacific Summer Water variability in the Arctic's Central Canada Basin. *Journal of Geophysical Research: Oceans*, **119**, 7523–7548.

Timmermans, M.-L. (2015) The impact of stored solar heat on Arctic sea ice growth. *Geophysical Research Letters*, **42**, 6399–6406.

Untersteiner, N. (1988) On the ice and heat balance in Fram Strait. *Journal of Geophysical Research: Oceans*, **93**, 527–531.

Wang, M. & Overland, J.E. (2012) An ice free summer Arctic within 30 years: An update from CMIP5 models. *Geophysical Research Letters*, **39**, L18501.

Winsor, P. & Björk, G. (2000) Polynya activity in the Arctic Ocean from 1958 to 1997. *Journal of Geophysical Research: Oceans*, **105**, 8789–8803.

Woodgate, R., Aagaard, K., Muench, R.D. et al. (2001) The Arctic Ocean Boundary Current along the Eurasian slope and the adjacent Lomonosov Ridge: Water mass properties, transports adn transformations from moored instruments. *Deep-Sea Research*, **48**, 1757–1792.

Woodgate, R., Weingartner, T. & Lindsay, R. (2010) The 2007 Bering Strait oceanic heat flux and anomalous Arctic sea-ice retreat. *Geophysical Research Letters*, **37**, L01602.

Zhang, J., Lindsay, R., Schweiger, A. & Steele, M. (2013) The impact of an intense summer cyclone on 2012 Arctic sea ice retreat. *Geophysical Research Letters*, **40**, 720–726.

CHAPTER 8

Oceanography and sea ice in the Southern Ocean

Michael P. Meredith[1] and Mark A. Brandon[2]

[1] *British Antarctic Survey, High Cross, Madingley Road, Cambridge, UK*
[2] *Department of Earth and Environmental Sciences, The Open University, Walton Hall, Milton Keynes, UK*

8.1 Introduction

The Arctic and the Antarctic are situated diametrically opposite each other on the planet, and their polar locations lead to them sharing some obvious similarities in climate and other characteristics. In some respects, however, their differences are more profound, and possibly the most significant of these is that the Arctic is fundamentally an ocean surrounded by land, whereas the Antarctic is a major continental landmass encircled by a vast and isolating ocean (Figure 8.1). This ocean – the Southern Ocean – has a profound influence on regional and global climate, on the glaciation of the Antarctic continent, and on biodiversity and ecosystem functioning.

The harsh environment and remoteness of Antarctica and the Southern Ocean were contributory factors in early Antarctic exploration being delayed compared with that in the Arctic, with the consequent scientific understanding that follows also lagging by comparison. It is thought that the first traveller who crossed the Arctic Circle and described their findings was the Greek Phytheas, around 325 BC, whereas Antarctic exploration progressed much more slowly, with James Cook crossing the Antarctic Circle on 17 January 1773. Scientific discovery in Antarctica progressed through the heroic age of exploration, with some early findings on the circulation of the Southern Ocean and the nature of the sea ice and icebergs that infest it. Often these studies were conducted for operational reasons, to assist in safe passage to, from and around Antarctica. Examples such as the sinking of Shackleton's ship *Endurance* in the Weddell Sea in 1915 demonstrate the need that these early explorers had for understanding of the behaviour of the Southern Ocean and its sea ice, and the limitations of that understanding.

It was the *Discovery Investigations* in the 1920s and 1930s that really marked the start of systematic scientific investigation into the Southern Ocean (e.g. Deacon, 1937). These were motivated by the recognition of a need to understand the functioning of the Southern Ocean ecosystem, some species of which (whales and seals) were commercially of immense value. These investigations included a large series of ship-based research expeditions (Figure 8.2), and, showing remarkable prescience, these early scientific investigations sought to understand the interdisciplinary functioning of the Southern Ocean, including ocean circulation, climate and sea ice, in order to better understand the life within it (Deacon, 1937, 1955).

Since the time of the *Discovery Investigations*, enormous progress has been made in understanding the Southern Ocean, and the role of sea ice in influencing its properties and determining its circulation is now better appreciated than ever. This understanding has been gained during a series of scientific experiments and long-term monitoring programmes, including the World Ocean Circulation Experiment (WOCE) in the 1990s (e.g. Ganachaud & Wunsch, 2000), which had a specific programmatic focus on the Southern Ocean (King, 2001).

Sea Ice, Third Edition. Edited by David N. Thomas.
© 2017 John Wiley & Sons, Ltd. Published 2017 by John Wiley & Sons, Ltd.

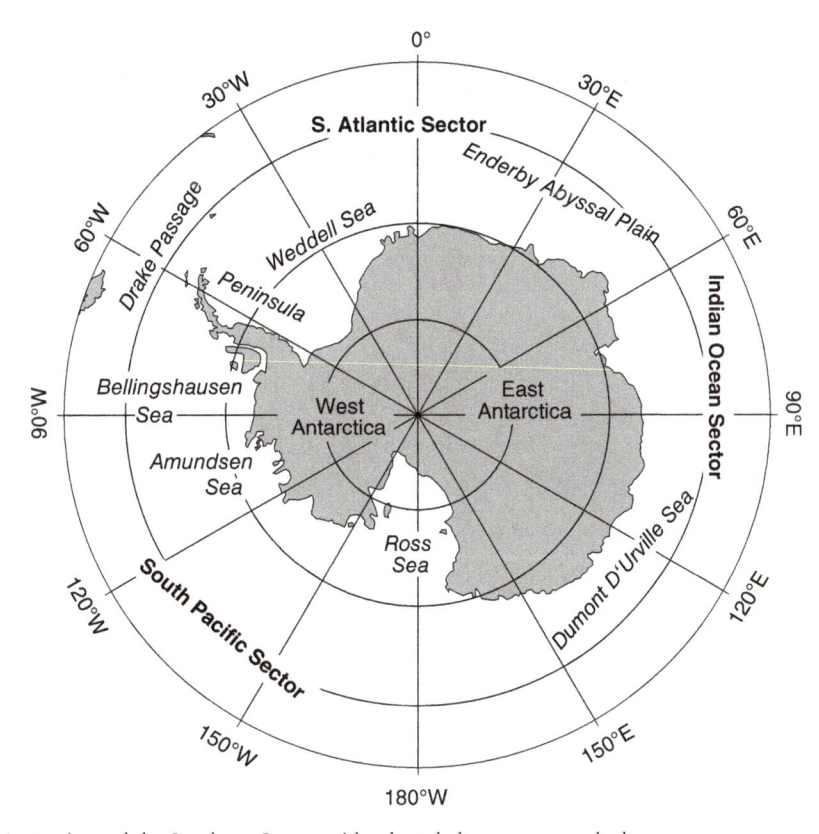

Figure 8.1 Map of Antarctica and the Southern Ocean, with selected place names marked.

Nonetheless, making progress in understanding the role of sea ice in structuring and controlling the Southern Ocean has remained challenging, in no small part due to the presence of the sea ice itself. Much of the research during WOCE and other programmes was conducted from research vessels, the vast majority of which are not icebreakers or ice-strengthened, and hence few of which could penetrate the Antarctic pack. There thus remains a dearth of direct observations concerning the interaction of the Southern Ocean with its seasonal and perennial sea ice cover, although with some very notable exceptions such as the landmark Ice Station Weddell expedition in 1992 (Muench & Gordon, 1995), and the ISPOL drift of the German ship R/V *Polarstern* in the Weddell Sea in 2006 (Hellmer et al., 2008).

Contemporary approaches to this problem are focusing on the progressively greater use of robotic instrumentation and autonomous (unmanned) ocean vehicles, some of which can be deployed into the ocean and navigated beneath the sea ice to collect data without the need for research vessels to attempt to access the pack directly. Great progress is being made, and significantly larger quantities of data are now being collected than ever before, enabling profound scientific advances across a range of disciplines. However, the ice-covered region of the Southern Ocean still remains the world's biggest data desert, and some of the most pressing scientific issues still require significantly more observational and research effort here to be fully addressed.

The purpose of this chapter is to provide an overview of some key aspects concerning how the sea ice around Antarctica shapes and influences the Southern Ocean's circulation and properties, how the ocean dictates the nature and behaviour of the sea ice, and collectively what the implications of these interactions are for larger-scale environmental issues. Without seeking to be exhaustive, we will highlight a few scientific areas of significant present activity, the current state of that research, and the emerging technologies that are

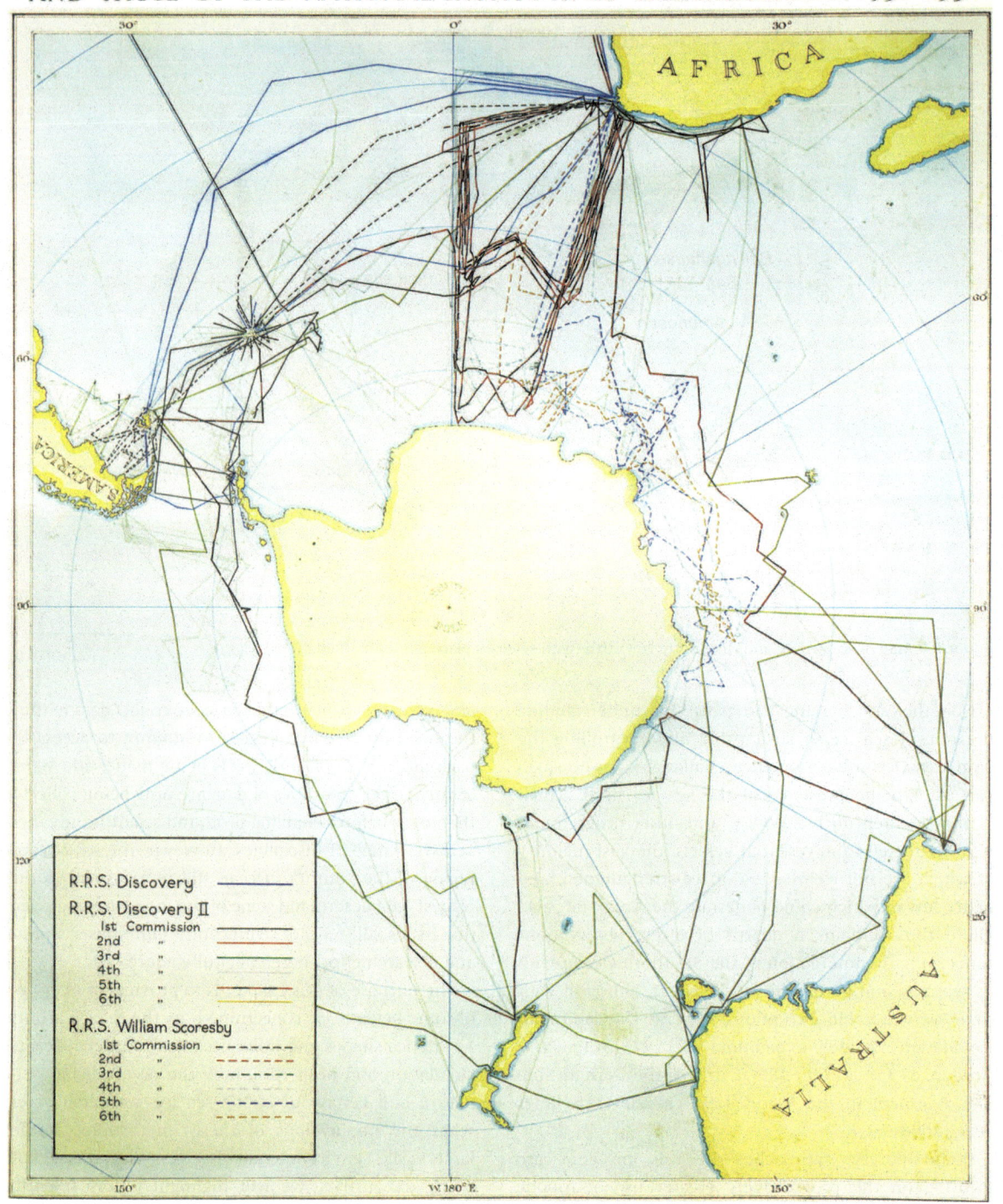

Figure 8.2 Map of the ship tracks undertaken during the *Discovery Investigations*, hand-drawn by E. Humphreys after the initial voyages of *Discovery* and *Discovery II*, and updated between the 1930s and 1950s. A monochrome reproduction of this appeared in Deacon (1955). Source: From Deacon, 1955. Reproduced with permission of John Wiley & Sons.

required to further these areas of investigation into the future.

8.2 Geographic and oceanographic setting of the Southern Ocean

The bathymetry that underlies the Southern Ocean is a mixture of comparatively smooth, oceanic abyssal plains of up to ~4000 m depth or deeper (e.g. the Weddell-Enderby Plain in the Atlantic sector), separated by much shallower and often convoluted ridge systems (Figure 8.3). The overall topographic configuration of the Southern Ocean is unique in the world, possessing a circumpolar channel that is open at all longitudes. This, combined with the strong westerly winds and buoyancy forcing that typifies the high latitudes, leads to the existence of the Antarctic Circumpolar Current (ACC; Figure 8.3; Rintoul et al., 2001).

The ACC is the largest current system in the world, continuously transporting approximately 130 Sv of water eastwards around Antarctica (1 Sv = 10^6 m^3 s^{-1}) (Meredith et al., 2011). It is a banded structure, consisting of relatively fast-moving jets separated by more quiescent zones of water. These jets coincide with oceanic fronts, namely (north to south), the

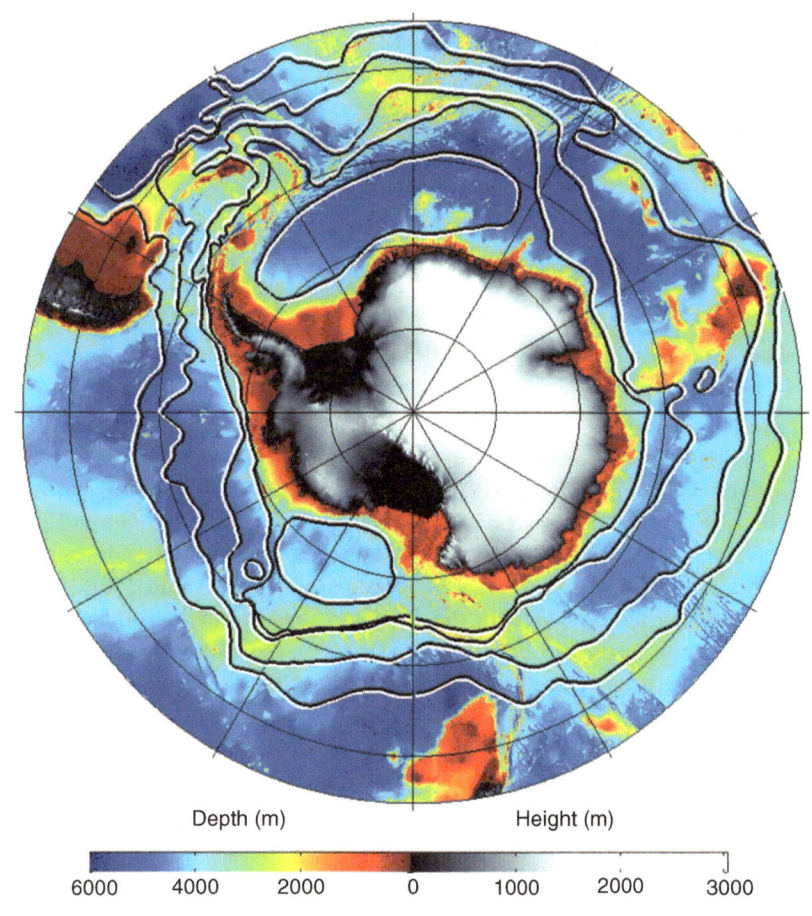

Figure 8.3 Bathymetry and topography of the Southern Ocean and Antarctica. Depths below sea level are coloured; heights above sea level are monochrome. Note the convoluted ridge systems, such as in and around Drake Passage (see Figure 8.1), separating wide expanses of oceanic abyssal plain. Also shown are schematic depictions of the fronts of the Antarctic Circumpolar Current (ACC) in the Southern Ocean, namely (north to south) the Subantarctic Front (SAF), the Polar Front (PF) and the Southern ACC Front (SACCF). South of the SACCF lies the Southern Boundary of the ACC. Frontal locations from Orsi et al. (1995). Poleward of the ACC lie sub-polar gyre systems, including the Weddell and Ross Gyres (marked).

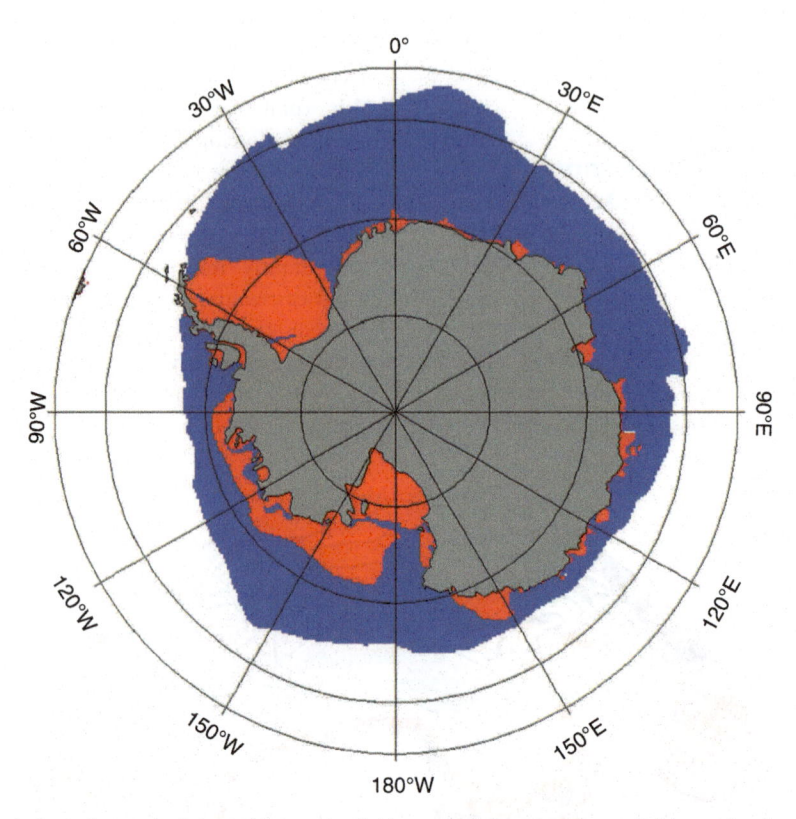

Figure 8.4 Climatological maxima and minima of the sea ice field around Antarctica, shown in blue and red, respectively.

Subantarctic Front, the Polar Front and the Southern ACC Front (Orsi et al., 1995). The southern edge of the ACC is marked by the Southern Boundary (Figure 8.3). Poleward of the ACC lie a series of sub-polar gyres, most notably in the Weddell Sea and Ross Sea (Figures 8.1 and 8.3). These transport a few tens of Sv cyclonically around the basins within which they reside, with the strongest parts of the circulation focused in boundary currents that lie adjacent to the periphery of the basins (Fahrbach et al., 1994).

The gyres are blanketed by sea ice in winter (Figure 8.4), which extends up to the southern part of the ACC during this season, with the sea ice edge being strongly steered by the path of the Southern Boundary (Figure 8.4, cf. Figure 8.3). The total ice coverage in the Southern Ocean reaches almost 19 million km² in the austral winter, compared with typically around 3 million km² in summer, a total range more than 1.5 times greater than the seasonal change in the Arctic. The perennial ice that remains in the Southern Ocean in winter typically inhabits the coastal regions and areas within the sub-polar gyres, most notably in the Weddell Sea (Figure 8.4).

In addition to the strong horizontal circulations in the Southern Ocean, and arguably of even more climatic importance, is a vigorous overturning circulation (Marshall & Speer, 2012). This is a key part of the global ocean overturning circulation, since the Southern Ocean is the main region within which deep waters are upwelled to the surface where they can exchange buoyancy with the atmosphere and be converted into both denser and lighter water masses (Rintoul et al., 2001). Sea ice is an important factor in the forcing of this overturning circulation and the formation of water masses, since it represents a key component of the buoyancy forcing in the Southern Ocean and a modulator of the wind stress imposed on the surface.

The overturning circulation in the Southern Ocean is shown schematically in Figure 8.5, where Circumpolar Deep Water (CDW; derived from the products of deep convection in the North Atlantic, and the most voluminous water mass of the ACC) is shown upwelling to the

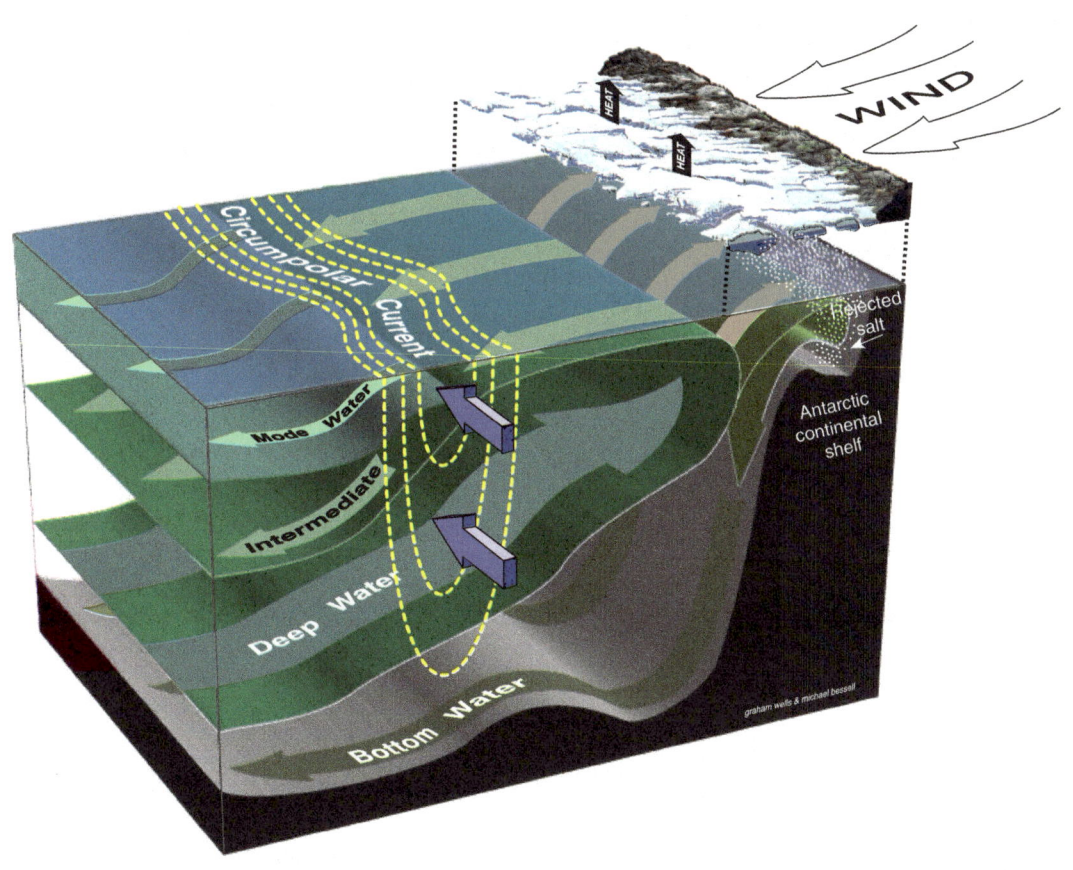

Figure 8.5 Schematic of the overturning circulation in the Southern Ocean. Circumpolar Deep Water upwells to the surface and is reprocessed into both lighter (mode and intermediate) waters and denser (bottom) waters. Source: From Rintoul, 2000. Reproduced with permission of the Royal Society of Tasmania, http://eprints.utas.edu.au/13587/.

surface in the vicinity of the ACC. One component of the CDW is converted by air–sea–ice interaction into lighter (mode and intermediate) water masses (McCartney, 1977), whilst a different component spreads southwards via the sub-polar gyres and becomes involved in the production of dense waters that sink to the seabed and spread out to fill much of the global abyss (Johnson, 2008). Such generalized views of Southern Ocean overturning are necessarily conceptual, and significant factors are not incorporated; nonetheless they remain useful guides to understanding.

Separating the ACC and the gyres from the continent itself are the shelf and slope regions of Antarctica. The Antarctic shelf is deeper than many shelf regions in the world, at typically a few hundred metres depth, and is dissected in many places by glacially scoured canyons. The waters on the Antarctic shelf are often subcategorized into those residing in a 'warm' sector (primarily the Bellingshausen and Amundsen seas; Figure 8.1), and those in a 'cold' sector (Figure 8.6; see also Clarke et al., 2009; Schmidtko et al., 2014). To leading order, this pattern is determined by the proximity of the ACC to the shelf. In the warm sector, the ACC lies immediately adjacent to the shelf, and hence warm waters can penetrate onto the shelf in relatively unmodified form, with the cross-shelf canyons being especially efficacious in enabling this transport (e.g. Martinson, 2011). In other sectors, the sub-polar gyres separate the warm waters of the ACC from the shelf regions, and hence any waters that can penetrate onto the shelf carry substantially less heat. This has a profound impact on ocean and atmospheric climate in

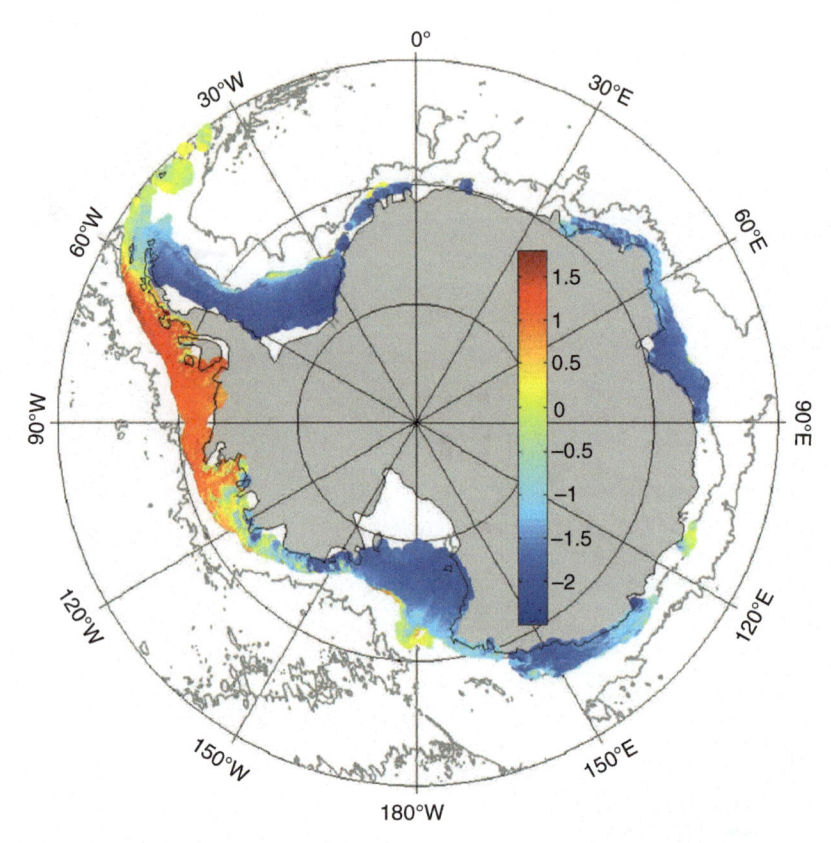

Figure 8.6 Bottom temperatures on the shelf around Antarctica. Note the existence of two primary regimes: a 'warm' sector in west Antarctica (the western Peninsula, Bellingshausen Sea and Amundsen Seas, between 60° and 120°W) and a 'cold' sector around the rest of the continent. The 1000 m and 4000 m depth contours are marked.

these regions, with significant consequences for sea ice production in the different sectors, as detailed below.

The overall water mass structure of the Southern Ocean is most easily comprehended in potential temperature–salinity space (Figure 8.7). The CDW of the ACC is notable for being a warm, saline water mass compared with waters further south. As CDW is the oceanic source for these more southerly waters, this illustrates that it loses heat and becomes freshened overall by air–sea–ice interactions during the transformation process. The upper layer water masses above CDW in the sub-polar region are Antarctic Surface Water (AASW, typically cool and fresh in summer) and winter water (WW, the summertime remnant of the previous winter's mixed layer, characterized by a temperature minimum at around 100 m depth). Filling the layers beneath CDW is Antarctic Bottom Water (AABW), the product of dense shelf water mixing with heavily modified CDW on the shelf and slope regions of Antarctica (Gill, 1973; Whitworth et al., 1998). The shelf waters have high salinity and low temperature, concomitant with their location on the surface melting–freezing line in potential temperature–salinity space (Figure 8.7).

North of the ACC, the upper ocean water mass structure differs greatly from that in the sub-polar region. Above the CDW (and its precursor North Atlantic Deep Water; NADW) lie Antarctic Intermediate Water (AAIW) and Subantarctic Mode Water (SAMW), formed by deep convection in winter and subduction of cold, fresh WW across the fronts of the ACC (Hanawa & Talley, 2001; McCartney, 1977). These waters are formed in specific sites circumpolarly, and their characteristics vary from basin to basin. AABW permeates the bottom layer here also, as part of its general northward spread in the abyss of the major ocean basins (Johnson, 2008).

Oceanography and sea ice in the Southern Ocean

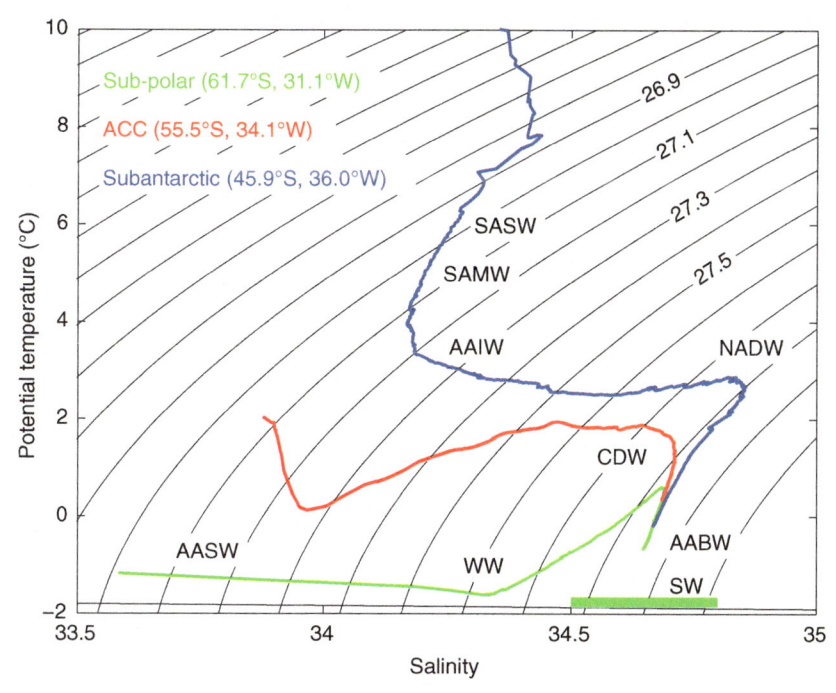

Figure 8.7 Sample profiles from different locations in the Atlantic sector of the Southern Ocean, specifically the Weddell Sea (green), the Antarctic Circumpolar Current (ACC, red) and the subantarctic South Atlantic (blue), in potential temperature–salinity space. Exact positions are as indicated. The melt–freeze line is shown along the bottom of the panel. Different water masses are marked, namely Antarctic Surface Water (AASW), Winter Water (WW), Subantarctic Surface Water (SASW), Antarctic Intermediate Water (AAIW), North Atlantic Deep Water (NADW), Circumpolar Deep Water (CDW), and Antarctic Bottom Water (AABW). The green rectangle denotes the properties of the freezing-point shelf water that provides the dense end member for AABW formation.

It is of particular note that whilst AAIW exists in the subantarctic zone and further north, it is renewed and freshened by upper-layer waters subducting from further south in the ACC (e.g. Naveira Garabato et al., 2009). This is the region where sea ice extent reaches its maximum in winter (Figure 8.4), and hence the upper ocean here is freshened seasonally as this ice melts. The seasonal renewal of AAIW has been demonstrated in the southwest Atlantic sector of the Southern Ocean, with seasonal inflation of the AASW and WW layers causing a pulse of new water to be injected into the AAIW layer (Evans et al., 2014). Precipitation and glacial melt will also contribute to the freshening as they are incorporated into the AASW and WW layers, but nonetheless this illustrates the importance of the sea ice around the Southern Ocean in determining the properties and formation rates of its interior water masses, even at comparatively low latitudes (see also Abernathey et al., 2016; Haumann et al., 2016).

8.3 Sea ice and dense water production in the Southern Ocean

8.3.1 Shelf waters as precursors of AABW

Of the major global water masses that form in the Southern Ocean, one that is notably strongly dependent on sea ice processes is AABW. Key to the formation of AABW is the production of dense waters on the Antarctic shelf, which can then mix and descend down the continental slope to ventilate and renew the ocean abyss (Figure 8.8). This is a very location-specific process, and traditionally it has been the Weddell Sea, Ross Sea and the ocean next to Adélie Land that have been recognized as the major locations where waters are formed that are sufficiently dense to lead to AABW production (Carmack & Foster, 1975; Gill, 1973; Jacobs, 2004).

There has been much renewed interest in AABW formation in recent years, due at least partly to the realization that this water mass is warming and freshening

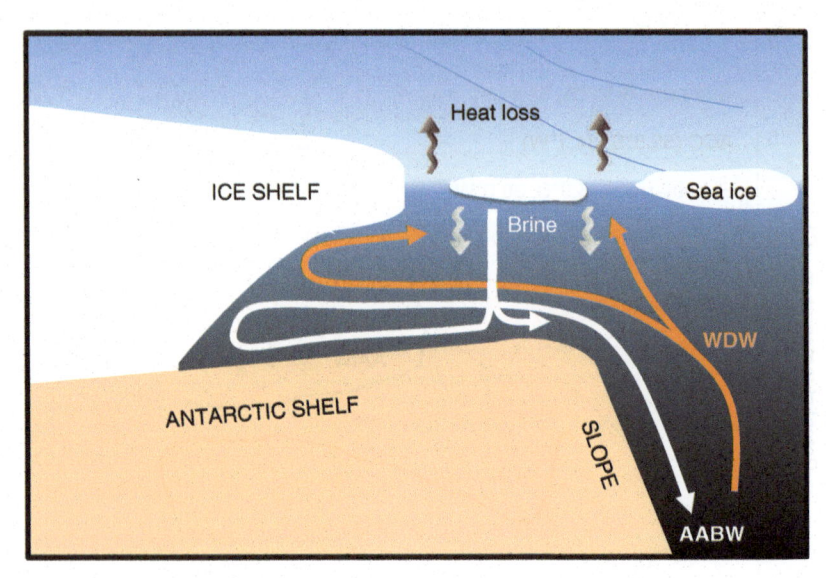

Figure 8.8 Schematic of the processes occurring adjacent to Antarctica that can lead to the production of dense shelf water, and its mixing and cascading to form Antarctic Bottom Water.

across large parts of the Southern Ocean and beyond, with consequences for the global heat budget, sea level rise, benthic biodiversity, and so on (Aoki et al., 2005; Johnson & Doney, 2006; Jullion et al., 2013; Meredith et al., 2008b). The freshening has been largely attributed to the extra injection of glacial melt from the Antarctic continent into the shelf waters that ultimately become incorporated into AABW (Jacobs et al., 2002; Rye et al., 2014), whilst the causes of the warming are still the subject of ongoing research.

The oceanic source for the dense shelf waters and AABW is ultimately CDW, which penetrates southward via the sub-polar gyres, becoming strongly modified en route (Fahrbach et al., 1994). Regionally, this modified water mass is often given a different name to CDW to highlight its modified nature, such as Warm Deep Water (WDW), or Weddell Deep Water in the Weddell Sea. Water from the deep ocean can access the Antarctic shelves in a number of places, with a range of processes (including seasonal wind forcing and fluxes associated with mesoscale eddies) believed to be important (Årthun et al., 2012; Thompson et al., 2014). On the shelf, a variety of shelf water types exist, often being quite fresh (and hence of comparatively low density) due to the input of precipitation and glacial melt. For the conversion of light shelf waters into dense waters, a range of complex processes are important (Figure 8.8),

including in some locations the penetration of water beneath the ice shelves that fringe Antarctica as the floating edge of the Antarctic ice sheet (Nicholls et al., 2009). Here, the inflowing water can change its salinity via interactions with the underside of the ice shelf, but it can also be cooled below the surface freezing point, with the pressure to which the water is subject allowing it to remain liquid. This super-cooled shelf water is commonly termed Ice Shelf Water (ISW), and is one of the shelf waters that can be incorporated into the AABW formation process in regions where significant ice shelves exist, such as the Weddell and Ross seas (e.g. Schlosser et al., 1990; Nicholls et al., 2009).

A key process in the densification of the shelf waters prior to their involvement in the production of AABW is the loss of buoyancy to the atmosphere (Renfrew et al., 2002). This is achieved by the loss of heat from shelf waters, which initially cools the waters to the freezing point, after which further heat loss leads to sea ice production. This process is most efficacious in winter, when the difference between ocean and atmosphere temperature is greatest, and once sea ice production commences, brine rejection into the ocean leads to the waters affected becoming progressively denser. In ocean regions with broad shelves, such as the Weddell and Ross seas and Adélie Land, substantial volumes of dense shelf waters can accumulate, having been exposed to sea ice freezing processes for several annual cycles (Jacobs, 2004). There are other regions where significant amounts of dense shelf waters can be produced, even in the absence of a broad shelf region, with the key factor being sufficiently high rates of sea ice production (Ohshima et al., 2013).

It is important to appreciate that, whilst sea ice formation and brine rejection are critical processes in the production of the dense shelf waters that ultimately feed into AABW, the major modification to CDW to become the denser AABW does not take the form of an increase in salinity. This is illustrated clearly in potential temperature – salinity space (Figure 8.7), where it can be seen that AABW is actually fresher than the oceanic source water CDW, and its greater density is due to its much lower temperature. In practice, CDW (in the form of WDW) provides a major source of salt to the Antarctic shelf regions, and dense water production requires loss of heat from these areas, and injection of salt from brine rejection to counter some of the freshening effects of precipitation and glacial melt.

It should also be noted that, whilst significant rates of sea ice production are required to form dense shelf waters and the precursors of AABW, the process of forming sea ice can itself restrict or halt this process, as a complete ice cover would prevent the further loss of ocean heat to the atmosphere. Gaps in the sea ice cover, where the ocean is exposed to the atmosphere even during the winter, are thus critical for the continuous production of sea ice and the creation of significant volumes of dense waters on the Antarctic shelves. Polynyas can be especially effective in this role.

8.3.2 Polynyas

The term polynya is used to denote an enclosed region within an area of ice cover that is persistently or recurrently free from ice, or shows lower ice concentrations than would typically be expected for that area (Chapter 7; Morales-Maqueda et al., 2004). Their sizes vary greatly and they can display significant variability in their occurrence and persistence. In this section, we will describe the importance of polynyas for the production of dense water around Antarctica (further information, and specifics of their importance in the Arctic, is given in Chapter 7).

There are two types of polynyas, namely latent-heat polynyas and sensible-heat polynyas (Figure 8.9). The latter of these are thermally driven, being created from an oceanic sensible heat flux in circumstances

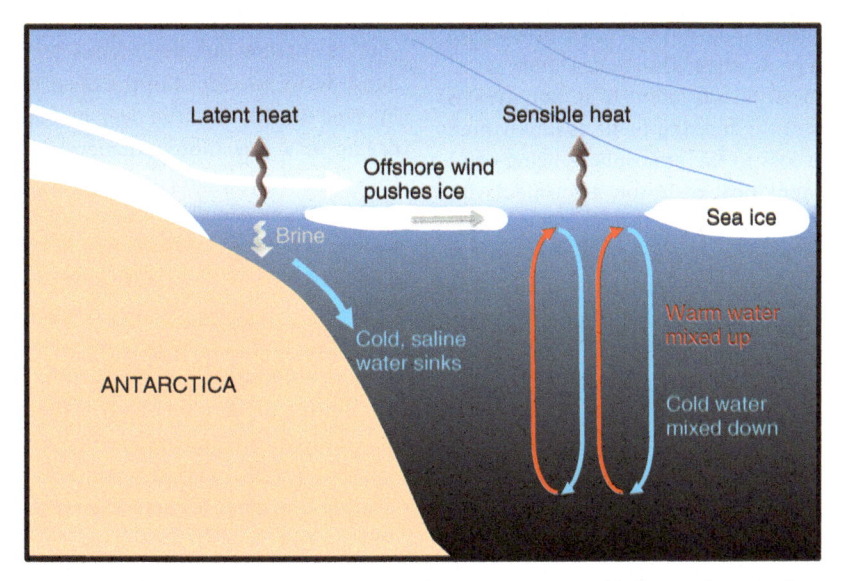

Figure 8.9 Schematic of two polynya types that occur in the Southern Ocean. A sensible-heat polynya is shown on the right, with upwelling of warm water acting to reduce ice production in the locality of the polynya, hence maintaining open water area and permitting heat loss to the atmosphere. A latent-heat polynya is shown on the left, with offshore winds from the Antarctic continent advecting ice away from the coast and maintaining a stretch of open water immediately adjacent to land. The loss of heat via this open water area can enable substantial ice production, brine rejection and the production of dense shelf waters.

where this flux is large enough to melt existing sea ice or restrict the formation of new ice (Martin, 2001). Typically, sensible-heat polynyas are characterized by low sea ice production rates and their dimensions show dependence on the scale of the warm water anomalies responsible for their formation. Upwelling regions can show tendencies to form sensible-heat polynyas, sometimes caused by interactions of ocean currents with submarine features such as seamounts. One example of this is a polynya that can form in the vicinity of Maud Rise, a seamount in the Weddell Sea (Bersch et al., 1992); however, there are comparatively few recurring sensible-heat polynyas in the Southern Ocean.

Of more significance to the production of dense shelf waters around Antarctica are latent-heat polynyas, which fringe the edge of the continent in winter (Kern, 2009). Latent-heat polynyas are mechanically driven, and in general form in the lee of islands, headlands, grounded icebergs and other obstacles that enable export of sea ice by winds or ocean currents without its replacement from upstream in the flow (Martin, 2001). Particularly effective at maintaining coastal latent-heat polynyas in the Antarctic are the katabatic winds, the powerful, gravity-driven flows of air that stream offshore from the continent, advecting the sea ice that has formed and exposing the ocean beneath to further air–sea fluxes and more sea ice production (Morales-Maqueda et al., 2004; Thorsten, 1998).

Surface water in Antarctic latent-heat polynyas is usually maintained at the freezing point during winter, with the huge heat loss to the atmosphere being compensated by the latent heat of fusion associated with new ice formation (Thorsten, 1998). Some polynyas can produce many metres of ice per year, which is very much more productive than the surrounding ice-covered areas. Because of the substantial volumes of sea ice that are produced, these coastal polynyas are often termed 'ice factories', and they play a key role in modifying shelf water properties (Thorsten, 1998; Tamura et al., 2008; Ohshima et al., 2013).

The spatial distribution of coastal polynyas around Antarctica is reflected in maps of sea ice production rates (e.g. Figure 8.10; Tamura et al., 2008). The locations of the significant latent-heat polynyas are clear, with associated high rates of ice production in the Ross Sea, Weddell Sea and along the fringes of East Antarctica. High rates of ice production are discernible in the Cape Darnley region, and it was shown recently that this area is a significant site for dense water production, with the water formed spreading around and into the Weddell Sea after descending from the shelf and becoming incorporated into the AABW mixture (Ohshima et al., 2013).

8.3.3 Deep convection in the open Southern Ocean

Whilst dense water formation in the Southern Ocean appears to be presently restricted to the shelf regions, there is evidence that this has not always been the case. In the mid-1970s, a large polynya feature was detected in the pack ice of the open Weddell Sea, some considerable distance from the shelves (Figure 8.11; see also Zwally & Gloersen, 1977). Dubbed the Weddell Polynya, it first appeared in 1974, and reappeared in the following two austral winters (Gordon, 1978).

The Weddell Polynya represented a very significant proportion of the open water area in the Weddell Sea at the times it existed, being around 0.3×10^6 km^2 in size. Consequently, it had the capacity to be a very significant contributor to the net winter heat loss from the ocean to the atmosphere, with consequences for dense water production. Unlike contemporary dense water production adjacent to Antarctica, where the water masses formed convect to the bottom of the shelves at a few hundred metres depth, any sufficiently dense water produced in the open Weddell Sea will be injected directly into the deep and abyssal layers of the ocean, at a few thousand metres depth (Figure 8.3), thus contributing directly to the reservoir of AABW in the world's ocean. The amount of deep water formed by the Weddell Polynya has been estimated to be approximately 1–3 Sv (Martinson et al., 1981).

The initial cause of the Weddell Polynya and its transience are still subject to some uncertainties, as there is not presently the opportunity to study it in full detail using modern techniques. However, it is believed that sea ice formation in the area was inhibited by ocean convection that injected warm waters from the ocean interior into the surface layer. Other, less persistent polynyas of smaller area have been observed in the vicinity, associated with anomalous upwelling of warm water caused by the interaction of ocean currents with Maud Rise (Bersch et al., 1992; Muench et al., 2001). However, the Weddell Polynya itself has not recurred since 1976.

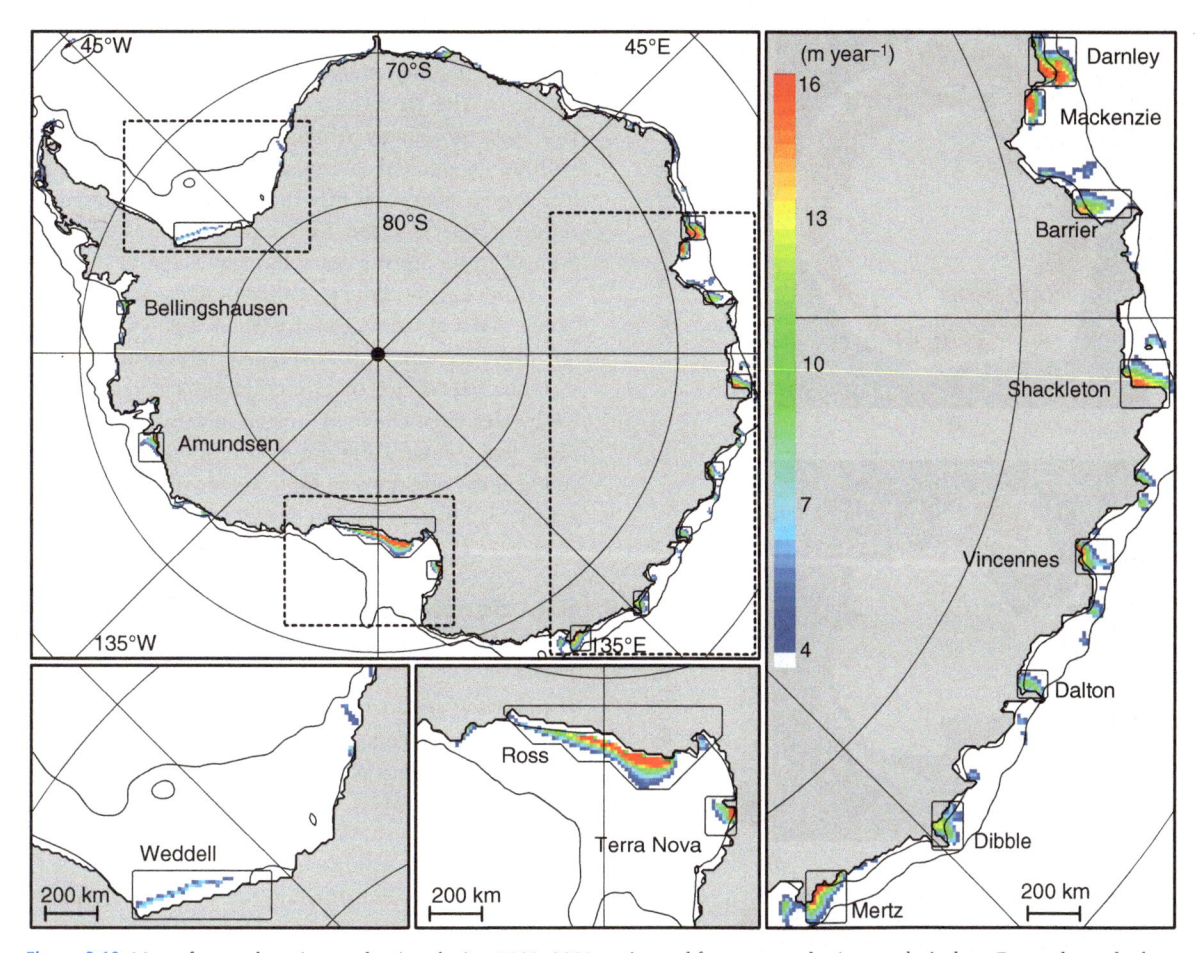

Figure 8.10 Map of annual sea ice production during 1992–2001, estimated from atmospheric reanalysis data. Zoomed panels show sea ice production in the Weddell Sea, Ross Sea and along the coast of East Antarctica. Source: From Tamura et al., 2008. Reproduced with permission from John Wiley & Sons.

There are numerous open and pressing research questions concerning the Weddell Polynya, including what triggered its original formation, what led to its disappearance, and when and how frequently it might recur. Recent investigations have highlighted the importance of the regional hydrological cycle on the stability of the ocean in this area. Ocean density depends predominantly on salinity at cold temperatures, and thus relatively small changes in the amount of freshwater injected to the surface ocean can have significant implications for upper ocean stratification and the propensity for deep convection to occur (Gordon et al., 2007). The hydrological cycle itself depends on a range of climatic factors, including coupled climate modes such as the El Niño-Southern Oscillation (ENSO) phenomenon, and the Southern Annular Mode (SAM; Thompson & Wallace, 2000). These operate at spatial scales from circumpolar to near-global, indicating the sensitivity that polynya activity and dense water production in the Weddell Sea can have to large-scale climatic changes.

Further to this, historical observations and model simulations were used recently to argue that the Weddell Polynya was previously more active than it has been in the satellite era, and that this decline in activity has been caused by human-induced changes in forcing (de Lavergne et al., 2014). These arguments were made on the basis of an observed freshening of the sub-polar Southern Ocean since the 1950s, which has

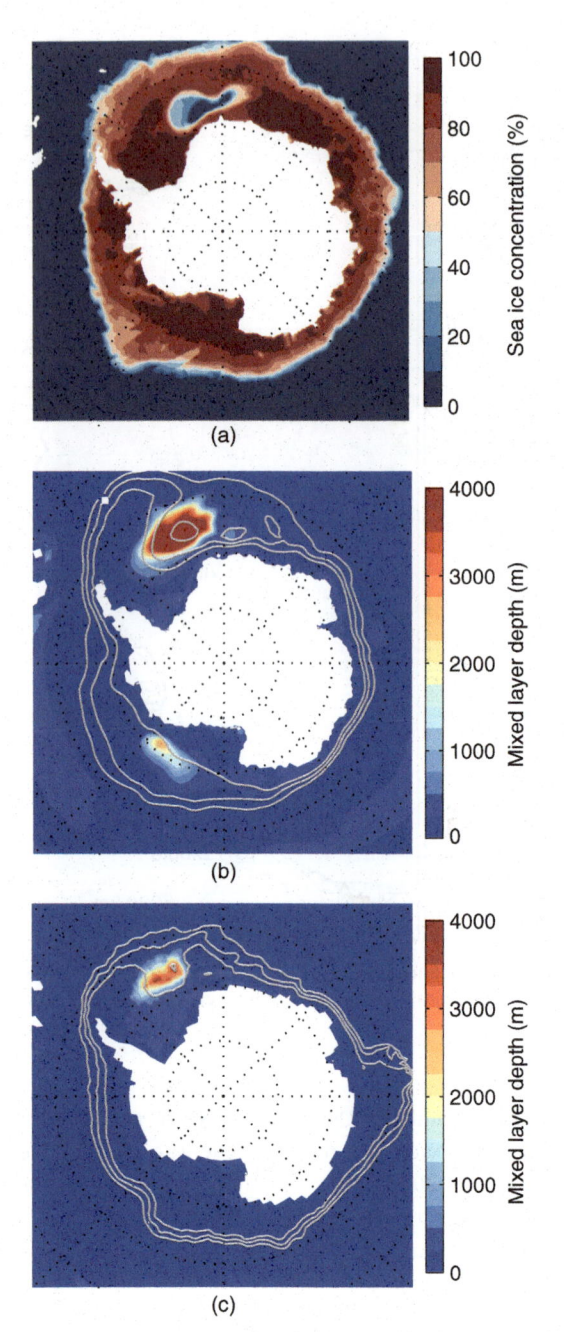

Figure 8.11 (a) Mean September sea ice cover for 1974–1976 from satellite measurements. Note the large polynya within the Weddell Sea pack ice during this period. (b, c) Simulated mixed layer depths using two difference climate models, illustrating the deep-reaching convection that occurred within the polynya. Source: From de Lavergne et al., 2014. Reproduced with permission from Macmillan Publishers Ltd.

increased the strength of the near-surface stratification, thus reducing the tendency for deep convective events to occur. The freshening is believed to derive from two sources, namely stronger precipitation associated with an accelerated hydrological cycle, and increasing injection of meltwater into the ocean from Antarctica's retreating glaciers and ice shelves. Looking forward, both of these processes are thought liable to intensify as anthropogenic climate change progresses, thus indicating that a recurrence of the Weddell Polynya may be less likely in future (de Lavergne et al., 2014), though it should be noted that interannual changes in the freshwater balance of this region associated with coupled climate modes and other natural fluctuations may still lead to conditions in some years where convective activity is possible (Gordon et al., 2007).

8.4 Seasonal cycles of ice and ocean on a 'warm' Antarctic shelf

The previous section outlined how sea ice plays a critical role in the production of the dense waters that feed the AABW layer in the global abyss. The shelf processes outlined occur exclusively in the 'cold' sector of Antarctica (Figure 8.6), but it would be misleading to imply that sea ice does not play a major role in influencing the ocean in the 'warm' (Amundsen/Bellingshausen Sea) sector also. Indeed, it is one of the most significant factors in controlling ocean circulation and properties here, and feedbacks from the ocean to the ice are very important in determining the nature and behaviour of the ice cover in this region (Stammerjohn et al., 2003; Venables & Meredith, 2014).

In this context, a great deal of useful information has been obtained from the western Antarctic Peninsula (wAP). This is the most accessible part of Antarctica, and hence has the greatest concentration of manned research stations. Their presence, combined with close proximity to the ocean, means that oceanographic and sea ice data can be collected here even during the winter months. Further, the wAP is of great scientific interest because it recently warmed more rapidly than any other region of the southern hemisphere, with annual-mean atmospheric temperatures at wAP research stations increasing by an average of 3.7 ± 1.6°C century^{-1} during the second half of the 20th century (Smith et al., 1996; Vaughan et al., 2003).

Oceanography and sea ice in the Southern Ocean

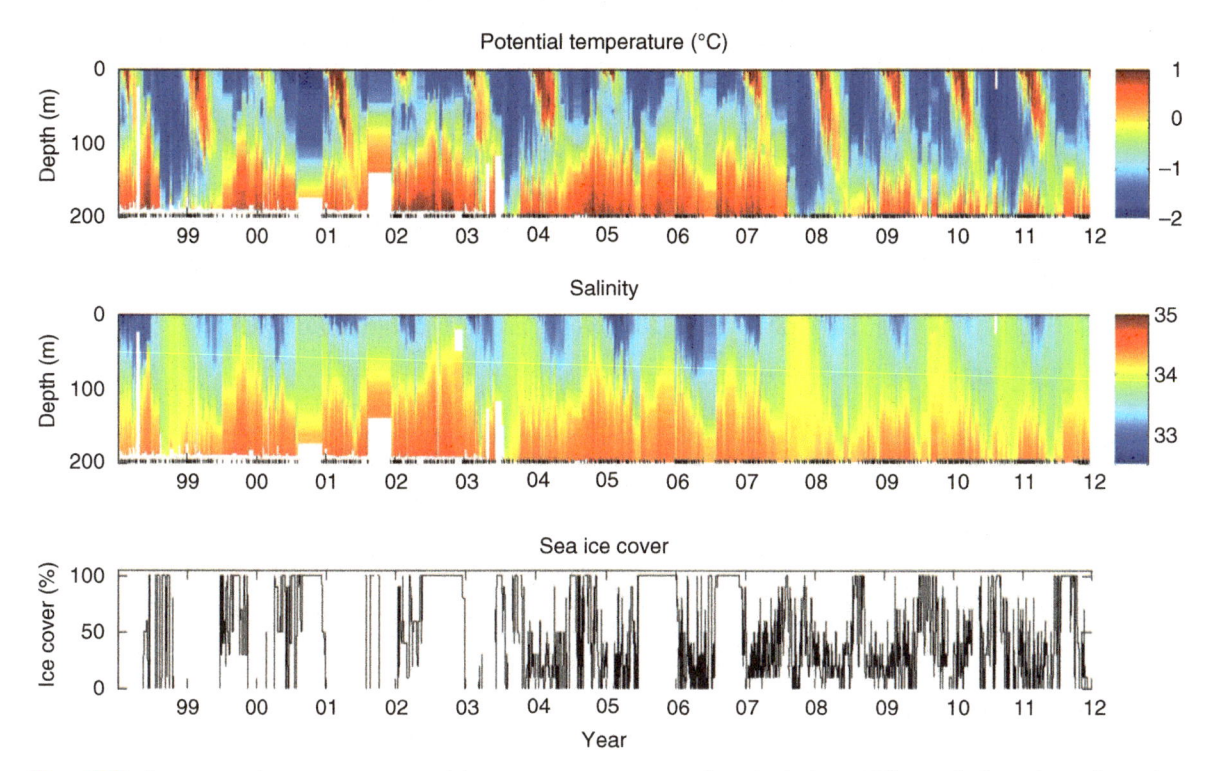

Figure 8.12 Sequences of upper-ocean potential temperature (upper panel) and salinity (middle panel) from the Rothera Time Series (RaTS) at the western Antarctic Peninsula. Note, in particular, the strong seasonality in both series, with interannual changes superposed, each of which is strongly coupled to shifts in the sea ice (lower panel).

Surface ocean temperatures at the wAP rose by more than 1°C during this same period (Meredith & King, 2005), with a concordant retreat of the majority of glaciers and an acceleration in their retreat rates (Cook et al., 2005). The causes of this rapid climatic change are not fully determined, but a wind-induced increase in the flow of CDW from the ACC onto the shelf has been implicated in a number of the changes observed (Martinson, 2011).

Many examples of progress in understanding the ocean–ice system at the wAP exist (e.g. Massom et al., 2008; Stammerjohn et al., 2008a; Turner et al., 2013), but here we present an illustrative case from a long, year-round time series in the nearshore waters on the Peninsula shelf. This dataset was collected adjacent to Rothera Research Station, approximately midway along the western side of the Antarctic Peninsula, and is appropriately called the Rothera Time Series (RaTS) (Clarke et al., 2008). This series includes full oceanographic data from all seasons, and as such is almost unique in the Antarctic context.

Strong seasonality in the upper ocean at the ice-influenced wAP is clearly visible in the data collected (Figure 8.12). The summer periods, with ocean temperatures above 0°C, are comparatively short, and interspersed with significantly longer periods during which the near-surface ocean is close to the freezing point (Figure 8.12; upper panel). This is a consequence of the energy budget of the ocean–sea ice system here, with significant thermal energy in early spring utilized to melt the residual sea ice from the previous winter, rather than warming the seawater; consequently the water temperature remains comparatively low during summer and rapidly reduces to freezing in autumn.

Salinity shows the same significant seasonality, with much fresher waters during spring and summer, interspersed with more saline upper-ocean characteristics during winter (Figure 8.12; middle panel). The phasing of this seasonality is closely tied to the seasonality of the sea ice itself, which exerts a strong, direct influence on ocean salinity (lower panel). This is most clearly manifested by the shift toward more saline water after

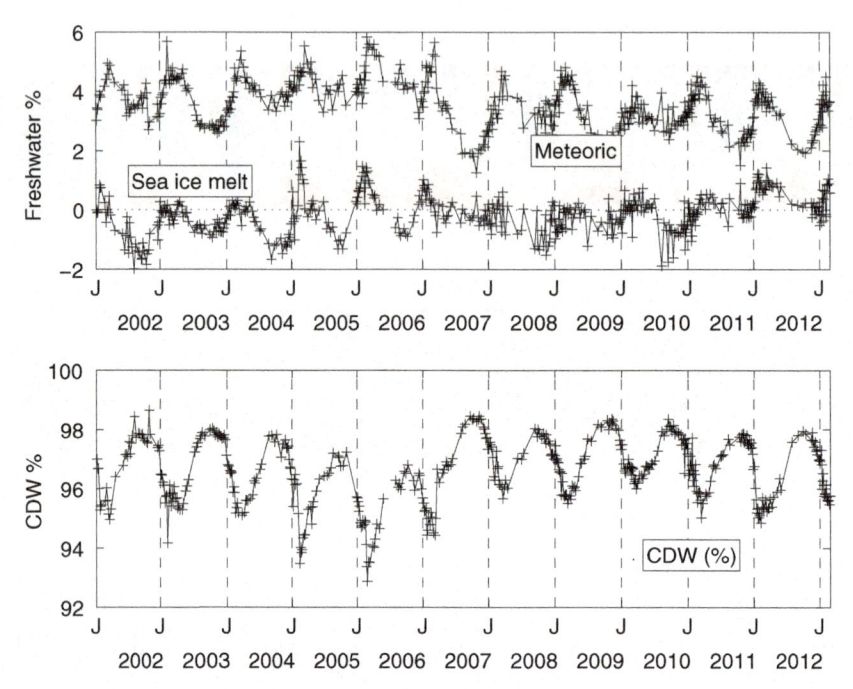

Figure 8.13 Sequences of sea ice melt and meteoric water (glacial discharge plus precipitation) at the western Antarctic Peninsula, as quantified in the near-surface ocean at Rothera Time Series (RaTS) using salinity and oxygen isotope data. Note that the sea ice melt contribution can be both positive and negative; this reflects the fact that sea ice can both melt into and form from seawater, whereas meteoric inputs can only freshen the seawater. Lower panel represents the contribution of saline Circumpolar Deep Water (CDW) to the overall budget. Note the strong seasonality in both freshwater components, and the general dominance of meteoric water over sea ice melt. Source: adapted from Meredith et al., 2013b.

the end of summer, which is caused by the sea ice production, mixed layer deepening and the entrainment of saline CDW from below. The consequent melt of this sea ice contributes to the subsequent freshening in spring and summer, but it is not the only contributor; other terms in the freshwater budget, such as glacial discharge and precipitation, are also important.

This can be illustrated more clearly, and the impact of sea ice on changing ocean salinity demonstrated more directly, by using other oceanographic freshwater tracers in addition to salinity. One very useful such tracer is the ratio of stable isotopes of oxygen in seawater, termed $\delta^{18}O$. The great utility of this ocean tracer is that, when measured alongside salinity, it allows the freshwater supplied to the ocean from sea ice melt to be quantified separately from that supplied by meteoric sources (here meaning the collective input from glaciers and precipitation) (Craig & Gordon, 1965).

When this technique is applied to water samples collected near the surface at Rothera, the seasonality in both the sea ice melt and meteoric freshwater is very clear (Figure 8.13; Meredith et al., 2008a, 2010). Interestingly, the seasonal signals are broadly comparable between the sea ice cycle and meteoric water cycle (about 2–3%, peak-to-peak), despite the growth and decay of sea ice around Antarctica often being conceived as the most significant seasonal signal on the planet. It is clear that, in this locality, other sources of freshwater contribute just as much to the seasonality in the oceanic freshwater budget.

It should be noted also that the meteoric freshwater term is always higher than the sea ice melt term. This is a consequence of the combined effect of glacial discharge and snow in the nearshore environment being dominant over the sea ice inputs. At other locations, further from the glacial sources and away from the orographically induced increases in precipitation caused by the Antarctic Peninsula mountains, the balance is likely to be different.

In the time series plots of freshwater contributions at Rothera (Figure 8.13), and in the sequence of hydrographic profiles (Figure 8.12), it is clear that there are some longer-term (interannual) changes in addition to the seasonal cycles. Some of the interannual changes relate to anomalous forcings associated with ENSO and the SAM, which are the major modes of coupled climatic variability that impact this part of the Southern Ocean (Stammerjohn et al., 2008b). Examples of this impact include the deep upper-ocean mixed layers observed in 1998, 2003 and 2010 (Figure 8.12), when anomalous atmospheric forcings led to perturbations in the sea ice field, and consequently modified the upper-ocean response. In particular, years with reduced sea ice coverage in winter (Figure 8.12; lower panel) are known to lead to deeper mixed layers, due to the combination of greater wind-induced upper ocean mixing and increased buoyancy loss. This reduced stratification can persist through to the following spring, creating feedbacks in the system via exchanges of heat between the atmosphere and ocean (Venables & Meredith, 2014), and with an impact on biological productivity (Venables et al., 2013).

There are also some manifestations of decadal variability in the data, including a decline in meteoric water prevalence (Figure 8.13), decreasing seasonality in the sea ice concentrations (Figure 8.13), and a progressive shift towards warmer and more saline upper ocean characteristics, with deeper mixed layers in winter (Figure 8.12). Whilst decadal variability cannot be not fully resolved in series of this length, these signals are known to be manifestations of longer-period climatic changes ongoing at the Antarctic Peninsula.

The very significant atmospheric warming that occurred recently at the Peninsula was commensurate with a rapid retreat of the sea ice (Stammerjohn et al., 2012; Turner et al., 2013). However, it is known that such changes in the sea ice field were not just a passive consequence of the atmospheric warming, but instead they played an active role in sustaining and accelerating the warming trend itself (Meredith & King, 2005). In practice, the ocean–ice–atmosphere system in this area of Antarctica was moved from one where relatively strong ice production in autumn and winter was balanced by relatively strong melt in spring and summer, to one with much weaker manifestations of these processes, and the role of the upper ocean and sea ice was to act as a positive feedback on this transition (Meredith & King, 2005). It is thus possible that if an atmospheric warming trend resumes at the Peninsula, the area will move from a predominantly Antarctic climate to a subantarctic one, becoming progressively free of sea ice in due course. The role of external forcings is already known to be significant (Stammerjohn et al., 2008b; Li et al., 2014;), and more research that reliably incorporates regional and global scales is needed to improve predictive skill in this regard.

8.5 The future

The Southern Ocean is disproportionately important in the functioning of the Earth system, being the prime location where deep waters in the global ocean are upwelled to the surface and converted into other water masses that sink and replenish the different limbs of the global ocean overturning circulation. It is a major regulator of planetary climate, it acts as a significant sink for anthropogenic carbon dioxide from the atmosphere, and it plays a strong role in controlling deglaciation of the Antarctic continent, with implications for sea level rise. It is also home to a unique ecosystem, within which some species are commercially exploited. Sea ice in the Southern Ocean influences each of these globally important functions, and understanding how the interactions of the ocean and ice occur is a high scientific priority.

Despite the importance attached to understanding the interactions of sea ice with the Southern Ocean, there is still a very great deal that remains undetermined. In no small part, this is due to the presence of sea ice itself, which presents a major obstacle to conventional (ship-based) fieldwork, creating significant difficulties in obtaining the coherent, sustained datasets that are vital in developing the process-based understanding and in testing model constructs. This difficulty is most challenging in winter, when the Southern Ocean is one of the harshest environments on the planet in which to conduct fieldwork, but nonetheless the winter season is when some of the key processes occur that make the Southern Ocean profoundly important on the planetary scale.

Ocean science has developed dramatically since the days when it was predominantly reliant on research vessels for data gathering, and this has opened up many opportunities to find solutions to the conundrum of how

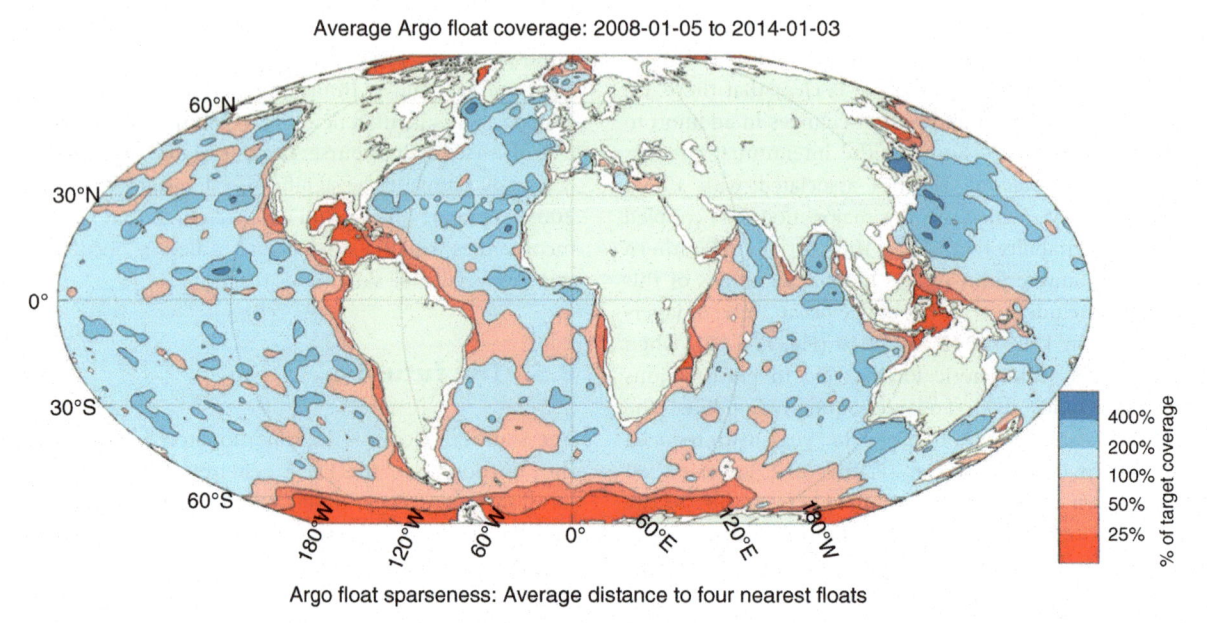

Figure 8.14 Distribution of Argo float coverage in the global ocean, for the period 2008 to 2014. Note the decline in coverage in major parts of the Southern Ocean, especially those typically covered by sea ice. Source: courtesy of JCOMMOPS/Argo.

to obtain sustained, systematic ocean data in the sea ice zone. One example of a technological development that has revolutionized ocean science is the development of the profiling float, which drifts passively at a prescribed 'parking depth' (typically 1000 m), and at intervals sinks then rises to the surface, collecting oceanographic data as it cycles. These data are transmitted to land stations by satellite when the float is at the surface. A network of more than 3000 such floats, called Argo, now exists in the world's oceans, providing real-time data on the upper 2000 m (Figure 8.14).

The Argo array has enabled many radical new insights into the functioning of the ocean within the global climate system, not least in accurately quantifying the oceanic uptake of heat due to global warming (currently >93% of the total extra heat in the Earth system). The distribution of Argo floats in the Southern Ocean is now very creditable in the predominantly ice-free regions (Figure 8.14); however, it declines markedly once the more ice-influenced areas are encountered. This is a consequence of the physical hazard that the ice presents to the float, damaging the float on impact as it ascends or crushing it while it is at the surface.

Various techniques have been developed to alleviate this issue, including equipping floats with ice-detection algorithms that enable them to estimate whether an ascent is likely to be impeded by sea ice at the surface and to abort any ascent accordingly (Klatt et al., 2007). When floats are unable to surface, they can record oceanographic data, but cannot obtain positional fixes via satellite. One ambitious project has circumvented this problem by deploying a network of sound sources in the Weddell Sea, and tracking the floats via underwater acoustics (Figure 8.15). These developments and other ongoing developments that include equipping the floats with biogeochemical sensors and extending their depth range will ensure the continued provision of vital data from the sub-sea-ice regions into the future.

Profiling floats are just one example of the innovative technology being used to better understand sea ice–ocean interactions around Antarctica. A further example is the equipping of marine mammals with miniaturized oceanographic sensors, which record profiles of ocean properties when the animal dives, and relay the data to land via satellite when the animal surfaces. Such sensors have been deployed widely on species such as elephant seals in the Southern Ocean; these can range for huge distances (thousands of km) and dive to great depths (down to ~1500 m), thus providing vast quantities of data that complement the

Oceanography and sea ice in the Southern Ocean   233

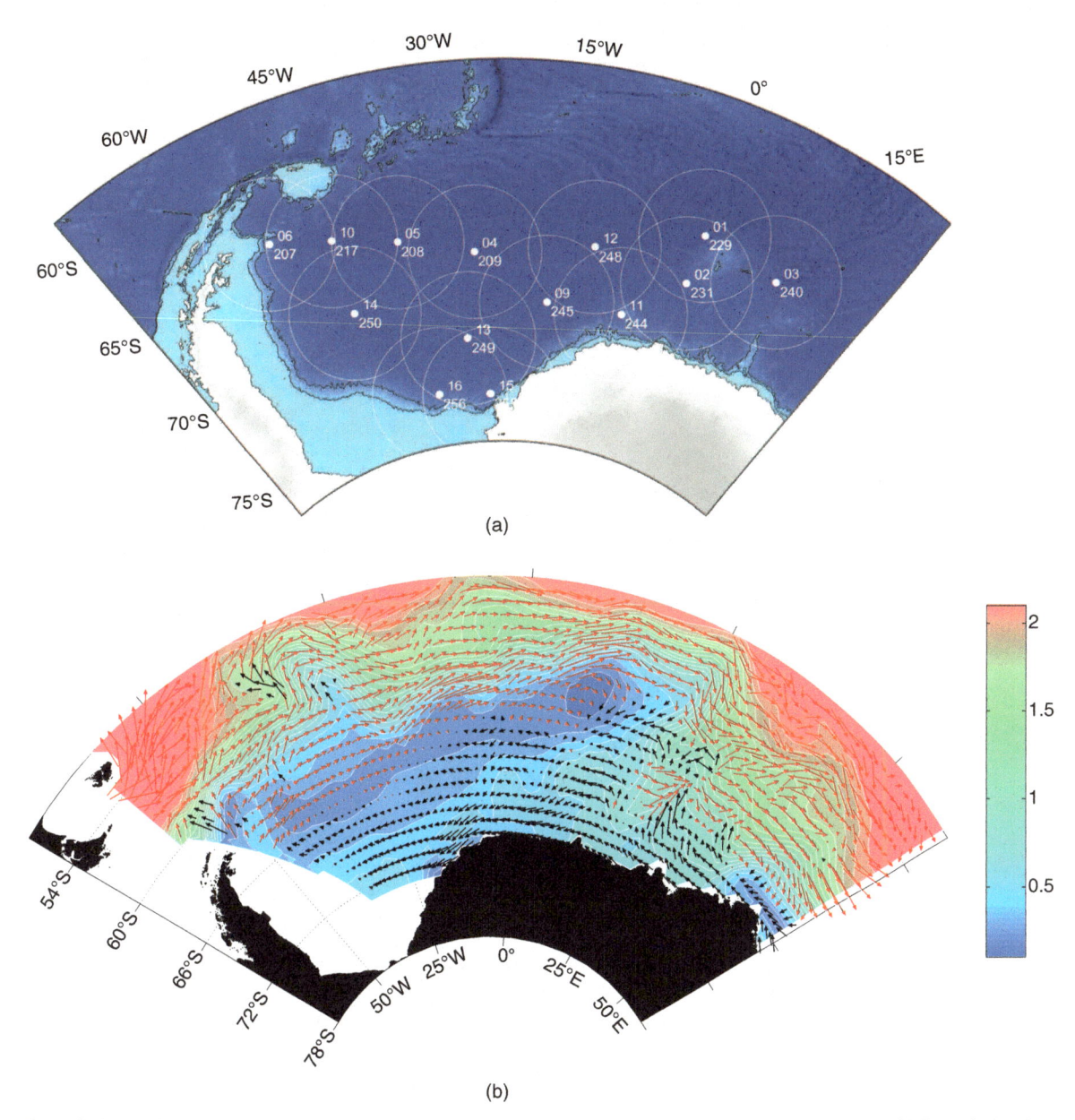

Figure 8.15 (a) The array of sound source moorings deployed by the Alfred Wegener Institute (AWI) in the Weddell Sea for tracking under-ice profiling floats (white circles denoting the area ensonified by each mooring); (b) map of the circulation and temperature at the drift depth of the floats from the data obtained. Source: Figures courtesy of Olaf Boebel and Olaf Klatt, AWI.

Argo float datasets (e.g. Boehme et al., 2008). The integrated Argo and seal datasets have been used to provide unprecedented detail on frontal structures in the Southern Ocean, and also to calculate sea ice production rates deep within the pack (Charrassin et al., 2008). Because the data are collected wherever the seals swim and forage, they are also inherently interdisciplinary and valuable for studies of animal behaviour and ecology as well as oceanography and ocean climate (Biuw et al., 2007).

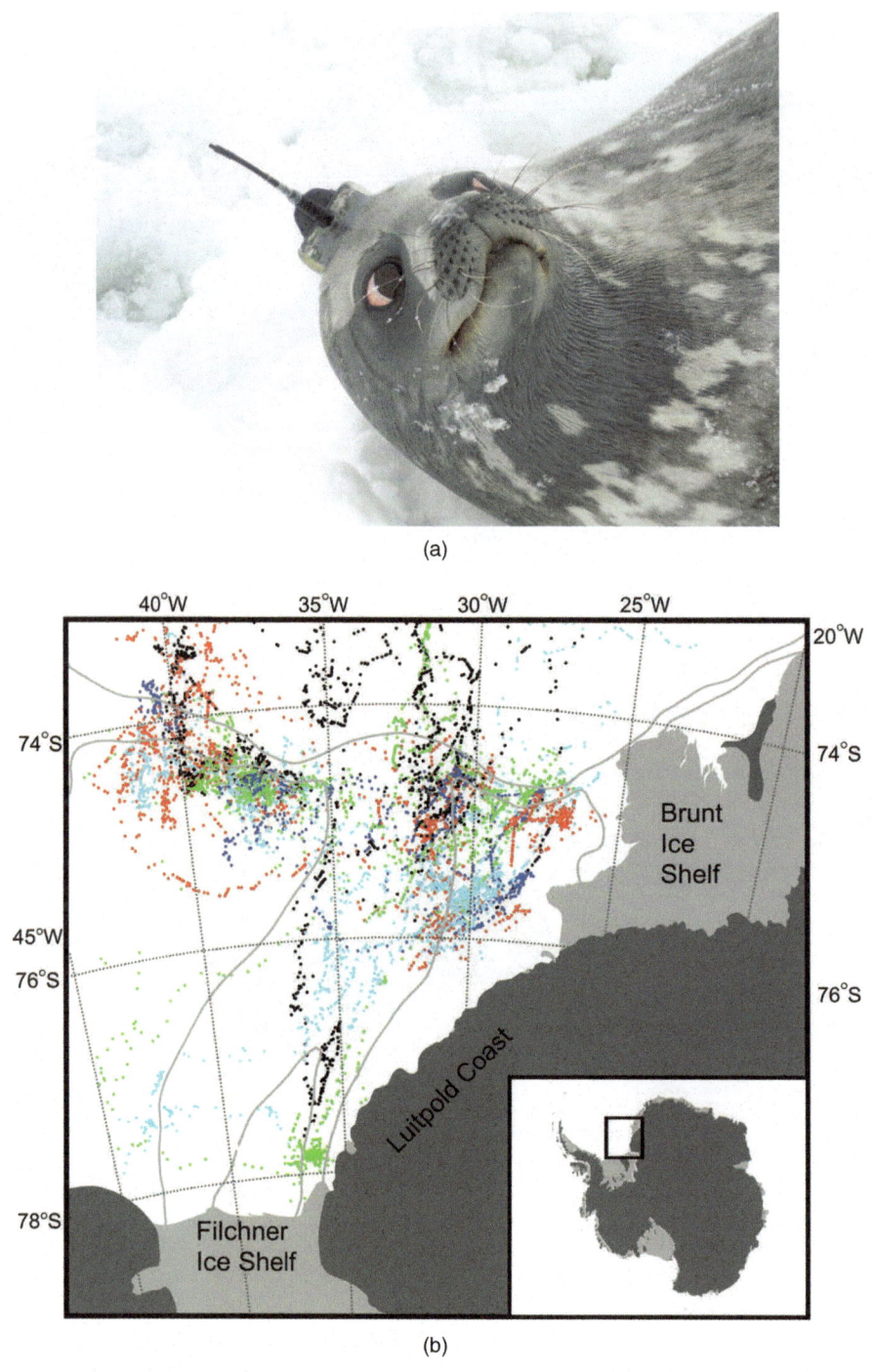

Figure 8.16 (a) A Weddell seal tagged with a miniaturized oceanographic sensor unit (photograph courtesy of Capt. Ralph Stevens); (b) spatial distribution of ocean profiles obtained from seals similarly tagged in 2011. Source: map by Keith Nicholls, adapted from Årthun et al., 2012 with permission; see also Årthun et al., 2012 and Nicholls et al., 2008.

Oceanography and sea ice in the Southern Ocean

In the context of sea ice, the great utility of this technique is that certain animal species venture into and exploit sea ice around Antarctica, including in areas from which it is extremely difficult or impossible to collect oceanographic data via other means. For example, Weddell seals differ from other seal species in that they do not move northwards in autumn and winter as the sea ice edge advances, but remain at high southern latitudes year-round, retaining vital access to the ocean by continually gnawing at holes in the ice to keep them open. Weddell seals tagged with miniaturized oceanographic sensors have produced some unique data that have given significant new insights into the wintertime interactions of ocean and sea ice in this region of the Southern Ocean (Nicholls et al., 2008; Figure 8.16).

Other techniques are continually being developed to help alleviate the dearth of data from the ice-infested regions of the Southern Ocean. Recent developments include ocean gliders, which are autonomous,

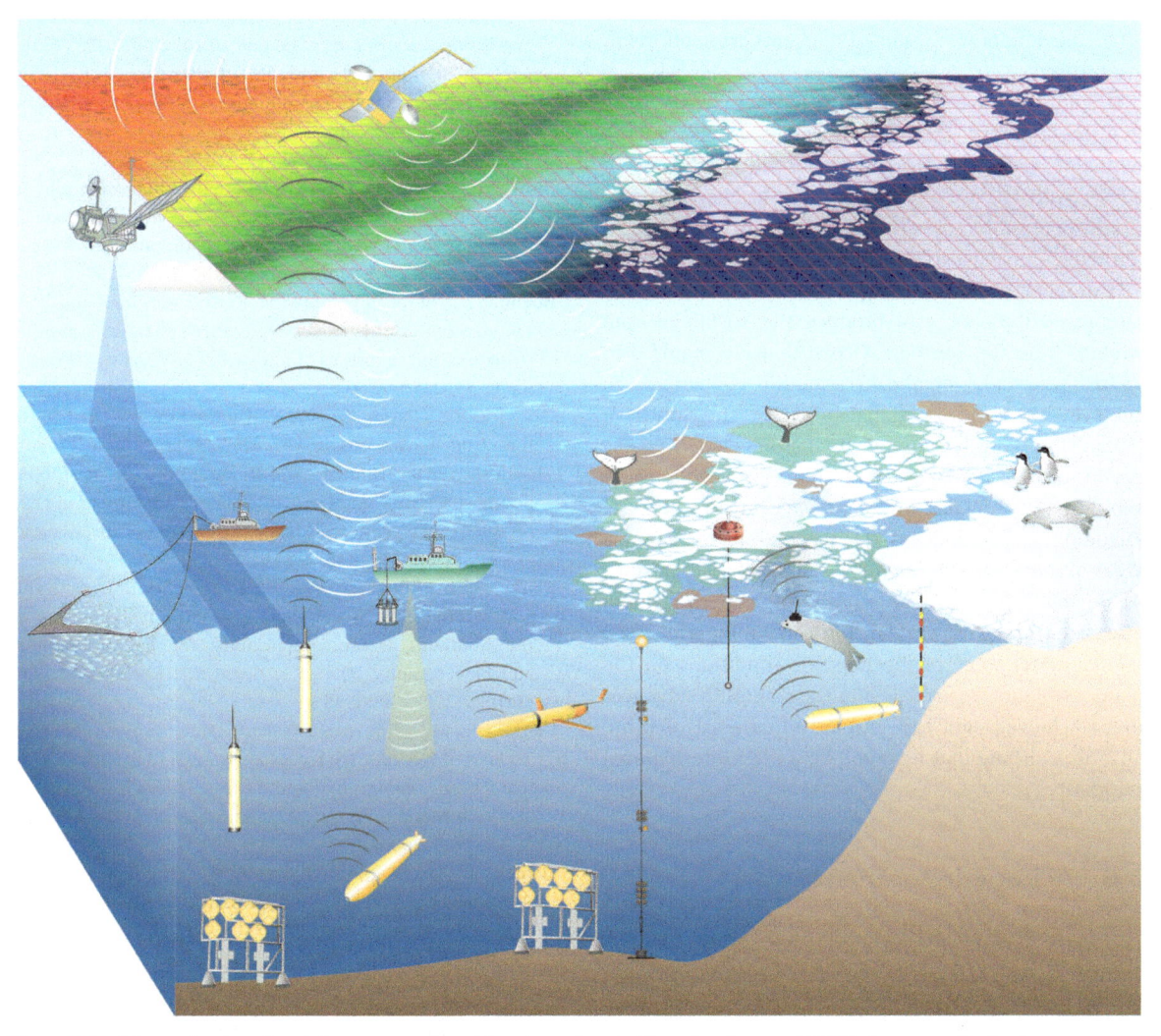

Figure 8.17 Schematic of a cyber-infrastructure-based Southern Ocean Observing System (SOOS), incorporating both autonomous and conventional platforms, but relying progressively more on the former over time. Source: Meredith et al., 2013. Reproduced with permission from Elsevier.

buoyancy-driven vehicles that can survey the upper 1000 m of the water column over large distances and long periods, with deployments of up to several months now possible. In the Arctic, deployments of such ocean gliders beneath the sea ice have been carried out, with navigation and control being made possible via an acoustic communications network. This offers great potential in the Southern Ocean also, for missions both beneath the perennial sea ice in summer and under the full pack ice in winter. Powered unmanned vehicles can deliver oceanographic data from beneath sea ice, and whilst their duration is typically much less than that of an ocean glider, this technology still promises to be a vital contribution to sustained data gathering efforts in the future.

Despite these technological advances, it seems likely that the Southern Ocean will be data-sparse compared with the rest of the world's oceans for some time to come, with the ice-covered regions of the Southern Ocean being the most affected. It thus behooves us to challenge ourselves to find the optimal combination of investments (ships, moorings, autonomous vehicles, satellite sensors, etc.) to produce the sustained data streams from the Southern Ocean that are required to address the most significant scientific challenges. To achieve this, an international initiative, the Southern Ocean Observing System (SOOS), has developed a strategy for how sustained ocean observations in this region should be coordinated and conducted into the future, including a specific focus on sea ice and the sub-ice areas of the ocean (Figure 8.17; Meredith et al., 2013a). Achieving this vision of a fully capable, sustained, integrated observing system that can function even in winter under some of the harshest conditions on the planet will not be trivial, but the scientific advances to be achieved, and the societal importance of those advances, dictate that it should be a high international priority.

8.6 Acknowledgements

The authors thank the numerous people who helped with the production of this chapter, including Hugh Venables, Pete Bucktrout, Andrew Gray, Steve Rintoul, Casimir de Lavergne, Eric Galbraith, Mathieu Belbeoch, Olaf Boebel, Olaf Klatt, Marius Årthun, Keith Nicholls, Ralph Stevens and Takeshi Tamura.

References

Abernathey, R.P., Cerovecki, I., Holland, P.R., Newsom, E., Mazloff, M. & Talley, L.D. (2016) Water-mass transformation by sea ice in the upper branch of the Southern Ocean overturning. *Nature Geoscience*, **9**, 596–601, doi: 10.1038.ngeo2749.

Aoki, S., Rintoul, S.R., Ushio, S., Watanabe, S. & Bindoff, N.L. (2005) Freshening of the Adélie Land Bottom Water near 140°E. *Geophysical Research Letters*, **32**, doi: 10.1029/2005GL024246.

Årthun, M., Nicholls, K.W., Makinson, K., Fedak, M.A. & Boehme, L. (2012) Seasonal inflow of warm water onto the southern Weddell Sea continental shelf, Antarctica. *Geophysical Research Letters*, **39**, doi: 10.1029/2012GL052856

Bersch, M., Becker, G.A., Frey, H. & Koltermann, K.P. (1992) Topographic effects of the Maud Rise on the stratification and circulation of the Weddell Gyre. *Deep-Sea Research Part* **1**, 39, 303–331.

Biuw, M., Boehme, L., Guinet, C. et al. (2007) Variations in behaviour and condition of a Southern Ocean top predator in relation to in situ oceanographic conditions. *Proceedings of the National Academy of Sciences USA*, **104**, 13705–13710.

Boehme, L., Thorpe, S.E., Biuw, M., Fedak, M. & Meredith, M.P. (2008) Monitoring Drake Passage with elephant seals: Frontal structures and snapshots of transport. *Limnology and Oceanography*, **53**, 2350–2360.

Carmack, E.C., Foster, T.D., 1975. On the flow of water out of the Weddell Sea. *Deep-Sea Research*, **22**, 711–724.

Charrassin, J.-B., Hindell, M., Rintoul, S.R. et al. (2008) Southern Ocean frontal structure and sea-ice formation rates revealed by elephant seals. *Proceedings of the National Academy of Sciences USA*, **105**, 11634–11639.

Clarke, A., Griffiths, H.J., Barnes, D., Meredith, M.P. & Grant, S.M. (2009) Spatial variation in seabed temperatures in the Southern Ocean: Implications for benthic ecology and biogeography. *Journal of Geophysical Research*, **114**, doi: 10.1029/2008JG000886

Clarke, A., Meredith, M.P., Wallace, M.I., Brandon, M.A. & Thomas, D.N. (2008) Seasonal and interannual variability in temperature, chlorophyll and macronutrients in Ryder Bay, northern Marguerite Bay, Antarctica. *Deep-Sea Research Part II*, **55**, 1988–2006.

Cook, A.J., Fox, A.J., Vaughan, D.G. & Ferrigno, J.G. (2005) Retreating glacier fronts on the Antarctic peninsula over the past half-century. *Science*, **308**, 541–544.

Craig, H. & Gordon, L. (1965) Deuterium and oxygen-18 variations in the ocean and the marine atmosphere. In: *Stable isotopes in Oceanographic Studies and Paleotemperatures* (Ed. E. Tongiorgio), pp. 9–130. Spoleto.

Deacon, G.E.R. (1937) *The Hydrology of the Southern Ocean*. Cambridge University Press, Cambridge.

Deacon, G.E.R. (1955) The Discovery Investigations in the Southern Ocean. *Transactions of the American Geophysical Union*, **36**, 877–880.

Evans, D.G., Zika, J.D., Naveira-Garabato, A.C. & Nurser, A.J.G. (2014) The imprint of Southern Ocean overturning on seasonal water mass variability in Drake Passage. *Journal of Geophysical Research*, **119**, 7987–8010.

Fahrbach, E., Rohardt, G., Schröder, M. & Strass, V. (1994) Transport and structure of the Weddell Gyre. *Annales Geophysicae*, **12**, 840–855.

Ganachaud, A. & Wunsch, C. (2000) Improved estimates of global ocean circulation, heat transport and mixing from hydrographic data. *Nature*, **408**, 453–457.

Gill, A.E. (1973) Circulation and bottom water production in the Weddell Sea. *Deep-Sea Research*, **20**, 111–140.

Gordon, A.L. (1978) Deep Antarctic convection west of Maud Rise. *Journal of Physical Oceanography*, **8**, 199–217.

Gordon, A.L., Visbeck, M. & Comiso, J. (2007) A possible link between the Weddell Polynya and the Southern Annular Mode. *Journal of Climate*, **20**, 2558–2571.

Hanawa, K. & Talley, L.D. (2001) Mode Waters. In: *Ocean Circulation and Climate* (Eds. G. Siedler, & J. Church), pp. 373–386. Academic Press.

Haumann, F.A., Gruber, N., Munnich, M., Frenger, I. & Kern, S. (2016) Sea-ice transport driving Southern Ocean salinity and its recent trends. *Nature*, **537**, 89–92, doi: 10.1038/nature19101.

Hellmer, H.H., Schroder, M., Haas, C., Dieckmann, G.S. & Spindler, M. (2008) The ISPOL Drift Experiment. *Deep-Sea Research Part II*, **55**, 913–917.

Jacobs, S.S. (2004) Bottom water production and its links with the thermohaline circulation. *Antarctic Science*, **16**, 427–437.

Jacobs, S.S., Giulivi, C.F. & Mele, P.A. (2002) Freshening of the Ross Sea during the late 20th century. *Science*, **297**, 386–389.

Johnson, G. (2008) Quantifying Antarctic bottom water and North Atlantic deep water volumes. *Journal of Geophysical Research*, **113**, doi: 10.1029/2007JC004477

Johnson, G.C. & Doney, S.C. (2006) Recent western South Atlantic bottom water warming. *Geophysical Research Letters*, **33**, doi: 10.1029/2006GL026769.

Jullion, L., Naveira-Garabato, A.C., Meredith, M.P., Holland, P.R., Courtois, P. & King, B.A. (2013) Decadal freshening of the Antarctic Bottom Water exported from the Weddell Sea. *Journal of Climate*, **26**, 8111–8125.

Kern, S. (2009) Wintertime Antarctic coastal polynya area: 1992–2008. *Geophysical Research Letters*, **36**, doi: 10.1029/2009GL038062

King, B.A. (2001) Introduction to special section: World Ocean Circulation Experiment: Southern Ocean results. *Journal of Geophysical Research*, **106**(C2), 2691.

Klatt, O., Boebel, O. & Fahrbach, E. (2007) A profiling float's sense of ice. *Journal of Atmospheric and Oceanic Technology*, **24**, 1301–1308.

de Lavergne, C., Palter, J.B., Galbraith, E.D., Bernadello, R. & Marinov, I. (2014) Cessation of deep convection in the open Southern Ocean under anthropogenic climate change. *Nature Climate Change*, **4**, 278–282.

Li, X., Holland, D.M., Gerber, E.P. & Yoo, C. (2014) Impacts of the north and tropical Atlantic Ocean on the Antarctic Peninsula and sea ice. *Nature*, **505**, 538–542.

Marshall, J. & Speer, K. (2012) Closure of the meridional overturning circulation through Southern Ocean upwelling. *Nature Geoscience*, **5**, 171–180.

Martin, S. (2001) Polynyas. In: *Encyclopedia of Ocean Sciences* (Eds. J.H. Steele, K.K. Turekian & S.A. Thorpe), pp. 2241–2247. Academic Press, San Diego.

Martinson, D.G. (2011) Transport of warm upper circumpolar deep water onto the Western Antarctic Peninsula Continental Shelf. *Ocean Science Discussions*, **8**, 2479–2502.

Martinson, D.G., Killworth, P.D. & Gordon, A.L. (1981) A Convective Model for the Weddell Polynya. *Journal of Physical Oceanography*, **11**, 466–488.

Massom, R.A., Stammerjohn, S.E., Lefebvre, W. et al. (2008) West Antarctic Peninsula sea ice in 2005: Extreme compaction and ice edge retreat due to strong anomaly with respect to climate. *Journal of Geophysical Research*, 113, doi: 10.1029/2007JC004239

McCartney, M.S. (1977) Subantarctic mode water. In: *A Voyage of Discovery* (Ed. M. Angel), pp. 103–119. Pergamon Press, Oxford.

Meredith, M., Brandon, M.A., Wallace, M.I. et al. (2008a) Variability in the freshwater balance of northern Marguerite Bay, Antarctic Peninsula: results from $\delta^{18}O$. *Deep-Sea Research Part II*, **55**, 309–322.

Meredith, M.P., Garabato, A.C.N., Gordon, A.L. & Johnson, G.C. (2008b) Evolution of the Deep and Bottom Waters of the Scotia Sea, Southern Ocean, 1995–2005. *Journal of Climate*, **21**, 3327–3343.

Meredith, M.P., Hibbert, A., Hogg, A.M. et al. (2011) Sustained monitoring of the Southern Ocean at Drake Passage: past achievements and future priorities. *Reviews of Geophysics*, **49**, doi: 10.1029/2010RG000348

Meredith, M.P. & King, J.C. (2005) Rapid climate change in the ocean to the west of the Antarctic Peninsula during the second half of the twentieth century. *Geophysical Research Letters*, **32**, doi:10.1029/2005GL024042.

Meredith, M.P., Schofield, O., Newman, L., Urban, E. & Sparrow, M.D. (2013a) The vision for a Southern Ocean observing system. *Current Opinion in Environmental Sustainability*, **5**, 306–313.

Meredith, M.P., Venables, H.J., Clarke, A. et al. (2013b) The freshwater system west of the Antarctic Peninsula: spatial and temporal changes. *Journal of Climate*, **26**, 1669–1684.

Meredith, M.P., Wallace, M.I., Stammerjohn, S.E. et al. (2010) Changes in the freshwater composition of the upper ocean west of the Antarctic Peninsula during the first decade of the 21st century. *Progress in Oceanography*, **87**, 127–143.

Morales-Maqueda, M.A., Willmott, A.J. & Biggs, N.R.T. (2004) Polynya dynamics: A review of observations and modelling. *Reviews of Geophysics*, **42**, RG1004, doi:10.1029/2002RG000116.

Muench, R.D. & Gordon, A.L. (1995) Circulation and transport of water along the western Weddell Sea margin. *Journal of Geophysical Research*, **100**, 18503–18515.

Muench, R.D., Morison, J.H., Padman, L. et al. (2001) Maud Rise revisited. *Journal of Geophysical Research*, **106**, 2423–2440.

Naveira Garabato, A.C., Jullion, L., Stevens, D.P., Heywood, K.J. & King, B.A. (2009) Variability of Subantarctic Mode Water and Antarctic Intermediate Water in Drake Passage during the late 20th and early 21st centuries. *Journal of Climate*, **13**, 3661–3688.

Nicholls, K.W., Boehme, L., Biuw, M. & Fedak, M.A. (2008) Wintertime ocean conditions over the southern Weddell Sea continental shelf, Antarctica. *Geophysical Research Letters*, **35**, doi: 10.1029/2008GL035742

Nicholls, K.W., Österhus, S., Makinson, K., Gammelsrod, T. & Fahrbach, E. (2009) Ice-ocean processes over the continental shelf of the southern Weddell Sea, Antarctica: a review. *Reviews of Geophysics*, **47**, RG3003, doi: 10.1029/2007RG000250

Ohshima, K.I., Fukamachi, Y., Williams, G.D. et al. (2013) Antarctic Bottom Water production by intense sea-ice formation in the Cape Darnley polynya. *Nature Geoscience*, **6**, 235–240.

Orsi, A.H., Whitworth, T. & Nowlin, W.D. (1995) On the meridional extent and fronts of the Antarctic Circumpolar Current. *Deep-Sea Research Part I*, **42**, 641–673.

Renfrew, I.A., King, J.C. & Markus, T. (2002) Coastal polynyas in the southern Weddell Sea: variability of the surface energy budget. *Journal of Geophysical Research*, **107**, 10.10129/12000JC000720.

Rintoul, S.R. (2000) Southern Ocean currents and climate. *Papers and Proceedings of the Royal Society of Tasmania*, **133**, 41–50.

Rintoul, S.R., Hughes, C. & Olbers, D. (2001) The Antarctic Circumpolar System. In: *Ocean Circulation and Climate* (Eds. G. Sielder, J. Church & J. Gould), pp. 271–302. Academic Press.

Rye, C.D., Garabato, A.C.N., Holland, P.R. et al. (2014) Rapid sea-level rise along the Antarctic margins in response to increased glacial discharge. *Nature Geoscience*, **7**, 732–735.

Schlosser, P., Bayer, R., Foldvik, A., Gammelsrod, T., Rohardt, G. & Munnich, K.O. (1990) Oxygen 18 and helium as tracers of Ice Shelf Water and water/ice interaction in the Weddell Sea. *Journal of Geophysical Research*, **95**, 3253–3263.

Schmidtko, S., Thompson, A.F. & Aoki, S. (2014) Multidecadal warming of Antarctic waters. *Science*, **346**, 1227–1231.

Smith, R.C., Stammerjohn, S.E. & Baker, K.S. (1996) Surface air temperature variations in the western Antarctic peninsula regions. In: *Foundations for Ecological Research West of the Antarctic Peninsula* (Eds. R.M. Ross, E.E. Hofmann & L.B. Quetin), pp. 105–121. American Geophysical Union, Washington, DC.

Stammerjohn, S.E., Drinkwater, M.R., Smith, R.C. & Liu, X. (2003) Ice-atmosphere interactions during sea-ice advance and retreat in the western Antarctic Peninsula region. *Journal of Geophysical Research Oceans*, **108**, 10.1029/2002JC001543.

Stammerjohn, S.E., Martinson, D.G., Smith, R.C. & Ianuzzi, R.A. (2008a) Sea ice in the western Antarctic Peninsula region: spatio-temporal variability from ecological and climate change perspectives. *Deep-Sea Research Part II*, **55**, 2041–2058.

Stammerjohn, S.E., Martinson, D.G., Smith, R.C., Yuan, X. & Rind, D. (2008b) Trends in Antarctic annual sea ice retreat and advance and their relation to El Ñino-Southern Oscillation and Southern Annular Mode variability. *Journal of Geophysical Research*, **113**, doi: 10.1029/2007JC004269

Stammerjohn, S.E., Massom, R., Rind, D. & Martinson, D. (2012) Regions of rapid sea ice change: An inter-hemispheric seasonal comparison. *Geophysical Research Letters*, **39**, L06501, doi: 10.1029/2012GL050874

Tamura, T., Ohshima, K.I. & Nihashi, S. (2008) Mapping sea ice production for Antarctic coastal polynyas. *Geophysical Research Letters*, **35**, doi: 10.1029/2007GL032903

Thompson, A.F., Heywood, K.J., Schmidtko, S. & Stewart, A.L. (2014) Eddy transport as a key component of the Antarctic overturning circulation. *Nature Geoscience*, **7**, 879–884.

Thompson, D.W.J. & Wallace, J.M. (2000) Annular modes in the extratropical circulation. Part I: Month-to-month variability. *Journal of Climate*, **13**, 1000–1016.

Thorsten, M. (1998) Ice formation in coastal polynyas in the Weddell Sea and their impact on oceanic salinity. *Antarctic Research Series*, **74**, 273–292.

Turner, J., Maksym, E., Phillips, A., Marshall, G.J. & Meredith, M.P. (2013) The impact of changes in sea ice advance on the large winter warming on the western Antarctic Peninsula. *International Journal of Climatology*, **33**, 852–861.

Vaughan, D.G., Marshall, G.J., Connolley, W.M. et al. (2003) Recent rapid regional climate warming on the Antarctic Peninsula. *Climatic Change*, **60**, 243–274.

Venables, H.J., Clarke, A. & Meredith, M.P. (2013) Wintertime controls on summer stratification and productivity at the western Antarctic Peninsula. *Limnology and Oceanography*, **58**, 1035–1047.

Venables, H.J. & Meredith, M.P. (2014) Feedbacks between ice cover, ocean stratification and heat content in Ryder Bay, western Antarctic Peninsula. *Journal of Geophysical Research*, **119**, 5323–5336.

Whitworth, T., Orsi, A.H., Kim, S.J., Nowlin, W.D. & Locarnini, R.A. (1998) Water masses and mixing near the Antarctic Slope Front. In: *Ocean, Ice and Atmosphere: Interactions at the Antarctic Continental Margin* (Eds. S.S. Jacobs & R.F. Weiss), pp. 1–27. American Geophysical Union, Washington, DC.

Zwally, H.J. & Gloersen, P. (1977) Passive microwave images of the polar regions and research applications. *Polar Record*, **18**, 431–450.

CHAPTER 9

Methods of satellite remote sensing of sea ice

Gunnar Spreen[1]* and Stefan Kern[2]

[1] *Norwegian Polar Institute, Tromsø, Norway*
[2] *University of Hamburg, Integrated Climate Data Centre – ICDC, Hamburg, Germany*

9.1 Introduction

Sea ice floats on the polar and sub-polar oceans. It covers on average between 17 million and 27 million square kilometres of the Earth (Parkinson, 2014). Such a vast area can only be monitored using satellite remote sensing. Every acquisition of information without physical contact can be called remote sensing, e.g. taking a photo. Hence, satellite remote sensing in the visible part of the electromagnetic (EM) spectrum can be compared with taking a photo from space. This chapter considers remote sensing over a wider range of the EM spectrum and will focus on the technical aspects of satellite sea ice remote sensing. The results are shown and interpreted in several other chapters in this book.

The Intergovernmental Panel on Climate Change (IPCC) Fifth Assessment Report (IPCC, 2013) gives an impression of how important remote sensing is for current sea ice research. In the executive summary of the chapter on 'Cryosphere', five important findings regarding changes in sea ice are highlighted. Four out of these five findings are based on satellite observations, namely since 1979: (1) Arctic sea ice extent decreased; (2) Arctic perennial and multi-year ice decreased; (3) the period of surface melt on Arctic perennial ice lengthened; and (4) Antarctic sea ice extent increased, however, with strong regional differences. Only the fifth point, the decrease of Arctic sea ice thickness, is based on a combination of submarine, airborne and satellite observations due to the short length of the satellite ice thickness record. Without satellites we would know much less and with lower confidence about sea ice and the climate in the polar regions.

9.1.1 Basic concepts and principles

Satellite sensors can obtain a vast number of observations compared with ground-based measurements. A single observation or image does not tell too much about the spatiotemporal development of the sea ice. One needs a series of images ideally of the same area at different times of the seasonal cycle of the sea ice. This is realized by using polar-orbiting satellites, which complete one orbit around the Earth in typically around 100 minutes. If the satellite sensor offers a swath width (Figure 9.1) of more than 1500 km, the complete global sea ice cover can be observed on a daily basis leaving only small, unobserved holes at the poles.

During polar night, taking a photo is difficult because the scene needs to be illuminated by sunlight. Also clouds may obscure the scene, rendering observations taken in the visible portion of the EM spectrum useless for sea ice research. Sensor technology has advanced very much from just taking a single photo, however. Satellite remote sensing of sea ice is carried out in the visible, infrared and microwave portions of the EM spectrum (Figure 9.2). This permits remote sensing throughout the polar night and mitigation of the influence of clouds and other atmospheric constituents.

Depending on the frequency range, one obtains a picture of the surface brightness which depends on solar illumination for the visible portion of the EM spectrum (Section 9.5). The measured intensity in the infrared and microwave frequency range depends on the capabilities of the sea ice to emit EM radiation as a function of its physical temperature and emissivity, which is determined by geophysical quantities such as salinity,

*Present address: Institute of Environmental Physics, University of Bremen, Otto-Hahn-Allee 1, D-28359 Bremen, Germany.

Sea Ice, Third Edition. Edited by David N. Thomas.
© 2017 John Wiley & Sons, Ltd. Published 2017 by John Wiley & Sons, Ltd.

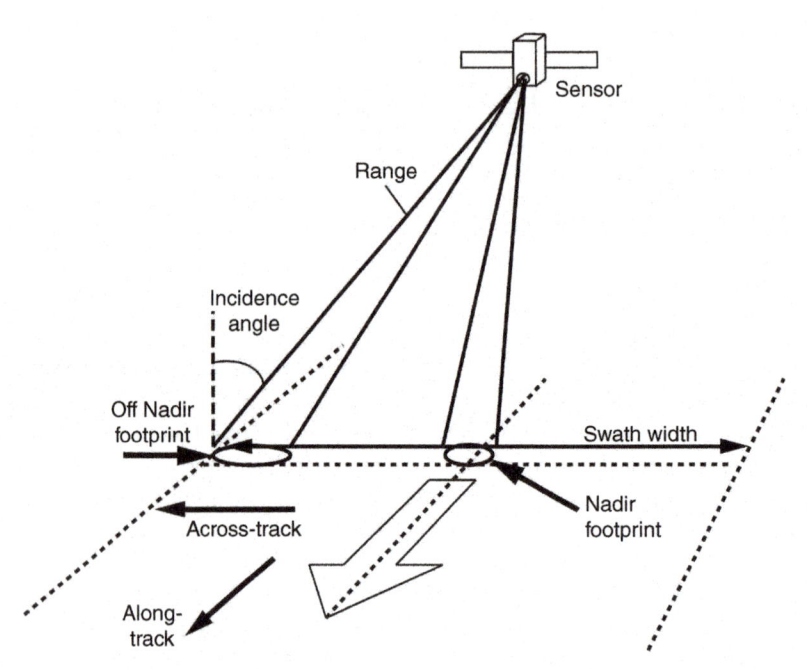

Figure 9.1 Schematic of a typical satellite viewing geometry and important definitions. Every sensor, however, has its specific geometry. For example, altimeters do not scan and only observe at nadir; while radiometers such as the special sensor microwave imager (SSM/I) scan elliptically (not along a line) under a constant incidence angle, have a constant footprint size but no nadir observations.

roughness, porosity and air content (Chapter 1). Such sensors are called (spectro-) radiometers (Section 9.2).

Other sensors actively emit EM radiation and measure the amount of this radiation scattered back to the sensor from the sea ice. This backscatter can be translated into an image of the backscattering properties of the sea ice. These properties depend on the internal and surface structure of the sea ice (see Chapter 1). Such sensors operate in the microwave frequency range and are called scatterometers (Section 9.2) and synthetic aperture radar (SAR; Section 9.3). Other sensors measure the runtime of an emitted EM pulse between the sensor and the surface. The runtime can be translated into the distance between the satellite sensor and the surface. Such sensors are called altimeters (Section 9.4).

Figure 9.1 shows a schematic of the typical geometry of how a satellite sensor sees the Earth. The size of the sensor footprint, i.e. the area illuminated or sensed by the sensor at the surface, determines the spatial resolution of the observations. Depending on the type of sensor, footprint sizes vary between a few metres (visible and SAR sensors) to tens of kilometres (radiometers). The swath width determines how much of the Earth's surface is observed during one satellite overflight, i.e. one orbit. The swath width varies between point measurements of a few tens of metres (laser altimeter) and about 2500 km (some optical sensors and radiometers). Depending on the swath width, it can take between 1 day and several months until the polar regions are covered by observations from a particular satellite sensor (compare 1 day of ICESat observations in Figure 9.4f with the 1 day coverage of other sensors in Figure 9.4a–d).

9.1.2 Short overview of important missions for sea ice remote sensing

The first images of sea ice from satellites reach as far back as 1964 (Meier et al., 2013). A milestone for sea ice remote sensing, however, was the launch of the Nimbus-7 satellite carrying the scanning multichannel microwave radiometer (SMMR) in October 1978. Measurements from this sensor together with its successors, the special sensor microwave imager (SSM/I) and the special sensor microwave imager/sounder (SSMIS) now form the basis of the more than 35-year-long time series of daily, global sea ice observations. The primary sea ice parameter obtained from these sensors is the sea ice area (Section 9.2.2), but other applications exist

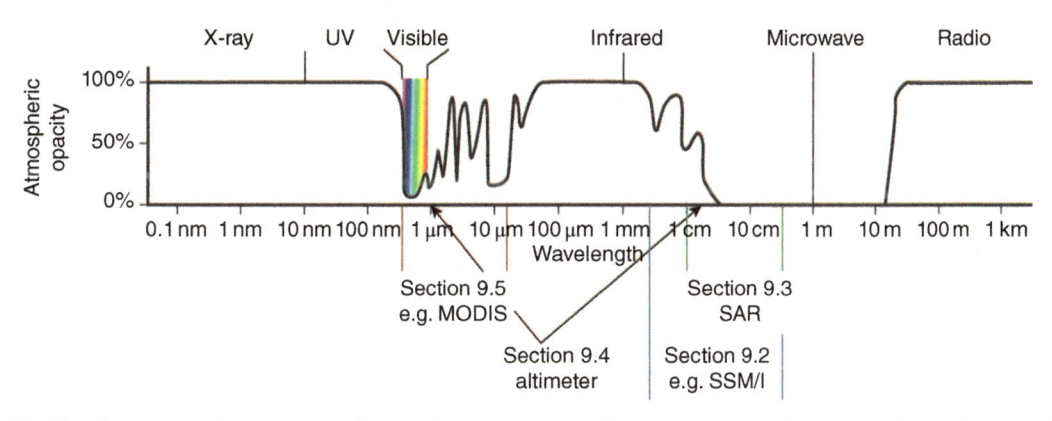

Figure 9.2 The electromagnetic spectrum and atmospheric opacity. Satellite remote sensing focuses on the window wavelength marked at the bottom with the sections in which the respective sensors are discussed.

(Sections 9.2.3–9.2.6). This sea ice area time series is one of the longest satellite records existing today and, as mentioned before, enabled us to monitor the unprecedented change in Arctic sea ice cover during recent decades. With a gap in observations between 1977 and 1978 and lower spatial and temporal resolution, this time series can even be extended back until December 1972, when the electrically scanning microwave radiometer (ESMR) on Nimbus-5 was launched.

Another milestone in sea ice remote sensing was reached when the first map of the Arctic Ocean sea ice thickness distribution from satellite altimetry was obtained (Laxon et al., 2003). Before this, only the spatial dimension of different sea ice variables could be obtained. The utilization of altimeters on the ERS1/2, Envisat, ICESat, and CryoSat-2 satellites for ice thickness retrieval added the third dimension to sea ice observations from space.

There are many more achievements in both satellite sensor technology and algorithm development, which cannot all be listed here. Some of the most important aspects will be discussed throughout the course of this chapter.

9.1.3 The EM spectrum

The most basic concept of satellite remote sensing is to measure EM radiation with a sensor in space (Campbell & Wynne, 2011). Figure 9.2 shows the EM spectrum and the wavelength ranges relevant for remote sensing of sea ice (including the sections they are discussed in). Sea ice remote sensing concentrates on the atmospheric windows in the visible, infrared and microwave portion of the EM spectrum. Only in these wavelength bands is the atmosphere (partly) transparent for EM radiation (Figure 9.2). There are two fundamentally different principles of remote sensing:

- Active sensors send out an EM signal themselves and record the reflected or backscattered radiation from the Earth's surface. Examples are the QuikSCAT scatterometer (Section 9.2) and all SAR sensors (Section 9.3). These measure the backscattered microwave radiation from the sea ice under typical incident angles between 20° and 55°. Other examples are the laser and radar altimeters (Section 9.4). These measure the EM signal runtime and thereby the distance between the sensor and the Earth's surface.

- The second group of sensors observes the Earth in a passive way without sending out a signal themselves. Here again two different principles can be discriminated: either the sensor measures the EM radiation from the Sun after it was reflected by the Earth's surface (this is the case for observations within the visible and near-infrared bandwidth and only possible in daylight; Section 9.5) or the sensor measures the EM radiation emitted by the Earth's surface itself. This is the case for the thermal-infrared and microwave wavelengths (Sections 9.2 and 9.5).

Figure 9.3 shows how these different measurement principles can be explained. Following Planck's law for black body radiators, the Earth is not emitting any radiation in the visible and near-infrared EM spectrum and therefore only reflected radiation from the Sun can be observed.

The Earth shows its maximum EM emission in the thermal-infrared spectrum, and for cloud-free conditions the physical Earth surface temperature can

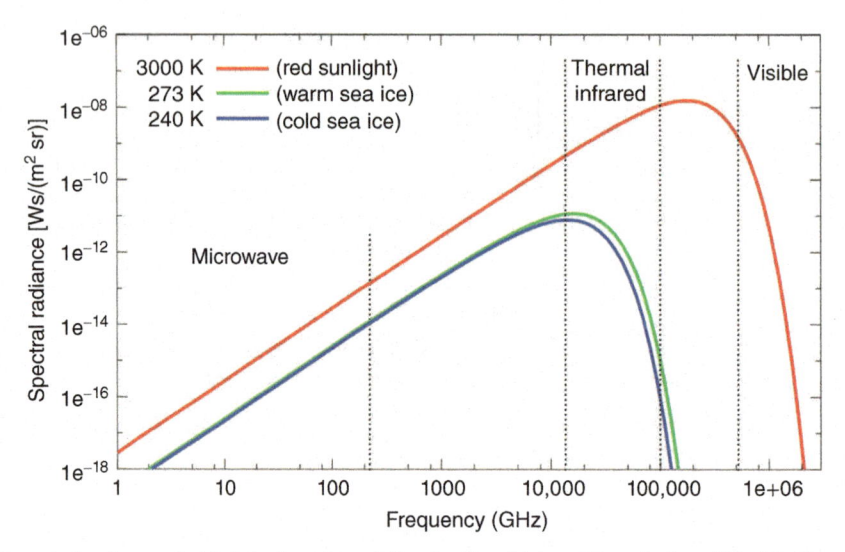

Figure 9.3 Emitted spectral radiance of a black body radiator following Planck's law. The radiation of black bodies with 240, 273 and 3000 K temperature is shown representing cold and warm sea ice surfaces, and the red part of the Sun's spectrum.

be measured in this EM range. The microwave EM range is well suited for spaceborne remote sensing, as microwave radiation can penetrate clouds. The emitted Earth radiation in the microwave range, however, is more than three orders of magnitudes lower than in the thermal infrared range (mind the exponential scale in Figure 9.3). For the radiation emitted by sea ice at microwave frequencies, the idealized black body radiator is not a good approximation. For the microwave brightness temperature, $T_B = \varepsilon\, T$, measured by a radiometer, the emissivity ε is not close to 1 (as it would be for a black body) but highly variable. Therefore radiometers mainly measure the emissivity ε in the microwave range and only to a second degree the physical temperature T.

9.1.4 Scattering mechanisms

How much energy is recorded by a satellite sensor is strongly determined by scattering of the EM waves in the atmosphere and on the ground. In the atmosphere aerosols, water vapour, and hydrometeors like rain, snow and cloud liquid water are the main sources of scatter. Depending on the ratio between EM wavelength λ and the radius r of these scattering particles, different scattering mechanisms have to be discriminated, e.g. Rayleigh scattering for $r \ll \lambda$ or Mie scattering for $r \approx \lambda/2\pi$. In addition to the size, the shape and dielectric properties of the particles also influence the scattering. Atmospheric scattering can be complex and is not discussed further here.

For EM scattering on the ground, two main processes can be discriminated: surface and volume scattering. Besides the dielectric properties of the surface material, the surface roughness is a key factor for the EM backscatter. A very flat or specular surface will only reflect EM radiation back to the sensor if the sensor's line of sight and the surface are perpendicular, as for a mirror. This is only the case for flat surfaces and nadir-looking sensors, or if the surface is tilted towards the sensor as is the case with mountains or, in our case, sea ice pressure ridges and water surface waves. Otherwise, for specular surfaces, all radiation will be reflected away from the sensor and the surface will have a very low backscatter value. Low backscatter is often displayed black or with dark colours in a satellite image. A rough surface will scatter a fraction of the incoming EM radiation back towards the sensor. In the case of a diffusely reflecting Lambertian surface, the backscatter value becomes independent of the incidence angle. If the surface exhibits a regular structure such as waves of a certain wavelength, it can cause positive interference of the incoming and backscattered EM radiation, the so-called Bragg scattering, which causes high backscatter values. For microwave frequencies, ripple waves on the water are of the order of the EM wavelength and can cause high Bragg backscatter for scatterometers and SAR sensors (Section 9.3; Jackson & Apel, 2004).

Electromagnetic waves penetrating into a medium can encounter volume scattering which increases the backscattered radiation. The penetration depth again depends on the dielectric properties and porosity of the medium. In the visible EM spectrum, the penetration depth is low for sea ice and snow and much higher for water. Hence surface scattering dominates for sea ice and snow. For water, as long as it is clear and scattering bodies like sediments or frazil ice are absent, both surface and volume scattering are low and it appears dark. In the microwave spectrum, the penetration depth into seawater is of the order of millimetres at most. Melting conditions with wet snow and melt ponds mask all retrievable information of the sea ice below. For sea ice the microwave penetration depth and volume scattering are very much related to the age of the ice. Young and first-year ice have a higher salinity and are less porous than the older multi-year ice. During the ice growth and especially during the summer melt, much of the brine is removed from the ice, leaving behind fresher ice with empty brine pockets and channels (Chapter 1). These pockets and channels cause a much higher backscatter value for multi-year than for first-year ice, as can be seen in Figure 9.4(e).

So far we have discussed the influence of scattering on satellite measurements. Most of these principles are very similar for microwave emission [compare Figure 9.4c (emission) and 9.4e (backscatter)].

Many more details about scattering mechanisms, microwave emission, and sea ice remote sensing in general can be found in Ulaby et al. (1982, 1986) and Lubin and Massom (2006).

9.2 Microwave radiometry and scatterometry

9.2.1 Theory/principles of measurements

Any material emits EM radiation as a function of its physical temperature and emissivity. The emissivity ε is a function of the dielectric properties of the material. Sea ice can be considered a mixture of ice crystals, brine solution and air. The sea ice dielectric properties are mainly determined by its salinity, temperature and porosity (Chapter 1). Snow, when dry, can be regarded as a mixture of ice crystals and air. Wet snow has liquid water as a third component (Chapters 1 and 3).

Microwave radiation penetrates snow and sea ice. Penetration depth is closely linked to ε and depends, like ε, on frequency and polarization. Radiation at, for example, 89 GHz, penetrates less into sea ice than radiation at, say, 19 GHz. EM radiation is reflected and refracted at strong dielectric property gradients such as the water–ice interface at the ice underside, the ice–air interface and the ice–snow interface. Within the ice and snow itself, the radiation is either attenuated as a function of salinity and liquid water content, or it is scattered at air inclusions, which each act as an ice–air interface. The higher the salinity and/or liquid water content in the sea ice and/or snow, the more radiation is attenuated. As a consequence, for example, the penetration depth is small and ε high for thin ice and first-year ice. The larger the concentration of air inclusions, the more radiation is scattered and the lower is ε, as is the case for multi-year ice. In addition to its high number of air inclusions, multi-year ice also has a low salinity, a prerequisite for enhanced scattering of microwave radiation within the ice.

Optimal for sea ice microwave remote sensing in terms of a minimum atmospheric influence are frequencies < 10 GHz. Sensor antenna aperture and antenna design for a satellite sensor limit the spatial resolution at such low frequencies to several 10 km. At higher frequencies, e.g. 89 GHz, the spatial resolution is only a few kilometres. This is illustrated in Figure 9.4, middle row (data: http://nsidc.org/data/amsre/). Advanced Microwave Scanning Radiometer – Earth (AMSR-E) brightness temperatures reveal many more details at 89 GHz (Figure 9.4c) than at 19 GHz (Figure 9.4d) – e.g. in the Fram Strait area. Areas of elevated brightness temperatures over open water (Figure 9.4; compare 89 with 19 GHz, Barents Sea) could be misinterpreted as sea ice but are actually caused by clouds and other atmospheric influences. Therefore satellite microwave remote sensing of sea ice is a trade-off between bearable atmospheric influence and a spatial resolution that is as fine as possible.

9.2.2 Sea ice area

Sea ice area and extent are computed from sea ice concentration (C): the sea ice area fraction of a known area, e.g. a grid cell. Brightness temperature (T_B) differences between open water and sea ice are utilized to compute

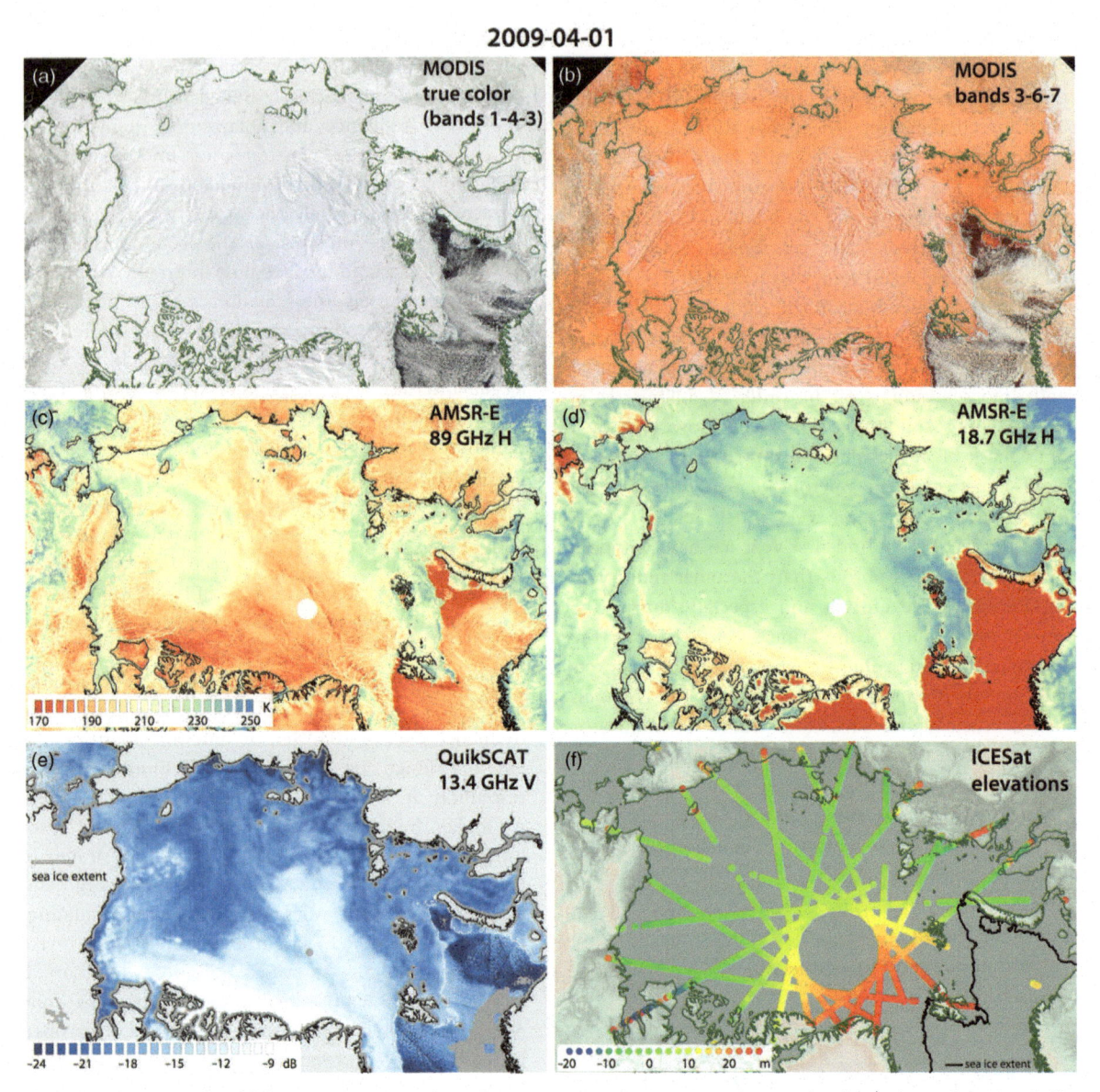

Figure 9.4 Example of how different satellite sensors see the Arctic sea ice cover on 1 April 2009. This figure shows the 'raw' satellite measurements, Figures 9.6 and 9.11 show derived geophysical sea ice quantities from these measurements. (a) MODIS visible bands (645, 555, 470 nm); (b) combined MODIS blue (470 nm) and short-wavelength infrared (SWIR) bands (1.6, 2.1 µm); (c, d) Advanced Microwave Scanning Radiometer – Earth (AMSR-E) microwave brightness temperatures at 89 GHz H (horizontally polarized) (c) and 19 GHz H (d); (e) backscatter from the QuikSCAT scatterometer at 13 GHz (K_u band) V (vertically polarized); (f) elevations (in reference to the Earth ellipsoid) from the ICESat GLAS laser altimeter. The ICESat footprint size is about 70 m and is increased here for better visibility by factor 1000 to 70 km. The white and grey discs located at the pole in images (c)–(f) denote the area of data missing due to the satellites' orbit inclination.

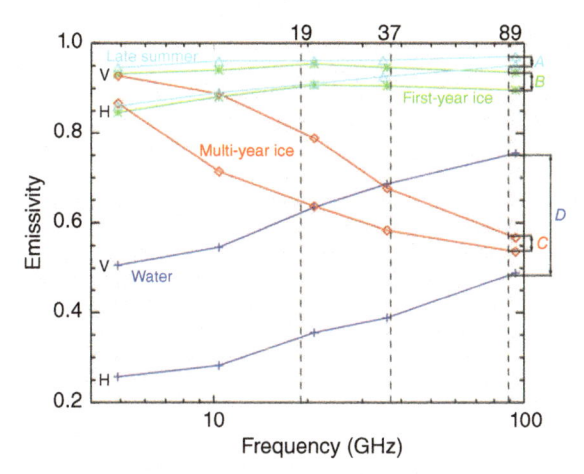

Figure 9.5 Observations of vertically (V) and horizontally (H) polarized emissivities ε of sea ice and sea water at an incident angle of 50° from two field campaigns. For frequencies typically used for sea ice concentration retrieval (19, 37, 89 GHz) polarization difference for water, D, is larger than for all sea ice types (A, B, C).

C. Usually the T_B of open water is smaller than the T_B of sea ice. Wind-induced roughening of the sea surface can increase open water emissivity ε, and hence the T_B, to the level of the sea ice T_B. Brightness temperatures measured over sea ice depend on its emissivity and physical temperature, but the T_B variability is dominated by the emissivity. Therefore T_B varies with polarization and frequency, as does ε. Figure 9.5 shows surface ε for different frequencies demonstrating that the polarization difference $P = T_{BV} - T_{BH}$ (V for vertical polarization, H for horizontal) can be used to discriminate water and sea ice. To remove the dependence on the physical temperature, the polarization ratio $PR = \frac{T_{BV} - T_{BH}}{T_{BV} + T_{BH}}$ is frequently used in sea ice algorithms.

Ivanova et al. (2014) and Andersen et al. (2006) give overviews of sea ice retrieval algorithms based on satellite microwave radiometry. Usually these algorithms combine T_B at two different polarizations (group I) or from two different frequencies (group II), or both, group III. The choice of the correct T_B combination is determined by a low sensitivity to atmospheric influence or to unwanted variations in ice and snow physical properties, a fine spatial resolution and a high accuracy.

Group I includes, for example, the COMISO Bootstrap algorithm, polarization mode (Comiso, 1986), based on 37 GHz T_B at vertical and horizontal polarization, and the ARTIST Sea Ice (ASI) algorithm (Kaleschke et al., 2001; Spreen et al., 2008), based on the near 90 GHz polarization difference. An example of ASI AMSR-E sea ice concentration is shown in Figure 9.6(b) (data: http://seaice.uni-bremen.de/amsr/). Group II includes, for example, the COMISO Bootstrap algorithm, frequency mode (Comiso, 1986), based on 19 and 37 GHz T_B, vertical polarization. Group III includes, for example, the NASA-Team (NT) algorithm (Cavalieri et al., 1984) or enhanced NASA-Team (NT2) algorithm (Markus & Cavalieri, 2000), based on T_B at 19 GHz, both polarizations, and 37 GHz, vertical polarization, plus the NT2 at near 90 GHz, both polarizations. The Bristol algorithm (Smith, 1996) also belongs to this group. A fourth group comprises hybrid approaches that combine two algorithms of the above-mentioned groups. The Eumetsat OSI-SAF algorithm is one such approach, combining the COMISO bootstrap algorithm, frequency mode, with the Bristol algorithm (Eastwood et al., 2012).

Sea ice extent can be derived also from radar backscatter from scatterometers like QuikSCAT (e.g. Remund & Long, 1999). The backscatter difference between open water and sea ice allows their discrimination and mapping of the ice-covered area. Sea ice extent based on scatterometry can be more accurate than the one obtained with microwave radiometry – particularly for an ice edge dominated by frazil and grease ice (Haarpaintner et al., 2004; Rivas et al., 2012) or melting, wet ice (e.g. Ozsoy-Cicek et al., 2011).

9.2.3 Sea ice motion

In order to derive sea ice motion, sea ice is tracked in a pair of consecutive overlapping satellite images. This section considers sea ice motion observed with coarse-resolution [i.e. O (10 km) or coarser] satellite imagery such as microwave radiometry or scatterometry. Small-scale sea ice motion derived from fine-resolution satellite imagery is described later.

Sea ice displacement can be observed with any kind of imaging satellite sensor as long as the sensor resolves surface features, which can be tracked over time. Features can be ice floe boundaries, deformation features such as ridges, different ice types and different surface properties, as, for example, determined by the snow cover.

The most common method to track surface features in consecutive images from scatterometry or microwave T_B is the maximum cross-correlation (MCC) method. For

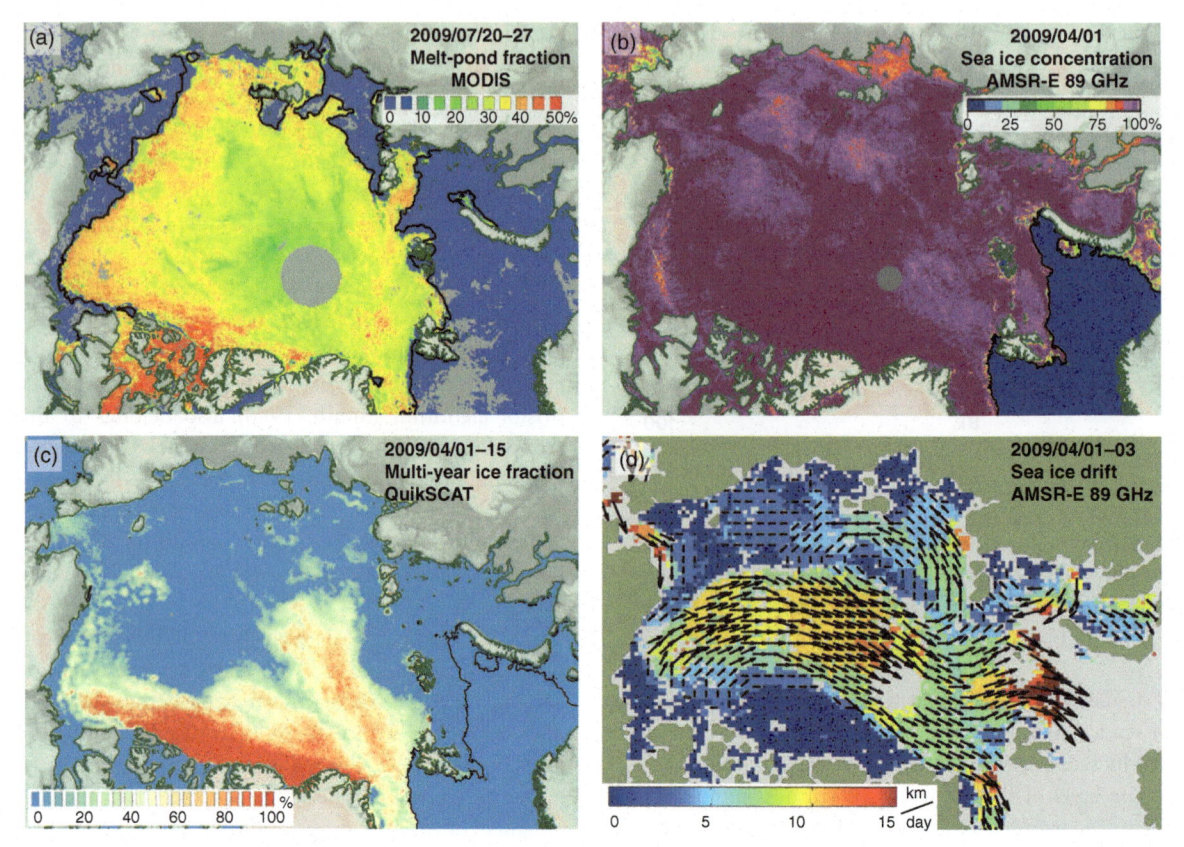

Figure 9.6 Example of geophysical sea ice quantities derived from satellite observations (see also the correspondent raw satellite observations in Figure 9.4). (a) Melt-pond fraction from an 8-day MODIS composite for 20–27 July 2009 (grid resolution 12.5 km); (b) sea ice concentration from the Advanced Microwave Scanning Radiometer – Earth (AMSR-E) 89 GHz channels on 1 April 2009 (grid resolution 6.25 km); (c) multi-year sea ice fraction from a 15-day QuikSCAT composite for 1–15 April 2009 (grid resolution 12.5 km); (d) sea ice drift from AMSR-E 89 GHz observations between 1 and 3 April 2009 (grid resolution 31.25 km; vectors show drift direction for every third grid cell with the vector length proportional to the drift speed).

this, a box of image pixels, e.g. 5 × 5, is selected in the first satellite image. This box has a characteristic radar backscatter or T_B distribution. A new box of a similar size is moved within a certain search window around the position of the original box in the second image. For each location of the new box, the cross-correlation between the radar backscatter or T_B distribution at the current position in the second image and the original box in the first image is computed. The location providing the highest (maximum) cross-correlation gives the best match between the box in the first and second images, respectively. The distance between the centres of the two boxes is the magnitude of the displacement vector. The location of the centres relative to each other gives the direction of the displacement vector. Sea ice drift speed is computed from the magnitude and the time difference between the two satellite image acquisitions. Typical ice drift speeds are 5–10 km day^{-1}, as can be seen in the example in Figure 9.6(d) using AMSR-E 89 GHz data as input (Girard-Ardhuin & Ezraty, 2012; data: http://cersat.ifremer.fr).

The correlation needs to be above a certain experience-based threshold. Too low thresholds could include spurious displacement vectors; too high thresholds could discard too many vectors and hence create unwanted data gaps. The correlation threshold should depend on the season because surface features can change, e.g. by melting, as do the distributions of radar backscatter and T_B (e.g. Girard-Ardhuin & Ezraty, 2012). During melting conditions sea ice motion retrieval is generally less reliable, if not even impossible, but alternatives exist (Kwok, 2008).

Usage of optical and infrared sensors is limited to clear-sky daylight cases. Still such imagery has been used for sea ice motion retrieval, e.g. from AVHRR imagery (Ninnis et al., 1986; Emery et al., 1991). Scatterometry (ESCAT, QuikSCAT or ASCAT) and microwave radiometry (SSM/I and AMSR-E/2) form the backbone of large-scale sea ice motion retrieval. Several sea ice drift products have been derived (e.g. Girard-Ardhuin & Ezraty, 2012; Kwok et al., 1998; Lavergne et al., 2010). Sumata et al. (2014) compare different drift products.

The grid resolution of a sea ice motion product is determined by the satellite channels used; it is of the order of 50 km or even coarser because several pixels need to be combined in a box (see earlier). This hampers retrieval of slow ice drift at daily temporal resolution. Also the sea ice motion is quantized as a function of the grid used, as shown and mitigated by Lavergne et al. (2010) introducing the continuous MCC.

9.2.4 Sea ice type

Radiometric and radar backscattering properties differ between different ice types as a function of their salinity, temperature, density, surface roughness and snow cover properties. Different properties can cause similar changes in T_B or radar backscatter. A good example for this is the gradient ratio (GR) between 37 and 19 GHz T_B at vertical polarization:

$$GR_{37,19} = \frac{T_B(37V) - T_B(19V)}{T_B(37V) + T_B(19V)} \quad (9.1)$$

In the Arctic the GR is used to estimate the multi-year ice fraction, e.g. with the NT algorithm (Cavalieri et al., 1984). Thin ice and first-year ice have a GR close to zero; the GR of multi-year ice is negative. Comiso (2012) used the difference in polarization at 37 GHz to discriminate first- and multi-year ice. In the Antarctic, multi-year ice, as is typical for the Arctic, is seldom found. Melt–refreeze cycles, however, can change snow properties such that its microwave signature is similar to that of multi-year ice.

Multi-year ice or coarse-grained metamorphous snow can be identified by higher radar backscatter values (Figure 9.4e; data from: http://cersat.ifremer.fr) because of the larger volume scattering contribution compared with more saline and substantially less porous first-year ice. Kwok (2004) used an empirical relationship to infer the Arctic multi-year ice fraction from QuikSCAT backscatter, as shown in Figure 9.6(c), and Lindell and Long (2016) derived and inter-compared the Arctic multi-year ice fraction derived from QuikSCAT and OSCAT-2 data.

Elevated radar backscatter is also found in the marginal ice zone, e.g. in the Barents Sea (Figure 9.4e). This could be due to a high fraction of pancake ice, which has high surface roughness and more floe edges and hence higher radar backscatter than other thin ice types or first-year ice (Dierking, 2001). Pancake ice is a likely source for the higher radar backscatter observed in the Antarctic marginal ice zone (Lange & Eicken, 1981). Alternatively, the observed elevated radar backscatter could be caused by strongly deformed ice with many ridges.

9.2.5 Thin sea ice thickness

The retrieval of the sea ice thickness in general is described in Section 9.4. Here we concentrate on thin ice. Approaches utilizing active microwave data (i.e. SAR; see Section 9.3) assume that the surface roughness of young and thin sea ice tends to increase with thickness and that the vertical salinity profile changes with thickness. The ratio of vertically to horizontally polarized radar backscatter (co-polarization ratio), at L- or C-band frequencies can be used empirically to infer the thickness of thin sea ice (e.g. Kwok et al., 1995; Nakamura et al., 2009). The maximum thickness to be retrieved ranges from 0.1 m (Kwok et al., 1995) to over 1 m (Nakamura et al., 2009). Most of these empirical approaches have not yet been applied widely.

Approaches utilizing microwave radiometry take advantage of the change in salinity with ice thickness. The approach using, for example, SSM/I or AMSR-E data is empirical and based on the ratio of the vertically to horizontally polarized T_B at 37 GHz (e.g. Martin et al., 2004, 2005). Even though limited to ice thickness values < 0.2 m, it has been developed further and applied regionally (e.g. Nihashi et al., 2009) and hemispherically (Tamura et al., 2008; Tamura & Ohshima, 2011). A sophisticated evaluation of the results is, however, hampered by limited access to thin sea ice regions to carry out *in situ* measurements.

A physically (Kaleschke et al., 2010) and empirically (Huntemann et al., 2014) based approach utilizing the sensitivity of L band T_B to salinity and temperature

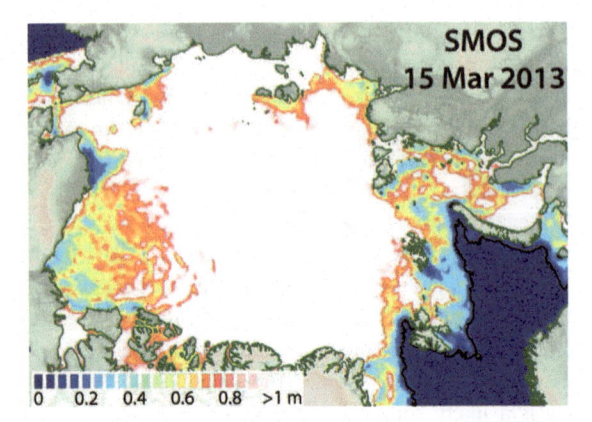

Figure 9.7 Thin sea ice thickness derived from the Soil Moisture and Ocean Salinity (SMOS) L-band radiometer. The thickness range is limited to 0–1 m and SMOS ice thickness is most accurate below about 0.5 m. The black line shows the sea ice extent for that day.

of sea ice has been developed for measurements by the Soil Moisture and Ocean Salinity (SMOS) sensor. These approaches enable the retrieval of ice thickness up to about 0.5 m independent of daylight and weather conditions for high ice concentration under freezing conditions; for ice of low salinity, as in the Baltic Sea, the maximum thickness to be retrieved is larger (Tian-Kunze et al., 2014). An example of SMOS thin sea ice thickness for 15 March 2013 is shown in Figure 9.7 (data: http://icdc.zmaw.de/l3c_smos_sit.html), which can be compared with the ice thickness from CryoSat-2 in Figure 9.11(b) (see later) for the same month (mind the different colour scales).

9.2.6 Snow thickness on sea ice

Snow depth S can be retrieved utilizing its empirical relationship to the gradient ratio (GR) between T_B at 37 and 19 GHz (see equation 9.11; Markus & Cavalieri, 1998). The retrieval is limited to $S < 50$ cm and to regions of first-year ice because the GR varies also with the multi-year ice fraction. The retrieval is influenced by grain size variations and snow wetness (Markus & Cavalieri, 1998) and underestimates snow depth for rough sea ice (e.g. Worby et al., 2008; Kern et al., 2011).

For thick sea ice, the insulating effect of a snow cover depends mainly on its depth and can be quantified by the change in ice surface temperature. Within a certain snow depth range, the increase in ice surface temperature as function of snow depth causes an increase in the brine volume fraction, which in turn changes the microwave emissivity at L-band frequencies. Maaß et al. (2013) exploited this to derive snow depth on thick sea ice from SMOS satellite microwave radiometry at L-band frequencies under cold conditions.

9.2.7 Melt onset

The seasonal evolution of melt and freezing conditions can be estimated using microwave radiometry and scatterometry. In the Arctic, the moisture increase in the snow during melt causes an increase in T_B and a decrease in radar backscatter compared with values before melt. For example, Markus et al. (2009) exploited this to map melt onset on Arctic sea ice. In the Antarctic, moisture increase in the snow is less pronounced. Instead, melt–refreeze cycles are very common and change snow grain size, which has an impact on T_B variability. Willmes et al. (2009) used this to map melt onset on Antarctic sea ice.

9.3 Synthetic aperture radar (SAR)

9.3.1 Theory/principles of measurements

One of the disadvantages of the radiometer and scatterometer measurements discussed in Section 9.2 is their low spatial resolution (Tables 9.1 and 9.2). The along-track (or azimuth) resolution ρ_a of an imaging radar depends on the range distance R (i.e. distance between the satellite and the observed object; see also Figure 9.1), EM wavelength λ, and length of the antenna l:

$$\rho_a = \frac{R\lambda}{l}$$

To achieve a resolution of $\rho_a = 10$ m for a C-band radar with wavelength $\lambda = 5$ cm and a range of 800 km (typical altitude of a satellite), one would need a 4-km-long antenna, which obviously is impractical. Here the advantage of SARs comes into play. A SAR uses the along-track movement of the satellite to 'simulate' a long antenna. Several measurements of the small, real antenna A_r on the satellite are used to construct the measurement of a synthetic, large antenna. Every object on the Earth's surface is covered several times by the large footprint of A_r, but under different incident angles. When an object enters the radar footprint in front of the nadir antenna position, its backscattered radar signal will be shifted to higher frequencies according

Methods of satellite remote sensing of sea ice 249

Table 9.1 List of commonly used sensors for radiometry in sea ice research (non-exclusive).

Name	Frequency (GHz)	Resolution (km)	Swath (km)	Incidence angle	Inclination	Operational	Polarization
ESMR	19.4	25–150	1280	0–50°	81° retrograde	1972–1976	H
SMMR	6.6, 10.7, 18.0, 21.0, 37.0	30–150	780	50°	99.15°	1978–1987	H and V
SSM/I; SSMIS	19.4, 22.2, 37.0, 85.5 (SSMIS: 91.7 + more)	15–50	1394 (SSMIS: 1707)	53°	98.9°	1987–2009 SSMIS: 2003–	H and V 22.2 GHz: V
AMSR;AMSR-E; AMSR2	6.9, 10.7, 18.7, 23.8, 36.5, 89.0 (AMSR/2: + more)	5–50	1450	55°	98.2°	2002–2011 AMSR2: 2012–	H and V
SMOS	1.4 (L band)	30–50	900	0–55°	98.44°	2009–	Full

ESMR, electrically scanning microwave radiometer; SMMR, scanning multichannel microwave radiometer; SSM/I, special sensor microwave imager; SSMIS, special sensor microwave imager/sounder; AMSR, Advanced Microwave Scanning Radiometer; SMOS, Soil Moisture and Ocean Salinity sensor; H, horizontal; V, vertical.

Table 9.2 List of commonly used sensors for scatterometry in sea ice research (non-exclusive).

Name	Frequency (GHz)	Resolution (km)	Swath (km)	Incidence angle	Inclination	Operational	Polarization
ESCAT (ERS-1 and 2)	5.3 (C-band)	50	500	18–59°	98.5°	1991–2011	V
QuikSCAT	13.4 (K_u-band)	25 x 6	1400–1800	46–54°	98.6°	1999–2009	H & V
Oceansat-2	13.5 (K_u-band)	50	1400–1840	49–58°	98.3°	2009–2014	H & V
ASCAT	5.3 (C-band)	50	500 (dual swath)	25–65°	98.7°	2006–	V

to the Doppler effect. If the same object is observed again when it leaves the footprint behind the antenna, its reflected signal will be Doppler-shifted to lower frequencies. Every object is observed several times within the footprint.

If the movement of the satellite in space is known with high accuracy, the synthetic aperture of a large antenna can be constructed using the different viewing geometries. The along-track resolution, ρ_{SAR}, of a SAR is, perhaps surprisingly, proportional to the real antenna length, l – i.e. the shorter the antenna, the higher the SAR resolution:

$$\rho_{SAR} = \frac{l}{2}$$

The resolution ρ_{SAR} is independent of wavelength λ and range distance R. The returned power, however, still depends on the emitted signal strength, which in turn also depends on the antenna size. To achieve a suitable signal-to-noise ratio, there is a lower limit to how small a SAR antenna can be. Typical SAR antenna lengths are between 5 and 15 m, offering a SAR resolution down to $\rho_{SAR} = 2$ m.

As for traditional radars, the across-track (slant range) resolution ρ_r depends on the pulse length τ of the SAR and the incidence angle θ:

$$\rho_r = \frac{c\tau}{2 \sin \theta}$$

where c is the speed of light. As for ρ_{SAR}, ρ_r is also independent of the λ and R. The radar bandwidth B, however, and therefore the pulse width $\tau = 1/B$ depend on λ.

These are only a few of the basic principles of SAR remote sensing. A more complete description can be found in Jackson and Apel (2004). An overview of the most commonly used SAR sensors for sea ice research is given in Table 9.3.

Figure 9.8 shows an example of the much higher resolution of an Envisat ASAR ScanSAR wide swath mode scene in comparison to AMSR-E radiometer observations at 89 GHz. The high resolution of 50–150 m for scenes of about 400–500 km in ScanSAR wide swath mode makes SAR data well suited for operational application. For national ice services (e.g. NIC USA, CIS

Table 9.3 List of commonly used synthetic aperture radar (SAR) sensors in sea ice research (non-exclusive).

Name	Frequency (GHz)	Resolution (m)	Swath (km)	Incidence angle	Inclination	Operational	Polarization
ERS-1 and 2	5.3 (C-band)	30	100	20–26°	98.5°	1991–2011	VV
Envisat ASAR	5.3 (C-band)	30–1000	100–400	15–45°	98.5°	2002–2012	Dual-polarimetric
Radarsat-1 & 2	5.3/5.4 (C-band)	3–100	18–500	10–60°	98.6°	1995–	HH (RS-1) full (RS-2)
Sentinel-1	5.4 (C-band)	5–100	80–400	19–47°	98.2°	2014–	Full
TerraSAR-X	9.6 (X-band)	1–40	5–200	15–60°	97.4°	2007–	Full
ALOS-1 & 2 PALSAR	1.3 (L-band)	3–100	20–350	8–70°	98.7° (AL-1) 97.9° (AL-2)	2006–2011 2014– (AL-2)	Full
COSMO-SkyMed	9.6 (X-band)	1–100	10–100	16–51°	97.9°	2007– Four satellites	Dual-polarimetric

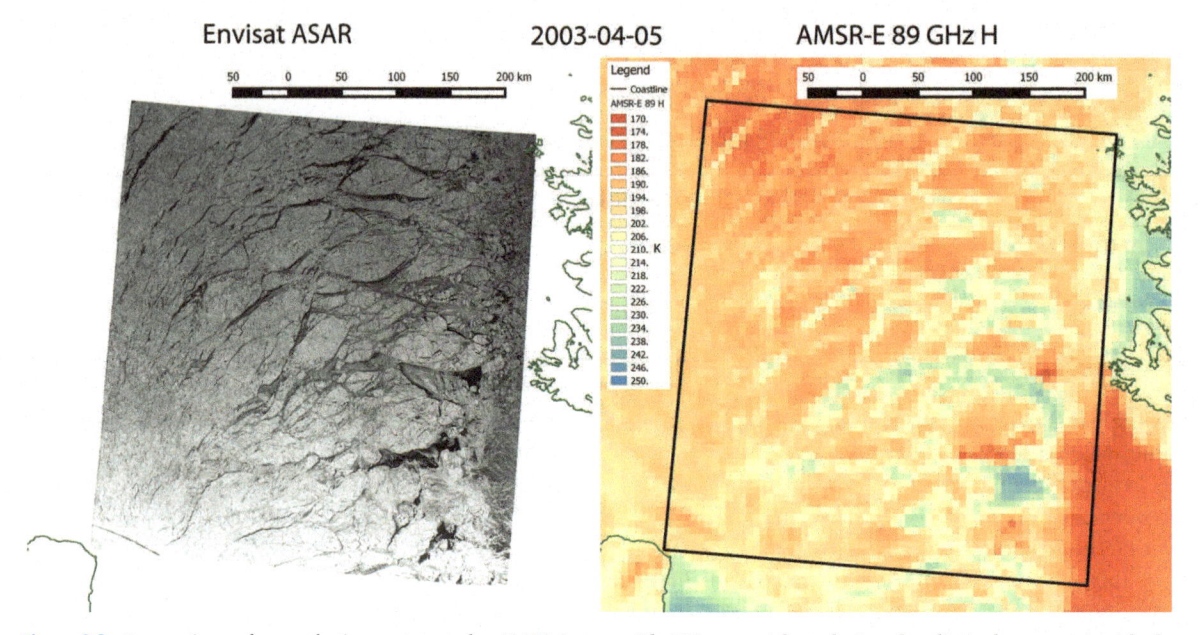

Figure 9.8 Comparison of a synthetic aperture radar (SAR) image with 150 m spatial resolution (but limited coverage) with the Advanced Microwave Scanning Radiometer – Earth (AMSR-E) radiometer observations at 89 GHz and about 5 km resolution on 5 April 2003. Similar features in the ice can be observed but with much different detail.

Canada, MET Norway, DMI Denmark), SAR data are the key information for deriving ice charts for their area of interest. Onstott & Shuchman (2004) and Dierking (2013) give a more extensive overview of SAR sea ice remote sensing.

9.3.2 Sea ice motion

If a consecutive series of SAR images is available for a given region, sea ice motion can be derived in a similar fashion to that described in Section 9.2.3 (e.g. Kwok et al., 1990; Komarov & Barber, 2014), but with a much higher spatial resolution. The SAR scene is divided in sub-regions. These are identified again in a subsequent scene using a cross-correlation method (MCC; see Section 9.2.3). The quality of the results depends on the time difference between the two SAR scenes, which should be as short as possible. The original cross-correlation method can be improved by several additions: a hierarchical approach where ice motion is first calculated with lower resolution patterns as a first guess before resolution is improved in several steps in a pyramid-like way. Correlations cannot only be calculated between patterns of the backscatter magnitudes in the space domain but also using a Fourier transform in the frequency domain (Komarov & Barber, 2014). If a pattern match is found its quality can be assessed calculating the back-trajectory from the second to the first scene.

The sea ice motion derived from SAR can have spatial resolutions finer than 10 km and, more importantly, uncertainties of the displacement vector are on the order of a few hundred metres (e.g. Hollands & Dierking, 2011). SAR sea ice motion is therefore well suited to observing small scale sea ice kinematic features caused by ice deformation. The Radarsat Geophysical Processor System (RGPS) (e.g. Kwok et al., 1990; Kwok, 2006) follows a grid of sea ice parcels over a complete season in a Lagrangian way. The initial grid distance is 10 km. The grid cells get deformed over time. Figure 9.9 shows the RGPS sea ice divergence, shear and vorticity for November 1999 (data: http://rkwok.jpl.nasa.gov/radarsat). Clearly linear kinematic features spanning several hundreds of kilometres over the Arctic Basin can be identified. Due to its high resolution and accuracy, SAR sea ice motion is ideally suited for such sea ice deformation, i.e. strain rate observations.

In the marginal ice zone where rotation of floes becomes more prevalent and the ice gets broken up, feature and object tracking approaches are commonly used and better suited than the cross-correlation method (e.g. Kwok et al., 1990).

9.3.3 Sea ice type

Sea ice types can be discriminated in SAR data in a similar manner to microwave radiometers and scatterometers (Section 9.2.4): multi-year ice shows a higher volume scattering than first-year ice (Section 9.1.4); deformed ice with ridges shows higher radar backscatter than level ice; and rough surfaces, e.g. pancake ice or frost flowers, or an increased number of floe edges for broken up ice increase the radar backscatter, etc. The much higher spatial resolution of SAR data, however, allows a much more detailed classification and a larger number of different surface types to be identified: in Figure 9.8 single floes and open and refrozen leads can be clearly identified in the ASAR data compared to the AMSR-E data. The high-resolution SAR modes (Table 9.3) allow identification of melt ponds or ridges. In single polarization SAR images, many ambiguities can exist between ice types and even wind roughened water and ice. Still ice type segmentation can be successfully applied and can even be used in an operational context (Ochilov & Clausi, 2012). The dual- and full-polarimetric modes of today's SAR sensors allow an even better segmentation of surface types (e.g. Moen et al., 2013), e.g. for ice versus water separation or identifying multi-year ice floes embedded in first-year ice. Different SAR frequencies have different strengths: deformed ice can be better discriminated at L-band than at C- and X-band frequencies, while the higher frequencies are better for multi-year versus first-year ice discrimination (e.g. Dierking, 2013).

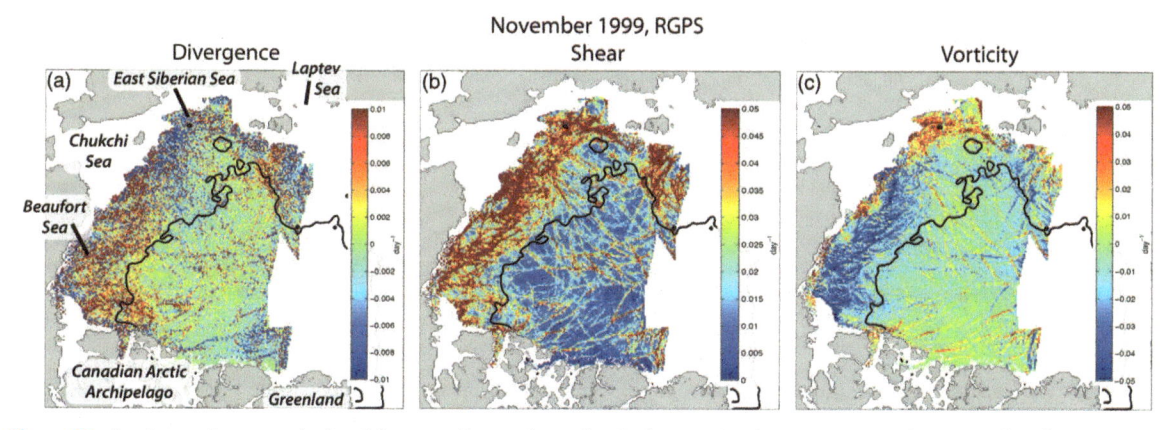

Figure 9.9 Sea ice strain rates calculated from sea ice motion using Radarsat-1 synthetic aperture radar (SAR) data for November 1999: (a) divergence; (b) shear; and (c) vorticity. The black line shows the multi-year sea ice extent from QuikSCAT data.

9.4 Altimetry

9.4.1 Theory/principles of measurements

Both spaceborne laser and radar altimeters can be used to retrieve the sea ice thickness I. The sea ice thickness is not measured directly but is inferred from the sea ice freeboard F, the part of the sea ice sticking out of the ocean. Radar altimeter data of sea ice have been available since the launch of ERS-1 in 1991 with the pulse-limited radar altimeter on board, covering the sea ice up to 82° latitude. ERS-1 was followed by ERS-2 and Envisat in similar orbits. The first spaceborne laser altimeter and the first altimeter covering the Arctic sea ice up to 86°N was the GLAS instrument on ICESat, operational from 2003 to 2009. The first radar altimeter using synthetic aperture processing and covering the area up to 88°N was CryoSat-2 launched in 2010.

Both laser and radar altimeters send out an EM pulse in nadir direction that is reflected back to the satellite by the Earth's surface. The amplitude of the returned power of the pulse is recorded within a short receiving time window. These digitized pulses are referred to as laser or radar waveforms and have a Gaussian, peaky shape for flat surfaces, such as leads, and a wider shape with a long tail for rough, scattering surfaces, such as sea ice. The location T_r of the first return signal related to the surface at nadir position, i.e. closest to the satellite, is now re-tracked from the waveform. Different waveform re-trackers for different surfaces and application exist (e.g. Zwally et al., 2002; Kurtz et al., 2014; Ricker et al., 2014). Typically, the waveform is re-tracked on the leading edge for sea ice surfaces and on the maximum peak for the smooth, Gaussian ocean or thin sea ice surfaces in leads (Laxon et al., 2013). From the travelling time Δt for the identified return point T_r and the speed of light c the distance $D = c\,\Delta t/2$ between the altimeter and the Earth's surface is calculated (see Figure 9.10). To get from the measured distance D to a surface elevation measurement E, the exact position of the altimeter, and thus the position of the satellite in space, has to be known with centimetre accuracy. The position of the satellite, h_{ellip}, is measured in reference to an Earth ellipsoid model. The surface elevation E is calculated as $E = h_{ellip} - D$ (an example for ICESat is shown in Figure 9.4f).

The shape of the recorded laser and radar waveforms also contains information about surface roughness and type (e.g. Zygmuntowska et al., 2013). The most important application of altimetry in sea ice research, however, is the retrieval of sea ice freeboard and thickness as, will be discussed in the next subsection.

9.4.2 Sea ice thickness

A prerequisite to obtaining the sea ice thickness I from altimeter measurements is the sea ice freeboard height F (an example for laser altimeters is shown in Figure 9.10). A major challenge in determining the freeboard F from altimetry is the identification of leads in between the ice floes. Leads with open water or covered by thin sea ice serve as reference points for the sea surface height (SSH). The difference between the SSH and the elevations E measured over the sea ice is the wanted

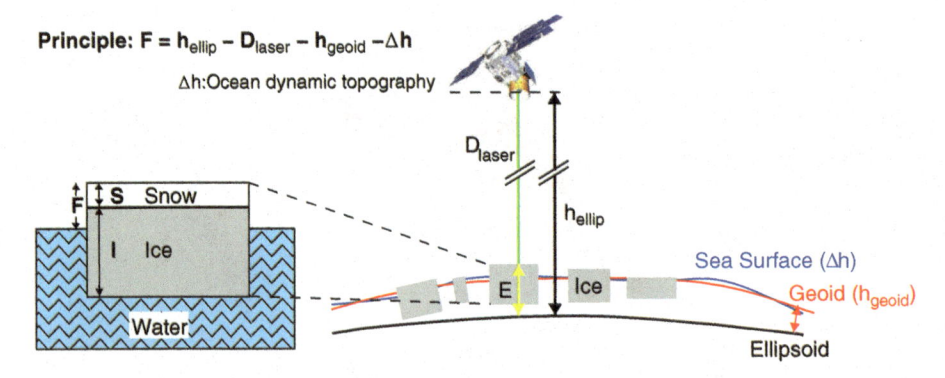

Figure 9.10 Schematic showing the principles of sea ice freeboard and thickness retrieval from satellite altimetry using the ICESat laser altimeter as an example. Left side: interrelation of sea ice freeboard, F, snow depth, S, and sea ice thickness, I. Right side: an artist's view of ICESat above the three involved surfaces: reference ellipsoid (black), geoid (red), and sea surface (blue). Figures are not to scale.

freeboard F:

$$F = E - h_{\text{geoid}} - \text{SSH}$$

where h_{geoid} is the Earth geoid height. The SSH varies in respect to the geoid according to the tides, ocean currents and atmospheric pressure. The SSH is not known with a high enough accuracy from other sources but has to be determined from the altimeter measurements themselves. Leads and thereby the SSH can be identified by the shape of the laser waveforms or the comparable lower elevation of the leads compared with the surrounding sea ice. The development of methods to determine the SSH in sea ice-covered regions is still an active field of research. One challenge for the SSH determination is the large footprint size, especially of radar altimeters. While ICESat's laser footprint has a diameter of about 70 m, CryoSat-2;s radar footprint is 380 m × 1650 m, and the footprint of all former radar altimeters is several km large (Table 9.4). In most cases leads are much smaller than that and have to be detected at sub-footprint size for accurate SSH retrieval. The specular reflection caused by leads, however, can be identified in the retrieved radar waveform even if the lead only covers a fraction of the footprint.

The freeboard heights can be converted to ice thickness under the assumption of isostatic balance of the ice floes in the water and knowledge of all involved densities ρ_x and snow depth S. According to Archimedes' principle, a body submerged in a liquid such as water will displace the volume of water equivalent to its own weight. If now the part of the body, in our case sea ice (with snow), sticking out of the water (the freeboard) is known, the mean sea ice thickness of the floe can be determined.

The signal of a laser altimeter such as the one on board ICESat is reflected at the snow surface on top of the sea ice. This example is shown in Figure 9.10 and the conversion between the snow freeboard, F_S, and ice thickness, I, can be described as:

$$I = F_S \frac{\rho_W}{\rho_W - \rho_I} + S \frac{\rho_S - \rho_W}{\rho_W - \rho_I} \quad \text{(for laser altimeter)} \quad (9.2)$$

Radar signals penetrate the snow either completely or partially. The signal of a radar altimeter therefore originates from either the sea ice surface or from somewhere in the snow pack. The exact origin of the radar altimeter return signal is often unknown (e.g. Willatt et al., 2010; Ricker et al., 2015). We will refer to the freeboard height measured by a radar altimeter as radar freeboard F_R that can be converted to ice thickness I by:

$$I = F_R \frac{\rho_W}{\rho_W - \rho_I} + S \frac{\rho_S}{\rho_W - \rho_I} \quad \text{(for radar altimeter)} \quad (9.3)$$

where S is the snow thickness above the radar return origin. From equations (9.2) and (9.3) it becomes obvious that accurate knowledge of snow depth S and the three densities ρ_x is needed and that the uncertainty of these variables will have a direct feedback on the accuracy of the retrieved ice thickness. Besides the uncertainty of the freeboard retrieval itself uncertainties in snow depth and ice density were identified as the most significant sources for the accuracy of the derived sea ice thickness (e.g. Giles et al., 2007; Kwok & Cunningham 2008; Kern & Spreen 2015; Ricker et al., 2015). For representative density estimates extended and repeated field in-situ measurements are of utmost importance. The same holds for a snow depth climatology. Snow depth, however, can also be estimated from airborne snow radar campaigns (e.g. Operation IceBridge; Kurtz et al., 2013) or satellite radiometers (see Section 9.2.6) but often with significant uncertainties or limited spatial coverage.

Table 9.4 List of radar and laser altimeters used for sea ice thickness retrieval.

Name	Frequency	Footprint	Inclination	Operational
ERS 1 and 2 radar altimeters	13.8 GHz (K_u-band)	16–20 km	98.5°	1991–2011
Envisat RA-2	13.575 GHz (K_u-band)	2–10 km	98.5°	2002–2012
CryoSat-2	13.575 GHz (K_u-band)	1650 m x 380 m	92°	2010–
SARAL/AltiKa	35.75 GHz (K_a-band)	8 km	98.5°	2013–
Sentinel-3 SRAL	13.575 (K_u-band)	2 km x 300 m	98.65°	2016–
ICESat	1064 nm (laser)	64 m	94°	2003–2009
ICESat-2	532 (laser)	10 m; 3 pairs of beams 3 km apart	94°	Launch 2017

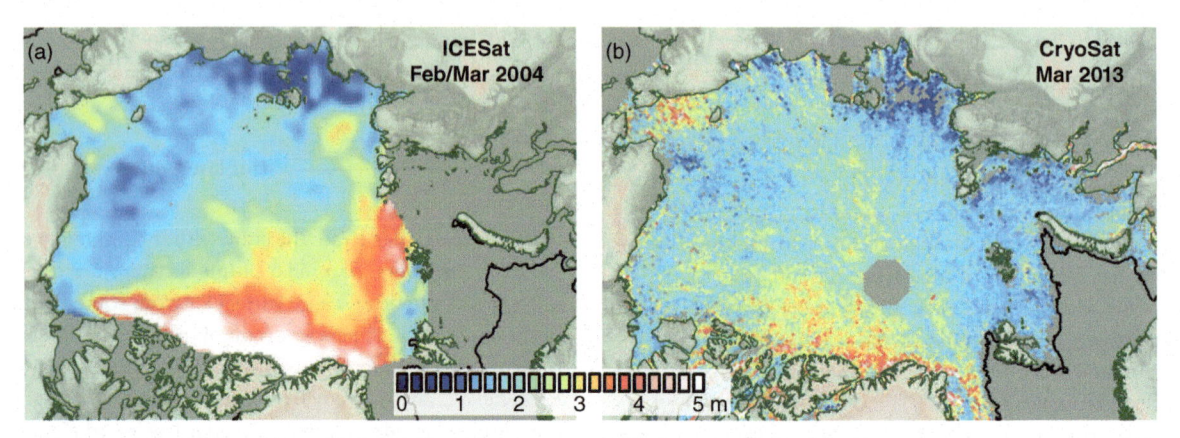

Figure 9.11 Sea ice thickness derived from satellite altimetry. (a) For February/March 2004 from the ICESat GLAS laser altimeter; (b) 9 years later for March 2013 from the CryoSat SIRAL radar altimeter. Both datasets are shown on a 25 km grid. The ICESat data are restricted to the Arctic Basin and interpolated for empty grid cells such as the pole hole. Black lines show sea ice extent for 1 March 2004 and 15 March 2013, respectively.

Figure 9.11 shows the monthly Arctic sea ice thickness for February/March 2004 from the ICESat laser altimeter (Kwok et al., 2009; data: http://rkwok.jpl.nasa.gov/icesat/) and for March 2013 from CryoSat-2 radar altimeter measurements (Ricker et al., 2014; data: Hendricks et al., 2013). The Arctic ice thickness decreased between these years (Laxon et al., 2013) but shows high interannual variation (Tilling et al., 2015). ICESat and CryoSat-2 observations can now be combined to retrieve a longer-term ice thickness record (Kwok & Cunningham, 2015).

The development of new sea ice thickness retrieval methods from satellite altimeters is ongoing. Improved sea ice thickness estimates can be expected from new methods, better validation, and new sensors such as those on ICESat-2 in the future.

9.5 Optical and thermal infrared imaging

9.5.1 Theory/principles of measurements

Sea ice reflects solar radiation to a much greater extent than open water – provided there is a large enough solar elevation angle. Hence sea ice has an average albedo > 0.6, whereas water has a low albedo, 0.07. The albedo usually increases with ice thickness for bare ice (Wadhams, 2000; Brandt et al., 2005). Snow on sea ice increases the albedo substantially, usually to > 0.8 (Brandt et al., 2005). Figure 9.4(a,b) show examples of MODIS visible and short-wavelength infrared (SWIR) channel composites (data: http://lance-modis.eosdis.nasa.gov). Satellite sensors measuring the surface reflectance in the optical frequency range of the EM spectrum can be used to obtain the surface albedo. Estimation of the melt-pond fraction (Rösel et al., 2012) and detection of polynyas and leads are typical applications of satellite optical remote sensing of sea ice. Optical satellite imagery proved valuable for the evaluation of sea ice concentration (e.g. Comiso & Steffen, 2001). An overview of commonly used sensors is given in Table 9.5.

Sea ice emits thermal radiation according to its infrared emissivity and surface temperature, $T_{Surface}$. Sensors measuring $T_{Surface}$ operate in the atmospheric window at wavelengths of 8–12 µm (Figure 9.2). At these wavelengths, the infrared emissivity of sea ice and snow is relatively constant around 0.98 (e.g. Rees & James, 1992). Typical application areas for satellite infrared imagery of sea ice are $T_{Surface}$ retrieval (Key and Haefliger, 1992; Hall et al., 2004), lead and polynya detection (e.g. Willmes & Heinemann, 2015, 2016) and thin ice thickness retrieval (Section 9.5.2).

Despite a requirement for daylight and clear-sky conditions, optical and infrared satellite imagery has also been used to derive sea ice concentration (Emery et al., 1994; Drüe & Heinemann 2004), and sea ice motion (e.g. Emery et al., 1991).

In the optical spectrum, the reflectivity of clouds is similar to the one of ice. Also, over ice the physical and thus infrared temperatures of clouds and the surface are similar for infrared data. Hence, even though sensors such as MODIS are equipped with numerous channels at

Table 9.5 List of commonly used optical and thermal infrared satellite sensors in sea ice research (non exclusive). If a parameter changed between earlier and current versions of the sensors both are given.

Name	Frequency range	Number of bands	Spatial resolution	Swath	Inclination	Operational
AVHRR	630 nm–12 µm	4–6	1090 m	2900 km	98°–99°	1978–
MODIS	405 nm–14.4 µm	36	250–1000 m	2330 km	98.2°	1999–
MERIS	390–1040 nm	15	300–1200 m	1150 km	98.4°	2002–2012
Landsat	480 nm–11.5 µm	7–8	80 m to 15–60 m	185 km	98°–99°	1972–
OLCI	400–1020 nm	21	300–1200 m	1270 km	98.65°	2016–

different wavelengths, cloud detection is still not optimal (e.g. Chan & Comiso, 2013). As shown in Figure 9.4(b), detection of some clouds over ice is possible with MODIS. Ice reflects in the blue channel 3 and absorbs in the SWIR infrared channels 6 and 7. Cloud droplets and crystals scatter in all three channels, which let them appear bright. The ocean absorbs in all three channels.

9.5.2 Thin sea ice thickness

The ice $T_{Surface}$ can be used to estimate the thickness h of thin sea ice. It is assumed that the heat flux through sea ice is balanced by the net surface heat flux F_{net}, comprising the turbulent latent F_L and sensible heat flux F_S, and the long- and short-wave radiative fluxes, F_{Long} and F_{Short}, at the ice surface. The upward F_{Long} can be computed from the ice $T_{Surface}$. The downward F_{Long} can be taken from numerical weather prediction models. During winter, F_{Short} can be neglected. Turbulent heat fluxes are usually computed via bulk formulas relating near surface vertical temperature (F_S) and moisture (F_L) gradients to wind speed and transfer coefficients of heat and moisture (e.g. Yu & Lindsay 2003; Ohshima et al., 2009). The heat flux through the sea ice is assumed to be linearly proportional to the temperature difference between the ice surface $T_{Surface}$ and the ice underside T_0. The proportionality coefficient is the conductive heat flux coefficient k. It is a function of sea ice salinity, temperature and thickness. This assumption holds for bare growing sea ice < 30–50 cm thick (Yu & Rothrock 1996; Drucker et al., 2003):

$$h = \frac{k(T_{surface} - T_0)}{F_{net}}$$

Snow on the thin ice disturbs the linearity assumption. Computation of the above-mentioned fluxes can be subject to large uncertainties. Observations of near-surface air temperature, humidity and wind speed are sparse in polar regions. Models or reanalyses often do not resolve the larger heat flux through thin ice (Willmes et al., 2011). For Antarctic coastal polynyas, models or reanalyses often fail to capture the influence of the Antarctic topography (Petrelli et al., 2008). The alternative to using model data is taking *in situ* observations at, for example, automatic weather stations (e.g. van Woert, 1999).

9.5.3 Melt-pond fraction

Surface properties of sea ice can be extremely variable – partly due to the snow cover and its interaction with the sea ice. A few of them were considered earlier, but there are others, such as grain size, wetness, melt-pond coverage and snow morphology. Here the focus is on melt ponds and the seasonal melt–refreeze cycle.

Melt ponds occur on Arctic sea ice as a result of snow and ice surface melt every summer. They can cover up to between 50% and 60% (90% in extreme cases) of the sea ice area (Fetterer & Untersteiner, 1998). Melt ponds can cause substantial biases in the summertime sea ice concentration computed from satellite microwave radiometry (Comiso & Kwok, 1996).

Melt ponds are typically smaller than 10 m² (Perovich et al., 2002). They cannot be detected by current satellite sensors pond by pond. More common are approaches estimating the contribution of the melt-pond fraction to the measured EM signal. Landsat, MODIS and MERIS have been used to identify melt ponds and to retrieve their aerial fraction (Markus et al., 2003; Tschudi et al., 2008; Istomina et al., 2015). Melt-pond cover fraction can be obtained, e.g. from 8-day composite clear-sky MODIS optical imagery using a spectral unmixing approach (Rösel et al., 2012; see the example in Figure 9.6a; data: http://icdc.cen.uni-hamburg.de). Reflectances from MODIS bands 1, 2 and 3 are combined utilizing the fact that reflectances of the surface types – melting snow or bare ice, and melt ponds – depend on the wavelength whereas the reflectance of open water between ice floes does not. By assuming that the reflectance measured at each band is composed of contributions from these surface types, the aerial fraction of the surface types is estimated. Rösel et al. (2012) used the approach to obtain Arctic-wide melt-pond cover fraction with an accuracy of about 5–10%.

9.6 Uncertainties and validation

This chapter has given an overview of the different methods used to derive sea ice quantities from satellite observations. All these methods contain several uncertainties, which will determine the quality of the final products. The estimation of uncertainties is an integral part of remote sensing of sea ice. Uncertainty ranges should always be published together with the satellite observations to allow sensible use of the products. In many cases, the estimation of the uncertainty is more challenging than the retrieval of the sea ice quantity itself. This section can only briefly summarize the different uncertainty sources and validation strategies for satellite data.

9.6.1 Sources of uncertainties

The first source of uncertainty is the accuracy of the measurement of the satellite sensor itself. Repeated measurements of the same target will give slightly different results caused by, for example, the sensor noise. This defines the precision of the measurement. In addition, a measurement system can have an offset or bias. Bias and precision together define the accuracy of the measurement. Constant biases are often uncritical as they can be easily corrected for. Some biases, however, change over time, e.g. those caused by degradation of a satellite sensor. This can introduce artificial trends in satellite time series, which are sometimes hard to discriminate from natural trends in the observed variable.

The satellite, in most cases, does not measure the wanted geophysical sea ice variable directly (see Sections 9.2–9.5): microwave brightness temperatures have to be converted to sea ice concentration, altimeter elevations have to be converted to ice freeboard and thickness, etc. This conversion, also called model, includes many error sources from which we list a few here:

- The algorithm used to map measurements to geophysical quantities contains errors or there simply is no direct analytical relationship between the two and the algorithm can only produce an approximation.
- The environmental conditions, e.g. the atmosphere, influence the measurement. The influence varies for different frequencies (e.g. clouds and haze in the optical and water vapour and ocean wind roughening in the microwave spectrum). In most cases, these environmental conditions are not well enough known and cause significant uncertainties.
- Variability within the satellite footprint. Satellite measurements integrate over the footprint area, which in many cases is several kilometres (see Tables 9.1–9.4). Within the footprint, different surface types of unknown distribution can exist. Assumptions are made, as for sea ice concentration, that the footprint is composed of the two distinct surface types, water and sea ice. In reality, the surface is much more complex (different ice types, calm and wind roughened water, etc.), which will cause uncertainty in the retrieval.

The combination of measurement accuracy and all other unknowns and errors determines the uncertainty of the derived sea ice variable. Some of these factors can be theoretically determined, but many cannot. In the end, however, validation experiments are needed to quantify the uncertainties of the satellite dataset.

9.6.2 Validation

In situ measurements are used to evaluate and validate satellite products. As the polar regions are sparsely populated and challenging to reach, *in situ* data in general are sparse. Dedicated validation campaigns by ships, helicopter and airplanes are conducted to collect satellite validation datasets. In many cases, detailed knowledge of the sea ice characteristics, as for example the ice stratigraphy or snow structure, is needed to more correctly interpret the satellite measurements and to quantify the uncertainties. On the other hand, the large spatial scales of satellite measurements demand a large number of measurements over larger distances. Therefore, a dual strategy is often used for satellite product validation: detailed measurements on the ice, typically done as ice stations from ships, deliver the required detail, and airborne observations from helicopters or airplanes bridge the gap between the small-scale ice station and often large-scale satellite measurements. One of the challenges is to estimate how many validation measurements within a satellite footprint are needed to represent the satellite measurement. The same is the case for the large-scale distribution (estimation of the covariance). It has to be estimated how many satellite measurements, i.e. footprints, have to be validated to make a sensible statement of the satellite product uncertainty on a hemispheric scale. The already-mentioned possible drift in time of satellite sensors further complicates the situation and demands repeated validation measurements throughout the lifetime of a satellite or a family of satellite sensors operated in parallel, such as SSM/I and SSMIS.

9.7 Acknowledgements

We gratefully acknowledge the provision of satellite datasets to produce the figures: MODIS, ICESat, QuikSCAT and RGPS data by NASA; Envisat ASAR, CryoSat-2 and SMOS data by ESA; AMSR-E data by JAXA and NSIDC; QuikSCAT data and sea ice motion by CERSAT/IFREMER; melt pond-fraction and SMOS ice thickness by ICDC/University of Hamburg; CryoSat-2 ice thickness by AWI; sea ice concentration by the University of Bremen. Parts of Figure 9.2 are courtesy of NASA.

References

Andersen, S., Tonboe, R., Kern, S. & Schyberg, H. (2006) Improved retrieval of sea ice total concentration from spaceborne passive microwave observations using numerical weather prediction model fields: An intercomparison of nine algorithms. *Remote Sensing of Environment*, **104**, 374–392.

Brandt, R.E., Warren, S.G., Worby, A.P. & Grenfell, T.C. (2005) Surface Albedo of the Antarctic sea ice zone. *Journal of Climate*, **18**, 3606–3622.

Campbell, J.B. & Wynne, R.H. (2011) *Introduction to Remote Sensing*, 5th edn. Guilford Press.

Cavalieri, D.J., Gloersen, P. & Campbell, W.J. (1984) Determination of sea ice parameters with the NIMBUS-7 SMMR. *Journal of Geophysical Research*, **89**, 5355–5369.

Chan, M.A. & Comiso, J.C. (2013) Arctic cloud characteristics as derived from MODIS, CALIPSO, and CloudSat. *Journal of Climate*, **26**, 3285–3306.

Comiso, J.C. (1986) Characteristics of Arctic winter sea ice from satellite multispectral microwave observations. *Journal of Geophysical Research*, **91**, 975–995.

Comiso, J.C. (2012) Large Decadal Decline of the Arctic Multiyear Ice Cover. *Journal of Climate*, **25**, 1176–1193.

Comiso, J.C. & Kwok, R. (1996) Surface and radiative characteristics of the summer Arctic sea ice cover from multisensor satellite observations. *Journal of Geophysical Research*, **101**, 28 397–28 416.

Comiso, J.C. & Steffen, K. (2001) Studies of Antarctic sea ice concentrations from satellite data and their applications. *Journal of Geophysical Research*, **106**, 31 361–31 385.

Dierking, W. (2001) Radar signatures of pancake and frazil ice – a review. CONVECTION Report No. 7, v01, October 2001, EU-Final Report for Contract EVK2-2000-0058.

Dierking, W. (2013) Sea ice monitoring by synthetic aperture radar. *Oceanography*, **26**, 100–111.

Drucker, R., Martin, S. & Moritz, R. (2003) Observations of ice thickness and frazil ice in the St. Lawrence Island polynya from satellite imagery, upward looking sonar, and salinity/temperature moorings. *Journal of Geophysical Research*, **108**, 3149.

Drüe, C. & Heinemann, G. (2004) High-resolution maps of the sea-ice concentration from MODIS satellite data. *Geophysical Research Letters*, **31**, L20403.

Eastwood, S., Larsen, K.R., Lavergne, T., Nielsen, E. & Tonboe, R. (2011) EUMETSAT OSI SAF Global Reprocessed Sea Ice Concentration dataset v1.1, Product User Manual v1.3. October 2011, SAF/OSI/CDOP/met.no/TEC/MA/138.

Emery, W.J., Fowler, C.W. & Maslanik, J. (1994) Arctic sea ice concentration from special sensor microwave imager and advanced very high resolution radiometer satellite data. *Journal of Geophysical Research*, **99**, 18 329–18 342.

Emery, W.J., Fowler, C.W., Hawkins, J. & Preller, R.H. (1991) Fram Strait satellite image-derived ice motions. *Journal of Geophysical Research*, **96**, 4751–4768.

Fetterer, F. & Untersteiner, N. (1998) Observations of melt ponds on Arctic sea ice. *Journal of Geophysical Research*, **103**(C11), 24821–24835.

Giles, K., Laxon, S., Wingham, D. et al. (2007) Combined airborne laser and radar altimeter measurements over the Fram Strait in May 2002. *Remote Sensing of Environment*, **111**, 182–194.

Girard-Ardhuin, F. & Ezraty, R. (2012) Enhanced Arctic sea ice drift estimation merging radiometer and scatterometer data. *IEEE Transactions Geoscience and Remote Sensing*, **50**, 2639–2648.

Haarpaintner, J., Tonboe, R.T., Long, D.G. & van Woert, M.L. (2004) Automatic detection and validity of the sea ice edge: An application of enhanced-resolution QuikScat/SeaWinds data. *IEEE Transactions Geoscience and Remote Sensing*, **42**, 1433–1443.

Hall, D.K., Key, J.R., Casey, K.A., Riggs, G.A. & Cavalieri, D.J. (2004) Sea ice surface temperature product from

MODIS. *IEEE Transactions Geoscience and Remote Sensing*, **42**, 1076–1087.

Hendricks, S., Ricker, R. & Helm, V. (2013) *AWI CryoSat-2 sea ice thickness*. From http://www.meereisportal.de/cryosat

Hollands, T. & Dierking, W. (2011) Performance of a multiscale correlation algorithm for the estimation of sea-ice drift from SAR images: initial results. *Annals of Glaciology*, **52**, 311–317.

Huntemann, M., Heygster, G., Kaleschke, L. et al. (2014) Empirical sea ice thickness retrieval during the freeze-up period from SMOS high incident angle observations. *The Cryosphere*, **8**, 439–451.

IPCC. (2013) *Climate Change 2013: The Physical Science Basis. Contribution of Working Group I to the Fifth Assessment Report of the Intergovernmental Panel on Climate Change* (Eds. T. Stocker et al.,) Cambridge University Press, Cambridge, UK.

Istomina, L., Heygster, G., Huntemann, M. et al. (2015) Melt pond fraction and spectral sea ice albedo retrieval from MERIS data – Part 2: Case studies and trends of sea ice albedo and melt ponds in the Arctic for years 2002–2011. *The Cryosphere*, **9**, 1567–1578.

Ivanova, N., Johannessen, O.M., Pedersen, L.T. & Tonboe, R.T. (2014) Retrieval of Arctic sea ice parameters by satellite passive microwave sensors: a comparison between eleven sea ice algorithms. *IEEE Transactions Geoscience and Remote Sensing*, **52**, 7233–7246.

Jackson, C. R. & Apel, J. R. (Eds.) (2004). *Synthetic Aperture Radar Marine User's Manual*. National Oceanic and Atmospheric Administration, US Dept. of Commerce, Washington, DC, USA.

Kaleschke, L., Lüpkes, C., Vihma, T. et al. (2001) SSM/I sea ice remote sensing for mesoscale ocean-atmosphere interaction analysis. *Canadian Journal of Remote Sensing*, **27**, 526–537.

Kaleschke, L., Maaß, N., Haas, C., Hendricks, S., Heygster, G. & Tonboe, R.T. (2010) A sea ice thickness retrieval model for 1.4 GHz radiometry and application to airborne measurements over low salinity sea ice. *The Cryosphere*, **4**, 583–592.

Kern, S., & Spreen, G. (2015) Uncertainties in Antarctic sea-ice thickness retrieval from ICESat. *Annals of Glaciology*, **56**, 107–119.

Kern, S., Ozsoy-Cicek, B., Willmes, S., Nicolaus, M., Haas, C. & Ackley, S. (2011) An intercomparison between AMSR-E snow-depth and satellite C- and Ku-band radar backscatter data for Antarctic sea ice. *Annals of Glaciology*, **52**, 279–290.

Key, J. & Haefliger, M. (1992) Arctic ice surface temperature retrieval from AHVRR thermal channels. *Journal of Geophysical Research*, **97**(D5), 5885–5893.

Komarov, A. & Barber, D. (2014) Sea Ice Motion Tracking From Sequential Dual-Polarization RADARSAT-2 Images. *IEEE Transactions Geoscience and Remote Sensing*, **52**, 121–136.

Kurtz, N.T., Farrell, S.L., Studinger, M. et al. (2013) Sea ice thickness, freeboard, and snow depth products from Operation IceBridge airborne data. *The Cryosphere*, **7**, 1035–1056.

Kurtz, N. T., Galin, N. & Studinger, M. (2014) An improved CryoSat-2 sea ice freeboard retrieval algorithm through the use of waveform fitting. *The Cryosphere*, **8**, 1217–1237.

Kwok, R. (2004) Annual cycles of multiyear sea ice coverage of the Arctic Ocean: 1999–2003. *Journal of Geophysical Research*, **109**, C11004.

Kwok, R. (2006) Contrasts in sea ice deformation and production in the Arctic seasonal and perennial ice zones. *Journal of Geophysical Research*, **111**, C11S22.

Kwok, R. (2008) Summer sea ice motion from the 18 GHz channel of AMSR-E and the exchange of sea ice between the Pacific and Atlantic sectors. *Geophysical Research Letters*, **35**, L03504.

Kwok, R. & Cunningham, G.F. (2008) ICESat over Arctic sea ice: Estimation of snow depth and ice thickness. *Journal of Geophysical Research*, **113**, C08010.

Kwok, R. & Cunningham, G.F. (2015) Variability of Arctic sea ice thickness and volume from CryoSat-2. *Philosophical Transactions of the Royal Society A*, **373**(2045).

Kwok, R., Nghiem, S.V., Yueh, S.H. & Huynh, D.D. (1995) Retrieval of thin ice thickness from multifrequency polarimetric SAR data. *Remote Sensing of Environment*, **51**, 361–374.

Kwok, R., Schweiger, A., Rothrock, D.A., Pang, S. & Kottmeier, C. (1998) Sea ice motion from satellite passive microwave imagery assessed with ERS SAR and buoy motions. *Journal of Geophysical Research*, **103**, 8191–8214.

Kwok, R., Cunningham, G.F., Wensnahan, M., Rigor, I., Zwally, H.J. & Yi, D. (2009) Thinning and volume loss of the Arctic Ocean sea ice cover: 2003–2008. *Journal of Geophysical Research*, **114**, C07005.

Kwok, R., Curlander, J., McConnel, R. & Pang, S.S. (1990) An ice-motion tracking system at the Alaska SAR Facility. *IEEE J. Oceanic Engineering*, **15**, 44–54.

Lange, M.A. & Eicken, H. (1981) Textural characteristics of sea ice and the major mechanisms of ice growth in the Weddell Sea. *Annals of Glaciology*, **15**, 210–215.

Lavergne, T., Eastwood, S., Teffah, Z., Schyberg, H. & Breivik, L.-A. (2010) Sea ice motion from low-resolution satellite sensors: An alternative method and its validation in the Arctic. *Journal of Geophysical Research*, **115**, C10032.

Laxon, S.W., Giles, K.A., Ridout, A.L. et al. (2013) CryoSat-2 estimates of Arctic sea ice thickness and volume. *Geophysical Research Letters*, **40**, 1–6.

Laxon, S., Peacock, N. & Smith, D. (2003) High interannual variability of sea ice thickness in the Arctic region. *Nature*, **425**, 947–950.

Lindell, D.B. & Long, D.G. (2016) Multiyear Arctic sea ice classification using OSCAT and QuikSCAT. *IEEE Transactions on Geoscience and Remote Sensing*, **54**, 167–175.

Lubin, D. & Massom, R.A. (2006) *Polar Remote Sensing. Volume 1: Atmosphere & Polar Oceans*. Praxis/Springer.

Maaß, N., Kaleschke, L., Tian-Kunze, X. & Drusch, M. (2013) Snow thickness retrieval over thick Arctic sea ice using SMOS satellite data. *The Cryosphere*, **7**, 1971–1989.

Markus, T. & Cavalieri, D.J. (1998) Snow depth distribution over sea ice in the Southern Ocean from satellite passive microwave data. In: *Antarctic Sea Ice: Physical Processes, Interactions, and Variability*. Antarctic Research Series (Ed. M.O. Jeffries). American Geophysical Union. Antarctic Research Series, **74**, 19–39.

Markus, T. & Cavalieri, D.J. (2000). An enhancement of the NASA Team sea ice algorithm. *IEEE Transactions Geoscience and Remote Sensing*, **38**, 1387–1398.

Markus, T., Cavalieri, D.J., Tschudi, M.A. & Ivanoff, A. (2003) Comparison of aerial video and Landsat 7 data over ponded sea ice. *Remote Sensing of Environment*, **86**, 458–469.

Markus, T., Stroeve, J.C. & Miller, J. (2009) Recent changes in Arctic sea ice melt onset, freezeup, and melt season length. *Journal of Geophysical Research*, **114**, C12024.

Martin, S., Drucker, R., Kwok, R. & Holt, B. (2004) Estimation of the thin ice thickness and heat flux for the Chukchi Sea Alaskan coast polynya from Special Sensor Microwave/Imager data, 1990–2001. *Journal of Geophysical Research*, **109**(C10), C10012.

Martin, S., Drucker, R., Kwok, R. & Holt, B. (2005) Improvements in the estimates of ice thickness and production in the Chukchi Sea polynyas derived from AMSR-E. *Geophysical Research Letters*, **32**, L05505.

Meier, W.N., Gallaher, D. & Campbell, G.G. (2013) New estimates of Arctic and Antarctic sea ice extent during September 1964 from recovered Nimbus I satellite imagery. *The Cryosphere*, **7**, 699–705.

Moen, M.-A. N., Doulgeris, A.P., Anfinsen, S.N. et al. (2013) Comparison of feature based segmentation of full polarimetric SAR satellite sea ice images with manually drawn ice charts. *The Cryosphere*, **7**, 1693–1705.

Nakamura, K., Wakabayashi, H., Naoki, K., Nishio, F., Moriyama, T. & Uratsuka, S. (2005) Observation of Sea-Ice Thickness in the Sea of Okhotsk by Using Dual-Frequency and Fully Polarimetric Airborne SAR (Pi-SAR) Data. *IEEE Transactions Geoscience and Remote Sensing*, **43**, 2460–2469.

Nakamura, K., Wakabayashi, H., Uto, S., Ushio, S. & Nishio, F. (2009). Observation of Sea-Ice Thickness Using ENVISAT Data From Lützow-Holm Bay, East Antarctica. *IEEE Transactions Geoscience and Remote Sensing Letters*, **6**, 277–281.

Nihashi, S., Ohshima, K.I., Tamura, T., Fukamachi, Y. & Saitoh, S. (2009) Thickness and production of sea ice in the Okhotsk Sea coastal polynyas from AMSR-E. *Journal of Geophysical Research*, **114**, C10025.

Ochilov, S. & Clausi, D. (2012) Operational SAR Sea-Ice Image Classification. *IEEE Transactions Geoscience and Remote Sensing*, **50**, 4397–4408.

Onstott, R.G. & Shuchman, R.A. (2004) SAR Measurement of Sea Ice. In: *Synthetic Aperture Radar Marine User's Manual* (Eds. C.R. Jackson & J.R. Apel), pp. 81–115. National Oceanic and Atmospheric Administration, US Dept. of Commerce, Washington, DC, USA.

Ozsoy-Cicek, B., Ackley, S.F., Worby, A., Xie, H. & Lieser, J. (2011) Antarctic sea-ice extents and concentrations: comparison of satellite and ship measurements from International Polar Year cruises. *Annals of Glaciology*, **52**, 318–326.

Parkinson, C. L. (2014) Global sea ice coverage from satellite data: Annual cycle and 35-yr trends. *Journal of Climate*, **27**, 9377–9382.

Perovich, D.K. & Polashenski, C. (2012) Albedo evolution of seasonal Arctic sea ice. *Geophysical Research Letters*, **39**, L08501.

Perovich, D.K., Tucker III, W.B. & Ligett, K.A. (2002) Aerial observations of the evolution of ice surface conditions during summer. *Journal of Geophysical Research*, **107**, 8048.

Petrelli, P., Bindoff, N.L. & Bergamasco, A. (2008) The sea ice dynamics of Terra Nova Bay and Ross Ice Shelf polynyas during a spring and winter simulation. *Journal of Geophysical Research*, **113**, C09003.

Rees, W.G. & James, S.P. (1992) Angular variation of the infrared emissivity of ice and water surfaces. *International Journal of Remote Sensing*, **13**, 2873–2886.

Remund, Q.P. & Long, D.G. (1999) Sea ice extent mapping using Ku-band scatterometer data. *Journal of Geophysical Research*, **104**, 11 515–11 527.

Ricker, R., Hendricks, S., Helm, V., Skourup, H. & Davidson, M. (2014) Sensitivity of CryoSat-2 Arctic sea-ice freeboard and thickness on radar-waveform interpretation. *The Cryosphere*, **8**, 1607–1622.

Ricker, R., Hendricks, S., Perovich, D.K., Helm, V. & Gerdes, R. (2015) Impact of snow accumulation on CryoSat-2 range retrievals over Arctic sea ice: An observational approach with buoy data. *Geophysical Research Letters*, **42**, 4447–4455.

Rivas, M.B., Verspeek, J., Verhoef, A. & Stoffelen, A. (2012) Bayesian sea ice detection with the advanced scatterometer ASCAT. *IEEE Transactions Geoscience and Remote Sensing*, **50**, 2649–2657.

Rösel, A. & Kaleschke, L. (2012) Exceptional melt pond occurrence in the years 2007 and 2011 on the Arctic sea ice revealed from MODIS satellite data. *Journal of Geophysical Research*, **117**(C5), C05018.

Rösel, A., Kaleschke, L. & Birnbaum, G. (2012) Melt ponds on Arctic sea ice determined from MODIS satellite data using an artificial neural network. *The Cryosphere*, **6**, 431–446.

Smith, D.M. (1996) Extraction of winter total sea-ice concentration in the Greenland and Barents Seas from SSM/I data. *International Journal of Remote Sensing*, **17**, 2625–2646.

Spreen, G., Kaleschke, L. & Heygster, G. (2008) Sea ice remote sensing using AMSR-E 89-GHz channels. *Journal of Geophysical Research*, **113**, C02S03.

Sumata, H., Lavergne, T., Girard-Ardhuin, F. et al. (2014) An Intercomparison of ice drift products in the Arctic Ocean for 2002–2006. *Journal of Geophysical Research*, **119**, 4887–4921.

Tamura, T. & Ohshima, K.I. (2011) Mapping of sea ice production in the Arctic coastal polynyas. *Journal of Geophysical Research*, **116**, C07030.

Tamura, T., Ohshima, K.I. & Nihashi, S. (2008) Mapping of sea ice production for Antarctic coastal polynyas. *Geophysical Research Letters*, **35**, L07606.

Tian-Kunze, X., Kaleschke, L., Maaß, N., Mäkynen, M., Serra, N. & Drusch, M. (2013) SMOS-derived thin sea ice thickness: algorithm baseline, product specifications and initial verification. *The Cryosphere*, **8**, 997–1018.

Tilling, R. L., Ridout, A., Shepherd, A. & Wingham, D. J. (2015) Increased Arctic sea ice volume after anomalously low melting in 2013. *Nature Geoscience*, **8**, 643–646.

Tschudi, M.A., Maslanik, J.A. & Perovich, D.K. (2008) Derivation of melt pond coverage on Arctic sea ice using MODIS observations. *Remote Sensing of Environment*, **112**, 2605–2614.

Ulaby, F.T., Moore, R.K. & Fung, A.K. (1982) *Microwave Remote Sensing, Active and Passive. Volume II: Radar Remote Sensing and Surface Scattering and Emission Theory.* Addison Wesley Publishing, London.

Ulaby, F.T., Moore, R.K. & Fung, A.K. (1986) *Microwave Remote Sensing, Active and Passive. Volume III: From Theory to Applications.* Addison Wesley Pubublishing, London.

Van Woert, M.L. (1999) Wintertime dynamics of Terra Nova Bay polynya. *Journal of Geophysical Research*, **104**, 7753–7769.

Wadhams, P. (2000) *Ice in the Ocean.* Gordon and Breach Science Publisher, London.

Willatt, R.C., Giles, K.A., Laxon, S.W., Stone-Drake, L. & Worby, A.P. (2010) Field investigations of Ku-band radar penetration into snow cover on Antarctic sea ice. *IEEE Transactions on Geoscience and Remote Sensing*, **48**, 365–372.

Willmes, S. & Heinemann, G. (2015) Pan-Arctic lead detection from MODIS thermal infrared imagery. *Annals of Glaciology*, **56**, 29–37.

Willmes, S. & Heinemann, G. (2016) Sea-ice wintertime lead frequencies and regional characteristics in the Arctic, 2003—2015. *Remote Sensing*, **8**, 4.

Willmes, S., Adams, S., Schröder, D. & Heinemann, G. (2011) Spatio-temporal variability of polynya dynamics and ice production in the Laptev Sea between winters of 1979/80 and 2007/08. *Polar Research*, **30**, 5971.

Willmes, S., Haas, C., Nicolaus, M. & Bareiss, J. (2009) Satellite microwave observations of the interannual variability of snowmelt on sea ice in the Southern Ocean. *Journal of Geophysical Research*, **114**, C03006.

Worby, A.P., Markus, T., Steer, A.D., Lytle, V.I. & Massom, R.A. (2008) Evaluation of AMSR-E snow depth product over East Antarctic sea ice using in situ measurements and aerial photography. *Journal of Geophysical Research*, **113**, C05S94.

Yu, Y. & Rothrock, D. A. (1996). Thin ice thickness from satellite thermal imagery. *Journal of Geophysical Research*, **101**, 25753–25766.

Yu, Y. & Lindsay, R.W. (2003) Comparison of thin ice thickness distributions derived from RADARSAT Geophysical Processor System and advanced very high resolution radiometer data sets. *Journal of Geophysical Research*, **108**, 3387.

Zwally, H.J., Schutz, B.E., Abdalati, W. et al. (2002) ICESat's laser measurements of polar ice, atmosphere, ocean, and land. *Journal of Geodynamics*, **34**, 405–445.

Zygmuntowska, M., Khvorostovsky, K., Helm, V. & Sandven, S. (2013) Waveform classification of airborne synthetic aperture radar altimeter over Arctic sea ice. *The Cryosphere*, **7**, 1315–1324.

CHAPTER 10

Gaining (and losing) Antarctic sea ice: variability, trends and mechanisms

Sharon Stammerjohn[1] and Ted Maksym[2]

[1] *Institute of Arctic and Alpine Research, University of Colorado, Boulder, CO, USA*
[2] *Department of Applied Ocean Physics and Engineering, Woods Hole Oceanographic Institution, Woods Hole, MA, USA*

10.1 Introduction

One of the greatest challenges facing the polar research community today is understanding why, over the past 36 years (1979–2014), Antarctic sea ice extent (spatially averaged) continues to increase in a warming world, while Arctic sea ice extent continues to decline (Chapter 11). But the Antarctic circumpolar average hides important details, as Antarctic sea ice is not increasing everywhere. In fact, there are several regions showing strong opposing regional trends, and those regions are changing at rates comparable to those observed in the Arctic (Cavalieri & Parkinson, 2012; Parkinson & Cavalieri, 2012; Stammerjohn et al., 2012). Nonetheless, over the last 3 years (2012–2014) there have been consecutive records broken in maximum Antarctic circumpolar sea ice extent, as observed by satellite since 1979 (Turner et al., 2013a; Reid & Massom, 2015; Reid et al., 2015). However, there was no single smoking gun behind these record maxima. Instead, for each of these 3 years (2012–2014), there were different regions and seasons contributing to the record sea ice extent (Reid & Massom, 2015).

There is one overriding factor that needs to be considered when contemplating Antarctic sea ice variability and change, especially when comparing Antarctic with Arctic sea ice changes: geography (Turner & Overland, 2009). The Antarctic consists of a vast glaciated polar continent (about 14 million km^2) surrounded by the Southern Ocean (Figure 10.1), a large portion of which becomes covered by seasonal sea ice (about 16 million km^2) plus a smaller fraction of sea ice that persists year-round (about 3 million km^2). Thus in winter, maximum sea ice extent is about 19 million km^2 (Figure 10.2, left column). This seasonal wax and wane of sea ice is more than a six-fold change in area of ice-covered ocean and is one of the largest seasonal signals on earth (Lieser et al., 2013).

The Arctic is the geographic opposite: a polar ocean (about 15 million km^2) surrounded by continents. Compared with the Antarctic, the Arctic Ocean has double the amount of sea ice that persists year-round (about 6 million km^2) and nearly half as much seasonal sea ice (about 9 million km^2; see Chapter 11, Figure 11.1 and Table 11.1), though these ratios are changing as Arctic summer sea ice continues to decline (Figure 11.2). Most of the sea ice-covered region in Antarctica encompasses lower latitudes (60–75°S) compared with the high-latitude (70–90°N) sea ice-covered polar ocean of the Arctic. Thus, changes in seasonal solar insolation, winds and related feedbacks will be markedly different between the polar regions. Finally, Antarctic sea ice is frequently exposed to some of the highest winds and waves that exist on Earth. Even at maximum winter sea ice extent, there are no protective land masses in the Southern Ocean for thousands of kilometres.

Since the ability to continuously monitor sea ice conditions from space starting in 1979, it has been known

---

Sea Ice, Third Edition. Edited by David N. Thomas.
© 2017 John Wiley & Sons, Ltd. Published 2017 by John Wiley & Sons, Ltd.

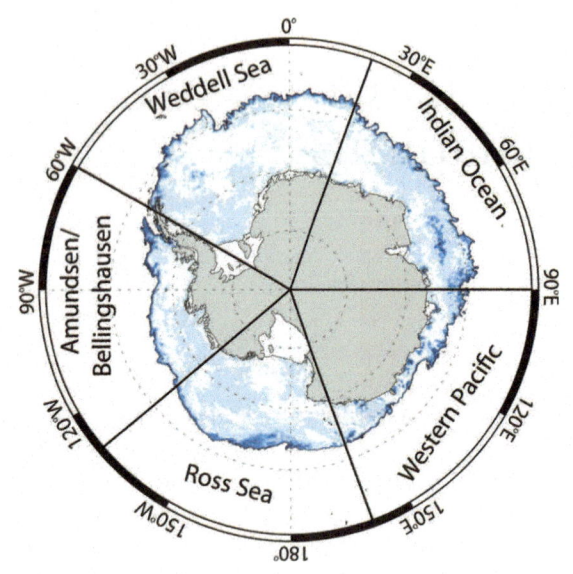

Figure 10.1 Map of Antarctica showing the boundaries of the five sectors used in this chapter. The sea ice map is an Advanced Microwave Scanning Radiometer – Earth image from 15 July 2010 showing features typical of the Antarctic sea ice cover. Darker blue represents low sea ice concentration, while white areas inside the ice edge represent 100% sea ice cover. The ice edge is not a smooth boundary, but consists of many meanders, varying spatially and temporally, as a result of the complex interactions of winds, waves and currents at the boundary between the ice cover and open ocean. A recurring feature during ice advance is the tongue offshore of the Antarctic Peninsula. The darker area in the central ice pack centred at 75°E consists of thinner sea ice and open water due to ice divergence in this area. Various coastal polynyas (Chapter 8), indicated by low sea ice concentration, can be seen along the coast in the Western Pacific sector. Sea ice data from Cavalieri et al. (2014).

that Antarctic sea ice exhibits high regional and seasonal variability (Zwally et al., 1983). Regional variability in winds, ice drift, clouds, air and ocean temperatures, precipitation and other freshwater inputs all contribute to the variability observed in seasonal sea ice extent, concentration and thickness. On daily timescales, the ice edge is highly variable due to frequent exposure to high winds and waves (Figure 10.1), and this influence can extend hundreds of kilometres into the pack ice (Kohout et al., 2014). Winds and waves at the ice edge during autumn can accelerate ice growth through rapid formation of pancake ice, while during spring, if winds and waves cause divergent conditions within the pack ice, they can accelerate ice melt by breaking up floes, decreasing ice coverage and increasing solar-warmed waters. In contrast, if winds cause convergent conditions (e.g. on-ice winds), then they can drive the ice edge poleward, cause rafting, ridging and deformation within the pack ice and increases in ice concentration (see Chapters 1 and 2 for more details on ice growth, deformation and melt-back processes).

Whereas wind forcing is typically the dominant control determining anomalous behaviour in the seasonal expansion and contraction of Antarctic sea ice (Massom et al., 2008; Stammerjohn et al., 2008; Holland & Kwok, 2012), air and ocean temperatures (Chapter 8) and precipitation (and hence snow depth), exert considerable control on sea ice growth in the Antarctic (Sturm & Massom, 2009). The deeper the snow cover, the more it insulates the surface from either freezing or warming temperatures, thus impeding ice growth in autumn/winter or ice melt in spring/summer, respectively. A further complication is relatively deep snow on thin ice, which depresses the ice surface below sea level, causing flooding, which may then subsequently freeze, producing a layer of snow ice (Ackley et al., 1990). Due to weak ocean stratification and the presence of warm deep waters, seasonal sea ice thickness is also limited by high ocean heat flux (Martinson & Iannuzzi, 1998; Chapter 8). Given these factors, thick Antarctic sea ice evolves either through rafting and ridging or from the top down through surface flooding, freezing and thickening, rather than the more usual way of thickening from the bottom up (Maksym et al., 2012).

Sea ice formation processes also vary considerably, depending on proximity to the coast or the continental shelf break. Much of the Antarctic coastline consists of floating ice shelves, glacier tongues, coastal polynyas (Nihashi & Ohshima, 2015) or pockets of thick perennial fast ice (Fraser et al., 2012). Strong and persistent low-level winds blow off the continent (e.g. 'katabatic' winds) and over the coastal areas, especially where there are strong confluences in the continental topography (Parish & Bromwich, 2007). Such winds have profound local effects, forming coastal latent-heat polynyas and causing high sea ice formation rates, which in turn cause high water mass transformation rates, the latter contributing to large-ranging effects on ocean circulation (Chapter 8). The width of the continental shelf also varies regionally, widest in the embayment areas, narrowest along East Antarctica, as does the regional proximity to the Antarctic Circumpolar Current (ACC) and its attendant warm (3–4°C above

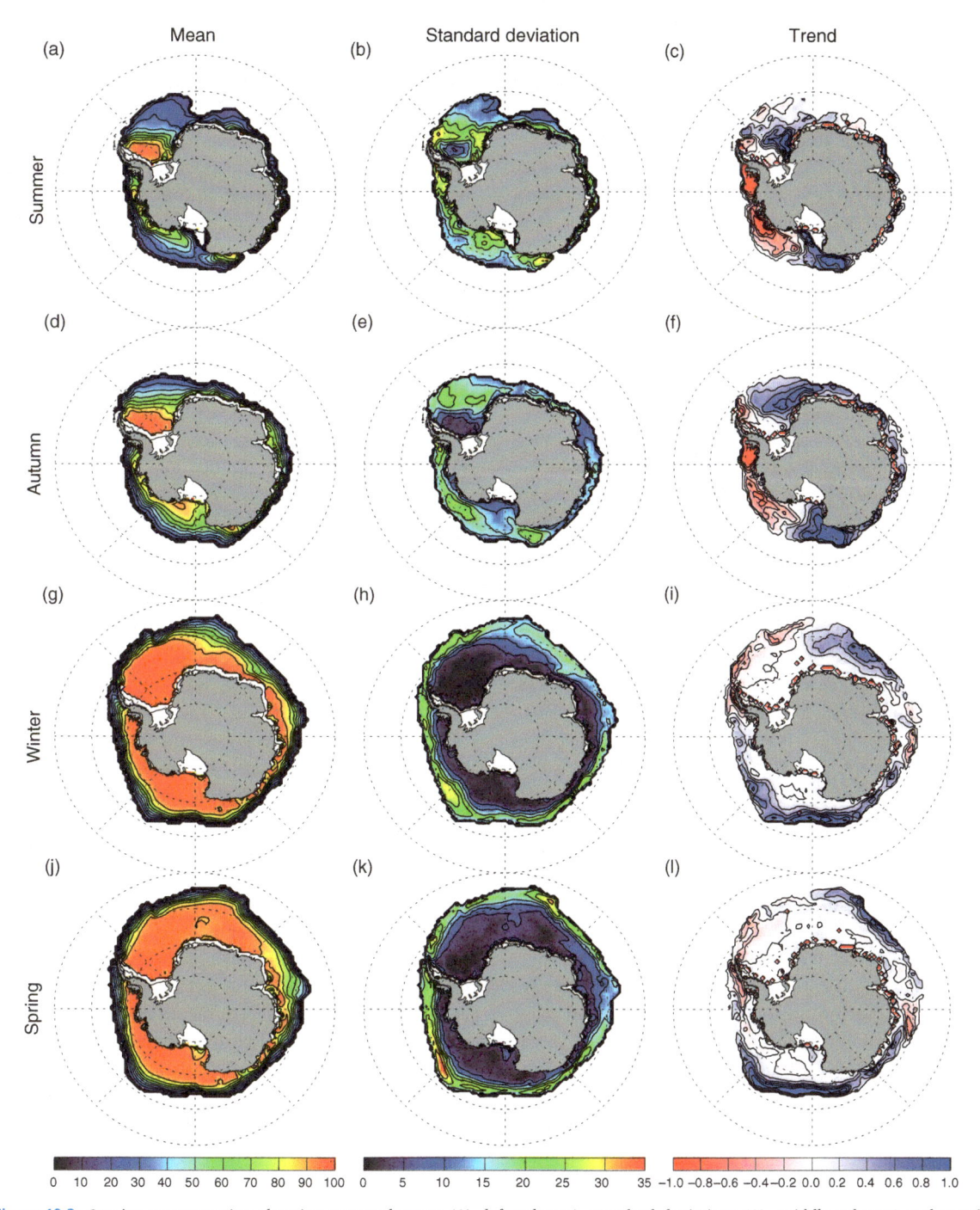

Gaining (and losing) Antarctic sea ice: variability, trends and mechanisms 263

Figure 10.2 Sea ice concentration showing seasonal mean (%, left column), standard deviations (%, middle column) and trends (% year^{-1}, right column) from November 1978 to December 2014 as derived from the Goddard Space Flight Center Bootstrap algorithm (Comiso, 2000). All values below 15% sea ice concentration are masked. Seasons are as follows: summer, December–January–February; autumn, March–April–May; winter, June–July–August; and spring, September–October–November. Contour intervals are indicated on the corresponding colour bars for each column.

freezing) Circumpolar Deep Water (CDW). All of these features have strong influences on sea ice formation processes and air–sea–ice interactions.

Regionally and seasonally, air–sea–ice interactions in the Antarctic are also quite sensitive to various modes of climate variability, e.g. the Southern Annular Mode (SAM) (e.g. Turner et al., 2009), El Niño-Southern Oscillation (ENSO) (e.g. Yuan, 2004), and the Atlantic Multidecadal Oscillation (Li et al., 2014) to name just a few. Thus, factors contributing to sea ice variability include a complex interplay between the atmosphere, ocean and cryosphere, each with different time and space scales of response and feedbacks (for an in-depth review, see Hobbs et al., 2016). This juxtaposition of both local and large-scale influences makes it all the more challenging to identify cause and effect and requires data and models at both the process-level and at large spatial scales and long timescales (Hobbs et al., 2015), resources that are developing and improving for the Antarctic sea ice community. We still need to make progress though as most climate models fail to capture the satellite-observed seasonal and regional variability (Turner et al., 2013c; Zunz et al., 2013), while also simulating net sea ice decreases instead of the observed increases over the satellite era (Arzel et al., 2006; Mahlstein et al., 2013). The inability of climate models to capture either the overall or regional behaviour of Antarctic sea ice is an indication that some processes and feedbacks are insufficiently represented or missing.

In this chapter, the large-scale behaviour of Antarctic sea ice is described, including regional and seasonal variability and trends in sea ice concentration, seasonal timing, ice motion and thickness. We also delve into air–sea–ice interactions, focusing on the effects of winds, waves and ocean feedbacks. This sets the stage for a discussion on ice–climate linkages and the perspective offered by the few historical observations that exist. To kick off these discussions, we first introduce the data sources and common approaches used to study large-scale sea ice variability.

10.2 Data sources and approaches

Satellite observations from coarse resolution (~25 km) passive microwave sensors, the scanning multichannel microwave radiometer/special sensor microwave imager/special sensor microwave imager/sounder (SMMR-SSMI-SSMIS) data (see Chapter 9 for details), provide a continuous record of (quasi) daily sea ice concentration from the late 1970s and are particularly useful for analysing regional and seasonal variability and change. Finer-scale features such as ice type, surface roughness and lead fraction can be identified and mapped with higher-resolution satellite data (e.g. synthetic aperture radar) on scales of metres to a few kilometres, but the high-resolution satellite data are limited in time and space. There has been a concerted effort to standardize and compile underway and on-ice ship-based sea ice observations, which provide valuable information on sea ice thickness in particular (see Section 10.4.1) and are useful for ground-truthing satellite observations. However, given that this chapter is focused on large-scale sea ice variability and trends, we will primarily rely on the long and continuous record provided by the passive microwave satellite data.

The SMMR-SSMI-SSMIS sea ice data are provided by the EOS Distributed Active Archive Center and can be electronically downloaded from the National Snow and Ice Data Center (NSIDC) at the University of Colorado in Boulder, Colorado (http://nsidc.org). There are several different algorithms for deriving sea ice concentration from the remotely sensed surface brightness temperature (see Chapter 9 and Ivanova et al. (2015) for an overview and inter-comparison). The sea ice data from NSIDC are provided on a nominal 25 × 25 km polar stereographic grid, and additional gridded information on latitude, longitude, grid cell size and land mask is available.

There are different biases in the satellite-derived sea ice concentrations depending on the algorithm used (Comiso et al., 1997; Eisenman et al., 2014; Ivanova et al., 2015). Although we discuss changes in sea ice concentration (e.g. with respect to winds), it must be acknowledged that different algorithms (and different versions of algorithms) give different estimates of sea ice concentration (Stroeve et al., 2016), depending on, for example, differences in empirically derived tie-points and whether those tie-points are applied regionally or seasonally. However, satellite-derived estimates of the nominal ice edge (which is typically defined by the 15% sea ice concentration threshold but other thresholds such as the 10% or 30% have also been used) have higher precision and less inter-algorithm differences (Steffen et al., 1992; Comiso et al., 1997) due to the large and less ambiguous contrast in emissivity between well-formed sea ice and open ocean. That said, thin ice, flooded ice and ponded sea ice are not well resolved by passive microwave radiometers. Therefore, ice edge detection by passive microwave radiometers may show offsets with respect to field observed ice edge locations (Worby & Comiso, 2004).

There are various ways to analyse and summarize the satellite-derived sea ice data. Maps of sea ice concentration, which is the percentage of ocean covered by sea ice, are useful for showing the strong regional variability that characterizes Antarctic sea ice. It is also convenient to distil the data into spatial averages of sea ice extent and sea ice area, which are, respectively, the total area inside the ice edge (of varying concentrations) and the total ice-covered area inside the ice edge, where each grid cell is multiplied by its concentration and then summed for the region of interest. Other ways to distil the satellite data into useful sea ice metrics include mapping in time the daily ice edge location over a full 'ice year', which nominally begins and ends during the mean summer minimum (February). This produces spatial maps of the timing of the autumn ice edge advance and spring retreat. By differencing the day of ice edge retreat from the day of ice edge advance, maps of the annual ice season duration (or ice season length) are produced (Parkinson, 1994, 2002, 2004; Stammerjohn et al., 2008, 2012). There are other useful approaches to analysing and summarizing sea ice data (e.g. Yuan & Martinson, 2001; Holland, 2014; Raphael & Hobbs, 2014; Reid & Massom, 2015), which the reader is encouraged to explore.

Additional data sets described in this chapter include ship- and satellite-based observations of sea ice thickness and satellite-derived sea ice motion. Methods and uncertainties of obtaining sea ice thickness and ice motion from satellite observations are described in Chapter 9. We also use and discuss numerically analysed meteorological reanalysis data in the examination of ice–atmosphere interactions. Because *in situ* meteorological data are sparse in the Southern Ocean, we are forced to rely on reanalysis data to study large-scale behaviour over the satellite era. Reanalysis data are also used to drive sea ice or ice–ocean coupled models, or they are used to validate climate model output. Unfortunately, these numerically analysed data are not well constrained in the Southern Ocean due to the sparsity of *in situ* data, so some care is necessary when drawing conclusions based on these data, particularly with respect to any trends prior to the satellite era. These issues are discussed in greater detail in Chapter 12.

10.3 Seasonal and regional variability and change

In this section we describe Antarctic sea ice using the following metrics: sea ice concentration, sea ice extent and seasonal timings. We discuss both seasonal and regional means, as well as variability and change.

10.3.1 The seasonal sea ice cover

Seasonal sea ice changes (Figure 10.2, left column; Figure 10.4a below) reflect in large part the seasonal changes in solar insolation. On average, the seasonal sea ice cover expands northwards from February to September. At maximum winter extent, sea ice is dispersed over about 19 million km^2. This frozen platform more than doubles the icy real estate occupied by Antarctica (about 14 million km^2). Then every austral spring, the seasonal sea ice cover begins to break up and melt, contracting southwards from September to February. Most of the sea ice melts, while some persists throughout the summer (about 3 million km^2) along the coast and in the southern embayments of the Weddell, Amundsen and Ross Sea regions.

Regional differences in sea ice concentration can be attributed, in part, to local geographical differences. The off-pole asymmetry of the Antarctic continent results in more poleward coastal latitudes (70–75°S) along most of West Antarctica, with winter sea ice extending ~700 km to 1500 km in the north–south direction. In contrast, the coast along East Antarctica between 0° and 90°E is at lower latitudes (65–70°S) and has slightly less extensive north–south winter sea ice. The lowest latitude coastline (equator-ward of 65°S) includes the northern half of the Antarctic Peninsula and a portion of East Antarctica between 90° and 135°E, the latter showing the narrowest band of seasonal sea ice (~250–750 km) (Massom et al., 2013). At the other extreme, the two largest and deepest embayments (in the Weddell and Ross Seas) are located poleward of 75°S and contain the most extensive winter sea ice coverage (1500–2000 km in the north–south direction). Finally, the north–south protrusion of the Antarctic Peninsula is geographically distinct and exhibits contrasting climates west to east, with mostly seasonal sea ice along the wind-exposed western side versus mostly perennial sea ice along the relatively quiescent east side. Mean seasonal and regional differences in the sea ice cover can also be attributed to mean winds, which will be discussed in Section 10.3.3, and to the mean circulation of sea ice, which will be discussed in Section 10.4.2.

Temporal variability in sea ice concentration (Figure 10.2, middle column) is highest at and inside the ice edge where concentrations range from 0% to ~75% (see also Simpkins et al., 2012). The width of this high-variability zone varies depending on the season and region. In summer and autumn, much of the sea ice zone in the West Antarctic region is characterized by high temporal variability. In winter and spring, the band of high variability follows the ice edge equator-wards and becomes more zonally distributed. This high-variability zone of the outer pack ice is characterized by strong surface gradients and frequent storms, and thus high winds and waves. Winds or surface currents can thermodynamically assist or hamper sea ice changes by bringing in colder or warmer air masses or surface waters, respectively. Winds, ocean currents and ocean swell also cause sea ice to drift, contributing to ice edge expansion or contraction, creating openings or closings within the pack ice. Inside the high-variability zone, the sea ice cover is more consolidated and is, on average, thicker, dampening the effect of winds and waves, but not necessarily removing those effects altogether. (This will be further explored in Section 10.4.4.)

Trends in seasonal sea ice concentration over 1979–2014 show remarkable seasonal and regional patterns (Figure 10.2, right column) (see also Simpkins et al., 2012; Holland, 2014; Turner et al., 2015; Hobbs et al., 2016). Trends are strongest in summer and autumn and show strong regional contrasts between increases in sea ice concentration in the western Weddell and Ross Sea regions and decreases in sea ice concentration in the Bellingshausen and Amundsen Sea region. In autumn, the East Antarctic sector also shows strong increases in sea ice concentration (see also Massom et al., 2013). In winter and spring, the magnitude of the trends and spatial contrasts weaken relative to those observed in summer and autumn. The areas of strong positive trends shift to the outer pack ice, with strongest trends found in the Indian Ocean (between ~20°E and 50°E) and in the Western Pacific Ocean and Ross Sea sectors (between ~100°E and ~135°W). Areas of negative trends include the west side of the Antarctic Peninsula, a thin outer band in the western Weddell Sea (between 60°W and 0°E) and a portion of East Antarctica (between ~90°E and 100°E).

10.3.2 The regional perspective

Additional insights into regional differences can be gained by creating regional summaries of sea ice extent. Here we define the regions based on traditional definitions (Gloersen et al., 1992; Zwally et al., 2002; Comiso, 2010; Parkinson & Cavalieri, 2012), which delineate by ocean basins: Weddell Sea (60°W to 20°E), Indian Ocean (20–90°E), Western Pacific Ocean (90–160°E), Ross Sea (160°E–130°W) and the Bellingshausen/Amundsen Sea (130–60°W). There are other ways to define regions that are not based on geography but on characteristic variability (Raphael & Hobbs, 2014) or change (Liu et al., 2004; Stammerjohn et al., 2008; Simpkins et al., 2012). Isolating regions based on their characteristic variability (seasonally and in frequency space) or change is a useful approach for investigating regionally varying ice–atmosphere–ocean interactions and connections to large-scale circulation modes, which operate on different time and space scales (Simpkins et al., 2013; Raphael & Hobbs, 2014). The reader is encouraged to consult these other approaches, but for this broad overview, we show the geographically delineated regions to allow comparison with previous studies.

To set the stage we first show the time series of circumpolar monthly sea ice extent (Figure 10.3a), which illustrates the strong seasonal swings between summer minimum and winter maximum, as well as the yearly variability in those two variables. The mean annual cycles for each region (Figure 10.4a) also indicate regional differences in the timing of ice expansion and contraction, as well as the magnitude and duration of winter maximum extent. These characteristics of the seasonal cycle will be explored in the next section, but for now we must remove the mean annual cycle to better reveal the monthly variability in sea ice extent. When we do this for each region, the regional differences in monthly variability and trends become clear (Figure 10.3b–g). Note, however, that as the regions do not have similar amounts of sea ice, the ranges in monthly anomalies are different for each region. Thus, for reference, dotted lines have been added to indicate the ±1 standard deviation range.

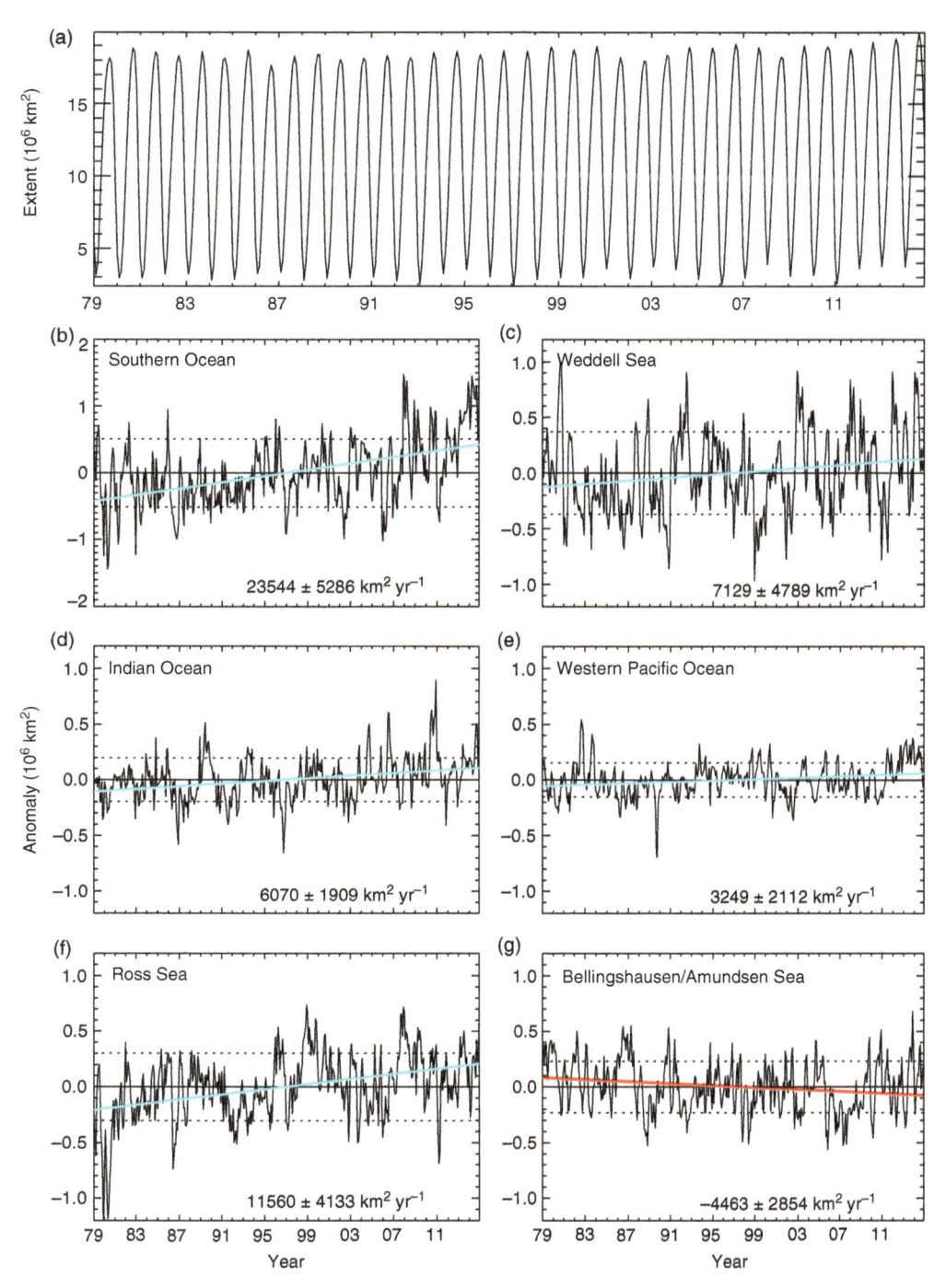

Gaining (and losing) Antarctic sea ice: variability, trends and mechanisms

Figure 10.3 (a) Monthly sea ice extent from January 1979 to December 2014 as derived from the Goddard Space Flight Center Bootstrap algorithm (Comiso, 2000). Monthly averages were computed from daily averages of sea ice extent. (b–g) Monthly sea ice extent anomalies over the same period shown for each region (as defined in Figure 10.1). The linear trend is also shown for each region (blue for positive, red for negative), and the solid and dotted black lines indicate zero and ± 1 standard deviation, respectively. The trends were computed on the annual means (of the daily data), and the standard error and significance levels are based on the effective degrees of freedom (Santer et al., 2000). The trends for the Southern Ocean, Indian Ocean and Ross Sea are significant at the > 99% level, while the other trends are significant at the > 90% level. (Adapted from Parkinson & Cavalieri, 2012).

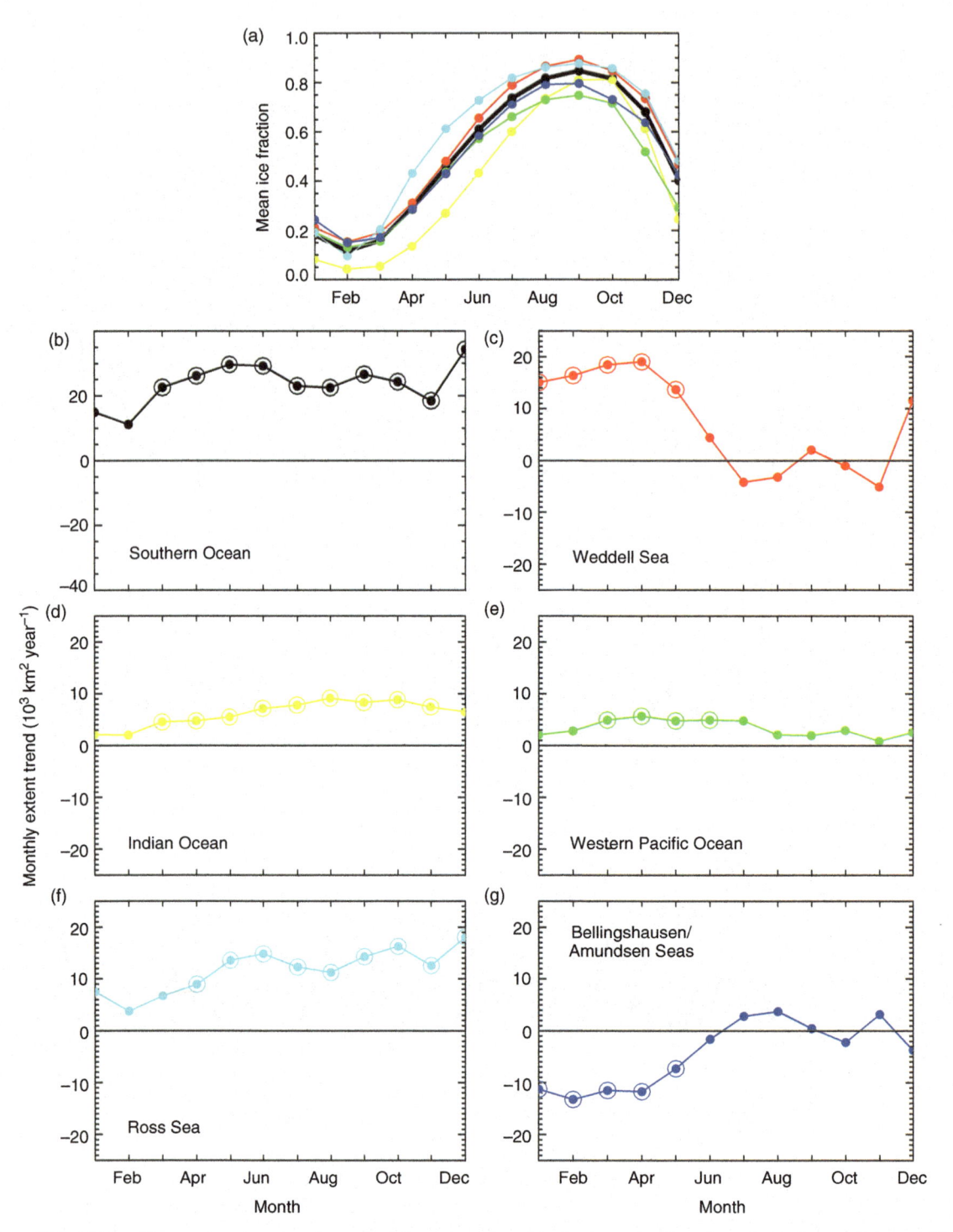

Figure 10.4 (a) Monthly mean sea ice extent for each of the regions shown in (b–g) matched by colour. (b–g) The trends in monthly sea ice extent for each region (adapted from Parkinson & Cavalieri, 2012; Holland, 2014). The zero line is added to highlight positive from negative trends, and those trends significant at the 95% level or higher are circled.

All regions but the Bellingshausen/Amundsen Sea show positive trends, the largest being in the Ross Sea. Thus, the circumpolar trend (Figure 10.3b) reflects contributions from all regions, and its magnitude has been increasing compared to previous reports (Parkinson & Cavalieri, 2012). Indeed new records for maximum sea ice extent have been consecutively broken over the last three years (Turner et al., 2013a; Reid & Massom, 2015; Reid et al., 2015). However, it is unclear whether this positive trend will continue, given that it is a product of large regional and seasonal variability. For additional perspective, the trend in annual averaged Antarctic sea ice extent over 1979–2014 is +23.5 ± 5.3 10^3 km^2 $year^{-1}$), which is almost half the magnitude of the Arctic annual mean negative trend of −52.6 ± 6.8 10^3 km^2 $year^{-1}$ (Chapter 11).

The relatively small regional trends observed in monthly sea ice extent (Figure 10.3) might appear to be at odds with the results shown in Figure 10.2 (right column). However, the regional sea ice extents involve spatial averaging, and their trends are computed over the entire monthly time series. Thus, strong or opposing seasonal trends will be muted or averaged out. When we plot the regional trends for each month (Figure 10.4 b–g) (Parkinson & Cavalieri, 2012; Holland, 2014), the results become more consistent with Figure 10.2 (right column) and further emphasize the regional and seasonal differences. We see strong positive trends during January to May in the Weddell Sea contrasted against strong negative trends in the Bellingshausen/Amundsen seas. In winter and spring the strongest trends in monthly sea ice extent are in the Ross Sea and Indian Ocean. In general, the circumpolar monthly trends are most similar to those in the Ross Sea, but with additional contributions from all other regions, particularly from the Weddell Sea during the first third of the year. Thus, we see clearly how different regions and different months contribute to the circumpolar trend (Holland, 2014).

10.3.3 Seasonal timing and feedbacks

Yet another way to view and analyse seasonal sea ice changes is to track the ice edge location in space through time, which allows us to create maps showing the timing of the ice edge advance in autumn, its retreat in spring, and the length or duration of its ice-covered season during winter (Figure 10.5, left column). When we do this, we see that the asymmetric seasonal cycle (Figure 10.4a) is readily captured in the timing of ice edge advance and retreat. On average, the autumn ice edge advance takes about 7–8 months to progress from near the coast to its most northward extent (note the contours on the mean maps are at 30-day intervals). In contrast, the spring ice edge retreat occurs over a much shorter interval, taking only about 4–5 months to return back to its summer minimum. This asymmetry has been a curiosity ever since early satellite images began capturing seasonal sea ice changes, especially since the Arctic region shows a more symmetric seasonal cycle.

It was soon suggested that this asymmetry in the Antarctic seasonal sea ice cycle was another reflection of its contrasting geography with the Arctic. The asymmetry is largely due to wind–ice interactions resulting from a twice-yearly (in autumn and spring) intensification and poleward contraction of the southern hemisphere atmospheric circumpolar trough (CPT) (van Loon, 1967; Van Loon, 1972; Walland & Simmonds, 1999). The CPT is a low-pressure belt located between 60° and 70°S formed by large numbers of cyclones that have either migrated from mid-latitudes or developed within the Antarctic coastal zone (Jones & Simmonds, 1993). The twice-yearly intensification and poleward contraction of the CPT is known as the semi-annual oscillation (SAO) (van Loon, 1967; Van Loon et al., 1993; Meehl et al., 1998; Simmonds & Jones, 1998) and is caused by different solar radiation input and absorption between the southern hemisphere's ice-dominated polar latitudes and ocean-dominated mid-latitudes. Consequently, the mid-to-high latitude mid-tropospheric temperature gradient is greatest during autumn and spring, thus causing a contraction and intensification of the CPT polewards. Subsequently, the mid-to-high latitude gradient weakens during summer and winter, but strengthens between low- and mid-latitudes, causing an expansion of the CPT equator-wards.

The seasonal strengthening and poleward intensification of the CPT has a distinct effect on the timing of the annual ice edge advance and retreat (Enomoto & Ohmura, 1990; Watkins & Simmonds, 1999; van den Broeke, 2000; Stammerjohn et al., 2003; Yuan & Li, 2008). The low-pressure minimum defining the CPT marks the boundary between westerlies to the north and easterlies to the south. Due to the Coriolis effect, the westerly winds impart a northward Ekman component on ice motion, whereas the easterly winds impart a southward component. As the CPT migrates south

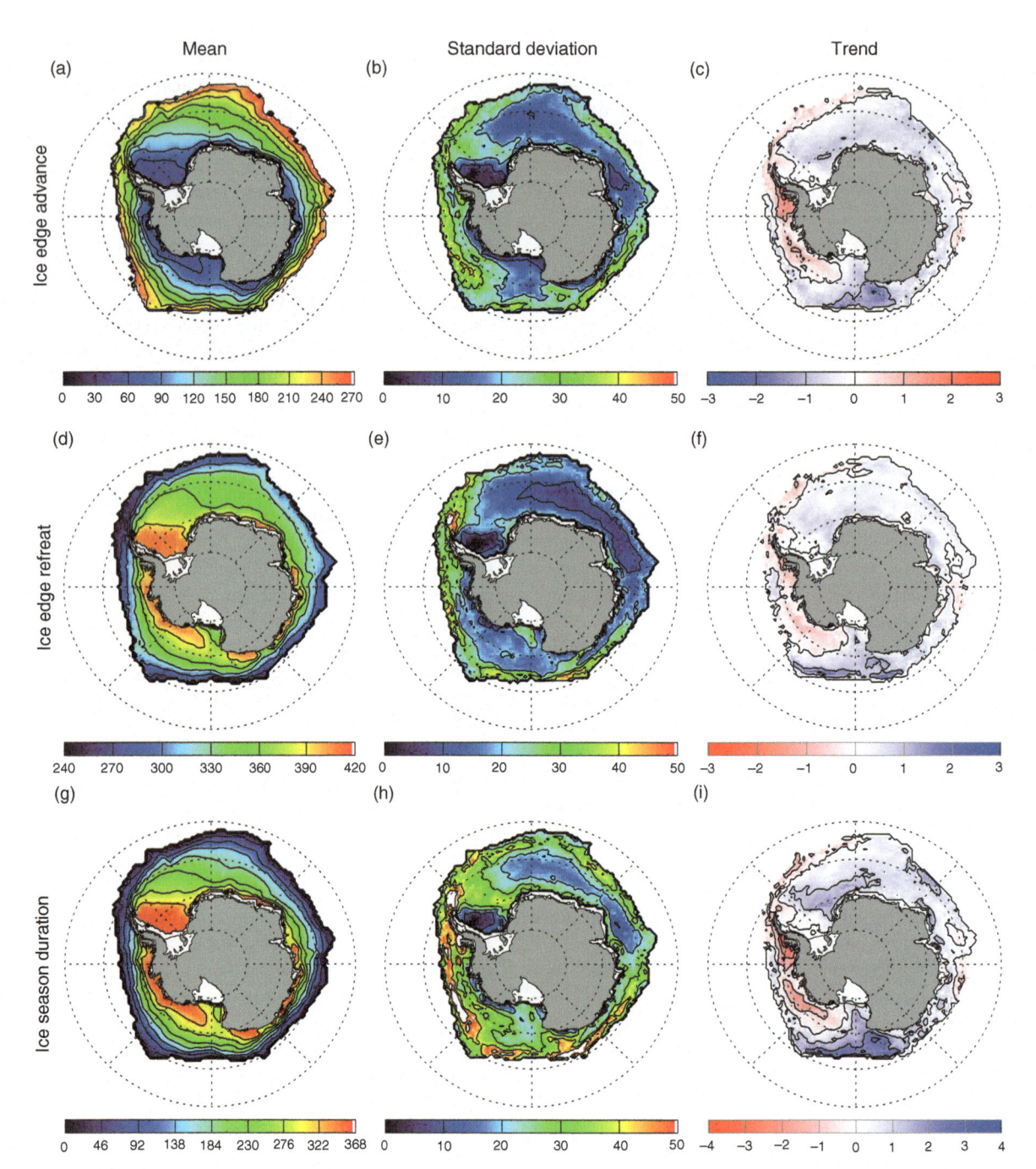

Figure 10.5 Seasonal timings in ice edge advance (year day, top row), ice edge retreat (year day, middle row) and ice season duration (days, bottom row), showing means (left column), standard deviations (middle column) and trends (days year^{-1}, right column; updated from Stammerjohn et al., 2012) over February 1979 to February 2014. The daily sea ice concentration data derived from the Goddard Space Flight Center Bootstrap algorithm (Comiso, 2000) were used to compute the seasonal timings (see Stammerjohn et al., 2008, for further details). Contour intervals are indicated on the corresponding color bars for each column.

towards the advancing ice edge in autumn, the advance is hampered by easterly winds and associated southward advection, thus moderating ice edge expansion. As the CPT weakens, the ice edge continues to advance and moves equator-ward of the CPT location. As winter progresses, the dynamical response to the divergence imparted by the westerly wind north of the CPT (but south of the ice edge) is now more sluggish due to the winter freezing conditions. As spring approaches, the retreat is marked first by a decrease in sea ice concentration inside the ice edge as the CPT (still located south of the ice edge) strengthens. This causes a divergence of sea ice (i.e. movement of sea ice towards the ice edge and the coast) that, in turn, creates open water areas that subsequently do not freeze due to increased insolation. During spring–summer the CPT moves north of the ice edge, thus facilitating a rapid retreat due to the southward Ekman forcing on the low sea ice concentrations (Enomoto & Ohmura, 1990; Watkins & Simmonds, 1999). Upward mixing of ocean heat due to this divergence also accelerates the melting of the thin ice cover and enhances the sea ice retreat (Gordon, 1981).

Although the SAO helps to explain the mean cycle of slow autumn advance and fast spring retreat, there is considerable yearly and regional variability caused by yearly and regional variability in storm frequency, intensity and duration (Simmonds & Keay, 2000; Simmonds, 2003), which also affect ice–ocean interactions (Chapter 8). Regional variability in the mean seasonal sea ice cycle, which in part reflects regional variability in seasonal storms (also see Raphael & Hobbs, 2014), is illustrated in Figures 10.4(a) and 10.5. It shows that the autumn ice edge advance is, on average, earliest in the Ross Sea and latest in the Indian Ocean, the spring ice retreat is earliest in the Western Pacific Ocean and latest in the Ross and Weddell Seas, and the winter maximum is earliest in the Bellingshausen/Amundsen seas and latest in the Indian Ocean. The spatial distribution of temporal variability (Figure 10.5, middle column) shows overall that the largest variability is in the more northern locations, particularly in the West Antarctic region. The low temporal variability in timing of ice edge advance and retreat in the central Weddell Sea and Indian Ocean is also noteworthy.

When we look at the spatial distribution of the trends in seasonal timings (Figure 10.5, right column), we are reminded of the seasonal trends in sea ice concentration. For trends in the ice edge advance, we see mostly negative trends in the day of advance, indicating an earlier advance in most regions, particularly in the western Ross Sea, where the advance is earlier by 1–2 days year^{-1}. In contrast, we see mostly positive trends in the day of advance in the Bellingshausen/Amundsen seas and parts of East Antarctica centred on 90°E, indicating a later ice edge advance, with the largest trends in delay of the ice edge advance occurring in the southern Bellingshausen Sea, at 1–2 days year^{-1}. The trends in ice edge retreat show similar spatial contrasts in the sign of the trend (later retreat everywhere but in the Bellingshausen/Amundsen seas and in the East Antarctic region centred on 90°E), but the trends are overall weaker (less than ±1 day year^{-1}), except along the northern ice edge in the Ross Sea where the ice edge retreat is later by 1–2 days year^{-1}.

Trends in annual ice season duration aggregate the trends just described, showing that the ice season is becoming longer everywhere but in the Bellingshausen/Amundsen seas and the East Antarctic region centred on 90°E. In the southern Bellingshausen Sea the ice season has become shorter by 2–3 days year^{-1} and in the western Ross Sea, longer by 2–3 days year^{-1}. When the change is totalled over the 35-year times series (1979/80–2013/14), we see that the ice season has become shorter or longer by more than 3 months in these two regions, respectively. Those are remarkably fast changes, as fast as observed in the high trending regions of the Arctic (Stammerjohn et al., 2012), and too fast for most marine life to adapt to a rapidly changing habitat (e.g. Ducklow et al., 2013).

Using these data we can also determine if there is any covariability between anomalies in the timing of ice edge advance and its subsequent retreat, and vice versa by correlating the de-trended time series of these two variables at each location. When we do this, we see weak correlations between the autumn ice edge advance and the subsequent spring retreat (i.e. over winter) (Figure 10.6a), but in contrast, we see strong correlations between the spring ice edge retreat and the subsequent autumn advance (i.e. over summer) (Figure 10.6b). This negative correlation indicates that, more often than not, an early (late) ice edge retreat in spring is followed by a late (early) ice edge advance the following autumn. The relationship between spring sea ice retreat and subsequent autumn advance is consistent with expected seasonal feedbacks at high latitudes: an earlier (later) spring retreat leads to

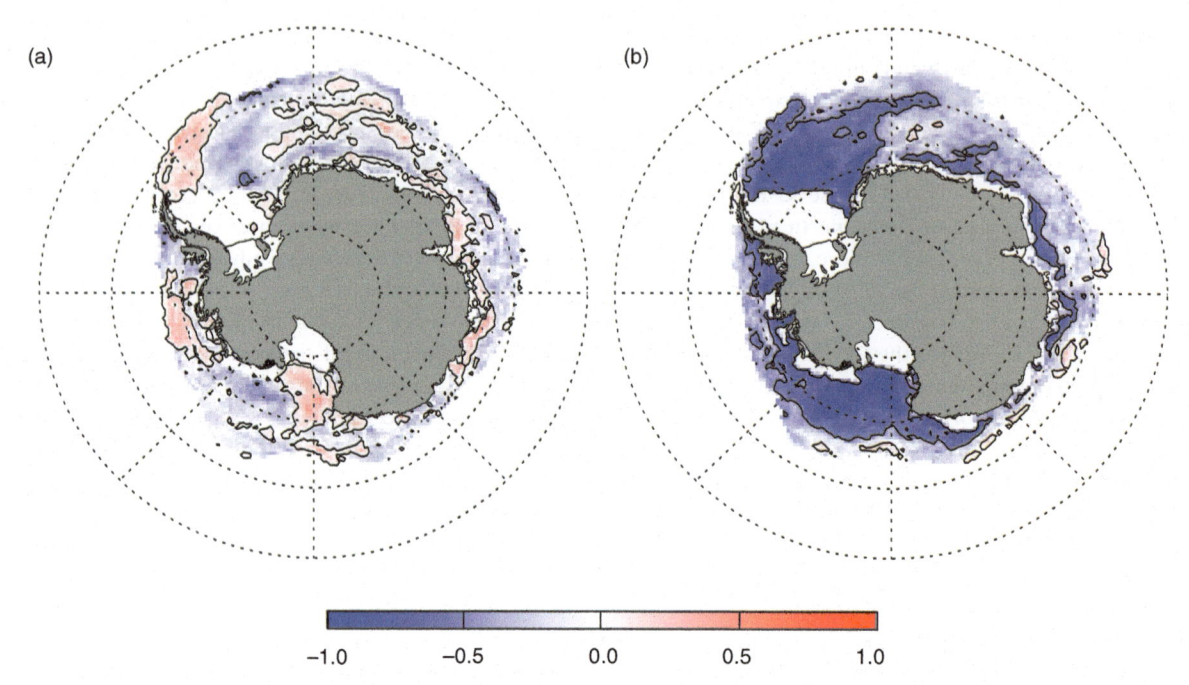

Figure 10.6 Temporal correlations between detrended anomalies in the autumn ice edge advance and the subsequent spring ice edge retreat (a), and the spring ice edge retreat and the subsequent autumn ice edge advance (b) over February 1979 to February 2014. Correlations were computed only for those pixels showing a day of advance and retreat for at least 20 of the 36 years. Contour intervals are indicated on the corresponding colour bars for each column.

increased (decreased) solar ocean warming. This addition (deficit) of solar-gained heat in the upper ocean must be removed before sea ice can grow, resulting in a later (earlier) sea ice advance in autumn (see also Nihashi & Ohshima, 2001; Holland, 2014).

10.4 Sea ice thickness, drift, winds and waves

Wind is the predominant factor driving sea ice motion, or drift, particularly in the Antarctic where exposure to the open Southern Ocean does not constrain the pack ice, and thus internal ice stresses provide less resistance to motion. This determines both the extent of sea ice and also the thickness, as divergence and convergence of the ice cover drive sea ice production and deformation (see Chapter 2). Thus, sea ice drift and ice thickness provide additional metrics for assessing air–sea–ice interactions and the impact of climate variability. In this section we discuss seasonal variability in ice drift and ice thickness and possible trends.

10.4.1 Seasonal and regional variability in ice thickness

Unlike sea ice extent, there has been no means to determine ice thickness from space until recently. And unlike the Arctic, there have been no widespread observations available from submarines or proxies for thick, multi-year sea ice such as ice age (see Chapter 11). *In situ* Antarctic sea ice thickness data are sparse, with very few observations prior to the late 1980s. Visual observations of sea ice from over 80 icebreaker expeditions have been compiled to provide a climatology of sea ice thickness and snow depth distribution (Figure 10.7d) (Worby et al., 2008), but the data are too sparse to provide useful information on interannual variability or on regional sea ice thickness in all seasons.

More recently, large-scale sea ice thickness has been estimated from NASA's ICESat LIDAR altimeter data available from 2003 to 2009 (Yi et al., 2011; Kurtz & Markus, 2012; Xie et al., 2013). Determination of sea ice thickness from altimeters is challenging in the Antarctic because the deep snow cover dominates the

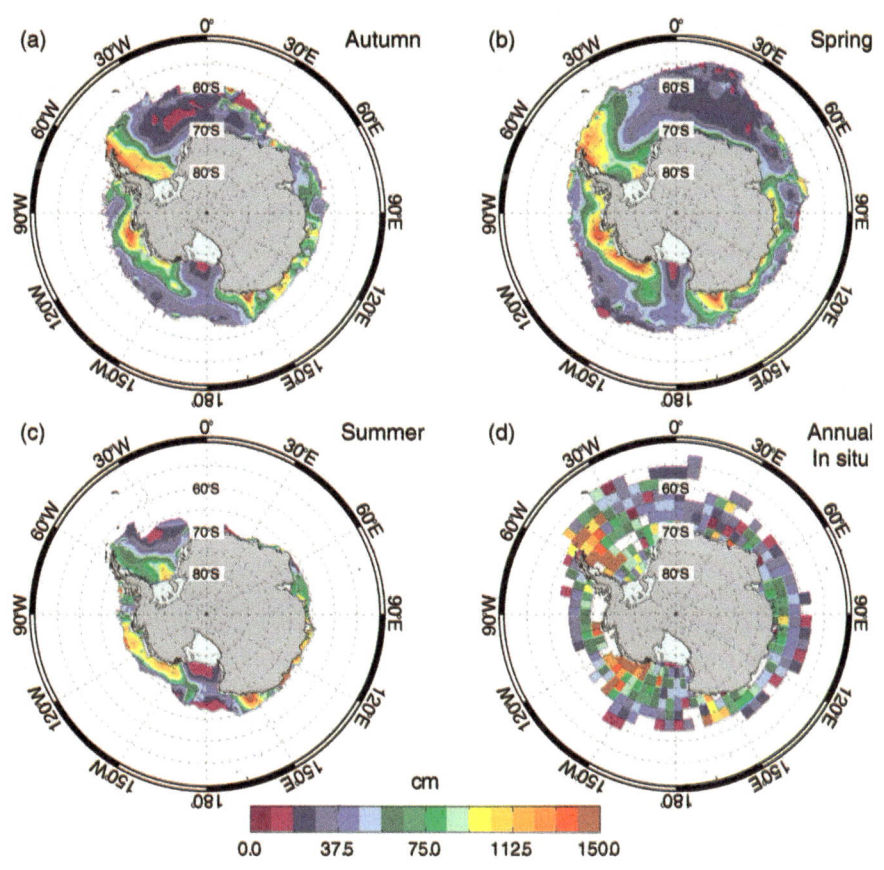

Figure 10.7 Average sea ice thickness estimates from ICESat for: (a) autumn 2004–06, (b) spring 2004–08, and (c) summer 2003–07 (adapted from Kurtz & Markus, 2012; see text for details). (d) A compilation of shipboard visual observations (adapted from Worby et al., 2008). Average ice thickness increases from autumn to summer, although average seasonal variability is small. The actual ice thickness may be somewhat greater than these estimates (see text). The thickest sea ice is found in areas where sea ice typically survives the summer melt. (ICESat data provided by N. Kurtz. Shipboard data from http://aspect.antarctica.gov.au/data.)

altimeter's measurement of freeboard (the height of the surface above the ocean). Reliable estimation of snow depth is difficult, so there is substantial uncertainty in the ice thickness estimates (Kern & Spreen, 2015) (see also Chapter 9). Kurtz and Markus (2012) provide the only circumpolar estimates of ice thickness from ICESat; these are illustrated in Figure 10.7. Although other studies suggest a substantially thicker ice cover (Worby et al., 2011; Yi et al., 2011; Xie et al., 2013; Kern & Spreen, 2015), the spatial patterns are similar.

Overall, the Antarctic pack ice is thin, with a mean thickness of less than 1 m and with broad regions of the outer pack ice at less than 50 cm from autumn to midwinter (Worby et al., 2008). Seasonal variation in areal average ice thickness is modest. Worby et al. (2008) report mean ice thickness of 0.68, 0.66, 0.81 and 1.17 m for autumn, winter, spring and summer, respectively. Kurtz and Markus (2012) show a smaller seasonal cycle; the differences are most probably due to regional gaps in the shipboard data and limitations in snow depth determination for the satellite data. Sea ice is thickest in summer, when the thinner, outer pack has melted, leaving thicker, often heavily deformed ice along the coast. Thin ice cover is maintained in the outer pack by a combination of ice divergence, which sustains a high proportion of new and young sea ice (Worby et al., 2008), and a high ocean heat flux which limits basal ice growth (Martinson & Iannuzzi, 1998).

Thicker sea ice is found predominantly along coastal regions, much of it multi-year sea ice. The extent of multi-year sea ice is low relative to the Arctic as the strong seasonal cycle in sea ice extent indicates. Most (~80%) multi-year sea ice is found in the western Weddell Sea, where there is a relatively slow northward drift and heavy deformation along the coast, known since the drift of the *Endurance* (Shackleton, 1920). Here, the low ocean heat flux (Martinson & Iannuzzi, 1998) helps to maintain a thick, perennial ice cover. Multi-year sea ice is also seen in coastal areas from the southern Bellingshausen Sea, across the Amundsen Sea, and extending into the eastern Ross Sea. Patches of thick multi-year sea ice are also found along the coast of East Antarctica, predominantly in the Western Pacific Ocean sector. In general, thick sea ice is found in areas of deep snow cover and heavy deformation.

The coastal areas in the Bellingshausen and Amundsen seas, and around the East Antarctic region have some of the highest snowfall rates over the Southern Ocean, which lead to thickening through snow ice production (Maksym & Markus, 2008). The southern Bellingshausen Sea also regularly experiences pronounced compaction episodes in winter due to strong northerly winds that can produce very thick, heavily deformed sea ice (Massom et al., 2006; Massom et al., 2008). This combination of deep snow and heavy deformation in these coastal areas leads to sea ice thicknesses averaging several metres thick (Haas, 1998; Kwok & Maksym, 2014; Williams et al., 2015).

10.4.2 Seasonal ice drift patterns

The spatial patterns of sea ice drift (Figure 10.8) explain much of the large-scale features of seasonal sea ice expansion (Figure 10.2) and ice thickness distribution (Figure 10.7). Ice drift patterns broadly follow prevailing wind patterns, as suggested by the sea level pressure contours overlaid on Figure 10.8, and in accord with observations that sea ice drift in the Antarctic can reasonably be approximated by free drift (Martinson & Wamser, 1990). Broadly, sea ice drift is characterized by an eastward drift in the outer pack (driven by prevailing westerly winds) and a westward drift around the coast (driven by coastal easterly winds). This pattern is broken by two prominent gyres – the Ross and Weddell gyres. The general pattern of northward sea ice drift is consistent with the seasonal expansion of the sea ice cover and the most northward extensions of the ice edge in the Ross and Weddell seas and Indian Ocean sectors (compare the winter and spring patterns of sea ice drift in Figure 10.8 with the mean winter and spring sea ice covers in Figure 10.2). While there are some seasonal variations in drift speed, there is little apparent seasonal variation in the patterns of ice drift, although the northward Ekman drift near the ice edge in spring as described in Section 10.3.3 is apparent.

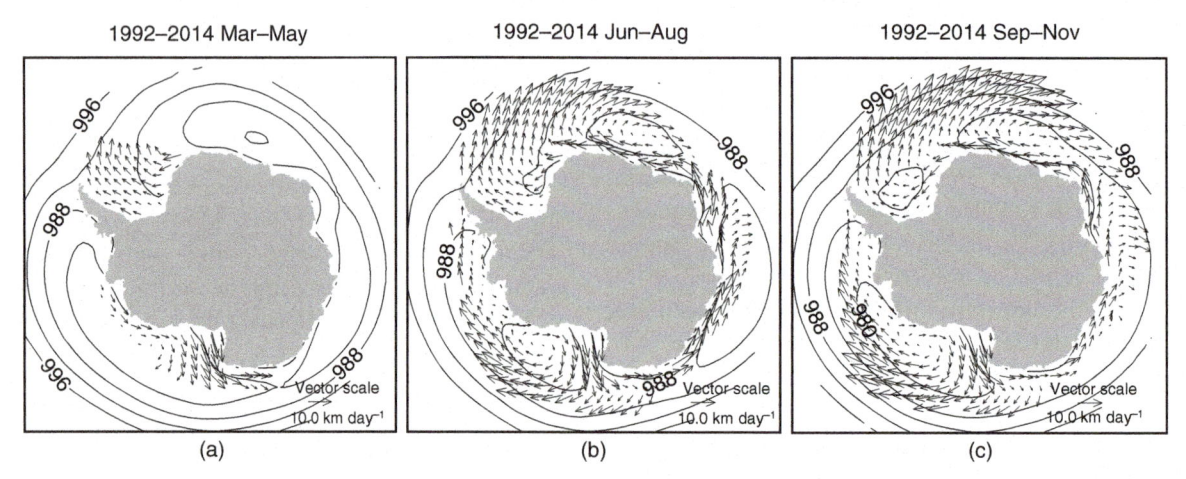

Figure 10.8 Average sea ice drift patterns for: (a) autumn, (b) winter, (c) spring. Drift is determined by maximum cross-correlation of successive images of microwave brightness temperature (Kwok et al., 1998; see also Chapter 9). Few drift vectors can be accurately determined near the ice edge due to rapidly changing conditions, or in summer, so no vectors are available in these cases. Mean sea level pressure contours are from ERA-interim reanalysis (http://data-portal.ecmwf.int/data/d/interim_daily/). (Figure provided by R. Kwok.)

The western limb of the Ross Gyre forms a strong outflow from the Ross ice shelf and Ross Sea polynya, leading to thinner sea ice in the south and along the 180° meridian. The northward drift diverts to the west off Oates Land, presumably supporting the existence of persistent, thick sea ice in that sector and westward to Terre Adelie. In the western Weddell, the sea ice drifts northwards to eventually melt at the northern ice edge (e.g. Massom, 1992). Through winter, the northern arm of the Weddell Gyre produces an ice tongue of thick sea ice, seen in the spring ice thickness map (Figure 10.7b). In some years this ice tongue extends well into the eastern Weddell Sea and persists well into summer even as sea ice to the south has retreated.

The band of thick sea ice stretching along the coast from the Bellingshausen Sea, across the Amundsen Sea, and into the eastern Ross Sea is sustained by a slight southerly component in the sea ice drift towards the coasts in the Bellingshausen and eastern Ross Seas. This promotes deformational thickening, and the slow westward drift then carries this thick ice eventually into the Ross Sea (see also Assmann et al., 2005). The strong drift along the East Antarctic coast probably promotes significant thickening through shear deformation (e.g. Williams et al., 2015).

10.4.3 Trends in winds, ice drift and ice thickness

There is evidence suggesting that westerly winds have intensified over the Southern Ocean since about the mid-1960s (Thompson & Solomon, 2002; Bracegirdle et al., 2008) in association with a strengthening of the SAM (Gong & Wang, 1999; Marshall, 2003; Thompson et al., 2011), which is defined by the difference in sea level pressure between mid- and high latitudes (these changes are described in more detail in Section 10.5). Over the sea ice cover, wind trends are more complex but there appears to be a concurrent deepening of the Amundsen Sea Low – a statistical low pressure centred over the Amundsen Sea (Turner et al., 2013d; 2015).

Trends in sea ice drift are highly correlated with trends in surface winds over much of the sea ice zone in winter (Figure 10.9a) and are consistent with a deepening of the Amundsen Sea Low, in particular (Figure 10.9b). Holland and Kwok (2012) further show that these trends in winds and sea ice drift coincide strongly with regions of increasing sea ice extent and concentration in the Ross Sea and eastern Weddell Sea and with areas of decreasing sea ice extent and concentration in the Bellingshausen Sea (Figure 10.9b).

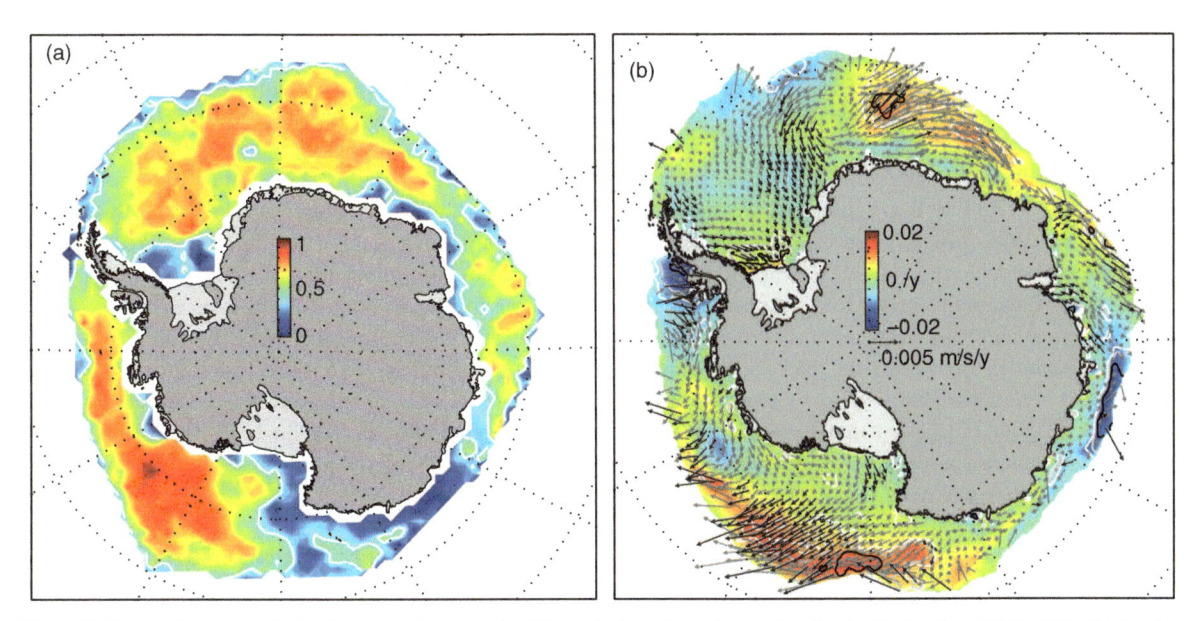

Figure 10.9 (a) Vector correlation between ERA-interim 10 m winds and sea ice motion for April–October, 1992–2010. (b) Sea ice motion trend vectors overlaid on sea ice concentration trends over the same period. Source: From Holland & Kwok, 2012, figure 3a and 1b, pages 3 and 2. Reproduced with permission from Macmillan Publishers Ltd.

However, not all regions show high correlations between surface winds and sea ice drift, such as the western Ross Sea and the eastern sector of the East Antarctic seasonal sea ice zone, where strong ocean currents may dominate ice motion variability (Holland & Kwok, 2012). As demonstrated by Holland and Kwok (2012) and Kimura and Wakatsuchi (2011), the sea ice concentration budget, at least for the inner pack ice, can be decomposed into dynamic and thermodynamic contributions, revealing regions where advective/divergent processes dominate the sea ice concentration changes (Pacific sector and Weddell Sea) versus thermodynamic processes elsewhere. Holland (2014) also points out that while wind trends clearly drive much of the observed trends in sea ice extent, there are ocean feedbacks, which play important roles in the seasonal sea ice trends (see also Section 10.3.3). The sea ice drift and wind trends also correspond well with modelled changes in sea ice thickness, as described next.

There is little information on the large-scale variability and trends of Antarctic ice thickness or snow depth, except for the 2003–2009 record from ICESat. This record is too short to identify any trends, particularly in the light of differences in the treatment of snow. Both Kurtz and Markus (2012) and Xie et al. (2013) suggest that mean sea ice thickness in the Bellingshausen Sea may have increased slightly over this period, while Yi et al. (2011) and Kurtz and Markus (2012) found opposing trends for the Weddell Sea, although none were significant at the 95% level. These trends are much smaller than the interannual variability in mean sea ice thickness, which can range as high as 40 cm. As this thickness variability is a significant fraction of the mean sea ice thickness, this suggests that ice volume may be more variable than sea ice extent. However, as much of the uncertainty in sea ice thickness estimates is due to uncertainty in snow depth (Kern & Spreen, 2015; see Chapter 9), the sea ice thickness variability may largely reflect variability in snow depth. Maksym and Markus (2008) suggested that snow depth on sea ice may have increased in most sectors between 1979 and 2001, except in the Amundsen and Bellingshausen seas where decreases were observed. Increases in snow depth are generally consistent with increases in cyclone frequency (Markus & Cavalieri, 2006).

Because of the short sea ice thickness record, models are the only means available to estimate possible long-term trends in Antarctic sea ice thickness. Several models suggest that Antarctic sea ice thickness may have been increasing over the past two to four decades at a rate comparable to or greater than the overall increase in sea ice extent (Massonnet et al., 2013; Holland et al., 2014; Zhang, 2014). These models generally show consistent regional trends in sea ice thickness, with increases in the Ross and southern Weddell seas, and a decrease in the Bellingshausen Sea (Figure 10.10b). Zhang (2014) attributes the overall thickening of the sea ice cover to intensifying winds driving increases in ice deformation (Figure 10.10a). The trends in sea ice drift and winds also correspond to modelled changes in ice thickness in the southern Bellingshausen Sea, where a corresponding increase in ridging production is consistent with increased northerly winds (e.g. Massom et al., 2006, 2008) and a westward transport of sea ice out of the southern Bellingshausen Sea (e.g. Assmann et al., 2005) that, in turn, contributed to the observed loss of summer sea ice in that sector (Jacobs & Comiso, 1997).

As described in Section 10.1, the ice edge is shaped not just by winds, but also by the interaction between waves and ice. For example, increased storms and waves may enhance sea ice retreat in spring. In autumn and winter, waves might enhance sea ice production through the pancake ice cycle (Chapter 1), or in some way retard sea ice advance. Kohout et al. (2014) suggest that waves may indeed play an important role. Where modelled significant wave heights north of the ice edge have increased, most regions show a corresponding reduction in sea ice extent (i.e. the ice edge has moved southwards) both during the autumn ice edge advance and spring retreat (Figure 10.11). Likewise, where modelled significant wave heights have decreased north of the ice edge, most regions show a corresponding increase in sea ice extent (i.e. the ice edge has moved northwards). As significant wave height is driven by winds and available fetch, this may mostly reflect the influence of changes in wind speed and direction at the ice edge, but it clearly illustrates the importance of regional forcing in driving regional sea ice variability.

10.5 Ice-climate connections

While winds and ice–ocean feedbacks are the largest proximate causes for forcing variability and change in the Antarctic sea ice cover, ultimately these changes are

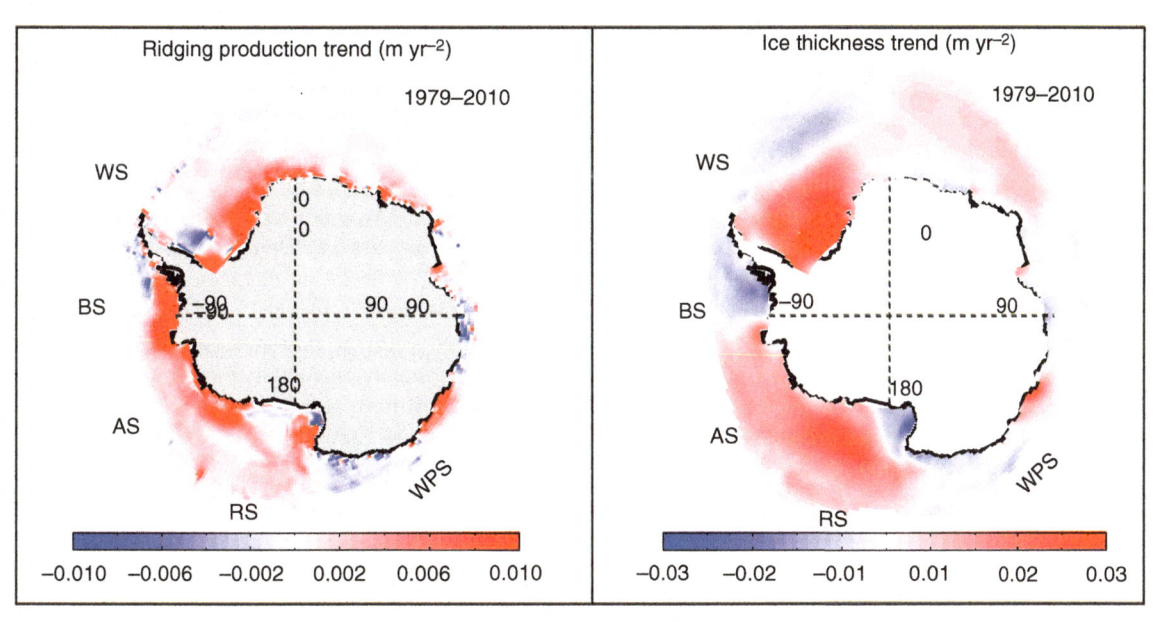

Gaining (and losing) Antarctic sea ice: variability, trends and mechanisms | 277

Figure 10.10 (a) Modelled trends in sea ice ridging production. (b) Modelled trends in sea ice thickness. RS, Ross Sea; WS, Weddell Sea; WPS, Western Pacific Sector; BS, Bellingshausen Sea; AS, Amundsen Sea. Source: From Zhang, 2012. Reproduced with permission of Jinlun Zhang.

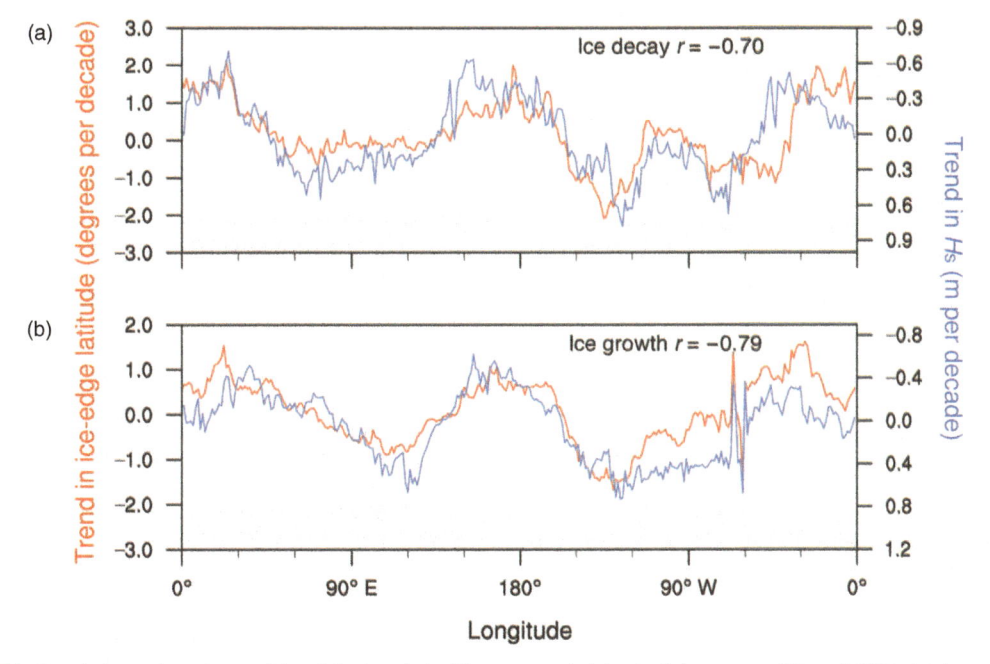

Figure 10.11 Trends in sea ice edge position (blue) and significant wave height (red) between 1979 and 2009. (a) Average trends during sea ice retreat (September–February). (b) Average trends during sea ice advance (March–August). H_S, ice thickness. Source: From Kohout et al., 2014, figure 4, page 3. Reproduced with permission from Macmillan Publishers Ltd.

driven by variability and change in large-scale climate. The climate of the Southern Ocean and Antarctica has undergone significant change over the satellite era. There has been a significant positive trend in SAM in summer and autumn (Gong & Wang, 1999; Marshall, 2003; Thompson et al., 2011), implying that westerly winds have also intensified (Thompson & Wallace, 2000). These changes have been linked to springtime decreases in stratospheric ozone and increases in greenhouse gases (e.g. Thompson et al., 2011), but other climate linkages may be involved as well (see later). Over the last 50+ years, surface air temperatures have increased over all of Antarctica (Steig et al., 2009), but increases have been most pronounced over West Antarctica (Steig et al., 2009; Bromwich et al., 2012; Nicolas & Bromwich, 2014), including the Antarctic Peninsula region (Turner et al., 2013b). There have been observed increases in precipitation (Turner et al., 2005; Thomas et al., 2008), ocean warming (Meredith & King, 2005; Jacobs, 2006; Gille, 2008; Meredith et al., 2008) and regional ocean freshening (Jacobs & Giulivi, 2010; Rye et al., 2014; see Chapter 8).

Various climate modes of variability affect the synoptic and large-scale atmospheric circulation at high southern latitudes, including the SAM (Gong & Wang, 1999; Thompson & Wallace, 2000; Thompson & Solomon, 2002) and the high-latitude atmospheric response to ENSO (Turner, 2004). Sea-level pressure (SLP) anomalies associated with these two modes of variability are shown in Figure 10.12. SAM variability is characterized by zonal atmospheric pressure anomalies of opposite sign between Antarctica and mid-latitudes, but it also includes a non-zonal or non-annular component in the Amundsen Sea region (centered approximately on 70°S and 120°W). Positive atmospheric pressure anomalies at mid-latitudes together with negative anomalies at high latitudes indicate a positive SAM, and vice-versa for a negative SAM. During a positive SAM, the circumpolar westerly winds strengthen over the Southern Ocean and the low pressure in the Amundsen Sea deepens, increasing warm northerly flow over the eastern Bellingshausen Sea and cold southerly flow over the Ross Sea.

A similar SLP pattern in the high latitude southeast Pacific is shown for ENSO, although its anomaly center is shifted to the northeast (~60°S, 100°W) and shows a wave-2 instead of a wave-1 structure. During La Niña, SLP anomalies in the southeast Pacific sector are negative, which strengthens warm northerly flow over the eastern Bellingshausen and western Weddell Seas and cold southerly flow over the eastern Ross Sea (and vice-versa for an El Niño).

Several studies have explored potential connections between Southern Ocean sea ice and SAM (Hall & Visbeck, 2002; Kwok & Comiso, 2002; Lefebvre et al., 2004; Liu et al., 2004; Lefebvre & Goosse, 2005; Turner et al., 2009). Lefebvre et al. (2004) and Turner et al. (2009) show that the regional sea ice response to SAM is associated with the non-annular spatial component of SAM variability, such that during a positive SAM the warm northerly winds along the eastern limb of the low-pressure centre contribute to negative sea ice anomalies in the eastern Bellingshausen and western Weddell seas, and vice versa for the cold southerly winds flowing over the Ross Sea. A somewhat similar sea ice variability pattern occurs in response to La Niña (and vice versa for negative SAM and El Niño).

Yuan (2004) provides a conceptualization of potential mechanisms proposed for the high-latitude ENSO teleconnection and sea ice response (see Figure 10.13). One of the primary features of this teleconnection is the weakening of the subtropical jet and a strengthening of the polar front jet in the South Pacific during La Niña events (Rind et al., 2001; Lachlan-Cope & Connolley, 2006) (Figure 10.13b). The strengthening of the polar front jet leads to more storms, warmer conditions and less sea ice in the southern Bellingshausen and western Weddell seas, but colder conditions and more sea ice in the Amundsen and Ross seas (in association with the spin-up of the quasi-stationary low-pressure cell in the Amundsen Sea). The opposite scenario applies to El Niño events (Figure 10.13a). This sea ice anomaly pattern between the Bellingshausen/Weddell seas and Amundsen/Ross seas during ENSO events is referred to as the Antarctic Dipole (ADP) (Yuan & Martinson, 2001).

However, there are other modes of climate variability, including zonal wave variability (Yuan & Li, 2008; Hobbs & Raphael, 2010; Raphael & Hobbs, 2014), and the leading mode of tropical and North Atlantic sea surface temperature variability, the Atlantic Multidecadal Oscillation (AMO) (Li et al., 2014). The AMO affects atmospheric circulation at high southern latitudes through similar teleconnections as the ENSO (e.g. Rossby wave propagation of atmospheric perturbations). An AMO influence is consistent with sea ice studies showing strong variance at decadal

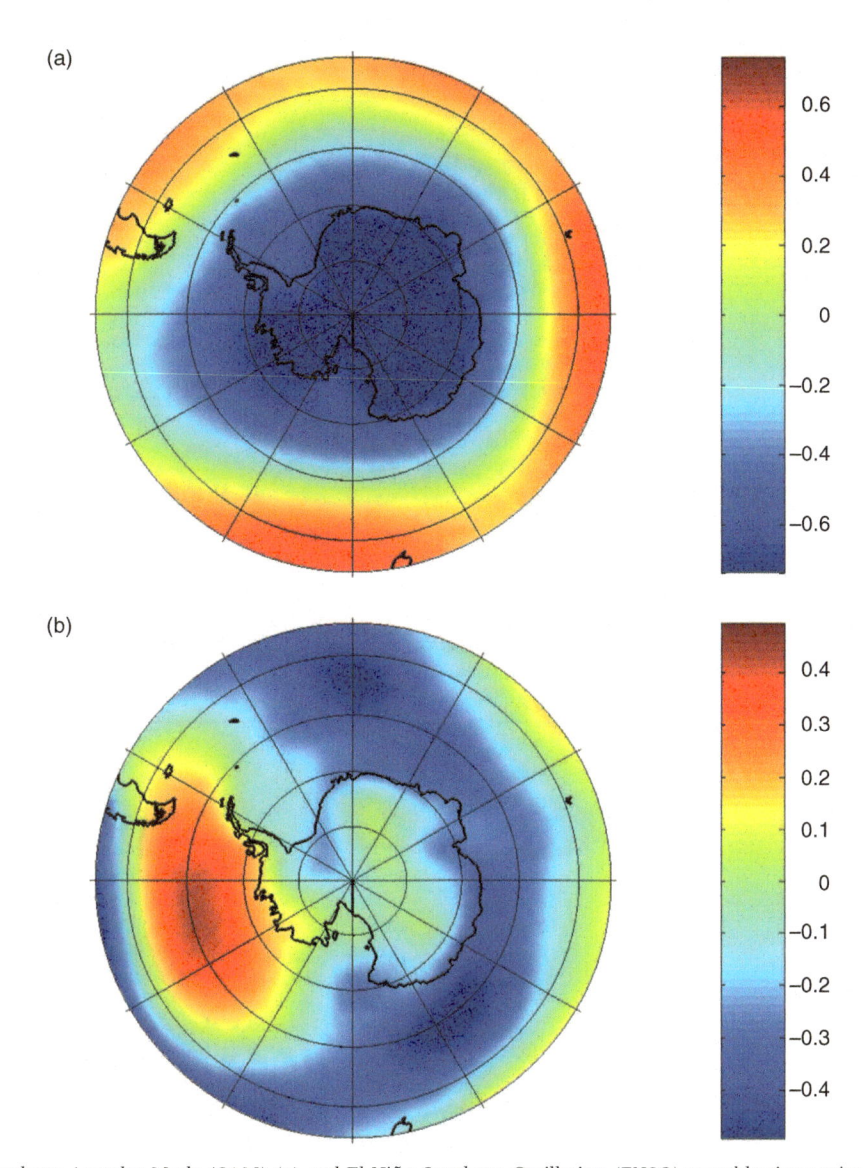

Figure 10.12 Southern Annular Mode (SAM) (a) and El Niño-Southern Oscillation (ENSO) monthly time series (1980–2004) (b) correlated with surface atmospheric pressure. The SAM index is from Marshall (2003; available at http://www.nerc-bas.ac.uk/icd/gjma/sam.html). The ENSO index is the Niño3.4 index of Cane et al. (1986). Sea-level pressure data are from the National Center of Environmental Prediction and National Center for Atmospheric Research Reanalysis project (Kalnay et al., 1996). (From Meredith et al., 2010, figure 3, p. 131. Reproduced with permission of Elsevier Limited.)

timescales in the Ross-Amundsen Sea sector, while an ENSO/SAM influence is consistent with sea ice studies showing strong variance at interannual timescales in the Bellingshausen/Amundsen seas (Simpkins et al., 2013; Raphael & Hobbs, 2014). Co-occurring climate modes can also amplify or dampen the regional atmospheric response. As described earlier, if a positive SAM co-occurs with La Niña (or a negative SAM with El Niño), this appears to anomalously intensify (weaken) the quasi-stationary atmospheric low-pressure system in the Amundsen Sea (Stammerjohn et al., 2008; Fogt et al., 2012; Turner et al., 2013d).

Connections between sea ice variability and these large-scale circulation modes or patterns have distinct

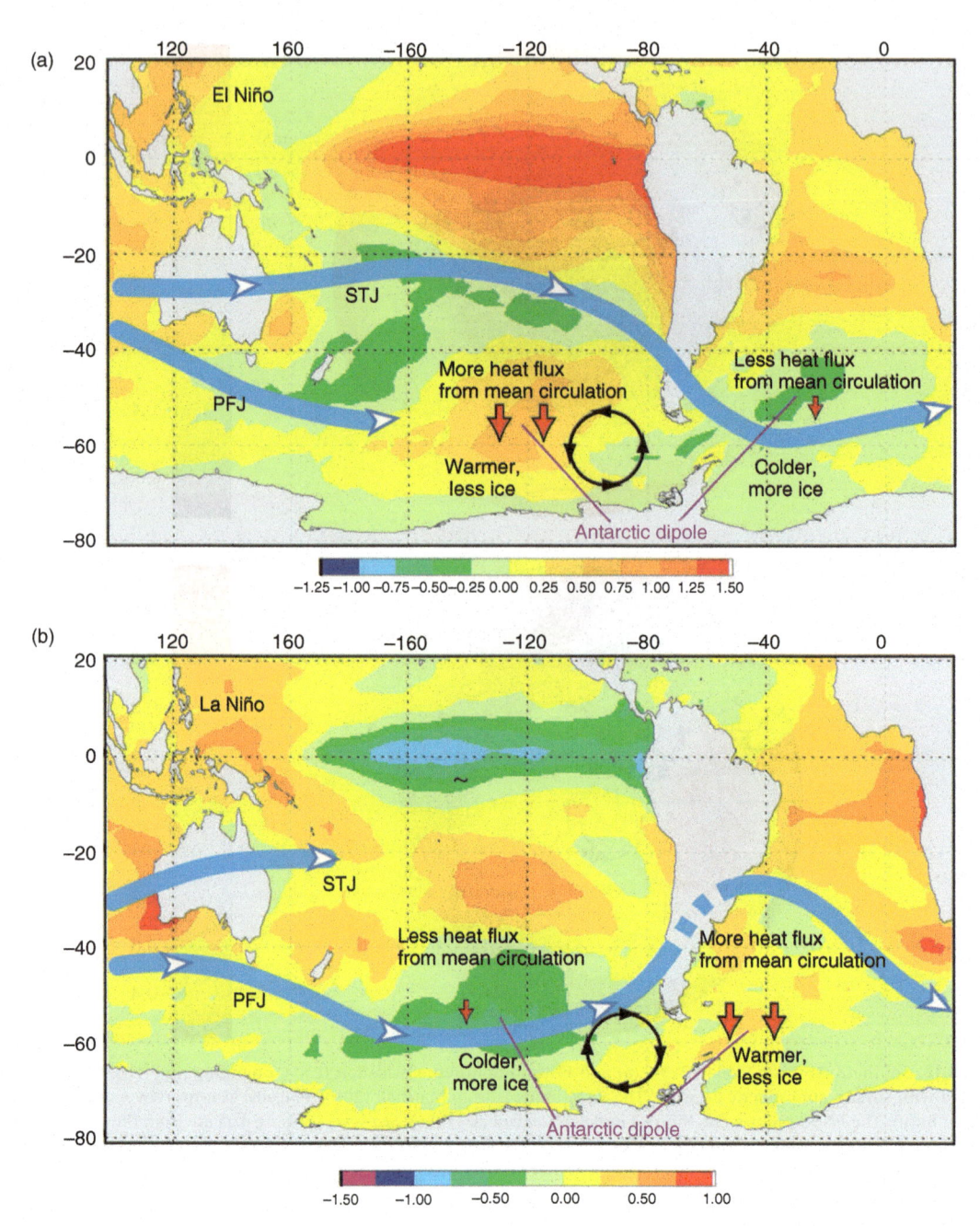

Figure 10.13 A schematic of El Niño (a) and La Niña (b) conditions showing sea surface temperature anomaly (°C) composites for each El Niño-Southern Oscillation (ENSO) phase. The positions of the subtropical jet (STJ) and polar front jet (PFJ) are shown for illustration, as well as the positions of high and low sea-level pressure anomalies, anomalous heat fluxes and the Antarctic dipole nodes for each phase. Source: From Yuan, 2004, figure 5, page 420. Reproduced with permission from Cambridge University Press.

seasonal and regional signatures (Raphael & Hobbs, 2014; Kohyama & Hartmann, 2016; Hobbs et al., 2016). For example, the high-latitude atmospheric response to ENSO variability is strongest in spring–summer (Fogt & Bromwich, 2006) and reveals an ADP pattern, with out-of-phase sea ice anomalies centred on the eastern Bellingshausen/western Weddell Sea sector and the eastern Ross/western Amundsen Sea sector (Figure 10.13; Yuan & Martinson, 2001; Yuan, 2004). The ENSO-related sea ice anomalies affect the spring–summer sea ice retreat in the dipole areas. Ocean thermal feedbacks can subsequently affect anomalies in the autumn sea ice advance, thus propagating the ENSO-related perturbation into the subsequent autumn–winter (Nihashi & Ohshima, 2001; Stammerjohn et al., 2012; Holland, 2014). The SAM has become more positive in summer and autumn (Marshall 2003, Abram et al., 2014), and connections between positive SAM-related atmospheric circulation anomalies and the autumn sea ice advance have been observed (Stammerjohn et al., 2008; Turner et al., 2009). Finally, the high-latitude response to Atlantic sea surface temperature warming is strongest in winter and appears to affect the Bellingshausen/Amundsen seas (Li et al., 2014).

There has been considerable attention focused on the mechanisms and possible causes of the net Antarctic sea ice increases as observed over the satellite era, particularly in response to the recent record high sea ice extents. Possible mechanisms for increased sea ice extent might include increased ocean stratification due to warmer air and sea surface temperatures (e.g. Zhang, 2007); increased surface freshening from increased precipitation (e.g. Liu & Curry, 2010) or glacial melt (Bintanja et al., 2013); and/or changes in winds in response to ozone depletion and greenhouse gas increases driving net sea ice extent increases (e.g. Turner et al., 2009; Haumann et al., 2014). For the latter, however, there is also the suggestion that the initial short-term response to ozone depletion is one of surface cooling and sea ice increase, followed by a longer-term warming and sea ice decrease (Marshall et al., 2014; Ferreira et al., 2015). Given that the net Antarctic sea ice increase is also the product of strong and opposing regional sea ice changes, it is very likely that there are several mechanisms and ice–climate connections (many of which have been described earlier) at work. Understanding the causes of the net increase necessarily requires an understanding of the stronger regional trends and the climate linkages that drive them (Hobbs et al., 2015). There is also the question of how the satellite record of Antarctic sea ice extent compares with any available historical records prior to 1979, a topic we explore next.

10.6 The historical perspective

The positive circumpolar sea ice extent trends discussed in the previousl section are in stark contrast to climate model predictions that show a modest decline over the satellite era (Maksym et al., 2012; Turner et al., 2013c; Hobbs et al., 2015). As the observed trends are also modest and, indeed, a consequence of large, contrasting regional trends that are themselves strongly influenced by various modes of climate variability, an important question is whether current trends are anomalous in a longer-term context. That a long-term, historically anomalous decline of sea ice is taking place in the Arctic is certain (Chapter 11); it is not yet clear how Antarctic sea ice has changed over the longer-term.

Prior to the continuous SMMR-SSMI-SSMIS passive microwave satellite record available since late 1978, there were a few satellite missions that provided passive microwave, visible and infrared imagery of the Antarctic sea ice cover. An early passive microwave sensor, the electrically scanning microwave radiometer (ESMR) on the Nimbus-5 satellite, provided imagery from December 1972 to March 1977 (see also Chapters 9 and 11). This instrument had only a single frequency radiometer and there was no overlap period with SMMR, so sea ice concentrations determined from ESMR are difficult to compare with those from SMMR. To bridge this gap, Cavalieri et al. (2003) used sea ice charts from the US Navy/ National Oceanic and Atmospheric Administration Joint Ice Center (now the National Ice Center) to produce an ice edge estimate that could be compared with both the ESMR and SMMR/SSMI data and provide a consistent ice extent record between 1973 and 2002. This analysis showed that overall sea ice extent decreased dramatically from 1973 to 1976, before a long, slow recovery over the course of the next few decades (Figure 10.14). The anomalously low ice extent for 1977 (when no passive microwave data were available) may indicate an issue with the accuracy of the ice charts. The decline over the ESMR period occurred in all seasons, but was strongly regional, with

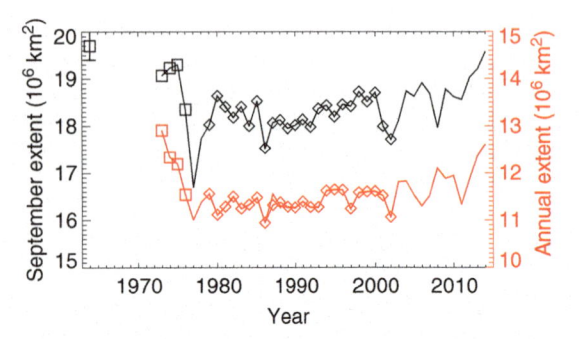

Figure 10.14 Sea ice extent from a merged electrically scanning microwave radiometer/scanning multichannel microwave radiometer/special sensor microwave/imager (ESMR-SMMR-SSMI) dataset between 1973 and 2002 (Cavalieri et al., 2003), updated with SSMI/special sensor microwave imager/sounder (SSMI-SSMIS) data for 2003–14. To maintain a consistent time series, the updated data were adjusted by the mean offset between the overlapping years (1979–02) in the two time series. September (black) and mean annual (red) sea ice extents show a decrease in ice extent in the mid-1970s, with a rebound occurring over the past several decades. Squares indicate ESMR years, and diamonds indicate SSMR/SSMI data from Cavalieri et al. (2003). There are few passive microwave data available for 1977 and 1978 (no symbols); therefore, sea ice extent in those years was based on ice charts from visible and infrared data. Sea ice extent for September 1964 estimated from Nimbus I visible imagery (Meier et al., 2013) is indicated by the single square with uncertainty bars.

the greatest reduction in ice extent occurring in the Weddell Sea, coinciding with the occurrence of the Weddell polynya from 1974 to 1976 (Zwally et al., 1983; see also Chapter 8).

Early comparisons with weather satellite imagery (Predoehl, 1966; Sissala et al., 1972; Streten, 1973) suggested that this decline was preceded by a comparable increase in sea ice extent in the 1960s (Zwally et al., 1983). A more reliable recent analysis of visible imagery from the Nimbus I and II missions show that the September sea ice extent in 1964 exceeded that found in every year of the passive microwave record (Meier et al., 2013; also shown in Figure 10.14), though was comparable to recent values. Early winter extents in 1964 and 1966 were also higher than in the modern record (Gallaher et al., 2014), although some caution is warranted as they also report an anomalously low sea ice extent in August 1966. This may reflect uncertainties in the interpretation of the imagery, but it may also indicate much greater variability in the recent past, as suggested by additional unpublished Nimbus imagery, which show years of both anomalously low and high sea ice extents. While there remains a need to produce a more consistent record from these early satellite data, they suggest that the increase in sea ice extent over the most recent 36 years (1979–2014) may be a reversal of a strong negative trend prior to 1979 that resulted from similarly high sea extents in the 1960s to early 1970s (at least in winter) followed by low sea ice extents in the mid-to-late 1970s.

Extending the record before the advent of satellite remote sensing is more challenging; the only indication of sea ice extent has been from occasional ship-based observations. Reports of ice edge position by explorers are too limited to determine if a substantial decrease in sea ice extent has occurred since the 18th and 19th centuries (Kukla & Garvin, 1981; Parkinson, 1990). More frequent observations are available from various expeditions between the 1930s and early 1960s, most notably from the British Discovery Investigations (Mackintosh, 1972; see Chapter 8). Comparisons between those records and the early satellite data have suggested that sea ice extent was significantly greater in the early 20th century (Kukla & Garvin, 1981; Rayner et al., 2003; Titchner & Rayner, 2014). However, care must be taken when comparing satellite and *in situ* observations, as the passive microwave ice edge is often 1–2° south of the true ice edge during the melt season (Worby & Comiso, 2004). Ackley et al. (2003) proposed that this difference could explain the apparent decline.

Because direct ship observations are so sparse in the pre-satellite period, proxy records may provide a better estimate of early 20th-century sea ice extent. De la Mare (1997, 2009) noted that whaling fleets operating between the 1930s and 1980s recorded the position of each catch, and that whales were typically taken close to the ice edge. Using the southernmost catch in each 10° longitude band, he suggested that the ice edge location during the melt season (October–April) shifted southwards by 1.89–2.80° latitude between the 1930s–1950s and 1970s–1980s. The reliability of these results has been questioned as there is a gap in the whaling records between the two periods (1930–1950 and 1970–1980), with a shift in species taken from predominantly blue whales in the early period to minke whales in the later period. There are other questions, too, including whether a whale-catch location near the ice edge is comparable to a passive microwave ice edge position, and regarding the substantial differences in the regional distribution of whale catches between the two periods (Ackley et al., 2003). However, de la Mare (2009) has shown that the whale-catch data are consistent with previous ship reports for the early period, and ice charts for the later period, although the decline based on early ship reports alone is a more modest 1.71° latitudinal shift (de la Mare, 2009).

The whaling data suggest that reductions have been greatest (as much as a 4° southerly shift) in the Weddell Sea and of a lesser degree in the adjacent Indian Ocean sector, with decreases less clear elsewhere (Cotte & Guinet, 2007; de la Mare, 2009). The decline in the Weddell Sea is consistent in sign with a century-long record of sea ice advance and retreat from the South Orkney Islands (Murphy et al., 2014), but the latter record suggests a much smaller decrease in sea ice extent (Abram et al., 2010). The apparent substantial decrease in the Weddell Sea may reflect changes in the eastward extent of the Weddell ice tongue, a remnant band of sea ice that extends eastwards across the northern Weddell Sea in summer, rather than a dramatic overall decrease in sea ice extent.

Proxy records from continental ice cores may also provide an indication of past sea ice extent (Abram et al., 2013). The most established of these is methanesulphonic acid (MSA), a by-product of primary productivity in surface waters. In many regions, MSA concentration in ice core records is positively correlated with sea ice extent, presumably resulting from spring blooms within the pack ice (Curran et al., 2003). In other areas, however, MSA may be negatively correlated with sea ice extent, such as in the Weddell Sea where it reflects increased cold offshore winds that limit MSA deposition (Abram et al., 2007), or in the Amundsen and Ross Sea regions, where it reflects productivity in coastal polynyas and summer absence of sea ice (Rhodes et al., 2009; Criscitiello et al., 2013). Curran et al. (2003) found a strong positive relationship between East Antarctic winter sea ice extent and MSA concentration in an ice core from Law Dome. Their data suggest a 20% decline in sea ice in the 80–140°E sector since the 1950s, corroborating the whaling and ship-based observations (although covering different seasons). They also observed large cyclic variability, with a period of about 11 years, which could mask long-term trends when looking at a shorter record.

In the Bellingshausen Sea, Abram et al. (2010) also show positive correlations between MSA in sea ice cores from the Antarctic Peninsula region and winter sea ice extent. These data suggest a long-term retreat in winter sea ice extent of about 0.7° over the last century in the Bellingshausen Sea, with the trend becoming more pronounced since the 1950s. This long-term trend is consistent with observed and inferred warming on the peninsula (Jones, 1990; Vaughan et al., 2003; Thomas et al., 2009). Using a different ice core proxy, Sinclair et al. (2014) suggest a modest decrease (~5%) in mean annual sea ice area in the Ross Sea between the 1950s and 1980s, which is comparable to the summer change reported for this region as derived from the whaling data (de la Mare, 2009).

Despite uncertainties in reconstructions of sea ice extent from historical ship records, whaling records and ice core proxies, all these data paint a similar picture – a significant decline in sea ice extent may have occurred between the 1930s–1950s and the 1980s, but with substantial regional variability in the magnitude and timing. In summary, the evidence suggests that the current positive trends in the Weddell and Ross Seas may be a partial recovery from earlier decreases, but the declines in the Bellingshausen are a continuation of a long-term retreat. Some of the records also exhibit significant variability (Curran et al., 2003; Abram et al., 2010), which may vary in periodicity (Murphy et al., 2014). A better understanding of this variability may aid in the interpretation of current trends and variability.

Given the dramatic sea ice declines suggested by several of these records, is there evidence for a substantial climate shift to support such a strong decrease? Here, evidence is equivocal, but several recent studies suggest that this longer context is important when considering potential modes and mechanisms of sea ice variability (Marshall et al., 2014; Ferreira et al., 2015; Hobbs et al., 2015; Kohyama & Hartmann, 2016; Hobbs et al., 2016). Consistent trends among sea ice extent, sea surface and air temperatures, winds, and sea level pressure show a reversal of surface climate trends between 1950–1978 and 1979–2011 (Fan et al., 2014). Thus, present-day sea ice trends may be partly a reflection of low-frequency climate variability. Links between sea ice spatial variability and low-frequency variability in the tropics (Li et al., 2014; Simpkins et al., 2014; Kohyama & Hartmann, 2016) also imply that current sea ice extent trends are plausibly a reversal from a previous downward trend. On the other hand, the SAM appears to have been increasing since the 1960s (Marshall,

2003), and is well correlated to the long-term warming on the Peninsula (Nicolas & Bromwich, 2014).

Because of the complex processes involving sea ice changes and feedbacks, and their regional connectivity to global climate change, considerable progress must still be made before we can understand past and present-day changes well enough to predict future changes with any confidence. There is considerable scope for improved understanding of sea ice variability in the recent past and its causes, and particularly promising are the new proxies from ice cores (Abram et al., 2013). Without a doubt, this is an area of active research requiring continued investigations, new ideas and new approaches.

10.7 Acknowledgements

The authors wish to thank the sea ice community involved in the creation of this book as well as the following people who generously provided material or inspiration for this chapter: Steve Ackley, Don Cavalieri, Joey Comiso, Ryan Fogt, Will Hobbs, Paul Holland, Alison Kohout, Nathan Kurtz, Ron Kwok, Rob Massom, Mike Meredith, Claire Parkinson, Marilyn Raphael, Phil Reid, Graham Simpkins, Tony Worby, Xiaojun Yuan, and Jinlun Zhang. Special thanks goes to David Thomas, who patiently encouraged us through several extended deadlines. Sharon Stammerjohn would also like to acknowledge support and contributions made possible through the National Science Foundation Office of Polar Programs Palmer Long-Term Ecological Project (ANT-1440435). Ted Maksym would like to acknowledge support from National Science Foundation grant ANT-1142075.

References

Abram, N.J., Wolff, E.W. & Curran, M.A. J. (2013) A review of sea ice proxy information from polar ice cores. *Quaternary Science Reviews*, **79**, 168–183.

Abram, N.J., Mulvaney, R., Wolff, E.W. & Mudelsee, M. (2007) Ice core records as sea ice proxies: an evaluation from the Weddell Sea region of Antarctica. *Journal of Geophysical Research*, **112**, D15101, doi: 10.1029/2006JD008139.

Abram, N.J., Thomas, E.R., McConnell, J.R., Mulvaney, R., Bracegirdle, T.J., Sime, L.C. & Aristarain, A.J. (2010) Ice core evidence for a 20th century decline of sea ice in the Bellingshausen Sea, Antarctica. *Journal of Geophysical Research*, **115**, D23101, doi: 10.1029/2010JD014644.

Ackley, S., Wadhams, P., Cosimo, J. & Worby, A. (2003) Decadal decrease of Antarctic sea ice extent inferred from whaling records revisited on the basis of historical and modern sea ice records. *Polar Research*, **22**, 19–25, doi: 10.1111/j.1751–8369.2003.tb00091.x.

Ackley, S.F., Lange, M.A. & Wadhams, P. (1990) Snow cover effects on Antarctic sea ice thickness, paper presented at Sea Ice Properties and Processes: *Proceedings of the W.F. Weeks Sea Ice Symposium held December 1988*. U.S. Army Cold Regions Research and Engineering Laboratory, San Francisco, CA.

Arzel, O., Fichefet, T. & Goosse, H. (2006) Sea ice evolution over the 20th and 21st centuries as simulated by current AOGCMs. *Ocean Modelling*, **12**, 401–415.

Assmann, K.M., Hellmer, H.H. & Jacobs, S.S. (2005) Amundsen Sea ice production and transport. *Journal of Geophysical Research*, **110**, C12013, doi: 10.1029/2004jc002797.

Bintanja, R., van Oldenborgh, G.J., Drijfhout, S.S., Wouters, B. & Katsman, C.A. (2013) Important role for ocean warming and increased ice-shelf melt in Antarctic sea-ice expansion. *Nature Geoscience*, **6**, 376–379.

Bracegirdle, T.J., Connolley, W.M. & Turner, J. (2008) Antarctic climate change over the twenty first century. *Journal of Geophysical Research*, **113**, D03103, doi: 10.1029/2007jd008933.

Bromwich, D.H., Nicolas, J.P., Monaghan, A.J. et al. (2012) Central West Antarctica among the most rapidly warming regions on Earth. *Nature Geoscience*, **6**, 139–145.

van den Broeke, M.R. (2000) The semi-annual oscillation and Antarctic Climate. Part 4: A note on sea ice cover in the Amundsen and Bellingshausen Seas. *International Journal of Climatology*, **20**, 455–462.

Cane, M.A., Zebiak, S.E. & Dolan, S.C. (1986) Experimental forecasts of El Niño. *Nature*, **322**, 827–832.

Cavalieri, D.J. & Parkinson, C.L. (2012) Arctic sea ice variability and trends, 1979–2010. *The Cryosphere*, **6**, 881–889.

Cavalieri, D.J., Parkinson, C.L. & Vinnikov, K.Y. (2003) 30-Year satellite record reveals contrasting Arctic and Antarctic decadal sea ice variability. *Geophysical Research Letters*, **30**, doi: 10.1029/2003GL018031.

Cavalieri, D.J., Markus, T. & Comiso, J.C. (2014) *AMSR-E/Aqua Daily L3 12.5 km Brightness Temperature, Sea Ice Concentration, & Snow Depth Polar Grids*, Version 3, edited, NASA National Snow and Ice Data Center Distributed Active Archive Center, 10.5067/AMSR-E/AE_SI12.003. Boulder, Colorado, USA.

Comiso, J. (2010) Variability and trends of the global sea ice cover. In: *Sea Ice*, 2nd edn (Eds. D.N. Thomas & G.S. Dieckmann), pp. 205–246. Wiley-Blackwell, Oxford.

Comiso, J.C. (2000) *Bootstrap sea ice concentrations from Nimbus-7 SMMR and DMSP SSM/I-SSMIS, version 2* [1978–2014 used], edited, National Snow and Ice Data Center, Boulder, CO. Digital media available online at http://nsidc.org/data/nsidc-0079.

Comiso, J.C., Cavalieri, D., Parkinson, C. & Gloersen, P. (1997) Passive microwave algorithms for sea ice concentration – a comparison of two techniques. *Remote Sensing of the Environment*, **60**, 357–384.

Cotte, C. & Guinet, C. (2007) Historical whaling records reveal major regional retreat of Antarctic sea ice. *Deep Sea Research Part I*, **54**, 243–252.

Criscitiello, A.S., Das, S.B., Evans, M.J., Frey, K.E., Conway, H., Joughin, I. et al. (2013) Ice sheet record of recent sea-ice behavior and polynya variability in the Amundsen Sea, West Antarctica. *Journal of Geophysical Research: Oceans*, **118**, doi: 10.1029/2012jc008077.

Curran, M.A. J., van Ommen, T.D., Morgan, V.I., Phillips, K.L. & Palmer, A.S. (2003) Ice core evidence for Antarctic sea ice decline since the 1950s. *Science*, **302**, 1203–1206.

Ducklow, H., Fraser, W., Meredith, M. et al. (2013) West Antarctic Peninsula: an ice-dependent coastal marine ecosystem in transition. *Oceanography*, **26**, 190–203.

Eisenman, I., Meier, W.N. & Norris, J.R. (2014) A spurious jump in the satellite record: is Antarctic sea ice really expanding? *The Cryosphere Discussions*, **8**, 273–288.

Enomoto, H. & Ohmura, A. (1990) Influences of atmospheric half-yearly cycle on the sea ice extent in the Antarctic. *Journal of Geophysical Research*, **95**, 9497–9511.

Fan, T., Deser, C. & Schneider, D.P. (2014) Recent Antarctic sea ice trends in the context of Southern Ocean surface climate variations since 1950. *Geophysical Research Letters*, **41**, doi: 10.1002/2014GL059239.

Ferreira, D., Marshall, J., Bitz, C.M., Solomon, S. & Plumb, A. (2015) Antarctic Ocean and Sea Ice Response to Ozone Depletion: A Two-Time-Scale Problem. *Journal of Climate*, **28**, 1206–1226.

Fogt, R.L. & Bromwich, D.H. (2006) Decadal variability of the ENSO teleconnection to the high latitude South Pacific governed by coupling with the Southern Annular Mode. *Journal of Climate*, **19**, 979–997.

Fogt, R.L., Wovrosh, A.J., Langen, R.A. & Simmonds, I. (2012) The characteristic variability and connection to the underlying synoptic activity of the Amundsen-Bellingshausen Seas Low. *Journal of Geophysical Research*, **117**, doi: 10.1029/2011jd017337.

Fraser, A.D., Massom, R.A., Michael, K.J., BK, G.-F. & Lieser, J.L. (2012) East Antarctic landfast sea ice distribution and variability, 2000–08. *Journal of Climate*, **25**, 1137–1156.

Gallaher, D.W., Campbell, G.G. & Meier, W.N. (2014) Anomalous variability in Antarctic sea ice extents during the 1960s with the use of Nimbus data. *IEEE Journal of Selected Topics in Applied Earth Observations and remote sensing*, **7**, 881–887.

Gille, S.T. (2008) Decadal-scale temperature trends in the Southern Hemisphere Ocean. *Journal of Climate*, **21**, 4749–4765.

Gloersen, P., Campbell, W.J., Cavalieri, D.J., Comiso, J.C., Parkinson, C.L. & Zwally, H.J. (1992) *Arctic and Antarctic Sea Ice, 1978–1987: Satellite Passive-Microwave Observations and Analysis*. National Aeronautics and Space Administration, Washington, DC.

Gong, D. & Wang, S. (1999) Definition of Antarctic oscillation index. *Geophysical Research Letters*, **26**, 459–462.

Gordon, A.L. (1981) Seasonality of southern ocean sea ice. *Journal of Geophysical Research*, **86**, 4193–4197.

Haas, C. (1998) Evaluation of ship-based electromagnetic-inductive thickness measurements of summer sea-ice in the Bellinghausen and Amundsen Seas, Antarctica. *Cold Regions Science and Technology*, **27**, 1–16.

Hall, A. & Visbeck, M. (2002) Synchronous variability in the Southern Hemisphere atmosphere, sea ice and ocean resulting from the Annular Mode. *Journal of Climate*, **15**, 3043–3057.

Haumann, F.A., Notz, D. & Schmidt, H. (2014) Anthropogenic influence on recent circulation-driven Antarctic sea ice changes. *Geophysical Research Letters*, **41**, doi: 10.1002/2014gl061659.

Hobbs, W.R. & Raphael, M.N. (2010) The Pacific zonal asymmetry and its influence on Southern Hemisphere sea ice variability. *Antarctic Science*, **22**, 559–571.

Hobbs, W.R., Bindoff, N.L. & Raphael, M.N. (2015) New perspectives on observed and simulated Antarctic sea ice extent trends using optimal fingerprinting techniques. *Journal of Climate*, **28**, 1543–1560.

Hobbs, W.R., Massom, R., Stammerjohn, S., Reid, P., Williams, G. & Meier, W. (2016) A review of recent changes in Southern Ocean sea ice, their drivers and forcings. *Global and Planetary Change*, **143**, doi: 10.1016/j.gloplacha.2016.06.008.

Holland, P.R. (2014) The seasonality of Antarctic sea ice trends. *Geophysical Research Letters*, **41**, doi: 10.1002/2014gl060172.

Holland, P.R. & Kwok, R. (2012) Wind-driven trends in Antarctic sea-ice drift. *Nature Geoscience*, **5**, 872–875.

Holland, P.R., Bruneau, N., Enright, C., Losch, M., Kurtz, N.T. & Kwok, R. (2014) Modeled Trends in Antarctic Sea Ice Thickness. *Journal of Climate*, **27**, 3784–3801.

Ivanova, N., Pedersen, L.T., Tonboe, R.T. et al. (2015) Inter-comparison and evaluation of sea ice algorithms: towards further identification of challenges and optimal approach using passive microwave observations. *The Cryosphere*, **9**, 1797–1817.

Jacobs, S. (2006) Observations of change in the Southern Ocean. *Philosophical transactions. Series A, Mathematical, Physical, and Engineering Sciences*, **364**, 1657–1681.

Jacobs, S.S. & Comiso, J.C. (1997) Climate variability in the Amundsen and Bellingshausen Seas. *Journal of Climate*, **10**, 697–709.

Jacobs, S.S. & Giulivi, C.F. (2010) Large multidecadal salinity trends near the Pacific–Antarctic Continental Margin. *Journal of Climate*, **23**, 4508–4524.

Jones, D.A. & Simmonds, I. (1993) A climatology of Southern Hemisphere extratropical cyclones. *Climate Dynamics*, **9**, 131–145.

Jones, P.D. (1990) Antarctic temperatures over the past century – a study of the early expedition record. *Journal of Climate*, **3**, 1193–1203.

Kalnay, E., Kanamitsu, M., Kistler, R. et al. (1996) The NCEP/NCAR 40-year reanalysis project. *Bulletin of American Meteorological Society*, **77**, 437–471.

Kern, S. & Spreen, G. (2015) Uncertainties in Antarctic sea-ice thickness retrieval from ICESat. *Annals of Glaciology*, **56**, 107–119.

Kimura, N., & Wakatsuchi, M. (2011) Large-scale processes governing the seasonal variability of the Antarctic sea ice. *Tellus A*, **63**, 828–840.

Kohout, A.L., Williams, M.J., Dean, S.M. & Meylan, M.H. (2014) Storm-induced sea-ice breakup and the implications for ice extent. *Nature*, **509**, 7502, 604–607.

Kohyama, T., & Hartmann, D.L. (2016) Antarctic sea ice response to weather and climate modes of variability. *Journal of Climate*, **29**, 721–741.

Kukla, G. & Garvin, J. (1981) Summer Ice and Carbon Dioxide. *Science*, **214**, 497–503.

Kurtz, N.T. & Markus, T. (2012) Satellite observations of Antarctic sea ice thickness and volume. *Journal of Geophysical Research*, **117**, C08025, doi: 10.1029/2012jc008141.

Kwok, R. & Comiso, J.C. (2002) Southern Ocean climate and sea ice anomalies associated with the Southern Oscillation. *Journal of Climate*, **15**, 487–501.

Kwok, R. & Maksym, T. (2014) Snow depth of the Weddell and Bellingshausen sea ice covers from IceBridge surveys in 2010 and 2011: An examination. *Journal of Geophysical Research: Oceans*, **119**, doi: 10.1002/2014jc009943.

Kwok, R., Schweiger, A., Rothrock, D.A., Pang, S. & Kottmeier, C. (1998) Sea ice motion from satellite passive microwave imagery assessed with ERS SAR and buoy motions. *Journal of Geophysical Research*, **103**, 8191–8214.

Lachlan-Cope, T. & Connolley, W. (2006) Teleconnections between the tropical Pacific and the Amundsen-Bellingshausen Sea: role of the El Niño/Southern Oscillation. *Journal of Geophysical Research*, **111**, doi: 10.1029/2005JD006386.

Lefebvre, W. & Goosse, H. (2005) Influence of the Southern Annular Mode on the sea ice-ocean system: the role of the thermal and mechanical forcing. *Ocean Sciences*, **1**, 145–157.

Lefebvre, W., Goosse, H., Timmermann, R. & Fichefet, T. (2004) Influence of the Southern Annular Mode on the sea ice-ocean system. *Journal of Geophysical Research*, **109**, doi: 10.1029/2004JC002403.

Li, X., Holland, D.M., Gerber, E.P. & Yoo, C. (2014) Impacts of the north and tropical Atlantic Ocean on the Antarctic Peninsula and sea ice. *Nature*, **505**, 538–542.

Lieser, J.L., Massom, R.A., Fraser, A.D. et al. (2013) *Position Analysis: Antarctic Sea Ice and Climate Change 2014*. Antarctic Climate & Ecosystems Cooperative Research Centre, Hobart, Tasmania, Australia.

Liu, J. & Curry, J.A. (2010) Accelerated warming of the Southern Ocean and its impacts on the hydrological cycle and sea ice. *Proceedings of the National Academy of Sciences USA*, **107**, 14987–14992.

Liu, J., Curry, J.A. & Martinson, D.G. (2004) Interpretation of recent Antarctic sea ice variability. *Geophysical Research Letters*, **31**, doi: 10.1029/2003GL018732.

van Loon, H. (1967) The half-yearly oscillations in middle and high southern latitudes and the coreless winter. *Journal of Atmospheric Science*, **24**, 472–486.

van Loon, H. (1972) Wind in the Southern Hemisphere. In: *Meteorology of the Southern Hemisphere*, pp. 87–100. American Meteorological Society.

van Loon, H., Kidson, J.W. & Mullan, A.B. (1993) Decadal variation of the annual cycle in the Australian dataset. *Journal of Climate*, **6**, 1227–1231.

Mackintosh, N. (1972) Life cycle of Antarctic krill in relation to ice and water conditions. *Discovery Reports*, **36**, 1–94.

Mahlstein, I., Gent, P.R. & Solomon, S. (2013) Historical Antarctic mean sea ice area, sea ice trends, and winds in CMIP5 simulations. *Journal of Geophysical Research*, **118**, doi: 10.1002/jgrd.50443.

Maksym, T. & Markus, T. (2008) Antarctic sea ice thickness and snow-to-ice conversion from atmospheric reanalysis and passive microwave snow depth. *Journal Geophysical Research*, **113**, C02S12, doi: 10.1029/2006JC004085.

Maksym, T., Stammerjohn, S., Ackley, S. & Massom, R. (2012) Antarctic Sea Ice – A Polar Opposite? *Oceanography*, **25**, 140–151.

de la Mare, W.K. (1997) Abrupt mid-twentieth century decline in Antarctic sea ice extent from whaling records. *Nature*, **389**, 57–60.

de la Mare, W.K. (2009) Changes in Antarctic sea ice extent from direct historical observations and whaling records. *Climate Change*, **92**, 461–493.

Markus, T. & Cavalieri, D.J. (2006) Interannual and regional variability of Southern Ocean snow on sea ice. *Journal of Glaciology*, **44**, 53–57.

Marshall, G.J. (2003) Trends in the Southern Annular Mode from observations and reanalyses. *Journal of Climate*, **16**, 4134–4143.

Marshall, J., Armour, K.C., Scott, J.R. et al. (2014) The ocean's role in polar climate change: asymmetric Arctic and Antarctic responses to greenhouse gas and ozone forcing. *Philosophical Transactions A Mathmatical Physical and Engineering Sciences* **372**, 1–17.

Martinson, D.G. & Wamser, C. (1990) Ice drift and momentum exchange in the winter Antarctic ice pack. *Journal of Geophysical Research*, **95**, 1741–1755.

Martinson, D.G. & Iannuzzi, R.A. (1998) Antarctic ocean-ice interaction: implications from ocean bulk property distributions in the Weddell gyre. In: *Antarctic Sea Ice: Physical Processes, Interactions and Variability* (Ed. M.O. Jeffries), pp. 243–271. American Geophysical Union, Washington, DC.

Massom, R.A. (1992) Observing the advection of sea ice in the Weddell Sea using buoy and satellite passive microwave data. *Journal of Geophysical Research*, **97**, 15559–15572.

Massom, R.A., Stammerjohn, S.E., Smith, R.C. et al. (2006) Extreme anomalous atmospheric circulation in the west Antarctic Peninsula region in austral spring and summer 2001/2, and its profound impact on sea ice and biota. *Journal of Climate*, **19**, 3544–3571.

Massom, R.A., Stammerjohn, S.E., Lefebvre, W. et al. (2008) West Antarctic Peninsula sea ice in 2005: Extreme compaction and ice edge retreat due to strong anomaly with respect to climate. *Journal of Geophysical Research*, **113**, doi: 10.1029/2007JC004239.

Massom, R.A., Reid, P., Stammerjohn, S., Raymond, B., Fraser, A. & Ushio, S. (2013) Change and variability in East Antarctic sea ice seasonality, 1979/80–2009/10. *PloS one*, **8**, e64756, doi: 10.1371/journal.pone.0064756.

Massonnet, F., Mathiot, P., Fichefet, T. et al. (2013) A model reconstruction of the Antarctic sea ice thickness and volume changes over 1980–2008 using data assimilation. *Ocean Modelling*, **64**, 67–75.

Meehl, G.A., Hurrell, J.W. & van Loon, H. (1998) A modulation of the mechanism of the semiannual oscillation in the Southern Hemisphere. *Tellus A*, **50**, 442–450.

Meier, W.N., Gallaher, D. & Campbell, G.G. (2013) New estimates of Arctic and Antarctic sea ice extent during September 1964 from recovered Nimbus I satellite imagery. *The Cryosphere*, **7**, 699–705.

Meredith, M.P. & King, J.C. (2005) Rapid climate change in the ocean west of the Antarctic Peninsula during the second half of the 20th century. *Geophysical Research Letters*, **32**, doi: 10.1029/2005GL024042.

Meredith, M.P., Murphy, E.J., Hawker, E.J., King, J.C. & Wallace, M.I. (2008) On the interannual variability of ocean temperatures around South Georgia, Southern Ocean: forcing by El Niño/Southern Oscillation and the Southern Annular Mode. *Deep-Sea Research Part II*, **55**, 2007 – 2022.

Murphy, E.J., Clarke, A., Abram, N.J. & Turner, J. (2014) Variability of sea-ice in the northern Weddell Sea during the 20th century. *Journal of Geophysical Research: Oceans*, **119**, doi: 10.1002/2013JC009511.

Nicolas, J.P. & Bromwich, D.H. (2014) New Reconstruction of Antarctic Near-Surface Temperatures: Multidecadal Trends and Reliability of Global Reanalyses. *Journal of Climate*, **27**, 8070–8093.

Nihashi, S. & Ohshima, K.I. (2001) Relationship between the sea ice cover in the retreat and advance seasons in the Antarctic Ocean. *Geophysical Research Letters*, **28**, doi: 10.1029/2001gl012842.

Nihashi, S., & Ohshima, K.I. (2015) Circumpolar mapping of Antarctic coastal polynyas and landfast sea ice: relationship and variability. *Journal of Climate*, **28**, 3650–3670.

Parish, T.R. & Bromwich, D.H. (2007) Reexamination of the near-surface airflow over the Antarctic continent and implications on atmospheric circulations at high southern latitudes. *Monthly Weather Review*, **135**, 1961–1973.

Parkinson, C.L. (1990) Search for the Little Ice Age in Southern Ocean sea-ice records. *Annals of Glaciology*, **14**, 221–225.

Parkinson, C.L. (1994) Spatial patterns in the length of the sea ice season in the Southern Ocean. *Journal of Geophysical Research*, **99**, 16327–16339.

Parkinson, C.L. (2002) Trends in the length of the Southern Ocean Sea Ice Season, 1979–1999. *Annals of Glaciology*, **34**, 435–440.

Parkinson, C.L. (2004) Southern Ocean sea ice and its wider linkages: insights revealed from models and observations. *Antarctic Science*, **16**, 387–400.

Parkinson, C.L. & Cavalieri, D.J. (2012) Antarctic sea ice variability and trends, 1979–2010. *The Cryosphere Discussions*, **6**, 931–956.

Predoehl, M.C. (1966) Antarctic Pack Ice—boundaries established from Nimbus 1 pictures. *Science*, **153**, 861–863.

Raphael, M.N. & Hobbs, W. (2014) The influence of the large-scale atmospheric circulation on Antarctic sea ice during ice advance and retreat seasons. *Geophysical Research Letters*, **41**, doi: 10.1002/2014gl060365.

Rayner, N.A., Parker, D.E., Horton, E.B. et al. (2003) Global analyses of sea surface temperature, sea ice, and night marine air temperature since the late nineteenth century. *Journal of Geophysical Research*, **108**, doi: 10.1029/2002JD002670.

Reid, P. & Massom, R.A. (2015) Successive Antarctic sea ice extent records during 2012, 2013, and 2014 [in "State of the Climate in 2014"]. *Bulletin of American Meteorological Society*, **96**, S163-S164.

Reid, P., Stammerjohn, S., Massom, R., Scambos, T. & Lieser, J. (2015) The record 2013 Southern Hemisphere sea-ice extent maximum. *Annals of Glaciology*, **56**, 99–106.

Rhodes, R.H., Bertler, N.A. N., Baker, J.A., Sneed, S.B., Oerter, H. & Arrigo, K.R. (2009) Sea ice variability and primary productivity in the Ross Sea, Antarctica, from methylsulphonate snow record. *Geophysical Research Letters*, **36**, L10704, doi: 10.1029/2009GL037311.

Rind, D., Chandler, M., Lerner, J., Martinson, D.G. & Yuan, X. (2001) Climate response to basin-specific changes in latitudinal temperature gradients and implications for sea ice variability. *Journal of Geophysical Research*, **106**, doi: 10.1029/2000JD900643.

Rye, C.D., Naveira Garabato, A.C., Holland, P.R. et al. (2014) Rapid sea-level rise along the Antarctic margins in response to increased glacial discharge. *Nature Geoscience*, **7**, 732–735.

Santer, B.D., Wigley, T.M. L., Boyle, J.S. et al. (2000) Statistical significance of trends and trend differences in layer-average atmospheric temperature time series. *Journal of Geophysical Research*, **105**, doi: 10.1029/1999jd901105.

Shackleton, E.H. (1920) *South: the Story of Shackleton's Last Expedition, 1914–1917*. Macmillan.

Simmonds, I. (2003) Modes of atmospheric variability over the Southern Ocean. *Journal of Geophysical Research*, **108**, doi: 10.1029/2000JC000542.

Simmonds, I. & Jones, D.A. (1998) The mean structure and temporal variability of the semiannual oscillation in the southern extratropics. *International Journal of Climatology*, **18**, 473–504.

Simmonds, I. & Keay, K. (2000) Mean southern hemisphere extratropical cyclone behavior in the 40-year NCEP-NCAR reanalysis. *Journal of Climate*, **13**, 873–883.

Simpkins, G.R., Ciasto, L.M., Thompson, D.W. J. & England, M.H. (2012) Seasonal relationships between large-scale climate variability and Antarctic sea ice concentration. *Journal of Climate*, **25**, 5451–5469.

Simpkins, G.R., Ciasto, L.M. & England, M.H. (2013) Observed variations in multidecadal Antarctic sea ice trends during 1979–2012. *Geophysical Research Letters*, **40**, doi: 10.1002/grl.50715.

Simpkins, G.R., McGregor, S., Taschetto, A.S., Ciasto, L.M. & England, M.H. (2014) Tropical connections to climatic change in the extratropical southern hemisphere: the role of Atlantic SST trends. *Journal of Climate*, **27**, 4923–4936.

Sinclair, K.E., Bertler, N.A., Bowen, M.M. & Arrigo, K.R. (2014) Twentieth century sea-ice trends in the Ross Sea from a high-resolution, coastal ice-core record. *Geophysical Research Letters*, **41**, doi: 10.1002/2014GL059821.

Sissala, J.F., Sabatini, R.R. & Ackerman, H.J. (1972) Nimbus satellite data for polar ice survey. *Polar Record*, **16**, 367–373.

Stammerjohn, S., Massom, R., Rind, D. & Martinson, D. (2012) Regions of rapid sea ice change: An inter-hemispheric seasonal comparison. *Geophysical Research Letters*, **39**, L06501, doi: 10.1029/2012gl050874.

Stammerjohn, S.E., Drinkwater, M.R., Smith, R.C. & Liu, X. (2003) Ice-atmosphere interactions during sea-ice advance and retreat in the western Antarctic Peninsula region. *Journal of Geophysical Research*, **108**, doi: 10.1029/2002JC001543.

Stammerjohn, S.E., Martinson, D.G., Smith, R.C., Yuan, X. & Rind, D. (2008) Trends in Antarctic annual sea ice retreat and advance and their relation to ENSO and Southern Annular Mode Variability. *Journal of Geophysical Research*, **113**, doi: 10.1029/2007JC004269.

Steffen, K., Key, J., Cavalieri, D.J. et al. (1992) The estimation of geophysical parameters using passive microwave algorithms. In: *Microwave Remote Sensing of Sea Ice* (Ed. F. Carsey), pp. 201–231, American Geophysical Union, Washington, DC.

Steig, E.J., Schneider, D.P., Rutherford, S.D., Mann, M.E., Comiso, J.C. & Shindell, D.T. (2009) Warming of the Antarctic ice-sheet surface since the 1957 International Geophysical Year. *Nature*, **457**, doi:10.1038/nature07669.

Streten, N.A. (1973) Satellite observations of the summer decay of the Antarctic sea-ice. *Archiv fur Meteorologie, Geophysik und Bioklimatologie, Serie A*, **A22**, 119–134.

Stroeve, J.C., Jenouvrier, S., Campbell, G.G., Barbraud, C. & Delord, K. (2016) Mapping and assessing variability in the Antarctic marginal ice zone, pack ice and coastal polynyas in two sea ice algorithms with implications on breeding success of snow petrels. *The Cryosphere*, **10**, doi:10.5194/tc-2016-26

Sturm, M. & Massom, R.A. (2009) Snow and sea ice. In: *Sea Ice*, 2nd edn (Eds. D. Thomas & G. Dieckmann), pp. 153–204. Oxford, Wiley-Blackwell.

Thomas, E.R., Marshall, G.J. & McConnell, J.R. (2008) A doubling in snow accumulation in the western Antarctic Peninsula since 1850. *Geophysical Research Letters*, **35**, L01706, doi: 10.1029/2007GL032529.

Thomas, E.R., Dennis, P.F., Bracegirdle, T.J. & Franzke, C. (2009) Ice core evidence for significant 100-year regional warming on the Antarctic Peninsula. *Geophysical Research Letters*, **36**, L20704, doi: 10.1029/2009GL040104.

Thompson, D.W. J. & Wallace, J.M. (2000) Annual modes in the extratropical circulation. Part I: month-to-month variability. *Journal of Climate*, **13**, 1000–1016.

Thompson, D.W. J. & Solomon, S. (2002) Interpretation of recent Southern Hemisphere climate change. *Science*, **296**, 895–899.

Thompson, D.W. J., Solomon, S., Kushner, P.J., England, M.H., Grise, K.M. & Karoly, D.J. (2011) Signatures of the Antarctic ozone hole in Southern Hemisphere surface climate change. *Nature Geoscience*, **4**, 741–749.

Titchner, H.A. & Rayner, N.A. (2014) The Met Office Hadley Centre sea ice and sea surface temperature data set, version 2: 1. Sea ice concentrations, *Journal of Geophysical Research*, **119**, doi: 10.1002/2013JD020316.

Turner, J. (2004) Review: the El Niño-Southern Oscillation and Antarctica. *International Journal of Climate*, **24**, 1–31.

Turner, J. & Overland, J. (2009) Contrasting climate change in the two polar regions. *Polar Research*, **28**,146–164.

Turner, J., Lachlan-Cope, T., Colwell, S. & Marshall, G.J. (2005) A positive trend in western Antarctic Peninsula precipitation over the last 50 years reflecting regional and Antarctic-wide atmospheric circulation changes. *Annals of Glaciology*, **41**, 85–91.

Turner, J., Comiso, J.C., Marshall, G. et al. (2009) Non-annular atmospheric circulation change induced by stratospheric ozone depletion and its role in the recent increase of Antarctic sea ice extent. *Geophysical Research Letters*, **36**, doi: 10.1029/2009GL037524.

Turner, J., Hosking, J.S., Phillips, T. & Marshall, G.J. (2013a) Temporal and spatial evolution of the Antarctic sea ice prior to the September 2012 record maximum extent. *Geophysical Research Letters*, **40**, doi: 10.1002/2013gl058371.

Turner, J., Maksym, T., Phillips, T., Marshall, G.J. & Meredith, M.P. (2013b) The impact of changes in sea ice advance on the large winter warming on the western Antarctic Peninsula. *International Journal of Climatology*, **33**, 852–861.

Turner, J., Bracegirdle, T.J., Phillips, T., Marshall, G.J. & Hosking, J.S. (2013c) An Initial Assessment of Antarctic Sea Ice Extent in the CMIP5 Models. *Journal of Climate*, **26**, 1473–1484.

Turner, J., Phillips, T., Hosking, J.S., Marshall, G.J. & Orr, A. (2013d) The Amundsen Sea low. *International Journal of Climatology*, **33**, 1818–1829.

Turner, J., Hosking, J.S., Bracegirdle, T.J., Marshall, G.J., & Phillips, T. (2015) Recent changes in Antarctic Sea Ice. *Philosophical Transactions of the Royal Society*, **373**, 1–13.

Vaughan, D.G., Marshall, G.J., Connolley, W.M. et al. (2003) Recent rapid regional climate warming on the Antarctic Peninsula. *Climatic Change*, **60**, 243–274.

Walland, D. & Simmonds, I. (1999) Baroclinicity, meridional temperature gradients, and the southern semiannual oscillation. *Journal of Climate*, **12**, 3376–3382.

Watkins, A.B. & Simmonds, I. (1999) A late spring surge in the open water of the Antarctic sea ice pack. *Geophysical Research Letters*, **26**, 1481–1484.

Williams, G., Maksym, T., Wilkinson, J. et al. (2015) Thick and deformed Antarctic sea ice mapped with autonomous underwater vehicles. *Nature Geoscience*, **8**, 61–67.

Worby, A.P. & Comiso, J.C. (2004) Studies of the Antarctic sea ice edge and ice extent from satellite and ship observations. *Remote Sensing of the Environment*, **92**, 98–111.

Worby, A.P., Geiger, C.A., Paget, M.J., Van Woert, M.L., Ackley, S.F. & DeLiberty, T.L. (2008) Thickness distribution of Antarctic sea ice. *Journal Geophysical Research*, **113**,C05S92, doi: 10.1029/2007JC004254.

Worby, A.P., Steer, A., Lieser, J.L. et al. (2011) Regional-scale sea-ice and snow thickness distributions from in situ and satellite measurements over East Antarctica during SIPEX 2007. *Deep-Sea Research Part II*, **58**, 1125–1136.

Xie, H., Tekeli, A.E., Ackley, S.F., Yi, D. & Zwally, H.J. (2013) Sea ice thickness estimations from ICESat Altimetry over the Bellingshausen and Amundsen Seas, 2003–2009. *Journal of Geophysical Research: Oceans*, **118**, doi: 10.1002/jgrc.20179.

Yi, D., Zwally, H.J. & Robbins, J.W. (2011) ICESat observations of seasonal and interannual variations of sea-ice freeboard and estimated thickness in the Weddell Sea (2003–2009). *Annals of Glaciology*, **52**, 43–51.

Yuan, X. (2004) ENSO-related impacts on Antarctic sea ice: a synthesis of phenomenon and mechanisms. *Antarctic Science*, **16**, 415–425.

Yuan, X. & Martinson, D.G. (2001) The Antarctic Dipole and its predictability. *Geophysical Research Letters*, **28**, doi: 10.1029/2001GL012969.

Yuan, X. & Li, C. (2008) Climate modes in southern high latitudes and their impacts on Antarctic sea ice. *Journal of Geophysical Research*, **113**,C06S91, doi: 10.1029/2006JC004067.

Zhang, J. (2007) Increasing Antarctic sea ice under warming atmospheric and oceanic conditions. *Journal of Climate*, **20**, 2515–2529.

Zhang, J. (2014) Modeling the impact of wind intensification on Antarctic sea ice volume. *Journal of Climate*, **27**,1, 202–214.

Zunz, V., Goosse, H. & Massonnet, F. (2013) How does internal variability influence the ability of CMIP5 models to reproduce the recent trend in Southern Ocean sea ice extent? *The Cryosphere*, **7**, 451–468.

Zwally, H.J., Comiso, J.C., Parkinson, C.L., Campbell, W.J., Carsey, F.D. & Gloersen, P. (1983) *Antarctic Sea Ice, 1973–1976: Satellite Passive-Microwave Observations*. National Aeronautics and Space Administration, Washington, DC.

Zwally, J.H., Comiso, J.C., Parkinson, C.L., Cavalieri, D.J. & Gloersen, P. (2002) Variability of Antarctic sea ice 1979–1998. *Journal of Geophysical Research*, **107**, doi: 10.1029/2000JC000733.

CHAPTER 11

Losing Arctic sea ice: observations of the recent decline and the long-term context

Walter N. Meier

Cryospheric Sciences Laboratory, NASA Goddard Space Flight Center, Greenbelt, MD, USA

11.1 Introduction

The Arctic is a rapidly changing environment and one of the most visible indicators of this change is sea ice. The Arctic sea ice cover has a significant seasonal cycle, reaching its maximum in late February or March and then declining through the spring and summer to a minimum sometime in September. During winter, the ice cover expands well southward of the Arctic Ocean, extending to the southern reaches of the Labrador Sea and covering much of the Sea of Okhotsk. Occasional sea ice is even found as far south as the Bohai Sea, at ~40° N latitude (Chapter 27). In summer, the ice retreats to within the bounds of the Arctic Ocean. While the sea ice seasonal cycle is substantial, it is smaller than Antarctic sea ice (Chapter 10) with a lower maximum and higher minimum extent. Of particular note, nearly half of the Arctic winter sea ice cover survives through the summer. Unlike in Antarctica, this means that Arctic sea ice has historically been dominated by perennial sea ice (also called multi-year ice – ice that has survived through one or more melt seasons). The older ice is generally thicker than seasonal sea ice (also called first-year ice – ice that has formed since the end of the previous summer melt season). Perennial ice is generally thick enough to survive summer melt and is primarily lost through advection out of the Arctic, mainly through Fram Strait along the east coast of Greenland (a smaller amount of ice exits through the Canadian Archipelago).

However, things are changing in the Arctic. Sea ice has shown a rapid decline in areal coverage in satellite data, and myriad satellite, *in situ* and proxy data indicate a substantial thinning and loss of old ice. While there has been an overall decline over the past four decades, it has become particularly pronounced in the last decade. These changes have suffused the entirety of the character of the ice cover. Melt onset is occurring earlier, resulting in lower surface albedo and enhanced absorption of solar energy. The younger, thinner ice is advected more efficiently by the winds and currents, leading to faster ice drift. The changes have been dramatic and rapid on a climate scale. However, caution is warranted as the Arctic sea ice experiences large interannual variability, and projections of future change, especially on a decadal scale, is a challenge. In this chapter, we review available data on Arctic sea ice, its long-term trends and recent changes in the ice cover.

11.2 Sea ice extent

The most common indicator of Arctic sea ice change is sea ice extent derived from passive microwave satellite imagery. Extent is the total area of a given region covered by ice of some minimum concentration (the most commonly used concentration threshold is 15%, though other values are sometimes used). Passive microwave data is optimal for estimating time series of sea ice extent because:

- there is generally a large contrast in the microwave emission between ice and water;
- there is a long record, starting in late 1978;

Sea Ice, Third Edition. Edited by David N. Thomas.
© 2017 John Wiley & Sons, Ltd. Published 2017 by John Wiley & Sons, Ltd.

- it has the ability to detect the surface during day and night conditions and through most weather (there is some influence from thick clouds, especially near the ice edge);
- it provides complete daily coverage of all sea ice-covered regions (coverage was once every other day for 1978 to mid-1987).

Sea ice area – simply the total surface area of ice (the area of a region covered by ice, weighted by concentration) – is another parameter that can be derived from concentration data for a time series. Area and extent generally show similar trends, but here we use extent because area is generally considered less accurate and less stable due to concentration biases during summer melt (liquid water on the ice surface is seen as reduced concentration and thus area estimates are biased low). Characteristics of passive microwave imagery and the derivation of sea ice concentrations and extent from it are discussed in Chapter 9. Further details can also be found in Steffen et al. (1992). Information on the data used in this section and a discussion of extent uncertainties is discussed in Section 11.2.3.

11.2.1 The modern passive microwave satellite record

The modern era of satellite remote sensing of sea ice extent began with the advent of multi-channel passive microwave radiometers in late 1978. This period yields the most complete, consistent and continuous record and thus there is the most confidence in the observed changes during this era. Overall, as of the date of this publication, there is more than a 36-year record (1979–2014) of extent. This provides a long enough record to reasonably investigate long-term trends and variability.

The story that the extent data tell is one of decline. All months have statistically significant declining trends in extent since 1979 (Table 11.1). The largest trends occur during summer, with the September trend of −13.3 % per decade (relative to the 1981–2010 average) having the largest magnitude. Trends during winter and early spring are smaller with the lowest magnitude of −2.3 % per decade in May. However, winter months also have less variability and thus all are statistically significant at >95% confidence level, as suggested by the two-standard deviation range in the trends (Table 11.1).

The spatial patterns of concentration trends reveal that the decline is Arctic-wide, with virtually all sectors showing negative trends in all seasons (Figure 11.1). The main exception is the Bering Sea during winter and spring. The regions of large trends are primarily at and near the ice edge, reflecting the declining trend ice extent. In the interior of the ice pack there is little

Table 11.1 Monthly and annual average Arctic sea ice extent estimates. Average extent for 1981–2010 and absolute and relative (to the 1981–2010 average) trends for 1979–2014 and 2001–2014. The two standard deviation range is provided in brackets next to each value. Data from the National Snow and Ice Data Center Sea Ice Index, Version 1 (Fetterer et al., 2002).

Month	1981–2010 Average (SD) extent (10^6 km^2)	1979–2014 Trend (SD) (km^2 year^{-1})	1979–2014 Trend (SD) (% per decade)	2001–2014 Trend (SD) (km^2 year^{-1})	2001–2014 Trend (SD) (% per decade)
January	14.56 (0.98)	−48 000 (5100)	−3.3 (0.5)	−48200 (27 100)	−3.3 (1.9)
February	15.35 (0.94)	−45 600 (8300)	−3.0 (0.5)	−55000 (31 600)	−3.6 (2.1)
March	15.49 (0.88)	−39 500 (8700)	−2.6 (0.6)	−35 900 (39 000)	−2.3 (2.5)
April	14.75 (0.89)	−35 200 (9400)	−2.4 (0.6)	−5000 (37 100)	−0.3 (2.5)
May	13.39 (0.84)	−30 800 (9800)	−2.3 (0.7)	−14600 (35 100)	−1.1 (2.6)
June	11.89 (0.91)	−42 400 (7200)	−3.6 (0.6)	−41600 (29 700)	−3.5 (2.5)
July	9.70 (1.41)	−72 000 (12 000)	−7.4 (1.2)	−113 400 (40 000)	−11.7 (4.1)
August	7.22 (1.51)	−74 400 (14 900)	−10.3 (2.1)	−111 600 (59 900)	−15.5 (8.3)
September	6.52 (1.81)	−86 700 (18 100)	−13.3 (2.8)	−137 300 (71 800)	−21.1 (11.0)
October	8.91 (1.39)	−61 500 (14 800)	−6.9 (1.7)	−93 500 (63 200)	−10.5 (7.1)
November	10.99 (1.19)	−51 300 (10200)	−4.7 (0.9)	−49 000 (33 600)	−4.5 (3.1)
December	13.07 (0.94)	−43 500 (6900)	−3.3 (0.5)	−39 800 (21 400)	−3.0 (1.6)
Annual average	11.82 (1.02)	−52 600 (6800)	−4.4 (0.6)	−62 100 (28 000)	−5.3 (2.4)

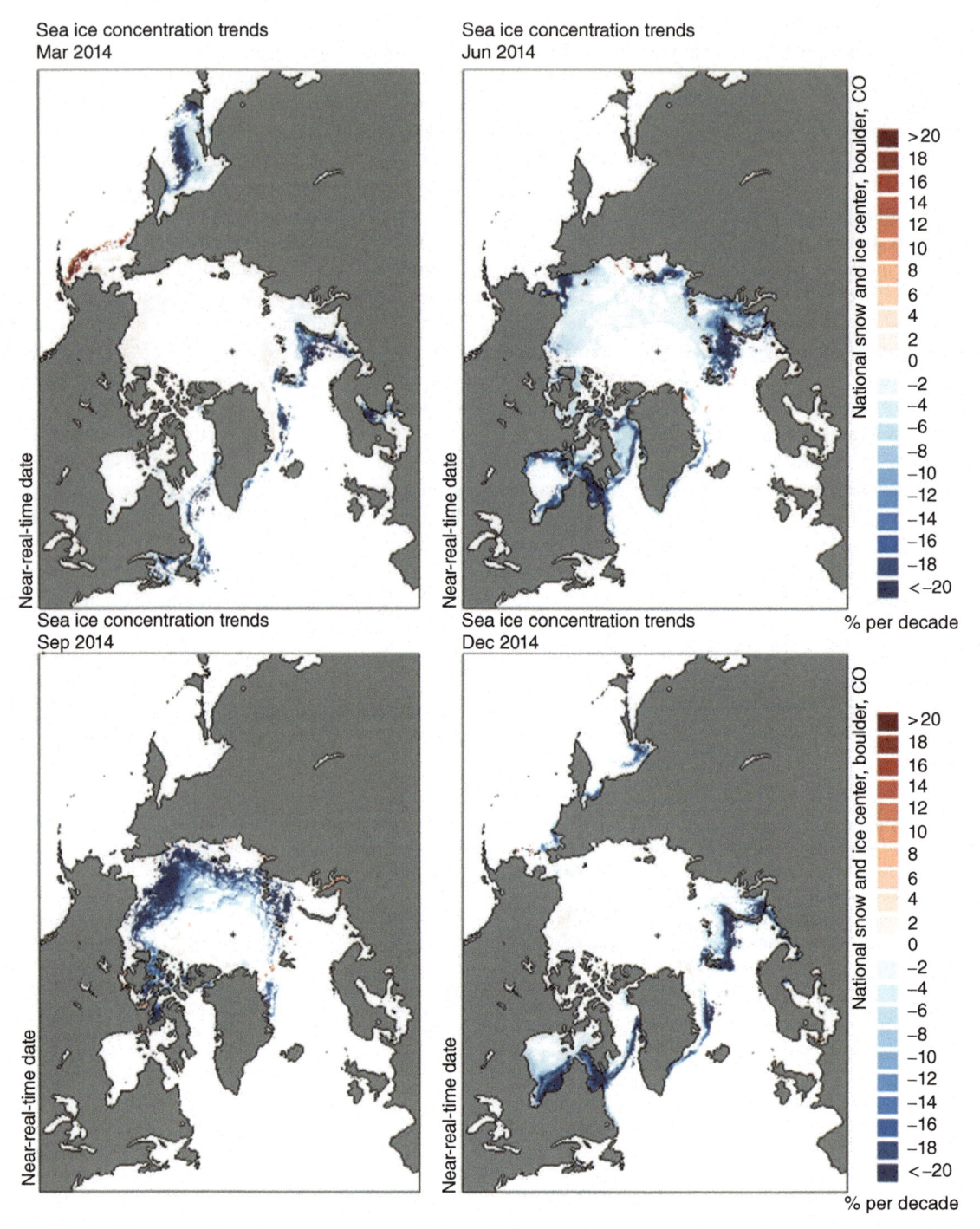

Figure 11.1 Spatial pattern of sea ice concentration trends for March, June, September and December through 2014. (From the National Snow and Ice Data Center Sea Ice Index, Version 1; Fetterer et al., 2002.)

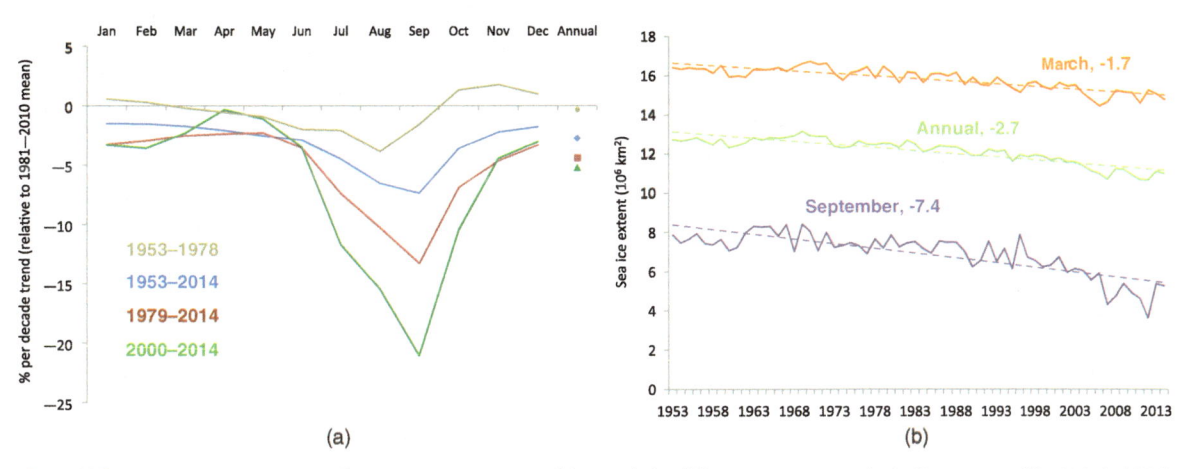

Figure 11.2 Sea ice extent time series for 1953–2014. (a) Monthly trends for different year ranges, including pre-satellite (1953–1978) and modern satellite (1979–2014). (b) Extent for 1953–2014 for March, September and the annual average; numbers in the legend are the % per decade trends over 1953–2014 (data updated from Meier et al., 2012.)

variability in the ice pack and concentration trends are small.

It is clear that the extent trends in most months have accelerated in the last 15 years, particularly during summer (Table 11.1, Figure 11.2). During this period, each September recorded extent well below the 1981–2010 average and record low monthly extents were set in September 2002, 2005, 2007 and 2012. The 2012 extent was 3.63×10^6 km^2, 44% below the 1981–2010 September average of 6.52×10^6 km^2. Even with a shorter record and higher trend standard deviations, many months were still statistically significant.

Care needs to be taken in interpreting these trend values and their implications. There is a strong correlation between decreasing sea ice and increasing Arctic surface temperatures, which have been linked to changing external forcing (i.e. increasing atmospheric concentration of greenhouse gases) (Notz&Marotzke, 2012). However, the statistical significance implied by the trend standard deviations does not in itself imply a change linked to external forcing. On shorter timescales, natural variability can dominate. For example, modelling studies indicate that even in the presence of increasing greenhouse gases, positive trends in sea ice extent over periods up to two decades during this century are not unrealistic (Kay et al., 2011).

11.2.2 Sea ice extent before the modern era

The large interannual variability of the sea ice cover and its implications for interpreting trends and the response of the ice cover to external forcing motivate the desire to lengthen the time series as much as possible. While a complete, continuous, consistent record is not available before the modern passive microwave record, other data sources are available. These include earlier satellite data and operational sea ice charts by international ice charting groups. An earlier single-channel passive microwave instrument [the electronically scanning microwave radiometer (ESMR) on the NASA Nimbus-7 satellite] that operated from 1972 to 1977 was merged with the modern passive microwave record to create a 30-year (1973–2002) record of extent (Cavalieri et al., 2003); however, this time series has not been updated is not consistent with modern 1979–present multi-channel passive microwave records, such as the one discussed in this chapter. Another earlier source of extent information is from a long time series of extent that has been compiled by the UK Hadley Centre (Rayner et al. 2003), based partly on a climatology compiled from ice charts by Walsh and Chapman (2001). However, there was not overlap between the satellite and pre-satellite data, resulting in a discontinuity that limits its usefulness for investigating trends and climatologies. The climatology and Hadley data extend through the 20th century, but climatology was used for many of the years before Arctic-wide ice charting began in 1953. A new version of the Hadley product, encompassing newly obtained sources of sea ice information and new methodologies, is now available (Titchner & Rayner, 2014), as is an updated Walsh and Chapman climatology (Walsh et al., 2015). Trends from both updated products have yet to be analysed in detail.

Meier et al. (2012) merged the original Hadley record, the ESMR data and the modern satellite record, adjusting the earlier pre-1979 data for consistency to create a 60+ year record of extent starting in 1953. This longer record, while not as consistent as the modern passive microwave product, provides a longer-term context to examine the recent downward trend (Figure 11.2). The pre-1979 record showed small negative trends (0 to −5% per decade) throughout much of the year with small positive trends during October–February. Overall, the 1953–2014 record has negative trends through all months, similar to the 1979–2014 data, but at smaller magnitudes.

Recently, other historical data, from the early Nimbus satellites, has been recovered. These data now encompass 1964, 1966, 1969 and 1970. Imagery was recovered, scanned and processed for geolocation and calibration. Then monthly mosaics were created from available clear sky imagery and the ice edge was contoured by manual inspection. Arctic estimates from August/September 1964 and 1966 have been published (Meier et al., 2013; Gallaher et al., 2014). These data obviously have much higher uncertainty than the passive microwave estimates, and consistency with the microwave record is not assured. Nonetheless they provide at least a qualitative context for the later data. The 1964 and 1966 data show Arctic extents that are on par with the 1981–2010 average, suggesting that summer Arctic sea ice was relatively stable prior to 1979, further supporting the Hadley results.

Before Arctic-wide ice charting began in 1953, sea ice extent information was much more limited. Charts of limited regions – e.g. the Canadian Archipelago, the Russian Arctic (Mahoney et al., 2008) date back to the early 1930s. There is even a compilation of ice edge locations in the Nordic seas as far back as 1750 CE (Divine & Dick, 2006, 2007). These data are useful in that they provide at least some information on parts of the Arctic. For example, the Russian Arctic charts indicate seasonal and regional variability on decadal scales, but the decline over the last 35 years is more consistent and widespread throughout the Russian Arctic. Much of these early data, including the Nimbus satellite records, have been retrieved through data rescue efforts (the so-called field of data archaeology). Such efforts are tedious and may be unsuccessful, but there is great value recovery of old data. There are probably still sea ice observations waiting in a file drawer to be discovered that will further illuminate our understanding of sea ice conditions before 1979.

Sea ice information from before the observational record can only be inferred from proxy records. Chemical and biological markers in sediment and ice cores, tree rings and drift wood show traces of the influence of the presence or absence of sea ice and these proxies can be used to infer sea ice conditions over hundreds and thousands of years before the present. Kinnard et al. (2011) analysed multiple proxies to investigate multi-decadal variability over the past 1450 years. They found that sea ice variability over much of the record was substantially linked to the advection of warm Atlantic water, but in recent decades a clear anthropogenic warming signal has emerged. Polyak et al. (2010) reviewed numerous palaeoclimate records and concluded that the recent change in sea ice extent was unprecedented over at least the past few thousand years.

11.2.3 A note on data sources and uncertainty

As noted in Chapter 9, several passive microwave sea ice concentration algorithms have been developed over the years. For consistency, all extent values and trends for 1979–2014 presented here are derived from the NASA Team algorithm (Cavalieri et al., 1984, 1999), developed at the NASA Goddard Space Flight Center. The source concentration data (Cavalieri et al., 1996) are distributed by the National Snow and Ice Data Center (NSIDC) and the extent values and concentration images discussed here are from the NSIDC Sea Ice Index, Version 1 (http://nsidc.org/data/seaice_index/) (Fetterer et al., 2002). The adjusted Hadley record (Meier et al., 2012) uses these data as well, with the 1953–1978 part of the record adjusted to be consistent with the Sea Ice Index values.

Other concentration products yield different total extents, and differences between products can be on the order of 500 000 km^2 (5–10%) (Kattsov et al., 2010). These differences are generally offsets due to how sensitive the algorithms are to ice in the marginal ice zone. As such, trends and variability differences between algorithms are much smaller and all algorithms generally yield the same conclusions. However, the differences can be important if trends or variability are small. This is the case in the Antarctic (Eisenman et al., 2014), but in the Arctic, where the decline trends are quite large, the algorithm differences are inconsequential except in the details.

Uncertainties in total extent and area are difficult to estimate because there is no reasonable validation data for such hemispheric parameters. Validation of algorithms is done via case-study intercomparisons with other satellite or airborne data in limited locations and time periods (e.g. Meier, 2005; Heinrichs et al., 2006; Andersen et al., 2007), but it is difficult to scale up such validations to estimate uncertainty in total hemispheric extents. The spread in estimates between different algorithms is one indication of uncertainty. Operational ice charts, produced by national ice charting services, such as the U.S. National Ice Center, provide potential validation because their charts are now created largely independent of passive microwave data, using primarily synthetic aperture radar (SAR) and visible/infrared imagery interpreted by trained analysts to provide the best ice cover estimate. However, such validation is limited by: potential subjectivity of the analyst, varying quality and quantity of data used in the chart, and different interpretations of the ice edge (operational charts will generally be more conservative and map ice to a completely ice-free line).

However, for investigating trends and variability, absolute uncertainty is not relevant. The key factor is consistency over time. And in this sense, the passive microwave record is optimal. Because the type of data is the same throughout, and even the sensors (and sensor channels) are nearly the same throughout, passive microwave imagers provide a consistent source. In addition, intercalibration is done between sensors to optimize consistency as newer sensors take over for earlier sensors (e.g. Cavalieri et al., 1999).

On the other hand, perfect intercalibration is not possible, particularly if the overlap periods are short, which has been the case for some passive microwave sensor transitions. Thus shorter intercalibration periods have relatively larger uncertainties than longer ones (e.g. Meier et al., 2011). This means that typically provided trend uncertainties (e.g. denoted by the standard deviation of trend lines and statistical significance) underestimate the real uncertainty. When trends are strongly significant, this is a minor issue, but if the trend significance is marginal, such intercalibration issues can be important (Eisenman et al., 2014).

Of course, uncertainties are much higher for extent estimates from the earlier satellite data (i.e. Nimbus), pre-satellite ice charts and proxy data. In many cases it is not possible to put realistic error bars on such estimates and effectively the information is qualitative. But as stated above, even though imperfect, such early data are precious because of the long-term context they provide.

11.3 Sea ice thickness

While extent provides a picture of the surface conditions, to really understand the overall condition of the ice, thickness information is needed. Unfortunately, unlike extent, sea ice thickness is much more difficult to observe. This is largely due to the fact that most satellite sensors observe surface characteristics and don't have the ability to 'see' into and through the ice. *In situ* measurements are very limited in the Arctic and even measurements from aircraft are sparse in space and time. It is only with the advent of satellite and airborne altimeters that large-scale estimates of thickness could be derived.

These began in the early 1990s with the launch of the ESA ERS-1 and ERS-2 radar altimeters. However, these had limited coverage in the Arctic and much of the ice cover was not imaged. The launch of the NASA ICESat laser altimeter in 2003 provided the first satellite-based coverage of the central Arctic. Unfortunately, instrument issues limited the data collection to twice per year and the instrument failed in 2009. Since then, NASA has continued observations in a limited manner using airborne platforms during Operation IceBridge. NASA plans to launch a follow-on satellite mission, ICESat-2, by 2017. In 2010, ESA launched CryoSat-2 (a replacement for the initial CryoSat, which was lost due to launch vehicle malfunction in 2005). These satellite and airborne data are now providing valuable thickness data,

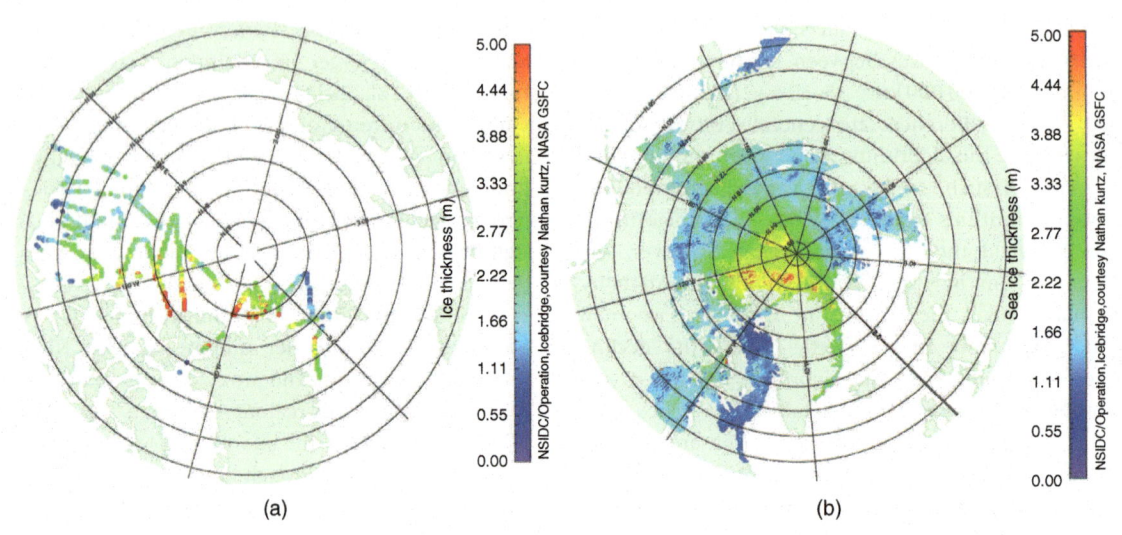

Figure 11.3 Sea ice thickness for April 2014 from: (a) IceBridge; (b) CryoSat-2. (Data and images from N. Kurtz, NASA Goddard; adapted from an image compiled by the National Snow and Ice Data Center.)

but there are still challenges to interpreting the data and many uncertainties (e.g. Kurtz et al., 2014) remain to be resolved (particularly snow cover, see Section 11.5). There are also challenges in integrating the data into a useful time series. An important goal of IceBridge is to provide intercalibration between ICESat, CryoSat-2 and ICESat-2 (Figure 11.3 for IceBridge and CryoSat-2 results). See Chapter 2 for a description of altimeter measurement techniques and their uncertainties.

Before the satellite era, the best source of thickness information has been derived from sea ice draft measurements by upward-looking sonars (ULSs) mounted on submarines or moored on the ocean floor. Submarine data, collected by the US and British Navy during the mid-1970s through to the late 1990s, were originally released as part of the joint US Russian Environmental Working Group in the mid-1990s. During the 1990s, dedicated science cruises were made on US submarines to collect data under the auspices of the Submarine Arctic Science programme (SCICEX). After a period of dormancy, the SCICEX programme was reinvigorated during the 2000s and is now regularly collecting data, though unlike the 1990s SCICEX programme, the cruises are not specifically scientifically oriented (i.e. collection is done as time permits within regular submarine operations).

The story that these measurements tell is one of thinning, particularly during the most recent years. A combined time series of submarine ULS and ICESat data indicate a thinning in the central Arctic from 3.64 m in 1980 to 1.89 m in 2008 (see Figure 2.11; Kwok & Rothrock, 2009). The thinning appeared to accelerate during the ICESat period (Kwok et al., 2009), though the record is too short to draw significant conclusions.

Thickness measurements from satellite have been carried forward now by CryoSat-2 (Figure 11.4) and 2012 showed distinct thinning from the 2004–2008 ICESat-derived average. However, 2012 was the record low extent and along with a rebound in minimum ice extent there are indications of increasing thickness in 2013 and 2014 (e.g. Kurtz et al., 2014). For example, an increase in total volume is seen in sea ice simulations from the University of Washington PIOMAS model (Zhang & Rothrock, 2003) that is due to larger extent and increased thickness. The model estimates are constrained by sea ice extent observations and have been found to compare well with CryoSat-2 thickness fields (Laxon et al., 2013). This points to the importance of natural variability in the Arctic sea ice system. Thickness generally responds more slowly than extent to variations in forcing and is thus considered a more 'stable' climate indicator. But the IceBridge and CryoSat-2 data are demonstrating that even thickness can have substantial year-to-year variability.

Estimate of Arctic sea ice thickness

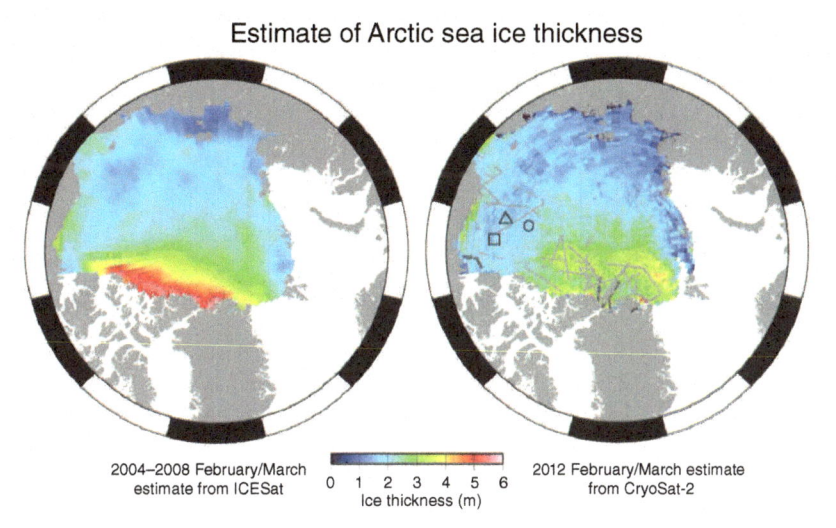

Figure 11.4 Sea ice thickness maps for February/March, 2004–2008 average from ICESat (left) and 2012 from CryoSat-2 (right) with location of validation data from upward-looking sonar moorings (triangle, circle and square) and airborne missions (grey and black lines). (From Meier et al., 2014, figure 4; adapted from Laxon et al., 2013.)

11.4 Sea ice type/age as a proxy for thickness

Although the record of direct thickness estimates is limited, there is a desire to have a long-term time series of thickness changes. While models can provide such a time series of estimates (e.g. PIOMAS), they are subject to potential biases in both the model physics and the forcings. Sea ice age is a useful, strictly observational proxy for thickness, because a general rule of thumb holds that older ice is thicker ice due to thermodynamic ice growth. As noted in Chapter 2, ice dynamics also plays an essential role, and ridging and rafting lead to a thicker ice cover than is achievable simply through thermodynamics, but the rule of thumb is generally valid at large scales.

Sea ice age can be estimated from microwave data due to the different emission and scattering properties between the higher salt content in younger ice and the 'fresher' older ice. It is feasible to discriminate between seasonal (first-year) and perennial (and to some degree between second-year and older multi-year) ice with either passive microwave (Comiso, 2012) or active scatterometer (e.g. Nghiem et al., 2007) imagery, at least during the winter period (during summer, meltwater on the surface inhibits the ice type determination).

Both methods show a downward trend in multi-year ice that is even faster than the overall sea ice extent. For example, Comiso (2012) calculated a −15.1% per decade trend in multi-year ice (defined therein as having survived more than two summers) extent. This loss of multi-year ice explains most of the decline in ice thickness, with first-year ice thickness remaining stable (Kwok et al., 2009).

Another approach to estimating ice thickness is through Lagrangian tracking of ice parcels. This has been done using a combination of passive microwave and visible/infrared imagery as well as *in situ* buoys (Maslanik et al., 2011). The advantage of this approach is that because it uses tracking (via cross-correlation feature matching for satellite imagery), the age can be estimated in any number of desired categories – in other words, beyond simply first-year and multi-year ice. This is useful because ice continues to thicken thermodynamically over several years. Also, more categories provide a longer time history of the evolution of the ice cover and thus more insights into the state of the ice cover.

The sea ice age data reveal a substantial loss in the oldest ice types (>4 years old) and a corresponding increasing in first-year ice (Figure 11.5). The oldest ice once comprised a substantial portion, ~2 × 10^6 km^2, of the

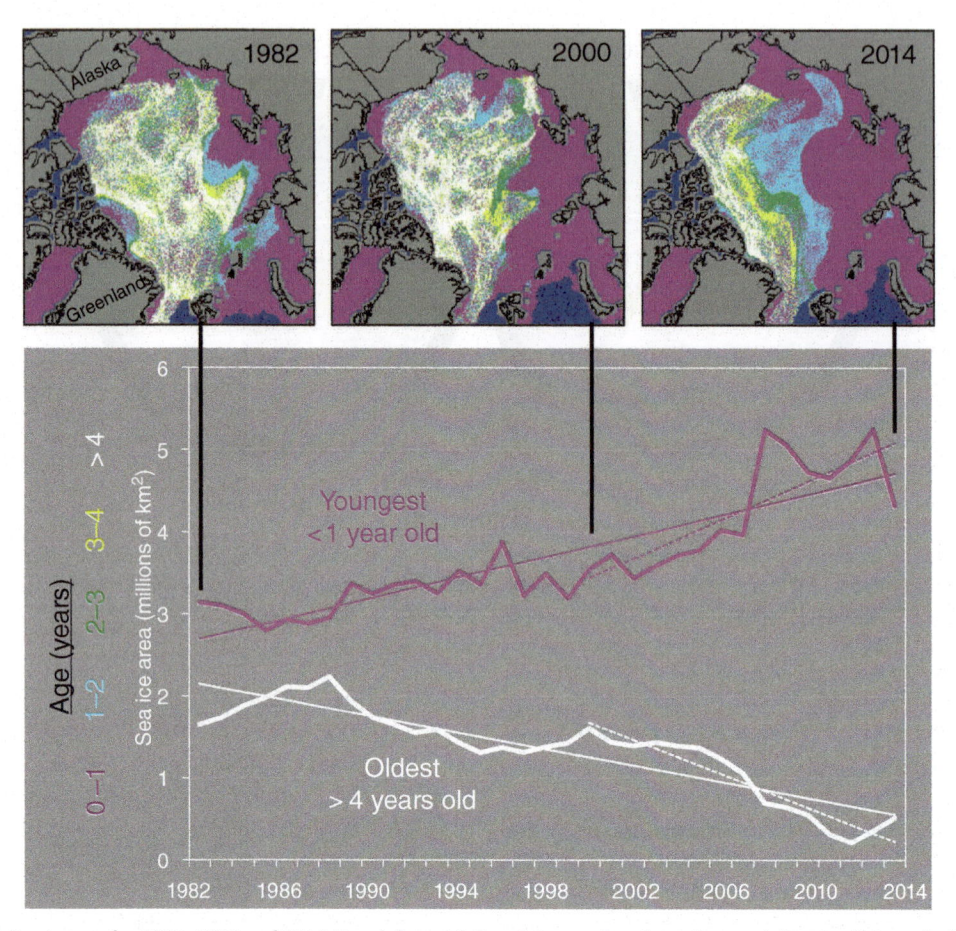

Figure 11.5 Sea ice age for 1982, 2000 and 2014 (top, left to right) and time series of age (bottom); for simplicity only first-year ice (in purple) and old ice (> 4 years old, in white) are shown. The colour key for the top images is provided to the right of the bottom time series plot. (Data updated from Maslanik et al., 2011, courtesy Mark Tschudi, University of Colorado.)

Arctic Ocean's winter ice cover. However, by 2012, the Arctic was virtually devoid of such ice. Since 2012, there has been a small rebound in old ice, a further indication of the increase in thickness and volume observed in the CryoSat-2 data and the PIOMAS model fields. Ice of intermediate age has remained relatively stable, though there is also a shift from 2- to 4-year-old ice to 1- to 2-year-old ice.

Much of the loss of older ice types is due to advection out of the Arctic through Fram Strait and the Canadian Archipelago that is not being replaced because less first-year ice is surviving the summer. However, the ice age data also reveal that in recent years there has been significant loss of multi-year ice through *in situ* melting within the Arctic, particularly in the Beaufort and Chukchi seas (Perovich et al., 2008; Hansen et al., 2013; Krishfield et al., 2014). This is probably related to warmer ocean waters from *in situ* solar heating of the Arctic Ocean during the summer (Steele et al., 2010) as well as inflow of warmer Pacific waters (Maslowski et al., 2012). Another factor may be that the ice cover is less consolidated with smaller floe sizes that are more vulnerable to lateral and bottom melt.

11.5 Snow depth on sea ice

An important component of the sea ice system is the overlying snow cover. Snow has a higher albedo, which reduces the absorption of solar energy during spring and summer and slows melt of the ice cover. During autumn and winter it acts as a blanket over the ice, limiting heat transfer between the underlying ocean and the atmosphere above; this slows ice growth. Snow also modifies reflected and emitted energy observed by satellite and airborne sensors, potentially increasing uncertainty. Of particular note is the effect of snow on altimeters. Laser altimeters (e.g. ICESat) measure the snow + ice freeboard, so knowledge of the snow cover is essential to derive total sea ice thickness (see Chapters 2 and 9). Radar altimeters (e.g. CryoSat-2) generally penetrate through the snow cover, but the weight of the snow on the ice affects the derivation of thickness from the freeboard retrievals. And, of course, changes in snow cover are a potentially important climate indicator. Unfortunately, snow data over sea ice are decidedly lacking. The most spatially complete data set is a 1954–1991 climatology record compiled from Russian ice stations and other *in situ* sources (Warren et al., 1999). A regional comparison of the Warren climatology with more recent airborne and *in situ* data suggests a thinning trend in the western Arctic of 30% or more, which is correlated with delayed autumn freeze-up in the region (Webster et al., 2014). However, the comparison is limited due to the sparse spatial and temporal coverage and the regional nature of the airborne and *in situ* data.

Satellite data can provide limited snow information. Passive microwave imagery can estimate snow over sea ice because snow modifies the surface emission, but only over seasonal ice (Markus & Cavalieri, 1999). The package of instruments aboard IceBridge flights provides an estimate snow depth by employing both radar and laser altimeters (the laser measure the top of the snow surface, the radar measuring the snow–ice interface). However, the IceBridge time span is short, and spatial and temporal coverage is limited (Kurtz & Farrell, 2011; Farrell et al., 2012). In the future, a constellation of a radar and laser altimeter could be used to routinely estimate snow depth over much of the Arctic. It is hoped that CryoSat-2 will still be operating when ICESat-2 is launched so that such a product could be developed. Thus, although snow cover is a critical component of the sea ice system, long-term comprehensive records are simply not available. This represents a significant gap in our observational capabilities, which, it is hoped, will eventually be filled.

11.6 Melt onset and freeze-up

The timing of melt onset and freeze-up of the sea ice cover is an important factor in the seasonal and interannual evolution of the ice cover. Earlier onset of melt and melt-pond formation reduces albedo and thus increases the amount of absorbed solar energy. Dry snow cover has an albedo of ~0.8, but begins dropping sharply with melt onset and the formation of melt ponds. A heavily ponded surface may have an albedo of 0.4 or less (Perovich & Polashenski, 2012). The amount of energy absorbed is quite sensitive to the timing of melt because the onset within the Arctic basin is near the time of the summer solstice, so even a few days' difference can substantially affect the total amount of solar energy absorbed over the summer (Stroeve et al., 2014).

Microwave emission is dependent on the phase state of water, so passive microwave sensors are very sensitive to the onset of melt. Thus there is a long time series (since 1979) of melt-onset dates (Markus et al., 2009; Wang et al., 2013; Bliss & Anderson, 2014), as well as freeze-up dates and total melt season length (Markus et al., 2009). These data show a clear trend of earlier melt and later freeze-up and an overall longer melt season (Stroeve et al., 2014).

Overall, these data indicate an Arctic average melt onset trend of 5 to 6 decade^{-1}, with regional trends up to 11 days per decade (Figure 11.6). Earlier melt correlates well with September sea ice extent because of the increased amount of absorbed solar energy. However, recent years (since 2000) do not show a trend in melt-onset date (Figure 11.6) even though extent has continued to decline. This suggests that other factors, such as ice age and ice motion, have become relatively more important.

11.7 Sea ice motion

As discussed in detail in Chapter 2, sea ice dynamics are a key component of the sea ice system. Convergent motion of the ice results in ice deformation, thickening the ice dynamically through ridging. Divergent motion results in leads and polynyas: open water areas that enhance energy and moisture fluxes that can modify the ocean and atmospheric boundary layer. As mentioned earlier, advection of the ice out of the Arctic through

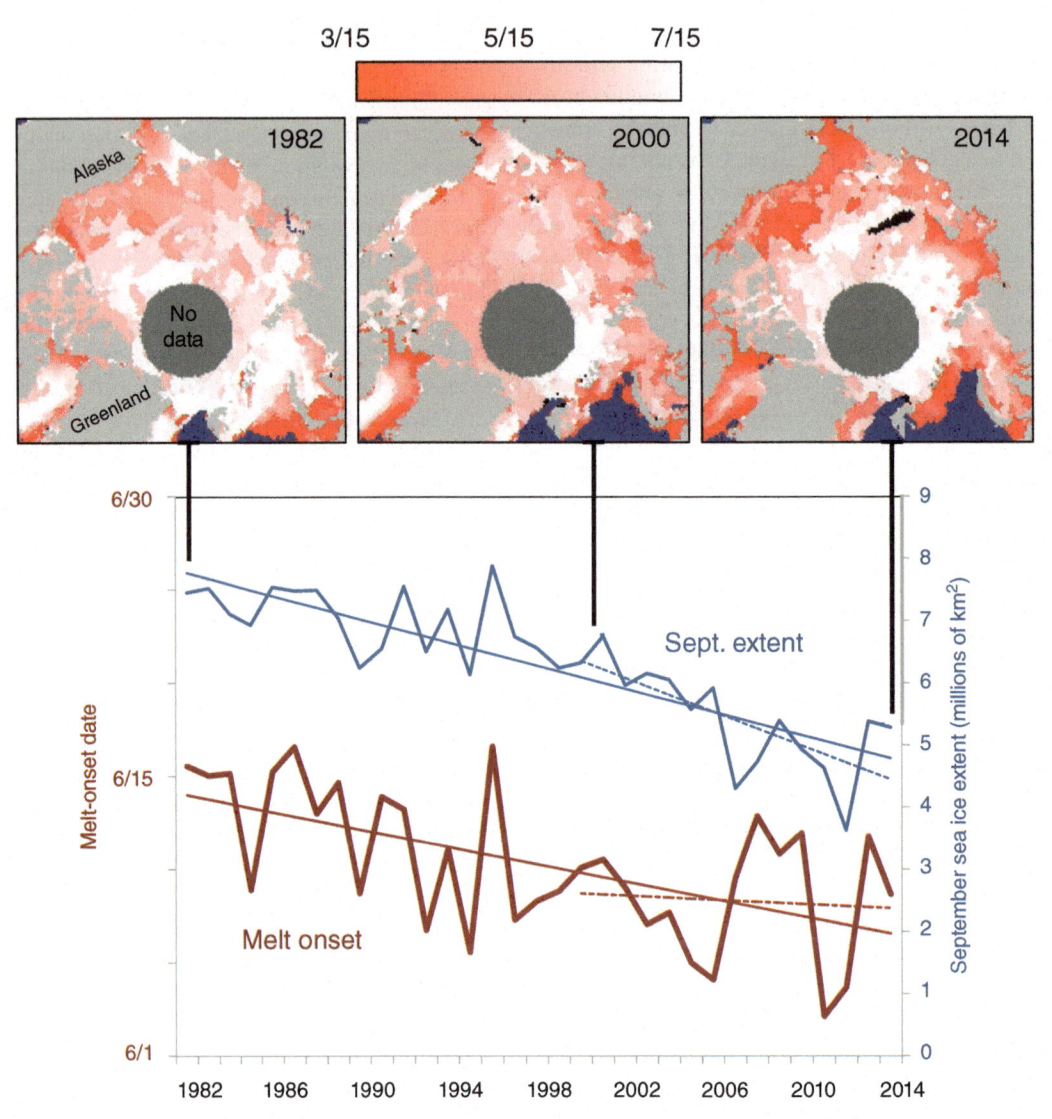

Figure 11.6 Melt-onset fields for 1982, 2000 and 2014 (top, left to right) and time series (bottom) of melt onset date (red) and September extent (blue) for 1982–2014. Dates are in month/day format. Trend lines are given for 1982–2014 and 2000–2014. (Data updated from Stroeve et al., 2014, courtesy T. Markus and J. Miller, NASA Goddard Space Flight Center.)

the Fram Strait and the Canadian Archipelago has been the primary sink for multi-year ice. Thus, ice motion has been an important component in the loss of older ice (more multi-year ice is advected out of the Arctic than is replaced).

Sea ice motion can be estimated by tracking of buoy positions and derived from satellite remote sensing imagery (primarily via passive microwave and scatterometer instruments) using cross-correlation feature tracking techniques (i.e. a feature in one image is matched in a subsequent image and motion is estimated by dividing the displacement by the time interval between images). Several sea ice motion products have been developed (Sumata et al., 2014) and while there are differences, overall the products compare reasonably well. Satellite retrievals of motion are limited during summer because of surface melt, so most analyses focus on fall through spring (e.g. October–May).

The ice motion data clearly show an increase in sea ice drift speed (e.g. Rampal et al., 2009; Spreen et al., 2011).

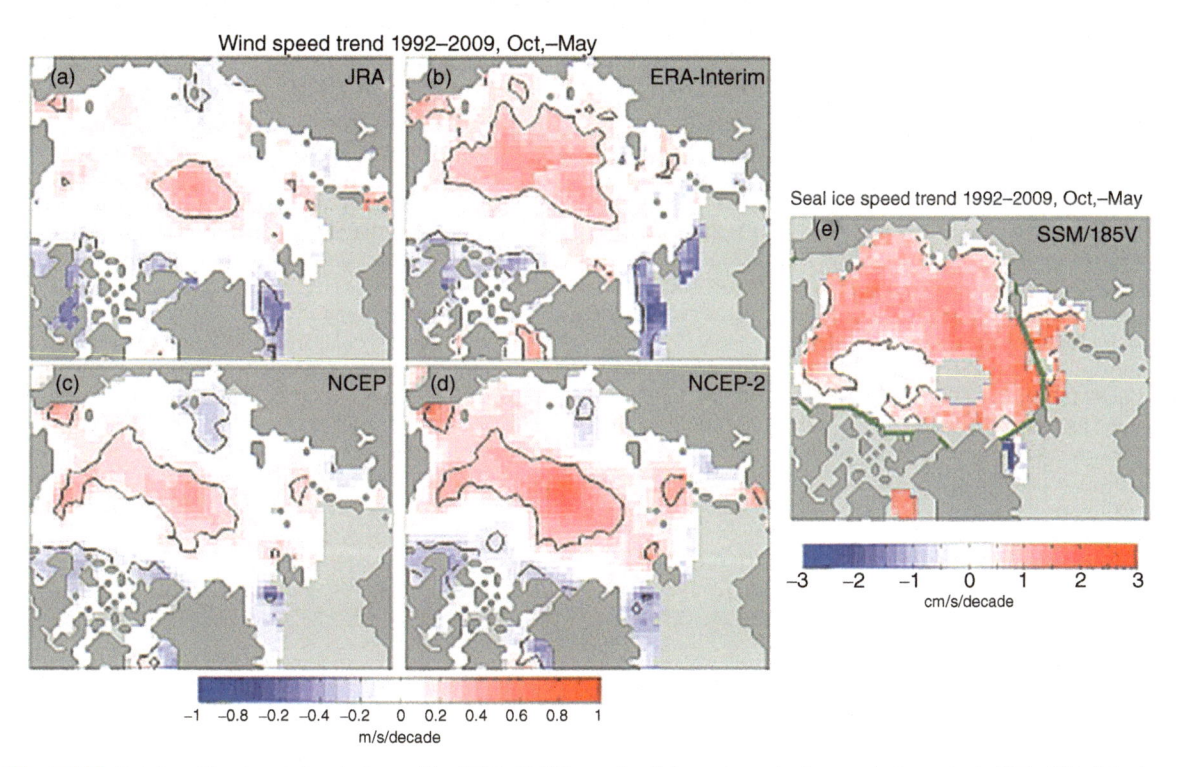

Figure 11.7 October–May trends in wind speed for 1992–2009 from: (a–d) four atmospheric reanalysis products (JRA, ERA-Interim, NCEP and NCEP-2); and (e) sea ice speed. Black contours outline statistically significant trends at the 99% confidence level. Source: From Spreen et al., 2011. Reproduced with permission of John Wiley & Sons.

Spreen et al. (2011) found an overall trend in speed of 10.6% per decade between 1992 and 2009, with some regions having increases up to 16%. Although wind is the major forcing mechanism on ice drift, there has not been a similar associated increase in winds (Figure 11.7). This means that the response of the ice to wind stress is changing. This is probably related to a thinner ice cover that has less resistance to the winds, as well as perhaps a less compact ice pack that drifts more freely.

References

Andersen, S., Tonboe, R., Kaleschke, L., Heygster, G. & Pedersen, L.T. (2007) Intercomparison of passive microwave sea ice concentration retrievals over the high-concentration Arctic sea ice, *Journal of Geophysical Research*, **112**, C08004, doi:10.1029/2006JC003543.

Bliss, A.C. & Anderson, M.R. (2014) Snowmelt onset over Arctic sea ice from passive microwave satellite data: 1979–2012. *The Cryosphere*, **8**, 2089–2100.

Cavalieri, D.J., Gloersen, P. & Campbell, W.J. (1984) Determination of Sea Ice Parameters with the NIMBUS-7 SMMR. *Journal of Geophysical Research*, **89**(D4), 5355–5369.

Cavalieri, D.J., Parkinson, C.L., Gloersen, P., Comiso, J.C. &Zwally, H.J. (1999) Deriving long-term time series of sea ice cover from satellite passive-microwave multisensor data sets. *Journal of Geophysical Research*, **104**, 15 803–15 814.

Cavalieri, D.J., Parkinson, C.L., Gloersen, P. &Zwally, H. (1996, updated yearly) *Sea Ice Concentrations from Nimbus-7 SMMR and DMSP SSM/I-SSMIS Passive Microwave Data*. Boulder, Colorado, USA. NASA DAAC at the National Snow and Ice Data Center.

Comiso, J.C. (2012) Large decadal decline of the Arctic multi-year ice cover. *Journal of Climate*, **25**, 1176–1193.

Divine, D.V. & Dick, C. (2006) Historical variability of sea ice edge position in the Nordic Seas. *Journal of Geophysical Research*, **111**(C1), C01001, 10.1029/2004JC002851.

Divine, D. V., Dick, C. (2007) *March through August ice edge positions in the Nordic Seas, 1750–2002*. National Snow and Ice Data Center, Boulder, CO, USA. doi.org/10.7265/N59884X1.

Eisenman, I., Meier, W.N. & Norris, J.R. (2014) A spurious jump in the satellite record: has Antarctic sea ice expansion been overestimated? *The Cryosphere*, **8**, 1289–1296.

Farrell, S.L., Kurtz, N., Connor, L.N. et al. (2012) A first assessment of IceBridge snow and ice thickness data over Arctic sea ice. *IEEE Transactions on Geoscience and Remote Sensing*, **50**, 2098–2111.

Fetterer, F., Knowles, K., Meier, W. &Savoie, M. (2002, updated daily) *Sea Ice Index*. National Snow and Ice Data Center, Boulder, CO, USA. Digital media.

Gallaher, D., Campbell, G.G. & Meier, W.N. (2014) Anomalous variability in Antarctic sea ice extents during the 1960s with the use of Nimbus data. *IEEE Journal of Selected Topics in Applied Earth Observations and Remote Sensing*, **3**, 881–887.

Hansen, E., Gerland, S., Granskog, M.A. et al. (2013) Thinning of Arctic sea ice observed in Fram Strait: 1990–2011. *Journal of Geophysical Research*, **118**, 5202–5221.

Heinrichs, J.F., Cavalieri, D.J., Markus, T. (2006) Assessment of the AMSR-E sea ice-concentration product at the ice edge using RADARSAT-1 and MODIS Imagery. *IEEE Transactions on Geoscience and Remote Sensing*, **44**, 3070–3080.

Kattsov, V.M., Ryabinin, V.E., Overland, J.E. et al. (2010) Arctic sea-ice change: a grand challenge of climate science. *Journal of Glaciology*, **56**, 1115–1121.

Kay, J.E., Holland, M.M. &Jahn, A. (2012) Inter-annual to multi-decadal Arctic sea ice trends in a warming world. *Geophysical Research Letters*, **38**, L15708, doi:101029/2011GL048008.

Kinnard, C., Zdanowicz, C.M., Fisher, D.A., Esaksson, E., de Vernal, A. & Thompson, L.G. (2011) Reconstructed changes in Arctic sea ice over the past 1,450 years, *Nature*, **24**, 509–512.

Krishfield, R.A., Proshutinsky, A., Tateyama, K. et al. (2014) Deterioration of perennial sea ice in the Beaufort Gyre from 2003 to 2012 and its impact on the oceanic freshwater cycle. *Journal of Geophysical Research*, **119**, 1271–1305.

Kurtz, N.T. & Farrell, S.L. (2011) Large-scale surveys of snow depth on Arctic sea ice from Operation IceBridge. *Geophysical Research Letters*, **38**, L20505, doi:10.1029/2011GL049216.

Kurtz, N., Galin, N. &Studinger, M. (2014) An improved CryoSat-2 sea ice freeboard retrieval algorithm through the use of waveform fitting, *The Cryosphere*, **8**, 1217–1237.

Kwok, R. &Rothrock, D.A. (2009) Decline in Arctic sea ice thickness from submarine and ICESat records: 1958–2008, *Geophysical Research Letters*, **36**, L15501, doi:10.1029/2009GL039035.

Kwok, R., Cunningham, G.F., Wensnahan, M., Rigor, I., Zwally, H.J. & Yi, D. (2009) Thinning and volume loss of the Arctic Ocean sea ice cover: 2003–2008, *Journal of Geophysical Research*, **114**, C07005, doi:10.1029/2009JC005312.

Laxon S.W., Giles, K.A., Ridout, A.L. et al. (2013) CryoSat-2 estimates of Arctic sea ice thickness and volume, *Geophysical Research Letters*, **40**, doi:10.1002/grl.50193.

Mahoney, A.R., Barry, R.G., Smolyanitsky, V. &Fetterer, F. (2008) Observed sea ice extent in the Russian Arctic, 1933–2006. *Journal of Geophysical Research*, **113**, C11005, doi:10.1029/2008JC004830.

Markus, T. &Cavalieri, D.J. (1998) Snow depth distribution over sea ice in the southern ocean from satellite passive microwave data. In: *Antarctic Sea Ice: Processes, Interactions & Variability* (Ed. M. Jeffries). *Antarctic Research Series*, **74**, 183–187.

Markus, T., Stroeve, J.C. & Miller, J. (2009) Recent changes in Arctic sea ice melt onset, freeze-up & melt season length. *Journal of Geophysical Research*, **114**, C12024, doi:10.1029/2009JC005436.

Maslanik, J., Stroeve, J., Fowler, C. & Emery, W. (2011) Distribution and trends in Arctic sea ice age through spring 2011, *Geophysical Research Letters*, **38**, L13502, doi:10.1029/2011GL047735.

Maslowski, W., Kinney, J.C., Higgins, M. & Roberts, A. (2012) The future of Arctic sea ice. *Annual Review of Earth and Planetary Sciences*, **40**, 625–654.

Meier, W.N. (2005) Comparison of passive microwave ice concentration algorithm retrievals with AVHRR imagery in the Arctic peripheral seas. *IEEE Transactions on Geoscience and Remote Sensing*, **43**, 1324–1337.

Meier, W.N., Khalsa, S.J.S. &Savoie, M.H. (2011) Intersensor calibration between F-13 SSM/I and F-17 SSMIS near-real-time sea ice estimates. *IEEE Transactions on Geoscience and Remote Sensing*, **49**, 3343–3349.

Meier, W.N., Stroeve, J., Barrett, A. &Fetterer, F. (2012) A simple approach to providing a more consistent Arctic sea ice extent timeseries from the 1950s to present. *The Cryosphere*, **6**, 1359–1368.

Meier, W.N., Gallaher, D. & Campbell, G.G. (2013) New estimates of Arctic and Antarctic sea ice extent from recovered Nimbus I satellite imagery. *The Cryosphere*, **7**, 699–705.

Meier, W.N., Hovelsrud, G., van Oort, B. et al. (2014) Arctic sea ice in transformation: A review of recent observed changes and impacts on biology and human activity, *Reviews of Geophysics*, 41, doi:10.1002/2013RG000431.

Nghiem, S.V., Rigor, I.G., Perovich, D.K., Clemente-Colon, P., Weatherly, J.W. & Neumann, G. (2007) Rapid reduction of Arctic perennial sea ice, *Geophysical Research Letters*, **34**, L19504, doi: 10.1029/2007GL031138.

Notz, D. &Marotzke, J. (2012) Observations reveal external driver for Arctic sea-ice retreat. *Geophysical Research Letters*, **39**, L08502, doi:10.1029/2012GL051094.

Perovich, D.K., Richter-Menge, J.A., Jones, K.F. & Light, B. (2008) Sunlight, water & ice: Extreme Arctic sea ice melt during the summer of 2007. *Geophysical Research Letters*, **35**, L11501, doi:10.1029/2008GL034007.

Perovich, D.K. &Polashenski, C. (2012) Albedo evolution of seasonal Arctic sea ice. *Geophysical Research Letters*, **39**, L08501, doi:10.1029/2012GL051432.

Polyak, L., Alley, R.B., Andrews, J.T. et al. (2010) History of sea ice in the Arctic. *Quaternary Science Reviews*, **29**, 1757–1778.

Rampal, P., Weiss, J. & Marsan, D. (2009) Positive trend in the mean speed and deformation rate of Arctic sea ice, 1979–2007. *Journal of Geophysical Research*, **114**, C05013, doi:10.1029/2008JC005066.

Rayner, N.A., Parker, D.E., Horton, E.B., Folland, C.K., Alexander, L.V. & Powell, D.P. (2003) Global analyses of sea surface temperature, sea ice & night marine air temperature since the late nineteenth century. *Journal of Geophysical Research*, **108**, 4407, doi:10.1029/2002JD002670.

Spreen, G., Kwok, R. &Menemenlis, D. (2011) Trends in Arctic sea ice drift and the role of wind forcing: 1992–2009. *Geophysical Research Letters*, **38**, L19501, doi:10.1029/2011GL048970.

Steele, M., Zhang, J. &Ermold, W. (2010) Mechanisms of summertime upper Arctic Ocean warming and the effect on sea ice melt. *Journal of Geophysical Research*, **115**, C11004, doi:10.1029/2009JC005849.

Steffen, K., Key, J., Cavalieri, D.J. et al. (1992) The estimation of geophysical parameters using passive microwave algorithms. In: *Microwave Remote Sensing of Sea Ice* (Ed. F.D. Carsey), pp. 201–231. American Geophysical Union, Washington, DC.

Stroeve, J.C., Markus, T., Boisvert, L., Miller, J. & Barrett, A. (2014) Changes in Arctic melt season and implications for sea ice loss. *Geophysical Research Letters*, **41**, doi:10.1002/2013GL058951.

Sumata, H., Lavergne, T., Girard-Ardhuin, F. et al. (2014) An intercomparison of Arctic ice drift products to deduce uncertainty estimates. *Journal of Geophysical Research-Oceans*, **119**, 4887–4921.

Titchner, H.A. & Rayner, N.A. (2014) The Met Office Hadley Centre sea ice and sea surface temperature data set, version 2: 1. Sea ice concentrations. *Journal of Geophysical Research*, **119**, doi:10.1002/2013JD020316.

Walsh, J.E. & Chapman, W.L. (2001) Twentieth-century sea ice variations from observational data. *Annals of Glaciology*, **33**, 444–448.

Walsh, J.E., Chapman, W.L., &Fetterer, F. (2015) *Gridded Monthly Sea Ice Extent and Concentration, 1850 Onward, Version 1*. National Snow and Ice Data Center (NSIDC), Boulder, CO, USA. http://dx.doi.org/10.7265/N5833PZ5.

Wang, L., Derksen, C., Brown, R. & Markus, T. (2013) Recent changes in pan-Arctic melt onset from satellite passive microwave measurements. *Geophysical Research Letters*, **40**, 1–7.

Warren, S.G., Rigor, I., Untersteiner, N. et al. (1999) Snow depth on Arctic sea ice. *Journal of Climate*, **12**, 1814–1829.

Webster, M.A., Rigor, I.G., Nghiem, S.V. et al. (2014) Interdecadal changes in snow depth on Arctic sea ice, *Journal of Geophysical Research*, **119**, doi:10.1002/2014JC009985.

Zhang, J.L. &Rothrock, D.A. (2003) Modeling global sea ice with a thickness and enthalpy distribution model in generalized curvilinear coordinates. *Monthly Weather Review*, **131**, 845–861.

CHAPTER 12

Sea ice in Earth system models

Dirk Notz[1] and Cecilia M. Bitz[2]

[1] Max Planck Institute for Meteorology, Hamburg, Germany
[2] Atmospheric Sciences Department, University of Washington, Seattle, WA, USA

12.1 Introduction

Modern Earth system models (ESMs) are the most advanced tools available to understand the functioning of the Earth's climate system and the interaction of its various components. An ESM is designed to simulate the physical, chemical and biological processes in the atmosphere, ocean, land and sea ice that govern the evolution of Earth's climate. Among the most important interactions that these models must represent is that of sea ice with both the atmosphere and the ocean. This importance derives from the fact that in polar regions, sea ice forms the interface between the ocean and the air, and as such controls to a large degree the exchange of heat, moisture and momentum between the two. Through this control, sea ice gives rise to a number of very effective feedbacks that can amplify or dampen the climate response to forcing in the Earth system.

In Section 12.2 of this chapter, we outline the numerical representation of sea ice processes in modern ESMs that govern the sea ice evolution as discussed in Chapters 1 and 2. In Section 12.3 we then describe how ESMs are used to provide insights into the evolution of sea ice–climate interactions and will outline some of the most important results. In particular, we describe what we have learned about the various feedbacks that drive the evolution of sea ice and polar climates, why sea ice in the Arctic and Antarctic has developed quite differently in the past few decades, and outline possible pathways of the future evolution of sea ice. The chapter concludes with a brief summary and outlook of the future of sea ice modelling in ESMs.

12.2 Representing sea ice in an Earth system model

12.2.1 The general setup of an ESM

Already in some of the earliest attempts to simulate the climate system of our Earth, sea ice was identified as a key component that one needs to consider in order for simulations of the Earth's climate system to be even remotely realistic (Budyko, 1969; Manabe & Stouffer, 1980). Since those early models, we have come a long way. Via the intermediate step of general circulation models (GCMs) that consisted of numerical representations of the physical processes that drive the interaction of atmosphere, ocean and sea ice, we have now reached the era of ESMs. In most ESMs the physical processes are even more advanced than in GCMs, with sea ice components that account for melt ponds and the latest sea ice dynamics, and ESMs generally also include biogeochemical interactions, e.g. through the exchange of CO_2 between land, ocean and atmosphere.

The general setup of ESMs is still very similar to those of the numerical exploration of the Earth's climate system in GCMs. The two main components of the

---

Sea Ice, Third Edition. Edited by David N. Thomas.
© 2017 John Wiley & Sons, Ltd. Published 2017 by John Wiley & Sons, Ltd.

climate system, namely the ocean and the atmosphere, are divided into grid cells, whose size in the horizontal has decreased from the several hundred kilometres, which was typical a few decades ago, to about 100 km in most cases today. However, there are many examples of integrations with finer grids, at a few tens of kilometres (e.g. Small et al., 2014) or even just a few kilometres in the most highly resolved modern ESMs. The trade-offs of utilizing higher horizontal resolution are often that run lengths are short; the number of ensemble members is one or very few; and vertical resolution and/or representation of processes are compromised. To avoid the polar singularity from converging meridians in the ocean at the North Pole, often the ocean grids are not simple spherical grids (see Figure 12.1), so the resolution cited for the ocean is often an average of the grid.

The numerical grid for computing atmospheric circulation is often on a spectral domain in which the variables are computed from an expansion in spherical harmonics, which also avoids the polar singularity. In this case, the resolution is commonly represented by a letter and number that indicate the truncation of the harmonics (i.e. the maximum resolved wavenumber). Even if spherical harmonics are employed, the atmosphere component must always compute 'physics' (radiation, clouds, etc.) on a spatial domain. In recent years, however, a number of modelling centres have started to develop atmospheric components that compute both dynamics and physics on the spatial domain, as these avoid spurious oscillations that arise from harmonic expansions and can be problematic in positive definite quantities (e.g. trace atmospheric constituents). Computation of dynamics on a grid in the spatial domain also tends to allow better use of modern massively parallel computer setups.

New spectral element methods (Evans et al., 2013) and alternative spatial discretizations (e.g. icosahedral-hexagons grids; Satoh et al., 2014) are being developed that aim to balance the attractive qualities of other schemes, while minimizing deficiencies. These methods can be applied to any ESM component and also permit desirable properties with selective grid refinement. Figure 12.2 shows an example of ocean and sea ice grids from the Model Prediction Across Scales (MPAS) model (Ringler et al., 2013). One panel shows a grid with increased resolution in the mid-latitudes of the North Atlantic, but the same technique could be used to increase resolution at the poles.

Both in the ocean and in the atmosphere, the physical laws that represent the transport of heat, momentum and mass are solved numerically on the underlying grid, which then allows one to simulate the time evolution of the system from any given initial condition. There are a variety of different timescales that typically govern the evolution of the ocean and atmosphere. The time step for dynamics in the ocean and atmosphere is usually tens of minutes or shorter, while the time step for atmospheric physics is usually about 1 hour. At the interface between the ocean and the atmosphere, the two systems need to exchange information, which occurs in

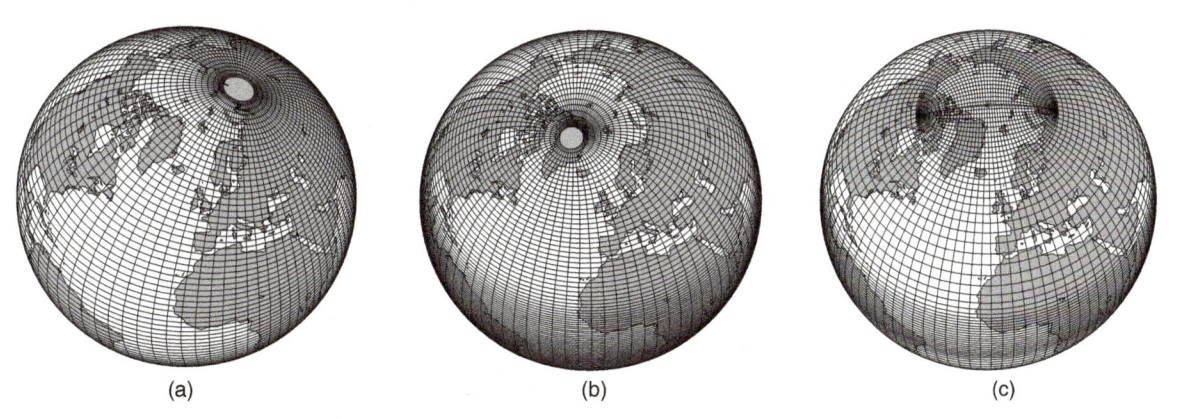

Figure 12.1 Examples of ocean and sea ice grids in Earth system models from CMIP5. Only every fourth line from the model grid is shown, so each cell as drawn encloses 16 grid cells in the model. The specific grids shown are: (a) a spherical mesh rotated so the grid North Pole is in Asia, from the INMCM4 model; (b) a non-spherical mesh in the northern hemisphere with the grid North Pole shifted into Greenland, from the CESM1 model; and (c) a tripole mesh, from the ACCESS1-3 model.

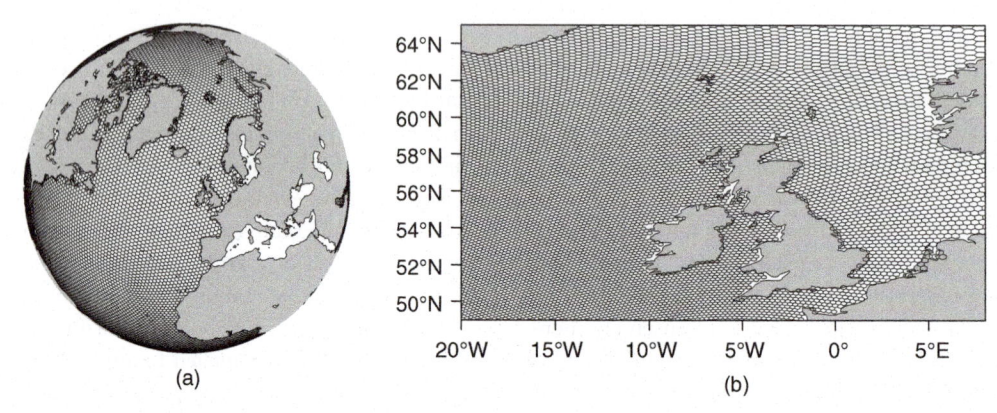

Figure 12.2 Examples of ocean and sea ice grids from the MPAS approach (Ringler et al., 2013): a uniform 120 km resolution mesh (a) and a multi-resolution mesh (b) varying from 75 km over most of the globe and reduced to 15 km near the southwest corner of the panel. Similar mesh refinement could be applied to the polar region. Source: Ringler et al., 2013. Reproduced with permission of Elsevier.

a special component of an ESM, called the coupler. This coupler receives and sends information about the surface state and fluxes exchanged across the surface between the atmospheric and ocean components. The interval at which such exchange of information occurs is called the coupling frequency, which may be as low as once per simulated day or could be hourly to allow ocean ecosystems to respond to the diurnal cycle and to drive ocean internal waves. While this general setup of an ESM is relatively straightforward, the existence of sea ice complicates the picture significantly. Because sea ice interacts simultaneously with both the ocean and the atmosphere, sea ice must exchange information with both systems simultaneously. Further, the coupling frequency is likely to be hourly between sea ice and atmosphere as the surface temperature of sea ice reacts immediately to changes in the incoming radiative fluxes, and the model would soon become numerically unstable if such reactions were delayed by several hours. Some ESMs also try to capture inertial oscillations of sea ice in the ocean (Roberts et al., 2015) and therefore the exchange of momentum flux between ocean and sea ice is approaching an hourly coupling frequency in some models. Clearly, the implementation of the sea ice component is a major design issue during the development of any ESM.

In sea ice components with a sub-grid-scale parameterization of the ice thickness distribution (see Section 12.2.4 and Figure 12.3), the coupling with the atmosphere must be considered separately for each thickness category. If we consider the open water as a zero-thickness category, then in principle any sea ice component that captures a variable sea ice fraction must also consider the coupling with the atmosphere separately over ice and open water. Because the sea ice fraction varies in time, a logical strategy is to control the entire atmosphere–surface exchange through the sea ice model component, even in areas that are not sea ice-covered at all. Hence, in a sense, the sea ice model component acts as an oceanic surface component, where the existence of sea ice in a particular grid cell modifies the magnitude of the variables exchanged with the atmosphere. With this strategy, it is most likely that the sea ice component is chosen to have the same grid as the ocean component, the coupling frequency between atmosphere and sea ice is equivalent to the atmosphere physics time step everywhere, and the coupling between sea ice and ocean is relatively simple. The coupler has the job of grid transformations between all variables exchanged between sea ice and atmosphere and managing the time at which components are run and variables are sent.

Not all ESMs operate this way, in part due to legacy codes that were developed when sea ice components were highly idealized. In some ESMs, the ocean is coupled directly to the atmosphere over the sea ice-free portion of the grid cell, in which case the ice model must send the ocean the sea ice fraction, and, in turn, the ocean must send the ice model any information about new ice growth over open water. It is also not uncommon for ESMs to incorporate a subset of the sea ice thermodynamics into the atmosphere component where it is computed on the atmosphere grid and on

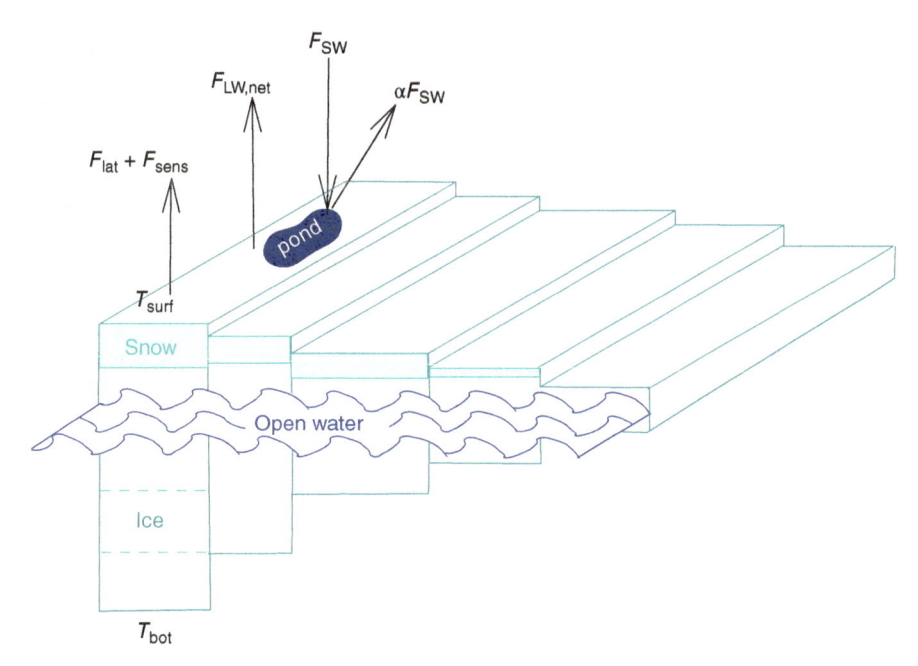

Figure 12.3 Schematic of sea ice processes within a grid cell of an Earth system model. The ice thickness distribution of sea ice is represented as a series of categories, each with a specific thickness range. Five categories with thickness greater than zero and another for open water are typical. The fractional ice coverage, ice thickness, snow depths, pond coverage, radiative transfer and surface energy balance are predicted for each category. Some models also treat the cycling of brine, aerosols, algae and chemistry within the ice and snow.

the atmosphere's physics time step. Then the rest of the sea ice processes lie within the ocean component and are solved on the ocean grid with the ocean time step.

The 'sea ice as ocean surface layer' approach requires all of the processes that relate to sea ice to exist in one truly independent sea ice component. In the next sections we describe the basic building blocks in this type of approach, derived from conservations of mass, energy and momentum. Note that we do not discuss how the transport of salt and biogeochemical tracers is implemented into modern sea ice models, which is instead described in detail in Chapter 20.

12.2.2 Sea ice thermodynamics

Sea ice interacts with the ocean and the atmosphere primarily through four of its main properties: its thickness, its areal coverage, its surface conditions, and temperature and salinity profiles. These are also the main state variables that evolve in response to the forcing that is external to the sea ice in any reasonable representation of an ESM. Though there are many details of how this interaction occurs in reality (see primarily Chapters 1 and 2), here we only briefly summarize how these interactions are typically represented in large-scale numerical models. We start with a description of thermodynamic processes and then later discuss the representation of changes in ice thickness and ice coverage through dynamic processes.

The first parameter that any thermodynamic model component needs to update is the surface temperature of the sea ice, T_{surf}. As discussed in Chapter 1, provided the temperature is below the surface melting point, T_{surf} can be calculated from the surface energy balance:

$$F_{cond} = F_{SW,net} + F_{LW,net} + F_{sens} + F_{lat} \qquad (12.1)$$

which describes that the heat flux into the ice, F_{cond}, is given by the sum of net short-wave radiation $F_{SW,net}$, net long-wave radiation $F_{LW,net}$, sensible heat flux F_{sens} and latent heat flux F_{lat}, including the latent heat exchange during surface melting or snow ice formation. Physically, each of these fluxes varies with surface temperature, and generally the surface temperature is obtained by iteratively solving the surface energy balance. In particular, since the outgoing long-wave radiation is proportional to T_{surf}^4, even a linear dependence of the other fluxes on surface temperature would still require one to solve a fourth-order polynomial to obtain the surface temperature. For numerical efficiency, one therefore often uses a Taylor's series expansion of those fluxes that depend non-linearly on T_{surf} and then only calculates the change in surface temperature relative to the previous time step (e.g. Winton, 2000). As a further simplification, one usually assumes that the surface albedo is given by the surface properties of the previous time step, which allows one to treat it as a constant for any given time step. This then allows for numerically efficient code. If the procedure for solving equation 1 for T_{surf} should result in a temperature above the surface melting point, then each flux in equation 1 is computed with T_{surf} set equal to the melting point and the balance in Equation (12.1) does not hold.

The conductive heat flux into the ice F_{cond} is represented with various degrees of complexity among modern sea ice models. Some models still use the very simple so-called zero-layer approach after Semtner (1976), where the total heat flux through the ice and snow is assumed to be vertically homogeneous, which in particular implies that the heat capacity of sea ice and snow is negligible. In these models, the heat flux is hence simply given as:

$$F_{cond} = \frac{T_{surf} - T_{bot}}{h_i/k_i + h_s/k_s}$$

Here, h_i and h_s denote the thickness of the snow and the ice, respectively, k_i and k_s are the heat conductivities of the two media, and T_{bot} is the bottom temperature. Because this approach neglects the heat capacity of the ice, in summer all excess energy goes into the thinning of the ice pack, which then causes an amplified seasonal cycle (Semtner, 1984).

More complex models, in contrast, numerically solve the transfer of heat through several layers of sea ice, which in particular allows one to capture the significant storage of heat within the ice itself. These models are typically formulated as enthalpy models, where the state variable that describes the sea ice in any given layer is given by the enthalpy of the layer, which in turn depends on the temperature of the ice and its solid fraction in each layer (Bitz & Lipscomb, 1999). Because the bulk salinity is usually held constant and thus known, the temperature and the solid fraction are uniquely determined by the enthalpy. The numerical solution of the heat-transfer equation then allows these models to update the entire temperature field within the sea ice at each time step, resulting in an energy-conserving time evolution of the energy that is stored within the ice. In even more realistic models, the salinity is a predicted quantity. Thus the heat conductivity that is used to transfer heat from one layer to another depends on the solid fraction given by temperature and bulk salinity, as described in Chapter 1.

Based on the update of the temperature field in the ice, it is then possible to calculate the freezing and melting of the ice in response to the net heat fluxes above the top $F_{net,atm}$ (equal to the sum of fluxes on the right hand side of equation 12.1) and below the bottom $F_{net,oc}$ surfaces. If the top surface is at the melting point, then the flux imbalance from $F_{cond} \neq F_{net,atm}$ changes the solid fraction at the surface and possibly an overall thinning of the ice $\partial h/\partial t$ at the top according to:

$$F_{cond} = F_{net,atm} + \rho L \left.\frac{\partial h}{\partial t}\right|_{top} \quad (12.2)$$

where ρL is the latent heat. Similarly at the bottom, the growth and melt of the ice are given by the bottom heat-flux balance:

$$f_{cond} = F_{net,oc} + \rho L \left.\frac{\partial h}{\partial t}\right|_{bot} \quad (12.3)$$

where f_{cond} is the heat conducted into the ice at the bottom. As with all numerical implementations of the equations discussed so far, a wide range of different approaches also exists to calculate the oceanic heat flux that provides heat to the bottom of the sea ice. In the simplest approach, the water temperature in any ocean grid cell that contains sea ice is forced to be at the freezing temperature of the water. All excess heat that might warm the water, e.g. by advection or heating from solar radiation, is received directly by the sea ice as $F_{net,oc}$. In more realistic methods, the oceanic heat flux is parameterized based on the friction velocity of the ocean and the temperature difference between the bottom of the ice and the far field water temperature. In the most advanced models, a three-equation approach is used, where the temperature at the bottom of the ice is determined from phase equilibrium including the turbulent transport of salt to the interface, whereas in simpler approaches only a two-equation approach is used, where the temperature at the ice–ocean interface is assumed to be equal to the freezing temperature of the far-field ocean water (Schmidt et al., 2004).

This short overview already demonstrates the wide complexity involved in representing the thermodynamic interaction of sea ice with both the ocean and the atmosphere. Because of space constraints, we cannot go into much further detail here regarding the detailed implementation of the various processes that govern the exchange of heat between the three media. This includes, for example, details of the various albedo parameterizations including melt-pond coverage, details of radiative transfer through the ice, the treatment of snow on the ice and details of the representation of sensible- and latent-heat fluxes and their relationship to surface properties of the ice. For a detailed description of some of these parameterizations, we refer the reader to some recent reviews and model documentation, including Hunke et al. (2010, 2011, 2015), Notz (2012), Bitz et al. (2012) and Vancoppenolle et al. (2012).

12.2.3 Sea ice dynamics

In addition to the thermodynamic processes discussed so far, numerical models of sea ice must also represent the movement of sea ice in response to external momentum transfer from wind and ocean currents. While these motions only redistribute the existing sea ice volume and hence, unlike thermodynamics, do not directly change the hemispheric total sea ice volume itself, sea ice dynamics has a very strong indirect impact on sea ice volume. For example, if sea ice divergence in a certain region causes the formation of open water, then in non-summer months significant ice growth can occur. Figure 12.4 illustrates this anti-correlation of the grid-cell mean sea ice thickness tendencies from thermodynamic and dynamic processes in the Antarctic fall. (Note that the grid-cell mean is an average over open water and sea ice-covered portions of the grid cell, so the grid-cell mean thickness is proportional to the sea ice volume in a grid cell.) In contrast, in summer, open water formation by divergence leads to enhanced absorption of solar radiation, and hence strongly increases the thermodynamic thinning of the surrounding ice. Hence, in numerical models the introduction of sea ice dynamics changes the overall ice volume significantly, where usually the increased melt caused by the export of sea ice into warmer regions and the additional absorption of solar radiation in divergent areas during summer dominates the annual mean response of the ice pack. The introduction of sea ice dynamics therefore generally causes an overall decrease in annual mean total sea ice volume (e.g. Holland et al., 2001).

The importance of sea ice dynamics for a realistic representation of sea ice was acknowledged for the earliest large-scale models of sea ice, leading during the 1970s to a rapid development of various approaches to implement sea ice dynamics into large-scale models. These approaches differ in particular in their ways of representing the mechanical response of sea ice to forcing by the atmosphere and ocean. Among others, such response can be divided into a plastic response, where the ice cover yields to a certain external stress, e.g. through the irreversible formation of ridges, into a viscous response, where the ice responds to an external stress similar to a viscous fluid, and an elastic response, where the ice regains its original state once the external stress disappears.

In the literature, various approaches have been combined to make a sea ice rheology for an individual model, such as the widely used viscous-plastic (VP) rheology of Hibler (1979). In this approach, the ice yields plastically to typical stresses but acts like a viscous fluid for small external stresses. The strength of the ice is parameterized in terms of the ice thickness and concentration in the individual grid cells.

Numerically, the implementation of any scheme describing sea ice dynamics is far more challenging than the implementation of sea ice thermodynamics, because dynamics transports sea ice properties among neighbouring grid cells. In contrast, thermodynamics is simply calculated within each grid cell, independent of any information in the surrounding cells. Furthermore, Hibler's VP rheology uses implicit numerical methods, which require large matrix inversions that pose challenges for numerical efficiency on modern massively parallel computers. To address this problem, a popular modern method introduces elastic waves to Hibler's VP rheology that only have a significant effect at timescales of minutes and shorter (Hunke & Dukowicz, 1997). Because of the elastic waves, this numerically more efficient method is referred to as an EVP rheology, even though it does not capture any true elastic response of the sea ice and reduces to the VP solution on timescales relevant to variations in the atmosphere and ocean forcing (see Feltham, 2008, for an extensive review).

Representing the behaviour of sea ice in response to some given external forcing is only one of the challenges in implementing sea ice dynamics into a

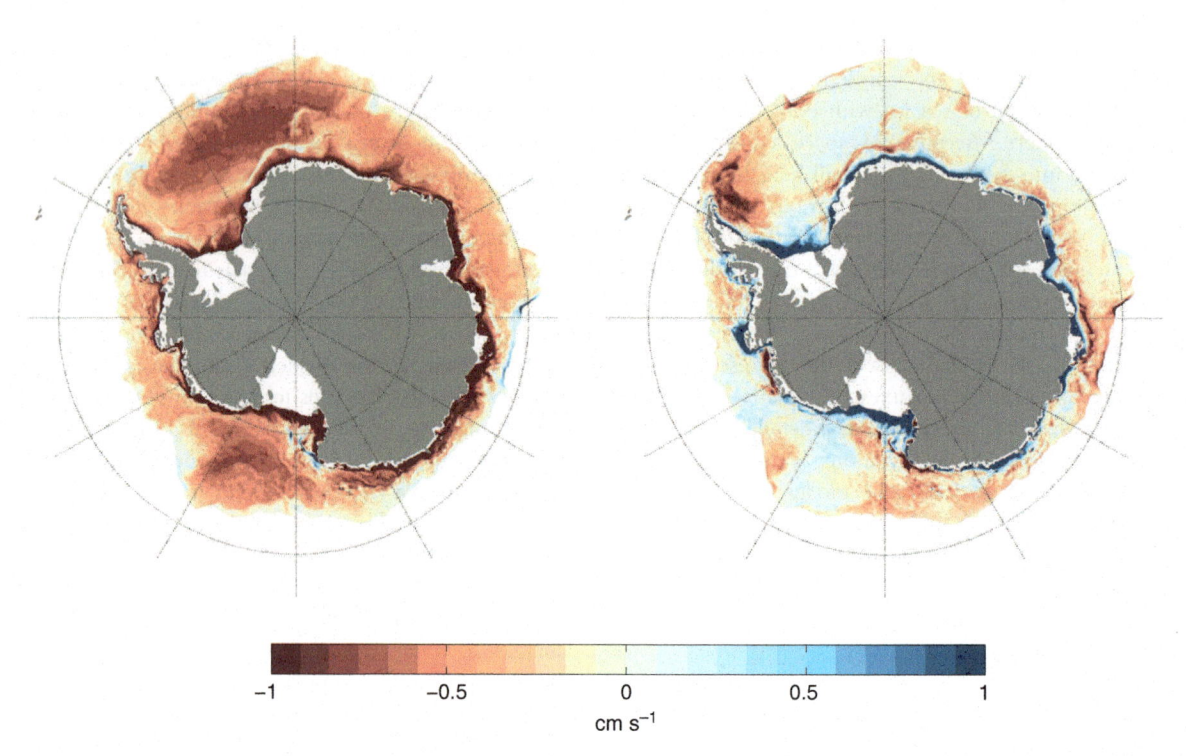

Figure 12.4 April–June sea ice grid-cell mean thickness tendencies (cm s^{-1}) from thermodynamic (left) and dynamic (right) processes in the CCSM3.5 at 0.1° resolution in the sea ice and ocean. Grid cells with sea ice concentration below 15% are omitted. There is a strong negative spatial correlation (R^2 = 0.84) between the two panels. Model output is from the last 35 years of a 150-year integration that is described in Kirtman et al. (2012).

numerical model. In general, the full force balance of sea ice that describes its change in movement according to Newton's second law is given by

$$m_A \frac{Dv}{Dt} = F_{\text{cor}} - m_A g \nabla H + \tau_a + \tau_w + \nabla \cdot \sigma$$

where the acceleration of a certain mass of sea ice m_A occurs in response to the Coriolis force, a gravitational force term that is related to the tilt of the sea surface, the atmospheric and oceanic drag and the divergence of the internal stress tensor. While the first two force terms are straightforward to calculate in a numerical model, the drag terms are very challenging to represent realistically (see Chapters 2 and 5). This is because for a realistic representation of these forces, the numerical model must realistically represent the roughness of the ice, the stratification of the ocean and the atmosphere, respectively, and hence the detailed profile of winds and currents very close to the ice interface.

This short overview again only gives a very rough outline of the challenges that any model developer faces when integrating sea ice dynamics into their model. For a more comprehensive overview of these challenges, see Leppäranta (2005) and Weiss (2013).

12.2.4 Sea ice thickness distribution

In any large-scale model, a large number of physical processes cannot be explicitly simulated but must be parameterized. This term describes any process in a large-scale model that occurs on scales that are too small to be explicitly represented by the numerical grid, and which hence must be represented by relating information that is available on the scale of the grid to the processes that occur on smaller length scales. In a sea ice context, one of the most important parameterizations relates to the distribution of ice thickness within a single grid cell. As discussed in detail in Chapter 2, the thickness of sea ice is spatially not uniform, but changes on all length scales from metres to the scale of the entire basin. While the changes in thickness that occur on spatial scales beyond the size of individual grid cells are directly represented by the sea ice module of any large-scale numerical model, the thickness changes that occur within an individual grid cell can only be parameterized.

Such parameterization is accomplished by introducing a certain number of ice thickness categories, each of which represents the areal fraction of sea ice of a given thickness. In the simplest approach, which is based on the model introduced by Hibler (1979), the grid cell is simply divided into two categories, namely a fraction of open water and a fraction of sea ice with a given ice thickness. At each time step, the volume of sea ice in that grid cell can change by the growth or melt of the existing ice and by growth of ice in the open water fraction. The total ice volume obtained as such is then assigned to a new areal coverage, which is obtained by a simplified approach that calculates any increase in areal coverage by simply calculating the area that the ice forming in open water would cover if it were redistributed to a collection thickness h_0. This parameterization then gives a new total areal coverage of all the ice in that grid cell, and the ice thickness for the next time step is obtained by dividing the total sea ice volume by the total sea ice covered area obtained such. During melting conditions, the decrease in areal sea ice coverage is obtained by assigning a certain percentage of the total loss of volume to the thinning of the sea ice and the remaining percentage to a loss of areal coverage. This *ad hoc* parameterization captures in a very simple, yet mass- and energy-conserving way, the change in areal coverage versus the change in sea ice thickness that occurs during melting and freezing.

In more advanced schemes (see Figure 12.3), in contrast, there are more than two categories in each grid cell, each of which represents the fractional coverage of a particular range of sea ice thickness (see, e.g. Thorndike et al., 1975). The thermodynamic change of sea ice thickness within each category is simply calculated using equations (12.2) and (12.3) with unique surface fluxes and melt/growth rates for each thickness category, so ice in thinner categories grows faster than the ice in thicker categories (also see Chapter 1). While these changes in sea ice thickness within the individual categories can be calculated based on first principles, the change in areal coverage of the individual ice thickness categories through mechanical redistribution and lateral freezing or melting must again be parameterized.

12.2.5 Tuning

Many of the processes that govern the evolution of sea ice in the real world must be parameterized in large-scale numerical models, since they occur on spatial scales that are too small to represent based on first principles on the coarse spatial resolution of the numerical grid. Such processes relate, for example, to the spatial distribution of sea ice thickness within individual grid cells that we just discussed, to the occurrence of melt ponds on the sea ice surface, to the proportion of lateral to vertical melt or to the integrated mechanical response of individual small ice floes to a given mechanical forcing. As these processes occur on scales that cannot be resolved, any parameterization trying to capture the bulk behaviour of a given process will contain individual parameters that cannot fully be determined from observations. This is because in the real world, the processes act on small spatial scales, whereas the parameterization can only capture the integrated large-scale response of the system. Because of this, any large-scale sea ice model (as well as any other ESM component) will contain a number of so-called tuning parameters, which need to be adjusted to obtain a realistic statistical representation of the smaller-scale processes on the large spatial scales of the numerical grid. This adjustment of the free parameters is referred to as tuning, and it is a necessary and routine operation in any ESM that normally only occurs during the model's development phase (Mauritsen et al., 2012).

Consider, for example, the parameterization of the sub-grid-ice-thickness distribution that we just discussed. The amount of lateral versus vertical melt in any such distribution depends in reality, among other variables, on the size of individual ice floes, on the amount of open water between these floes, on the roughness of the ice floes, on the detailed wave field that governs small-scale oceanic heat transport, and on many other properties of the ice pack that cannot be represented by a large-scale numerical model. Therefore, the division of melting between a decrease of ice thickness and a decrease of the ice areal coverage can only be parameterized, and the developer of the model will have some liberty regarding how to represent this division. As the parameterization is only a simplified representation of the real world, there is no overall 'correct' way to distribute the heat, and the model developer can set this number such that the ice cover as simulated at the large scale looks most realistic. For example, if the sea ice model has too little areal coverage of summer sea ice, the amount of lateral melt can be decreased, at the expense of then having more thinning of the ice during summer.

In practice, because of the computational expense of running an ESM, the tuning of the sea ice model component will only ever involve the adjustment of very few parameters. In particular, such adjustment will never be able to fully compensate for missing physical realism of those processes that can be resolved at the grid scale. For example, the adjustment of surface roughness will change the velocity of the ice in response to surface winds, but it cannot improve an unrealistic large-scale drift pattern. Similarly, while an adjustment of the lateral versus vertical proportion of heat in the model can affect the overall volume of sea ice, it will not be able to compensate for an unrealistic large-scale distribution of sea ice thickness.

While some tuning is unavoidable, generally when resolution is increased and/or more processes are included in a model, there are fewer tuning parameters. For example, when the number of ice thickness categories was increased so that the distribution of thin ice was better resolved within a grid cell, Bitz et al. (2001) found no need to partition an unrealistically high portion of melting to lateral melt because thin ice categories could melt away, thereby raising the open water fraction entirely through vertical melt. Thus model advancements ought to reduce the range parameters need to be tuned or possibly even reduce the number of tunable parameters.

Finally, we should also note that tuning of the sea ice model component usually aims for matching a long-term time mean tuning target, such as pre-industrial sea ice volume or the mean winter sea ice thickness, but cannot be used to match, for example, the long-term evolution of sea ice in response to changes in the external forcing.

Having described briefly how sea ice is represented in large-scale models, we now turn to the question of how these large-scale models can be used to actually learn something about the behaviour of sea ice. After some general remarks, we will discuss, in particular, insights from large-scale model studies on sea ice-related feedbacks, on the drivers of the ongoing changes in the Arctic and Antarctic, and on the possible future evolution of sea ice.

12.3 Learning from ESMs

12.3.1 Evaluating the usefulness of ESMs

Before we start examining some of the findings from modelling sea ice in ESMs, it seems useful to generally examine how such insights are gained and, in particular, how the relevance of any result from an ESM can meaningfully relate to the functioning of the real climate system of our planet (see Notz, 2015, for a detailed discussion). In doing so, we first note that *per se*, an ESM is nothing but a tool that has been developed to answer certain questions. This implies, for example, that a climate model can be neither verified nor falsified, similar to the fact that a screwdriver cannot be verified or falsified either. One can, however, possibly establish the usefulness of a climate model to answer a question at hand. Further, a given climate model might be found useful for understanding why sea ice has increased in the Southern Ocean but might not allow one to determine if the Northern Sea Route will be navigable next year.

The evaluation of a model generally aims to establish how much a given simulation might indeed represent the relevant aspects of the real climate system. However, for a number of reasons, such comparison is not straightforward. Most importantly, the Earth's climate system is a chaotic one, which is why for any given forcing, the observed trajectory of any observable is just one manifestation of an infinite number of possible trajectories. No numerical model can hence ever be expected to precisely simulate the observed evolution of any observable, and a mismatch between a simulation and the observations might simply be caused by internal variability.

A useful analogue in this respect might be a model that simulates the casting of a die: if three sequential simulations of the model results in a 3, a 5 and another 3, the model is by no means proven 'wrong' if casting a real die results in a 6. In the same way, a climate model that simulates, for example, a slower loss of Arctic sea ice than has been observed cannot per se be considered wrong. Indeed, a model that on average agrees with the observed decline might be a rather poor model. Similarly, if our model that simulates the casting of a standard die gives a 4, a 6 and an 8, it is a very poor model, even though on average it agrees with the 'observed' 6 of the real die in our example.

Such internal variability is just one reason why it can be difficult to prove or disprove the usefulness of a climate model for answering specific questions about the real world, but there are others. For example, observational uncertainty is an additional hindrance of a robust model evaluation. In the case of sea ice, the longest time series of large-scale satellite observations are those of sea ice concentration based on the measurement of the passive microwave signature of the surface, giving a continuous record since the 1970s. Unfortunately, however, the passive microwave signal that satellites measure is not directly related to the amount of sea ice in any given area. The signal is distorted as it travels through the atmosphere and snow cover. In addition, it is difficult to differentiate passive microwave emissions from thin sea ice or sea ice covered by melt ponds from the surrounding open water. For all these reasons, there is significant uncertainty in establishing the sea ice concentration from an analysis of the passive microwave signature of the surface. In particular, in summer, different algorithms give, for the same passive microwave signal, differences in sea ice concentration that can locally reach 20% and more (Meier & Notz, 2010). This hinders, for example, a robust evaluation of the sea ice concentration patterns of Arctic sea ice during summer, where some models and the Bootstrap algorithm suggest a high fraction of high-concentration sea ice, while other models and the NASA-Team algorithm suggest a modest fraction of high-concentration sea ice (see Figure 12.5).

Another difficulty is that it usually is not clear if a high-quality simulation of the present-day sea ice cover in an ESM also implies that that model has realistic simulations of the future sea ice cover. This uncertainty arises for a number of reasons. First, the physical processes

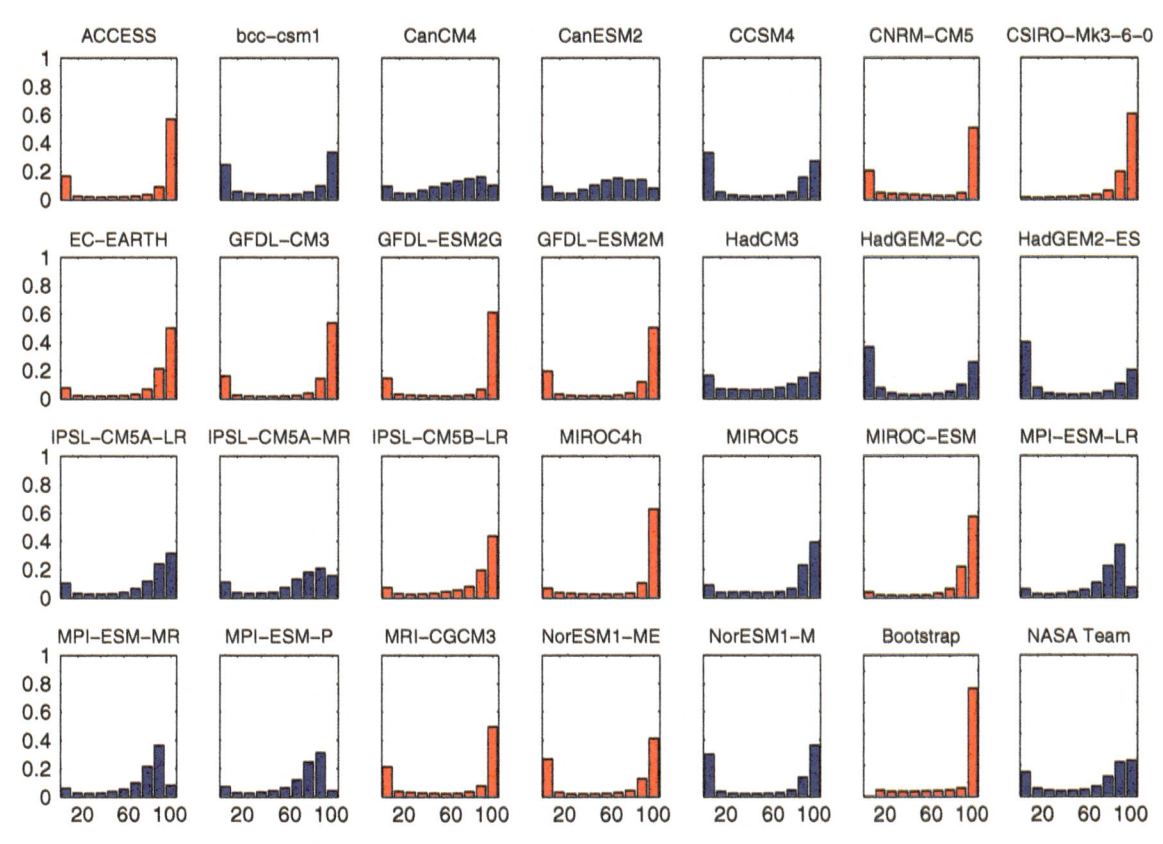

Figure 12.5 Histogram of 1979–2005 September sea ice concentration in all grid cells with at least 0.1 % sea ice concentration in CMIP5 model simulations. The numbers on the *x*-axis denote the upper limit of each bar: e.g. 20 denotes the concentration range 10–20%. Red panels denote histograms with a compact sea ice cover, while blue panels denote histograms with a loose sea ice cover. For models with multiple simulations, the ensemble mean is shown. The last two panels show two different satellite retrievals. Source: Notz, 2014.

that drive the evolution of sea ice might change as the ice cover and the climate state as a whole are changing. Second, a model might have been tuned to match the observable in question during the observational period, which does not necessarily imply that this model will be particularly good in modelling the future. And, last but not least, the quality of the prediction of any aspect of the sea ice cover in a model is primarily determined by the evolution of the simulated forcing by the atmosphere and the ocean, which hence must be evaluated too.

This last aspect indeed often makes it very difficult to establish whether any given shortcoming of a simulated sea ice cover is related to a shortcoming of the sea ice model itself, or simply a consequence of unrealistic simulated forcing of that sea ice model by the other components of the ESM. Indeed, even a relatively crude sea ice model will give a much more realistic simulation of sea ice evolution if forced by a high-quality ocean and atmosphere model than would a very-high quality sea ice model that is exposed to unrealistic forcing.

These difficulties might make it seem almost impossible ever to establish the usefulness of a climate model to answer a specific research question. Luckily, in reality such model evaluation is quite possible, particularly as the number of observables on which a robust model evaluation can be based is ever increasing, as more and more high-quality remote sensing and *in situ* data become available.

Nevertheless, model evaluation remains challenging, which is why currently those insights gained from model simulations are obviously considered most reliable that are fully consistent with the observational record (see Stroeve & Notz, 2015). In the following sections, we present some of these findings, which in particular underpin the fact that many of the key aspects of large-scale sea ice evolution are realistically represented by modern ESMs.

12.3.2 The importance of feedbacks

One of the major types of processes that an ESM must represent realistically is sea ice-related feedbacks (see Figure 12.6). Such feedbacks are important because they

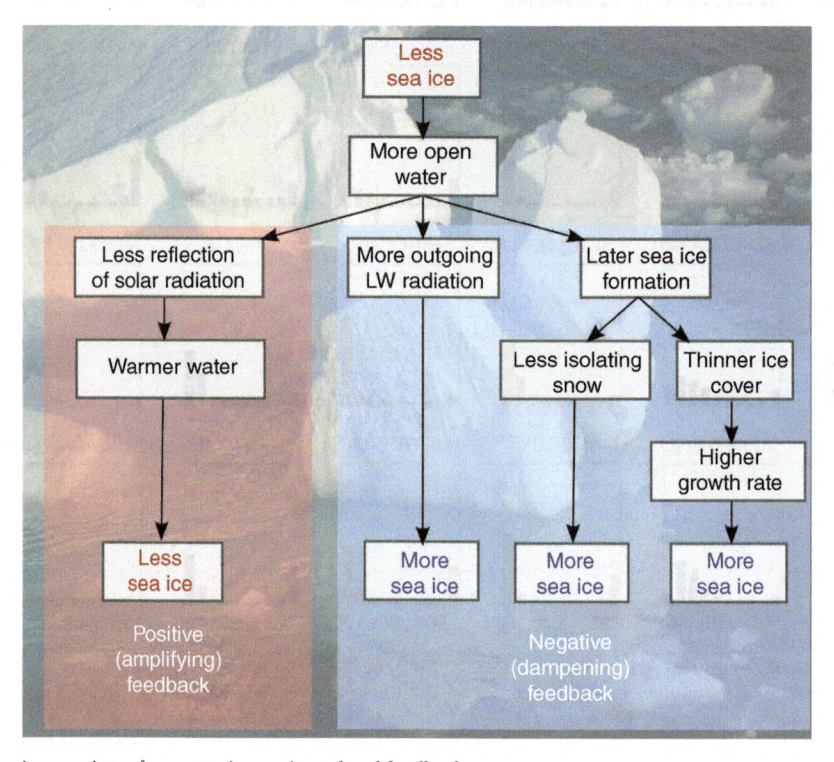

Figure 12.6 Schematic overview of some major sea ice-related feedbacks.

will either dampen or amplify any response to forced change. We consider feedbacks from two perspectives: first, in terms of how feedbacks directly influence the sea ice state; and second, in terms of feedbacks that affect radiative fluxes at the top of the atmosphere and the larger climate system more generally.

Sea ice-related feedbacks and their impact on the sea ice itself

Feedbacks that directly amplify sea ice change to forcing external to the sea ice are particularly important, because they can lead to so-called tipping points beyond which any change in the sea ice cover becomes dominated by internal self-acceleration. The most well known of such feedbacks is the ice–albedo feedback, which is related to the fact that the albedo of sea ice is much higher than that of the surrounding water. Hence, if the sea ice coverage decreases, more open water becomes exposed and more solar radiation is taken up by the system. This can then lead to even more sea ice loss, causing an acceleration of the initial decrease even if the external forcing remains constant.

The existence of such positive feedbacks related to changes in the sea ice cover has given rise to some speculation that the ongoing retreat of Arctic sea ice might already be unstoppable. However, this is probably not the case, as one can see from examining both the observational record and related simulations with large-scale models. In the observational record, it is striking that the ice cover always recovered again after a significant decrease from one summer to the next (Notz & Marotzke, 2012). Such behaviour is inconsistent with a dominating impact of amplifying feedbacks, such as the ice–albedo feedback, because positive feedbacks should lead to an even larger loss after a year with substantial sea ice retreat. Large-scale model simulations show the same behaviour, in that the sea ice year-to-year change (e.g. the running difference of one September mean to the next) has a negative autocorrelation at 1-year lag.

Dedicated model studies have further established that negative (i.e. stabilizing) feedbacks dominate the total response of sea ice to exposed changes. This implies that the evolution of the sea ice cover will remain tightly coupled to the ongoing change in the external forcing, as described later in this Chapter (see, e.g., related studies by Eisenman & Wettlaufer, 2009; Armour et al., 2011; Tietsche et al., 2011; Li et al., 2013). These modelling studies have allowed us not only to confirm the robustness of the observed sea ice behaviour for its future evolution, but also, in particular, to understand which feedbacks contribute to stabilizing the ice cover. One of the most important stabilizing feedbacks is related to the change in the outgoing long-wave radiation as the sea ice cover decreases. The open water that is then exposed to the atmosphere is much warmer than the surrounding sea ice cover, and in winter, in particular, causes a strong increase in the outgoing long-wave radiation. As analysed by Tietsche et al. (2011), the increase in outgoing long-wave radiation during winter very effectively compensates for the increased uptake of solar radiation that is caused by a retreating ice cover during summer. A second feedback that very effectively stabilizes the ice cover is related to the fact that thin ice grows much faster than thick ice (see Chapter 2 and Bitz & Roe, 2004). This is related to the fact that the thin ice more effectively transports heat from the ocean to the atmosphere, allowing for a significant gain in ice mass after a particularly strong thinning during summer. Third, the snow cover will be thinner on ice that formed later in the season (Hezel et al., 2012), which again stabilizes the ice cover: With less insulating snow covering the ice surface, the ice during winter is more directly exposed to the cold atmosphere, again causing a significant increase in sea ice growth rates (see Chapter 3 and Notz, 2009).

For the Antarctic, existing model studies have also found that any externally exposed change in sea ice coverage does not become self-accelerating (Armour et al., 2011; Li et al., 2013). However, the response time of Antarctic sea ice to a cooling of the climate system is much longer than that of Arctic sea ice, which then causes a certain decoupling of the sea ice state from the external forcing on timescales of several centuries. This is related to the fact that sea ice coverage in the Southern Ocean is tightly coupled to the properties of the underlying ocean, which might take several hundred years to adjust to a given change in the external forcing. However, these findings are much less robust than those related to Arctic sea ice, because current ESMs still have significant difficulties in simulating the observed distribution of water masses in the Southern Ocean and its possible response to changes in the external forcing (Sallée et al., 2013; Stössel et al., 2014).

Sea ice-related effects on radiative feedbacks

The large-scale climate response to climate forcing is traditionally viewed from a top-of-atmosphere (TOA) perspective. Even though the polar regions are often cloudy, the influence of the sea ice on surface reflectivity is unmistakable at the TOA, and hence the presence of sea ice reduces the net absorbed short-wave radiation at the top of the atmosphere in the polar region. Sea ice also alters the basic thermal state of the lower atmosphere by insulating the atmosphere from the ocean. As a result, the polar winter atmosphere tends to have a deep temperature inversion, while the polar summer atmosphere tends to have a surface temperature near 0°C. Before we explain how these characteristics influence large-scale climate change, we first describe the basics of climate feedbacks.

Climate forcing causes a change in the TOA fluxes, which alone causes a TOA flux imbalance of ΔR_f. Assuming the climate is stable, the surface temperature response ΔT acts to return the net TOA flux back into balance. The sign convention used is that a warming ($\Delta T > 0$) results from a positive climate forcing ($\Delta R_f > 0$). During the transient phase of the climate response (while the surface climate is still changing), there is yet a radiative imbalance at the TOA:

$$\Delta R = \Delta R_f + \lambda \Delta T$$

where λ is the net radiative feedback. A radiative feedback can be viewed as a measure of the inefficiency of climate processes at fluxing heat out the TOA for a given warming. For example, the ice–albedo feedback is so inefficient that it actually does the opposite: warming causes an increase in absorbed short-wave radiation, so the ice–albedo feedback is positive. Because the Earth's climate is stable, λ is negative, but as we shall see, only in the global mean.

The net radiative feedback can be broken into different components:

$$\lambda = \lambda_p + \lambda_{LR} + \lambda_\alpha + \lambda_w + \lambda_c$$

where the components in order are the Planck feedback, lapse-rate feedback, surface albedo feedback, water-vapour feedback and cloud feedback. Using the kernel method to evaluate feedbacks (Soden & Held, 2006; Soden et al., 2008) applied to an ensemble of ESMs, we find that the first three components have a substantially greater magnitude in the polar regions compared with other latitudes. Figure 12.7 shows the net feedback and these three key components. Here we see that there is net positive radiative feedback in latitudes with sea ice. However, none of these simulations exhibits runaway warming at any latitude, so clearly small regions with net positive feedback do not cause instability. Further, we see that the albedo feedback is not the only factor that causes polar amplification. The lapse-rate and Planck feedbacks are just as important (see also Pithan & Mauritsen, 2014). The lapse-rate feedback is positive because polar regions have strong temperature inversions, which trap surface warming and yield little change at the TOA fluxes. The Planck feedback, while always negative, is less so in the polar regions because it depends on temperature, and polar temperatures are low.

12.3.3 The ongoing changes in the Arctic and Antarctic sea ice

Having just described sea ice-related feedbacks and found no particularly strong hemispheric asymmetries, it is natural to wonder what then is the driving mechanism behind the observed opposing evolution of Arctic and Antarctic sea ice: While Arctic sea ice has been retreating rapidly in recent decades (Chapter 11), Antarctic sea ice has shown a slight increase (Chapter 10). Here we bring a modelling perspective to this question, and, unlike previous chapters, consider both hemispheres at once (Figure 12.8).

Turning first to the Arctic, after having established that the net effect of sea ice-related feedbacks does not cause a self-acceleration of forced changes in the sea ice cover, only two possibilities remain. First, the observed changes could simply be caused by large internal fluctuations of the climate system, and second, these changes could be caused by a response to climate forcing (such as the decrease in outgoing long-wave radiation at the top of the atmosphere from an increase in greenhouse gases in the atmosphere). Of these two possibilities, both the observational record and a large number of related modelling studies point to climate forcing as the main driver of the observed sea ice evolution. This is because internal variability as manifested in long-term observational records of sea ice from ships and airplanes (Notz & Marotzke, 2012) is too small to explain the observed retreat, as is the internal variability that is simulated by climate models in their so-called pre-industrial control simulations where the external

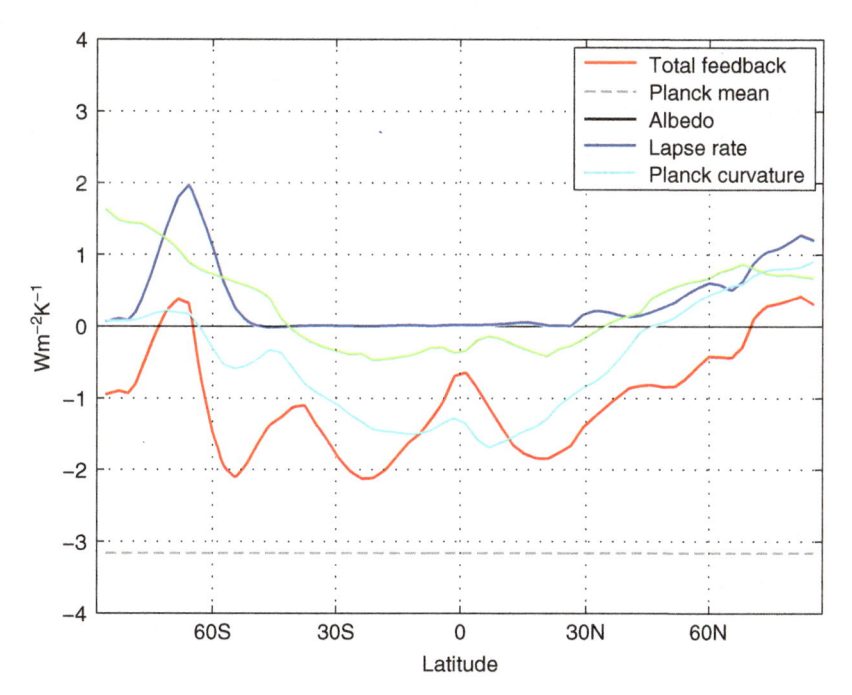

Figure 12.7 Radiative feedbacks in the multi-model ensemble mean of CMIP5 models computed using the method of Soden and Held (2006) and Soden et al. (2008) applied to the climate response 140 years after quadrupling CO_2 instantly relative to a pre-industrial climate. The Planck feedback is shown as an overall global mean and its 'curvature', which is the departure from the global mean at each latitude. Here, based on Armour et al. (2013), in a minor yet important departure from the original method, the feedbacks at each latitude are normalized by the temperature response of the latitudinal mean rather than the global mean. This normalization reduces the pole-to-equator gradient of the feedbacks, and hence our conclusions about the relative strength of feedbacks in the polar regions are conservative compared with the original normalization procedure.

forcing is held constant (Kay et al., 2011). Hence, while internal variability might have contributed to the observed loss, it is too small to explain the very rapid, large-scale loss of Arctic sea ice in recent decades.

This then implies that the increase in climate forcing is by far the most likely driver of the observed retreat of Arctic sea ice. Out of all possible contributors to this climate forcing, the observed increase of atmospheric greenhouse gas concentrations caused by anthropogenic emissions is most probably the main driver. This finding is again both consistent with the observational record and with large-scale ESM simulations, which both show a striking, largely linear relationship between the increase in atmospheric CO_2 concentration and Arctic sea ice extent over the past few decades (Mahlstein & Knutti, 2012; Notz & Marotzke, 2012). No other climate forcing agent shows a similarly robust relationship, and the agreement of model simulations with the observational record strongly suggests that anthropogenic CO_2 emissions are indeed the main driver of the observed changes in Arctic sea ice. This finding is robust even if many of the models show a slower retreat than has been observed, as the mismatch between the observed and the simulated trends is largely explicable by internal variability, as discussed earlier (also see Notz, 2015).

In the Antarctic, in contrast, passive microwave satellite records indicate that sea ice has slightly increased over the past few decades. While some studies have claimed that the increase is statistically significant, we currently have too little understanding of the long-term memory of the system and the uncertainty of the measurement to establish statistical significance. This is related to the fact that standard tests of significance assume that the individual observations of sea ice coverage over the years are not related to each other and are Gaussian-distributed. However, the evolution of both Arctic and Antarctic sea ice is obviously partly driven by both short-term and long-term memory of the previous state of the system, rendering standard tests of significance of only little use (e.g. Cohn & Lins, 2005).

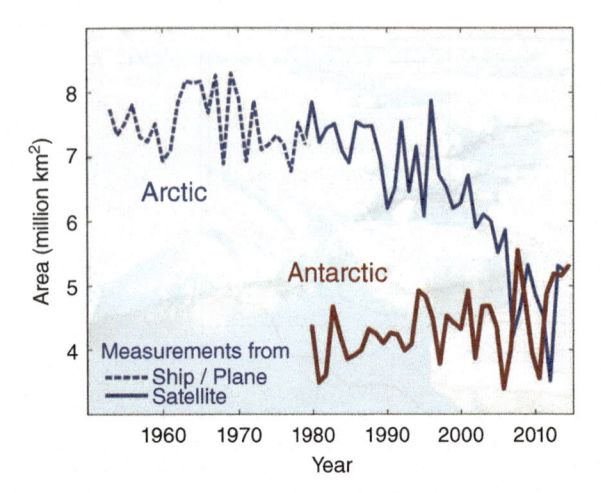

Figure 12.8 Observed evolution of Arctic and Antarctic sea ice during September (1953–1978: HadISST, 1979–2015: NSIDC Bootstrap satellite retrievals).

Problems with the passive microwave record and long-term memory aside, Antarctic sea ice is clearly not decreasing significantly, and it is a valid question as to why, particularly as most ESMs simulate a substantial decrease of Antarctic sea ice over the past few decades (Turner et al., 2012). One obvious candidate for this mismatch is again internal variability, and indeed the observed increase has been found to be of a similar magnitude to the multi-decadal increase in Antarctic sea ice in isolated seasons (Mahlstein et al., 2013; Polvani & Smith, 2013). However, when the spatial patterns are taken into account, the observed dipole trend pattern in the Ross Sea and Amundsen-Bellingshausen Seas falls outside the range of natural variability in the models (Hobbs et al., 2015). Furthermore, if indeed internal variability were the sole driver of the observed change, and if the models simulated the correct amplitude and pattern of such internal variability, Antarctic sea ice should increase in many of the ESM simulations. This, however, is not the case, suggesting that either the models underestimate large-scale internal variability of the system, or they fail to correctly simulate the response of the system to changes in the external forcing.

To shed some light on this question, a number of studies have examined the possible causes of the observed increase in Antarctic sea ice. Currently, it seems most likely that there are multiple causes but that a major contributor is the large-scale atmospheric circulation, which can affect the sea ice cover through both related changes in the wind and precipitation. This assumption is based on the fact that the observed overall increase is related to a year-round expansion in sea ice coverage in the Ross Sea. There, the offshore winds have clearly increased over the past few years in response to a deepening of a dominating low-pressure system, and the pattern of changes in the wind forcing shows a very clear connection to the observed changes in the sea ice coverage (e.g. Holland & Kwok, 2012; Haumann et al., 2014). Such deepening of the dominating low-pressure system is largely absent in most ESM simulations (see Figure 12.9), hinting at issues with the atmospheric forcing as a possible contributor to the mismatch between the observed and the modelled evolution of Antarctic sea ice. However, the observed wind changes have not changed in a commensurate way to explain the sea ice expansion in the Weddell Sea and Indian Ocean sectors of the Southern Ocean (see Figures 12.9 and 10.2).

12.3.4 The future evolution of sea ice

Let us now finally turn to the future evolution of sea ice. Obviously, such future evolution can only be robustly estimated by using ESMs that incorporate possible pathways or 'scenarios' of future climate forcings. For any such scenario, different simulations will result in different possible future trajectories of the sea ice cover, both because the models differ in their physical realism and because internal variability always causes some spread in the possible trajectories (see Figure 12.10). Hence, regarding the future evolution of sea ice, uncertainty arises from three distinctly different sources. First, uncertainty arises because we do not know the future climate-forcing scenario, most notably the future evolution of anthropogenic greenhouse gas emissions, as aerosols become a smaller relative forcing factor in the future. Second, uncertainty arises because internal variability caused by the chaotic nature of the climate system is present. And third, uncertainty arises because of differences in the realism of the various models.

These sources of uncertainty affect the usefulness of simulations of future sea ice evolution in very different ways, depending on the timescale of interest. On short timescales, ranging from a few months to a few years, uncertainty in the climate forcing is likely to be small, and the model's skill in simulating the short-term future evolution of sea ice is largely limited

Sea ice in Earth system models   **319**

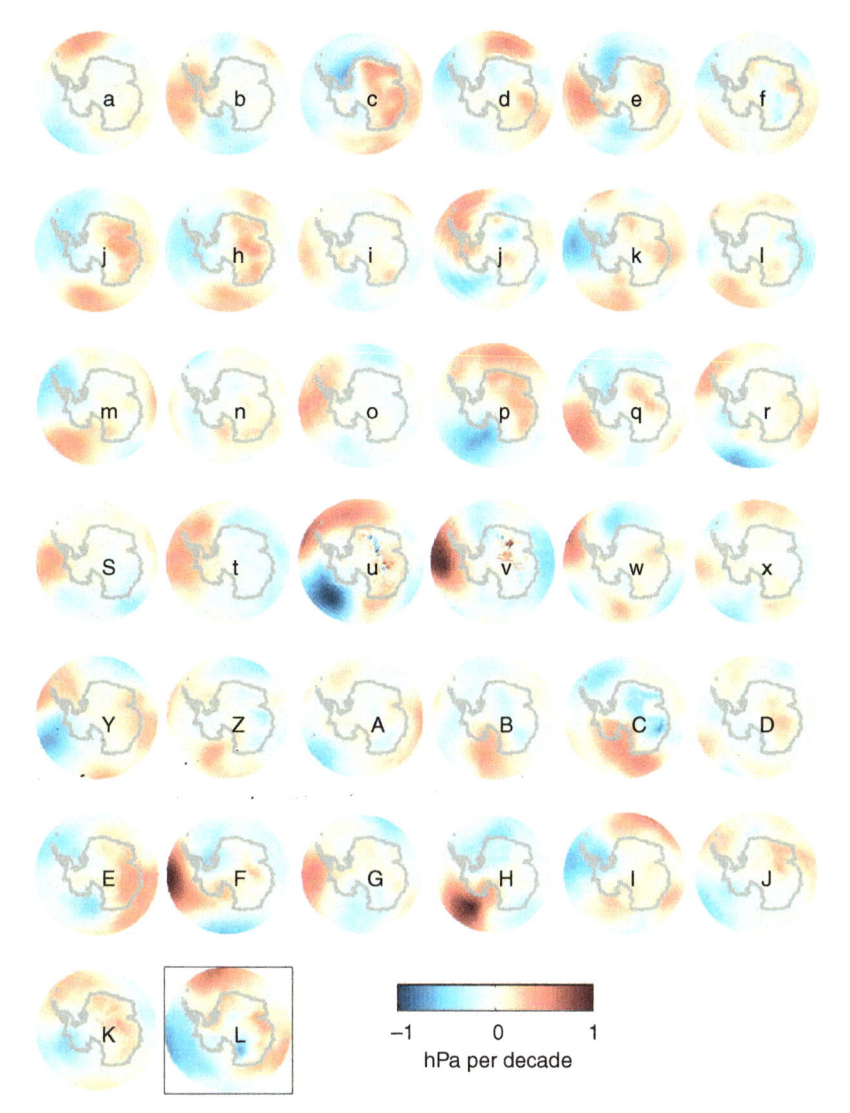

Figure 12.9 Sea level pressure trends (in hPa per decade) of annual means from 1979 to 2013 from 37 different CMIP5 models (a–z, A–K) and from ERA interim reanalysis (boxed and labelled L). The latitude-mean trend has been removed at each latitude to emphasize the meridional circulation. The reanalysis has a significant decrease in the Amundsen Sea low (see Haumann et al., 2014, for discussion of significance). The rare model that resembles the reanalysis appears to be random, which supports the view that the atmospheric meridional wind trends are likely to be unrelated to climate forcing in reality and in the models. Only the first ensemble member of each CMIP5 model is shown, using historical runs through 2005 and RCP8.5 thereafter. Annual means are computed for December to November. The CMIP5 models in this figure are ACCESS1-0 (a), ACCESS1-3 (b), BCC-CSM1-1-M (c), BCC-CSM1-1 (d), BNU-ESM (e), CANESM2 (f), CCSM4 (g), CESM1-BGC (h), CESM1-CAM5 (i), CMCC-CESM (j), CMCC-CM (k), CMCC-CMS (l), CNRM-CM5 (m), CSIRO-MK3-6-0 (n), FGOALS-G2 (o), FGOALS-S2 (p), FIO-ESM (q), GFDL-CM3 (r), GFDL-ESM2G (s), GFDL-ESM2M (t), GISS-E2-H (u), GISS-E2-R (v), HADGEM2-AO (w), HADGEM2-CC (x), HADGEM2-ES (y), INMCM4 (z), IPSL-CM5A-LR (A), IPSL-CM5A-MR (B), IPSL-CM5B-LR (C), MIROC-ESM-CHEM (D), MIROC-ESM (E), MIROC5 (F), MPI-ESM-LR (G), MPI-ESM-MR (H), MRI-CGCM3 (I), NORESM1-M (J) and NORESM1-ME (K). Source: Haumann et al., 2014. Reproduced with permission of John Wiley & Sons.

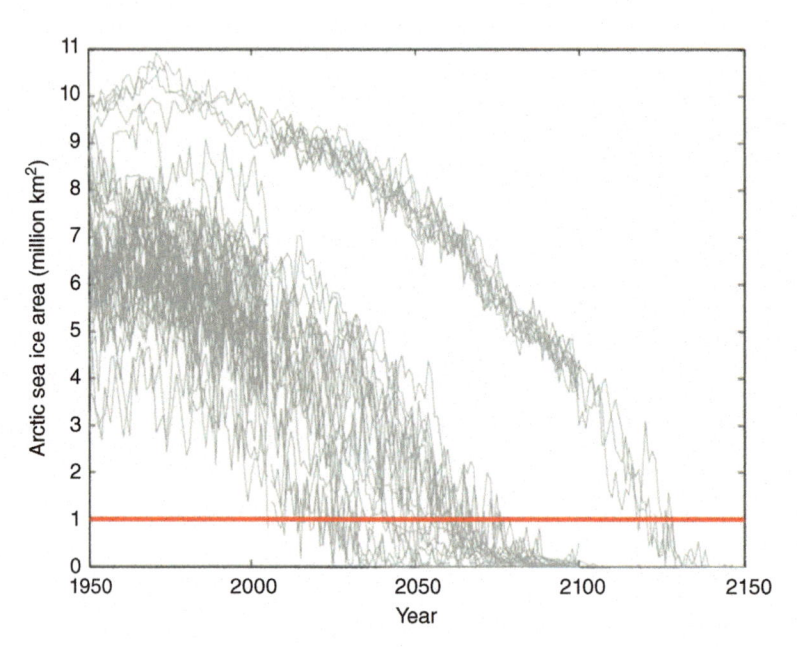

Figure 12.10 Possible future evolution of Arctic September sea ice as simulated in the CMIP5 model ensemble for the scenario representative concentration pathway (RCP) 8.5. This scenario assumes that by the end of this century, anthropogenic activities have increased the radiative forcing of the Earth's climate system by 8.5 W m^{-2}. The red line denotes 1 million km^2, which is often taken as the definition of a near-ice free Arctic Ocean.

by internal variability and model shortcomings. No study has yet systematically evaluated the role of model performance on short-term sea ice predictability, but numerous studies have used the perfect model approach to estimate predictability from the other two sources by using one simulation of a given model as a substitute for the real world, to test if this estimate of the truth can be robustly predicted by the same model for slightly different initial conditions (see, e.g., Tietsche et al., 2014; Guemas et al., 2014). Using this approach, Blanchard-Wrigglesworth (2011) showed that sea ice predictability is controlled by internal variability for the first few years and climate forcing only takes over after about 3 years (see Figure 12.11). In the more important test, when models are initialized with observations and retrospective forecasts are evaluated against the real world, the forecast horizon for pan-Arctic sea ice extent is only about a half-year or so (e.g. Chevallier et al., 2013; Msadek et al., 2014). The issues of short-term initialized forecasts underperforming compared with perfect-model studies might be partly related to unrealistic initialization of the models or an actual shortcoming of the models to simulate the processes that control the internal variability of short-term sea ice evolution.

On longer timescales, in contrast, the uncertainty of the long-term evolution of the external forcing dominates the uncertainty range of future sea ice evolution. To estimate this uncertainty, a number of different possible scenarios of future greenhouse gas emissions are used in the external future forcing of ESM simulations. Commonly used scenarios go under the name of representative concentration pathways (RCPs), followed by a specific number that describes the additional radiative forcing that is caused by anthropogenic emissions of long-lived greenhouse gases. In the RCP 2.6 scenario, which describes a future pathway of greenhouse gas emissions that is based on a rapidly decreasing use of fossil fuels, Arctic and Antarctic sea ice at the end of this century are in most model simulations comparable to the sea ice state that we observe today. In contrast, in the scenarios RCP 4.5 and RCP 8.5, which assume only little reduction in greenhouse gas emissions, the sea ice coverage both in the Arctic and in the Antarctic decrease further, with many models showing for the RCP 8.5 scenario a total loss of Arctic summer sea ice within this century (see Figure 12.10).

The fact that the long-term future sea ice evolution is closely related to the long-term evolution of greenhouse

Sea ice in Earth system models   321

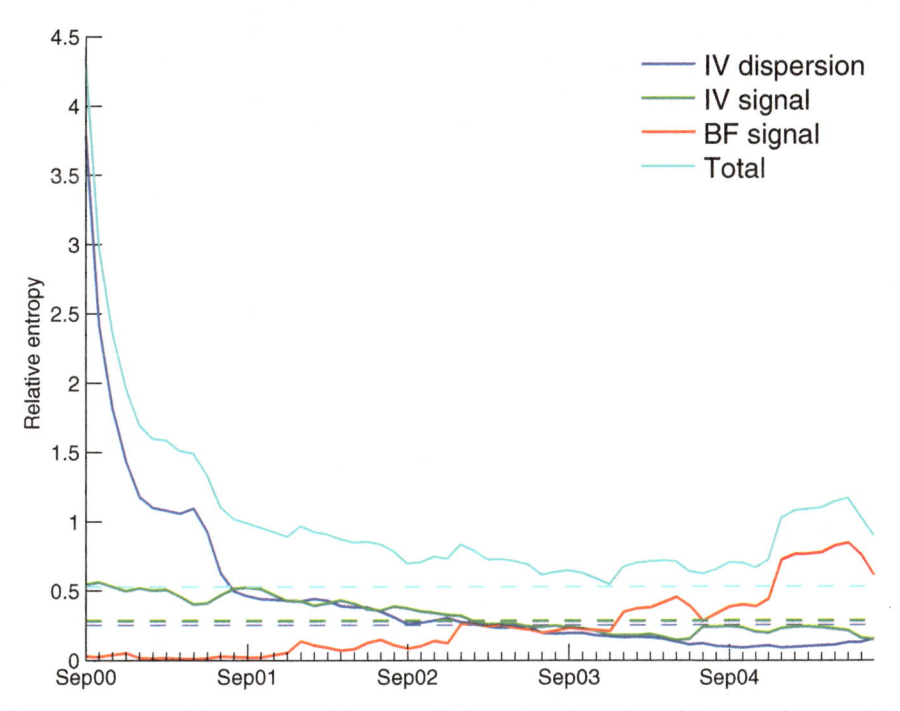

Figure 12.11 Relative entropy, a unitless measure of forecast skill, for pan-Arctic sea ice volume in perfect model simulations with the CCSM4 beginning in September of year 2000. The total (cyan) is broken into contributions due to the response to climate forcing (red; BF, background forced), the response to a particular initial condition (since any given initial condition differs from the 'background' trend due to climate forcing) (green; IV, initial value), and the dispersion (i.e. spreading of the forecast ensemble) after being initialized (blue). The skill is high when the relative entropy is far above the level of null hypothesis (dashed lines). The total is high at first because the forecast is essentially equal to the initial condition, but decays as the ensemble spread grows. The response to the particular value of the initial condition offers continuous predictability for several years, and sometime during the third year, the climate forcing becomes a larger factor. The dispersion component is only small yet significant for the first few years. This figure is redrafted from Blanchard-Wrigglesworth et al. (2011), who provided a detailed discussion of the null hypothesis. Source: Blanchard-Wrigglesworth et al., 2011. Reproduced with permission of John Wiley & Sons.

gas emissions is consistent with our understanding that these emissions have also been driving much of the observed evolution of Arctic sea ice in recent years. This finding is therefore consistent with the observational record and thus very robust. Uncertainties remain, however, as to when precisely, for example, the Arctic Ocean might have lost its summer sea ice in the RCP 8.5 scenario, with model estimates ranging from 2005 until not within this century. To narrow down this range, a number of studies have tried to provide estimates that are based on a subset of the models, which narrows down the uncertainty range to a few decades around mid-century (e.g. Massonnet et al., 2012; Wang & Overland, 2012). However, the approach of sub-selecting models faces all the difficulties that we described in Section 12.3.1, and it is currently not clear to which degree such sub-selection gives more credible results than the range provided by all models (Stroeve & Notz, 2015).

A different approach to narrowing the uncertainty range of future sea ice evolution is based on the use of so-called emergent constraints. This term describes specific observables that have been found to be closely related to the future evolution either of the same or of a different observable. For example, model simulations have shown relationships between the evolution of global mean temperature and Arctic sea ice extent (Winton, 2011; Mahlstein & Knutti, 2012), between the trend in Arctic sea ice and the percentage of remaining ice during the period 2021–2040 (Boe et al., 2009; Collins et al., 2012), and between the amount of Arctic sea ice during the observational period and the sea ice extent during the period 2018–2022 (Liu et al., 2013). These relationships, if proven to be robust, then allow one to obtain an estimate of the most likely future trajectory of Arctic sea ice evolution based on the observational record. Such analysis again points towards a loss of Arctic summer sea ice in the course of this century for emission scenarios that do not involve a rapid reduction of anthropogenic greenhouse gas emissions.

12.4 Summary and outlook

We have come a long way in simulating sea ice in modern ESMs. The models capture the basic conservation of mass, energy and momentum, and have become increasingly skilful in simulating the processes and feedbacks that are important for predicting the response of sea ice to natural and anthropogenic climate forcing. Despite such progress, the spread of sea ice characteristics in ESM simulations has not narrowed much over the last decade (Massonnet et al., 2012; Stroeve et al., 2012), for which there are three main explanations. First, part of the spread simply reflects the fact that the evolution of a chaotic system such as the Earth's climate system can never be fully captured by any given model. Second, the spread is related to the fact that the various sea ice models are exposed to different forcing from the atmosphere and the ocean in any ESM, and the quality of this forcing crucially affects the quality of the sea ice simulations. And third, the spread is related to the fact that the complexity and realism of the sea ice components in ESMs vary a great deal across models.

To address, and ideally to minimize, this third reason for the large spread in sea ice simulations, the sea ice research community is striving for ever more realistic models of sea ice. Realistic models are developed through an increased understanding of the ways in which sea ice functions, as obtained usually from combining observations with insights gained from dedicated modelling studies. Hence, progress is required on both the reliability and coverage of observational records, as well as our understanding of some of the fundamental processes that govern sea ice evolution. Such understanding will then allow us to incorporate an ever more realistic representation of key processes into our models.

One example of such development, which allows one to remove *ad hoc* parameterizations from the models, is a method to simulate the evolution of sea ice floe size (Zhang et al., 2015). Based on this method, one can represent the differentiation of melting between lateral melting and thinning of the ice floes on more physical grounds than is possible by other existing parameterizations. The fact that this method requires a realistic representation of ocean waves is one example of how one often has to develop the various components of an ESM in parallel to allow for a more realistic representation of processes that act across different components of the ESM.

Another very active area of development relates to sea ice dynamics, with a number of groups aiming to treat sea ice as an anisotropic material, which then allows one to predict lead orientation and to use those for a more realistic simulation of the anisotropic distribution of sea ice strength. A number of studies have explored this type of dynamics, but only recently has an anisotropic scheme been implemented and tested in a model that is complete in other ways so that it could be used in an ESM (see Tsamados et al., 2013).

Regarding sea ice thermodynamics, recent developments relate to a more realistic treatment of snow (Lecomte et al., 2015), to the representation of light-absorbing aerosols and melt ponds on sea ice (Holland et al., 2012; Lecomte et al., 2015), and to methods that actually simulate the temporal evolution of sea ice salinity (Turner et al., 2013; Griewank & Notz, 2013).

Finally, we anticipate that it won't be long before studies will use grid refinement (Figure 12.2b) to enhance the resolution locally at one or both of the poles, to better resolve the small-scale processes in the sea ice and in its interaction with the atmosphere and ocean. Such a strategy allows an ESM to replicate the benefits of a regional model, while avoiding lateral boundary conditions.

Such ongoing model development allows us to obtain simulations from ESMs that capture ever more of the complexity of the real climate system of our planet – including one of its most beautiful components, namely sea ice.

12.5 Acknowledgements

We thank Nicole Feldl for computing the kernel feedbacks shown in Figure 12.7. We acknowledge the World

Climate Research Programme's Working Group on Coupled Modelling, which is responsible for CMIP, and we thank the climate modelling groups (listed in the caption of Figure 12.9) for producing and making available their model output. For CMIP the U.S. Department of Energy's Program for Climate Model Diagnosis and Intercomparison provides coordinating support and led development of software infrastructure in partnership with the Global Organization for Earth System Science Portals. Dirk Notz is grateful for his funding through a Max Planck Research Fellowship. Cecilia M. Bitz is grateful for financial support to write this chapter from the Office of Naval Research grant N000141310793, from the National Science Foundation through grant PLR-1341497, and from the National Aeronautical and Space Agency grant NNX14AIO3G.

References

Armour, K.C., Bitz, C.M. & Roe, G.H. (2013) Time-varying climate sensitivity from regional feedbacks. *Journal of Climate*, **26**, 4518–5434.

Armour, K., Eisenman, I., Blanchard-Wrigglesworth, E., McCusker, K. & Bitz, C.M. (2011) The reversibility of sea ice loss in a state-of-the-art climate model. *Geophysical Research Letters*, **38**, L16705, doi:10.1029/2011GL048739.

Bitz, C., Holland, M., Eby, M. & Weaver, A. (2001) Simulating the ice-thickness distribution in a coupled climate model. *Journal of Geophysical Research*, **106**, 2441–2464.

Bitz, C., Ridley, J., Holland, M. & Cattle, H. (2012) Global climate models and 20th and 21st century arctic climate change. In: *Arctic Climate Change: The ACSYS Decade and Beyond* (Eds. P. Lemke & H.-W. Jacobi), pp. 405–436. Springer, Dordrecht, the Netherlands.

Bitz, C. & Roe, G. (2004) A mechanism for the high rate of sea ice thinning in the Arctic Ocean. *Journal Climate*, **17**, 3623–3632.

Bitz, C.M. & Lipscomb, W.H. (1999) An energy-conserving thermodynamic model of sea ice. *Journal of Geophysical Research*, **104**, 15 669–15 677.

Blanchard-Wrigglesworth, E., Bitz, C.M. & Holland, M.M. (2011) Influence of initial conditions and boundary forcing on predictability in the Arctic. *Geophysical Research Letters*, **38**, L18503, doi:10.1029/2011GL048807.

Boe, J., Hall, A. & Qu, X. (2009) September sea-ice cover in the Arctic Ocean projected to vanish by 2100. *Nature Geoscience*, **2**, 341–343.

Budyko, M.I. (1969) The effect of solar radiation variations on the climate of the earth. *Tellus*, **21**, 611–619.

Chevallier, M., Salas y Melia, D., Voldoire, A., Deque, M. & Garric, G. (2013) Seasonal forecasts of the pan-Arctic sea ice extent using a GCM-based seasonal prediction system. *Journal of Climate*, **26**, 6092–6104.

Cohn, T.A. & Lins, H.F. (2005) Nature's style: Naturally trendy. *Geophysical Research Letters*, **32**, L23402, doi:10.1029/2005GL024,476.

Collins, M., Chandler, R.E., Cox, P.M., Huthnance, J.M., Rougier, J. & Stephenson, D.B. (2012) Quantifying future climate change. *Nature Climate Change*, **2**, 403–409.

Eisenman, I. & Wettlaufer, J.S. (2009) Nonlinear threshold behavior during the loss of Arctic sea ice. *Proceedings of the National Academy of Science USA*, **106**, 28–32.

Evans, K.J., Lauritzen, P.H., Misra, S.K., Neale, R.B., Taylor, M.A. & Tribbia, J.J. (2013) AMIP simulations with the CAM4 spectral element dynamical core. *Journal of Climate*, **26**, 689–709.

Feltham, D.L. (2008) Sea ice rheology. *Annual Reviews of Fluid Mechanics*, **40**, 91–112.

Griewank, P.J. & Notz, D. (2013), Insights into brine dynamics and sea ice desalination from a 1-D model study of gravity drainage. *Journal Geophysical Research*, **118**, 3370–3386.

Guemas, V., Blanchard-Wrigglesworth, E., Chevallier, M. et al. (2014) A review on Arctic sea ice predictability and prediction on seasonal-to-decadal timescales. *Quarterly Journal of the Royal Meteorological Society*, doi:10.1002/qj.2401.

Haumann, F.A., Notz, D. & Schmidt, H. (2014) Anthropogenic influence on recent circulation-driven antarctic sea ice changes. *Geophysical Research Letters*, **41**, 8429–9437.

Hezel, P.J., Zhang, X., Bitz, C.M., Kelly, B.P., & Massonnet, F. (2012) Projected decline in spring snow depth on Arctic sea ice caused by progressively later autumn open ocean freeze-up this century. *Geophysical Research Letters*, **39**, L17505, doi:10.1029/2012GL052794.

Hibler, W.D. (1979) A dynamic thermodynamic sea ice model. *Journal of Physical Oceanography*, **9**, 815–846.

Hobbs, W.R., Bindoff, N.L. & Raphael, M.N. (2015) New perspectives on observed and simulated Antarctic sea ice extent trends using optimal fingerprinting techniques. *Journal of Climate*, **28**, 1543–1560.

Holland, M.M., Bitz, C.M. & Weaver, A.J. (2001) The influence of sea ice physics on simulations of climate change, *Journal Geophysical Research*, **106**, 19639–19655.

Holland, M.M., Bailey, D.A., Briegleb, B.P., Light, B. & Hunke, E. (2012) Improved sea ice shortwave radiation physics in CCSM4: the impact of melt ponds and aerosols on Arctic sea ice. *Journal of Climate*, **25**, 1413–30.

Holland, P.R. & Kwok, R. (2012) Wind-driven trends in Antarctic sea-ice drift. *Nature Geoscience*, **5**, 872–875.

Hunke, E. & Dukowicz, J. (1997) An elastic-viscous-plastic model for sea ice dynamics. *Journal of Physical Oceanography*, **27**, 1849–1867.

Hunke, E.C., Lipscomb, W.H. & Turner, A.K. (2010) Sea-ice models for climate study: retrospective and new directions. *Journal of Glaciology*, **56**, 1162–1172.

Hunke, E.C., Lipscomb, W.H., Turner, A.K. & Elliott, N.J. S. (2015) CICE: the Los Alamos sea ice model documentation and software users manual. *Technical Report LA-CC- 06–012*. Los Alamos National Laboratory, Los Alamos, US.

Hunke, E.C., Notz, D., Turner, A.K. & Vancoppenolle, M. (2011) The multiphase physics of sea ice: a review for model developers. *The Cryosphere*, **5**, 989–1009.

Kay, J.E., Holland, M.M. & Jahn, A. (2011) Inter-annual to multi-decadal arctic sea ice extent trends in a warming world. *Geophysical Research Letters*, **38**, L15708.

Kirtman, B.P., Bitz, C., Bryan, F. et al. (2012) Impact of ocean model resolution on CCSM climate simulations. *Climate Dynamics*, **39**, 1303–1328.

Lepparänta, M. (2005) *The Drift of Sea Ice*. Springer-Verlag, Berlin, Germany.

Lecomte, O., T. Fichefet, D. Flocco, D. Schroeder & M. Vancoppenolle (2015) Interactions between wind-blown snow redistribution and melt ponds in a coupled ocean–sea ice model. *Ocean Modelling*, **87**, 67–80.

Li, C., Notz, D., Tietsche, S. & Marotzke, J. (2013) The transient versus the equilibrium response of sea ice to global warming. *Journal of Climate*, **26**, 5624–5636.

Liu, J., Song, M., Horton, R.M. & Hu, Y. (2013) Reducing spread in climate model projections of a September ice-free Arctic. *Proceedings of the National Academy of Science USA*, **110**, 12571–12576.

Mahlstein, I., Gent, P.R. & Solomon, S. (2013) Historical Antarctic mean sea ice area, sea ice trends, and winds in cmip5 simulations. *Journal of Geophysical Research*, **118**, 5105–5110.

Mahlstein, I. & Knutti, R. (2012) September arctic sea ice predicted to disappear near 2°C global warming above present. *Journal of Geophysical Research*, **117**, D06104, doi 10.1029/2011JD016709.

Manabe, S. & Stouffer, R.J. (1980) Sensitivity of a global climate model to an increase of CO_2 concentration in the atmosphere. *Journal of Geophysical Research*, **85**, 5529–5554.

Massonnet, F., Fichefet, T., Goosse, H. et al. (2012) Constraining projections of summer arctic sea ice. *The Cryosphere*, **6**, 1383–1394.

Mauritsen, T., Stevens, B., Roeckner, E. et al. (2012) Tuning the climate of a global model. *Journal of Advances in Modeling Earth Systems*, **4**, M00A01, doi: 10.1029/2012MS000154

Meier, W. & Notz, D. (2010) *A note on the accuracy and reliability of satellite-derived passive microwave estimates of sea-ice extent*. CliC Arctic sea ice working group consensus document. World Climate Research Program.

Msadek, R., Vecchi, G.A., Winton, M. & Gudgel, R.G. (2014) Importance of initial conditions in seasonal predictions of Arctic sea ice extent. *Geophysical Research Letters*, **41**, 5208–5215.

Notz, D. (2009) The future of ice sheets and sea ice: between reversible retreat and unstoppable loss. *Proceedings of the National Academy of Science USA*, **106**, 20590–20595.

Notz, D. (2012) Challenges in simulating sea ice in earth system models. *WIREs Climate Change*, **3**, 509–526.

Notz, D. (2014) Sea-ice extent and its trend provide limited metrics of model performance. *The Cryosphere*, **8**, 229–243.

Notz, D. (2015) How well must climate models agree with observations? *Philosophical Transactions of the Royal Society Series A*, doi: 10.1098/rsta.2014.0164

Notz, D. & Marotzke, J. (2012) Observations reveal external driver for arctic sea-ice retreat. *Geophysical Research Letters*, **39**, L051094, doi: 10.1029/2012GL051094.

Pithan, F. & Mauritsen, T. (2014) Arctic amplification dominated by temperature feedbacks in contemporary climate models, *Nature Geosciences*, **7**, 181–84.

Polvani, L.M. & Smith, K.L. (2013) Can natural variability explain observed antarctic sea ice trends? New modeling evidence from CMIP5. *Geophysical Research Letters*, **40**, 3195–3199.

Ringler, T., Petersen, M., Higdon, R.L., Jacobsen, D., Jones, P. & Maltrud, M. (2013) A multi-resolution approach to global ocean modeling. *Ocean Modelling*, **69**, 211–232.

Roberts, A., Craig, A., Maslowski, W. et al. (2015) Simulating transient ice-ocean Ekman transport in the regional arctic system model and the community earth system model. *Annals of Glaciology*, **56**, 211–228.

Sallée, J.-B., Shuckburgh, E., Bruneau, N. et al. (2013), Assessment of Southern Ocean water mass circulation and characteristics in CMIP5 models: historical bias and forcing response, *Journal of Geophysical Research*, **118**, 1830–1844.

Satoh, M., Tomita, H., Yashiro, H. et al. (2014) The non-hydrostatic icosahedral atmospheric model: description and development. *Progress in Earth and Planetary Science*, **1**:18, doi:10.1186/s40645-014-0018-1.

Schmidt, G.A., Bitz, C.M., Mikolajewicz, U. & Tremblay, L. (2004) Ice-ocean boundary conditions for coupled models. *Ocean Modelling*, **7**, 59–74.

Semtner, A.J. (1976) A model for the thermodynamic growth of sea ice in numerical investigations of climate. *Journal of Physical Oceanography*, **6**, 379–389.

Semtner, A.J. (1984) On modelling the seasonal thermodynamic cycle of sea ice in studies of climatic change. *Climatic Change*, **6**, 27–37.

Small, R.J., Bacmeister, J., Bailey, D. et al. (2014) A new synoptic scale resolving global climate simulation using the community earth system model. *Journal of Advances in Modeling Earth Systems*, doi: 10.1002/2014MS000363.

Soden, B.J., & I.M. Held (2006) An assessment of climate feedbacks in coupled ocean–atmosphere models. *Journal of Climate*, **19**, 3354–60.

Soden, B.J., Held, I.M., Colman, R., Shell, K.M., Kiehl, J.T. & Shields, C.A. (2008), Quantifying climate feedbacks using radiative kernels. *Journal of Climate*, **21**, 3504–20.

Stössel, A., Notz, D., Haumann, F.A., Haak, H., Jungclaus, J. & Mikolajewicz, U. (2014) Controlling high-latitude southern ocean convection in climate models. *Ocean Modelling*, **86**, 58–75.

Stroeve, J.C., Kattsov, V., Barrett, A. et al. (2012) Trends in Arctic sea ice extent from CMIP5, CMIP3 and observations. *Geophysical Research Letters*, **39**, L16502. doi:10.1029/2012GL052676.

Stroeve, J. & Notz, D. (2015) Insights on past and future sea-ice evolution from combining observations and models. *Global Planetary Change*, **135**, 119–132.

Thorndike, A.S., Rothrock, D.A., Maykut, G.A. & Colony, R. (1975) The thickness distribution of sea ice. *Journal of Geophysical Research*, **80**, 4501–4513.

Tietsche, S., Notz, D., Jungclaus, J.H. & Marotzke, J. (2011) Recovery mechanisms of Arctic summer sea ice. *Geophysical Research Letters*, **38**, L02707, doi: 10.1029/2010GL045698.

Tietsche S., Day, J.J., Guemas, V. et al. (2014) Seasonal to interannual Arctic sea-ice predictability in current global climate models. *Geophysical Research Letters*, **41**, 1035–1043.

Tsamados, M., D.L. Feltham, & A.V. Wilchinsky (2013), Impact of a new anisotropic rheology on simulations of Arctic sea ice. *Journal of Geophysical Research*, **118**: 91–107.

Turner, J., Bracegirdle, T.J., Phillips, T., Marshall, G.J. & Hosking, J.S. (2012) An initial assessment of Antarctic sea ice extent in the CMIP5 models. *Journal of Climate*, **26**, 1473–84.

Turner, A.K., Hunke, E.C., & Bitz, C.M., 2013: Two modes of sea-ice gravity drainage: A parameterization for large-scale modeling. *Journal of Geophysical Research*, **118**, 2279–2294.

Vancoppenolle, M., Bouillon, S., Fichefet, T. et al. (2012) The Louvain-la-neuve sea ice model. *Techical Report*. Université catholique de Louvain.

Wang, M. & Overland, J.E. (2012) A sea ice free summer arctic within 30years: An update from CMIP5 models. *Geophysical Research Letters*, **39**, L18501, doi: 10.1029/2012GL052868.

Weiss, J. (2013) *Drift, Deformation, and Fracture of Sea Ice*. Springer Briefs in Earth Sciences. Springer, Dordrecht, the Netherlands.

Winton, M. (2000) A reformulated three-layer sea ice model. *Journal of Atmosphere and Ocean Technology*, **17**, 525–531.

Winton, M. (2011) Do climate models underestimate the sensitivity of Northern Hemisphere sea ice cover? *Journal of Climate*, **24**, 3924–3934.

Zhang, J., Schweiger, A., Steele, M. & Stern H. (2015) Sea ice floe size distribution in the marginal ice zone: theory and numerical Experiments. *Journal of Geophysical Research*, **120**, 3484–98.

CHAPTER 13

Sea ice as a habitat for Bacteria, Archaea and viruses

Jody W. Deming[1] and R. Eric Collins[2]

[1] School of Oceanography, University of Washington, Seattle, WA, USA
[2] School of Fisheries and Ocean Sciences, University of Alaska Fairbanks, AK, USA

13.1 Introduction and short history

The history of understanding sea ice as a habitat for the smallest forms of life is not a long one. It has lagged behind the early recognition of algal blooms in sea ice (reviewed by Horner, 1985), observable by the naked eye as pigmented bands in the ice. Documenting the habitation of sea ice by single-celled akaryotes[1] – members of the phylogenetic domains of Bacteria and Archaea, typically no larger than a micrometer wide in nature – can be seen as having taken a methods-driven pathway.

Efforts began with culturing approaches in the 1960s, using melted samples of both Arctic and Antarctic sea ice as inocula (reviewed by Baross & Morita, 1978; Sullivan & Palmisano, 1984) and media that selected primarily for heterotrophic bacteria. Such work gave way in the 1970s to microbial community analyses, with the advent of epifluorescence microscopy and application of radioisotopes to track organic carbon consumption. Initial studies on Arctic sea ice revealed bacterial densities and rates of heterotrophic activity higher in (melted) sea ice than in underlying seawater (Horner & Alexander, 1972; Kaneko et al., 1978; Griffiths et al., 1978). Antarctic expeditions in the 1980s defined a decade when sea ice bacteria were studied intensively for their heterotrophic responses to the ice-algal bloom and competitive use and recycling of nutrients (Garrison et al., 1986; Kottmeier et al., 1987). A common theme to emerge was the prevalence of physical associations between bacteria and algae, and with particulate matter in general, in sea ice (Grossi et al., 1984; reviewed by Stewart & Fritsen, 2004). Seawater bacteria were considered to incorporate into growing sea ice primarily as riders attached to algae and other particles (Garrison et al., 1983; Ackley & Sullivan, 1994). The successional stages of bacteria, at least where it was possible to follow them, were thus tracked in relation to the presence of ice algae (Delille et al., 1995; reviewed by Kaartokallio et al., 2008).

In the 1990s, several groups took a renewed interest in culturing heterotrophic bacteria from sea ice, this time to acquire uniquely cold-adapted strains and explore their ecological, evolutionary and applied significance (Delille, 1992; Helmke & Weyland, 1995; Bowman et al., 1997; Gosink et al., 1997, 1998; Junge et al., 1998; Nichols et al., 1999). In the same time frame, the presence of bacterial viruses[2] in sea ice, at even higher densities than their hosts, was finally discovered (Maranger et al., 1994). The detection of Archaea in sea ice was not confirmed until this century, when archaeal gene-based techniques were applied, first to Arctic winter sea ice (Junge et al., 2004; Collins & Deming, 2010) and then to Antarctic sea ice (Cowie et al., 2011). In terms of abundance, the archaeal contribution to the total microbial community in sea ice has appeared small (<7%), regardless of latitude or season, yet the significance of their potential ecological functions should not be overlooked.

Evaluating bacterial diversity in sea ice based on the 16S rRNA gene came late to the study of sea ice, relative to other environments. It began with analysis of a collection of bacteria cultured from Antarctic sea ice (Bowman et al., 1997), an approach that could

---

Sea Ice, Third Edition. Edited by David N. Thomas.
© 2017 John Wiley & Sons, Ltd. Published 2017 by John Wiley & Sons, Ltd.

only reveal limited (heterotrophic) diversity, but was followed by a series of phylogenetic studies targeting natural bacterial assemblages in the ice that could begin to address truer *in situ* diversity. In this series, the developed methodologies of the day were applied: clone libraries generated from extracts of sea ice DNA to identify the dominant inhabitants in sea ice; fluorescent *in situ* hybridization (FISH) probes applied to quantify selected bacterial groups microscopically; and DNA fingerprints generated via a number of methods, including temperature and denaturing gradient gel electrophoresis (TGGE and DGGE, respectively), terminal restriction fragment length polymorphisms (T-RFLP), and amplified ribosomal intergenic spacer analysis (ARISA), to compare communities across time and/or space. As a result, questions on microbial diversity, biogeography, seasonal succession, and even responses to hydrocarbon spills in sea ice were addressed in ways that had not been possible before (Staley & Gosink, 1999; Brown & Bowman, 2001; Junge et al., 2002, 2004; Brinkmeyer et al., 2003, 2004; Gerdes et al., 2005; Kaartokallio et al., 2008; Collins et al., 2010). Other technologies and approaches were developed in the 2000s specifically for the study of sea ice, including microelectrodes that could be frozen into the ice to record oxygen changes (Mock et al., 2002) and the clever use of under-ice cameras (Rysgaard et al., 2001). These developments contributed to the detection of anoxic zones within the ice, where levels of microbial denitrification approached those classically detected in marine sediments (Rysgaard & Glud, 2004), giving new microbial meaning to the sea ice ecosystem as the inverted benthos or 'upside-down counterpart of the bottom life' (Mohr & Tibbs, 1963, p. 246).

All along the way, ideas and methods from disciplines other than microbiology were brought to bear on the study of sea ice bacteria and their viruses. Biochemists uncovered features of bacterial enzymes, lipids and other cellular constituents that make life in ice possible (reviewed by Feller & Gerday, 2003; D'Amico et al., 2006) and that have encouraged the pursuit of sea ice bacteria and their gene products for biotechnology (e.g., Mancuso Nichols et al., 2005; Cavicchioli et al., 2011; Bowman & Deming, 2014). Geophysicists helped to push the observational scale of sea ice to the micrometer level to reveal the *in situ* habitats of microbes in unmelted sea ice (see Chapter 1) and the bacteria within them (see images in Junge et al., 2001).

Molecular biologists obtained the first whole-genome sequences for bacterial species known to reside in sea ice, *Colwellia psychrerythraea* and *Psychroflexus torquis*, from which a list of attributes reflecting and enabling life in ice were deduced (Methé et al., 2005; Bowman, 2008; Feng et al., 2014). More recently, evidence that viruses may function not only to control bacterial populations in sea ice (Wells & Deming, 2006a) but also to mediate lateral gene transfer in the cold has emerged (Raymond & Kim, 2012; Colangelo-Lillis & Deming, 2013; Collins & Deming, 2013; DeMaere et al., 2013; Bowman, 2014; Feng et al., 2014).

Astrobiologists continue to ask if targeted studies of sea ice and its inhabitants can contribute to understanding the possible habitability of extraterrestrial ices (Deming & Eicken, 2007; Firth et al., 2015; Nadeau et al., 2015; Wallace et al., 2015), while those examining scientific and philosophical questions on the definition, origin and evolution of life have not missed the relevance of saline ice formation as a process that concentrates elements and biopolymers into liquid-filled pore spaces that then become biochemical reactors in the cold (Kanavarioti et al., 2001; Vajda & Hollosi, 2001; Menor-Salván & Marín-Yaseli, 2012; Attwater et al., 2013).

In the following sections, we revisit some material presented earlier (Deming, 2010) to highlight key aspects of approaches to studying bacteria and viruses in sea ice and what has been learned from them. We also present a new synthesis of the available phylogenetic information on bacteria in sea ice and address recent findings on their potential carbon and nitrogen cycling functions in sea ice, using metagenomics, stable isotopes, modelling and *in situ* experimental approaches. Microbial interactions with each other and with viruses in sea ice, from lateral gene transfer to synergistic survival strategies, figure prominently. The physical properties of sea ice during its growth and demise, the latter receiving new interest in a warming climate, continue to provide a fundamental understanding of this unique habitat and to inform on how the smallest of its inhabitants survive and function within the ice and impact the ocean and atmosphere.

13.2 Sampling issues

13.2.1 The melting dilemma

Sampling sea ice to investigate its akaryotic inhabitants has typically relied upon methods developed for the larger eukaryotic inhabitants of the ice, from algae and other protists to macrofauna. For sea ice > 10 cm thick, an ice core is taken and sectioned vertically on site, using sterile technique to the extent possible, then transported in sterile containers and later melted for analysis. Thinner or unconsolidated sea ice has been collected using study-specific techniques to examine bacteria in relation to the ice formation process in autumn (e.g., Grossmann & Dieckmann, 1994; Gradinger & Ikävalko, 1998). For collecting features on the surface of young ice, including expelled brines, frost flowers and the saline snow layer that develops after snowfall, sterile spatulas borrowed from snow studies are used (Bowman & Deming, 2010; Ewert et al., 2013; Barber et al., 2014). Regardless of the ice type, melting the sample has been standard and convenient: all methodologies and analyses developed for pelagic bacteria can be applied to a melted sample. Unfortunately, as the sea ice melts, organisms in the liquid brine phase of the ice are exposed to much fresher salinities and thus to osmotic shock and potential lysis (cell loss), despite cellular strategies for tolerating natural fluctuations in salinity (Ewert & Deming, 2014; Firth et al., 2016). If the ice is also allowed to warm during melting, cold-adapted microbes can succumb to thermal shock.

The melting dilemma was recognized years ago for fragile flagellates and ciliates (Garrison & Buck, 1986). An accepted solution has been to melt the ice into a measured volume of seawater that is first filtered to remove pelagic microbes. The brine within the ice still freshens as the ice melts, but the drop in salinity and consequent cell losses are minimized. This seawater-melting approach was also adopted for the study of akaryotic cells; losses during melting are minimized or avoided, unless the temperature of the original ice is below −10°C, when more than half of the population can be lost (Ewert et al., 2013). At such low temperatures in upper winter sea ice, brine salinities are much higher than seawater salinity, so that melting into seawater represents a significant salinity shock (e.g., from 21% salt in −20°C brines to 3.5% salt in −1°C seawater). A cell-protective solution to melting winter sea ice is to melt into a sterile brine solution at a salinity pre-calculated to yield an isohaline melt (Junge et al., 2004; Collins et al., 2008; Ewert et al., 2013). Although viruses are not subject to osmotic shock and only decay very slowly in subzero brine solutions (Wells & Deming, 2006c), the study of viruses in sea ice also requires special attention to the melting solution, either to account for viruses present in the solution (Maranger et al., 1994) or to remove them prior to its use (Gowing et al., 2002). Even with measures to avoid cell loss due to osmotic shock or viral complications, the melting of sea ice alters the physical associations within a brine inclusion that Bacteria, Archaea and viruses may have had with each other or with larger organisms, particulate debris, gelatinous material, or the ice itself (Dolev et al., 2016). Isohaline melts appear to leave many of these associations intact (Figure 13.1a).

An alternative approach to melting sea ice has been to collect the brine phase directly. The brine in warm, unconsolidated or thin ice will drain from it immediately

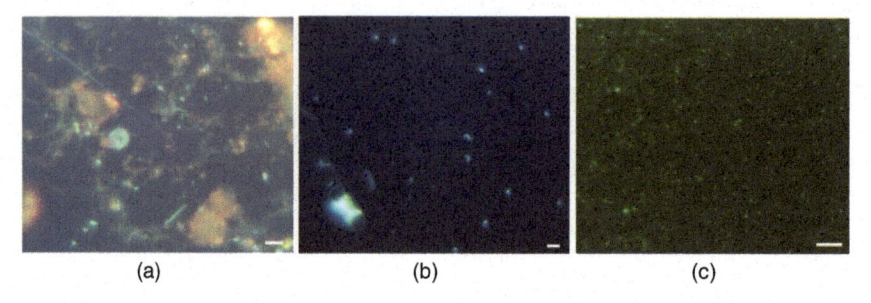

Figure 13.1 Examples of epifluorescence micrographs of sea ice samples prepared for qualitative and quantitative observations of bacteria and viruses. (a) An isohaline melt of an interior section from an ice core, after staining with the DNA-specific stain DAPI and concentrated on a 0.2 μm filter, shows qualitative associations of bacteria (blue) with larger eukaryotes (blue), particulate detritus (orange) and strands of exopolymeric material. (b) Brine collected by sackhole drainage, similarly stained and filtered, shows bacteria ready for counting. (c) A 0.2 μm filtrate of sackhole brine, stained with SYBR Gold, shows viral particles too dense to count. Samples were taken in March, from Kanajorsuit Fjord (a) or Kobbefjord (b, c) near Nuuk, Greenland, when *in situ* brine temperatures were −3 to −6°C. Scale bars indicate 3 μm (micrographs by S. Carpenter).

upon recovery, so that collecting a quantitative or representative sample is challenging. If the ice is thicker, sections of it can be centrifuged at sub-zero temperature to collect the brine phase, but this method is rarely used in microbial studies: the collection volume is limited, the match between *in situ* and centrifugation temperature is usually suboptimal, and the efficiency of recovering organisms in the brine is unpredictable. Instead, the common practice for thick ice involves drilling partway into the ice and leaving the empty (covered) hole to accumulate brine at its base for later sampling. The amount of brine collected by this 'sackhole' approach is temperature- and time-dependent; it is not an effective means to collect brine from very cold horizons of winter sea ice. In other seasons or warmer depths in the ice, where the temperature is near or above the threshold for bulk ice permeability (−5°C; Golden et al., 1998), the approach yields sufficient volumes over a reasonable period of time (e.g., ~500 mL in an hour at −5°C) for most microbial studies. Although neither the locational origin of the drained brine nor the percentage of cells left behind in the ice can be known with precision, studies of sackhole brines have shown that substantial numbers of bacteria (Figure 13.1b) and viruses (Figure 13.1c) exist in the brine inclusions of sea ice.

13.2.2 Unmelted ice

The study of microbes in colder sea ice, at temperatures that prevent brine drainage, was advanced by developing means to examine unmelted ice sections under a microscope (Junge et al., 2001). A freezer room equipped with ice-sectioning tools and an epifluorescence microscope, factory-modified to operate at sub-zero temperatures, was required for this approach, but a cold stage on a microscope at room temperature, as used for glacial ice (Mader et al., 2006), may be adaptable to sea ice. The freezer-room setup yielded images of bacteria and diatoms within undisturbed brine-filled ice pores examined at −15°C. Autofluorescing chloroplasts allowed diatoms to be seen with transmitted light, but the DNA-specific stain 4′,6-diamidino-2-phenylindole (DAPI), prepared in an isothermal-isohaline solution and diffused into the brine network of a microtomed ice section, was required to observe bacteria (Junge et al., 2001). The vast majority (95%) of bacteria were determined to be present in the liquid phase of the ice, where they were often observed in aggregates or associated with particulate debris and algal cells, validating the importance of such associations often observed in melted samples (Figure 13.1a). Some of the bacteria appeared to adhere to the face of an ice crystal framing the inhabited pore, though by what mechanism was not clear (Junge et al., 2001). Although surface-associated bacteria are well studied in other environments, and eukaryotic inhabitants of sea ice have long been identified as surface-seeking or 'thigmotactic' (Mohr & Tibbs, 1963, p. 246), the study of bacterial thigmotaxy in sea ice remains in its infancy, despite evidence that purified proteins and other compounds produced by bacteria have ice-affine properties (Dolev et al., 2016; Section 13.5).

Other stains have also been applied to unmelted sea ice samples. The use of Alcian Blue, which binds to the complex organic polymers known as extracellular polysaccharide substances (EPS), revealed that virtually every pore space within natural sea ice contains EPS, even when no organism is visible in the pore (Krembs et al., 2002, 2011). Some EPS derives from source waters prior to freezing, but these exopolymers are also produced within the ice by both sea ice algae (copiously in algal bands) and bacteria (to a lesser degree, but throughout the ice). Initially understood to serve as cryoprotectants (Krembs et al., 2002; Marx et al., 2009), new information indicates additional roles for EPS (Section 13.5 and Chapter 17).

The stain 5-cyano-2,3-ditolyltetrazolium chloride (CTC), which allows detection of oxygen-respiring bacteria after an incubation period, was first applied to melted summer sea ice, revealing higher percentages of active cells in the ice than in underlying seawater (Junge et al., 2002). When applied to isothermal-isohaline melts of winter sea ice, CTC staining showed that active bacteria were associated with particulate matter captured on a 3 μm filter; in the coldest ice examined (−20°C), the only active cells were those associated with this particle size fraction (Junge et al., 2004). Meiners et al. (2008) then applied CTC to unmelted spring and summer sea ice sections that were returned to ice-core holes for *in situ* incubation, mimicking an *in situ* primary production method developed earlier (Mock & Gradinger, 1999). Activity was stopped in the ice sections by fixative in the melting solution, after which DAPI and Alcian Blue stains were applied to subsamples of the melted ice. Using all three stains in this way, which bypasses the need for a freezer-adapted microscope, Meiners et al.

(2008) were able to characterize gel-like particles of EPS not only as the particulate matter with which bacteria were associating physically but also as hotspots of *in situ* bacterial activity.

The stains available for detecting viruses or bacteriophage in seawater have not been applied to unmelted sea ice; based in part on sack-hole brine collections (Figure 13.1c), viruses are assumed to partition into the brine phase along with their microbial hosts, EPS and other 'impurities' in seawater. Nor have fluorescently labelled gene probes, effective at quantifying phylogenetic types of Bacteria in melted ice samples (Brinkmeyer et al., 2004), been used in unmelted ice. Application of a fluorescently labelled protein substrate analogue, however, to unmelted winter sea ice indicated extracellular enzyme activity down to −18°C (Deming, 2007). Renewed efforts to use unmelted sea ice to measure total microbial autotrophy and heterotrophy in sea ice show that rates determined in parallel melted samples underestimate the *in situ* activity (Sogaard et al., 2010). We continue to risk the accuracy of our conclusions about bacterial activity in sea ice by relying upon melted samples and thus the behaviour of bacterial populations diluted from their original concentrations in the brine phase of the ice, even if protected against osmotic and thermal losses during melting.

13.2.3 Scaling

Confirmation that almost all akaryotes inhabit the liquid phase of sea ice raises the issue of scaling. When bacterial numbers from melted samples are scaled to brine volume, the resulting densities necessarily exceed those scaled to ice volume. The discrepancy is greater the colder the ice (and thus the smaller the brine volume fraction). For research questions that concern bacterial interactions with each other or with other organisms, viruses, organic compounds or inorganic salts also partitioned into brine, scaling to brine volume yields more significant relationships between bacterial and other parameters than scaling to ice volume (Junge et al., 2004; Torstensson et al., 2015). Scaling viruses to brine volume indicates the high contact rates between viruses and hosts that are possible within the brines of very cold sea ice (Wells & Deming, 2006a).

Comparing biological results between different studies of sea ice is made difficult by lack of consistency in scaling approaches, particularly when details required to convert to consistent units are not provided. For example, the algal-rich bottom of spring and summer ice has frequently been the horizon targeted for study, yet a fundamental difference between akaryotic and eukaryotic sea ice inhabitants is that bacteria are present in significant numbers throughout the ice, regardless of ice age, thickness or season (Deming, 2010). The depth-integrated biomass of bacteria in an ice core can exceed algal (and other) biomass not only during the cold dark winter (Collins et al., 2008), when shrinking pore space and encroaching ice crystals limit sea ice inhabitants to the smallest organisms (Krembs et al., 2000, 2002), but also during the algal bloom season (Gradinger et al., 1999). To enable comparison across studies, where the goal concerns microbial contributions to the ecosystem or to biogeochemical cycling, the most useful approach may be scaling parameters to the depth-specific volume of the sea ice sampled (Deming & Eicken, 2007).

13.3 Abundance and distribution

13.3.1 Ubiquity, dominance

Unlike other inhabitants of sea ice, bacteria have been found in significant numbers in every ice sample that has been examined for them. Reported densities, scaled to volume of ice, range from 4×10^3 mL^{-1} in upper horizons of Arctic winter sea ice (Collins et al., 2008) to nearly 3×10^7 mL^{-1} in summer bottom ice following the ice-algal bloom, in both Arctic and Antarctic sea ice (Krembs & Engel, 2001; Thomas et al., 2001). When scaled to brine volume, the maximum density reported is 2×10^8 bacteria mL^{-1} (Collins et al., 2008). Regardless of scaling, these upper end-points rival or exceed the maxima found in most aquatic environments, at any latitude but including surface seawater in the autumn prior to freezing, where bacterial densities range from 3×10^5 to 2×10^6 mL^{-1} (Gradinger & Ikävalko, 1998; Riedel et al., 2007; Collins & Deming, 2011b).

Although only a handful of sea ice samples have been examined to quantify viruses microscopically, bacteria are always accompanied by bacteriophage throughout the Earth's environments, including other extreme ones (Ortmann & Suttle, 2005), and almost always in higher numbers than their hosts. A typical virus to bacteria ratio in aquatic environments is about 10, but the ratios in sea ice can range over several orders of magnitude, with the highest ratios in young ice (Collins & Deming,

2011b), implying a need to study viral production rates before and after freeze-up. Scaled to volume of ice sampled, the densities of bacteriophage in sea ice range from 2×10^4 mL^{-1} in surface brine skim (Barber et al., 2014) to 1.3×10^8 mL^{-1} in bottom ice during the spring ice-algal bloom (Maranger et al., 1994). Densities are higher when scaled to brine volume. The larger and less well-studied eukaryotic viruses usually account for less than 10% of the total number of viruses in sea ice (Gowing et al., 2002; Gowing, 2003).

13.3.2 Enrichment

Planktonic organisms are incorporated into growing sea ice in numbers disproportionate to those in surface seawater. The proposed mechanism involves the physical scavenging of organisms by frazil ice crystals that form in underlying waters and rise to concentrate at the surface during sea ice formation (Garrison et al., 1983; Ackley & Sullivan, 1994). Most observations indicate that bacteria are physically enriched in new sea ice (Grossmann, 1994; Gradinger & Ikävalko, 1998; Riedel et al., 2007; Collins & Deming, 2011a), with evidence that an adjustment period to ice conditions is needed before growth can occur (Grossmann & Dieckmann, 1994; Eronen-Rasimus et al., 2015). The bacterial enrichment factor for ice over seawater is typically 5–10, but ranges from 2 to 54 (Sullivan & Palmisano, 1984; Gradinger & Ikävalko, 1998; Riedel et al., 2007; Collins & Deming 2011a).

Because particle size strongly influences entrainment, with larger particles entraining preferentially into sea ice, the enrichment of bacteria in new ice has been explained by their attachment to larger particles, particularly algal cells with a 'sticky' exterior coating (Grossmann & Dieckmann, 1994; Gradinger & Ikävalko, 1998). Organic detritus, sediment grains and aggregates, which also partition into the liquid phase of the ice matrix (Stierle & Eicken, 2002), can deliver large numbers of attached bacteria into the ice (Junge et al., 2004; Deming & Eicken, 2007). In both melted and unmelted sea ice, DNA-stained bacteria have been observed attached to organic debris and embedded in gelatinous exopolymer particles (Section 13.2; Figure 13.1a).

Whether specific types of bacteria entrain into sea ice more readily than others has not been tested experimentally, but a study of Baltic Sea ice shows that the bacterial community in newly-formed pancake ice was nearly indistinguishable from the source seawater (Eronen-Rasimus et al., 2015). As the ice consolidated and thickened, however, strong changes in community composition emerged, suggesting that retention in the ice during brine expulsion may be as critical to shaping the sea ice community as initial entrainment. While only a small fraction of bacteria in seawater may be attached to particulate matter, 50% or more of the bacteria in sea ice are attached to surfaces (Junge et al., 2004). Bacteria that associate physically with algae or produce their own EPS (Mancuso Nichols et al., 2005; Marx et al., 2009) may be able to influence their own fates by biasing their entrainment and retention in the ice. In support of this hypothesis are findings that some sea ice bacteria produce proteins and external coatings that have an affinity for ice as a mineral surface (Section 13.5). The possible role that swimming may play in the bacterial colonization of sea ice or their subsequent location within it is under consideration (Junge et al., 2003; Wallace et al., 2015; Nadeau et al., 2016).

Selective entrainment of Archaea into sea ice is unexplored, although the ones entrained clearly persist through winter (Collins et al., 2010). Viruses of both akaryotes and eukaryotes appear to be enriched into sea ice during its formation, by factors of 10–100 (Maranger et al., 1994; Gowing et al., 2004; Collins & Deming, 2011b). Interpreting these viral results as physical enrichment is complicated by the possibility that freezing stress induces lysogenic phage and subsequent production of viruses in young ice or that viral production ongoing within hosts in seawater is simply completed (new viruses released) in the ice.

13.3.3 Losses

The reverse process – the loss of akaryotes and viruses from sea ice – can take several forms, both at the bottom and at the top of the growing ice cover. Physical losses with downward brine expulsion during ice formation do not appear to offset net gains in the consolidated ice, but the onset of ice melt, when algal biomass is released from bottom ice into the water column (Krembs & Engel, 2001), can be expected to release substantial numbers of akaryotes and viruses as well. Quantifying that process has been a challenge (Riedel et al., 2006). At the top of the ice, physical losses of bacteria and viruses may occur through the winter, as brine and its contents are expelled upwards due to pressure gradients in the thickening ice. As with the entrainment process, organisms that are transported upwards do not appear to act as conservative components of the brine. Recent work on snow-covered sea ice in late winter indicates that a significant fraction of the bacterial community in the ice is either physically retained in the ice through the season, again due to ice-affine cellular coatings and pore-clogging EPS (Krembs et al., 2011), or else succumbs (physically lyses) to more extreme conditions when expelled upwards (Ewert et al., 2013; Ewert & Deming, 2014). Cell losses due to viral infection have been considered, but direct evidence is not available (Collins et al., 2008; Barber et al., 2014). In the warmer seasons, when snowmelts and melt ponds form on the surface of the ice, the discrepancy between bacterial densities in the ice and in the melt ponds suggests that many sea ice bacteria succumb to osmotic shock (Brinkmeyer et al., 2004). The fate of Archaea and viruses in melt ponds is unknown.

Prior to snow deposition, the surface of new sea ice can develop a thin layer of liquid brine, often called 'brine skim', as brine is expelled upwards. When the temperature gradient between atmosphere and new ice is steep enough and other favourable conditions exist (moisture, nucleation points), frost flowers will grow on the surface of the new ice, where they can become remarkably salty (> 120 ppt) as they wick brine upwards from the brine skim (Barber et al., 2014). These delicate ice crystal structures have been the subject of multidisciplinary study in recent years, including examining them for microbial content (Barber et al., 2014). The discovery of elevated bacterial densities (scaled to total sample volume) in saline frost flowers, compared with underlying sea ice (Bowman & Deming, 2010), led to further studies that now provide a unique data set for considering the transport of bacteria as components of brine, from source seawater to features on the new ice surface, over a short period of time.

Here we use dilution curves for this purpose (as in Barber et al., 2014), where the property of interest (bacterial density) is plotted against sample salinity to reveal whether or not bacteria behave conservatively; deviations from a 1:1 line based on starting seawater values indicate non-conservative behaviour. Bacterial losses due to retention in the ice matrix or to cell lysis (e.g., by osmotic shock or viral infection) plot below the dilution curve, while gains due to taxis (directed motility) or growth (new cell production) plot above the line. In most cases, results of such analyses indicate cell losses. For new sea ice that grew rapidly (hours) in March in an experimental opening in the ice cover of an oligotrophic Greenland fjord, bacterial densities in brine skim and frost flowers were lower than expected, although the implied losses are minimal (Figure 13.2a). By comparison, bacterial densities in sackhole brines collected in March from older ice, also in an oligotrophic Greenland fjord, fall well below expected densities (red line and data in Figure 13.2a). Other measures showed these communities to be active, so the apparent losses can be attributed to inefficient drainage from the surrounding ice matrix (rather than cell lysis). For new sea ice in April in the coastal Chukchi Sea, the bacterial losses were generally greater, even though some data fall on the dilution curve (Figure 13.2b). These results support phylogenetic (Bowman et al., 2013; Barber et al., 2014) and metagenomic (Bowman et al., 2014) evidence of a selective process at work in determining the composition of bacterial communities in frost flowers. For new sea ice in the central Arctic Ocean in September, where the frost flowers were less saline (Bowman & Deming, 2010), substantial bacterial gains were apparent, suggesting bacterial growth or an accumulation mechanism in frost flowers not yet understood. Overall, this analysis indicates that bacteria behave non-conservatively, or dynamically, in sea ice brines and that local conditions determine gains or losses.

13.4 Seasonal dynamics

13.4.1 Autumn

As sea ice forms in autumn, the bacteria entrained from seawater appear to suffer a temporary reduction in activity, with resurgence in cellular production once the ice has consolidated. Some bacteria may not adjust to ice sufficiently to survive or flourish within it, but the overall success of the adjustment is reflected in rates of ice-bacterial production that exceed ice-algal production (Grossman & Dieckmann, 1994). In recent experimental treatments of sea ice derived from North Sea water, the addition of dissolved organic matter dramatically shortened the adjustment time and increased the production rates of bacteria in the ice (Eronen-Rasimus et al., 2014). A shift in the balance towards sea ice bacteria at this stage is consistent with

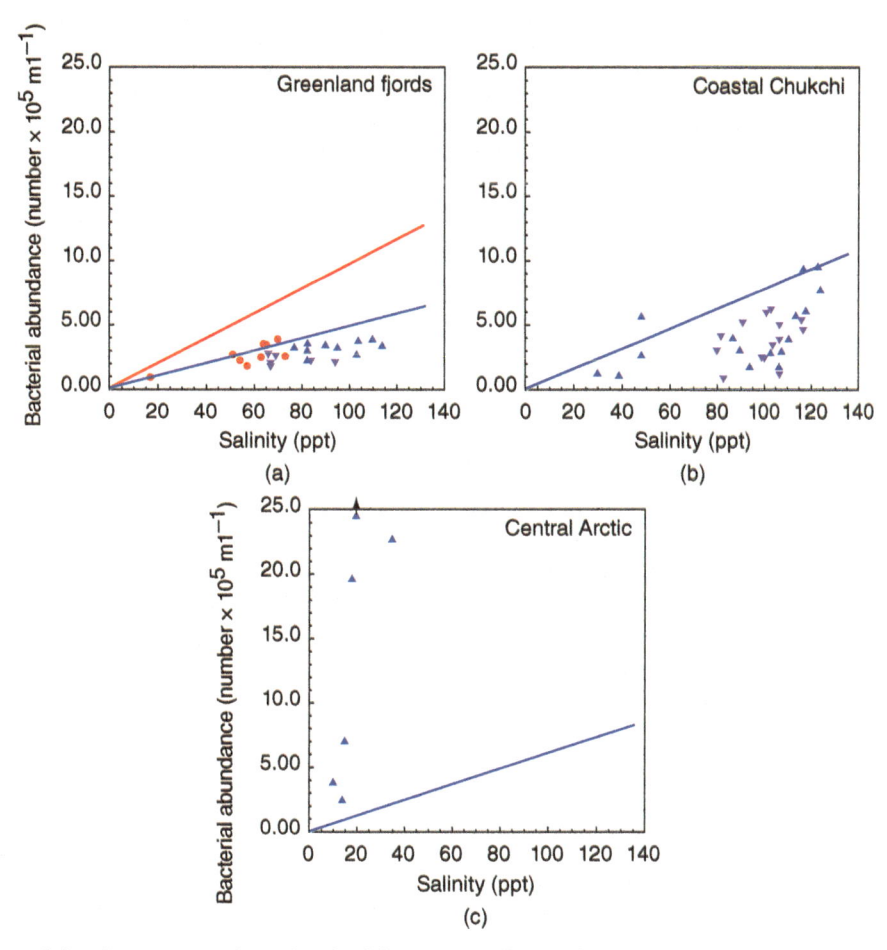

Figure 13.2 Bacterial abundance versus salinity data for different types of sea ice brine samples. Red, brines collected in sackholes in first-year sea ice; purple, brines expelled onto the surface of young sea ice and collected as brine skim; blue, brines wicked into frost flowers growing on the surface of young sea ice and collected as saline frost flowers. Solid lines indicate dilution curves calculated from seawater values as if bacteria were transported passively 1:1 with salt; data above or below the lines indicate, respectively, possible bacterial gains (growth or taxis) or losses (cell lysis or retention in the ice matrix). Data from sea ice in Greenland fjords (a) are derive from Barber et al. (2014) and Firth et al. (2016); data from coastal sea ice in the Chukchi Sea (b) and from central Arctic sea ice (c) are from Bowman & Deming (2010). [Panel (c) is modified from Bowman & Deming, 2010. Reproduced with permission of John Wiley & Sons.]

their access to elevated concentrations of dissolved organic matter in the ice and with the onset of light limitation for the entrained algae. It is also consistent with observations that other limits on bacterial production, set by protistan grazing and viral infection, do not appear to increase in young sea ice relative to those in seawater. For heterotrophic protists, the reductions in habitable space within the sea ice may limit their relative grazing impacts on ice bacteria, except near the ice–water interface (Krembs et al., 2000; Niemi et al., 2011), relegating more important grazing impacts to warmer ice seasons.

The limited amount of information available on bacterial, archaeal or viral dynamics in autumn sea ice, however, leaves many questions. For example, what happens to the viral–host relationship when seawater bacteria are newly encased in sea ice and experience temporarily stressful conditions is not known. Reports of some very high virus-to-bacteria ratios in autumn sea ice samples (100–1000; Gowing et al., 2004; Collins &

Deming, 2011b) suggest that bulk measurements may be masking important activities occurring at the scale of the organism.

13.4.2 Winter

As autumn gives way to winter, temperatures drop, especially in the upper horizons of Arctic sea ice (Collins et al., 2008; Ewert & Deming, 2014). Once below the permeability threshold of the ice (about −5°C; Golden et al., 1998), continued decreases in temperature reduce the fraction of the ice that remains liquid. Reduction of liquid volume means that bacteria and viruses, along with sea salts, become increasingly concentrated in these brine inclusions, as do any other particulate, gelatinous or dissolved organic and inorganic components. Bacterivorous protists are not known to inhabit the restricted brine space of winter sea ice in sufficient numbers to alter the abundance of their bacterial prey (Krembs et al., 2000, 2002). Bacteriophage, however, are present in winter ice, have high contact rates with their potential hosts due to the brine-concentrating factor, and appear capable of reducing bacterial numbers in sea ice brines (at −12°C and 16% salt, the most extreme conditions tested; Wells & Deming, 2006a).

The winter conundrum for heterotrophic bacteria is that food availability increases in the pores they inhabit just as other parameters (temperature, brine salinity and, potentially, viral infection) become limiting. Predictably, the minima for various measures of bacterial activity, including reductions in total bacterial abundance over time (Delille et al., 1995; Collins et al., 2008), are observed in the coldest horizons of the ice during the winter season, whether in the Arctic Ocean where conditions tend to be more severe (Junge et al., 2004; Ewert et al., 2013), in Antarctic seas (Helmke & Weyland, 1995) or in the Baltic Sea under milder conditions (Kaartokallio et al., 2008).

Despite these minima under severe conditions, bacteria remain active in the cold horizons of winter sea ice. The observed production of viruses (Wells & Deming, 2006a) and EPS under winter ice conditions (Krembs et al., 2002; Collins et al., 2008; Marx et al., 2009) both require active bacterial cells (eukaryotes being largely absent in upper winter ice). In one study of upper winter ice and its overlying saline snow layer, 85% of the bacteria were observed as functional by 'live/dead' staining (Ewert et al., 2013). In another, up to 86% were maintaining protein synthesis (as indicated by flurorescent 16S rRNA probes), while up to 4% were respiring enough oxygen to be detected by CTC stain at temperatures as low as −20°C and corresponding brine salinities as high as 209 ppt (Junge et al., 2004). Protection against osmotic shock and cell lysis, as the concentrations of inorganic salts increase to such levels, also requires active cells with functioning membrane transport systems to regulate the ionic balance between the interior of the cell and its briny surroundings; organic solutes (called compatible solutes) are taken up by bacteria in this process (Collins & Deming, 2013; Ewert & Deming 2014; Firth et al., 2016; Section 13.5). In warmer bottom layers bathed by seawater, rates of bacterial production equal or exceed rates of primary production measured in the same ice, giving winter sea ice its reputation for net heterotrophy (Kottmeier & Sullivan, 1987; Kottmeier et al., 1987; Helmke & Weyland, 1995; Niemi et al., 2011). Winter is the season for reduced activity, but the bacterial community that persists through the winter remains active in many ways.

13.4.3 Spring

The transition from winter to spring sees a reversal of the winter trends with the onset of renewed heterotrophic growth and production. The organic substrates to support this resurgence derive from both old and new resources. Bacterial increases near the seawater interface, observed in both Arctic and Antarctic bottom sea ice, are attributed to the onset of the ice-algal bloom and its early release of new dissolved organic matter (see Chapter 17). In upper regions of the ice, where algae are not blooming or even present (unless by seawater flooding; Arrigo et al., 2014), bacterial increases are attributed to consumption of existing dissolved organic carbon and possibly EPS, concentrated in the ice during autumn formation or produced during winter for cryoprotection. With the winter condundrum relaxing (temperature and salinity becoming milder), heterotrophic bacterial activity can thus increase throughout the ice without the provision of new resources. A new hypothesis invokes the remineralization of nitrogen-rich compatible solutes, stored intracellularly during winter for osmoprotection but no longer required for that purpose as warming freshens the liquid phase of sea ice (Collins & Deming, 2013; Firth et al., 2016).

Studies of viruses during the winter–spring transition in sea ice are not available, but at the start of the ice-algal bloom, densities of bacterial viruses in bottom ice are already high (9×10^7 mL^{-1} ice; Maranger et al., 1994), and much higher than densities reported for underlying seawater. Both bacteria and viruses then follow the progress of the ice-algal bloom, increasing in abundance and activity as chlorophyll concentrations rise. Other measures of cellular activity (by CTC stain) support this idea, given the increase in active bacteria in spring sea ice (up to 38% in EPS aggregates within the ice; Meiners et al., 2008) compared with winter sea ice (up to 4%; Junge et al., 2004). Whereas low rates of viral production may explain some bacterial losses in winter sea ice, viral production and host death simply accompany the more robust bacterial production that occurs in spring sea ice, as new dissolved substrates become plentiful. Bacterivorous protists may be the more important agents of bacterial mortality during the warmer seasons (Różańska et al., 2008; Piwosz et al., 2013). How organic substrates 'stored' from autumn may influence bacterial–viral dynamics remains to be determined.

Concentrated and labile dissolved organic matter derived largely from ice algae ensure active and robust bacterial assemblages during the bloom season (Smith & Clement 1990; Grossmann & Dieckmann 1994; Helmke & Weyland 1995), with production rates on the order of 20–30 mg C m^{-3} day^{-1}. Although these rates are considerably higher than reported autumn–winter rates, bacterial production typically represents less than 10% of primary production during spring and summer (Grossi et al., 1984; Kottmeier et al., 1987; Kottmeier & Sullivan, 1987; Smith & Clement, 1990; Grossmann & Dieckmann, 1994). Integrating measurements over the full thickness of the ice, rather than only the bottom layers, reveals a dominance of bacterial production over primary production in many cases (Gradinger & Zhang, 1997). Spring sea ice may be net heterotrophic in spite of intensive algal blooms at its base. Important advances in this regard come from measurements of bacterial respiration and growth efficiency in spring sea ice, which move beyond bacterial production assays and show that bacterial respiration may be higher than previously realized, especially in warmer ice (Nguyen & Maranger, 2011). The contribution of bacterial respiration to CO_2 fluxes to and from the ice is not yet clear, given the complex inorganic chemistry of sea ice brines (Miller et al., 2011; Rysgaard et al., 2011) and that these various rates, all based on melted ice samples, may not accurately reflect what is happening *in situ* (Sogaard et al., 2010).

Heterotrophic bacteria are clearly involved in nitrogen cycling in sea ice, and particularly in regenerating nitrogen in the form of ammonia, potentially aiding a jumpstart, enhancement or extension of the spring bloom season. Close associations between bacteria and protists in the ice probably contribute to efficient nitrogen recycling and transfer (Torstensson et al., 2015). Ammonium may be produced heterotrophically, even apart from algal associations, based on remineralization of organic compounds stored extracellularly in ice pores and intracellularly in bacteria. EPS concentrations in sea ice have been observed to decrease as ammonia increased (Riedel et al., 2007), while recent work shows that both psychrophilic bacteria and sea ice bacterial communities degrade nitrogen-rich organic solutes when they are no longer needed as compatible solutes for salt stress (Collins & Deming, 2013; Firth et al., 2016; Section 13.5). The general fate of nitrogen in sea ice has come under scrutiny as a result of the discovery of zones of anoxia in sea ice that support both denitrification and anaerobic ammonia oxidation or 'anammox' (Chapter 18; Rysgaard & Glud, 2004; Rysgaard et al., 2008). In some coastal ice formations, ice-algal blooms stimulate enough heterotrophic bacterial activity that oxygen is fully consumed, even prior to peak algal activity, leaving pockets of anoxia in the ice. Because bacterial denitrification and anammox in such pockets represent nitrogen losses to the system, these processes can be expected to influence nutrient availability for ice algae and thus impact overall primary productivity. The recent use of nitrogen isotopes to track nitrogen cycling pathways in sea ice indicates that nitrification, the production of nitrate from ammonia, is more important in spring sea ice than previously imagined (Fripiat et al., 2014; 2015; Firth et al., 2016), and may counterbalance nitrogen losses from the system. Genes for nitrogen fixation have been detected in sea ice (Diez et al., 2012), but rates to indicate any nitrogen gains to sea ice via this pathway await future work.

13.4.4 Summer

The physical demise of sea ice begins in summer, including the melting of the bottom ice layer that can be rich in organic compounds, especially EPS, and organisms, from bacteria and algae to various grazers, including viruses (Krembs & Engel, 2001; Riedel et al., 2006; Meiners et al., 2008; Chapter 17). By late summer, the physical demise of the ice is well under way, even as it remains microbially active. Rates of bacterial production in 'crack pools' of the disintegrating ice cover, believed to represent the 'climax' community leaving the ice (Gleitz et al., 1996), rank among the highest rates detected for the sea ice environment; they also tend to exceed concurrent rates of primary production. Still intact portions of the sea ice cover can support sizeable snowmelt ponds on its surface, of considerable importance to the physics and albedo of the ice. Though virtual death traps for sea ice organisms due to osmotic shock, these melt ponds eventually develop distinctive microbial ecosystems supported by freshwater algae (Brinkmeyer et al., 2004), in which bacterial production can surpass all other sea ice-related habitats with rates up to $1 g C m^{-3} day^{-1}$ (Kottmeier & Sullivan, 1987, 1990; Kottmeier et al., 1987; Grossmann & Dieckmann, 1994). The fate of this new biomass includes drainage to underlying seawater via cracks in the ice cover, as well as percolation into the increasingly porous ice itself, complicating evaluations of sea ice bacterial dynamics at the end of the annual cycle.

13.4.5 Multi-year ice

Sea ice that persists through the summer takes on physical properties that differ from the first-year ice addressed in the preceding sections. The summer season brings the same surface snowmelt ponds, freshwater flushing through the ice, and general desalination due to gravity drainage, yet the ice persists and grows new bottom layers during the cold seasons, becoming thicker than first-year sea ice (Chapter 1). The resulting mix of old and new ice and altered brine content translates to a correspondingly mixed bacterial community (Archaea fall below detection limits in multi-year ice; viruses have not been studied.) Bacterial abundance and diversity have been examined in multi-year ice (Gradinger & Zhang, 1999; Brinkmeyer et al., 2003), including with advanced DNA sequencing technology (Bowman et al., 2012; Hatam et al., 2014; Section 13.5), but little is known of bacterial dynamics within it, as sampling such thick sea ice or readily finding it in the warming Arctic Ocean is a challenge.

13.5 Diversity

13.5.1 From cultures to phylogeny

The study of bacterial diversity in sea ice began with culturing work, where the focus was on the physiology of heterotrophs and not phylogeny. The general understanding today is that sea ice strongly selects for cold-adapted bacteria (Bowman et al., 1997; Junge et al., 2002; Helmke & Weyland, 2004) and, consequently, for cold-active viruses (Borriss et al., 2003, 2007; Wells & Deming, 2006a,b). Culturable sea ice populations are often dominated by 'true' psychrophiles that fail to grow at 20°C and express lower temperature growth optima and minima than other bacteria (according to the definition of Morita, 1975). How psychrophiles tolerate the wide range of salinities experienced in sea ice appears to be species-specific (Ewert & Deming, 2014).

When the heterotrophic bacteria cultured from sea ice were finally identified phylogenetically by 16S rRNA gene sequence, representatives from specific genera turned up repeatedly in both Arctic and Antarctic sea ice. These taxa, primarily Gammaproteobacteria, Alphaproteobacteria and Bacteroidetes, were again found to be dominant when culture-independent methods such as clone library sequencing of environmental DNA came to the fore in the 1990s. The most common genera found in summer sea ice are of the orders Alteromonadales and Flavobacteriales: *Pseudoalteromonas*, *Colwellia*, *Shewanella*, *Flavobacterium* and *Polaribacter* (Table 13.1). Other bacterial groups, including Betaproteobacteria, Actinobacteria and Firmicutes, were also identified from sea ice but at lower frequency (Gosink & Staley, 1995; Bowman et al., 1997; Brown & Bowman, 2001; Brinkmeyer et al., 2003).

Perhaps not unexpectedly, given an early focus on spring and summer sea ice and the long observed physical association of sea ice bacteria with algal cells, many of the dominant sea ice bacteria in culture fall within lineages generally associated with marine algae (Gosink & Staley, 1995; Bowman et al., 1997; Junge et al., 2002; Brinkmeyer et al., 2003). The availability of algae-derived organic substrates for heterotrophic bacteria in sea ice also explains the high percentage of the total number of bacteria that can be cultured (up to 60%; Junge et al., 2002) and identified microscopically by fluorescent gene probes (~95%; Brinkmeyer et al., 2003), making summer sea ice a unique microbial

Table 13.1 Most commonly identified bacterial and archaeal taxa in sea ice phylogenetic studies. The column 'Studies' includes both published articles and unpublished datasets available in public databases. The column 'Clones' is scaled to the number of clone library sequences in public databases for each genus. The column 'Reads' is the total number of next-generation sequencing reads from MGP138 and MGP6099 (see Table 13.2) hitting each genus.

Phylum	Order	Genus	Studies	Clones	Reads	Public genomes Available	Public genomes In progress	Public genomes From sea ice
Gammaproteobacteria	Alteromonadales	Pseudoalteromonas	22	317	47 932	55	43	6
		Colwellia	21	133	15 684	10	3	1
		Shewanella	20	107	336 546	43	25	0
		Glaciecola	12	75	62	15	0	3
		Marinobacter	12	41	12 661	12	7	0
	Oceanospirillales	Marinomonas	13	75	72 583	10	0	0
		Halomonas	9	75	11 856	29	21	0
	Pseudomonadales	Psychrobacter	13	84	6 992	20	9	0
		Pseudomonas	13	69	239 292	758	359	0
Bacteroidetes	Flavobacteriales	Flavobacterium	17	46	868 456	48	52	0
		Polaribacter	15	95	36 039	3	7	0
Alphaproteobacteria	Rhodobacterales	Octadecabacter	11	34	10 069	2	0	2
		Sulfitobacter	9	46	17 854	16	2	0
	Rickettsiales	SAR11 clade	4	103	283 295	29	28	0
Thaumarchaeaota	Nitrosopumilales	Marine Group I	3	100	668 093	7	2	0

habitat relative to other natural aquatic environments (Junge et al., 2002).

Early results of phylogenetic studies indicated the existence of bacterial sea ice specialists and raised questions about biogeography and endemism (Staley & Gosink, 1999). Did the sea ice environment in general select for similar species at both poles, or were the Arctic and Antarctic sufficiently different and geographically removed from each other to have fostered endemism? Although the definition of bacterial species continues to be debated, there is little evidence for sea ice endemism at the common 'species' threshold of 97% 16S rRNA similarity. Highly similar 16S rRNA gene sequences have been found for many bacteria in sea ice from both polar regions (Brown & Bowman, 2001; Brinkmeyer et al., 2003; Vollmers et al., 2013), while two named species, *Colwellia psychrerythraea* and *Polaribacter irgensii*, are known from both Arctic and Antarctic sea ice (see Section 13.5.3).

An analysis of available phylogenetic data (Figure 13.3) shows that spring and summer sea ice communities are overrepresented in the literature and in sequence databases, which can be attributed to logistical constraints that have limited access to sea ice in the colder seasons when the role of sea ice algae is also limited.

Though the bacterial genera considered to dominate sea ice communities are thus found abundantly and nearly ubiquitously in spring and summer sea ice, they appear to be rare in winter sea ice. Only representatives of the Flavobacteriales, common in spring and summer sea ice, were also identified in a season-long study of Arctic winter ice (Collins et al., 2010). Within the Alteromonadales, the clades that dominate in spring and summer ice were absent from the winter ice (Collins et al., 2010), replaced instead by oligotrophic marine Gammaproteobacteria (e.g., clade OM182) apparently sourced from autumn seawater.

Archaea, well known in polar waters (DeLong et al., 1994; Bano et al., 2004), are present in Arctic winter ice and persist through the severity of that season (Collins et al., 2010). Their absence from spring and summer sea ice has suggested that the winter–spring transition, with the onset of algal productivity, so strongly favours heterotrophic Bacteria that Archaea fade toward undetectability (Brinkmeyer et al., 2003). The more recent detection of Archaea in spring and summer Antarctic sea ice, where the dominant members were closely related to an archaeal ammonia-oxidizer (Cowie et al., 2011), suggests that archaeal diversity and function in sea ice merits more study.

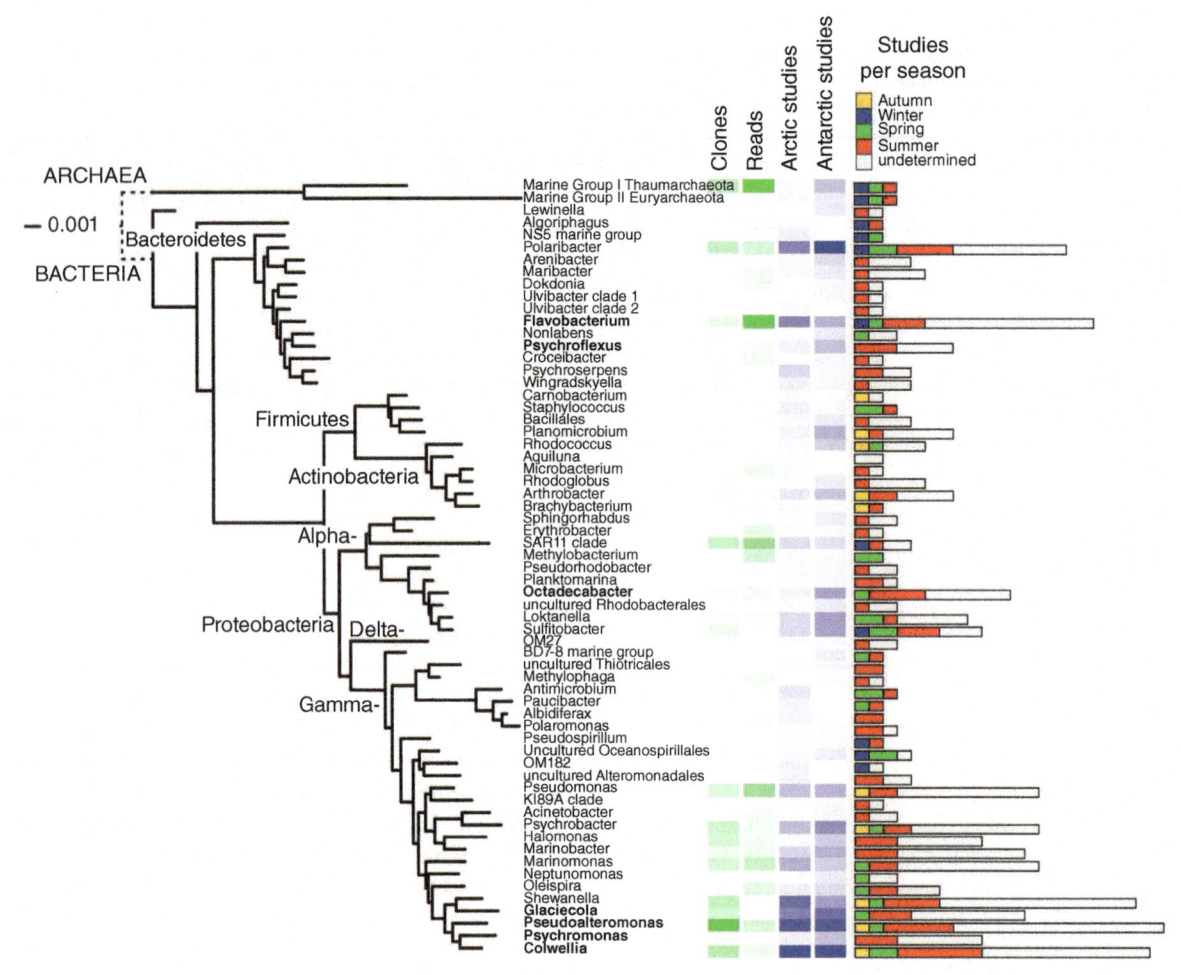

Figure 13.3 Representative sea ice Bacteria and Archaea on a maximum likelihood phylogenetic tree generated from 16S ribosomal RNA gene sequences. Only genera with at least two sequences, detected in different studies, were included on the tree. Genera in bold type include sea ice representatives for which genome sequences are available. The column 'Clones' is scaled to the number of clone library sequences in public databases for each genus (range 2–317). The column 'Reads' is scaled to the square root of the total number of next-generation sequencing reads from MGP138 and MGP6099 (see Table 13.2) hitting each genus (range 5–868 456). The columns 'Arctic studies' and 'Antarctic studies' are scaled to the number of studies in each location where the genus was identified (range 0–17). The column 'Studies per season' is a bar chart showing the number of studies where each genus was identified, colour-coded by sampling season when available (range 2–22) (data from references are available at http://github.com/rec3141/pubs/tree/master/Thomas2015).

13.5.2 Temporal and spatial diversity

The use of comparative phylogenetic techniques to study bacterial succession in sea ice is in its infancy. Only in short-lived, relatively warm and thin, Baltic Sea ice are community DNA fingerprints available to evaluate succession throughout the lifetime of an ice cover (Kaartokallio et al., 2008); the patterns predictably follow the ice-algal bloom. Succession in polar sea ice, by contrast, can only be pieced together from the results of separate seasonal studies. For example, between autumn ice formation and a sampling plan initiated in winter, the nature of the bacterial community by DNA fingerprinting best resembled the bacterioplankton community in autumn seawater (Collins et al., 2010). For the rest of the sea ice cycle, influenced by the ice-algal bloom, the familiar suite of heterotrophic bacteria appear, with some additional uncultured phylotypes (Table 13.1; Brown & Bowman, 2001; Brinkmeyer et al., 2003).

The largely freshwater melt ponds that form on the surface of sea ice in late summer present sea ice diversity in a different light. Unlike communities within the sea ice, these melt-pond communities of Bacteria (no Archaea have been detected) at the surface of the ice are dominated by Betaproteobacteria (Table 13.1), considered to be physiologically adapted to and characteristic of freshwater habitats. Gram-positive, often spore-forming species (Firmicutes), also rarely detected in the interior of the ice, sometimes comprise 10% of the bacterial community in melt ponds (Brinkmeyer et al., 2004). The net result is the appearance of an estuarine community, where Betaproteobacteria dominate freshwater zones and Alphaproteobacteria and Gammaproteobacteria prevail in more saline zones of summer melt ponds (Brinkmeyer et al., 2004). Aerial delivery explains the colonization of melt ponds by freshwater and spore-forming (desiccation-resistant) bacteria, while penetration of melt-pond fluids into sea ice explains their detection as components of the ice community, particularly in multi-year ice (Hatam et al., 2014).

Comparative phylogenetic studies focused on surface features of new sea ice show that the process of ice formation exerts a selective force on the bacteria that expel upwards to populate brine skim and frost flowers on the surface of the ice (as also shown by tracking bacterial density as a function of salinity; Figure 13.1). Community composition differs between study sites, but within-site comparison of surface and underlying ice communities indicates that frost flowers select for distinctive communities. In the case of new ice in the coastal Chukchi Sea, frost flowers (sampled in bulk with brine skim) selected for members of the marine Rhizobiales, about which little is known (Bowman et al., 2013, 2014); in the case of new ice in an oligotrophic Greenland fjord, frost flowers and brine skim selected for clades of the oligotrophic Pelagibacteria (SAR 11 clade) and members of the genus *Nitrospina* known for nitrification (Barber et al., 2014). The fate of these surface-ice communities is not clear, but the options include aerial dispersal, potentially colonizing distant locations (Bowman & Deming, 2010), or incorporation into a saline snow layer on the surface of the growing ice, where the majority (85%) of the bacteria present appear viable by 'live/dead' staining (Ewert et al., 2013).

13.5.3 Step change in methodology

The methodology used to evaluate the diversity of sea ice microbial communities has undergone a step change over the past decade. As culture-based methods transitioned to culture-independent methods, the depth of our view into sea ice communities increased exponentially. Besides environmental clone library sequencing, DNA fingerprinting techniques such as DGGE and T-RFLP were used widely in the 2000s. In reviewing 76 studies on sea ice phylogenetic diversity, we found that about 60% included isolation of pure cultures, 30% included sequence libraries from environmental cloning, and 15% utilized DNA fingerprinting (full reference list available at http://github.com/rec3141/Thomas2015).

More recently, a shift towards 'next-generation' DNA sequencing technologies (NGS) has increased the scale of environmental sampling of microbial diversity by a factor of a million, while also rapidly increasing our ability to investigate functional genes and, consequently, genetic and putative biochemical and biogeochemical pathways (Bowman, 2015). Only a few published studies (Bowman et al., 2012, 2013, 2014; Hatam et al., 2014) and a handful of public (but unpublished) sea ice environmental datasets include results from NGS technologies (Table 13.2), including amplicon sequencing ('metabarcoding') and shotgun metagenomics, but not transcriptomics. The scale of sampling is so greatly increased with these methods that thousands of samples of sea ice will have been analysed by the end of this decade, leading to the discovery of much new diversity as different sea ice seasons, formations and features are examined.

In the near future a comprehensive catalogue of the phylogenetic diversity of microorganisms in sea ice will exist, but understanding the mechanisms of microbial adaptation to this environment requires the functional gene diversity to be explored as well. The primary methods used today for functional genomics are metagenomic analysis and complete genome sequencing of isolated strains of microorganisms. As of 2015, few metagenomes have been published from sea ice (e.g., Bowman et al., 2014), though many unpublished sequences are in the pipeline. Likewise, only a handful of genome sequences have been completed for Bacteria

Table 13.2 Next-generation DNA sequencing studies of sea ice microbial communities.

Study type	Location	Technology	Data (Mbp)	Accession	Reference
Amplicon	Arctic multi-year sea ice (North Pole)	454 GS FLX	12	SRP006990	Bowman et al. (2012)
Amplicon	Arctic sea ice (Ellef Ringnes Island)	Illumina	1 100	MGP138	H. Findlay (unpubl. data)
Amplicon	Arctic sea ice and frost flowers (Young Sound)	454 GS FLX	49	PRJNA239386	Barber et al. (2014)
Amplicon	Antarctic sea ice (McMurdo Sound)	454 GS FLX Titanium	552	PRJNA50699	JCVI (unpubl. data)
Amplicon	Arctic multi-year sea ice (Ellesmere Island)	454 GS FLX Titanium	23	PRJNA247832	Hatam et al. (2014)
Shotgun metagenome	Arctic sea ice and frost flowers (Point Barrow)	Illumina	3 100	MGP6099	Bowman et al. (2013, 2014)
Shotgun virome	Arctic sea ice (Point Barrow)	454 GS FLX Titanium	289	PRJNA47457	J. Deming (unpubl. data)

from sea ice (Table 13.3), and none for Archaea. Of the 10 genera most commonly found in sea ice, only four include genome sequences from sea ice isolates: *Colwellia*, *Pseudoalteromonas*, *Glaciecola* and *Octadecabacter*. As most of the completed genome sequences are for Gammaproteobacteria, future genome-sequencing efforts targeting clades of Bacteroidetes, Alphaproteobacteria and Archaea will provide better coverage of the dominant phylotypes found in environmental sequences libraries. In particular, sea ice isolates of *Flavobacterium* and *Polaribacter* among the Bacteroidetes, and *Pelagibacter*, *Loktanella* and *Sulfitobacter* among the Alphaproteobacteria, would be valuable additions to research on the genomic basis of adaptation to sea ice as an extreme environment.

Recent genome sequencing efforts have led to revisiting the question of endemism in sea ice at the strain and even gene level. In the case of *Octadecabacter arcticus* and *Octadecabacter antarcticus*, they share > 99% 16S rRNA similarity, yet < 80% of their gene content (Vollmers et al., 2013). High rates of rearrangement in these genomes indicate a strong influence of mobile genetic elements on their evolution, a signature that has been observed in genomes from other sea ice bacteria (Bowman, 2008, 2014; Feng et al., 2014).

13.5.4 Viral diversity

Several studies of the size range and morphological diversity of viruses in sea ice are available (Maranger et al., 1994; Gowing et al., 2002, 2004; Gowing, 2003), indicating the overwhelming predominance of bacterial viruses, but the phylogenetic diversity of viruses in sea ice is not known. In only two studies have bacterial viruses been obtained in culture against cold-adapted bacterial hosts known from polar sea ice (Borriss et al., 2003; Wells & Deming, 2006b). From these efforts, as well as recent work on subarctic ice (Senčilo et al., 2014), only the presence of common, tailed, double-stranded DNA viruses, classified as members of the *Siphoviridae* and *Myoviridae*, can be ascribed to sea ice. Two of the genomes of the isolated phage from polar ice have since been sequenced (Borriss et al., 2007; Colangelo-Lillis & Deming, 2012). Genomic analysis of the most cold-active of these viruses, Colwelliaphage 9A, and earlier experimental work with it (Wells & Deming, 2006a) suggest that cold-active phage may have a broader host range than 'mesophilic' phage, which are understood to be species-specific. If so, then low phage diversity in sea ice may be balanced by broad infectivity. Given the importance of viruses in the evolution and dynamics of sea ice bacteria, and potentially of eukaryotic organisms, direct analyses of viral DNA extracted from sea ice is needed to move beyond what can be learned from isolated phage in culture.

13.6 Adaptations

13.6.1 Temperature and salinity

Bacterial and archaeal adaptations that enable life in sea ice necessarily include molecular means to function at low temperature (reviewed by D'Amico et al., 2006; Bowman, 2008; De Maayer et al., 2014). Universal hallmarks of cold adaptation include modifications of the amino acid composition of the proteome, which confer flexibility to proteins for their enzymatic functions (and usually increased thermolability as a trade-off),

Table 13.3 Bacterial strains isolated from sea ice for which public genome sequences are available.

Taxon	Isolated from	Sequencing status	Proteins	BioProject	Reference
Colwellia SLW05[a]	Antarctic sea ice melt puddle (West Antarctic Peninsula)	Reads	nd	PRJNA74001	10.1111/j.1574-6941.2007.00345.x
Flavobacterium frigoris PS1	Antarctic sea ice algal assemblage (McMurdo Sound)	Contigs	3590	PRJNA156765	10.1371/journal.pone.0035968
Glaciecola pallidula ACAM 615	Antarctic sea ice (Taynaya Bay)	Scaffolds	4036	PRJNA178727	10.1111/1462-2920.12318
Glaciecola psychrophila 170	Arctic sea ice (Canada Basin)	Complete	5618	PRJNA174842	10.1128/genomeA.00199-13
Glaciecola punicea ACAM 611	Antarctic sea ice (Prydz Bay)	Contigs	2883	PRJNA157053	10.1111/1462-2920.12318
Octadecabacter antarcticus 307	Antarctic sea ice (McMurdo Sound)	Complete	4492	PRJNA54701	10.1371/journal.pone.0063422
Octadecabacter arcticus 238	Arctic sea ice (Beaufort Sea)	Complete	4683	PRJNA19331	10.1371/journal.pone.0063422
Pseudoalteromonas sp. BSi20311	Arctic sea ice (Beaufort Sea)	Contigs	3676	PRJDA74505	10.1128/JB.06427-11
Pseudoalteromonas sp. BSi20429	Arctic sea ice (Beaufort Sea)	Contigs	4030	PRJDA74507	10.1128/JB.06427-11
Pseudoalteromonas sp. BSi20439	Arctic sea ice (Beaufort Sea)	Contigs	3612	PRJDA74509	10.1128/JB.06427-11
Pseudoalteromonas sp. BSi20480	Arctic sea ice (Beaufort Sea)	Contigs	3967	PRJDA74513	10.1128/JB.06427-11
Pseudoalteromonas sp. BSi20495	Arctic sea ice (Beaufort Sea)	Contigs	4365	PRJDA74515	10.1128/JB.06427-11
Pseudoalteromonas sp. BSi20652	Arctic sea ice (Beaufort Sea)	Contigs	4085	PRJDA74291	10.1128/JB.06427-11
Psychroflexus torquis ATCC 700755	Antarctic sea ice-algal assemblage (Prydz Bay)	Complete	3526	PRJNA13542	10.1093/gbe/evt209
Psychromonas ingrahamii 37	Arctic sea ice (Point Barrow)	Complete	3545	PRJNA16187	10.1186/1471-2164-9-210

[a] The strain 34H of Colwellia psychrerythraea, a bacterial species isolated from Arctic sediments and since found in both Arctic and Antarctic sea ice, has also been sequenced (Methé et al., 2005).

and adjustments to the lipid composition of cellular membranes, which maintain fluidity for their transport functions in the cold. Without compositional changes, enzymes and membranes become 'rigid' at low temperatures and lose functionality; with them, bacteria become competitive in the cold, even if absolute rates of chemical reactions and diffusive transport slow as temperatures drop (and viscosity rises).

Comparative studies of the whole-genome sequences and other traits of cultured strains of C. psychrerythraea and P. torquis, two bacterial species known to reside in sea ice, reveal additional traits that contribute to success at sub-zero temperatures (Methé et al., 2005; Bowman, 2008; Feng et al., 2014). For example, both organisms show evidence of storing various intracellular energy reserves, which may counter the slowing of reactions involved in nutrient uptake across membranes at sub-zero temperatures. These reserves take the form of large polymers such as glycogen, polyhydroxybutyrate, polyphosphates or neutral lipid globules, but also protein-like polymers (Bowman, 2008) and even monomers (Collins & Deming, 2013; Firth et al., 2016) that would provide a reserve of organic nitrogen.

Of course, bacteria in sea ice experience not only sub-zero temperatures but also simultaneous extremes in salinity; they must tolerate both seasonal and diurnal fluctuations in these extremes (Ewert & Deming, 2014). The hallmark defence against salt or osmotic shock is to produce or accumulate within the cell what are called compatible solutes – typically sugars, polyols and amino acids and their derivatives, such as betaine. Both *C. psychrerythraea* and *P. torquis* carry genes involved in the production or transport of compatible solutes, particularly betaine; *P. torquis* also appears capable of using the sugar trehalose for the dual purpose of osmo- and cryoprotection (Bowman, 2008). By adjusting intracellular concentrations of these compounds, an organism can balance the influx or efflux of salts that occurs when external conditions shift; at high intracellular concentrations of these organic solutes, the freezing point may also be depressed so that potentially damaging ice crystals do not form. Compatible solutes were so named because, unlike salts, they can be accumulated to high levels within the cell without compromising its macromolecular functions. Required ionic interactions can be stabilized by compatible solutes as they interact with the hydration sphere around enzymes, nucleic acids and other macromolecules during osmotic shifts.

A comparative analysis of compatible solute systems encoded by the genomes of *C. psychrerythraea* and another psychrophilic Gammaproteobacterium, *Psychrobacter arcticus*, isolated from permafrost, suggests subtle but potentially important diversity in strategies for tolerating salinity shifts (Figure 13.4). In general, *C. psychrerythraea*, representing its class of sea ice bacteria, appears more invested in the use

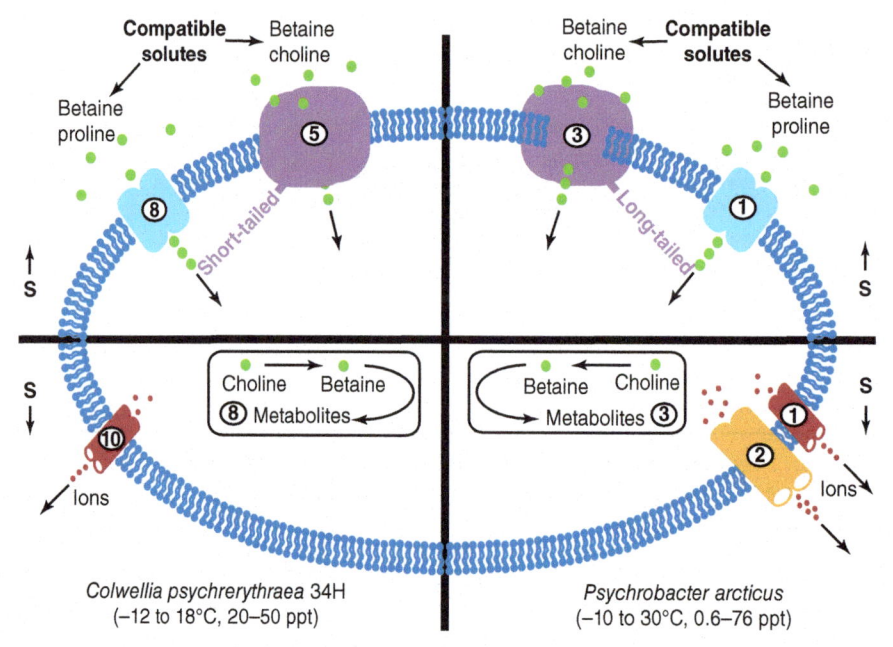

Figure 13.4 Simplified schematic of a composite cold-adapted bacterium showing strategies for tolerating salinity changes driven by temperature changes in sea ice. *Colwellia psychrerythraea* 34H is represented on the left half; *Psychrobacter arctica* on the right half. Cellular response to an increase in external salinity (S) is shown in the upper half; and response to a decrease in S in the lower half. When S increases, cells use transport proteins (aqua and purple shapes) to import compatible solutes (e.g., betaine, choline, proline; green dots) and maintain cellular integrity and function; when S decreases, cells rapidly export salt ions (red dots) for the same purpose, through large-conductance (orange tubes) and small-conductance (red tubes) mechanosensitive ion channels, and then metabolize stored compatible solutes (interior boxes) no longer needed for osmotolerance. Circled numbers indicate the known number of copies, in the genomes of each organism, for genes putatively involved in each of these steps (based on data in Ewert, 2013; Ewert & Deming, 2014).

of compatible solutes, as it carries more identifiable genes for this purpose than *P. arcticus*. In the event of a salinity upshift, both organisms appear capable of transporting proline, betaine and choline (a precursor to betaine), but *C. psychrerythraea* encodes for twice as many transporter proteins. In the case of a salinity downshift, *C. psychrerythraea* appears to control salt release through small-conductance mechanosensitive ion channels and to metabolize its store of compatible solutes, presumably for energy and nitrogen. *P. arcticus* has similar but fewer of these genes; it also encodes for large-conductance mechanosensitive ion channels (that release ions more rapidly) not detected in the *C. psychrerythraea* genome. An intriguing feature of some of the compatible solute transporters encoded by *C. psychrerythraea* is the short-tailed C-terminus (Figure 13.4), which in other bacteria functions as a temperature sensor in the membrane; the comparable genes in *P. arcticus* encode for transporters with a long-tailed C-terminus, known to function as an osmosensor (Ziegler et al., 2010). Although such genome analyses are speculative and need functional verification (as in Firth et al., 2016), we can hypothesize that *C. psychrerythraea* strain 34H is uniquely adapted to surviving in sea ice: its ability to sense a temperature change is directly linked to its ability to respond to a salinity shift. It may not outcompete other sea ice bacteria adapted for growth at higher salinities (Ewert & Deming, 2014), but it appears to have 'genetic memory' of inhabiting sea ice.

Related to adaptations to salinity stress is the discovery that numerous sea ice bacteria, including *P. torquis* (Feng et al., 2013), *O. arcticus* and *O. antarcticus* (Vollmers et al., 2013), carry genes for proteorhodopsin, which are expressed at all depths in sea ice (Koh et al., 2010). Proteorhodopsins are membrane-bound proteins that function as light-driven proton pumps, harvesting supplemental energy for the cell. Work with *P. torquis* shows a direct link between light-stimulated growth and elevated salinity (Feng et al., 2013); higher brine salinity and greater light availability characterize upper sea ice, where *P. torquis* may have the competitive edge.

13.6.2 Extracellular interactions

Partitioned into the brine-filled pores of sea ice, bacteria and other components from seawater experience higher contact rates with each other and with surfaces, including the ice itself, than in seawater (Section 13.2). Adaptations reflecting this higher degree of interaction include the production of antifreeze proteins, cold-active enzymes and EPS. All microorganisms present in a brine inclusion can benefit from these released products of an individual, including many other metabolic products (even if not yet explored in sea ice), which helps to explain the intimate associations often observed between bacteria and sea ice algae (Stewart & Fritsen, 2004).

Antifreeze proteins, a structurally diverse group, include proteins that bind to ice crystals and others that inhibit the recrystallization of ice. The former alter ice morphology, which is thought to favourably modify the habitat for bacteria; the latter prevent formation of large ice crystals that could damage membranes. Not all sea ice bacterial genomes examined encode for antifreeze proteins, so either the trait is not required for inhabiting ice or new antifreeze proteins are yet to be discovered. *P. torquis* carries a gene that codes for a large putative cell surface protein suspected of antifreeze activity, while *Colwellia* strain SLW05, a close relative of *C. psychrerythraea* isolated from Antarctic sea ice, encodes a small antifreeze protein that both binds ice crystals and prevents recrystallization (Raymond et al., 2007). The *Colwellia* protein appears to have arrived in the genome via horizontal gene transfer (Raymond & Kim, 2012), an evolutionary story that may well have been repeated in other sea ice microorganisms (Bayer-Giraldi et al., 2010; Kiko, 2010). Experiments with exudates from *C. psychrerythraea* indicated the presence of protein-containing polysaccharides that are preferentially entrained into sea ice (Ewert & Deming, 2011). Subsequent analysis of the capsular EPS that coats the exterior of this bacterium revealed a novel polysaccharide unusual for its decoration with threonine in a pattern that imparts the ability to inhibit ice recrystallization (Carillo et al., 2014).

Extracellular enzymes released by cold-adapted bacteria, including *C. psychrerythraea*, follow established fundamentals of cold activity in an enzyme (Feller & Gerday, 2003; D'Amico et al., 2006). In an unconventional twist, the cell-free proteases of *C. psychrerythraea* were stabilized against thermal stress when allowed to interact with the complex polymers of EPS produced by the same organism (Huston et al., 2004). Various enzyme–EPS interactions are expected in sea ice brines, but little is known of such interactions *in situ*. Extracellular enzyme activities in sea ice are known, however, for their unusually low-temperature optima (Huston et al., 2000; Deming, 2007).

The prevalence of EPS throughout Arctic and Antarctic sea ice is attributed to production by microalgae either before or after their entrainment into the ice (Krembs et al., 2002; Underwood et al., 2013). These gelatinous exopolymers alter the microstructure of the ice, increasing the habitable pore space (Krembs et al., 2011), and act as cryoprotection, maintaining a liquid environment for both eukaryotes and akaryotes and buffering against ice-crystal damage (Krembs et al., 2002; Aslam et al., 2012). The production of EPS by individual bacteria, however, also represents an important adaptation to surviving in ice, particularly in upper horizons of winter ice where ice algae are virtually absent, yet increases in EPS are observed (Krembs et al., 2002; Collins et al., 2008). These increases, best explained as bacterial EPS production, appeared to minimize the observed losses in bacterial numbers through the winter (from a 50% loss to an undetectable loss in some ice horizons; Collins et al., 2008). The bulk amount of bacterially produced EPS in winter sea ice may pale compared with EPS production by algae in spring ice, but the ability of an individual bacterium to modify its immediate surroundings with its own EPS may be critical to its survival. That sea ice bacteria in culture produce EPS, often in copious amounts, and in response to multiple stressors is well established (Mancuso Nichols et al., 2005; Marx et al., 2009). Like extracellular enzymes, the production and release of excessive amounts of EPS can be triggered by a drop in temperature (or salinity), as occurs for both *P. torquis* (Bowman, 2008) and *C. psychrerythraea* (Marx et al., 2009). In the latter case, the specific trigger appears not to be a function of dropping temperature, but the event of becoming encased in ice.

13.6.3 Mechanisms of adaptation

The vertical inheritance of beneficial gene mutations from parent cells is the mechanism traditionally invoked to explain acquisition of new traits that make an organism better adapted to (more competitive in) its environment. Much of what is known about cold adaptation in proteins, for example, has derived from studies using site-directed mutagenesis and other genetic approaches based on the vertical inheritance of modified genes. The alternative mechanism is to receive beneficial genes via horizontal transfer. This mechanism occurs in the laboratory via the uptake of free DNA from the medium (transformation), the transfer of DNA between physically connected bacteria (conjugation), and the viral introduction of bacterial DNA from a former host into the genome of its new host (transduction). In the latter case, if an environmental or biological trigger does not cause the virus to enter its lytic cycle (to the demise of the host), the viral DNA and bacterial gene(s) brought with it can pass benignly (vertically) to new generations of bacteria, with the virus eventually losing its capacity to become lytic again.

The advent of genomics has brought ample evidence that virally mediated gene transfer has been an important evolutionary mechanism for bacteria (Ochman et al., 2000). For sea ice bacteria, genomic evidence for gene transfer via transduction and other horizontal mechanisms is abundant and accumulating (Methé et al., 2005; Bowman, 2008; Collins & Deming, 2013; Vollmers et al., 2013; Feng et al., 2014; Møller et al., 2014). For example, the genome of *P. torquis* is rich in insertion elements, including transposases and their remnants, retroviral integrases and phage-like genes, all suggesting a strong influence of viruses in its evolution (Bowman, 2008; Feng et al., 2014). In particular, the genome locus for EPS production is flanked by sequences suggestive of an assembly process that involves a series of horizontal gene transfer events. Genes for putative ice-binding proteins and other ice-adaptive traits also appear to be the products of horizontal gene transfer.

Observed concentrations of free DNA and viruses, scaled to sea ice brine volume, indicate that these agents of lateral gene transfer surround bacteria encased in autumn and winter sea ice (Wells & Deming, 2006a; Collins & Deming, 2011a,b). Model calculations yield virus–bacteria and extracellular DNA–bacteria contact rates that are hundreds of times higher in these sea ice brines than in underlying seawater (Wells & Deming, 2006a). Where eukaryotes are also present, cross-domain gene transfers may be possible, as suggested by studies of ice-binding proteins (Raymond & Kim, 2012). In spite of its sometimes extreme thermal and saline conditions, the interior brine habitats of sea ice may provide dynamic settings for the evolution of ice adaptations (Figure 13.5) and for cold adaptation in general.

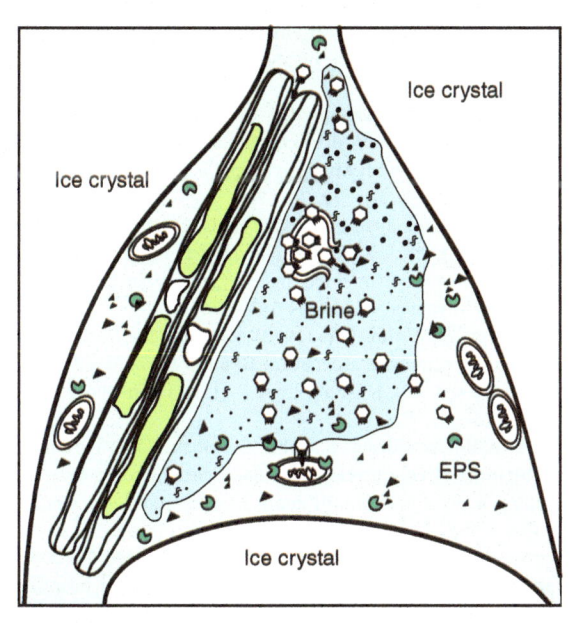

Figure 13.5 Idealized schematic of the contents of a brine inclusion at the juncture of three ice crystals in sea ice. Bacteria and a pair of eukaryotic algae (diatoms) are shown embedded in exopolymer gels (EPS) lining the pore space, with dissolved organic substrates (triangles) concentrated in the interior and extracellular enzymes (green) hydrolysing the substrates. Also shown are cold-active viral enzymes penetrating EPS and successful viral infection of, reproduction within and lysis of a host bacterium, releasing free DNA and new viruses, potential agents of horizontal gene transfer between bacteria or across domains (bacteria–diatom).

13.7 Frontiers

Much has been learned about sea ice Bacteria over the short history of their study; their abundance, distribution, dynamics, diversity and adaptations are reasonably well understood compared with the more recently discovered Archaea in sea ice. While virtually everything remains to be learned about Archaea in ice, the grounds for advancing knowledge of sea ice Bacteria are still fertile, and accompanied by a sense of urgency. The continued loss of sea ice in the Arctic, for example, makes the role of sea ice bacteria in various elemental and gas cycles important to understanding feedbacks to the climate system and inputs to the ecosystem. The climate-active gas, dimethylsulfide, usually considered produced in ice-free waters by planktonic microorganisms, was recently found in very high concentrations in sea ice, cycling at unexpected rates in a medium previously thought to be uninvolved in gas exchange between ocean and atmosphere (Asher et al., 2011; Chapter 18). The role of sea ice bacteria in fluxes of carbon dioxide to and from sea ice remains unclear, even as the fluxes themselves gain attention (Miller et al., 2011; Rysgaard et al., 2011). The biological fate of toxic mercury in polar regions may depend on bacterial processes only recently shown to operate in sea ice brines (Møller et al., 2011, 2014). Converging lines of evidence, from stable isotope and phylogenetic studies, suggest that the production of nitrate via nitrification is more important in sea ice than previously recognized (Barber et al., 2014; Fripiat et al., 2014, 2015). The increase in new ice formation in the changing Arctic Ocean may be providing a selective force for nitrifying bacteria (Barber et al., 2014).

The diminishing sea ice cover in the Arctic is opening the ocean to increased industrial and navigational activity, including offshore oil drilling. Though studies to date are limited, the heterotrophic bacteria of sea ice in both polar regions appear poised to confront the concentrated input of a non-algal source of organic compounds, in the form of an oil spill which could be expected to pool on or under sea ice. The bacterial community has been observed to shift, by both culturing and non-culturing methods, towards familiar representatives of the Gammaproteobacteria – species of *Marinobacter*, *Shewanella* and *Pseudomonas*, with the apparent potential to degrade hydrocarbons, although that remains to be verified (Gerdes et al., 2005). Analyses of the available genomes from cold-adapted bacteria, including from sea ice, reveal that genes involved in hydrolysis of alkane, a major component of oil spills, encode for cold-active enzymes (Bowman & Deming, 2014). *C. psychrerythraea*, one of the first microbial responders in the oil plume that developed in cold deep waters of the Gulf of Mexico following the Deepwater Horizon disaster (and the same species as found in polar sea ice), produces EPS that has oil-dispersant properties at low temperatures (Deming, unpublished). Much societally relevant research on sea ice bacteria clearly remains to be pursued.

The microbial habitation of the Earth's sea ice continues to inspire astrobiological pursuits, using sea ice as an analogous extreme environment or instrument test-bed (Wallace et al., 2015; Nadeau et al., 2016). Viruses in sea ice, however, can be considered the newest frontier. The genomic evidence of horizontal gene transfer in sea ice bacteria promises that the study of viruses, other agents of horizontal gene transfer, and more bacterial (and anticipated archaeal) genomes from sea ice will yield a better understanding of the evolution of cold-adapted proteins and enzymes, of fundamental or applied importance, and cold-adapted life in general. The freezing of polar waters annually, and over geological time, represents a significant planetary forcing for such evolution, given the astronomical numbers of Bacteria, Archaea and viruses that pass through this frozen gauntlet.

Endnotes

1 In this chapter, we use 'akaryotes' (microorganisms without nuclei) according to Penny et al. (2014), recognizing that 'prokaryotes' can be a misleading term when the direction of evolution between particular prokaryotes and eukaryotes remains ambiguous. We also use 'bacteria' as a general term for members of the akaryotic domains of Bacteria and Archaea when a more specific designation cannot be given.
2 Viruses that infect bacteria are also called 'bacteriophage' and are typically < 110 nm in diameter; larger size classes of viruses infect eukaryotic cells.

References

Ackley, S.F. & Sullivan, C.W. (1994) Physical controls on the development and characteristics of Antarctic sea ice biological communities – a review and synthesis. *Deep-Sea Research*, **41**, 1583–1604.

Arrigo, K.R., Brown, Z.W. & Mills, M.M. (2014) Sea ice algal biomass and physiology in the Amundsen Sea, Antarctica. *Elementa: Science of the Anthropocene*, **2**, 000028, doi: 10.12952/journal.elementa.000028

Asher, E.C., Dacey, J.W., Mills, M.M., Arrigo, K.R. & Tortell, P.D. (2011) High concentrations and turnover rates of DMS, DMSP and DMSO in Antarctic sea ice. *Geophysical Research Letters*, **38**, doi:10.1029/2011GL049712

Aslam, S.N., Cresswell-Maynard, T., Thomas, D.N. & Underwood, G.J. (2011) Production and characterization of the intra- and extracellular carbohydrates and polymeric substances (EPS) of three sea-ice diatom species, and evidence for a cryoprotective role for EPS. *Journal of Phycology*, **48**, 1494–1509.

Attwater, J., Wochner, A. & Holliger, P. (2013) In-ice evolution of RNA polymerase ribozyme activity. *Nature Chemistry*, **5**, 1011–1018.

Bano, N., Ruffin, S., Ransom, B. & Hollibaugh, J.T. (2004) Phylogenetic composition of Arctic Ocean archaeal assemblages and comparison with Antarctic assemblages. *Applied and Environmental Microbiology*, **70**, 781–789.

Barber, D.G., Ehn, J.K., Pućko, M. et al. (2014). Frost flowers on young Arctic sea ice: The climatic, chemical, and microbial significance of an emerging ice type. *Journal of Geophysical Research: Atmospheres*, **119**, 11 593–11 612.

Baross, J.A. & Morita, R.Y. (1978) Microbial life at low temperatures: ecological aspects. In: *Microbial Life in Extreme Environments* (Ed. D.J. Kushner), pp. 7–91. Academic Press, New York.

Bayer-Giraldi, M., Uhlig, C., John, U., Mock, T. & Valentin, K. (2010) Antifreeze proteins in polar sea ice diatoms: diversity and gene expression in the genus *Fragilariopsis*. *Environmental Microbiology*, **12**, 1041–1052.

Borriss, M., Helmke, E. Hanschke, R. & Schweder, T. (2003) Isolation and characterization of marine psychrophilic phage-host systems from Arctic sea ice. *Extremophiles*, **7**, 377–384.

Borriss, M., Lombardot, T., Glöckner, F.O., Becher, D., Albrecht, D. & Schweder, T. (2007) Genome and proteome characterization of the psychrophilic *Flavobacterium* bacteriophage 11b. *Extremophiles*, **11**, 95–104.

Bowman, J.P. (2008) Genomic analysis of psychrophilic prokaryotes. In: *Psychrophiles: from Biodiversity to Biotechnology* (Eds. R. Margesin, F. Schinner, J.-C. Marx & C. Gerday), pp. 265–284. Springer-Verlag, Berlin.

Bowman, J.P., McCammon, S.A., Brown, M.V., Nichols, D.S. & McMeekin, T.A. (1997) Diversity and association of psychrophilic bacteria in Antarctic sea ice. *Applied and Environmental Microbiology*, **63**, 3068–3078.

Bowman, J.S. (2014) *Life in the Cold Biosphere: The Ecology of Psychrophile Communities, Genomes, and Genes*, Doctoral dissertation, University of Washington, Seattle.

Bowman, J.S. (2015). The relationship between sea ice bacterial community structure and biogeochemistry: A synthesis of current knowledge and known unknowns. *Elementa: Science of the Anthropocene*, **3**, 000072.

Bowman, J.S. & Deming, J.W. (2010) Elevated bacterial abundance and exopolymers in saline frost flowers and implications for atmospheric chemistry and microbial dispersal. *Geophysical Research Letters*, **37**, L13501, doi:10.1029/2010GL043020

Bowman, J.S. & Deming, J.W. (2014) Alkane hydroxylase genes in psychrophile genomes and the potential for cold active catalysis. *BMC Genomics* **15**, 1120.

Bowman, J.S., Rasmussen, S., Blom, N., Deming, J.W., Rysgaard, S. & Sicheritz-Ponten, T. (2012) Microbial community structure of Arctic multiyear sea ice and surface seawater by 454 sequencing of the 16S RNA gene. *The ISME Journal*, **6**, 11–20.

Bowman, J.S., Larose, C., Vogel, T.M. & Deming, J.W. (2013) Selective occurrence of Rhizobiales in frost flowers on the surface of young sea ice near Barrow, Alaska and distribution in the polar marine rare biosphere. *Environmental Microbiology Reports*, **5**, 575–582.

Bowman, J.S., Berthiaume, C.T., Armbrust, E.V. & Deming, J.W. (2014) The genetic potential for key biogeochemical processes in Arctic frost flowers and young sea ice revealed by metagenomic analysis. *FEMS Microbiology Ecology*, **89**, 376–387.

Brinkmeyer, R., Knittel, K., Jugens, J., Weyland, H., Amann, R. & Helmke, E. (2003) Diversity and structure of bacterial communities in Arctic versus Antarctic pack ice. *Applied and Environmental Microbiology*, **69**, 6610–6619.

Brinkmeyer, R., Glockner, F.-O. Helmke, E. & Amann, R. (2004) Predominance of β-proteobacteria in summer melt pools on Arctic pack ice. *Limnology and Oceanography*, **49**, 1013–1021.

Brown, M. & Bowman J.P. (2001) A molecular phylogenetic survey of sea-ice microbial communities (SIMCO). *FEMS Microbiology Ecology*, **35**, 267–275.

Carillo, S., Casillo, A., Pieretti, G. et al. (2015) A unique capsular polysaccharide structure from the psychrophilic marine bacterium *Colwellia psychrerythraea* 34H that mimicks antifreeze (glyco) proteins. *Journal of the American Chemical Society*, **137**, 179–189.

Cavicchioli, R., Charlton, T., Ertan, H., Mohd Omar, S., Siddiqui, K.S. & Williams, T.J. (2011) Biotechnological uses of enzymes from psychrophiles. *Microbial Biotechnology*, **4**, 449–460.

Colangelo-Lillis, J.R. & Deming, J.W. (2013) Genomic analysis of cold-active *Colwelliaphage* 9A and psychrophilic phage-host interactions. *Extremophiles*, **17**, 99–114.

Collins, R.E. & Deming, J.W. (2011a). Abundant dissolved genetic material in Arctic sea ice Part I: Extracellular DNA. *Polar Biology*, **34**, 1819–1830.

Collins, R.E. & Deming, J.W. (2011b). Abundant dissolved genetic material in Arctic sea ice Part II: Viral dynamics during autumn freeze-up. *Polar Biology*, **34**, 1831–1841.

Collins, R.E. & Deming, J.W. (2013) Identification of an inter-Order lateral gene transfer event enabling the catabolism of compatible solutes by *Colwellia* spp. *Extremophiles*, **17**, 601–610.

Collins, R.E., Carpenter S.D. & Deming J.W. (2008) Spatial heterogeneity and temporal dynamics of particles, bacteria, and pEPS in Arctic winter sea ice. *Journal of Marine Systems*, **74**, 902–917.

Collins, R.E., Rocap, G. & Deming, J.W. (2010) Persistence of bacterial and archaeal communities in sea ice through an Arctic winter. *Environmental Microbiology* **12**, 1828–1841.

Cowie, R.O., Maas, E.W. & Ryan, K.G. (2011) Archaeal diversity revealed in Antarctic sea ice. *Antarctic Science-Institutional Subscription*, **23**, 531–536.

D'Amico, S., Collins, T., Marx, J.-C., Feller, G. & Gerday, C. (2006) Psychrophilic microorganisms: challenges for life. *EMBO Reports*, **7**, 385–389.

Delille, D. (1992) Marine bacterioplankton at the Weddell Sea ice edge, distribution of psychrophilic and psychrotrophic populations. *Polar Biology*, **12**, 205–210.

Delille, D., Fiala, M. & Rosiers, C. (1995) Seasonal changes in phytoplankton and bacterioplankton distribution at the ice-water interface in the Antarctic neritic area. *Marine Ecology Progress Series*, **123**, 225–233.

DeLong, E.F., Wu, K.Y., Prézelin, B.B. & Jovine, R.V. (1994) High abundance of Archaea in Antarctic marine picoplankton. *Nature*, **371**, 695–697.

De Maayer, P., Anderson, D., Cary, C. & Cowan, D.A. (2014) Some like it cold: understanding the survival strategies of psychrophiles. *EMBO Reports*, **15**, 508–517.

Deming, J.W. (2007) Life in ice formations at very cold temperatures. In: *Physiology and Biochemistry of Extremophiles* (Eds. C. Gerday & N. Glansdorff), pp. 133–145. ASM Press, Washington, DC.

Deming, J.W. (2010) Sea ice bacteria and viruses. In: *Sea Ice – An Introduction to its Physics, Chemistry, Biology and Geology*, Second Edition (Eds. D.N. Thomas & G.S. Dieckmann), pp. 247–282. Blackwell Science Ltd, Oxford.

Deming, J.W. & Eicken, H. (2007) Life in ice. In: *Planets and Life: The Emerging Science of Astrobiology* (Eds. W.T. Sullivan & J.A. Baross), pp. 292–312. Cambridge University Press.

Diez, B., Bergman, B., Pedrós-Alió, C., Anto, M. & Snoeijs, P. (2012) High cyanobacterial nifH gene diversity in Arctic seawater and sea ice brine. *Environmental Microbiology Reports*, **4**, 360–366.

Dolev, M.B., Bernheim, R., Guo, S., Davies, P.L., & Braslavsky, I. (2016) Putting life on ice: bacteria that bind to frozen water. *Journal of the Royal Society Interface*, 20160210, http://dx.doi.org/10.1098/rsif.2016.0210.

Eronen-Rasimus, E., Kaartokallio, H., Lyra, C., Autio, R., Kuosa, H., Dieckmann, G.S. & Thomas, D.N. (2014) Bacterial community dynamics and activity in relation to dissolved organic matter availability during sea-ice formation in a mesocosm experiment. *MicrobiologyOpen*, **3**, 139–156.

Ewert, M. (2013). *Microbial Challenges and Solutions to Inhabiting the Dynamic Architecture of Saline Ice Formations*. Doctoral dissertation, University of Washington, Seattle.

Ewert, M. & Deming, J.W. (2011) Selective retention in saline ice of extracellular polysaccharides produced by the cold-adapted marine bacterium *Colwellia psychrerythraea* strain 34H. *Annals of Glaciology*, **52**, 111–117.

Ewert, M. & Deming, J.W. (2014) Bacterial responses to fluctuations and extremes in temperature and brine salinity at the surface of Arctic winter sea ice. *FEMS Microbiology Ecology* **89**, 476–489.

Ewert, M. Carpenter, S.D., Colangelo-Lillis, J. & Deming, J.W. (2013) Bacterial and extracellular polysaccharide content of brine-wetted snow over Arctic winter first-year sea ice. *Journal of Geophysical Research Oceans*, **118**, 726–735.

Feller, G. & Gerday, C. (2003) Psychrophilic enzymes: Hot topics in cold adaptation. *Nature Reviews*, **1**, 200–208.

Feng, S., Powell, S.M., Wilson, R. & Bowman, J.P. (2013) Light-stimulated growth of proteorhodopsin-bearing sea-ice psychrophile *Psychroflexus torquis* is salinity dependent. *The ISME Journal*, **7**, 2206–2213.

Feng, S., Powell, S.M., Wilson, R. & Bowman, J.P. (2014) Extensive gene acquisition in the extremely psychrophilic bacterial species *Psychroflexus torquis* and the link to sea-ice ecosystem specialism. *Genome Biology and Evolution*, **6**, 133–148.

Firth, E., Carpenter, S.D., Jorgensen, H., Collins, R.E. & Deming, J.W. (2016) Bacterial use of choline to tolerate salinity shifts in sea ice brines. *Elementa: Science of the Anthropocene* 4, 000120, doi:10.12952/journal.elementa.000120.

Fripiat, F., Sigman, D.M., Fawcett, S.E., Rafter, P.A., Weigand, M.A. & Tison, J.L. (2014) New insights into sea ice nitrogen biogeochemical dynamics from the nitrogen isotopes. *Global Biogeochemical Cycles*, **28**, 115–130.

Fripiat, F., Sigman, D.M., Massé, G. & Tison, J.L. (2015) High turnover rates indicated by changes in the fixed N forms and their stable isotopes in Antarctic landfast sea ice. *Journal of Geophysical Research: Oceans*, **120**, 3079–3097.

Garrison, D.L. & Buck, K.R. (1986) Organism losses during ice melting: a serious bias in sea ice community studies. *Polar Biology*, **6**, 237–239.

Garrison, D.L., Ackley, S.F. & Buck, K.R. (1983) A physical mechanism for establishing algal populations in frazil ice. *Nature*, **306**, 363–365.

Garrison, D.L., Sullivan, C.W. & Ackley, S.F. (1986) Sea ice microbial communities in Antarctica. *BioScience*, **36**, 243–250.

Gerdes, B., Brinkmeyer, R., Dieckmann, G. & Helmke, E. (2005) Influence of crude oil on changes of bacterial communities in Arctic sea-ice. *FEMS Microbiology Ecology*, **53**, 129–139,

Gleitz, M., Grossmann, S., Scharek, R. & Smetacek, V. (1996) Ecology of diatom and bacterial assemblages in water associated with melting summer sea ice in the Weddell Sea, Antarctica. *Antarctic Science*, **8**, 135–146.

Golden, K.M., Ackley, S.F. & Lytle, V.I. (1998) The percolation phase transition in sea ice. *Science*, **282**, 2238–2241.

Gosink, J.J., Herwig, R.P. & Staley, J.T. (1997) *Octadecabacter arcticus* gen. nov., and *O. antarcticus*, sp. nov., non-pigmented, psychrophilic gas vacuolate bacteria form polar sea ice and water. *Systematic and Applied Mcirobiology*, **20**, 356–365.

Gosink J.J., Woese, C.R. & Staley, J.T. (1998) *Polaribacter* gen. nov., with three new species, *P. irgensii*, sp. nov., *P. franzmannii* sp. nov., and *P. filamentus*, sp. nov., gas vacuolate polar marine bacteria of the Cytophaga/Flavobacterium/Bacteroides Group and reclassification of '*Flectobacillus glomeratus*' as *Polaribacter glomeratus*. *International Journal of Systematic Bacteriology*, **48**, 223–235.

Gowing, M.M. (2003) Large viruses and infected microeukaryotes in Ross Sea summer pack ice habitats. *Marine Biology*, **142**, 1029–1040.

Gowing, M.M., Riggs, B.E., Garrison, D.L., Gibson, A.H. & Jeffries, M.O. (2002) Large viruses in Ross Sea late autumn pack ice Habitats. *Marine Ecology Progress Series*, **241**, 1–11.

Gowing, M.M., Garrison, D.L., Gibson, A.H., Krupp, J.M., Jeffries, M.O. & Fritsen, C.H. (2004) Bacterial and viral abundance in Ross Sea summer pack ice communities. *Marine Ecology Progress Series*, **279**, 3–12.

Gradinger, R. & Ikävalko, J. (1998) Organism incorporation into newly forming Arctic sea ice in the Greenland Sea. *Journal of Plankton Research*, **20**, 871–886.

Gradinger, R. & Zhang, Q. (1997) Vertical distribution of bacteria in Arctic sea ice from the Barents and Laptev Seas. *Polar Biology*, **17**, 448–454.

Gradinger, R., Friedrich, C. & Spindler, M. (1999) Abundance, biomass and composition of the sea ice biota of the Greenland Sea pack ice. *Deep-Sea Research Part II*, **46**, 1457–1472.

Griffiths, R.P., Hayasaka, S.S., McNamara, T.M. & Morita, R.Y. (1978) Relative microbial activity and bacterial concentrations in water and sediment samples taken in the Beaufort Sea. *Canadian Journal of Microbiology*, **24**, 1217–1226.

Grossi, S.M., Kottmeier, S.T. & Sullivan, C.W. (1984) Sea ice microbial communities. III. Seasonal abundance of microalgae and associated bacteria, McMurdo Sound, Antarctica. *Microbial Ecology*, **10**, 231–242.

Grossmann, S. (1994) Bacterial activity in sea ice and open water of the Weddell Sea, Antarctica: a microautoradiographic study. *Microbial Ecology*, **28**, 1–18.

Grossmann, S. & Dieckmann, G.S. (1994) Bacterial standing stock, activity, and carbon production during formation and growth of sea ice in the Weddell Sea, Antarctica. *Applied and Environmental Microbiology*, **60**, 2746–2753.

Hatam, I., Charchuk, R., Lange, B., Beckers, J., Haas, C. & Lanoil, B. (2014) Distinct bacterial assemblages reside at different depths in Arctic multiyear sea ice. *FEMS Microbiology Ecology*, **90**, 115–125.

Helmke, E. & Weyland, H. (1995) Bacteria in sea ice and underlying water of the Eastern Weddell Sea in midwinter. *Marine Ecology Progress Series*, **117**, 269–287.

Helmke, E. & Weyland, H. (2004) Psychrophilic versus psychrotolerant bacteria – occurrence and significance in polar and temperate marine habitats. *Cellular and Molecular Biology*, **50**, 553–561.

Horner, R. (1985) Ecology of sea ice microalgae. In: *Sea Ice Biota* (Ed. R.A. Horner), pp. 83–103. CRC Press, Boca Raton, FL.

Horner, R. & Alexander, V. (1972) Algal populations in arctic sea ice: an investigation of heterotrophy. *Limnology and Oceanography*, **17**, 454–458.

Huston, A.L., Krieger-Brockett, B.B. & Deming, J.W. (2000) Remarkably low temperature optima for extracellular enzyme activity from Arctic bacteria and sea ice. *Environmental Microbiology*, **2**, 383–388.

Huston, A.L., Methé, B. & Deming, J.W. (2004) Purification, characterization and sequencing of an extracellular cold-active aminopeptidase produced by marine psychrophile *Colwellia psychrerythraea* strain 34H. *Applied and Environmental Microbiology*, **70**, 3321–3328.

Junge, K., Eicken, H. & Deming, J.W. (2003) Motility of *Colwellia psychrerythraea* strain 34H at subzero temperatures. *Applied and Environmental Microbiology*, **69**, 4282–4284.

Junge, K., Eicken, H. & Deming, J.W. (2004) Bacterial activity at −2 to −20°C in Arctic wintertime sea ice. *Applied and Environmental Microbiology*, **70**, 550–557.

Junge, K., Gosink, J.J., Hoppe, H.G. & Staley, J.T. (1998) *Arthrobacter*, *Brachybacterium* and *Planococcus* isolates identified from Antarctic sea ice brine – description of *Planococcus mcmeekinii*. *Systematic and Applied Microbiology*, **21**, 306–314.

Junge, K., Imhoff, J.F., Staley, J.T. & Deming, J.W. (2002) Phylogenetic diversity of numerically important bacteria in Arctic sea ice. *Microbial Ecology*, **43**, 315–328.

Junge, K., Krembs, C., Deming, J., Stierle, A. & Eicken, H. (2001) A microscopic approach to investigate bacteria under in-situ conditions in sea-ice samples. *Annals of Glaciology*, **33**, 304–310.

Kaartokallio, H., Tuomainen, J., Kuosa, H., Kuparinen, J., Martikainen, P. & Servomaa, K. (2008) Succession of sea-ice bacterial communities in the Baltic Sea fast ice. *Polar Biology*, **31**, 783–793.

Kanavarioti, A., Monnard, P.-A. & Deamer, D.W. (2001) Eutectic phases in ice facilitate nonenzymatic nucleic acid synthesis. *Astrobiology*, **1**, 271–281.

Kaneko, T., Roubal, G. & Atlas, R.M. (1978) Bacterial populations in the Beaufort Sea. *Arctic*, **31**, 97–107.

Kiko, R. (2010) Acquisition of freeze protection in a sea-ice crustacean through horizontal gene transfer?. *Polar Biology*, **33**, 543–556.

Koh, E.Y., Atamna-Ismaeel, N., Martin, A. et al. (2010) Proteorhodopsin-bearing bacteria in Antarctic sea ice. *Applied and Environmental Microbiology*, **76**, 5918–5925.

Kottmeier, S.T. & Sullivan, C.W. (1987) Late winter primary production and bacterial production in sea ice and seawater west of the Antarctic Peninsula. *Marine Ecology Progress Series*, **36**, 287–298.

Kottmeier, S.T. & Sullivan, C.W. (1990) Bacterial biomass and production in pack ice of Antarctic marginal ice edge zones. *Deep-Sea Research*, **37**, 1311–1330.

Kottmeier, S.T., McGrath-Grossi, S. & Sullivan, C.W. (1987) Sea ice microbial communities. VIII. Bacterial production in annual sea ice of McMurdo Sound, Antarctica. *Marine Ecology Progress Series*, **35**, 175–186.

Krembs, C. & Engel, A. (2001) Abundance and variability of microorganisms and transparent exopolymer particles across the ice–water interface of melting first-year sea ice in the Laptev Sea (Arctic). *Marine Biology*, **138**, 173–185.

Krembs, C., Eicken, H. & Deming, J.W. (2011). Exopolymer alteration of physical properties of sea ice and implications for ice habitability and biogeochemistry in a warmer Arctic. *Proceedings of the US National Academy of Sciences*, **108**, 3653–3658.

Krembs, C., Gradinger, R. & Spindler, M. (2000) Implications of brine channel geometry and surface area for the interaction of sympagic organisms in Arctic sea ice. *Journal of Experimental Marine Biology and Ecology*, **243**, 55–80.

Krembs, C., Deming, J.W. & Eicken, H. (2008) Effects of exopolymeric substances on sea-ice microstructure and salt retention. *Marine Ecology Progress Series (submitted)*.

Krembs, C., Deming, J.W., Junge, K. & Eicken, H. (2002) High concentrations of exopolymeric substances in wintertime sea ice: Implications for the polar ocean carbon cycle and cryoprotection of diatoms. *Deep-Sea Research Part I*, **49**, 2163–2181.

Mader, H.M., Pettitt, M.E., Wadham, J.L., Wolff, E.W. & Parkes, R.J. (2006) Subsurface ice as a microbial habitat. *Geology*, **34**, 169–172.

Mancuso Nichols, C., Guezennec, J. & Bowman, J.P. (2005) Bacterial exopolysaccharides from extreme marine environments with special consideration of the Southern Ocean, sea ice, and deep-sea hydrothermal vents: a review. *Marine Biotechnology*, **7**, 253–271.

Maranger, R., Bird, D.F. & Juniper, K. (1994) Viral and bacterial dynamics in Arctic sea ice during the spring algal bloom near Resolute, N.W.T., Canada. *Marine Ecology Progress Series*, **111**, 121–127.

Marx, J.G., Carpenter, S.D. & Deming, J.W. (2009) Production of cryoprotectant extracellular polysaccharide substances (EPS) by the marine psychrophilic bacterium *Colwellia psychrerythraea* strain 34H under extreme conditions. *Canadian Journal of Microbiology*, **55**, 63–72.

Meiners, K., Krembs, C. & Gradinger, R. (2008) Exopolymer particles: microbial hotspots of enhanced bacterial activity in Arctic fast ice (Chukchi Sea). *Aquatic Microbial Ecology*, **52**, 195–207.

Menor-Salván, C. & Marín-Yaseli, M.R. (2012) Prebiotic chemistry in eutectic solutions at the water–ice matrix. *Chemical Society Reviews*, **41**, 5404–5415.

Methé, B.A., Nelson, K.E., Deming, J.W. et al. (2005) The psychrophilic lifestyle as revealed by the genome sequence of *Colwellia psychrerythraea* 34H through genomic and proteomic analyses. *Proceedings of the National Academy of Sciences USA*, **102**, 10913–10918.

Miller, L.A., Papakyriakou, T.N., Collins, R.E. et al. (2011) Carbon dynamics in sea ice: A winter flux time series. *Journal of Geophysical Research: Oceans (1978–2012)*, **116**, C02028, doi:10.1029/2009JC006058

Mock, T. & Gradinger, R. (1999) Determination of Arctic ice algal production with a new in situ incubation technique. *Marine Ecology Progress Series*, **177**, 15–26

Mock, T., Dieckmann, G.S., Haas, C. et al. (2002) Micro-optodes in sea ice: a new approach to investigate oxygen dynamics during sea ice formation. *Aquatic Microbial Ecology*, **29**, 297–306

Mohr, J.L. & Tibbs, J. (1963) Ecology of ice substrates. *Proceedings of the Arctic Basin Symposium October 1962*, pp. 245–248. The Arctic Institute of North America, Tidewater Publishing, Centreville, MD.

Møller, A.K., Barkay, T., Al-Soud, W.A., Sørensen, S.J., Skov, H. & Kroer, N. (2011) Diversity and characterization of mercury-resistant bacteria in snow, freshwater and sea-ice brine from the High Arctic. *FEMS Microbiology Ecology*, **75**, 390–401.

Møller, A.K., Barkay, T., Hansen, M.A. et al. (2014) Mercuric reductase genes (merA) and mercury resistance plasmids in High Arctic snow, freshwater and sea-ice brine. *FEMS Microbiology Ecology*, **87**, 52–63.

Morita, R.Y. (1975) Psychrophilic bacteria. *Bacteriological Reviews*, **39**, 144–167.

Nadeau, J., Lindensmith, C., Deming, J.W., Fernandez, V.I. & Stocker, R. (2015) Bacterial morphology and motility as biosignatures for outer planet missions. *Astrobiology Journal* doi: 10.1089/ast.2015.1376.

Nguyen, D. & Maranger, R. (2011) Respiration and bacterial carbon dynamics in Arctic sea ice. *Polar Biology*, **34**, 1843–1855.

Nichols, D., Bowman, J.P., Sanderson, K. et al. (1999) Developments with Antarctic microorganisms: culture collections, bioactivity screening, taxonomy, PUFA production and cold-adapted enzymes. *Current Opinion in Biotechnology*, **10**, 240–246.

Niemi, A., Michel, C., Hille, K. & Poulin, M. (2011) Protist assemblages in winter sea ice: setting the stage for the spring ice algal bloom. *Polar Biology*, **34**, 1803–1817.

Ochman, H., Lawrence, J.G. & Grolsman, E.A. (2000) Lateral gene transfer and the nature of bacterial innovation. *Nature*, **405**, 299–304.

Ortmann, A.C. & Suttle, C.A. (2005) High abundances of viruses in a deep-sea hydrothermal vent system indicates viral mediated microbial mortality. *Deep-Sea Research Part I*, **52**, 1515–1527.

Piwosz, K., Wiktor, J.M., Niemi, A., Tatarek, A. & Michel, C. (2013) Mesoscale distribution and functional diversity of picoeukaryotes in the first-year sea ice of the Canadian Arctic. *The ISME Journal*, **7**, 1461–1471.

Raymond, J.A., Fritsen, C. & Shen, K. (2007) An ice-binding protein from an Antarctic sea ice bacterium. *FEMS Microbiology Ecology*, **61**, 214–221.

Riedel, A., Michel, C. & Gosselin, M. (2006) Seasonal study of sea-ice exopolymeric substances on the Mackenzie shelf: implications for transport of sea-ice bacteria and algae. *Aquatic Microbial Ecology*, **45**, 195–206.

Riedel, A., Michel, C., Gosselin, M. & LeBlanc, B. (2007) Enrichment of nutrients, exopolymeric substances and microorganisms in newly formed sea ice on the Mackenzie shelf. *Marine Ecology Progress Series*, **342**, 55–67.

Różańska, M., Poulin, M. & Gosselin, M. (2008) Protist entrapment in newly formed sea ice in the Coastal Arctic Ocean. *Journal of Marine Systems*, **74**, 887–901.

Rysgaard, S. & Glud, R.N. (2004) Anaerobic N_2 production in Arctic sea ice. *Limnology and Oceanography*, **49**, 86–94.

Rysgaard, S., Kuhl, M., Glud, R.N. & Hansen, J.W. (2001) Biomass, production and horizontal patchiness of sea ice algae in a high-Arctic fjord (Young Sound, NE Greenland). *Marine Ecology Progress Series*, **223**, 15–26.

Rysgaard, S. Glud, R.N., Sejr, M.K., Blicher, M.E. & Stahl, H.J. (2008) Denitrification activity and oxygen dynamics in Arctic sea ice. *Polar Biology*, **31**, 527–537.

Rysgaard, S., Bendtsen, J., Delille, B. et al. (2011) Sea ice contribution to the air–sea CO_2 exchange in the Arctic and Southern Oceans. *Tellus B*, **63**, 823–830.

Senčilo, A., Luhtanen, A.-M., Saarijärvi, M., Bamford, D.H. and Roine, E. (2014) Cold-active bacteriophages from the Baltic Sea ice have diverse genomes and virus–host interactions. *Environmental Microbiology*, doi: 10.1111/1462-2920.12611

Smith, R.E.H. & Clement, P. (1990) Heterotrophic activity and bacterial productivity in assemblages of microbes from sea ice in the high Arctic. *Polar Biology*, **10**, 351–357.

Søgaard, D.H., Kristensen, M., Rysgaard, S., Glud, R.N., Hansen, P.J. & Hilligsøe, K.M. (2010) Autotrophic and heterotrophic activity in Arctic first-year sea ice: seasonal study from Malene Bight, SW Greenland. *Marine Ecology Progress Series*, **419**, 31–45.

Staley, J.T. & Gosink, J.J. (1999) Poles apart: biodiversity and biogeography of sea ice bacteria. *Annual Reviews in Microbiology*, **53**, 189–215.

Stewart, F.J. & Fritsen, C.H. (2004) Bacteria-algal relationships in Antarctic sea ice. *Antarctic Science*, **16**, 143–156.

Stierle, A.P. & Eicken, H. (2002) Sediment inclusions in Alaskan coastal sea ice: spatial distribution, interannual variability, and entrainment requirements. *Arctic Antarctic and Alpine Research*, **34**, 465–476.

Sullivan, C.W. & Palmisano, A.C. (1981) Sea ice microbial communities in McMurdo Sound, Antarctica. *Antarctic Journal US*, **16**, 126-127.

Sullivan, C.W. & Palmisano, A.C. (1984) Sea ice microbial communities: distribution, abundance, and diversity of ice bacteria in McMurdo Sound, Antarctica in 1980. *Applied Environmental Microbiology*, **47**, 788–795.

Sullivan, C.W., Palmisano, A.C., Kottmeier, S. & Moe, R. (1982) Development of the sea ice microbial community (SIMCO) in McMurdo Sound, Antarctica. *Antarctic Journal US*, **17**, 155-157.

Thomas, D.N., Engbrodt, R., Giannelli, V. et al. (2001) Dissolved organic matter in Antarctic sea ice. *Annals of Glaciology*, **33**, 297–303.

Torstensson, A., Dinasquet, J., Chierci, M., Fransson, A., Riemann, L. & Wulff, A. (2015) Physiochemical control of bacterial and protist community composition and diversity in Antarctic sea ice. *Environmental Microbiology*, doi:10.1111/1462-2920.12865.

Underwood, G.J., Aslam, S.N., Michel, C. et al. (2013) Broad-scale predictability of carbohydrates and exopolymers in Antarctic and Arctic sea ice. *Proceedings of the National Academy of Sciences*, **110**, 15734–15739.

Vajda, T. & Hollosi, M. (2001) Cryo-bioorganic chemistry: freezing effect on stereoselection of L- and DL-leucine cooligomerization in aqueous solution. *Cellular & Molecular Life Science*, **58**, 343–346.

Vollmers, J., Voget, S., Dietrich, S., Gollnow, K. et al. (2013) Poles apart: Arctic and Antarctic *Octadecabacter* strains share high genome plasticity and a new type of xanthorhodopsin, *PLoS ONE*, **8**, e63422, doi:10.1371/journal.pone.0063422

Wallace, J.K., Deming, J.W., Rider, S. et al. (2015) A simple, robust common-mode implementation of a digital holographic microscope. *Optics Express* **23**, doi:10.1364/OE.23.0173675.

Wells, L.E. & Deming, J.W. (2006a) Modelled and measured dynamics of viruses in Arctic winter sea-ice brines. *Environmental Microbiology*, **8**, 1115–1121.

Wells, L.E. & Deming, J.W. (2006b) Characterization of a cold-active bacteriophage on two psychrophilic marine hosts. *Aquatic Microbial Ecology*, **45**, 15–29.

Wells, L.E. & Deming, J.W. (2006c) Effects of temperature, salinity and clay particles on inactivation and decay of cold-active marine Bacteriophage 9A. *Aquatic Microbial Ecology*, **45**, 31–39.

Ziegler, C., Bremer, E. & Krämer, R. (2010) The BCCT family of carriers: from physiology to crystal structure. *Molecular Microbiology*, **78**, 13–34.

CHAPTER 14

Sea ice as a habitat for primary producers

Kevin R. Arrigo

Department of Earth System Science, Stanford University, Stanford, CA, USA

14.1 Introduction

The extensive pack and fast ice that forms in both Antarctic and Arctic regions provides a unique habitat for polar microbial assemblages (see the recent review by Arrigo 2014). Algal communities, in particular, are known to flourish within the distinct microhabitats that are created when sea ice forms and ages. The primary advantage afforded by sea ice is that it provides a platform from which sea ice algae can remain suspended in the upper ocean, where light and nutrients are sufficient for net growth. These autotrophic organisms play a critical role in polar marine ecology in both the Arctic and the Antarctic. For example, although rates of primary production (photosynthetic carbon assimilation) by sea ice algae are generally low compared with their phytoplankton counterparts, they are often virtually the sole source of fixed carbon for higher trophic levels in ice-covered waters. In fact, sea ice algae have been shown to sustain a wide variety of organisms (Chapters 15 and 16), including krill, through the winter months when other sources of food are lacking.

14.2 Sea ice as a habitat

In summer, the Arctic is covered by approximately 4.5–9.0 million km^2 of persistent multi-year ice, compared with only 3.5 million km^2 of multi-year ice in the Antarctic. During the winter, the Arctic adds 7–12 million km^2 of annual ice, bringing the total sea ice extent to approximately 16 million km^2. The Antarctic, in contrast, produces more than twice as much first-year ice as the Arctic, averaging 15.5 million km^2 year^{-1} of new ice, and has a greater total sea ice extent of 19 million km^2. Taken together, sea ice in the Arctic and Antarctic comprises one of the largest ecosystems on Earth.

Sea ice microalgae flourish in those regions of the ice floe that are most tightly coupled to the underlying seawater. This is because one of the primary factors controlling the growth of algae in sea ice is access to nutrients (Chapter 17). Except for those areas where snow cover is extremely thick, light is usually sufficient for net photosynthesis during the polar spring and summer (Chapter 4). Sea ice habitats are often characterized by steep gradients in temperature, salinity, light and nutrient concentration. The greatest fraction of sea ice microalgae often reside in the bottom 20 cm of the ice sheet where environmental conditions are generally stable and more favourable for growth (Figure 14.1). Bottom ice communities form in the skeletal layer and extend upwards as far as 0.2 m, their upward distribution generally being limited by nutrient availability and high brine salinity characteristic of the sea ice interior when temperatures are low (Arrigo & Sullivan 1992).

Under certain conditions, microalgae may also be found in internal layers (Figure 14.1), where they are often subjected to large environmental fluctuations (Lizotte & Sullivan, 1991). For example, brines with salinities as high as 173 psu and temperatures as low as −16°C have been collected from the upper 1.0–1.5 m of the sea ice in McMurdo Sound (Kottmeier & Sullivan 1988), high enough to prevent detectable metabolic activity in most sea ice algae (Grant & Horner, 1976; Arrigo & Sullivan, 1992). Internal communities are

---

Sea Ice, Third Edition. Edited by David N. Thomas.
© 2017 John Wiley & Sons, Ltd. Published 2017 by John Wiley & Sons, Ltd.

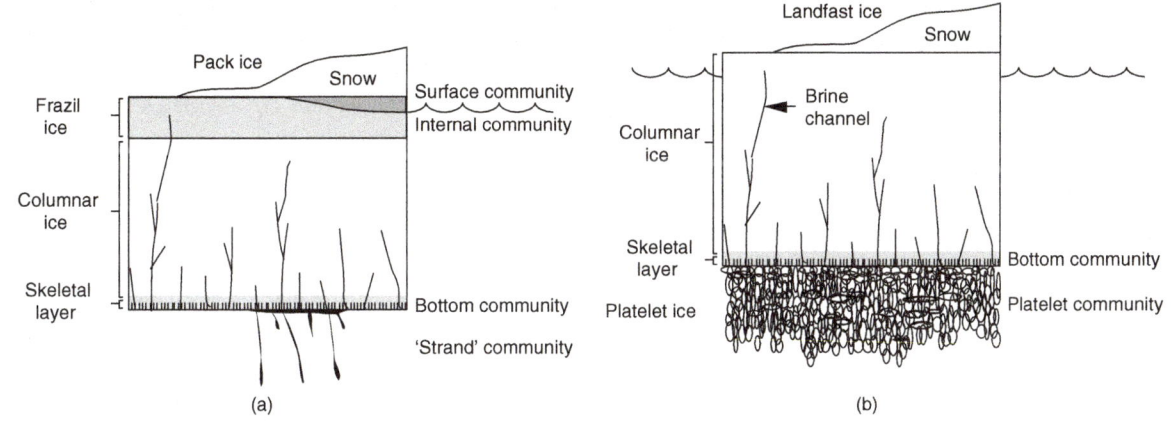

Figure 14.1 Highly idealized schematic illustration of pack ice (a) and landfast ice ecosystems (b) in the Arctic and Antarctic, showing the location of the major ice algal communities. Obviously, whether a given community is found in a particular location is highly variable and dependent on the sea ice structure and formation history.

generally associated with a frazil ice layer or can be found at the freeboard level where seawater can infiltrate the ice floe (Kattner et al., 2004). These communities are seeded by the particles scavenged during frazil ice formation and are especially dependent upon nutrient availability and salinity. They are more common in the Antarctic than the Arctic.

Although the surface and near-surface communities often have adequate light for growth, the availability of nutrients for these communities is often restricted (Figure 14.1). Measurements of salinity in the cavity of the freeboard layer indicate that this layer is infiltrated by seawater that provides the nutrients required by the algal community (Kattner et al., 2004). However, this supply of nutrients depends on the porosity of the surrounding ice. For the algal communities growing at the snow–ice interface, surface flooding, caused by snow loading and submersion of the ice pack, is an important source of nutrients (Saenz & Arrigo, 2012). Such surface flooding occurs over 15–30% of the ice pack in Antarctica (Wadhams et al., 1987). Although snow cover has the negative effect of reducing the amount of light available for algal growth, it is also responsible for providing nutrients to the surface community (Arrigo et al., 1997).

Less common sea ice assemblages include those that grow in a 'strand' layer just beneath the sea ice (Figure 14.1a), and within a sub-ice platelet layer (Figure 14.1b). Mat and strand communities, where algae are loosely attached to the underside of the sea ice but extend well into the water column, are found mainly in Arctic regions (Johnsen & Hegseth, 1991). Conversely, platelet ice, a semi-consolidated layer of ice ranging from a few centimetres to several metres in thickness, is commonly observed beneath sea ice in regions adjacent to floating ice shelves in the Antarctic where 45% of the continental margin is associated with an ice shelf (Bindschadler, 1990; Kipfstuhl, 1991). Platelet ice is the most porous of all sea ice types, being composed of approximately 20% ice and 80% seawater by volume, and harbours some of the largest concentrations of sea ice algae found anywhere on Earth (Arrigo et al., 1995).

14.3 Algae inhabiting sea ice

14.3.1 Biodiversity of photosynthetic organisms

The organisms responsible for photosynthetic production in sea ice are nearly always diatoms, though blooms of other algae have been observed. Extensive reviews of the dominant species were assembled by Horner (1985) and Garrison (1991). These reviews and more recent studies imply that diatoms make up most (>90%) of the photosynthetic organism diversity, which probably exceeds 500 species. However, improved methods in taxonomy have increased recognition of flagellate species (Gast et al., 2006). Overall, species diversity appears to be higher in the Arctic than in the Antarctic.

Possible explanations for higher Arctic diversity include the greater extension south along continental shelves of the Atlantic and Pacific Oceans, and the greater potential for terrestrial species introductions (via rivers, winds and migrating animals).

Among the diatoms, species with a pennate form (longer in one axis) are most common (e.g. species of the genera *Nitzschia*, *Fragilariopsis* and *Navicula*), and many form chains of cells (Figure 14.2). Unicellular forms tend to be large, and thus, as either cells or chains, the sea ice diatoms make up large food particles (>20 μm) compared with oceanic phytoplankton. Blooms of pennate diatoms are very common in bottom ice, but have also been noted in surface ice (Whitaker & Richardson, 1980; Lizotte & Sullivan, 1992b), infiltration communities (Garrison, 1991) and platelet ice (Arrigo et al., 1995). In the Arctic, species of *Fragilaria*, *Cylindrotheca* and *Achnanthes* are relatively common, with some species common with riverine and benthic habitats. In the Antarctic, large species of *Amphiprora*, *Pinnularia*, *Pleurosigma*, *Synedra* and *Tropidoneis* are commonly reported, especially in landfast ice.

Centric diatoms are also common in sea ice, but only dominate species composition under certain circumstances. In recently formed sea ice, it is possible for centric diatoms that had dominated the water column (e.g. autumn phytoplankton blooms of *Thalassiosira* or *Cheatoceros* species) to be entrained into the sea ice. Further growth by these planktonic species may be limited within the sea ice. In the Antarctic, dense blooms of centric diatoms (species of *Porosira* and *Thalassiosira*) have been observed to grow in platelet ice habitats (Lizotte & Sullivan, 1992b; Smetacek et al., 1992).

A few species of diatoms are associated with sea ice as attached sub-ice forms with extensive extracellular matrix material that can extend centimetres to metres below the ice bottom. In the Arctic, colonies of the centric diatom *Melosira arctica* can reach lengths of several metres below the ice (Melnikov, 1997; Ambrose et al., 2005). In the Antarctic, the most common strand-forming species are from the genus *Berkeleya*, which only reach lengths of up to 15 cm (McConville & Wetherbee, 1983).

Flagellate species of algae also have been reported for sea ice, especially in surface ice and ponds of trapped seawater or meltwater. Prymnesiophytes (*Phaeocystis* spp.), dinoflagellates (e.g. *Gymnodinium* spp.), prasinophytes (species of *Mantoniella* and *Pyramimonas*), chlorophytes (species of *Monoraphidium* and *Chlamydomonas*), chrysophytes and cryptophytes are reported from both poles. Silicoflagellates (species of *Dictyocha*) have been reported in Arctic sea ice.

Abundance and biomass

Algae cell concentrations in sea ice vary by up to six orders of magnitude (from $<10^4$ to $>10^9$ cells L^{-1}), ranging from low values typical of oceanic waters to some of the highest values recorded for any aquatic environment. Low abundance is more typical of newly formed ice or the upper ice column where environmental conditions are extreme. The highest concentrations are found at ice–seawater interfaces, ice-enclosed pockets of seawater (e.g. platelet layers, brash ice and crack pools) and surface ponds.

The chlorophyll *a* (Chl*a*) biomass in sea ice varies by geographic region, ice type and over seasons, as summarized for the Antarctic by Meiners et al. (2012). To explore these variations, the maximum accumulations of Chl*a* recorded in sea ice have been tabulated (Tables 14.1 and 14.2). Volumetric Chl*a* concentrations range from 3 to 800 mg m^{-3} (μg L^{-1}) in the Arctic and 3–10 100 mg m^{-3} in the Antarctic. Given that sea ice may start with only minimal algal biomass from seawater (e.g. 0.01 mg m^{-3}), the extreme values represent changes of five to seven orders of magnitude. To accumulate this biomass from algal growth via cell division would require 16–20 generations. Most of the reported biomass data are from first-year ice, so this

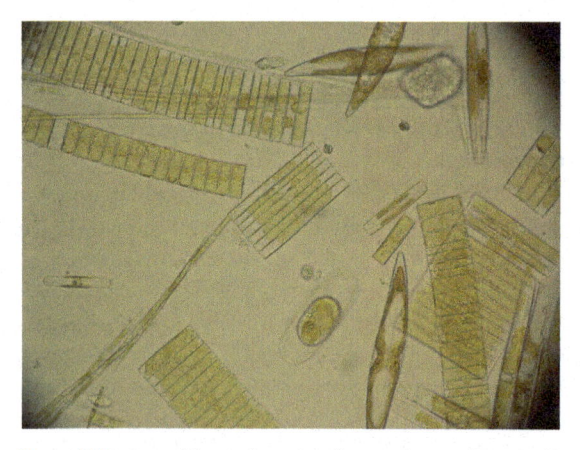

Figure 14.2 Assemblage of sea ice diatoms from a bottom ice layer of sea ice from the Western Weddell Sea. Source: photograph by Jacqueline Stefels.

Table 14.1 Maximum algal biomass as chlorophyll *a* (Chl*a*) reported for independent studies of Arctic sea ice.

Ice type	Region	Season	mg m^{-3}	mg m^{-2}	References
Bottom	Canadian Arctic Archipelago	Spring	704	14, 23, 55, 80, 89, 110, 130, 140, 160, 260, 300, 320, 340	Apollonia (1965); Dunbar & Acreman (1980); Smith et al. (1987, 1988); Cota & Horne (1989); Cota et al. (1990); Smith & Herman (1991); Herman et al. (1993); Maranger et al. (1994); Michel et al. (1996); Suzuki et al. (1997); Welch & Bergmann (1989); Apollonia et al. (2002); Lavoie et al. (2005)
	Hudson Bay	Spring	>800	21, 40, 170	Gosselin et al. (1986); Legendre et al. (1987); Welch et al. (1991); Monti et al. (1996)
	Labrador Sea	Spring		5, 8	Grainger (1977); Hsiao (1980)
	Baffin Bay	Spring		25, 56	Nozais et al. (2001); Michel et al. (2002)
	Chukchi Sea Chukchi/Beaufort	Summer Spring	123 48	304	Ambrose et al. (2005) Gradinger (2009)
	Beaufort Sea Darnley Bay Barrow	Spring–summer Spring Spring Winter	427, 600, 711 2.6 ~100 696 ~3	24, 26, 30, 31, 64 1.7 13.6	Meguro et al. (1967); Alexander (1974); Horner (1976); Alexander & Chapman (1981); Horner & Schrader (1982); Jin et al. (2006); Riedel et al. (2006 Mundy et al. (2012) Lee et al. (2010) Manes & Gradinger (2009) Manes & Gradinger (2009)
	Sea of Okhotsk	Winter	20, 153	6, 8, 30, 36, 119	Kudoh et al. (1997); Robineau et al. (1997); Asami and Imada (2001); McMinn et al. (2008)
	White Sea	Spring	35		Krell et al. (2003)
	Baltic Sea	Winter spring	6, 15, 17, 19, >60	3	Meiners et al. (2002); Granskog et al. (2003); Kaartokallio (2004); Steffens et al. (2006); Kaartokallio et al. (2007)
	Barents Sea	Spring	84	35.4	Mock & Gradinger (1999) McMinn and Hegseth (2007)
	Spitzbergen Svalbard	Spring–summer Spring	3, 67	23.4, 92	Dobryzn and Tatur (2003); Schunemann & Werner (2005) Leu et al. (2011)
	Greenland Sea	Spring	15, 30, 260	2	Buck et al. (1998); Mock & Gradinger (1999); Rysgaard et al. (2001)
Surface Pack	Bering Sea			70	Niebauer et al. (1981
	Central Arctic Ocean	Summer–autumn	9	7, 15	Gosselin et al. (1997); Gradinger (1999); Melnikov et al. (2002)
	Labrador Sea	Spring	191		Irwin (1990)
	Sea of Okhotsk	Winter	3		McMinn et al. (2008)
	Laptev Sea	Autumn	17	2	Gradinger & Zhang (1997)
	Barents Sea	Summer	17	1	Gradinger & Zhang (1997)
	Greenland Sea West Greenland	Summer Spring	57 2.6	3	Gradinger et al. (1999) Mikkelsen et al. (2008
	Fram Strait	Autumn	17		Meiners et al. (2003
Mean			160	75	
Median			35	31	

Table 14.2 Maximum algal biomass as chlorophyll *a* reported for independent studies of Antarctic sea ice.

Ice type	Region	Season	mg m^{-3}	mg m^{-2}	References
Bottom	McMurdo Sound	Spring	656, > 1000, 10100	9, 170, 170, 173, 252, 294, 309, 378	Sullivan et al. (1982); Palmisano & Sullivan (1983); Palmisano et al. (1985, 1988); Grossi et al. (1987); Lizotte & Sullivan (1992b); Arrigo et al. (1995); Trenerry et al. (2002); Ryan et al. (2006)
	Terra Nova Bay	Spring	430, 2480		Guglielmo et al. (2000, 2004)
	Cape Hallett	Spring		36	Ryan et al. (2006)
	Adelie Coast	Spring	24, 3100	>500	Riaux-Gobin et al. (2000, 2005)
	Prydz Bay	Spring		15, 96, 110	McConville et al. (1985); Archer et al. (1996); McMinn & Ashworth (1998)
	Lutzow-Holm Bay	Spring	> 1000, 2980, 5320	35, 125	Hoshiai (1977, 1981); Watanabe & Satoh (1987)
	Lutzow-Holm Bay	Autumn	> 829, 944	65	Hoshiai (1977, 1981)
	Weddell Sea	Summer–autumn	540	9	Dieckmann et al. (1990); Lizotte & Sullivan (1991)
Platelet	McMurdo Sound	Spring	132, 250, 430, > 6000	76, 125, 164, 1076, 1090	Bunt (1963, 1968); Bunt & Lee (1969, 1970); Grossi et al. (1987); Lizotte & Sullivan (1992b); Arrigo et al. (1993b, 1995)
	Terra Nova Bay	Spring	360		Guglielmo et al. (2004)
	Weddell Sea	Spring	36	20	Smetacek et al. (1992)
Internal	McMurdo Sound	Summer		10	Stoecker et al. (2000)
	Prydz Bay	Spring		6	Archer et al. (1996)
	Weddell Sea	Summer–autumn	4, 5, 10, 440	1, 1, 10, 29	Ackley et al. (1979); Garrison & Buck (1982); Clarke & Ackley (1984); Lizotte & Sullivan (1991); Hegseth & von Quillfeldt (2002); Kattner et al. (2004)
Surface	McMurdo Sound	Spring	730	51	Lizotte & Sullivan (1992b); Robinson et al. (1997)
	Lutzow-Holm Bay	Summer	670	97	Meguro (1962)
	Weddell Sea	Winter–spring	43	80	Clarke & Ackley (1984); Kottmeier & Sullivan (1990)
	Palmer Peninsula	Summer	407	117	Burkholder & Mandelli (1965)
	South Orkney Is.	Winter	7500	244	Whitaker & Richardson (1980)
New	Weddell Sea	Summer–autumn	5, 6, 27	4	Garrison et al. (1983); Lizotte & Sullivan (1991); Mock (2002)

(continued)

Table 14.2 (*Continued*)

Ice type	Region	Season	mg m^{-3}	mg m^{-2}	References
Pack	Weddell Sea	Winter	14, 15, 19, 23	16, 18, 23, 29, 53	Quetin & Ross (1988); Kottmeier et al. (1987); Schnack-Schiel (1987); Augstein et al. (1991); Garrison & Close (1993)
		Spring	15, 23, 27	18, 19, 35, 47	Ainley & Sullivan (1984); Schnack-Schiel (1987); Hempel (1989); Gradinger (1999)
		Summer	3, 59, 202	73, 77, 99, 453	Hempel (1985); Bathmann et al. (1992); Miller & Grobe (1996); Gradinger (1999)
		Autumn	10, 16, 37	11, 19, 30, 32, 38	Garrison & Buck (1982); Spindler et al. (1993); Fritsen et al. (1994); Lemke (1994); Gradinger (1999)
	Bransfield Strait	Winter	1300		Lizotte et al. (1998)
	Bellingshausen–Amundsen Sea	Summer–autumn	230, 400	60 72.2	Thomas et al. (1998); Meiners et al. (2004) Arrigo et al. (2014)
	Amundsen Sea	Spring–summer	110		
	Ross Sea	Spring	18	11	Arrigo et al. 2003
		Summer–autumn	84, 97, 1456	12	Fritsen et al. (2001); Garrison et al. (2003); Gowing (2003)
	Prydz Bay Casey Station	Spring Spring		8 3.8, 21.6	McMinn et al. (2007) McMinn et al. (2012)
Mean			974	120	
Median			170	43	

growth is restricted to a single growth season of a few months. The accumulated biomass per unit area ranges from 1 to 340 mg m^{-2} in the Arctic and <1 to 1090 mg m^{-2} in the Antarctic.

Across the independent studies of algal biomass accumulation (Tables 14.1 and 14.2), the median values for the Arctic are much lower (35 mg Chl*a* m^{-3} and 31 Chl*a* mg m^{-2}; $n > 30$) than for the Antarctic (170 and 43 mg Chl*a* m^{-2} $n > 50$). This may be attributed to major oceanographic differences, such as higher surface macronutrient availability in Antarctic seas, or the lower annual light availability at extreme polar latitudes in the Arctic. However, the differences also may be due to sampling bias, as much of the work in the Antarctic has focused on high biomass platelet and surface ice communities.

Concentrations of sea ice algae change over the seasons, particularly in first-year ice, increasing from winter through a spring bloom, and possibly continuing into summer or subsequent blooms in summer and autumn (Meiners et al., 2012). Low-latitude sites may only have sea ice during winter. Chl*a* accumulations in the Baltic Sea, Labrador Sea and Sea of Okhotsk in the north are approximately an order of magnitude lower than those at higher latitude sites (Table 14.1). The South Orkney Islands hold a similar position in the south, but the surface ice diatom blooms recorded there in the 1970s are amongst the highest biomass ever recorded for sea ice (Table 14.2). Most sea ice studies have been conducted during spring at higher latitudes, when ice is a relatively stable (growing or melting slowly) substrate for algal communities. Spring also may begin as a season with relatively low grazer pressure, if invertebrate grazers have an annual life cycle timed to the accumulation and release of ice algae (Chapters 15 and 16). In general, summer and autumn biomass levels are lower than the spring peaks. Pack ice of the Weddell Sea is the only ice type–region combination for which numerous observations have been made in all seasons; summer appears to be the peak season in this high latitude basin (Table 14.2).

14.4 Primary production

Primary production via photosynthesis (as distinct from chemosynthesis) is a complex suite of processes that includes light harvesting, electron transport and carbon fixation, each with different sensitivities to environmental conditions and cellular controls. The environmental conditions that have been studied thus far include light intensity and spectral quality, temperature, salinity and nutrient depletion. In particular, salinity, light and nutrient concentrations have been shown to be important factors governing the growth of ice algae, while variability in pH, CO_2 concentration and UV radiation appears to have little effect (Ryan et al., 2002; McMinn et al., 2003; McMinn et al., 2014).

14.4.1 Measured primary production rates

Owing to the difficulty of the measurement, estimates of primary production in sea ice, even indirect ones, are relatively rare. The available data suggest that daily rates of primary production in both the Arctic and Antarctic are highly variable, ranging from 0.2 to 463 mg C m^{-2} day^{-1} and 0.5 to 1250 mg C m^{-2} day^{-1}, respectively (Tables 14.3 and 14.4). By far the highest rates have been measured in the platelet ice layer of the Antarctic, where both Grossi et al. (1987) and Arrigo et al. (1995) measured maximum primary production rates of approximately 1200 mg C m^{-2} day^{-1}. Ignoring the platelet layer, maximum production rates in the Arctic exceed those in the Antarctic by two- to three-fold. In both regions, daily production is greatest in the bottom ice community.

Annual rates of sea ice primary production in the Arctic and Antarctic are similar in magnitude, ranging from 0.001 to 23 g C m^{-2} year^{-1} and from 0.3 to 38 g C m^{-2} year^{-1}, respectively (Tables 14.3 and 14.4). These values are consistent with biomass accumulation data from the two regions that represent a minimum estimate of annual production. For example, assuming a carbon:Chl$a$ ratio for sea ice diatoms of 40, then annual production estimated from maximum spring/summer Chl$a$ abundance ranges from 0.2-12 g C m^{-2} year^{-1} for the Arctic and from 0.04-44 g C m^{-2} year^{-1} for the Antarctic. Ignoring production in the platelet ice reduces the range for the Antarctic to 0.04-20 g C m^{-2} year^{-1}. Despite the small sample sizes, it is still probably fair to conclude that even in the most productive sea ice habitats, annual production is below 50 g C m^{-2} year^{-1}, an amount similar in magnitude to the oligotrophic central gyres of the open oceans, although produced in a small fraction of the time.

Table 14.3 Primary production (mg C m^{-2}) rates reported for Arctic sea ice.

Ice type	Region	Season	Daily (mg C m^{-2} day^{-1})	Annual (g C m^{-2} year^{-1})	References
Bottom	Canadian Arctic Archipelago	Spring	21–463	3–23	Smith et al. (1988)
			20–157	2–14	Smith & Herman (1991)
				5	Bergmann et al. (1991)
	Baffin Bay	Spring	26–317		Nozais et al. (2001)
		Spring	2–150		Michel et al. (2002)
				5	Horner (1976)
	Beaufort Sea			0.001	Horner & Schrader (1982)
	Barents Sea	Spring	3–7		Mock & Gradinger (1999)
			4.9–55	5	Hegseth (1998) McMinn & Hegseth (2007)
	Greenland Sea	Spring	0.2		Rysgaard et al. (2001)
Pack	Central Arctic Ocean	Summer	0.5–310	15	Gosselin et al. (1997)
	Labrador Sea	Spring	58		Irwin (1990)
	Davis Strait	Spring	0.03–2.4		Booth (1984)
	Bering Sea West Greenland Chukchi/Beaufort	Spring Spring Spring	2.2–4.8 0.7 4–30	0.78 0.4–2.7	McRoy & Goering (1974) Mikkelsen et al. (2008) Gradinger (2009)

Table 14.4 Primary production (mg C m^{-2}) rates reported for Antarctic sea ice.

Ice type	Region	Season	Daily (mg C m^{-2} day^{-1})	Annual (g C m^{-2} year^{-1})	References
Bottom	McMurdo Sound	Spring	0.5–85	0.3–3.4	Grossi et al. (1987)
				1–6	Bunt & Lee (1970)
	Prydz Bay	Spring	140	4.6	Archer et al. (1996)
Platelet	McMurdo Sound	Spring	2–1250	4–38	Grossi et al. (1987)
			200–1200		Arrigo et al. (1995)
Internal	McMurdo Sound	Summer	0.5–12		Stoecker et al. (2000)
New	Weddell Sea	Autumn	0.02–0.25		Mock (2002)
Pack	Bransfield Strait	Winter	42–60		Kottmeier & Sullivan (1987)
bottom	Casey Station	Spring	103		McMinn et al. (2012)

14.4.2 Measured photosynthetic rates

More common than estimates of primary production are estimates of photosynthetic rate, which when combined with measured light distributions in the ice, can be used to calculate primary production. Measurements of P_m^* exhibit some interesting patterns that, when combined with biomass accumulation patterns, provide some insight into the factors responsible for regulating primary production in sea ice.

Unlike the hemispheric pattern exhibited by Chl$a$ accumulation in sea ice, whereby values in the Antarctic are substantially higher than in the Arctic, data on P_m^* display no such bias, although the data set available for evaluation is somewhat small, particularly in the Arctic (Tables 14.5 and 14.6). While these two observations may seem initially to be inconsistent, they can be explained in part by the different habitats that dominate each region. A major reason for the difference in biomass accumulation between the two poles is that platelet ice, which contains so much more algal biomass than any other habitat, is restricted to the Antarctic. Ignoring the contribution by platelet ice brings algal biomass accumulation in the two hemispheres more closely in line.

An obvious pattern in P_m^* that does emerge from the data are the consistently higher values measured in pack ice relative to landfast ice in both the Antarctic (Table 14.6; Lizotte & Sullivan 1992a) and possibly the Arctic (Table 14.5). Values of P_m^* for the fast ice are consistently below 2 mg C mg^{-1} Chl$a$ hour^{-1} and often much lower. In contrast, estimates of P_m^* for the pack ice can be as high as > 8 mg C mg^{-1} Chl$a$ hour^{-1}. The differences in P_m^* between these two habitats can probably be attributed to variation in sea ice thickness. Pack ice tends to be much thinner than landfast ice in both polar regions and, as a result, light levels are substantially higher. Because the P_m^* of algae is highly dependent upon previous light history, fast ice, with its reduced irradiance, exhibits lower values for P_m^* than that of the pack.

The fact that photosynthetic rates are greater in the pack ice is in apparent conflict with the observation that landfast ice accumulates more algal biomass than pack ice. These observations can be reconciled only if losses of algal biomass from the pack ice are greater than from landfast ice. This provides strong, albeit indirect, evidence that grazing by zooplankton on sea ice algae is greater in pack ice than in the landfast ice. Although quantitative data with which to compare relative grazing effort in pack and landfast ice are lacking, structural characteristics of the two habitats support a higher grazing rate within pack ice. Pack ice tends to be composed of a mixture of congelation and frazil ice whereas landfast ice is predominantly congelation ice. As a result, pack ice tends to be more porous, and therefore more susceptible to grazers. Consistent with this notion, there have been numerous reports of herbivores such as copepods, amphipods, protists, etc., actively grazing within pack ice in both the Arctic and the Antarctic (Chapters 15 and 16). These organisms appear to be less common in the more consolidated congelation ice type characteristic of landfast ice until much later in the season when the ice begins to melt and the brine volume increases dramatically (Stoecker et al., 2000). Losses due to sinking of algal material must also be considered, but because pack ice has a higher fraction of surface and internal

Table 14.5 Photosynthesis rates per unit chlorophyll *a* (Chla) biomass (mg C mg^{-1} Chl*a* hour^{-1}) rates reported for Arctic sea ice.

Ice type	Region	Season	Rate (mg C mg^{-1} Chla hour^{-1})	References
Bottom	Canadian Arctic Archipelago	Spring	0.03–0.08	Cota (1985)
			0.01–0.33	Bates & Cota (1986)
			0.60–1.17	Smith et al. (1988)
			0.20–4.9	Cota & Horne (1989)
			0.01–0.60	Cota & Smith (1991)
			0.08–0.41	Smith and Herman (1991)
			0.10–0.20	Herman et al. (1993)
	Hudson Bay	Spring	0.10–1.2	Gosselin et al. (1985)
			1.8–5.2	Gosselin et al. (1985)
	Davis Strait	Spring	0.10–1.2	Cota & Smith (1991)
	Beaufort Sea	Spring	0.01–0.38	Cota & Smith (1991)
	Baltic Sea	Winter	0.90	Kaartokallio et al. (2007)
	Barents Sea	Spring	0.16–0.24	Johnsen & Hegseth (1991)
			0.05–0.90	Mock & Gradinger (1999)
			0.02–0.20	McMinn & Hegseth (2007)
	Greenland Sea	Spring	0.01–0.40	Mock & Gradinger (1999)
Pack	Labrador Sea	Spring	1.4–3.1	Irwin (1990)
Strand	Barents Sea	Summer	0.03–0.09	Johnsen & Hegseth (1991)

communities that tend to be more resistant to sinking, differences in sinking are unlikely to be important. In addition, fast ice, at least in the Antarctic, is often associated with an under-ice platelet layer that would tend to further minimize losses from the ice due to sinking and grazing.

Therefore, pack ice may represent the more important sea ice habitat in terms of providing a food source for upper trophic level organisms. Having photosynthetic rates as much as four-fold higher than landfast ice and accumulating only half the algal biomass (with little expected difference in sinking losses) imply that a much larger fraction of the primary production in pack ice is being consumed by higher trophic levels. The importance of these differences in food web structure is further magnified by the fact that pack is so much more prevalent than fast ice in both the Arctic and the Antarctic.

14.4.3 Large-scale estimates of sea ice primary production

Measurements made in nearshore sea ice led to early annual primary production estimates of 1 to 6 g C m^{-2} year^{-1} (Bunt & Lee, 1970; Horner, 1976), suggesting that sea ice production was several orders of magnitude lower than that being reported for polar phytoplankton. As a result, the contribution of primary production in sea ice was dismissed as insignificant. However, further investigation suggested that rates of production in sea ice could be very high (Palmisano & Sullivan, 1983; Welch & Bergman, 1989; Arrigo et al., 1995), fuelling renewed interest in quantifying the contribution by sea ice to total primary production in both the Arctic and Antarctic regions.

Historical estimates of sea ice primary production for the entire Arctic vary by over an order of magnitude, from 6 Tg C year^{-1} (Subba Rau & Platt, 1984) to as high as 73 Tg C year^{-1} (Legendre et al., 1992), although this latter value is on the high end of the range estimated by those authors (Table 14.7). In the Antarctic, total production in sea ice was estimated to be somewhat higher than in the Arctic, in the range of 63–70 Tg C year^{-1} (Legendre et al., 1992). However, all of these estimates were based on a very limited number of field observations that were extrapolated over the entire area of the ice pack. Given the recognition that the number of *in situ* estimates of primary production in sea ice is likely to remain small for the foreseeable future, attempts were

Table 14.6 Photosynthesis rates per unit chlorophyll *a* (Chla) biomass (mg C mg^{-1} Chla hour^{-1}) rates reported for Antarctic sea ice.

Ice type	Region	Season	Rate (mg C mg^{-1} Chla hour^{-1})	References
Bottom	McMurdo Sound	Spring	0.06–0.32	Palmisano & Sullivan (1985)
			0.05–0.22	Palmisano et al. (1985)
			0.11–0.83	Palmisano et al. (1987)
			0.17–1.9	Grossi et al. (1987)
			0.3–3.0	Kottmeier & Sullivan (1988)
			0.04–0.38	Cota & Sullivan (1990)
			0.1	Arrigo & Sullivan (1992)
			0.02–0.05	McMinn et al. (1999)
			0.29–2.0	McMinn et al. (2000)
	Terra Nova Bay	Spring	0.11–0.12	Guglielmo et al. (2000)
			0.09–0.24	Lazzara et al. (2007)
	Prydz bay	Spring	0.01–0.05	McMinn & Ashworth (1998)
Platelet	McMurdo Sound	Spring	0.15	Palmisano & Sullivan (1985)
			0.10–0.76	Grossi et al. (1987)
			0.09–0.28	Lizotte & Sullivan (1991)
			0.05–0.20	Arrigo et al. (1993b)
			0.04–0.21	Robinson et al. (1995)
Surface	McMurdo Sound	Spring	3.9 ± 1.5	Palmisano & Sullivan (1985)
			0.40–0.42	Lizotte et al. (1991)
			0.50–1.2	Robinson et al. (1997)
	Bransfield Strait	Summer	2.6–2.7	Burkholder & Mandelli (1965)
New	Bransfield Strait	Winter	0.12–1.4	Lizotte & Sullivan (1991)
	Weddell Sea	Autumn	0.02–1.2	Mock (2002)
Pack	Ross Sea	Spring	0.41–0.81	Arrigo et al. (2003)
	Prydz Bay	Spring	0.05–5.4	McMinn et al. (2007)
	Weddell Sea	Autumn–winter	0.17–4.6	Lizotte & Sullivan (1991)
			0.04–8.6	Lizotte & Sullivan (1992a)
	Bransfield Strait	Winter	0.11–8.6	Lizotte & Sullivan (1991)
	Casey Station	Spring–summer	0.12–1.4	Lizotte & Sullivan (1992a)
	Amundsen Sea	Spring–summer	0.15–3.77	McMinn et al. (2010) Arrigo et al. (2014)
			0.07–3.2	

made to use ecosystem models as a tool to estimate rates of primary production in sea ice.

The first ecosystem model capable of providing estimates of algal primary production in sea ice was developed for the Antarctic and described in Arrigo et al. (1993a) and Arrigo and Sullivan (1994). Since then, these models have grown in both number and complexity and include sea ice ecosystems in the Antarctic and the Arctic regions (Chapter 20; Jin et al., 2006; Silbert et al., 2011; Saenz & Arrigo, 2012; Tedesco et al., 2012). Although initial models were only one-dimensional (vertical) representations of the sea ice ecosystem, later versions were expanded to three dimensions, providing large-scale estimates of the seasonal cycle of sea ice primary production over both the entire Arctic and Antarctic sea ice zones (Arrigo et al., 1997, 1998; Deal et al., 2011; Dupont, 2012; Jin et al., 2012; Saenz & Arrigo, 2015).

Table 14.7 Total annual primary production in sea ice.

	Method	Rate (Tg C year^{-1})	References
Pan–Arctic	Model	10.1[a]	Deal et al. (2011)
	Model	36[b]	Dupont (2012)
	Model	21.3	Jin et al. (2012)
	Field data	9–73	Legendre et al. (1992)
	Field data	6	Subba Rao & Platt (1984)
Pan–Antarctic	Model	35.7	Arrigo (1997)
	Model	23.7	Saenz & Arrigo (2014)
	Field data	63–70	Legendre et al. (1992)

[a] 4.2 Tg C year^{-1} for the Bering Sea was subtracted from a pan-Arctic production rate of 15.1 Tg C year^{-1}.
[b] Assuming a net primary production rate of 3 g C m^{-2} over an ice covered area of 12 million km^2

Model estimates of annual primary production for Arctic sea ice fall within the wide range of earlier estimates that were based on field data (Table 14.7), varying from 10 to 36 Tg C year^{-1} (Deal et al., 2011; Dupont, 2012; Jin et al., 2012). Estimates of total sea ice primary production in the Antarctic are similar to those of the Arctic, ranging from 23.7 to 35.7 Tg C year^{-1} (Arrigo et al., 1997; Seanz & Arrigo, 2014). This similarity is surprising given the larger extent of sea ice in the Antarctic (19 million km^2) than in the Arctic (15 million km^2) and demonstrates that per-metre rates of production in the Antarctic are lower than those in the Arctic, perhaps due to a thicker snow cover.

These estimates allow an assessment of the relative contribution of sea ice microalgae to total primary production in both the Arctic and the Antarctic. Recent large-scale estimates of annual primary production by pelagic phytoplankton in the Arctic range from 329 (Sakshaug, 2003) to 350–500 Tg C year^{-1} (Pabi et al., 2008), the latter estimate being based on satellite observations of surface Chl*a* over the entire Arctic between 1998 and 2006. This estimate by Pabi et al. (2008) also includes interannual variability in productivity resulting from recent losses of Arctic sea ice (e.g. Arrigo et al. 2008a; Chapter 11). These annual production estimates for the pelagic environment, coupled with similar estimates for the sea ice, suggest that Arctic sea ice contributes approximately 2–10% of total primary production in Arctic waters (ice+water column), in agreement with previous estimates of 7.5% (Dupont, 2012) and 5% (Jin et al., 2012) that were based on models that include both sea ice and pelagic primary production. This contribution is substantially less than the value of 4–26% estimated by Legendre et al. (1992). The difference is primarily due to the fact that model-based estimates of sea ice production are lower, and satellite-based estimates of pelagic production are higher, than previous field-based extrapolations used by Legendre et al. (1992).

In Antarctic waters (south of 50°S, total area = 43 million km^2), estimates of phytoplankton primary production range from 1949 ± 70.1 Tg C year^{-1} (Arrigo et al., 2008b) to 2900 Tg C year^{-1} (Moore and Abbott, 2000). Assuming that sea ice algae contribute an additional 23–70 Tg C year^{-1} (Table 14.7), then sea ice ecosystems account for approximately 1–3% of total annual production in Antarctic waters. If this analysis is restricted to the 19 million km^2 of the Southern Ocean that are ice-covered at least part of the year, where annual primary production in the water column is estimated to range from 224 to 912 Tg C year^{-1} (Legendre et al., 1992; Moore and Abbott, 2000; Arrigo et al., 2008b), then sea ice would account for 2–24% of the total primary production in the sea ice zone.

Although the fraction of annual primary production contributed by sea ice algae in both the Arctic and Antarctic is relatively low, there are times of the year when its importance is relatively greater. This is because the seasonal cycles of production in the sea ice and the water column are offset from each other, with peak sea ice production preceding peak water column production by 1–4 months (Figure 14.3). Consequently, sea ice accounts for ~18% of total Arctic primary production in the month of April and about 12% in May, well in excess of the annual mean of 2–10%. Similarly, sea ice accounts for 4% of total Antarctic production south of 50°S in November, much greater than the annual mean of 1%. Therefore, the presence of sea ice algal communities in polar waters extends the season of food availability for zooplankton grazers by up to a few months each year (Tremblay et al., 2008; Dupont, 2012; Jin et al., 2012).

In the future, sea ice is expected to decrease in extent and thickness as the climate warms (Wang & Overland, 2009). In the near term, thinner ice may favour higher per-metre rates of sea ice primary production. However, this could be offset by increased precipitation

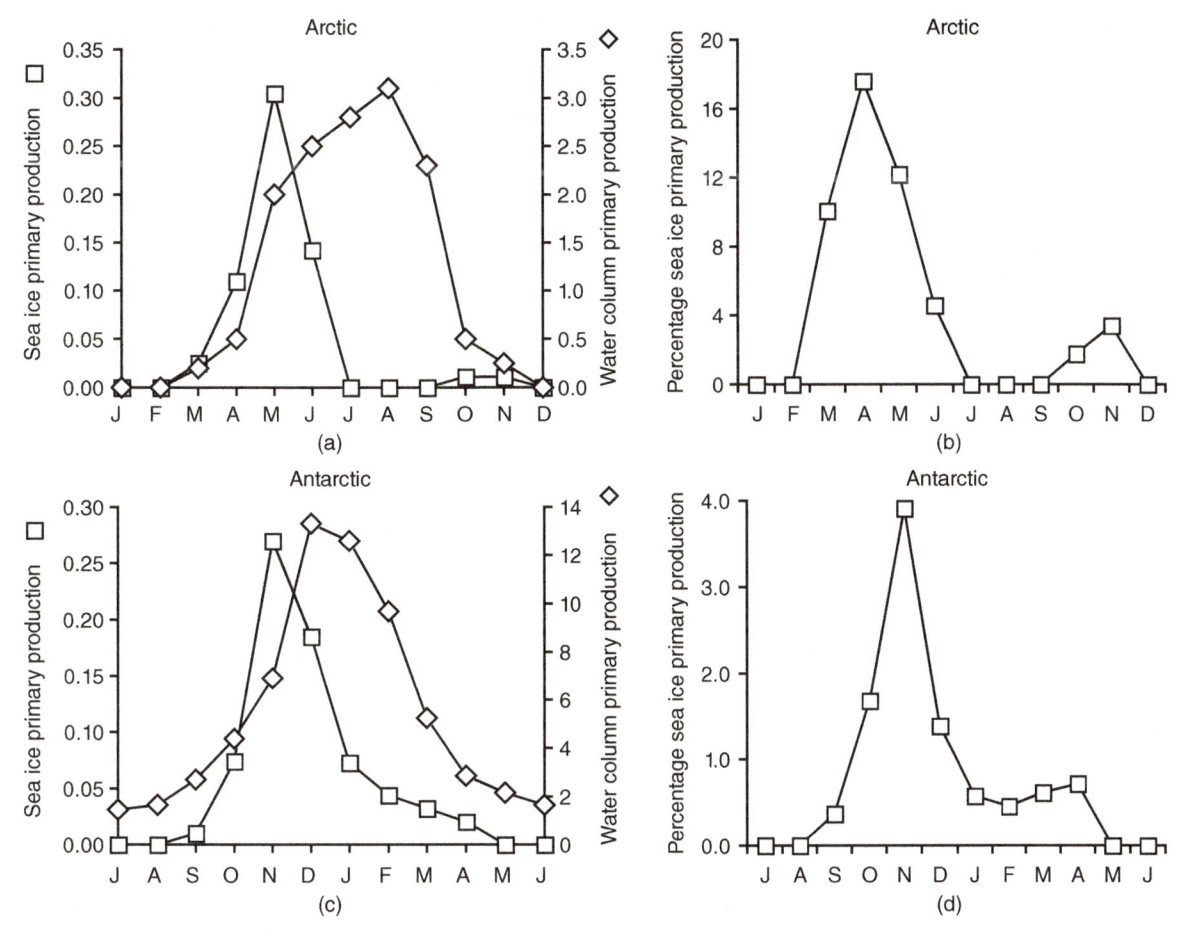

Figure 14.3 The seasonal cycle of spatially integrated primary productivity (Tg C day^{-1}) for the Arctic (a, b) and the Antarctic (c, d) in both the sea ice and the water column. Data for sea ice in (a) are from Deal et al. (2011), Jun et al. (2011) and Dupont (2012). Data for the water column in (a) are from Pabi et al. (2008). Data for sea ice in (c) are from Arrigo et al. (1997) and Saenz and Arrigo (2014). Data for the water column in (c) are from Arrigo et al. (2008b). Also shown are the seasonal cycles of sea ice primary production as a percentage of total production (sea ice + water column) for both the Arctic (b) and Antarctic (d).

associated with an accelerated hydrological cycle and by a reduction in sea ice extent (Chapter 11). Recent evidence does suggest that in areas where sea ice extent has declined, primary production in the pelagic regions of polar oceans has increased and is likely to continue to do so (Arrigo et al., 2008a, 2011, 2014; Zhang et al., 2010; Jin et al., 2012; Belanger et al., 2013), further reducing the proportion of annual production attributable to sea ice algae. However, if surface water stratification in polar waters intensifies along with the reduction in sea ice cover, the associated reduction in nutrient renewal could limit or reverse this increase in pelagic production (Walsh et al., 2005; Lavoie et al., 2010), thereby increasing the relative importance of sea ice algae.

References

Ackley, S.F., Buck, K.A. & Taguchi, S. (1979) Standing crop of algae in the sea ice of the Weddell Sea region. *Deep-Sea Research*, **26**, 269–281.

Ainley, D. & Sullivan, C.W. (1984) AMIEREZ 1983: A summary of activities on board the R/V Mellville and USCGC Westwind. *Antarctic Journal of the United States*, **19**, 177–179.

Alexander, V. (1974) Primary productivity regimes of the nearshore Beaufort Sea, with reference to potential roles of ice biota. In: *The coast and shelf of the Beaufort Sea* (Eds. J.C.

Reed & J.E. Sater), pp. 609–632. Arctic Institute of North America, Arlington, Va.

Alexander, V. & Chapman, T. (1981) The role of epontic algal communities in Bering Sea ice. In: *The Eastern Bering Sea Shelf: Oceanography and Resources*, vol.**2** (Eds. D.W. Hood & J.A. Calder), pp. 773–780. Univ. Washington Press, Seattle.

Ambrose, W.G., Jr., von Quillfeldt, C., Clough, L.M., Tilney, P.V. R., & Tucker, T. (2005) The sub-ice algal community in the Chukchi Sea: Large- and small-scale patterns of abundance based on images from a remotely operated vehicle. *Polar Biology*, **28**, 784–795.

Apollonio, S. (1965) Chlorophyll in Arctic sea ice. *Arctic*, **18**, 118–122.

Apollonia, S., Pennington, M. & Cota, G.F. (2002) Stimulation of phytoplankton photosynthesis by bottom-ice extracts in the Arctic. *Polar Biology*, **25**, 350–354.

Archer, S.D., Leakey, R.J. G., Burkill, P.H., Sleigh, M.A., & Appleby, C.J. (1996) Microbial ecology of sea ice at a coastal Antarctic site; community composition; biomass and temporal change. *Marine Ecology Progress Series*, **135**, 179–195.

Arrigo, K.R. (2014) Sea ice ecosystems. *Annual Reviews of Marine Science*, **6**, 13.1–13.29.

Arrigo, K.R. & Sullivan, C.W. (1992) The influence of salinity and temperature covariation on the photophysiological characteristics of Antarctic sea ice microalgae. *Journal of Phycology* **28**, 746–756.

Arrigo, K.R. & Sullivan, C.W. (1994) A high resolution bio-optical model of microalgal growth: Tests using sea ice algal community time series data. *Limnology and Oceanography*, **39**, 609–631.

Arrigo, K.R. & van Dijken, G.L. (2011) Secular trends in Arctic Ocean net primary production. *Journal of Geophysical Research*, **116**, C09011, doi:10.1029/2011JC007151.

Arrigo, K.R. & van Dijken, G.L. (2015) Continued increases in Arctic Ocean primary production. *Progress in Oceanography*, **136**, 60–70.

Arrigo, K.R., Kremer, J.N. & Sullivan, C.W. (1993a) A simulated Antarctic fast-ice ecosystem. *Journal of Geophysical Research*, **98**, 6929–6946.

Arrigo, K.R., Robinson, D.H. & Sullivan, C.W. (1993b) A high resolution study of the platelet ice ecosystem in McMurdo Sound, Antarctica: Photosynthetic and bio-optical characteristics of a dense microalgal bloom. *Marine Ecology Progress Series*, **98**, 173–185.

Arrigo, K.R., Dieckmann, G., Gosselin, M., Robinson, D.H., Fritsen, C.H. & Sullivan, C.W. (1995) A high resolution study of the platelet ice ecosystem in McMurdo Sound, Antarctica: biomass, nutrient, and production profiles within a dense microalgal bloom. *Marine Ecology Progress Series*, **127**, 255–268.

Arrigo, K.R., Lizotte, M.P., Worthen, D.L., Dixon, P., & Dieckmann, G. (1997) Primary production in Antarctic sea ice. *Science*, **276**, 394–397.

Arrigo, K.R., Worthen, D.L., Dixon, P. & Lizotte, M.P. (1998) Primary productivity of near surface communities within Antarctic pack ice. In *Antarctic Sea Ice Biological Processes, Interactions, and Variability* (Eds. M.P. Lizotte & K.R. Arrigo). Antarctic Research Series, **73**, 23–43.

Arrigo, K.R., Robinson, D.H., Dunbar, R.B., Leventer, A.R., & Lizotte, M.P. (2003) Physical control of chlorophyll *a*, POC, and PON distributions in the pack ice of the Ross Sea, Antarctica. *Journal of Geophysical Research*, **108**, 3316, 10.1029/2001JC001138.

Arrigo, K.R., G.L. van Dijken & S. Pabi (2008a) The impact of a shrinking Arctic ice cover on marine primary production. *Geophysical Research Letters*, **35**, L19603, doi:10.1029/2008GL035028.

Arrigo, K.R., G.L. van Dijken & S. Bushinsky (2008b) Primary Production in the Southern Ocean, 1997–2006. *Journal of Geophysical Research*, **113**. C08004, doi:10.1029/2007JC004551.

Arrigo, K.R., Brown, Z.W., Mills, M.M., Maldonado, M.T., & Tortell, P.T. (2014) Sea ice algal biomass and physiology in the Amundsen Sea, Antarctica. *Elementa Science of the Anthropocene*, **2**, doi: 10.12952/journal.elementa.000028

Asami, H. & Imada, K. (2001) Ice algae and phytoplankton in the late ice-covered season in Notoro Ko lagoon, Hokkaido. *Polar Bioscience*, **14**, 24–32.

Augstein, E., Bagriantsev, N. & Schenke, H-W. (1991) Die expedition Antarktis VIII/1–2, 1989 mit der Winter Weddell Gyre study der Forschungsschiffe Polarstern and Akademik Federov. *Berichte zum Polarfurschung*, **84**.

Bates, S.S. & Cota, G.F. (1986) Fluorescence induction and photosynthetic responses of arctic algae to sample treatment and salinity. *Journal of Phycology*, **22**, 421–429.

Bathmann, U., Schulz-Baldes, M., Farbach, E., Smetacek, V. & Hubberten, H-W. (1992) Die expedition Antarktis IX/1–4 des Forschungsschiffes Polarstern. *Berichte zum Polarfurschung*, 100 pp.

Bélanger, S., Babin, M., and Tremblay, J.É. 2013. Increasing cloudiness in Arctic damps the increase in phytoplankton primary production due to sea ice receding. *Biogeosciences*, doi:10.5194/bg-10-4087-2013.

Bergmann, M.A., Welch, H.E., Butler-Walker, J.E. & Silfred, T.D. (1991) Ice algal photosynthesis at Resolute and Saqvaqjuac in the Canadian Arctic, *Journal of Marine Systems*, **2**, 43–52.

Bindschadler, R. (1990) SeaRISE: A multidisciplinary research initiative to predict rapid changes in global sea level caused by collapse of marine ice sheets. *NASA Conference Publication 3075*.

Booth, J.A. (1984) The epontic algal community of the ice edge zone and its significance to the Davis Strait ecosystem. *Arctic*, **37**, 234–243.

Buck, K.R., Nielsen, T.G., Hansen, B.W., Gastrup-Hansen, D., & Thomsen, H. (1998) Infiltration phyto- and protozooplankton assemblages in the annual sea ice of Disko Island, West Greenland, spring 1996. *Polar Biology*, **20**, 377–381.

Bunt, J.S. 1963. Microbiology of Antarctic sea-ice - diatoms of Antarctic sea-ice as agents of primary production. *Nature*, **199**, 1255–1257.

Bunt, J.S. (1968) Probing the ecosystem. In: *Advances in Microbiology of the Sea*, Vol. **1** (Eds. M.R. Droop & E.J.F. Wood), pp 215–222. Academic Press, New York.

Bunt, J.S. & Lee, C.C. (1969) Observations within and beneath Antarctic sea ice in McMurdo Sound and the Weddell Sea, 1967–1968: Methods and data. *Institute of Marine Science, University of Miami Publication* **69-1**, 1–12. Institute of Marine Science of the University of Miami, Miami, Florida.

Bunt, J.S. & Lee, C.C. (1970) Seasonal primary production in Antarctic sea ice at McMurdo Sound in 1967. *Journal of Marine Research*, **28**, 304–320.

Burkholder, P.R. & Mandelli, E.F. (1965) Productivity of microalgae in Antarctic sea ice. *Science*, **149**, 872–874.

Clarke, D.B. & Ackley, S.F. (1984) Sea ice structure and biological activity in the Antarctic marginal ice zone. *Journal of Geophysical Research*, **89**, 2087–2095.

Cota, G.F. (1985) Photoadaptation of high arctic ice algae. *Nature*, **315**, 219–222.

Cota, G.F. & Horne, E.P. W. (1989) Physical control of Arctic ice algal production. *Marine Ecology Progress Series*, **52**, 111–121.

Cota, G.F. & Sullivan, C.W. (1990) Photoadaptation, growth and production of bottom ice algae in the Antarctic. *Journal of Phycology*, **26**, 399–411.

Cota, G.F. & Smith, R.E. H. (1991) Ecology of bottom ice algae II: dynamics, distributions and productivity. *Journal of Marine Systems*, **2**, 279–295.

Cota, G.F., Anning, J.L., Harris, L.R., Harrison, W.G. & Smith, R.E. H. (1990) Impact of ice algae on inorganic nutrients in seawater and sea ice in Barrow Strait, N.W.T., *Canada, during spring Canadian Journal of Fisheries and Aquatic Sciences*, **47**, 1402–1415.

Deal, C., Jin, M., Elliott, S., Hunke, E., Maltrud, M. & Jeffery, N. (2011) Large-scale modeling of primary production and ice algal biomass within arctic sea ice in 1992. *Journal of Geophysical Research*, **116**, C07004, doi:10.1029/2010JC006409

Dieckmann, G., Sullivan, C.W. & Garrison, D. (1990) Seasonal standing crop of ice algae in pack ice of the Weddell Sea, Antarctica. *Eos, Transactions of the American Geophysical Union*, **71**, 79.

Dobrzyn, P. & Tatur, A. (2003) Algal pigments in fast ice and under-ice water in an Arctic fjord. *Sarsia*, **88**, 291–296.

Dunbar, M.J. & Acreman, J.C. (1980) Standing crops and species composition of diatoms in sea ice from Robeson Channel to the Gulf of St. Lawrence. *Ophelia*, **19**, 61–72.

Dupont, F. (2012), Impact of sea-ice biology on overall primary production in a biophysical model of the pan-Arctic Ocean, *Journal of Geophysical Research*, **117**, C00D17, doi:10.1029/2011JC006983.

Fritsen, C.H., Lytle, V.I., Ackley, S.F. & Sullivan, C.W. (1994) Autumn bloom of Antarctic pack-ice algae. *Science*, **266**, 782–784.

Fritsen, C.H., Coale, S., Neenan, D., Gibson, A. & Garrison, D. (2001). Biomass, production and microhabitat characteristics near the freeboard of ice floes in the Ross Sea during the austral summer. *Annals of Glaciology*, **33**, 280–286.

Garrison, D.L. (1991) Antarctic sea ice biota. *American Zoologist*, **31**, 17–33.

Garrison, D.L. & Buck, K.R. (1982) Sea-ice algae in the Weddell Sea. I. Biomass distribution and the physical environment. *Eos, Transactions of the American Geophysical Union*, **63**, 47.

Garrison, D.L. & Close, A.R. (1993) Winter ecology of the sea-ice biota in Weddell Sea pack ice. *Marine Ecology Progress Series*, **96**, 17–31.

Garrison, D.L., Ackley, S.F. & Buck, K.R. (1983) A physical mechanism for establishing algal populations in frazil ice. *Nature*, **306**, 363–365.

Garrison, D.L., Jeffries, M.O., Gibson, A. et al. (2003) Development of sea ice microbial communities during autumn ice formation in the Ross Sea. *Marine Ecology Progress Series*, **259**, 1–15.

Gast. R. J., Moran, D.M., Beaudoin, D.J., Blythe, J.N., Dennett, M.R. & Caron, D.A. (2006) Abundance of a novel dinoflagellate phylotype in the Ross Sea, Antarctica. *Journal of Phycology*, **42**, 233–422.

Gosselin, M., Legendre, L., Demers, S. & Ingram, R.G. (1985) Responses of sea-ice microalgae to climatic and fortnightly tidal energy inputs (Manitounuk Sound, Hudson Bay). *Canadian Journal of Fisheries and Aquatic Sciences*, **42**, 999–1006.

Gosselin, M., Legendre, L., Therriault, J.C., Demers, S. & Rochet, M. (1986) Physical control of the horizontal patchiness of sea-ice microalgae. *Marine Ecology-Progress Series*, **9**, 289–298

Gosselin, M., Levasseur, M., Wheeler, P.A., Horner, R.A. & Booth, B.C. (1997). New measurements of phytoplankton and ice algal production in the Arctic Ocean. *Deep-Sea Research II*, **44**, 1623–1644.

Gowing, M.M. (2003) Large viruses and infected microeukaryotes in Ross Sea summer pack ice habitats. *Marine Biology*, **142**, 1029–1040.

Gradinger, R. (2009) Sea-ice algae: Major contributors to primary production and algal biomass in the Chukchi and Beaufort Seas during May/June 2002. *Deep-Sea Research, Part II*, **56**, 1201–1212.

Gradinger, R. & Zhang, Q. (1997) Vertical distribution of bacteria in Arctic sea ice from the Barents and Laptev Seas. *Polar Biology*, **17**, 448–454.

Gradinger, R., Friedrich, C. & Spindler, M. (1999) Abundance, biomass and composition of the sea ice biota of the Greenland Sea pack ice. *Deep-Sea Research, Part II*, **46**, 1457–1472.

Grainger, E.H. (1977) The annual nutrient cycle in sea ice. In: *Polar Oceans* (Ed. M.J. Dunbar), pp. 285–299. Arctic Institute of North America, Calgary, Canada.

Granskog, M.A., Kaartokallio, H. & Shirasawa, K. (2003) Nutrient status of Baltic Sea ice: evidence for control by snow-ice formation, ice permeability, and ice algae. *Journal of Geophysical Research*, **108**(C8), 9-1-9.

Grant, W.S. & Horner, R.A. (1976) Growth responses to salinity variation in 4 Arctic ice diatoms. *Journal of Phycology*, **12**, 180–185.

Grossi, S.M., Kottmeier, S.T., Moe, R.L., Taylor, G.T., & Sullivan, C.W. (1987) Sea ice microbial communities. VI. Growth and primary production in bottom ice under graded snow cover. *Marine Ecology Progress Series*. **35**, 153–164.

Guglielmo, L., Carrada, G.C., Catalano, G. et al. (2000) Structural and functional properties of sympagic communities in the annual sea ice at Terra Nova Bay (Ross Sea, Antarctica). *Polar Biology*, **23**, 137–146.

Guglielmo, L., Carrada, G.C., Catalano, G. et al. (2004) Biogeochemistry and algal communities in the annual sea ice at Terra Nova Bay (Ross Sea, Antarctica). *Chemistry and Ecology*. **20**, S43-S55.

Hegseth, E.N. (1998) Primary production of the northern Barents Sea. *Polar Research*, **17**, 113–123.

Hegseth, E.N. & von Quillfeldt, C.H. (2002) Low phytoplankton biomass and ice algal blooms in the Weddell Sea during the ice-filled summer of 1997. *Antarctic Science*, **14**, 231–243.

Hempel, G. (1985) Die expedition Antarktis III/3 mit F.S. Polarstern 1984/85. *Berichte zum Polarfurschung*, 25.

Hempel, I. (1989) The expedition Antarctic VII/1 and 2 (EPOS 1) of R.V. Polarstern in 1988/89. *Berichte zum Polarfurschung*, 62.

Herman, A.W., Knox, D.F., Conrad, J. & Mitchell, M.R. (1993) Instruments for measuring subice algal profiles and productivity in situ. *Canadian Journal of Fisheries and Aquatic Sciences*, **50**, 359–369.

Horner, R.A. (1976) Sea ice organisms. *Oceanography and Marine Biology, An Annual Review*, **14**, 167–182.

Horner, R.A. (Ed.) (1985) *Sea Ice Biota*. CRC Press, Boca Raton, FL.

Horner, R.A. & Schrader, G.C. (1982) Relative contributions of ice algae, phytoplankton, and benthic microalgae to primary production in nearshore regions of the Beaufort Sea. *Arctic*, **35**, 485–503.

Hoshiai, T. (1977) Seasonal change of ice communities in the sea ice near Syowa Station, Antarctica. In: *Polar oceans* (Ed. by M.J. Dunbar.), pp. 307–317. Arctic Institute of North America, Calgary, Canada.

Hoshiai, T. (1981) Proliferation of ice algae in the Syowa Station area, Antarctica. *Memoirs of the National Institute of Polar Research, Series E*, **34**, 1–12.

Hsiao, S.I. C. (1980) Quantitative composition, distribution, community structure and standing stock of sea ice microalgae in the Canadian Arctic. *Arctic*, **33**, 768–793.

Irwin, B.D. (1990) Primary production of ice algae on a seasonally-ice-covered, continental shelf. *Polar Biology*, **10**, 247–254.

Jin, M., Deal, C.J., Wang, J., Shin, K., Tanaka, N., Whitledge, T.E., Lee, S.H. & Gradinger, R.R. (2006) Controls of the landfast ice-ocean ecosystem offshore Barrow, Alaska. *Annals of Glaciology*, **44**, 63–72.

Jin, M., Deal, C.J., Lee, S.H., Elliott, S., Hunke, E., Maltrud, M. & Jeffery, N. (2012) Investigation of Arctic sea ice and ocean primary production for the period 1992–2007 using a 3-D global ice–ocean ecosystem model. *Deep-Sea Research Part II*, **81–84**, 28–35.

Johnsen, J. & Hegseth, E.N. (1991) Photoadaptation of sea ice microalgae in the Barents Sea. *Polar Biology*, **11**, 179–184.

Kaartokallio, H. (2004) Food web components, and physical and chemical properties of Baltic Sea ice. *Marine Ecology Progress Series*, **273**, 49–63.

Kaartokallio, H., Kuosa, H., Thomas, D.N., Granskog, M.A. & Kivi, K. (2007) Biomass, composition and activity of organism assemblages along a salinity gradient in sea ice subjected to river discharge in the Baltic Sea. *Polar Biology*, **30**, 183–197.

Kattner, G., Thomas, D.N., Haas, C., Kennedy, H. & Dieckmann, G.S. (2004) Surface ice and gap layers in Antarctic sea ice: highly productive habitats. *Marine Ecology Progress Series*, **277**, 1–12.

Kipfstuhl, J. (1991) On the formation of underwater ice and the growth and energy budget of the sea ice in Atka Bay, Antarctica. *Reports on Polar Research*, **85**, 1–89.

Kottmeier, S.T. & Sullivan, C.W. (1988) Sea ice microbial communities. IX. Effects of temperature and salinity on metabolism and growth of autotrophs and heterotrophs. *Polar Biology*, **8**, 293–304.

Kottmeier, S.T. & Sullivan, C.W. (1990) Bacterial biomass and bacterial production in pack ice of Antarctic marginal ice edge zones. *Deep-Sea Research*, **37**, 1311–1330.

Kottmeier, S.T., Grossi, S.M., & Sullivan, C.W. (1987) Sea ice microbial communities .8. Bacterial production in annual sea ice of McMurdo Sound, Antarctica. *Marine Ecology Progress Series*, **35**, 175–186.

Krell, A., Ummenhofer, C., Kattner, G. et al. (2003) The biology and chemistry of land fast ice in the White Sea, Russia - A comparison of winter and spring conditions. *Polar Biology*, **26**, 707–719.

Kudoh, S., Robineau, B., Suzuki, Y., Fujiyoshi, Y. & Takahashi, M. (1997) Photosynthetic acclimation and the estimation of temperate ice algal primary production in Saroma-ko Lagoon, Japan. *Journal of Marine Systems*, **11**, 93–109.

Lavoie, D., Denman, K. & Michel, C. (2005) Modeling ice algal growth and decline in a seasonally ice-covered region of the Arctic (Resolute Passage, Canadian Archipelago). *Journal of Geophysical Research*, **110**, C11009, doi:10.1029/2005JC002922.

Lavoie, D., Denman, K.L. & Macdonald, R.W. (2010), Effects of future climate change on primary productivity and export fluxes in the Beaufort Sea. *Journal of Geophysical Research*, **115**, C04018, doi:10.1029/2009JC005493.

Lazzara, L., Nardello, I., Ermanni, C., Mangoni, O. & Saggiomo, V. (2007) Light environment and seasonal dynamics of microalgae in the annual sea ice at Terra Nova Bay, Ross Sea, Antarctica. *Antarctic Science*, **19**, 83–92.

Lee, S.H., Jin, M. & Whitledge, T.E. (2010) Comparison of bottom sea-ice algal characteristics from coastal and offshore regions in the Arctic Ocean. *Polar Biology*, **33**, 1331–1337.

Legendre, L., Demers, S. & Gosselin, M. (1987) Chlorophyll and photosynthetic efficiency of size-fractionated sea-ice microalgae (Hudson-Bay, Canadian Arctic). *Marine Ecology Progress Series*, **40**, 199–203

Legendre, L, Ackley, S.F., Dieckmann, G.S. et al. (1992) Ecology of sea ice biota: 2. Global significance. *Polar Biology*, **12**, 429–444.

Lemke, P. (1994) Die expedition Antarktis X/4 mit F.S. Polarstern 1992. *Berichte zum Polarfurschung*, **140**.

Leu, E., Søreide, J.E., Hessen, D.O., Falk-Petersen, S., & Berge, J. (2011) Consequences of changing sea-ice cover for primary and secondary producers in the European Arctic shelf seas: Timing, quantity, and quality. *Progress in Oceanography*, **90**, 18–22.

Lizotte, M.P. & Sullivan, C.W. (1991) Photosynthesis-irradiance relationships in microalgae associated with Antarctic pack ice: evidence for *in situ* activity. *Marine Ecology Progress Series*, **71**, 175–184.

Lizotte, M.P. & Sullivan, C.W. (1992a) Photosynthetic capacity in microalgae associated with Antarctic pack ice. *Polar Biology*, **12**, 497–502.

Lizotte, M.P. & Sullivan, C.W. (1992b) Biochemical composition and photosynthate distribution in sea ice algae of McMurdo Sound: evidence for nutrient deficiencies during the bloom. *Antarctic Science*, **4**, 23–30.

Lizotte, M.P., Robinson, D.H. & Sullivan, C.W. (1998) Algal pigment signatures in Antarctic sea ice. In: Lizotte, M.P. & Arrigo, K.R. (Eds.) *Antarctic Sea Ice: Biological Processes, Interactions and Variability. Antarctic Research Series*, **73**, 93–106.

Manes, S.S. & Gradinger, R. (2009) Small scale vertical gradients of Arctic ice algal photophysiological properties. *Photosynthesis Research*, **102**, 53–66.

Maranger, R., Bird, D.F. & Juniper, S.K. (1994) Viral and bacterial dynamics in Arctic sea ice during the spring algal bloom near Resolute, N.W.T., Canada. *Marine Ecology Progress Series*, **111**, 121–127.

McConville, M.J. & Wetherbee, R. (1983) The bottom-ice microalgal community from annual ice in the inshore waters of East Antarctica. *Journal of Phycology*, **19**, 431–439.

McConville, M.J., Mitchell, C. & Wetherbee, R. (1985) Patterns of carbon assimilation in a microalgal community from annual sea ice, East Antarctica. *Polar Biology*, **4**, 135–141.

McMinn, A. & Ashworth, C. (1998) The use of oxygen microelectrodes to determine the net production by an Antarctic sea ice algal community. *Antarctic Science*, **10**, 39–44.

McMinn, A. & Hegseth, E.N. (2007) Sea ice primary productivity in the northern Barents Sea, spring 2004. *Polar Biology*, **30**, 289–294.

McMinn A., Ashworth, C. & Ryan, K. (1999) Growth and productivity of Antarctic Sea ice algae under PAR and UV irradiances, *Botanica Marina*, **42**, 401–407.

McMinn, A., Ashworth, C. & Ryan, K.G. (2000) In situ net primary productivity of an Antarctic fast ice bottom algal community. *Aquatic Microbial Ecology*, **21**, 177–185.

McMinn, A., Ryan, K. & Gademann, R. (2003) Diurnal changes in photosynthesis of Antarctic fast ice algal communities determined by pulse amplitude modulation fluorometry. *Marine Biology*, **143**, 359–367.

McMinn, A., Ryan, K.G., Ralph, P.J. & Pankowski, A. (2007) Spring sea ice photosynthesis, primary productivity and biomass distribution in eastern Antarctica, 2002–2004. *Marine Biology*, **151**, 985–995.

McMinn, A., Hattori, H., Hirawake, T. & Iwamoto, A. (2008) Preliminary investigation of Okhotsk Sea ice algae; taxonomic composition and photosynthetic activity. *Polar Biology*, **31**, 1011–1015.

McMinn, A., Pankowskii, A. Ashworth, C., Bhagooli, R., Ralph, P. & Ryan, K. (2010) In situ net primary productivity and photosynthesis of Antarctic sea ice algal, phytoplankton and benthic algal communities. *Marine Biology*, **157**, 1345–1356.

McMinn, A. Ashworth, C., Bhagooli, R. et al. (2012) Antarctic coastal microalgal primary production and photosynthesis. *Marine Biology*, **159**, 2827–2837.

McMinn, A., Muller, M.N., Martin, A. & Ryan, K.G. (2014) The response of Antarctic sea ice algae to changes in pH and CO_2. *PLoS ONE*, **9**, e86984, doi:10.1371/journal.pone.0086984

McRoy, C.P. & Goering, J.J. (1974) The influence of ice on the primary productivity of the Bering Sea. In: *Oceanography of the Bering Sea with Emphasis on Renewable Resources* (Ed. D.W. Hood & E.J. Kelley), pp. 403–421. Institute of Marine Science, University of Alaska, Fairbanks, Alaska.

Meguro, H. (1962) Plankton ice in the Antarctic Ocean. *Antarctic Record*, **14**, 1192–1199.

Meguro, H., Ito, K. & Fukushima, H. (1967) Ice flora (bottom type): a mechanism of primary production in polar seas and the growth of diatoms in sea ice. *Arctic*, **20**, 114–133.

Meiners, K., Fehling, J., Granskog, M.A. & Spindler, M. (2002) Abundance, biomass and composition of biota in Baltic Sea ice and underlying water (March 2000). *Polar Biology*, **25**, 761–770.

Meiners, K., Gradinger, R., Fehling, J., Civitarese, G. & Spindler, M. (2003) Vertical distribution of exopolymer particles in sea ice of the Fram Strait (Arctic) during autumn. *Marine Ecology Progress Series*, **248**, 1–13.

Meiners, K., Brinkmeyer, R., Granskog, M.A. & Lindfors, A. (2004) Abundance, size distribution and bacterial colonization of exopolymer particles in Antarctic sea ice (Bellingshausen Sea). *Aquatic Microbial Ecology*, **35**, 283–296.

Meiners, K.M., Vancoppenolle, M., Thanassekos, S. et al. (2012) Chlorophyll *a* in Antarctic sea ice from historical ice core data. *Geophys. Res. Lett.*, **39**, L21602, doi:10.1029/2012GL053478.

Melnikov, I.A. (1997) *The Arctic Sea Ice Ecosystem*. Gordon and Breach Science Publishers, Amsterdam,.

Melnikov, I.A., Kolosova, E.G., Welch, H.E. & Zhitina, L.S. (2002) Sea ice biological communities and nutrient dynamics in the Canada Basin of the Arctic Ocean. *Deep-Sea Research, Part I*, **49**, 1623–49.

Michel, C., Legendre, L., Ingram, R.G., Gosselin, M. & Levasseur, M. (1996) Carbon budget of sea-ice algae in spring: evidence of a significant transfer to the zooplankton grazers. *Journal of Geophysical Research*, **101**, 18345–18360.

Michel, C., Gosselin, M. & Nozais, C. (2002). Preferential sinking export of biogenic silica during the spring and summer in the North Water Polynya (northern Baffin Bay): temperature or biological control. *Journal of Geophysical Research*, **107**, 1-1 to 1-14.

Mikkelsen, D.M., Rusgaard, S. & Glud, R.N. (2008) Microalgal composition and primary production in Arctic sea ice: a seasonal study from Kobbefjord (Kangerluarsunnguaq), West Greenland. *Marine Ecology Progress Series*, **368**, 65–74.

Miller, H. & Grobe, H. (1996) Die expedition Antarktis X1/3 mit F.S. Polarstern 1994. *Berichte zum Polarfursching*, 188.

Mock, T. (2002) In situ primary production in young Antarctic sea ice. *Hydrobiologia*, **470**, 127–132.

Mock, T. & Gradinger, R. (1999) Determination of Arctic ice algal production with a new *in situ* incubation technique, *Marine Ecology Progress Series*, **177**, 15–26.

Monti, D., Legendre, L., Therriault, J.C. & Demers, S. (1996) Horizontal distribution of sea-ice microalgae: Environmental control and spatial processes (southeastern Hudson Bay, Canada). *Marine Ecology Progress Series*, **133**, 229–240.

Moore, J. K & Abbott, M.R. (2000) Phytoplankton chlorophyll distributions and primary production in the Southern Ocean. *Journal of Geophysical Research*, **105**, 28 709–28 722.

Mundy, C.J., Gosselin, M., Ehn, J.K. et al. (2012) Characteristics of two distinct high-light acclimated algal communities during advanced stages of sea ice melt. *Polar Biology*, **34**, 1869–1886.

Niebauer, H.J., Roberts, J. & Royer, T.C. (1981) Shelf break circulation in the northern Gulf of Alaska, *Journal of Geophysical Research*, **86**, 4231–4242.

Nozais, C., Gosselin, M., Michel, C. & Tita, G. (2001) Abundance, biomass, composition and grazing impact of the sea-ice meiofauna in the North Water, northern Baffin Bay *Marine Ecology Progress Series*, **217**, 235–250.

Pabi, S., van Dijken, G.L. & K.R. Arrigo. (2008) Primary Production in the Arctic Ocean, 1998–2006. *Journal of Geophysical Research*. **113**, C08005, doi:10.1029/2007JC004578.

Palmisano A.C. & Sullivan, C.W. (1983) Sea ice microbial communities (SIMCO). I. Distribution, abundance, and primary production of ice microalgae in McMurdo Sound, Antarctica in 1980. *Polar Biology*, **2**, 171–177.

Palmisano, A.C. & Sullivan, C.W. (1985) Pathways of photosynthetic carbon assimilation in sea-ice microalgae from McMurdo Sound, Antarctica. *Limnology and Oceanography*, **30**, 674–678.

Palmisano, A.C., SooHoo, J.B. & Sullivan, C.W. (1985) Photosynthesis-irradiance relationships in sea ice microalgae from McMurdo Sound, Antarctica. *Journal of Phycology*, **21**, 341–346.

Palmisano, A.C., SooHoo, J.B. & Sullivan, C.W. (1987) Effects of four environmental variables on photosynthesis-irradiance relation-ships in Antarctic sea ice microalgae, *Marine Biology*, **94**, 299–306.

Quetin, L. & Ross, R. (1988) Summary of WINCRUISE II to the Antarctic Peninsula during June and July, 1987. *Antarctic Journal of the United States*, **23**, 149–151.

Riaux-Gobin, C., Treguer, P., Poulin, M. & Vetion, G. (2000) Nutrients, algal biomass and communities in land-fast ice and seawater off Adelie Land (Antarctica). *Antarctic Science*, **12**, 160–171.

Riaux-Gobin, C., Treguer, P., Dieckmann, G., Maria, E., Vetion, G. & Poulin, M. (2005) Land-fast ice off Adelie Land (Antarctica): short-term variations in nutrients and chlorophyll just before ice break-up. *Journal of Marine Systems*, **55**, 235–248.

Riedel, A., Michel, C. & Gosselin, M. (2006) Seasonal study of sea-ice exopolymeric substances on the Mackenzie shelf: implications for transport of sea-ice bacteria and algae. *Aquatic Microbial Ecology*, **45**, 195–206.

Robineau, B., Legendre, L., Kishino, M. & Kudoh, S. (1997) Horizontal heterogeneity of microalgal biomass in the first- year sea ice of Saroma-ko Lagoon (Hokkaido, Japan). *Journal of Marine Systems*, **11**, 81–91.

Robinson, D.H., Arrigo, K.R., Iturriaga, R. & Sullivan, C.W. (1995) Adaptation to low irradiance and restricted spectral distribution by Antarctic microalgae from under-ice habitats. *Journal of Phycology*, **31**, 508–520.

Robinson, D.H., Kolber, Z., & Sullivan, C.W. (1997) Photophysiology and photoacclimation in surface sea ice algae from McMurdo Sound, Antarctica. *Marine Ecology Progress Series*, **147**, 243–256.

Ryan, K.G., McMinn, A., Mitchell, K.A. & Trenerry, L. (2002) Mycosporine-like amino acids in Antarctic Sea ice algae, and their response to UVB radiation. *Zeitschrift fuer Naturforschung Section C Journal of Biosciences*, **57**, 471–477.

Ryan, K.G., Hegseth, E.N., Martin, A. et al. (2006) Comparison of the microalgal community within fast ice at two sites along the Ross Sea coast, Antarctica. *Antarctic Science*, **18**, 583–594.

Rysgaard, S., Kuhl, M., Glud, R.N., & Hansen, J.W. (2001) Biomass, production and horizontal patchiness of sea ice algae in a high-Arctic fjord (Young Sound, NE Greenland). *Marine Ecology Progress Series*, **223**, 15–26.

Saenz, B.T. & Arrigo, K.R. (2012) Simulation of a sea ice ecosystem using a hybrid model for slush layer desalination. *Journal of Geophysical Research*, **117**, C05007, doi:10.1029/2011JC007544

Saenz, B.T. & Arrigo, K.R. (2014) Primary production in Antarctic sea ice during 2005–2006 from a sea ice state estimate. *Journal of Geophysical Research*, **119**, 3645–3678.

Sakshaug, E. (2003) Primary and secondary production in the Arctic Seas. In: *The Organic Carbon Cycle in the Arctic Ocean* (Eds. R. Stein & R.W. Macdonald), pp. 57–81. Springer-Verlag, Berlin.

Schnack-Schiel, S. (1987) The winter expedition of RV Polarstern to the Antarctic (ANT V/1–3). *Berichte Reports on Polar Research*, **39**.

Schunemann, H. & Werner, I. (2005) Seasonal variations in distribution patterns of sympagic meiofauna in Arctic pack ice. *Marine Biology*, **146**, 1091–1102.

Sibert, V., Zakardjian, B., Gosselin, M., Starr, M., Senneville, S. & LeClainche, Y. (2011) 3D bio-physical model of the sympagic and planktonic productions in the Hudson Bay system. *Journal of Marine Systems*, **88**, 401–422.

Smetacek, V., Scharek, R. Gordon, L.I. et al. (1992) Early spring phytoplankton blooms in ice platelet layers of the southern Weddell Sea, Antarctica. *Deep-Sea Research*, **39**, 153–168.

Smith, R.E.H. & Herman, A.W. (1991) Productivity of sea ice algae: in situ vs incubator methods. *Journal of Marine Systems*, **2**, 97–110.

Smith, R.E. H., Clement, P., Cota, G. & Li, W.K. W. (1987) Intracellular photosynthate allocation and the control of Arctic marine ice algal production. *Journal of Phycology*, **23**, 124–132.

Smith, R.E. H., Anning, J., Clement, P. & Cota, G. (1988) Abundance and production of ice algae in Resolute Passage, Canadian Arctic. *Marine Ecology Progress Series*, **48**, 251–263.

Spindler, M., Dieckmann, G. & Thomas, D.N. (1993) Die expedition Antarktis X/3 mit F.S. Polarstern. *Berichte zum Polarfurschung*, **121**, 122 pp.

Steffens, M., Granskog, M.A., Kaartokallio, H. et al. (2006) Spatial variation of biogeochemical properties of landfast ice in the Gulf of Bothnia, Baltic Sea. *Annals of Glaciology*, **44**, 80–87.

Stoecker, D.K., Gustafson, D.E., Baier, C.T. & Black, M.M. D. (2000) Primary production in the upper sea ice. *Aquatic Microbial Ecology*, **21**, 275–287.

Subba Rau, D.V. & Platt, T. (1984) Primary production of Arctic waters. *Polar Biology*, **3**, 191–201.

Sullivan, C.W., Palmisano, A.C., Kottmeier, S.T. & Moe, R. (1982) Development of the sea ice microbial community in McMurdo Sound. *Antarctic Journal of the United States*, **17**, 155–157.

Suzuki, Y., Kudoh, S. & Takahashi, M. (1997) Photosynthetic and respiratory characteristics of an Arctic ice algal community living in low light and low temperature conditions *Journal of Marine Systems*, **11**, 111–121.

Tedesco, L., Vichi, M. & Thomas, D.N. (2012) Process studies on the ecological coupling between sea ice algae and phytoplankton. *Ecological Modeling*, **226**, 120–138.

Thomas, D.N., Lara, R.J., Haas, C. et al. (1998) Biological soup within decaying summer sea ice in the Amundsen Sea, Antarctica. In: *Antarctic Sea Ice: Biological Processes, Interactions and Variability* (Eds. M.P. Lizotte & K.R. Arrigo). American Geophysical Union, *Antarctic Research Series*, **73**, 161–171.

Tremblay, J.E., Simpson, K.G., Martin, J. et al. (2008) Vertical stability and the annual dynamics of nutrients and chlorophyll fluorescence in the coastal, southeast Beaufort Sea. *Journal of Geophysical Research*, **113**, C07S90.

Trenerry, L.J., McMinn, A., & Ryan, K.G. (2002) In situ oxygen microelectrode measurements of bottom-ice algal production in McMurdo Sound, Antarctica. *Polar Biology*, **25**, 72–80.

Wadhams, P., Lange, M.A. & Ackley, S.F. (1987) The ice thickness distribution across the Atlantic sector of the Antarctic Ocean in midwinter. *Journal of Geophysical Research*, **92**(C13), 14535–14552.

Wang, M., & Overland, J.E. (2009) A sea ice free summer Arctic within 30 years. *Geophysical Research Letters*, **36**, L07502, http://dx.doi.org/10.1029/2009GL037820.

Watanabe, K., & Satoh, H. (1987) Seasonal variations of ice algal standing crop near Syowa Station, East Antarctica, in 1983/84. *Bulletin of Plankton Society of Japan*, **34**, 143–164.

Welch, H.E. & Bergmann, M.A. (1989) Seasonal development of ice algae and its prediction from environmental factors near Resolute, N.W.T., Canada. *Canadian Journal of Fisheries and Aquatic Sciences*, **46**, 1793–1804.

Welch, H.E., Bergmann, M.A., Siferd, T.D. & Amarualik, P.S. (1991) Seasonal development of ice algae near Chesterfield inlet, NWT, Canada. *Canadian Journal of Fisheries and Aquatic Sciences*, **48**, 2395–2402.

Whitaker, T.M. & Richardson, M.G. (1980) Morphology and chemical composition of a natural population of an ice-associated Antarctic diatom *Navicula glaciei*. *Journal of Phycology*, **16**, 250–257.

Zhang, J., Spitz, Y.H., Steele, M., Ashjian, C., Campbell, R., Berline, L. & Matrai, P. (2010) Modeling the impact of declining sea ice on the Arctic marine planktonic ecosystem, *Journal of Geophysical Research*, **115**, C10015, doi:10.1029/2009JC005387.

CHAPTER 15

Sea ice as a habitat for micrograzers

David A. Caron,[1] Rebecca J. Gast[2] and Marie-Ève Garneau[1]

[1] Department of Biological Sciences, University of Southern California, Los Angeles, CA, USA
[2] Department of Biology, Woods Hole Oceanographic Institution, Woods Hole, MA, USA

15.1 Introduction

Sea ice makes up a physically complex, geographically extensive, but often seasonally ephemeral biome on Earth. Despite the extremely harsh environmental conditions under which it forms and exists for much of the year, sea ice can serve as a suitable, even favourable, habitat for dense assemblages of microorganisms. Our knowledge of the existence of high abundances of microalgae (largely diatoms) in sea ice spans at least a century, but for many years it remained unknown whether these massive accumulations were composed of metabolically active or merely inactive cells brought together through physical processes associated with ice formation. Work began approximately a quarter century ago started to fill this void in our knowledge of sea ice microbiota.

Algal populations and their attendant bacterial assemblages were initially believed to exist largely in the absence of grazing mortality from herbivorous and bacterivorous organisms. Our perception of these microbial assemblages has changed, however, and it is now clear that sea ice supports highly enriched, taxonomically diverse and trophically active microbial consumers in Arctic and Antarctic environments (Laurion et al., 1995; Sime-Ngando et al., 1997a; Thomas & Dieckmann, 2002; Kaartokallio, 2004; Riedel et al., 2007).

Biological aspects of sea ice bacteria and microalgae have been detailed in Chapters 13 and 14, respectively, of this book. The present chapter focuses on the microbial consumers that characterize these unique marine habitats. Within the complex microbial communities of sea ice, the major consumers of bacterial and algal biomass are single-celled eukaryotic microorganisms displaying heterotrophic ability, usually phagotrophy, which is the engulfment of food particles. The term 'protozoa' has traditionally been employed and is still commonly used to describe these species, but wholesale revision of the evolutionary relationships among eukaryotic taxa throughout the past decade (Adl et al., 2012), and greater understanding of the complex nutritional modes exhibited by single-celled eukaryotes have called the accuracy of this term into question. In particular, the traditional distinction between phototrophic protists (i.e. unicellular algae) and their heterotrophic counterparts (protozoa) presupposes that phototrophy and heterotrophy are mutually exclusive nutritional modes, but they are not. Technically, the word 'protozoa' adequately describes truly heterotrophic protists, but it does not take into account the fact that phagotrophy is a common behaviour among many microscopic algae, nor does it recognize that some apparently photosynthetic species of protists are in fact kleptoplastidic, i.e. heterotrophs that consume photosynthetic prey and retain the chloroplasts in a functional state. Due to the existence of these mixotrophic species, the expression 'phagotrophic protists' provides a more accurate description of the many protistan species employing heterotrophic nutrition, regardless of the presence or absence of chloroplasts. These deficiencies having been noted, the term 'protozoa' will be used synonymously with 'heterotrophic protists' (which includes mixotrophic species) in this text for the sake of brevity.

Scientific understanding of the diversity, abundances and trophic activities of sea ice microconsumers has

Sea Ice, Third Edition. Edited by David N. Thomas.
© 2017 John Wiley & Sons, Ltd. Published 2017 by John Wiley & Sons, Ltd.

only emerged within the past few decades, and much remains to be learned. Although their ecology is still not as thoroughly described or investigated as that of many photosynthetic taxa, bacteria and even some metazoa from polar ecosystems, a basic understanding of the breadth of their abundances and activities now exists. Protozoan populations in sea ice reduce bacterial and algal abundances, remineralize major nutrients contained in their biomass, and constitute additional sources of nutrition for metazoa within or associated with the ice. The available information, complemented by a wealth of knowledge regarding the ecology of heterotrophic microbial eukaryotes from temperate and tropical regions, is beginning to clarify the pivotal role played by protozoa in polar marine ecosystems.

15.2 Origins and fates of sea ice protists

The processes giving rise to sea ice protozoan assemblages presumably are the same for phytoplankton, bacteria and invertebrates. Microbial populations can be either entrained into ice platelets formed in subsurface waters, trapped at the surface during the formation of frazil ice, or invade existing ice through brine channels or fissures (Garrison et al., 1986; Petrich & Eicken, 2015). These processes may selectively concentrate certain microbial taxa because of size selection or other characteristics (e.g. pennate diatoms may attach to ice, while foraminifera possess sticky pseudopodia that may result in their scavenging by ice crystals). Thus, the initial colonizing assemblages in sea ice may represent a 'biased' sampling of microorganisms in the water column (Eicken, 1992).

Microbes that are incorporated into the ice experience chemical and physical characteristics that are vastly different from the surrounding seawater environment. The structure of the ice microbial community is therefore dictated by the growth and mortality of species contained within the initial colonizing assemblage (or invading from the water column), under selective pressures that differ greatly from those in the surrounding water. These selective pressures result in a divergence in community composition from that in the water as the ice ages. For this reason, protistan assemblages in newly formed sea ice have been shown to be similar to assemblages in the water column initially (Gradinger & Ikavalko, 1998; Garrison et al., 2003), but this similarity decreases as the assemblages adapt to conditions and food availability in the ice (Różanska et al., 2008).

Demonstrations of different dominant taxa in sea ice and the water column underlying or adjacent to the ice have been documented using traditional and, more recently, genetic techniques (Figure 15.1). DNA fragment analysis, gene sequencing and fluorescent *in situ* hybridization (FISH) approaches have corroborated such differences in drifting and landfast ice, underlying water and overlying snow in Antarctica, the Arctic and the Baltic Sea (Gast et al., 2004, 2006; Bachy et al., 2011; Terrado et al., 2011; Majaneva et al., 2012; Comeau et al., 2013; Piwosz et al., 2013). For example, microbial eukaryotic assemblages in ice samples collected from different sites in the Antarctic were more similar to each other than to the assemblages from water samples collected at the same sites as the ice samples (Gast et al., 2004). Similarly, a microscopical study of sea ice protistan assemblages collected from different locations in the Ross Sea, Antarctica, showed a relatively high degree of similarity in major biomass components, presumably indicating the dominance of specific taxa that were more adapted to existence within the sea ice than in the water (Garrison et al., 2005). Similarities in protistan assemblages within ice from freshwater ecosystems have even been noted between Arctic and Antarctic samples (Jungblut et al., 2012). Heterotrophic protists were observed in all ice samples, often at significant abundances.

These similarities in ice communities notwithstanding, vertical gradients of environmental conditions within sea ice can create microhabitats that differ significantly in their physico-chemical characteristics (Figure 15.2). Snow cover on ice dramatically affects light penetration, an important variable controlling algal biomass within and especially at the bottom of the ice (Grossi & Sullivan, 1985; Gosselin et al., 1986). Gradients of temperature and salinity resulting from seawater infiltration and/or melting of snow and ice produce small-scale spatial heterogeneity in the ice. These factors can lead to dramatic differences in the composition and biomass of the microbial communities growing there. Brine channels, compression ridges, meltwater (slush) at the snow-ice interface, and the bottom of the ice are often microhabitats of exceptionally high microbial biomass, and at times high diversity (Figure 15.2), as are meltwater ponds on sea

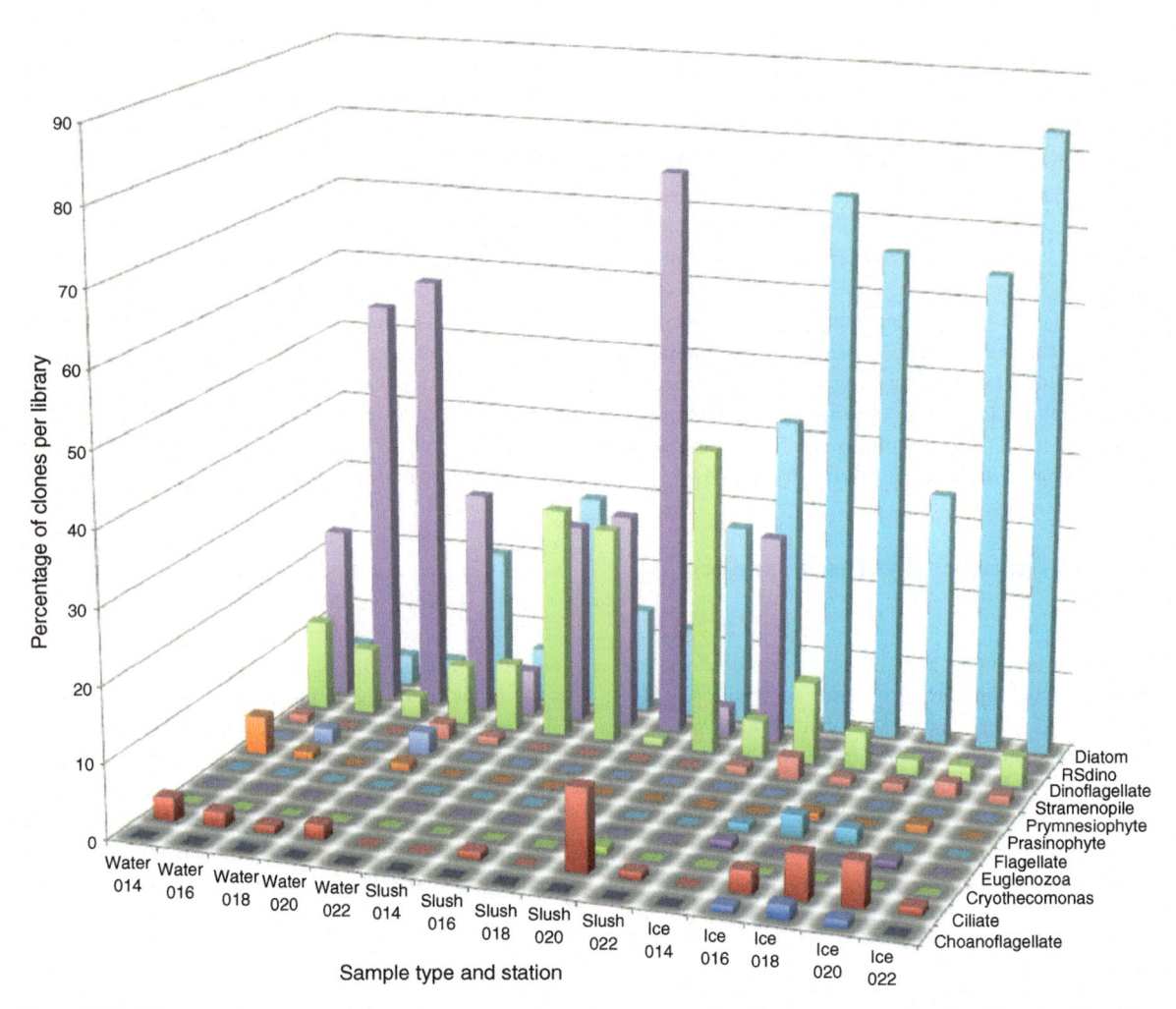

Figure 15.1 Differences in protistan community structure in sea ice, water and meltwater communities (slush) on ice floes from the Ross Sea, Antarctica, as determined from small subunit ribosomal RNA gene sequence libraries. Note the strong dominance of diatoms among the photosynthetic protists within sea ice, but also differences in the abundances of dinoflagellates (including RS dino, which is a kleptoplastidic form) and ciliates within the ice, slush and water.

ice in the Arctic (Figure 15.3) (Vincent, 2010; Mundy et al., 2011). Dramatic differences in the dominant taxa within these microhabitats can exist over small vertical scales due to snow cover, nutrient supply and the presence/absence of specific consumer populations (Figure 15.2g–k). However, we still know very little about the growth and trophic activities of protozoa within the physico-chemical gradients in sea ice (Kaartokallio et al., 2007). Even basic information such as which taxa are metabolically active in sea ice is only now becoming available (Majaneva et al., 2012).

Spatial (horizontal) variability in sea ice protozoan communities are influenced by variations in the algal assemblages in the ice (Gosselin et al., 1986; Laurion et al., 1995), which in turn are affected by wind-driven advection of pack ice (Garrison et al., 2003). Vertical variability in sea ice algal communities (Mundy et al., 2011) as well as interannual variability in their biomass have also been reported (Fritsen et al., 2008). Heterotrophic protists often mirror these spatial distributions, or exhibit a lagged response to increases or decreases in the abundance of phototrophic protists and

Sea ice as a habitat for micrograzers

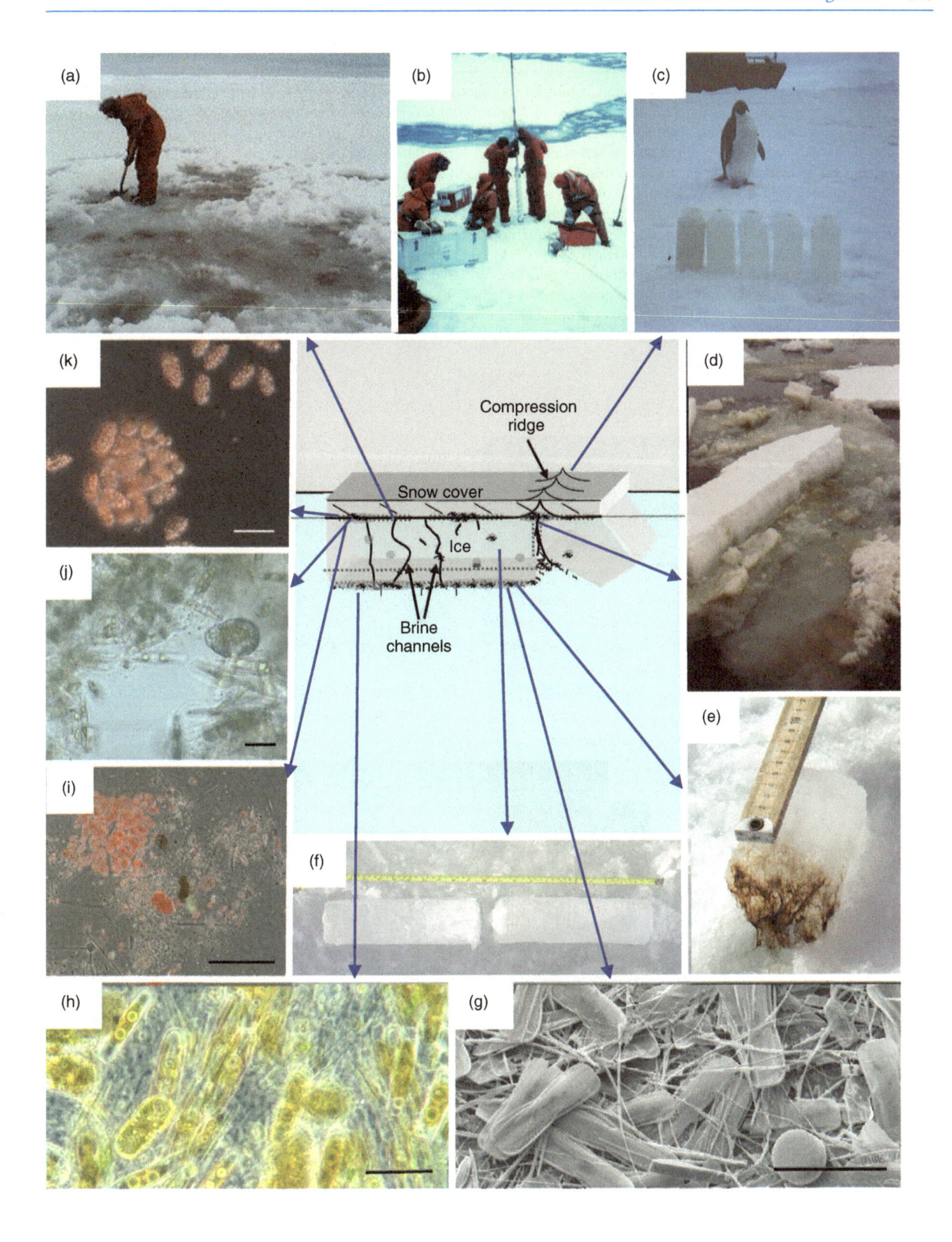

374 Chapter 15

Figure 15.2 A schematic of sea ice microhabitats colonized by protozoa (centre), and the macroscopic manifestations and dominant taxa of these microhabitats (surrounding pictures). (a) Removal of several centimetres of snow reveals dense assemblages of microorganisms (coloured patches) at the tops of brine channels on Antarctic pack ice. (b) Collecting ice core samples from pack ice. (c) Samples of slush (meltwater and infiltration seawater) collected across a ridge compression on pack ice. High concentrations of microorganisms (brown colour in left-hand bottles) found in fissures in the ice decrease rapidly away from the axis of the ridge. (d) Pack ice overturned by a ship reveals densely coloured microbial communities of the snow–ice meltwater microhabitat. (e) Dense mat of diatoms and associated bacteria and protozoa from the bottom of a core of landfast ice in the Ross Sea, Antarctica. (f) A core of pack ice showing low microbial biomass (no visible colour). (g, h) Scanning electron micrograph (g) and transmitted light micrograph (h) of diatoms from the bottom of Antarctic sea ice. (i, j, k) Light micrographs showing the dominance of different taxa in the different meltwater microhabitats. These communities may contain large amounts of phototrophic and heterotrophic populations (bacteria and protozoa) and debris (i), and be dominated by diatoms (j) or phototrophic dinoflagellates (k). The red colour in (i) and (k) is due to chlorophyll fluorescence by phototrophic protists, predominantly dinoflagellates. Scale bars are 100 µm (g, h, i) or 20 µm (j, k).

Figure 15.3 Meltwater ponds on sea ice (a) are a common feature of the Arctic during summer (Cape Parry, Northwest Territories, Canada). These ponds allow greater light penetration than snow-covered ice, enhancing primary production in the underlying ice and water. Gross similarities between sea ice microbial communities from the Ross Sea, Antarctica (Figure 2i,j) and sea ice from the Arctic Ocean in the Chukchi Sea (b, c) are exemplified by the complex assortment of photosynthetic and heterotrophic protists, bacteria and detritus. Scale bars in (b) and (c) are 100 µm.

bacteria (Sime-Ngando et al., 1997a, 1999; Riedel et al., 2007; Eddie et al., 2010). Seasonal successions within sea ice appear to be similar to classical patterns observed for water column microbes (Parsons et al., 1984; Strom, 2000), but there are few complete data sets because of the logistical issues of obtaining such time series in polar environments.

The Antarctic Marine Ecosystem Research in the Ice Edge Zone (AMERIEZ) programme remains one of the most complete data sets for examining temporal relationships among sea ice microbial communities (Garrison & Buck, 1989). This study revealed that large quantities of phototrophic protists during spring and early summer gave way to higher abundances of bacteria as organic substrate concentrations increased due to primary production. Heterotrophic protistan abundances increased during summer as prey proliferated (bacteria and algae) and temperature rose. The accumulation of phototrophic biomass observed prior to the build-up of substantial numbers of heterotrophs may be explained in part by the differential effect of temperature on phototrophic and heterotrophic processes (see Section 15.4.2) (Rose & Caron, 2007, 2013).

The physical structure of the ice is presumably a major determinant for the survival and growth of heterotrophic protists, because food acquisition for these species is dependent on movement and prey capture. The physical dimensions of brine channels in sea ice have been examined using water-soluble resin (Weissenberger et al., 1992). These studies indicated a typical diameter of 200 µm for many of the channels, creating highly convoluted, complex microhabitats of narrow passageways and high surface area. These 'ice veins' are probably microhabitats for the survival and growth of sea ice microbes (Price, 2003), and possibly are the reason why many sea ice protozoa display morphological or behavioural features that are more characteristic of protozoa from sediment than from planktonic ecosystems (Figure 15.6d,e,g,i,l and 15.8a,b,g – see later). Active microbes in ice appear to be relegated to the brine, and in liquid water that is maintained through the production of ice-retarding substances (Janech et al., 2006; Uhlig et al., 2015).

The potential influence of the physical structure of sea ice on microbial colonization has been examined experimentally (Krembs et al., 2000). These researchers estimated that 6–41% of the brine channel surface area may be covered by microorganisms. The interstitial spaces in ice appear to be important determinants of microbial community structure and biomass through the exclusion of some consumers, and controls on the renewal of nutrients or organic substrates.

Sea ice microbiota have been hypothesized to serve as 'seed populations' for spring phytoplankton blooms in polar ecosystems (Kuosa et al., 1992; Michel et al., 1993). A similar role could be hypothesized for protozoa. Sea ice protozoa contribute to microbial grazing pressure in the water column upon release from melting sea ice during spring/summer thaws. As noted above, however, many of the protozoa within sea ice are adapted to consuming prey that are associated with surfaces, or present in microhabitats at highly elevated abundances relative to abundances in the plankton. It is therefore unclear if many protozoa released from ice could capture sufficient prey to support their growth.

Aggregations of microbes in sea ice may also contribute to the vertical flux of carbon and other elements in coastal ecosystems. Sime-Ngando et al (1997a) reported that net secondary production by heterotrophic protists in sea ice was two to four times greater than that reported for sea ice bacteria, but how that carbon is used remains a question. Protozoan biomass in sea ice would be expected to be at its maximum during late summer when ice melt rate is maximal, and thus the contribution of protozoan biomass and protozoan waste materials could be a significant source of sinking particles during that period.

Sea ice protozoa probably serve as prey for metazoa either within the ice or in the surrounding water but we know virtually nothing about this process. In the water column of the south-central polynya of the Ross Sea, *Phaeocystis antarctica* dominates phytoplankton biomass during austral summer (Smith et al., 1998). This alga is a poor food source for some metazoa, but protozoa that consume the alga directly, or consume bacterial biomass supported by the breakdown of *P. antarctica* and other algal detritus, may be important intermediates in the food chain between primary production and zooplankton in this region (Caron et al., 1997). Whether protozoa in sea ice occupy a similar niche between sea ice algae and metazoa is presently unknown.

Some protozoa trapped during ice formation may encyst (flagellates and dinoflagellates) or form resting spores (diatoms) (Garrison & Buck, 1989; Buck et al., 1992; Stoecker et al., 1992, 1997; Ikävalko & Gradinger, 1997; Okolodkov, 1998; Montresor et al., 1999). Many

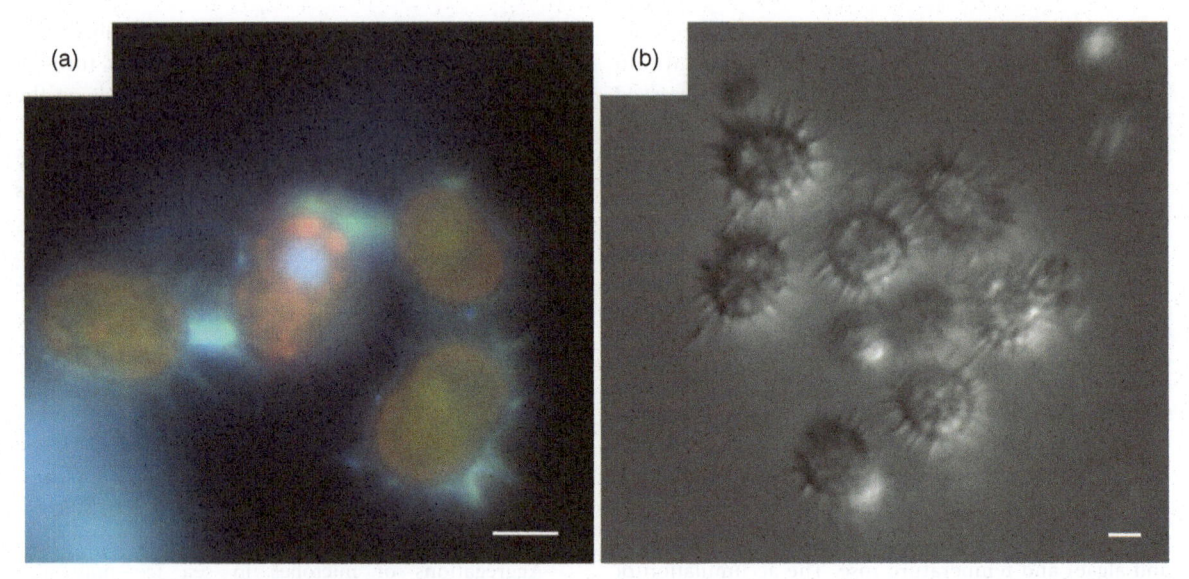

Figure 15.4 Epifluorescence micrograph (a) and transmitted light micrograph (b) of spined cysts of a phototrophic dinoflagellate (probably *Polarella glacialis*) from sea ice of the Ross Sea, Antarctica. The red colour in (a) is due to the chlorophyll fluorescence, and the whitish blue colour is from the fluorochrome DAPI. Scale bars are 10 µm.

photosynthetic protists occur as dormant forms in sea ice (Figure 15.4). The incorporation of dinoflagellate and chrysophyte cysts (i.e. hypnozygotes and statocysts, respectively) into sea ice may allow their retention near surface waters during the winter until daylight is sufficient for vegetative growth. Such a cycle has been documented for some phototrophic protistan taxa in the landfast ice of McMurdo Sound, Antarctica (Stoecker et al., 1997, 1998). Alternatively, cyst formation by some protistan taxa may enable overwintering in sea ice for species whose vegetative life stages take place in the water column during the spring–autumn period.

One additional, potential fate of sea ice protozoa that has recently emerged is infection and lysis by viruses. Viral infection of some heterotrophic protists in pack ice habitats of the Ross Sea has been demonstrated (Gowing, 2003). The high abundances and constrained mobility of microbial eukaryotes may make sea ice a particularly suitable condition for viral infection and transmission.

15.3 Diversity and abundances

The unique physical and chemical conditions of sea ice create a variety of extreme conditions for the microbial assemblages inhabiting it; salinities can range from highly saline brine to nearly fresh water; exceptionally low temperatures can occur in the upper portions of sea ice to temperatures near those of water just below the ice; light intensities vary vertically from surface melt ponds (Figure 15.3a) to ice bottom, and seasonally from very high and constant during summer to continuous darkness. There is little doubt that the protistan species in sea ice have adapted to these harsh conditions, although it is presently unclear if these adaptations are strain-specific or reflect the presence of large numbers of unique, endemic species in these polar microhabitats.

Sea ice often has conspicuous assemblages of photosynthetic protists, but protozoa also contribute significantly to its communities. Visible discolorations can be the result of high abundances of diatoms, dinoflagellates, *Phaeocystis* (the latter primarily in Antarctic waters) and other taxa. The magnitude of these accumulations on the bottom of some regions of fast ice can be remarkable (Michel et al., 2002) (Figure 15.2e,g,h). These exceptional situations aside, Archer et al. (1996b) noted that sea ice protozoa contributed up to nearly 20% of the total integrated microbial biomass at a coastal Antarctic site and up to approximately 10% of the biomass attached to the bottom of the ice. Similar proportions of phototrophic and heterotrophic biomass in Antarctic sea ice were observed by Garrison & Buck (1991) and Garrison et al. (2005).

Euglenoid flagellates, heterotrophic dinoflagellates, ciliates and 'nanoheterotrophs' (heterotrophic flagellates 2–20 µm in size) constituted major components of the heterotrophic protistan assemblages in the latter studies, and pack ice and landfast ice assemblages usually differed in species composition. Euglenoid flagellates alone (Figure 15.5d) contributed up to 20% of the biomass in coastal Antarctic sea ice, while heterotrophic dinoflagellates such as *Protoperidinium* (Figure 15.5a) were also abundant (Archer et al., 1996a). In Greenland Sea pack ice, heterotrophic flagellates alone constituted an average of 20% of the total microbial biomass (Gradinger et al., 1999).

These and other comparisons of protists in Arctic and Antarctic sea ice have indicated that relative abundances of heterotrophic and phototrophic species can be comparable (Garrison & Buck, 1989; Buck et al., 1998; Różanska et al., 2008) but, as noted earlier, the relationship varies with season. Seasonal shifts in the proportion of phototrophic and heterotrophic microbial biomass are also typical of assemblages in the water column of polar ecosystems (Dennett et al., 2001; Lovejoy et al., 2002; Terrado et al., 2011). Absolute abundances of both phototrophic and heterotrophic protists in sea ice (cells per volume of melted sea ice) are often one to two orders of magnitude higher than in the water column (cells per volume of seawater) (Garrison, 1991; Davidson & Marchant, 1992; Garrison et al., 1993; Garrison & Close, 1993; Sime-Ngando et al., 1997a,b; Vaqué et al., 2004; Fonda Umani et al., 2005). These latter studies did not consider the contribution of mixotrophic algae to heterotrophic processes, but one recent study indicates that species known to be mixotrophic were a dominant component of the chloroplast-bearing protists in Arctic sea ice after several months of polar night (Bachy et al., 2011).

Estimates of the total species richness of protozoa in sea ice do not yet exist. Nevertheless, a wide diversity of protozoa has been documented from high-latitude ecosystems and specifically from sea ice (Palmisano & Garrison, 1993; Garrison et al., 2005; Scott & Marchant, 2005; Bluhm et al., 2011; Poulin et al., 2011). Over the last few decades, extensive taxonomic or ecological analyses have been published for specific protozoan groups such as ciliates (Petz et al., 1995) or choanoflagellates (Marchant, 1985; Marchant & Perrin, 1990). These studies, using traditional approaches of microscopy and culture, have concluded that polar ecosystems are characterized by high protozoan diversity (Vørs, 1993; Petz et al., 1995). The description of a number of previously undescribed species of protozoa from sea ice appears to indicate that these microhabitats support unique protozoan assemblages and endemic species (Song & Wilbert, 2000).

Genetic approaches for characterizing and studying protistan species have begun to strongly complement traditional microscopical approaches, and thereby expand our knowledge of the breadth of microbial diversity in sea ice as they have in other ecosystems (Vaulot et al., 2008; Caron et al., 2012). Genetic methods have the advantage that they can characterize species composition across a wide taxonomic range of organisms. Traditional, morphology-based taxonomies of protists are highly taxon-specific, and their application is dependent on considerable expertise. Molecular approaches are able to discriminate small protists that often exhibit few distinctive morphological features (Slapeta et al., 2006; Stoeck et al., 2008), such as the marine stramenopile (MAST) cells (Massana et al., 2004) that were subsequently found in Arctic seas as well (Lovejoy et al., 2006). It is not yet possible to easily distinguish some protozoa (heterotrophic protists) from phototrophic or mixotrophic protists using DNA sequence information alone but, as they improve, genetic analyses will become indispensible for characterizing the phylogenetically diverse protistan assemblages present in sea ice habitats.

15.3.1 Ciliates

Ciliated protozoa constitute a conspicuous component of the protozoan assemblages colonizing sea ice (Figure 15.6). Morphological studies have documented a very wide diversity of taxa from ice habitats of this monophyletic protistan lineage (Stoecker et al., 1993; Petz et al., 1995; Lynn, 2008).

A number of the ciliates associated with sea ice are also common inhabitants of the water column at the same locations (Garrison, 1991). For example, several tintinnid species (Figure 15.6b,c,f,j,k) and strombidiid ciliates are typical filter-feeding 'planktonic' forms whose presence in ice presumably indicates the existence of significant quantities of water within some ice microhabitats. *Mesodinium rubrum*, a kleptoplastidic

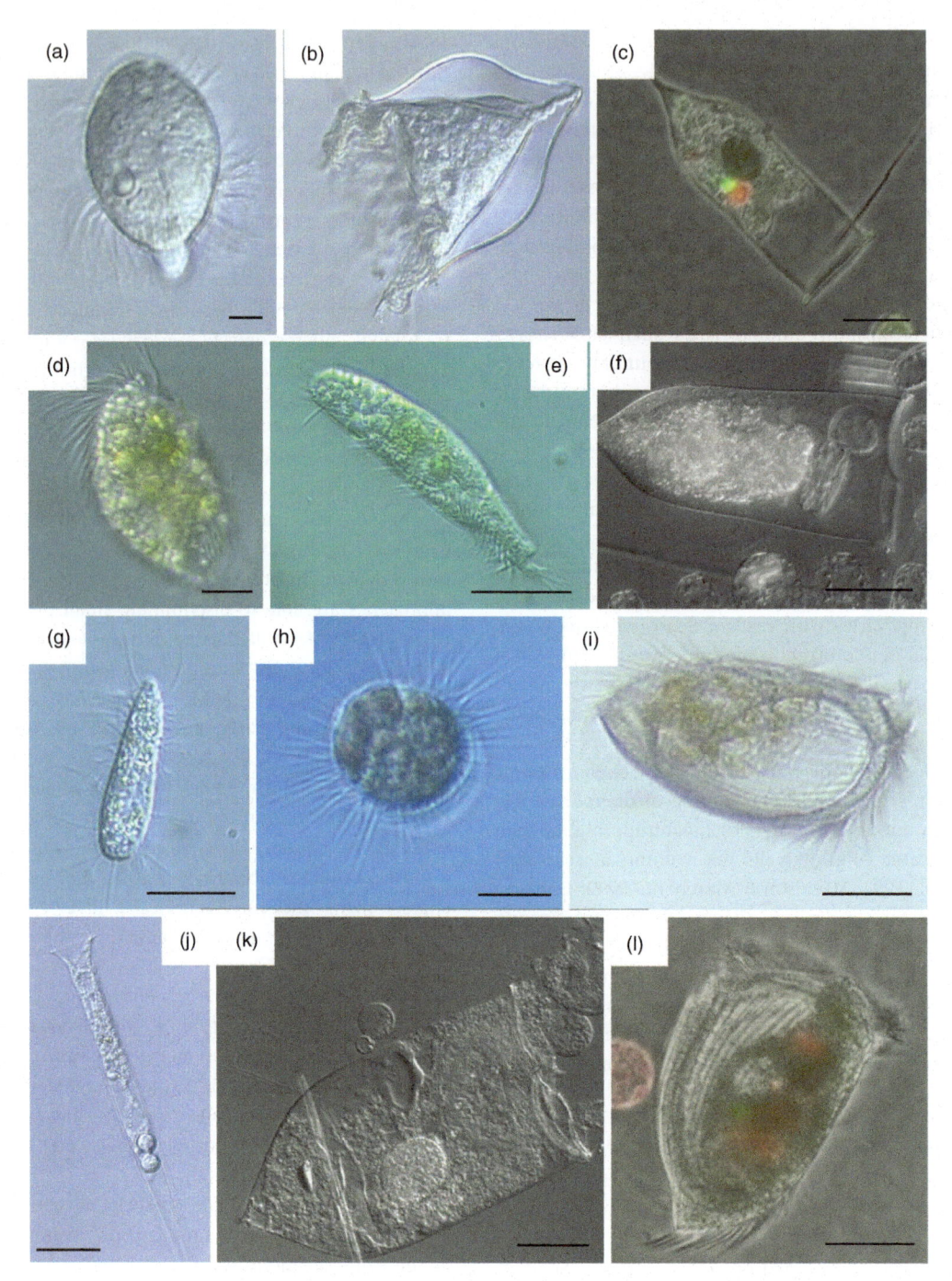

Figure 15.5 Ciliates of sea ice microhabitats in the Ross Sea, Antarctica. (a) *Didinium*, a predatory ciliate. (b, c, f, j, k) Species of tintinnids often encountered in the meltwater assemblages at the ice–snow interface of pack ice are also common in the plankton. (d) An unidentified hypotrich ciliate possessing many food vacuoles filled with algal prey. (e) A heterotrich species with elongated shape characteristic of benthic ciliates. (g) A scuticociliate that displays bacterivory in culture. (h) *Mesodinium* sp., a chloroplast-retaining ciliate. (i, l) Algal prey in food vacuoles in two large ciliates, possibly *Chlamydonella* sp., are evidence of herbivory in sea ice. The red colour in (l) is chlorophyll fluorescence from ingested algae. Scale bars are 10 (a, g), 20 (b, c, d, f, h, k) and 40 μm (e, i, j, l).

Sea ice as a habitat for micrograzers 379

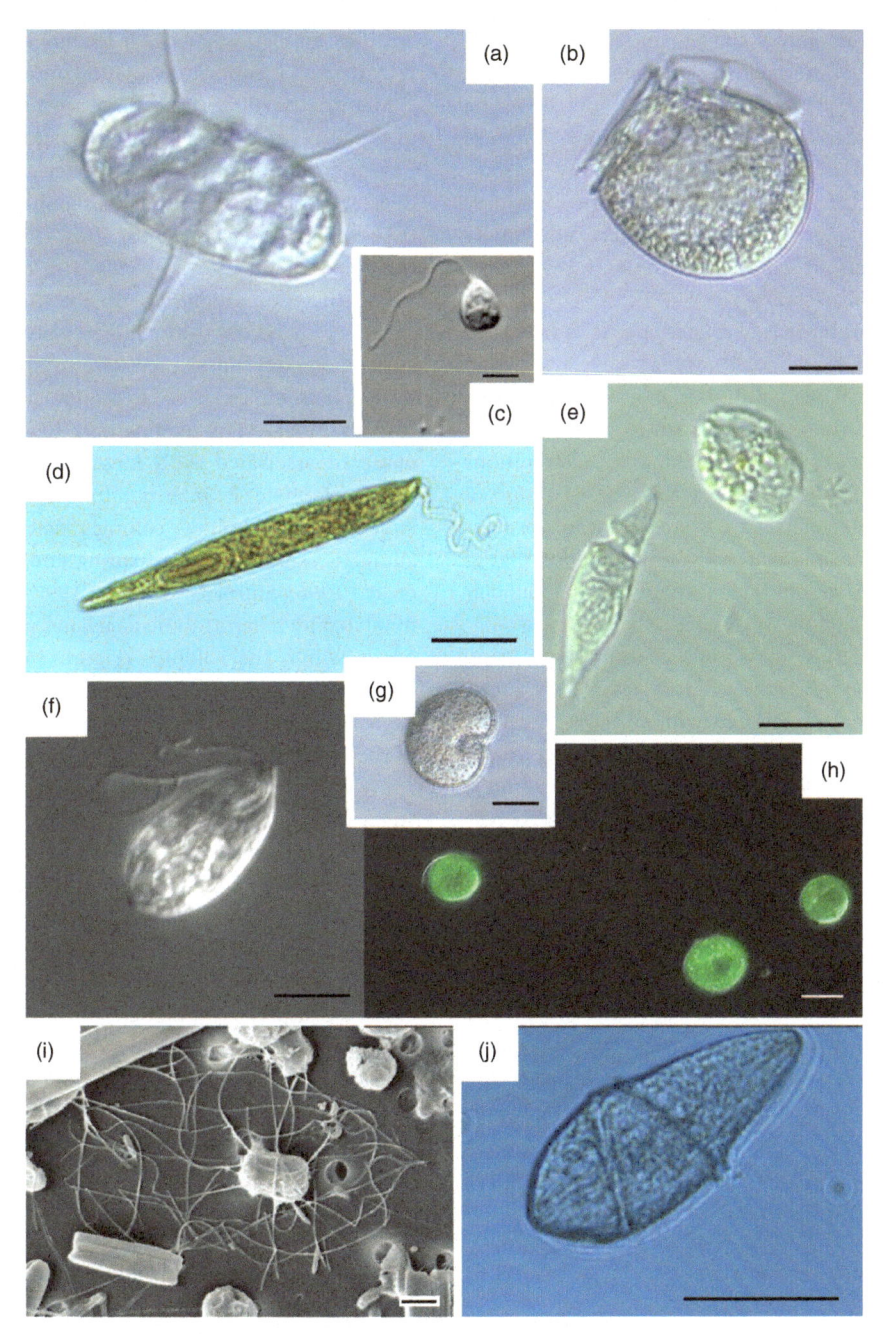

Figure 15.6 Heterotrophic flagellates are taxonomically diverse and often numerically dominant within sea ice. Heterotrophic dinoflagellates (a, b, e, g, h, j) such as *Protoperidinium* (a), *Dinophysis* (b), *Katodinium* (e, left), *Protoperidinium* (g) and *Gyrodinium* (j) may be major consumers of phytoplankton in the ice. (h) An unidentified dinoflagellate displaying a distinctive apple-green fluorescence with blue light excitation. Many heterotrophic dinoflagellates exhibit this fluorescent signal, making them very conspicuous when viewed by epifluorescence microscopy. (c) *Pseudobodo* sp., a small bacterivorous species. (d) A heterotrophic euglenoid flagellate stained with Lugol's iodine to visualize the cell and flagellum. (f) An unidentified flagellate cultured from the Ross Sea, Antarctica. (i) Scanning electron micrograph of the choanoflagellate *Diaphanoeca* sp. showing the cell and the lattice of the lorica (thin structure of bars surrounding the cell). Choanoflagellates are often extremely abundant in the plankton of high-latitude ecosystems, and are also common in sea ice microbial communities. Scale bars are 5 (c, f, i), 20 (a, b, d, e, g, h) or 50 μm (j).

(i.e. chloroplast-retaining) ciliate, can sometimes comprise up to 25% of the ciliate assemblage in sea ice, and over a third of the plankton assemblage in neighbouring waters (Sime-Ngando et al., 1997a). Brine channels, the slush habitats that develop at the ice–snow interface of summer sea ice, and melt ponds (Figure 15.3) may provide environmental conditions that adequately mimic conditions in many planktonic (i.e. non-ice) ecosystems in the ocean.

The global distributions of ciliates appear to vary with species. Several species of tintinnids and strombidiid taxa (e.g. *Strombidium*) are abundant in the marine plankton worldwide (Pierce & Turner, 1992), while other species have only Arctic or Antarctic distributions (Dolan & Pierce, 2012). Petz et al. (2007) conducted taxonomic studies of ciliates from freshwater ecosystems in the Arctic and Antarctic and concluded that at least some of the species (i.e. morphotypes) had bipolar distributions.

In contrast to truly 'planktonic' morphotypes, many of the ciliate species described from sea ice are forms that are commonly associated with surfaces, the benthos or eutrophic conditions that are not typical of oceanic plankton (Figure 15.6d,e,g,i,l). For example, species of *Lacrymaria* and *Condylostoma* are large ciliates often observed within dense accumulations of microbial cells and debris. These species possess highly elongated, flexible bodies that facilitate movement and prey capture in confined spaces. Spirotrich, heterotrich and hypotrich ciliate species that possess fused ciliary structures such as membranelles and cirri are mobile along surfaces where they consume surface-associated prey. Such taxa have been commonly observed in the pack ice in the Ross Sea (Figure 15.6d,e,i,l), and in landfast ice of Baffin Bay (Michel et al., 2002), but are rarely observed in the plankton. Many of these latter species show evidence of a herbivorous mode of nutrition (Figure 15.6c,d,e,g,h,i,k,l). Other species, such as *Chlamydonella*, consume fairly large algal prey such as pennate diatoms (Petz et al., 1995).

Oligohymenophorean ciliates comprise a group of predominantly bacterivorous species that are often found in sea ice (Petz et al., 1995; Song & Wilbert, 2000; Scott et al., 2001). *Pseudocohnilembus* sp. isolated from Arctic sea ice preferentially ingested bacteria-sized particles when offered particles of different size (Scott et al., 2001). These species are capable of rapid growth, but they are best adapted for feeding at relatively high prey (bacteria) concentrations (Figure 15.6g).

Many of the ciliates noted here are rarely encountered in the water column. Prey abundances presumably are not sufficiently dense, or the pelagic habitat not conducive to the development of significant populations of these large, robust ciliates. Thus, it is unclear where the 'seed' populations of ciliate assemblages in the ice originate, although it is possible that low numbers rafted on particles through the water column serve as an inoculum but are too rare to appear commonly in water samples. One exception is the existence of substantial ciliate populations that are frequently observed associated with large algal colonies of the prymnesiophyte *P. antarctica* in the Ross Sea. These large (>1 mm) hollow colonies reach extraordinary abundances during austral spring and summer (Smith et al., 1996; Smith & Gordon, 1997), and are often inhabited by heterotrophic flagellates and ciliates living on or within the colonies (Figure 15.7). These large structures may provide a mechanism for inoculating sea ice with particle-associated ciliate taxa.

The morphological and behavioural adaptations of many of the ice ciliates imply that sea ice microhabitats create a 'false benthos' in polar marine waters that are analogous to microhabitats created by large detrital aggregates in pelagic environments, so-called 'marine snow' (Caron et al., 1986). Marine snow particles are colonized by ciliated protozoa equipped for movement on surfaces and in confined spaces. Similarly, the ciliates of sea ice display morphologies that presumably reflect a strong selection for forms capable of existence in physically complex, nutritionally rich microhabitats.

15.3.2 Heterotrophic and mixotrophic flagellates

Heterotrophic flagellates comprise a phylogenetically diverse group of protistan taxa with representatives from no fewer than six different protistan supergroups (Simpson & Roger, 2004; Burki et al., 2007) (Figure 15.5). Among these disparate taxa, our knowledge regarding species richness is probably most complete for the heterotrophic dinoflagellates because they often display high abundances, large size and distinctive morphologies. Species of *Protoperidinium*, *Gymnodinium*, *Gyrodinium* and *Polykrikos* are members of sea ice communities (Stoecker et al., 1993; Archer

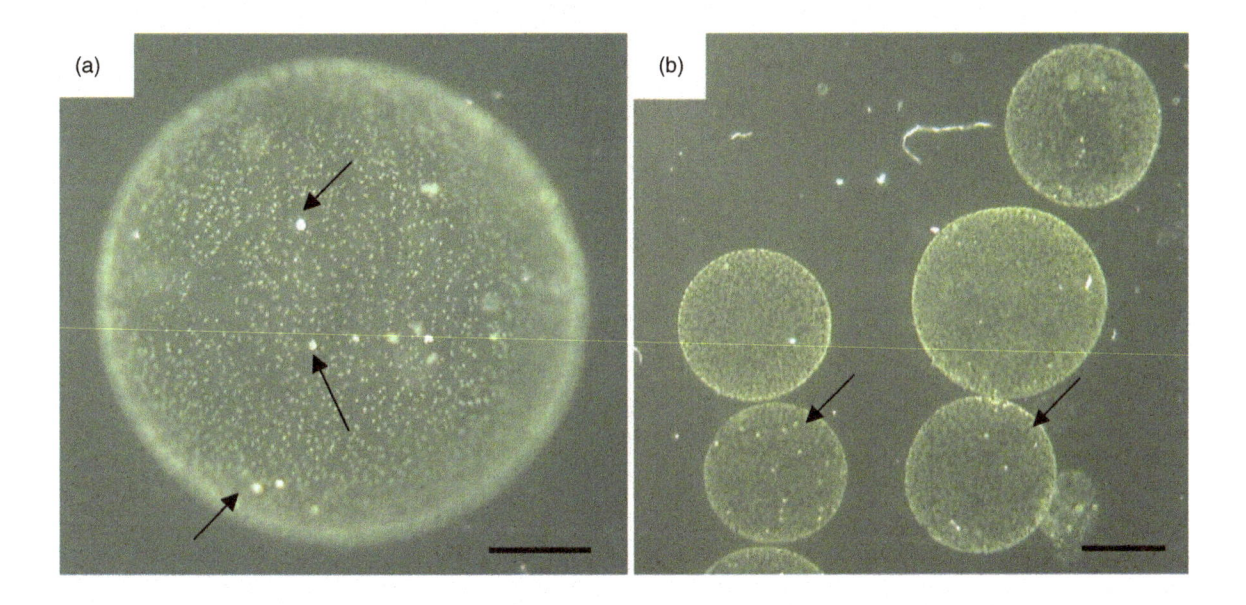

Figure 15.7 Colonization of *Phaeocystis antarctica* colonies by protozoa in the Ross Sea, Antarctica. Individual algal cells are visible in a single colony (a) as minute green dots. Arrows point to several heterotrophic protists associated with the colony. (b) Several colonies from a water sample. The interior of two of the colonies (arrows) are colonized with several heterotrophic protists each (white dots). Scale bars are 200 (a) and 500 (b) μm.

et al., 1996a; Michel et al., 2002), and are large enough to be easily visualized using a dissecting microscope.

Heterotrophic dinoflagellates (including kleptoplastidic taxa) are important herbivores in polar planktonic marine ecosystems (Sherr & Sherr, 2007, 2009; Sherr et al. 2009, 2013). Species inhabiting sea ice presumably exhibit similar trophic activities as their planktonic counterparts, although quantitative estimates of their grazing impact in ice microhabitats are virtually nonexistent. Many of these species feed using delicate pseudopodial projections to surround prey that are often larger than themselves (Jacobson & Anderson, 1986, 1996; Strom, 1991). Because sample handling and preparation for microscopy easily disrupt these feeding structures, it is probable that mixotrophic behaviour by dinoflagellates in sea ice has been severely underestimated. One dominant dinoflagellate from the Ross Sea has been shown to prey on *P. antarctica* and retain its chloroplasts for several months (Gast et al., 2007). Species of *Dinophysis* (Figure 15.5b) also capture and retain chloroplasts from their prey, the ciliate *Mesodinium rubrum*, which in turn have stolen chloroplasts from its cryptophyte prey (Park et al., 2006).

Choanoflagellates are frequent and abundant components of the heterotrophic plankton of Arctic and Antarctic coastal waters (Buck, 1981; Marchant, 1985; Caron et al., 1997; Sherr et al., 1997; Tong et al., 1997; Lovejoy et al., 2006; Poulin et al., 2011). Blooms of these minute (usually < 10 μm) bacterivorous species often appear in the upper water column following the summer maxima in bacterial abundances, and are also found in sea ice in the Arctic (Ikävalko & Gradinger, 1997) and Antarctic (Garrison et al., 2005) (Figure 15.5i). Our knowledge of their diversity is aided by the intricate and taxonomically informative skeletal structures produced by these species. Together with *Cryothecomonas*-like cercozoa, choanoflagellates dominated the heterotrophic flagellate assemblages in the ice communities of Saroma-ko lagoon in Japan (Sime-Ngando et al., 1997b).

Cercozoan flagellates of the genus *Cryothecomonas* are also ubiquitous heterotrophic flagellates of Antarctic (Garrison & Buck, 1989; Thomsen et al., 1991; Garrison

& Close, 1993; Stoecker et al., 1993; Gast et al., 2006), and Arctic sea ice (Ikävalko & Gradinger, 1997; Bachy et al., 2011; Thaler & Lovejoy, 2012, 2015). These species appear to be particularly well adapted for survival and growth at extreme temperatures.

Much less information is known regarding the many small, morphologically-nondescript groups of heterotrophic flagellates in sea ice. These species contribute significantly to bacterial mortality in aquatic ecosystems worldwide (Strom, 2000; Sherr & Sherr, 2002), but their overall abundance and diversity are not well known. This is due to the relative absence of taxonomically informative morphological features for most of these species (the choanoflagellates are an exception because they possess distinctive collar structures and loricae) and the increasing tendency to incorporate physiological information into the species concept (Boenigk, 2008). In any event, chrysomonad (especially *Paraphysomonas* spp.), bodonid (Figure 15.5c) and heterotrophic euglenoid flagellates (Figure 15.7d) are commonly observed in enrichment cultures established from samples of sea ice.

15.3.3 Amoeboid forms

Amoeboid protozoan species are a collection of phylogenetically diverse taxa that span a range from some of the least to the most striking protozoan forms observed in sea ice (Figure 15.8). Gymnamoebae or 'naked' amoebae (i.e. lacking skeletal structures) and heliozoa are easily cultured from sea ice samples. However, these species are only occasionally observed directly by microscopy (Garrison & Buck, 1989; Moran et al., 2007) because they are extremely difficult to distinguish when they are entangled with debris and other microbial taxa in natural samples of sea ice communities (Figure 15.8a,b,g). Amoebae are not unique to sea ice but are present in non-ice environments around Antarctica and in marine Arctic waters (Vørs, 1993; Mayes et al., 1997; Tong et al., 1997; Lovejoy et al., 2002).

Larger, much more noticeable amoeboid protozoa in sea ice include the foraminifera, acantharia and radiolaria. The foraminifera bear calcium carbonate tests, and at least one species, *Neogloboquadrina pachyderma*, is a consistent inhabitant of sea ice where it can attain significant abundances (Figure 15.8e). Acantharia also appear in sea ice, but infrequently (Figure 15.8c,d,f,i). Species identifications are based on their distinctive skeletons of strontium sulphate. These structures dissolve rapidly in most preservatives, and can thus be easily overlooked in preserved samples. The taxonomy of these species is difficult, and as a result, their overall diversity and phylogeny are still poorly known (Decelle et al., 2012). Most of these species possess symbiotic algae that are held within the cytoplasm of the host and which supplement the nutrition of these phagotrophic protists. Acantharian host–symbiont

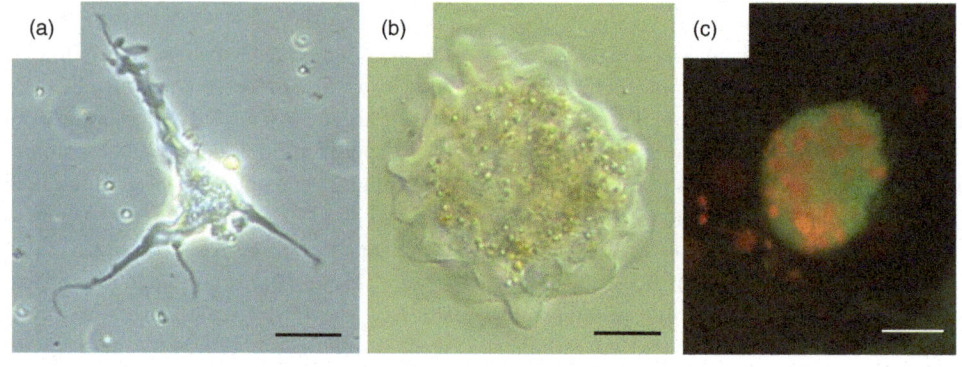

Figure 15.8 Amoeboid protozoa of the plankton and sea ice of the Ross Sea, Antarctica. Gymnamoebae (a, b, g) are inconspicuous, particle-associated protozoa. Acantharia (c, d, f, i) occur in the water column but are also observed in slush microhabitats in apparently good physiological condition. Red-fluorescing structures in a specimen observed by epifluorescence microscopy (c) indicate the presence of endocellular symbiotic algae in the same specimen examined by transmitted light microscopy (f). Endosymbionts are also present as yellowish-brown areas in the light micrographs of specimens in (d) and (i). (e) The 'planktonic' foraminifer *Neogloboquadrina pachyderma* is often present in sea ice. (h, j, k) Heliozoa and radiolaria. The unusual heliozoan *Sticholonche* sp. (h) and phaeodarian radiolaria (j, k) are occasionally observed in sea ice, but are more commonly found at the pack ice edge. (k) *Protocystis* sp. Scale bars are 10 (a, b, g, i), 20 (d, h, k), 50 (c, f, j) and 100 μm (e).

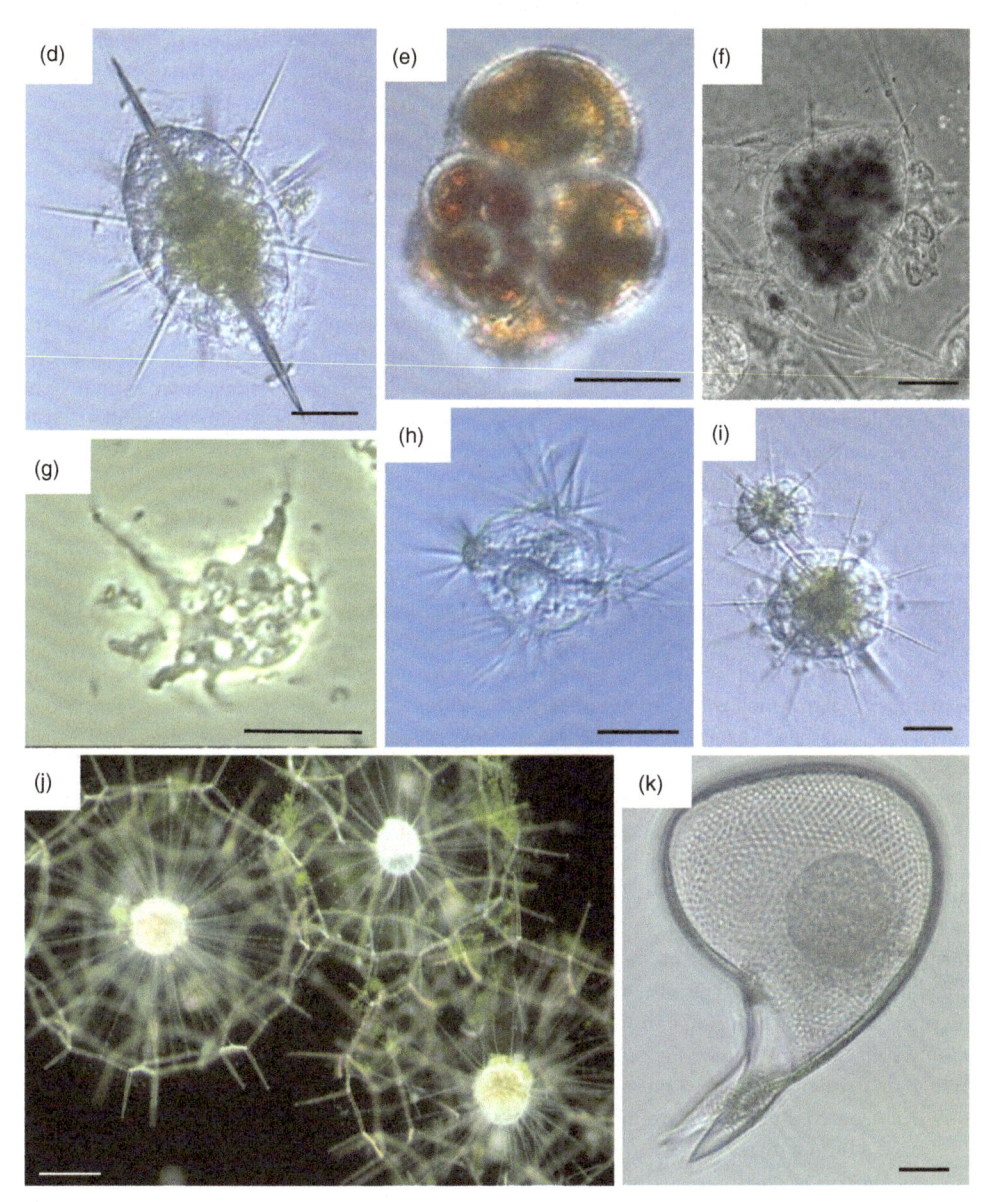

Figure 15.8 (*Continued*)

associations are abundant in low-latitude ecosystems where the acantharia can contribute significantly to primary productivity and to the vertical distribution of strontium (Michaels, 1991; Caron et al., 1995). It is unclear if the acantharia associated with sea ice habitats are actively growing and reproducing, but specimens in apparently good physiological condition can be recovered from melted sea ice (Caron and Gast, personal observation; Figure 15.8c,d,f,i).

Phaeodarian radiolarians (Figure 15.8j,k) are common in the ice edge regions of Antarctic coastal seas where they can reach abundances comparable to other larger sarcodine protozoa, such as foraminifera and polycystine radiolaria (Gowing, 1989; Gowing & Garrison, 1992). These populations are omnivorous in the water column, consuming prey ranging from bacteria to algae and other protozoa, based on analyses of food vacuole contents. Radiolarians have also been found at the ice edge in Arctic waters, where juveniles increase in importance as the radiolarian community gets closer to the ice (Swanberg & Eide, 1992). As for other larger amoeboid protozoa, it is not clear whether they are metabolically active within sea ice microhabitats.

15.4 Ecology and biogeochemistry of sea ice protozoa

15.4.1 Trophic behaviours (herbivory, bacterivory and mixotrophy)

Many of the trophic activities ascribed to protozoa in sea ice have been derived from observations of food vacuole contents of recently collected specimens (e.g. Figure 15.6d,i,l). Some information has also been obtained from experimental work conducted on sea ice microbes thawed into seawater (Moorthi et al., 2009), but the results of such studies are difficult to interpret due to possible artifacts of removing microbes from the spatially structured environment of the ice. Observations and experiments performed on samples collected from the water column near sea ice have also been extrapolated to provide information on the probable trophic activity of these species in ice microhabitats. As noted earlier, several species found in the water column may also be encountered in sea ice, so there is reason to believe that many of the same processes take place in both habitats.

Direct measurements of the grazing activities of phagotrophic protists in sea ice typically require the introduction of marker cells for tracing and quantifying consumption, which in turn causes drastic changes in the physical matrix of the ice. Generally, these experiments have been conducted by carefully melting sea ice in a large volume of seawater to minimize osmotic shock when the ice melts, and then applying routine protocols for the measurement of bacterivory or herbivory (Caron, 2000). The physical constraints imposed by sea ice, the very low temperatures that can be experienced in sea ice during winter, and the considerable dilution that microbial populations undergo during thawing and experimental manipulations may alter the rates, and perhaps the very nature, of the trophic interactions between heterotrophic protists and potential prey populations.

As a consequence, there have been very few quantitative measurements of herbivory in sea ice, and virtually no work on this topic until approximately the last decade. Indeed, some early reports hypothesized that algal growth took place largely in the absence of grazing pressure. This hypothesis is not surprising given that chlorophyll concentrations in excess of $1000\,\mu g\,l^{-1}$ have been observed (Chapter 14), and primary production by sea ice algae can constitute a significant fraction of total system productivity (Arrigo & Thomas, 2004). However, the high abundances of heterotrophic protists in some samples of sea ice are circumstantial evidence that significant herbivory and/or bacterivory occur there. Strong correlations such as that between ciliate abundances and those of pennate diatoms in Arctic sea ice also imply a predator–prey relationship (Gradinger et al., 1992). Anecdotal information that grazing takes place is further provided by shifts in the proportion of phototrophic to heterotrophic or mixotrophic protists in sea ice during summer and autumn (Gradinger et al., 1992; Stoecker et al., 1993; Garrison et al., 2003, 2005; Bachy et al., 2011).

As noted earlier, heterotrophic dinoflagellates are important herbivores in high-latitude ecosystems (Archer et al., 1996a; Becquevort, 1997; Levinsen & Nielsen, 2002; Sherr & Sherr, 2007). These species are particularly well suited for consuming large diatoms that often dominate coastal polar seas. Dinoflagellate species also reach high abundances in sea ice habitats, and leave tangible evidence of herbivory. Buck et al. (1990) reported an abundant ($>10^5\,L^{-1}$), heterotrophic dinoflagellate that produced 'faecal pellets' containing large numbers of diatom frustules. These faecal pellets are commonly encountered in sea ice samples (Figure 15.9), indicating that the species producing these pellets are probably important consumers of sea ice diatoms (compare figure 15.9 with figure 1E in Nöthig & Bodungen, 1989). As noted above, at least one dinoflagellate common in sea ice also plays a role as a consumer of the prymnesiophyte *P. antarctica*, from which it steals and maintains its chloroplasts (Gast et al., 2007; Torstensson et al., 2015).

Observations of the microbial communities of landfast ice in McMurdo Sound, Antarctica, indicate the potential importance of flagellate taxa other than dinoflagellates as herbivores in sea ice. Stoecker et al. (1993) noted high numbers of cysts of phototrophic protists in sea ice during austral winter (Figure 15.4), many of which excysted during austral spring as day length became sufficient for vegetative growth.

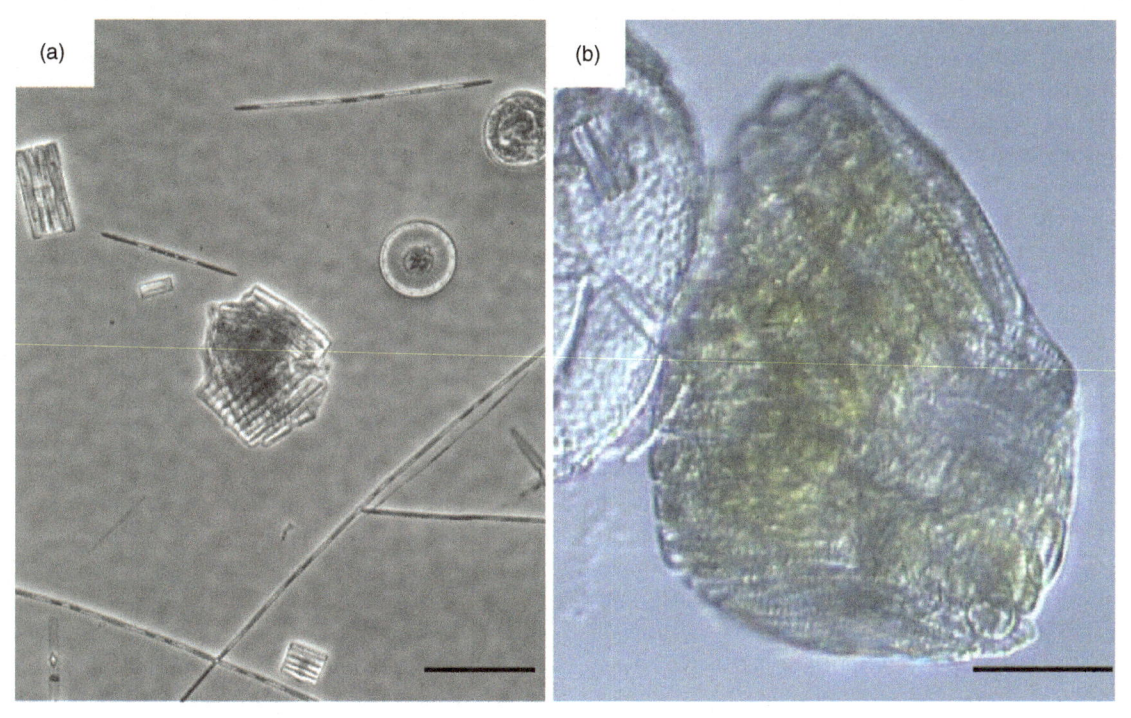

Figure 15.9 Light micrographs of the faecal pellets of heterotrophic dinoflagellates. These membrane-bound pellets are often encountered in samples of sea ice containing high abundances of diatoms. Scale bars are 100 (a) and 50 μm (b).

The appearance of large populations of heterotrophic protists, in particular the cercozoan *Cryothecomonas*, coincided with high abundances of the phototrophic protists that were presumed its prey (Stoecker et al., 1997; Stoecker et al., 1998). *Cryothecomonas* spp. also use pseudopodia to feed on large photosynthetic prey such as diatoms (Schnepf & Kühn, 2000).

Large ciliates are also potentially important herbivores within sea ice (note ingested prey in Figure 15.6), although based on the structure of the oral apparatus, many of these species are better adapted for the consumption of non-diatom prey (Dolan, 2010). For example, many of the species in the dense diatom assemblages at the bottom of landfast ice may not be suitable prey for *Strombidium* spp., which typically consume small algae and bacteria (Michel et al., 2002).

Bacterivory in sea ice is probably dominated by a variety of heterotrophic flagellates (Leakey et al., 1996; Becquevort, 1997; Vaqué et al., 2004). Gradinger et al. (1992) noted a ratio of bacteria to heterotrophic flagellates in Arctic sea ice of approximately 1000:1, a relationship typical of pelagic ecosystems where these assemblages exhibit strong predator:prey coupling (Sanders et al., 1992). Choanoflagellates appear to be particularly abundant bacterivores in polar ecosystems (Becquevort, 1997; Caron et al., 1997). However, these minute bacterivores are adapted for feeding on small planktonic bacteria, while many of the bacteria associated with sea ice are attached and filamentous. It is not clear how bacterivores adapted for feeding on free-living bacteria are effective at feeding on sea ice bacteria. Thus, the primary consumers of these ice-attached bacteria are not well known.

Measurements of bacterial ingestion in sea ice are rare. Laurion et al. (1995) and Sime-Ngando et al. (1999) examined bacterivory in Arctic ice and sea ice from a Japanese lagoon, respectively. These studies employed fluorescently labelled bacteria to examine bacterial grazing by small protozoa. Both studies noted that bacterivory was an important trophic activity among small ice protozoa, but further work is needed to better define and quantify this trophic connection.

Omnivory is suspected but unconfirmed in many sea ice protozoa, although some feeding specialists undoubtedly exist. For example, the predatory ciliate *Didinium* (Figure 15.6a) is a predator of other ciliates.

Yet many strains of sea ice protozoa cultured in the laboratory will consume a mixture of algae, bacteria and other small protozoa. Phaeodarian radiolaria are particularly well known for their omnivorous diets (Gowing, 1986, 1989; Nöthig & Gowing, 1991; Gowing & Garrison, 1992). Specimens collected from polar waters often possess numerous food vacuoles that contain the remains of a wide variety of prey. Foraminifera, polycystine radiolaria and acantharia are also known to be feeding generalists in the water column (Caron & Swanberg, 1990; Swanberg & Caron, 1991), but resolving the trophic activities of these larger amoeboid protozoa in sea ice will require further study.

Appreciation of the importance of mixotrophic nutrition by protozoa in Arctic and Antarctic food webs is recent, but expanding rapidly. This broad category of behaviours encompasses phototrophic algae that are capable of ingesting food particles, heterotrophic protists (predominantly ciliates and dinoflagellates) that consume algae and retain the functional chloroplasts of their prey, and symbiont-bearing heterotrophic protists, including many of the radiolaria, acantharia and foraminifera.

The extent of mixotrophic nutrition by phototrophic protists in polar regions is a relatively recent discovery, but is now being recognized as a common ecological strategy in both Arctic and Antarctic ecosystems (Moorthi et al., 2009; Bachy et al., 2011; Sanders & Gast, 2011). A surprising finding of these studies is that very small algae that were previously believed to be exclusively phototrophic appear capable of consuming small bacteria. For example, it has recently been demonstrated that the minute prasinophytes *Pyramimonas gelidicola* and *Micromonas* sp. are both capable of bacterial ingestion (Bell & Laybourn-Parry, 2003; McKie-Krisberg & Sanders, 2014). These species are common in water and sea ice samples from the Antarctic and Arctic, respectively. The contribution of mixotrophic activity by these species to bacterivory and to their own algal nutrition is presently unknown.

Kleptoplastidic ciliates, including *Strombidium* spp. and *Mesodinium rubrum*, have been observed in the brine channels of Antarctic and Arctic sea ice (Stoecker et al., 1993; Archer et al., 1996b; Michel et al., 2002). *M. rubrum* was often the most frequently occurring ciliate species in these studies. These chloroplast-retaining ciliates were also common in such disparate locales as the pack ice of the Ross Sea and Weddell Sea, and in sea ice of Saroma-ko lagoon (Japan) and Baffin Bay (Garrison & Buck, 1989; Michel et al., 2002; Garrison et al., 2005). The documentation of kleptoplastidy in a novel heterotrophic dinoflagellate of the Ross Sea indicates our rudimentary knowledge regarding the importance of this nutritional strategy in polar microbial ecology (Gast et al., 2007). The potential importance of this behaviour is evidenced by the fact that DNA sequences of this kleptoplastidic, heterotrophic dinoflagellate dominated samples collected from the water column and the microbial communities, as well as by microscopic observation of the dinoflagellate in the slush at the snow–ice interface of Ross Sea pack ice (Figure 15.2i,k).

The presence of symbiont-bearing acantharia in sea ice has been noted previously (Figure 15.8c,d,f,i). These symbiont–host associations are exceptionally abundant in the water column at various tropical and subtropical locations where they can form a significant fraction of the total primary production (Caron et al., 1995). High-latitude regions appear to have substantially smaller contributions of these associations, but this strategy still appears to be important for the autecology of some polar taxa.

15.4.2 Biogeochemical processes and rates

Protozoa play significant biogeochemical roles in all natural ecosystems on Earth. Their trophic activities result in the repackaging of (usually) smaller prey, such as bacteria, small algae and small protozoa, into larger cells that serve as food for other protozoa and metazoa. In the process, a percentage of the prey biomass is released as dissolved and particulate organic compounds that serve as substrates for bacterial growth, or are remineralized completely to inorganic nutrients that fuel primary productivity (Sherr et al., 2007). These species undoubtedly play similar roles in sea ice, but our understanding of the extent of their activities is only beginning to develop, and few models of energy or elemental flow within sea ice communities have thus far been constructed. Vézina et al. (1997) developed a model to examine carbon flow through the sea ice microbial community of Resolute Passage, Canadian High Arctic. The approach used changes in standing stocks of microbial assemblages to infer carbon flow. An important role of herbivory by flagellates and ciliates was indicated by the results of their modelling effort.

While modelling, field observations and experimental studies indicate an important role for protozoa in sea ice, the fate of carbon and energy entering the microbial food webs in these extremely cold environments is less clear. Temperature directly affects the metabolic rates of microorganisms, and might also change the eventual fate of microbial biomass through its influence on growth rates and growth efficiencies of the microconsumers. There is no question that many species of protozoa can survive and grow at low polar *in situ* temperatures. However, it is probable that most sea ice protozoa do not exhibit true physiological compensation, i.e. maintain constant physiological rates in the face of changing temperature (Peck, 2002). Studies with Antarctic ciliates did not indicate any significant degree of physiological compensation; growth at low temperatures was substantially slower than growth at higher temperatures (Fenchel & Lee, 1972; Lee & Fenchel, 1972). Low temperature has also been reported to greatly depress the growth rates of Antarctic amoebae (Mayes et al., 1997).

Rose and Caron (2007) summarized information on the growth rates of phototrophic and heterotrophic protists from a wide variety of studies, and concluded that the effect of temperature on maximal potential growth rates under optimal conditions (abundant prey or nutrients) is not the same for phototrophic and heterotrophic protistan species. Maximal growth rates for heterotrophic protists (both bacterivorous and herbivorous) decrease more rapidly with decreasing temperature than maximal growth rates for phototrophic protists. Differences between maximal potential growth rates of phototrophs and heterotrophs are most acute at the temperatures experienced by sea ice. Thus, the dominance of total microbial biomass by dense algal accumulations in sea ice during spring may be due, in part, to more rapid algal growth, whereas protozoan growth is constrained by extremely low temperatures. Net growth rates of heterotrophic protists in Antarctic sea ice estimated on changes in abundances in the ice have indicated relatively slow rates, 0.005–0.5 day^{-1} (Laurion et al., 1995; Archer et al., 1996b; Levinsen et al., 2000), in general agreement with the overall constraint of low temperature on metabolic rate.

Studies of the rates of bacterivory and herbivory in sea ice and in coastal polar waters also support the controlling effect of temperature on overall protozoan rate processes. Decreasing water temperature was shown to significantly decrease microzooplankton (predominantly protistan) herbivory when this process was compared over a large range of environmental temperatures (see summary and references in Caron et al., 2000). Experiments to characterize bacterivory in sea ice (Laurion et al., 1995; Sime-Ngando et al., 1999) also have often yielded low rates (i.e. a few bacteria per protistan consumer per hour). These grazing rates are commensurate with rates expected from other ecosystems with low *in situ* temperatures (Berninger et al., 1991; Leakey et al., 1996; Becquevort, 1997; Caron et al., 1997; Vaqué et al., 2004), but there are presently too few measurements to accurately determine the impact of sample handling in these experiments on the rates of bacterivory in sea ice specifically.

There are conflicting views at present regarding the direction and magnitude of the effect of temperature on the overall fate of microbial biomass consumed by protozoa in polar ecosystems. Growth at low temperature has been reported to increase (Choi & Peters, 1992), dramatically decrease (Mayes et al., 1997) or leave unaffected (Rose et al., 2008) the gross growth efficiencies of protozoa. These differing results are probably due, at least in part, to the fact that measurements of gross efficiency are difficult to make, and are particularly exacerbated by potential artifacts at low environmental temperature (Rose et al., 2013). Nonetheless, the effect of low temperature on protozoan growth efficiency has important implications for energy flow and elemental recycling in sea ice. Low growth efficiencies imply that protozoa act primarily as agents for the decomposition of organic matter and the remineralization of nutrients contained in their microbial prey. On the other hand, high growth efficiencies would suggest a much more important role for protozoa as intermediate steps in the food webs of sea ice biological communities. Resolution of these apparently conflicting observations awaits further study.

15.5 Conclusions and future directions

Studies of sea ice protozoan assemblages during the past few decades provide a basic understanding of the major taxa occurring in sea ice and their contributions to total microbial biomass. Many important trophic interactions involving these species have been observed and characterized, derived from observations of protozoa in sea ice samples, extrapolated from our knowledge of the activities of the same or similar species in the water column, or established from observations and experimental studies on cultured species.

Major gaps in our knowledge remain, particularly regarding the physiological rates (feeding, growth and mortality rates) of sea ice microbes in intact ice microhabitats. Evidence of feeding activity, such as food vacuole contents and feeding behaviour in laboratory cultures, provide information on the types of prey consumed by protozoa, but it is unclear how this information extrapolates to actual rates of consumption and growth in the ice matrix. Similarly, information on the growth efficiencies of protozoa at environmentally relevant temperatures must be obtained for species that are ecologically important in ice microhabitats in order to establish their impact on decomposition, nutrient remineralization and trophic transfer. These goals are not without significant methodological hurdles and potential artifacts, but this information is key to understanding the overall biogeochemical importance of protozoa to sea ice microbial community structure and function.

Studies on protozoan physiological adaptations, species diversity and abundances within sea ice microhabitats will further improve our understanding of community structure and function in polar ecosystems, and the diverse roles that protozoa play in oceanic biogeochemical cycles. Isolation and culture of these species will provide genetic fodder for biotechnological exploitation of their unique physiological capabilities (Caron et al., 2012). Additionally, information from sea ice protistan species living at the cold extreme of environmental temperature will help to address difficult issues relating to the morphology-based species concept of protists, the existence of cryptic species, and the distribution of these taxa on our planet. Future studies of their diversity and ecology will undoubtedly continue to incorporate genetic approaches, but these findings must be interpreted in conjunction with traditional approaches of microscopy, culture, experimentation and field-based ecological observations.

Acknowledgements

The authors thank Julie M. Rose, Dawn M. Moran, Mark R. Dennett, Peter D. Countway, Astrid Schnetzer, Stephanie Moorthi, Robert W. Sanders and Paige Connell for contributing some of the micrographs. The authors are grateful for field support by the captains, crew and marine technical staff of the RVIB *Nathaniel B. Palmer* and the USCGC *Healy*. Preparation of the chapter was supported by NSF grants OPP-9714299, OPP-0125437 and OPP-0542456, and PLR-1341362.

References

Adl, S.M., Simpson, A.G.B., Lane, C.E. et al. (2012) The revised classification of eukaryotes. *Journal of Eukaryotic Microbiology*, **59**, 429–514.

Archer, S.D., Leakey, R.J.G., Burkill, P.H. & Sleigh, M.A. (1996a) Microbial dynamics in coastal waters of east Antarctica: herbivory by heterotrophic dinoflagellates. *Marine Ecology Progress Series*, **139**, 239–255.

Archer, S.D., Leakey, R.J.G., Burkill, P.H., Sleigh, M.A. & Appleby, C.J. (1996b) Microbial ecology of sea ice at a coastal Antarctic site: Community composition, biomass and temporal change. *Marine Ecology Progress Series*, **135**, 179–195.

Arrigo, K.R. & Thomas, D.N. (2004) Large scale importance of sea ice biology in the Southern Ocean. *Antarctic Science*, **16**, 471–486.

Bachy, C., López-García, P., Vereshchaka, A. & Moreira, D. (2011) Diversity and vertical distribution of microbial eukaryotes in the snow, sea ice and seawater near the North Pole at the dnd of the polar night. *Frontiers in Microbiology*, **2**, 106.

Becquevort, S. (1997) Nanoprotozooplankton in the Atlantic sector of the Southern Ocean during early spring: biomass and feeding activities. *Deep-Sea Research II*, **44**, 355–373.

Bell, E.M. & Laybourn-Parry, J. (2003) Mixotrophy in the Antarctic phytoflagellate, *Pyramimonas gelidicola* (Chlorophyta: Prymnesiophyceae). *Journal of Phycology*, **39**, 644–649.

Berninger, U.-G., Caron, D.A., Sanders, R.W. & Finlay, B.J. (1991) Heterotrophic flagellates of planktonic communities, their characteristics and methods of study. In: *The biology of free-living heterotrophic flagellates*, special volume **45** (Eds. D.J .& Patterson & J. Larsen), pp. 39–56. Clarendon Press, Oxford.

Bluhm, B.A., Gebruk, A.V., Gradinger, R. et al. (2011) Arctic marine biodiversity: An update of species richness and examples of biodiversity change. *Oceanography*, **24**, 232–248.

Boenigk, J. (2008) The past and present classification problem with nanoflagellates exemplied by the genus *Monas*. *Protist*, **159**, 319–337.

Buck, K. (1981) A study of choanoflagellates (Acanthoecidae) from the Weddell Sea, including a description of *Diaphanoeca multiannulata n.sp. Journal of Protozoology*, **28**, 47–54.

Buck, K.R., Bolt, P.A., Bentham, W.N. & Garrison, D.L. (1992) A dinoflagellate cyst from Antarctic sea ice. *Journal of Phycology*, **28**, 15–18.

Buck, K.R., Nielsen, T.G., Hansen, B.W., Gastrup-Hansen, D. & Thomsen, H.A. (1998) infiltration phyto- and protozooplankton assemblages in the annual sea ice of Disko Island, West Greenland, spring 1996. *Polar Biology*, **20**, 377–381.

Buck, K.R., Bolt, P.A., & Garrison, D.L. (1990) Phagotrophy and fecal pellet production by an athecate dinoflagellate in Antarctic sea ice. *Marine Ecology Progress Series*, **60**, 75–84.

Burki, F., Shalchian-Tabrizi, K., Skjaeveland, Å., Nikolaev, S.I., Jakobsen, K.S. & Pawlowski, J. (2007) Phylogenomics reshuffles the eukaryotic supergroups. *PLoS One*, **8**, E790.

Caron, D.A. (2000) Protistan herbivory and bacterivory. In: Marine Microbiology: Methods in Microbiology, Vol. 30 (Ed. J. Paul), pp. 289–315. San Diego: Academic Press Inc.

Caron, D.A., Countway, P.D., Jones, A.C., Kim, D.Y. & Schnetzer, A. (2012) Marine protistan diversity. *Annual Review of Marine Science*, **4**, 467–493.

Caron, D.A., Davis, P.G., Madin, L.P. & Sieburth, J.M. (1986) Enrichment of microbial populations in macroaggregates (marine snow) from the surface waters of the North Atlantic. *Journal Of Marine Research*, **44**, 543–565.

Caron, D.A., Dennett, M.R., Lonsdale, D.J., Moran, D.M. & Shalapyonok, L. (2000) Microzooplankton herbivory in the Ross Sea, Antarctica. *Deep-Sea Research*, **47**, 15–16.

Caron, D.A., Lonsdale, D.J. & Dennett, M.R. (1997) Bacterivory and herbivory play key roles in fate of Ross Sea production. *Antarctic Journal of the United States*, **32**, 81–83.

Caron, D.A., Michaels, A.F., Swanberg, N.R. & Howse, F.A. (1995) Primary productivity by symbiont-bearing planktonic sarcodines (Acantharia, Radiolaria, Foraminifera) in surface waters near Bermuda. *Journal of Plankton Research*, **17**, 103–129.

Caron, D.A. & Swanberg, N.R. (1990) The ecology of planktonic sarcodines. *Reviews in Aquatic Sciences*, **3**, 147–180.

Choi, J.W. & Peters, F. (1992) Effects of temperature on two psychrophilic ecotypes of a heterotrophic nanoflagellate, *Paraphysomonas imperforata*. *Applied and Environmental Microbiology*, **58**, 593–599.

Comeau, A.M., Philippe, B., Thaler, M., Gosselin, M., Poulin, M. & Lovejoy, C. (2013) Protists in Arctic drift and land-fast sea ice. *Journal of Phycology*, **49**, 229–240.

Davidson, A.T. & Marchant, H.J. (1992) Protist abundance and carbon concentration during a *Phaeocystis*-dominated bloom at an Antarctic coastal site. *Polar Biology*, **12**, 387–395.

Decelle, J., Suzuki, N., Mahé, F., de Vargas, C. & F., N. (2012) Molecular phylogeny and morphological evolution of the Acantharia (Radiolaria). *Protist*, **163**, 435–450.

Dennett, M.R., Mathot, S., Caron, D.A., Smith, W.O. & Lonsdale, D.J. (2001) Abundance and distribution of phototrophic and heterotrophic nano- and microplankton in the southern Ross Sea. *Deep-Sea Research Part Ii-Topical Studies in Oceanography*, **48**, 4019–4037.

Dolan, J.R. (2010) Morphology and ecology in tintinnid ciliates of the marine plankton: correlates of lorica dimensions. *Acta Protozoologica*, **49**, 235–244.

Dolan, J.R. & Pierce, R.W. (2012) Diversity and distributions of tintinnids. In: *The Biology and Ecology of Tintinnid Ciliates: Models for Marine Plankton* (Eds. J.R. Dolan et al.), pp. 214–243. John Wiley & Sons Ltd, New York.

Eddie, B., Juhl, A., Krembs, C., Baysinger, C. & Neuer, S. (2010) Effect of environmental variables on eukaryotic microbial community structure of land-fast Arctic sea ice. *Environmental Microbiology*, **12**, 797–809.

Eicken, H. (1992) The role of sea ice in structuring Antarctic ecosystems. *Polar Biology*, **12**, 3–13.

Fenchel, T. & Lee, C.C. (1972) Studies on ciliates associated with sea ice from Antarctica: I. The nature of the fauna. *Archiv für Protistenkunde*, **114**, 231–236.

Fonda Umani, S., Monti, M., Bergamasco, A. et al. (2005) Plankton community structure and dynamics versus physical structure from Terra Nova Bay to Ross Ice Shelf (Antarctica). *Journal of Marine Systems*, **55**, 31–46.

Fritsen, C.H., Memmmott, J. & Stewart, F.J. (2008) Inter-annual dynamics and micro-algal biomass in winter pack ice of Marguerite Bay, Antarctica. *Deep-Sea Research Part II*, **55**, 2059–2067.

Garrison, D.L. (1991) An overview of the abundance and role of protozooplankton in Antarctic waters. *Journal of Marine Systems*, **2**, 317–331.

Garrison, D.L., Ackley, S.F. & Buck, K.R. (1986) A physical mechanism for establishing algal populations in frazil ice. *Nature*, **306**, 363–365.

Garrison, D.L. & Buck, K.R. (1989) The biota of Antarctic pack ice in the Weddell Sea and Antarctic peninsula regions. *Polar Biology*, **10**, 211–219.

Garrison, D.L. & Buck, K.R. (1991) Surface-layer sea ice assemblages in Antarctic pack ice during the austral spring: environmental conditions, primary production and community structure. *Marine Ecology Progress Series*, **75**, 161–172.

Garrison, D.L., Buck, K.R. & Gowing, M.M. (1993) Winter plankton assemblage in the ice edge zone of the Weddell and Scotia Seas: composition, biomass and spatial distribution. *Deep-Sea Research*, **40**, 311–338.

Garrison, D.L. & Close, A.R. (1993) Winter ecology of the sea ice biota in Weddell Sea pack ice. *Marine Ecology Progress Series*, **96**, 17–31.

Garrison, D.L., Gibson, A., Coale, S.L. et al. (2005) Sea-ice microbial communities in the Ross Sea: autumn and summer biota. *Marine Ecology Progress Series*, **300**, 39–52.

Garrison, D.L., Jeffries, M.O., Gibson, A. et al. (2003) Development of sea ice microbial communities during autumn ice formation in the Ross Sea. *Marine Ecology Progress Series*, **259**, 1–15.

Gast, R.J., Dennett, M.R. & Caron, D.A. (2004) Characterization of protistan assemblages in the Ross Sea, Antarctica by denaturing gradient gel electrophoresis. *Applied and Environmental Microbiology*, **70**, 2028–2037.

Gast, R.J., Moran, D.M., Beaudoin, D.J., Blythe, J.N., Dennett, M.R. & Caron, D.A. (2006) Abundance of a novel dinoflagellate phylotype in the Ross Sea, Antarctica. *Journal of Phycology*, **42**, 233–242.

Gast, R.J., Moran, D.M., Dennett, M.R. & Caron, D.A. (2007) Kleptoplasty in an Antarctic dinoflagellate: caught in evolutionary transition? *Environmental Microbiology*, **9**, 39–45.

Gosselin, M., Legendre, L., Therriault, J.-C., Demers, S. & Rochet, M. (1986) Physical control of the horizontal patchiness of sea-ice microalgae. *Marine Ecology Progress Series*, **29**, 289–298.

Gowing, M.M. (1986) Trophic biology of phaeodarian radiolarians and flux of living radiolarians in the upper 2000 m of the North Pacific central gyre. *Deep-Sea Research*, **33**, 655–674.

Gowing, M.M. (1989) Abundance and feeding ecology of Antarctic phaeodarian radiolarians. *Marine Biology*, **103**, 107–118.

Gowing, M.M. (2003) Large viruses and infected microeukaryotes in Ross Sea summer pack ice habitats. *Marine Biology*, **142**, 1029–1040.

Gowing, M.M. & Garrison, D.L. (1992) Abundance and feeding ecology of larger protozooplankton in the ice edge zone of the Weddell and Scotia Seas during the austral winter. *Deep-Sea Research*, **39**, 893–919.

Gradinger, R., Friedrich, C. & Spindler, M. (1999) Abundance, biomass and composition of the sea ice biota of the Greenland Sea pack ice. *Deep-Sea Research II*, **46**, 1457–1472.

Gradinger, R. & Ikavalko, J. (1998) Organism incorporation into newly forming Arctic sea ice in the Greenland Sea. *Journal of Plankton Research*, **20**, 871–886.

Gradinger, R., Spindler, M. & Weissenberger, J. (1992) On the structure and development of Arctic pack ice communities in Fram Strait: a multivariate approach. *Polar Biology*, **12**, 727–733.

Grossi, S.M. & Sullivan, C.W. (1985) Sea ice microbial communities.V. The vertical zonation of diatoms in an Antarctic fast ice community. *Journal of Phycology*, **21**, 401–409.

Ikävalko, J. & Gradinger, R. (1997) Flagellates and heliozoans in the Greenland Sea ice studied alive using light microscopy. *Polar Biology*, **17**, 473–481.

Jacobson, D.M. & Anderson, D.M. (1986) Thecate heterotrophic dinoflagellates: feeding behavior and mechanisms. *Journal of Phycology*, **22**, 249–258.

Jacobson, D.M. & Anderson, D.M. (1996) Widespread phagocytosis of ciliates and other protists by marine mixotrophic and heterotrophic thecate dinoflagellates. *Journal of Phycology*, **32**, 279–285.

Janech, M.G., Krell, A., Mock, T., Kang, J.S. & Raymond, J.A. (2006) Ice-binding proteins from sea ice diatoms (Bacillariophyceae). *Journal of Phycology*, **42**, 410–416.

Jungblut, A.D., Vincent, W.F. & Lovejoy, C. (2012) Eukaryotes in Arctic and Antarctic cyanobacterial mats. *FEMS Microbiology Ecology*, **82**, 416–428.

Kaartokallio, H. (2004) Food web components, and physical and chemical properties of Baltic sea ice. *Marine Ecology Progress Series*, **273**, 49–63.

Kaartokallio, H., Kuosa, H., Thomas, D.N., Granskog, M. & Kivi, K. (2007) Biomass, composition and activity of organism assemblages along a salinity gradient in sea ice subjected to river discharge in the Baltic Sea. *Polar Biology*, **30**, 183–197.

Krembs, C., Gradinger, R. & Spindler, M. (2000) Implications of brine channel geometry and surface area for the interaction of sympagic organisms in Arctic sea ice. *Journal of Experimental Marine Biology and Ecology*, **243**, 55–80.

Kuosa, H., Borrman, B., Kivi, K. & Brandini, F. (1992) Effects fo Antarctic sea ice biota on seeding as studied in aquarium experiments. *Polar Biology*, **12**, 333–339.

Laurion, I., Demers, S. & Vézina, A.F. (1995) The microbial food web associated with the ice algal assemblage: biomass and bacterivory of nanoflagellate protozoans in Resolute Passage (High Canadian Arctic). *Marine Ecology Progress Series*, **120**, 77–87.

Leakey, R.J.G., Archer, S.D. & Grey, J. (1996) Microbial dynamics in coastal waters of East Antarctica: bacterial production and nanoflagellate bacterivory. *Marine Ecology Progress Series*, **142**, 3–17.

Lee, C.C. & Fenchel, T. (1972) Studies on ciliates associated with sea ice from Antarctica. II. Temperature responses and tolerances in ciliates from antarctic, temperate and tropical habitats. *Archiv für Protistenkunde*, **114**, 237–244.

Levinsen, H., Gissel Nielsen, T. & Winding Hansen, B. (2000) Annual succession of marine pelagic protozoans in Disko Bay, West Greenland, with emphasis on winter dynamics. *Marine Ecology Progress Series*, **206**, 119–134.

Levinsen, H. & Nielsen, T.G. (2002) The trophic role of marine pelagic ciliates and heterotrophic dinoflagellates in arctic and temperate coastal ecosystems: A cross-latitude comparison. *Limnology and Oceanography*, **47**.

Lovejoy, C., Legendre, L., Martineau, M.J., Bâcle, J. & von Quillfeldt, C.-H.H. (2002) Distribution of phytoplankton and other protists in the North Water. *Deep Sea Research Part II: Topical Studies in Oceanography*, **49**, 5027–5047.

Lovejoy, C., Massana, R. & Pedros-Alio, C. (2006) Diversity and distribution of marine microbial eukaryotes in the Arctic Ocean and adjacent seas. *Applied and Environmental Microbiology*, **72**, 3085–3095.

Lynn, D.H. (2008) *The Ciliated Protozoa: Characterization, Classification, and Guide to the Literature*. Springer, New York.

Majaneva, M., Rintala, J.-M., Piisilä, M., Fewer, D. & Blomster, J. (2012) Comparison of wintertime eukaryotic community from sea ice and open water in the Baltic Sea, based on sequencing of the 18S rRNA gene. *Polar Biology*, **35**, 875–889.

Marchant, H.J. (1985) Choanoflagellates in the Antarctic marine food chain. In: Antarctic Nutrient Cycles and Food Webs. (Eds. W.R. Siegfried et al.), pp. 271–276. Springer-Verlag, Berlin.

Marchant, H.J. & Perrin, R. (1990) Seasonal variation in abundance and species composition of choanoflagellates (Acanthoecideae) at Antarctic coastal sites. *Polar Biology*, **10**, 499–505.

Massana, R., Castresana, J., Balagué, V. et al. (2004) Phylogenetic and ecological analysis of novel marine stramenopiles. *Applied and Environmental Microbiology*, **70**, 3528–3534.

Mayes, D.F., Rogerson, A., Marchant, H. & Laybourn-Parry, J. (1997) Growth and consumption rates of bacterivorous Antarctic naked marine amoebae. *Marine Ecology Progress Series*, **160**, 101–108.

McKie-Krisberg, Z.M. & Sanders, R.W. (2014) Phagotrophy by the picoeukaryotic green alga *Micromonas*: implications for Arctic Oceans. *The ISME Journal*, **8**, 1953–1961.

Michaels, A.F. (1991) Acantharian abundance and symbiont productivity at the VERTEX seasonal station. *Journal of Plankton Research*, **13**, 399–418.

Michel, C., Gissel Nielsen, T.G., Nozais, C. & Gosselin, M. (2002) Significance of sedimentation and grazing by ice micro- and meiofauna for carbon cycling in annual sea ice (northern Baffin Bay). *Aquatic Microbial Ecology*, **30**, 57–68.

Michel, C., Legendre, L., Therriault, J.C., Demers, S. & Vandevelde, T. (1993) Springtime coupling between ice algal and phytoplankton assemblages in southeastern Hudson Bay, Canadian Arctic. *Polar Biology*, **13**, 441–449.

Montresor, M., Procaccini, G. & Stoecker, D.K. (1999) *Polarella glacialis*, gen. nov. sp. nov. (Dinophyceae): Suessiaceae are still alive! *Journal of Phycology*, **35**, 186–197.

Moorthi, S.D., Caron, D.A., Gast, R.J. & Sanders, R.W. (2009) Mixotrophy: a widespread and important ecological strategy for planktonic and sea-ice nanoflagellates in the Ross Sea, Antarctica. *Aquatic Microbial Ecology*, **54**, 269–277.

Moran, D.M., Anderson, O.R., Dennett, M.R., Caron, D.A. & Gast, R.J. (2007) A description of seven Antarctic marine Gymnamoebae including a new subspecies, two new species and a new genus: *Neoparamoeba aestuarina antarctica* n. subsp., *Platyamoeba oblongata* n. sp., *Platyamoeba contorta* n. sp. and *Vermistella antarctica* n. gen.n. sp. *Journal of Eukaryotic Microbiology*, **54**, 169–183.

Mundy, C.J., Gosselin, M., Ehn, J.K. et al. (2011) Characteristics of two distinct high-light acclimated algal communities during advanced stages of sea ice melt. *Polar Biology*, **34**, 1869–1886.

Nöthig, E.-M. & Bodungen, B.V. (1989) Occurrence and vertical flux of faecal pellets of probably protozoan origin in the southeastern Weddell Sea (Antarctica). *Marine Ecology Progress Series*, **56**, 281–289.

Nöthig, E.-M. & Gowing, M.M. (1991) Late winter abundance and distribution of phaeodarian radiolarians, other large protozooplankton and copepod nauplii in the Weddell Sea, Antarctica. *Marine Biology*, **111**, 473–484.

Okolodkov, Y.B. (1998) A check list of dinoflagellates recorded from the Russian arctic seas. *Sarsia*, 83.

Palmisano, A.C. & Garrison, D.L. (1993) Microorganisms in antarctic sea ice. In: *Antarctic Microbiology* (Ed. E.I. Friedmann), pp. 167–218. Wiley-Liss, New York.

Park, M.G., Kim, S., Kim, H.S., Myung, G., Kang, Y.G. & Yih, W. (2006) First successful culture of the marine dinoflagellate *Dinophysis acuminata*. *Aquatic Microbial Ecology*, **45**, 101–106.

Parsons, T.R., Takahashi, M. & Hargrave, B. (1984) *Biological Oceanographic Processes*. Pergamon Press, Elmsford.

Peck, L.S. (2002) Ecophysiology of Antarctic marine ectotherms: limits of life. *Polar Biology*, **25**, 31–40.

Petrich, C. & Eicken, H. (2015) Growth structure & properties of sea ice. In: *Sea Ice* (Ed. D.N. Thomas). John Wiledy & Sons, Ltd., West Sussex, in press.

Petz, W., Song, W. & Wilbert, N. (1995) Taxonomy and ecology of the ciliate fauna (Protozoa, Ciliophora) in the endopagial and pelagial of the Weddel Sea, Antarctica. *Stapfia*, **40**.

Petz, W., Valonesi, A., Schiftner, U., Quesada, A. & Ellis-Evans, J.C. (2007) Ciliate biogeography in Antarctic and Arctic freshwater ecosystems: endemism or global distribution of species? *FEMS Microbiology Ecology*, **59**, 396–408.

Pierce, R.W. & Turner, J.T. (1992) Ecology of planktonic ciliates in marine food webs. *Reviews in Aquatic Sciences*, **6**, 139–181.

Piwosz, K., Wiktor, J.M., Niemi, A., Tatarek, A. & Michel, C. (2013) Mesoscale distribution and functional diversity of picoeukaryotes in the first-year sea ice of the Canadian Arctic. *The ISME Journal*, **7**, 1461–1471.

Poulin, M., Daugbjerg, N., Gradinger, R., Ilyash, L., Ratkova, T. & von Quillfeldt, C. (2011) The pan-Arctic biodiversity of marine pelagic and sea-ice unicellular eukaryotes: a first-attempt assessment. *Marine Biodiversity*, **41**, 13–28.

Price, P.B. (2003) Life in solid ice. In: *Astronomical Society of the Pacific*, Vol. **213** (Eds. R. Norris & F. Stootman), pp. 363–366 IAU Symposium Series.

Riedel, A., Michel, C. & Gosselin, M. (2007) Grazing of large-sized bacteria by sea-ice heterotrophic protists on the Mackenzie Shelf during the winter-spring transition. *Aquatic Microbial Ecology*, **50**, 25–38.

Rose, J., Fitzpatrick, E., Wang, A., Gast, R. & Caron, D. (2013) Low temperature constrains growth rates but not short-term ingestion rates of Antarctic ciliates. *Polar Biology*, **36**, 645–659.

Rose, J.M. & Caron, D.A. (2007) Does low temperature constrain the growth rates of heterotrophic protists? Evidence and implications for algal blooms in cold water. *Limnology and Oceanography*, **52**, 886–895.

Rose, J.M., Vora, N.M. & Caron, D.A. (2008) Effect of temperature and prey type on nutrient regeneration by an Antarctic bacterivorous protist. *Microbial Ecology*, **56**, 101–111.

Rózanska, M., Poulin, M. & Gosselin, M. (2008) Protist entrapment in newly formed sea ice in the coastal Arctic Ocean. *Journal of Marine Systems*, **74**, 887–901.

Sanders, R.W., Caron, D.A. & Berninger, U.-G. (1992) Relationships between bacteria and heterotrophic nanoplankton in marine and fresh water: an inter-ecosystem comparison. *Marine Ecology Progress Series*, **86**, 1–14.

Sanders, R.W. & Gast, R.J. (2011) Bacterivory by phototrophic picoplankton and nanoplankton in Arctic waters. *FEMS Microbiology Ecology*, **82**, 242–253.

Schnepf, E. & Kühn, S.F. (2000) Food uptake and fine structure of *Cryothecomonas longipes* sp. nov., a marine nanoflagellate incertae sedis feeding phagotrophically on large diatoms. *Helgoland Marine Research*, **54**, 18–32.

Scott, F.J., Davidson, A.T. & Marchant, H.J. (2001) Grazing by the Antarctic sea-ice ciliate *Pseudocohnilembus*. *Polar Biology*, **24**, 127–131.

Scott, F.J. & Marchant, H.J. (2005) *Antarctic Marine Protists*. Goanna Print, Canberra.

Sherr, B.F., Sherr, E.B., Caron, D.A., Vaulot, D. & Worden, A.Z. (2007) A sea of microbes: oceanic protists. *Oceanography*, **20**, 102–106.

Sherr, E. & Sherr, B. (2002) Significance of predation by protists in aquatic microbial food webs. *Antonie van Leeuwenhoek*, **81**, 293–308.

Sherr, E.B. & Sherr, B.F. (2007) Heterotrophic dinoflagellates: a signficant component of microzooplankton biomass and major grazers of diatoms in the sea. *Marine Ecology Progress Series*, **352**, 187–197.

Sherr, E.B. & Sherr, B.F. (2009) Capacity of herbivorous protists to control initiation and development of mass phytoplankton blooms. *Aquatic Microbial Ecology*, **57**, 253–262.

Sherr, E.B., Sherr, B.F. & Fessenden, L. (1997) Heterotrophic protists in the Central Arctic Ocean. *Deep-Sea Research II*, **44**, 1665–1682.

Sherr, E.B., Sherr, B.F. & Hartz, A.J. (2009) Microzooplankton grazing impact in the Western Arctic Ocean. *Deep Sea Research Part II: Topical Studies in Oceanography*, **56**, 1264–1273.

Sherr, E.B., Sherr, B.F. & Ross, C. (2013) Microzooplankton grazing impact in the Bering Sea during spring sea ice conditions. *Deep Sea Research Part II: Topical Studies in Oceanography*, **94**, 57–67.

Sime-Ngando, T., Demers, S. & Juniper, S.K. (1999) Protozoan bacterivory in the ice and water column of a cold temperate lagoon. *Microbial Ecology*, **37**, 95–106.

Sime-Ngando, T., Gosselin, M., Juniper, S.K. & Levasseur, M. (1997a) Changes in phagotrophic microprotists (20–200 μm) during the spring bloom, Canadian Arctic Archipelago. *Journal of Marine Systems*, **11**, 163–172.

Sime-Ngando, T., Juniper, S.K. & Demers, S. (1997b) Ice-brine and planktonic microheterotrophs from Saroma-ko Lagoon, Hokkaido (Japan): quantitative importance and trophodynamics. *Journal of Marine Systems*, **11**, 149–161.

Simpson, A.G.B. & Roger, A.J. (2004) The real 'kingdoms' of eukaryotes. *Current Biology*, **14**, R693–696.

Slapeta, J., López-García, P. & Moreira, D. (2006) Global dispersal and ancient cryptic species in the smallest marine eukaryotes. *Molecular Biology And Evolution*, **23**, 23–29.

Smith, W.O., Jr.,, Carlson, C.A., Ducklow, H.W. & Hansell, D.A. (1998) Growth dynamics of *Phaeocystis antarctica*-dominated plankton assemblages from the Ross Sea. *Marine Ecology Progress Series*, **168**, 229–244.

Smith, W.O., Jr., & Gordon, L.I. (1997) Hyperproductivity of the Ross Sea (Antarctica) polynya during austral spring. *Geophysical Research Letters*, **24**, 233–236.

Smith, W.O.J., Nelson, D.M., DiTullio, G.R. & Leventer, A.R. (1996) Temporal and spatial patterns in the Ross Sea: phytoplankton biomass, elemental composition productivity and growth rates. *Journal of Geophysical Research*, **101**, 18,455–418,465.

Song, W. & Wilbert, N. (2000) Ciliates from Antarctic sea ice. *Polar Biology*, **23**, 212–222.

Stoeck, T., Jost, S. & Boenigk, J. (2008) Multigene phylogenies of clonal *Spumella*-like strains, a cryptic heterotrophic nanoflagellate, isolated from different geographical regions. *International Journal of Systematic and Evolutionary Microbiology*, **58**, 716–724.

Stoecker, D.K., Buck, K.R. & Putt, M. (1992) Changes in the sea-ice brine community during the spring-summer transition, McMurdo Sound, Antarctica. 1. Photosynthetic protists. *Marine Ecology Progress Series*, **84**, 265–278.

Stoecker, D.K., Buck, K.R. & Putt, M. (1993) Changes in the sea-ice brine community during the spring-summer transition, McMurdo Sound, Antarctica. 2. Phagotrophic protists. *Marine Ecology Progress Series*, **95**, 103–113.

Stoecker, D.K., Gustafson, D.E., Black, M.M.D. & Baier, C.T. (1998) Population dynamics of microalgae in the upper land-fast sea ice at a snow-free location. *Journal of Phycology*, **34**, 60–69.

Stoecker, D.K., Gustafson, D.E., Merrell, J.R., Black, M.M.D. & Baier, C.T. (1997) Excystment and growth of chrysophytes and dinoflagellates at low temperatures and high salinities in Antarctic sea-ice. *Journal of Phycology*, **33**, 585–595.

Strom, S.L. (1991) Growth and grazing rates of the herbivorous dinoflagellate *Gymnodinium* sp. from the open subarctic Pacific Ocean. *Marine Ecology Progress Series*, **78**, 103–113.

Strom, S.L. (2000) Bacterivory: Interactions between bacteria and their grazers. In: *Microbial Ecology of the Oceans* (Ed. D.L. Kirchman) , pp. 351–386. Wiley-Liss, Inc., New York.

Swanberg, N.R. & Caron, D.A. (1991) Patterns of sarcodine feeding in epipelagic oceanic plankton. *Journal of Plankton Research*, **13**, 287–312.

Swanberg, N.R. & Eide, L.K. (1992) The radiolarian fauna at the ice edge in the Greenland Sea during summer. *Journal Of Marine Research*, **50**, 297–320.

Terrado, R., Medrinal, E., Dasilva, C., Thaler, M., Vincent, W. & Lovejoy, C. (2011) Protist community composition during spring in an Arctic flaw lead polynya. *Polar Biology*, **34**, 1901–1914.

Thaler, M. & Lovejoy, C. (2012) Distribution and diversity of a protist predator *Cryothecomonas* (Cercozoa) in Arctic marine waters. *Journal of Eukaryotic Microbiology*, **59**, 291–299.

Thaler, M. & Lovejoy, C. (2015) Biogeography of heterotrophic flagellate populations indicates the presence of generalist and specialist taxa in the Arctic Ocean. *Applied and Environmental Microbiology* **81**, 2137–2148.

Thomas, D.N. & Dieckmann, G.S. (2002) Antarctic sea ice – a habitat for extremophiles. *Science*, **295**, 641–644.

Thomsen, H.A., Buck, K.R., P.A., B. & Garrison, D.L. (1991) Fine structure and biology of *Cryothecomonas* gen. nov. (Protista incertae sedis) from the ice biota. *Canadian Journal of Zoology*, **69**, 1048–1070.

Tong, S., Vørs, N. & Patterson, D.J. (1997) Heterotrophic flagellates, centrohelid heliozoa and filose amoebae from marine and freshwater sites in the Antarctic. *Polar Biology*, **18**, 91–106.

Torstensson, A., Dinasquet, J., Chierici, M., Fransson, A., Riemann, L. & A. Wulff. (2015) Physicochemical control of bacterial and protist community composition and diversity in Antarctic sea ice. *Environmental Microbiology* **17**, 3869–3881.

Uhlig, C., Kilpert, F., Frickenhaus, S. et al. (2015) In situ expression of eukaryotic ice-binding proteins in microbial communities of Arctic and Antarctic sea ice. *ISME Journal*, **9**, 2537–2540.

Vaqué, D., Agustí, S. & Duarte, C.M. (2004) Response of bacterial grazing rates to experimental manipulation of an Antarctic coastal nanoflagellate community. *Aquatic Microbial Ecology*, **36**, 41–52.

Vaulot, D., Eikrem, W., Viprey, M. & Moreau, H. (2008) The diversity of small eukaryotic phytoplankton (≤3 µm) in marine ecosystems. *FEMS Microbiology Reviews*, **32**, 792–820.

Vézina, A.F., Demers, S., Laurion, I., Sime-Ngando, T., Kim Juniper, S. & Devine, L. (1997) Carbon flows through the microbial food web of first-year ice in Resolute Passage (Canadian High Arctic). *Journal of Marine Systems*, **11**, 173–189.

Vincent, W.F. (2010) Microbial ecosystem responses to rapid climate change in the Arctic. *The ISME Journal*, **4**, 1087–1090.

Vørs, N. (1993) Heterotrophic amoebae, flagellates and heliozoa from Arctic marine waters (North West Territories, Canada and West Greenland). *Polar Biology*, **13**, 113–126.

Weissenberger, J., Kieckmann, G.S., Gradinger, R. & Spindler, M. (1992) Sea ice: a cast technique to examine and analyse brine pockets and channel structure. *Limnology and Oceanography*, **37**, 179–183.

CHAPTER 16

Sea ice as a habitat for macrograzers

Bodil A Bluhm,[1] Kerrie M. Swadling[2] and Rolf Gradinger[1]

[1] *Institute of Arctic and Marine Biology, UiT – The Arctic University of Norway, Tromsø, Norway*
[2] *Antarctic Climate & Ecosystems Cooperative Research Centre and Institute for Marine and Antarctic Studies, Hobart, Australia*

16.1 Environmental setting and relevant forcing factors for ice macrograzers

Sea ice provides a wide range of microhabitats for associated fauna due to seasonal and regional variations in its structure, physical and chemical gradients, and interactions with pelagic and benthic biota. It is this interplay of ice structural, physical and biological factors (e.g. food availability) that shapes the seasonal occurrence of sea ice fauna. The term 'macrograzers' is used here for the ice-associated metazoan fauna community both inside the brine channel system (hereafter 'meiofauna') and at the ice–water boundary layer (hereafter 'under-ice fauna', or 'macrofauna' when referring to larger sizes), and includes taxa beyond algal grazers.

Biota in and under the ice varies across spatial and temporal scales in response to variations in environmental conditions at those scales. On the small scale of micro- to millimetres, the spatial dimensions of the sea ice brine channel network limit the maximum size of meiofauna (Krembs et al., 2000). Diameters of the brine network vary from <1 µm to many mm, with temperature-dependent size distribution across the vertical thickness of the sea ice. Environmental conditions within the brine channels also change with seasonal cooling or warming, including brine volume and its salinity (Chapter 1). Salinity ranges span freshwater conditions in melt ponds on top of Arctic multi-year ice (MYI), while brine salinities in older parts of pressure ridges can exceed 120 in winter. The ice underside provides a highly variable and heterogeneous habitat at the ice–water interface at the scale of centimetres to metres (Hop et al., 2000; Arndt & Swadling, 2006). Seasonally, the physical conditions in this boundary layer may change from fully marine conditions during most of the ice growth period to brackish conditions during ice melt, at which time freshwater may accumulate under the sea ice, causing osmotic stress to marine biota. In this under-ice habitat, sea ice pressure ridges, formed by mechanical break-up of ice floes and in constant evolution because of ice drift and friction, provide gaps and spatial niches where larger biota, including fish, find refuge from predatory birds and seals in both the Arctic and Antarctic (Smetacek et al., 1990; Hop et al., 2000; Gradinger et al., 2010).

On yet larger regional scales, the age of ice floes and their geographic location influence biotic patterns. In the central Arctic Ocean, which used to be dominated by MYI floes that survived the seasonal summer ice melt for several years to decades, sea ice fauna can reproduce and complete their life cycles associated with the ice habitat, allowing for the evolution of sea ice endemic taxa (Arndt & Swadling, 2006). Water depths of several thousand metres in these regions exclude exchange of organisms with the sea floor but allow for interactions with the pelagic fauna. The other extreme is coastal Arctic fast ice in shallow waters (<20 m) on the shelves which remains attached to land during the entire ice-covered period. Under such conditions, no ice-endemic fauna has evolved due to the absence of the habitat during the ice-free period. There, ice fauna partition their life cycles between the sea ice and the water column and/or sea floor and include larvae and juveniles of benthic taxa. The seasonal pack or drift ice in Antarctica lies in the middle between both extremes,

---

Sea Ice, Third Edition. Edited by David N. Thomas.
© 2017 John Wiley & Sons, Ltd. Published 2017 by John Wiley & Sons, Ltd.

as the generally great water depths on the shelf limit exchanges with benthic communities and ice biota often has to bridge open water conditions for several months per year.

The sampling of sea ice fauna is generally still based on technologies that have been successfully used for decades, including sea ice coring with subsequent melting and microscopic examinations for meiofauna, plus optical tools (e.g. remotely operated vehicles, divers) for collection and identification of macrofauna (Gradinger & Bluhm, 2009). Over recent years, new techniques, including metagenomics, have been applied to determine sympagic diversity (Collins et al., 2010) and new sampling tools such as the Surface and Under-ice Trawl (SUIT) were developed to collect macrofauna (Flores et al., 2011). Studies on sea ice fauna will continue to require fieldwork, as no remote sensing tools are yet available for their determination, although long-term deployment of acoustic and optical instrumentation attached to ice-tethered platforms are under development.

16.2 Diversity, distribution and abundance

Sea ice is inhabited by only a few ice-endemic species, while the majority of taxa are also found in the water column and/or at the sea floor (Arndt & Swadling, 2006; Gradinger et al., 2010). A few dozen ice meiofaunal species are known from each polar region (Figure 16.1–16.3), with poor taxonomic resolution for several taxa. While some taxa are common in both polar regions, rotifers and nematodes are more common in Arctic sea ice, and copepods more prominent in Antarctic sea ice (Figures 16.1 and 16.2, Table 16.1; Gradinger, 1999; Arndt & Swadling, 2006). Cnidarians and nemerteans are so far only known from Arctic sea ice (Piraino et al. 2008; Marquardt et al., 2011), while ctenophores and nudibranchs may be limited to the brine channel system of Antarctic pack ice (Kiko et al., 2008a). The definition and therefore diversity of 'under-ice' metazoans varies by the extent to which the upper water column is included, but species richness is generally higher than it is within the ice proper, with amphipods, ctenophores, pteropods, euphausiids and large copepods linked to the sea ice via direct and indirect feeding (Purcell et al. 2010; Jia et al., 2016). At least 40 species are known from that habitat (Arndt & Swadling, 2006; Flores et al., 2011), although there are only very few fish species (Mecklenburg et al., 2011; Vacchi et al., 2012).

Many phyla are represented in the sea ice habitats, but few species are described from each, with the exception of Arthropoda. Cnidarians (Figures 16.1 and 16.2; Table 16.1) are represented, typically in very low densities, by one species of hydroid in the Arctic (*Sympagohydra tuuli*; Piraino et al., 2008) and two species of ctenophore in the Antarctic, *Callianira antarctica* (Kiko et al., 2008a) and another possibly from the genus *Euplokamis* (Kramer et al., 2011). Several species of ctenophores and siphonophores can be abundant or at least present in the upper ~20 m immediately under Arctic sea ice (Figure 16.1, Table 16.1; Purcell et al., 2010). Eight rotifer species are recorded from Arctic sea ice so far, with *Synchaeta hyperborea* and *Synchaeta tamara* being most common, although few studies have identified this group to species level (Friedrich & De Smet, 2000). Abundance can be regionally very high, as in the Bering Sea, but is negligible in some other regions. No representatives of this group have been described for Antarctic sea ice. Nematodes can dominate meiofaunal communities at peak abundances exceeding 200 000 individuals m^{-2} (Gradinger et al., 2009). Several nematode species are endemic to Arctic sea ice, namely *Teristus melnikovii*, *Cryonema crassum* and *C. tenue*, and additional, undescribed species have also been recorded (Table 16.1; Tchesunov & Riemann, 1995; Riemann & Sime-Ngando, 1997).

Only one observation reports the monhysteroid *Geomonhystera glaciei* and three unidentified nematodes from Antarctic ice (Blome & Riemann, 1999). Acoela and Platyhelminthes are often dominant members of the Arctic brine channel-inhabiting meiofaunal community, where they can occur at densities exceeding 15,000 individuals m^{-2} in spring (Friedrich & Hendelberg, 2001). In Antarctic pack ice, platyhelminthes, of the acoel and rhabditophor types, occur in abundances up to 133 000 individuals m^{-2} in the Weddell Sea, and up to 4500 individuals m^{-2} in the Indian Ocean sector (Kramer et al., 2011). To date, no detailed taxonomic study is available for Antarctic and Arctic Platyhelminthes, but in the Arctic they may represent at least a handful of species (K. Dilliplaine, personal communication).

Figure 16.1 Antarctic ice-associated fauna. (a) *Euphausia superba* (4 cm); (b) *Trematomus* sp.; (c) *Oithona similis*; (d) *Stephos longipes*; (e) *Drescheriella glacialis*; (f) *Pleuragramma antarcticum* eggs; (g) Platyhelminthes; (h) *Tergipes antarcticus*; (i) *Paralabidocera antarctica* copepodid; (j) *Neogloboquadrina pachyderma*. [Photographs by: Cheryl Hopcroft (a; University of Alaska Fairbanks), Claire Davies (b, c, e; CSIRO, Australia), Jan Michels (d; University of Kiel, Germany), Noelia Estévez Calvar ((f; IMAS, XXIX Campaign PNRA (Programma Nazionale di Ricerche in Antartide) RAISE Project (Ricerche integrate sulla ecologia dell'Antarctic Silverfish nel Mare di Ross)), Maike Kramer (g), Rainer Kiko (h; University of Kiel), Sarah Payne (I; University of Tasmania), Hannes Grobe (j; AWI).]

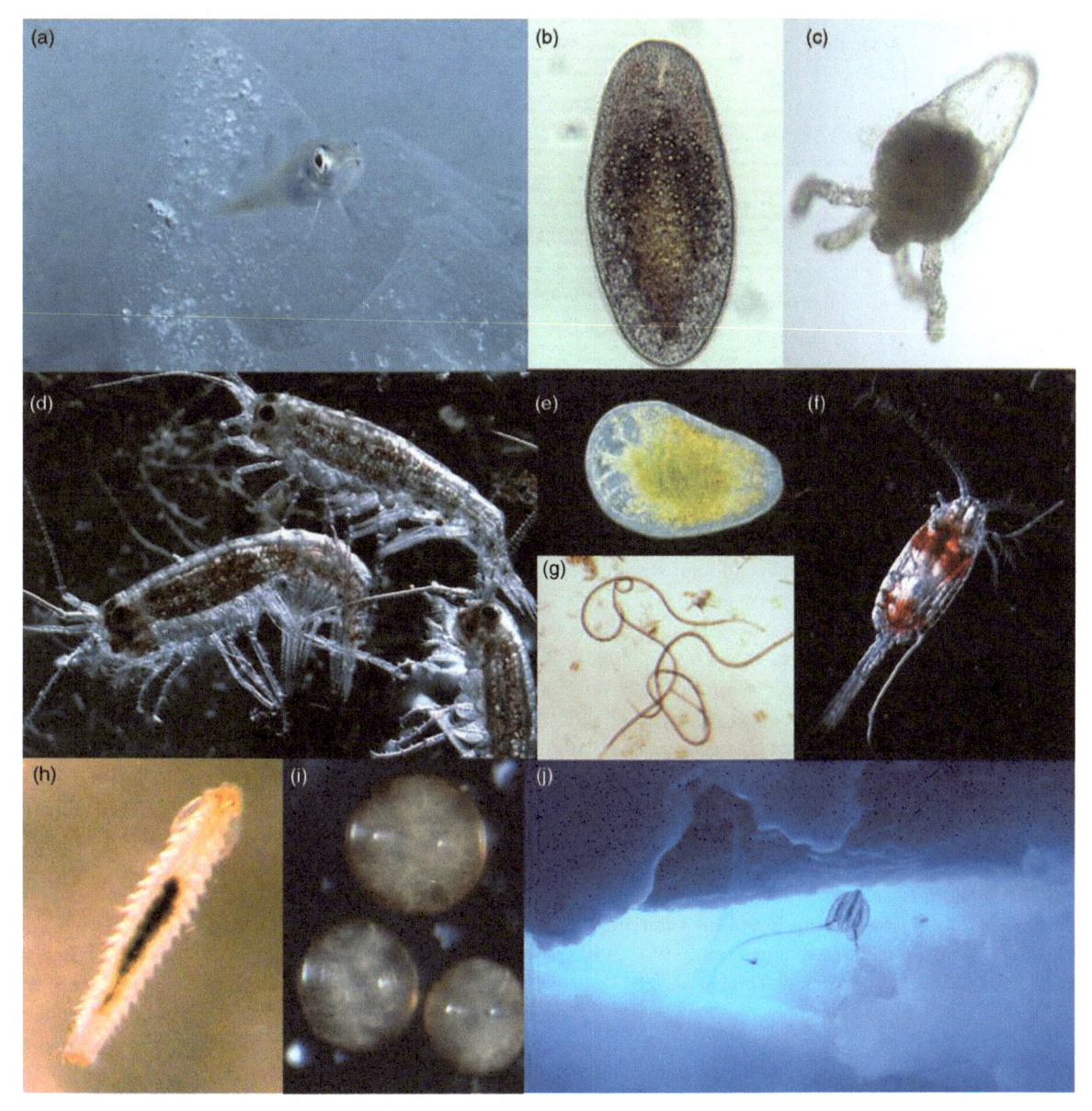

Figure 16.2 Arctic ice-associated fauna. (a) *Boreogadus saida*; (b) *Acoel flatworm*; (c) *Sympagohydra tuuli*; (d) *Apherusa glacialis*; (e) acoel flatworm; (f) *Pseudocalanus* sp.; (g) nematodes; (h) *Scolelepis squamata*; (i) Platyhelminthes eggs; (j) *Mertensia* sp. [Photographs by Bodil Bluhm (d, f, g, h), Kyle Dilliplaine (b, e, i), Rolf Gradinger (c), and Katrin Iken (j; all Univ. of Alaska Fairbanks).]

In spring, larvae and juveniles of benthic communities, including polychaetes and gastropods, inhabit both the coastal Arctic fast ice brine channel system and ice underside (Table 16.1; Conover et al., 1990; Carey, 1992; Gradinger et al., 2009). The seasonally very abundant (>100 000 individuals m^{-2}) spionid polychaete *Scolelepis squamata* is an example from the coastal fast ice in the Chukchi and Beaufort Seas (Figures 16.2 and 16.4), and the pelagic polychaete *Tomopteris carpenteri* has been reported from the sea ice–water boundary layer. Meroplanktonic larvae are less common in Antarctic sea ice (Andriashev, 1968), although echinoderm larvae were found during the winter in the pack ice of the Indian Ocean sector (Wallis et al., 2015). Decapod,

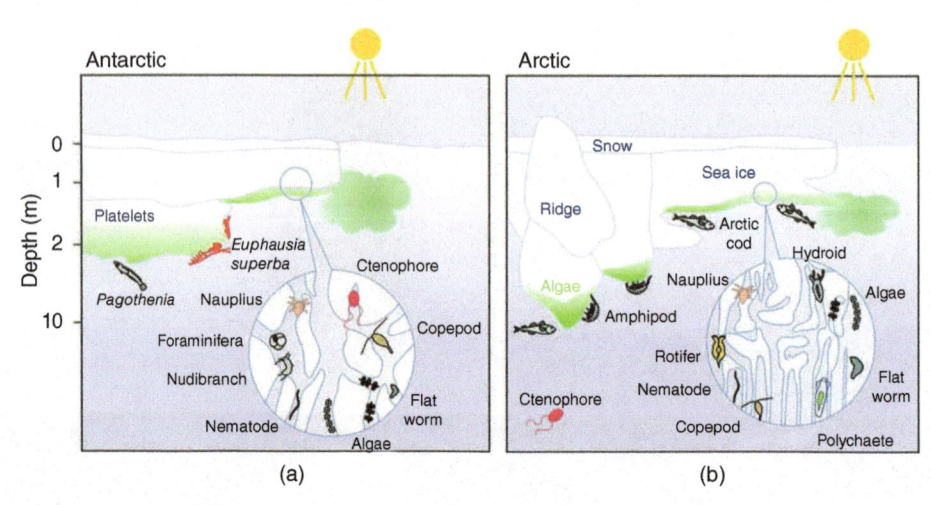

Figure 16.3 Schematic depiction of dominant ice habitats in (a) Antarctica and (b) the Arctic, and associated sympagic meio- and macrofauna. Expanded circles show the meiofauna and algae that inhabit the brine channel systems in sea ice.

cephalopod and fish larvae have been found within the top 2 m below Antarctic ice (Flores et al., 2011).

Crustaceans are the most species-rich sympagic macrograzers (Table 16.1; Figures 16.1–16.3 and 16.5). Three copepod species dominate the Antarctic ice community: the harpacticoid *Drescheriella glacialis* and the calanoids *Paralabidocera antarctica* and *Stephos longipes*. Several other harpacticoids occur sporadically, either in low numbers or in pulses: *Idomene antarctica*, *Ectinosoma* sp., *Harpacticus furcifer* and *Diarthrodes* cf. *lilacinus* (Dahms et al., 1990). The calanoids *Ctenocalanus citer* and *Metridia gerlachei*, and the cyclopoids *Oncaea curvata* and *Oithona similis* are occasionally found in Antarctic sea ice in very low numbers (Swadling et al., 1997, 2001; Guglielmo et al., 2007; Schnack-Schiel et al., 2008). In the Arctic, the cyclopoid copepods *Cyclopina gracilis* and *C. schneideri*, as well as the harpacticoids *Harpacticus* spp., *Halectinosoma* sp. and *Tisbe furcata* are common inside sea ice (Carey & Montagna, 1982; Grainger, 1991). Common pelagic calanoids such as *Acartia longiremis*, *Pseudocalanus* sp., *Calanus glacialis* and the cyclopoid *Oithona similis* occasionally reach noteworthy densities in coastal fast ice (Carey & Montagna, 1982). Other rarely recorded taxa include mysids (Menshenina & Melnikov, 1995). An extensive review of all crustacean species associated with Arctic and Antarctic sea ice is given by Arndt and Swadling (2006).

Crustaceans also dominate under-ice communities, primarily euphausiids in Antarctica (O'Brien, 1987; Nicol, 2006) and gammarid amphipods in the Arctic (Figures 16.1–16.3; Beuchel & Lønne, 2002; Macnaughton et al., 2007; David et al., 2015). The euphausiid species *Euphausia superba* and, to a much lesser extent, *Euphausia crystallorophias* occur in dense swarms at the ice–water interface, feeding on ice algal biomass, particularly in winter and early spring, when phytoplankton biomass is still low (O'Brien, 1987; Marschall, 1988). *E. superba* occurs directly at the ice underside in densities between 1 and 100 individuals m^{-2} or in free-swimming aggregations in the water column directly beneath the sea ice (O'Brien, 1987). *Thysanoessa raschii* can be abundant under the ice in the Bering Sea (Gradinger et al., unpublished). Gammarid amphipods occur in variable abundances of < 5 to > 100 individuals m^{-2} underneath Arctic sea ice, except in the Bering Sea (summary tables in Hop et al., 2000 & Arndt & Swadling, 2006). Dominant species in offshore pack ice include *Apherusa glacialis*, *Gammarus wilkitzkii*, *Onisimus nanseni* and *O. glacialis*, while nearshore areas are inhabited by *O. litoralis*, *Gammarus setosus*, *Gammaracantus loricatus* and others (Table 16.1; summarized by Arndt and Swadling, 2006). The hyperiid *Parathemisto libellula* occurs regularly but mostly in low densities at the undersides of Arctic ice (Gulliksen & Lønne, 1991; Melnikov, 1997). In Antarctica, gammarid amphipods also occur in locally very high abundances exceeding 10 000 individuals m^{-2} at the under-ice surface in both shallow and deep waters (Andriashev, 1968; Richardson

Table 16.1 Sea ice-associated macrogazers

	Arctic	Antarctic
Taxon (in ice)		
Cnidaria	*Sympagohydra tuuli*	
Ctenophora		*Callianira antarctica*
Acoelomorpha	Several unidentified species	
Platyhelminthes	Several unidentified species	several unidentified species
Nemertea	Several unidentified species	
Nematoda	*Theristus melnikovii, Cryonema tenue, C. crassum*	Monhysteroidea
Rotifera	*Synchaeta hyperborea, Encentrum graingeri*	
Gastropoda	Unidentified larvae	*Tergipes antarcticus*
Polychaeta	*Scolelepis squamata* (juveniles), larval stages of other species	
Calanoid copepods	*Pseudocalanus* sp.	*Paralabidocera antarctica, Stephos longipes*
Harpacticoid copepods	*Halectinosoma* sp., *Harpacticus superflexus, Tisbe furcata, Microsetella norvegica*	e.g. *Drescheriella glacialis, Idomene antarctica, Harpacticus furcifer*
Cyclopoid copepods	*Cyclopina gracilis, C. schneideri, Oithona similis*	*Oncaea curvata, Oithona similis*
Amphipoda	Gammarid amphipod juveniles	
Echinodermata		Meroplanktonic stages of echinoderms
Taxon (under ice)		
Ctenophora	*Mertensia ovum, Bolinopsis infundibulum, Beroe cucumis*	
Gastropoda	Meroplanktonic juvenile stages	
Polychaeta	Meroplanktonic juvenile stages	
Others	Meroplanktonic stages of bivalves, echinoderms	Meroplanktonic stages of echinoderms
Calanoid copepods	*Pseudocalanus* sp., *Calanus glacialis*	*Stephos longipes, Paralabidocera antarctica*
Harpacticoid copepods	*Halectinosoma finmarchicum, Tisbe* spp., *Harpacticus superflexus*	*Drescheriella glacialis, Ectinosoma* sp., *Idomene antarctica, Diarthrodes* cf. *lilacinus*
Cyclopoid copepods	*Cyclopina schneideri, Oithona similis, Oncaea borealis*	*Oithona* sp., *Oncaea* sp.
Amphipoda	*Aperusa glacialis, Gammarus wilkitzkii, Onisimus nanseni, O. glacialis, O. litoralis, Gammaracantus loricatus, Ischyocerus anguipes, Weyprechtia pinguis, Eusirus holmii*	*Paramoera walkeri, Cheirimedon fougneri, Pontogoneia antarctica, Eusirus antarcticus, E. laticarpus, Orchomene plebs*
Euphausids	*Thysanoessa raschii*	*Euphausia superba, E. crystallorophias*
Fishes	*Boreogadus saida, Arctogadus glacialis*	*Pagothenia* (=*Trematomus*) *borchgrevinki, Dissostichus mawsoni*

& Whitaker, 1979; Gulliksen & Lønne, 1991; Günther et al., 1999). At least a dozen species of amphipods, both gammarid and hyperiid, have been recorded from under-ice hauls in the Weddell and Lazarev Seas, with *Eusirus antarcticus* and *E. laticarpus* most numerous and strongly ice-associated (Krapp et al., 2008; Flores et al., 2011).

Copepods can be highly concentrated directly under the ice in both polar regions. The calanoids *Stephos longipes* and *Paralabidocera antarctica* and the harpacticoids *Drescheriella glacialis, Ectinosoma* sp., *Idomene antarctica* and *Diarthrodes* cf. *lilacinus* are abundant in the Antarctic under-ice layer (Table 16.1; Schnack-Schiel et al., 1998; Swadling et al., 2001; Kiko et al., 2008a). Other copepods including the calanoid *Ctenocalanus citer*, the cyclopoids *Oithona similis, Pseudocyclopina belgicae* and *Oncaea curvata* occur occasionally near the ice–water interface (Günther et al., 1999; Swadling et al., 2001; Kiko et al., 2008a). In the Arctic, the calanoids *Pseudocalanus* sp., *Calanus glacialis* and various nauplii were collected in high concentrations at the ice-water interface (Conover et al., 1986). The small calanoid *Acartia longiremis* and cyclopoid *Oithona similis* occur in variable numbers, although regularly, at the ice–water interface (Runge & Ingram, 1988; Conover et al., 1990). Total abundance of Arctic sub-ice fauna is on the order of 200–30,000 individuals m^{-3} (Werner & Martinez-Arbizu, 1999).

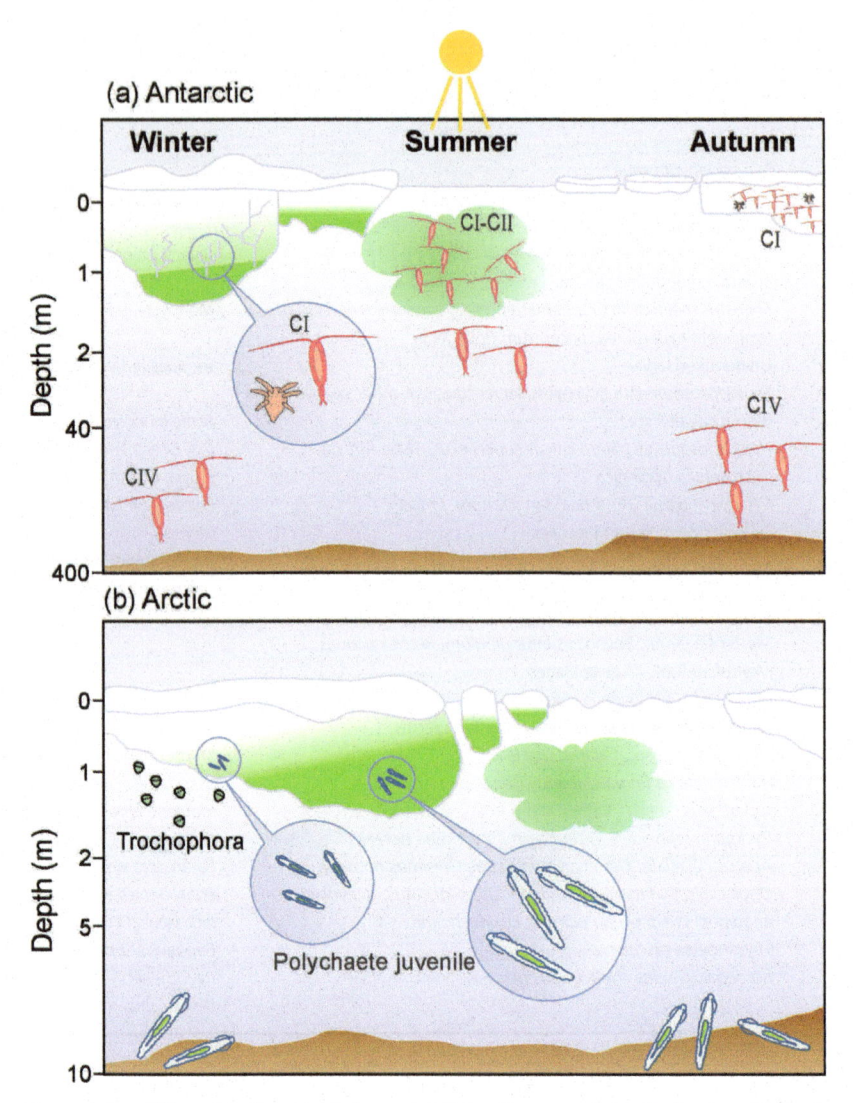

Figure 16.4 Schematic examples of life cycles of sympagic fauna. (a) Antarctic copepod, *Stephos longipes*; (b) Arctic polychaete, *Scolelpis squamata*. For a description, see text. Source: Schnack-Schiel et al., 1995. Reproduced with permission of Elsevier.

Sympagic fish fauna is not diverse. In the Arctic, young year classes of two fish species, *Boreogadus saida* and *Arctogadus glacialis* live in close association with the ice underside (Carey, 1985; Melnikov, 1997; Hop & Gjøsæater, 2013). *B. saida* at times utilize cavities and narrow wedges of sea water along the edges of melting ice floes (Carey, 1985; Gradinger & Bluhm, 2004), whereas older fish tend to inhabit the water column and sea floor (Logerwell et al., 2011). School sizes of *B. saida* associated with sea ice tends to be far smaller (groups of one to a few dozens; Gradinger & Bluhm, 2004) than schools recorded from open water (summarized in Hop & Gjøsæter, 2013), although larger swarms also occur (Melnikov & Chernova, 2013). In the Antarctic, swarms of young and adult broadhead fish, *Pagothenia borchgrevinki* and young giantfish *Dissostichus mawsoni* live immediately under the ice (Andriashev, 1968; Goutte et al., 2014), and large numbers of the eggs of *Pleuragramma antarcticum* (Antarctic silverfish) have been found in platelet ice layers in Terra Nova Bay (Vacchi et al. 2012).

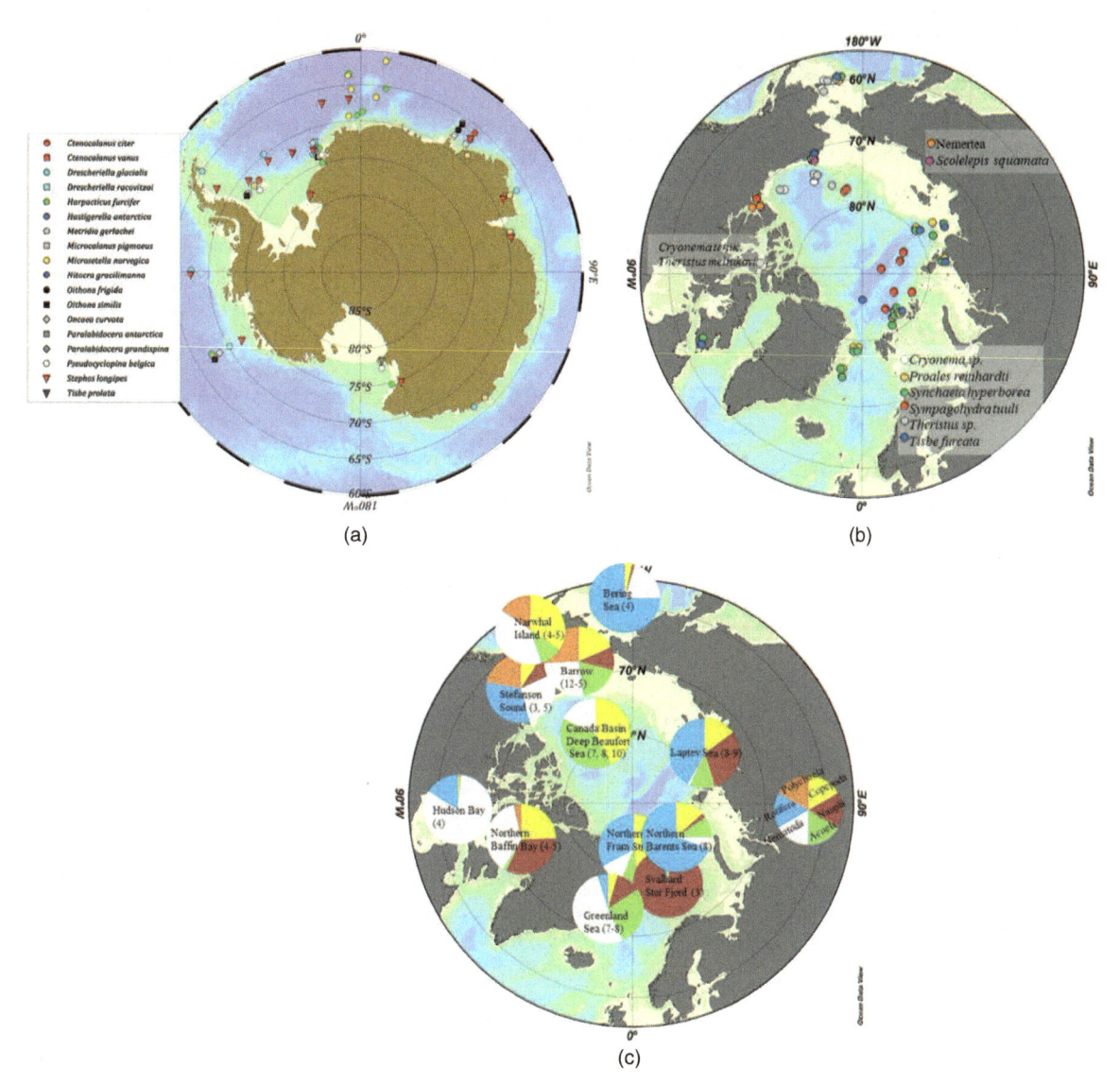

Figure 16.5 Distribution of some sympagic macrograzers in the Antarctic (a) and Arctic sea ice (b), and relative composition of Arctic ice meiofauna (c). The focus is on crustaceans in and under sea ice in Antarctica, and on ice meiofauna taxa inside the Arctic brine channel system. Numbers in parentheses indicate the month of sampling.

Abundance and distribution of ice-associated biota vary on scales from days to years (Hoshiai, et al. 1987), and from less than one meter (Spindler & Dieckmann, 1986) to many kilometres, and up to basin wide scales (Garrison & Close, 1993; Swadling et al. 1997). Densities can vary on several orders of magnitude from <10 to >300,000 individuals m^{-2} inside the brine channels and from 0 to >10,000 individuals m^{-2} under the ice (Figure 16.6; Gradinger et al. 2009). Swadling et al. (1997) examined the effects of four spatial scales on the distribution of fast ice biota. The kilometre scale contributed significant variation to the variables measured, while intermediate scales of hundreds of meters and tens of meters did not. Partitioning of variance components revealed that between 50 and 100 % of the total variance came from residual or unexplained

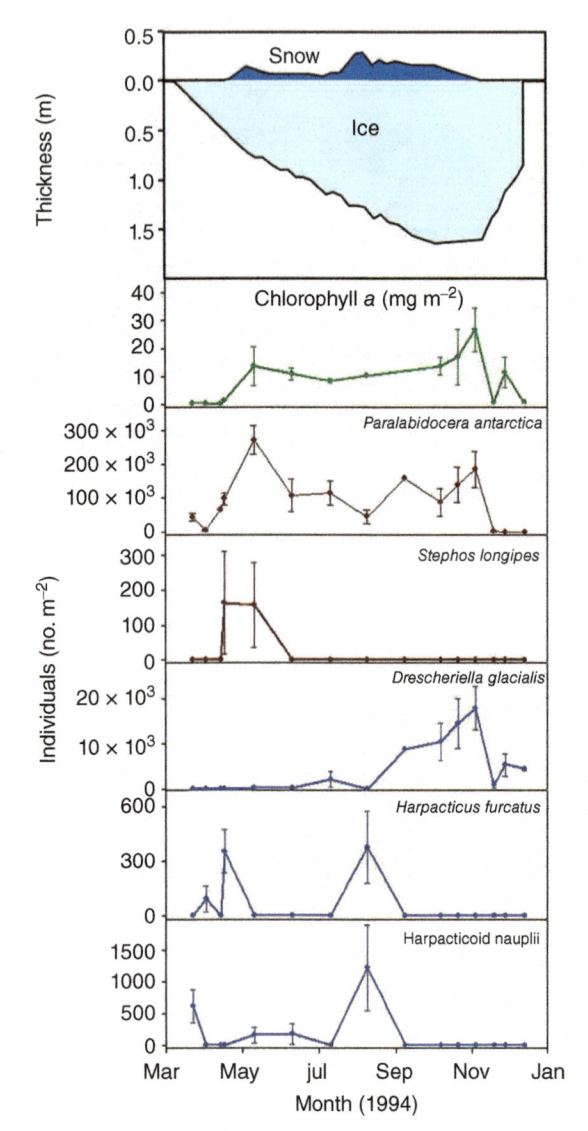

Figure 16.6 Seasonality in the abundance of Antarctic sympagic copepods, chlorophyll concentration in sea ice and ice thickness over an annual cycle. Data were collected throughout 1994 at Davis Station in the Indian Ocean sector of the Southern Ocean. Abundances show the mean (± SE) of five sea ice cores (from Swadling, 1998).

variance, thus highlighting the fact that horizontal patchiness of sympagic biota varies as much at scales of less than one meter as it does at scales of several kilometers. Part of the spatial availability is explained by the physical factors outlined in section 1, and by food availability, which again varies on scales of meters (e.g. related to light; Campbell et al., 2014) to kilometers (e.g. related to nutrient concentrations; Gradinger, 2009), with associated metazoan density varying by several orders of magnitude (Gradinger et al., 2009). For macrofauna, the vertical variability on scales of several meters has been attributed to avoidance of low salinity melt water directly under MYI during the summer melt, when higher abundances of amphipods and meiofauna were found at greater depths in pressure ridges (Gradinger et al., 2010).

16.3 Adaptations to sea ice habitat

Sympagic macrograzers spend part or all of their lives in association with sea ice. They reach this habitat through entrapment during frazil ice formation (Garrison et al., 1983), platelet and anchor ice formation in Antarctica (Dayton et al., 1969; Dieckmann et al., 1986), attachment to larger mobile fauna (Riemann & Sime-Ngando, 1997), and lateral and/or vertical, active migration (Berge et al., 2012) which can be aided by positive phototaxis (Gradinger et al. 2009). In seasonal ice, sympagic fauna is released to the water column and/or sea floor during ice melt, while some permanent inhabitants remain with MYI or pressure ridges throughout the summer. Adaptations to the habitat are related to life cycles, and physiological and other biological traits.

16.3.1 Life cycles

Few sympagic mainly Arctic species complete their entire life cycles in or under sea ice and can be considered ice endemic or autochthonous. Other taxa either enter the ice for a specific life stage, or complete their reproductive cycles in sea ice, but get released during ice melt, or opportunistically use the sea ice during the productive season. Most sympagic macrograzers at both poles are in this latter category.

In Antarctica, the life cycles of three dominant copepods, *Paralabiocera antarctia*, *Stephos longipes* and *Drescheriella glacialis*, are well understood (Figure 16.4). *P. antarctica* recruits from the sediments into the sea ice as naupliar stages I and II, and enters newly forming ice at the end of summer where it remains in the lower part of the ice with environmental conditions resembling those of the sub-ice environment. Unusually for a calanoid copepod, *P. antarctica* undergoes a long overwintering period as late naupliar stages, possibly as a response to the size restrictions placed on them by the width of the brine channels (Swadling et al., 2004). Metametaphosis through the copepodid stages takes place in spring and the adults mate and reproduce in the sub-ice habitat before ice breaks out, and eggs rest in the sediments during the periods of open water (Swadling et al., 2004). In contrast to this univoltine life cycle, *Drescheriella glacialis* displays a multivoltine life cycle, with several reproduction events per year. It has a close association with the sea ice, but must at times survive in the open water column before it can recolonize the ice (Schnack-Schiel et al., 1998; Swadling, 2001). Several developmental stages are often observed at one time and the females carry egg sacs, which probably help in the recolonization process as the nauplii are not strong swimmers (Dahms et al., 1991). Other harpacticoid copepods appear to have similar life cycles and are often found in the upper portion of the ice. This is supported both by their small size and their ability to withstand large fluctuations in salinity and temperature. *Stephos longipes* has a looser association with sea ice; the adults breed in the ice to release eggs and larvae, but older stages are found deeper in the water column (Kurbjeweit et al. 1993).

The Antarctic nudibranch *Tergipes antarcticus* can complete its full life cycle in/at the sea ice as eggs, veliger larvae, juveniles and adults have all been sampled from sea ice (Kiko et al., 2008b). The species is small (≤4 mm) with reduced appendages, both of which enable it to inhabit the brine system. Developmental time from egg to larva takes about one month, with subsequent larval development ~3-6 weeks before hatching. Eggs and larvae can endure temperatures <-5 °C and corresponding calculated salinities >100 and may occur throughout the sea ice. Hardly anything is known about life cycles of other Antarctic meiofauna. Arctic species with full life cycles in/at the sea ice include the gammarid amphipods *Gammarus wilkitzkii*, *Onisimus nanseni* and *O. glacialis* (Poltermann et al., 2000; Beuchel & Lønne, 2002). *Apherusa glacialis* is often included, but seasonal vertical migration has recently been suggested (Berge et al., 2012). Juvenile ice amphipods of all species are abundant in spring and summer (Hop et al., 2011), but adult longevity varies. Life cycle studies suggest a 2-year life cycle for *A. glacialis* (Beuchel & Lønne, 2002), 3 years for *O. litoralis* (Weslawski et al., 2000), 3–4 years for *O. glacialis*

(Arndt & Beuchel, 2006) and at least 5–6 years for *G. wilkitzkii* (Poltermann et al. 2000). Full life cycles can also be completed in Arctic sea ice by nematodes, Platyhelminths and Acoela, as eggs were frequently observed releasing juveniles in melted ice cores (Bluhm & Gradinger, personal observation; Figure 16.2i) and adult stages of the same taxa can be present in large densities throughout the spring.

The Antarctic amphipods *Paramoera walkeri* and *Pontogeneia antarctica* (Rakusa-Suszczewski, 1972; Richardson & Whitaker, 1979) migrate from the seabed to the newly forming sea ice to release their hatchlings, probably to protect the offspring from pelagic predators. In Arctic coastal fast ice, hatchlings of the gammarid *Onisimus litoralis* occasionally inhabit the largest brine pockets during the ice-algal spring bloom. Similarly, juvenile stages of the polychaete *Scolelpis squamata* utilize the sea ice for feeding in coastal fast ice, while trochophores are primarily found in the water column and adults inhabit the sea floor (Figure 16.4b). The most abundant and best studied sympagic crustacean is Antarctic krill, *Euphausia superba* (Figure 16.1a), found throughout the Southern Ocean, although not uniformly distributed, with over 50% of the biomass found in the southwest Atlantic sector (Atkinson et al., 2004). Sea ice is important habitat for the winter survival of larval krill (e.g. Atkinson et al., 2004; Meyer et al., 2009). Larval krill are unable to store lipid and hence need to feed throughout the winter months (Meyer et al., 2009; O'Brien et al., 2011). Food is scarce in the water column, so sea ice-associated biota are important for larval krill survival (reviewed in Meyer, 2012). Recent surveys using SUITs in the Lazarev Sea have documented an almost year-round association of Antarctic krill with the under-ice habitat, including post-larval krill in winter (Flores et al., 2012).

16.3.2 Physiology

Adaptations to low temperatures and high brine salinities are prerequisites for the survival of sympagic organisms in sea ice and influence sympagic diversity (Gradinger & Schnack-Schiel, 1998). Many invertebrates and some vertebrates avoid freezing by super-cooling, whereby the body fluids are cooled below the equilibrium freezing point without freezing, or they have the potential to tolerate ice formation in their tissues (Aarset, 1991; Waller et al., 2006). Examples include the Arctic amphipod *Gammarus oceanicus* and juveniles of the Antarctic krill *E. superba*; the krill larvae were able to survive freezing temperatures as low as −4 °C at a salinity of 69 (Aarset & Torres, 1989; Aarset, 1991). *Apherusa glacialis* had a super-cooling (i.e. lowering the temperature of a liquid below its freezing point without it becoming a solid) point of −7.9°C (Kiko et al., 2008c), compared with −9.0 °C for *E. superba* (Aarset & Torres, 1989). The Arctic *Gammarus wilkitzkii* and *Onisimus glacialis* and the Antarctic *E. antarcticus* do not tolerate freezing in solid sea ice (Aarset & Torres, 1989). Fish, including the Antarctic notothenioids *Pagothenia borchgrevinki* and *Dissostichus mawsoni*, use macromolecular antifreeze, specifically proteins or glycoproteins (AFGPs), as found in the Arctic gadids *Boreogadus saida* and *Arctogadus glacialis* (DeVries, 1971, 1997; Praebel & Ramlov, 2005). This antifreeze lowers the hysteresis freezing point to −2 °C or lower (Praebel & Ramlov, 2005).

In Antarctic fish, antifreeze glycopeptides are synthesized year-round, but *P. borchgrevinki* decreases its AFGP concentrations in response to warm acclimation (Jin & DeVries, 2006). Arctic fish living at summer temperatures well above freezing produce such compounds only during winter (DeVries, 1997). Thermal hysteresis is also present in some polar invertebrates, including the sympagic nudibranch *T. antarcticus* (Kiko et al., 2008b) and copepod *Stephos longipes* (Kiko, 2008), while *Paralabidocera antarctica* lacks that capability (Kiko, 2008). The difference may be explained by habitat use: *P. antarctica* occurs preferentially in the lowermost parts of the sea ice where temperature and salinities are similar to those of the open water (Kiko, 2008), whereas *S. longipes* occurs throughout the sea ice in all habitats, including refrozen gaps and surface layers (Schnack-Schiel et al., 2001; Kiko et al., 2008a).

Temperature and salinity extremes inside sea ice are related and can therefore potentially act on organisms at the same time. Salinity exceeds 60 in the brine channels during cold temperatures and drops below 2 at the ice–water interface during periods of summer ice melt. The Arctic sympagic amphipods *Gammarus wilkitzkii*, *Onisimus glacialis* and *Apherusa glacialis* are hyperosmotic regulators at low salinities (<30) and regulate extracellular concentrations of sodium and chloride in the haemolymph. At higher salinities (>30) they are osmoconform over the tolerated salinity range (~34–60) (Aarset, 1991; Kiko et al., 2008c). Ice meiofaunal tolerance to high salinities varies between taxa but often exceeds that of zooplankton (Gradinger, 2001). In experiments, Arctic nematodes tolerated salinities up to 80, Acoela tolerated salinities up to 60 (with deformations occurring at 70), while juvenile polychaetes *Scolelepis squamata* survived a salinity of 50, but died at a salinity of 60 (Kaufmann et al., unpubl. data). Antarctic sea ice flat worms survived salinities up to 75, while pelagic copepods only tolerated salinities of up to 34 (Gradinger & Schnack Schiel, 1998). The Antarctic amphipod *E. antarcticus* and juveniles of *E. superba* are osmoconformers with narrower salinity tolerances (26–40 and 25–45, respectively; Aarset & Torres, 1989), although other observations suggest *E. superba* may also tolerate lower salinities (Schnack-Schiel et al., 2001).

In addition to temperature and salinity tolerance, some sympagic macrograzers are prepared to produce and store large energy reserves that carry them through the food-scarce winter period. Accumulation of large lipid depots, specifically wax esters, is typical of some ice-associated and pelagic crustaceans of high latitudes (Lee et al., 2006), including *E. superba*, which at the onset of winter has lipid contents exceeding 35% of their dry mass (Atkinson et al., 2002). Ice-associated crustaceans and fish feeding during the dark season store mostly triacylglycerols rather than wax esters. These include the Antarctic species *Stephos longipes* (Schnack-Schiel et al., 1995), *Paralabidocera antarctica* (Swadling et al., 2000), *Euphausia superba* (Falk-Petersen et al., 2000) and *Pagothenia borchgrevinki* (Clarke et al., 1984), and the Arctic species, *Apherusa glacialis*, *Gammarus wilkitzkii*, *Onisimus nanseni* and *Boreogadus saida* (Scott et al., 1999).

16.4 Food web and carbon cycle

16.4.1 Trophic connections

While the phenology of ice algal blooms, their release from the melting ice and fate through pelagic grazing and/or sedimentation to the benthos have been directly quantified, the contribution of sea ice macrograzers to the carbon cycle remains poorly studied. Typical zooplankton approaches such as serial dilution experiments or addition or removal of grazers to/from natural communities are not feasible in the *in situ* ice habitat, so that we largely rely on indirect measures to understand the connectivity, carbon flux and trophic connections within the ice beyond primary production. Sympagic food web relationships have been investigated by gut contents, and a range of biomarkers are suitable to track ice-derived production based on the following underlying mechanisms: During times of high algal production, ice-derived particular organic matter (POM) has enriched carbon stable isotope values compared with pelagic POM because the preferred light carbon isotope can become limiting inside the sea ice brine channel system (Wang et al. 2014), and can, therefore, be used as a tracer. Similarly, profiles and stable isotopic signatures of fatty acids characteristic for certain sea ice algae can be separately tracked through the food web (Budge et al., 2008; Wang et al., 2014). Also, highly branched isoprenoid (HBI) markers characteristic for ice algae, specifically the ice proxy IP25, have been used to track sympagic diatoms in the Arctic (Brown & Belt, 2012), and certain other HBIs have been quantified from various Antarctic algae (Goutte et al., 2013). Recently, genetic markers were used to study the importance of ice algae in under-ice zooplankton (Durbin & Casas, 2013). A drawback of the biomarker experiments remains the often unknown temporal variability of the marker signal in the potential food sources (e.g. Gradinger et al., 2009) and the poorly known turnover times in the fauna after exposure to a prey shift (Kaufman et al., 2008). Experimental studies have been limited to simulated-*in-situ* experiments on Antarctic krill (Marschall, 1988), Arctic amphipod grazing (Werner, 1997), and the growth response of sea ice-inhabiting polychaete juveniles to varying concentrations of ice algal food (McConnell et al., 2012), while true *in situ* experiments do not exist.

There is wide agreement that sympagic meiofauna is mainly herbivorous and grows much faster at the high food concentrations found in the ice in the spring compared with feeding at the much lower concurrent phytoplankton concentrations (McConnell et al., 2012). In the Canadian Arctic, 13 of 16 sea ice meiofauna taxa were entirely herbivorous and grazed on 26 genera of algae (Grainger & Hsiao, 1990). Similarly, Acoela, harpacticoid copepods and polychaete juveniles from coastal fast ice show bright green stomachs in spring (Bluhm & Gradinger, personal observation), and are mainly herbivorous based on their isotopic signatures (Gradinger & Bluhm, unpubl. data). Feeding experiments also confirmed herbivory for the under-ice amphipod *Apherusa glacialis* (Werner, 1997). In the Bering Sea, ice algae may contribute substantial fractions to the nutrition of pelagic crustaceans (*Calanus glacialis/marshallae*, *Thysanoessa rashii*) and benthic biota up to highest trophic levels as revealed by stable isotope (Iken et al., unpubl. data; Wang et al., 2014) and DNA investigations (Durbin & Casas, 2013). Females of the dominant Arctic grazing copepod *Calanus glacialis* used the ice algal bloom to support early maturation and reproduction in Svalbard (Søreide et al., 2010). Food switches can occur with season as observed for the opportunistic Arctic nearshore amphipod *Onisimus litoralis*, which feeds on the spring ice algal bloom in spring and uses other food sources for the rest of the year (Carey & Boudrias, 1987; Gradinger & Bluhm, 2010).

Antarctic stable isotopes studies demonstrated dietary plasticity in *Euphausia superba* furciliae between years, being herbivores when feeding on sea ice biota, and having a more heterotrophic diet when feeding from both the sea ice and the water column. *Limacina helicina*, *Oithona* spp., ostracods and amphipods relied heavily on sea ice biota, while post-larval *E. superba* and *Thysanoessa macrura* consumed both plankton and ice biota (Figure 16.7). The separation of $\delta^{13}C$ values through the mesozooplankton food web confirmed that the sea ice and water column served as important dietary sources for the under-ice zooplankton community during the winter–spring transition. Herbivorous copepods tended to obtain food from the water column even when they lived under the ice cover. At the top of this food web, carnivores, including chaetognaths and copepods, showed carbon isotopic signals that were similar to the water POM (Jia et al., 2016).

Sea ice algae are not the only food source for sea ice metazoans. Currently we know very little about the potential use of dissolved organic material, bacteria and protists other than diatoms in the nutrition of sea ice fauna, but such food sources have been suggested for sympagic nematodes (Tchesunov & Riemann, 1995) and rotifers (e.g. Kaartokallio, 2004). Most omnivorous or predatory taxa are larger sympagic organisms. The exception in the Arctic brine channel system is the hydroid *Sympagohydra tuuli*, whose prey include nauplii and rotifers (Siebert et al., 2009). The ice-endemic nematode *Cryonema crassum* may also feed on smaller nematodes. In the under-ice habitat, adult *Gammarus wilkitzkii* preferably ingest calanoid copepods (Werner,

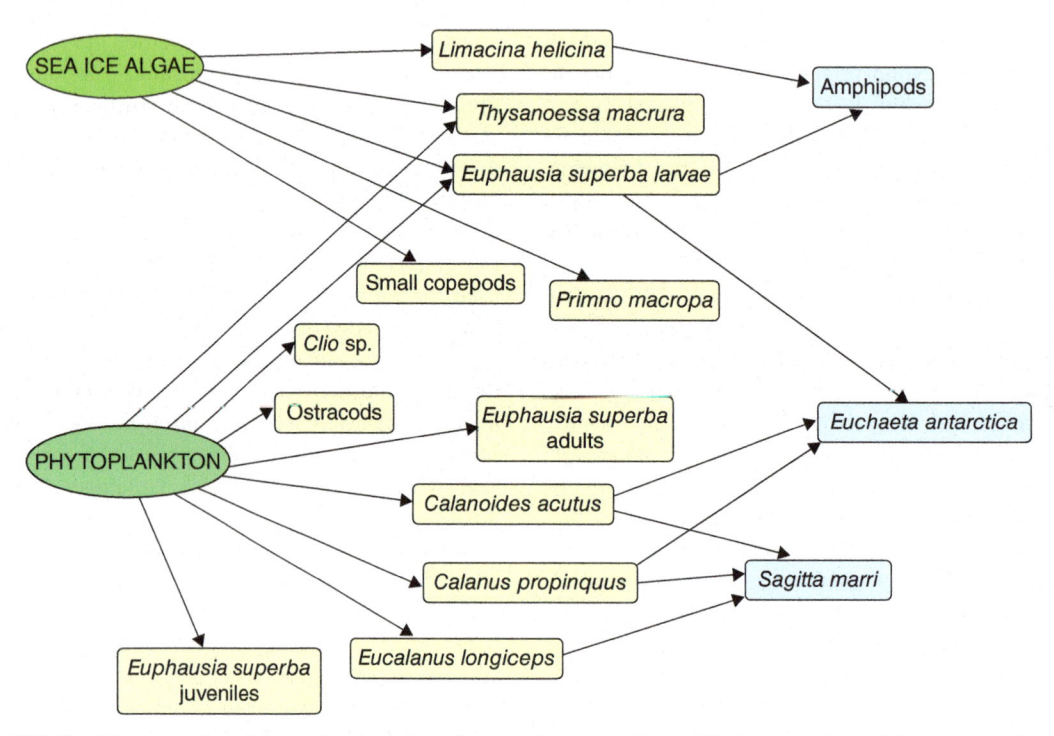

Figure 16.7 Trophic connection of Antarctic sympagic and pelagic fauna based on stable isotope values of these taxa and potential food sources. Schematics show food webs from the Indian sector of the Southern Ocean during the winter–spring transition in: (a) 2007; (b) 2012. Source: From Jia et al., 2016. Reproduced with permission of Elsevier.

1997; Scott et al., 1999) while stable isotopes suggest that young *G. wilkitzkii* may favour primary producers (Iken et al., 2005). Antarctic *P. walkeri* at times ingest newly hatched *P. antarctica*. Fatty acid signatures of young *B. saida* (Scott et al., 1999) point towards largely pelagic carbon sources (Graham et al., 2014), while earlier studies also identified ice-associated amphipods. Antarctic *Pagothenia borchgrevinki* consume a range of sympagic and planktonic metazoans, including pelagic snails, copepods, hyperiid and gammarid amphipods, chaetognaths, euphausiids and even smaller fish, with proportions and prey items varying with the size of the fish (Eastman & DeVries, 1985; Foster et al., 1985). Immediately underneath Arctic sea ice, the ctenophores *Mertensia ovum* and *Bolinopsis infundibulum* can consume copepods, amphipods and pteropods at very high rates (Purcell et al., 2010) and the Antarctic ctenophore *Callianira antarctica* preys on krill under sea ice (Hammer & Hammer, 2000; Scolardi et al., 2006). *Onisimus nanseni* and *O. glacialis* are omnivorous, as documented by stomach content analysis and experiments (Werner, 1997; Poltermann, 1997).

16.5 Grazing impact and contribution to sea ice carbon cycle

To date, all estimates of grazing impacts of sea ice meiofauna are based on indirect estimates, using allometric equations (e.g. Gradinger, 1999; Nozais et al., 2001), while direct measurements under *in situ* conditions are missing. These studies indicate no top-down control of ice algal biomass, with ingestion rates typically below 10% of the available ice algal biomass during the spring bloom (Gradinger et al., 1999, 2005; Nozais et al., 2001; Michel et al., 2002). Currently, meiofauna is also assumed to have no significance in the dynamics of dissolved organic material within the ice, where the interplay between algae and bacteria might be dominating (e.g. Zhou et al., 2014), although some indications exist for direct DOM uptake specifically by sea ice nematodes (Tchesunov & Riemann, 1995). The combined biomass of ice primary producers, meiofauna and microbial communities becomes available to under-ice inhabitants during ice melt. In shallow fast ice, the contribution of meiofauna to the remaining particulate organic matter ranged from 4% to 70% depending on season and locations (Gradinger & Bluhm, unpubl. data).

The characteristics of ice-derived feeds available for macrofauna at the ice–water interface are modified by the degree of grazing, nutritional value and year-to-year variability in the timing of the release from the ice. While direct grazing on sea ice algae is documented for Arctic herbivorous amphipods (see earlier), pelagic copepods (Runge et al., 1991) and krill (*Thysanoessa raschii*; Bering Sea only), the only estimates for grazing impact are based on experiments and suggest summer grazing rates for amphipods to be even less than sea ice meiofauna under pack ice conditions (2.6% day^{-1}; Werner, 1997). Grazing rates may be higher in coastal fast ice areas where benthic amphipods also exploit the sea ice algal biomass and can occur in very high abundances (Carey & Boudrias, 1987; Nygård, 2011).

The vertical flux of faecal pellets produced by sympagic macrofauna can be an important contribution to the food web. Faecal pellets may contain about 20% organic carbon in their dry mass and may amount to a daily release of 2% of the Arctic ice-based organic carbon (Werner, 2000). In nearshore waters, daily flux rates of ice-related amphipod faecal pellets could exceed 1500 pellets m^{-2} day^{-1} (Carey et al., 1987). Such pellets typically contribute the majority of the export from the ice prior to ice melt (Michel et al., 2002), with fast sinking rates of over 100 m day^{-1} linking sea ice to the biota in shallow Arctic seas, but also to the deep sea, weeks prior to the sinking of ice algae released during periods of ice melt (e.g. Boetius et al., 2013). However, macrograzer feeding and associated release of sea ice carbon is a minor contribution compared with the release of all ice-produced carbon during complete sea ice melt (Michel et al., 2002).

16.6 Change and trends

Biological responses to the ongoing changes in the sea ice regimes are gradually emerging for pelagic and benthic realms (Montes-Hugo et al., 2009, Wassmann et al., 2011; Fossheim et al., 2015), while predictions for the sea ice biota are difficult due to the lack of time series in the Arctic and Antarctic (with few notable exceptions). In the Arctic, the most drastic sea ice-mediated changes for macrograzers are related to: (a) the availability of habitat due to the decline in MYI and shift to a primarily seasonally ice-covered Arctic Ocean (Comiso, 2012); (b) the change in the phenology of food availability in the form of algal blooms (Adyrna et al., 2014); and (c) the arrival of potentially resource-competing species from sub-polar areas (Cheung et al., 2009). Intuitively, one might predict that (a) and (c) combined would result in a decline of Arctic sympagic populations, with (a) mostly acting on species tied to MYI, and (c) mostly affecting species at the ice margins where the invasions are highest. On the contrary, thinner ice, a larger marginal ice zone and a longer growing season perhaps associated with (b) (Arrigo & van Dijken, 2011) may result in better food conditions for macrograzers inhabiting first-year ice. In the Antarctic, the focus of attention has been on sea ice variability, with declines in the Bellingshausen, Amundsen and Scotia seas (Schmidt et al., 2012), but increases in other areas such as the Ross Sea, Weddell Sea and Indian Ocean (Stammerjohn et al., 2012) and related consequences, especially for krill populations (Flores et al., 2012; Steinberg et al. 2015). For ice-associated commercially fished species such as *Boreogadus saida* and *Euphausia superba*, sources of population trends are even more complex to determine. So what is the evidence for change in Arctic and Antarctic ice fauna?

There is indeed evidence that the decline in MYI cover may affect associated sympagic species. In the Arctic, sympagic ice amphipod abundance and biomass in the Svalbard and Fram Strait region have declined between the 1980s and the 2000s (Hop et al., 2013). Both the low population size of the amphipod *Gammarus wilkitzkii* (Figure 16.2) in that time series in the last decade and the virtual absence of ice meiofauna in the Beaufort Gyre during the Surface Heat Budget of the Arctic drift compared with earlier studies (Melnikov et al., 2001) have been interpreted as a consequence of the reduction in MYI in that region. *G. wilkitzkii* is a long-lived species (5–7 years) that prefers the MYI habitat (Lønne & Gulliksen, 1991). Future refugia may remain for MYI-associated species in the Canadian Arctic and north of Greenland where MYI has persisted (Lange et al., 2015), and at pressure ridges that survive the summer melt. *G. wilkitzkii* and other ice fauna appear to concentrate at ridges in summer from where they can potentially recolonize adjacent newly forming ice in the autumn and winter (Hop et al., 2000;

Gradinger et al., 2010). Another adaptive strategy to a seasonally ice-covered Arctic has been proposed for *Apherusa glacialis*, which may descend to deep water in the Atlantic Arctic during ice-free periods where it would encounter currents that transport it northward again before ascending to the upper ocean during ice formation, thereby avoiding export from the Arctic (Berge et al., 2012).

While both MYI and overall ice extent have been shrinking in the Arctic, certain sea ice habitats may actually be increasing in area and temporal availability. Arctic summer melt ponds have increased in frequency and are occurring earlier in the year than in previous decades (Nicolaus et al., 2012). Meio- and macrofaunal communities similar to those found in the bottom layers of sea ice also occur in these ponds (Kramer & Kiko, 2011), while small flagellated species with limited primary productivity dominate the algal communities (Lee et al., 2012). The apparently now widespread occurrence and locally high abundance of large ice algal aggregates (1–15 cm diameter) below melting Arctic sea ice north of Svalbard (Assmy et al., 2013) and on the upper surface of sea ice in the Canada Basin (Gradinger & Bluhm, personal observation) may provide a dense food source for ice-associated fauna. In the Svalbard study, high concentrations of degraded chlorophyll, high abundance of ciliates and ice amphipods suggest an active food web coupled to those aggregates.

Evaluating spatial and temporal trends of *Euphausia superba* in the light of environmental change is now achievable due to time series collected from regions of the Antarctic Peninsula [e.g. North Atlantic Peninsula (Loeb & Santora, 2015) and the Long Term Ecological Research program at Palmer Station (PAL-LTER; Steinberg et al., 2015)]. When Southern Annular Mode (SAM) conditions are negative, cold southerly winds flow across the Peninsula, resulting in increased sea ice duration and extent during winter and delayed retreat in spring (Saba et al., 2014). These conditions favour increased phytoplankton growth, which, in turn, promotes the winter survival of larval krill, the stage that is probably the most susceptible to warming temperatures. Conversely, during positive SAM conditions, when warmer northwesterly winds flow across the Peninsula, sea ice duration and extent are reduced, leading to reductions in water column stability, and thereby decreasing the biomass of diatoms and negatively impacting on krill recruitment. The increasing trends in positive SAM conditions have contributed to Antarctic warming, though every 4–6 years negative SAM dominates, which is critical for the continued resetting of ice conditions and, consequently, for krill recruitment. A situation where the duration of positive SAM is longer than the life cycle of krill could result in major decreases in the krill population (Saba et al., 2014), with potentially negative effects for higher trophic levels. While the direct effects of warming will probably impact krill, it appears that a decline in summer chlorophyll concentration could have a more direct influence on krill recruitment (Hill et al., 2013). Whether the krill are genotypically or phenotypically plastic enough to cope with the changes is not known. The krill genome is large and is not yet described fully; however, efforts are under way to characterize transcriptomes (Meyer et al., 2015). An additional stressor includes increasing levels of CO_2 with possible consequences of stock collapse within the next 250 years (Kawaguchi et al., 2013), and possible effects of long-term fisheries (Nicol et al., 2011), that necessitate implementation of ecosystem-based management practices designed to prevent fishery-induced irreversible changes in krill productivity or ecological relationships within Southern Ocean food webs (Hill, et al., 2013).

The potential for species invasions is highest for high latitudes (Cheung et al., 2009), and indeed northward range expansions of sub-polar/boreal species have been reported from the Arctic inflow shelves (Fossheim et al., 2015). We suggest that the chances for newcomers in the polar areas to become associated with sea ice are highest for species with similar biological traits as the ones currently inhabiting sea ice. For example, benthic species with meroplanktonic larvae/juveniles inhabiting shallow seas could adopt a similar strategy as the spionid *Scolelepis squamata* which occurs both outside of and in the Arctic. Potential newly incoming opportunistic species able to exploit resources in different habitats, such as the nearshore amphipod *Onisimus litoralis*, may take advantage of ice algal blooms while present (Gradinger & Bluhm, 2010). Finally, species already occurring in sea ice of sub-polar or boreal areas, such as rotifers prominent in Baltic Sea ice (Meiners et al., 2002), have already been documented as capable of thriving under similar conditions experienced by polar sympagic fauna.

16.7 Summary and outlook

Sea ice in the Arctic and Antarctic houses abundant and phylogenetically diverse life, ranging from microbes to mammals, many of which have specialized life cycles and physiological adaptations to their unique habitat. The macrograzers presented in this chapter inhabiting the brine channel system and/or utilizing the under-ice habitat are less well studied than ice algae, with a few notable exceptions. While seasonally and spatially variable in abundance and composition, they serve as a trophic link connecting primary producers and the microbial loop within the ice to higher trophic levels that inhabit the pelagic and benthic realms. The best-studied sympagic taxa include amphipods and the fish *Boreogadus saida* in the Arctic, and krill, *Euphausia superba*, in the Antarctic, because of both their central roles in the food web and their potential or realized commercial relevance. Increases in harvest rate, as discussed for *B. saida* in areas becoming more easily accessible for vessel traffic, will affect food webs in the future.

Climate-mediated changes in the physical sea ice system in both polar regions are monitored on pan-Arctic and pan-Antarctic scales (Chapters 10 and 11). With regard to sympagic biota, evidence is accumulating that biological and biochemical changes are following those physical changes with regard to primary production (Chapter 14), the carbon, and macrograzer abundance and diversity. Some of these changes are unexpected, such as the decrease in food quality of sea ice algae with increasing light availability (Leu et al., 2010). Most biotic components of sea ice are not monitored regularly – with the notable exceptions of Antarctic krill stocks and subarctic *B. saida* in the Barents Sea – so changes may remain undetected. We therefore encourage a focus on monitoring and building time series plus recovery of historic data by funding agencies and researchers in the future. Some efforts in that direction are under way, such as the work by the sea ice group in the Arctic Circumpolar Biodiversity Monitoring Program and the recently published CAML/SCAR-MarBIN Biogeographic Atlas of the Southern Ocean.

Clearly, warming climate is not the only pressure affecting the sea ice ecosystem; rather, cumulative impacts will increase as shipping, oil and gas exploration, fisheries, tourism and other industries expand towards the poles (Tin et al., 2014) and can affect all marine life, including sympagic taxa. Studies document acute and chronic effects of petroleum compounds through oil contamination in metazoans (for sympagic taxa: Hatlen et al., 2009; Geraudie et al., 2014) and comparatively low rates of oil degradation by bacteria at low temperatures (McFarlin et al., 2014). Even with the predicted loss of Arctic summer sea ice and the changes already measured in the Western Antarctic Peninsula, winter sea ice will remain for the foreseeable future in both polar seas. However, the changing phenology of ice extent and algal production could lead to mismatch situations between life cycles of ice macrograzers and ice algal food availability. In-depth knowledge of the role of sympagic biota in the Arctic and Antarctic food webs is therefore critical, as neither climate warming nor contamination risks associated with increasing human footprint in polar areas can be argued away by industry, policy makers or carbon trade quotas.

16.8 Acknowledgements

Schematic figures were prepared by Patricia Kimber with assistance by Malin Daase. Editorial assistance was provided by Mette Kaufman. We thank the data providers to Figure 16.5 for their contributions. The editor is thanked for his encouragement and patience.

References

Aarset, A.V. (1991) The ecophysiology of under-ice fauna. *Polar Research*, **10**, 309–324.

Aarset, A.V. & Torres, J.J. (1989) Cold resistance and metabolic responses to salinity variations in the amphipod *Eusirus antarcticus* and the krill *Euphausia superba*. *Polar Biology*, **9**, 491–497.

Andriashev, A.P. (1968) The problem of the life community associated with the Antarctic fast ice. In: *Symposium on Antarctic Oceanography* (Ed. R.I. Currie), pp. 147–155. Scott Polar Research Institute, Cambridge.

Ardyna, M., Babin, M., Gosselin, M., Devred, E., Rainville, L., & Tremblay, J. É. (2014). Recent Arctic Ocean sea ice loss triggers novel fall phytoplankton blooms. *Geophysical Research Letters*, **41**, 6207–6212.

Arndt, C.E. & Beuchel, F. (2006) Life history and population dynamics of the Arctic sympagic amphipods *Onisimus nanseni* Sars and *O. glacialis* Sars (Gammaridea: Lysianassidae). *Polar Biology*, **29**, 239–248.

Arndt, C.E & Swadling, K.M. (2006) Crustacea in Arctic and Antarctic sea ice: distribution, diet and life history strategies. *Advances in Marine Biology*, **51**, 197–315.

Arrigo, K.R. & van Dijken, G.L. (2011) Secular trends in Arctic Ocean net primary production. *Journal of Geophysical Research: Oceans (1978–2012)*, **116**(C9).

Assmy, P., Ehn, J.K., Fernández-Méndez, M. et al (2013). Floating ice-algal aggregates below melting Arctic sea ice. *PloS one*, **8**(10), e76599.

Atkinson, A., Meyer, B., Stübing, D., Hagen, W., Schmidt, K. & Bathmann, U.V. (2002) Feeding and energy budgets of Antarctic krill *Euphausia superba* at the onset of winter – II. Juveniles and adults. *Limnology and Oceanography*, **47**, 953–966.

Atkinson, A., Siegel, V., Pakhomov, E., & Rothery, P. (2004). Long-term decline in krill stock and increase in salps within the Southern Ocean. *Nature*, **432**, 100–103.

Berge, J., Varpe, Ø., Moline, M.A. et al. (2012) Retention of ice-associated amphipods: possible consequences for an ice-free Arctic Ocean. *Biology Letters*, **8**, 1012–1015.

Beuchel, F. & Lønne, O. (2002) Population dynamics of the sympagic amphipods *Gammarus wilkitzkii* and *Apherusa glacialis* in sea ice north of Svalbard. *Polar Biology*, **25**, 241–250.

Blome, D. & Riemann, F. (1999) Antarctic sea ice nematodes, with description of *Geomonhystera glaciei* sp. nov. (Monhysteridae). *Mitteilungen aus dem Hamburgischen Zoologischen Museum und Institut*, **96**, 15–20.

Boetius, A., Albrecht, S., Bakker, K. et al. (2013). Export of algal biomass from the melting Arctic sea ice. *Science*, **339**, 1430–1432.

Brown, T.A. & Belt, S.T. (2012) Identification of the sea ice diatom biomarker IP25 in Arctic benthic macrofauna: direct evidence for a sea ice diatom diet in Arctic heterotrophs. *Polar Biology*, **35**, 131–137.

Budge, S.M., Wooller, M.J., Springer, A.M., Iverson, S.J., McRoy, C.P. & Divoky, G.J. (2008) Tracing carbon flow in an arctic marine food web using fatty acid-stable isotope analysis. *Oecologia*, **157**, 117–129.

Campbell, K., Mundy, C.J., Barber, D.G. & Gosselin, M. (2014) Characterizing the sea ice algae chlorophyll a–snow depth relationship over Arctic spring melt using transmitted irradiance. *Journal of Marine Systems*, **147**, 76–84.

Carey, A.G., Jr, (1985) Marine ice fauna. In: *Sea Ice Biota* (Ed. R.A. Horner), pp. 173–190. CRC Press, Boca Raton, Florida.

Carey, A.G., Jr, (1987). Particle flux beneath fast ice in the shallow southwestern Beaufort Sea, Arctic Ocean. *Marine Ecology Progress Series*, **40**, 247-257.

Carey, A.G., Jr., (1992) The ice fauna in the shallow southwestern Beaufort Sea, Arctic Ocean. *Journal of Marine Systems*, **3**, 225–236.

Carey, A.G.J. & Boudrias, M.A. (1987) Feeding ecology of *Pseudalibrotus* (=*Onisimus*) *litoralis* Krøyer (Crustacea:Amphipoda) on the Beaufort Sea inner continental shelf. *Polar Biology*, **8**, 29–33.

Carey, A.G. & Montagna, P.A. (1982) Arctic sea ice formation assemblage: first approach to description and source of the underice meiofauna. *Marine Ecology Progress Series*, **8**, 1–8.

Cheung, W.W., Lam, V.W., Sarmiento, J.L., Kearney, K., Watson, R. & Pauly, D. (2009) Projecting global marine biodiversity impacts under climate change scenarios. *Fish and Fisheries*, **10**, 235–251.

Clarke, A., Doherty, N., DeVries, A.L. & Eastman, J.T. (1984) Lipid content and composition of three species of Antarctic fish in relation to buoyancy. *Polar Biology*, **3**, 77–83.

Collins, R.E., Rocap, G. & Deming, J.W. (2010) Persistence of bacterial and archaeal communities in sea ice through an Arctic winter. *Environmental microbiology*, **12**, 1828–1841.

Comiso, J.C. (2012) Large decadal decline of the Arctic multi-year ice cover. *Journal of Climate*, **25**, 1176–1193.

Conover, R.J, Cota, G.F., Harrison, W.G., Horne, E.P.W. & Smith, R.E.H. (1990) Ice/water interactions and their effect on biological oceanography in the Arctic Archipelago. In: *Canada's Missing Dimension: Science and History in the Canadian Arctic Islands*, (Ed. C.R. Harington), pp. 204–228. Canadian Museum of Nature, Ottowa.

Conover, R.J., Herman, A.W., Prinsenberg, S.J. & Harris, L.R. (1986) Distribution of and feeding by the copepod *Pseudocalanus* under fast ice during the Arctic spring. *Science*, **232**, 1245–1247.

Dahms, H. U., Bergmans, M., & Schminke, H. K. (1990). Distribution and adaptations of sea ice inhabiting Harpacticoida (Crustacea, Copepoda) of the Weddell Sea (Antarctica). *Marine Ecology*, **11**, 207–226.

David, C., Lange, B., Rabe, B., & Flores, H. (2015). Community structure of under-ice fauna in the Eurasian central Arctic Ocean in relation to environmental properties of sea-ice habitats. *Marine Ecology-Progress Series*, **522**, 15–32.

Dayton, P.K., Robilliard, G.A. & DeVries, A.L. (1969) Anchor ice formation in McMurdo Sound, Antarctica, and its biological effects. *Science*, **163**, 273–274.

DeVries, A.L. (1971) Glycoproteins as biological antifreeze agents in Antarctic fishes. *Science*, **172**, 1152–1155.

DeVries, A.L. (1997) The role of antifreeze proteins in survival of Antarctic fishes in freezing environments. In: *Antarctic Communities. Species, Structure and Survival* (Eds. B. Battaglia, J. Valencia & D.W.H. Walton), pp. 202–208. Cambridge University Press, Cambridge.

Dieckmann, G.S., Rohardt, G., Hellmer, H. & Kipfstuhl, J. (1986) The occurrence of ice platelets at 250 m near the Filchner Ice Shelf and its significance to sea ice biology. *Deep-Sea Research*, **33**, 141–148.

Durbin, E. G., & Casas, M. C. (2013). Early reproduction by *Calanus glacialis* in the Northern Bering Sea: the role of ice algae as revealed by molecular analysis. *Journal of Plankton Research*, **0**, 1–19.

Eastman, J. T., & DeVries, A. L. (1985). Adaptations for cryopelagic life in the Antarctic notothenioid fish *Pagothenia borchgrevinki*. *Polar Biology*, **4**, 45–52.

Falk-Petersen, S., Hagen, W., Kattner, G., Clarke, A. & Sargent, J. (2000) Lipids, trophic relationships, and biodiversity in Arctic and Antarctic krill. *Canadian Journal of Fisheries and Aquatic Sciences*, **57** (Suppl. 3), 178–191.

Flores, H., van Franeker, J.A., Cisewski, B. et al. (2011) Macrofauna under sea ice and in the open surface layer of the Lazarev Sea, Southern Ocean. *Deep Sea Research Part II*, **58**, 1948–1961.

Flores, H., Van Franeker, J.A., Siegel, V. et al. (2012) The association of Antarctic krill *Euphausia superba* with the under-ice habitat. *PloS one*, **7**, e31775.

Fossheim, M., Primicerio, R., Johannesen, E. et al. (2015) Recent warming leads to a rapid borealization of fish communities in the Arctic. *Nature Climate Change*, doi:10.1038/nclimate2647.

Foster, B. A., Cargill, J. M., & Montgomery, J. C. (1987). Planktivory in *Pagothenia borchgrevinki* (Pisces: Nototheniidae) in McMurdo Sound, Antarctica. *Polar Biology*, **8**, 49–54.

Friedrich, C. & De Smet, W.H. (2000) The rotifer fauna of Arctic sea ice from the Barents Sea, Laptev Sea and Greenland Sea. *Hydrobiologia*, **432**, 73–90.

Friedrich, C. & Hendelberg, J. (2001) On the ecology of Acoela living in the Arctic sea ice. *Belgian Journal of Zoology*, **131**, 213–216.

Garrison, D.L., Ackley, S.F. & Buck, K.R. (1983) A physical mechanism for establishing algal populations in frazil ice. *Nature*, **306**, 363–365.

Garrison, D.L., & Close, A.R. (1993) Winter ecology of the sea ice biota in Weddell Sea pack ice. *Marine Ecology Progress Series*, **96**, 17–31.

Geraudie, P., Nahrgang, J., Forget-Leray, J., Minier, C. & Camus, L. (2014) In vivo effects of environmental concentrations of produced water on the reproductive function of polar cod (*Boreogadus saida*). *Journal of Toxicology and Environmental Health, Part A*, **77**, 557–573.

Goutte, A., Cherel, Y., Houssais, M.N. et al. (2013) Diatom-specific highly branched isoprenoids as biomarkers in Antarctic consumers. *PloS one*, **8**, e56504.

Goutte, A., Cherel, Y., Ozouf-Costaz, C., Robineau, C., Lanshere, J. & Massé, G. (2014) Contribution of sea ice organic matter in the diet of Antarctic fishes: a diatom-specific highly branched isoprenoid approach. *Polar biology*, **37**, 903–910.

Gradinger, R. (1999) Integrated abundance and biomass of sympagic meiofauna in Arctic and Antarctic pack ice. *Polar Biology*, **22**, 169–177.

Gradinger, R. (2001) Adaptation of Arctic and Antarctic ice metazoan to their habitat. *Zoology* **104**, 339–345.

Gradinger, R. (2009) Sea-ice algae: Major contributors to primary production and algal biomass in the Chukchi and Beaufort Seas during May/June 2002. *Deep Sea Research Part II: Topical Studies in Oceanography*, **56**(17), 1201–1212.

Gradinger, R. & Bluhm, B. (2009) Assessment of the abundance and diversity of sea ice biota. *Field Techniques for Sea-Ice Research*, 283.

Gradinger, R., Bluhm, B.A. (2010) Timing of ice algal grazing by the Arctic nearshore benthic amphipod *Onisimus litoralis*. *Arctic* **63**, 355–358.

Gradinger, R., Kaufman, M.R. & Bluhm, B.A. (2009) Pivotal role of sea ice sediments in the seasonal development of near-shore Arctic fast ice biota. *Marine Ecology Progress Series*, **394**, 49–63.

Gradinger, R., Bluhm, B. & Iken, K. (2010) Arctic sea-ice ridges—Safe havens for sea-ice fauna during periods of extreme ice melt? *Deep Sea Research Part II*, **57**, 86–95.

Gradinger, R. & Schnack-Schiel, S.B. (1998) Potential effect of ice formation on Antarctic pelagic copepods: salinity induced mortality of *Calanus propinquus* and *Metridia gerlachei* in comparison to sympagic acoel turbellarians. *Polar Biology*, **20**, 139–142.

Gradinger, R.R. & Bluhm, B.A. (2004) *In situ* observations on the distribution and behavior of amphipods and Arctic cod (*Boreogadus saida*) under the sea ice of the high Arctic Canadian Basin. *Polar Biology*, **27**, 595–603.

Gradinger, R., Friedrich, C. & Spindler, M. (1999) Abundance, biomass and composition of the sea ice biota of the Greenland Sea pack ice. *Deep-Sea Research*, **46**, 1457–1472.

Gradinger, R.R., Meiners, K., Plumley, G., Zhang, Q. & Bluhm, B.A. (2005) Abundance and composition of the sea-ice meiofauna in off-shore pack ice of the Beaufort Gyre in summer 2002 and 2003. *Polar Biology*, **28**, 171–181.

Graham, C., Oxtoby, L., Wang, S. W., Budge, S. M., & Wooller, M. J. (2014). Sourcing fatty acids to juvenile polar cod (*Boreogadus saida*) in the Beaufort Sea using compound-specific stable carbon isotope analyses. *Polar Biology*, **37**, 697–705.

Grainger, E.H. (1991) Exploitation of arctic sea ice by epibenthic copepods. *Marine Ecology Progress Series*, **77**, 119–124.

Grainger, E.H. & Hsiao, S.I.C. (1990) Trophic relationship of the sea ice meiofauna in Frobisher Bay, Arctic Canada. *Polar Biology*, **10**, 283–292.

Guglielmo, L., Zagami, G., Saggiimo, V., Catalano, G. & Granata, A. (2007) Copepods in spring annual sea ice at Terra Nova Bay (Ross Sea, Antarctica). *Polar Biology*, **30**, 747–758.

Gulliksen, B. & Lønne, O.J. (1991) Sea ice macrofauna in the Antarctic and the Arctic. *Journal of Marine Systems*, **2**, 53–61.

Günther, S., George, K.H. & Gleitz, M. (1999) High sympagic metazoan abundance in platelet layers at Drescher-Inlet, Weddell Sea, Antarctica. *Polar Biology*, **22**, 82–89.

Hammer, W.M. & Hammer, P.P. (2000) Behavior of Antarctic krill (*Euphausia superba*): schooling, foraging, and antipredatory behavior. *Canadian Journal of Fishery and Aquatic Science*, **57** (Suppl. 3), 192–202.

Hatlen, K., Camus, L., Berge, J., Olsen, G.H. & Baussant, T. (2009) Biological effects of water soluble fraction of crude oil on the Arctic sea ice amphipod *Gammarus wilkitzkii*. *Chemistry and Ecology*, **25**, 151–162.

Hill, S.L., Phillips, T. & Atkinson, A. (2013) Potential climate change effects on the habitat of Antarctic Krill in the Weddell quadrant of the Southern Ocean. *PLoS One*, **8**, e72246, doi:10.1371/journal.pone.0072246.

Hop, H., Poltermann, M., Lønne, O.J., Falk-Petersen, S., Kornes R. & Budgell, W.P. (2000) Ice amphipod distribution relative to ice density and under-ice topography in the northern Barents Sea. *Polar Biology*, **23**, 357–367.

Hop, H., Bluhm, B.A., Daase, M., Gradinger, R. & Poulin, M. (2013) Arctic sea ice biota. In: *Arctic Report Card 2013* (Eds. M.O. Jeffries, J.A. Richter-Menge & J.E. Overland). NOAA, MD, USA.

Hop, H., Mundy, C.J., Gosselin, M., Rossnagel, A. L. & Barber, D.G. (2011) Zooplankton boom and ice amphipod bust below melting sea ice in the Amundsen Gulf, Arctic Canada. *Polar Biology*, **34**, 1947–1958.

Hop, H. & Gjøsæter, H. (2013) Polar cod (*Boreogadus saida*) and capelin (*Mallotus villosus*) as key species in marine food webs of the Arctic and the Barents Sea. *Marine Biology Research*, **9**, 878–894.

Hoshiai, T., Tanimura, A. & Watanaba, K. (1987) Ice algae as food of an Antarctic ice-associated copepod, *Paralabidocera antarctica* (I.C. Thompson). *Proceedings of the NIPR Symposium on Polar Biology*, **1**, 105–111.

Iken, K., Bluhm, B. A., & Gradinger, R. (2005). Food web structure in the high Arctic Canada Basin: evidence from $\delta 13C$ and $\delta 15N$ analysis. *Polar Biology*, **28**(3), 238–249.

Jia Z., Swadling K.M., Meiners K.M., Kawaguchi S. & Virtue P. (2016) *The zooplankton food web under East Antarctic pack ice – a stable isotope study. Deep-Sea Research II*, in press.

Jin, Y., & DeVries, A. L. (2006). Antifreeze glycoprotein levels in Antarctic notothenioid fishes inhabiting different thermal environments and the effect of warm acclimation. *Comparative Biochemistry and Physiology Part B: Biochemistry and Molecular Biology*, **144**, 290–300.

Kaartokallio, H. (2004). Food web components, and physical and chemical properties of Baltic Sea ice. *Marine Ecology Progress Series*, **273**, 49–63.

Kaufman, M. R., Gradinger, R. R., Bluhm, B. A., & O'Brien, D. M. (2008). Using stable isotopes to assess carbon and nitrogen turnover in the Arctic sympagic amphipod *Onisimus litoralis*. *Oecologia*, **158**, 11–22.

Kawaguchi, S., Ishida, A., King, R. et al. (2013). Risk maps for Antarctic krill under projected Southern Ocean acidification. *Nature Climate Change*, **3**, 843–847.

Kiko, R. (2008) Ecophysiology of Antarctic sea-ice meiofauna. Dissertation. Kiel University. 116p.

Kiko, R., Michels, J., Mizdalski, E., Schnack-Schiel, S.B. & Werner, I. (2008a) Living conditions, abundance and composition of the metazoan fauna in surface and sub-ice layers in pack ice of the western Weddell Sea during late spring. *Deep-Sea Research Part II*, **55**, 1000–1014.

Kiko, R., Kramer, M., Spindler, M. & Werner, I. (2008b) *Tergipes antarcticus* (Gastropoda, Nudibranchia): distribution, life cycle, morphology, anatomy and adaptation of the first mollusk known to live in Antarctic sea ice. *Polar Biology*, **31**, 1383–1395.

Kiko, R., Werner, I. & Wittmann, A. (2008c) Osmotic and ionic regulation in response to salinity variations and cold resistance in the Arctic under-ice amphipod *Apherusa glacialis*. *Polar Biology*, **32**, 393–398.

Kramer, M. & Kiko, R. (2011) Brackish meltponds on Arctic sea ice—a new habitat for marine metazoans. *Polar Biology*, **34**, 603–608.

Kramer, M., Swadling, K.M., Meiners, K.M. et al. (2011) Antarctic sympagic meiofauna in winter: comparing diversity, abundance and biomass between perennially and seasonally ice-covered regions. *Deep-Sea Research Part II*, **58**, 1062–1074.

Krapp, R.H., Berge, J., Flores, H., Gulliksen, B. & Werner, I. (2008) Sympagic occurrence of eusirid and lysianassoid amphipods under Antarctic pack ice. *Deep-Sea Research Part II*, **55**, 1015–1023.

Krembs, C., Gradinger, R., & Spindler, M. (2000). Implications of brine channel geometry and surface area for the interaction of sympagic organisms in Arctic sea ice. *Journal of Experimental Marine Biology and Ecology*, **243**, 55–80.

Kurbjeweit, F., Gradinger, R. & Weissenberger, J. (1993) The life cycle of *Stephos longipes* – an example for cryopelagic coupling in the Weddell Sea (Antarctica). *Marine Ecology Progress Series*, **98**, 255–262.

Lange, B.A., Michel, C., Beckers, J.F. et al. (2015) Comparing springtime ice-algal chlorophyll *a* and physical properties of multi-year and first-year sea ice from the Lincoln Sea. *Plos One*, **10**, doi: 10.1371/journal.pone.0122418. eCollection 2015.

Lee, R.F., Hagen, W. & Kattner, G. (2006) Lipid storage in marine zooplankton. *Marine Ecology Progress Series*, **307**, 273–306.

Lee, S.H., Stockwell, D., Joo, H-.M., Son, Y.B., Kang, C-.K. & Whitledge, T. (2012) Phytoplankton production from melting ponds on Arctic sea ice. *Journal of Geophysical Research: Oceans (1978–2012)*, 117, C4.

Leu, E., Wiktor, J., Søreide, J.E., Berge, J. & Falk-Petersen, S. (2010) Increased irradiance reduces food quality of sea ice algae. *Marine Ecology Progress Series*, **411**, 49–60.

Loeb, V.J. & Santora, J.A. (2015) Climate variability and spatiotemporal dynamics of five Southern Ocean krill species. *Progress in Oceanography*, **134**, 93–122.

Lonne, O.J., & Gulliksen, B. (1991) Sympagic macro-fauna from multiyear sea-ice near Svalbard. *Polar Biology*, **11**(7), 471–477.

Logerwell, E., Rand, K. & Weingartner, T.J. (2011) Oceanographic characteristics of the habitat of benthic fish and invertebrates in the Beaufort Sea. *Polar Biology*, **34**, 1783–1796.

Macnaughton, M.O., Thormar, J. & Berge, J. (2007) Sympagic amphipods in the Arctic pack ice: redescriptions of *Eusirus holmii* Hansen, 1887 and *Pleusymtes karstensi* (Barnard, 1959). *Polar Biology*, **30**, 1013–1025.

Marquardt, M., Kramer, M., Carnat, G. & Werner, I. (2011) Vertical distribution of sympagic meiofauna in sea ice in the Canadian Beaufort Sea. *Polar biology*, **34**, 1887–1900.

Marschall, P. (1988) The overwintering strategy of Antarctic krill under the pack ice of the Weddell Sea. *Polar Biology*, **9**, 129–135.

McConnell, B., Gradinger, R., Iken, K. & Bluhm, B.A. (2012) Growth rates of arctic juvenile *Scolelepis squamata* (Polychaeta: Spionidae) isolated from Chukchi Sea fast ice. *Polar biology*, **35**, 1487–1494.

McFarlin, K.M., Prince, R.C., Perkins, R. & Leigh, M.B. (2014) Biodegradation of dispersed oil in arctic seawater at-1 C. *PloS One*, **9**, e84297.

Mecklenburg, C.W., Møller, P.R. & Steinke, D. (2011) Biodiversity of arctic marine fishes: taxonomy and zoogeography. *Marine Biodiversity*, **41**, 109–140.

Meiners, K., Fehling, J., Granskog, M. A. & Spindler, M. (2002) Abundance, biomass and composition of biota in Baltic sea ice and underlying water (March 2000). *Polar Biology*, **25**, 761–770.

Melnikov, I.A. (1997) *The Arctic Sea Ice Ecosystem*. Gordon and Breach Science Publishers, Amsterdam.

Melnikov, I.A., Zhitina, L.S. & Kolosova, H.G. (2001) The Arctic sea ice biological communities in recent environmental changes (scientific note). *Memoirs of National Institute of Polar Research. Special issue*, **54**, 409416.

Melnikov, I.A. & Chernova, N.V. (2013) Characteristics of under-ice swarming of polar cod *Boreogadus saida* (Gadidae) in the central Arctic ocean. *Journal of Ichthyology*, **53**, 7–15.

Menshenina, L.L. & Melnikov, I.A. (1995) Under-ice zooplankton of the western Weddell Sea. *Proceedings of the NIPR Symposium on Polar Biology*, **8**, 126–138.'

Meyer, B. (2012) The overwintering of Antarctic krill, *Euphausia superba*, from an ecophysiological perspective. *Polar Biology*, **35**, 15–37.

Meyer, B., Fuentes, V., Guerra, C. et al. (2009) Physiology, growth and development of larval krill *Euphausia superba* in autumn and winter in the Lazarev Sea, Antarctica. *Limnology and Oceanography*, **54**, 1595–1614.

Meyer, B., Martini, P., Biscontin, A. et al. (2015) Pyrosequencing and *de novo* assembly of Antarctic krill (*Euphausia superba*) transcriptome to study the adaptability of krill to climate-induced environmental changes. *Molecular Ecology Resources*, doi: 10.1111/1755-0998.12408.

Michel, C., Nielsen, T.G., Nozais, C. & Gosselin, M. (2002) Significance of sedimentation and grazing by ice micro- and meiofauna for carbon cycling in annual sea ice (northern Baffin Bay). *Aquatic Microbial Ecology*, **30**, 57–68.

Montes-Hugo, M., Doney, S.C., Ducklow, H.W. et al. (2009) Recent changes in phytoplankton communities associated with rapid regional climate change along the Western Antarctic Peninsula. *Science*, **323**, 1470–1473.

Nicol, S. (2006) Krill, currents, and sea ice: *Euphausia superba* and its changing environment. *BioScience*, **56**, 111–120.

Nicol, S., Foster, J., & Kawaguchi, S. (2011) The fishery for Antarctic krill – recent developments. *Fish and Fisheries*, **13**, 30–40.

Nicolaus, M., Katlein, C., Maslanik, J. & Hendricks, S. (2012) Changes in Arctic sea ice result in increasing light transmittance and absorption. *Geophysical Research Letters*, **39**,

Nozais, C., Gosselin, M., Michel, C. & Tita, G. (2001) Abundance, biomass, composition and grazing impact of the sea-ice meiofauna in the North Water, northern Baffin Bay. *Marine Ecology Progress Series*, **217**, 235–250.

Nygård, H. (2011) Scavenging amphipods in the high Arctic. PhD thesis. Arctic University of Norway, Tromsø.

O'Brien, D.P. (1987) Direct observations of the behaviour of *Euphausia superba* and *Euphausia crystallorophias* (Crustacea: Euphausiacea) under pack ice during the Antarctic spring. *Journal of Crustacean Biology*, **7**, 437–448.

Piraino, S., Bluhm, B.A., Gradinger, R. & Boero, F. (2008) *Sympagohydra tuuli* gen. nov. et sp. nov. (Cnidaria: Hydrozoa) a cool hydroid from the Arctic sea ice. *Journal of the Marine Biological Association of the U.K.*, **88**, doi:10.1017/S0025315408002166.

Poltermann, M. (1997) Biologische und ökologische Untersuchungen zur kryopelagischen Amphipodenfauna des arktischen Meereises. *Berichte zur Polarforschung*, **225**, 1–170. (In German)

Poltermann, M., Hop, H. & Falk-Petersen, S. (2000) Life under Arctic sea ice - reproduction strategies of two sympagic (ice-associated) amphipod species, *Gammarus wilkitzkii* and *Apherusa glacialis*. *Marine Biology*, **136**, 913–920.

Praebel, K. & Ramlov, H. (2005) Antifreeze activity in the gastrointestinal fluids of *Arctogadus glacialis* (Peters 1874) is dependent on food type. *Journal of Experimental Biology*, **208**, 2609–2613.

Purcell, J.E., Hopcroft, R.R., Kosobokova, K.N. & Whitledge, T.E. (2010) Distribution, abundance, and predation effects of epipelagic ctenophores and jellyfish in the western Arctic Ocean. *Deep Sea Research Part II*, **57**, 127–135.

Rakusa-Suszczewski, S. (1972) The biology of *Paramoera walkeri* Stebbing (Amphipoda) and the Antarctic sub-fast ice community. *Polish Archives of Hydrobiology*, **19**, 11–36.

Richardson, M.G. & Whitaker, T.M. (1979) An Antarctic fast-ice food chain: observations on the interaction on the amphipod *Pontogeneia antarctica* Chevreux with ice-associated micro-algae. *British Antarctic Survey Bulletin*, **47**, 107–115.

Riemann, F. & Sime-Ngando, T. (1997) Note on sea-ice nematodes (Monhysteroidea) from Resolute Passage, Canadian High Arctic. *Polar Biology*, **18**, 70–75.

Runge, J.A. & Ingram, R.G. (1988) Underice grazing by planktonic, calanoid copepods in relation to a bloom of ice microalgae in southeastern Hudson Bay. *Limnology and Oceanography*, **33**, 280–286.

Runge, J.A., Therriault, J.C., Legendre, L., Ingram, R.G. & Demers, S. (1991) Coupling between ice microalgal productivity and the pelagic, metazoan food web in southeastern Hudson Bay: a synthesis of results. *Polar Research*, **10**, 325–338.

Saba, G.K., Fraser, W.R., Saba, V.S. et al. (2014) Winter and spring controls on the summer food web of the coastal West Antarctic Peninsula. *Nature Communications*, **5**, 4318, doi: 10.1038/ncomms5318.

Schmidt, K., Atkinson, A., Venables, H. J., & Pond, D. W. (2012). Early spawning of Antarctic krill in the Scotia Sea is fueled by "superfluous" feeding on non-ice associated phytoplankton blooms. *Deep Sea Research Part II: Topical Studies in Oceanography*, **59**, 159–172.

Schnack-Schiel, S.B., Haas, C., Michels, J. et al. (2008) Copepods in sea ice of the western Weddell Sea during austral spring 2004. *Deep-Sea Research Part II*, **55**, 1056–1067.

Schnack-Schiel, S.B., Thomas, D.N., Dahms, H.U., Haas, C. & Mizdalski, E. (1998) Copepods in Antarctic Sea ice. In: *Antarctic Sea Ice Biological Processes, Interactions and Variability*, (Eds. M.P. Lizotte & K. Arrigo), pp. 173–182. American Geophysical Union, Washington, D.C. Antarctic Research Series, **73**.

Schnack-Schiel, S.B., Thomas, D., Dieckmann, G.S. et al. (1995) Life cycle strategy of the Antarctic calanoid copepod *Stephos longipes*. *Progress in Oceanography*, **36**, 45–75.

Schnack-Schiel, S.B., Thomas, D.N., Haas, C., Dieckmann, G.S. & Alheit, R. (2001) On the occurrence of the copepods *Stephos longipes* (Calanoida) and *Drescheriella glacialis* (Harpacticoida) in summer sea ice in the Weddell Sea. *Antarctic Science*, **13**, 150–157.

Scolardi, K.M., Daly, K.L., Pakhomov, E.A. & Torres, J.J. (2006) Feeding ecology and metabolism of the Antarctic cydippid ctenophore *Callianira antarctica*. *Marine Ecology Progress Series*, **317**, 111–126.

Scott, C.L., Falk-Petersen, S., Sargent, J.R., Hop, H., Lønne, O.J. & Poltermann, M. (1999) Lipid and trophic interactions of ice fauna and pelagic zooplankton in the marginal zone of the Barents Sea. *Polar Biology*, **21**, 65–70.

Siebert, S., Anton-Erxleben, F., Kiko, R. & Kramer, M. (2009) *Sympagohydra tuuli* (Cnidaria, Hydrozoa): first report from sea ice of the central Arctic Ocean and insights into histology, reproduction and locomotion. *Marine Biology*, **156**, 541–554.

Smetacek, V., Scharek, R. & Nöthig, E-M. (1990) Seasonal and regional variation in the pelagial and its relationship to the life history cycle of krill. In *Antarctic Ecosystems*, pp. 103–114. Springer, Berlin.

Søreide, J.E., Leu, E.V.A., Berge, J., Graeve, M. & Falk-Petersen, S.T.I.G. (2010) Timing of blooms, algal food quality and *Calanus glacialis* reproduction and growth in a changing Arctic. *Global Change Biology*, **16**, 3154–3163.

Spindler, M., & Dieckmann, G. S. (1986). Distribution and abundance of the planktic foraminifer *Neogloboquadrina pachyderma* in sea ice of the Weddell Sea (Antarctica). *Polar Biology*, **5**, 185–191.

Stammerjohn, S. E., Massom, R., Rind, D. & Martinson, D. (2012) Regions of rapid sea ice change: An inter-hemispheric seasonal comparison. *Geophysical Research Letters*, **39**, L06501.

Steinberg, D.K., Ruck, K.E., Gleiber, M.R. et al. (2015) Long-term (1993–2013) changes in macrozooplankton off the Western Antarctic Peninsula. *Deep Sea Research Part I: Oceanographic Research Papers*, **101**, 54–70.

Swadling, K.M. (1998) Influence of seasonal ice formation on the life cycle strategies of Antarctic copepods. PhD Thesis. University of Tasmania, Australia.

Swadling, K.M. (2001) Population structure of two Antarctic ice-associated copepods. *Drescheriella glacialis* and *Paralabidocera antarctica*, in winter sea ice. *Marine Biology*, **129**, 597–603.

Swadling, K.M., Gibson, J.A.E., Ritz, D.A. & Nichols, P.D. (1997) Horizontal patchiness in sympagic organisms of the Antarctic fast ice. *Antarctic Science*, **9**, 399–406.

Swadling, K.M., McKinnon, A.D., De'ath, G. & Gibson, J.A.E. (2004) Life cycle plasticity and differential growth and development in marine and lacustrine populations of an Antarctic copepod. *Limnology and Oceanography*, **49**, 644–655.

Swadling, K.M., McPhee, A.D. & McMinn, A. (2001) Spatial distribution of copepods in fast ice of eastern Antarctica. *Polar Bioscience*, **13**, 55–65.

Swadling, K.M., Nichols, P.D., Gibson, J.A.E. & Ritz, D.A. (2000) Role of lipid in the life cycles of ice-dependent and ice-independent populations of the copepod *Paralabidocera antarctica*. *Marine Ecology Progress Series*, **208**, 171–182.

Tchesunov, A.V. & Riemann, F. (1995) Arctic sea ice nematodes (Monhysteroidea) with descriptions of *Cryonema crassum* gen. n. sp. n. and *C. tenue* sp. n. *Nematologica*, **41**, 35–50.

Tin, T., Lamers, M., Liggett, D., Maher, P.T. & Hughes, K.A. (2014) Setting the scene: Human activities, environmental impacts and governance arrangements in Antarctica." In: *Antarctic Futures*. (Eds. T. Tin et al.), pp. 1–24. Springer, the Netherlands.

Vacchi, M., Koubbi, P., Ghigliotti, L. & Pisano, E. (2012) Sea-ice interactions with polar fish: focus on the antarctic silverfish life history. *Adaptation and Evolution in Marine Environments, Volume 1*, pp. 51–73. Springer, Berlin.

Waller, C., Worland, M., Convey, P. & Barnes, D. (2006) Ecophysiological strategies of Antarctic intertidal invertebrates faced with freezing stress. *Polar Biology*, **29**, 1077–1083.

Wallis, J.R., Swadling, K.M., Everitt, J.D. et al.(2015) Zooplankton abundance and biomass size spectra in the East Antarctic sea-ice zone during the winter-spring transition. *Deep-Sea Research Part II*, doi.org/10.1016/j.dsr2.2015.10.002.

Wang, S.W., Budge, S.M., Gradinger, R.R., Iken, K. & Wooller, M.J. (2014) Fatty acid and stable isotope characteristics of sea ice and pelagic particulate organic matter in the Bering Sea: tools for estimating sea ice algal contribution to Arctic food web production. *Oecologia*, **174**, 699–712.

Wassmann, P., Duarte, C.M., Agusti, S. & Sejr, M.K. (2011) Footprints of climate change in the Arctic marine ecosystem. *Global Change Biology*, **17**, 1235–1249.

Werner, I. (1997) Grazing of Arctic under-ice amphipods on sea-ice algae. *Marine Ecology Progress Series*, **160**, 93–99.

Werner, I. (2000) Faecal pellet production by Arctic under-ice amphipods – transfer of organic matter through the ice/water interface. *Hydrobiologia*, **426**, 89–96.

Werner, I. & Martinez-Arbizu, P.M. (1999) The sub-ice fauna of the Laptev Sea and the adjacent Arctic Ocean in summer 1995. *Polar Biology*, **21**, 71–79.

Weslawski, J.M., Opalinski, K. & Legezynska, J. (2000) Life cycle and production of *Onisimus litoralis* (Crustacea Amphipoda): The key species in the Arctic soft sediment littoral. *Polskie Archiwum Hydrobiologii*, **47**, 585–596.

Zhou, J., Delille, B., Kaartokallio, H. et al. (2014) Physical and bacterial controls on inorganic nutrients and dissolved organic carbon during a sea ice growth and decay experiment. *Marine Chemistry*, **166**, 59–69.

CHAPTER 17

Dynamics of nutrients, dissolved organic matter and exopolymers in sea ice

Klaus M. Meiners[1] and Christine Michel[2]

[1] *Australian Antarctic Division and Antarctic Climate and Ecosystems Cooperative Research Centre, University of Tasmania, Australia*
[2] *Freshwater Institute, Fisheries and Oceans Canada, Winnipeg, Manitoba, Canada*

17.1 Introduction

Sea ice is not a static medium. It provides a seasonal to multi-year reservoir for particulates and solutes, and supports various ecosystem processes that cycle nutrients and organic matter, driving polar marine biogeochemistry (Vancoppenolle et al., 2013a). Sea ice consists of a dynamic matrix of pure ice, brine-filled interstices and gas bubbles (Chapter 1). Coupled physical, chemical and biological processes affect the pools as well as the cycling and fate of dissolved nutrients and dissolved organic matter (DOM) in sea ice. In general, initial concentrations of solutes in sea ice are controlled by the composition of the water from which the sea ice forms. Therefore, solutes initially behave conservatively with salinity. Boundary fluxes, both across ice–ocean as well as atmosphere–snow/ice interfaces, additionally affect solute concentrations in sea ice (Nomura et al., 2010; Vancoppenolle et al., 2010; Saenz & Arrigo, 2012; Figure 17.1). Dissolved inorganic and organic nutrient pools are then modified within the ice, and may become depleted or enriched when compared with salinity (Gleitz et al., 1996; Thomas et al., 2010). Upon ice melt, dissolved and particulate nutrients and organic matter are released into the water column where they can impact on pelagic processes, or contribute to particle export (Riebesell et al., 1991; Norman et al., 2011; Boetius et al., 2013; Lannuzel et al., 2013; Niemi et al., 2014).

In contrast to classical concepts of conservative incorporation of dissolved constituents into sea ice, recent research on dissolved organic carbon and so-called exopolymeric substances (EPS) have shown preferential incorporation of these substances into newly forming sea ice in the field and laboratory (Riedel et al., 2007a; Ewert & Deming, 2011; Müller et al., 2013; Zhou et al., 2014, 2016), and also demonstrated their effect on physical sea ice properties with implications for sea ice biogeochemistry and habitability (Krembs & Deming, 2008; Krembs et al., 2011). In addition, advancing from simple salt rejection and constant diffusive transport theory, new concepts for the exchange of solutes across the ice–ocean interface and solute transport between sea ice layers with different porosity, temperature and brine salinity have been developed. These include new measurements of sea ice permeability thresholds and the application of mushy-layer and thermohaline convection theory (Notz & Worster, 2009; Hunke et al., 2011; Loose et al., 2011). In turn, these new physical concepts have significantly improved and informed emerging sea ice biogeochemical and primary production studies and modelling efforts (Saenz & Arrigo, 2012, 2014; Tedesco & Vichi, 2014; Chapters 14 and 20). A number of recent summary articles have highlighted the importance of sea ice biogeochemistry and ecology for polar marine systems (Loose et al., 2011; Underwood et al., 2013; Vancoppenolle et al., 2013a; Arrigo, 2014; Meier et al., 2014) and also provide recommendations for best-practice sampling methodologies (Eicken & Salganek, 2010; Miller et al., 2015).

Building on these previous reviews, this chapter aims to provide an overview of the current understanding of the sources and sinks of biologically important inorganic and organic solutes in sea ice. Our focus will be on physical, chemical and biological processes that influence the pools of dissolved inorganic macronutrients, the micronutrient iron and DOM in sea ice. In addition, we discuss the dynamics and role of sea ice EPS,

---

Sea Ice, Third Edition. Edited by David N. Thomas.
© 2017 John Wiley & Sons, Ltd. Published 2017 by John Wiley & Sons, Ltd.

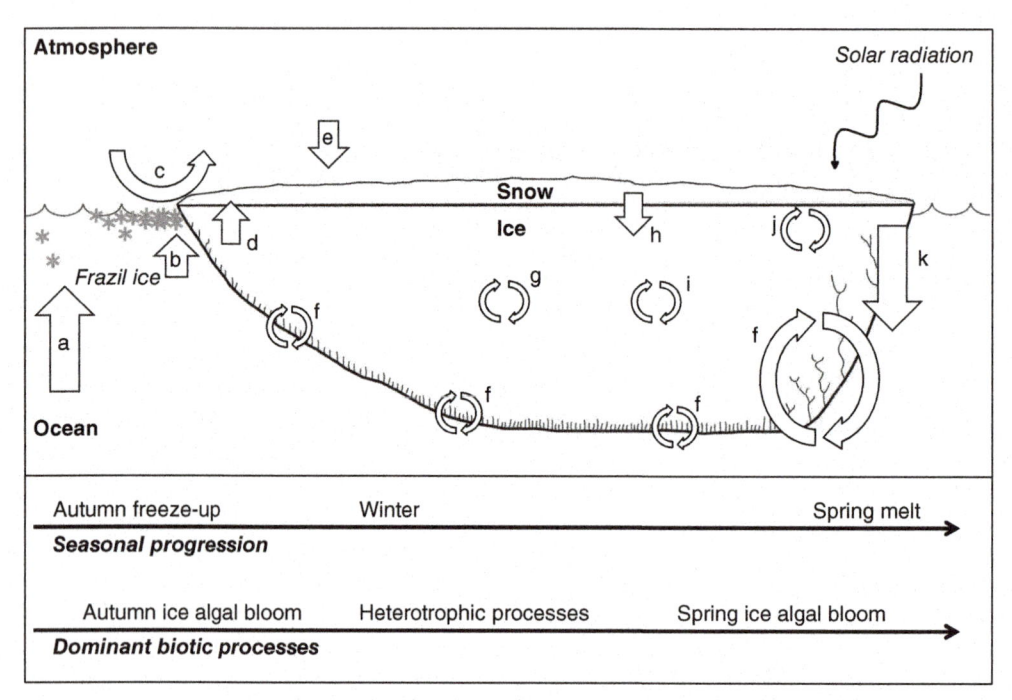

Figure 17.1 Schematic of the sea ice growth, melt cycle and processes affecting dissolved and particulate matter pools in sea ice. a, scavenging of lithogenic and biogenic material by frazil crystals; b, sieving of particulates by newly forming and young sea ice due to wave-field pumping and convection; c, aerosol formation due to wind erosion and outgassing; d, brine expulsion; e, atmospheric deposition; f, brine convection in permeable sea ice layers; g, chemical/physical solubilization of lithogenic/biogenic particulate matter; h, snow ice formation; i, freeze-concentration and coagulation of dissolved organic precursor materials into gels; j, photochemical processes affecting both organic and inorganic solute pools; and k, melt water flushing. For details see text.

including exopolymeric gels and particles. The chapter includes a general overview of processes occurring in Arctic, Antarctic and lower-latitude sea ice. Finally, it provides a brief outlook for sea ice biogeochemical and ecosystem research needs in response to an increasingly altered climate. Current and future changes in sea ice environments (Chapters 10–12) will be accompanied by significant changes to biogeochemical cycles, highlighting the need for an improved understanding of the coupled physical–chemical–biological processes of ice-covered oceans in both hemispheres.

17.2 Definitions and biogeochemical importance of solutes in sea ice

17.2.1 Macronutrients

In marine science, the term 'macronutrients' generally refers to the inorganic chemical elements that algae take up in the largest quantities. These include dissolved inorganic nitrate (NO_3^-), nitrite (NO_2^-), ammonium (NH_4^+), phosphate (PO_4^{3-}) and silicic acid ($Si(OH)_4$). Bulk concentrations of nutrients in sea ice are generally low when compared with seawater (Vancoppenolle et al., 2013a), and in the absence of biological uptake or remineralization, are proportional to salinity and more or less follow so-called theoretical dilution lines (Figure 17.2). Nutrient concentrations in sea ice may limit ice algal growth at times of high ice algal productivity and/or restricted external input. Observations and modelling studies suggest that ice algal growth in both hemispheres can be limited by nitrate as well as silicic acid (Lavoie et al., 2005; Gradinger, 2009; Saenz & Arrigo, 2014). Silicic acid limitation particularly affects sea ice diatoms, often the dominant taxonomic group in algal communities. Sea ice communities in the brackish northern Baltic Sea have been shown to be phosphate-limited, with total seasonal ice-associated primary production strongly influenced by the nutrient concentration of the water from which the sea ice

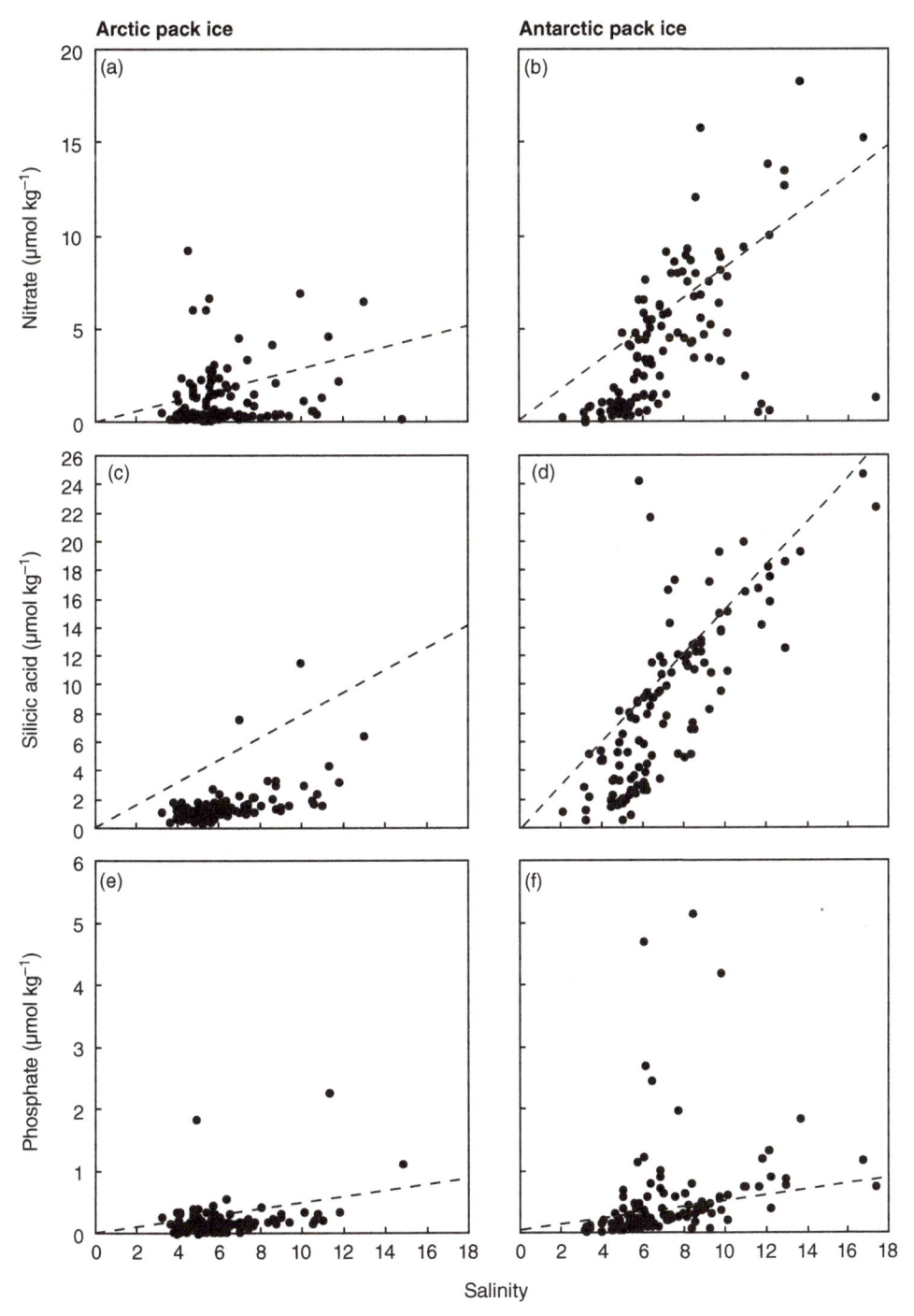

Figure 17.2 Concentrations of dissolved nutrients versus bulk ice salinity in Arctic (a, c, e) and Antarctic (b, d, f) pack ice. Dashed lines (theoretical dilution lines; for details see text) indicate nutrient concentrations to be expected from mean surface water values if the respective nutrient behaved conservatively with salinity. Arctic data are from the Beaufort Sea (Circumpolar Flaw Lead study, March 2007–June 2008, C. Michel, unpubl. data); Antarctic data were collected in September–October 2007 off East Antarctica (Meiners et al., 2011).

formed (Chapter 27). Nutrient limitation affects ice algal physiology and metabolism, impacting carbon, lipid and fatty acid content of cells, with associated effects on the fate of ice algae due to changes in aggregation potential, sinking rates and food quality for herbivores (e.g. Riebesell et al., 1991; Leu et al., 2010).

17.2.2 Dissolved iron

Iron is a key limiting micronutrient controlling phytoplankton production in the ocean, in particular in so-called high-nutrient–low-chlorophyll areas where macronutrients are high but phytoplankton biomass is low due to iron limitation (Boyd et al., 2007). Dissolved iron in Antarctic sea ice can be enriched by one to three orders of magnitude relative to iron-deplete seawater (Lannuzel et al., 2007, 2010; van der Merwe et al., 2011; de Jong et al., 2013), making sea ice an important iron reservoir. The release of this micronutrient during ice melt in spring – in combination with increasing light levels and surface ocean meltwater stratification – is thought to contribute to the development of Antarctic ice-edge blooms. While initial studies focused on the atmospheric supply of iron, i.e. by snow deposition (Edwards & Sedwick, 2001), more recent research has shown that marine iron sources are more important (Lannuzel et al., 2007, 2010; van der Merwe et al., 2009; de Jong et al., 2013).

Studies from the northern hemisphere have reported high dissolved iron concentrations in sea ice in the Bering Sea as well as in the Sea of Okhotsk, where iron released from melting sea ice may also significantly contribute to phytoplankton production (Aguilar-Islas et al., 2008; Tovar-Sánchez et al., 2010; Kanna et al., 2014). Iron is released from melting sea ice together with organic ligands, which may affect its residence time and bioavailability in the water column (van der Merwe et al., 2009). Although sea ice algae generally show large cell sizes (Arrigo et al., 2010), indicating a high iron demand (Sunda & Huntsman, 1995), ice algal communities are generally not considered to be iron-limited due to overall high dissolved iron concentrations within the sea ice brine channel network (Pankowski & McMinn, 2008; Lannuzel et al., 2014).

17.2.3 Dissolved organic matter

Dissolved organic matter is a diverse pool of complex organic compounds that is defined to be smaller than 0.2 μm. In practice, DOM is often operationally defined as material that passes through glass-fibre filters with a nominal pore size of 0.7 μm (Thomas et al., 2010). Sea ice DOM pools consist of carbohydrates, amino acids as well as more complex substances such as humic acids. Sea ice DOM can also contain high levels of dissolved organosulphur compounds, including the climate-relevant dimethylsulphoniopropionate (Chapter 18). DOM remains poorly characterized at the molecular level and can range in size from monomers to large polymers. Molecular size can be indicative of the bioavailability of compounds and DOM is often grouped into biolabile and refractory fractions. Sea ice DOM includes material that has been produced *in situ* through biological release from organisms living in sea ice (autochthonous; Underwood et al., 2010; Aslam et al., 2012a) as well as material trapped during sea ice formation (allochthonous; Stedmon et al., 2007). The former can be orders of magnitude higher than the latter and can reach concentrations of up to 2.5 mmol L^{-1} (Thomas et al., 2001; Riedel et al., 2008). Understanding the incorporation of DOM into sea ice, as well as its fate, is an active area of research and current studies indicate that some DOM can be enriched into newly forming sea ice relative to salinity (Müller et al., 2011, 2013; Zhou et al., 2014, 2016).

Dissolved organic matter is most commonly measured as dissolved organic carbon (DOC) and dissolved organic nitrogen (DON) (Thomas et al., 2010). Sea ice DOC concentrations are generally enriched when compared with the seawater from which the ice has formed (Figure 17.3). In Antarctic sea ice, DOC and DON concentrations are generally correlated, although DOC/DON ratios are both relatively high and highly variable (Thomas et al., 2001; Norman et al., 2011). Even higher DOC/DON ratios, increasing with total DOC concentration, have been reported from Arctic sea ice (Thomas et al., 2001, 2010) and are probably a result of DOC-rich riverine supply. The high carbon content of sea ice DOM has also been attributed to high concentrations of biologically produced sea ice carbohydrates, which may be a result of high concentrations of sea ice EPS (Underwood et al., 2010, 2013). Recent experimental results show that the release of sea ice DOM at the time of ice melt can stimulate prokaryotic activity in surface water microbial communities (Niemi et al., 2014). Importantly, DOM can be chromophoric, and so-called coloured DOM (CDOM) absorbs light at the ultraviolet radiation (UVR) and photosynthetically active radiation

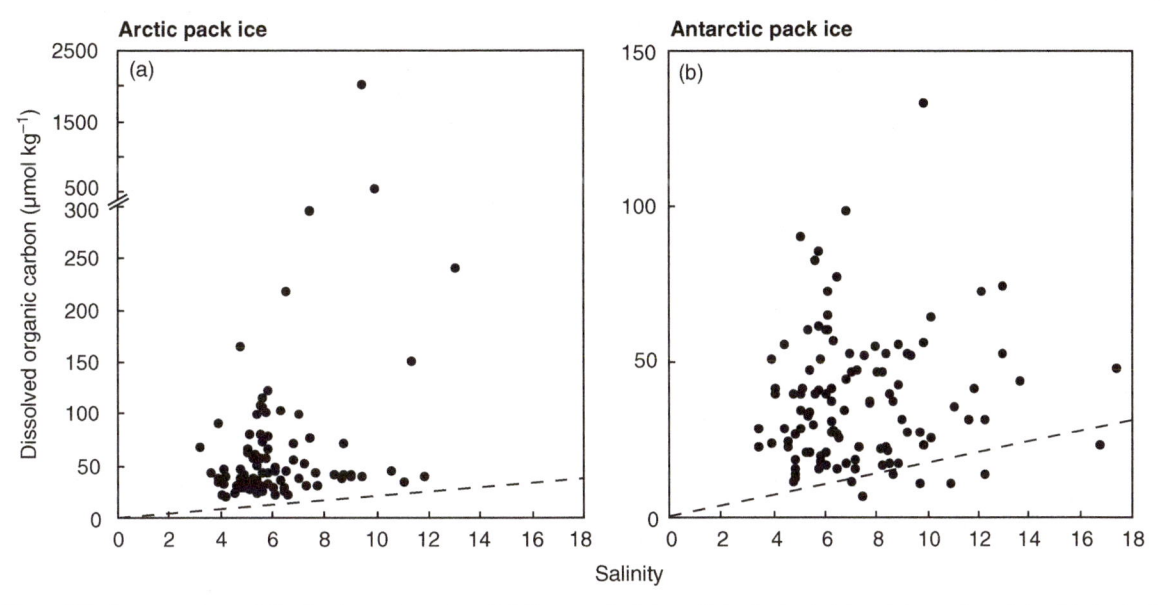

Figure 17.3 Concentrations of dissolved organic carbon (DOC) versus bulk ice salinity in Arctic (a) and Antarctic pack ice (b). Dashed lines (theoretical dilution lines; for details see text) indicate concentrations to be expected from mean surface water values if DOC behaved conservatively with salinity. Note the different y-axes and axis break. Arctic data are from the Beaufort Sea (Circumpolar Flaw Lead study, March 2007–June 2008, C. Michel, unpubl. data); Antarctic data were collected off East Antarctica (Norman et al., 2011).

(PAR) wavelength bands, therefore affecting the energy budget of sea ice as well as the exposure of ice-associated organism to UVR and PAR (Belzile et al., 2000; Uusikivi et al., 2010).

Upon ice melt, Antarctic sea ice-derived CDOM has been suggested to affect the optical properties and phytoplankton production of surface waters via UV protection and attenuation of visible light (Norman et al., 2011). This effect might be somewhat reduced in the Arctic Ocean, due to already high DOM concentrations in surface waters from riverine sources. There is growing evidence that sea ice CDOM is strongly affected by photochemical processes, and this may be particularly true in the surface layers of sea ice floes exposed to high irradiances (Uusikivi et al., 2010; Beine et al., 2012). Photochemical degradation of DOM can lead to direct remineralization of DOM to CO and CO_2, but also to the formation of biolabile organic compounds that can fuel sea ice microbial food webs (Thomas et al., 2010; Norman et al., 2011). DOM cycling in sea ice needs to be considered as a function of physico-chemical, in particular photochemical, and microbial processes acting in parallel.

17.2.4 Exopolymeric substances

Marine and ice-associated microbial EPS exist in a dynamic equilibrium and along a size continuum from dissolved polysaccharides to complex gel-like and particulate EPS (retained on 0.2–0.4 um filters) that can form aggregates at the millimetre to centimetre scale (Krembs et al., 2011; Underwood et al., 2013). EPS are produced by algae and bacteria and can differ widely in chemical composition and structure (Aslam et al., 2016). However, in general, EPS are rich in carbohydrates, resulting in a high carbon:nitrogen ratio of the material. Exopolymers are characterized by polysaccharides that have carbon backbones of high molecular weight, typically carrying carboxylic acid groups such as uronic acids, which give EPS a predominantly poly-anionic nature. Exopolymeric gels can self-assemble from dissolved precursors and the process is facilitated by divalent cations like calcium and magnesium ions that can cross-link negatively charged groups (Verdugo, 2012).

Exopolymeric substances can also contain other chemical substances, including sulphates, lipids, proteins and nucleic acids that may strongly affect sea ice EPS chemical properties (Mancuso Nichols et al., 2005;

Ewert & Deming, 2011). The use of different analytical methods to quantify EPS in sea ice makes comparisons difficult, but they coherently indicate that EPS contribute a significant fraction to sea ice carbon pools in both the Arctic and Antarctica (Meiners et al., 2003, 2004; Riedel et al., 2006; Collins et al., 2008). Moreover, recent evidence of a consistent contribution of uronic acids to dissolved EPS across a variety of ice types, seasons and locations in both hemispheres points to broad-scale predictability of carbohydrates in sea ice (Underwood et al., 2013).

Exopolymeric substances play multiple ecological roles in serving marine bacteria and diatoms in adhesion to surfaces, locomotion, supporting sequestration of dissolved nutrients and providing a protective buffer against unfavourable environmental changes such as shifts in ionic, osmotic, pH, desiccation and toxic-metal conditions. In sea ice, EPS may serve particularly as protectant against hyper- and hypo-osmotic conditions as well as against mechanical damage from growing ice crystals (Krembs & Deming, 2008).

The stickiness of EPS affects incorporation of organisms into newly forming sea ice and also their retention within the sea ice during ice melt, thereby affecting cryopelagic coupling processes (Meiners et al., 2003; Riedel et al., 2006; Juhl et al., 2011). Krembs et al. (2011) highlighted the role of EPS in altering physical properties of sea ice. These authors showed that high concentrations of EPS affect sea ice bulk salinity and sea ice brine pocket complexity by EPS-clogging of brine channel micropores, reducing brine drainage (Figure 17.4). This increases sea ice habitable pore space, and the EPS coatings also maintain an aqueous environment around cells that may facilitate retention of excrete compounds, including extracellular enzymes and ice-binding proteins (Ewert & Deming, 2013). The overall result of EPS enrichment is a sea ice structure and brine channel environment that is more favourable for microbial communities. Information on the fate of sea ice EPS remains poor, although particulate EPS in sea ice can be considered as a hotspot for biogeochemical cycling, e.g. by harbouring high numbers of bacteria that may contribute to its degradation and by serving as a carbon source for heterotrophic protists (Meiners et al., 2004, 2008; Riedel et al., 2007b).

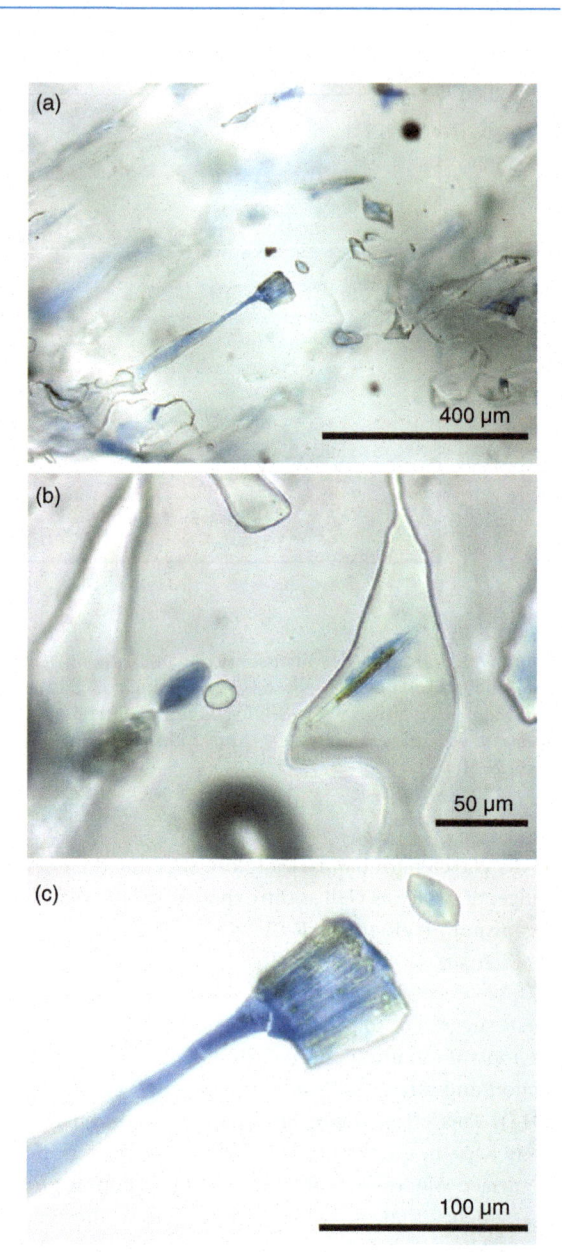

Figure 17.4 Photomicrographs of sea ice thin sections containing biological material. Images show sea ice pore structure (a), sea ice diatoms (a–c) and exopolymeric gels (stained with Alcian Blue). Note the plug-like exopolymeric substance matrix (darker blue colours) in (c). Source: Images provided by C. Krembs. Image (c) from Krembs et al., 2011. [Reproduced with permission from the Proceedings of the National Academy of Science (USA)].

17.3 Physical sources and sinks

17.3.1 Incorporation and exchange of dissolved solutes

Salt and inorganic solute concentrations in sea ice are generally much lower than in the seawater from which the ice has formed. Salts and solutes are mainly rejected from ice crystals and therefore become concentrated in interstitial brines. A notable exception is NH_4^+ which can be trapped within ice crystals during freezing (Weeks, 2010) and/or can also be biologically produced in newly formed sea ice (Riedel et al., 2007a). Neglecting gas inclusions and precipitation of some salts, sea ice remains a mixture of solid ice and liquid salty brines during all stages of growth and melt. Assuming equilibrium conditions, brine salinity is a function of ice temperature and brine volume is a function of temperature and sea ice bulk salinity (Chapter 1). The brine volume of sea ice controls its permeability, i.e. its ability for fluid transport (Golden et al., 1998). Above a brine volume fraction of 0.05 (volume$_{brine}$/volume$_{ice}$), brine pockets coalesce and form brine channels and sea ice with a columnar structure, i.e. ice formed by congelation growth, becomes permeable for fluid transport. The ~0.05 brine volume fraction corresponds to a bulk ice salinity of 5 and temperature of −5°C ('the rule of fives'; Golden et al., 1998). It has been suggested that the permeability of granular ice, formed by frazil ice accumulation, may be smaller than the permeability of columnar ice at a given porosity (Golden et al., 1998; Saenz & Arrigo, 2012). This can be explained by a more random, less aligned distribution of brine pockets and channels in granular sea ice (Chapter 1).

Classically five processes have been suggested to contribute to salt and solute release from the sea ice cover: initial solute rejection at the ice–water freezing interface; diffusion of solutes; brine expulsion; gravity drainage; and flushing (Notz & Worster, 2009; Hunke et al., 2011). New laboratory experiments and field research in conjunction with improved understanding of the multi-phase behaviour of sea ice through application of mushy layer theory (Feltham et al., 2006; Notz & Worster, 2009; Loose et al., 2011) have now identified gravity drainage and flushing by meltwater as the key drivers of salt loss (Notz & Worster, 2009; Hunke et al., 2011) and these must therefore also be recognized as the major physical processes controlling macronutrients in sea ice.

Gravity drainage results from atmospheric cooling of sea ice. The temperature decrease towards the ice surface increases brine salinity, leading to an unstable density gradient in brine columns, which causes convective overturning in connected brine channels and, ultimately, the release of high saline brine from permeable sea ice layers into the water column. The lost brine is replaced by less saline and generally nutrient-replete seawater. This resupply of nutrients is considered to be important for high algal accumulation in sea ice bottom and interior layers during spring when the sea ice warms and permeability increases, and also during autumn when surface cooling destabilizes brine stratification in an existing and generally porous sea ice cover allowing convective overturning. Convective resupply of iron has been hypothesized to transport dissolved iron into the bottom layers of sea ice where it is enriched into the particulate fraction due to ice algal uptake, i.e. bioaccumulation (van der Merwe et al., 2009). Vancoppenolle et al. (2010) have captured these processes in detail in a model combining brine convection and silicic acid uptake by an ice algal community.

Forced convection, i.e. convection induced by under-ice currents, may additionally enhance transport of nutrients into the ice and may be a key driver of high biomass accumulations observed in many landfast ice regimes open to costal currents and tidal influence (Cota et al., 1987; Lavoie et al., 2005; Vancoppenolle et al., 2013a). Overturning convection of sea ice brines throughout the entire ice column can also be the result of the refreezing of so-called surface slush layers (Vancoppenolle et al., 2010; Saenz & Arrigo, 2012). These layers are a significant and widespread feature of Antarctic sea ice and originate from submersion of the snow–ice interface due to snow-loading or result from melting of near-surface sea ice due to warmer summertime surface temperatures, amongst other processes (Ackley et al., 2008). Surface slush layers can be enriched in nutrients supplied from seawater, and these nutrients can be supplied to sea ice interior layers by convection induced by refreezing. In contrast, surface melting of snow, slush and ice results in the flushing out of nutrients due to meltwater percolation through the sea ice. The percolation of freshwater through the sea ice results in a loss of solute-rich brine and is primarily observed during advanced melt stages, i.e. in highly permeable sea ice (Petrich & Eicken, 2010; Chapter 1). Surface melt is a key feature of Arctic sea ice, where melt ponds can cover a significant area of the ice surface (Polashenski et al., 2012).

While nutrients trapped into the ice during formation provide an initial pool, brine convection controlled by sea ice porosity and brine density gradients is important in providing new, i.e. non-recycled, nutrients to sea ice microbial communities. The capacity of sea ice layers for gravity drainage can be quantified in a mushy-layer Rayleigh number which expresses the ratio between negative buoyancy in brines and dissipation (Vancoppenolle et al., 2013b). This concept can be applied to determine convectively active regions in the sea ice cover (Vancoppenolle et al., 2013b; Zhou et al., 2013, 2014, 2016).

17.3.2 Enrichment of solutes in newly forming sea ice

A small number of studies have observed enrichment of inorganic solutes into new sea ice formed from natural seawater, with non-conservative enrichment of specific trace metals as well as macronutrients (Granskog et al., 2004; Zhou et al., 2014, 2016). This enrichment may be attributed to biogeochemical processes that do not completely stop during ice formation as well as undefined physico-chemical processes. Various mechanisms for iron enrichment in sea ice have since been discussed. They are dependent on the mode and timing of ice formation, ice cover duration and sea ice carbon content and ice algal biomass (van der Merwe et al., 2009, 2011; Lannuzel et al., 2010; de Jong et al., 2013).

Several studies have shown that DOM and EPS can be enriched in newly forming artificial and natural sea ice, and suggest that incorporation is related to physico-chemical processes (Giannelli et al., 2001; Riedel et al., 2007a; Ewert & Deming, 2011). Recent field and laboratory evidence show that freezing can modify the quality of DOM. Examples of such are an enrichment of chromo- and fluorophoric DOM and changes in the size distribution of EPS and DOM (Aslam et al., 2012b; Müller et al. 2011, 2013). Ewert and Deming (2011) demonstrated that EPS produced by the cold-adapted marine gammaproteobacterium, *Colwellia psychrerythraea*, a species found in cold sea water and sea ice, is enriched in sea ice produced experimentally. The retention of the *Colwellia*-produced EPS was attributed to a heat-labile fraction, suggesting that the responsible component is a protein with ice-affinity properties. Antarctic sea ice bacterial EPS has been shown to contain a high fraction of protein (Mancuso Nichols et al., 2005). So-called ice binding proteins that attach to and change the structure of ice crystals are produced by sea ice bacteria as well as sea ice diatoms (Janech et al., 2006; Raymond et al., 2007).

In addition to the direct binding of molecules to ice crystals, other processes for selective retention of organic solutes in sea ice have been suggested. These include coagulation of DOM, facilitated by increased cation and dissolved precursor concentrations in brines. Spontaneous self-assembly and coagulation of gels has been reported from marine environments and is influenced by cation concentrations (e.g. Verdugo, 2012), but – to the best of our knowledge – has not yet been observed in sea ice. Orellana et al. (2011) demonstrated that in the high Arctic, marine gels (<1 μm in diameter) have unique physico-chemical characteristics that originate in the organic material produced by ice algae and ice-associated surface plankton. The polymers in this dissolved organic pool assembled faster and with higher micro-gel yields than at lower latitudes. After initial coagulation of DOM, small colloids and aggregates may further serve as sites for selective binding of additional DOM through hydrogen and ionic bonds (Verdugo, 2012; Müller et al., 2013).

In summary, the mechanisms for non-conservative uptake of biogeochemically active solutes into sea ice remain largely unresolved. They must be considered as a result of complex interactions of biological activity, brine dynamics and physico-chemical processes affecting constituents in solution and at ice crystal interfaces. Quantitatively and qualitatively differential solute loading of parent sea waters with DOC and nutrients is expected to influence bulk concentrations of these constituents in young sea ice, both along onshore–offshore gradients in the Arctic and, in particular, in between hemispheres. As a result, there could be distinct biogeochemical cycling of matter in Arctic and Antarctic sea ice, but comparative studies remain scarce (Underwood et al., 2013; Fripiat et al., 2014a; Zhou et al., 2014, 2016).

17.3.3 Dissolution of particulate precursors

Particulates including biogenic (algae and detritus) and lithogenic material can be incorporated into sea ice through various processes, including nucleation where marine particles serve as ice-condensation nuclei; scavenging of particles, also referred to as suspension freezing, by frazil ice crystals rising through the water column; and sieving of particle-containing seawater through layers of newly forming or established sea ice, trapping particles into the sea ice matrix (Ackley & Sullivan, 1994; Spindler, 1994; Gradinger & Ikävalko, 1998). Once embedded in the sea ice, particulate matter is exposed to high salinities as well as mechanical stress through ice crystal growth, potentially enhancing particle solubilization. Incorporation of algal material from autumn blooms into sea ice and entrainment of resuspended sediments are therefore considered to affect not only the particulate but also the dissolved nutrient loading of sea ice. These processes are considered to be particularly relevant for particulate and dissolved iron accumulation in Antarctic landfast ice (Grotti et al., 2005; de Jong et al., 2013) and are also important in offshore pack ice (Lannuzel et al., 2010).

High concentrations of dissolved macronutrients [NO_2^-, PO_4^{3-}, $Si(OH)_4$] have been reported in association with organic-rich sediment layers from sea ice in the Sea of Okhotsk, indicating strong remineralization of particulate matter in association with 'turbid' sea ice (Nomura et al., 2010). The layers were depleted in NO_3^-, suggesting that they may also serve as sites for denitrification (Nomura et al., 2010). The understanding of the mechanisms and quantitative importance of remineralization processes of allochtonous particulate material in sea ice remains poor. However, the widespread occurrence of sediment inclusions in Arctic sea ice (Eicken et al., 2000) may significantly affect the spatiotemporal distribution and cycling of dissolved macronutrients and trace metals in the Arctic Ocean (Measures, 1999). Due to low sediment loading of Antarctic sea ice (Chapter 1), the process is assumed to be less important in Southern Ocean environments. There, the generally dynamic conditions during sea ice formation result in a large contribution of frazil ice to Antarctic sea ice (Chapter 1), indicating a high potential for scavenging of biogenic particles and ice-associated remineralization (e.g. Lannuzel et al., 2010).

17.3.4 Sea ice surface processes and atmospheric deposition

Frost flowers, due to their unique growth process and chemical composition, provide an important link in the ocean–ice–atmosphere system in polar regions (Bowman & Deming, 2010; Douglas et al., 2012). Brine expulsion can transport highly saline brine to the sea ice surface, where expelled brines may contribute to the formation of frost flowers, i.e. vapour-deposited crystals that wick brines from the sea ice surface. Precipitation of snow results in the wicking of brine into surface snow layers, resulting in so-called brine-wetted snow. Incorporation of brine EPS and other solutes into frost flowers and brine-wetted snow has been reported from laboratory experiments and Arctic first-year sea ice (Bowman & Deming, 2010; Aslam et al., 2012b; Ewert et al., 2013). Material enriched in frost flowers and snow can be transported into the atmosphere by wind (Shaw et al., 2010). This part of the 'marine' aerosol pool may then affect atmospheric chemistry and cloud properties or, after redistribution, be again deposited into sea ice or open water surface layers (Alvarez-Aviles et al., 2008; Bowman & Deming, 2010; Orellana et al., 2011).

The deposition of atmospheric aerosols can also provide a nutrient source to sea ice. Granskog et al. (2003) found that nitrogen-rich snow metamorphosing into snow-ice provides an important nitrogen pool in Baltic Sea ice. High nitrate and ammonium concentrations have also been reported in snow ice in the Sea of Okhotsk (Nomura et al., 2010).

In the Southern Ocean, deposition of aerosol iron can contribute to the trace metal load of pack ice (Edwards & Sedwick, 2001), although marine sources are considered to be more important for the total (vertically integrated) iron content of Antarctic pack ice (Lannuzel et al., 2007, 2010). Studies on Antarctic landfast sea ice indicate that deposition of iron-rich dust from ice-free regions can be an important source to the particulate iron load of coastal sea ice (e.g. de Jong et al., 2013). Importantly, and as a function of geographic location, atmosphere-derived nutrients accumulate in sea ice during its entire growth season and can be redistributed through wind events or sea ice drift, but are generally released to the water column during a relatively short melt period in a localized area. This collation of material over time and space, followed by a more focused release, constitutes a sea ice 'funnelling' effect. Iron released from sea ice in this way has been suggested to be an important driver of productivity in the Antarctic marginal ice zone (Lancelot et al., 2009).

17.4 Chemical sources and sinks

High and variable concentrations of solutes in sea ice brines provide a dynamic environment for chemical processes. Concentrations are controlled by freeze-concentration and melting cycles over a range of temporal and spatial scales. Chemical processes within the brine-filled interstices, including photochemical processes, constitute key drivers in sea ice nutrient and organic matter cycling (Thomas et al., 2010). Chemical reactions associated with the polycrystalline ice matrix, i.e. occurring on ice grain surfaces and in veins, nodes and pre-melted layers, may also impact bulk concentrations of biogeochemical solutes, but are not the focus of this chapter. For more details on ice microphysics and their impacts on chemical processes in environmental ice, we refer to the recent reviews by Wettlaufer (2010) and McNeill et al. (2012) (and citations therein).

17.4.1 Effects of freeze-concentration and salt precipitation

Hyper-saline brines generated during the freezing of seawater undergo changes with respect to their mineral-liquid thermodynamic equilibrium and are conducive to sequential mineral precipitation (Marion, 2001), e.g. calcium carbonate as ikaite (Papadimitriou et al., 2007, 2012; Dieckmann et al., 2008). The freezing process also leads to changes in ion concentration and composition and increases encounter rates of molecules. In the absence of biology, sea ice brine pH evolution is considered to be primarily controlled by the carbonate system, i.e. the brine CO_2 concentration, which in turn is affected by freeze concentration, outgassing, ikaite precipitation and brine drainage (Miller et al., 2011; Papadimitriou et al., 2012; Rysgaard et al., 2013). Direct measurements of the pH of sea ice are scarce, but indicate strong positive and negative deviations from seawater values due to combined abiotic (as discussed earlier) as well as biotic processes (photosynthesis, respiration), which are influenced by the age of the ice as well as the vertical position of microhabitats within the ice cover (Gleitz et al., 1995; Hare et al., 2013). While only limited information is available for sea ice, pH changes have been shown to strongly influence marine biogeochemical processes in various ways, including effects on micro- and macronutrient availability and algal physiology (Shi et al., 2009; McMinn et al., 2014).

Changes in pH as well as the concentrations of cations affect the solubility of DOM and EPS and phase transitions among the dissolved, colloidal or particulate phases (Verdugo, 2012; Ewert & Deming, 2013). The extent to which reversible gel self-assembly might occur in sea ice brine channels, where precursor concentrations are high and pH and cation concentrations are conducive to self-coagulation, has not yet been studied. An initial study on natural ice cores kept frozen for several months indicated limited self-assembly (Krembs et al., 2002), but both Arctic and Antarctic sea ice harbour high concentrations of EPS aggregates with diameters > 2 μm, suggesting the potential for chemically driven changes in the hydration phases of micro-gels and EPS aggregation in addition to biological production The closely linked changes in brine temperature, salinity and volume are therefore expected to have both direct and indirect effects on the composition and fractionation of organic matter pools in sea ice (Meiners et al., 2003; Ewert & Deming, 2013). As DOM pools also include chelating agents that influence the bioavailability of metals, in particular iron, freezing as well as dilution of brines may additionally affect micronutrient availability in sea ice and, during ice melt, in ice-associated surface waters.

17.4.2 Photochemical degradation processes

Photochemical transformation of dissolved organic and inorganic matter, driven in particular by high-energy radiation in the UVR and blue wavelength ranges, can degrade sea ice dissolved matter and lead to the formation of reactive products, such as radicals (Thomas et al., 2010; Norman et al., 2011). Photodegradation of DOM can produce labile nitrogen, mainly ammonium (photoammonification), as well as phosphate, which may contribute to biological production in marine environments (Stedmon et al., 2007, 2011; Xie et al., 2012). Significant photochemical production of carbon monoxide and carbon dioxide from DOM has been observed in Arctic first-year sea ice (Xie & Gosselin, 2005) and OH^- radical production due to oxidation of hydrogen peroxide (H_2O_2) and nitrate has been reported in Antarctic sea ice (King et al., 2005).

Increased absorption in the UVR wavelength range of EPS associated with bacterial communities in the surface layer of Arctic winter sea ice suggests that EPS may also be susceptible to photochemical processes (Ewert & Deming, 2013). High concentrations of DOM that strongly absorbs UV radiation have been reported from Arctic, Antarctic and Baltic sea ice (Belzile et al., 2000; Uusikivi et al., 2010; Norman et al., 2011), indicating that significant photo-oxidation of the material may occur. High concentrations of CDOM in sea ice surface layers can also significantly modulate UVR exposure of ice algal communities and under-ice phytoplankton. As both sea ice and snow effectively attenuate UVR and PAR, photochemical processes most likely occur in sea ice layers exposed to high and increasing irradiance levels, e.g. surface ice layers with thin and melting snow covers during spring. Other factors influencing photochemical reactions in sea ice include ambient precursor concentrations and composition (e.g., DON:DOC ratios) as well as environmental controls such as pH, iron and oxygen concentrations, as discussed earlier (e.g. Xie et al., 2012).

Improved understanding of the photochemical processes affecting CDOM pools in sea ice is tightly linked to gaining a better knowledge of CDOM quality, in particular its molecular composition. The recent application of DOM fluorescence spectroscopy to sea ice environments has proven a powerful tool for quantifying different sea ice DOM fractions and characterizing their individual dynamics (Stedmon et al., 2011; Müller et al., 2011, 2013). The combination of classical DOC and DON measurements via combustion and chemical analysis, CDOM absorption measurements and fluorescence spectroscopy may provide a way forward to tease out the complex interplay of physical, chemical and biological sources and sinks of DOM and CDOM in sea ice habitats. Time-series experiments as well as spatial studies are needed to characterize the behaviour of DOM and CDOM during the seasonal ageing of sea ice and to characterize different DOM sources, e.g., riverine *versus* open ocean, and their impact on organic matter cycling in sea ice.

17.5 Biological cycling of nutrients, dissolved carbon and EPS

Sea ice communities consist of organisms that use different energy sources and comprise multiple trophic levels, including chemoautotrophic bacteria and archaea; photosynthetic bacteria and algae; and heterotrophic bacteria, archaea, protists, fungi and metazoans, which are all involved in the biogeochemical cycling of nutrients, dissolved organic carbon and EPS (Thomas & Dieckmann, 2010; Chapters 13–16). Major biological processes in the cycling of matter between the dissolved and particulate phase include nutrient uptake by algae and bacteria, excretion of intracellular compounds and extracellular remineralization of particulate or DOM through release of exoenzymes. Complete remineralization of organic compounds is frequently mediated by several organisms or bacterial groups with different metabolic pathways, often requiring diverse environmental conditions, such as aerobic and anaerobic micro-zones, for existence and growth (Rysgaard & Glud, 2004; Ewert & Deming, 2013; Fripiat et al., 2014a).

17.5.1 The role of bacteria

The elemental composition of substrates that bacteria grow on differs from that of bacterial biomass, so heterotrophic bacteria need the capacity to both take up and release inorganic nutrients in order to maintain steady-state elemental composition. The critical role of heterotrophic bacteria in sea ice environments is considered to be recycling, i.e. providing a biological source of inorganic nutrients through remineralization of organic matter (Riedel et al., 2007b). For example, high accumulations of PO_4^{3-} in older sea ice have been linked to bacterial remineralization (Thomas et al., 2010). Remineralization rates of silicate in sea ice are generally believed to be slow due to low temperatures (Thomas et al., 2010), although recent evidence of high silica dissolution:production ratios (range 0.4–0.9) from Si isotope measurements in Arctic and Antarctic sea ice suggests otherwise (Fripiat et al., 2014b).

High bacterial biomass and respiration rates occurring in the semi-confined sea ice environment may result in the development of anaerobic micro-zones (Rysgaard & Glud, 2004; Glud et al., 2014). These conditions are favourable for nitrogen cycling through NO_3^- to NO_2^- reduction and potential further reduction to N_2 (denitrification), as well as anaerobic ammonium oxidation (anammox). Bacterial groups involved in these processes have been identified in Antarctic sea ice (Staley & Gosink, 1999), and denitrification, i.e. significant losses of fixed nitrogen through production of molecular nitrogen (N_2), has been suggested for both polar and temperate regions (Rysgaard et al,. 2008; Arrigo et al., 2014; Chapter 27).

Using a stable isotope approach, Fripiat et al. (2014a) suggest that nitrification, an aerobic process, may provide an important source of oxidized nitrogen (NO_3^-) for primary production in Antarctic pack ice. The first step of nitrification, e.g. NH_4^+ to NO_2^- oxidation, has been measured in Antarctic fast ice (Priscu et al., 1990), and nitrification has also been inferred from significant NO_2^- accumulations in sea ice (e.g. Arrigo et al., 1995). This indicates that some of the NO_3^-, generally assumed a 'new' nutrient, may in fact be a 'regenerated' nutrient in sea ice (Fripiat et al., 2014a). In summary, there is now strong evidence for complex nitrogen cycling in sea ice including N assimilation, N remineralization, denitrification and nitrification (Rysgaard & Glud, 2004). However, the relative importance of these processes, their spatial distribution across sea ice micro-zones, as well as their seasonal patterns still remain poorly constrained (Rysgaard et al., 2008; Fripiat et al., 2014a).

Bacterial uptake of DOM produced by ice algae is a key process in sea ice microbal food webs and considered the main pathway for the transfer of DOM into the particulate fraction. Bacterial biomass in turn serves as a food source for heterotrophic protists that link microbial to metazoan food webs. Seasonal shifts in carbon cycling also take place, as exemplified by an Arctic first-year carbon budget showing that DOC and/or EPS consumption is required to supplement bacterivory to meet heterotrophic carbon requirements during, but not prior to, the ice algal bloom period (Riedel et al., 2007b). Bacterial uptake of DOM combined with extracellular breakdown of dissolved matter may also mediate the release of essential micro- and macronutrients as well as vitamins, providing positive feedback for ice algal primary production. Limited information on viruses in sea ice indicates high viral abundance and high virus–bacteria contact rates (Wells & Deming, 2006; Deming, 2010). Research in pelagic environments suggests that DOC released by viral lysis of bacterial host populations can serve as a substrate for the growth of non-infected bacterial populations. This effect may be enhanced in the semi-enclosed brine channel system of sea ice, highlighting the potential role of viruses for sea ice–brine biogeochemical cycling, although DOC standing stocks are generally high and the relative contribution of lysate DOC might be small. Unravelling bacteria- and virus-mediated DOC dynamics in sea ice remains a major challenge (Deming, 2010; Chapter 13).

To our knowledge, no *in situ* bacterial EPS production rates have been measured in sea ice habitats so far. Crude estimates, calculated through a combination of cell-specific rates of temperate organisms corrected for temperature effects and abundance data from sea ice habitats, indicate that in sea ice, bacterial EPS production is generally low compared with algal EPS production (Meiners et al., 2003, 2004). However, in distinct habitats, e.g. surface horizons in winter sea ice, bacteria may provide the dominant source of sea ice-associated EPS (Collins et al., 2008). Temperature has been shown to affect not only the yield of EPS production of an Antarctic sea ice bacterial isolate of *Pseudoalteromonas*, but also the uronic acid content of the produced material (Mancuso Nichols et al., 2005). Furthermore, the EPS produced by *Pseudoalteromonas* is poly-anionic and may bind cations such as trace metals (Mancuso Nichols et al., 2005). EPS produced by sea ice bacteria may therefore affect trace metal availability (van der Merwe et al., 2009), but interactions remain poorly understood. Sea ice EPS may also serve as substrate for bacterial growth (Meiners et al., 2004, 2008) and have been shown to increase NH_4^+ production, probably a result of grazing by heterotrophic protists (Riedel et al., 2007b).

17.5.2 The role of algae

Ice algae utilize carbon dioxide and inorganic nutrients contained in sea ice brine and are the most important producers of autochtonous DOC and EPS in sea ice systems. However, they can also take up DOC and EPS under both light and dark conditions and may contribute to the degradation of DOM (Rivkin & Putt, 1987; Palmisano & Garrison, 1993), but information on ice algal facultative heterotrophy remains extremely sparse.

Highest ice algal biomass generally occurs in microenvironments that are coupled to the underlying water column and exhibit stable growth conditions, e.g. bottom horizons and porous freeboard layers that are easily infiltrated with seawater (Arrigo et al., 2010; Meiners et al., 2012). Growth of ice algae requires acquisition of C, N and P and a host of micronutrients such as iron, as well as Si for diatoms. Ice algae can serve as a reservoir for nutrients, and biomass-rich bottom ice layers have been described as a 'capacitor' for nutrients due to their ability to seasonally bio-accumulate macronutrients as well as iron in particulate biogenic fractions contained in sea ice (van der Merwe et al., 2009; Vancoppenolle et al., 2010; Noble et al., 2013). Key limiting nutrients for sea ice algal communities are NO_3^- and $Si(OH)_4$, and some studies have shown positive relationships between sea ice and under-ice water nutrient concentration and algal biomass (Rózanska et al., 2009).

Nutrient stress affects algal physiology and metabolism and is often accompanied by a release of DOC from the cells by both passive (leakage) and active processes (exudation). A strong increase in the production of extracellular organic carbon fractions (both DOC and EPS) by sea ice algae under environmental stress has been reported (Aslam et al., 2012a; Ugalde et al., 2013). Moreover, field studies have shown strong relationships between the distribution of ice algae and EPS concentrations, with EPS quality and complexity influenced by environmental conditions such as ice temperature and brine salinity (Underwood et al., 2010, 2013). These findings are underpinned by laboratory experiments indicating that the sea ice diatom *Fragilariopsis cylindrus* responds to freezing, high-saline growth conditions and raised pH with an increase in the production of EPS, and that EPS yields of ice algae vary as function of growth rate (Aslam et al., 2012a; Ugalde et al., 2013). Coupled physical (e.g. temperature, salinity) and chemical drivers (e.g. macronutrient concentration, pH) are therefore assumed to regulate the extracellular carbon (both DOC and EPS) production of ice algae.

17.5.3 Heterotrophic protists and metazoans

The principal biogeochemical role of brine channel-dwelling heterotrophic protists and metazoans (for details see Chapter 16) is ingestion of particulate organic carbon (e.g. particulate EPS, bacteria and algae) and remineralization of inorganic nutrients through excretion. The limited number of studies on sea ice microbial food webs indicates the existence of a functional microbial loop, with bacterial standing stocks generally linked to high DOC and EPS concentrations, and grazed upon, but generally not controlled, by heterotrophic protists (Riedel et al., 2007b, 2008). Estimates of the grazing impact and carbon demand of in-ice herbivores are low, constituting only a few percent of ice algal primary production (Gradinger, 1999; Nozais et al., 2001; Michel et al., 2002), with most of the organic carbon produced in sea ice released during ice melt, thereby contributing primarily to pelagic food webs and particle export (Michel et al., 2006; Juul-Pedersen et al., 2008). While the overall importance of heterotrophic protists and metazoans for carbon transfer is estimated to be small, they may contribute significantly to nutrient – in particular nitrogen – cycling, but information remains sparse. Sea ice food web dynamics remain challenging to study and must be considered as distinctly different from those observed in the pelagic realm due to the spatial complexity of the brine channel habitat, including attachment and association of organisms with internal surfaces, e.g. brine channel walls and particle surfaces, as well as clogging of brine channels by EPS (Krembs et al., 2000; Riedel et al., 2007b; Deming, 2010; Ewert & Deming, 2013).

17.6 Summary and future challenges

We have provided a brief overview on the various sources, sinks and processes involved in the cycling of a selected suite of biogeochemically relevant solutes in sea ice. We highlighted the fact that sea ice not only provides an important reservoir for macro- and micronutrients, DOC and EPS in the coupled ocean–sea ice–atmosphere systems of both hemispheres, but importantly it also serves as a dynamic interface in the cycling of these substances.

Cross-disciplinary field and laboratory (including large-scale ice tank) studies that simultaneously measure physical, chemical and biological properties and provide the basis for integrated analyses have been – and will be – key in furthering our understanding of sea ice biogeochemical processes. Recent development and application of new *in situ* (micro-) sensors, robust and automated sea ice observatories, and advances in under-ice platforms and vehicle technology provide promising new tools for measuring various sea ice properties (and their variability) on large scales. Multidisciplinary approaches on increasing temporal (seasonally to multi-year) and spatial (10s of metres to 100s of kilometres) scales, and across different ice regimes, are needed to improve the representation of biogeochemical processes in numerical sea ice models. Refined identification and parameterization of physical drivers of sea ice biogeochemical processes are particularly needed to improve sea ice primary production models and predict future trends in polar marine ecosystem function. This in turn will improve assessments of the impacts of changing sea ice conditions on polar marine systems and beyond.

17.7 Acknowledgements

Many colleagues and friends have contributed to our knowledge on sea ice biogeochemistry. They are too numerous to list here, but we would like especially to thank Christopher Krembs, Delphine Lannuzel, C.J. Mundy, Andrea Niemi and Martin Vancoppenolle for stimulating discussions. We are also indebted to the captains and crews of the research vessels and personnel of field stations that supported our work. Klaus Meiners was supported by the Australian Cooperative Research Centres Programme through the Antarctic Climate and Ecosystems Cooperative Research Centre and Australian Antarctic Science projects 2767 and 4073. Christine Michel was supported by grants from the International Governance Strategy, Fisheries and Oceans Canada, the Canadian International Polar Year Federal Program, and the Natural Sciences and Engineering Research Council of Canada through IPY Funds and a Discovery Grant. This work is a contribution to the Marine Productivity Laboratory, Fisheries and Oceans Canada.

References

Ackley, S.F. & Sullivan, C.W. (1994) Physical controls on the development and characteristics of Antarctic sea ice biological communities – a review and synthesis. *Deep Sea Research Part I: Oceanographic Research Papers*, **41**, 1583–1604.

Ackley, S.F., Lewis, M.J., Fritsen, C.H. & Xie, H. (2008) Internal melting in Antarctic sea ice: Development of 'gap layers'. *Geophysical Research Letters*, **35**, L11503, 10.1029/2008GL033644.

Aguilar-Islas, A.M., Rember, R.D., Mordy, C.W. & Wu, J. (2008) Sea ice-derived dissolved iron and its potential influence on the spring algal bloom in the Bering Sea. *Geophysical Research Letters*, **35**, L24601, 10.1029/2008GL035736.

Alvarez-Aviles, L., Simpson, W.R., Douglas, T.A., Sturm, M., Perovich, D. & Domine, F. (2008) Frost flower chemical composition during growth and its implications for aerosol production and bromine activation. *Journal of Geophysical Research*, **113**, D213064, 10.1029/2008JD010277.

Arrigo, K.R. (2014) Sea ice ecosystems. *Annual Reviews of Marine Science*, **6**, 439–467.

Arrigo, K.R., Dieckmann, G., Gosselin, M., Robinson, D.H., Fritsen, C.H. & Sullivan, C.W. (1995) High-resolution study of the platelet ice ecosystem in McMurdo Sound, Antarctica – biomass, nutrient, and production profiles within a dense microalgal bloom. *Marine Ecology Progress Series*, **127**, 255–268.

Arrigo, K.R., Mock T., Lizotte, M.P. (2010) Primary producers and sea ice. In: *Sea Ice*, 2nd ed (Eds. D.N. Thomas & G.S. Dieckmann), pp. 283–325. Wiley-Blackwell, Oxford, UK.

Aslam, S.N., Cresswell-Maynard, T., Thomas, D.N. & Underwood, G.J.C. (2012a) Production and characterization of the intra-and extracellular carbohydrates and polymeric substances (EPS) of three sea-ice diatom species, and evidence for a cryoprotective role for EPS. *Journal of Phycology*, **48**, 1494–1509.

Aslam, S.N., Michel, C., Niemi, A., & Underwood, G.J.C. (2016) Patterns and drivers of carbohydrate budgets in ice algal assemblages from first year Arctic sea ice. *Limnology and Oceanography* **61**, 919–937.

Aslam, S.N., Underwood, G.J.C., Kaartokallio, H. et al. (2012b) Dissolved extracellular polymeric substances (dEPS) dynamics and bacterial growth during sea ice formation in an ice tank study. *Polar Biology*, **35**, 661–676.

Beine, H., Anastasio, C., Domine, F. et al. (2012) Soluble chromophores in marine snow, seawater, sea ice and frost flowers near Barrow, Alaska. *Journal of Geophysical Research*, **117**, D00R15, 10.1029/2011JD016650.

Belzile, C., Johannessen, S.C., Gosselin, M., Demers, S. & Miller, W.L. (2000) Ultraviolet attenuation by dissolved and particulate constituents of first-year ice during late spring in an Arctic polynya. *Limnology and Oceanography*, **45**, 1265–1273.

Boetius, A., Albrecht, S., Bakker, K. et al. (2013) Export of algal biomass from the melting Arctic sea ice. *Science*, **339**, 1430–1432.

Bowman, J.S. & Deming, J.W. (2010) Elevated bacterial abundance and exopolymers in saline frost flowers and implications for atmospheric chemistry and microbial dispersal. *Geophysical Research Letters*, **37**, L13501, 10.1029/2010GL043020.

Boyd, P.W., Jickells, T., Law, C.S. et al. (2007) Mesoscale iron enrichment experiments 1993–2005: Synthesis and future directions. *Science*, **315**, 612–617.

Collins, R.E., Carpenter, S.D. & Deming, J.W. (2008) Spatial heterogeneity and temporal dynamics of particles, bacteria, and pEPS in Arctic winter sea ice. *Journal of Marine Systems*, **74**, 902–917.

Cota, G.F., Prinsenberg, S.J., Bennett, E.B. et al. (1987) Nutrient fluxes during extended blooms of Arctic ice algae. *Journal of Geophysical Research*, **92**, 1951–1962.

de Jong, J., Schoemann, V., Maricq, N. et al. (2013) Iron in land-fast sea ice of McMurdo Sound derived from sediment resuspension and wind-blown dust attributes to primary productivity in the Ross Sea, Antarctica. *Marine Chemistry*, **157**, 24–40.

Dieckmann, G.S., Nehrke, G., Papadimitriou, S. et al. (2008) Calcium carbonate as ikaite crystals in Antarctic sea ice, *Geophysical Research Letters*, **35**, L08501, 10.1029/2008GL033540.

Douglas, T.A., Domine, F., Barret, M. et al. (2012) Frost flowers growing in the Arctic ocean-atmosphere–sea ice–snow interface: 1. Chemical composition, *Journal of Geophysical Research*, **117**, D00R09, 10.1029/2011JD016460.

Edwards, R., & Sedwick, P. (2001) Iron in East Antarctic snow: Implications for atmospheric iron deposition and algal production in Antarctic waters. *Geophysical Research Letters*, **28**, 3907–3910.

Eicken, H. & Salganek, M. (Eds.) (2010) *Field Techniques for Sea-ice Research*. University of Alaska Press, Fairbanks.

Eicken, H., Kolatschek, J., Freitag, F., Lindemann, F., Kassens, H. & Dmitrenko, I. (2000) A key source area and constraints on entrainment for basinscale sediment transport by Arctic sea ice. *Geophysical Research Letters*, **27**, 1919–1922.

Ewert, M. & Deming, J.W. (2011) Selective retention in saline ice of extracellular polysaccharides produced by the cold-adapted marine bacterium *Colwellia psychrerythraea* strain 34H. *Annals of Glaciology*, **52**, 111–117.

Ewert, M. & Deming, J.W. (2013) Sea ice microorganisms: environmental constraints and extracellular responses. *Biology*, **2**, 603–628.

Ewert, M., Carpenter, S.D., Colangelo-Lillis, J. & Deming, J.W. (2013) Bacterial and extracellular polysaccharide content of brine-wetted snow over Arctic winter first-year sea ice. *Journal of Geophysical Research*, **118**, 726–735.

Feltham, D.L., Untersteiner, N., Wettlaufer, J.S. & Worster, M.G. (2006) Sea ice is a mushy layer. *Geophysical Research Letters*, **33**, L14501, 10.1029/2006GL026290.

Fripiat, F., Sigman, D.M., Fawcett, S.E., Rafter, P.A., Weigand, M.A. & Tison, J.-L. (2014a) New insights into sea ice nitrogen biogeochemical dynamics from the nitrogen isotopes. *Global Biogeochemical Cycles*, **28**, 115–130.

Fripiat, F., Tison, J.-L., André, L., Notz, D., & Delille, B. (2014b) Biogenic silica recycling in sea ice inferred from Si-isotopes: constraints from Arctic winter first-year sea ice. *Biogeochemistry*, **119**, 5–33.

Giannelli, V., Thomas, D.N., Haas, C., Kattner, G., Kennedy, H., & Dieckmann, G.S. (2001) Behaviour of dissolved organic matter and inorganic nutrients during experimental sea-ice formation. *Annals of Glaciology*, **33**, 317–321.

Gleitz, M., Kukert, H., Riebesell, U. & Dieckmann, G.S. (1996) Carbon acquisition and growth of Antarctic sea ice diatoms in closed bottle incubations. *Marine Ecology Progress Series*, **135**, 169–177.

Gleitz, M., Thomas, D.N., Dieckmann, G.S. & Millero, F.J. (1995) Comparison of summer and winter inorganic carbon, oxygen and nutrient concentrations in Antarctic sea ice brine. *Marine Chemistry*, **51**, 81–91.

Glud, R.N., Rysgaard, S., Turner, G., McGinnis, D.F. & Leakey, R.J.G. (2014) Biological- and physical-induced oxygen dynamics in melting sea ice of the Fram Strait. *Limnology and Oceanography*, **59**, 1097–1111.

Golden, K.M., Ackley, S.F & Lytle, V.I. (1998) The percolation phase transition in sea ice. *Science*, **282**, 2238–2241.

Gradinger, R. (1999) Integrated abundance and biomass of sympagic meiofauna in Arctic and Antarctic pack ice. *Polar Biology*, **22**, 169–177.

Gradinger, R. (2009) Sea-ice algae: Major contributors to primary production and algal biomass in the Chukchi and Beaufort Seas during May/June 2002. *Deep Sea Research Part II: Topical Studies in Oceanography*, **56**, 1201–1212.

Gradinger, R. & Ikävalko, J. (1998) Organism incorporation into newly forming Arctic sea ice in the Greenland Sea. *Journal of Plankton Research*, **20**, 871–886.

Granskog, M.A., Kaartokallio, H. & Shirasawa, K. (2003) Nutrient status of Baltic Sea ice: Evidence for control by snow-ice formation, ice permeability, and ice algae. *Journal of Geophysical Research*, **108**, 3253, 10.1029/2002JC001386.

Granskog, M.A., Virkkunen, K., Thomas, D.N., Ehn, J., Kola, H. & Martma, T. (2004) Chemical properties of brackish water ice in the Bothnian Bay, the Baltic Sea. *Journal of Glaciology*, **50**, 292–302, 10.3189/172756504781830079.

Grotti, M., Soggia, F., Ianni, C. & Frache, R. (2005) Trace metals distributions in coastal sea ice of Terra Nova Bay, Ross Sea, Antarctica. *Antarctic Science*, **17**, 289–300.

Hare, A.A., Wang, F., Barber, D. Geilfus, N.-X., Galley, R.J & Rysgaard, S. (2013) pH evolution in sea ice grown at an outdoor experimental facility. *Marine Chemistry*, **154**, 46–54.

Hunke, E.C., Notz, D., Turner, A.K, Vancoppenolle, M. (2011) The multiphase physics of sea ice: a review for model developers. *The Cryosphere*, **5**, 989–1009.

Janech, M.G., Krell, A., Mock, T., Kang, J.-S. & Raymond, J.A. (2006) Ice-binding proteins from sea ice diatoms (Bacillariophyceae). *Journal of Phycology*, **42**, 410–416.

Juhl, A.R., Krembs, C. & Meiners, K.M. (2011) Seasonal development and differential retention of ice algae and other organic fractions in first-year Arctic sea ice. *Marine Ecology Progress Series*, **436**, 1–16.

Juul-Pedersen, T., Michel, C., Gosselin, M. & Seuthe, L. (2008) Seasonal changes in the sinking export of particulate material under first-year sea ice on the Mackenzie Shelf (western Canadian Arctic). *Marine Ecology Progress Series*, **353**, 13–25.

Kanna, N., Toyota, T. & Nishioka, J. (2014) Iron and macro-nutrient concentrations in sea ice and their impact on the nutritional status of surface waters in the southern Okhotsk Sea. *Progress in Oceanography*, **126**, 44–57.

King, M.D., France, J.L., Fisher, F.N. & Beine, H.J. (2005) Measurement and modelling of UV radiation penetration and photolysis rates of nitrate and hydrogen peroxide in Antarctic sea ice: an estimate of the production rate of hydroxyl radicals in first-year sea ice. *Journal of Photochemistry and Photobiology*, **176**, 39e49, 10.1016/j.jphotochem.2005.08.032.

Krembs, C. & Deming, J.W. (2008) The role of exopolymers in microbial adaptation to sea ice. In: *Psychrophiles: from biodiversity to biotechnology*, (Eds. R. Margesin et al.), pp. 247–264. Springer Verlag GmbH, Berlin Heidelberg.

Krembs, C., Eicken, H., Junge, K. & Deming, J.W. (2002) High concentrations of exopolymeric substances in Arctic winter sea ice: implications for the polar ocean carbon cycle and cryoprotection of diatoms. *Deep Sea Research Part I: Oceanographic Research Papers*, **49**, 2163–2181.

Krembs, C., Eicken, H. & Deming, J.W. (2011) Exopolymer alteration of physical properties of sea ice and implications for ice habitability and biogeochemistry in a warmer Arctic. *Proceedings of the National Academy of Sciences*, **108**, 3653–3658.

Krembs, C., Gradinger, R. & Spindler, M. (2000). Implications of brine channel geometry and surface area for the interaction of sympagic organisms in Arctic sea ice. *Journal of Experimental Marine Biology and Ecology*, **243**, 55–80.

Lancelot, C., de Montety, A., Goosse, H. et al. (2009) Spatial distribution of the iron supply to phytoplankton in the Southern Ocean: a model study. *Biogeosciences*, **6**, 2861–2878.

Lannuzel, D., Schoemann, V., de Jong, J. et al. (2010) Distribution of dissolved iron in Antarctic sea ice: Spatial, seasonal, and inter-annual variability. *Journal of Geophysical Research*, **115**, G03022, 10.1029/2009JG001031.

Lannuzel, D., Schoemann, V., de Jong, J., Tison, J.-L. & Chou, L. (2007) Distribution and biogeochemical behaviour of iron in the East Antarctic sea ice. *Marine Chemistry*, **106**, 18–32.

Lannuzel, D., Schoemann, V., Dumont, I. et al. (2013) Effect of melting Antarctic sea ice on the fate of microbial communities studied in microcosms. *Polar Biology*, **36**, 1438–1497.

Lannuzel, D., Chever, F., van der Merwe, P. et al. (2014) Iron biogeochemistry in Antarctic pack ice during SIPEX-2. *Deep Sea Research Part II: Topical Studies in Oceanography*, doi:10.1016/j.dsr2.2014.12.003.

Lavoie, D., Denman, K. & Michel, C. (2005) Modeling ice algal growth and decline in a seasonally ice-covered region of the Arctic (Resolute Passage, Canadian Archipelago). *Journal of Geophysical Research*, **110**, C11009, 10.1029/2005JC002922

Leu, E., Wiktor, J., Søreide, J.E., Berge, J. & Falk-Petersen, S. (2010) Increased irradiance reduces food quality of sea ice algae. *Marine Ecology Progress Series*, **411**, 49–60.

Loose, B., Miller, L.A., Elliott, S. & Papakyriakou, T. (2011) Sea ice biogeochemistry and material transport across the frozen interface. *Oceanography*, **24**, 202–218.

Mancuso Nichols, C., Bowman, J.P., & Guezennec, J. (2005) Effects of incubation temperature on growth and production of exopolysaccharides by an Antarctic sea ice bacterium grown in batch culture. *Applied and Environmental Microbiology*, **71**, 3519–3523.

Marion, G.M. (2001) Carbonate mineral solubility at low temperatures in the Na-K-Mg-Ca-H-Cl-SO_4-OH-HCO_3-CO_3-CO_2-H_2O system. *Geochimica et Cosmochimica Acta*, **65**, 1883–1896.

McMinn, A., Müller, M.N., Martin, A., & Ryan, K.G. (2014). The response of Antarctic sea ice algae to changes in pH and CO_2. *PloS ONE*, **9**, e86984.

McNeill, V.F., Grannas, A.M., Abbatt, J.P.D. et al. (2012) Organics in environmental ices: sources, chemistry, and impacts. *Atmospheric Chemistry and Physics*, **12**, 9653–9678.

Measures, C. I. (1999) The role of entrained sediments in sea ice in the distribution of aluminium and iron in the surface waters of the Arctic Ocean, *Marine Chemistry*, **68**, 59–70, doi:10.1016/S0304-4203(99)00065-1.

Meier, W.N., Hoverlsrud, G.K., van Oort, B.E.H. et al. (2014) Arctic sea ice in transformation: A review of recent observed changes and impacts on biology and human activity. *Reviews of Geophysics*, 51, 10.1002/2013RG000431.

Meiners, K., Brinkmeyer, R., Granskog, M.A. & Lindfors, A. (2004) Abundance, size distribution and bacterial colonization of exopolymer particles in Antarctic sea ice (Bellingshausen Sea). *Aquatic Microbial Ecology*, **35**, 283–296.

Meiners, K., Gradinger, R., Fehling, J., Civitarese, G & Spindler, M. (2003) Vertical distribution of exopolymer particles in sea ice of the Fram Strait (Arctic) during autumn. *Marine Ecology Progress Series*, **248**, 1–13.

Meiners, K., Krembs, C., Gradinger, R. (2008) Exopolymer particles: microbial hotspots of enhanced bacterial activity in Arctic fast ice (Chukchi Sea). *Aquatic Microbial Ecology*, **52**, 195–207.

Meiners, K. M., Vancoppenolle, M., Thanassekos, S. et al. (2012) Chlorophyll a in Antarctic sea ice from historical ice core data. *Geophysical Research Letters*, **39**, L21602, 10.1029/2012GL053478.

Michel, C., Ingram, R.G. & Harris, L.R. (2006) Variability in oceanographic and ecological processes in the Canadian Arctic Archipelago. *Progress in Oceanography*, **71**, 379–401.

Michel, C., Nielsen, T.G., Gosselin, M. & Nozais, C. (2002) Significance of sedimentation and grazing by ice micro- and meiofauna for carbon cycling in annual sea ice (Northern Baffin Bay). *Aquatic Microbial Ecology*, **30**, 57–68.

Miller, L.A., Fripiat, F., Else, B.G.T. et al. (2015) Methods for biogeochemical studies of sea ice: the state of the art, caveats, and recommendations. *Elementa Science of the Anthropocene*, **3**: 000038. doi: 10.12952/journal.elementa.000038.

Miller, L.A., Papakyriakou, T.N., Collins, R.E. et al. (2011) Carbon dynamics in sea ice: A winter flux time series. *Journal of Geophysical Research*, **116**, C02028, 10.1029/2009JC006058.

Müller, S., Vähätalo, A.V., Granskog, M.A., Autio, R. & Kaartokallio, H. (2011) Behaviour of dissolved organic matter during formation of natural and artificially grown Baltic Sea ice. *Annals of Glaciology*, **52**, 233–241.

Müller, S., Vähätalo, A.V, Stedmon, C.A. et al. (2013) Selective incorporation of dissolved organic matter (DOM) during sea ice formation. *Marine Chemistry*, **155**, 148–157.

Niemi, A., Meisterhans, G. & Michel, C. (2014) Response of under-ice prokaryotes to experimental sea-ice DOM enrichment. *Aquatic Microbial Ecology*, **73**, 17–28.

Noble, A. E., Moran, D. M., Allen, A. E. & Saito, M.A. (2013) Dissolved and particulate trace metal micronutrients under the McMurdo Sound seasonal sea ice: basal sea ice communities as a capacitor for iron. *Frontiers in Chemistry*, 1, 10.3389/fchem.2013.00025.

Nomura, D., Nishioka, J., Granskog, M.A. et al. (2010) Nutrient distributions associated with snow and sediment-laden layers in sea ice of the southern Sea of Okhotsk. *Marine Chemistry*, **119**, 1–8.

Norman, L., Thomas, D.N., Stedmon, C.A. et al. (2011) The characteristics of dissolved organic matter (DOM) and chromophoric dissolved organic matter (CDOM) in Antarctic sea ice. *Deep Sea Research Part II: Topical Studies in Oceanography*, **58**, 1075–1091.

Notz, D. & Worster, M.G. (2009) Desalination processes of sea ice revisited. *Journal of Geophysical Research*, **114**, C05006, 10.1029/2008JC004885.

Nozais, C., Gosselin, M., Michel, C., Tita, G. (2001) Abundance, biomass, composition and grazing impact of the sea ice meiofauna in the NorthWater, Northern Baffin Bay. *Marine Ecology Progress Series*, **217**, 235–250.

Orellana, M.V., Matrai, P.A., Leck, C., Rauschenberg, C.D., Lee, A.M. & Coz, E. (2011) Marine microgels as a source of cloud condensation nuclei in the high Arctic. *Proceedings of the National Academy of Sciences*, **108**, 13 612–13 617.

Palmisano, A.C. & Garrison, D.L. (1993) Microorganisms in Antarctic sea ice. In: *Antarctic microbiology* (Ed. Friedmann, E.I.), pp. 167–218. Wiley-Liss Publishers, New York.

Pankowski, A. & McMinn, A. (2008) Ferredoxin and flavodoxin in eastern Antarctica pack ice. *Polar Biology*, **31**, 1153–1165.

Papadimitriou, S., Kennedy, H., Norman, L., Kennedy, D.P., Dieckmann, G.S. & Thomas, D.N. (2012) The effect of biological activity, $CaCO_3$ mineral dynamics, and CO_2 degassing in the inorganic carbon cycle in sea ice in late winter-early spring in the Weddell Sea, Antarctica. *Journal of Geophysical Research*, **117**, C08011, 10.1029/2012JC008058.

Papadimitriou, S., Thomas, D.N., Kennedy, H. et al. (2007) Biogeochemical composition of natural sea ice brines from the Weddell Sea during early austral summer. *Limnology and Oceanography*, **52**, 1809–1823.

Petrich, C. & Eicken, H. (2010) Growth, structure and properties of sea ice. In: *Sea Ice*, 2nd edn (Eds. D.N. Thomas & G.S. Dieckmann), pp. 23–77. Wiley-Blackwell, Oxford, UK.

Polashenski, C., Perovich, D. & Courville, Z. (2012) The mechanisms of sea ice melt pond formation and evolution. *Journal of Geophysical Research*, **117**, C01001, 10.1029/2011JC007231.

Priscu, J.C, Downes, M.T., Priscu, L.R., Palmisano, A.C & Sullivan, C.W. (1990) Dynamics of ammonium oxidizer activity and nitrous oxide (N_2O) within and beneath Antarctic sea ice. *Marine Ecology Progress Series*, **62**, 37–46.

Raymond, J.A., Fritsen C. & Shen, K. (2007) An ice-binding protein from an Antarctic sea ice bacterium. *FEMS Microbiology Ecology*, **61**, 214–221.

Riebesell, U., Schloss, I. & Smetacek, V. (1991) Aggregation of algae released from melting sea ice: implications for seeding and sedimentation. *Polar Biology*, **11**, 239–248.

Riedel, A., Michel, C. & Gosselin, M. (2006) Seasonal study of sea-ice exopolymeric substances on the Mackenzie shelf: implications for transport of sea-ice bacteria and algae. *Aquatic Microbial Ecology*, **45**, 195–206.

Riedel, A., Michel, C., Gosselin, M. & LeBlanc, B. (2007a) Enrichment of nutrients, exopolymeric substances and microorganisms in newly formed sea ice on the Mackenzie shelf. *Marine Ecology Progress Series*, **342**, 55–67.

Riedel, A., C. Michel C, & Gosselin, M. (2007b) Grazing of large-sized bacteria by sea-ice heterotrophic protists on the Mackenzie shelf during the winter-spring transition. *Aquatic Microbial Ecology*, **50**, 25–38.

Riedel, A., Michel C, Gosselin M., and LeBlanc, B. (2008) Winter spring dynamics in sea-ice carbon cycling on the Mackenzie shelf, Canadian Arctic. *Journal of Marine Systems*, **74**, 918–932.

Rivkin, R.B. & Putt, M. (1987) Heterotrophy and photoheterotrophy by Antarctic microalgae: light-dependent incorporation of amino acids and glucose. *Journal of Phycology*, **23**, 442–452.

Rózanska, M., Gosselin, M., Poulin, M., Wiktor, J.M. & Michel, C. (2009) Influence of environmental factors on the development of bottom ice protist communities during the winter-spring transition. *Marine Ecology Progress Series*, **386**, 43–59.

Rysgaard, S. & Glud, R.N. (2004) Anaerobic N_2 production in Arctic sea ice. *Limnology and Oceanography*, **49**, 86–94.

Rysgaard, S., Glud, R. N., Sejr, M. K., Blicher, M. E. & Stahl, H. J. (2008). Denitrification activity and oxygen dynamics in Arctic sea ice. *Polar Biology*, **31**, 527–537.

Rysgaard, S., Søgaard, D.H., Cooper, M. et al. (2013) Ikaite crystal distribution in winter sea ice and implications for CO_2 system dynamics. *Cryosphere*, **7**, 707–718.

Saenz, B.T. & Arrigo, K.R. (2012) Simulation of a sea ice ecosystem using a hybrid model for slush layer desalination. *Journal of Geophysical Research*, **117**, C05007, 10.1029/2011JC007544.

Saenz, B.T. & Arrigo, K.R. (2014) Annual primary production in Antarctic sea ice during 2005-2006 from a sea ice state estimate. *Journal of Geophysical Research*, **119**, 3645–3678.

Shaw, P.M., Russell, L.M., Jefferson, A. & Quinn, P.K. (2010) Arctic organic aerosol measurements show particles from mixed combustion in spring haze and from frost flowers in winter. *Geophysical Research Letters*, **37**, L10803, 10.1029/2010GL042831.

Shi, D., Xu, Y. & Morel, F.M.M. (2009) Effects of the pH/pCO_2 control method on medium chemistry and phytoplankton growth. *Biogeosciences*, **6**, 1199–1207.

Spindler, M. (1994) Notes on the biology of sea ice in the Arctic and Antarctic. *Polar Biology*, **14**, 319–324.

Staley, J.T. & Gosink, J.J. (1999) Poles apart: biodiversity and biogeography of sea ice bacteria. *Annual Review of Microbiology*, **53**, 189–215.

Stedmon, C.A., Thomas, D.N., Granskog, M., Kaartokallio, H., Papadimitriou, S., Kuosa, K. (2007) Characteristics of dissolved organic matter in Baltic coastal sea ice: allochthonous or autochthonous origins? *Environmental Science & Technology*, **41**, 7273–7279.

Stedmon, C.A., Thomas, D.N, Papadimitriou, S., Granskog, M.A., Dieckmann, G.S. (2011) Using fluorescence to characterize dissolved organic matter in Antarctic sea ice brines. *Journal of Geophysical Research*, **116**, G03027, 10.1029/2011JG001716.

Sunda, W.G. & Huntsman, S.A. (1995) Iron uptake and growth limitation in oceanic and coastal phytoplankton. *Marine Chemistry*, **50**, 189–206.

Tedesco, L. & Vichi, M. (2014) Sea Ice Biogeochemistry: A Guide for Modellers. *PLoS ONE*, **9**, e89217, 10.1371/journal.pone.0089217.

Thomas, D.N. & Dieckmann, G.S. (Eds.) (2010) *Sea Ice*, 2nd edn. Wiley-Blackwell, Oxford, UK.

Thomas, D.N., Kattner, G., Engbrodt, R. et al. (2001) Dissolved organic matter in Antarctic sea ice. *Annals of Glaciology*, **33**, 297–303.

Thomas, D.N., Papadimitriou, S., Michel, C. (2010) Biogeochemistry of Sea Ice. In: *Sea Ice*, 2nd edn (Eds. D.N. Thomas & G.S. Dieckmann), pp. 425–468, Wiley-Blackwell, Oxford.

Tovar-Sánchez, A., Duarte, C.M., Alonso, J.C., Lacorte, S., Tauler, R. & Galbán-Malagón, C. (2010) Impacts of metals and nutrients released from melting multiyear Arctic sea ice. *Journal of Geophysical Research*, **115**, C07003, 10.1029/2009JC005685.

Ugalde, S.C., Meiners, K.M., Davidson, A.T., Westwood, K.J. & McMinn, A. (2013) Photosynthetic carbon allocation of an Antarctic sea ice diatom (*Fragilariopsis cylindrus*). *Journal of Experimental Marine Biology and Ecology*, **446**, 228–235.

Underwood, G.J.C., Aslam, S.N., Michel, C. et al. (2013) Broad-scale predictability of carbohydrates and exopolymers in Antarctic and Arctic sea ice. *Proceedings of the National Academy of Sciences*, **110**, 15 734–15 739.

Underwood, G.J.C., Fietz, S., Papadimitriou, S., Thomas, D.N., Dieckmann, G.S. (2010) Distribution and composition of dissolved extracellular polymeric substances (EPS) in Antarctic sea ice. *Marine Ecology Progress Series*, **404**, 1–19.

Uusikivi, J., Vähätalo, A.V., Granskog, M.A., Sommaruga, R. (2010) Contribution of mycosporine-like amino acids and colored dissolved and particulate matter to sea ice optical properties and ultraviolet attenuation. *Limnology and Oceanography*, **55**, 703–713.

Vancoppenolle, M., Goosse, H., De Montety, A., Fichefet, T., Tremblay, B. & Tison, J.-L. (2010) Modeling brine and nutrient dynamics in Antarctic sea ice: The case of dissolved silica. *Journal of Geophysical Research*, **115**, C02005, 10.1029/2009JC005369.

Vancoppenolle, M., Meiners, K.M., Michel, C. et al. (2013a) Role of sea ice in global biogeochemical cycles: Emerging views and challenges, *Quaternary Science Reviews*, **79**, 207–230.

Vancoppenolle, M., Notz, D., Vivier, F. et al. (2013b) Technical Note: On the use of the mushy-layer Rayleigh number for the interpretation of sea-ice-core data. *The Cryosphere Discussion*, **7**, 3209–3230.

van der Merwe, P., Lannuzel, D., Bowie, A.R., Mancuso Nichols, C.A. & Meiners, K.M. (2011) Iron fractionation in pack and fast ice in East Antarctica: Temporal decoupling between the release of dissolved and particulate iron during spring melt. *Deep Sea Research Part II: Topical Studies in Oceanography*, **58**, 1222–1236.

van der Merwe, P., Lannuzel, D., Mancuso Nichols, C.A. et al. (2009) Biogeochemical observations during the winter–spring transition in East Antarctic sea ice: evidence of iron and exopolysaccharide controls. *Marine Chemistry*, **115**, 163–175.

Verdugo, P. (2012) Marine microgels. *Annual Review of Marine Science*, **4**, 375–400.

Weeks, W.F. (2010) *On Sea Ice*. University of Alaska Press, Fairbanks.

Wells, L.E. & Deming, J.W. (2006). Modelled and measured dynamics of viruses in Arctic winter sea-ice brines. *Environmental Microbiology*, **8**, 1115–1121.

Wettlaufer, J.S. (2010) Sea ice and astrobiology. In: *Sea Ice*, 2nd edn (Eds. D.N. Thomas & G.S. Dieckmann), pp. 579–594, Wiley-Blackwell, Oxford.

Xie, H. & Gosselin, M. (2005) Photoproduction of carbon monoxide in first-year sea ice in Franklin Bay, southeastern Beaufort Sea. *Geophysical Research Letters*, **32**, L12606, 10.1029/2005GL022803.

Xie, H., Bélanger, S., Song, G., Benner, R. et al. (2012) Photoproduction of ammonium in the southeastern Beaufort Sea and its biogeochemical implications. *Biogeosciences*, **9**, 3047–3061.

Zhou, J., Delille, B., Eicken, H. et al. (2013) Physical and biogeochemical properties in landfast sea ice (Barrow, Alaska): Insights on brine and gas dynamics across seasons. *Journal of Geophysical Research*, **118**, 3172–3189.

Zhou, J., Delille, B., Kaartokallio, H., Kattner, G. et al. (2014) Physical and bacterial controls on inorganic nutrients and dissolved organic carbon during a sea ice growth and decay experiment. *Marine Chemistry*, **166**, 59–69.

Zhou, J., Kotovitch, M., Kaartokallio, H. et al. (2016) The impact of dissolved organic carbon and bacterial respiration on pCO_2 in experimental sea ice. *Progress in Oceanography*, **141**, 153–167.

CHAPTER 18

Gases in sea ice

Jean-Louis Tison,[1] Bruno Delille[2] and Stathys Papadimitriou[3]

[1] *Laboratoire de Glaciologie, Université Libre de Bruxelles, Bruxelles, Belgium*
[2] *Astrophysics, Geophysics and Oceanography Department, Université de Liège, Liège, Belgium*
[3] *School of Ocean Sciences, Bangor University, Menai Bridge, Anglesey, UK*

18.1 Introduction

18.1.1 Environmental and climatic role of gases in sea ice

The interest of the scientific community in gases in sea ice dates back to the early 1960s, with the pioneering work of Japanese scientists (Miyake & Matsuo, 1963; Matsuo & Miyake, 1966). Sea ice acquires its initial gas content from the pool of dissolved gases in the surface ocean. This is controlled by exchange of gaseous components of the atmosphere with the surface oceanic water, biological activity and inorganic reactions, such as photochemical oxidation and ionization. Much like the dissolved ions in seawater, dissolved gases will also be subject to solute rejection and physical concentration in the residual brine that forms during seawater freezing and is trapped – the small part of it that escapes gravity drainage to the underlying ocean – as inclusions in thermal equilibrium with the ice crystal matrix in sea ice. Again like the dissolved ionic species, the concentrated dissolved gases will be further subject to biological and abiotic reactions, including most prominently the aqueous–gaseous phase transition via gas bubble nucleation controlled by gas solubility, a thermodynamic property of gas interaction with electrolyte solutions. Such reactions will modulate the concentration of the dissolved gases in the sea ice brines away from the effect of the physical concentration mechanism.

The gas content and gas composition of sea ice have become a subject of intense investigation by the glaciological and oceanographic communities on account of their potential feedback into the climate system via the influence of gas bubbles on the sea ice albedo due to their specific optical properties (see Chapter 4) and on account of exchange of greenhouse gases with the atmosphere. Reports of exchange of greenhouse gases between sea ice and the atmosphere in the last decade have defied the traditional view of sea ice as an inert and impermeable barrier for gas exchange and have raised interest in gas dynamics within sea ice. The investigation of gases is also an important component in biogeochemical studies of sea ice on account of gas transformations by biological activity (sympagic primary and secondary production) and physical-chemical processes (brine concentration–dilution and degassing–gas solvation). The diversity of ice types within the sea ice cover of polar and sub-polar surface waters (snow ice, superimposed ice, frazil ice, columnar ice, platelet ice – see Chapter 1) and the versatility of its surface properties in the course of its growth–decay cycle (bare ice, fresh snow cover, wet snow cover, melt ponds) result in a large range of total gas content (gaseous in gas bubbles plus dissolved in sea ice brine) and also in the variable fractional contribution of each of the individual gaseous species to the total gas content of sea ice. This high temporal and spatial variability in the phase distribution of gas species in sea ice is conveyed on the albedo of its surface, which is still a major challenge for climate modelling. Further, the amount of light that will penetrate the sea ice cover and reach the ice–water interface will also be quite variable, with potential implications for the photosynthetic activity and light sensitivity of the sympagic algal communities. Also, the density of sea ice will vary considerably, with known complications for the development of the algorithms used to derive sea ice thickness information from satellite laser

---

Sea Ice, Third Edition. Edited by David N. Thomas.
© 2017 John Wiley & Sons, Ltd. Published 2017 by John Wiley & Sons, Ltd.

altimetry (see Chapters 2 and 9). In addition, the state of the gas content of sea ice (dissolved vs gaseous) will affect the modality of gas transfer within sea ice and, therefore, the efficiency of gas exchange between the ocean and the atmosphere in sea ice-covered polar waters. The very dynamic thermal regime of the sea ice cover will control both the distribution of the gas content in sea ice, the phase transition of individual gas species and the permeability of the medium. Finally, because sea ice hosts an array of physical–chemical reactions and is also a rich ecosystem, it does not simply operate as a passive stratum between the ocean and the atmosphere, but as a reactor, particularly in the context of gas mass balance in the surface of the polar oceans. As a result, sea ice has only very recently been recognized as a potential key player in climate-relevant gas fluxes to and from the atmosphere, with still unquantified global consequences in a warming climate.

18.1.2 Dissolved versus gaseous state

Ice crystals cannot accommodate gas molecules in their lattice at atmospheric pressure (Hobbs, 1974), but only at the higher pressures commonly encountered in ice sheets in the form of gas clathrates (e.g. Pauer et al., 1999; Kipfstuhl et al., 2001). As opposed to meteoric and lake (fresh water) ice, sea ice contains a non-negligible amount of gases in the dissolved and gaseous state in, respectively, brine pockets and channels and in gas bubbles. The heterogeneous 'mushy layer' character of the sea ice cover complicates sampling and the understanding of gas distribution and phase changes in the temporally dynamic course of thermodynamic and biogeochemical reactions in the medium. It is therefore crucial to clarify in reports the measured gas component, as well as the method and the units (Figure 18.1). Depending on the sampling technique (i.e. sackhole brines and intact, or crushed, or melted bulk sea ice), quantities and units will differ. For example, total gas content is typically expressed in $mL_{air\ STP}\ L_{brine}^{-1}$ or $mL_{air\ STP}\ kg_{ice}^{-1}$ and, occasionally, in $cm^3\ L_{ice}^{-1}$, gas mixing ratios are expressed in volume percent (%) or parts per million (ppm), gas concentrations in $\mu mol\ L_{brine}^{-1}$, $\mu mol\ kg_{brine}^{-1}$ or $\mu mol\ kg_{ice}^{-1}$, and partial pressure or fugacity in μatm. In some cases, the determined gas quantities will only refer to the dissolved phase, while in others they will only refer to the gaseous phase and in others still to both.

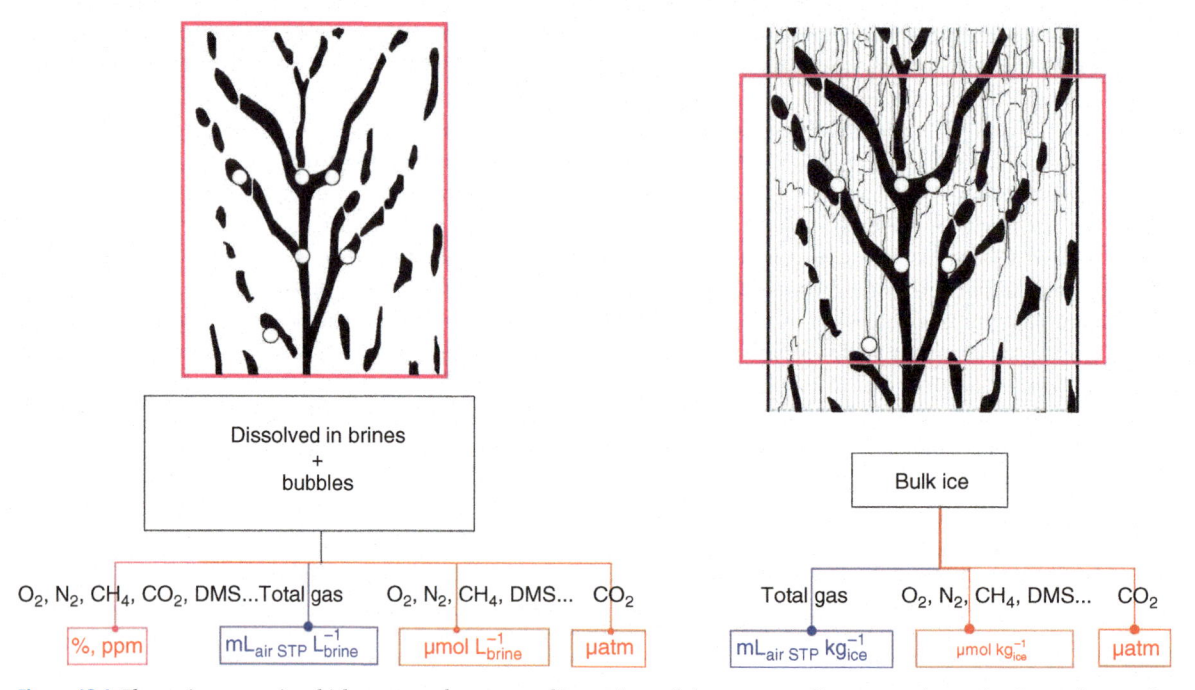

Figure 18.1 The various ways in which gases can be measured in sea ice and the corresponding appropriate units. Depending on the sampling technique, the measurements may refer to either the dissolved phase exclusively or the gaseous phase, or both.

18.1.3 The concept of gas solubility in sea ice

The property of thermodynamic gas solubility describes the maximum amount of gas that can be dissolved at gas–solution equilibrium as a function of temperature and total pressure, corrected for non-ideality. The thermodynamic equilibrium solubility is described by the Henry's Law constant (K_H), which, in the case of the biologically reactive CO_2, is expressed as

$$K_H = \frac{\alpha_{CO_2(aq)}}{f_{CO_2(g)}} = \frac{m_{CO_2(aq)} \gamma_{CO_2(aq)}}{\gamma_{CO_2(g)} pCO_2} \quad (18.1)$$

In the above equation, $\alpha_{CO_2(aq)}$ is the activity of CO_2 in solution at gas–solution equilibrium, $f_{CO_2(g)}$ is the fugacity of gaseous CO_2, $m_{CO_2(aq)}$ is the molal concentration of CO_2 in solution at gas–solution equilibrium (in mol $kg_{H_2O}^{-1}$), $\gamma_{CO_2(aq)}$ and $\gamma_{CO_2(g)}$ are activity coefficients of dissolved and gaseous CO_2, respectively, and pCO_2 is the partial pressure of gaseous CO_2 (in atm) (Plummer & Busenberg, 1982). In complex electrolyte solutions, such as seawater and sea ice brines, the determination of the parameters on the activity scale for thermodynamic calculations is challenging and the equilibrium CO_2 solubility is defined on the conventionally measured concentration scale as follows:

$$K_o = \frac{M_{CO_2(aq)}}{f_{CO_2(g)}} \quad (18.2)$$

where $M_{CO_2(aq)}$ is the molar CO_2 concentration in solution (on a volumetric or gravimetric unit basis) at gas–solution equilibrium and $f_{CO_2(g)}$ as described earlier (Weiss 1974). The fugacity of CO_2 in solution can be measured with standardized oceanographic techniques (Pierrot et al., 2007); when it is equivalent to that in the gas phase, the solution and gas phase are said to be at thermodynamic equilibrium with respect to CO_2, the solution is gas-saturated and the equilibrium $M_{CO_2(aq)}$ can be computed from equation (18.2). When the gas fugacity in solution is more or less than that in the gas phase, the solution is said to be super-saturated or undersaturated, respectively, with respect to the gaseous component. The change in the equilibrium CO_2 solubility is described at standard atmospheric pressure by the empirical salinity–temperature (S–T) function of Weiss (1974) with a minimum temperature of applicability at −10°C. The same concepts hold and a similar function at standard atmospheric pressure has been derived for the equilibrium solubility of dissolved molecular oxygen (O_2) (Garcia & Gordon, 1992), nitrogen (N_2), argon (Ar), neon (Ne) (Hamme & Emerson, 2004) and methane (CH_4) (Wiesenburg & Guinasso, 1979).

The coupled variations in brine salinity and temperature accompanying the structural changes during the seasonal formation–decay cycle of sea ice have direct bearing on the solubility, ionization equilibria and biological reactions of gases. The determination of the physical–chemical changes in gas solubility and ionization equilibria in the brine inclusions rely on the extrapolation (still with indeterminate errors today) of the empirical S–T functions that describe such reactions and were derived from the available determinations at standard oceanographic conditions (above-zero temperature; hyposaline, saline and mildly hypersaline solutions). Several investigators (Tison et al., 2010; Brabant, 2012; Zhou et al., 2013, 2014b, Crabeck et al., 2014a,b) have used the concept of gas solubility in their interpretation of measurements of the total gas content in sea ice with respect to phase distribution. In those instances, the total gas content of a sea ice section was compared with a thermodynamic concentration derived from the saturation concentration of the gas in the brine at thermal equilibrium with the ice, calculated as described above, at the relevant S–T conditions of the ice section and its brine inclusions and further weighted by the brine volume fraction (porosity) in the ice section. Excess total gas content relative to the porosity-weighted brine saturation value has been considered as a first-order estimate of the gaseous gas component in sea ice because of the brine–atmosphere equilibrium assumption implicit in this type of computation. However, in many instances of direct determination of dissolved biophilic gases in sackhole brines, brine–air disequilibrium has been observed (see later sections for details). A more robust technique is clearly required to determine reliably the gaseous gas component separately from the dissolved gas component in bulk sea ice.

18.2 The total gas content in sea ice

Tsurikov (1979) provided the first extensive discussion of the processes responsible for the presence and fate of gases, especially gas bubbles, in sea ice. Nine different processes were invoked, which can be divided into

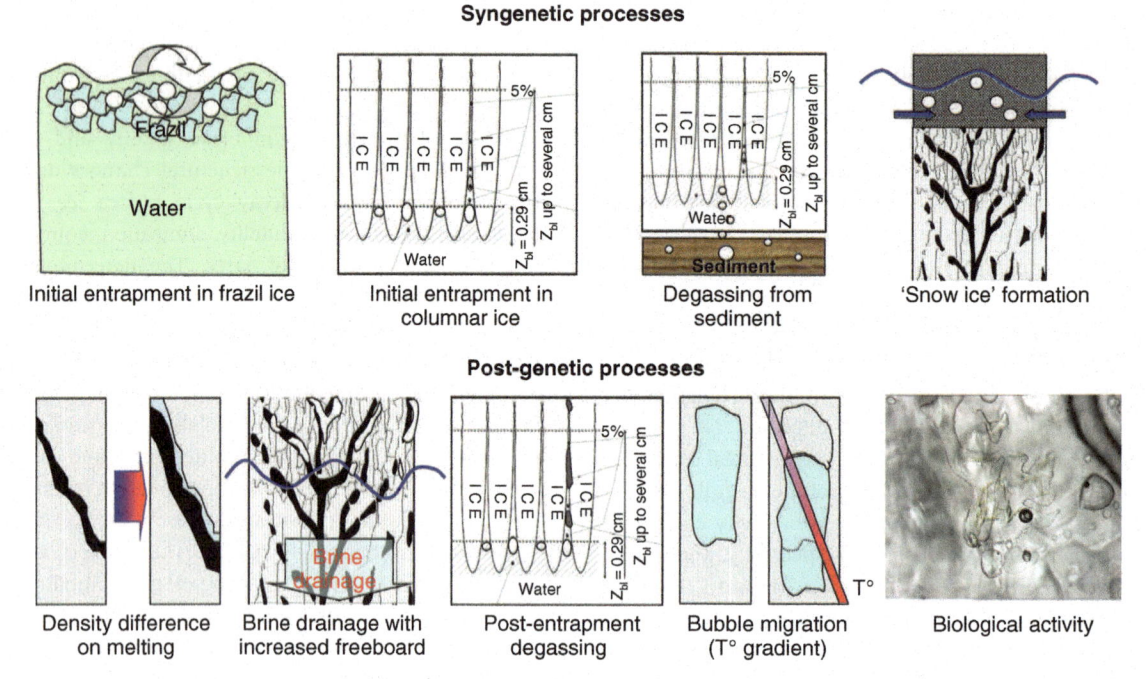

Figure 18.2 Schematic summary of the nine processes suggested by Tsurikov (1979) for inclusion of dissolved gases and air bubbles within sea ice.

syngenetic and post-genetic processes (Figure 18.2). Syngenetic processes encompass:
- initial entrapment in wind- and wave-induced frazil ice;
- initial entrapment in columnar ice;
- degassing from sediments in sea ice above shallow continental shelf areas, typically in the Arctic;
- snow ice formation, when refreezing sea water traps the air initially present within the flooded snow cover.

Post-genetic processes include:
- ice to water density difference during internal melting;
- brine drainage under increasing freeboard as the ice grows;
- internal degassing during thermal change;
- the relatively slow process of gas bubble migration along temperature gradients;
- biological activity (metabolism).

There has been a fair amount of theoretical and empirical work done on impurity (dissolved salts, gases, particles) inclusions in sea ice (see Chapter 1). Total gas content has not yet been measured directly in sea ice brines. Equations (1.6)–(1.9) in Chapter 1, derived from the work of Cox and Weeks (1983), provide a semi-empirical way of calculating the relative volume (V_a/V) of total gas (air) in sea ice. However, this requires knowledge of the density of bulk sea ice, a difficult measurement. Three techniques have been adopted or developed to measure the total gas content of bulk sea ice: gas extraction with a Toepler pump (Tison et al., 2002); ice melting in artificial sea water (Rysgaard & Glud, 2004); and summing individual concentrations of the major gases (O_2, N_2, Ar) measured by gas chromatography. The first technique was developed for gas measurements in meteoric ice for paleoclimatic reconstructions (Raynaud et al., 1988; Martinerie et al., 1994). Table 18.1 summarizes the range of total gas content measured with this technique in artificial sea ice and in natural sea ice from the Arctic and the Antarctic. Most of the available total gas content determinations (Figure 18.3) are less than 24 mL kg_{ice}^{-1}, which is the amount of air dissolved in air-saturated seawater at 1 atm. This value, therefore, represents the maximum amount of total gas in sea ice if the total gas content of the initial sea water is conserved throughout the

Table 18.1 Range of total gas content measured in artificial and natural Arctic and Antarctic sea ice.

Cruise	Period	Method	Total gas content min (mL$_{air\ STP}$ kg$_{ice}^{-1}$)	max (mL$_{air\ STP}$ kg$_{ice}^{-1}$)	
Japanese Antarctic Research Expedition Vicinity of Showa Base, Antarctica	February 1959 to February 1962	Toepler	2.0	21.4	Matsuo & Miyake (1966)
Interice III HSVA, Hamburg	18 days	Toepler	3.5	18.0	Tison et al. (2002)
ARISE Antarctica (*Aurora Australis*)	September–October 2003	Toepler	4.0	17.0	Tison et al. (unpubl. data)
ISPOL Weddell Sea Antarctica (*Polarstern*)	November–December 2004	Toepler	6.0	23.0	Tison et al. (unpubl. data)
Young Sound Fjord Greenland Arctic	May–June 2002	Melting in artificial seawater (bubble content)	0	90	Rysgaard & Glud (2004)
Barrow Alaska, USA Arctic	February–June 2009	Toepler	2.4	29.2	Zhou et al. (unpubl. data)

freezing and decay process and is shown as a dotted line in the graphs of Figure 18.3. Total gas content above this value indicates air accumulation in sea ice via syngenetic incorporation of air during initial freeze-up as wave-induced air entrainment in frazil ice, post-genetic gas bubble migration within the ice cover, postgenetic air incorporation during snow ice and superimposed ice formation, or postgenetic air incorporation during sea ice decay, when draining brines can be replaced by air in sea ice layers above the freeboard level. Another potential source of excess total gas is the syngenetic entrapment of gas bubbles rising from the sediment floor in shallow depth areas. The extremely high values for the total gas content (up to 80 cm^3 L$_{ice}^{-1}$; Figure 18.3) in sea ice from Young Sound, Greenland, and the maximum value (28 mL$_{air\ STP}$ kg$_{ice}^{-1}$; Figure 18.3) observed at Barrow, Alaska, both occur in the upper layers of decaying first-year sea ice, suggesting origin from replacement of drained brine by air, upward gas bubble migration, or superimposed ice formation (Rysgaard & Glud, 2004; Zhou et al., 2013). Apart from these exceptions, the available observations are consistent with the early measurements (2.0–21.4 mL$_{air\ STP}$ kg$_{ice}^{-1}$) of Matsuo and Miyake (1966) and with a mean air content estimate derived from the Cox and Weeks (1983) semi-empirical equations using a bulk ice salinity of 6 and a mean sea ice density of 0.93.

Based on the above and the more extensive work of Zhou et al. (2014a,b) (Figure 18.3), the main features of the total gas content in first-year sea ice can be summarized as follows:

- Total gas content in excess of that for air-saturated seawater typically appears to be confined in the uppermost sections of ice cores, with the remainder of sea ice layers exhibiting a relative deficit and, hence, a syngenetic or post-genetic loss of the air content of the initial air-saturated seawater from the sea ice.
- All values observed in bottom sea ice are close to that controlled by the solubility of air in sea ice, suggesting initial entrapment of air-saturated sea water in sea ice with its total gas content effectively cropped and modulated by changes in the solubility of the individual atmospheric gases due to salinity and temperature changes in the ice–brine system further up in the ice cover.
- The total gas content is well above that of gas-saturated sea ice throughout the sea ice column above the bottom layer in winter and spring, suggesting that a significant fraction of gases is in the gaseous phase, assuming no brine–gas disequilibrium.

438 Chapter 18

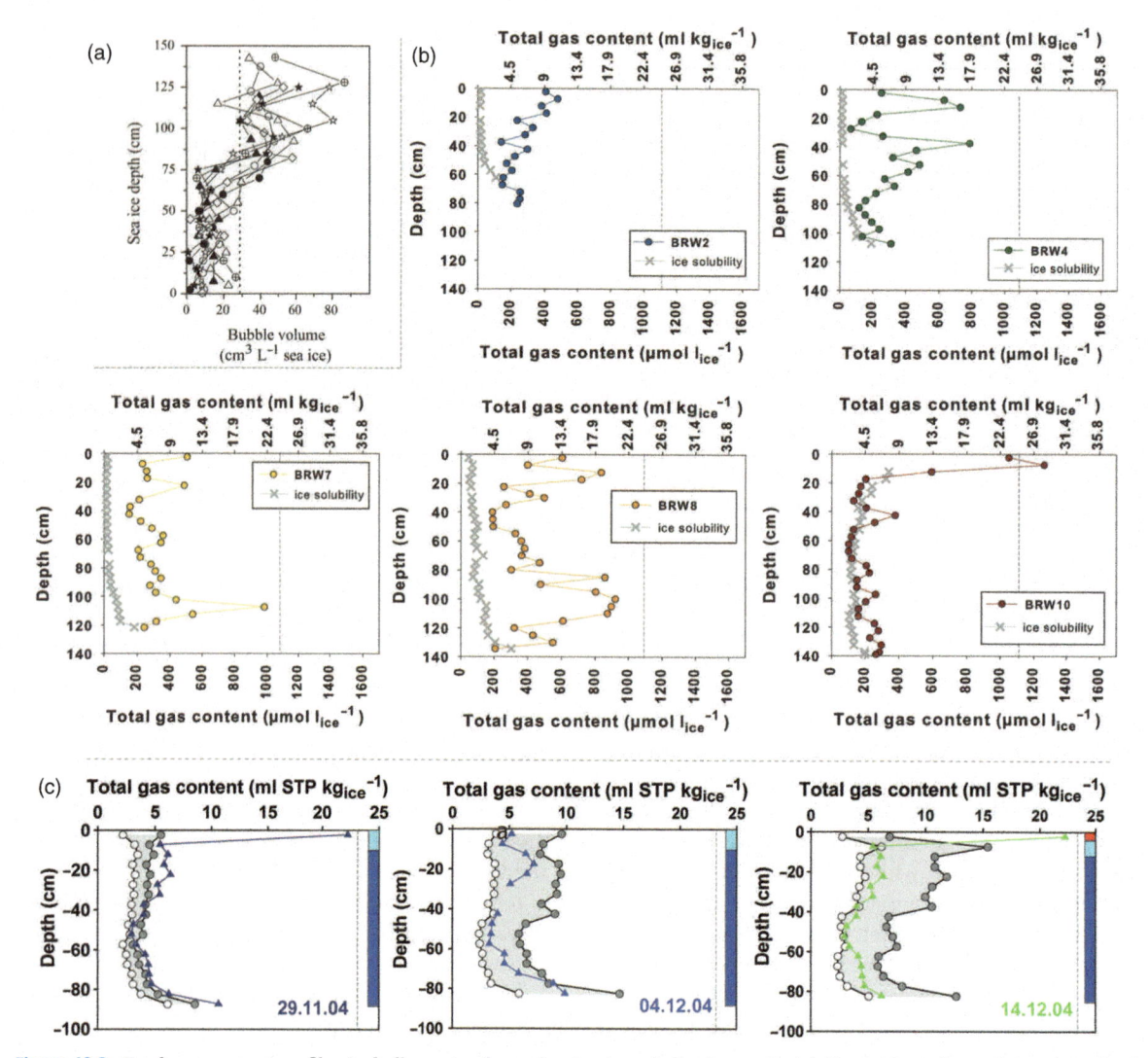

Figure 18.3 Total gas content profiles in bulk sea ice from the Arctic and the Antarctic: (a) Young Sound, northeast Greenland (28 May–12 June 2002) (Rysgaard & Glud, 2004); (b) Barrow Point, Alaska (3 February–7 June 2009) (Zhou, unpubl. data); and (c) drifting ice floe (Ice Station Polarstern, ISPOL), western Weddell Sea, Antarctica (29 November–30 December 2004) (Tison & Brabant, unpubl. data). Dotted lines represent estimates for bubble volume/total gas content from closed system freezing of sea water. In (c), two curves are plotted for the ice solubility with super-saturation factors of 1.0 (white circles) and 2.2 (grey circles), respectively. Colour bars in (c) represent ice types: red, superimposed ice; cyan, frazil ice and blue, columnar ice. In (a) and (b), the ice structure is predominantly columnar ice. BRW2 = 03.02.09; BRW4 = 31.03.09; BRW7 = 10.04.09; BRW8 = 08.05.09 and BRW10 = 05.06.09. Source: From Rysgaard & Glud, 2004. Reproduced with permission of John Wiley & Sons.

- The total gas content drops to the solubility value in sea ice in summer, suggesting loss of the gaseous phase from the sea ice, with the potential exception of the surface layers.

18.3 The gas composition in sea ice

The measurement of gas composition in sea ice is a relatively young discipline. It has been building up on

Table 18.2 Techniques for measuring concentrations of individual gas species in sea ice (see text for details and relevant investigations).

Gas	Bulk ice Destructive	Bulk ice Non-destructive	Brine Direct	Brine Indirect
O_2	Dry crushing, GC; Melting (Tedlar bags or syringe), GC or Winkler	Electrodes; Optodes	Winkler	
N_2	Dry crushing, GC		GC and mass spectrometry	
Ar	Dry crushing, GC			
CH_4	Melting–refreezing, GC; Melting (Tedlar bags), purge and trap to GC		Purge and trap to GC	
CO_2	Dry crushing, GC; Melting–refreezing, GC	Bulk sea ice sample equilibration with gas mixture of known CO_2 content	Li-Cor in portable sea ice equilibrator system	Computed from DIC and TA
N_2O	Melting (Tedlar bags or syringe) purge and trap to GC		Purge and trap to GC	
DMS	Dry crushing, purge and trap GC; Melting in filtered seawater, purge and trap GC; Melting in filtered seawater, purge and trap to PTRMS		Purge and trap to GC	
Halocarbons			Purge and trap to GC	

GC, gas chromatography; DIC, dissolved inorganic carbon; TA, total alkalinity; DMS, dimethylsulphide.

the experience of glaciologists and oceanographers, with the complications underlined before and related to the heterogeneous nature of sea ice as exemplified by the 'mushy layer' theory (Wettlaufer et al., 1997). Measurements of the concentration of individual gases have been conducted either in whole ice core segments with accurately constrained vertical resolution along the ice column or in sackhole brines with the caveat of the crude spatial resolution afforded by this sampling protocol. Analytical techniques are summarized in Table 18.2. More details on challenges in sea ice sampling for physical and biogeochemical investigations can be found in Eicken et al. (2009) and Miller et al. (2015). However, as mentioned earlier, it should be kept in mind that different techniques provide different outputs (e.g. dissolved gases only, or the gas content of gas bubbles only, or the gas content of both phases) and that extreme caution should be taken when reporting and comparing results from different protocols. Below, we examine the existing information on individual gases in sea ice.

18.3.1 Argon: an inert gas as proxy for physical processes

As discussed earlier, the total gas content of individual gases and their phase transition and distribution with temperature and salt content in sea ice evolve under the combined control of syngenetic and post-genetic physical processes and biogeochemical reactions. In order to improve our understanding and modelling capability of the evolution of gas properties in the sea ice cover and its impact on gas fluxes across the seawater–ice and

440 Chapter 18

ice–air interfaces, it is worthwhile to attempt separating the impact of the physical and biogeochemical drivers. In that respect, argon (Ar), as an inert gas, can be used to examine the role of physical processes (temperature and salinity changes and related mechanisms, such as brine dilution and concentration and brine transport) on gas properties (concentration, phase changes and fluxes) in sea ice. This inert gas has never been measured directly in sea ice brines. The available Ar data in sea ice have been collected using dry-crushing of whole ice segments and gas chromatography (Table 18.2), which has been used by glaciologists for meteoric ice and, in sea ice, collects both the dissolved and the gaseous phase (Raynaud et al., 1982; Tison et al., 2002; Zhou et al., 2013; Zhou et al., 2014b). Early work (Mock et al., 2002; Tison et al., 2002; Rysgaard & Glud, 2004) has suggested that gas compounds in sea ice should be present predominantly in the gaseous phase in the form of bubbles (see Section 18.2) and that, therefore, buoyancy-driven redistribution should be an important process in the determination of gas profiles in the medium. These studies were, however, based on short-term experiments or field campaigns, not encompassing the full growth-decay cycle of sea ice.

In a more recent investigation (January–June 2009; Zhou et al., 2013) of the total gas content and individual gases in landfast sea ice at Barrow, Alaska, bulk sea ice Ar concentrations were found to be consistent with those determined in the pioneering study of gases in sea ice by Matsuo and Miyake (1966). As briefly discussed below, the study identified three stages of brine dynamics across the annual cycle of sea ice growth and decay based on brine salinity, brine volume fraction and porous-medium Rayleigh number (Ra): bottom layer convection, full-depth convection and brine stratification (Figure 18.4). The Ar profiles from

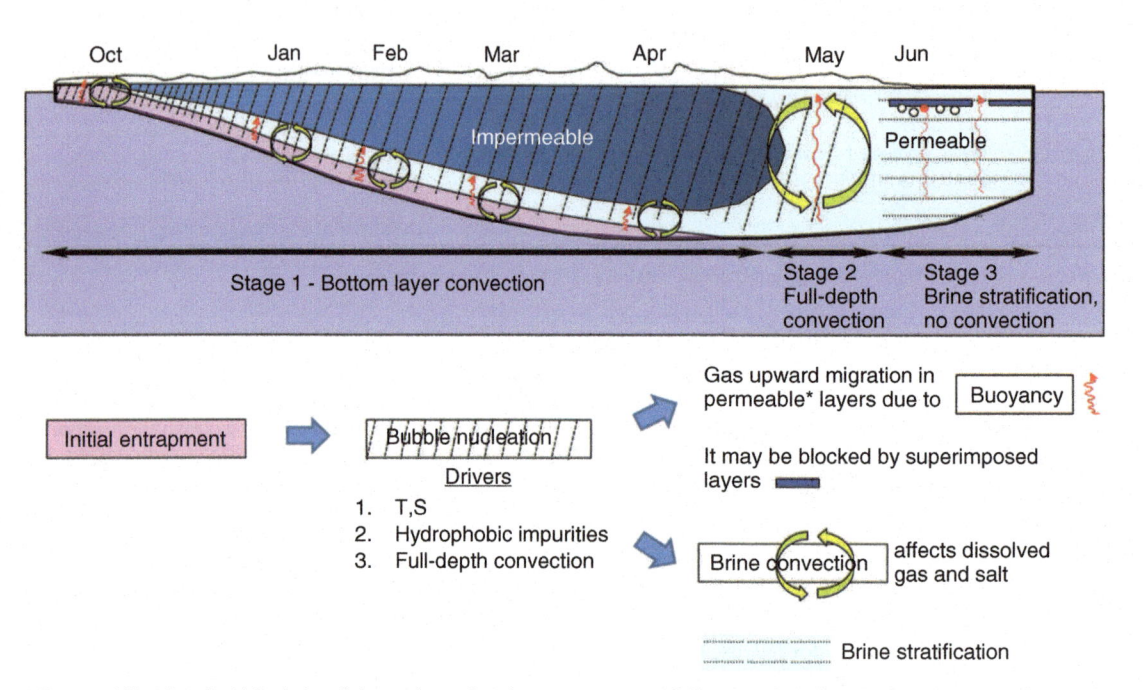

Figure 18.4 Schematic view of gas entrapment and evolution in sea ice through the three stages of brine dynamics (Zhou et al., 2013). After initial entrapment, changes in temperature (T) and salinity (S), the presence of hydrophobic impurities and full-depth convection contribute to gas bubble nucleation. Gas bubbles could then migrate upward in permeable layers due to their buoyancy, while gas dissolved in brine would migrate along with the dissolved salt by brine convection. Diffusion and the formation of superimposed ice layers may also control gas content in permeable sea ice. (From Zhou et al., 2013. Reproduced with permission of John Wiley & Sons.)

* Permeability threshold for brine: brine volume fraction > 5%; permeability threshold for gas: brine volume fraction >7.5%

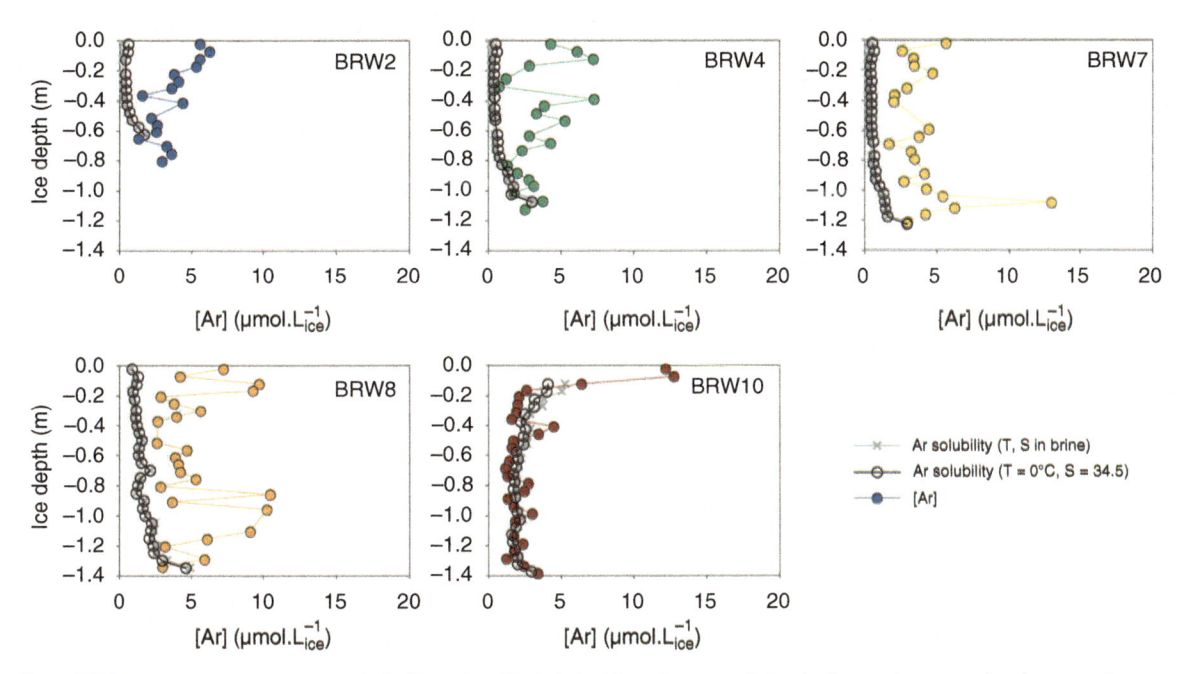

Figure 18.5 Argon concentration ([Ar]) in bulk sea ice (filled circles) from Barrow, Alaska, in discrete ice cores taken between January and June 2009 (from Zhou et al., 2013) and the Ar solubility limit set by the porosity-weighted concentration in the brine at air–brine equilibrium at *in situ* ice temperature (T) and brine salinity (S) (crosses), or at T and S fixed at 0°C and 34.5, respectively (open circles). Excess [Ar] relative to the solubility limit indicates a first-order estimate of the gaseous Ar content in gas bubbles in sea ice, while [Ar] at the solubility limit suggests only the presence of dissolved gas in brine. Station dates as in Figure 18.3. Source: From Zhou et al., 2013. Reproduced with permission of John Wiley & Sons.

this study (Figure 18.5) and the very similar, also concurrent, profiles of total gas content (Figure 18.3) show the dominant control of physical processes (temperature- and salinity-driven solubility changes in brines and degassing or air entrainment in the upper layers) on the gas distribution and phase transition with depth in sea ice. Specifically, during the ice growth period (winter–spring), a portion of the dissolved gas in the brines appeared to transfer to gas bubbles as evidenced in the total Ar super-saturation by Zhou et al. (2013), which ranged between 2200% and 3800% and was consistent with the super-saturation levels at which Killawee et al. (1998) observed nitrogen bubble formation.

First results of model simulations (Moreau et al., 2014) have indicated at least 50% underestimation of the total Ar content in sea ice if gas bubble formation is neglected. Further, the highest Ar super-saturation was found at the base of the transitional layer between granular and columnar ice where optical observations yielded high gas bubble content. The bulk sea ice Ar profiles in that study had the following salient features:
- excess total Ar relative to saturation, except in the bottom sea ice sections at all times and throughout the ice in summer (BRW10; Figure 18.5);
- temporal increase at the base of the surficial transition zone between granular and columnar ice, not observed in any of the other investigated parameters (bulk salinity, nutrients);
- total Ar accumulation during convective brine drainage in the winter–spring transition phase (BRW8; Figure 18.5);
- Ar decrease to saturation in the warm sea ice in summer (BRW 10; Figure 18.5).

The investigators suggested bubble nucleation and migration along the ice column as the main drivers for the observed depth distribution of total Ar in bulk sea ice. Further, given Ar inertness towards biological reactions and using it as a proxy for the physical–chemical reactions initiated by the dynamic temperature and salinity changes in sea ice, the investigators concluded that the seasonal spatial and phase distribution of gases in sea ice should be constrained by initial incorporation of gases dissolved in seawater at the ice–water interface and subsequent transition cycles from the dissolved to the gaseous phase and vice versa, controlled by solubility changes driven by the salinity and temperature changes in the brine–ice system and relative buoyancy.

However, total Ar should remain constant with depth in sea ice in a closed system and equal to that observed in the bottom layer of sea ice (~ 3 $\mu mol\ L_{ice}^{-1}$), where incorporation of sea water Ar (about 17 $\mu mol\ L^{-1}$) occurs during freezing and ice growth. The increasing deviation of the measured Ar from this value moving upwards in the sea ice column (Figure 18.5), therefore, requires the addition of Ar in an open system. Two processes can contribute to this: first, on syngenetic and post-genetic nucleation; gas bubbles may rise in the ice column due to their buoyancy (Frank et al., 2007) until their progress is barred by reductions in ice permeability. In this vein, Zhou et al. (2013) demonstrated that the permeability threshold for gaseous Ar transport in sea ice is larger than the 5% value proposed for brine transport (Golden et al., 1998; Chapter 1) and probably lies between 7.5% and 10%. Second, in an open system (i.e. permeable ice), brine replacement by Ar-richer seawater via convection will increase the total Ar content of sea ice. Repeated convective events may then effectively raise bulk ice Ar concentration through successive cooling and bubble nucleation and migration cycles. Zhou et al. (2013) noted that these Ar enrichment processes will dominate the winter growth period with its strong temperature gradients in sea ice.

Following Jones et al. (1999), Zhou et al. (2013) also argued that bubble nucleation can also occur at low super-saturation (i.e. in the latest growth stages) provided the occurrence of gas cavities in the sea ice structure or of convection-induced local fluctuations in the gas-super-saturated liquid. Further, Killawee et al. (1998), Tison et al. (2002) and Verbeke et al. (2002) showed that large contrasts in critical super-saturation levels for bubble nucleation can occur at the ice–water interface, depending on small-scale hydrodynamic conditions. The decoupling between Ar and CH_4 profiles observed in sea ice by Zhou et al. (2014a) is consistent with excess methane input from the underlying sediments in the shallow environment of Barrow, Alaska.

The distribution with depth and phase transition in sea ice of biologically and chemically reactive gas species is controlled by ionization (CO_2) and biological reactions (N_2, O_2, CO_2), which will be outlined below. However, the solubility- and permeability-driven depth and phase distribution of biologically and chemically inert Ar outlined above should also underlie the reactive gas species in sea ice.

18.3.2 Nitrogen: the dominant gas species

As gases in sea ice primarily originate in the pool of dissolved gases in seawater or, on occasion, from the atmosphere, molecular nitrogen (N_2) is the dominant gas species in sea ice. Until now, N_2 concentrations in sea ice have been obtained using the dry-crushing technique coupled with gas chromatography (Table 18.2), while no direct observations from brines are available as yet. Figure 18.6 shows the N_2 concentration profiles at the long-term Barrow stations in Alaska studied by Zhou et al. (2014b), which mimic those of Ar in Figure 18.5, suggesting that N_2 might also be considered as an inert gas, faithfully tracking physical processes only. Zhou et al. (2014b) suggest that this is indeed primarily the case by showing an excellent linear relationship between the two gas species in their samples (Figure 18.6). A similar rationale is followed by Crabeck et al. (2014a) for spring ice samples from the Kapisillit Fjord in southwest Greenland. However, Glud et al. (2002) suggest that molecular oxygen (O_2) depletion in brines by strong dilution in thawing sea ice can create favourable conditions for anaerobic bacterial growth. In this event, anoxic bacterial metabolism leading to sulphate reduction and N_2 regeneration via denitrification and anaerobic ammonium oxidation (anammox) could occur within the ice matrix. Petri and Imhoff (2001) reported the presence of anoxygenic phototrophic purple sulphur bacteria in ice, and Kaartokallio (2001) reported evidence of denitrification in first-year ice, both incidents in the Baltic Sea, indicating the occurrence of O_2-deficient or anoxic zones within the ice. Rysgaard and Glud (2004) later quantified denitrification and anammox in Arctic first-year sea ice. Biological activity in the form of denitrification and anammox will then affect the N_2 content in sea ice in similar circumstances. Although it is fascinating that sea ice harbours anaerobic processes within the top metre or so, over many millions of square kilometres of the surface ocean, the effect of these processes should not be

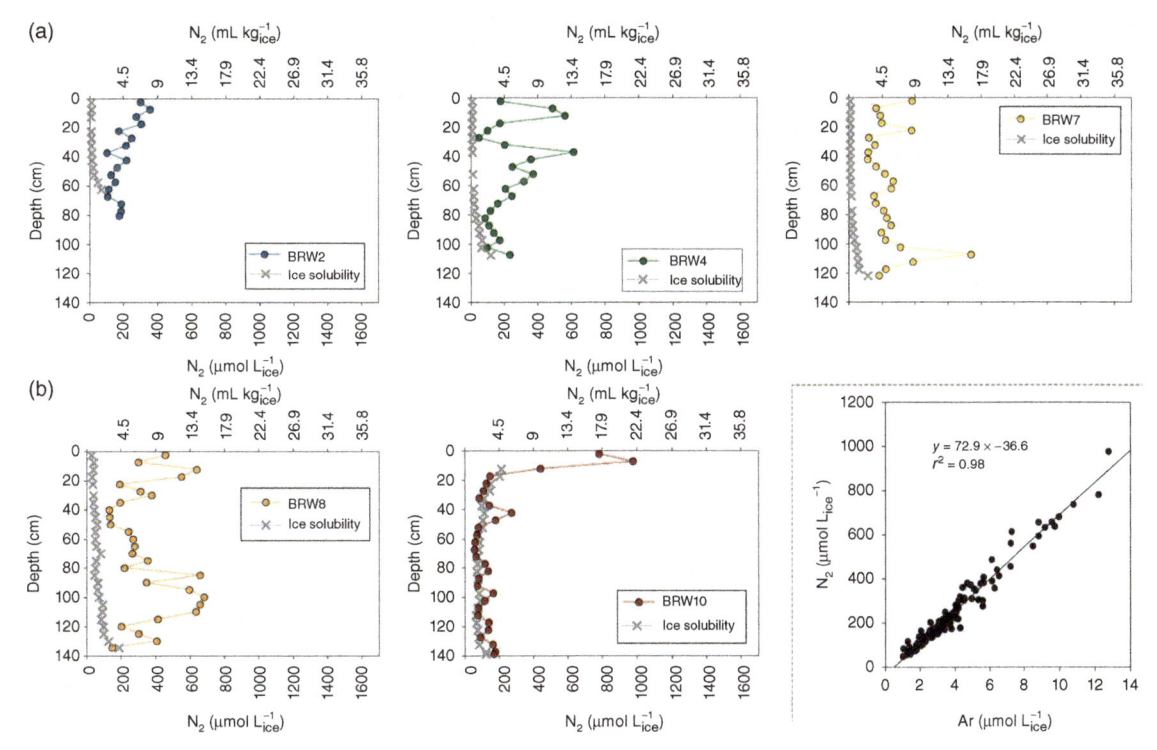

Figure 18.6 (a) The concentration of molecular nitrogen ([N$_2$]) in bulk sea ice (filled circles) from Barrow, Alaska, in discrete ice cores taken between January and June 2009 (from Zhou et al., 2013, 2014a,b) and the N$_2$ solubility limit set by the porosity-weighted concentration in the brine at air–brine equilibrium and *in situ* ice temperature (T) and brine salinity (S) (crosses). Excess [N$_2$] relative to the solubility limit indicates a first-order estimate of the gaseous N$_2$ content in gas bubbles in sea ice, while [N$_2$] at the solubility limit suggests only the presence of dissolved gas in brine. (b) Relationship between [Ar] and [N$_2$] in the sea ice from Barrow, Alaska (data from Zhou et al., 2014b) Station dates as in Figure 18.3.

overemphasized (Zhou et al., 2014b). Compared with the average bulk sea ice N$_2$ concentration (248 μmol L^{-1}) measured at Barrow, the maximum denitrification rate reported by Rysgaard et al. (2008) (22.6 nmol N$_2$ L^{-1} day^{-1}) suggests that anoxic bacterial metabolism is too slow in sea ice to have a measurable impact on its N$_2$ distribution. In the next section, we discuss how N$_2$ and Ar can be used as reference (inert) gases to elucidate the impact of biological activity on O$_2$ concentrations in sea ice.

18.3.3 Oxygen: a tracer for net community production

Molecular oxygen (O$_2$) is one of the most investigated gases in sea ice because of its link to biological activity. It has been measured using a variety of techniques (Table 18.2) in bulk sea ice and in sackhole and centrifuged brines. As for Ar and N$_2$, dry-crushing collects both the oxygen dissolved in brines and that present in gas bubbles (Tison et al., 2002; Zhou et al., 2014b). Melting in Tedlar bags or syringes is coupled with gas chromatographic analysis of the gas bubbles collected from the system and Winkler titration on the artificial seawater (with known initial O$_2$ content) in which the sample is melted (Rysgaard et al., 2004, 2008). Automated Winkler titration has also been used frequently for the determination of dissolved oxygen in brines from sackholes (Gleitz et al., 1995; Papadimitriou et al., 2007; Meiners et al., 2009). McMinn and Asworth (1998), Kühl et al. (2001), Rysgaard et al. (2001), Trenerry et al. (2002) and others successfully introduced O$_2$ microelectrodes in sea ice ecology studies. These studies were conducted either in the diffusive boundary layer between sea ice and seawater with remote-controlled devices (McMinn et al. 2000; Trenerry et al. 2002) or with a tripod and divers (Kühl et al. 2001; Rysgaard et al. 2001). Oxygen micro-optodes trapped into the growing sea ice have also been used by Mock et al. (2002) with encouraging results.

Analysis in sackhole brines, typically by Winkler titration, is the only sampling protocol so far that yields the dissolved O_2 concentration in sea ice but with the caveat of limited information on its depth distribution in the whole of the sea ice column (Gleitz et al., 1995; Papadimitriou et al., 2007; Meiners et al., 2009). Given that sackholes are usually shallow (20–60 cm) through the top of the sea ice, such observations describe the dissolved O_2 status mostly in the upper layers of sea ice (Papadimitriou et al., 2007).

The dissolved O_2 concentration in sea ice brines will be modified by physical concentration and dilution in the freezing and warming periods, respectively, of the system, degassing and gaseous O_2 solvation depending on deviation from thermodynamic solubility and contact with gas bubbles and biological activity, with photosynthesis causing accumulation of dissolved O_2 in a closed brine system and vice versa for respiration. Extrapolation of the empirical S–T function that describes O_2 solubility in seawater (Garcia & Gordon, 1992) to sea ice brine S–T conditions yields a decreasing O_2 concentration in brine in equilibrium with air (air-saturated brine) (Figure 18.7a). The physical concentration–dilution effect, illustrated in Figure 18.7(a) for the freezing of air-saturated surface oceanic water of salinity 35, results in increasing O_2 super-saturation with decreasing temperature during air-saturated seawater freezing and sea ice cooling and vice versa during warming. Measurements made in artificial sea ice indicated that the brine released from sea ice tended to be super-saturated with O_2, as expected from the physical concentration mechanism, whereas melting bulk ice was under-saturated with respect to O_2 (Glud et al., 2002).

Available dissolved O_2 observations in sea ice brines (Gleitz et al., 1995; Papadimitriou et al., 2007, 2012) have been collected by opportunistic sampling from different regions of the seasonal sea ice zone of the Weddell Sea, Antarctica, and cover the period from the austral spring–summer transition to mid-autumn (Figure 18.7). These observations show a tendency for decreasing deviation from thermodynamic solubility with decreasing temperature, consistent with the dominance of the physical–chemical controls on the dissolved O_2 concentration due to super-saturation in the brine and ensuing transfer to the gaseous phase (brine degassing) in the coldest brines collected right after the period of minimal biological activity and insolation in austral winter. At the other end of the temperature spectrum, warm brines exhibited increasing excess dissolved O_2 with increasing temperature, away from the physical concentration–dilution curve, with maximal hyperoxia in the warmest and most productive period in austral summer (Figure 18.7), a condition seldom found in other marine systems (Raven et al., 1994). Whether it is net respiration or net photosynthetic activity that controls the dissolved O_2 concentration in sea ice brines requires additional biogeochemical evidence, typically in the form of macronutrient concentrations and estimates of the concentration of dissolved carbon dioxide [CO_2(aq)] in the same brines. The available macronutrient data sets (Figure 18.7b,c) bear the mark of net photosynthesis in the hyperoxic brines in summer in the trend of increasing oxygen super-saturation with increasing net phosphate and silicic acid deficits in the brine. The macronutrient deficits are used as indication of net utilization by sympagic primary producers, with silicic acid being a proxy for the activity of siliciferous diatoms, one of the two major micro-algal taxa to be found in sea ice. Hyperoxia is quite common in predominantly autotrophic sea ice habitats (Gleitz et al., 1995; Kennedy et al., 2002; Papadimitriou et al., 2007).

The total O_2 concentration profiles in bulk sea ice shown in Figure 18.8 were derived from the same ice cores as those of total gas content, Ar and N_2 discussed earlier (Figures 18.3, 18.5 and 18.6, respectively; Tison et al., 2002; Zhou et al., 2014b). Again, the similarities amongst gases are striking, further illustrating the dominant control on the total gas content of the temperature–salinity–solubility-controlled gas bubble nucleation and phase transition (degassing) of the gases dissolved in internal brine into the gaseous phase of the sea ice. A salient discrepancy, however, is the relative excess in total O_2, especially in the lower half of the sea ice (BRW10). The latter represents the balance between oxygen production from primary producers (dominant in this case) and respiration from the whole microbial community, i.e. net community production (NCP). Zhou et al. (2014b) compared the temporal changes of standing stocks of total Ar, N_2 and O_2 in bulk sea ice and showed loss during ice decay and, more specifically, higher loss for the inert gases than for O_2, the latter

Gases in sea ice   445

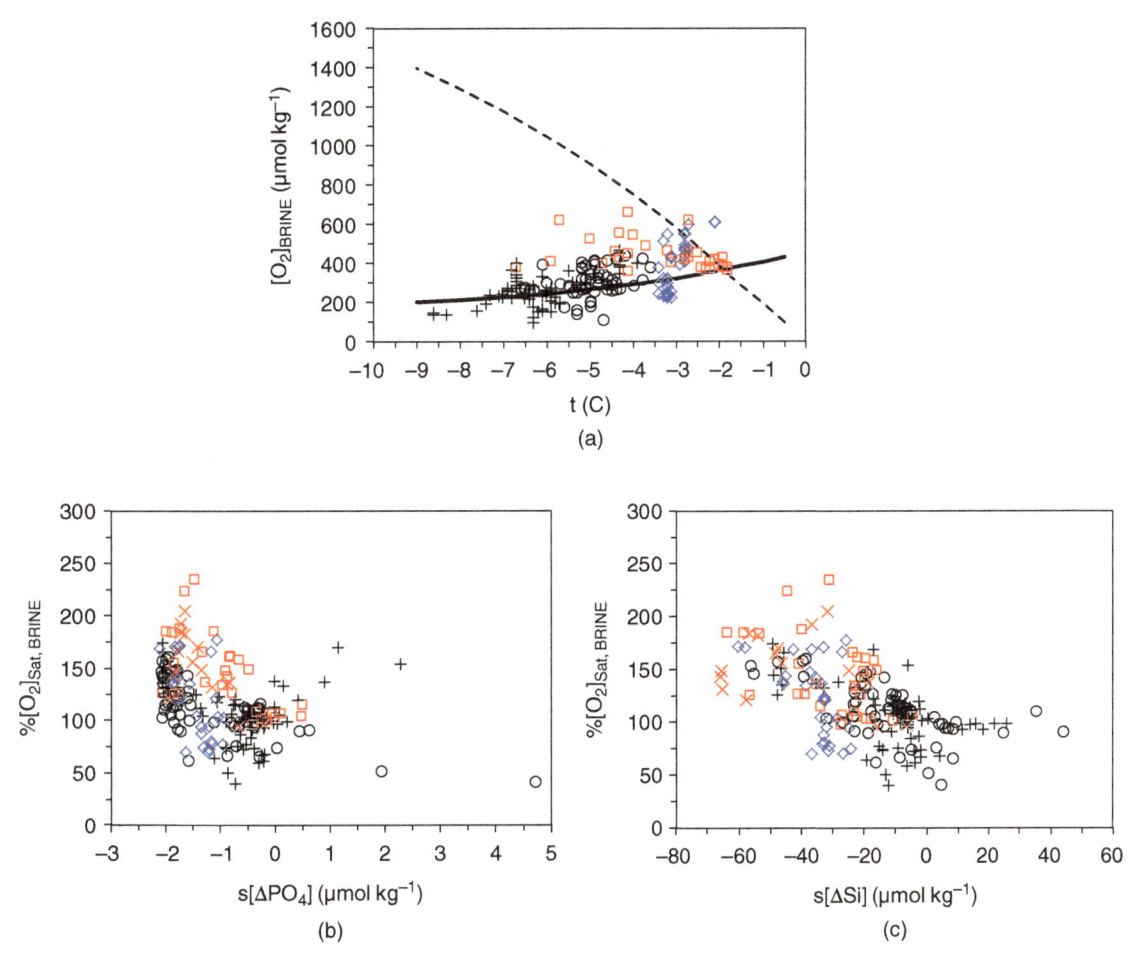

Figure 18.7 Observations in sackhole brines: (a) Dissolved molecular oxygen (O_2) concentration as a function of brine temperature, (b) percent dissolved O_2 saturation vs deficit of the salinity-normalized (S = 35) soluble reactive phosphorus (phosphate) concentration (s[ΔPO_4]) relative to that in surface oceanic water (2.11 μmol kg^{-1}), and (c) percent dissolved O_2 saturation vs deficit of the salinity-normalized (S = 35) silicic acid concentration (s[ΔSi]) relative to that in surface oceanic water (69 μmol kg^{-1}). The data were collected in the sea ice zone of the western Weddell Sea, Antarctica, during the Winter Weddell Outflow Study (WWOS) in September 2006 (circles) and October 2006 (crosses), and during the Ice Station Polarstern study (ISPOL) in December 2004 (diamonds), as well as in the sea ice zone of the southern Weddell Sea during the ANT IX/3 expedition in January – March 1991 (x) (no in situ temperature available) and during the ANT X/3 expedition in April – May 1992 (squares). The WWOS data (including the surface seawater observations) were reported in Papadimitriou et al. (2012), except for O_2 and silicic acid (D. N. Thomas, unpublished data). The ISPOL data were reported in Papadimitriou et al. (2007), while the ANT IX/3 and ANT X/3 data were reported in Gleitz et al. (1995). The solid line in panel (a) indicates the trend in dissolved O_2 concentration at air-saturation, calculated as a function of salinity and temperature from the equation in Garcia & Gordon (1992) at the freezing point in concentrated and diluted sea icebrine. The dashed line in panel (a) indicates the effect of physical concentration and dilutionon the dissolved O_2concentration in sea ice brines (conservative with respect to brine salinity).For this, air-saturated surface oceanic water of salinity 35 was considered as the freezing medium, with its dissolved O_2 concentration at air saturation [367 μmol kg^{-1}, Garcia & Gordon (1992)] calculated at the freezing point (UNESCO, 1983) and the conservative dissolved O_2 concentration in the resulting brine calculated as a linear function of brine salinity. The salinity-temperature relationship in sea ice brines (see Chapter 18.1) was used in the determination of both curves in panel (a).

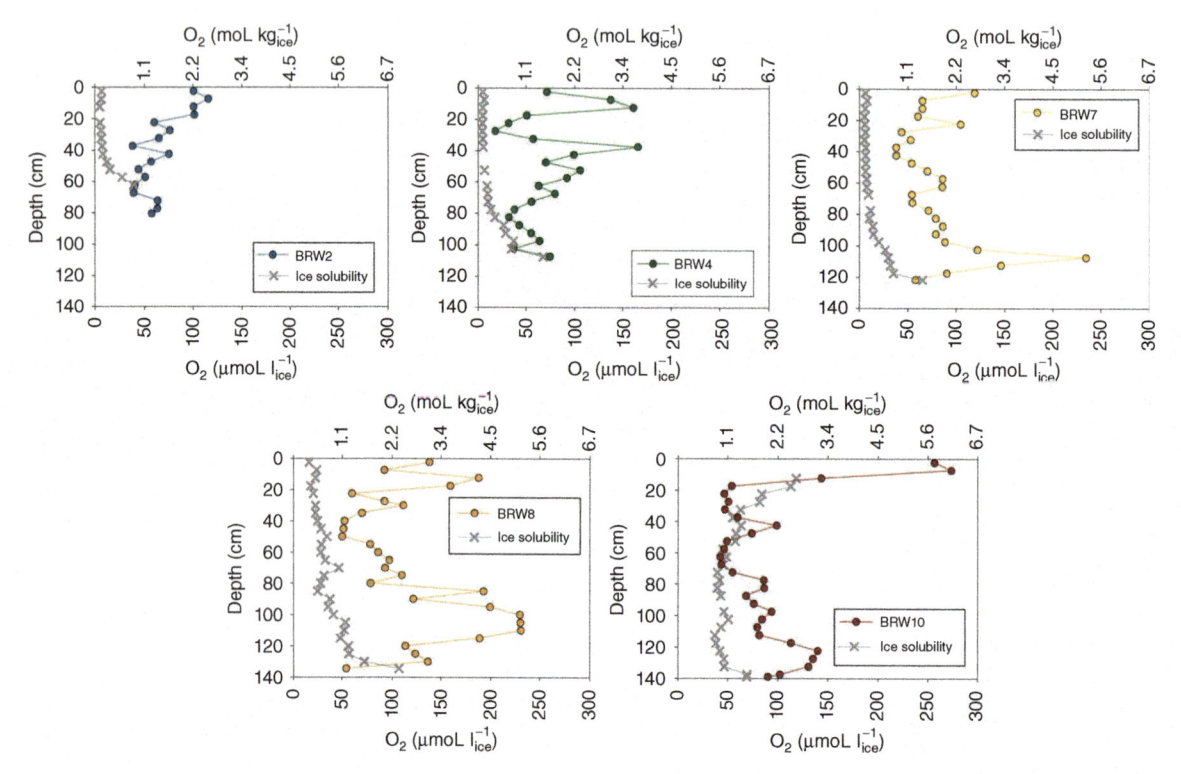

Figure 18.8 Molecular oxygen concentration ([O_2]) in bulk sea ice (filled circles) from Barrow, Alaska, in discrete ice cores taken between January and June 2009 (Zhou et al. 2014b) and the O_2 solubility limit set by the porosity-weighted concentration in the brine at air-brine equilibrium and in situ ice temperature (T) and brine salinity (S) (crosses). Excess [O_2] relative to the solubility limit indicates a first order estimate of the gaseous O_2 content in gas bubbles in sea ice, while [O_2] at the solubility limit suggests only the presence of dissolved gas in brine Station dates as in Figure 18.3.

indicating the control of in situ O_2 production in brines from phototrophic activity over solubility changes, phase transition and gas bubble buoyancy.

Biological activity can have strong influence on the seasonal O_2 content of sea ice (Figure 18.7), but the use of total O_2 profiles in bulk sea ice for quantification of NCP in sea ice is complicated by the overprint of physical-chemical processes. Zhou et al. (2014b) have recently used the molar O_2 to Ar ratio (O_2/Ar) for this purpose following the rationale of oceanographic studies (Cassar et al., 2009; Castro-Morales et al., 2013; Hendricks et al., 2004; Reuer et al., 2007). Because denitrification and anammox rates are negligible compared with the total N_2 concentrations in bulk sea ice (see above), Zhou et al. (2014b) also considered N_2 as an inert gas and used O_2/N_2 along with O_2/Ar in their analysis. In Figure 18.9, the O_2 profiles at Barrow, Alaska, are compared with chlorophyll a (Chl$a$) and phaeopigments, O_2/Ar and O_2/N_2. The O_2/Ar and O_2/N_2 in the atmosphere (O_2/Ar = 22.50 and O_2/N_2 = 0.27) and in seawater (O_2/Ar = 13.65 and O_2/N_2 = 0.51) were used as reference ratios indicating the range of values in sea ice when the gas content is controlled only by physical–chemical processes (concentration–dilution, degassing and solvation). These values and the range they constrain for each ratio are referred to as abiotic. The available O_2/N_2 observations prior to the growing season of sympagic microalgae (0.27–0.41; Figure 18.9) are within the abiotic range and are consistent with those obtained from sea ice growth experiments in abiotic conditions (0.37–0.45; Killawee et al., 1998), or with negligible bacterial activity (0.32–0.44; Tison et al., 2002). During the growing season (BRW10 in Figure 18.9), the O_2/N_2 increased beyond the abiotic range to maximum values (1.47) as a result of NCP production of O_2. The O_2/Ar values show the same trend

(Figure 18.9) but with a better sensitivity and smaller uncertainty as demonstrated by Zhou et al. (2014b).

A quantitative estimate of the NCP in sea ice can be derived from the equations used in seawater studies, namely the deviation of O_2/Ar from saturation [$\Delta(O_2/Ar)$; equation 18.3], also referred to as biological O_2 super-saturation (Castro-Morales et al., 2013) and the O_2 concentration associated with the *in situ* NCP ([O_2]$_{bio}$) (equation 18.4):

$$\Delta(O_2/Ar) = \frac{[O_2]/[Ar]}{[O_2]_{eq}/[Ar]_{eq}} - 1 \quad (18.3)$$

$$[O_2]_{bio} = [O_2]_{eq}\Delta(O_2/Ar), \quad (18.4)$$

In the above equations, [O_2]$_{eq}$ and [Ar]$_{eq}$ are, respectively, the equilibrium concentrations of O_2 and Ar in ice at air–brine saturation. Multiplying the [O_2]$_{bio}$ obtained from equation (18.4) (in mol O_2 L_{ice}^{-1}) by the thickness of the sampled ice layers yields an estimate of the depth-integrated [O_2]$_{bio}$ (in mol O_2 m^{-2}). Assuming $O_2/C = 1.43$ for the biological production (Glud et al., 2002), an estimate of NCP can be obtained (in g C m^{-2}). In sea ice, however, the background gas concentration, on which changes due to biological activity are superimposed, may be higher than the *in situ* gas solubility expressed in the [O_2]$_{eq}$/[Ar]$_{eq}$ in the above equations. This is due to the effect of the

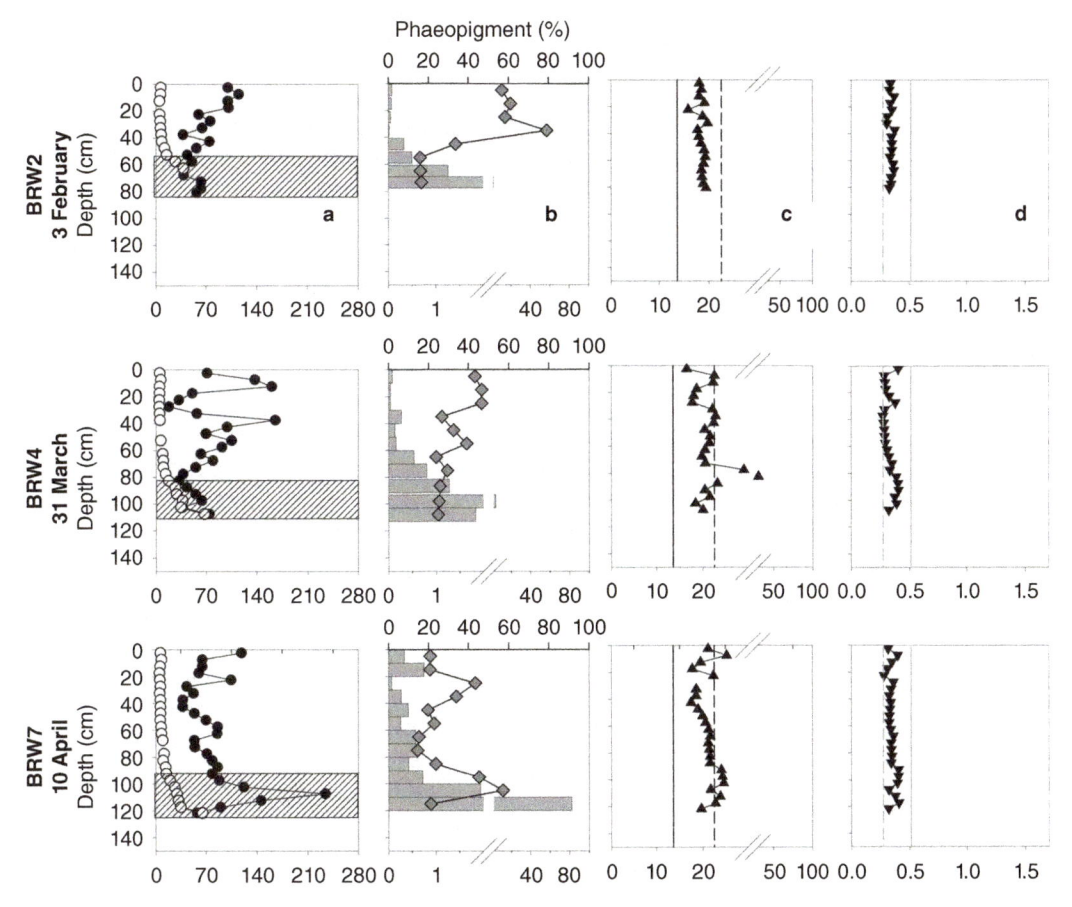

Figure 18.9 (a) Molecular oxygen concentration ([O_2]) in bulk sea ice (filled circles) and the O_2 solubility limit set by the porosity-weighted concentration in the brine at air-brine equilibrium and *in situ* ice temperature (T) and brine salinity (S) (open circles). The hatched areas refer to brine-permeable sea ice layers (brine volume fraction > 5%). (b) Chlorophyll *a* (Chl*a*) concentrations (horizontal bars) with a break at 2 μg L_{ice}^{-1} and percentage of phaeopigments (diamonds). (c) O_2/Ar in bulk sea ice (triangles) with a break at 25, in seawater (vertical solid line) and in the atmosphere (vertical dashed line). (d) O_2/N_2 in bulk sea ice (triangles), in seawater (vertical solid line) and in the atmosphere (vertical dashed line) (data from Zhou et al., 2014b).

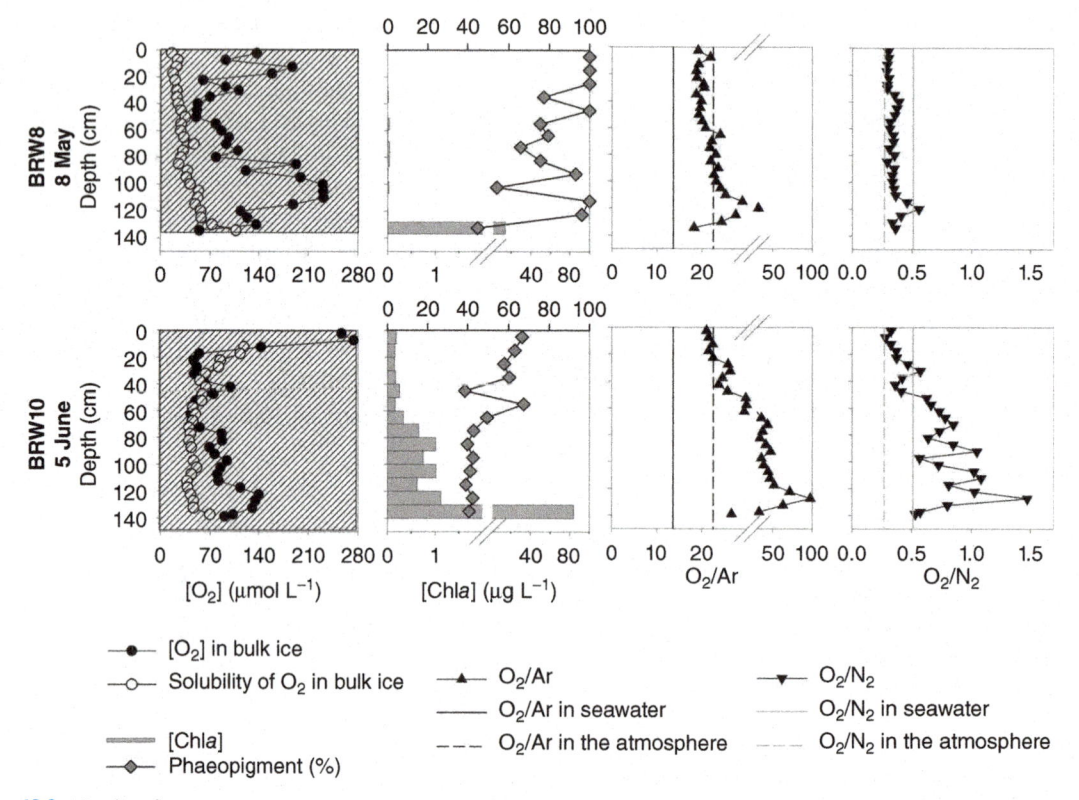

Figure 18.9 (*Continued*)

physical–chemical processes on gas concentration in the sea ice system discussed earlier in this section. For example, Ar super-saturation approached 500% for a brine volume fraction of 5% (transition from permeable to impermeable layers at Barrow in 2009) in comparison with the 1% of super-saturation reported in seawater studies (Hamme & Severinghaus, 2007). This is a major challenge, to which Zhou et al. (2014b) only give preliminary answers.

Microelectrodes have enabled non-destructive measurements of the O_2 distribution within loose layers of microalgae just beneath the ice cover and within the lowermost millimetre of the skeletal layer of sea ice and have allowed the elucidation of the O_2 saturation state of the brine (Figure 18.10). Because of their delicate design, microelectrodes have never been deployed directly into brine channels. As underlined by Mock et al. (2002), these instruments have limitations related to the measuring principle of the sensors, their stability and their construction (Klimant et al. 1997). Unlike electrodes, optodes do not consume O_2 (Klimant et al., 1995) and are also able to measure gaseous O_2 but have their own constraints, as they have to be frozen into the sea ice cover at the onset of the freezing period. Also, it is never certain where the sensor tip is located within the ice–bubble–brine system (Figure 18.11a). Comparison of O_2 measurements made with optodes and gas chromatography in twin cores (C. Krembs and J.-L. Tison, unpubl. data) has shown that optodes systematically underestimated the O_2 concentration in bulk sea ice past the solubility limit of the system (ca. 50 μmol O_2 L_{ice}^{-1}; Figure 18.11b). Optode-derived O_2 profiles from the INTERICE III experiment at HSVA (Hamburg, Germany) were promising (Figure 18.11c; Mock et al., 2002). The difference between the total and dissolved O_2 clearly demonstrates the dominance of the gaseous phase in growing young sea ice. Discrepancies were noted, however, between the optode measurements and measurements in brines using the Winkler titration method. Finally, the use of O_2 eddy correlation systems at the ice–water interface (Long et al., 2012) is a recent approach that allows the integrated NCP

Gases in sea ice 449

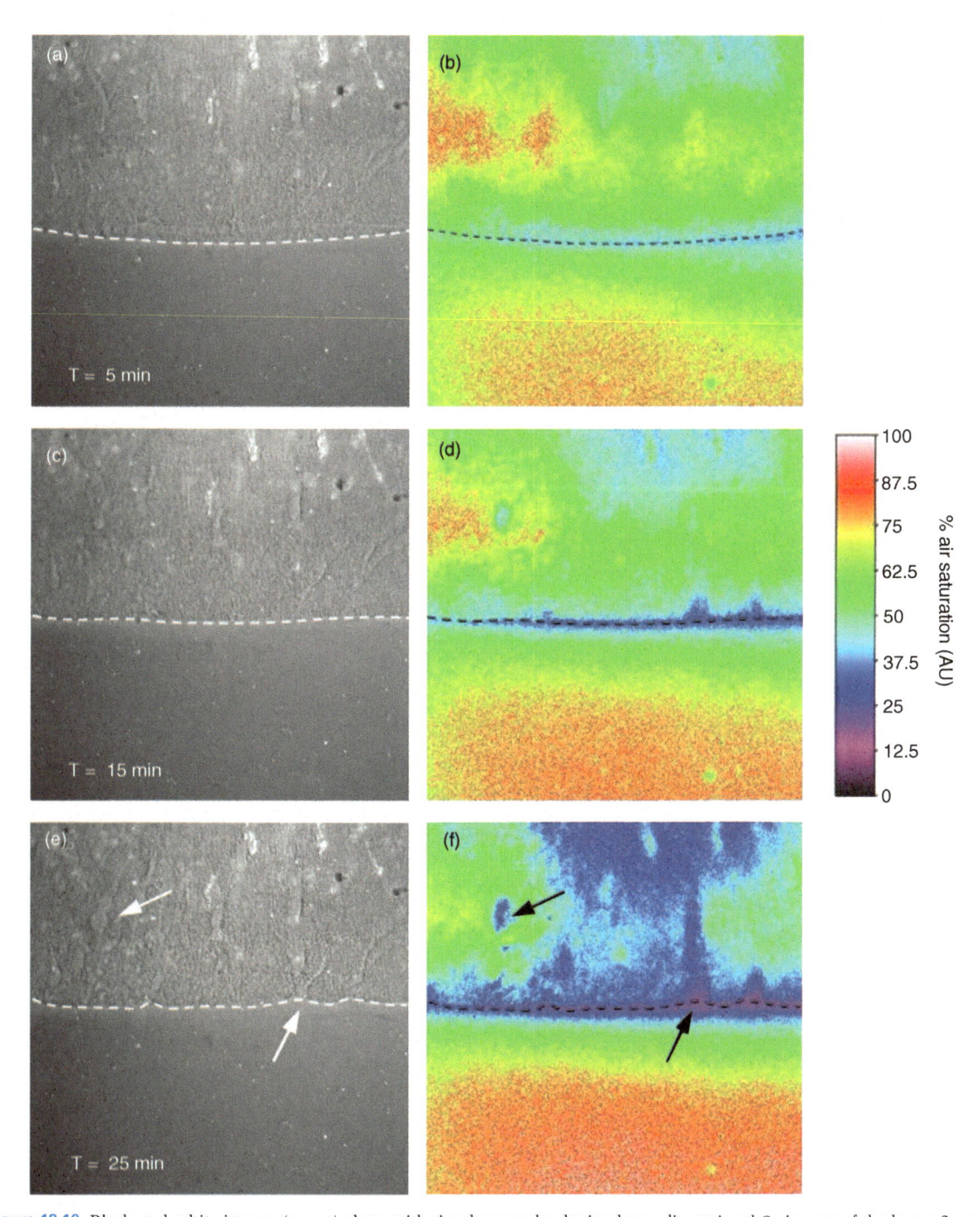

Figure 18.10 Black-and-white images (a, c, e) along with simultaneously obtained two-dimensional O_2 images of the lower 3 cm of a sea ice core during ice melt. The photos represent a time series after the temperature of the cold plate was increased from −18 to 0°C (at time 0 min). The heterogeneous distribution of O_2 within the sea ice matrix is apparent, and gradual thawing induces O_2 depletion in brine enclosures and channels (see arrows). Because the sensor was temperature-sensitive and there was no information on the micro-scale distribution of temperature, the O_2 images cannot be fully calibrated. An offset of 1°C per °C would lead to a 2.4% overestimation of the O_2 saturation. Scale: images across are 45 mm. Source: From Rysgaard et al., 2008. Reproduced with permission from Springer Science + Business Media.

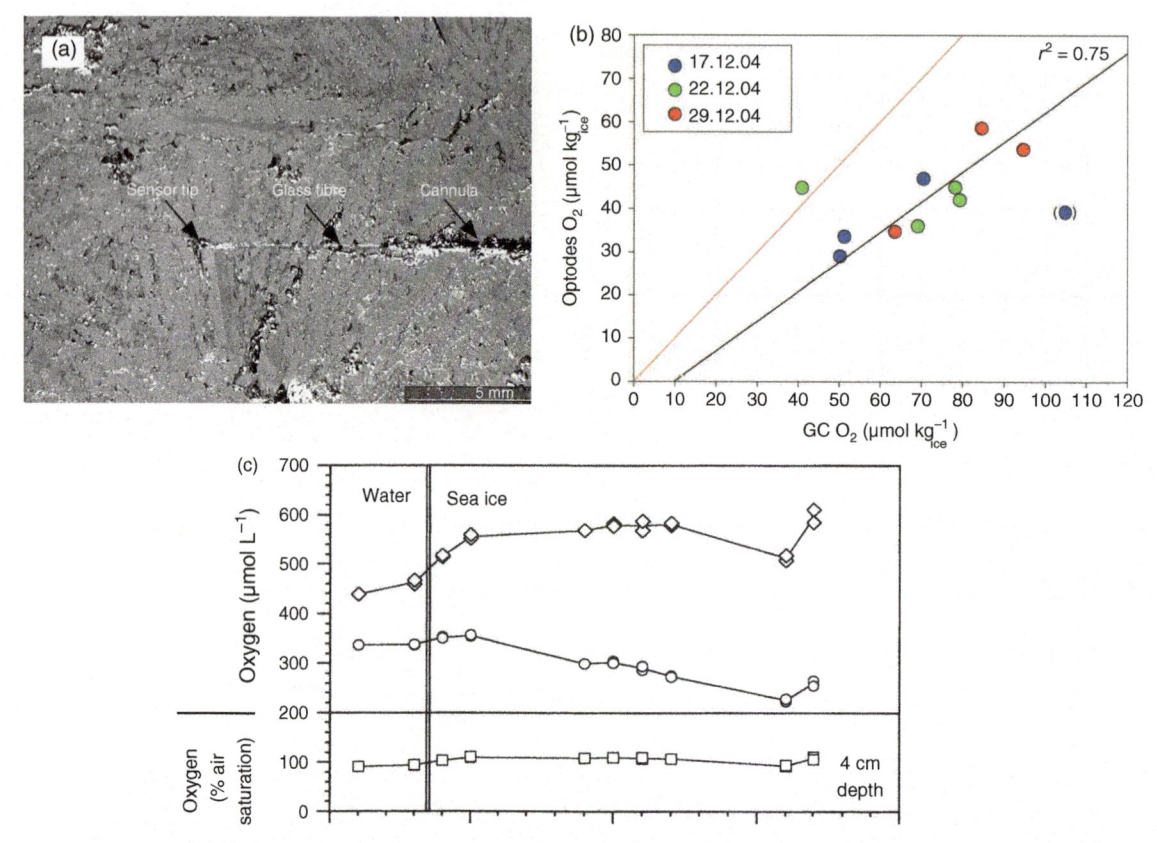

Figure 18.11 Measurements of molecular oxygen concentrations in sea ice with optodes: (a) setting of an optode in sea ice (Mock et al., 2002); (b) comparison of optode and gas chromatography (GC) measurements in twin cores during the Ice Station Polarstern (ISPOL) study in the western Weddell Sea, Antarctica (November–December 2004), with the red line representing the 1:1 relationship and the bracketed observation representing potential contamination from an air crack in the ice. The optode measurements were corrected for temperature but not for salinity, to potentially represent both dissolved and gaseous oxygen (Mock et al., 2002; A. Krell and J.-L. Tison, unpubl. data); (c) temporal change in O_2 as percentage air saturation (squares), dissolved oxygen concentration (circles) and total oxygen concentration (diamonds) in water and at 4 cm depth in artificial sea ice during the Interice III tank experiment at HSVA in Hamburg, Germany. Source: From Mock et al., 2002. Used under CC-BY-3.0 http://creativecommons.org/licenses/by/3.0/.

to be measured *in situ* for long periods (given that no strong stratification develops under the ice) and with a larger footprint than previous methods. However, this method only captures fluxes of dissolved O_2 and does not account for potential fluxes of gaseous O_2, as does the method by Mock et al. (2002).

18.3.4 Carbon dioxide and the carbonate system: the all-time favourites

Carbon dioxide in sea ice has been determined using several techniques in both the brine inclusions and the bulk sea ice (Table 18.2). It has been quite common to determine the dissolved CO_2 in sea ice brines recovered from sackholes either as a solute [CO_2(aq)] or as equivalent gas partial pressure (pCO_2) or equivalent (non-ideal) gas fugacity (f_{CO_2}) (for definitions, see equations 18.1 and 18.2). This has been routinely done indirectly by solving the equations that describe the thermodynamic equilibria of the oceanic CO_2 system using as input values direct measurements of the total dissolved inorganic carbon (DIC) and another parameter of the CO_2 system confidently measurable in sea ice brines, typically total alkalinity (TA) (see below for details). Other measurement techniques have emerged in recent years, especially in order to overcome the computation limitations of the indirect technique and the sample resolution limitation of the sackhole sampling technique. These techniques are brine equilibration with SIES (Delille et al., 2007; Geilfus et al., 2012a; Delille et al., 2014), peepers (Miller et al., 2011) and whole sea ice segment equilibration with a gas mixture of known CO_2 content (Geilfus et al., 2012b; Crabeck et al., 2014b).

The indirect technique for the determination of dissolved CO_2 in sea ice brines exploits the thermodynamics of CO_2 interaction with water. CO_2 is subject to fast ionization upon solvation in water. The solvation and its reverse process of degassing and the ionization of CO_2 in aqueous media are all controlled by fast equilibrium reactions amongst the gaseous CO_2 [$CO_2(g)$], the dissolved CO_2 [$CO_2(aq)$] and the bicarbonate (HCO_3^-) and carbonate (CO_3^{2-}) ions that result from the rapid dissociation of carbonic acid (H_2CO_3), the reaction product of dissolved CO_2 with water. The overall equilibrium reaction pathway is described by the equation (Millero, 1995):

$$CO_2(g) \leftrightarrow CO_2(aq) \qquad (18.5a)$$

$$CO_2(aq) + H_2O \leftrightarrow H_2CO_3 \leftrightarrow H^+ + HCO_3^- \qquad (18.5b)$$

$$HCO_3^- \leftrightarrow H^+ + CO_3^{2-} \qquad (18.5c)$$

These reactions are characterized by stoichiometric equilibrium constants that depend on salinity and temperature, including K_0 (equation 18.2) which describes the equilibrium between $CO_2(g)$ and $CO_2(aq)$ in equation (18.5a). The first and second dissociation constants of H_2CO_3 that describe the ionization steps in equation (18.5c) have been characterized as functions of these state variables for estuarine waters, sea water and mildly hypersaline oceanic waters ($S \leq 50$) and above-zero temperatures (Dickson & Millero, 1987; Millero, 1995; Millero et al., 2006). The sum of the dissolved CO_2 species is the parameter measured as DIC reliably and most frequently in oceanographic and, specifically, sea ice studies.

The ionized species of carbonic acid are also central in the charge balance of aquatic systems as electrolyte solutions, determined by potentiometric acid titration as TA. Total alkalinity, therefore, includes all known weak acids and bases in the oceanic environment, i.e. the carbonic, boric, silicic and phosphoric acids, ammonium and sulphide, each governed by equilibrium dissociation reactions and state variable functions for their equilibrium constants as outlined above for H_2CO_3 and collectively referred to as the oceanic CO_2 system (Wolf-Gladrow et al., 2007). TA and DIC are two of the four parameters of the CO_2 system that can be measured reliably in standard oceanographic temperature and salinity conditions, the other two being pH and f_{CO_2}. The determination of any two of the four measurable parameters (DIC, TA, pH and f_{CO_2}) of the CO_2 system is adequate for the complete determination of its components via iterative solution of the system of the equilibrium equations that describe it. Application of this technique has been central in the investigation of the CO_2 system in sea ice brines until recently (Gleitz et al., 1995; Papadimitriou et al., 2004; Delille et al., 2007; Papadimitriou et al., 2007) but only gives first-order estimates of the concentration of dissolved CO_2 and its ionic species in sea ice brines because it requires extrapolation of the existing oceanographic functions to the hypersaline and sub-zero temperature conditions of the brine–ice system with large uncertainties in the output values of the individual components of the sea ice CO_2 system (Brown et al., 2014). The result of these indirect determinations of CO_2 in sea ice brines have indicated an extremely variable parameter in space and time, ranging from the severely depleted CO_2 of sackhole brines in summer sea ice (Gleitz et al. 1995; Delille et al., 2007; Papadimitriou et al., 2007) to the highly concentrated values in late winter and early spring sea ice brines (from the data set in Papadimitriou et al., 2012; Figure 18.12a).

As with all seawater solutes, the CO_2 in the sea ice will be affected by physical concentration–dilution in the brine–ice system during the sea ice growth–decay cycle, further modulated by the change in the solubility of CO_2 and the equilibria in the CO_2 system with the temperature and salinity changes in the system. The net result of the influence of these fundamental physical–chemical reactions is shown in Figure 18.12, with physical concentration resulting in brine f_{CO_2} increase in a closed system with respect to gas exchange and vice versa for dilution. Degassing during sea ice formation and growth will tend to depress the brine f_{CO_2} towards local atmospheric equilibrium. In addition to the physical–chemical processes, the CO_2 in sea ice is influenced strongly by biological activity, with net CO_2 drawdown by primary production resulting in brine f_{CO_2} depression (Gleitz et al., 1995; Kennedy et al., 2002; Delille et al., 2007; Papadimitriou et al., 2007; Geilfus et al., 2012a; Delille et al. 2014) and net CO_2

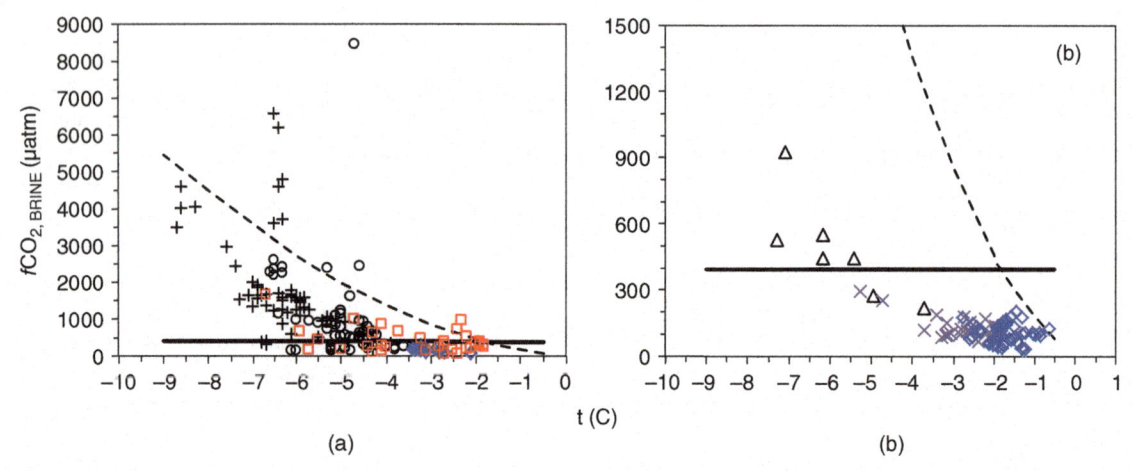

Figure 18.12 The fugacity of carbon dioxide (fCO_2) in sackhole brines In Antarctic sea ice as a function of brine temperature. The data in panel (a) are indirect determinations and were computed at in situ salinity and temperature from DIC and TA measurements for the ISPOL (December 2004, diamonds) and WWOS studies (September 2006, circles; October 2006, crosses), and from the CO_2(aq) data derived from pH and TA measurements in Gleitz et al. (1995) (April – May 1992, squares). The data in panel (b) are from direct measurements [Aurora, September – October 2003 (triangles); SIMBA, September – October 2007, (x); ISPOL, December 2004 (diamonds)]. Note the 6-fold reduction in the y-axis scale in panel (b). The WWOS measurements have been reported in Papadimitriou et al. (2012), while the ISPOL measurements have been reported in Papadimitriou et al. (2007) and Delille et al. (2014) (direct fCO_2 measurements). Delille et al. (2014) have also reported the direct fCO_2 measurements obtained during the SIMBA and Aurora studies. The solid line in panels (a) and (b) indicates the trend in fCO_2 in brine at atmospheric equilibrium, computed as inPierrotet at. (2009) for water-vapour-saturated air above brine at the freezing point and 1 atm total pressure based on the annual mean atmospheric mole fraction of CO_2 in dry air for 2014 (www.esrl.noaa.gov/gmd/ccgg/trends/). The dashed line in panels (a) and (b) indicates the effect of physical concentration and dilution on the brine fCO_2, determined by solving the CO_2 system for conservative (with respect to brine salinity) DIC and TA in a closed sea ice system resulting from the freezing of surface sea water (SSW) of salinity 35 with $[TA]_{SSW}$ = 2369 μmol kg^{-1}, and $[DIC]_{SSW}$ = 2264 μmol kg^{-1} (Papadimitriou et al., 2012). A linear salinity dependence was assumed for the concentration in the brine of TA, DIC, and the boric, phosphoric, and silicic acids relative to surface seawater, while the oceanographic salinity-temperature functions that describe the equilibria of the marine CO_2 system were extrapolated to below-zero temperatures and high salinities. Specifically for the first and second dissociation constants of carbonic acid, the salinity-temperature functions of Dickson and Millero (1987) were used from their refit of the Mehrbach et al. (1973) data. More details of this type of computation are given in Papadimitriou et al. (2014).The salinity-temperature relationship in sea ice brines (see Chapter 18.1) was used in in the determination of both curves.

production by respiration by all sea ice biota resulting in brine f_{CO_2} increase. Finally, the CO_2 is intimately linked to the solid $CaCO_3$ precipitation–dissolution cycle in sea ice brines (Papadimitriou et al., 2013, 2014; Delille et al., 2014), which is dominated by the hexahydrate polymorph of $CaCO_3$, ikaite (Dieckmann et al., 2008; Fischer et al., 2013). Carbon mass balance from DIC, $\delta^{13}C_{DIC}$, TA and inorganic nutrients has indicated the importance of primary production, $CaCO_3$ precipitation and degassing in the CO_2 system in cold brines from both natural and artificial sea ice (Papadimitriou et al., 2004, 2012; Munro et al., 2010; Søgaard et al., 2013; Delille et al., 2014; Moreau et al., 2015).

Direct measurements of brine f_{CO_2} have begun to emerge in the last decade (Delille et al., 2007, 2014; Geilfus et al., 2012b; Miller et al., 2014) with the standard oceanographic technique of equilibration of a volume of brine with a closed gas loop and non-dispersive infrared detection of CO_2 in the gas loop (Pierrot et al., 2009; Delille et al., 2007) (Figure 18.12b), while Miller et al. (2011) used silicone exchange chambers (Owens, 2008). In the latter technique, gas-permeable silicon chambers (peepers) were frozen into sea ice. The silicon membrane of the peepers is permeable to CO_2, allowing CO_2 equilibrium between the interior of the chamber and the surrounding sea ice. Sampling was conducted via tubes connecting the chambers to the ice surface and the pCO_2 of the chamber was measured by gas chromatography (Miller et al., 2011) or non-dispersive infrared detection (Loose et al., 2011b). Geilfus et al. (2012b) proposed a technique to measure f_{CO_2} in bulk sea ice in order to achieve high vertical resolution profiles. While bulk sea ice and direct brine f_{CO_2} show similar vertical and seasonal distribution patterns, there are some differences in the results of these two techniques (Geilfus et al. 2014) due to the fact that direct brine f_{CO_2} measurements derive only from the mobile brines that can seep through the inter-linked brine channels exposed to the sackhole, while bulk sea ice f_{CO_2} also includes the CO_2 in gas bubbles and in brines located in pockets isolated from the channel network and which are thus immune to transport. Brine f_{CO_2} is therefore more affected by physical processes, such as change in atmospheric temperature and transport, than bulk f_{CO_2} (Geilfus et al., 2014). The high vertical resolution offered by bulk sea ice f_{CO_2} measurements also allows robust assessment of the air–ice f_{CO_2} gradient that can drive air–ice CO_2 fluxes, as outlined in the relevant section 18.5.

18.3.5 Methane: the rising star

Methane (CH_4) is the second largest, atmospherically well-mixed greenhouse gas contributor to global warming (IPCC, 2013). The global ocean currently accounts for 3% of the global tropospheric CH_4 input (Kirschke et al., 2013), but this contribution could increase considering the potential CH_4 emission from the destabilized 'Arctic carbon hyper pool' associated with climate warming (Semiletov et al., 2012). The Arctic carbon hyper pool refers to the huge content of organic carbon buried inland and within the sedimentary basin of the Arctic Ocean (Semiletov et al., 2012). The latter includes pools such as seabed sediments, subsea permafrost, taliks (e.g. thaw bulbs) and gas hydrates (Shakhova et al., 2010; Semiletov et al., 2012). Thawing subsea permafrost promotes the development of taliks, which are favourable environments for modern methanogenesis and also provide migration pathways for CH_4 from seabed deposits (Semiletov et al., 2012). The produced CH_4 could then be transported from destabilized seabed sediments and gas hydrates in the seabed to the ocean surface, resulting in high CH_4 concentrations (super-saturation) in the oceanic waters [e.g. Shakhova et al. (2010) for the East Siberian Arctic Shelf (ESAS)].

High CH_4 concentrations in surface waters and the related release of CH_4 to the atmosphere are not limited to the ESAS (Kort et al., 2012; He et al., 2013 and references therein) and so this phenomenon is not a general feature over the continental shelf of the Arctic Ocean alone. Away from shelves, super-saturation of CH_4 was also observed in the oceanic waters of the central Arctic and was ascribed to aerobic production in the water column via degradation of methylated compounds, such as dimethylsulfoniopropionate (DMSP, $C_5H_{10}O_2S$; Damm et al., 2010). This suggests that riverine input, respiration and seepage from sediments are not the only sources of the Arctic Ocean CH_4 load. As a result, the role of sea ice in the fluxes of CH_4 in the Arctic Ocean deserves attention. Sea ice is not only a DMSP-rich environment (see below) but also a shield that prevents rapid gas escape from the water column. In addition, CH_4 oxidation can be enhanced in under-ice sea water or in sea ice. However, to date, only three studies have reported CH_4 concentrations in sea ice (Crabeck et al., 2014b; Shakhova et al., 2010; Zhou et al., 2014b) and only two studies have reported air–ice CH_4 fluxes (He et al., 2013; Kort et al., 2012). Shakhova et al. (2010) observed strikingly large bubbles with up to 11 400 parts per million by volume (ppmv) CH_4 in sea ice over the ESAS. Such bubbles are an indication of significant ebullitive fluxes of CH_4 from sediments. On the other hand, Zhou et al. (2014b) and Crabeck et al. (2014b) reported much lower concentrations, below 105 ppmv, in landfast ice over the Chukchi shelf and in a west Greenland fjord, respectively. The CH_4 concentrations reported by Zhou et al. (2014b) and Crabeck et al. (2014b) were of the same order of magnitude, ranging from 1.8 to 17.2 nmol L_{ice}^{-1}, and were generally lower than the concentration in under-ice sea water. However, the CH_4 concentrations in the brine were always above the atmospheric concentration of methane, so that sea ice could act as a source of CH_4 to the atmosphere. Nevertheless, He et al. (2013) observed both positive (from the ice to the atmosphere) and negative CH_4 fluxes in summer. They proposed that CH_4 consumption via photochemical and biochemical oxidation in sea ice might explain the negative fluxes. However, the comprehensive investigation by Zhou et al. (2014b) of the temporal change of the CH_4 concentration in sea ice from ice formation to ice melt suggested that the impact of sympagic biological activity on CH_4 is rather minor compared with the impact of the physical processes in sea ice related to the seasonal salinity and temperature changes in the system. Bubbles were present in the Chukchi Sea but were much smaller than over the ESAS and Zhou et al. (2014b) proposed that these bubbles should have formed within the sea ice as a result of solubility changes rather than being transported from the underlying sediments. In either case, bubbles promote vertical migration of CH_4 through the ice and subsequent release to the atmosphere when sea ice permeability increases.

18.3.6 Nitrous oxide: the great unknown

Nitrous oxide (N_2O) is a long-lived greenhouse gas naturally present in the atmosphere, with rising concentration in the industrial era. Nitrous oxide has a global warming potential (GWP) of around 300 (Forster et al., 2007) in comparison with a GWP for CO_2 of 1, contributing 6% to radiative forcing of the anthropogenic long-lived greenhouse gases (IPCC, 2013). Nitrous oxide is also described as the dominant ozone-depleting substance emitted in the 21st century (Ravishankara et al., 2009).

There are still large uncertainties and gaps to date in the understanding of the cycle of this compound in sea ice. Very few studies present N_2O measurements in sea ice (Gosink, 1980; Randall et al., 2012). Randall et al. (2012) reported N_2O concentrations in bulk sea ice ranging from 3 to 8 nM in the bottom 10 cm of the ice and noted that these concentrations were higher than would be expected from conservative physical concentration of the N_2O dissolved in seawater, suggesting internal biological N_2O production. The main chemical processes involved in the N_2O cycle in the aquatic environment are denitrification and nitrification. Denitrification reduces fixed inorganic nitrogen compounds to N_2 via an N_2O reaction step and can thus act as a sink or a source of N_2O depending on the O_2 concentration. In strictly anaerobic conditions, N_2O is removed during denitrification. However, denitrification can also occur in the presence of O_2 at trace concentrations (<0.2 mg L^{-1}; Seitzinger et al., 2006), and, in these conditions, there is a large net N_2O production. Kaartokallio (2001) and Rysgaard et al. (2008) reported the occurrence of denitrification in O_2-depleted microenvironments in Arctic sea ice. However, large concentrations of oxygen, and even hyperoxia, in sea ice brines have also regularly been reported, as mentioned earlier (Crabeck et al., 2014a; Delille et al., 2007; Gleitz et al., 1995; Papadimitriou et al., 2007; Zhou et al., 2014a). Nitrous oxide is also a by-product of nitrification during which ammonium is oxidized to nitrate by nitrifying bacteria. It has been estimated that nitrification supplies up to 70% of the nitrate assimilated by sympagic microalgae in Antarctic sea ice (Fripiat et al., 2014; Priscu et al., 1990). Hence, the production of N_2O during nitrification in productive sea ice habitats can be significant.

Taking into account N_2O saturation, Randall et al. (2012) suggested that Arctic sea ice formation and melt have the potential to generate sea-to-air and air-to-sea N_2O fluxes, respectively. Furthermore, because of the N_2O super-saturation of its brine inclusions (Randall et al., 2012), Arctic sea ice should act as a source of N_2O to the atmosphere during the period between the freezing and melting phases of the system. This is consistent with the enrichment of N_2O observed in the subnivean over landfast ice in the Antarctic by Gosink (1980), which indicated N_2O transfer from the ice to the atmosphere.

18.3.7 Dimethylsulphide: a key to the global sulphur cycle?

Dimethylsulphide (DMS, $C_2H_6S_2$) is a gaseous species derived from DMSP (see earlier), an intracellular metabolite synthesized by macro- and microalgae (including the sympagic algae). DMSP has been suggested to serve as an osmoregulator and cryoprotectant (Karsten et al., 1996; Stefels, 2000), an antioxidant (Sunda et al., 2002) and, potentially, an alternative to N-containing osmolytes under nitrogen limitation. DMS has attracted much attention since the late 1980s, because of its potential role in feedbacks of the global climate (the CLAW hypothesis; Charlson et al., 1987). Indeed, after ventilation to the atmosphere, DMS is quickly oxidized to various sulphur-containing compounds, including SO_2. The latter can be further oxidized to sulphuric acid, which can then nucleate or condense on existing particles to produce sulphate aerosols. The sulphate aerosols alter the radiative properties of the atmosphere directly by scattering of the incoming solar radiation, and indirectly by forming cloud condensation nuclei that increase the reflectivity of clouds. Although the global impact of DMS release to the atmosphere has been seriously questioned in recent years (Quinn & Bates, 2011), the current thinking is that it might still play a crucial role in climate regulation in the polar regions (Sharma et al., 1999, Krüger & Grassl, 2011, Chang et al., 2011; Lana et al., 2012; Levasseur, 2013; Tunved et al., 2013).

The oceanic DMSP cycle is extremely complex, involving both particulate and dissolved DMSP, DMS and dimethylsulphoxide (DMSO, C_2H_6OS), with two-way conversions mediated by algae, bacteria, grazers and viral infections (reviewed by Stefels et al., 2007). It is therefore still debated which fraction of the DMSP initially produced in the ocean ends up as climate-active atmospheric DMS.

The concentration of DMSP within sea ice organisms can be orders of magnitude higher than that measured in the open ocean water, which is typically up to a few nanomolar in the Southern Ocean and in the under-ice seawater (Kirst et al., 1991; DiTullio et al., 1998; Trevena et al., 2000, 2003; Trevena & Jones, 2006; Tison et al., 2010), with a World Ocean mean [DMSP] of 43 nM (Kettle et al., 1999). Typically, DMSP and related compounds (DMS, DMSO) in sea ice are measured in melted bulk ice segments. However, ice melting, even in filtered seawater, results in salinity stress, which potentially affects the physiology of sympagic algae and may even cause cell lysis (Garrison & Buck, 1986; cf. Thomas et al., 1998). Upon cell lysis, intracellular metabolites, such as DMSP, are released from the cell, with the released cellular DMSP potentially degraded to DMS by exo-enzymes (Brabant et al., 2011). Therefore, some investigators have adopted the dry-crushing protocol for the analysis of DMS, DMSP and DMSO in sea ice (Tison et al., 2010; Stefels et al., 2012). Stefels et al. (2012) provide a clear illustration of the potential bias in this protocol (Figure 18.13) by comparing measurements obtained with the two techniques from twin samples from the same core and conclude that, if relative contributions of DMS, DMSP and, by extension, DMSO are to be discussed using the ice-melting technique, the measurements need to be complemented by a sulphur isotopic analysis to trace the relative importance of conversion pathways. Due to these complications, until very recently, most of the sea ice publications only reported the concentration of the combined DMSP and DMS pool. Further, due to the difficulties in separating the particulate DMSP (DMSPp) from the dissolved DMSP (DMSPd), total DMSP (DMSPt) is generally reported. Table 18.3 gives a summary of the concentration range and mean DMS, DMSP and DMSO, combining all relevant field studies in the Arctic and the Antarctic to date. These numbers should, however, be considered with caution, as they integrate very different vertical sampling resolutions and periods of the year.

With the exception of the Circumpolar Flaw Lead Study – International Polar Year (CFL-IPY; Carnat, 2014; see later), mostly the lowermost 2–3 cm of ice cores have been investigated for DMSP and its derivatives in Arctic sea ice studies. Record bulk sea ice DMSPt concentrations between 12 000 and 15 000 nM were detected in these layers by Levasseur et al. (1994), Uzuka (2003) and Galindo et al. (2014). Galindo et al. (2014) recently provided the only available DMSPd concentrations in sea ice to date, ranging widely from 0.6 to 6110 nM. Comparing these concentrations with those in the under-ice sea water, these investigators estimated that the under-ice DMSPd pool represented

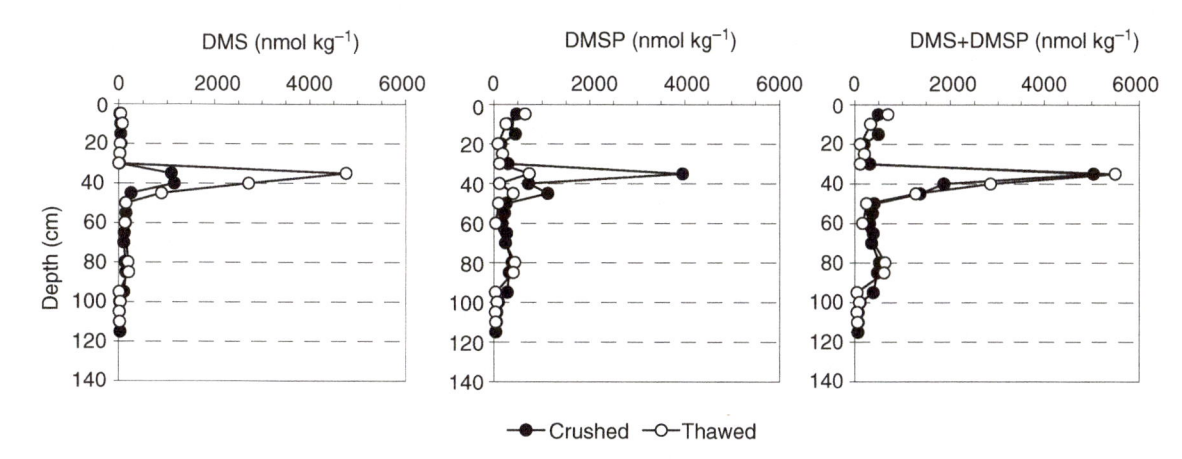

Figure 18.13 Comparison of crushing and thawing of ice samples taken from the same core (sampled on 14 November 2004) after 12 days of storage at −30°C: (A) DMS profile, (B) DMSP profile, (C) DMS+DMSP profile. Source: From Stefels et al., 2012. Reproduced with permission from Elsevier.

Table 18.3 Concentration range and mean of dimethylsulphide (DMS), dimethylsulfoniopropionate (DMSP) and dimethylsulphoxide (DMSO) in Arctic and Antarctic sea ice. The values correspond to different seasons and depths in sea ice (see original references for details).

	DMS (nM) Arctic	DMS (nM) Antarctic	DMSP (nM) Arctic	DMSP (nM) Antarctic	DMSO (nM) Arctic	DMSO (nM) Antarctic
Minimum	<0.3	<0.3–0.5	<0.3–500 (DMSPp = 9.8, DMSPd = 0.6)	<0.3–81	1.3–1.9	6.0–6.6
Mean	3–14	8–58	30–325	40–409	8–13	33–78
Maximum	95 – 769	75–1,430	987–15,000 (DMSPp = 15 082, DMSPd = 6,110)	193–11 352	102–5428	289–1202

DMSPp, particulate DMSP; DMSPd, dissolved DMSP.
Data were derived from Levasseur et al. (1994), Lee et al. (2001), Uzuka (2003), Gali and Simo (2010), Levasseur (2013), Galindo et al. (2014), Galindo (2014) and Carnat (2014) for the Arctic; and from Kirst et al. (1991), Turner et al. (1995), DiTullio et al. (1998), Curran et al. (1998), Curran and Jones (2000), Trevena et al. (2000), Trevena et al. (2003), Gambaro et al. (2004), Trevena and Jones (2006), Zemmelink et al. (2008a), Tison et al. (2010) and Carnat et al. (2014) for the Antarctic.

only 13% of the DMSPd lost from the ice through brine drainage, which suggested rapid microbial consumption. Galindo (2014) has further shown that microbial consumption of DMSPd lost from the ice results in rapid conversion to DMS in the water column. The DMSPt concentrations determined at 5 cm resolution in the entire ice column throughout the sea ice growth–decay cycle during the CFL-IPY study were more modest, with a mean concentration of 30 nM and a concentration range from 0.3 to 1636 nM (Carnat, 2014). The CFL-IPY study has also provided the only comprehensive data set available to date for DMS, DMSP and DMSO in Arctic sea ice from measurements on the same samples, with mean ice core DMS concentrations between 3 and 14 nM and a range from less than 0.3 nM (detection limit) to 769 nM. Lee et al. (2001) reported the first DMSO data from Arctic sea ice, with a mean concentration of 13.7 nM and a concentration range from 1.3 to 102 nM (particulate fraction only). Carnat (2014) considerably extended the data set in the Amundsen Gulf (mean concentration, 7.9 nM in autumn–winter to 112.6 nM in spring–summer; concentration range, 1.5 – 5427.9 nM). A typical Arctic vertical profile of DMSP concentration in first-year pack ice in spring shows a small concentration peak in surface sea ice and a much larger concentration peak in bottom sea ice, with lower concentrations or no DMSP in the intervening sea ice layers. Further, the DMSP concentrations increase from close to the detection limit in March to maximum concentrations at the end of April and decrease again to near the detection limit in May–June.

There have been relatively more studies of the distribution of DMS, DMSP and DMSO in sea ice in the Antarctic, and these have typically focused chiefly on DMSPt and the spring–summer season (Kirst et al., 1991; Turner et al., 1995; Curran et al., 1998; DiTullio et al., 1998; Curran & Jones, 2000; Trevena et al., 2000; Trevena et al., 2003; Gambaro et al., 2004; Tison et al., 2010; Carnat et al., 2014; Carnat, 2014). Mean DMSP concentrations ranged between 40 and 409 nM, with an overall concentration range of less than 0.3 nM (detection limit) to 3900 nM. The concentrations of DMS in sea ice have been measured by Trevena and Jones (2006), Tison et al. (2010), Carnat et al. (2014) and Carnat (2014), with mean concentrations of between 7 and 60 nM and a cross-study concentration range of less than 0.3 to 1430 nM. Finally, using a new measurement technique (Brabant et al., 2011), Tison et al. (2010), Carnat et al. (2014) and Carnat (2014) also determined DMSO in Antarctic sea ice (mean concentration = 33–78 nM; concentration range = 6–1202 nM). These investigations showed that the relative contribution of DMSO to the total of these sulphur compounds was much lower in Antarctic than in Arctic sea ice for reasons still to be discovered. Asher et al. (2011) determined DMS (mean concentration = 33.2 nmolS L^{-1}, maximum concentration = 277 nmolS L^{-1}),

DMSP (mean concentration = 305 nmolS L^{-1}, maximum concentration = 2990 nmolS L^{-1}) and DMSO (mean concentration = 138 nmolS L^{-1}, maximum concentration = 471 nmolS L^{-1}) in sea ice brines.

The concentrations of these three sulphur species were generally highest in brines collected in the near surface ice, while the DMSPt and DMSO distribution with depth in the sea ice explained much (66%) of the vertical variability. Indeed, unlike Arctic sea ice, the vertical distribution of DMSPt and DMS concentrations does not follow a unique pattern in Antarctic sea ice. Maximum concentrations have been found either in surface ice (Gambaro et al., 2004) or in interior ice (Trevena et al., 2003), or sometimes in bottom ice (Tison et al., 2010), reflecting the primary influence of the (past or present) distribution of the sympagic community in Antarctic sea ice as opposed to their distribution in Arctic sea ice, which is dominated by bottom communities. A comparison between thin young ice and thicker first-year ice in the Antarctic also suggested that concentrations could be higher in the former. Finally, Zemmelink et al. (2008) reported DMSPt concentrations as high as 11 352 nmolS L^{-1} in the surface layer of second-year ice in early austral summer (November–December 2004) in the Weddell Sea, Antarctica.

Controls on the concentrations of DMS, DMSP and DMSO and their temporal change within the sea ice cover are manifold: community structure, temperature, salinity, light and nutrient supply. Some of these controls are illustrated in Figure 18.14, which shows the temporal change in the concentrations of DMSP and DMS within a single drifting sea ice floe investigated during the Ice Station Polarstern (ISPOL) field campaign in the western Weddell Sea, Antarctica, in December 2004 (Tison et al., 2010). During the investigation, the sea ice switched from a convective brine drainage regime to a diffusion-controlled stratified brine network (Figure 18.14a). The Chl*a* concentrations showed a small surface maximum (a few µg L^{-1}) and a larger bottom maximum (30 µg L^{-1}) during the study period (Figure 18.14b). The surface community was dominated by *Phaeocystis* sp., while the bottom community consisted of pennate diatoms (Figure 18.14c). The initial DMSP profile (Figure 18.14d) showed two maxima associated with the Chl*a* maxima. The upper and weaker maximum resulted from production by a sparse population of DMSP-efficient *Phaeocystis* sp., which has cellular DMSP to Chl*a* ratios (DMSP/Chl*a*) as high as 3200 nmol DMSP µg^{-1} Chl*a* (Trevena et al., 2002, 2003). The lower and stronger maximum resulted from a large number of much less DMSP-efficient diatoms with cellular DMSP/Chl*a* = 1.5–7.6 nmol DMSP µg^{-1} Chl*a* (Gambaro et al., 2004). This illustrates the control of the community structure on DMSP and has also been documented in Arctic sea ice (Galindo et al., 2014). Further, the decrease in DMSP with time in the upper 40 cm suggests that, as the brine salinity decreased during sea ice warming, DMSP was converted to DMS and acrylic acid by the algae, as a response to changes in the environmental conditions (primarily temperature and salinity). However, no concurrent increase in DMS (Figure 18.14) was found in the upper layer during the second half of the period. Instead, a regular DMS gradient was observed from the negligible concentrations at the surface to the concentration maxima at the bottom of the sea ice, which Tison et al. (2010) interpreted as the result of diffusive loss to the atmosphere within the now stratified and increasingly permeable brine channel system, showing again the important role of the physical parameters of the sea ice cover.

In the bottom layers, nutrient accessibility fosters primary production and, therefore, DMSP production, as well as bacterial and viral activity, with an equally large amount of DMS being produced. DMSP is indeed broken down to DMS and acrylic acid through the action of the algal and bacterial enzyme DMSP-lyase and also from grazing by protozoans and metazoans, as well as viral infection. Furthermore, DMSP breaks down to DMS in alkaline conditions, and so the metabolism-associated increase of the brine pH to values up to 10 (Gleitz et al., 1995) may facilitate this chemical DMSP conversion to DMS in productive sea ice layers. Interestingly, the DMS:DMSP ratio increased with time in the Chl*a*-rich bottom layer of the ISPOL ice floe, suggesting that degradation takes over production of DMSP as the ice decays. As seen in Figure 18.14, convective brine drainage during the first half of the study period resulted in a DMSP and DMS loss from the ice, commensurate with a gain of these two compounds in the underlying surface sea water. As underlined above, how much of this is actually released as DMS to the atmosphere still remains conjectural. Similarly, in the Arctic (Michel et al., 2002; Juul-Pedersen et al., 2008; Galindo et al., 2014), a large input of DMSPp from the sea ice into surface oceanic waters is expected

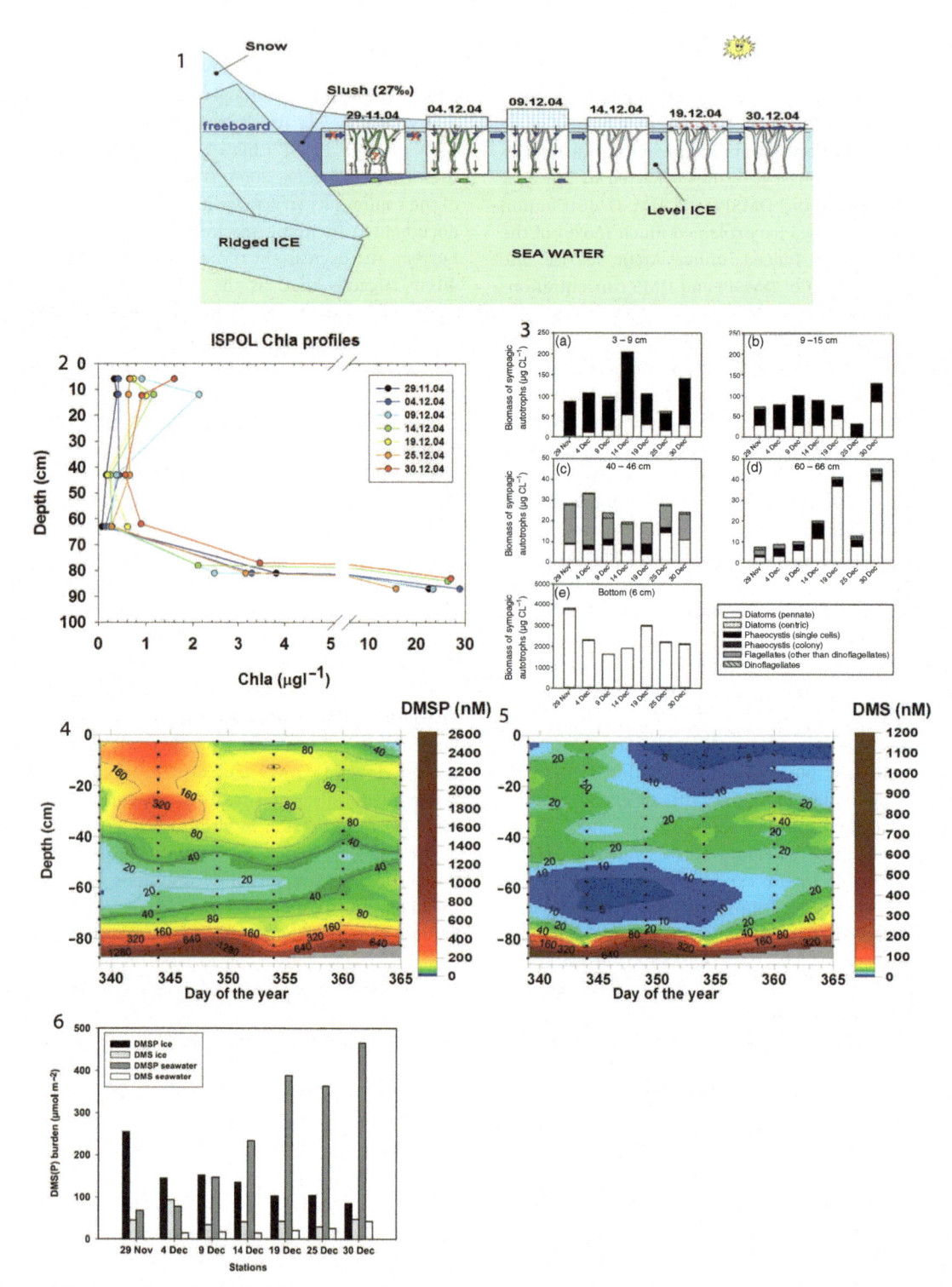

Figure 18.14 Controls on dimethylsulphide (DMS), dimethylsulfoniopropionate (DMSP) and dimethylsulphoxide (DMSO) dynamics at Ice Station Polarstern (ISPOL, December 2004, western Weddell Sea, Antarctica). (1) Schematic of the dynamics of brine; (2) chlorophyll *a* (Chl*a*) profiles; (3) biological community structure; (4) total DMSP (DMSP$_t$); (5) DMS; (6) DMSP and DMS load in sea ice and under-ice waters. Source: From Tison et al., 2010. Reproduced with permission from John Wiley & Sons.

to occur during the melt period and at ice edges. Increased heterotrophic bacterial activity and grazing at the end of ice algal blooms (Haecky & Andersson, 1999; Kaartokallio, 2004) and the fact that periods and sites of ice ablation are usually associated with high grazing activity point to an enhanced release of DMSP in surface waters and, potentially, of DMS to the atmosphere. Whether DMSP is mainly transferred from the sea ice to the surface seawater in particulate or dissolved form at the time of melt and how much bacterial DMSP-lyase activity takes place in the sea ice versus surface waters both warrant closer investigation.

Light is also an important factor, not only because it controls primary production, but also because it can affect DMSP synthesis and can foster conversions within the sea ice sulphur cycle. Hefu and Kirst (1997) showed that ultraviolet radiation significantly reduced the production of DMSP by *Phaeocystis antarctica*, a prymnesiophyte and prolific DMSP producer, while at the same time they observed an increase in the conversion rate of DMSP to DMS and acrylic acid. Recently, Carnat et al. (2014) discussed the potential conversion of DMS to DMSO under high light input in the upper layers of the sea ice cover.

18.3.8 Halogenated volatile compounds: a tight link to atmospheric chemistry

Reactive halogen species are responsible for ozone depletion events (ODEs) in the polar tropospheric boundary layer (reviewed by Simpson et al., 2007). In addition, reactive halogen species drive rapid oxidation of DMS (von Glasow & Crutzen, 2004) and elemental mercury (Lindberg et al., 2002) and change the oxidizing capacity of the troposphere by increasing the NO_2 to NO ratio and decreasing the HO_2 to OH ratio (Saiz-Lopez et al., 2007). Strong relationships between ODE and sea ice exposure have been reported (Gilman et al., 2010; Jacobi et al., 2010). Large concentrations of two reactive halogen species, bromine monoxide (BrO) and iodine monoxide (IO), have been observed over and near sea ice (Schönhardt et al., 2008, 2012). It has been proposed that sea ice, frost flowers and saline snow can affect the atmospheric halogen load in two ways: rapid release of Br_2 involved in the so-called 'bromine explosion' that is responsible for ODE events (reviewed by Abbatt et al., 2012); and release of halocarbons, the precursors of reactive halogen species (Abbatt et al., 2012; Granfors et al., 2013a,b).

As reviewed by Simpson et al. (2007), bromine explosion can be described by the following simplified reaction sequence:

$$HOBr + Br^- + H^+ \xrightarrow{mp} H_2O + Br_2 \quad (18.6a)$$

$$Br_2 + h\nu \rightarrow 2Br \quad (18.6b)$$

$$Br + O_3 \rightarrow BrO + O_2 \quad (18.6c)$$

$$BrO + HO_2 \rightarrow HOBr + O_2 \quad (18.6d)$$

Hypobromous acid (HOBr) is highly soluble and can dissolve in brine skim or in the disordered interface of sea ice, in frost flowers and in snow grains. The oxidation of dissolved bromide Br^- by HOBr produces Br_2 (equation 18.6a), which is released into the atmosphere. The reaction is autocatalytic, with the reactive halogen product acting as catalyst, further speeding up the reaction (Simpson et al., 2007). The conditions and location where the reaction (equation 18.6a) takes place are still debatable (Abbatt et al., 2012) but, according to Pratt et al. (2013), snow grains with acidic surface layer enriched in Br^- compared with Cl^- are suitable candidates, as such conditions are encountered in the topmost centimetres of snowpack over sea ice and tundra.

Sea ice contains halocarbons that are precursors of reactive halogen species. Microalgae are the main producers of halocarbons in the ocean (Saiz-Lopez et al., 2007). Sympagic microalgae appear to produce iodocarbons (CH_3I, C_2H_5I, CH_2ClI, $1-C_3H_7I$, CH_2BrI) and also $CHBrCl_2$ (Granfors et al., 2013b), while bacterial production of halocarbons has not been precluded (Granfors et al., 2013a). This biological production fuels exchange with the surrounding environment. In addition, large concentrations of abiotically produced bromocarbons, including the long-lived bromoform ($CHBr_3$) and bromodichloromethane ($CHBrCl_2$), have been observed in the top 10 cm of Antarctic ice in summer (Granfors et al., 2013b) and winter (Granfors, 2014). Sea ice production of $CHBr_3$ in winter drives a significant air–ice $CHBr_3$ flux, estimated to be 50 nmol m^{-2} $hour^{-1}$ during the polar night (Granfors, 2014).

18.4 Gas incorporation and transport in sea ice

18.4.1 Initial entrapment: Boundary layer versus mushy layer theory

A suite of physical processes affects dissolved gases in the ocean–sea-ice–air system and their effect was evident in the temporal changes of Ar concentrations in sea ice as outlined early in section 18.3.1. These processes are solute concentration and dilution via heat exchange and concomitant temperature change, gas solubility changes in concert with ice temperature and brine salinity changes, gas bubble nucleation and gaseous–aqueous phase transition, brine convection and buoyancy-driven upward gas bubble migration. The traditional approach to explain and model the temporal changes of solute profiles in sea ice (Weeks & Ackley, 1986; Eicken, 1998) combines a phase of 'initial salt entrapment', based on the boundary layer theory developed for metal solidification, with a semi-empirical relationship for ice desalination initially developed by Cox and Weeks (1988) (see Chapter 1). Although only strictly valid for a planar freezing interface, the boundary layer theory was found to give satisfactory results for the dependency of initial salt entrapment on the ice growth rate, the diffusive boundary layer thickness and the diffusion coefficients of the solutes. This theory predicts solute (including dissolved gases) enrichment of the residual liquid during freezing at the ice–water interface as a result of the balance between rejection from the ice matrix and diffusion away from the boundary layer towards the main water reservoir.

In the late 1990s, a new theoretical approach integrating both initial entrapment and desalination, known as the mushy layer theory, was put forward by several investigators (Wettlaufer et al., 1997; Notz et al., 2005, 2009; Feltham et al., 2006; Notz &Worster, 2008). This theory suggests that, during ice growth, the salinity field is continuous across the ice–ocean interface, with no immediate solute segregation at the advancing freezing front (Notz et al., 2005). This theory has been confirmed by precise *in situ* measurements of temperature and resistivity (Notz &Worster, 2008), but it is recognized that solute segregation might occur in the bottom 'skeletal layer', just above the ice–water interface (Notz & Worster, 2009). The latter could partly explain the evolution of gas ratios in that cooling permeable layer (Tison et al., 2002; Brabant, 2013). The good agreement between observed gas concentrations in bottom ice and estimates based on solubility lends some support to the mushy layer theory. However, it should be kept in mind that molecular diffusion might be active in brine channels in the permeable layers of sea ice or at the ice–water interface and also during the 'diffusion mode' periods when the potential for convective flow, as quantified by the Rayleigh number, is minimal. Also, during sea ice decay, as the brine system develops a stable density structure, with density decreasing upward, diffusion might become a dominant transport mechanism for dissolved gases.

18.4.2 Diffusivity in sea ice

Little is currently known about the diffusion coefficient of gases in sea ice. It is a challenging measurement to perform in this very heterogeneous medium, in which gases can migrate in both the gaseous and dissolved phases. Gosink et al. (1976) provided pioneering '*in situ*' experimental diffusion coefficient estimates of 0.9×10^{-5} cm^2 s^{-1} for sulphur hexafluoride (SF$_6$) and 40×10^{-5} cm^2 s^{-1} for CO$_2$ at temperatures between –12 and –5°C (Table 18.4). More than 30 years later, Loose et al. (2011a) provided the first laboratory measurements of the gas diffusion coefficients in bulk sea ice of O$_2$ (D_{O_2}) and SF$_6$ (D_{SF_6}) through columnar sea ice of constant thickness and for ice surface temperatures between –4 and –12°C. On average, D_{SF6} was 13×10^{-5} cm^2 s^{-1} (±40%) and D_{O_2} was 3.9×10^{-5} cm^2 s^{-1} (Table 18.4). These are the same order of magnitude as diffusion coefficients in seawater, although clearly higher. They also compare well with the average diffusion coefficient measured by Shaw et al. (2011) for volatile organic iodinated compounds, $D_{VOIC} = 5 \times 10^{-5}$ cm^2 s^{-1} at –3°C. Loose et al. (2011a) therefore proposed a model combining distinct gaseous and aqueous diffusive pathways through the ice, with preferential partitioning of SF6 to the gas phase explaining its faster diffusivity. Recently, taking the opportunity of the natural occurrence of a gas maximum in the decaying sea ice cover at Kapisilit (southwest Greenland), Crabeck et al. (2014a) used the diffusive decrease of that maximum over time to provide a new set of *in situ* diffusivity estimates for Ar, O$_2$ and N$_2$ (Table 18.4). These estimates are similar to those obtained by Loose et al. (2011a), but are closer to the diffusion coefficients in seawater. This is understandable as solute diffusion (as dissolved gas) should prevail in decaying sea ice, from which most of

Table 18.4 Gas diffusivity as diffusion coefficient (D) in sea ice, seawater and air (after Crabeck et al., 2014).

Study	Temperature (°C)	Brine volume	D_{Ar}	D_{O_2}	D_{N_2}	D_{SF_6}	D_{CO_2}
Gas diffusion coefficient in sea ice ($\times 10^{-5}$ cm^2 s^{-1})							
Gosink et al. (1976)	−15 to −7					0.9	40
Loose et al. (2011)	−12 to −4	6 to 8		3.9 (±41%)		13 (±40%)	
Crabeck et al.(2014a)	−5 to −2	4 to 8	1.54–1.76 (±17%)	1.55–1.74 (±9%)	2.49 (±11%)		
Gas diffusion in water (10^{-5} cm^2 s^{-1})							
Broecker & Peng (1974)	0		0.88	1.17	0.95		
Stauffer et al. (1985)	0			2.08	1.61		
Wise & Houghton (1966)	+10		1.7	1.7	1.8		
Gas diffusion in air of the order of 10^{-1} cm^2 s^{-1}							

the gaseous component as internal gas bubbles would probably have escaped already. These diffusivities should provide an adequate reference for modelling attempts to describe gas transport in sea ice (e.g. Moreau et al., 2014; Chapter 20).

18.5 Air–sea ice–ocean gas fluxes

For decades, sea ice has been considered an inert and impermeable barrier for gas exchange between the ocean and the atmosphere. However, as discussed earlier, sea ice is not an inert environment for gases; several physical and biogeochemical processes alter the concentration and drive the phase transition and movement of gas species in sea ice, which, moreover, is not an impermeable barrier as attested by recent reports of a wide range of air–ice CO_2 fluxes in ice-covered regions of the ocean. However, the processes that drive or control gas fluxes in sea ice, as well as their magnitude and significance to atmospheric and oceanic chemistry, are not fully understood.

Five different approaches are currently used to determine gas fluxes in sea ice-covered oceanic systems: measurements using tracers (SF_6, volatile organic iodinated compounds, ^{222}Rn to ^{226}Rn ratio), the chamber and eddy correlation techniques, a simple box model based on the inorganic carbon content of sea ice and a one-dimensional halo-thermodynamic sea ice model that includes gas physics and carbon biogeochemistry. All these techniques have yielded results that have led to contrasting and somewhat puzzling conclusions. Use of an intentionally introduced tracer in laboratory experiments (Loose et al., 2011a; Shaw et al., 2011) and measurement of the ^{222}Rn to ^{226}Rn ratio in under-ice water (Rutgers van der Loeff et al., 2014) have suggested that ventilation of the surface mixed oceanic layer by diffusion through sea ice may be negligible (Loose et al., 2011a) compared with air–sea fluxes in ice-free conditions (open water flux) and that diffusion of gases is weak in sea ice and across the ice–air interface. Ice–air CO_2 fluxes measured with the chamber technique over bare sea ice ranged from −5.2 (negative flux signifies gas invasion in the sea ice) to +9.9 mmol m^{-2} day^{-1} (positive flux signifies gas evasion from the sea ice; Table 18.5).

These fluxes are relatively modest compared with ice-free air–sea exchange over productive or turbulent surface oceanic waters. However, these fluxes can be significant when taking into account the areal extent of the sea ice cover in both polar oceanic regions. Chamber CO_2 flux measurements from three different cruises over Antarctic pack ice and their scaling up to the entire Antarctic ice pack by Delille et al. (2014) (Table 18.5) have suggested that Antarctic sea ice could act as a sink of 0.029 Pg C of atmospheric CO_2 in spring and summer. A CO_2 sink of this magnitude is equivalent to 58% of the oceanic CO_2 sink in the Southern Ocean. Such an assessment is supported by simple box models based on pCO_2, DIC and TA (Delille et al., 2014; Rysgaard et al., 2011; Table 18.6).

However, the atmospheric CO_2 uptake observed during ice melt is mainly driven from the CO_2 depletion

Table 18.5 Air–ice CO_2 fluxes (adapted from Geilfus, 2011). A positive flux indicates flux from sea ice to the atmosphere and vice versa for negative fluxes.

Sea ice conditions	Period of year	Location	Method	CO_2 flux (mmol m^{-2} day^{-1})	Integration period	Reference
Landfast ice (with melt ponds)	June	Barrow, Alaska	EC–open path[a]	−69 to +36	Hourly average	Semiletov et al. (2004)
Slush over pack ice	November–December	Weddell Sea	EC–open path[a]	−30 to +6.8	30 min integration	Zemmelink et al. (2006)
Landfast ice	May–June	Arctic Archipelago	EC–open path	−259 to +86	Hourly average	Papakyriakou & Miller (2011)
Landfast ice	December–June	Beaufort Sea	EC–open path	−1123 to +346	Hourly average	Miller et al. (2011)
Landfast ice	December–June	Beaufort Sea	EC–open path	−60 to +74	Daily average	Miller et al. (2011)
Grease ice	November	Beaufort Sea	EC–open path	−156	Daily average	Else et al. (2011)
Landfast ice (growth phase) First year (<40 cm)	November	Amundsen Gulf	EC–open path	20	Daily average	Else et al. (2011)
Landfast ice	December	Amundsen Gulf	EC–open path	36	3-day average	Else et al. (2011)
Pack ice	November–January	Amundsen Gulf	EC–open path	−74 to +50	Daily average	Else et al. (2011)
Ice and upwind leads	November–January	Amundsen Gulf	EC–open path	−2322 to +181	Hourly average	Else et al. (2011)
Landfast ice	April	West Greenland	EC–open path	−115 to +130	Hourly average	Sorensen et al. (2014)
Landfast ice	March–April	Barrow, Alaska	EC–closed path	−70 to +120	30 min integration	Heinesch et al. (2009)
Artificial sea ice (growth phase)			Chamber	+0.4 to +0.9	Hourly integration	Nomura et al. (2006)
Landfast ice	February–March	Hokkaido Bay Japan	Chamber	−3.6 to +1.0	40 min integration	Nomura et al. (2006)
Landfast ice First year (<40 cm)	April	Barrow, Alaska	Chamber	+4.2 to +9.9	20 min integration	Geilfus et al. (2013)
Pack ice First year	April–May	Svalbard Arctic Ocean	Chamber	−0.6 to +0.1		Nomura et al. (2013)
Landfast ice	May	Barrow, Alaska	Chamber	−1.0 to +0.7	40 min integration	Nomura et al. (2010a)
Landfast and pack ice First year	April–June	Beaufort Sea	Chamber	−1.49 to +2.10	20 min integration	Geilfus et al. (2012a)
Melt pond	June	Barrow, Alaska	Chamber	−51 to −4		Semiletov et al. (2004)
Pack ice	October	Western Antarctica	Chamber	−1.1 to +1.9	20 min integration	Delille et al. (2014)
Pack ice	October	Bellingshausen Sea	Chamber	−2.9 to −0.4	20 min integration	Delille et al. (2014)
Pack ice	November–December	Weddell Sea	Chamber	−5.2 to −1.8	20 min integration	Delille et al. (2014)
Landfast ice (multi-year)	January–February	Lützow-Holm Bay, Antarctica	Chamber	−0.3 to +0.1		Nomura et al. (2013)

[a]Not corrected for open-path CO_2 analyser at cold temperature. EC, eddy correlation.

Table 18.6 Net aerially and temporally integrated CO_2 air–ice fluxes in the Arctic and Antarctic sea-ice-covered oceanic waters (in Pg C units).

	Arctic	Antarctic	Reference
Net CO_2 air–ice flux in spring and summer derived from a simple box model involving bulk sea ice total alkalinity (TA) and dissolved inorganic carbon (DIC) without internal $CaCO_3$ cycle in sea ice (TA:DIC = 1.06)	−0.014	−0.019	Rysgaard et al. (2011)
Net CO_2 air–ice flux in spring and summer derived from a simple box model involving bulk sea ice TA and DIC with the internal $CaCO_3$ cycle in sea ice (TA:DIC = 1.80)	−0.031	−0.052	Rysgaard et al. (2011)
Net CO_2 air–ice flux in spring and summer derived from a simple box model involving TA and pCO_2 in sea ice brines		−0.024	Delille et al. (2014)
Net CO_2 air–ice flux in spring and summer derived from scaling up *in situ* chamber flux measurements using the NEMO-LIM3 model		−0.029	Delille et al. (2014)
CO_2 air–sea (open water) flux in polar oceans estimated from the air–sea pCO_2 gradient and the air–sea gas transfer rate parameterized as a function of wind speed	−0.154	−0.045	Takahashi et al. (2009)

pCO_2, partial pressure of gaseous CO_2.

in sea ice that develops in autumn and winter. Carbon dioxide is expelled from the ice to both the underlying water (like other solutes during ice growth) and the atmosphere. However, because CO_2 can be in the gaseous phase of sea ice (Crabeck et al., 2014b; Loose et al., 2011a; Zhou et al., 2013), upwards expulsion of gaseous CO_2 might potentially be higher than solute expulsion due to the buoyancy of gas bubbles. Furthermore, the expulsion of CO_2 is enhanced by precipitation of $CaCO_3$ which produces CO_2 in autumn and winter (Rysgaard et al., 2007; Delille et al., 2014). Atmospheric CO_2 uptake by sea ice in spring and summer is therefore counterbalanced by autumn and winter release of CO_2 [see 'first-year thin landfast ice (<40 cm)' and 'artificial growing ice' in Table 18.5 as an indication of the autumn and winter fluxes from sea ice to the atmosphere]. The significance of the autumnal and winter release of CO_2 from sea ice is still to be estimated. In addition, the CO_2 expelled in the under-ice oceanic water either can be transported with the draining dense brines to the deep ocean, where it can be effectively sequestered, or it can remain in the surface oceanic mixed layer, where it can be ventilated back to the atmosphere in summer. The effectiveness of CO_2 expulsion from sea ice into the under-ice ocean for sequestration therefore depends on how deep in the oceanic water column the CO_2-rich brines draining from sea ice are transported (Rysgaard et al., 2007, 2011).

The eddy covariance technique is often preferred to the chamber technique for the measurement of CO_2 fluxes in terrestrial ecosystems (Wang et al., 2013; Riederer et al., 2014) unless suitable conditions for the eddy covariance technique cannot be fulfilled, such as a stable concentration of atmospheric CO_2 for 0.5–1 hour during measurement, flat and homogeneous terrain and turbulent exchange. In particular, atmospheric turbulence cannot be reproduced inside the chamber and it is difficult to account for spatial heterogeneity (see Riederer et al., 2014 for further discussion). However, while both techniques agree reasonably well in terrestrial ecosystems, strikingly, there is one to two orders of magnitude difference between chamber and eddy covariance measurements over sea ice (Table 18.5). The eddy covariance CO_2 flux measurements over sea ice are somewhat at the higher end of what is commonly observed over open (ice-free) oceanic water. The measurements of Semiletov et al. (2004) and Zemmelink et al. (2006) have not been properly corrected for the open-path CO_2 analyser in cold temperatures used in these investigations (Burba et al., 2008). All the other measurements, however, converge towards large CO_2 fluxes (Heinesch et al., 2009; Else et al., 2011; Miller et al., 2011; Papakyriakou & Miller, 2011; Sørensen et al., 2014). Sea ice can be seen as an ideal environment for micro-meteorological measurement of CO_2 fluxes (Loose & Schlosser, 2011a) because the surface is relatively smooth and level, especially for landfast ice. In addition, measurements carried out with two different types of CO_2 analysers [open and closed path; see Garbe et al. (2014) for further details] are in agreement (Table 18.5) and so there seems to be little room for analytical bias. However, the large CO_2 fluxes out of sea ice observed in winter cannot easily be explained considering the low diffusion rates at low temperatures and the low and finite DIC content of the bulk sea ice (Moreau et al., 2015). The CO_2 fluxes observed in winter may therefore indicate a direct connection between the under-ice oceanic water and the atmosphere through micro-fractures or cracks in the sea ice cover, because only the CO_2 reservoir in the under-ice oceanic water could sustain such large fluxes out of sea ice in winter. Still, taking into account the apparent low surface extent of fractures and cracks, especially in landfast ice, and also the wind shield provided by the ice cover, the higher end of the *in situ* CO_2 flux observations remain difficult to explain despite the fact that shear and convection promote air–sea exchange in ice-covered areas (Loose et al., 2014). Hence, it remains an open question which processes are responsible for the elevated CO_2 fluxes observed over ice using the eddy correlation technique. Taking into account the magnitude of the fluxes, this question clearly deserves further attention.

Focusing only on measurements carried out with the chamber technique, the observed seasonal pattern of air–ice CO_2 flux indicates that sea ice acts as a source of CO_2 to the atmosphere during ice growth as long as the sea ice is permeable to gas transport. As the sea ice becomes impermeable, air–ice CO_2 fluxes are likely to be hampered. In spring, when the sea ice becomes permeable to gas transport again, the sea ice acts first as a transient source as a result of CO_2 super-saturation in brine. As the sea ice system warms up in the transition from spring to summer, the internal pCO_2 decreases below air saturation and the sea ice then shifts from a source to a sink for atmospheric CO_2 (Nomura et al., 2010a; Geilfus et al., 2012a; Delille et al., 2014). The magnitude of the fluxes is altered by flooding (Delille et al., 2014; Nomura et al., 2013), changes in snow structure (Nomura et al., 2010b, 2013) and the occurrence of melt ponds (Geilfus et al., 2012a).

The flux of other gases at the air–ice interface is poorly documented. Measurements of air–ice CH_4 fluxes were carried out using the chamber method along a transect from Alaska to near the North Pole in summer (He et al., 2013). The CH_4 flux ranged from −0.059 to +0.048 mmol m^{-2} day^{-1} with an average of −0.000 27 mmol m^{-2} day^{-1}. This study was somewhat inconclusive regarding the overall direction of the flux and the driving processes. Zemmelink et al. (2008b) reported measurements by relaxed eddy accumulation of the DMS flux in summer from multi-year sea ice in the western Weddell Sea, Antarctica. DMS-rich sea ice and slush released DMS to the atmosphere, with a DMS flux, averaged over a month, of 11 µmol m^{-2} day^{-1} and a maximum of 30 µmol m^{-2} day^{-1}. Based on these measurements, Zemmelink et al. (2008b) assessed that the multi-year ice of the western Weddell Sea released 14.5 GgS to the atmosphere during the period of observation (December), accounting for 6–19% of the yearly emission of DMS by the Southern Ocean between 50° and 70°S.

With the exception of the study by Loose et al. (2011a) mentioned earlier, direct measurement of gas fluxes at the ice–water interface have received less attention, but recent measurements of O_2 fluxes with the eddy correlation technique (Long et al., 2012) present potentially significant progress in measuring sea ice community production.

18.6 Revisiting Tsurikov's classification

As a conclusion to this chapter, it seems appropriate to summarize the progress that has been made in the recent decades in our understanding of the suite of processes and reactions that regulate gas concentrations in sea ice, thereby providing a useful update to the pioneering classification from Tsurikov (1979) (Figure 18.2) presented at the beginning of this chapter. This has been done in Table 18.7, where processes are organized in four major categories (entrapment, transport, nucleation and dissolution, and biogeochemical processes) and examples of relevant studies are given.

Table 18.7 Review of processes regulating gas concentrations in sea ice (from Zhou, 2014)

Process	Example(s)
Entrapment	
Entrapment of dissolved gases	Tsurikov (1979), Tison et al. (2002)
Entrapment of gas bubbles (e.g. methane)	Tsurikov (1979), Perovic & Gow (1996), Shakhova et al. (2010), Zhou et al. (2014a)
Air replacing downward moving brine in permeable sea ice pores above freeboard	Tsurikov (1979), Perovich & Gow (1996)
Superimposed ice formation – air entrapment in snow melt water	Zhou et al. (2013)
Snow ice formation –entrapment in snow of air and gases dissolved in sea water	Tsurikov (1979)
Transport	
Dissolved gas transport with brine movement	Moreau et al. (2014)
Air–ice and ice–ocean exchange	Delille et al. (2014), Loose et al. (2011), Nomura et al. (2010)
Buoyant migration of gas bubbles	Moreau et al. (2014), Zhou et al. (2013)
Gas movement with upward brine percolation	Geilfus et al. (2013)
Nucleation and dissolution	
Brine concentration (cooling) and dilution (warming)	Zhou et al. (2013)
Changes in gas solubility	Zhou et al. (2013, 2014a)
Pressure changes (e.g. void formation due to internal ice melting)	Tsurikov (1979), Light et al. (2003), Zhou et al. (2014b)
Biogeochemical processes	
Biological activity (primary and secondary production and respiration)	Zhou et al. (2014b) for O_2; Carnat et al. (2014) for DMS; Delille et al. (2007) for CO_2; Rysgaard & Glud (2004) and Zhou et al. (2014a) for N_2; Randall et al. (2012) for N_2O
$CaCO_3$ precipitation and dissolution for CO_2	Delille et al. (2007), Geilfus et al. (2013), Papadimitriou et al. (2013)

References

Abbatt, J.P.D., Thomas, J.L., Abrahamsson, K. et al. (2012) Halogen activation via interactions with environmental ice and snow in the polar lower troposphere and other regions. *Atmospheric Chemistry and Physics*, **12**, 14, 6237–6271.

Asher, E.C., Dacey, J.W.H., Mills, M.M., Arrigo, K.R. & Tortell, P.D. (2011) High concentrations and turnover rates of DMS, DMSP and DMSO in Antarctic sea ice. *Geophysical Research Letters*, **38**, L23609, doi: 10.1029/2011GL049712.

Brabant, F., El Amri, S. & Tison, J.-L. (2011) A robust approach for the determination of dimethylsulfoxide in sea ice. *Limnology and Oceanography: Methods*, **9**, 261–274.

Brabant, F. (2012) *Physical and biogeochemical controls on the DMS/P/O cycle in Antarctic sea ice*. Ph.D. Thesis, Université Libre de Bruxelles, Bruxelles.

Broecker, W.S. & Peng, T.H. (1974) Gas exchange rates between air and sea, *Tellus* **26**, 21–35.

Brown, K.A., Miller, L.A., Davelaar, M., Francois, R. & Tortell, P.D. (2014) Over-determination of the carbonate system in natural sea ice brine and assessment of carbonic acid dissociation constants under low temperature, high salinity conditions. *Marine Chemistry*, **165**, 36–45.

Burba, G., McDermitt, D.K., Grelle, A., Anderson, D.J. & Xu, L. (2008) Addressing the influence of instrument surface heat exchange on the measurements of CO_2 flux from open-path gas analyzers. *Global Change Biology*, **14**, 8, 1854–1876.

Carnat, G. (2014). Towards an understanding of the physical and biological controls on the cycling of dimethylsulfide (DMS) in Arctic and Antarctic sea ice, PhD, Department of Environment and Geography, University of Manitoba, Winnipeg, Manitoba, Canada.

Carnat, G., Zhou, J., Papakyriakou, T. et al.(2014) Physical and biological controls on DMS,P dynamics in ice shelf-influenced fast ice during a winter-spring and a spring-summer transitions. *Journal of Geophysical Research: Oceans*, **119**, 5, 2882–2905.

Cassar, N., Barnett, B., Bender, M.L., Kaiser, J., Hamme, R.C. & Tilbrook, B. (2009) Continuous high-frequency dissolved O_2/Ar measurements by equilibrator inlet mass spectrometry (EIMS). *Analytical Chemistry*, **81**, 5, 1855–1864.

Castro-Morales, K., Cassar, N., Shoosmith, D.R. & Kaiser, J. (2013) Biological production in the Bellingshausen Sea from oxygen-to-argon ratios and oxygen triple isotopes. *Biogeosciences*, **10**, 2273–2291.

Chang, R.Y.-W., Sjostedt, S.J., Pierce, J.R. et al. (2011) Relating atmospheric and oceanic DMS levels to particle nucleation events in the Canadian Arctic. *Journal of Geophysical Research: Atmospheres*, **116**, D17, 10.1029/2011JD015926.

Charlson, R.J., Lovelock, J.E., Andreae, M.O. & Warren, S G. (1987) Oceanic phytoplankton, atmospheric sulphur, cloud albedo and climate. *Nature*, **326**, 655–661.

Cox, G.F.N. & Weeks, W.F. (1983) Equations for determining the gas and brine volumes in sea-ice samples. *Journal of Glaciology*, **29**, 102, 306–316.

Cox, G.F.N. & Weeks, W.F. (1988) Numerical simulations of the profile properties of undeformed first-year sea ice during the growth season. *Journal of Geophysical Research: Oceans*, **93**, C10, 12449–12460.

Crabeck, O., Delille, B., Rysgaard, S. et al. (2014a) First "in situ" determination of gas transport coefficient (D_{O_2}, D_{Ar}, D_{N_2}) from bulk gas concentration measurements (O_2, N_2, Ar) in natural sea ice. *Journal of Geophysical Research: Oceans*, **119**, 6655–6668.

Crabeck, O., Delille, B., Thomas, D.N., Geilfus, N.X., Rysgaard, S. & Tison, J.L. (2014b), CO_2 and CH_4 in sea ice from a subarctic fjord. *Biogeosciences*, **11**, 4047–4083.

Curran, M.A.J., Jones, G.B. & Burton, H. (1998) Spatial distribution of dimethylsulfide and dimethylsulfoniopropionate in the Australasian sector of the Southern Ocean. *Journal of Geophysical Research*, **103** (D13), 16677–16689.

Curran, M.A.J. & Jones, G.B. (2000) Dimethyl sulfide in the Southern Ocean: seasonality and flux. *Journal of Geophysical Research*, **105** (D16), 20451–20459.

Damm, E., Helmke, E., Thoms, S. et al. (2010) Methane production in aerobic oligotrophic surface water in the central Arctic Ocean. *Biogeosciences*, **7**, 1099–1108.

Delille, B., Jourdain, B., Borges, A.V., Tison, J.-L. & Delille, D. (2007) Biogas (CO_2, O_2, dimethylsulfide) dynamics in spring Antarctic fast ice. *Limnology and Oceanography*, **52**, 1367–1379.

Delille, B., Vancoppenolle, M., Geilfus, N.-X. et al. (2014) Southern Ocean CO_2 sink: the contribution of the sea ice. *Journal of Geophysical Research: Oceans*, **119**, 6340–6355.

Dickson, A.G. & Millero, F.J. (1987) A comparison of the equilibrium constants for the dissociation of carbonic acid in seawater media. *Deep-Sea Research Part I*, **34**, 1733–1743.

Dieckmann, G.S., Nehrke, G., Papadimitriou, S. et al. (2008) Calcium carbonate as ikaite crystals in Antarctic sea ice. *Geophysical Research Letters*, **35**, L08501, doi:10.1029/2008GL033540.

DiTullio, G.R., Garrison, D.L. & Mathot, S. (1998) Dimethylsulfoniopropionate in sea ice algae from the Ross Sea polynya. *Antarctic Sea Ice Biological Processes, Interactions & Variability, Antarctic Series*, **73**, 139–146.

Eicken, H. (1998) Factors determining microstructure, salinity and stable-isotope composition of Antarctic sea ice: Deriving modes and rates of ice growth in the Weddell Sea. In: *Antarctic Sea Ice Physical Processes, Interactions and Variability* (Ed. M.O. Jeffries), pp. 89–122. American Geophysical Union, Washington.

Eicken, H., Gradinger, R., Salganek, M., Shirasawa, K., Perovich, D. & Lepparanta, M. (Ed.) (2009) *Field Techniques for Sea Ice Research*. University of Alaska Press, Fairbanks.

Else, B.G.T., Papakyriakou, T.N., Galley, R.J., Drennan, W.M., Miller, L.A. & Thomas, H., (2011) Wintertime CO_2 fluxes in an Arctic polynya using eddy covariance: Evidence for enhanced air-sea gas transfer during ice formation. *Journal of Geophysical Research: Oceans*, **116**, C00G03, doi:10.1029/2010JC006760.

Feltham, D.L., Untersteiner, N., Wettlaufer, J.S. & Worster, M.G. (2006) Sea ice is mushy layer. *Geophysical Research Letters*, **33**, L14501, 10.1029/2006GL026290.

Fischer, M., Thomas, D.N., Krell, A. et al. (2013) Quantification of ikaite in Antarctic sea ice. *Antarctic Science*, **25**, 421–432.

Forster, P., Ramaswamy, V., Artaxo, P. et al. (2007) Changes in Atmospheric Constituents and in Radiative Forcing. In: *Climate Change 2007: The Physical Science Basis. Contribution of Working Group I to the Fourth Assessment Report of the Intergovernmental Panel on Climate Change* (Eds. S. Solomon et al.), Cambridge University Press, Cambridge, USA.

Frank, X., Dietrich, N., Wu, J., Barraud, R. & Li, H.Z. (2007) Bubble nucleation and growth in fluids. *Chemical Engineering Science*, **62**, 7090–7097.

Fripiat, F., Sigman, D.M., Fawcett, S.E. et al. (2014) New insights into sea ice nitrogen biogeochemical dynamics from the nitrogen isotopes. *Global Biogeochemical Cycles*, **28**, 115–130.

Gali M. & Simo, R. (2010). Occurrence and cycling of dimethylated sulphur compounds in the Arctic during summer receding ice edge, *Marine Chemistry*, **122**, 105–117.

Galindo, V. (2014) Dynamique du dimethylsulfonopropionate (DMSP) produit par les assemblages d'algues sympagiques et planctoniques lors de la fonte printanière dans l'Archipel Arctique Canadien. *PhD Thesis*. Université de laval, Québec, Canada.

Galindo, V., Levasseur, M., Mundy, C.J. et al. (2014) Biological and physical processes influencing sea ice, under-ice algae & dimethylsulfoniopropionate during spring in the Canadian Arctic Archipelago. *Journal of Geophysical Research Oceans*, **119**, 10.1002/2013JC009497.

Gambaro, A., Moret, I., Piazza, R. et al. (2004) Temporal evolution of DMS and DMSP in Antarctic Coastal Sea water. *International Journal of Environmental Analytical Chemistry*, **84**, 401–412.

Garbe, C., Rutgersson, A., Boutin, J. et al. (2014) Transfer Across the Air-Sea Interface. In: *Ocean-Atmosphere Interactions of Gases and Particles. Springer Earth System Sciences* (Eds. P.S. Liss & M.T. Johnson), pp. 55–112. Springer Berlin Heidelberg.

Garcia, H.E and Gordon, L.I. (1992) Oxygen solubility in seawater: better fitting equations. *Limnology and Oceanography*, **37**, 1307–1312.

Garrison, D.L. & Buck, K.R. (1986) Organism losses during ice melting: a serious bias in sea ice community studies. *Polar Biology*, **6**, 237–239.

Geilfus, N.-X., Carnat, G., Papakyriakou, T. et al. (2012a) Dynamics of pCO$_2$ and related air-ice CO$_2$ fluxes in the Arctic coastal zone (Amundsen Gulf, Beaufort Sea). *Journal of Geophysical Research: Oceans*, **117**, C00G10, doi:10.1029/2011JC007118.

Geilfus, N.-X., Delille, B., Verbeke, V. & Tison, J.L. (2012b). Towards a method for high vertical resolution measurements of the partial pressure of CO$_2$ within bulk sea ice. *Journal of Glaciology*, **58**, 287–300.

Geilfus, N.-X. (2011). *Inorganic carbon dynamics in coastal arctic sea ice and related air–ice CO$_2$ exchanges*. PhD Thesis. Université de Liège, Liège.

Geilfus, N.X., Tison, J.L., Ackley, S.F. et al. (2014) Sea ice pCO$_2$ dynamics and air–ice CO$_2$ fluxes during the Sea Ice Mass Balance in the Antarctic (SIMBA) experiment – Bellingshausen Sea, Antarctica. *The Cryosphere*, **8**, 2395–2407.

Gilman, J.B., Burkhart, J.F., Lerner, B.M. et al. (2010) Ozone variability and halogen oxidation within the Arctic and sub-Arctic springtime boundary layer. *Atmospheric Chemistry and Physics*, **10**, 10223–10236.

von Glasow, R. & Crutzen, P.J. (2004) Model study of multiphase DMS oxidation with a focus on halogens. *Atmospheric Chemistry and Physics*, **4**, 589–608.

Gleitz, M., Rutgers van der Loeff, M., Thomas, D.N., Dieckmann, G.S. & Millero, F.J. (1995) Comparison of summer and winter inorganic carbon, oxygen and nutrient concentrations in Antarctic sea ice brines. *Marine Chemistry*, **51**, 81–91.

Glud, R.N., Rysgaard, S. & Kühl, M. (2002) A laboratory study in O$_2$ dynamics and photosynthesis in ice algal communities: quantification by microsensors, O$_2$ exchange rates, ^{14}C incubations and a PAM fluorometer. *Aquatic Microbial Ecology*, **27**, 301–311.

Gosink, T.A., Pearson, J.G. & Kelley, J.J. (1976) Gas movement through sea ice. *Nature*, **263**, 41–42.

Gosink, T.A. (1980). Atmospheric trace gases in association with sea ice. *Antarctic journal of the United States*, **15**, 82–83.

Granfors, A. (2014) *Biogenic Halocarbons in Polar Sea Ice*. PhD Thesis. University of Gothenburg, http://hdl.handle.net/2077/35507.

Granfors, A., Andersson, M., Chierici, M. et al. (2013a) Biogenic halocarbons in young Arctic sea ice and frost flowers. *Marine Chemistry*, **155**, 124–134.

Granfors, A., Karlsson, A., Mattsson, E., Smith, W.O. & Abrahamsson, K. (2013b) Contribution of sea ice in the Southern Ocean to the cycling of volatile halogenated organic compounds. *Geophysical Research Letters*, **40**, 3950–3955.

Haecky, P. & Anderson, A. (1999) Primary and bacterial production in sea ice in the northern Baltic Sea. *Aquatic Microbial Ecology*, **20**, 107–118.

Hamme, R.C. & Emerson, S.R. (2004) The solubility of neon, nitrogen and argon in distilled water and seawater. *Deep-Sea Research*, **51**, 1517–1528.

Hamme, R.C. & Severinghaus, J.P. (2007) Trace gas disequilibria during deep-water formation. *Deep Sea Research Part I: Oceanographic Research Papers*, **54**, 939–950.

He, X., Sun, L., Xie, Z. et al. (2013) Sea ice in the Arctic Ocean: role of shielding and consumption of methane. *Atmospheric Environment*, **67**, 8–13.

Hefu, Y. & Kirst, G.O. (1997) Effect of UV-radiation on DMSP content and DMS formation of Phaeocystis antarctica. *Polar Biology*, **18**, 402–409.

Heinesch, B., Yernaux, M., Aubinet, M. et al. (2009) Measuring air-ice CO$_2$ fluxes in the Arctic. *FluxLetter, the Newsletter of FLUXNET*, **2**, 9–10.

Hendricks, M.B., Bender, M.L. & Barnett, B.A. (2004) Net and gross O$_2$ production in the southern ocean from measurements of biological O$_2$ saturation and its triple isotope composition. *Deep Sea Research Part I: Oceanographic Research Papers*, **51**, 1541–1561.

Hobbs, P.V. (Ed.) (1974) *Ice Physics*. Clarendon Press, Oxford.

Intergovernmental Panel on Climate Change (IPCC) (2013) *Climate Change 2013: The Physical Science Basis. Contribution of Working Group I to the Fifth Assessment Report of the Intergovernmental Panel on Climate Change* (Eds. Stocker, T.F. et al.) Cambridge University Press, Cambridge.

Jacobi, H.-W., Morin, S. & Bottenheim, J.W. (2010) Observation of widespread depletion of ozone in the springtime boundary layer of the central Arctic linked to mesoscale synoptic conditions. *Journal of Geophysical Research: Atmospheres*, **115**, D17, D17302.

Jones, S.F., Evans, G.M. & Galvin, K.P. (1999) Bubble nucleation from gas cavities – a review. *Advances in Colloid and Interface Science*, **80**, 27–50.

Juul-Pedersen, T., Michel, C., Gosselin, M. & Seuthe, L. (2008) Seasonal changes in the sinking export of particulate material under first-year sea ice on the Mackenzie Shelf (western Canadian Arctic). *Marine Ecology Progress Series*, **353**, 13–25.

Kaartokallio, H. (2001) Evidence for active microbial nitrogen transformations in sea ice (Gulf of Bothnia, Baltic Sea) in midwinter. *Polar Biology*, **24**, 21–28.

Kaartokallio, H. (2004) Food web components & physical and chemical properties of Baltic Sea ice. *Marine Ecology Progress Series*, **273**, 49–63.

Karsten, U., Kück, K., Vogt, C. & Kirst, G.O. (1996) Dimethylsulfoniopropionate production in phototrophic organisms and its physiological function as a cryoprotectant. In: *Biological and Environmental Chemistry of DMSP and Related Sulfonium Compounds* (Ed. R. P. Kiene), pp. 143–153. Plenum Press, New York.

Kennedy, H., Thomas, D.N., Kattner, G., Haas, C. & Dieckmann, G.S. (2002) Particulate organic matter in Antarctic summer sea ice: concentration and stable isotopic composition. *Marine Ecology Progress Series*, **238**, 1–13.

Kettle, A.J., Andreae, M.O., Amoroux, D. et al. (1999) A global database of sea surface dimethylsulfide (DMS) measurements and a procedure to predict sea surface DMS as a function of latitude, longitude & month. *Global Biogeochemical Cycles*, **13**, 399–444.

Killawee, J.A., Fairchild, I.J., Tison, J.-L., Janssens, L. & Lorrain, R. (1998) Segregation of solutes and gases in experimental freezing of dilute solutions: Implications for natural glacial systems. *Geochimica et Cosmochimica Acta*, **62**, 3637–3655.

Kipfstuhl, S., Pauer, F., Kuhs, W.F and Shoji, H. (2001) Air bubbles and clathrate hydrates in the transition zone of the NGRIP deep ice core. *Geophysical Research Letters*, **28**, 591–594.

Kirschke, S., Bousquet, P., Ciais, P. et al. (2013) Three decades of global methane sources and sinks. *Nature Geosciences*, **6**, 813–823.

Kirst, G.O., Thiel, C., Wolff, H., Nothnagel, J., Wanzek, M. & Ulmke, R. (1991) Dimethylsufoniopropionate (DMSP) in ice-algae and its possible biological role. *Marine Chemistry*, **35**, 381–388.

Klimant, I., Meyer, V. & Kühl, M. (1995) Fiber-optic oxygen microsensors, a new tool in aquatic biology. *Limnology and Oceanography*, **40**, 1159–1165.

Klimant, I., Kühl, M., Glud, R.N. & Holst, G. (1997) Optical measurement of oxygen and temperature in microscale: strategies and biological applications. *Sensors Actuators B*, **38–39**, 29–37.

Kort, E.A., Wofsy, S.C., Daube, B.C. et al. (2012) Atmospheric observations of Arctic Ocean methane emissions up to 82° north. *Nature Geoscience*, **5**, 318–321.

Krüger, O. & Grassl, H. (2011) Southern Ocean phytoplankton increases cloud albedo and reduces precipitation. *Geophysical Research Letters*, **38**, L08809, 10.1029/2011GL047116.

Kühl, M., Glud, R.N., Borum, J., Roberts, R. & Rysgaard, S. (2002) Photosynthetic performance of surface associated algae below sea ice as measured with a pulse amplitude modulated (PAM) fluorometer and O_2 microsensors. *Marine Ecology Progress Series*, **223**, 1–14.

Lana, A., Simó, R., Vallina, S.M. & Dachs, J. (2012) Potential for a biogenic influence on cloud microphysics over the ocean: a correlation study with satellite-derived data. *Atmospheric Chemistry and Physics*, **12**, 7977–7993.

Lee, P.A., de Mora, S.J., Gosselin, M. et al. (2001) Particulate dimethylsulfoxide in Arctic sea-ice algal communities: the cryoprotectant hypothesis revisited. *Journal of Phycology*, **37**, 488–499.

Levasseur, M. (2013) Impact of Arctic meltdown on the microbial cycling of sulphur. *Nature Geoscience*, **6**, 691–698,

Levasseur, M., Gosselin, M. & Michaud, S. (1994) A new source of dimethylsulfide (DMS) for the arctic atmosphere - ice diatoms. *Marine Biology*, **121**, 381–387.

Lindberg, S.E., Brooks, S., Lin, C.J. et al. (2002) Dynamic oxidation of gaseous mercury in the Arctic troposphere at polar sunrise. *Environmental Science & Technology*, **36**, 1245–1256.

Long, M.H., Koopmans, D., Berg, P., Rysgaard, S., Glud, R.N., Søgaard, D.H. (2012) Oxygen exchange and ice melt measured at the ice-water interface by eddy correlation. *Biogeosciences*, **9**, 1957–1967.

Loose, B. & Schlosser, P. (2011) Sea ice and its effect on CO_2 flux between the atmosphere and the Southern Ocean interior. *Journal of Geophysical Research: Oceans*, **116**, C11, C11019.

Loose, B., Schlosser, P., Perovich, D. et al. (2011a) Gas diffusion through columnar laboratory sea ice: implications for mixed-layer ventilation of CO_2 in the seasonal ice zone. *Tellus B*, **63**, 23–39.

Loose, B., Miller, L.A., Elliott, S. & Papakyriakou, T. (2011b) Sea ice biogeochemistry and material transport across the frozen interface. *Oceanography*, **24**, 202–218.

Loose, B., McGillis, W.R., Perovich, D., Zappa, C.J. & Schlosser, P. (2014) A parameter model of gas exchange for the seasonal sea ice zone. *Ocean Sciences*, **10**, 17–28.

Martinerie, P., Lipenkov, V.Y., Raynaud, D., Chappellaz, J., Barkov, N.I. & Lorius, C. (1994) Air content paleo-record in the Vostok ice core (Antarctica): A mixed record of climatic and glaciological parameters. *Journal of Geophysical Research*, **99**, 10565–10576.

Matsuo, S. & Miyake, Y. (1966) Gas composition in ice samples from Antarctica. *Journal of Geophysical Research*, **71**, 5235–5241.

McMinn, A. & Ashworth, C. (1998) Use of oxygen microelectrodes to determine the net production by an Antarctic sea ice algal community. *Antarctic Science*, **10**, 39–44.

McMinn, A., Ashworth, C. & Ryan, K.G. (2000) In situ net primary productivity of an Antarctic fast ice bottom algal community. *Aquatic Microbial Ecology*, **21**, 177–185.

Mehrbach, C., Culberson, C.H., Hawley, J.E., Pytkowicz, R.M. (1973) Measurement of the apparent dissociation constants of carbonic acid in seawater at atmospheric pressure. *Limnology and Oceanography* **18**, 897–907.

Meiners K.M., Papadimitriou, S., Thomas, D.N., Norman, L. & Dieckmann, G.S. (2009) Biogeochemical conditions and ice algal photosynthetic parameters in Weddell Sea ice during early spring, *Polar Biology*, **32**, 1055–1065.

Michel, C., Nielsen, T.G., Nozias, C. & Gosselin, M. (2002). Significance of sedimentation and grazing by ice micro- and meiofauna for carbon cycling in annual sea ice (Northern Baffin Bay). *Aquatic Microbial Ecology*, **30**, 57–68.

Miller, L.A., Papakyriakou, T.N., Collins, R.E. et al. (2011) Carbon dynamics in sea ice: A winter flux time series. *Journal of Geophysical Research*, **116**, C02028, doi:10.1029/2009JC006058.

Miller, L.A., Fripiat, F., Else, B.G.T. et al.(2015) Methods for Biogeochemical Studies of Sea Ice: The State of the Art, Caveats & Recommendations. *Elementa: Science of the Anhropocene*, **3**, 000038, doi: 10.12952/journal.elementa.000038.

Millero, F.J., Graham, T.B., Huang, F., Bustos-Serrano, H. & Pierrot, D. (2006) Dissociation constants of carbonic acid in seawater as a function of salinity and temperature. *Marine Chemistry*, **100**, 80–94.

Millero, F.J. (1995) Thermodynamics of the carbon dioxide system in the ocean. *Geochimica et Cosmochimica Acta*, **59**, 661–677.

Miyake, Y. & Matsuo, S. (1963) A role of sea ice and sea water in the Antarctic on the carbon dioxide cycle in the atmosphere. *Papers in Meteorology and Geophysics*, **14**, 120–125.

Mock, T., Dieckmann, G.S., Haas, C. et al. (2002) Micro-optodes in sea ice: a new approach to investigate oxygen dynamics during sea ice formation. *Aquatic Microbial Ecology*, **29**, 297–306.

Moreau, S., Vancoppenolle, M., Zhou, J., Tison, J.-L., Delille, B. & Goosse, H. (2014) Modelling argon dynamics in first-year sea ice. *Ocean Modelling*, **73**, 1–18.

Moreau, S., Vancoppenolle, M., Delille, B. et al. (2015) Drivers of inorganic carbon dynamics in first-year sea ice: a model study. *Journal of Geophysical Research*, **120**, doi:10.1002/2014JC010388

Munro, D.R., Dunbar, R.B., Mucciarone, D.A., Arrigo, K.R. & Long, M.C. (2010) Stable isotopic composition of dissolved inorganic carbon and particulate organic carbon in sea ice from the Ross Sea, Antarctica. *Journal of Geophysical Research*, **115**, C09005, doi:10.1029/2009JC005661.

Nomura, D., Eicken, H., Gradinger, R. & Shirasawa, K. (2010a). Rapid Physically driven inversion of the air-sea ice CO_2 flux in the seasonal landfast ice off Barrow, Alaska after onset of surface melt. *Continental Shelf Research*, **30**, 1998–2004.

Nomura, D., Yoshikawa-Inoue, H., Toyota, T. & Shirasawa, K. (2010b). Effects of snow, snowmelting and refreezing processes on air-sea-ice CO_2 flux. *Journal of Glaciology*, **56**, 262–270.

Nomura, D., Granskog, M., Assmy, P., Simizu, D. & Hashida, G. (2013) Arctic and Antarctic sea ice acts as a sink for atmospheric CO_2 during periods of snow melt and surface flooding. *Journal of Geophysical Research: Oceans*, **118**, 6511–6524.

Nomura D, Yoshikawa-Inoue H, Toyota T (2006) The effect of sea-ice growth on air-sea CO_2 flux in a tank experiment. *Tellus, Series B: Chemical and Physical Meteorology*, **58**, 418–426.

Notz, D., Wettlaufer, J.S. & Worster, M.G. (2005) A non-destructive method for measuring the salinity and solid fraction of growing sea ice in situ. *Journal of Glaciology*, **51**, 159–166.

Notz, D. & Worster, M.G. (2008) In situ measurements of the evolution of young sea ice. *Journal of Geophysical Research: Oceans*, **113**, C3, C03001.

Notz, D. & Worster, M.G. (2009) Desalination processes of sea ice revisited. *Journal of Geophysical Research: Oceans*, **114**, C5, C05006.

Owens, O.C. (2008) *Wintertime measurements of pCO_2 in Arctic landfast sea ice*. MSc Thesis. University of Manitoba, Winnipeg, Manitoba, Canada.

Papadimitriou, S., Kennedy, H., Kennedy, P. & Thomas, D.N. (2014) Kinetics of ikaite precipitation and dissolution in seawater-derived brines at sub-zero temperatures to 265 K. *Geochimica et Cosmochimica Acta*, **140**, 199–211.

Papadimitriou, S., Kennedy, H., Kennedy, P. & Thomas, D.N. (2013) Ikaite solubility in seawater-derived brines at 1 atm and sub-zero temperatures to 265 K. *Geochimica et Cosmochimica Acta*, **109**, 241–253.

Papadimitriou, S., Kennedy, H., Norman, L., Kennedy, D.P. & Thomas, D.N. (2012) The effect of biological activity, $CaCO_3$ mineral dynamics & CO_2 degassing in the inorganic carbon cycle in sea ice in late winter-early spring in the Weddell Sea, Antarctica. *Journal of Geophysical Research*, **117**, C08011, doi:10.1029/2012JC008058.

Papadimitriou, S., Thomas, D.N., Kennedy, H. et al. (2007) Biogeochemical composition of natural sea ice brines from the Weddell Sea during early austral summer. *Limnology and Oceanography*, **52**, 1809–1823.

Papadimitriou, S., Kenned, y H., Kattner, G., Dieckmann, G.S. & Thomas, D.N. (2004) Experimental evidence for carbonate precipitation and CO_2 degassing during sea ice formation. *Geochimica et Cosmochimica Acta*, **68**, 1749–1761.

Papakyriakou, T.N. & Miller, L.A. (2011) Springtime CO_2 exchange over seasonal sea ice in the Canadian Arctic Archipelago. *Annals of Glaciology*, **52**, 215–224.

Pauer, F., Kipfstuhl, S., Kuhs, W.F. & Shoji, H. (1999) Air clathrate crystals from the GRIP deep ice core a number-, size- & shape-distribution study. *Journal of Glaciology*, **45**, 22–30.

Pierrot, D., Neill, C., Sullivan, K. et al. (2009) Recommendations for autonomous underway pCO_2 measuring systems and data-reduction routines. *Deep-Sea Research Part II*, **56**, 512–522.

Plummer, L.N. & Busenberg, E. (1982) The solubilities of calcite, aragonite and vaterite in CO_2-H_2O solutions between 0 and 90°C & an evaluation of the aqueous model for the system $CaCO_3$-CO_2-H_2O. *Geochimica et Cosmochimica Acta*, **46**,1011–1040.

Pratt, K.A., Custard, K.D., Shepson, P.B. et al. (2013) Photochemical production of molecular bromine in Arctic surface snowpacks. *Nature Geosciences*, **6**, 351–356.

Priscu, J.C., Downes, M.T., Priscu, L.R., Palmisano, A.C. & Siullivan, C.W. (1990) Dynamics of ammonium oxidizer activity and nitrous oxide (N_2O) within and beneath Antarctic sea ice. *Marine Ecology Progress Series*, **62**, 37–46.

Quinn, P.K. & Bates, T.S. (2011) The case against climate regulation via oceanic phytoplankton sulphur emissions. *Nature*, **480**, 51–56.

Randall, K., Scarratt, M., Levasseur, M., Michaud, S., Xie, H. & Gosselin, M. (2012) First measurements of nitrous oxide in Arctic sea ice. *Journal of Geophysical Research*, **117**, C00G15.

Raven J.A., Johnston A.M., Parsons R. & Kübler J.E. (1994) The influence of natural and experimental high O_2 concentrations on O_2-evolving phototrophs. *Biological Reviews*, **69**, 61–94.

Ravishankara, A., Daniel, J.S. & Portmann, R.W., 2009. Nitrous oxide (N_2O): the dominant ozone-depleting substance emitted in the 21st century. *Science*, **326**, 123–125.

Raynaud, D., Chappellaz, J., Barnola, J.M., Korotkevich, Y.S. & Lorius, C. (1988) Climatic and CH_4 cycle implications of glacial-interglacial CH_4 change in the Vostok ice core. *Nature*, **333**, 655–657.

Raynaud, D., Delmas, R., Ascencio, J.M. & Legrand, M. (1982) Gas extraction from polar ice cores: a critical issue for studying the evolution of atmospheric CO_2 and ice-sheet surface elevation. *Annals of Glaciology*, **3**, 265–268.

Reuer, M.K., Barnett, B.A., Bender, M.L., Falkowski, P.G. & Hendricks, M.B. (2007) New estimates of Southern Ocean biological production rates from O_2/Ar ratios and the triple isotope composition of O_2. *Deep Sea Research Part I*, **54**, 951–974.

Riederer, M., Serafimovich, A. & Foken, T. (2014) Net ecosystem CO2 exchange measurements by the closed chamber method and the eddy covariance technique and their dependence on atmospheric conditions. *Atmospheric Measurement Techniques*, **7**, 1057–1064.

Rutgers van der Loeff, M.M., Cassar, N., Nicolaus, M., Rabe, B. & Stimac, I. (2014). The influence of sea ice cover on air-sea gas exchange estimated with radon-222 profiles. *Journal of Geophysical Research: Oceans*, **119**, 2735–2751.

Rysgaard, S., Kuhl, M., Glud, R.N. & Hansen, J.W. (2001) Biomass, production and horizontal patchiness of sea ice algae in a high-Arctic fjord (Young Sound, NE Greenland). *Marine Ecology Progress Series*, **223**, 15–26.

Rysgaard, S. & Glud, R.N. (2004) Anaerobic N_2 production in Arctic sea ice. *Limnology and Oceanography*, **49**, 86–94.

Rysgaard, S., Glud, R.N., Sejr, M.K., Bendtsen, J. & Christensen, P.B. (2007) Inorganic carbon transport during sea ice growth and decay: a carbon pump in polar seas. *Journal of Geophysical Research-Oceans*, **112**, C03016, doi:10.1029/2006JC003572.

Rysgaard, S., Glud, R.N., Sejr, M.K., Blicher, M.E. & Stahl, H.J. (2008) Denitrification activity and oxygen dynamics in Arctic sea ice. *Polar Biology*, **31**, 527–537.

Rysgaard, S., Bendtsen, J., Delille, B. et al (2011) Sea ice contribution to the air–sea CO_2 exchange in the Arctic and Southern Oceans. *Tellus B*, **63**, 823–830.

Saiz-Lopez, A., Mahajan, A.S., Salmon, R.A. et al. (2007) Boundary layer halogens in coastal Antarctica. *Science*, **317**, 348–351.

Schönhardt, A., Begoin, M., Richter, A. et al. (2012) Simultaneous satellite observations of IO and BrO over Antarctica. *Atmospheric Chemistry and Physics*, **12**, 6565–6580.

Schönhardt, A., Richter, A., Wittrock, F. et al. (2008). Observations of iodine monoxide columns from satellite. *Atmospheric Chemistry and Physics*, **8**, 637–653.

Seitzinger, S., Harrison, J.A., Böhlke, J. et al. (2006) Denitrification across landscapes and waterscapes: a synthesis. *Ecological Applications*, **16**, 2064–2090.

Semiletov, I., Shakhova, N., Sergienko, V.I., Pipko, I. & Dudarev, O. (2012) On carbon transport and fate in the East Siberian Arctic land–shelf–atmosphere system. *Environmental Research Letters*, **7**, 015201, doi:10.1088/1748-9326/7/1/015201

Semiletov, I., Makshtas, A., Akasofu, S.I. & Andreas, E.L. (2004) Atmospheric CO_2 balance: The role of Arctic sea ice. *Geophysical Research Letters*, **31**, L05121, doi: 10.1029/2003GL017996.

Shakhova, N., Semiletov, I. & Gustafsson, O. (2010) Methane from the East Siberian Arctic Shelf Response. *Science*, **329**, 1147–1148.

Sharma, S., Barrie, L.A., Hastie, D.R. & Kelly, C. (1999) Dimethyl sulfide emissions to the atmosphere from lakes of the Canadian boreal region. *Journal of Geophysical Research*, **104**, 11585–11592.

Shaw, M.D., Carpenter, L.J., Baeza-Romero, M.T. & Jackson, A.V. (2011) Thermal evolution of diffusive transport of atmospheric halocarbons through artificial sea–ice. *Atmospheric Environment*, **45**, 6393–6402.

Simpson, W.R., von Glasow, R., Riedel, K. et al. (2007) Halogens and their role in polar boundary-layer ozone depletion. *Atmospheric Chemistry and Physics*, **7**, 4375–4418.

Sørensen, L.L., Jensen, B., Glud, R.N. et al. (2014) Parameterization of atmosphere–surface exchange of CO_2 over sea ice. *The Cryosphere*, **8**, 853–866.

Stefels, J. (2000) Physiological aspects of the production and conversion of DMSP in marine algae and higher plants. *Journal of Sea Research*, **43**, 183–197.

Stefels, J., Steinke, M., Turner, S., Malin, G. & Belviso, S. (2007) Environmental Constraints on the Production and Removal of the Climatically Active Gas Dimethylsulphide (DMS) and Implications for Ecosystem Modelling. *Biogeochemistry*, **83**, 245–275.

Stefels, J., Carnat, G., Dacey, J.W.H., Goossens, T., Elzenga, J.T.M. & Tison, J.-L. (2012) The analysis of dimethylsulfide and dimethylsulfoniopropionate in sea ice: Dry-crushing and melting using stable isotope additions. *Marine Chemistry*, **128–129**, 34–43.

Sunda, W., Kieber, D.J., Kiene, R.P. & Huntsman, S. (2002) An antioxidant function for DMSP and DMS in marine algae. *Nature*, **418**, 317–320.

Takahashi, T., Sutherland, S.C., Wanninkhof, R. et al. (2009) Climatological mean and decadal change in surface ocean pCO_2 & net sea-air CO_2 flux over the global oceans. *Deep Sea Research Part II*, **56**, 554–577.

Thomas, D.N., Lara, R.J., Haas, C. et al.(1998) Biological soup within decaying summer sea ice in the Amundsen Sea, Antarctica. In: *Antarctic Sea Ice: Biological Processes, Interactions and Variability* (Eds.M.P. Lizotte & K.R. Arrigo). American Geophysical Union, Antarctic Research Series,**73**, 161–171.

Tison, J.-L., Haas, C., Gowing, M.M., Sleewaegen, S. & Bernard, A. (2002) Tank study of physico-chemical controls on gas content and composition during growth of young sea ice. *Journal of Glaciology*, **48**, 267–278.

Tison, J.-L., Brabant, F., Dumont, I. & Stefels, J. (2010) High resolution DMS and DMSP time series profiles in decaying summer first-year sea ice at ISPOL (Western Weddell Sea, Antarctica). *Journal of Geophysical Research: Biogeochemistry*, **115**, G04044, 10.1029/2010JG001427.

Trenerry, L.J., McMinn, A. & Ryan, K.G. (2002) In situ oxygen microelectrode measurements of the bottom-ice algal production in McMurdo Sound, Antarctica. *Polar Biology*, **25**, 72–80.

Trevena, A.J., Jones, G.B., Wright, S.W. & van den Enden, R.L. (2000) Profiles of DMSP, algal pigments, nutrients and salinity in pack ice from eastern Antarctica. *Journal of Sea Research*, **43**, 265–273.

Trevena, A.J., Jones, G.B., Wright, S.W. & van den Enden, R.L. (2003) Profiles of dimethylsulphoniopropionate (DMSP), algal pigments, nutrients and salinity in the fast ice of Prydz Bay, Antarctica. *Journal of Geophysical Research*, **108**, doi: 10.1029/2002JC001369.

Trevena, A.J. & Jones, G.B. (2006) Dimethylsulphide and dimethylsulphoniopropionate in Antarctic sea ice and their release during sea ice melting. *Marine Chemistry*, **98**, 210–222.

Tsurikov, V.L. (1979) The formation and composition of the gas content of sea ice. *Journal of Glaciology*, **22**, 67–81.

Tunved, P., Ström, J. & Krejci, R. (2013) Arctic aerosol life cycle: linking aerosol size distribution observed between 2000 and 2010 with air mass transport and precipitation at Zeppelin station, Ny-Alesund, Svalbard. *Atmospheric Chemistry and Physics*, **13**, 3643–3660.

Turner, S.M., Nightingale, P.D., Broadgate, W. & Liss, P.S. (1995) The distribution of dimethyl sulphide and dimethylsulphoniopropionate in Antarctic waters and sea ice. *Deep Sea Research Part II*, **42**, 1059–1080.

UNESCO (1983) Algorithms for computation of fundamental properties of seawater. *Unesco Techical Papers in Marine Science*, **44**, 1–53.

Uzuka, N. (2003) A Time Series Observation of DMSP Production in the Fast Ice Zone Near Barrow. *Tôhoku Geophysical Journal*, **36**, 439–442.

Verbeke, V., Lorrain, R., Johnsen, S.J. & Tison, J.-L. (2002) A multiple-step deformation history of basal ice from the Dye3 (Greenland) core: new insights from the CO_2 and CH_4 content. *Annals of Glaciology*, **35**, 231–236.

Wang, K., Liu, C., Zheng, X. et al. (2013) Comparison between eddy covariance and automatic chamber techniques for measuring net ecosystem exchange of carbon dioxide in cotton and wheat fields. *Biogeosciences*, **10**, 6865–6877.

Weeks, W.F. & Ackley, S.F. (1986) The growth, structure and properties of sea ice. In: *The Geophysics of Sea Ice* (Ed. N. Untersteiner), pp. 9–164. Plenum Press, New York.

Weiss, R.S. (1974) Carbon dioxide in water and seawater: The solubility of a non-ideal gas. *Marine Chemistry*, **2**, 203–215.

Wettlaufer, J.S., Worster, M.G. & Huppert, H.E. (1997) The phase evolution of young sea ice. *Geophysical Research Letters*, **24**, 1251–1254.

Wiesenburg, D.A. & Guinasso Jr.,, N.L. (1979) Equilibrium solubilities of methane, carbon monoxide & hydrogen in water and sea water. *Journal of Chemical and Engineering Data*, **24**, 356–360.

Wolf-Gladrow, D.A., Zeebe, R.E., Klaas, C., Körtzinger, A. & Dickson, A.G. (2007) Total Alkalinity: the explicit conservative expression and its application to biogeochemical processes. *Marine Chemistry*, **106**, 287–300.

Zemmelink, H.J., Delille, B., Tison, J.L., Hintsa, E.J., Houghton, L. & Dacey, J.W.H. (2006) CO_2 deposition over the multi-year ice of the western Weddell Sea. *Geophysical Research Letters*, **33**, L13606, doi:10.1029/2006gl026320.

Zemmelink, H.J., Dacey, J.W.H., Houghton, L., Hintsa, E.J. & Liss, P.S. (2008) Dimethylsulfide emissions over the multi-year ice of the western Weddell Sea. *Geophysical Research Letters*, **35**, L06603.

Zhou, J., Delille, B., Eicken, H., Vancoppenolle, M. et al. (2013) Physical and biogeochemical properties in landfast sea ice (Barrow, Alaska): Insights on brine and gas dynamics across seasons. *Journal of Geophysical Research: Oceans*, **118**, 3172–3189.

Zhou, J., Tison, J.-L., Carnat, G., Geilfus, N.-X. & Delille, B. (2014a) Physical controls on the storage of methane in landfast sea ice. *The Cryosphere*, **8**, 1019–1029.

Zhou, J., Delille, B., Brabant, F. & Tison, J.-L. (2014b) Insights into oxygen transport and net community production in sea ice from oxygen, nitrogen and argon concentrations. *Biogeosciences*, **11**, 5007–5020.

Zhou, J. (2014) The physical and biological controls on the distribution of gases and solutes in sea ice from ice growth to ice decay. PhD Thesis. Université Libre de Bruxelles and Université de Liège, Bruxelles, http://hdl.handle.net/2268/174416.

CHAPTER 19

Transport and transformation of contaminants in sea ice

Feiyue Wang,[1,2] Monika Pućko[1,2] and Gary Stern[1]
[1] Centre for Earth Observation Science, University of Manitoba, Winnipeg, Canada
[2] Department of Environment and Geography, University of Manitoba, Winnipeg, Canada

19.1 Introduction

The presence of chemical contaminants in sea ice is not unexpected, yet the processes governing their transport and transformation in the sea ice environment and their implications for the polar marine ecosystem have only recently started to be appreciated. Since the onset of the Anthropocene in the late 18th century (Crutzen, 2002), and especially after World War II, many toxic substances have been released to the environment in large quantities in industrialized areas of the world, and made their way to the remote polar environments through long-range atmospheric, oceanic, riverine and biological transport (Macdonald et al., 2000, 2005). Although the role of sea ice as a transport medium for contaminants has been recognized since at least the beginning of the 1990s [e.g. Pfirman et al., 1995; Macdonald & Bewers, 1996; Arctic Monitoring and Assessment Programme (AMAP), 1998], detailed studies on the behaviour of contaminants in sea ice and across the ocean–sea ice–atmosphere (OSA) interface only began in the mid-2000s (e.g. Douglas et al., 2005; Pućko et al., 2010b; Chaulk et al., 2011).

As a dynamic and porous 'lid' between the atmosphere and the ocean, sea ice acts as both a temporary storage and an effective transporter for contaminants, moving them in space (vertically between the atmosphere and the ocean, and laterally as ice drifts) and in time (seasonal storage and release via freeze and melt).

Sea ice plays a profound role in concentrating and diluting exposures to chemical contaminants through a number of recently characterized processes. A variety of chemical changes can also take place in snowpack, sea ice and melt ponds, resulting in chemical transformation of the contaminants. Furthermore, evidence is mounting that through uptake by sea ice communities and via melting, sea ice may provide an important pathway for contaminants to enter the polar marine ecosystem food web.

This chapter discusses major processes governing the transport and transformation of contaminants in sea ice and across the OSA interface, and their implications for contaminant uptake and bioaccumulation in the polar marine ecosystem under a changing climate. Our discussion will be focused on the Arctic sea ice, as studies on contaminants in the Antarctic sea ice environment are scarce. An essentially Mediterranean sea surrounded by large land mass with greater human population and industrial activities, the Arctic also receives a higher input of contaminants than the Antarctic (Poland et al., 2003; Bargagli, 2008). Although some of the processes and implications are relevant to the Antarctic, caution is advised when extrapolating from one polar region to the other. Despite their obvious similarities, the sea ice environments in the Arctic and Antarctic differ greatly in their structure, dynamics, ecology and projected change (Chapters 1 and 10).

Sea Ice, Third Edition. Edited by David N. Thomas.
© 2017 John Wiley & Sons, Ltd. Published 2017 by John Wiley & Sons, Ltd.

19.2 Contaminants in the Arctic sea ice

A contaminant in the Arctic marine ecosystem refers to either a naturally occurring substance (e.g. a metal) that is present at a concentration clearly above what is considered normal ('baseline') or an anthropogenically produced substance (e.g. an organochlorine pesticide) that is readily detectable (Macdonald & Bewers, 1996). Depending on its concentration and speciation (i.e. the distribution of a contaminant among its chemical forms), the presence of a contaminant in an environmental medium may or may not result in any detectable adverse effects; where such effects are evident, the contaminant is often referred to as a pollutant (Macdonald & Bewers, 1996).

A variety of contaminants have been detected in the abiotic and biotic compartments of the Arctic, including trace metals (e.g. mercury, cadmium, lead), persistent organic pollutants (POPs; e.g. organochlorine, organobromine and organofluorine compounds), polycyclic aromatic hydrocarbons (PAHs) and artificial radionuclides (e.g. ^{137}Cs) (AMAP, 1998, 2010, 2011; Macdonald et al., 2000, 2005). Of particular concern are mercury (Hg) and POPs that reach the Arctic via long-range transport from other parts of the globe and can accumulate in the Arctic marine ecosystem to concentrations that pose a toxicological risk to wildlife and humans. Production and usage of some POPs [e.g. dichloro-diphenyl-trichloroethane (DDT), polychlorinated biphenyls (PCBs), hexachlorocyclohexane (HCH)] and Hg have been banned or are being increasingly placed under control due to international conventions, but large amounts remain in the environment from past use (hence 'legacy contaminants'). Many other contaminants are still being used for various purposes [e.g. current-use pesticides (CUPs) such as chlorothalonil] and are thus continuously transported to the Arctic (AMAP, 2010).

All the contaminants that have been detected in the Arctic are expected to be present, sometimes even concentrated, in the sea ice environment. For instance, Arctic sea ice has been found to contain concentrations of microplastics (small fragments of plastic debris) that are several orders of magnitude greater than in highly contaminated surface seawaters (Obbard et al., 2014). Our discussion will, however, focus primarily on three representative examples of contaminants whose behaviour in the sea ice environment has been studied to some details: Hg as a trace metal, HCH (α- and γ-isomers) as an organochlorine POP, and petrogenic PAHs that are associated with crude oil. Basic physical–chemical properties of Hg, HCH and naphthalene (as an example PAH) are summarized in Table 19.1. Typical concentrations of Hg and HCH in the Arctic sea ice environment are shown in Tables 19.2 and 19.3, respectively.

19.3 Pathways of contaminants entering sea ice

Contaminants are transported to the Arctic marine environment via long-range atmospheric transport, oceanic circulation, riverine and coastal input, or biological transport (Figure 19.1). During ice formation and growth, these contaminants can enter sea ice from the underlying seawater and shelf sediments via freeze rejection and particle entrapment ('bottom up'), or from the overlying snowpack and atmosphere via dry and wet deposition ('top down').

19.3.1 Freeze rejection

As sea ice starts forming in the fall, contaminants dissolved in seawater undergo freeze rejection from ice crystal lattice due to constraints on size or electric charge, in a similar way to major ions such as Na$^+$ and Cl$^-$ (Chapter 1). A fraction of the rejected contaminants is lost to the underlying seawater, with the rest being retained in brine pockets. The concentrations and distribution of conservative, dissolved contaminants in growing sea ice are thus expected to follow closely that of salinity, as demonstrated in Figure 19.2 based on recent data on HCH (Pućko et al., 2010b) and Hg (Chaulk et al., 2011) in the first-year ice (FYI) in the Beaufort Sea. Both α- and γ-HCH concentrations in the bulk ice, 0.13–0.44 and 0.02–0.07 ng L^{-1}, respectively, increased with increasing salinity and were lower than those detected in the underlying seawater (~1 and ~0.1 ng L^{-1} for α- and γ-isomer, respectively) which had a higher salinity (Figure 19.2a,b). For the more water-soluble α-HCH isomer (that is present in seawater primarily in the dissolved phase), 70% of its concentration variance in the ice could be explained by salinity (Figure 19.2a). In comparison, γ-HCH is both water soluble and particle-reactive, and hence

Table 19.1 Basic physical–chemical properties (at 25°C and 1 atm) of mercury, hexachlorocyclohexane (HCH) and naphthalene. For mercury, three species are shown: elemental mercury (Hg⁰), inorganic Hg^{II} (using HgCl$_2$ as an example) and methylmercury (using CH$_3$HgCl as an example). For HCH, two isomers, α-HCH and γ-HCH, are shown. Naphthalene is shown as an example polycyclic aromatic hydrocarbon (PAH) in crude oil.

	Mercury			HCH		Naphthalene
	Hg⁰	HgCl$_2$	CH$_3$HgCl	α-HCH (Two enantiomers)	γ-HCH	
Structure	Hg⁰	Cl–Hg–Cl	H$_3$C–Hg–Cl			
Molecular mass (g mol^{-1})	200.6	271.5	250.1	290.8	290.8	128.2
Melting point (°C)	–39[a]	277[a]	167[a]	158[b]	114[b]	80[c]
Boiling point (°C)	357[a]	303[a]	–	288	323	218
Aqueous solubility (mol kg^{-1})	3.0 × 10^{-7}[d]	0.27[d]	4.0 × 10^{-4}[e]	3.3 × 10^{-4}[b]	2.5 × 10^{-4}[b]	8.1 × 10^{-4}[c]
Vapour pressure ρ (Pa)	0.26[f]	0.017[g]	1.8[g]	0.25[b]	0.076[b]	38[c]
Henry's Law constant (Pa m³ mol^{-1})	790[h]	9.3 × 10^{-5}[i]	4.6 × 10^{-3}[j]	0.74[b]	0.31[b]	50[c]
Octanol-water distribution coefficient K$_{OW}$	4.2[k]	3.3[k]	1.7[k]	8.6 × 10^{3}[b]	6.8 × 10^{3}[b]	2.5 × 10^{3}[c]
Octanol-air distribution coefficient K$_{OA}$	12[l]	1.3 × 10^{8}[l]	9.6 × 10^{2}[l]	2.9 × 10^{7}[b]	5.5 × 10^{7}[b]	1.5 × 10^{5}[c]
Half-life in the Arctic Ocean, t$_{1/2}$ (yr)				6: (+)- α-HCH [m]; 23: (-)- α-HCH [m]	19[n]	
Current sources	Yes	Yes	Converted from inorganic Hg	No	No	Yes

[a] Schroeder & Munthe (1998);
[b] Xiao et al. (2004);
[c] Ma et al. (2010);
[d] Clever et al. (1985);
[e] NRC (2000);
[f] Huber et al. (2006);
[g] Lu & Schroeder (2004);
[h] Andersson et al. (2008);
[i] Sommar et al. (2000);
[j] Iverfeldt & Lindqvist (1982);
[k] Mason et al. (1996);
[l] Calculated from aqueous solubility, vapour pressure and K$_{OW}$;
[m] Harner et al. (1999).

salinity alone could only explain ~40% of the variance in its concentrations (Figure 19.2b). In the case of Hg, which is highly particle-reactive (Figure 19.2c), although there is also an increasing relationship with salinity (Figure 19.3c), its concentrations in the bulk ice (0.1–4 ng L^{-1}) were similar to or greater than in seawater from which the ice had formed (~0.2 ng L^{-1}), due to the dominant role of particle entrapment (see later). In either case, as brine volume in sea ice is typically small, this results in highly elevated HCH and Hg concentrations in sea ice brine (Pućko et al., 2010b; Chaulk et al., 2011).

Table 19.2 Concentration ranges of total mercury (Hg$_T$) and methylated mercury (MeHg) measured in bulk sea ice, brine, frost flowers and surface hoar in the Arctic, along with concentrations in surface seawater, deposited and blowing snow, and diamond dust.

	Hg$_T$ (ng L^{-1})	MeHg (ng L^{-1})	Study area
FYI – bulk	0.1–4[a]	–	Amundsen Gulf
FYI – brine	3–70[a]	–	Amundsen Gulf
FYI – frost flowers	14–190[b]	–	Barrow
FYI – surface hoar	420–820[b]	–	Barrow
	50–700[c]		Barrow
FYI – melt ponds	2.1–17.1[d]	–	Amundsen Gulf
MYI – bulk	0.1–12[a,e]	<0.02–0.5[e]	Beaufort Sea and McClure Strait
Surface seawater	0.1–0.3[a]	–	Amundsen Gulf
	0.2–0.6[f]	<0.01–0.06[f]	Beaufort Sea
	1.5–4[b]	–	Barrow
Snow – deposited	1–150[a]	–	Amundsen Gulf
	50–110[b]	–	Barrow
Snow – blowing	10–1500[c]	–	Barrow
Diamond dust	100–1100[c]	–	Barrow

FYI, first-year ice; MYI, multi-year ice.
[a] Chaulk et al. (2011);
[b] Douglas et al. (2005);
[c] Douglas et al. (2008);
[d] Chaulk (2011);
[e] Beattie et al. (2014);
[f] Wang et al. (2012).

Table 19.3 Concentration ranges of α- and γ-hexachlorocyclohexane (α- and γ-HCH) measured in bulk sea ice, brine, and melt ponds in the Arctic along with concentrations in surface seawater and snow.

	α-HCH (ng L^{-1})	EF	Study area	γ-HCH (ng L^{-1})	Study area
FYI – bulk	0.13–0.44[a]	0.26–0.55[a]	Amundsen Gulf	0.02–0.07[a]	Amundsen Gulf
FYI – brine	1.83–5.32[b]	0.45–0.46[c]	Amundsen Gulf	0.22–0.57[b]	Amundsen Gulf
MYI – bulk	0.12–0.18[a]	0.27–0.46[a]	Beaufort Sea	0.03–0.04[a]	Beaufort Sea
	0.05–0.11[d]	0.46–0.52[d]	Beaufort Sea		
MYI – melt ponds	0.22–0.83[d]	0.48–0.52[d]	Beaufort Sea		
Surface seawater	0.82–2.00[a]	0.36–0.47[a]	Amundsen Gulf	0.07–0.17[a]	
	0.88–1.14[b]	–	Amundsen Gulf	0.09–0.13[b]	Amundsen Gulf
Snow - deposited	0.20–0.50[c]	0.46–0.52[c]	Amundsen Gulf	0.04–0.10[c]	Amundsen Gulf

EF, enantiomer fraction; FYI, first-year ice; MYI, multi-year ice.
[a] Pućko et al. (2010b);
[b] Pućko et al. (2010a);
[c] Pućko et al. (2011);
[d] Pućko et al. (2012).

During ice formation, incorporation of some small inorganic contaminants, such as fluoride, ammonium ions and some gases, in the ice crystal matrix is possible (Chapter 1). Molecular dynamic simulations also suggest that while PAHs such as naphthalene and phenanthrene in super-cooled aqueous solution are displaced to the air–ice interface upon freezing, benzene, the simplest single-ring aromatic hydrocarbon, can

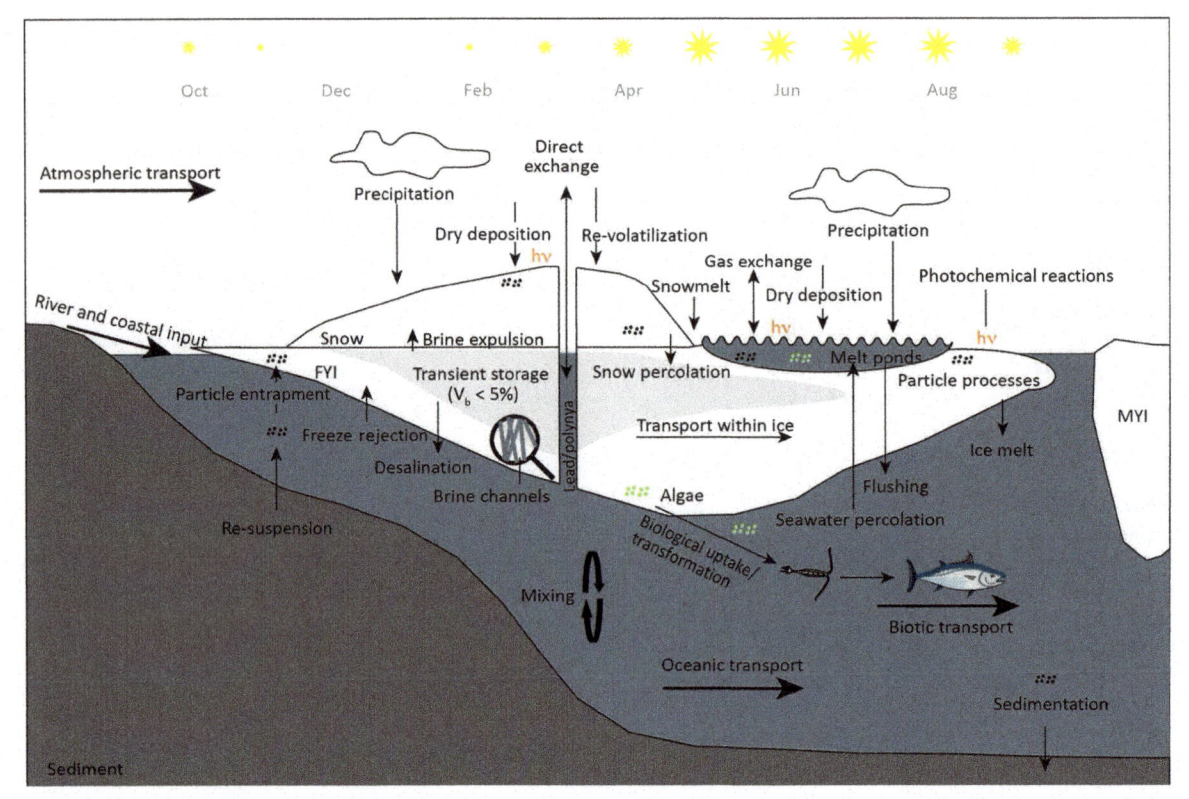

Figure 19.1 Major processes determining contaminant concentrations and bioaccumulation across the ocean–sea ice–atmosphere interface. FYI, first-year ice; MYI, multi-year ice; V_b, brine volume fraction; hν, solar radiation.

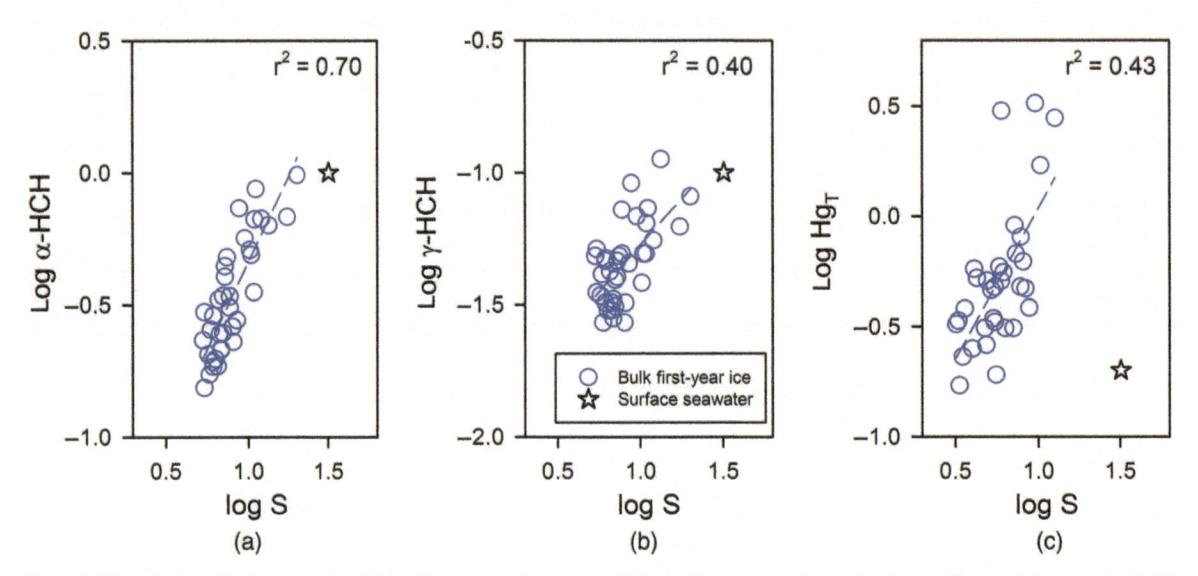

Figure 19.2 Relationship between the bulk salinity (S) of sea ice and the bulk concentrations of α-hexachlorocyclohexane (α-HCH) (ng L^{-1}) (a), γ-HCH (ng L^{-1}) (b) and total mercury (Hg$_T$, ng L^{-1}) (c) in first-year sea ice sampled from the Amundsen Gulf and Beaufort Sea. The graphs are plotted based on the data from Pućko et al. (2010b) (a, b) and Chaulk et al. (2011) (c).

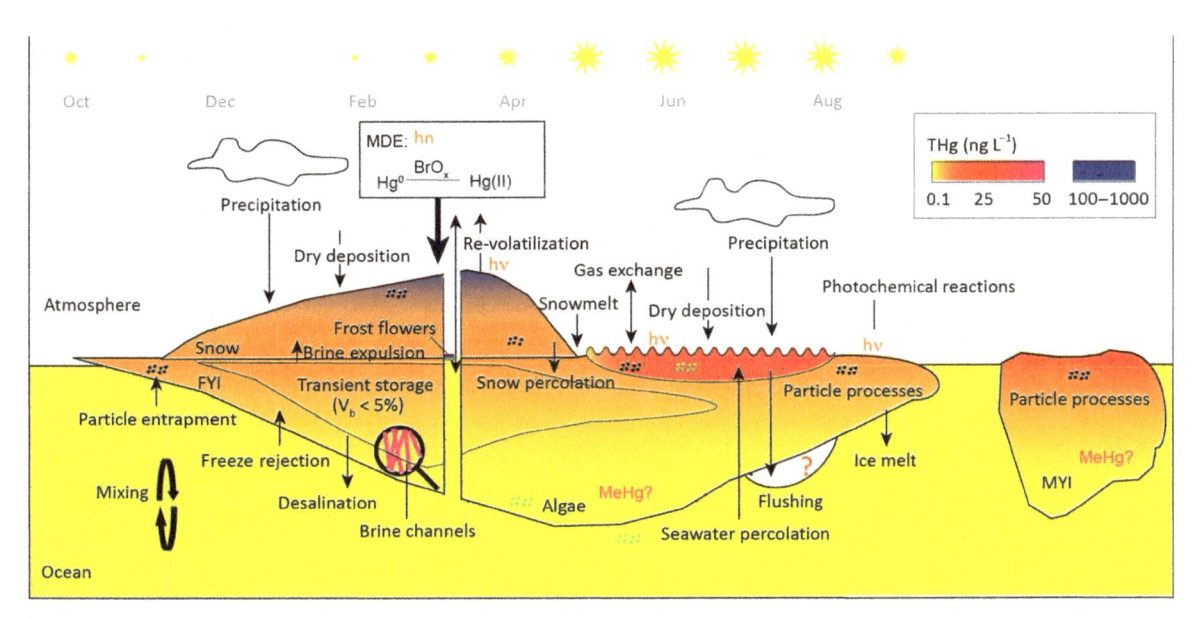

Figure 19.3 A simplistic schematic of mercury cycling in the Arctic sea ice environment. Colours denote approximate concentrations. Hg_T, total mercury; MeHg, methylmercury; MDE, mercury depletion events; BrO_x, reactive bromine species ($x = 0, 1$); FYI, first-year ice; MYI, multi-year ice; V_b, brine volume fraction.

become trapped inside the ice lattice due to its smaller size (Liyana-Arachchi et al., 2012).

19.3.2 Particle entrapment

Another mechanism for contaminants to enter sea ice is via particle entrapment. Aquatic particles not only aid in the initial nucleation of frazil ice crystals, but can also be trapped during the process of suspension freezing. Suspension freezing occurs in shallow shelf waters with a water depth typically < 50 m, where turbulent scavenging and filtration of suspended sediment by rising ice crystals promote the entrainment of particles in newly forming ice (Reimnitz et al., 1993; Dethleff & Kuhlmann, 2009). As a result, while concentrations of particles in the Arctic sea ice are generally very low (a few mg L^{-1}), turbid sea ice has a particle concentration typically ranging from 50 to 500 mg L^{-1} and can be as high as 3000 mg L^{-1} (Dethleff & Kuhlmann, 2009). Suspension freezing preferentially enriches fine-grained silt and clay minerals from bottom sediments to sea ice (Dethleff & Kuhlmann, 2009). As many trace metals and organic contaminants have high affinity to these fine-grained sediment, this provides an effective mechanism for particle-reactive contaminants to enter sea ice from shelf sediments (Pfirman et al., 1995),

resulting in higher concentrations in the bulk ice than in seawater (e.g. Figure 19.2c). Once in sea ice, particles are less prone to undergoing brine movement, enriching particulate contaminants (e.g. Hg; Chaulk, 2011) in the topmost layer of the ice, especially in multi-year ice (MYI) where soluble contaminants may be lost during recurrent melt–thaw cycles (Beattie et al., 2014) (see Figure 19.4a,b).

19.3.3 Wet and dry deposition

Airborne contaminants can enter the sea ice environment via wet and dry deposition. Snow is an excellent scavenger of gas-phase and particle-bound contaminants from the atmosphere. Scavenging of particle-bound contaminants can occur during the ice-forming process in the mid-air when the particles act as cloud condensation nuclei or ice-forming nuclei, or during the snow-falling process when snowflakes act as a particle filter (Wania et al., 1998; Grannas et al., 2013). Scavenging of gas-phase contaminants occurs on the snowflake surface via adsorption, a process that is enhanced by large surface areas of snowflakes and by cold temperatures (Grannas et al., 2013). Even though total annual snowfall over the Arctic Ocean is rather small (~300 mm $year^{-1}$), due to its desert-like conditions

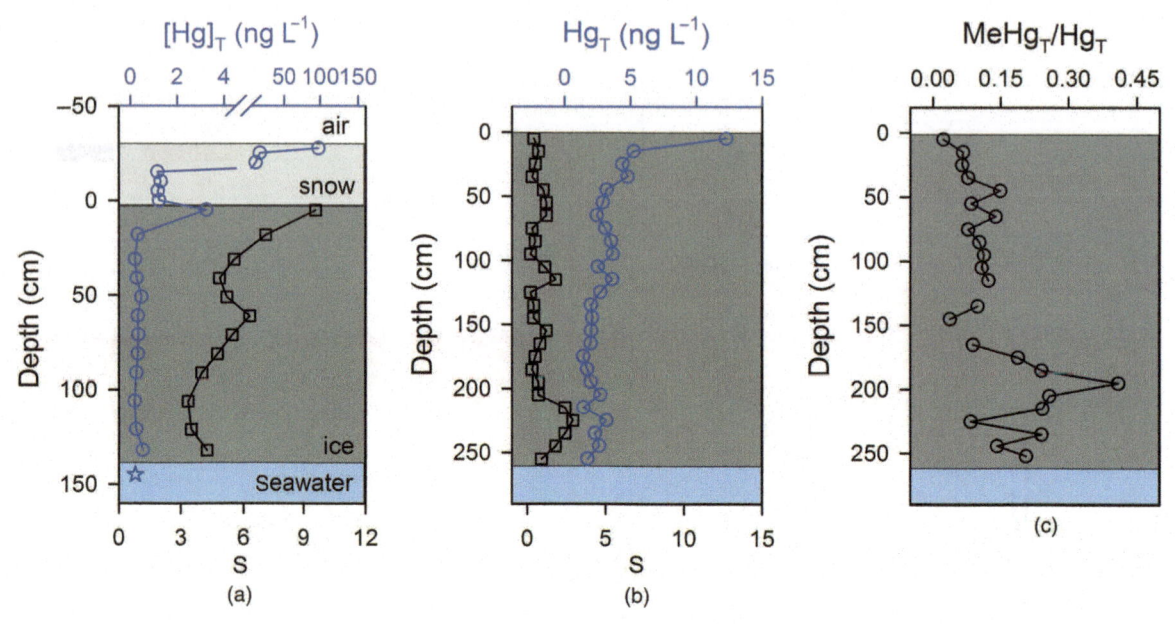

Figure 19.4 Mercury distribution profiles in the Arctic sea ice environment. (a) First-year ice (FYI): Total mercury (Hg$_T$) and salinity (S) distribution profiles across the ocean–sea ice–snow–air interface at a drifting FYI site in the Amundsen Gulf sampled in March 2008 (based on the data from Chaulk et al., 2011). Note the highly elevated Hg$_T$ at the surface of snow due to the occurrence of mercury depletion events (MDEs) in the air. (b, c) Multi-year sea ice (MYI): Hg$_T$, S and the ratio of total methylmercury (MeHg$_T$) to Hg$_T$ profiles across an MYI floe in McClure Strait sampled in September 2014 (based on the data from Beattie et al., 2014). Note the elevated MeHg$_T$/Hg$_T$ ratio in the bottom 70 cm of the ice.

(Chapter 3), the flux of contaminants associated with snowfall can be considerable, especially in the case of Hg during springtime atmospheric mercury depletion events (see Section 19.5.1). It should be noted that only a small fraction of the snow-scavenged contaminants enters the ocean directly via open leads or polynyas; the vast majority falls onto snowpack overlying the sea ice, where it undergoes various post-depositional transport and transformation processes before being released to sea ice and/or seawater, or re-volatilized back to the atmosphere.

Dry deposition of contaminants also occurs to snowpack, bare sea ice surface, open leads and melt ponds. Biomass burning in Asia and other industrial activities, for instance, produce springtime Arctic haze when concentrations of aerosols can reach well above 2 μg m^{-3} (compared with < 0.5 μg m^{-3} in summer) (Shaw et al., 2010), transporting particle-bound contaminants from faraway sources to the Arctic sea ice environment. Also, greater concentrations of aerosols originating primarily from frost flowers can occur during winter (Shaw et al., 2010).

Melt ponds present a unique opportunity for atmospheric dry deposition of organic contaminants (Pućko et al., 2012, 2014b). Melt ponds are relatively shallow and well-mixed water reservoirs developed on the surface of the ice during summer melt and can last for about 30–50 days (Eicken et al., 2002). They receive not only additional contaminant input from melting snowpack and precipitation, but also potentially large quantities of atmospherically deposited contaminants due to their large surface area-to-volume ratio. The latter is especially important for contaminants with relatively high Henry's Law constants (i.e. will be more readily deposited to the melt-pond water). For instance, melt-pond waters have been found to contain potentially up to 10 times higher concentrations of endosulfan (a POP banned under the Stockholm Convention) and chlorothalonil (a CUP) (Pućko et al., 2014b), and up to 100 times greater concentration of Hg than the surface seawater (Chaulk, 2011).

19.4 Transport of contaminants across the OSA interface

19.4.1 Vertical transport

The presence of a sea ice cover at the surface of the ocean inevitably slows down the exchange of contaminants between the atmosphere and the ocean. Such exchange may become essentially absent in midwinter when the ice permeability is too low to allow fluid transport (Chapter 1). However, exchange across the ice–water interface and across the ice–snow/air interface is never stopped due to greater permeability in the warmer, lowermost part of the ice and due to brine expulsion into the overlying snow layer, respectively (Figure 19.1). When the ice reaches a thickness of a few centimetres and cools, some of the brine freezes, creating internal pressure resulting in formation of a thin, surface quasi-liquid layer (QLL) as the excess volume is being pushed up (Perovich & Richter-Menge, 1994; Martin et al., 1995). The brine expelled from the ice to the snow will redistribute contaminants, and the significance of this process differs between contaminants of different partitioning behaviour in ice, brine and snow. For instance, brine expulsion has been shown to be capable of increasing α- and γ-HCH concentrations in snow by up to 50% (Pućko et al., 2011).

Transport across the entire OSA interface becomes possible in spring when the brine volume fractions (V_b) of snow and ice increase due to warmer temperatures. This leads to contaminant percolation from snow to sea ice, redistribution within the ice column and, eventually, flushing out of the ice to the underlying seawater (Figure 19.1). During summer melt, the meltwater flushes out not only dissolved chemicals but also particulate species. Depending on the contaminant physical–chemical properties, some contaminants will be enriched in the meltwater early during melt (e.g. relatively water-soluble chemicals such as α-HCH), end of melt (e.g. particle-reactive chemicals such as Hg), or both (e.g. contaminants that are both water-soluble and particle-reactive such as γ-HCH) (Grannas et al., 2013).

Direct exchange between the atmosphere and the ocean can occur even in midwinter via open leads and polynyas. As gas-phase (e.g. Hg^0) and semi-volatile (e.g. HCH) chemicals tend to oversaturate under winter pack ice, greater ocean-to-atmosphere fluxes of these contaminants can be expected wherever there is open water. On the other hand, open water in the otherwise frozen ocean also provides a rapid and effective pathway for airborne contaminants to enter the ocean via dry deposition, direct snowfall and blowing snow. The absolute values of these fluxes are probably small due to small surface areas of leads and polynyas (~1% of the total pack ice area in the central Arctic Ocean; Chapter 1); however, their role in contaminant transport into the marine ecosystem cannot be overlooked, as polynyas are important habitats where birds and mammals tend to overwinter (Chapters 22 and 24) and where primary production is the highest in spring (Chapter 14).

19.4.2 Lateral transport

Continental shelves and coastal polynyas are areas with the greatest rates of ice production (Iwamoto et al., 2014). Contaminants incorporated in the ice from shelf sediments can then be exported laterally to other regions via ice drifts. There are two large-scale, distinct ice drift trajectories in the Arctic Ocean (Chapter 7) that could move ice-borne contaminants over long distances. The Transpolar Drift moves ice from the eastern Arctic to the central Arctic and exits to the North Atlantic, whereas the Beaufort Gyre recirculates ice in the Western Arctic in a clockwise fashion connecting the Chukchi, East Siberian and Beaufort Seas (Figure 7.2). Assuming some of the ice survives the summer melt (i.e. MYI), modelling studies have shown that sea ice can transport contaminants from potential sources in the Kara and Laptev Seas to the Central Arctic Ocean and eventually to the Fram Strait within 2–5 years, and from the East Siberian, Chukchi and Beaufort Seas to the Fram Strait within 6–11 years (Pavlov et al., 2004; Hutchings & Rigor, 2012). As the ice drifts within the Arctic Ocean, the melt–thaw cycle tends to ablate the ice surface in summer and add ice to the bottom of multi-year floes in winter. This process releases soluble contaminants along with brine, but moves particles and associated contaminants to the ice surface (Macdonald & Bewers, 1996). These particle-bound contaminants will ultimately be released at locations where the ice completely melts (e.g. in the Greenland Sea).

Due to the relatively small volume of sea ice, net transport of contaminants via ice drift may not be significant when compared with other transport pathways such as atmospheric, oceanic, riverine transport and sedimentation. Ice export has been shown to account for < 1.5% of the total loss of α-HCH (Macdonald et al., 2000) and < 4% of Hg (Outridge et al., 2008) from the Arctic Ocean. However, sea ice motion could act as the main pathway of long-distance transport of oil should an oil spill occur in the Arctic Ocean (see Section 19.7.3).

19.4.3 Seasonal storage and transport

Sea ice transports contaminants not only in space, but also in time due to annual freeze and melt cycles. After the initial few weeks of ice life cycle, the majority of the ice column will be cold enough to become impermeable to fluid transport. Using brine volume fraction $V_b = 5\%$ as the lower threshold for fluid transport within sea ice (Chapter 1), in winter only the lowermost part of the Arctic sea ice, around one-third of the total ice thickness, is permeable (Vancoppenolle et al., 2013) (Figure 19.2). Therefore, around two-thirds of the winter sea ice essentially becomes a transient storage for contaminants with distributions reflecting the ice growth history, e.g. location of formation, growth rate and meteorological conditions (Pućko et al., 2011). As the temperature warms up, the permeability increases and the ice-borne contaminants start to be released into seawater or the atmosphere. Through these annual freeze and melt cycles, sea ice accumulates and stores contaminants during fall and winter and then releases them completely (in FYI) or partially (in MYI) in spring and summer in a pulsed fashion at locations close to (e.g. landfast ice) or far away from (e.g. MYI) the source region.

19.5 Transformation of contaminants across the OSA Interface

Sea ice not only acts as a physical storage and transporter for contaminants, it also provides a unique environment where a variety of chemical processes take place, resulting in changes in chemical forms, and thus fate and effects of these contaminants. The mechanisms and pathways of these cryospheric chemical processes at sub-zero temperatures and high ionic strengths remain poorly characterized, but the ones that are kinetically fast appear to be photolytically initiated or microbially mediated.

19.5.1 Photochemical processes

Photochemical processes occur in the sea ice environment wherever the sunlight irradiates (Chapter 4). Surface snow, open lead and melt ponds are locations with the highest solar irradiance, but photochemical reactions could take place across the entire OSA interface, depending on the thickness and structure of snowpack and ice column (Xu et al., 2016).

Some contaminants in the sea ice environment may undergo direct photochemical reactions (e.g. photoreduction of Hg^{II}), while others react with reactive species that are photolytically produced in the sea ice environment. The latter is perhaps best demonstrated by the occurrence of bromine explosion events (BEEs) (Simpson et al., 2007). Following polar sunrise, episodic BEEs occur when concentrations of reactive bromine species (e.g. Br, BrO, HOBr) increase exponentially in the marine boundary layer. Their rapid reactions with ozone result in concurrent ozone depletion events (ODEs) that can sometimes completely deplete ozone in the boundary layer (Simpson et al., 2007). Detailed mechanisms leading to the occurrence of BEEs are yet to be fully understood (Abbatt et al., 2012), but recent studies suggest that the upper, sunlit layer of snow pack overlying sea ice is the most likely location where bromine activation from bromide occurs (Pratt et al., 2013; Xu et al., 2016). In addition to reactive bromine species, other reactive halogenated species (e.g. Cl, I) (Abbatt et al., 2012; Liao et al., 2014) and reactive oxygen species (e.g. O_3, H_2O_2, OH, NO_x) (Grannas et al., 2013) can also be produced.

The occurrence of photolytically produced reactive halogenated species, reactive oxygen species, along with the depletion of ozone, significantly changes the oxidative conditions of the Arctic marine boundary layer, affecting chemical transformation of many inorganic and organic contaminants. For instance, the oxidation of gaseous elemental Hg (Hg^0) is typically very slow in the troposphere, allowing it to be transported over long distances. During BEEs, however, Hg^0 in the air is rapidly oxidized by reactive bromine species, producing oxidized Hg (Hg^{II}) which is readily removed from the air via wet and dry deposition. This results in the so-called mercury depletion events (MDEs) that accompany BEEs and ODEs (Schroeder et al., 1998; Simpson et al., 2007). MDEs provide a rapid and efficient mechanism for depositing atmospheric Hg to the sea ice environment. Once in the snowpack, Hg^{II} can undergo rapid photoreduction and part of it is re-volatilized back to the atmosphere in the form of Hg^0.

Many organic contaminants react readily with reactive oxygen species produced from snow or ice (Grannas et al., 2013). While photolysis of at least some aromatic compounds at saline ice surfaces seems to follow similar kinetics as in liquid saline water (Kahan et al., 2014), photochemical reactions at pure or freshwater ice surfaces (e.g. fresh snow) may occur differently from those at the liquid water surfaces, both in the reaction rate and in the end products (Grannas et al., 2013; Kahan et al., 2014). For instance, photochemical degradation of two organochlorine POPs (aldrin and dieldrin) (Rowland et al., 2011) and several PAHs (Kahan & Donaldson, 2007) has been shown to be more efficient when the aqueous solution is frozen. The reason for such differences remains a mystery. It could be due to enrichment of the contaminant in the QLL at the ice surface, with the resulting heterogeneous reactions exhibiting faster kinetics than their homogeneous counterparts (Liyana-Arachchi et al., 2012; Grannas et al., 2013). In the case of PAHs, the enhanced photolysis kinetics at the ice surfaces could be due to increased absorption of solar radiation, as aromatic species undergo extensive self-association at the ice surfaces, resulting in red shifts in their absorption spectra (Kahan & Donaldson, 2007; Kahan et al., 2014). Furthermore, water molecules act as nucleophiles in aqueous solutions but not in ice (Grannas et al., 2013).

19.5.2 Microbially mediated redox reactions

During spring thaw, oxygen depletion may develop in the bottom ice layer due to heterotrophic activity and melting of deoxygenated ice crystals (Rysgaard & Glud, 2004), creating favourable conditions for anaerobic processes such as denitrification to take place (Chapters 13 and 18). This seasonal occurrence of anaerobic processes in sea ice makes it possible for anaerobic transformation of sea ice-borne contaminants at the surface of the ocean. For instance, one of the key processes in Hg biogeochemistry is methylation of inorganic Hg, a process that is primarily mediated by sulphate- or iron-reducing bacteria (Driscoll et al., 2013). Although to date none of these bacteria have been identified in sea ice (Chapter 13), elevated methylmercury (MeHg) concentrations have recently been reported in the mid-to-bottom layer of MYI during the summer melt season (see Figure 19.4), suggesting the possible occurrence of microbially mediated Hg methylation in sea ice (Beattie et al., 2014).

19.6 Biological uptake of sea ice-borne contaminants

Upon melting of sea ice, ice-borne contaminants are released to the surface ocean. Even though the overall flux may not be very large, this could represent an important pathway for uptake of contaminants into the marine food web, as the period of melting coincides with the period of maximum primary productivity in the surface ocean (Macdonald & Bewers, 1996). Contaminants bound to sinking particles from snow and ice melt could be taken up by benthic organisms. Biological uptake of ice-borne contaminants also occurs within the ice-associated communities long before the ice melts. Contaminants in sea ice brine, highly concentrated due to freeze rejection (Pućko et al., 2010b; Chaulk et al., 2011), can be directly taken up by sea ice-associated biota, such as microbial communities (Chapter 13), ice algae (Chapter 14), or under-ice amphipods (Chapters 15 and 16). Furthermore, organisms such as fishes, seals, polar bears and birds that depend on sea ice for habitat, feeding or spawning (Chapters 16, 21, 22, 23, & 24) are also exposed to sea-ice borne contaminants.

Of particular importance are sea ice algae, which contribute 10–60% of the annual primary production in seasonally ice-covered seas (Chapter 14). Burt et al. (2013) examined Hg bioaccumulation by a FYI algal community in the Beaufort Sea during the spring of 2008. Mercury concentrations in particles at the bottom ice ranged from 4 to 22 µg g^{-1} (dry weight) and were limited by the amount of Hg available for uptake when the spring bloom commenced. The same authors also found that Hg in ice algae originated from a combination of brine and seawater, while atmospheric MDEs did not appear to have significantly contributed as a source in a coupled manner.

19.7 Case studies

19.7.1 Mercury in sea ice

Mercury is one of the primary contaminants of concern in the Arctic; it biomagnifies in the marine food web and its methylated form MeHg is a known developmental neurotoxin. Highly elevated Hg concentrations have been detected in marine mammals such as seals, whales and polar bears throughout the circumpolar Arctic (AMAP, 2011), raising health concerns of these animals and of the northern indigenous people who consume these animal tissues as part of their traditional diet (AMAP, 2009). Mercury is a naturally occurring element in the Earth's crust, but its release to the environment has been greatly disturbed and augmented by human activities such as fossil fuel combustion and mining. Present-day global emission of Hg to the atmosphere is estimated at ~7500 t year^{-1}, only < 10% of which is considered to be from natural geogenic emissions; the remaining 90% is attributed to primary anthropogenic emissions (~30%) and re-emissions of previously deposited anthropogenic and natural Hg (~60%) (Pirrone et al., 2010; Driscoll et al., 2013). Even though anthropogenic emissions of Hg are gradually placed under control, especially with the recent successful negotiation of the Minamata Convention on Mercury, re-emissions and cycling from the large reservoir of historic 'legacy' Hg will continue to take place in the Arctic environment. The present-day standing stock of Hg in the upper 200 m of the Arctic Ocean is estimated to be ~950 t (~8000 t in the entire Arctic Ocean), with a total annual input of ~200 t year^{-1} (Outridge et al., 2008). In general, Hg enters the Arctic Ocean via atmospheric deposition (48%), ocean currents (23%), coastal erosion (23%) and riverine input (6%). The main routes of removal of Hg from the Arctic include sedimentation (59%), ocean currents (37%) and ice export (4%).

In its elemental form, Hg^0 is highly volatile (Table 19.1). The volatility of Hg^0 and its generally low reactivity in the atmosphere are responsible for its long-range atmospheric transport to remote regions such as the Arctic. Background atmospheric Hg^0 concentrations in the Arctic typically vary within a narrow range of 1.5–1.7 ng m^{-3} except during MDEs (Fisher et al., 2013). Following Hg^0 oxidation to Hg^{II} (see HgCl$_2$ as an example compound in Table 19.1), atmospheric Hg can be readily deposited onto the surface environment. Such oxidation is typically slow, resulting in Hg^0 being the dominant atmospheric Hg species. Once in the ocean, inorganic Hg^{II} can be methylated to form MeHg (see CH$_3$HgCl as an example compound in Table 19.1), which biomagnifies in the marine food web and can cross the blood–brain barrier, posing neurological and other health risks to animals at higher trophic levels, including humans (Driscoll et al., 2013). Methylation of Hg in the aquatic environment is generally thought to be primarily an anoxic microbial process by some species of sulphate- or iron-reducing bacteria (King et al., 2000; Fleming et al., 2006), commonly abundant in shelf sediments. However, the recent discovery of a sub-surface peak in MeHg concentration in Arctic seawater, at a depth with low but considerable concentrations of dissolved oxygen (Wang et al., 2012), suggests that MeHg can be produced in the oxic water column either within reducing microenvironments such as zooplankton guts (Pućko et al., 2014a) or via a new methylation pathway.

Extensive studies in the past two decades have shown that the sea ice environment plays a pivotal role in Hg transport, transformation and biological uptake in the Arctic Ocean (Figures 19.3 and 19.4). One of the unique processes occurring to Hg across the OSA interface is the springtime episodic MDEs. As mentioned in Section 19.5.1, during MDEs, Hg^0 in the marine boundary layer is rapidly oxidized by reactive halogen species that are photolytically activated, most likely from the upper layer of the snowpack. MDEs are thought to have been enhanced by convective forcing from leads, which brings air with undepleted Hg^0 for renewed oxidation (Moore et al., 2014). As a result, frost flowers, surface hoar and diamond dust in the proximity of large leads with convective plumes have been reported to contain very high concentrations of Hg (14–190, 50–820 and 100–1100 ng L^{-1}, respectively), which are two to four orders of magnitude higher than in the surface ocean (Table 19.2; Figure 19.3). Deposition of MDE-derived Hg^{II} also results in very high Hg concentrations in snow (up to 1500 ng L^{-1}; Table 19.2, Figure 19.4). However, ~75% of Hg^{II} deposited to Arctic snowpacks is re-volatilized back to the atmosphere following rapid photoreduction to Hg^0 (Outridge et al., 2008; Durnford et al., 2012). The rest can enter the ocean via leaching through sea ice or during melt. Very limited data on melt ponds overlying FYI showed a Hg concentration range of 2.1–17.1 ng L^{-1} (Chaulk, 2011) (Table 19.2).

Sea ice incorporates Hg primarily from seawater and sediments. Total Hg concentrations in FYI sampled from the Beaufort Sea were generally low (~0.5 ng L^{-1}), only slightly higher than that in surface seawater (~0.2 ng L^{-1}), except in the topmost layer where concentrations could reach up to 4 ng L^{-1} (Figure 19.4), which can be explained by the combined effect of particle entrapment and freeze rejection (Chaulk et al., 2011). The influence of atmospheric Hg on sea ice appears to be limited, as new ice grown during MDEs does not contain much higher Hg than the underlying seawater (Chaulk et al., 2011). The exceptions are frost flowers that scavenge Hg from the air during kinetic crystal growth from the vapour phase (Douglas et al., 2005, 2008) and during the melt season when snow mercury percolates through ice (Chaulk et al., 2011).

Surface enrichment of Hg becomes even more profound in MYI, with bulk concentrations to be as high as 12 ng L^{-1} in the Beaufort Sea (Table 19.2; Figure 19.4) (Beattie et al., 2014). As dissolved Hg may have been leached out from the ice after recurrent desalination and meltwater flushing, the surface of MYI retains particles and associated particulate Hg from the air and melt ponds over its lifetime, resulting in the surface enrichment of particulate Hg. Mercury methylation in MYI has also been recently proposed (Beattieet al., 2014); MeHg concentrations as high as 0.5 ng L^{-1} have been measured in the mid-to-bottom layer of the ice (Table 19.2), representing up to 40% of the total Hg concentration in the ice (Figure 19.4). If this is ubiquitous in the Arctic Ocean, MYI could be an important 'hotspot' for *in situ* Hg methylation, which has not been previously accounted for. The mechanism for this MeHg production has yet to be elucidated, but methylation by anaerobic bacteria is possible as anoxic conditions are known to develop in the bottom layer of ice during spring melt (see Section 19.5.2).

19.7.2 HCH in sea ice

Hexachlorocyclohexane is an organochlorine insecticide that was used heavily worldwide from the 1940s. HCH was introduced to the environment in two major formulations: technical HCH which consists of five stable geometric isomers (55–80% α-HCH, 5–14% β-HCH, 8–15% γ-HCH, 2–16% δ-HCH and 3–5% ε-HCH) and lindane (>90% of γ-HCH, the only isomer with insecticidal properties) (Li & Macdonald, 2005). Technical HCH was the dominating contributor of HCH to the environment until the late 1970s, and thereafter lindane became the major source of this contaminant. Since 2009, HCH use has been banned worldwide under the Stockholm Convention on POPs. Structures of α- and γ-HCH, the two most abundant HCH isomers in the Arctic, are shown in Table 19.1.

In 2000, the estimated standing stock of HCH in the upper 200 m of the Arctic Ocean was ~3000 t, with total annual inputs of 186 t of α-HCH and 83 t of γ-HCH (Macdonald et al., 2000). α- and γ-HCH entered the Arctic Ocean via ocean currents (58% and 39%, respectively), atmospheric deposition (28% and 12%, respectively) and riverine input (13% and 49%, respectively) (Macdonald et al., 2000). The main routes of removal of α- and γ-HCH from the Arctic are water outflow (49% and 58% respectively), degradation (43% and 34%, respectively), volatilization (7% for both isomers) and ice export (1% for both isomers) (Macdonald et al., 2000). Microbial degradation accounts for ~85% and hydrolysis for ~15% of the total degradation of HCH in the Arctic Ocean (Harner et al., 1999). Currently, the major removal of HCHs is taking place by bacterial degradation, and complete elimination is expected within the next two decades (Li et al., 2004; Pućko et al., 2013).

Typical concentration ranges of HCH in the Arctic sea ice are shown in Table 19.3. Figure 19.5 provides a simplistic schematic of its cycling in the Arctic sea ice environment. Concentrations of HCH measured in the Arctic snow are comparable with those recorded in the bulk FYI (0.20–0.50 vs 0.13–0.44 and 0.04–0.10 vs 0.02–0.07 ng L^{-1} for α- and γ-isomer, respectively). Meteorological conditions after snowfall are known to have a major impact on the dynamics of HCH in the snowpack. Approximately 40% of HCH from fresh snow were shown to be lost within a day after snowfall under windy conditions (Pućko et al., 2011), due to the rapid reduction in snow surface area which then leads to increased snow density and reduced snow capacity to hold sorbed chemicals effectively (i.e. sintering) (Burniston et al., 2007).

In spring, HCH in snow and ice is leached out and eventually flushed out to the underlying seawater. The more water-soluble α-HCH will be amplified in the meltwater early during melt, whereas γ-HCH will be amplified at both early and final stages of melt, as it is partially dissolved in the aqueous meltwater and partially sorbed to particulate matter (Grannas et al., 2013). α-HCH concentrations in melt ponds on MYI were measured at 0.22–0.83 ng L^{-1}, a relatively high value compared with thick FYI (Table 19.3; Figure 19.5), and atmospheric dry deposition and percolation of seawater into the ponds once they thaw were identified as the main processes responsible for the observed high values (Pućko et al., 2012). α-HCH concentrations in bulk MYI were found to be variable with relatively high values in some parts of the ice cores, 0.05–0.18 ng L^{-1}, and significant elevations at the surface, due to

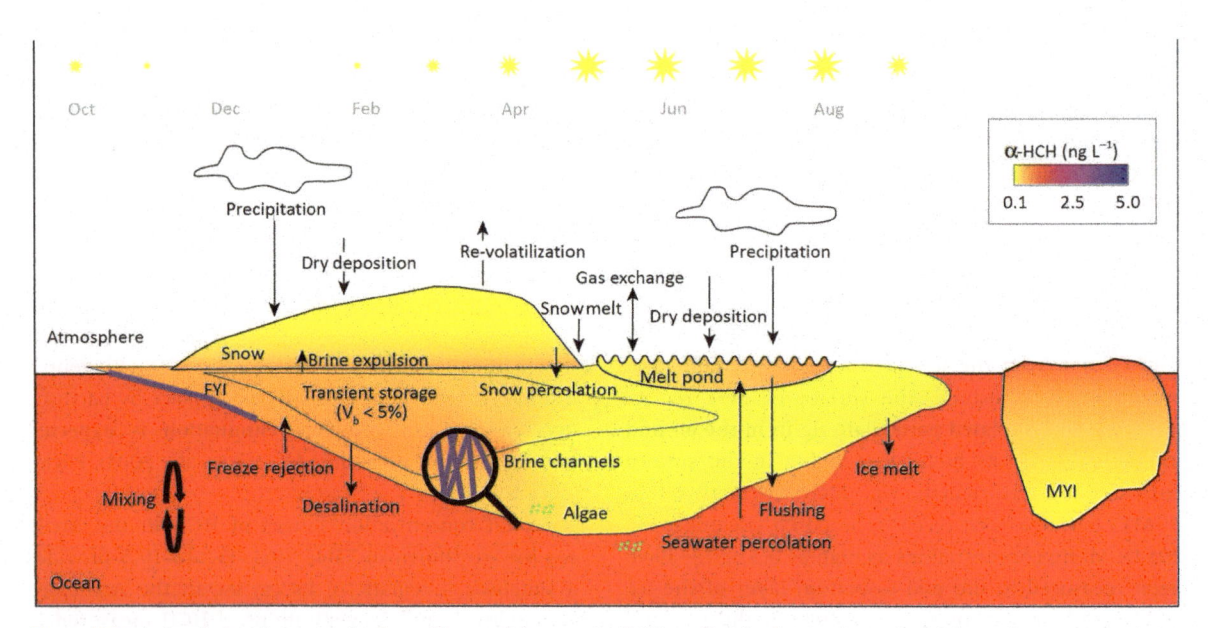

Figure 19.5 A simplistic schematic of α-hexachlorocyclohexane (α-HCH) cycling in the Arctic sea ice. Colours denote approximate concentrations. FYI, first-year ice; MYI, multi-year ice; V_b, brine volume fraction.

air–meltwater gas exchange (Pućko et al., 2012, 2014b), as discussed earlier. As α-HCH is present primarily in the dissolved fraction, its cycling in MYI presented by Pućko et al. (2012) does not include the particulate component.

α-Hexachlorocyclohexane is the only chiral HCH isomer that forms optical isomers (enantiomers), termed (+) and (−) α-HCH, which are non-superimposable mirror images of each other. α-HCH was released to the environment as a racemic mixture (1:1 ratio of enantiomers), but undergoes enantiodepletion when enantioselective processes come into play (Müller & Kohler, 2004). As a result, a change in enantiomer fraction (EF):

$$EF = (+)\alpha\text{-HCH}/[\text{sum of } (+) \text{ and } (-)\alpha\text{-HCH}] \quad (19.1)$$

has been used to study pathways and fate of this contaminant in the Arctic sea ice under the same fundamental principle used in studies of soil–water–air systems: only enzymatic processes such as microbial degradation, biological uptake or elimination are known to change the racemic α-HCH EF (0.5) of the initially applied mixture (Faller et al., 1991; Hühnerfuss et al., 1993). In the Canadian Arctic where the majority of studies on contaminants in sea ice took place, degradation of (+) α-HCH is found to be favoured in seawater and the EFs in surface seawater under ice-covered conditions range from 0.36 to 0.47 (Table 19.3). The air over the sea ice, on the other hand, was reported to have racemic composition of α-HCH (EF = 0.50 ± 0.01) (Wong et al., 2011).

Figure 19.6 presents a simplistic schematic diagram depicting Arctic sea ice processes that affect enantiomeric composition of α-HCH as traced by EF analysis. In general, FYI in the Beaufort Sea region has an α-HCH EF of ~0.47, reflecting the typical enantiomeric signature of seawater from which it had formed (Pućko et al., 2010b). In spring, however, α-HCH EF in FYI can be as high as >0.5 or as low as 0.26 at the bottom of the ice, reflecting snowmelt and biological uptake by ice algae or degradation by ice microbes, respectively (Table 19.3). α-HCH EF of snow reflects the signature of the atmosphere (~0.5), but can decrease to ~0.47 close to the ice surface as a result of brine expulsion. Initial EFs of the surface melt in spring are close to the values in sea ice and increase gradually as a result of dry atmospheric deposition until the ponds thaw and the surface seawater starts percolating into the ponds and bringing the EFs back down to ~0.47. Thus, melt-pond EF has been shown to be a good tracer of the melt pond evolution stage (Pućko et al., 2012). In late spring-early summer, atmospheric EFs were used as tracers of sea ice break-up, which can decrease the racemic EFs in the air down to ~0.48 (Wong et al., 2011). α-HCH EF distribution in MYI (racemic in the surface and

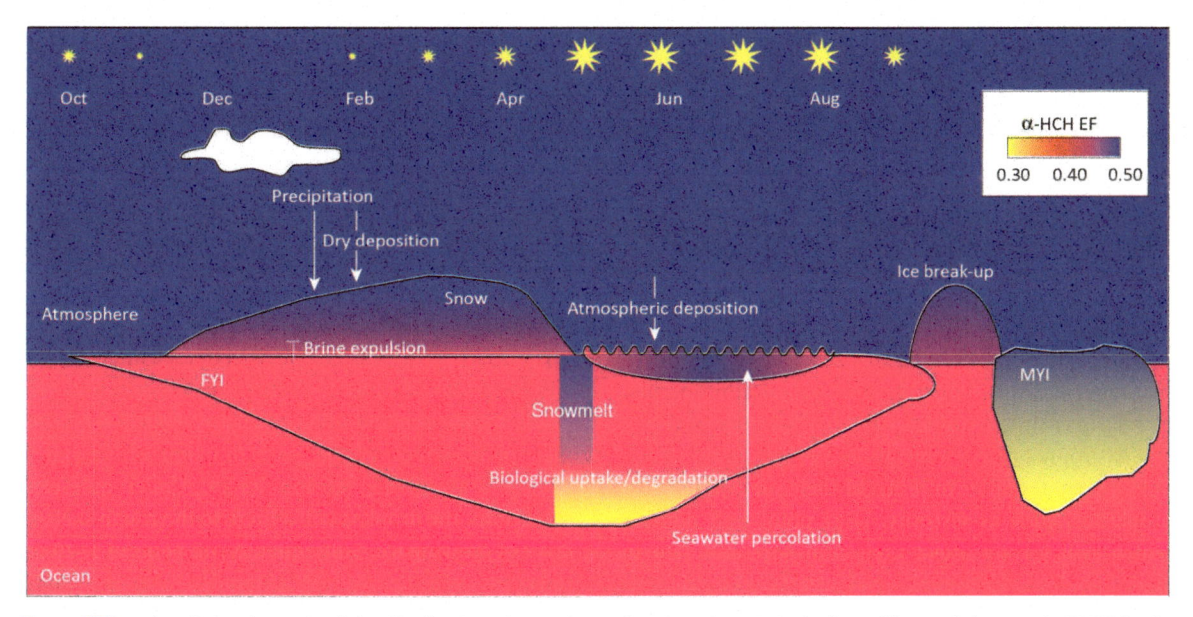

Figure 19.6 A simplistic schematic of the distribution of enantiomer fractionation (EF) of α-hexachlorocyclohexane (α-HCH) in the Arctic sea ice environment and the associated processes traced by EF. Colours denote approximate EF values. FYI, first-year ice; MYI, multi-year ice.

highly depleted in (+)-α-HCH in the middle to bottom parts of the ice) reflects spring processes, namely melt pond α-HCH enrichment of atmospheric origin and bottom-ice depletion due to biological activity in sea ice (ice algae and bacteria blooms) (Pućko et al., 2010a).

19.7.3 Crude oil and PAHs in sea ice

About 13% of the world's undiscovered oil is expected to be found north of the Arctic circle, mostly offshore under < 500 m of water (Gautier et al., 2009). Production drilling has been established in the offshore Alaskan Beaufort Sea since the 1990s, and the first ice-resistant oil platform has been operating in the Russian Pechora Sea since 2013. Offshore oil and gas production may release petrogenic hydrocarbons to the environment through chronic oil seeps, accidental oil spills (e.g. tanker accidents, burst pipelines under the seafloor), and produced water and drill cutting discharge from oil installations (Bakke et al., 2013). The Arctic may be exceptionally vulnerable to oil spill for several reasons. Forming or growing sea ice will probably encapsulate oil, transport it potentially undetected, and release it at the surface (surface pooling) upon melt away from the spill zone (Fingas & Hollebone, 2003). Seasonal or perennial presence of sea ice may greatly impair the detection and recovery response (Payne et al., 1991). Low temperatures in the Arctic waters generally inhibit bacterial degradation (Leahy & Colwell, 1990). Furthermore, typical recovery techniques developed for warmer regions are hard to apply in sea ice (Payne et al., 1991).

Parent (unsubstituted) and alkylated PAHs constitute a relatively small fraction of crude oil (up to ~1.5%) (Pampanin & Sydnes, 2013); however, they are of particular environmental significance due to their potential to form carcinogenic and mutagenic diols and epoxides that react with DNA (Shimada & Fujii-Kuriyama, 2004). Sixteen parent PAHs are on the U.S. Environmental Protection Agency's Priority Pollutants list, among which the most abundant in crude oil are naphthalene, phenanthrene, fluorene, acenapthene and chrysene (Payne et al., 1991; Pampanin & Sydnes, 2013) (Figure 19.7). Alkylated homologues of the most abundant parent PAHs such as naphthalene and phenanthrene are also of interest as their toxicity is higher and their abundance may be greater, particularly in weathered oil, compared with parent compounds (Black et al., 1983; Peterson et al., 2003).

The cycling and fate of crude oil in sea ice are especially complex as the mixture contains chemicals

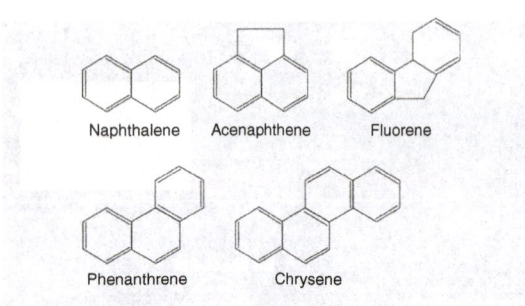

Figure 19.7 Structures of the most abundant parent polycyclic aromatic hydrocarbons in crude oil.

that differ greatly in their physical–chemical properties, including volatility, chemical and biological persistence, and hydrophobicity (Pampanin & Sydnes, 2013). Additionally, crudes differ in terms of their specific physical properties, such as viscosity, pour point and density, which will greatly affect their transport and fate in ice-infested waters (Payne et al., 1991). Considerable research has been done to understand the behavioural mechanisms of oil spilled in freezing waters using test tanks as well as controlled environmental spill experiments (Fingas & Hollebone, 2003); however, only a few quantitative studies exist (Payne et al., 1991; Faksness & Brandvik, 2008a,b). A summary of mechanistic behaviour of oil spilled in ice-infested waters is presented in Figure 19.8. As quantitative studies have been focused primarily on PAHs, our discussion will be limited to this class of compounds only.

If oil is released into water under freezing conditions of active ice growth, lower-molecular-weight PAHs, which are more water-soluble, may sink to the bottom with the brine generated during new ice formation, and may persist there in highly toxic dissolved and bioavailable form for up to several months (Payne et al., 1991). If the suspended sediment concentration in the water exceeds 100 mg L^{-1}, emulsified oil may adhere to particles and sink to the bottom and subsequently cause acute and long-term chronic toxic effects in benthic organisms (Payne et al., 1991). Oil released under the ice will form small droplets, < 1 cm in diameter, which will rise to the surface and upon contact with the ice bottom will spread concentrically as a function of buoyancy, viscous and surface tension forces. Under quiescent conditions, the droplets will adhere to the rough skeletal layer of growing ice. Subsequently, oil pools of 1–20 cm thickness will be formed in under-ice depressions, which will then become completely encapsulated by growing sea ice within hours to days. Ocean currents of > 20 cm s^{-1} may strip the oil from under-ice depressions and redistribute it. Throughout winter, water-soluble fractions of crude oil (mainly naphthalene and its alkylated homologues) can reach concentrations in the surrounding ice layers of up to 4 μg L^{-1}, which may cause acute toxicity to ice-associated organisms (Faksness & Brandvik, 2008a,b). The encapsulated oil will remain in place throughout winter in its primarily

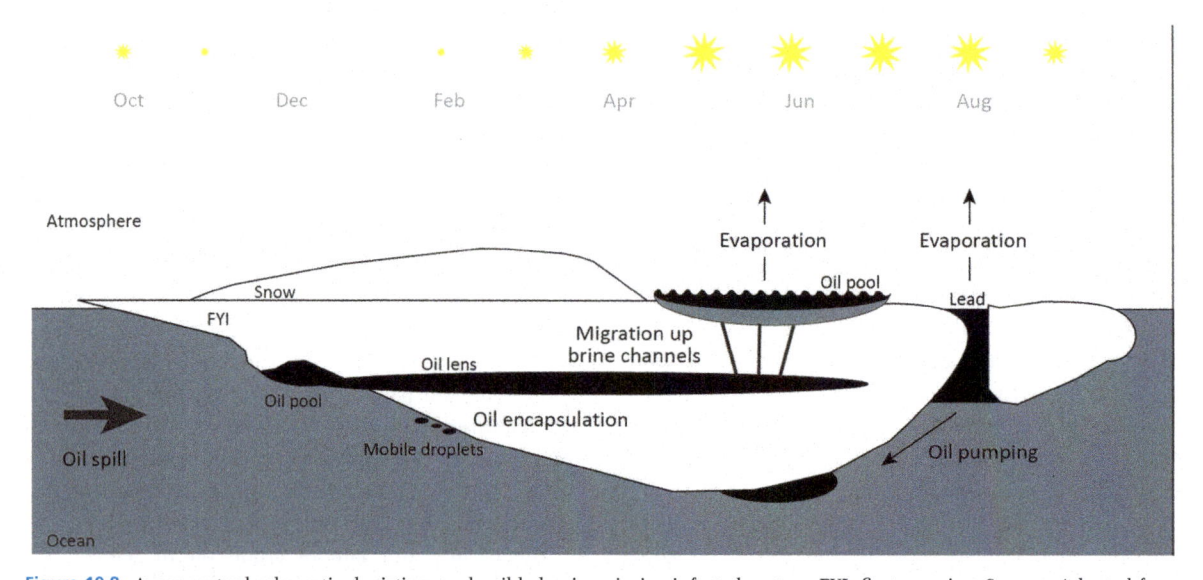

Figure 19.8 A conceptual schematic depicting crude oil behaviour in ice-infested waters. FYI, first-year ice. Source: Adapted from AMAP, 1998.

unweathered form (Payne et al., 1991), and will be released at the ice surface in spring by a combination of two processes: vertical rise of oil through the brine channels and ablation of ice surface down to the oil lens (Payne et al., 1991; Fingas & Hollebone, 2003).

The encapsulated oil will be released in the next melt season from FYI, but can persist in the MYI for up to five years. Oil exposed to the air will gradually weather due to natural physical and chemical processes such as spreading, evaporation, dispersion, emulsification, dissolution, photo-oxidation, adsorption to particles, sinking, ingestion by biota and biodegradation (Payne et al., 1991). When agitated, oil on surface seawater forms water-in-oil emulsions (up to 70% of water volumetric content after just a few hours of turbulence) with greatly increased viscosity and greatly decreased ignitability and ability to be dispersed by typical spill dispersants (Payne et al., 1991). As a result, spill clean-up strategies in spring ice with oil pooling, such as pumping, burning or use of dispersing agents, must be applied immediately after the oil surfaces in order to be effective; this is a rather unrealistic expectation given the difficulties in detecting and tracking encapsulated oil in sea ice (Payne et al., 1991). In heavy ice conditions (ice concentration >90%), low temperatures and reduced spreading and wave action significantly slow down weathering processes, resulting in reduced emulsification and evaporation of oil (Brandvik & Faksness, 2009). Thus, dense ice conditions may extend the operational time window for contingency methods application (Brandvik & Faksness, 2009).

Virtually all existing scientific efforts with regards to oil spills in sea ice did not address the issue of how particles in ice affect the pathways and fate of petrogenic PAHs. This question is of importance as PAHs with higher molecular weight tend to be bound to particles (Yunker & Macdonald, 1995). Furthermore, sediment resuspension and ice entrainment in the shallow Beaufort Sea and Russian shelf were recently shown to represent a significant term in the western Arctic Ocean sediment budget and are expected to further increase in the future (Eicken et al., 2005). Major gaps in knowledge on the pathways and fate of water-soluble fraction (WSF) and particle-associated fraction (PAF) of PAH in the Arctic sea ice are depicted in Figure 19.9. Specific relationships among the ice physical and thermodynamic state and PAH encapsulation, vertical transport, surface pooling, fractionation between water and particles, and weathering are yet to be understood. Finally, the question of which known or novel remediation techniques should be applied in ice-infested waters in order to minimize the harm to the aquatic environment remains to be answered.

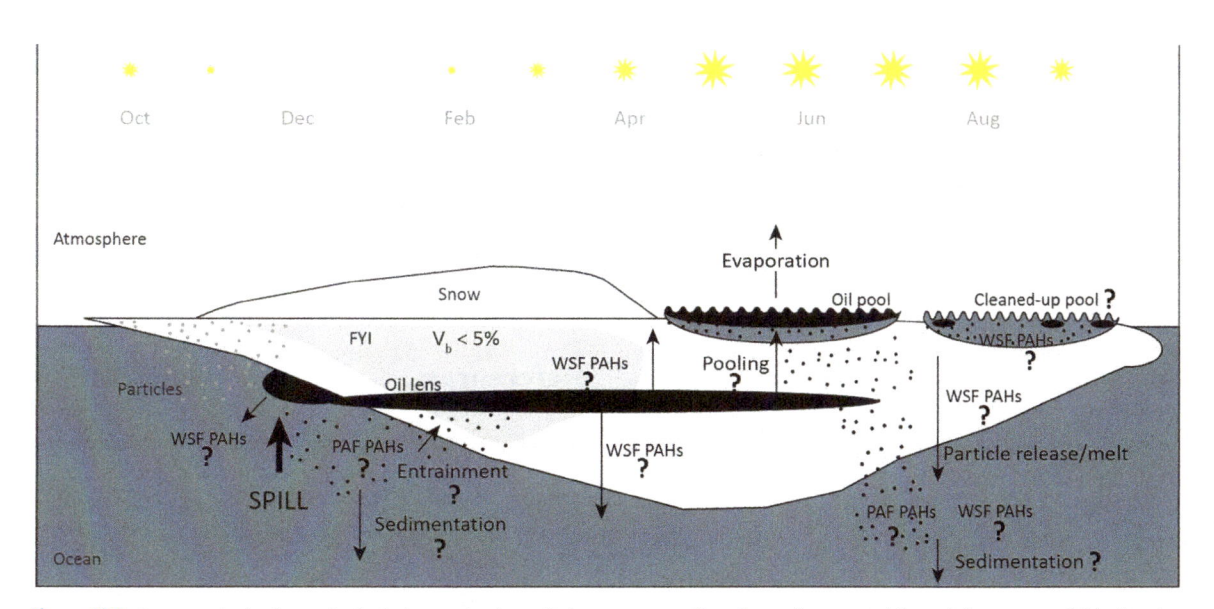

Figure 19.9 A conceptual schematic depicting major knowledge gaps regarding the pathways and fate of the water-soluble fraction (WSF) and particle-associated fraction (PAF) of polycyclic aromatic hydrocarbons (PAHs) in the Arctic sea ice. FYI, first-year ice; V_b, brine volume fraction.

19.8 Implications of a changing sea ice environment

The Arctic Ocean icescape has been undergoing dramatic changes in recent decades (Chapter 11). Rapid melting of MYI has resulted in an unprecedented reduction in the summer sea ice extent, which has declined by about 14% per decade, or 40% overall, from 1979 to 2013 (Stroeve et al., 2014). Accompanying the shrink in aerial coverage of MYI is an overall thinning of the ice pack, decreasing about 52% from ~3.64 m in 1980 to ~1.75 m in 2008 (Kwok & Rothrock, 2009). The reduction of the summer ice extent resulted in more open water present in the Arctic Ocean in the autumn, leading, in turn, to increased FYI production in the marginal ice zone and significant shifts in overall ice production areas from continental shelves to the deeper basin (Iwamoto et al., 2014).

Given the crucial role of sea ice in the transport, transformation and biological uptake of contaminants, a rapidly changing sea ice environment in the Arctic Ocean will have a major impact on the fate and effects of many contaminants in the polar marine ecosystem. This is particularly important for contaminants such as Hg whose bioaccumulation in the Arctic Ocean is postulated to be increasingly controlled by climate-induced changes in biogeochemical processes rather than by changes in external emission sources (Wang et al., 2010). For instance, one of the major differences between FYI and MYI is the bulk salinity. As concentrations of contaminants in ice tend to increase with salinity due to freeze rejection, replacing 'fresher' MYI with more saline FYI will probably result in sea ice brine concentrating more contaminants from seawater over winter. This will increase the exposure of ice-associated organisms such as ice algae in spring, as well as the exposure of under-ice and water-column communities during melt due to more concentrated re-release of contaminants to the ocean at a biologically important time of the year (Grannas et al., 2013). More saline brine in younger ice may also enhance downward migration of contaminants from the overlying snowpack to the ice (Grannas et al., 2013).

However, projecting the magnitude or even the direction of how a changing sea ice environment would affect contaminant loading to the Arctic marine ecosystem is impossible at this time, as our understanding of how contaminants move and change within this complex and dynamic Arctic cryospheric system is only in its infancy. Considerable research on Hg and HCH pathways and processes within Arctic sea ice in recent years, as discussed in this chapter, should be seen as a necessary first step in advancing our knowledge of those complicated yet important issues and in defining future research directions. For instance, there are a few outstanding findings that call for immediate attention of the scientific community and warrant further studies:

1 highly elevated concentrations of contaminants reported in sea ice brine and associated risk of increased exposures to ice, under-ice and water-column biota;
2 the possibility of enhanced contaminant loading to melt ponds over summer from melting snow and dry atmospheric deposition and the associated risk of increased exposures to biota after brine drainage;
3 the potential of *in situ* Hg methylation within the ice column; and,
4 the role of sea ice in transport and transformation of crude oil and associated PAHs.

19.9 Acknowledgements

Most of the studies on sea ice contaminants discussed in this chapter were carried out aboard the Canadian Research Icebreaker *CCGS Amundsen* in the Canadian Arctic as part of the Circumpolar Flaw Lead (CFL) System Study (2007–2009), an International Polar Year (IPY) project funded by the Canadian IPY program, and as part of ArcticNet, a Network of Centers of Excellence programme in Canada. We are indebted to all our colleagues and the entire crew of the *CCGS Amundsen* for their stimulating discussions, assistance and friendship during these studies. We thank Mark Hanson for his comments on an earlier draft of this chapter.

References

Abbatt, J.P.D., Thomas, J.L., Abrahamsson, K. et al. (2012) Halogen activation via interactions with environmental ice and snow in the polar lower troposphere and other regions. *Atmospheric Chemistry and Physics*, **12**: 6237–6271.

AMAP. (1998) *AMAP Assessment Report: Arctic Pollution Issues*. Arctic Monitoring and Assessment Programme, Oslo.

AMAP. (2009) *AMAP Assessment 2009: Human Health in the Arctic*. Arctic Monitoring and Assessment Program, Oslo.

AMAP. (2010) AMAP Assessment 2009 - Persistent Organic Pollutants (POPs) in the Arctic. *Science of the Total Environment*, **408**, 2851–3051.

AMAP. (2011) *AMAP Assessment 2011: Mercury in the Arctic*. Arctic Monitoring and Assessment Program, Oslo.

Andersson, M.E., Gardfeldt, K., Wangberg, I. & Stromberg, D. (2008) Determination of Henry's Law constant for elemental mercury. *Chemosphere*, **73**, 587–592.

Bakke, T., Klungsøyr, J. & Sanni, S. (2013) Environmental impacts of produced water and drilling waste discharges from the Norwegian offshore petroleum industry. *Marine Environmental Research*, **92**, 154–169.

Bargagli, R. (2008) Environmental contamination in Antarctic ecosystems. *Science of the Total Environment*, **400**: 212–226.

Beattie, S.A., Armstrong, D., Chaulk, A., Comte, J., Gosselin, M. & Wang F. (2014) Total and methylated mercury in arctic multiyear sea ice. *Environmental Science and Technology*, **48**, 5575–5582.

Black, J.A., Birge, W.J., Westerman, A.G. & Francis, P.C. (1983) Comparative aquatic toxicology of aromatic hydrocarbons. *Fundamental and Applied Toxicology*, **3**, 353–358.

Brandvik, P.J. & Faksness, L.-G. (2009) Weathering processes in Arctic oil spills: Meso-scale experiments with different ice conditions. *Cold Regions Science and Technology*, **55** 160–166.

Burniston, D.A., Strachan, W.M.J., Hoff, J.T. & Wania, F. (2007) Changes in surface area and concentrations of semivolatile organic contaminants in aging snow. *Environmental Science and Technology*, **41**, 4932–4937.

Burt, A., Wang, F., Pućko, M. et al. (2013) Mercury uptake within an ice algal community during the spring bloom in first-year Arctic sea ice. *Journal of Geophysical Research Oceans*, **118**, 4746–4754.

Chaulk, A., Stern, G.A., Armstrong, D., Barber, D. & Wang, F. (2011) Mercury distribution and transport across the ocean-sea-ice-atmosphere interface in the Arctic Ocean. *Environmental Science and Technology*, **45**, 1866–1872.

Chaulk, A. (2011) Distribution and partitioning of mercury in the Arctic cryosphere: Transport across snow-sea ice-water interface in the Western Arctic Ocean. *MSc. Thesis*. Department of Environment and Geography. University of Manitoba, Winnipeg.

Clever, H.L., Johnson, S.A. & Derrick, M.E. (1985) The solubility of mercury and some sparingly soluble mercury salts in water and aqueous electrolyte solutions. *Journal of Physical and Chemical Reference Data*, **14**, 631–680.

Crutzen, P.J. (2002) Geology of mankind. *Nature*, **415**, 23.

Dethleff, D. & Kuhlmann, G. (2009) Entrainment of fine-grained surface deposits into new ice in the southwestern Kara Sea, Siberian Arctic. *Continental Shelf Research*, **29**, 691–701.

Douglas, T.A., Sturm, M., Simpson, W.R. et al. (2008) Influence of snow and ice crystal formation and accumulation on mercury deposition to the Arctic. *Environmental Science and Technology*, **42**, 1542–1551.

Douglas, T.A., Sturm, M., Simpson, W.R., Brooks, S., Lindberg, S.E. & Perovich, D.K. (2005) Elevated mercury measured in snow and frost flowers near Arctic sea ice leads. *Geophysical Research Letters*, 32, doi:10.1029/2004GL022132.

Driscoll, C.T., Mason, R.P., Chan, H.M., Jacob, D.J. & Pirrone, N. (2013) Mercury as a global pollutant: Sources, pathways and effects. *Environmental Scienceand Technology*, **47**, 4967–4983.

Durnford, D.A., Dastoor, A.P., Steen, A.O. et al. (2012) How relevant is the deposition of mercury onto snowpacks? - Part 1: A statistical study on the impact of environmental factors. *Atmospheric Chemistry and Physics*, **12**, 9221–9249.

Eicken, H., Gradinger, R., Gaylord, A., Mahoney, A., Rigor, I. & Melling, H. (2005) Sediment transport by sea ice in the Chukchi and Beaufort Seas: Increasing importance due to changing ice conditions? *Deep-Sea ResearchPart II*, **52**, 3281–3302.

Eicken, H., Krouse, H., Kadko, D. & Perovich, D. (2002) Tracer studies of pathways and rates of meltwater transport through Arctic summer sea ice. *Journal of Geophysical Research*, **107**(C10), 8046, doi:10.1029/2000JC000583.

Faksness, L.-G. & Brandvik, P.J. (2008a) Distribution of water soluble components from Arctic marine oil spills – A combined laboratory and field study. *Cold Regions Science and Technology*, **54**, 97–105.

Faksness, L.-G. & Brandvik, P.J. (2008b) Distribution of water soluble components from oil encapsulated in Arctic sea ice: Summary of three field seasons. *Cold Regions Science and Technology*, **54**, 106–114.

Faller, J., Hühnerfuss, H., König, W.A., Krebber, R. & Ludwig, P. (1991) Do marine bacteria degrade α-hexachlorocyclohexane stereoselectively? *Environmental Science and Technology*, **25**: 676–678.

Fingas, M.F. & Hollebone, B.P. (2003) Review of behaviour of oil in freezing environments. *Marine Pollution Bulletin*, **47**, 333–340.

Fleming, E.J., Mack, E.E., Green, P.G. & Nelson, D.C.. Mercury methylation from unexpected sources: molybdateinhibited freshwater sediments and iron-reducing bacterium. *Applied and Environmenal Microbiology* 2006; **72**: 457–464.

Fisher, J.A., Jacob, D.J., Soerensen, A.L. et al. (2013) Factors driving mercury variability in the Arctic atmosphere and ocean over the past 30 years. *Global Biogeochemical Cycles*, **27**, 1226–1235.

Gautier, D.L., Bird, K.J., Charpentier, R.R. et al. (2009) Assessment of undiscovered oil and gas in the Arctic. *Science*, **324**, 1175–1179.

Grannas, A.M., Bogdal, C., Hageman, K.J. et al. (2013) The role of the global cryosphere in the fate of organic contaminants. *Atmospheric Chemistry and Physics*, **13**, 3271–3305.

Harner, T., Kylin, H., Bidleman, T.F. & Strachan, W.M.J. (1999) Removal of α and γ-hexachlorocyclohexanes and enantiomers of α-hexachlorocyclohexane in the eastern Arctic Ocean. *Environmental Science and Technology*, **33**, 1157–1164.

Huber, M.L., Laesecke, A. & Friend, D.G. (2006) Correlation for the vapor pressure of mercury. *Industrial and Engineering Chemistry Research*, **45**, 7351–7361.

Hühnerfuss, H., Faller, J., Kallenborn, R. et al. (1993) Enantioselective and non-enantioselective degradation of organic pollutants in the marine ecosystem. *Chirality*, **5**, 393–399.

Hutchings, J.K. & Rigor, I.G. (2012) Role of ice dynamics in anomalous ice conditions in the Beaufort Sea during 2006 and 2007. *Journal of Geophysical Research Oceans*, **117**, C00E04, doi:10.1029/2011JC007182.

Iverfeldt, A. & Lindqvist, O. (1982) Distribution equilibrium of methyl mercury chloride between water and air. *Atmospheric Environment*, **16**, 2917–2925.

Iwamoto, K., Ohshima, K.I. & Tamura, T. (2014) Improved mapping of sea ice production in the Arctic Ocean using AMSR-E thin ice thickness algorithm. *Journal of Geophysical Research – Oceans*, **119**, 3574–3594.

Kahan, T.F. & Donaldson, D.J. (2007) Photolysis of polycyclic aromatic hydrocarbons on water and ice surfaces. *Journal of Physical Chemistry A*, **111**, 1277–1285.

Kahan, T.F., Wren, S.N. & Donaldson, D.J. (2014) A pinch of salt is all it takes: chemistry at the frozen water surface. *Accounts of Chemical Research*, **47**, 1587–1594.

King, J.K., Kostka, J.E., Frischer, M.E. & Saunders, F.M. (2000) Sulfate-reducing bacteria methylate mercury at variable rates in pure culture and in marine sediments. *Applied and Environmenal Microbiology*, **66**, 2430–2437.

Kwok, R. & Rothrock, D.A. (2009) Decline in Arctic sea ice thickness from submarine and ICESat records: 1958–2008. *Geophysical Research Letters*, **36**, L15501, doi:10.1029/2009GL039035.

Leahy, J.G & Colwell, R.R. (1990) Microbial degradation of hydrocarbons in the environment. *Microbiological Reviews*, **54**, 305–315.

Li, Y.F. & Macdonald, R.W. (2005) Sources and pathways of selected organochlorine pesticides to the Arctic and the effect of pathway divergence on HCH trends in biota: a review. *Science of the Total Environment*, **342**, 87–106.

Li, Y.F., Macdonald, R.W., Ma, J.M., Hung, H. & Venkatesh, S. (2004) Historical α-HCH budget in the Arctic Ocean: the Arctic Mass Balance Box Model (AMBBM). *Science of the Total Environment*, **324**, 115–139.

Liao, J., Huey, L.G., Liu, Z. et al. (2014) High levels of molecular chlorine in the Arctic atmosphere. *Nature Geoscience*, **27**, 91–94.

Liyana-Arachchi, T.P., Valsaraj, K.T. & Hung, F.R. (2012) Ice growth from supercooled aqueous solutions of benzene, naphthalene, and phenanthrene. *Journal of Physical Chemistry A*, **116**, 8539–8546.

Lu, J.Y. & Schroeder, W.H. (2004) Annual time-series of total filterable atmospheric mercury concentrations in the Arctic. *Tellus Series B-Chemical and Physical Meteorology*, **56**, 213–222.

Ma, Y.-G., Lei, Y.D., Xiao, H., Wania, F. & Wang, W-H. (2010) Critical review and recommended values for the physical-chemical property data of 15 polycyclic aromatic hydrocarbons at 25°C. *Journal of Chemical and Engineering Data*, **55**, 819–825.

Macdonald, R. & Bewers, J.M. (1996) Contaminants in the arctic marine environment: priorities for protection. *ICES Journal of Marine Science*, **53**, 537–563.

Macdonald, R.W., Barrie, L.A., Bidleman, T.F. et al. (2000). Contaminants in the Canadian Arctic: 5 years of progress in understanding sources, occurrence and pathways. *Science of the Total Environment*, **254**, 93–234.

Macdonald, R.W., Harner, T. & Fyfe, J. (2005). Recent climate change in the Arctic and its impact on contaminant pathways and interpretation of temporal trend data. *Science of the Total Environment*, **342**, 5–86.

Martin, S., Drucker, R. & Fort, M.A. (1995) Laboratory study of frost flower growth on the surface of young sea ice. *Journal of Geophysical Research*, **100**, 7027–7036.

Mason, R.P., Reinfelder, J.R. & Morel, F.M.M. (1996) Uptake, toxicity, and trophic transfer of mercury in a coastal diatom. *Environmental Science and Technology*, **30**, 1835–1845.

Moore, C.W., Obrist, D., Steffen, A. et al. (2014) Convective forcing of mercury and ozone in the Arctic boundary layer induced by leads in sea ice. *Nature*, **506**, 81–84.

Müller, T.A. & Kohler, H.-P.E. (2004) Chirality of pollutants – effects on metabolism and fate. *Applied Microbiology and Biotechnology*, **64**, 300–316.

NRC (2000) *Toxicological Effects of Methylmercury*. National Research Council, Washington.

Obbard, R.W., Sadri, S., Wong, Y.Q., Khitun, A.A., Baker, I. & Thompson, R.C. (2014) Global warming releases microplastic legacy frozen in Arctic Sea ice. *Earth's Future*, **2**, 315–320.

Outridge, P.M., Macdonald, R.W., Wang, F., Stern, G.A. & Dastoor, A.P. (2008) A mass balance inventory of mercury in the Arctic Ocean. *Environmental Chemistry*, **5**, 89–111.

Pampanin, D.M. & Sydnes, M.O. (2013) Polycyclic aromatic hydrocarbons a constituent of petroleum: Presence and influence in the aquatic environment. In: *Hydrocarbon* (Eds. V. Kutcherov & A. Kolesnikov), pp. 83–118. InTech, Croatia.

Pavlov, V., Pavlova, O. & Korsnes, R. (2004) Sea ice fluxes and drift trajectories from potential pollution sources, computed with a statistical sea ice model of the Arctic Ocean. *Journal of Marine Systems*, **48**: 133–157.

Payne, J.R., McNabb, G.D. Jr., & Clayton, J.R. Jr., (1991) Oil-weathering behavior in Arctic environments. *Polar Research*, **10**, 631–662.

Perovich, D.K. & Richter-Menge, J.A. (1994) Surface characteristics of lead ice. *Journal of Geophysical Research: Oceans*, **99**, 16341–16350.

Peterson, C.H., Rice, S.D., Short, J.W. et al. (2003) Long-term ecosystem response to the Exxon Valdez oil spill. *Science*, **302**, 2082–2086.

Pfirman, S., Eicken, H., Bauch, D. & Weeks, W.F. (1995) The potential transport of pollutants by Arctic sea ice. *Science of the Total Environment*, **159**, 129–146.

Pirrone, N., Cinnirella, S., Feng, X. et al. (2010) Global mercury emissions to the atmosphere from anthropogenic and natural sources. *Atmospheric Chemistry and Physics*, **10**, 5951–5964.

Poland, J.S., Riddle, M.J. & Zeeb B.A. (2003) Contaminants in the Arctic and the Antarctic: a comparison of sources, impacts, and remediation options. *Polar Record*, **39**, 369–383.

Pratt, K.A., Custard, K.D., Shepson, P.B. et al. (2013) Photochemical production of molecular bromine in Arctic surface snowpacks. *Nature Geoscience*, **6**, 351–356.

Pućko, M., Stern, GA., Macdonald, R. & Barber, D. (2010a) α- and γ-hexachlorocyclohexane (HCH) measurements in the brine fraction of sea ice in the Canadian High Arctic using a sump-hole technique. *Environmental Science and Technology*, **44**, 9258–9264.

Pućko, M., Stern, G.A., Barber, D.G., Macdonald, R.W. & Rosenberg, B. (2010b) The International Polar Year (IPY) Circumpolar Flaw Lead (CFL) System Study: the importance of brine processes for α- and γ-hexachlorocyclohexane accumulation/rejection in the sea ice. *Atmosphere and Ocean*, **48**, 244–262.

Pućko, M., Stern, G.A., Macdonald, R.W., Rosenberg, B. & Barber D.G. (2011) The influence of the atmosphere-snow-ice-ocean interactions on the levels of hexachlorocyclohexanes in the Arctic cryosphere. *Journal of Geophysical Research Oceans*, **116**, C02035, doi:10.1029/2010JC006614.

Pućko, M., Stern, G.A., Barber, D.G., Macdonald, R.W., Warner, K.A. & Fuchs, C. (2012) Mechanisms and implications of alpha-HCH enrichment in melt pond water on Arctic sea ice. *Environmental Science and Technology*, **46**, 11862–11869.

Pućko, M., Stern, G.A., Macdonald, R.W., Barber, D.G., Rosenberg, B. & Walkusz, W. (2013) When will alpha-HCH disappear from the western Arctic Ocean? *Journal of Marine Systems*, **127**, 88–100.

Pućko, M., Burt, A., Walkusz, W. et al. (2014a) Transformation of mercury at the bottom of the Arctic food web: An overlooked puzzle in the mercury exposure narrative. *Environmental Science and Technology* **48**, 7280–7288.

Pućko, M., Stern, G.A,. Macdonald, R.W. et al. (2015) The delivery of contaminants to the Arctic food web: Why sea ice matters. *Science of the Total Environment*, **506**, 444–452.

Reimnitz, E., Barnes, P.W. & Weber, W.S. (1993) Particulate matter in pack ice of the Beaufort Gyre. *Journal of Glaciology*, **39**, 186–198.

Rowland, G.A., Bausch, A.R. & Grannas, A.M. (2011) Photochemical processing of aldrin and dieldrin in frozen aqueous solutions under arctic field conditions. *Environmental Pollution*, **159**, 1076–1084.

Rysgaard, S. & Glud, R.N. (2004) Anaerobic N_2 production in Arctic sea ice. *Limnology and Oceanography*, **49**, 86–94.

Schroeder, W.H., Anlauf, K.G., Barrie, L.A. et al. (1998) Arctic springtime depletion of mercury. *Nature*, **394**, 331–332.

Schroeder, W.H. & Munthe, J. (1998) Atmospheric mercury – an overview. *Atmospheric Environment*, **32**, 809–822.

Shaw, P.M., Russell, L.M., Jefferson, A. & Quinn, P.K. (2010) Arctic organic aerosol measurements show particles from mixed combustion in spring haze and from frost flowers in winter. *Geophysical Research Letters*, **37**, L10803, doi:10.1029/2010GL042831.

Shimada, T. & Fujii-Kuriyama, Y. (2004) Metabolic activation of polycyclic aromatic hydrocarbons to carcinogens by cytochromes P450 1A1 and 1B1. *Cancer Science*, **95**, 1–6.

Simpson, W.R., von Glasow, R., Riedel, K. et al. (2007) Halogens and their role in polar boundary-layer ozone depletion. *Atmospheric Chemistry and Physics*, **7**, 4375–4418.

Sommar, J., Lindqvist, O. & Stromberg, D. (2000) Distribution equilibrium of mercury (II) chloride between water and air applied to flue gas scrubbing. *Journal of the Air and Waste Management Association*, **50**, 1663–1666.

Stroeve, J., Hamilton, L.C., Bitz, C.M. & Blanchard-Wrigglesworth, E. (2014) Predicting September sea ice: ensemble skill of the SEARCH Sea Ice Outlook 2008–2013. *Geophysical Research Letters*, **41**, 2411–2418.

Vancoppenolle, M., Meiners, K.M., Michel, C. et al. (2013) Role of sea ice in global biogeochemical cycles: emerging views and challenges. *Quaternary Science Reviews*, **79**, 207–230.

Wang, F., Macdonald, R., Armstrong, D. & Stern, G. (2012) Total and methylated mercury in the Beaufort Sea: The role of local and recent organic remineralization. *Environmental Scienceand Technology*, **46**, 11821–11828.

Wang, F., Macdonald, R.W, Stern, G.A. & Outridge, P.M. (2010) When noise becomes the signal: Chemical contamination of aquatic ecosystems under a changing climate. *Marine Pollution Bulletin*, **60**, 1633–1635.

Wania, F., Hoff, J.T., Jia, C.Q. & Mackay, D. (1998) The effects of snow and ice on the environmental behavior of hydrophobic organic chemicals. *Environmental Pollution*, **102**, 25–41.

Wong, F., Jantunen, L.M., Pućko, M., Papakyriakou, T., Staebler, R.M., Stern, G.A. & Bidleman, T.F. (2011) Air-water exchange of anthropogenic and natural organohalogens on International Polar Year (IPY) expeditions in the Canadian Arctic. *Environmental Science and Technology*, **45**, 876–881.

Xiao, H., Li, N. & Wania, F. (2004) Compilation, evaluation, and selection of physical-chemical property data for α-, β- and γ-hexachlorocyclohexane. *Journal of Chemical and Engineering Data*, **49**, 173–185.

Xu, W., Tenuta, M. & Wang, F. (2016) Bromide and chloride distribution across the snow-sea ice-pocean interface: A comparative study between an Arctic coastal marine site and an experimental sea ice mesocosm. *Journal of Geophysical Research: Oceans*, **121**, doi:10.1002/2015JC011409.

Yunker, M.B. & Macdonald, R.W. (1995) Composition and origins of polycyclic aromatic hydrocarbons in the Mackenzie River and on the Beaufort Sea shelf. *Arctic* **48**, 118–129.

CHAPTER 20

Numerical models of sea ice biogeochemistry

Martin Vancoppenolle[1] and Letizia Tedesco[2]

[1] *LOCEAN-IPSL, Sorbonne Universités (UPMC Paris 6), CNRS/IRD/MNHN, Paris, France*
[2] *Finnish Environment Institute (SYKE), Helsinki, Finland*

20.1 Introduction

Observations over the past 30 years reveal active biological and chemical processes in the sea ice zone. Large changes in the polar oceans and their sea ice cover are observed and projected for the future. These changes in the sea ice could affect the biogeochemical cycles and marine ecosystems, yet by what means, to what extent, and with what consequences on the Earth system are not well understood. This is motivating the scientific community to implement model representations of sea ice impact on biogeochemical (BGC) processes.

There are several sea ice-related BGC processes that could be important globally (for reviews, see Chapters 13–19; Loose et al., 2011; Vancoppenolle et al., 2013). These have been considered only in a few modelling studies, dedicated to:

- understanding mechanisms, in particular those driving the seasonality of microbial ecosystems in sea ice (e.g. Arrigo et al., 1993; Lavoie et al., 2005; Jin et al., 2006; Tedesco et al., 2010; Saenz & Arrigo, 2012);
- computing the budget of a chemical element in sea ice at a given location (e.g. Moreau et al., 2015);
- quantifying large-scale quantities, in particular biomass and primary production for the entire sea ice packs of the Arctic (e.g. Deal et al., 2011; Sibert et al., 2011; Dupont 2012) and the Antarctic (Arrigo et al., 1997; Saenz & Arrigo 2014);
- projecting the future, e.g. the primary productivity of polar marine ecosystems (Lavoie et al. 2010; Tedesco et al. 2012);
- studying the large-scale impact of particular processes, such as the impact of specific pathways of air–ice–sea carbon exchange on the global carbon cycle (Steiner et al. 2013; Delille et al. 2014); the effect of iron uptake and release during sea ice growth and melt on pelagic productivity (Lancelot et al. 2009; Wang et al. 2014)' and the impact of sulphur cycle processes (Elliott et al. 2012).

The sea ice BGC models and parameterizations used in these studies formalize some selected physical and BGC processes in a computer code. At present, the few existing models are highly simplified, Earth system models (ESMs) do not include explicit sea ice BGC processes and there is no consensus on a 'standard' sea ice BGC model – if this concept has a meaning at all. This situation is changing, as several research groups in the world work towards improved model representations of the impact of sea ice processes and marine BGC cycles.

In this context, the present chapter focuses on a currently active research topic. We first briefly describe the main modelling approaches used so far and their scientific applications (Section 20.2). Then, we focus on the one-dimensional (1D) models of microbial communities in sea ice. We choose this example because it has attracted the most attention so far. We describe how such models work and how modelling decisions are made (Section 20.3). In a simple case study (Section 20.4), we then illustrate which factors control simulated ice algae, from modelling choices to environmental parameters. The last section (Section 20.5) is dedicated to future research directions.

20.2 The use of numerical models for sea ice biogeochemical studies

The type of model used and the processes represented are guided by the scientific questions to be answered and

Sea Ice, Third Edition. Edited by David N. Thomas.
© 2017 John Wiley & Sons, Ltd. Published 2017 by John Wiley & Sons, Ltd.

by the spatial and temporal scales of interest. Indeed, the sea ice environment is highly variable at multiple spatial scales (see Chapter 2), in particular ice thickness and snow depth, which models need to consider. Besides, there are many physical and BGC processes and many ways to represent them in a sea ice model. A series of decisions on the modelling approach must therefore be made, depending on the problem that is being studied.

Sea ice models traditionally separate vertical and horizontal processes. The vertical processes include ice thermodynamics, brine and gas dynamics, radiative transfer and BGC processes. Sea ice growth (frazil, congelation and snow ice) incorporates oceanic material into the ice, whereas melting (lateral, surface, basal) releases it back to the ocean. The temperature changes seasonally, which in turn drastically affects the size and salinity of liquid brine inclusions (see Chapter 1). Brines are small and highly saline when the ice is cold, and become fresher, larger and more connected as the temperature increases. Temperature and brine salinity variations have important chemical (ionic strength) and biological (physiology) consequences, some of which will be discussed later. The brine network, if connected, provides pathways for vertical fluid transport. Brine can exchange gas with the atmosphere. Solar radiation, necessary for photosynthesis, interacts with snow, ice, brine, gas inclusions, algae and organic matter. Finally, many active BGC processes, such as photosynthesis, respiration, nutrient remineralization and ikaite precipitation, occur in the sea ice. By contrast, horizontal processes are purely physical. They encompass the transport and deformation processes induced by the horizontal ice motion, which transports ice floes and their load of BGC material horizontally. In addition, horizontal convergence and shear deform ice floes through ridging and rafting, which redistributes tracers vertically. Seawater and its BGC content are trapped into newly formed ridges.

Numerical models with sea ice BGC (Figure 20.1) can be broadly categorized into ESMs, large-scale ice models and sea ice process models. These models are different yet not independent, often sharing common modules. Each model type is, as explained in the following, dedicated to specific spatiotemporal scales and has its own advantages and shortcomings.

20.2.1 Earth system models

Earth system models (Flato et al., 2013) are the large numerical codes used for the large-scale Earth system research, including Intergovernmental Panel on Climate Change studies. They provide means to study the role of sea ice processes in the climate and global carbon cycle context. They presently include large-scale three-dimensional (3D) representations of atmosphere, ocean, sea ice, atmospheric chemistry, aerosols and

	Earth system models	Ice–ocean models	Sea ice process models
Components	Atmosphere Ocean + sea ice Atmospheric chemistry/aerosols Land/ocean BGC	Prescribed atmosphere (Ocean) + Sea ice (Ocean BGC) (Sea ice BGC)	(Ocean) + sea ice + (Ocean BGC) + (sea ice BGC)
Coverage resolution	3D–global 0.25–2°	3D–global/regional 0.05–2°	1D–local n.a.
Applications	Climate, carbon cycle	Ocean, sea ice and BGC dynamics	Sea ice processes
References	Lengaigne et al. (2009), Steiner et al. (2013)	Sibert et al. (2011), Deal et al. (2011) Dupont et al. (2012), Elliott et al. (2012) Jin et al. (2012), Delille et al. (2014), Saenz & Arrigo (2014).	Arrigo et al. (1993), Zeebe et al. (1996), Lavoie et al. (2005), Jin et al. (2006), Nishi & Tabeta (2008), Vancoppenolle et al. (2010), Lavoie et al. (2010), Tedesco et al. (2010), Pogson et al. (2011), Saenz & Arrigo (2012), Tedesco et al. (2012), Tedesco & Vichi (2014).

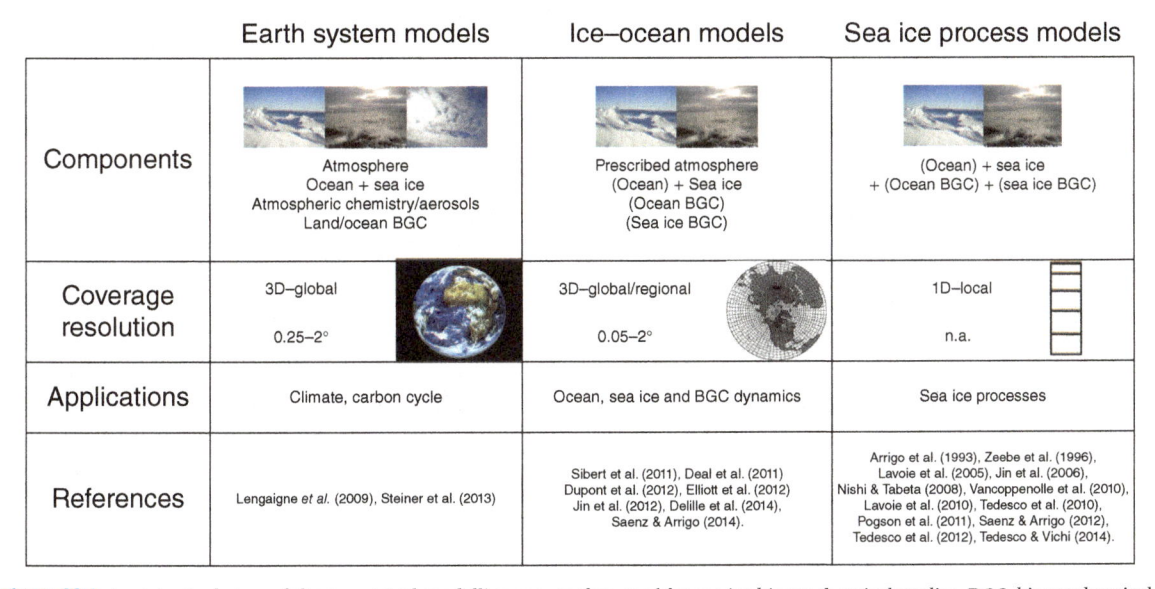

Figure 20.1 A contextual view of the numerical modelling approaches used for sea ice biogeochemical studies. BGC, biogeochemical.

land/ocean BGC. Because of their computer cost, they can only afford to cover the whole globe at resolutions of ~1°. They are not able to resolve detailed processes. In addition, there are model biases. Because the atmosphere is chaotic, ensemble simulations are required. The sea ice component of ESMs includes selected horizontal and vertical processes (see Chapter 12). The ocean surface is divided into grid cells. For each grid cell, the sea ice can be divided into ice categories that discretize the sub-grid-scale sea ice thickness distribution. In each thickness category, the snow–ice system is divided in vertical layers. BGC tracers can therefore possibly be defined for every grid cell, thickness category and vertical layer. Steiner et al. (2013) studied the impact of air–sea CO_2 exchange through sea ice cracks and small leads on the large-scale ocean–atmosphere carbon exchange, using an ESM with modified air–sea fluxes in ice-covered regions. Lengaigne et al. (2009) studied the feedback between Arctic sea ice retreat and enhanced marine productivity in the 21st century.

20.2.2 Large-scale sea ice models

Standalone configurations of ESMs, here referred to as large-scale sea ice models, can be run globally or regionally at spatial resolutions up to 0.1°, and even finer resolutions in the most advanced cases. This is because they are forced by a prescribed atmospheric state. These models can be used to study the large-scale sea ice and ocean dynamics, but miss some important ice–atmosphere or ice–ocean feedbacks. In addition, they are subject to forcing errors that can be large.

Deal et al. (2011) introduced a representation of sea ice microbial processes in a large-scale standalone sea ice model and calculated the gross primary production (GPP) due to Arctic sea ice algae to be ~15 TgC year^{-1} for 1992. Delille et al. (2014) introduce a representation of air–ice CO_2 exchange in a 3D ice–ocean model and evaluate the contribution of the Antarctic sea ice zone in spring (1997–2007) to represent a net sink for atmospheric CO_2 of ~29 TgC year^{-1}.

20.2.3 Sea ice process models

The representation of vertical processes in large-scale sea ice models is based on 1D sea ice process models. Such models are used to study processes and compute the budget of a given chemical element in a specific place. As they only consider vertical processes and hence do not model horizontal variations, they are computationally less expensive to run than 3D models. They divide the snow–ice system into a series of vertical layers that exchange heat and tracers with an underlying ocean component. The most attention in sea ice BGC studies has been given to microbial ecosystems (ice algae and nutrients). Presently, research teams work towards closing the carbon budget, including ice–atmosphere exchanges, carbonate chemistry in brine and calcium carbonate precipitation. Other elements (e.g. iron, sulphur) are yet to be considered. Process models are computationally inexpensive and can have as many details as needed. However, as they ignore large-scale processes and interactions with the rest of the Earth system, the scope of the conclusions drawn from such models is limited.

Arrigo et al. (1993) showed that the seasonal light limitation and inhibition of photosynthesis due to high brine salinities explain the spatiotemporal dynamics of sea ice algae in Antarctic fast ice. Tedesco et al. (2012) studied acclimation strategies for ice algae and phytoplankton. Moreau et al. (2015) attempt to quantify the contribution of the different processes that shape the carbon budget in sea ice.

20.3 One-dimensional modelling of sea ice microbial communities

In this section, we detail the framework and process formulations used in 1D models of microbial sea ice communities, which simply and elegantly shows the development of numerical sea ice BGC models.

The representation of physical processes in sea ice models originates from studies spanning the last 40 years. By comparison with physics, the mathematical laws describing biological processes are less well known and more empirical: they cannot be derived from first principles, their range of validity is limited and the associated uncertainties are large. Most of the concepts underlying model representations of microbial sea ice ecosystems originate from open ocean studies. Such representations involve many assumptions and simplifications, which correspond to the limits of our experimental understanding.

One-dimensional models of microbial sea ice communities were the first natural outcome of merging the open ocean microbial ecosystem models with the

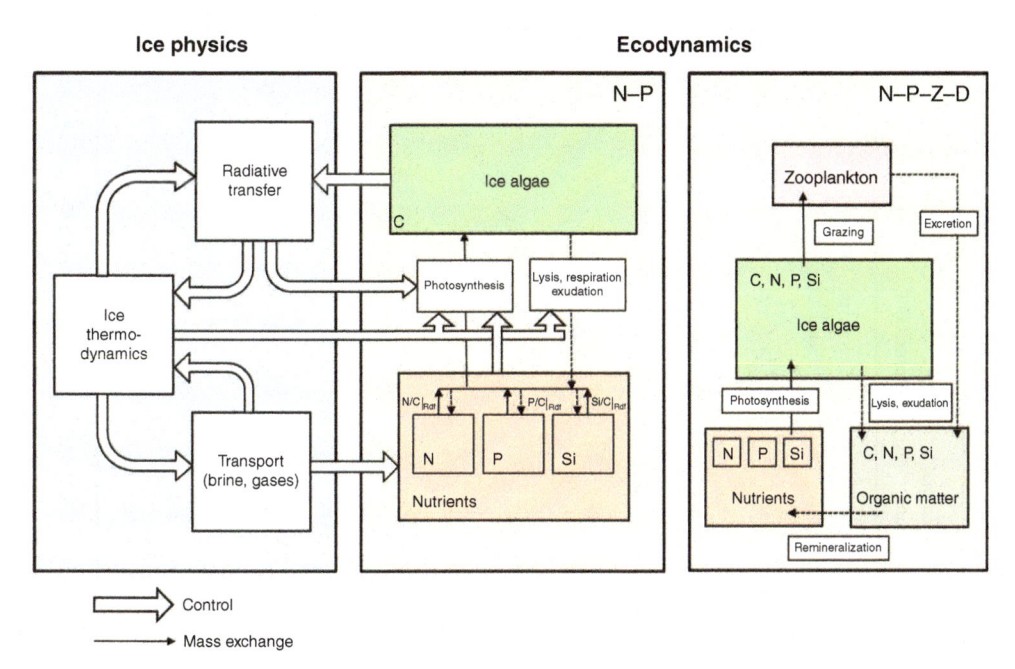

Figure 20.2 General structure of a one-dimensional biophysical model in sea ice, with ice physics on the left and ecodynamics on the right. The simple N-P (centre) and the more complex N-P-Z-D (right) approaches are depicted.

1D physical sea ice models. This approach was warranted, as ice algae behave as strongly shade-adapted phytoplankton (Cota & Smith 1991). One-dimensional sea ice models have a typical structure, as depicted in Figure 20.2. Vertical (physical and BGC) processes can have various representations (Table 20.1) detailed in this section.

20.3.1 The 1D approach

The fundamental structure of 1D sea ice models is based on the observation by Maykut & Untersteiner (1971; hereafter MU71) that the exchanges of heat and tracers are essentially vertical. Hence, if dynamic processes are neglected, the growth and decay of sea ice can be represented assuming an infinite horizontally homogeneous slab interacting with the atmosphere and the ocean. In practice, such setups can be seen as the numerical equivalent of a time-series observation from landfast ice or a drifting station on pack ice.

However, even for such experiments, 1D models are an idealization that is rarely achieved in practice. This is because, first, ice and snow are highly variable at the sub-floe scale (Thorndike et al. 1975) and second, the processes that control microbial communities, in particular radiative transfer (e.g. Katlein et al. 2014) and brine dynamics (Eicken et al., 2002; Wells et al. 2012) are fundamentally three-dimensional, which supposedly explains the strongly observed ice algae patchiness (Rysgaard et al. 2001). Arguably, 1D models are able to represent the larger (floe) scale processes for a given ice thickness. This is presently not warranted because our understanding of microbial communities at such scales is limited.

The development of a model involves a series of choices for the modeller with regard to:
1 tracers;
2 the vertical discretization;
3 the connection with physical processes;
4 BGC source and sink processes.

20.3.2 Tracer framework

The representation of biogeochemistry in sea ice models is based on a series of markers of biogeochemical activity (tracers), whose abundance is quantified by bulk concentration C (./m^3). In 1D models, C is a function of time t and depth in the ice z. The components of the microbial ecosystem are represented using a series of 'reservoirs', each characterized by a specific tracer (see Figure 20.2). Which tracers are used depends on the problem of interest. For the modelling of microbial

Table 20.1 Sea ice biogeochemical models and their components.

References	Ice physics	Transport	Radiation	Grid for ecodynamics	Functional groups	Cell quotas/Chl:C	Limiting elements	Ocean
Arrigo et al. (1993)	1D SM 0L	~ dh/dt	N bands; BL	Multi-layer	3N-1P	RFD / constant	N, P, Si	n.a.
Arrigo et al. (1997)	3D* SM 0L			Multi-layer		RFD / diagnostic		n.a.
Saenz & Arrigo (2012)	1D EC ML	~ dS/dt	32 bands; DE	Multi-layer	3N-1P-1D	RFD / constant	N, P, Si	n.a.
Saenz & Arrigo (2014)	3D*EC ML							
Lavoie et al. (2005)	1D SM 3L	Tide-dependent convection	1 band; BL	1L bottom	1N-1P	RFD / constant	Si	n.a.
Pogson et al. (2011)	1D EC 3L							n.a.
Lavoie et al. (2010)	3D SM 3L							1D
Dupont (2012)								3D
Nishi & Tabeta (2008)	3D MU ML	Growth/melt, diffusion, convection	1 band; BL	1L bottom	2N-1P-1Z-2D	Quota / n.a.	N, Si, C	3D
Jin et al. (2006)	Prescribed	~ dh/dt	1 band; BL	1L bottom	3N-1P	RFD / constant	N, Si	1D
Deal et al. (2011); Jin et al. (2012)	3D EC ML		4 bands; DE					3D
Elliott et al. (2012)	3D EC ML				3N-1P-1Z-1D		N, Si, C, S	
Sibert et al. (2011)	3D SM 3L	n.a.	1 band; BL	1L static	1N-1P-1Z	RFD / diagnostic	N	3D
Tedesco et al. (2010), Tedesco et al. (2012)	1D SM 0L	Growth/melt	1 band; BL	1L bottom (dynamic)	4N-2P-2D	RFD / prognostic	N, P, Si, C	1D
Tedesco & Vichi (2014)					1N-1P-2D		Si, C	
Vancoppenolle et al. (2010)	1D EC ML	Mushy-layer	1 band; BL	Multi-layer	1N, prescribed PP	RFD / n.a.	Si	n.a.

SM, Semtner (1976) model; MU, Maykut & Untersteiner (1971) model; EC, energy-conserving model (Bitz & Lipscomb, 1999); 0L/3L/ML, 0, 3 and multi-layer; BL, Beer–Lambert law; DE, Delta-Eddington; N , nutrient; P, algae; D, detritus; Z, grazers; B, bacteria; RFD, prescribed Redfieldian cell quotas; Quota, varying cell quota.

communities in sea ice, model tracers include at least one ice algal species and one nutrient. More complex models can include more algal species and nutrients, organic matter and grazers as well.

The computation of the time evolution of BGC tracer concentrations by models is based on formulations of physical (φ) and BGC source and sink (β) processes, solving equations of the form:

$$\frac{\partial C}{\partial t} = \frac{\partial C}{\partial t}\bigg|_{\varphi} + \frac{\partial C}{\partial t}\bigg|_{\beta}. \qquad (20.1)$$

Physical processes include the tracer uptake from/ release to seawater associated with growth and melt, as well as transport, where the mode of transport depends on the tracer of interest.

Dissolved tracers (e.g. macronutrients) are fully dissolved in brine and undergo transport due to brine motion. Hence for these tracers the brine concentration C^{br} must be introduced, using:

$$C = eC^{br}, \qquad (20.2)$$

where e is the brine fraction, diagnosed from model temperature and bulk salinity at depth z. At first order, dissolved tracers follow salt and their bulk concentration is typically much smaller the concentration in seawater C^w. However, as e is generally smaller than 10%, C^{br} can be much higher than C and C^w. In recently introduced sea ice tracer schemes the impact of brine motion on C^{br} is calculated, and then C is retrieved (Vancoppenolle et al., 2010; Jeffery et al., 2011; Saenz & Arrigo 2012).

Particulate tracers cannot be dissolved in brine, and hence their concentration does not directly relate to brine characteristics. Regarding vertical motion, particulate tracers can remain sealed in the sea ice microstructure (e.g. ice diatoms), move passively following brine motion (e.g. some iron particulates), move actively (e.g. flagellates), or stick on the ice crystals or on other impurities (as suggested for dissolved iron on exopolymeric substances).

For some applications, dissolved gases (e.g. oxygen) have to be included. Those are specific because they not only dilute in brine but also form bubbles: the contraction of brine inclusions associated with decreasing temperature can increase their concentration above saturation (see Chapter 18). For such gases, the bulk concentration in gas bubbles has to be included (C^{bub}): $C = eC^{br} + C^{bub}$ (Zhou et al., 2013; Moreau et al. 2014). In 1D models of microbial communities, nutrients are transported with brine motion, whereas ice algae do not move vertically. Integrating bulk concentration give (standing) stocks $C(./m^2)$:

$$M = \int_0^{h_i} C\, dz \qquad (20.3)$$

20.3.3 Vertical discretization

The next important modelling choice regards the vertical discretization for tracers. Three standard approaches are available from the literature (see Figure 20.3) regarding model biologically active layers (BALs), yet they have not formally been compared with each other.

Prescribed BAL

The simplest models prescribe the location of microbial communities in a layer near the ice–ocean interface, based on the assumption that most of the biomass lies in this 'skeletal' layer (e.g. Lavoie et al. 2005; Jin et al. 2006; Deal et al. 2011; Dupont et al. 2012). The thickness of this layer ranges from $h_{BAL} = 2$–5 cm in the different models. This approach is argued to be appropriate for the Arctic where bottom algal communities dominate.

Dynamic BAL

In this second type of BAL, h_{BAL} changes with time: the location of the upper limit of the BAL is the 5% brine fraction contour (Tedesco et al. 2010). This discretization is justified by the fact that the brine network is in connection with the ocean for $e > 5\%$, and hence models of the first type could miss a significant part of the production.

Multi-layer BAL

The third type of discretization is that every layer of the physical component is biologically active (Arrigo et al., 1993; Vancoppenolle et al., 2010; Saenz & Arrigo 2014). This approach is the most general and is supposed to be more adapted to the Antarctic because of the significant contribution of surface and internal communities there (Meiners et al. 2012).

20.3.4 Physical processes

The relevant environmental parameters for microbial ecosystems are ice thickness (h_i), snow depth (h_s), temperature (T), brine fraction (e), brine salinity (S_{br}), brine transport and photosynthetically active radiation (PAR). All are computed in the model physical component.

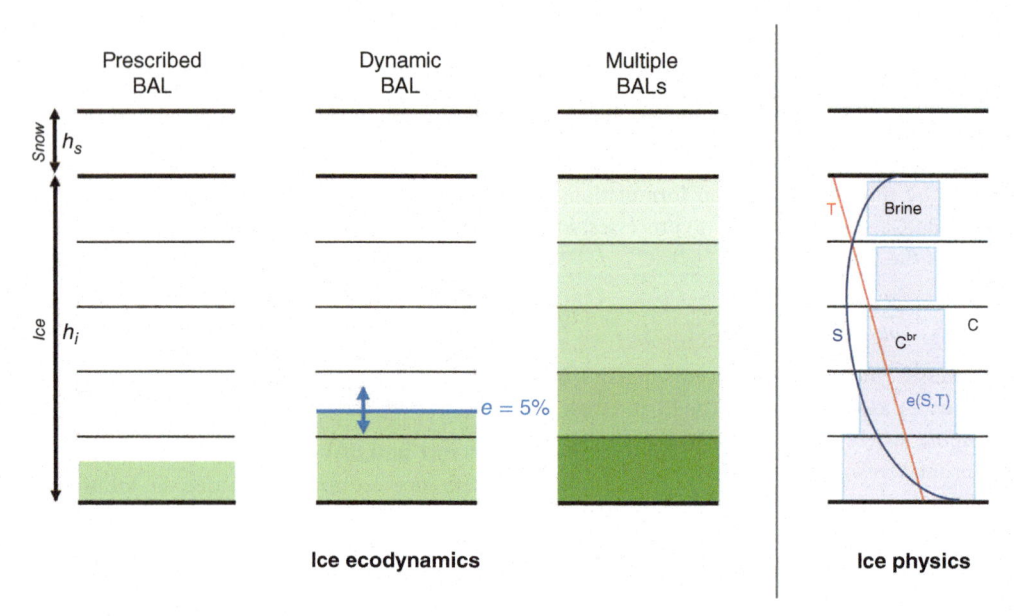

Figure 20.3 The vertical setup of model grids for sea ice algal dynamics and ice physics. Options for algal dynamics are prescribed biologically active layer (BAL), dynamic BAL, or multi-layer BALs. h_i, ice thickness; h_s, snow depth; T, temperature; e, brine fraction; S, bulk salinity; C^{br}, brine concentration; C, bulk concentration.

Ice thermodynamics

The physical component of 1D models follows the generic framework introduced by MU71. The sea ice column is made of ice (divided into N layers) covered by snow and characterized by vertical temperature and often bulk salinity (S) profiles (see Figure 20.3, right panel). Temperature is updated by solving the heat equation (see Chapter 12). Changes in ice thickness and snow depth are derived from the interfacial heat budgets, computed from oceanic, inner conductive and atmospheric fluxes (front-tracking models). Most current thermodynamic formulations in sea ice models are based on the principles laid by MU71. The most popular MU71 reformulations are the zero-layer and three-layer Semtner (1976) models and the multi-layer approach of Bitz and Lipscomb (1999).

Brine dynamics

The original MU71 model prescribes the salinity profile, but salinity changes over time due to brine dynamics. Brine dynamics matter to BGC tracers in sea ice, as they control brine concentrations in dissolved tracers as well as transport (Reeburgh, 1984; Vancoppenolle et al., 2010). Cox and Weeks (1988) introduced the first sea ice desalination scheme based on empirical relationships. However, it is hard to derive consistent transport rates from such empirical functions. The proper theoretical framework to handle brine dynamics derives from the mushy-layer theory (Worster 1992). Sea ice constitutes an example of a mushy layer, in the sense that it is a two-phase, two-component, reactive porous medium (Notz & Worster 2009). Based on a weighted decomposition between the solid and liquid phases, the mushy-layer theory provides consistent formulations for heat, salt and tracer conservation in sea ice (Hunke et al. 2011). Recently, several groups used mushy-layer theories to address the problem of brine dynamics in 1D sea ice models (Vancoppenolle et al., 2010; Turner et al., 2013; Griewank & Notz, 2013; Jeffery & Hunke, 2014; Rees Jones & Worster, 2014), some with sea ice biogeochemistry as a target. The different groups propose conceptually similar approaches, based on the following principles:

1. Changes in ice salinity are derived from a salt transport equation with parameterizations for natural brine convection (gravity drainage), brine percolation (flushing) and flooding, of the form (Hunke et al. 2011):

$$\frac{\partial S}{\partial t} = -w\frac{\partial S^{br}}{\partial z} + \frac{\partial}{\partial z}\left(\overline{D}\frac{\partial S^{br}}{\partial z}\right), \qquad (20.4)$$

where w is the net brine flow due to vertical transport induced by flushing and flooding and $\overline{D}$ is an effective diffusivity, designed to represent the effect of brine convection (or gravity drainage). Brine convection is active where and when the ice is permeable and the brine density structure is unstable, which is achieved in practice for surface temperature significantly lower than the seawater freezing point. Flushing is active during the surface melting period. Flooding of the snow with brine and/or seawater is active for snow depths large enough to depress the snow–ice interface below sea level.

2. Brine inclusions are assumed in thermal equilibrium with the pure ice, hence brine salinity (S_{br}) and fraction (e) can be diagnosed from S and T. At thermal equilibrium, brine inclusions adjust their salinity to the local freezing point, given by the liquidus relation. The linear liquidus $T^{liq} = -\mu S^{br}$ ($\mu = 0.054°C$ kg g^{-1}; Assur, 1958) represents the energetics of the system well (Notz 2005) but largely overestimates S^{br} at low temperature. As this may have undesirable biological or chemical consequences, a third-order fit of Assur (1958) data (see Notz 2005) is better suited in biogeochemical models:

$$S^{br} = -21.4T - 0.886T^2 - 0.0170T^3. \quad (20.5)$$

Brine fraction is well approximated using the relation (Bitz & Lipscomb 1999):

$$e = -\frac{\mu S}{T}. \quad (20.6)$$

The sea ice thermal properties depend either directly on e (more recent models) or implicitly on S and T (MU71; Bitz & Lipscomb 1999).

The various brine transport formulations (Vancoppenolle et al., 2010; Griewank & Notz 2013; Turner et al., 2013; Jeffery & Hunke 2014) differ in their details. At this stage, the different formulations have not been compared with observations or each other systematically, which currently hinders an assessment of their individual advantages and disadvantages. It is likely that inter-model differences in ice thickness and temperature would be small, whereas those in bulk salinity and brine transport could be larger. The ongoing scientific debate should clarify this and better constrain model formulations in the near future.

Tracer transport

Brine dynamics provide the basis for the transport of dissolved tracers, including nutrients. Hence, the physical term for such tracers is of the following form (Vancoppenolle et al. 2010):

$$\left.\frac{\partial C}{\partial t}\right|_\varphi = -w\frac{\partial C^{br}}{\partial z} + \frac{\partial}{\partial z}\left(\overline{D}\frac{\partial C^{br}}{\partial z}\right). \quad (20.7)$$

In this manner, nutrients would perfectly follow salt if there were no BGC source and sink term.

Radiative transfer

A radiative transfer module is needed for the computation of two important fields: the ice radiative heating term $R(z)$, required in the heat equation; and (ii) PAR, required to evaluate photosynthesis. Radiative transfer in sea ice is governed by two processes: absorption by ice, algae and impurities (e.g. algae, soot, sediments); and scattering, the redirection of light into a different direction by brine and gas inclusions. Scattering has no wavelength dependence, whereas absorption is strongly wavelength-dependent. In the ocean, scattering is low, hence the radiation essentially goes downwards. By contrast, radiation in sea ice is dominated by scattering (Jin et al. 1994). Chapter 4 deals with the details of the interactions between light and sea ice. Here, we only describe how it is currently represented in models.

A detailed explicit account of radiative transfer is practically impossible in sea ice models because they do not resolve the entire microstructure of sea ice floes. Hence, parameterizations must be used. Currently, two types of parameterizations of radiative transfer in sea ice are used:

- the two-level Beer–Lambert (BL) law, assuming exponential attenuation of radiation with depth (MU71), combined with *ad hoc* parameterizations of the surface albedo (e.g. Shine & Henderson-Sellers 1985);
- the Delta-Eddington approximation approach, using observed values of inherent optical properties (attenuation and scattering coefficients) to compute radiative transfer (e.g. Briegleb & Light 2007; Saenz & Arrigo 2012). In such models, the albedo is diagnosed and not prescribed.

With the exception of surface albedo, the radiative transfer schemes in sea ice models are rarely evaluated using observations.

The two-layer BL law

The original parameterization of radiative transfer introduced by MU71 decomposes the snow–ice system into two layers: a thin surface scattering layer with thickness $h_o = 10$ cm, absorbing most of the incoming short-wave radiation and another layer where the remaining light is attenuated (Figure 20.4). F_{sw}, the incoming solar irradiance in W m^{-2}, covering the whole solar spectrum (250–2500 nm), is given as a model forcing, and most is reflected (αF_{sw}), where α is a prescribed or computed broadband albedo. Then a prescribed fraction i_o of the net flux, $I_o = i_o(1 - \alpha)F_{sw}$, is transmitted below the surface, while the rest contributes to the surface energy budget. i_o is often called the surface transmission parameter. The transmitted downwelling irradiance attenuates with depth:

$$I(z) = I_o e^{-\kappa z}, \quad (20.8)$$

where κ is the attenuation coefficient. In the original MU71 paper, κ was set to 1.5 m^{-1} in the ice, i_o was set to 0.17 for snow-free ice and zero for snow-covered ice, which practically precludes photosynthesis as long as snow is present. Hence the original MU71 scheme requires adjustments for a minimalist modelling of ice algal photosynthesis. Another issue is that the MU71 scheme neglects the attenuation of light by ice algae and detritus.

Numerical values or parameterizations for κ, i_o and h_o are required for both ice and snow. These parameters could depend on the state of ice and snow and the light properties (direct/diffuse). Light et al. (2008) synthesized parameter values for sea ice in the visible range (see their table 4). For snow, there are no established values at this stage and various *ad hoc* values have been proposed (Vancoppenolle et al., 2010; Pogson et al., 2011).

The absorption of radiation by ice algae and organic matter is strongly dependent on wavelength and increases with the concentration of pigments (e.g. Perovich et al. 1993; Ehn et al. 2008; Fritsen et al. 2011). Based on such studies, Arrigo et al. (1993) suggested that an attenuation coefficient of the form:

$$\kappa = \kappa_i + (a^* C_{Chl} + a_{det}), \quad (20.9)$$

where κ_i is the attenuation coefficient due to ice and brine, a^* is the spectral specific absorption coefficient of ice algae, C_{chl} the bulk chlorophyll (Chl)

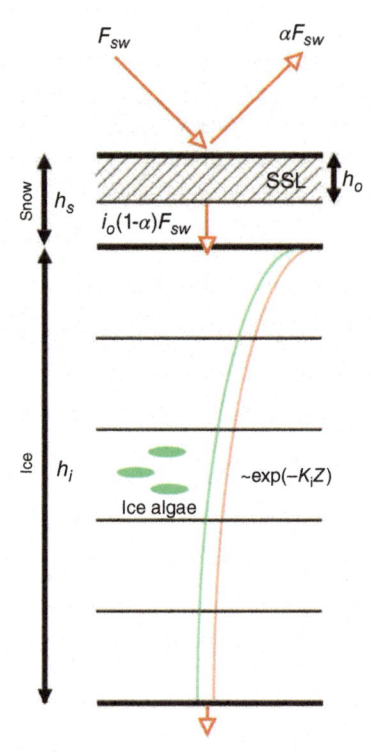

Figure 20.4 Schematics of radiative transfer in sea ice following the two-layer Beer–Lambert law formulation. SSL, surface scattering layer; h_i, ice thickness; h_s, snow depth; α, broadband albedo; F_{SW}, incoming solar irradiance; i_o, surface transmission parameter; κ_i, attenuation coefficient due to ice and brine; z, depth in the ice.

concentration, and a_{det} represents the contribution of particular organic matter. For broadband parameterizations (no spectral dependence), the mean of the specific absorption coefficients can be used. The mean of $a^*(\lambda)$ of Arrigo et al. (1991) is $a^* = 0.008$ [m^{-1} (mg Chl$a$ m^{-3})$^{-1}$], whereas Lavoie et al. (2005) use $a^* = 0.02$ [m^{-1} (mg Chl$a$ m^{-3})$^{-1}$].

The value of a^* is important in models because it determines the levels of self-shading and algal radiative heating in the model. Both effects are small in most cases, except for large Chl concentrations (Figure 20.5). Self-shading occurs once algae higher up in the ice absorb so much light that they effectively shade their counterparts below. An efficient shading (assumed here as 30%) occurs only once the Chl$a$ stock is large enough, i.e. near 15 (45) mg Chl$a$ m^{-2} for the Lavoie (Arrigo) value of a^* (Figure 20.5a). Hence, in models, self-shading is generally weak but important:

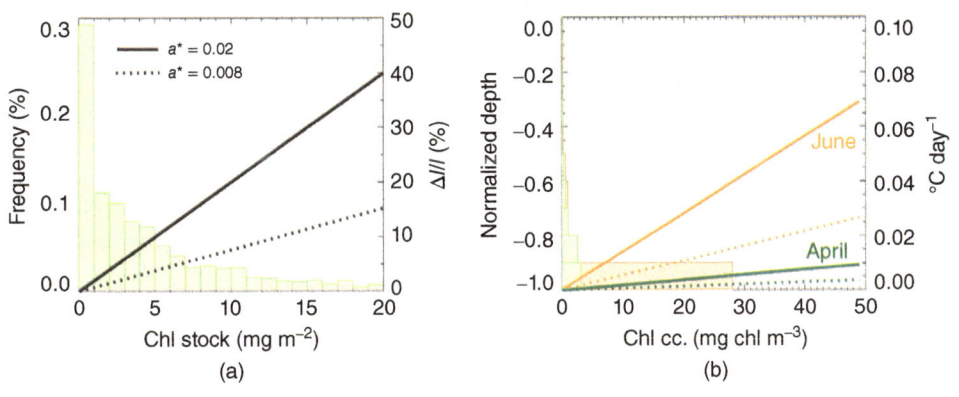

Figure 20.5 (a) Relative change in downwelling irradiance ($\Delta I/I$, %; lines) versus chlorophyll (Chl) stock (mg m^{-2}), for spectral specific absorption coefficients $a^* = 0.008$ (Arrigo et al., 1991) and $a^* = 0.02$ m^{-1} (mg Chl$a$ m^{-2})$^{-1}$ (Lavoie et al., 2005). For context, the discrete probability density function of Chl$a$ stock in Antarctic pack ice (histograms, ASPECT-BIO data, 1300 cores; Meiners et al., 2012) is also depicted. (b) Algal-induced radiative heating $\partial T/\partial t = R^a/\rho c_p$ (°C day^{-1}; lines) versus Chl$a$ concentration (mg m^{-3}), computed using the two a^* values, two values for the prescribed transmitted solar fluxes ($I = 2$ W m^{-2}, April; $I = 15$ W m^{-2}, June), the sea ice density $\rho = 917$ kg m^{-3} and specific heat $c_p = 20\ 152$ J kg^{-1} °C^{-1} (salinity, $S = 4$ g kg^{-1} and temperature, $T = -2$°C). For context, the histograms show two normalized vertical Chl$a$ profiles (histograms) in early April and June from landfast sea ice near Point Barrow, Alaska. Source: Zhou et al., 2013.

self-shading is a safeguard efficiently hampering the development of ice algae beyond large Chl$a$ stocks, i.e. above a few tenths of mg m^{-2}. This analysis excludes detrital matter, which would enhance this effect.

The algal radiative heating term reads:

$$R^a(z) = I(z)C_{chl}(z)a^* \quad (20.10)$$

The contribution of ice algae to radiative heating increases at both large irradiances and Chl$a$, as shown in Figure 20.5(b). Simple computations suggest that ice algal heating of ~0.05°C day^{-1} require significant levels of penetrating radiation in the ice (> 10 W m^{-2}), which occur only once the ice is free of snow. Such heating rates can be viewed as unlikely, as they also require rare Chl concentrations of at least 30 mg m^{-3}, for the larger value $a^* = 0.02$ [m^{-1} (mg Chl$a$ m^{-3})$^{-1}$], and even higher concentrations for the lower value of a^*. Hence, algal heating would be efficient close to the summer solstice and for large Chl$a$ concentrations. Those simple arguments are supported by a more in-depth simulation-based analysis (Zeebe et al. 1996).

Assuming that snow and the surface-scattering layer efficiently attenuate non-visible light (Perovich 1996), the computation of PAR at each level can practically assume that penetrating solar radiation is only in the visible wavelengths. The conversion of irradiance to PAR is done using PAR(z) = 4.91I(z), where the factor 4.91 µE J^{-1} gives the conversion from J to µE for radiation in the visible range (see Vancoppenolle et al. 2011).

The Delta-Eddington approach

Schemes based on the Delta-Eddington approximation for the scattering phase function quite likely outperform Beer's law in terms of precision (Briegleb & Light, 2007; Saenz & Arrigo 2012). These schemes account for multiple scattering effects on direct and diffuse radiation. Computations are based on inherent optical properties (scattering, absorption, asymmetry parameters) for different classes of materials (ice/snow), accounting for the effects of various contributors (i.e. ice, brine, particles, algae). Saenz & Arrigo (2012) use 31 spectral bands for PAR, five ice and snow classes, and include the effects of Chl$a$ and particulate organic matter (POC). The implementation of Briegleb & Light (2007) in the CICE model (Hunke et al. 2013) uses four spectral bands in the whole solar spectrum and accounts for melt ponds, dust, algae and aerosols. How such schemes affect the computation of photosynthesis is yet to be evaluated.

20.3.5 Biological processes

In this section, we will explain the basic principles underlying model representations of microbial ecosystem processes in sea ice. These mostly derive from ocean formulations, which are extensively described in recent textbooks (Sarmiento & Gruber, 2006; Williams & Follows 2011).

Structure of the microbial ecosystem

The structure of microbial ecosystems is represented using a certain number of boxes, or tracers (Figure 20.2). The structure can vary in complexity, depending on authors and their scientific questioning (see Table 20.1).

N-P models

The simplest models are so-called N-P models, including two types of reservoirs: nutrients and phytoplankton. Meant to represent the growth of ice algae and their nutrient uptake, these models can be used to estimate GPP in sea ice (e.g. Deal et al. 2011; Dupont 2012). The elemental composition of ice algae and organics is typically fixed, through prescribed Redfield C:N:Si:P ratios and fixed Chl:C ratios, which implies that only a single ice algal compartment is needed, which reduces model complexity and computational cost.

N-P-Z/D models

N-P models are simple to implement, but their scope is limited, mostly because they neglect detrital organic matter (D) and heterotrophic (Z) processes. N-P-Z/D models include grazers (Z) and/or detrital products (D) (see Figure 20.2). They are useful when remineralization and export have to be properly estimated. N-P-Z/D model structures have been used in sea ice modelling studies (see Table 20.1), but the contribution of the Z/D pools to the quality of the simulation remains unevaluated in the case of sea ice.

Quota models versus Redfield–Monod models

Prescribing elemental ratios neglects photo-acclimation and simplifies algal physiology: carbon assimilation and nutrient uptake are intrinsically coupled, whereas these are distinct processes in reality (e.g. Droop 1983; Geider et al., 1998; Flynn, 2010). Using a varying Chl:C ratio seems particularly indicated in sea ice, as ice algae experience large changes from very low to mild irradiances (Tedesco et al. 2012). Most models prescribe cell stoichiometry, and about half of the models prescribe the Chl:C ratio. It seems clear that acclimation and variable stoichiometry systematically improve the realism of models in the marine realm (Ayata et al. 2013). As far as sea ice is concerned, knowing in which cases acclimation and variable stoichiometry are necessary should be further studied.

Biogeochemical processes

Depending on the model structure, many biological and chemical processes can be incorporated in models. Here, we will discuss the most important processes, in the framework of a very simple example, that of an N-P model with one algal group (concentration C_C, mmol C m^{-3}), a single nutrient (concentration C_N, mmol N m^{-3}), and fixed algal stoichiometry r_C^N (mol mol^{-1}). For such a model, the two model equations read:

$$\left.\frac{\partial C_C}{\partial t}\right|_\beta = (\mu - \lambda)C_C, \quad (20.11)$$

$$\left.\frac{\partial C_N}{\partial t}\right|_\beta = r_C^N(-\mu + f\lambda)C_C, \quad (20.12)$$

where μ and λ represent the rates (s^{-1}) of photosynthesis and carbon loss (cell lysis, exudation, respiration, grazing). f is the remineralized fraction of ice algal loss (if different from 1, the model does not conserve nutrient mass).

Organic matter synthesis

Photosynthesis is the creation of new organic matter by microorganisms. During the process, light is converted into chemical energy, stored into organic molecules – formed from CO_2, water and minerals – whereas oxygen is released. The photosynthetic rate is the amount (mass or moles) of organic carbon synthesized per unit time. Pigments (mainly Chl*a*) are the central part of the photosynthetic apparatus. They are specialized organic molecules, whose electrons get excited due to the absorption of light at given wavelengths. Model representations of photosynthesis are based on the following assumptions:

1 The photosynthetic rate is directly proportional to the abundance of pigments.
2 At constant temperature and sufficient resources, the photosynthetic rate per weight of Chl initially increases with irradiance at low irradiances, and stabilizes for larger irradiance values (Jassby & Platt 1976).
3 Temperature sets an upper limit to the rate of photosynthesis per weight of Chl, and this limit can be predicted (Eppley 1972).

4 Any departure of brine salinity from that of seawater affects the photosynthetic rate per weight of Chl; brine salinity and temperature effects are decoupled (Arrigo & Sullivan, 1992; Ralph et al. 2007).
5 The main limiting resources for algal growth are macronutrients, and their uptake by algae follows the hyperbolic enzymatic kinetics (Morel 1987).

The photosynthetic rate (s^{-1}) is assumed to depart from its maximum value due to non-optimal environmental conditions:

$$\mu = \mu_m L_{PAR} L_N f_S f_T \quad (20.13)$$

μ_m is the maximum photosynthetic rate (s^{-1}), convertible into a maximum number of cell doublings day^{-1}: $d_m = \ln(2)/\mu_m$. The individual parameters describe the dependence of carbon assimilation on temperature (f_T), brine salinity (f_S), light (L_{PAR}) and extracellular nutrients (L_N). These will be now further discussed.

Light limitation (L_{PAR})

The response of photosynthesis to light has been widely studied. Based on ~200 light-saturation experiments on coastal phytoplankton, Jassby & Platt (1976) retained, among different formulations:

$$L_{PAR} = \tanh\left(\frac{\alpha\, PAR}{P_m}\right). \quad (20.14)$$

Other formulations exist and, among these, the exponential and the inverse square-root only slightly differ from equation (20.14). The so-called photosynthetic efficiency α [gC (g Chl$a$ hour µE m^{-2} s^{-1})$^{-1}$] is the increase of carbon assimilation (h^{-1}) per g Chl per unit increase of PAR. It indicates how rapidly ice algae reach the light-saturated specific photosynthetic rate P_m [gC (g Chl$a$ hour)$^{-1}$]. Acclimation to low light slightly increases α and strongly reduces P_m, which implies a net reduction of the photo-adaptive index (roughly the comfort level of irradiance) $E_k = P_m/\alpha$ (e.g. Cota & Smith, 1991). The maximum photosynthetic rate can be re-expressed using (Geider et al. 1998):

$$P_m = \mu_m f_S f_T L_N / r_C^{Chl}, \quad (20.15)$$

This introduces the Chl:C ratio, r_C^{Chl} (g g^{-1}).

Various strategies to express photosynthetic parameters P_m and α have been used in sea ice models. Prescribing E_k from observational P_m and α values is the simplest approach (Lavoie et al. 2005). It is formally equivalent to prescribe r_C^{Chl} and implies the absence of photo-acclimation. Because acclimation efficiently adjusts r_C^{Chl} (Michel et al. 1988), one is forced to readjust E_k at different sites (e.g. thin vs deep snow). Including acclimation can be achieved using diagnostic or prognostic approaches. The diagnostic approach empirically relates r_C^{Chl} to PAR and nutrient limitation status (Arrigo & Sullivan, 1994; Sibert et al. 2011). For instance, Aumont et al. (2003), based on Doney et al. (1996), propose:

$$r_C^{Chl} = \left[r_C^{Chl}|_{max} - (r_C^{Chl}|_{max} - r_C^{Chl}|_{min})\, min\left(\frac{PAR}{E^*}, 1\right)\right] L_N. \quad (20.16)$$

0.01 and 0.05 g g^{-1} are the prescribed minimum and maximum values of r_C^{Chl} (Cloern et al. 1995); E^* is the PAR level (µE m^{-2} s^{-1}) where r_C^{Chl} is assumed minimum. This equation is justified by (i) the roughly linear dependence of r_C^{Chl} on irradiance (Sakshaug et al. 1989); (ii) the preferential nutrient uptake compared with Chl synthesis under limited resources. However, equation (20.16) is poorly constrained by sea ice observations, in particular E^*. Using a diagnostic r_C^{Chl} drastically reduces light limitation at low irradiance (Figure 20.6a). The prognostic approach relies on a specific Chl compartment (Tedesco et al. 2010), with its own synthesis and loss terms. Compared with a prognostic r_C^{Chl}, the diagnostic approach was found sufficient for marine biogeochemical models as long as timescales larger than a few days are considered (Flynn 2003).

Nutrient limitation (L_N)

In Redfield–Monod models, the limitation of photosynthesis by nutrient availability is often assumed to follow a hyperbolic (Michaelis–Menten) form (Monod 1949; see Figure 20.6b):

$$L_N = \frac{C_N^{br}}{k_N + C_N^{br}}. \quad (20.17)$$

If several limiting nutrients are considered, L_N can be taken as the minimum of several such factors. The hyperbole corresponds to the steady-state reaction rate of an enzymatic reaction, where a substrate (N) is transferred through a product (ice algae), by an enzyme (porter site for nutrient). The half-saturation constant (k_N), for which the growth rate is half its maximum value, is typically derived from long-term laboratory cultures. Sarthou et al. (2005) propose $k_N = 1.6$ µM,

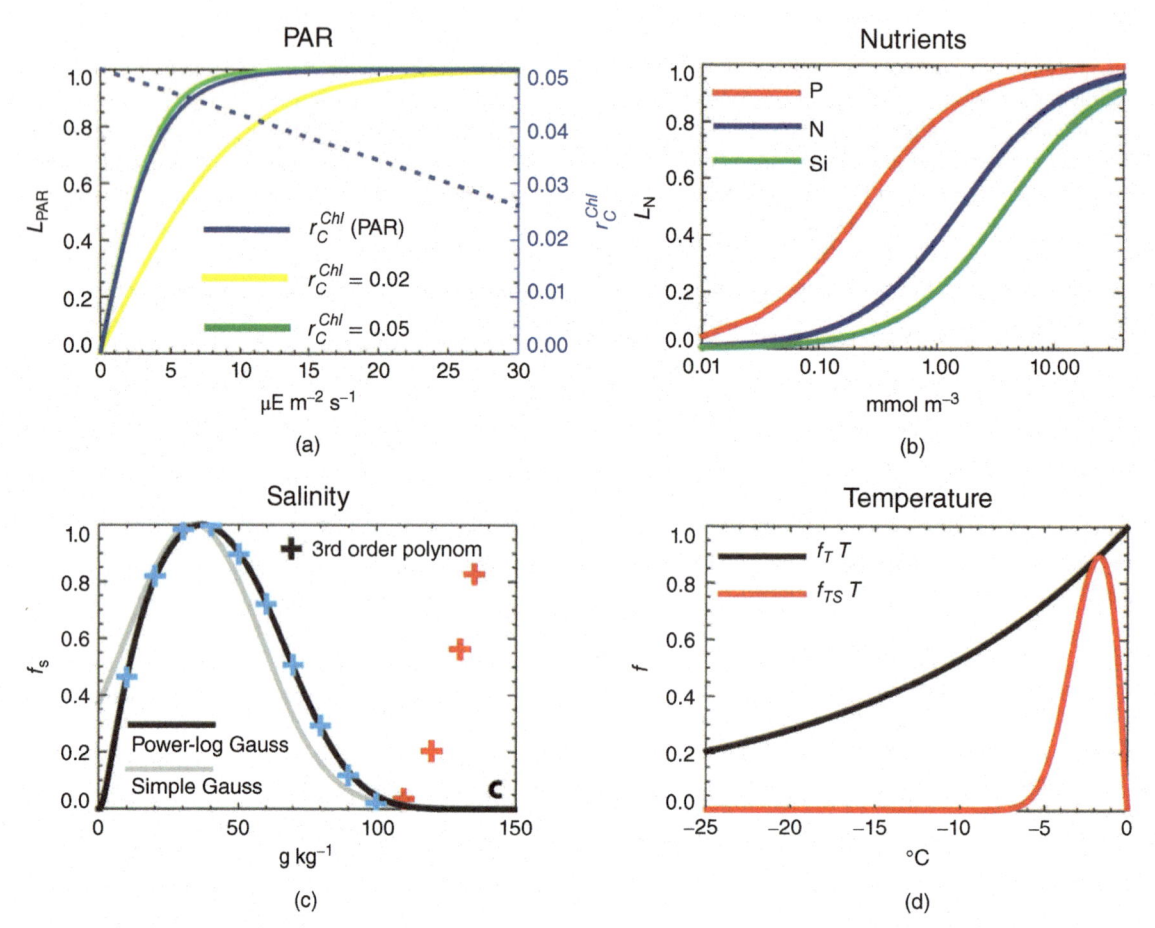

Figure 20.6 Classical formulations of the multiplicative factors for carbon assimilation. (a) Light limitation factor (*tanh*; Jassby & Platt, 1976), using $\mu = 2.10^{-5}$ s^{-1}, $\alpha = 1.0e{-}4$ g C (g Chl*a* h µE m^{-2} s^{-1})$^{-1}$ and various options for r_C^{Chl} (prescribed low, prescribed high, and interactive with $E^* = 50$ µE m^{-2} s^{-1}). (b) Extracellular nutrient limitation factor for N, P, Si (Monod, 1949) using half-saturation constants typical of marine diatoms (Sarthou et al., 2005). (c) Brine salinity factor, using the third-order polynomial fit (from Arrigo, 2003, symbols) and the simple and power-log Gaussian fits. (d) Temperature factor for sole temperature effects (f_T; Eppley, 1972) and combined temperature and brine salinity effects f_{TS}.

$k_{Si} = 3.9$ µM and $k_P = 0.24$ µM based on a compilation of marine diatoms data.

In cold ice, brine concentrations in nutrients remain much higher than half-saturation constants, where the brine network is ventilated (e.g. near the ice base in winter and spring), and therefore such ice nutrients are hardly limiting. However, because the brine fraction is ≪ 1, the stocks of nutrients are quickly exhausted if the brine network is not ventilated (e.g. upper in the ice). This formulation of nutrient limitation assumes that nutrient uptake and carbon assimilation are coupled, which is true only at steady state, but rarely achieved in reality (Droop, 1983; Flynn 2010).

Salinity dependence (f_S)

Incubations of ice algae at different salinities clearly indicate that ice algae reduce their growth rate when exposed to salinities that depart from standard seawater values (e.g. Grant & Horner, 1976; Arrigo & Sullivan, 1992; Ralph et al. 2007). The salinity dependence varies among different species. Arrigo (2003) proposes a third-order polynomial fit, based on ice algal samples taken at McMurdo Sound (Arrigo & Sullivan 1992). We propose an alternative power-log Gaussian fit to their data:

$$f_S(S_{br}) = exp[-(2.16 - 8.3 \times 10^{-5} \cdot S_{br}^{2.11} - 0.55 \, ln(Sbr))^2]. \tag{20.18}$$

This fit is as accurate as the original one, with two advantages: it guarantees values strictly within [0,1] and behaves nicely beyond 100 g kg^{-1} range (Figure 20.6c). A regular Gaussian fit would be more flexible, but cannot handle the asymmetry of the distribution. Brine salinity effects are strongest within the ice in winter, when the ice is cold. Including brine salinity effects is necessary in multi-layer models. It is not required for the models that prescribe bottom communities, as the temperature and salinity near the ice base are close to seawater values.

Temperature dependence (f_T)

The standard formulation for the temperature dependence of the ice algal specific growth rate comes from a compilation of laboratory culture data of various phytoplankton species (Eppley, 1972; black curve in Figure 20.6d):

$$f_T = e^{r_g T_c} \tag{20.19}$$

where T_c is the temperature in °C and $r_g = 0.0633$ °C^{-1}. An implicit assumption is that all coefficients correspond to a reference of 0°C. Other formulations exist; departing from equation (20.19) by the use of specific exponents (e.g. Q_{10}) or a different reference temperature: all are equivalent to equation (20.19).

In practice, the effects of temperature and brine salinity on growth rate are coupled. This is because the salinity of brine is related to temperature through the local thermal equilibrium (equation 20.5). The combined temperature and salinity effects define a window of temperature tolerance of ice algae defined by $f_{TS} = f_T f_S > 0.5$, which for the ice algae studied by Arrigo & Sullivan (1992) corresponds to the [−3.5, −0.5°C] interval (Figure 20.6d).

Carbon losses

Compared with photosynthesis, carbon losses by sea ice microalgae typically receive much less attention in the scientific literature. They should in principle include detailed formulations for cell lysis, exudation, respiration and grazing (see, e.g., Nishi & Tabeta 2008; Tedesco et al. 2010). With little experimental background to constrain such formulations, the latter are somewhat speculative and vary among authors. Representing these processes is, however, required to correctly distribute the chemical elements among the different model pools (i.e. algae, detritus, dissolved species). In practice, carbon losses by lysis, exudation and respiration are formulated as calibrated fractions of the gross production rate, photosynthesis, or the ice algal carbon content. The corresponding proportionality factors are either constant or functions of the environmental factors. In many instances, those terms are grouped together as a single loss term (e.g. Lavoie et al. 2005).

The contribution of heterotrophic grazing to biogeochemical cycling in sea ice is not well documented, but can be expected to be small due to the small size of brine inclusions. In a landfast ice study in the Candadian Archipelago, Michel et al. (2002) found that very little ice algal production was channelled through the meio- and microfauna within the ice. If no Z compartment is included, the grazing rate can be parameterized as a function of the GPP rate (e.g. Arrigo & Sullivan, 1994; Lavoie et al., 2005) or ignored. If a Z compartment is included in the model, grazing is typically chosen proportional to the grazer population (Sibert et al., 2011). Models including a grazing term have more carbon losses and recycling, with large associated uncertainties.

Nutrient uptake and remineralization

In Redfield models, nutrient uptake depends directly on production and the cell C-nutrient quota. Quota models (e.g. Tedesco et al., 2010) decouple nutrient uptake and carbon assimilation. Remineralization in N-P models can only be related to ice algal carbon loss, which ignores the important role of organic matter degradation for the resupply of chemical elements in sea ice, as indicated by recent observation-based analyses (Fripiat et al., 2014a,b). The sympagic organic matter pool increases in size throughout the season, which intensifies the regeneration loop and supports regenerated production late in the season. Released during sea ice melt in the water column, organic matter synthesized in the sea ice fuels heterotrophs, remineralization in the water column and export to depth, in proportions that vary from one study to another (Michel et al., 1996; Renaud et al., 2007). Models including a D pool can represent these processes and formulate nutrient recycling as a function of the detrital pool should better estimate the timing and magnitude of nutrient recycling.

20.4 Selected model simulations

Now that we have covered the basic processes that are typically included in sea ice models of microbial sea ice communities, we show how the environmental factors control the seasonality of model sea ice algae. To this end, a few idealized 1-year simulations performed with a multi-layer 1D physical sea ice model coupled with a simple N-P ice algal module are analysed in the following sections.

20.4.1 Short model description

The model includes most of the mechanisms described in the left and central boxes of Figure 20.2. The physical component of the model has ice thermodynamics, brine transport and radiative transfer. Ice thermodynamics follow the energy-conserving approach of Bitz & Lipscomb (1999). Brine dynamics include natural brine convection and percolation (Vancoppenolle et al., 2010; equation 20.4). Radiative transfer simply describes the exponential attenuation of the solar radiation that penetrates through the snow–ice system using the two-level BL approach (equation 20.8), assuming that penetrating radiation is purely visible. Solar radiation attenuation depends on Chl$a$ concentration (Equation 20.9).

The ice algal component is a simple N-P module with one ice algal group, and three dissolved macronutrients (N, P and Si). Ice algae have a prescribed Redfield stoichiometry, typical of marine diatoms (C:N:Si:P = 73:10:6:1; Sarthou et al., 2005). Photosynthesis is limited by light (Jassby & Platt, 1976; equations 20.14 and 20.15) and nutrients (equation 20.17), and accounts for the effects of brine salinity (Arrigo & Sullivan, 1992; equation 20.18) and temperature (Eppley, 1972; equation 20.19). r_C^{Chl} directly changes with irradiance (equation 20.16). Ice algal carbon loss is a temperature-dependent fraction of ice algal carbon content. Ice algae and nutrients are considered as passive tracers. Nutrients are dissolved in brine, hence their bulk concentration changes due to brine dynamics (equation 20.7), whereas their brine concentration is retrieved from bulk concentration (equation 20.2). Ice algae are considered as particulate tracers, not moving with brine. All tracers are trapped in forming ice at their ocean concentration when ice forms and are rejected to the ocean at their basal ice concentration when it melts.

20.4.2 Model setup

The model simulations are designed to reproduce the main features of undeformed first-year Antarctic pack ice, at the location of Ice Station Polarstern (ISPOL). ISPOL was a drift station experiment in the western Weddell Sea taking place within 67–68.5°S and 54–56°W, from 27 November 2004 to 3 January 2005. This particular setup was chosen because of the availability of thoroughly analysed time-series data: physical and biogeochemical sea ice properties were measured over seven ice stations (Tison et al. 2008) in December. Other undeformed sea ice sites could also have been chosen, but their broad characteristics would probably not qualitatively differ from that of ISPOL.

Simulations cover an annual cycle, from 14 February 2004 to 13 February 2005. The initial date was diagnosed as the onset of freezing from atmospheric data. Atmospheric forcing derives from a combination of reanalysis, *in situ* observations and formulae. Seawater nutrient and Chl$a$ concentrations (C^w) are prescribed following ISPOL observations (Zemmelink et al. 2008). Ice characteristics are initialized using thin ice values: h_i = 10 cm, h_s = 2 cm, S^{ini} = 13 g kg^{-1}. Bulk ice nutrient concentrations (C^{ini}) are initialized assuming that their ratio to seawater concentrations (C^w) is the same as that of salt: C^{ini} = $C^w \cdot S^{ini}/S^w$, where S^w = 34 g kg^{-1} is a reference seawater salinity in the ISPOL region.

Three model parameters were calibrated: the maximum photosynthetic rate μ_m = 0.86 day^{-1}, the carbon loss rate λ = 0.15 day^{-1} and the photosynthetic efficiency α = 0.09 g C (g Chl$a$ hour)$^{-1}$ (µE m^{-2} s^{-1})$^{-1}$, in order to get values of Chl and nutrients in reasonable agreement with observations.

20.4.3 Simulated environmental characteristics and ice algal development

The simulated seasonality and vertical variations of the environmental factors (PAR, temperature, brine salinity and phosphate, the limiting nutrient in this model setup), the corresponding multiplication factors as well as bulk algae carbon (C_C) and Chl concentrations of ice algae are depicted in Figure 20.7.

The simulated sea ice grows from mid-February to early October when bottom melt starts. There is little

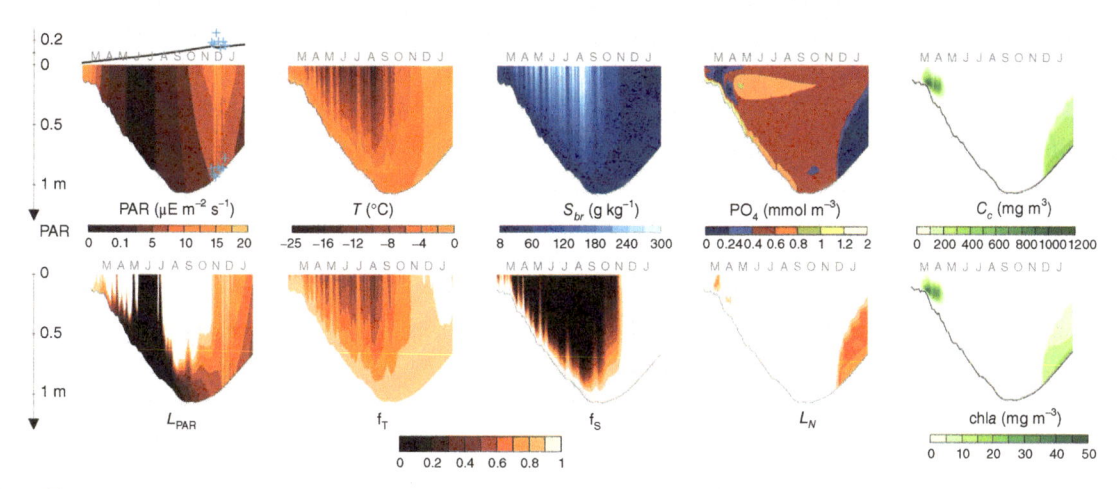

Figure 20.7 An example simulation of the vertical distribution of sea ice characteristics at the location of Ice Station Polarstern (ISPOL) drift station, over a complete year, using the one-dimensional sea ice model described in the text. Vertical axis is depth (m), horizontal axis is time (months). The contoured fields are: (upper row) photosynthetically active radiation (PAR), temperature (T), brine salinity (S_{br}), bulk concentration of phosphate (PO_4) and carbon content of ice algae (C_C); (lower row) the multiplication factors for ice algal growth rate, associated with PAR (L_{PAR}), temperature (f_T), brine salinity (f_S), nutrients (L_N) and bulk chlorophyll (Chl$a$) concentration. On the PAR plot, the simulated snow depth (black line) and observed snow depth and ice thickness (blue crosses) are depicted.

surface melting, and it occurs in December and January only. The PAR transmitted under snow is typically < 15 µE m^{-2} s^{-1}, and the polar night lasts from early May to late August. The average temperature is below −5°C from mid-April to mid November.

There are two main ice algal development periods: one in autumn, and the other starting in late spring and lasting until the end of the simulation in mid-February 2015. The autumn algal development is reasonable, for instance, compared with a young ice experiment during the Ice Station Weddell (Melnikov 1998): it starts in early March, lasts about 1 month, until biomass reaches ∼ 100 mg C m^{-2}, with a maximum Chl ∼ 30 mg m^{-3}. Over the active growth period, GPP is ∼ 750 mg C m^{-2} (8.4 ± 8.6 mg C m^{-2} day^{-1}). Algal growth is sustained by suitable environmental conditions: sufficient light, relatively warm ice, mild brine salinities and plentiful nutrients. Nutrients can be supplied from seawater through entrapment during growth and brine convection and then recycled within the ice. Two-thirds of autumn GPP is recycled in the presented simulation. Algal growth stops in mid-April, triggered by average brine salinities above 90 g kg^{-1}, which occurs when the temperature drops below −5°C.

Ice algal growth resumes by mid-November, which seems rather late in the season, but cannot be confirmed or invalidated using on-site observations, spanning December only. Ice algal biomass reaches ∼ 150 mg C m^{-2} by the end of the simulation. GPP is ∼ 2200 mg C m^{-2} (24 ± 12.8 mg C m^{-2} day^{-1}) for the whole spring and summer activity period. Algal development starts once the mean temperature is > −5°C, which relieves the inhibiting effect of brine salinity on growth. The ice algae develop the most rapidly at the base of the ice, where brine salinity is the closest to seawater values and progresses upwards later on. Because of stable brine stratification promoted by nearly isothermal ice, there is little nutrient supply in summer, and hence most of the production is regenerated (90%).

The simulated characteristics of the ice algae compare reasonably with ISPOL observations, in terms of Chl$a$ (Figure 20.8). In terms of nutrients, a precise model–observation agreement is difficult to reach. In our case, there is too much nitrate and too little phosphate early on. The robustness of this result should be investigated further. This could be due to an improper representation of model processes (in particular, brine dynamics and the remineralization of detrital organic matter) or to systematic measurement and/or sampling errors.

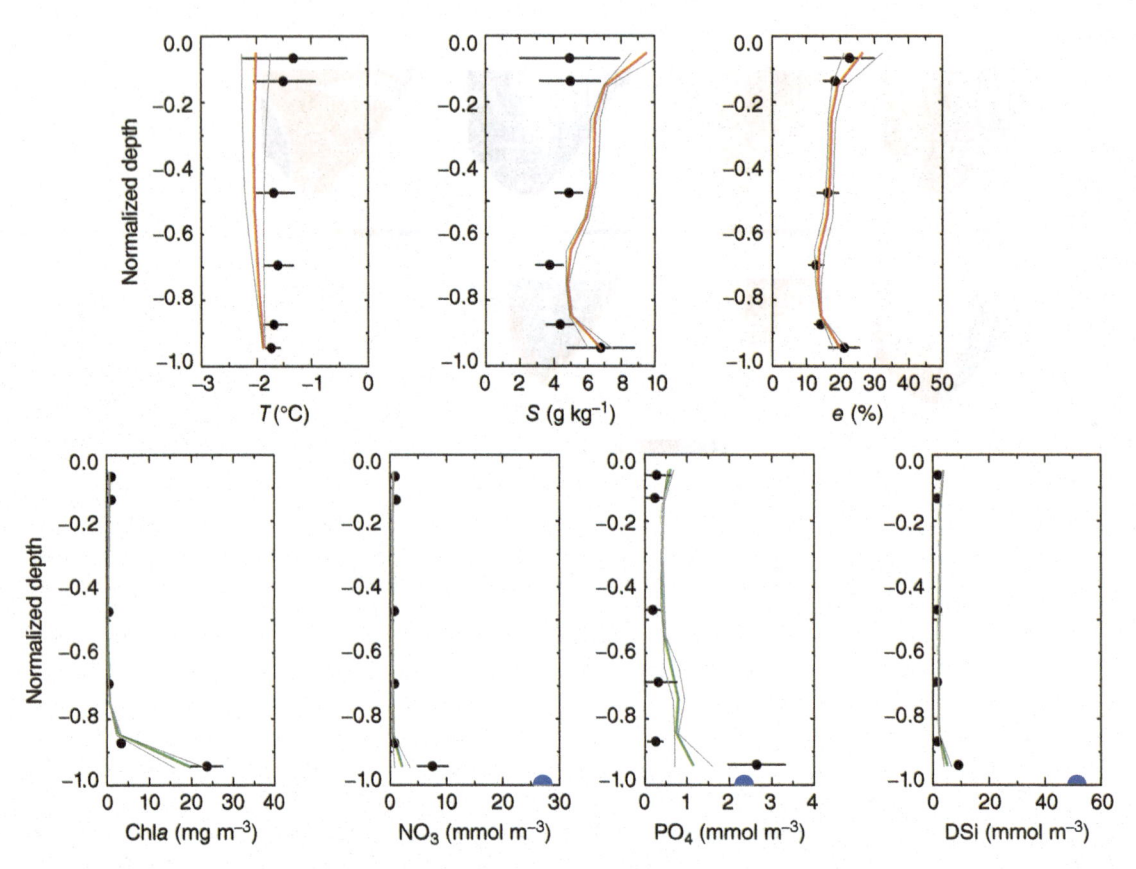

Figure 20.8 Mean normalized vertical profiles in sea ice at Ice Station Polarstern (ISPOL) drift station Tison et al., 2008: observations in sea ice (black circles) and seawater (blue circles); vs simulations (lines), mean (red/green) and ± 1 standard deviation (grey). Standard deviations are computed at each normalized depth in the ice, using the seven station values (obs) and using the daily values from the simulation time series over the ISPOL drift station period (model). T, temperature; S, salinity; e, brine fraction; Chl$a$, chlorophyll a; DSi, dissolved silica.

20.4.4 Response of ice algae to changes in the formulation of ice algal growth

To illustrate how the simulated ice algal growth responds to the model processes, the results of a series of simulations corresponding to specific combinations of multiplicative factors are depicted in Figure 20.9 (vertical Chl$a$ structure) and Figure 20.10 (carbon and dissolved nitrogen stocks).

No photosynthesis: $\mu = 0$

Without photosynthesis, ice algae obviously cannot grow. All nutrients follow salinity, and the evolution of nutrient stocks (mmol m^{-2}) mostly reflect changes in ice thickness.

Nutrient limitation only: $\mu = \mu(N)$ (Figure 20.9a)

If only nutrients limit growth, carbon stocks reach up to > 1000 mg m^{-2}. Nutrients are efficiently recycled due to large biomasses upper in the ice. Yet the net nutrient algal uptake is such that the stock of dissolved nutrients (see Figure 20.10b for dissolved N) is always smaller than with $\mu = 0$. Nutrient concentrations in brine at the bottommost sea ice layer often prove smaller than in seawater, and hence brine convection involves a net ocean–ice nutrient transport. This $\mu(N)$ run shows that nutrients hardly limit algal growth in the model if they are available in seawater: they can be pumped into the ice thanks to brine convection during the growth period.

in the upper part of the ice. There, brine circulation is not efficient and nutrients are effectively trapped. This run indicates that nutrient recycling from the ice algae growing in autumn increases the nutrient stocks available for the spring–summer growth period.

Light limitation

Adding light limitation (Figure 20.10c) – $\mu = \mu(N, \text{PAR})$ – adds a strong seasonality to the system; ice algae cannot grow during the dark period. The build-up of biomass is much more reasonable in comparison to what is observed in nature (~ 200 mg C m^{-2}). Recycling of nutrients from decaying algae grown in fall is active, but much less than with $\mu(N, T, S_{br})$.

Combination of effects

Only the combination of all effects (Figure 20.10d) – $\mu = \mu(N, T, S_{br}, \text{PAR})$ – gives a seasonal and vertical structure that resembles observed situations. In this simulation, there are two periods of higher algal activity (autumn and late spring), and very little activity in winter. The autumn algal development spans the full ice depth, whereas the one in spring starts from the ice base and expands towards the surface. This picture is broadly consistent with observations (Meiners et al., 2012). The details of this structure are, however, quite sensitive to model parameters, and the derived gross and net primary production (NPP) as well. For instance, the annual GPP ranges over 3.01 ± 0.65 g C m^{-2}, if $\pm 10\%$ changes are applied to μ_m, λ and α.

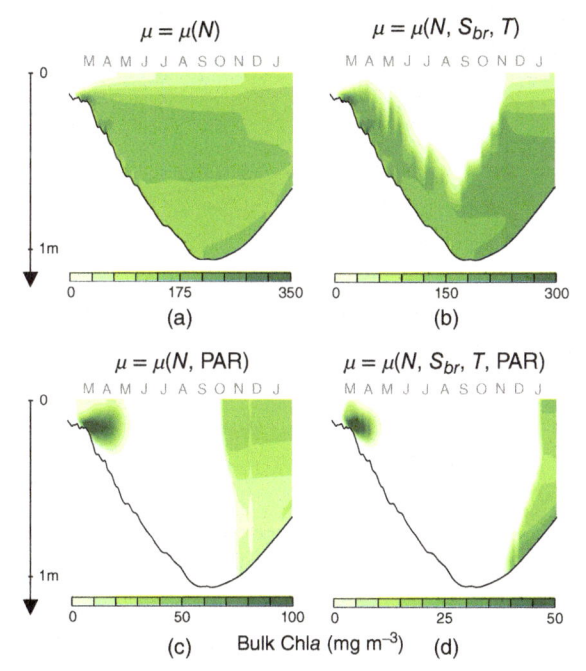

Figure 20.9 Simulated evolution in time of the vertical profile of bulk chlorophyll (Chl$a$) concentration in sea ice, over one full year of simulation at the Ice Station Polarstern (ISPOL) location. Each plot corresponds to a specific idealized model setup – ice algal growth rate function of: (a) nutrients only; (b) nutrients, brine salinity and temperature; (c) nutrients and light; (d) nutrients, brine salinity, temperature and light. To better isolate the effects of limitations, constant loss rate λ (no temperature dependence) and prescribed Chl:C ratio, $r_C^{Chl} = 0.05$ g g^{-1}, are used. μ, rate of photosynthesis; S, bulk ice salinity; T, temperature; S_{br}, brine salinity; PAR, photosynthetically active radiation.

Brine salinity and temperature effects

Adding brine salinity and temperature effects (Figure 20.9b) – $\mu = \mu(N, T, S_{br})$ – mostly introduces a vertical structure, with more algae near the ice base than near the surface. In particular, the upper part of the ice becomes unsuitable for algal growth from May to November, because of large brine salinities there. The build-up of ice algal carbon stock is still very large, but not as much as for $\mu = \mu(N)$, because the algal growth rate with $\mu(N, T, S_{br})$ is always smaller than in the $\mu(N)$ case, in particular during winter. During the initial algal development in March and April, nutrient stocks are lower than with $\mu = 0$ because of intense algal uptake. However, they become larger later on, because of the recycling of nutrients from decaying ice algae

20.5 Discussion and outlook

The modelling of sea ice biogeochemical processes has various applications: understand processes that are difficult to observe, close the budget of chemical elements, upscale the results of local observations, perform projections for the future, and study impacts on other compartments of the Earth system. Sea ice biogeochemical models have mostly been used to study sea ice algae at local and large scales; other aspects have received little consideration to date.

20.5.1 Qualitative understanding of the main ice algal growth drivers

The basic concept used in all sea ice algae models is that ice algae accumulate carbon at a rate that is

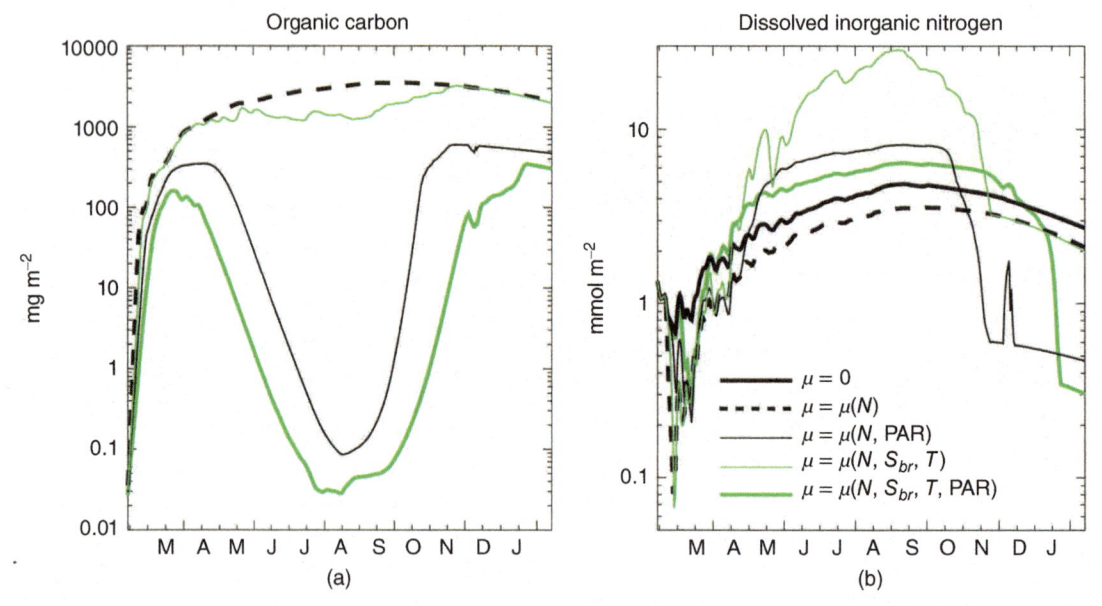

Figure 20.10 As Figure 20.9, but for the simulated evolution in time of stocks of: (a) ice algal carbon; and (b) dissolved inorganic nitrogen. Each curve corresponds to a specific idealized model setup – ice algal growth rate function of: nutrients only; nutrients, brine salinity and temperature; nutrients and light; nutrients, brine salinity, temperature and light. A constant loss rate and a prescribed chlorophyll:C ratio are used. μ, rate of photosynthesis; S, salinity; T, temperature; S_{br}, brine salinity; PAR, photosynthetically active radiation.

limited by the characteristics of the environment (light, temperature, brine salinity, nutrients) and lose carbon at a prescribed rate. With such a model, two bloom periods emerge: autumn and spring–summer.

The seasonality and vertical distribution of ice algae is driven by light and temperature variations in the ice (via brine salinity). The succession of clear and dark periods introduces the observed seasonality of ice algae. The vertical structure, most pronounced in summer, with more algae at the ice base, is due to brine salinity characterized by a strong winter upward gradient, which drastically limits winter algae upper in the ice. Brine dynamics can replenish depleted nutrients in the ice with ocean nutrients only when brine convection is active.

20.5.2 Role of model structure and complexity

Sea ice algae models vary widely in terms of structure and complexity. The impact of technical choices regarding the details of the model on simulated ice algal complexity is relatively unexplored. The impact of the representation of transport, radiation, vertical grid, ecosystem model structure and acclimation is probably important, but it is unexplored. Major unknowns are the role of detrital organic matter, the emergence of surface communities, and the comparison between models that prescribe production in a BAL of prescribed thickness near the ice base and multi-layer models.

20.5.3 Limitations of the predictive capacity of models

If the qualitative functioning of ice algal ecosystems is relatively well reproduced in models, their quantitative performance is doubtable, for several reasons.

1 The conceptual understanding from observations of sea ice microbial communities is limited. Time-series observations, integrating at least several ecosystem parameters over a sufficiently long period, are only few in number (Melnikov, 1998; Lavoie et al., 2005; Tison et al. 2008; Tedesco et al., 2010; Zhou et al. 2013). Protocols to measure many of the chemical elements, compounds and biogeochemical processes relevant to Earth system now exist but they are not really inter-compared (Miller et al., 2015). Measurement errors are difficult to control, and sampling issues in such a heterogeneous environment are

Table 20.2 Estimates of primary production (TgC year^{-1}) in the polar oceans.

	Primary production (TgC year^{-1})	Method	Source
Arctic sea ice	9–73	Observation-based	Legendre et al. (1992)
	15.1	Model (1992)	Deal et al. (2011)
	21.3	Model (1998–2007)	Jin et al. (2012)
	43	Model (1980–1999)	Dupont (2012)
Arctic Ocean (>66°N)	993	Satellite net PP	Hill et al. (2013)
	441–585	Satellite net PP	Arrigo & van Dijken (2011)
Antarctic sea ice	63–70	Observation–based	Legendre et al. (1992)
	35.7	Model (1989–1990)	Arrigo et al. (1997)
	23.7	Model (2005–2006)	Saenz & Arrigo (2014)
Southern Ocean (>50°S; sea ice zone)	1949; 180	Satellite net PP	Arrigo et al. (2008)

of primary importance. Spatial variability at the metre-scale for both physical and biogeochemical parameters can be substantial (e.g. Chl, PO$_4$; Zhou et al., 2013). The only large-scale sea ice biogeochemical data set (Meiners et al., 2012) includes 1300 vertical profiles of Chl*a* from Antarctic sea ice cores, which is small compared with comparable marine databases. Spatial variability is not explicitly accounted for in models, whereas observed fields are highly variable spatially. Besides, model evaluation is often limited to observations of Chl standing stock (m^{-2}) and concentration (m^{-3}), whereas important parameters, such as light, nutrients and POC are often overlooked, mostly because of the paucity of integrated time-series observations. This means the models are rarely supported by many observations, which casts doubt on their ability to reproduce a large number of observations. A major challenge for the future is to evaluate models with more observations and from several locations.

2 The environmental factors cannot always be well constrained in models, in particular light and brine dynamics, of large importance for ice algae. The light propagating through the heterogeneous snow–ice system was until recently difficult to measure repeatedly (Nicolaus et al., 2013) and therefore has rarely been evaluated in models (except in the study of Lavoie et al., 2005). The characteristics of the brine flow in a dynamic mushy layer are better understood now (e.g. Wells et al., 2011), but its characteristics as simulated by new model parameterizations (Vancoppenolle et al., 2010; Griewank & Notz, 2013; Turner et al., 2013; Jeffery & Hunke, 2014) are not yet fully evaluated.

3 Uncertainties in the model formulations of biological processes are large. First, parameterizations are empirical, hence they can work only for what they have been evaluated. Second, the physiological characteristics (nutrient affinity, photosynthetic parameters, osmotic tolerance) need to be *a priori* specified for the model, whereas they are highly variable in nature. Third, models miss or misrepresent some important processes (e.g. detrital organic matter, ice–ocean coupling).

We now turn to the impact of these model uncertainties on the large-scale sea ice primary production estimates. Specifically, these are GPP estimates (see Table 20.2), computed by integrating the ice algal growth rates (excluding losses) over the sea ice domain (thickness and extent) and over a year. These numbers are typically compared with NPP in the Arctic and Southern Ocean estimates (including respiration). Arctic sea ice model GPP (15.1–43 TgC year^{-1}, N = 3) is within simple observation-based computations (9–73 TgC year^{-1}; Legendre et al., 1992). Arctic sea ice GPP is small (<15%) compared with oceanic NPP above the Arctic Circle (455–993 TgC year^{-1}), which can be attributed to a thin BAL combined with a short production period. In the Southern Ocean sea ice zone, sea ice model GPP is 23.7–35.7 TgC year^{-1} (N = 2), which is less than observation-based computations (63–70 TgC year^{-1}). Antarctic sea ice GPP could be, at most, 40% of oceanic NPP, which is significant. There are only a few models, and the uncertainties on model processes are significant; hence, the confidence in sea ice GPP estimates is relatively small. However, it would be hard to imagine how these estimates could increase hugely if the models are wrong, as nutrients are ultimately limiting and production is restricted in space and time.

20.5.4 Perspectives

Important future research directions for sea ice BGC modelling, in the context of contemporary scientific questioning on the role of sea ice within the global biogeochemical cycles, are:

1 To evaluate ice algae models with more observations, including all aspects of the system (algae, organic matter, nutrients, light and brine), in terms of stocks and concentrations.
2 To increase the scope of models towards more processes that we think are relevant processes to complete the coupling of sea ice with marine biogeochemistry, such as inorganic carbon dynamics (e.g. Moreau et al., 2015), iron processes, sulphur.
3 To understand what are the salient features of the elemental cycles in sea ice, and simplify models for their implementation in ESMs.

20.6 Acknowledgements

The model simulations were run with the LIM1D model. The authors would like to acknowledge the contributors to LIM1D (T. Fichefet, H. Goosse, S. Moreau, O. Lecomte, C.M. Bitz, C. Lancelot, J.-L. Tison, B. Delille) as well as the people behind the Biogeochemical Exchange Processes at the Sea-Ice Intefaces (BEPSII) SCOR Working Group 140. Seb Moreau, Dirk Notz, Milos Zivadinovic and Gurvan Madec are acknowledged for helpful comments on an earlier version of this text.

References

Arrigo, K.R. (2003) Primary production in sea ice. In: *Sea ice: An Introduction to its Physics, Biology, Chemistry and Geology*, (Eds. D.N. Thomas & G.S. Dieckmann), pp. 143–183. Blackwell Science Ltd, Oxford, UK.

Arrigo, K.R., Kremer, J.N. & Sullivan, C.W. (1993) A simulated Antarctic fast ice ecosystem. *Journal of Geophysical Research*, **98**, C4, 6929–6946, doi: 10.1029/93JC00141.

Arrigo, K.R. & Sullivan, C.W. (1992) The influence of salinity and temperature covariation on the photophysiological characteristics of Antarctic sea ice microalgae. *Journal of Phycology*, **28**, 746–756.

Arrigo, K.R. & Sullivan, C.W. (1994) A high-resolution bio-optical model of microalgal growth: Tests using sea-ice algal community time-series data. *Limnology and Oceanography*, **39**, 3, 609–631.

Arrigo, K.R., Sullivan, C.W. & Kremer, J.N. (1991) A bio-optical model of Antarctic sea ice. *Journal of Geophysical Research*, **96**, C6, 10 581–10 592.

Arrigo, K.R., Worthen, D.L., Lizotte, M.P., Dixon, P. & Dieckmann, G. (1997) Primary production in Antarctic sea ice. *Science*, **276**, 5311, 394–397.

Arrigo, K.R. & van Dijken, G.L. (2011) Secular trends in Arctic Ocean net primary production. *Journal of Geophysical Research*, **116**, C09011, 10.1029/2011JC007151.

Arrigo, K.R., van Dijken, G.L. & Bushinsky, S. (2008) Primary production in the Southern Ocean, 1997–2006. *Journal of Geophysical Research*, **113**, C08004, 10.1029/2007JC004551.

Assur, A. (1958) Composition of sea ice and its tensile strength. In: *Arctic Sea Ice, Conference held at Easton, Maryland, February 24–27, 1958*, pp. 106–138. Publications of National Research Council, **598**, Washington, DC.

Aumont, O., Maier-Reimer, E., Blain, S. & Monfray, P. (2003) An ecosystem model of the global ocean including Fe, Si, P colimitations. *Global Biogeochemical Cycles*, **17**, 1060, 10.1029/2001GB001745.

Ayata, S.-D., Lévy, M., Aumont, O. et al. (2013) Phytoplankton growth formulation in marine ecosystem models : Should we take into account photo-acclimation and variable stoichiometry in oligotrophic areas ? *Journal of Marine Systems*, **125**, 29–40.

Bitz, C.M. & Lipscomb, W.H. (1999) An energy-conserving thermodynamic model of sea ice. *Journal of Geophysical Research*, **104**, C7, 15 669–15 677.

Briegleb, B.P. & Light, B. (2007) A Delta-Eddington multiple scattering parameterization for solar radiation in the sea ice component of the Community Climate System Model. *NCAR Tech. Note NCAR/TN-472+STR*. National Center for Atmospheric Research, Boulder, CO.

Cloern, J.E., Grenz, C. & Vidergar-Lucas, L. (1995) An empirical model of the phytoplankton chlorophyll:carbon ratio – the conversion factor between productivity and growth rate. *Limnology and Oceanography*, **40**, 1313–1321.

Cota, G.F. & Smith, R.E.H. (1991) Ecology of bottom ice algae: III. Comparative physiology. *Journal of Marine Systems*, **2**, 297–315.

Cox, G.F.N. & Weeks, W.F. (1988) Numerical simulations of the profile properties of undeformed first-year sea ice during growth season. *Journal of Geophysical Research*, **93**, 12 449–12 460.

Deal, C., Jin, M., Elliott, S., Hunke, E., Maltrud, M. & Jeffery, N. (2011) Large-scale modelling of primary production and ice algal biomass within Arctic sea ice in 1992. *Journal of Geophysical Research*, **116**, C07004, 10.1029/2010JC006409.

Delille, B., Vancoppenolle, M., Geilfus, N.-X. et al. (2014) Southern Ocean CO2 sink: The contribution of the sea ice, *Journal of Geophysical Research*, **119**, 6340–6355.

Doney, S.C., Glover, D.M. & Najjar, R.G. (1996) A new coupled one-dimensional biological-physical model for the upper ocean: applications to the JGOFS Bermuda Atlantic Time-series Study (BATS) site. *Deep Sea Research II*, **43**, 2–3, 591–624.

Droop, M.R. (1983) 25 years of algal growth kinetics, a personal view. *Botanica Marina*, **26**, 3, 99–112.

Dupont, F. (2012) Impact of sea-ice biology on overall primary production in a biophysical model of the pan-Arctic Ocean, *Journal of Geophysical Research*, **117**, C00D17, 10.1029/2011JC006983.

Eicken, H., Krouse, H.R., Kadko, D. & Perovich, D.K. (2002) Tracer studies of pathways and rates of meltwater transport through Arctic summer sea ice. *Journal of Geophysical Research*, **107**, C10, 8046, 10.1029/2000JC000583.

Ehn, J.K., Mundy, C.J. & Barber, D.G. (2008) Bio-optical and structural properties inferred from irradiance measurements within the bottommost layers in an Arctic landfast sea ice cover. *Journal of Geophysical Research*, **113**, C03S03, 10.1029/2007JC004194.

Elliott, S., Deal, C., Humphries, G. et al. (2012) Pan-Arctic simulation of coupled nutrient-sulfur cycling due to sea ice biology: Preliminary results. *Journal of Geophysical Research*, **117**, G01016, 10.1029/2011JG001649.

Eppley, R.W. (1972) Temperature and phytoplankton growth in the sea. *Fisheries Bulletin*, **70**, 1063–1085.

Flato, G., Marotzke, J., Abiodun, B. et al. (2013) Evaluation of Climate Models. In: *Climate Change 2013: The Physical Science Basis. Contribution of Working Group I to the Fifth Assessment Report of the Intergovernmental Panel on Climate Change*, (Eds. T.F. Stocker et al.), pp. 741–882. Cambridge University Press, Cambridge, UK.

Flynn, K.J. (2003) Do we need complex mechanistic photoacclimation models for phytoplankton? *Limnology and Oceanography*, **48**, 6, 2243–2249.

Flynn, K.J. (2010) Ecological modelling in a sea of variable stoichiometry: dysfunctionality and the legacy of Redfield and Monod. *Progress in Oceanography*, **84**, 52–65.

Fripiat, F., Sigman, D.M., Fawcett, S.E., Rafter, P.A., Weigand, M.A. & Tison, J.-L. (2014a) New insights into sea ice nitrogen biogeochemical dynamics from the nitrogen isotopes, *Global Biogeochemical Cycles*, **28**, 115–130.

Fripiat, F., Tison, J.-L., André, L., Notz, D.K. & Delille, B. (2014b) Biogenic silica recycling in sea ice inferred from Si-isotopes: constraints from Arctic winter first-year sea ice. *Biogeochemistry*, **119**, 1, 25–33.

Fritsen, C.H., Wirthlin, E.D., Momberg, D.K., Lewis, M.J. & Ackley, S.F. (2011) Bio-optical properties of Antarctic pack ice in the early austral spring. *Deep Sea Research, Part II*, **58**, 1052–1061.

Geider, R.J., MacIntyre, H.L. & Kana, T.M. (1998) A dynamic regulatory model of phytoplanktonic acclimation to light, nutrients and temperature. *Limnology and Oceanography*, **43**, 679–694.

Grant, W.S. & Horner, R.A. (1976) Growth responses to salinity variation in four Arctic ice diatoms. *Journal of Phycology*, **12**, 180–185.

Griewank, P.J. & Notz, D. (2013) Insights into brine dynamics and sea-ice desalination from a 1D model study of gravity drainage. *Journal of Geophysical Research*, **118**, 3370–3386.

Hill, V.J., Matrai, P.A., Olson, E. et al. (2013) Synthesis of integrated primary production in the Arctic Ocean: II. In situ and remotely sensed estimates. *Progress in Oceanography*, **110**, 107–125.

Hunke, E.C., Notz, D., Turner, A.K. & Vancoppenolle, M. (2011) The multiphase physics of sea ice: a review for model developers. *Cryosphere*, **5**, 989–1009.

Hunke, E.C., Lipscomb, W.H., Turner, A.K., Jeffery, N. & Elliott, S. (2013) CICE: the Los Alamos Sea Ice Model, Documentation and Software User's Manual, version 5.0. *Technical Report LA-CC-06-012*. Los Alamos National Laboratory, Los Alamos, New Mexico.

Jassby, A.D. & Platt, T. (1976) Mathematical formulation of the relationship between photosynthesis and light for phytoplankton. *Limnology and Oceanography*, **21**, 540–547.

Jeffery, N. & Hunke, E.C. (2014) Modelling the winter-spring transition of first-year ice in the western Weddell Sea. *Journal of Geophysical Research*, **119**, 5891–5920.

Jeffery, N., Hunke, E.C. & Elliott, S.M. (2011) Modelling the transport of passive tracers in sea ice. *Journal of Geophysical Research*, **116**, C07020, 10.1029/2010JC006527.

Jin, M., Deal, C.J., Wang, J. et al. (2006) Controls of the landfast ice-ocean ecosystem offshore Barrow, Alaska. *Annals of Glaciology*, **44**, 63–72.

Jin, M., Deal, C.J., Lee, S. et al. (2012) Investigation of Arctic sea ice and ocean primary production for the period 1992 to 2007 using a 3-D global ice-ocean ecosystem model. *Deep Sea Research Part II*, **81–84**, 28–35.

Jin, Z., Stamnes, K., Weeks, W.F. & Tsay, S.-C. (1994) The effect of sea ice on the solar energy budget in the atmosphere-sea ice-ocean system: A model study. *Journal of Geophysical Research*, **99**, C2, 25281–25294.

Katlein, C., Nicolaus, M. & Petrich, C. (2014) The anisotropic scattering coefficient of sea ice, *Journal of Geophysical Research*, **119**, 842–855.

Lancelot C., de Montety, A., Goosse, H. et al. (2009) Spatial distribution of the iron supply to phytoplankton in the Southern Ocean: a model study. *Biogeosciences*, **6**, 2861–2878.

Lavoie, D., Denman, K. & Michel, C. (2005) Modelling ice algal growth and decline in a seasonally ice-covered region of the Arctic (Resolute Passage, Canadian Archipelago). *Journal of Geophysical Research*, **110**, C11009, 10.1029/2005JC002922.

Lavoie, D., Denman, K.L. & Macdonald, R.W. (2010) Effects of future climate change on primary productivity and export fluxes in the Beaufort Sea. *Journal of Geophysical Research*, **115**, C04018, 10.1029/2009JC005493.

Legendre, L., Ackley, S.F., Dieckmann, G.S. et al. (1992), Ecology of sea ice biota: Part 2. Global significance. *Polar Biology*, **12**, 3, 429–444.

Lengaigne, M., Madec, G., Bopp, L., Menkes, C., Aumont, O. & Cadule, P. (2009) Bio-physical feedbacks in the Arctic Ocean using an Earth system model. *Geophysical Research Letters*, **36**, L21602, 10.1029/2009GL040145.

Light, B., Grenfell, T.C. & Perovich, D.K. (2008) Transmission and absorption of solar radiation by Arctic sea ice during the melt season. *Journal of Geophysical Research*, **113**, C03023, 10.1029/2006JC003977.

Loose, B., Miller, L.A., Elliott, S. & Papakyriakou, T. (2011) Sea ice biogeochemistry and material transport across the frozen interface. *Oceanography*, **24**, 202–218.

Maykut, G.A. & Untersteiner, N. (1971) Some results from a time-dependent thermodynamic model of sea ice. *Journal of Geophysical Research*, **76**, 6, 1550–1575.

Meiners, K.M., Vancoppenolle, M., Thanassekos, S. et al. (2012) Chlorophyll a in Antarctic sea ice from historical ice core data. *Geophysical Research Letters*, **39**, L21602, 10.1029/2012GL053478.

Melnikov, I.A. (1998) Winter production of sea ice algae in the western Weddell Sea. *Journal of Marine Systems*, **17**, 195–205.

Michel, C., Legendre, L., Demers, S. & Therriault, J.-C. (1988) Photoadaptation of sea-ice microalgae in springtime: photosynthesis and carboxylating enzymes. *Marine Ecology Progress Series*, **50**, 177–185.

Michel, C., Legendre, L., Ingram, R.G., Gosselin, M. & Levasseur, M. (1996) Carbon budget of sea-ice algae under first-year ice in spring: evidence of a significant transfer to zooplankton grazers. *Journal of Geophysical Research*, **101**, 18 345–18 360.

Michel, C., Nielsen T.G., Nozais, C. & Gosselin, M. (2002) Significance of sedimentation and grazing by ice micro- and meiofauna for carbon cycling in annual sea ice (northern Baffin Bay). *Aquatic Microbial Ecology*, **30**, 57–68.

Miller, L.A., Fripiat, F., Else, B.G.T. et al. (2015) Methods for biogeochemical studies of sea ice: where we are and where we are going. *Elementa: Science of the Anthropocene*, **3**, 000038, 10.12952/journal.elementa.000038

Monod, J. (1949) The growth of bacterial cultures. *Annual Review of Microbiology*, **3**, 371–394.

Moreau, S., Vancoppenolle, M., Delille, B. et al. (2015) Inorganic Carbon dynamics in first-year sea ice: a model study. *Journal of Geophysical Research*, **120**, 10.1002/2015JC010388.

Moreau, S., Vancoppenolle, M., Zhou, J., Tison, J.-L., Delille, B. & Goosse, H. (2014) Modelling argon dynamics in first-year sea ice. *Ocean Modelling*, **73**, 1–18.

Morel, F.M.M. (1987) Kinetics of nutrient uptake and growth in phytoplankton. *Journal of Phycology*, **23**, 137–150.

Nicolaus, M., Petrich, C., Hudson, S.R. & Granskog, M.A. (2013) Variability of light transmission through Arctic land-fast sea ice during spring. *Cryosphere*, **7**, 977–986.

Nishi, Y. & Tabeta, S. (2008) Relation of material exchange between sea ice and water to a coupled ice-ocean ecosystem at the Hokkaido coastal region of the Okhotsk Sea. *Journal of Geophysical Research*, **113**, C01003, 10.1029/2006JC004077.

Notz, D.K. (2005) Thermodynamic and fluid-dynamical processes in sea ice. *PhD. Thesis*. University of Cambridge, Cambridge, UK.

Notz, D.K. & Worster, M.G. (2009) Desalination processes of sea ice revisited. *Journal of Geophysical Research*, **114**, C05006, 10.1029/2008JC004885.

Perovich, D.K., Cota, G.F., Maykut, G.A. & Grenfell, T.C. (1993) Bio-optical observations of first-year Arctic sea ice. *Geophysical Research Letters*, **20**, 1052–1062.

Pogson, L., Tremblay, B., Lavoie, D., Michel, C. & Vancoppenolle, M. (2011) Development and validation of a one-dimensional snow-ice algae model against observations in Resolute Passage, Canadian Arctic Archipelago. *Journal of Geophysical Research*, **116**, C04010, 10.1029/2010JC006119.

Ralph, P.J., Ryan, K.G., Martin, A. & Fenton, G. (2007) Melting out of sea ice causes greater photosynthetic stress in algae than freezing in. *Journal of Phycology*, **43**, 948–956.

Reeburgh, W.S. (1984) Fluxes associated with brine motion in growing sea ice. *Polar Biology*, **3**, 29–33.

Rees Jones, D.W. & Worster, M.G. (2014) A physically based parameterization of gravity drainage for sea-ice modelling. *Journal of Geophysical Research*, **119**, 5599–5621.

Renaud, P.E., Riedel, A., Michel, C. et al. (2007) Seasonal variations in the benthic community oxygen demand: a response to an ice algal bloom in the Beaufort Sea, Canadian Arctic. *Journal of Marine Systems*, **67**, 1–12.

Rysgaard, S., Kühl, M., Glud, R.N. & Hansen, J.W. (2001) Biomass, production and horizontal patchiness of sea ice algae in a high-Arctic fjord (Young Sound, NE Greenland). *Marine Ecology Progress Series*, **223**, 15–26.

Saenz, B.T. & Arrigo, K.R. (2012) Simulation of a sea ice ecosystem using a hybrid model for slush layer desalination. *Journal of Geophysical Research*, **117**, C05007, 10.1029/2011JC007544.

Saenz, B.T. & Arrigo, K.R. (2014), Annual primary production in Antarctic sea ice during 2005–2006 from a sea ice state estimate. *Journal of Geophysical Research*, **119**, 3645–3678.

Sakshaug, E., Andresen, K. & Kiefer, D.A. (1989) A steady state description of growth and light absorption in the marine planktonic diatom Skeletonema costatum. *Limnology and Oceanography*, **34**, 198–205.

Sarmiento, J.L. & Gruber, N. (2006) *Ocean biogeochemical dynamics*. Princeton University Press.

Sarthou, G., Timmermans, K.R., Blain, S. & Tréguer, P. (2005) Growth, physiology and fate of diatoms in the ocean: a review. *Journal of Sea Research*, **53**, 25–42.

Semtner, A.J. (1976) A model for the thermodynamic growth of sea ice in numerical investigations of climate. *Journal of Physical Oceanography*, **6**, 379–389.

Shine, K.P. & Henderson-Sellers, A. (1985) The sensitivity of a thermodynamic sea ice model to changes in surface albedo parameterization. *Journal of Geophysical Research*, **90**, 2243–2250.

Sibert, V., Zakardjian, B., Gosselin, M., Starr, M., Senneville, S. & LeClainche, Y. (2011) 3D bio-physical model of the sympagic and planktonic productions in the Hudson Bay system. *Journal of Marine Systems*, **88**, 401–422.

Steiner, N.S., Lee, W.G., Christian, J.R. (2013) Enhanced gas fluxes in small sea ice leads and cracks: Effects on CO_2 exchange and ocean acidification, *Journal of Geophysical Research*, **118**, 10.1002/jgrc.20100.

Tedesco, L. & Vichi, M. (2014) Sea Ice Biogeochemistry: A guide for modellers. *PlosOne*, **9**, e89217, 10.1371/journal.pone.0089217.

Tedesco, L., Vichi, M., Haapala, J. & Stipa, T. (2010) A dynamic Biologically Active Layer for numerical studies of the sea ice ecosystem. *Ocean Modelling*, **35**, 89–104.

Tedesco, L., Vichi, M. & Thomas, D.N. (2012) Process studies on the ecological coupling between sea ice algae and phytoplankton. *Ecological Modelling*, **226**, 120–138.

Thorndike, A.S., Rothrock, D.A., Maykut, G.A. & Colony, R. (1975) The thickness distribution of sea ice. *Journal of Geophysical Research*, **80**, 4501–4513.

Tison, J.-L., Worby, A.P., Delille, B. et al. (2008) Temporal evolution of decaying summer first-year sea ice in the western Weddell Sea, Antarctica. *Deep Sea Research II*, **55**, 975–987.

Turner, A.K., Hunke, E.C. & Bitz, C.M. (2013) Two modes of sea-ice gravity drainage: A parameterization for large-scale modelling. *Journal of Geophysical Research*, **118**, 10.1002/jgrc.20171.

Vancoppenolle, M., Goosse, H., de Montety, A., Fichefet, T., Tremblay, B. & Tison, J.-L. (2010) Modelling brine and nutrient dynamics in Antarctic sea ice : the case of dissolved silica. *Journal of Geophysical Research*, **115**, C2, C02005, 10.1029/2009JC005369.

Vancoppenolle, M., Timmermann, R., Ackley, S.F., Fichefet, T., Goosse, H., Heile, P., Leonard, K.C., Lieser, I., Nicolaus, M., Papakyriakou, T. & Tison, J.-L. (2011) Assessment of radiation forcing data sets for large-scale sea ice models in the Southern Ocean. *Deep Sea Research Part II*, **58**, 1237–1249.

Vancoppenolle, M., Meiners, K., Michel, C. et al. (2013) Role of sea ice in Global Biogeochemical. Cycles: emerging views and challenges. *Quaternary Science Reviews*, **79**, 207–230.

Wang, S., Baily, D., Lindsay, K., Moore, J.K. & Holland, M. (2014) Impact of sea ice on the marine iron cycle and phytoplankton productivity. *Biogeosciences*, **11**, 4713–4731.

Wells, A.J., Wettlaufer, J.S. & Orszag, S.A. (2011) Brine fluxes from growing sea ice. *Geophysical Research Letters*, **38**, L04501, 10.1029/2010GL046288.

Williams, R.G. & Follows, M.J. (2011) *Ocean Dynamics and the Carbon Cycle*. Cambridge University Press, Cambridge, UK.

Worster, M.G. (1992) The dynamics of mushy layers. In: *Interactive Dynamics of Convection and Solidification* (Eds. S.H. Davis et al.), pp. 113–138. NATO ASI Series.

Zeebe, R.E., Eicken, H., Robinson, D.H., Wolf-Gladrow, D. & Dieckmann, G.S. (1996) Modelling the heating and melting of sea ice through light absorption by microalgae. *Journal of Geophysical Research*, **101**, 1163–1181.

Zemmelink, H., Houghton, L., Dacey, J.W.H. et al. (2008) Stratification and the distribution of phytoplankton, nutrients, inorganic carbon, and sulfur in the surface waters of Weddell sea leads. *Deep Sea Research II*, **55**, 988–999.

Zhou, J., Delille, B., Geilfus, N.-X. et al. (2013) Physical and biogeochemical properties in landfast sea ice (Barrow, Alaska): insights for brine and gas dynamics. *Journal of Geophysical Research*, **118**, 10.1002/jgrc.20232.

CHAPTER 21

Arctic marine mammals and sea ice

Kristin L. Laidre[1] and Eric V. Regehr[2]

[1] Polar Science Center, Applied Physics Laboratory and School of Aquatic and Fishery Sciences, University of Washington, Seattle, WA, USA
[2] Marine Mammals Management, US Fish and Wildlife Service, Anchorage, AK, USA

21.1 Introduction

There are 11 species of ice-dependent or ice-associated Arctic marine mammals. Of these, the following seven species are endemic to the Arctic and are dependent on, or highly associated with, sea ice for all or part of the year: the narwhal (*Monodon monoceros*), beluga (*Delphinapterus leucas*), bowhead whale (*Balaena mysticetus*), ringed seal (*Pusa hispida*), bearded seal (*Erignathus barbatus*), walrus (*Odobenus rosmarus*) and polar bear (*Ursus maritimus*). The following four species of ice seals depend on sea ice for whelping in the southern Arctic in the spring, but are generally pelagic or use subarctic waters for the rest of the year: the spotted seal (*Phoca largha*), ribbon seal (*Phoca fasciata*), harp seal (*Pagophilus groenlandicus*) and hooded seal (*Cystophora cristata*). Additionally, there are approximately 24 other species of marine mammals that use the Arctic seasonally, primarily as a habitat for foraging in the late spring and summer.

21.2 Ecology, distribution, and population structure of ice-dependent or ice-associated Arctic marine mammals

21.2.1 Cetaceans

The three species of Arctic cetaceans have different ecology, patterns of distribution and population structure. The narwhal is a deep-diving odontocete (toothed whale) that follows the formation and retreat of annual sea ice. The narwhal range is confined to the eastern Canadian high Arctic and around West and East Greenland, Svalbard and Franz Joseph Land (Koski & Davis, 1994; Innes et al., 2002; Heide-Jørgensen et al., 2013). The other Arctic odontocete, the beluga whale, is a generalist feeder that inhabits diverse Arctic and subarctic habitats. Belugas have a circumpolar distribution and occur in discrete subpopulations, generally defined by summering areas (Richard et al., 2001; Innes et al., 2002; Palsbøll et al., 2002). Finally, the bowhead whale is the only baleen whale to inhabit the Arctic year-round, foraging primarily on copepods and amphipods. Bowheads have a circumpolar distribution and move between high-Arctic waters in summer and low-Arctic waters in winter (George et al., 1989; Heide-Jørgensen et al., 2006).

Narwhal

Narwhals spend approximately 2 months each summer in ice-free bays and fjords of the high Arctic. In autumn they migrate to overwintering areas that tend to be deep, offshore, ice-covered habitats along the continental slope (Innes et al., 2002; Laidre et al., 2003; Laidre & Heide-Jørgensen, 2011). Narwhals are highly adapted to winter habitats with limited open water (Laidre et al., 2004a,2004b,2004c). No other cetacean uses areas with such dense sea-ice cover for up to 6 months each winter (Figure 21.1). The male narwhal has an elongated spiral tooth, or tusk, which serves as a secondary sexual

---

Sea Ice, Third Edition. Edited by David N. Thomas.
© 2017 John Wiley & Sons, Ltd. Published 2017 by John Wiley & Sons, Ltd.

Figure 21.1 Narwhals are perhaps the most ice-adapted whale, capable of using areas with > 98% ice cover for 6 months of the year. Source: Photograph by Kristin Laidre.

characteristic (Kelley et al., 2014). Calving occurs in spring and most feeding takes place in winter (Laidre et al., 2003; Laidre & Heide-Jørgensen, 2005). The worldwide population of narwhals is approximately 100 000 animals (Laidre et al., 2015) divided into several subpopulations based on summering location (Heide-Jørgensen et al., 2013). The narwhals that summer in the Canadian High Arctic and East Baffin Island number at least 70 000 animals (Innes *et al.*, 2002). In these areas, individual subpopulation sizes range between approximately 5000 and 30 000 animals (Richard et al., 2010). Some areas in Canada contain unsurveyed aggregations that probably consist of small numbers of whales. In West Greenland, the primary subpopulations are Inglefield Bredning and Melville Bay (Heide-Jørgensen et al., 2010). There is at least one subpopulation in East Greenland.

Beluga whale

Beluga whales occupy estuaries, continental shelf and slope waters, and deep ocean basins in conditions of open water, loose ice and heavy pack ice. Belugas are born dark grey and gradually lighten with maturity to the white colour of adults. Belugas feed on a wide variety of prey, including pelagic and benthic fish, squid and amphipods. Some belugas undertake large-scale annual migrations between summering and wintering sites, while others remain in the same area year-round, shifting offshore only when excluded from coastal habitat by the formation of fast ice. Beluga abundance worldwide is estimated to be at least 150 000 animals divided into approximately 19 subpopulations based on summering ground location (Laidre et al., 2015). Large populations of beluga whales are thought to occur in the Eastern Bering Sea, the Eastern Beaufort Sea and Western Hudson Bay. Very little is known about population abundance of belugas in the Russian sector of the Arctic.

Bowhead whale

Bowhead whales summer in the Arctic and migrate to subarctic areas in winter (Figure 21.2). The species often inhabits polynyas and the marginal ice zone (MIZ) in winter and early spring (Moore & Reeves, 1993), and can easily move through extensive areas of nearly solid sea ice cover (Heide-Jørgensen et al., 2006). Bowheads are the Arctic's largest and most zooplankton-dependent predator. Their diet has been well described based on the stomach contents of subsistence-harvested whales. Bowheads feed throughout the water column, including near or on the bottom, and eat mostly a variety of pelagic and epibenthic crustaceans (Lowry, 1993; Lowry et al., 2004). Copepods (primarily *Calanus* spp.) and euphausiids are often the most important prey (Laidre et al., 2007).

Bowhead whales number approximately 25 000 animals worldwide (Laidre et al., 2015). There are four subpopulations: Bering-Chukchi-Beaufort Seas, Eastern Canada-West Greenland, Svalbard-Barents Sea and the Okhotsk Sea. The largest fraction of the global population is located in the Bering-Chukchi-Beaufort seas, with approximately 17 000 whales (Givens et al., 2013). The Svalbard-Barents Sea and the Sea of Okhotsk subpopulations are small (Boertman et al., 2009; Wiig et al., 2010; Ivaschenko & Clapham, 2009).

Figure 21.2 Bowhead whales are the longest-lived mammal on earth, reaching ages of over 200 years, and represent an important subsistence resource for native people throughout the Arctic. Several bowhead populations have increased since cessation of commercial whaling. Source: Photograph by Kristin Laidre.

21.2.2 Pinnipeds

Of the three true Arctic pinnipeds, ringed seals are pelagic feeders that have a circumpolar distribution, inhabiting seasonally or permanently ice-covered areas from the North Pole to the low Arctic, including some lake and river systems. Bearded seals are primarily benthic feeders and also have a circumpolar distribution that is more restricted to Arctic waters over the continental shelf. The walrus has a discontinuous circumpolar distribution and is found in two recognized subspecies: the Atlantic walrus (*O. r. rosmarus*), distributed from the eastern Canadian Arctic to the Kara Sea; and the Pacific walrus (*O. r. divergens*), distributed from Mys Shelagskyi in Siberia to Barter Island in Alaska, and throughout the Bering Sea to the south. It is unclear whether the Laptev walrus (*O. r. laptevi*), which is confined to the Laptev Sea region, is part of the Pacific walrus subspecies (Lindquist et al., 2009). Of the four subarctic ice seals, the spotted seal and ribbon seal are found in the Pacific Arctic and peripheral seas (Burns, 1981; Boveng et al., 2008; Boveng et al., 2009). Conversely, the harp seal and hooded seal are confined to the Atlantic Arctic and are widely distributed throughout the North Atlantic and Arctic Ocean shelf and seas (Lavigne & Kovacs, 1988).

Ringed seal

The ringed seal occupies the largest geographic range and diversity of habitats of all the Arctic seals. The primary breeding habitat of the ringed seal is landfast ice over the continental shelf along coasts, bays and interisland channels (Smith & Hammill, 1981), yet they may also use drifting pack ice both nearshore and offshore (Finley et al, 1983; Wiig et al., 1999). Ringed seals are able to inhabit the fast ice between freeze-up and breakup because they maintain breathing holes by continuing to abrade thick ice with the heavy claws of their foreflippers (McLaren, 1958). In late spring, ringed seals give birth and nurse their pups in under-snow lairs excavated above the breathing holes (Smith & Stirling, 1975). These birth lairs are a key life history feature for the species, providing protection from predators and shelter for wet pups (Smith et al., 1991; Lydersen et al., 1992). During the open-water season (i.e. between sea-ice breakup in early summer to freeze-up in the autumn) ringed seals remain pelagic (Smith, 1987; Harwood & Stirling, 1992). Ringed seals' diet is diverse and varies seasonally and regionally, including large zooplankton, epibenthic and under-ice crustaceans, and pelagic and demersal fishes (Lowry et al., 1980a; Weslawski et al., 1994; Wathne et al., 2000). Throughout much of the Arctic ringed seals are the primary prey of polar bears (Stirling, 2002).

Ringed seal abundance is estimated to be in the low millions (Laidre et al., 2015), although few data are available on regional subpopulation sizes. The species occurs at lower densities in multi-year ice of the high Arctic than in annual ice areas, probably because biological productivity is lower and it is more difficult to maintain breathing holes in the thicker ice. Ringed seals are divided into five recognized subspecies: Arctic ringed seal (*P. h. hispida*), Baltic Sea ringed seal (*P. h. botnica*), Lake Ladoga ringed seal (*P. h. ladogensis*), Lake Saimaa ringed seal (*P. h. saimensis*) and the Okhotsk ringed seal (*P. h. ochotensis*).

Bearded seal

Bearded seals are widely distributed throughout the circumpolar Arctic, mainly over the shallow waters of the continental shelf and usually in association with moving ice or shore leads and polynyas (Burns, 1970). They are thought to be mainly pelagic during the summer and autumn, although they may remain in or near the sea ice year-round (Figure 21.3). Bearded seals are predominantly benthic feeders and take shrimp, clams, crabs, other benthic invertebrates and fishes (Lowry et al., 1980b; Antonelis et al., 1994; Hjelset et al., 1999). Less is known about their diet in deep offshore areas. The worldwide abundance of bearded seals is unknown but may be in the range of 500 000 to 750 000 animals (Laidre et al., 2015). There are two subspecies of bearded seal: the Atlantic subspecies (*E. b. barbatus*) occurs from the central Canadian Arctic east to the central Eurasian Arctic; and the Pacific subspecies (*E. b. nauticus*) occurs from the Laptev Sea east to the central Canadian Arctic, including the Sea of Okhotsk (Rice, 1998).

Walrus

Walruses are specialized feeders that target benthic invertebrates, primarily mollusks, in shallow waters (Fay, 1982; Born et al., 2003). Walruses depend on a suitable substrate (i.e. sea ice or land) close to foraging grounds for resting. They often overwinter in areas with polynyas that provide open water and access to benthic food resources (Stirling, 1997). Walruses occasionally

Figure 21.3 Bearded seals use their whiskers to locate prey on the ocean floor, including clams, squid, and fish. While the smaller ringed seal is the primary prey of polar bears, bearded seals can also be an important food source for large male bears. Source: © Jenny E. Ross/LifeOnThinIce.org

predate on seals (Lowry & Fay, 1984; Gjertz & Wiig, 1992; Born et al., 1995), especially in deep water where they do not have access to benthic prey. Walruses breed aquatically from January through March in their wintering areas near polynyas or in areas of drifting pack ice (Fay, 1982).

Pacific walruses spend the winter in the Bering Sea where they use ice floes to haul out over the relatively shallow continental shelf. As ice cover recedes in spring, juvenile and adult female walruses move northward with the ice into the Chukchi, east Siberian and Beaufort seas (Fay, 1982). Most adult males remain farther south and spend the summer hauled out on terrestrial sites along the Russian coast of the Bering and Chukchi seas and a few sites on the Alaskan coast, from which they presumably have access to suitable foraging areas. In the autumn, females and juveniles move southward to rejoin the males for the winter. The size of the Pacific walrus subpopulation has been estimated to be at least 129 000 animals (Speckman et al., 2011).

Atlantic walruses use both sea ice and terrestrial haulouts as a hub for feeding excursions on shallow nearshore banks with substantial bivalve mollusk production (Born et al., 1994). In summer, walruses of both sexes and all age-classes often leave their ice-based haulouts for terrestrial haulouts, although many terrestrial haulouts (especially in Greenland and Svalbard) have been abandoned due to excessive hunting (Born et al., 1995; Gjertz & Wiig, 1994). In East Greenland, formation of sea ice in winter forces walruses to leave terrestrial haulouts and move offshore to waters >100 m deep (Born, 2005). Total abundance of Atlantic walruses is probably about 20 000 animals, comprising at least nine separate stocks (Laidre et al., 2015).

Spotted seals

Spotted seals occur in the North Pacific and peripheral areas including the Sea of Japan, Okhotsk, Bering, Chukchi and Beaufort seas (Shaughnessey & Fay, 1977; Boveng et al., 2009). Spotted seals give birth to and care for their pups near the southern edge of seasonal pack ice, over the relatively shallow continental shelf (Burns, 1970). As the sea ice melts in the summer, some spotted seals use coastal haulouts in the Bering Sea region, while others follow the retreating ice northward and use coastal haulouts in the Chukchi and western Beaufort seas (Lowry et al., 1998). Haulouts occur in predictable locations, often near lagoon systems that may provide opportunities to escape from predators such as humans and bears (Frost et al., 1993). Spotted seals show considerable flexibility in the concentrations of sea ice they use (Lowry et al., 2000), yet appear most commonly near the ice edge and prefer relatively small floes (Simpkins et al., 2003). They move southward with advancing sea ice in October, and in winter and spring use a broad band of sea ice extending up to 300 km north of the ice edge in the eastern Bering Sea. They have a diverse diet that varies regionally and exploits both pelagic and benthic communities including fishes, shrimp and other crustaceans, and octopus (Bukhtiyarov et al., 1984). Worldwide abundance of spotted seals is poorly known. Boveng et al., (2009) concluded that there are likely to be at least 100 000 spotted seals in the Bering Sea (including the seasonal inhabitants of the Chukchi Sea), 100 000 animals in the Sea of Okhotsk, and about 3,300 animals in the Yellow Sea and Sea of Japan.

Ribbon seals

Ribbon seals occur only in the North Pacific and peripheral seas (i.e. the Okhotsk, Bering, Chukchi and Beaufort seas) and use the ice edge or MIZ in late winter through spring where they give birth, care for pups and moult (Burns, 1981). Ribbon seals may adopt a largely pelagic lifestyle in summer and autumn. In the Bering Sea, ribbon seals feed on many of the same species of fish and invertebrates as spotted seals (Frost & Lowry, 1990). In the Okhotsk Sea, juveniles feed on euphausiids and shrimp and adults feed on mostly pollock (Fedoseev, 2002). Worldwide abundance of ribbon seals is poorly known and likely exceeds 200 000 animals (Boveng et al., 2008). There are two main breeding areas for ribbon seals, one in the Sea of Okhotsk and one in the Bering Sea, although to date there is no evidence of discrete subpopulation structure.

Harp seals

Harp seals have broad habitat preferences yet depend on sea ice in specific locations for whelping and lactation, which occurs over a period of about 12 days just prior to breakup in the spring (Lavigne & Kovacs, 1988) (Figure 21.4). Mating takes place in the water at the same time, and moulting occurs on ice floes shortly thereafter. At some times of the year, harp seals lead a pelagic lifestyle and do not require sea ice or land for hauling out. In summer and early autumn, harp seals range widely throughout the North Atlantic moving north into ice-free Arctic seas in Baffin Bay, West or East Greenland, and the interisland channels of the Canadian high Arctic (Sergeant, 1991). They are also known to range widely into subarctic and temperate waters near the Faroe Islands and the Barents and Norwegian seas. Throughout their range, harp seals tend to prefer waters over the continental shelf, often feeding at depths less than a few hundred meters. Their diet is varied and includes capelin, mysids, shrimp and euphausids (Lydersen et al., 1991; Sergeant, 1991; Hammill et al., 2005).

Harp seals are the most abundant pinniped species in the Northern Hemisphere, numbering approximately nine million animals (Laidre et al., 2015). Harp seals are divided into three recognized subpopulations that inhabit the following regions: the Labrador and Newfoundland coasts and in the Gulf of Saint Lawrence, East Greenland north of Jan Mayen and the White Sea (Lavigne & Kovacs, 1988).

Hooded seals

Hooded seals depend on stable ice floes for parturition, similar to harp seals (Figure 21.5). In late March, hooded seal pups are born in an advanced developmental stage and are weaned in approximately 4 days (Bowen et al., 1985), after which they are left alone on the ice for up to several weeks. When the pups first enter the water to feed, they prey on krill and other invertebrates until their swimming and diving skills are sufficient to capture fish (Hammill & Stenson, 2000). Female hooded seals mate immediately after weaning their pups, and subsequently moult on the pack ice in June (Lavigne & Kovacs, 1988). In the open-water season, hooded seals are widely distributed throughout subarctic waters, although they occasionally occur along the European coast and the east coast of the United States (Lavigne & Kovacs, 1988). They are deep divers and can reach depths of 1000 m when foraging for benthic prey such as Greenland halibut (Folkow et al., 1999).

Hooded seal subpopulation size in the northwest Atlantic has been estimated to be just over half a million animals, based on pup counts. In the northeast Atlantic, hooded seal subpopulation size was estimated to be 82 000 animals in 2007 (Salberg *et al.*, 2008; ICES, 2008).

Figure 21.4 Harp seals are the most abundant pinniped in the northern hemisphere, numbering approximately nine million animals. In the spring, harp seals need sea ice as a platform to give birth, nurse and moult. Source: © Jenny E. Ross/LifeOnThinIce.org

Figure 21.5 The hooded seal can be recognized by an elastic nasal cavity located at the top of its head. This feature, only found in males, is used for acoustic signalling and communication. Source: © Jenny E. Ross/LifeOnThinIce.org

21.2.3 Polar bears

Polar bears occur throughout ice-covered areas of the circumpolar Arctic, especially on annual sea ice over the continental shelf and in the inter-island channels of various archipelagos. Polar bears depend on sea ice as a platform for hunting, seasonal movements, mating and sometimes maternal denning (Amstrup, 2003). In the polar basin and adjacent areas, some polar bears remain on multi-year sea ice in the summer and autumn while others spend the open-water season on land. In more southerly areas (i.e. Hudson Bay, Foxe Basin, Baffin Bay and Davis Strait) the annual ice melts completely and all bears spend up to several months on land until freeze-up allows them to return to the ice (e.g. Stirling & Parkinson, 2006). While on land, polar bears are largely cut off from their marine mammal prey (Regehr et al., 2007). In the winter, only pregnant females that are in maternal dens hibernate. Bears of all ages and sex classes may seek shelter in temporary dens in drifted snow for up to several weeks to escape periods of intense cold or inclement weather or when seals are less accessible (Ferguson et al., 2000). Ringed seals are the primary prey of polar bears throughout most of their range. The most important feeding time is in the spring after naïve seal pups are weaned and have peak fat stores of about 50% fat by wet weight (Stirling & McEwan, 1975). Polar bears are opportunistic hunters and will occasionally take belugas, narwhals, walrus, bearded seals, harbour seals (*Phoca vitulina*), reindeer (*Rangifer tarandus*) and birds or bird eggs (Thiemann et al., 2008).

Worldwide abundance of polar bears is approximately 26 000 animals divided into 19 recognized subpopulations (Figure 21.6; Obbard et al., 2010; Wiig et al., 2015). Genetic analysis indicates considerable gene

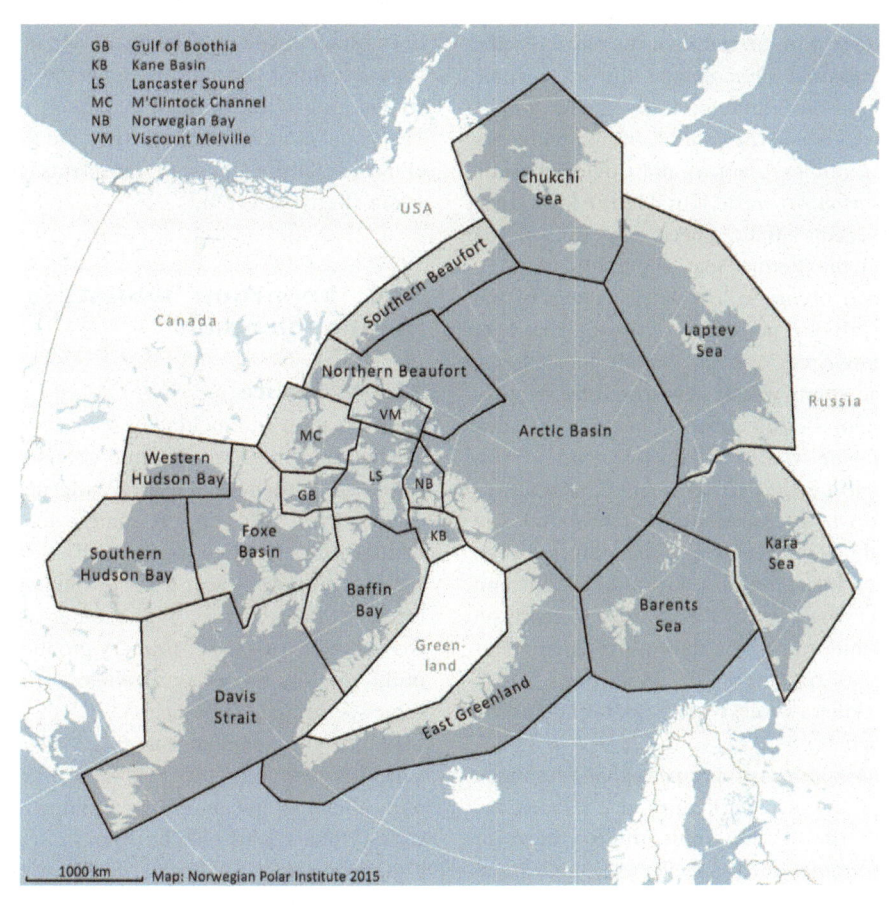

Figure 21.6 Polar bears are divided into 19 subpopulations with a worldwide abundance of approximately 20 000–25 000 animals. The current status of polar bears varies among subpopulations, although nearly all bears face the long-term threat of sea ice loss due to climate change.

flow between subpopulations, with four higher-level clusters corresponding roughly to current ecological and oceanographic factors (Peacock et al., 2015).

21.3 Marine mammals that seasonally occupy Arctic waters

There are 19 species of cetaceans that seasonally occupy low- or high-Arctic waters, primarily in the summer to feed. The North Atlantic right whale (*Eubalaena glacialis*) uses low-Arctic waters of East Greenland. The Northern bottlenose whale (*Hyperoodon ampullatus*) is found in the low-Arctic waters of the Atlantic, including Davis Strait, Baffin Bay, East Greenland and the Barents Sea. The white-beaked dolphin (*Lagenorhynchus albirostris*), long-finned pilot whale (*Globicephala melas*) and Atlantic white-sided dolphin (*Lagenorhynchus acutis*) are all found in the Atlantic Arctic. In the summer, the North Pacific right whale (*Eubalaena japonica*) and the grey whale (*Eschrichtius robustus*) are confined to the Pacific Arctic. The Baird's beaked whale (*Berardius bairdii*), Stejneger's beaked whale (*Berardius stejnegeri*) and Cuvier's beaked whale (*Ziphius cavirostris*) are all found in the low-Arctic waters of the Pacific Arctic, specifically in the Sea of Okhotsk and the Bering Sea. The Dall's porpoise (*Phocoenoides dalli*) occurs in low-Arctic waters of the Pacific, while harbour porpoises (*Phocoena phocoena*) are found in low-Arctic waters of both the Atlantic and Pacific. The sperm whale (*Physeter catodon*) may be found in the low-Arctic waters of both the Atlantic and Pacific. The blue whale (*Balaenoptera musculus*), fin whale (*Balaenoptera physalus*), sei whale (*Balaenoptera borealis*), minke whale (*Balaenoptera acutorostrata*) and humpback whale (*Megaptera navaeangliae*) are found in both the low and high Arctic, in both the Atlantic and Pacific. The killer whale (*Orcinus orca*) can visit nearly all areas of the circumpolar Arctic during the open-water season (Figure 21.7). Several of these species (e.g. balaenopterids, killer whales) are also found in the southern hemisphere.

There are four species of pinnipeds that seasonally use the Arctic. These include the grey seal (*Halichoerus grypus*), found in the Atlantic Arctic, the Northern fur seal (*Callorhinus ursinus*) and the Steller's sea lion (*Eumetopias jubatus*), found in the Pacific low Arctic within the Sea of Okhotsk and Bering Sea (Priblof Islands), and the harbour seal, which has a circumpolar distribution

Figure 21.7 Killer whales and other subarctic marine mammals migrate into Arctic waters each summer to feed. Sea ice loss is causing some of these species to arrive earlier and stay longer. Source: © Jenny E. Ross/LifeOnThinIce.org

across low-Arctic waters, with one population living in the high Arctic near Svalbard. Finally, the sea otter (*Enhydra lutris*) extends peripherally into low-Arctic waters around the Pribilof Islands in the Bering Sea. This population was extirpated by the Russian fur trade and re-established through transplants in the 1950s and 1960s (Kenyon, 1969).

21.4 Important habitats for Arctic marine mammals

21.4.1 Sea ice

The formation and melting of sea ice are defining characteristics of the Arctic marine ecosystem, influencing nearly all aspects of life for Arctic marine mammals (Figure 21.8). Most Arctic marine mammals rely on annual sea ice, in the form of landfast (fast) ice or floating pack ice. Multi-year ice is less important because it forms a thicker barrier at the water–air interface and is associated with lower primary productivity, although multi-year ice serves as a refuge during summer for some species (e.g. polar bears).

Ringed and bearded seals forage below fast ice and pack ice, and rely on it as a platform for hauling out, whelping and moulting. In late winter and early spring, ringed seals give birth and nurse their pups in under-snow lairs excavated above the breathing holes in the ice (Smith & Stirling, 1975). For walruses, sea ice provides access to offshore feeding areas, isolation from terrestrial predators and hunting pressure, a

Arctic marine mammals and sea ice   523

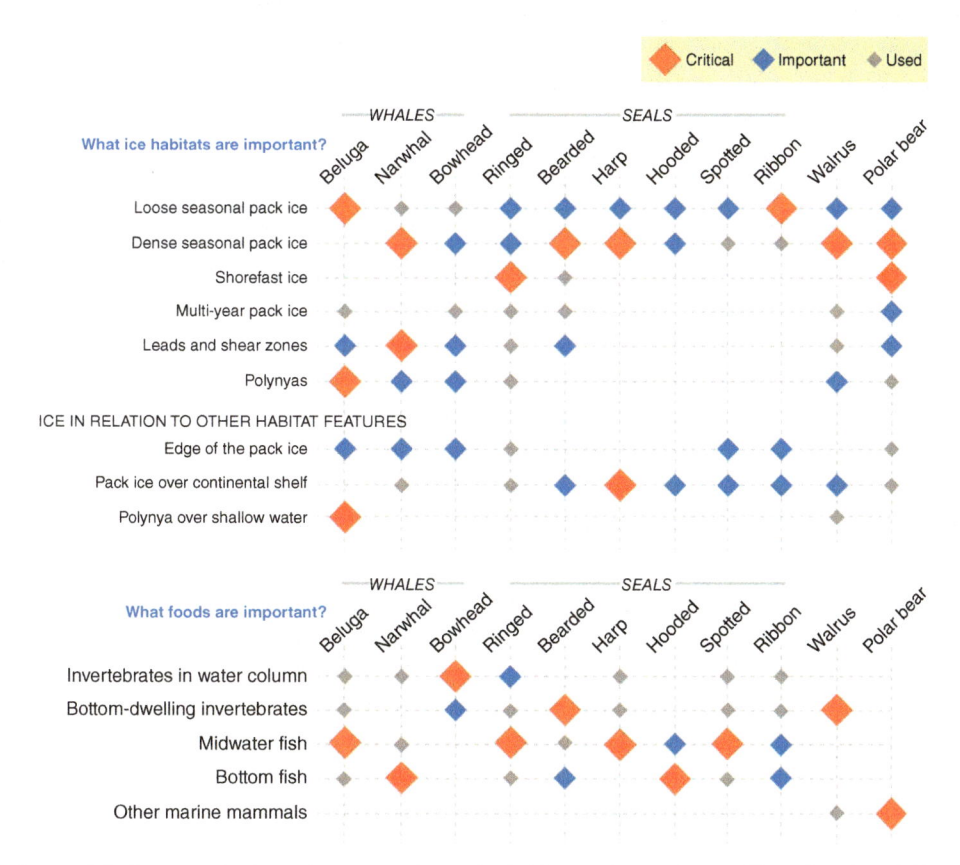

Figure 21.8 Connectivity between Arctic marine mammals, sea ice habitats and prey. Source: Figure based on Laidre et al. (2008) adapted by Eamer et al. (2013). Reproduced with permission from Joan Eamer and CAFF.

platform for giving birth and caring for young, and shelter from stormy seas (Fay, 1982; Ray et al., 2006; Ovsyanikov et al., 2007). All subarctic ice seals rely on pack ice at the southern limit of the ice extent in spring for whelping, lactation, resting, moulting and rearing pups (Johnston et al., 2005; Kovacs & Lydersen, 2008; Stenson & Hammill, 2014).

Arctic cetaceans are well-adapted for life in ice-covered waters. They may forage under the ice or use it to avoid predators, and can breathe by breaking through thin ice or surfacing between floes. Heavier sea ice is often a barrier for cetaceans, excluding them from underlying marine areas they might otherwise access for feeding and movements.

The formation and retreat of annual sea ice have cascading ecological impacts (Figure 21.9). Interactions between sea ice extent, breakup and solar radiation constrain and control the onset and pattern of primary production (Bluhm & Gradinger, 2008). This applies particularly to the MIZ where the ice edge retreats northwards, exposing the waters to sunlight and creating conditions necessary for a primary production bloom fuelled by the winter store of nutrients (Bluhm & Gradinger, 2008). A production bloom thus slowly sweeps across the area previously covered with annual sea ice. This relatively slow, geographically widespread and spatially variable transfer of production to the higher trophic levels of the food web is predominantly mediated by zooplankton (Pershing et al., 2004), although in some regions of the Arctic, production passes through the water column to the benthos (Hunt et al., 2002). The MIZ is a critical seasonal habitat and supports relatively high concentrations of Arctic marine mammals.

21.4.2 Polynyas

In some parts of the Arctic, polynyas (areas of open water surrounded by sea ice) may last throughout the

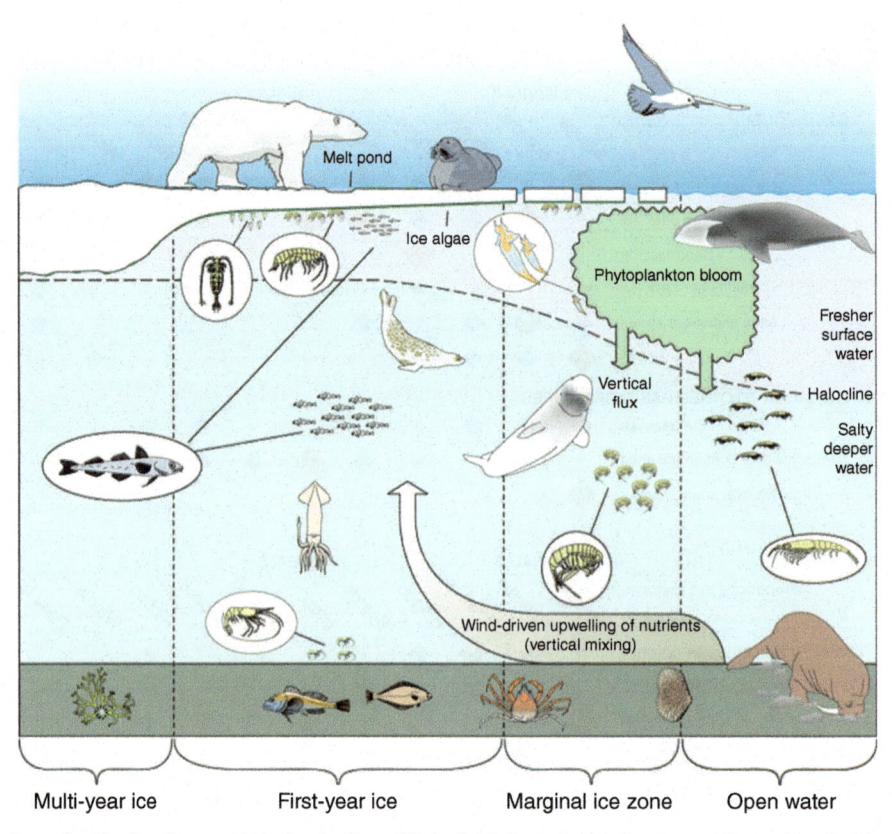

Figure 21.9 Arctic marine food webs are relatively simple and linked tightly to the sea ice. Source: Reproduced with permission from Audun Igesund and the Norwegian Polar Institute.

winter and serve an important ecological function. Polynyas vary in size from hundreds of square metres to hundreds of square kilometres. Some polynyas occur as unique events, while others, known as recurring polynyas, develop at the same time and place each year. This allows marine mammals to seasonally depend on them for feeding and overwintering (Stirling, 1997; Stirling, 1980). Recurring polynyas result from persistent upwelling of warmer water, unidirectional winds, tidal currents, or a combination of these factors (Stirling & Cleator, 1981). Larger polynyas such as the North Water Polynya are capable of having a substantial positive influence on productivity (Odate et al., 2002; Ringuette et al., 2002), often rivalling that of the MIZ and attracting large numbers of marine mammals. For example, a majority of the beluga population summering in northern Canada overwinters in the North Water Polynya (Richard et al., 1998a,1998b). The Northeast Water Polynya in East Greenland is an important wintering area for walruses (Born, 2005) and the Saint Lawrence Island Polynya in the Bering Sea hosts thousands of seals, walruses, bowhead whales and belugas every winter (Simpkins et al., 2003). Smaller recurrent polynyas are also biologically important (Stirling, 1997; Figure 21.10).

21.4.3 Glacial ice

Glacial (freshwater) ice provides important habitat for some Arctic marine mammals (Lydersen et al., 2014). It can provide an ice substrate in summer when annual ice has disappeared. Glacial fronts tend to be areas of high biological productivity, even outside of the spring primary production bloom, due to the provision of nutrients from flowing subglacial waters. Ringed seals, polar bears (Freitas et al., 2012), narwhals and belugas (Lydersen et al., 2001; Freitas et al., 2008; Heide-Jørgensen et al., 2010) congregate in front of tidewater glacial outlets in Greenland and Svalbard (Figure 21.11).

Figure 21.10 Distribution of polynyas in the circumpolar Arctic. Due to warming temperatures, and altered sea ice conditions and circulation patterns, when and where some polynyas occur are changing. Source: Meltofte, H. (Ed) (2013) Status and trends in Arctic biodiversity: Synthesis. Arctic Council ministerial meeting, Kiruna, Sweden, May 2013.

Figure 21.11 Glacial ice can provide important habitat for Arctic marine mammals, especially in the summer when annual sea ice may not be available. Source: © Jenny E. Ross/LifeOnThinIce.org

21.4.4 Non-ice habitats

Other important aspects of marine habitat include oceanographic and topographic features largely independent of sea ice. Ocean depth and the structure of the ocean floor can steer major currents on the shelf, slope and basin, which directly influences densities of zooplankton and fish and, therefore, areas where marine mammals forage. Shallow continental shelves constitute a large portion of the Arctic and tend to be areas where nutrients are more easily mixed in the water column, thus promoting phytoplankton production. Topographic features such as canyons, shelf breaks (including the continental shelf), ridges and plateaus often attract or funnel prey into specific areas where they may be easily targeted by marine mammals (Moore et al., 2000; Laidre et al., 2004a).

Dynamic oceanographic features such as sea surface temperatures, chlorophyll *a* concentrations, eddies, gyres or currents also influence densities of prey and are important in structuring Arctic habitats for marine mammals (Wilson et al., 2014). The coastline and land features may also serve as important seasonal habitats, especially for walrus and polar bears in some regions (Gleason & Rode, 2009).

21.5 Conservation and management challenges

The loss of sea ice due to climate change represents the greatest long-term threat to most Arctic marine mammals, particularly the ice-associated pinnipeds and polar bears (Figure 21.12). Spring is especially important for several species because it coincides with reproduction and critical feeding periods. Sea ice declines over the past several decades have been clearly documented (Perovich & Richter-Menge, 2009; IPCC, 2013) and most projections indicate a nearly ice-free Arctic in summer in the near future (Overland & Wang, 2013). Such declines may result in the following types of habitat loss for ice-associated or ice-dependent species: reduction in total habitat area (i.e. reduced ice cover); habitat fragmentation (i.e. discontinuous pack ice); habitat deterioration (i.e. thinner sea ice, increased rainfall, reduced primary production); and decreased temporal availability of key habitats (Laidre et al., 2008).

Declines in reproduction and survival of ringed seals have been linked to variations in their sea ice habitat, including early or late sea ice breakup, or unusually light or heavy sea ice conditions (Kingsley & Byers, 1998; Harwood et al., 2000). While some ringed seal pups are born on pack ice (Wiig et al., 1999), in many areas landfast ice with sufficient snow cover is required to build lairs that provide protection from the elements and from polar bear predation (McLaren, 1958; Burns, 1970; Hammill & Smith, 1991). Warmer conditions without snow can make lairs impossible to build, and rain events can destroy lairs (Stirling & Smith, 2004; Hezel et al., 2012). Although diminishing snow cover and sea ice represent broad concerns for ringed seals (Kelly et al., 2010), some subpopulations currently appear to be doing well despite sea ice loss (Quakenbush et al., 2011).

Figure 21.12 In some areas of the Arctic, longer ice-free seasons have led to polar bears consuming more birds and bird eggs while on land. Terrestrial food sources are not expected to replace the polar bear's preferred diet of energy-rich seals. Source: © Jenny E. Ross/LifeOnThinIce.org.

Loss of sea ice is also considered the primary threat to walruses, especially in the Pacific where suitable strata for whelping and breeding, and access to offshore feeding areas, will be reduced (Jay et al., 2014). In recent summers Pacific walruses have been hauling out on land in large numbers along the Alaskan and Chukotkan coasts of the Chukchi Sea (Garlich-Miller et al., 2011), which has been attributed to retreat of pack ice beyond the continental shelf (Figure 21.13). In contrast, reduced sea ice cover may increase feeding opportunities for Atlantic walruses (Born, 2005) because they have historically hauled out on land.

In the case of subarctic ice seals, Johnston et al. (2012) found that harp seal mortality in the Gulf of

Figure 21.13 Increasing numbers of Pacific walruses have hauled out on land in recent years as the pack ice retreats beyond the continental shelf. Source: © Jenny E. Ross/LifeOnThinIce.org.

St Lawrence was positively correlated with lighter ice cover and lower values of the North Atlantic Oscillation index. Rosing-Asvid (2008) reported unprecedented numbers of harp seals along the ice edge of central West Greenland in late January and early February, including pregnant females with fetuses close to birth weight, suggesting that harp seals may be whelping in new areas. Large herds of harp seals also have been seen on the east coast of Svalbard during winter (Kovacs et al., 2011).

Declines in sea ice habitat are considered the most significant long-term threat to polar bears (Durner et al., 2009; Obbard et al., 2010). In western Hudson Bay and the southern Beaufort Sea, sea ice loss has been associated with declines in body condition, reproduction, survival and population size (Stirling et al., 1999; Regehr et al., 2007, 2010; Rode et al., 2010; Bromaghin et al., 2015). However, in some other parts of the Arctic, polar bear subpopulations are apparently either productive or stable despite sea ice loss (Obbard et al., 2007; Stirling et al., 2011; Rode et al., 2014). For example, increased polar bear survival and abundance have been associated with large increases in harp seal abundance in Davis Strait (Peacock et al., 2013). Fundamentally, polar bears depend on sea ice for many aspects of their life history, including foraging, travelling, mating and, in some areas, constructing maternal dens (Derocher et al., 2004). Thus, the long-term effects of sea ice loss on polar bears will largely be negative, although when and where such effects occur is expected to vary throughout the Arctic.

Sea ice loss has the potential to bring polar bears into increasing contact with humans. As the ice recedes or breaks up earlier in the spring, the critical on-ice foraging period for polar bears is shortened (Figure 21.14). In regions of the Arctic where polar bears spend the ice-retreat season on land, this can lead to larger numbers of polar bears that arrive in relatively poor nutritional condition and stay on land for longer periods (Stirling et al., 1999). It can also lead to polar bears appearing in areas they had not previously used (Gleason & Rode, 2009), local perceptions of over-abundant bears (Dowsley & Wenzel, 2009), and increasing conflicts with humans (Towns et al., 2009). In some regions, such conflicts are a concern to both human safety and polar bear conservation, requiring community-based solutions that may involve reducing human–bear interactions and using non-lethal deterrents (Peacock et al., 2011).

Figure 21.14 Continued loss of Arctic sea ice is expected to lead to increased competition for ice-dependent species. Source: © Jenny E. Ross/LifeOnThinIce.org.

Figure 21.15 Native people of the Arctic have lived with and harvested polar bears for millennia. Their knowledge and participation are important to managing this species, as well as other Arctic marine mammals, in an uncertain future. Source: © Jenny E. Ross/LifeOnThinIce.org.

The impacts of sea ice loss on Arctic cetaceans are currently unclear and variable. Sea ice loss has the potential to open new habitats (Heide-Jørgensen et al., 2011), increase primary productivity (Zhang et al., 2010), and increase the duration of the production season, which could increase foraging opportunities for baleen whales (Moore & Laidre, 2006; Laidre et al., 2010). In West Greenland, beluga whales have shifted farther offshore as the sea ice edge recedes earlier in the spring (Heide-Jørgensen et al., 2010). In 2010 two satellite-tagged bowhead whales, one from West Greenland and one from Northern Alaska, entered the Northwest Passage from opposite directions and spent approximately 10 days in the same area (Heide-Jørgensen et al., 2011). This is the first time distributional overlap between the Bering–Chukchi–Beaufort and the Eastern Canada-West Greenland bowhead whale populations has been documented, although these populations were connected at some point in history.

Fundamentally, the spatial and temporal dynamics of Arctic sea ice cover influence energy fluxes, levels of biological production and ultimately the survival of species at higher trophic levels, although many of the ecological consequences of declining sea ice remain poorly understood (Post et al., 2013). Some aspects of changing ice dynamics may have unforeseen impacts at the population level (Laidre et al., 2008; Kovacs et al., 2011). For example, irregular freeze-up patterns may have negative effects for ice-associated Arctic cetaceans, such as beluga whales and narwhals, which are susceptible to ice entrapments (Laidre et al., 2011).

The impacts of sea ice loss on seasonal Arctic marine mammals are likely positive, given the potential for southerly species to extend their ranges and occupy previously ice-covered areas (e.g. Heide-Jorgensen et al., 2011). In the Canadian high Arctic, killer whales have been reported to be expanding their range northwards during the ice-free period (Higdon & Ferguson, 2009), increasing predation pressure and competition in some areas. Observations of some subarctic baleen whales (fin and blue) father north may indicate that these species are expanding their ranges as well (Simon et al., 2010; Kovacs et al., 2011; Moore et al., 2011).

In addition to sea ice loss, increasing anthropogenic activity in a warming Arctic may pose conservation challenges or threats to Arctic marine mammals (Laidre et al., 2008). Such activities include expanding oil, gas and mineral development (seismic exploration, drilling, habitat degradation), growing vessel traffic due to reduced sea ice opening new shipping routes (Reeves et al., 2013), and increased interest in tourism. The northward expansion of fisheries (McBride et al., 2014) may lead to increased incidental mortality and injury caused by ship strikes or entanglement in fishing gear. Furthermore, accumulation of environmental contaminants has raised concern about biological effects in some species and food safety among people that rely on Arctic marine mammals for food. Because many Arctic marine mammals migrate long distances and range across international borders, coordinated scientific monitoring (Vongraven et al., 2012) and the development of new international agreements will be key to long-term conservation in a changing Arctic.

Marine mammals represent an important nutritional, cultural and, in some cases, economic resource for native people throughout the Arctic (Chapter 25; Hovelsrud et al., 2008). Currently, approximately 78% of Arctic marine mammal populations support a legal subsistence harvest (Laidre et al., 2015). Most subsistence harvests are relatively closely monitored by local, national and international organizations and appear to be sustainable, although there are exceptions (Klein, 2005). Harvest strategies designed to remove a certain fraction of the population each year (as opposed to removing a constant level of animals) are relatively robust, if updated periodically with new information, even under conditions of changing environmental carrying capacity (Regehr et al., 2015). Broadly, the potential for over-harvest becomes an increasing concern if harvests are poorly monitored, basic population data are lacking, a system is not in place to adjust harvest based on current conditions, there are economic incentives for harvest not subject to conservation controls, or populations are small and isolated. Throughout much of the Arctic, implementation of co-management systems that involve both hunters and wildlife managers (local, regional or national) is key to obtaining local support for harvest regulations, thus balancing the long-term conservation of Arctic marine mammals with improved responsiveness to local practices and subsistence needs (Huntington & Moore, 2008).

References

Amstrup, S.C. (2003) Polar Bear (*Ursus maritimus*). In: *Mammals of North America: Biology, Management, and Conservation.* (Eds. G.A. Feldhamer, B.C. Thompson & J.A. Chapman), pp. 587–610. John Hopkins University Press, Baltimore, Maryland, USA.

Antonelis, G.A., Melin, S.R. & Bukhitiyarov, Y.A. (1994) Early spring-feeding-habits of bearded seals (*Erignathus barbatus*) in the central Bering Sea, 1981. *Arctic* **47**, 74–79.

Boertmann, D., Merkel, F. & Durinik, J. (2009) Bowhead whales in East Greenland, summers 2006–2008. *Polar Biology*, **32**, 1805–1809.

Born, E.W. (2005) An assessment of the effects of hunting and climate on walruses in Greenland. Natural History Museum. PhD Thesis. Oslo, University of Oslo.

Born, E.W., Heide-Jørgensen, M.P. & Davis, R.A. (1994) The Atlantic walrus (*Odobenus rosmarus rosmarus*) in West Greenland. *Meddelelser om Grønland Bioscience*, **40**, 1–33.

Born, E.W., Gjertz, I. & Reeves, R.R. (1995) Population assessment of Atlantic walrus. *Norsk Polarinstitutt Meddelelser*, **138**, 1–100.

Born, E.W., Rysgaard, S., Ehlme, G., Sejr, M., Acquarone, M. & Levermann, N. (2003) Underwater observations of foraging free-living Atlantic walruses (*Odobenus rosmarus rosmarus*) and estimates of their food consumption. *Polar Biology*, **26**, 348–357.

Boveng, P.L., Bengtson, J.L., Buckley, T.W. et al. (2008) Status review of the ribbon seal (*Histriophoca fasciata*). U.S. Department of Commerce, NOAA Technical Memorandum NMFS-AFSC-191.

Boveng, P.L., Bengtson, J.L., Buckley, T.W. et al. (2009) Status review of the spotted seal (*Phoca largha*). U.S. Department of Commerce, NOAA Technical Memorandum NMFS-AFSC-200.

Bowen, W.D., Oftedal, O.T. & Bowness, D.J. (1985) Birth to weaning in four days: remarkable growth in the hooded seal, *Cystophora cristata*. *Canadian Journal of Zoology*, **63**, 2841–2846.

Bromaghin, J.F., McDonald, T.L., Stirling, I. et al. (2015) Polar bear population dynamics in the southern Beaufort Sea during a period of sea ice decline. Ecological Applications, in press. 10.1890/14-1129.1

Bukhtiyarov, Y.A., Frost, K.J. & Lowry, L.F. (1984) New information on the foods of the spotted seal, *Phoca largha*, in the Bering Sea in spring, pp. 55–59 In: *Soviet-American Cooperative Research On Marine Mammals*, Vol. **1**. (Eds. F.H. Fay & G.A. Fedoseev) U.S. Department of Commerce, NOAA, Technical Report NMFS 12.

Burns, J.J. (1970) Remarks on the distribution and natural history of pagophilic pinnipeds in the Bering and Chukchi Seas. *Journal of Mammalogy*, **51**, 445–454.

Burns, J.J. (1981) Ribbon seal-*Phoca fasciata*. In: Handbook of Marine Mammals, vol. 2. Seals (Eds. S.H. Ridgway & R.J. Harrison), pp. 89–109. Academic Press, New York.

Derocher, A.E., Lunn, N.J. & Stirling, I. (2004) Polar bears in a warming climate. *Integrative and Comparative Biology*, **44**, 163–176.

Durner, G.M., Douglas, D.C., Nielson, R.M. et al. (2009) Predicting 21st-century polar bear habitat distribution from global climate models. *Ecological Monographs*, **79**, 25–58.

Eamer, J., J., Donaldson, G.M., Gaston, A.J. et al. (2013) Life linked to ice: A guide to sea-ice-associated biodiversity in this time of rapid change. CAFF Assessment Series No. 10. Conservation of Arctic Flora and Fauna, Iceland.

Fay, F.H. (1974) The role of ice in the ecology of marine mammals of the Bering Sea. In: *Oceanography of the Bering Sea* (Eds. D.W. Hood & E.J. Kelley), pp. 383–399. Institute of Marine Science, University of Alaska, Fairbanks.

Fay, F. (1982) Ecology and biology of the Pacific walrus, *Odobenus rosmarus divergence*, Illiger. North American Fauna **74**.

Fedoseev, G. (2002) Ribbon seal (*Histriophoca fasciata*). In: *Encyclopedia of Marine Mammals* (Eds. W.P. Perrin, B. Würsig & J.G.M. Thewissen), pp. 1027–1030. Academic Press, San Diego, USA.

Ferguson, S.H., Taylor, M.K., Rosing-Asvid, A., Born, E.W. & Messier, F. (2000) Relationship between denning of polar bears and sea-ice conditions. *Journal of Mammalogy*, **81**, 1118–1127.

Finley, K.J., Miller, G.M., Davis, R.A. & Koski, W.R. (1983) A distinctive large breeding population of ringed seals (*Phoca hispida*) inhabiting the Baffin Bay pack ice. *Arctic*, **36**, 162–173.

Folkow, L.P. & Blix, A.S. (1999) Diving behaviour of hooded seals (*Cystophora cristata*) in the Greenland and Norwegian Seas. *Polar Biology*, **22**, 61–74.

Freitas, C., Kovacs, K.M., Andersen, M., Aars, J. et al., (2012) Importance of fast ice and glacier fronts for female polar bears and their cubs during spring in Svalbard, Norway. *Marine Ecology Progress Series*, **447**, 289–304.

Freitas C., Kovacs, K.M., Ims, R.A., Fedak, M.A. & Lydersen, C. (2008) Ringed seal post-moulting movement tactics and habitat selection. *Oecologia*, **155**, 193–204.

Frost, K.J. & Lowry, L.F. (1990.) Feeding of ribbon seals (*Phoca fasciata*) in the Bering Sea in spring. *Canadian Journal of Zoology*, **58**, 1601–1607.

Frost, K.J., Lowry, L.F. & Carroll, G. (1993) Beluga whale and spotted seal use of a coastal lagoon system in the northeastern Chukchi Sea. *Arctic*, **46**, 8–16.

Garlich-Miller, J.L., MacCracken, J.G., Snyder, J. et al. (2011) Status review of the Pacific walrus (*Odobenus rosmarus divergens*). *U.S. Fish and Wildlife Service, Marine Mammals Management*, January 2011, Anchorage, AK.

George, J.C., Clark, C., Carroll, G.M. & Ellison, W.T. (1989) Observations on the ice-breaking and ice navigation behavior of migrating bowhead whales (*Balaena mysticetus*) near Point Barrow, Alaska, spring 1985. *Arctic*, **42**, 24–30.

Givens, G.H., Edmondson, S.L., George, J.C. et al. (2013) Estimate of 2011 abundance of the Bering-Chukchi-Beaufort Seas bowhead whale population. *Paper SC/65a/BRG01*. International Whaling Commission, Cambridge, England.

Gjertz, I. & Wiig, Ø. (1992) Feeding of walrus *Odobenus rosmarus* at Svalbard. *Polar Record*, **28**, 57–59.

Gleason, J.S. & Rode, K.D. (2009) Polar bear distribution and habitat association reflect long-term changes in fall sea-ice conditions in the Alaskan Beaufort Sea. *Arctic*, **62**, 405 - 417.

Hammill, M.O. & Smith, T.G. (1991) The role of predation in the ecology of ringed seal in Barrow strait, Northwest Territories, Canada. *Marine Mammal Science*, **7**, 123–135.

Harwood, L.A., Smith, T.G. & Melling, H. (2000) Variation in reproduction and body condition of the ringed seal (*Phoca hipsida*) in the Western Prince Albert sound, NT, Canada, as assessed through a harvest-based sampling program. *Arctic*, **53**(4), 422–431.

Hammill, M.O. & Stenson, G.B. (2000) Estimated prey consumption by harp seals (*Phoca groenlandica*), grey seals (*Halichoerus grypus*), harbour seals (*Phoca vitulina*) and hooded seals (Cystophora cristata). *Journal of Northwest Atlantic Fishery Science*, **26**, 1–23.

Hammill, M.O., Lesage, V. & Carter, P. (2005) What do harp seals eat? Comparing diet composition from different compartments of the digestive tract with diets estimated from stable isotope ratios. *Canadian Journal of Zoology*, **833**, 1365–1372.

Harwood, L. & Stirling, I. (1992) Distribution of ringed seals in the southeastern Beaufort Sea during late summer. *Canadian Journal of Zoology*, **70**, 891–900.

Heide-Jørgensen M.P., Laidre, K.L., Borchers, D., Marques, T.A., Stern, H. & Simon, M.J. (2010) The effect of sea ice loss on beluga whales (*Delphinapterus leucas*) in West Greenland. *Polar Research*, **29**, 198–208.

Heide-Jørgensen M.P., Laidre, K.L., Burt, M.L. et al. (2010) Abundance of narwhals (*Monodon monoceros* L.) on the hunting grounds in Greenland. *Journal of Mammalogy*, **91**, 1135–1151.

Heide-Jørgensen, M.P., Laidre, K.L., Quakenbush, L.T. & Citta, J. (2011) Northwest Passage opens for bowhead whales. *Biology Letters*, **3**, 577–580

Heide-Jorgensen, M.P., Laidre, K.L., Jensen, M.V., Dueck, L. & Postma, L.D. (2006) Dissolving stock discreteness with satellite tracking: Bowhead whales in Baffin Bay. *Marine Mammal Science* **22**, 34–45.

Heide-Jørgensen, M.P., Richard, P.R., Dietz, R. & Laidre, K.L. (2013) A metapopulation model for narwhals. *Animal Conservation*, **16**, 331–343. doi:10.1111/acv.12000

Hezel, P.J., Zhang, X., Bitz, C.M., Kelly, B.P. & Massonnet, F. (2012), Projected decline in spring snow depth on Arctic sea ice caused by progressively later autumn open ocean freeze-up this century, *Geophysical Research Letters*, **39**, L17505, doi:10.1029/2012GL052794.

Higdon, J.W. & Ferguson, S.H. (2009) Loss of Arctic sea ice causing punctuated change in sightings of killer whales (*Orcinus orca*) over the past century. *Ecological Applications*, **19**(5), 1365–1375.

Hjelset, A.M., Andersen, M., Gjertz, I., Lydersen, C. & Gulliksen, B. (1999) Feeding habits of bearded seals (*Erignathous barbatus*) from the Svalbard area, Norway. *Polar Biology*, **21**, 186–193.

Hovelsrud, G.K., Mckenna, M. & Huntington, H.P. (2008). Marine Mammal Harvests and Other Interactions With Humans. *Ecological Applications*, **18**, S135-S147.

Hunt, G.L., Stabeno, P., Walters, G. et al. (2002) Climate change and control of the southeastern Bering Sea pelagic ecosystem. *Deep Sea Research II*, **49**, 5821–5853.

Hunter, C.M., Caswell, H., Runge, M.C., Regehr, E.V., Amstrup, S.C. & Stirling, I. (2010) Climate change threatens polar bear populations: a stochastic demographic analysis. *Ecology*, **91**, 2883–2897.

Huntington, H.P. & Moore, S.E. (2008) Assessing the impacts of climate change on Arctic marine mammals. *Ecological Applications*, **18**, S1-S2.

International Council for the Exploration of the Sea (ICES) (2008) Report of the Joint ICES/NAFO Working Group on Harp and Hooded Seals, 27–30 August 2008, Tromsø, Norway. ICES CM 2008/ACOM 17.

Innes, S., Heide-Jørgensen, M.P., Laake, J.L. et al. (2002) Surveys of belugas and narwhals in the Canadian high Arctic in 1996. *NAMMCO Scientific Publications*, **4**, 147–190.

IPCC (2013) Summary for Policymakers. Climate Change 2013: The Physical Science Basis. *Contribution of Working Group I to the Fifth Assessment Report of the Intergovernmental Panel on Climate Change*. Cambridge University Press, Cambridge, United Kingdom and New York, NY, USA.

Ivaschenko, Y. & Clapham, P. (2009) Bowhead whales *Balaena mysticetus* in the Okhotsk Sea. *Mammal Review*. doi: 10.1111/j.1365-2907.2009.00152.x

Jay, C.V., Grebmeier, J.M., Fischbach, A.S. et al., (2014) Pacific Walrus (*Odobenus rosmarus divergens*) Resource Selection in the Northern Bering Sea. *PLoS ONE*, **9**, e93035. doi:10.1371/journal.pone.0093035

Johnston, D.W., Bowers, M.T., Friedlaender, A.S. & Lavigne, D.M. (2012) The effects of climate change on harp seals (*Pagophilus groenlandicus*). *PLoS ONE*, **7**, 1–8.

Johnston, D.W., Friedlaender, A.S., Torres, L.G. & Lavigne, D.M. (2005) Variation in sea ice cover on the east coast of Canada from1969 to 2002: climate variability and implications for harp and hooded seals. *Climate Research*, **29**, 209–222.

Kelly, B.P., Bentson, J.L., Boveng, P.L. et al. (2010) Status Review of the Ringed Seal (*Phoca hispida*). NOAA Technical Memorandum NMFS-AFSC **212**:i-xiv, 1–250.

Kelley, T.C., Stewart, R.E. A., Yurkowski, D.J., Ryan, A. & Ferguson, S. (2014) Mating ecology of beluga (*Delphinapterus leucas*) and narwhal (*Monodon monoceros*) as estimated by reproductive tract metrics. *Marine Mammal Science*. DOI: 10.1111/mms.12165

Kenyon, K.W. (1969) The sea otter in the Eastern Pacific Ocean. *North American Fauna*, **68**, 1 – 352.

Kingsley, M.C.S. & Byers, T.J. (1998) Failure in reproduction of ringed seals (*Phoca hispida*) in Amunsden Gulf, Nortwest Territories in 1984–1987. In: *Ringed Seals in the North Atlantic* (Eds. M.P. Heide-Jorgensen & C. Lydersen), pp. 197–210. The North Atlantic Marine Mammal Commission, Tromso.

Klein, D.R. (2005) Management and conservation of wildlife in a changing arctic environment. *Arctic Climate Impact Assessment*, pp. 597–648. Cambridge: Cambridge University Press.

Koski, W.R. & Davis, R.A. (1994) Distribution and numbers of narwhals (*Monodon monoceros*) in Baffin Bay and Davis Strait. *Meddelelser om Grønland, Bioscience*, **39**, 15–40.

Kovacs, K.M. & Lydersen, C. (2008) Climate change impacts on seals and whales in the North Atlantic Arctic and adjacent shelf seas. *Science Progress*, **91**, 117–150.

Kovacs, K.M., Lydersen, C., Overland, J.E. & Moore, S.E. (2011) Impacts of changing sea-ice conditions on Arctic marine mammals. *Marine Biodiversity*, **41**, 181–194 DOI 10.1007/s12526-010-0061-0

Laidre, K.L., Stern, H., Kovacs, K.M. et al. (2015) A circumpolar assessment of Arctic marine mammals and sea ice loss, with conservation recommendations for the 21st century. *Conservation Biology*, **29**, 724–737.

Laidre, K.L. & Heide-Jørgensen, M.P. (2011) Life in the lead: Extreme densities of narwhals in the offshore pack ice. *Marine Ecology Progress Series*, **423**, 269–278.

Laidre K.L., Heide-Jørgensen, M.P., Stern, H. & Richard P. (2012) Unusual sea ice entrapments and delayed autumn ice-up timing reinforce narwhal vulnerability to climate change. *Polar Biology*, **35**, 149–154.

Laidre, K.L., Heide-Jørgensen, M.P., Heagerty, P., Cossio, A., Bergstrom, B. & Simon, M. (2010) Spatial associations between large baleen whales and their prey in West Greenland. *Marine Ecology Progress Series*, **402**, 269–284.

Laidre, K.L., Stirling, I., Lowry, L.F., Wiig, Ø., Heide-Jørgensen, M.P. & Ferguson, S.H. (2008) Quantifying the sensitivity of Arctic marine mammals to climate-induced habitat change. *Ecological Applications*, **18**, 97–125.

Laidre, K.L., Heide-Jørgensen, M.P. & Nielsen, T.G. (2007) The role of the bowhead whale as a predator in West Greenland. *Marine Ecology Progress Series*, **346**, 285–297.

Laidre, K.L. & Heide-Jørgensen, M.P. (2005) Winter feeding intensity of narwhals (*Monodon monoceros*). *Marine Mammal Science*, **21**, 45–57.

Laidre, K.L., Heide-Jørgensen, M.P., Logdson, M.L. et al. (2004a) Seasonal narwhal habitat associations in the high Arctic. *Marine Biology*, **145**, 821–831.

Laidre, K.L., Heide-Jørgensen, M.P., Jørgensen, O.A. & Treble, M.A. (2004b) Deep ocean predation by a high Arctic cetacean. *ICES Journal of Marine Science*, **61**, 430–440.

Laidre, K.L., Heide-Jørgensen, M.P., Logsdon, M.L., Hobbs, R.C., Dietz, R. & VanBlaricom, G.R. (2004c) Fractal analysis of narwhal space use patterns. *Zoology*, **107**, 3–11.

Laidre, K.L., Heide-Jørgensen, M.P., Dietz, R., Hobbs, R.C. & Jørgensen, O.A. (2003) Deep-diving by narwhals, *Monodon monoceros*: differences in foraging behavior between wintering areas? *Marine Ecology Progress Series*, **261**, 269–281.

Lavigne, D.M. & Kovacs, K.M. (1988) *Harps and Hoods: Ice-breeding Seals of the Northwest Atlantic*. University of Waterloo Press, Waterloo, Ontario, Canada.

Lindqvist, C., Bachmann, L., Andersen, L.W. Et al. (2009) The Laptev Sea walrus *Odobenus rosmarus laptevi*: an enigma revisited. *Zoologica Scripta*, **38**, 113–127.

Lowry, L.F. (1993) Foods and feeding ecology. In: *The Bowhead Whale* (Eds. J.J. Burns, J.J. Montague, & C.J. Cowles), pp. 201–283. Society for Marine Mammalogy, Special Publication No. 2. Allen Press, Lawrence, Kansas USA.

Lowry, L.F. & Fay, F.H. (1984) Seal eating by walruses in the Bering and Chukchi Seas. *Polar Biology*, **3**, 11–18.

Lowry, L.F., Frost, K.J. & Burns, J.J. (1980a) Variability in the diet of ringed seals, *Phoca hispida*, in Alaska. *Canadian Journal of Fisheries and Aquatic Sciences*, **37**, 2254–2261.

Lowry, L.F., Frost, K.J. & Burns, J.J. (1980b) Feeding of bearded seals in the Bering and Chukchi Seas and trophic interaction with Pacific walruses. *Arctic*, **33**, 330–342.

Lowry, L.F., Frost, K.J., Davis, R., DeMaster, D.P. & Suydam, R.S. (1998) Movements and behavior of satellite-tagged spotted seals (*Phoca largha*) in the Bering and Chukchi Seas. *Polar Biology*, **19**, 221–230.

Lowry, L.F., Burkanov, V.N., Frost, K.J. et al.. (2000) Habitat use and habitat selection by spotted seals (*Phoca largha*) in the Bering Sea. *Canadian Journal of Zoology*, **78**, 1959–1971.

Lowry, L.F., Sheffield, G. & George, G.C. (2004) Bowhead whale feeding in the Alaska Beaufort Sea, based on stomach content analysis. *Journal of Cetacean Research and Management*, **6**, 215–223.

Lydersen, C., Angantyr, L.A., Wiig, Ø. & Øritsland, T. (1991) Feeding habits of Northeast Atlantic harp seals (*Phoca groenlandica*) along the summer ice edge of the Barents Sea. *Canadian Journal of Fisheries and Aquatic Science*, **48**, 2180–2183.

Lydersen, C., Martin, A.R., Kovacs, K.M. & Gjertz, I. (2001) Summer and autumn movements of white whales *Delphinapterus leucas* in Svalbard, Norway. *Marine Ecology Progress Series*, **219**, 265–274.

Lydersen, C., Hammill, M.O. & Ryg, M. (1992) Water flux and mass gain during lactation in free living ringed seal (*Phoca hispida*) pups. *Journal of Zoology (London)*, **228**, 361–369.

Lydersen, C., Assmy, P., Falk-Petersen, S. et al.(2014) The importance of tidewater glaciers for marine mammals and seabirds in Svalbard, Norway. *Journal of Marine Systems*, **129**, 452–471.

McBride, M.M., Dalpadado, P., Drinkwater, K.F. et al. (2014) Krill, climate, and contrasting future scenarios for Arctic and Antarctic fisheries. *Ices Journal of Marine Science*, **71**, 1934–1955.

McLaren, I.A. (1958) The biology of the ringed seal, *Phoca hispida*, in the eastern Canadian Arctic. *Fisheries Research Board of Canada Bulletin*, **118**, 1–97.

Molnár, P.K., Derocher, A.E., Thiemann, G.W. & Lewis, M.A. (2010) Predicting survival, reproduction and abundance of polar bears under climate change. *Biological Conservation*, **143**(7), 1612–1622.

Moore, S.E. & Reeves, R.R. (1993) Distribution and movement. In: *The Bowhead Whale* (Eds. J.J. Burns, J.J. Montague, & C.J. Cowles), pp. 313–386. Society for Marine Mammalogy, Special Publication No. 2. Allen Press, Lawrence, Kansas USA.

Moore, S.E., DeMaster, D.P. & Dayton, P.K. (2000) Cetacean habitat selection in the Alaskan Arctic during summer and autumn. *Arctic*, **53**, 432–447.

Moore, S.E. & Laidre, K.L. (2006) Analysis of sea ice trends scaled to habitats used by bowhead whales in the western Arctic. *Ecological Applications*, **16**, 932–944.

Moore, S.E., Stafford, K.M., Melling, H. et al. (2011) Comparing marine mammal acoustic habitats in Atlantic and Pacific sectors of the High Arctic: year-long records from Fram Strait and the Chukchi Plateau. *Polar Biology*, **35**, 475–480.

Obbard, M.E., Thiemann, G.W., Peacock, E. & DeBruyn, T.D. (Eds) (2010) Proceedings of the 15th Working Meeting of the IUCN/SSC Polar Bear Specialist Group, 29 June-3 July

2009, Copenhagen, Denmark. Occasional Paper of the IUCN Species Survival Commission, 43.

Odate, T., Hirawake, T., Kudoh, S., Klein, B., LeBlanc, B. & Fukuchi, M. (2002) Temporal and spatial patterns in the surface water biomass of phytoplankton in the North Water. *Deep-Sea Research*, **49**, 4947–4958.

Overland, J.E. & Wang, M. (2013) When will the summer Arctic be nearly sea ice free? *Geophysical Research Letters*, **40**, 2097–2101.

Ovsyanikov, N.G., Menyushina, I.E. & Bezrukov, A.V. (2007) Unusual walrus mortality at Wrangel Island in 2007. 2007 Field Report, *Wrangel Island State Nature Reserve*, 1–4.

Palsbøll P., Heide-Jørgensen, M.P. & Berubé, M. (2002) Analysis of mitochondrial control region nucleotide sequences from Baffin Bay belugas, *Delphinapterus leucas*: detecting pods or sub-populations? *NAMMCO Scientific Publications*, **4**, 39–50.

Peacock, E., Derocher, A.E., Thiemann, G.W. & Stirling, I. (2011) Conservation and management of Canada's polar bears in a changing Arctic. *Canadian Journal of Zoology*, **89**, 371–385.

Peacock E., Sonsthagen, S.A., Obbard, M.E., Boltunov, A., Regehr E.V. et al. (2015) Implications of the circumpolar genetic structure of polar bears for their conservation in a rapidly warming Arctic. *PLoS ONE*. doi:10.1371/journal.pone.0112021

Peacock, E., Taylor, M.K., Laake, J. & Stirling, I. (2013) Population ecology of polar bears in Davis Strait, Canada and Greenland. *Journal of Wildlife Management*, **77**, 463–476.

Perovich, D.K. & Richter-Menge, J.A. (2009) Loss of sea ice in the Arctic. *Annual Review of Marine Science*, **1**, 417–441.

Pershing, A.J., Greene, C.H., Planque, B. & Fromentin, J-M. (2004) The influences of climate variability on North Atlantic zooplankton populations. In: *Marine Ecosystems and Climate Variation* (Eds. N.C. Stenseth et al.), pp. 59–69. Oxford University Press, Oxford England.

Post, E., Bhatt, U.S., Bitz, C.M. et al. (2013) Ecological consequences of sea-ice decline. *Science*, **341**, 519–524.

Quakenbush, L., Citta, J. & Crawford, J. (2011) Biology of the Ringed Seal (*Phoca hispida*) in Alaska, 1960–2010. *Final Report to the National Marine Fisheries Service*. Fairbanks, Alaska, USA.

Reeves, R.R., Ewins, P.J., Agbayani, S. et al. (2014) Distribution of endemic cetaceans in relation to hydrocarbon development and commercial shipping in a warming Arctic. *Marine Policy*, **44**, 375–389.

Regehr, E.V., Lunn, N.J., Amstrup, S.C. & Stirling, I. (2007) Effects of earlier sea-ice breakup on survival and population size of polar bears in western Hudson Bay. *Journal of Wildlife Management*, **71**, 2673–2683.

Regehr, E.V., Hunter, C.M., Caswell, H., Amstrup, S.C. & Stirling, I. (2010) Survival and breeding of polar bears in the southern Beaufort Sea in relation to sea-ice. *Journal of Animal Ecology*, **79**, 117–127.

Regehr, E.V., Wilson, R.R., Rode, K.D. & Runge, M.C. (2015) Resilience and risk-A demographic model to inform conservation planning for polar bears.U.S. Geological Survey Open-File Report, 56 pp.

Rice, D.W. (1998) Marine mammals of the world: systematics and distribution. *Society for Marine Mammalogy, Special Publication*, **4**, 1–230.

Richard, P.R., Orr, J. R, Dietz, R. & Dueck, L. (1998a) Sightings of belugas and other marine mammals in the North Water, late March 1993. *Arctic*, **51**, 1–4.

Richard, P.R., Heide-Jørgensen, M.P. & St. Aubin, D. (1998b) Fall movements of belugas (*Delphinapterus leucas*) with satellite-linked transmitters in Lancaster Sound. *Arctic*, **51**, 5–16.

Richard, P.R., Heide-Jørgensen, M.P., Or, J., Dietz, R. & Smith, T.G. (2001) Summer and autumn movements and habitat use by belugas in the Canadian high Arctic and adjacent waters. *Arctic*, **54**, 207–222.

Ringuette, M., Forture, L., Fortier, M. et al. (2002) Advanced recruitment and accelerated population development in Arctic calanoid copepods of the North Water. *Deep-Sea Research Part II*, **49**, 5801–5100.

Rode, K.D., Amstrup, S.C. & Regehr, E.V. (2010) Reduced body size and cub recruitment in polar bears associated with sea-ice decline. *Ecological Applications*, **20**, 768–782.

Rode, K.D., Regehr, E.V., Douglas, D.C. et al. (2014) Variation in the response of an Arctic top predator experiencing habitat loss: feeding and reproductive ecology of two polar bear populations. *Global Change Biology*, **20**, 76–88.

Rosing-Asvid, A. (2008) A new harp seal whelping ground near South Greenland. *Marine Mammal Science*, **24**(3), 730–736.

Salberg A.-B., Haug, T. & Nilssen, K.T. (2008) Estimation of hooded seal (*Cystophora cristata*) pup production in the Greenland Sea pack ice during the 2005 whelping season. *Polar Biology*, **31**, 867–878.

Sergeant, D.E. (1991) Harp seals, man and ice. *Canadian Special Publications of Fisheries and Aquatic Science No. 114*.

Shaughnessy, P.D. & Fay, F.H. (1977) A review of the taxonomy and nomenclature of North Pacific harbour seals. *Journal of Zoology (London)*, **182**, 385–419.

Simon, M.J., Stafford, K., Beedholm, K., Lee, C.M. & Madsen, P.T. (2010) Singing behavior of fin whales in the Davis Strait with implications for mating, migration and foraging. *Journal of the Acoustical Society of America*, **128**, 3200–3210.

Simpkins, M.A., Hiruki-Raring, L.M., Sheffield, G., Grebmeier, J.M. & Bengtson, J.L. (2003) Habitat selection by ice-associated pinnipeds near St. Lawrence Island, Alaska in March 2001. *Polar Biology*, **26**, 577–586.

Smith, T. (1987) The ringed seal, *Phoca hispida*, of the Canadian western Arctic. *Canadian Bulletin of Fisheries and Aquatic Science*, 216.

Smith, T.G. & Stirling, I. 1975. The breeding habitat of the ringed seal (*Phoca hispida*): the birth lair and associated structures. *Canadian Journal of Zoology*, **53**, 1297–1305.

Smith, T.G. & Hammill, M.O. (1981) The ecology of the ringed seal, (*Phoca hispida*) in its fast ice breeding habitat. *Canadian Journal of Zoology*, **59**, 966–981.

Smith, T.G. Hammill, M.O., & Taugbol, G. (1991) A review of the developmental, behavioural, and physiological adaptations of the ringed seal, *Phoca hispida*, to life in the arctic winter. *Arctic*, **44**, 124–141.

Speckman, S.G., Chernook, V.I., Burn, D.M. et al. (2011) Results and evaluation of a survey to estimated Pacific walrus population size, 2006. *Marine Mammal Science*, **27**, 51–553.

Stenson, G.B. & Hammill, M.O. Can ice breeding seals adapt to habitat loss in a time of climate change? *ICES Journal of Marine Science*, **71**, 1977–1986.

Stirling, I. (1980) The biological importance of polynyas in the Canadian Arctic. *Arctic*, **33**, 303–315.

Stirling, I. (1997) The importance of polynyas, ice edges, and leads to marine mammals and birds. *Journal of Marine Systems*, **10**, 9–21.

Stirling, I. (2002) Polar bears and seals in the eastern Beaufort Sea and Amundsen Gulf: A synthesis of population trends and ecological relationships over three decades. *Arctic*, **55**, 59–76.

Stirling, I. & McEwan, E.H. (1975) The caloric value of whole ringed seals (*Phoca hispida*) in relation to polar bear (*Ursus maritimus*) ecology and hunting behavior. *Canadian Journal of Zoology*, **53**, 1021–1026.

Stirling, I. & Cleator, H. (1981) Polynyas in the Canadian Arctic. *Canadian Wildlife Service Occasional Paper*, **45**, 1–73.

Stirling, I. & Smith, T.G. (2004) Implications of warm temperatures and an unusual rain event for the survival of ringed seals on the coast of Southeastern Baffin Island. *Arctic*, **57**, 59–67.

Stirling, I. & Parkinson, C.L. (2006) Possible effects of climate warming on selected populations of polar bears (*Ursus maritimus*) in the Canadian Arctic. *Arctic*, **59**, 261–275.

Thiemann, G.W., Iverson, S.J. & Stirling, I. (2008) Polar bear diets and Arctic marine food webs: insights from fatty acid analysis. *Ecological Monographs*, **78**, 591–613.

Towns, L., Derocher, A.E., Stirling, I., Lunn, N.J. and Hedman, D. (2009) Spatial and temporal patterns of problem polar bears in Churchill, Manitoba. *Polar Biology*, **32**, 1529–1537.

Vongraven D., Aars J., Amstrup S. et al. (2012) A circumpolar monitoring framework for polar bears. *Ursus Monograph Series*, **5**, 1–66.

Wathne, J.A., Haug, T. & Lydersen, C. (2000) Prey preference and niche overlap of ringed seals *Phoca hispida* and harp seals *Phoca groenlandica* in the Barents Sea. *Marine Ecology Progress Series*, **194**, 233–239.

Weslawski, M.J., Ryg, M., Smith, T.G. & Øritsland, N.A. (1994) Diet of ringed seals (*Phoca hispida*) in a Fjord of West Svalbard. *Arctic*, **47**, 109–114.

Wiig, Ø., Amstrup, S., Atwood, T., Laidre, K., Lunn, N., Obbard, M., Regehr, E. & Thiemann, G. (2015). *Ursus maritimus*. The IUCN Red List of Threatened Species 2015: e.T22823A14871490.

Wiig, Ø., Derocher, A.E. & Belikov, S.E. (1999) Ringed seal (*Phoca hispida*) breeding in the drifting pack ice of the Barents Sea. *Marine Mammal Science*, **15**, 595–598.

Wiig, Ø., Bachmann, L., Øien, N., Kovacs, K.M. & Lydersen, C. (2010) Observations of bowhead whales (*Balaena mysticetus*) in the Svalbard area 1940–2009. *Polar Biology*, **33**, 979–984.

Wilson, R.R., Horne, J.S., Rode, K.D., Regehr, E.V. & Durner, G.M. (2014) Identifying polar bear resource selection patterns to inform offshore development in a dynamic and changing Arctic. *Ecosphere*, **5**, 1–24.

Zhang, J., Spitz, Y.H., Steele, M. et al. (2010) Modeling the impact of declining sea ice on the Arctic marine planktonic ecosystem. *Journal of Geophysical Research*, **115**, C10015.

CHAPTER 22

Antarctic marine mammals and sea ice

Marthán N. Bester,[1] Horst Bornemann[2] and Trevor McIntyre[1]

[1] Mammal Research Institute, Department of Zoology and Entomology, University of Pretoria, Hatfield, South Africa
[2] Alfred-Wegener-Institut, Helmholtz-Zentrum für Polar- und Meeresforschung, Bremerhaven, Germany

22.1 Pinnipedia

Within the Southern Ocean there are four phocid species, the crabeater seal (*Lobodon carcinophaga*), the leopard seal (*Hydrurga leptonyx*), the Ross seal (*Ommatophoca rossii*) and the Weddell seal (*Leptonychotes weddellii*), that are intimately tied to the circumpolar sea ice. By contrast, a fifth phocid, the southern elephant seal (*Mirounga leonina*), and only one of several otariid species, the Antarctic fur seal (*Arctocephalus gazella*), breed almost exclusively on Antarctic and subantarctic islands and migrate seasonally to the sea ice to forage. Information on pinnipeds was gleaned primarily from Laws (1993), Reeves et al. (2002), Berta et al. (2006), Fontaine (2007), Siniff et al. (2008), Perrin et al. (2009), Kovacs et al. (2012), Southwell et al. (2012), Ropert-Coudert et al. (2013) and Hund (2014).

22.1.1 Antarctic fur seal

The Antarctic fur seal, *Arctocephalus gazella*, breeds exclusively on Southern Ocean islands mostly south of the Antarctic Polar Front (APF) (Gentry & Kooyman, 1986). At three island groups situated just north of the APF (Prince Edward Islands, Iles Crozet and Macquarie Island) they occur in sympatry with the related subantarctic fur seal, *Arctocephalus tropicalis*, with limited hybridization (Reeves et al., 2002; Perrin et al., 2009). Large populations breed primarily on South Georgia, with numerous smaller populations also on Bouvet, Heard, McDonald, Kerguelen, South Shetland, South Sandwich and the South Orkney islands. The genus name is from the Greek *arktos* (a bear) and *kephale* (a head that has a bear-like appearance) and *gazella* from being first described from a specimen collected by the German vessel S.M.S. *Gazelle* in 1874. Antarctic fur seal females grow to ~1.45 m long and weigh up to 50 kg. Adult males are considerably larger, at most 2 m and weighing 230 kg. Females may live up to ~23 years and mature sexually at ~2–5 years, pupping after a gestation period of about a year. Males may survive to 14 years and mature sexually at ~3–4 years although they only mature socially at ~7 years when they start holding harems on breeding colony beaches. These beaches vary from sand, shingle and cobbles to vegetated areas, depending on the island locality. A Southern Ocean-wide estimate of numbers is around two to three million.

Antarctic fur seal males are heavily built, with a short, broad and blunt snout compared with other southern fur seals, with long facial vibrissae. Uniformly dark brown to charcoal, the neck and chest are silvery-grey with longer guard hairs on the mane. Females and juveniles are grey dorsally with creamy throat and chest, with a light blaze on the flanks. Pups are black at birth, but about one in every 1000 pups born at South Georgia is leucistic, i.e. they are characterized by reduced pigmentation, resulting in dark-eyed anomalously white animals which are not albinos.

The age of first pupping is 3 years, and all females have pupped by age of 6 years, rarely producing more than one pup on occasion. The breeding (pupping) season extends from late November to late December, with the peak pupping date in the range 4–8 December. Births are highly synchronized and 90% occur within ~21 days. The species is polygynous, with a 'harem' male:female ratio of about 1:10 on crowded beaches at the peak of female numbers present. Males are fiercely territorial during the breeding season when they mate

---

Sea Ice, Third Edition. Edited by David N. Thomas.
© 2017 John Wiley & Sons, Ltd. Published 2017 by John Wiley & Sons, Ltd.

with mothers within about ~7 days after the birth of a pup. Subsequently, females undertake foraging trips of about 1 week long, punctuated by periods of ~3 days ashore to attend their pups, although overnight trips also take place. Pups are weaned at 4 months old, after moulting the black pelage before finally departing to the sea. Moulting of older animals is gradual and completed by the time they become pelagic. Females then remain continually at sea between breeding seasons, although some return to their breeding localities for short periods during winter. Male Antarctic fur seals appear to undertake a southward migration during the post-breeding period.

Males dive deeper (~354 m) and longer (~9 min) than females (~208 m and ~4 min) which are nocturnal foragers, employing short, shallow dives throughout the night (~1 minute to ~15 m depth, rarely > 126 m). Deeper, longer dives occur towards dawn and dusk as they presumably follow vertically migrating prey. Myctophid fish, cephalopods, krill, pelagic icefish and nototheniids are the usual prey, although there is substantial geographic variability, as well as variation within areas and seasons. Males also kill and eat Gentoo, king and macaroni penguins at some islands.

Single individuals are sighted in Antarctica, on the southern fringes of South America, and they also turn up temporarily on temperate islands during spring, sometimes in groups (e.g. at Gough Island), amongst subantarctic fur seals. During their pelagic phase, individuals in the high latitudes haul out on ice floes even in winter, although mainly on pancake floes in the outer fringes of the ice (Figure 22.1a). However, observations remain scarce, and even the most recent Antarctic seal census survey, the international Antarctic Pack Ice Seals (APIS) programme, disregarded potential counts of land breeding seal species such as fur seals and southern elephant seals in the ice.

Threats

Exploited for their pelts in the 18th and 19th centuries, the species has not been commercially exploited since the early 1900s. From virtual extinction, Antarctic fur seals recovered exponentially under protection. Main predators are killer whales *Orcinus orca* and leopard seals *Hydrurga leptonyx*, especially where the populations are smaller. Their foraging range overlaps fisheries activities, especially long-lining and trawling, and individuals frequently become entangled in lines, hooks and strapping, which may result in death. Predictions about probable sensitivity to climate change-related alterations in habitat suggest that the distribution and abundance of the ice-tolerant Antarctic fur seal will not be negatively influenced by changes in pack ice characteristics, but the links between global climate change and predator responses are being explored (e.g. Forcada et al., 2005). Population size and distribution may, however, be influenced most immediately by changes in their food resources due to factors other than climate. Significant sea ice loss, leading to open water and the possible resultant expansion of the krill fishery in the future, may play a role.

Conservation status

The Antarctic fur seal is listed as 'Lower Risk, Least Concern' on the IUCN Red List and is protected under legislation of the various states that have sovereignty of island sites; within the Antarctic Treaty Area it is protected under the Protocol on Environmental Protection to the Antarctic Treaty, although it is no longer designated a specially protected species. It is fully protected under the Convention for the Conservation of Antarctic Seals (CCAS) and the Convention on the Conservation of Antarctic Marine Living Resources (CCAMLR).

22.1.2 Crabeater seal

The crabeater seal (*Lobodon carcinophaga*) is one of four phocid pinnipeds that lives exclusively in the southern hemisphere with breeding populations confined to the circumpolar pack ice of Antarctica (Bester, 2014a). The genus name is from the Greek *lobos* (meaning lobe) plus *odons* (tooth), and refers to the conspicuously lobed postcanine teeth, while the species is from the Greek *karkinos* (crab) , plus *phagein* (to eat), a mistaken reference to its diet. Crabeater seals grow to about 2.65 m long, rarely 277 cm, and around 210 kg, rarely up to 410 kg. Females are usually larger than males. Maximum longevity is 40 years, but usually they live for 20–25 years, and sexual maturity is thought to be attained at 4–5 years of age. Counting ice-seal seal numbers is problematic, but a Southern Ocean-wide estimate of crabeater seals is somewhere between seven and 15 million, making crabeater seals the most abundant seals anywhere.

Crabeater seals have a relatively slim, lithe and streamlined body, with an elongated snout, slightly tip-tilted or pig-like due to the dorsally positioned

Figure 22.1 Antarctic pinnipeds. (a) Antarctic fur seal (photograph by Jan Andries van Franeker, IMARES); (b) crabeater seal (photograph by Ryan Reisinger, MRI); (c) leopard seal (photograph by Ryan Reisinger, MRI); (d) Ross seal (photograph by Jan Andries van Franeker, IMARES); (e) southern elephant seals (photograph by H. Bornemann, AWI); (f) Weddell seal. Source: (photograph by Chris Oosthuizen, MRI).

nostrils (Figure 22.1b). The head is small relative to the body. The canines are moderately sized with the post-canines sieve-like with elaborate four to five blunt cusps each, and interlocking when the jaw is closed. The interlocking teeth reflect their ability to filter crustaceans out of the water column. The short pelage is usually silvery-grey to golden or creamy white (faded), sometimes brown with the back darker than the belly.

There are reticulated chocolate brown markings and fleckings on shoulders, sides and flanks, shading into predominantly dark hind and fore flippers and head. They have a high incidence of obvious scarring on their bodies, mostly caused by leopard seal attacks.

Crabeater seal pups are born from October to early November, perhaps into December, with the exact timing varying by region. Mating has never been observed and presumably takes place in the water in pack ice habitats from December to early January, within approximately 4 days after females wean their single pups following about 2–3 weeks of nursing. However, exact timing may be dependent on the region of the Antarctic, and the pupping season probably varies with location. Crabeater seals moult through January and February, sometimes extending into March. The mean dive depth by crabeater seals ranges from 40 to 140 m, usually less than 5 minutes in duration, but they do sometimes dive as deep as 713 m. Diving patterns vary seasonally, with a clear preference for diving during darkness and hauling out during daylight in summer and autumn. Overall, seals dive deeper (average 92 m, range 6–713 m) and longer (average 5.26 min, range 0.2–23.6 min) in winter, hauling out during the night rather than the day (Burns et al., 2004). Diurnal or diel patterns in the diving behaviour of crabeater seals are perhaps in response to diel vertical migrations of prey and they also showed seasonal shifts in foraging patterns consistent with foraging on vertically migrating prey. They are believed to feed primarily on Antarctic krill, but also to eat fish and cephalopods when krill is not available.

Crabeater seals occur all around the Antarctic continent and are most abundant in pack ice habitats, rarely frequenting open waters. Tending to stay over the continental shelf during the breeding season, they become more widespread in the outer pack ice during summer and autumn. Crabeater seals usually occur as single animals or in small groups, or sometimes in larger groups of up to 28 individuals, and occasionally in large aggregations on fast ice in bays or straits. They often swim directionally in small groups, although larger groups of up to 500 individuals have been observed. The seals move with the pack ice as it expands and contracts seasonally. Crabeater seals have been satellite-tracked over several thousand kilometres, and occasionally haul out at subantarctic islands, but rarely further north along the southern fringes of the continents. Dispersed individuals may also wander onto the Antarctic continent, far up glaciers and into dry inland valleys.

Threats

Although they were harvested commercially twice during the previous century, the sealing ventures were economically unsuccessful, and none are planned or likely to be considered. Crabeater seals rely on certain specific sea ice characteristics to complete their annual cycle, and climate change-related alterations in habitat and constriction of seasonal pack ice will likely affect them most. As the major krill consumer, recent rapid environmental change which led to sea ice loss and an associated decline in, and less predictable, krill biomass is likely to have detrimental effects. Furthermore, population size and distribution may be altered through changes in food web dynamics. Sea ice loss also reduces suitable breeding and resting habitat, protection from predators and increases the distance to areas that concentrate prey. Significant sea ice loss also leads to open water and possible expansion for the krill fishery in the future. These factors may combine to potentially reduce food for pack-ice seals in general, and crabeater seals in particular. High uncertainty in krill and seal stock trends and in their environmental drivers requires a precautionary management of the krill fishery, in the absence of survey data to support management based on specific conservation objectives for pack ice seals (Forcada et al., 2012).

Conservation status

The crabeater seal is listed as 'Lower Risk, Least Concern' on the IUCN Red List and is protected in its range under the Protocol on Environmental Protection to the Antarctic Treaty. It is also protected under the CCAS, Annex I of which provides for commercial harvests of limited numbers, as do the CCAMLR, Article II.

22.1.3 Leopard seal

The leopard seal (*Hydrurga leptonyx*) lives exclusively in the southern hemisphere with breeding populations confined to the circumpolar pack ice of Antarctica (Bester, 2014b). The genus name is from the Greek *hudor* (meaning water) and (possibly) suffix *ourgos* (a worker in), while the species is from the Greek *leptos* (small, slender) and *onux* (claw). Leopard seals grow to about 4.5 m long and weigh up to 600 kg, with females being larger than males. Maximum longevity is possibly up to 25 years, and sexual maturity is thought to be attained at 4–5 years of age. Counting leopard seal numbers is fraught with problems, but a Southern Ocean-wide estimate of leopard seals is somewhere from 220 000 to 440 000.

Leopard seals have relatively large, very long and slender bodies with large 'shoulders' and disproportionally large 'reptilian' heads with long snouts, marked neck constriction and flanks tailing off (Figure 22.1c). The gape is noticeably large and the nostrils point upwards rather than forwards as in other Antarctic phocids. The teeth are all large, incisors and canine teeth are pointed with the post-canines recurved pointed, each with main crown and two lateral cusps, and interlocking when the jaw is closed. The robust teeth reflect their mainly carnivorous diet, although they are adapted for eating a variety of prey, as well as filtering crustaceans out of the water column. The short pelage is usually very dark (black or dark grey) dorsally and silvery ventrally, liberally spotted with light and dark grey, and black spots with a relatively sharp line of demarcation along the sides.

The little information on the reproductive biology of leopard seals suggests that pups are born from October to mid-November. Mating occurs from December to early January, evidently in the water, at about the same time as females wean their pups. Females are thought to nurse their pups for a month, or perhaps less. However, the exact timing may be dependent on the region of the Antarctic, and the pupping season probably varies with location. Leopard seals moult through January and February and are primarily shallow divers, with most dives to depths of 10–50 m and only occasionally 200 m or more. They feed on a variety of prey, including fish, cephalopods, crustaceans, penguins and other seals, but diet composition varies between seasons and regions. These seals can be exceptionally aggressive towards human divers with several known attacks and one fatality.

Although relatively little is known of the species' distribution, abundance, life history and basic natural history, they occur all around the Antarctic continent and are most abundant in pack ice and fast ice habitats especially along the Antarctic Peninsula. Leopard seals are primarily solitary but may congregate around large penguin colonies. The seals sometimes move north as the pack ice expands and stay in open water before moving south back to the ice edge while others do not show such pronounced north–south movement. Northward movement is consistent with numerous observations of leopard seals hauling out at subantarctic islands in winter, and year-round on some islands to the south of the APF. Vagrants also occur at all continents abutting the Southern Ocean.

Threats

There has been essentially no commercial harvest of the species and none are planned or likely to be considered. The non-aggregating nature and remote breeding habitat of leopard seals shelter them from most potential direct interactions with human activities. The apparent solitary behaviour may also reduce direct interactions with commercial fishing activities. Perhaps the most important threat is loss of breeding habitat accompanying ocean climate warming and constriction of seasonal pack ice. The effects of changes in sea ice on leopard seal foraging seem unlikely to be detrimental as they often forage at considerable distances from the continent, beyond the sea ice, and thus may be preying on species not particularly linked with the pack ice ecosystem. On the other hand, population size and distribution may be altered through changes in food web dynamics and their non-breeding season foraging north of pack ice zones may overlap with southern elephant seals, fur seals and other migratory subantarctic marine vertebrates.

Conservation status

The leopard seal is listed as 'Lower Risk, Least Concern' on the IUCN Red List and is protected in its range under the Protocol on Environmental Protection to the Antarctic Treaty. It is also protected under the CCAS, Annex I of which provides for commercial harvests of limited numbers, as does the CCAMLR, Article II.

22.1.4 Ross seal

The Ross seal (*Ommatophoca rossii*) lives exclusively in the southern hemisphere with breeding populations confined to the circumpolar pack ice of Antarctica (Bester & Hofmeyr, 2007; Bester, 2014c). The species was named after Sir James Clark Ross, who collected two of these seals in 1840 during his voyage into the Ross Sea. The genus name is from the Greek *omma* (meaning eye), highlighting its large size. Ross seals grow to about 2–2.5 m long and up to 216 kg. Estimated longevity is 21 years and age of sexual maturity is 3–4 years for males and 2–7 years for females. Although counting Ross seal numbers is fraught with problems, a Southern Ocean-wide estimate of 130 000–220 000 is not impossible.

Ross seals have relatively small but robust bodies with short, broad heads (Figure 22.1d). The eyes are noticeably large and forward-pointing, reflecting adaptations to their deep-diving and foraging habits. The teeth are all small and the post-canines are simple without shearing or grinding structure. The canine teeth are very sharply conical and adapted for catching squid, which seems to be the primary prey. The short pelage is dark brown dorsally and cream or tan ventrally, with several dark stripes radiating down the throat from the mouth and some spotting along the boundary between the counter-shaded dorsal-ventral pattern.

The little information on the reproductive biology of Ross seals suggest that pups are born from mid-October through November. Mating may occur soon thereafter in December and early January. One definitive study showed the nursing period to be 13 days in mid-November, with mating occurring at the end, or shortly after the nursing period, evidently in the last week of November. However, the exact timing is dependent on the region of the Antarctic and the pupping season probably varies with location.

Ross seals moult from late December, but especially through late January to early February. Just after the moult in February, adults head north and stay pelagic in the area south of the APF, until October when they go south into the pack ice. Throughout the year they make about 100 dives a day, mostly to a depth of 100–300 m, the deepest dive on record being 792 m. Most dives, outside the breeding and moulting period, last for 5–15 minutes. This diving behaviour is consistent with feeding on mid-water Antarctic silverfish, squid, and to some extent krill (*Euphausia superba*), when in the pack ice, and myctophid fish and several species of squid, when in the open ocean.

Although relatively little is known of the species' distribution, abundance, life history and basic natural history, they may range all around the Antarctic continent, although areas of higher density appear to be in the Ross Sea, the King Haakon VII Sea and possibly parts of the western Weddell Sea. Most sightings of Ross seals have been of solitary seals, but small groups and aggregations occasionally occur. Although adult Ross seals give birth and mate in remote and inaccessible areas of pack ice, and forage in open water far north of seasonal pack ice after they finish the moult, immature and non-breeding seals may spend an entire year or more in pelagic habitats. Unsurprisingly, then, vagrants occur at several subantarctic islands, New Zealand and Australia.

Threats

There has been essentially no commercial harvest of the species and none are planned or likely to be considered. The non-aggregating nature and remote breeding habitat of Ross seals shelter them from virtually all potential direct interactions with human activities. The apparent solitary behaviour and broad distribution of non-breeding seals may also reduce direct interactions with commercial fishing activities. Perhaps the most important threat is loss of breeding habitat accompanying ocean climate warming and constriction of seasonal pack ice. The effects of changes in sea ice on Ross seal foraging seem unlikely to be detrimental as they often forage at considerable distances from the continent, beyond the sea ice, and thus may be preying on species not particularly linked with the pack ice ecosystem. However, population size and distribution may be altered through changes in food web dynamics and their non-breeding season foraging habitats north of pack ice zones may overlap with southern elephant seals and other migratory subantarctic marine vertebrates.

Conservation status

The Ross seal is listed as 'Lower Risk, Least Concern' on the IUCN Red List and is protected in its range as Specially Protected Species under Annex II to the Protocol on Environmental Protection of the Antarctic Treaty. It is also protected under the CCAS.

22.1.5 Southern elephant seal

The southern elephant seal (*Mirounga leonina*) breeds almost exclusively on Southern Ocean islands, with one continental population at the Valdés Peninsula in Argentina, and occasionally on seasonal pack ice near Signy Island, South Orkney Islands (Laws, 1956; Le Boeuf & Laws, 1994). Large populations breed on South Georgia, Iles Kerguelen, Heard Island and Macquarie Island, with numerous smaller populations primarily at the Prince Edward, Crozet, South Shetland, South Sandwich and the South Orkney islands

(Bester, 2014d). The genus name is from *miouroung*, the Australian aboriginal name for the species, and *leonina* from Latin *leoninus* (lion-like), probably referring to both size and roar. Elephant seal females may grow to about 2.8 m long and weigh up to 900 kg. Adult males are considerably larger at maximally 5 m and 5000 kg, a remarkable sexual dimorphism (Figure 22.1e). They live to 23+ years and females mature sexually at about 2–6 years, males at about 4–5 years, although males only mature socially at 6–8 years, the majority of breeding bulls being 9–12 years old. The age at social maturity is related to the intensity of competition within differently sized populations. A Southern Ocean-wide estimate of numbers is around 600 000.

Southern elephant seals are heavily built, with an inconspicuous neck and large thorax. After the first 3 years, males are already larger than females. The adult male has a conspicuous proboscis that enlarges during the breeding season. The males especially have large canines with small peg-like post-canines. The short pelage is typically dark grey, lighter ventrally, fading to various shades of brown. They have no superimposed pattern of spots or other markings, but adults have scarring around the neck and chest, adult females a lighter yoke around the neck from bites during mating.

The southern elephant seal females aged 3–23+ years give birth to single pups from late September to early November, singly but usually within aggregations of two to 1600 females (the largest on record, in October 1985, at Heard Island) depending on location. Pups nurse for a period of 3 weeks, to late November, but the timing of events depends on latitude and is usually delayed at higher latitudes. A few days before the pups wean, females mate on land with dominant males ('beachmasters') that usually allow assistant beachmasters into the harem when it exceeds about 60 females. After the breeding season, adults depart to sea, only to return for a drastic moult on land where the entire epidermis is shed in patches from December to March with younger age classes starting already in November. Elephant seals dive deeply, feeding primarily in the mid-water regions of the water column at depths of around 200–600 m with dive durations of about 20–30 minutes, and to about 2133 m lasting 120 minutes. Elephant seals spend up to 90% of their time at sea, diving with surface intervals of approximately 2–3 minutes. They typically exhibit diel diving patterns, with shallower dives at night and deeper dives during the day, probably as a response to vertically migrating prey. Myctophid fish and cephalopods are the usual prey, although there appears to be substantial geographic variability as well as considerable variability within areas and seasons.

Elephant seals primarily use beaches for breeding and vegetated areas inland from breeding beaches during the month-long moult haulout. There is a predictable pattern of movement in elephant seals. Repeated deployments of satellite transmitters on individuals belonging to the population at Marion Island (Prince Edward Islands) over the past decades documented a post-breeding pelagic period of about 2–3 months in females (approximately late October to mid-January) and 6 months for males when they remain in a reasonably well-circumscribed area, e.g. mostly within 1500 km from the island, while during the 8-month post-moulting period, the feeding area is some 2000–3100 km distant at the APF and farther south over the Antarctic continental shelf. These statistics vary with different populations and the bathymetry of foraging areas, and depend on whether the animals use a pelagic or benthic diving strategy. Elephant seals tracked from a number of breeding populations appear to particularly exploit the marginal sea ice zone and areas of increased productivity associated with the ice edge, whilst evidently avoiding areas of persistent ice cover. However, multiple tracking studies illustrated the ability of elephant seals to also exploit areas of high sea ice concentrations. These seem to mostly be limited to seals tracked from high-latitude breeding colonies, such as those located around the Antarctic Peninsula and some of the southernmost islands. Furthermore, seals from some populations (notably the small population from the Prince Edward Islands, southern Indian Ocean) rarely travel as far south as the ice edge, and undertake foraging trips covering a wide, ice-free area mostly south of the Subantarctic Front. Single individuals are commonly sighted on the southern fringes of Australia, New Zealand, South Africa and South America, primarily in summer.

Threats

Although they were exploited for their oil in the 18th and 19th centuries, the depleted populations subsequently recovered under protection. There then followed a period of decline from the 1950s, at different rates at different places and periods. A subsequent stabilization, and a currently detectable increase followed despite continued predation by their main predator, the killer whale *Orcinus orca*, particularly where the populations are smaller. Their foraging ranges overlap with fisheries activities, especially long-lining for Patagonian toothfish, and individuals frequently become entangled in lines and hooks, which may result in death. The distribution and abundance of the ice-tolerant southern elephant seal will be the least negatively influenced by changes in pack ice characteristics. In fact, inferences based on genetic data (de Bruyn et al., 2009, 2014) suggest that a now-extinct breeding population on the Victoria Land Coast in the Ross Sea was apparently founded when retreating ice released breeding habitat about 8000 years before present (YBP). The new population grew to approximately an order of magnitude greater abundance than the source population at Macquarie Island, and then declined to extinction when the ice returned approximately 1000 YBP. Population size and distribution may also be influenced most immediately by changes in their food resources due to factors other than climate (see the case study in Section 22.3.1).

Conservation status

The southern elephant seal is listed as 'Lower Risk, Least Concern' on the IUCN Red List and is protected under legislation of the various states that have sovereignty of island sites; within the Antarctic Treat Area it is protected under the Protocol on Environmental Protection to the Antarctic Treaty. It is also fully protected under the CCAS and the CCAMLR.

22.1.6 Weddell Seal

The Weddell seal (*Leptonychotes weddellii*) lives exclusively in the southern hemisphere, with breeding populations largely confined to the circumpolar fast ice and some pack ice habitats of Antarctica (Bester, 2014e). Small populations also breed on South Georgia, the South Sandwich Islands, the South Shetland Islands and the South Orkney Islands. The genus name is from the Greek *leptos* (small, slender), and *onux* (claw) plus suffix *-otes* (denoting possession). It refers to small claws on hind flippers, while the species name *weddellii* is named after James Weddell, who commanded the British sealing expedition, 1822–24, which penetrated to the head of the Weddell Sea. Weddell seals may grow to about 3.3 m long and weigh up to 550 kg. Females may be slightly larger than males. They live to about 22 years, and females mature sexually at about 3–6 years, males at about 7–8 years. Counting ice-seal seal numbers is difficult, but a continent-wide estimate of Weddell seals is approximately 800 000.

Weddell seals have a large, heavy 'barrel-shaped' body with a relatively small head, moderate snout and no distinct neck. The incisors and canines are moderately sized and often greatly worn. The upper, outer incisors and the canines are larger and procumbent for ice sawing and the cheek teeth are blunt with a central prominent point on the molars with a smaller point behind. The seals maintain breathing holes in the fast ice using their teeth. The short pelage is typically dark, slightly lighter ventrally, mottled with large darker and lighter patches, with white patches predominant ventrally.

Weddell seals give birth to single pups, mostly within breeding aggregations of up to 50 females, from late September to early November depending on location. Pups nurse for periods of 7–8 weeks, up to December. After the pups wean, females mate in the water with dominant males that defend underwater territories ('maritories') in fast ice habitats. Adults moult from December to March. Weddell seals dive deeply, feeding primarily in the mid-water regions of the water column at depths of around 100–300 m, and to about 600 m for up to 82 minutes. They may also exhibit diurnal feeding patterns within two depth layers (0–160 m and 340–450 m) as a response to vertically migrating prey. Fish (mainly Antarctic silverfish) including large Antarctic cod, followed by cephalopods and crustaceans, are the usual prey, although there appears to be substantial geographic variability as well as considerable variability within areas and seasons.

Weddell seals primarily use fast ice and nearby pack ice habitats close to the coast (Figure 22.1f). Adult Weddell seals seem reluctant to leave the fast ice where they bred until March or April and then mostly remain within 50–100 km of their summer breeding colonies, although some seals move longer distances and spend long periods in heavy winter pack ice. From bio-acoustic investigations, it is inferred that those males that occupy maritories year-round might have an advantage over non-territorial males or males that move away in winter, in that they are already resident when females arrive at the breeding sites.

In East Antarctica they forage offshore within pack ice in winter for periods of up to 30 days, but return to the stable fast ice to haul out. There is no predictable migration in Weddell seals, and they often aggregate in larger groups in fast ice habitats in early summer when breeding. They may form large colonies of many hundreds on inshore fast ice at this time, often in hummocked ice and near tide cracks. In pack ice they are usually found singly on large smooth floes and haul out on beaches or fast ice during summer, usually singly or in small scattered groups, occasionally in close aggregations of up to 60. Single individuals are occasionally sighted on subantarctic islands, and rarely on the southern fringes of Australia, New Zealand and South America.

Threats

Although they were taken near a number of research stations to feed sled dogs from the 1950s through the early 1980s, and a small-scale experimental harvest was conducted in the 1980s, none are planned or likely to be considered. The distribution and abundance of Weddell seals will probably be negatively affected by changes in the extent, persistence and type of annual sea ice as a result of climate change. Population size and distribution may be altered through changes in food web dynamics. The effects of changes in sea ice on Weddell seals foraging seems likely to be detrimental as they forage on species that are linked with the pack ice ecosystem.

Conservation status

The Weddell seal is listed as 'Lower Risk, Least Concern' on the IUCN Red List and is protected in its range under the Protocol on Environmental Protection to the Antarctic Treaty. It is also protected under the CCAS, Annex I of which provides for commercial harvests of limited numbers, as does the CCAMLR, Article II.

22.2 Cetacea

Antarctic cetaceans can be divided ecologically among those associated with fast ice, pack ice or the open ocean (Boyd, 2009). The Antarctic minke whale (*Balaenoptera bonaerensis*), southern bottlenose whale (*Hyperoodon planifrons*), Antarctic blue whale (*Balaenoptera musculus intermedia*) and killer whale (*Orca orcinus*) are the only cetaceans that are regularly associated with the Antarctic sea ice zone, also during winter. Sperm whales (*Physeter macrocephalus*) and Arnoux's beaked whales (*Berardius arnuxii*) are the only other cetaceans that are often found at the ice edge, and even penetrate it, during the austral summer. These are the six cetacean species that we are concerned with here. Those that are relatively plastic in their movement patterns and more flexible in their habitat utilization may be taking advantage of the receding sea ice in spring due to climate change to utilize new habitats for feeding. Information on cetaceans was gleaned primarily from Stewardson (1997), Rice (1998), Reeves et al. (2002), Berta et al. (2006), Best (2007), Fontaine (2007), Perrin et al. (2009), Parsons (2013), Ropert-Coudert et al. (2013) and Bester (2014f). Our South African bias in describing the aforementioned cetacean species shows through heavy reliance on details presented by Best (2007). We disregard humpback whales (*Megaptera novaeangliae*) and fin whales (*Balaenoptera physalus*), both cosmopolitan cetaceans that have not been sighted near the ice limit in the most extreme latitudes, although continuous recordings from the PerenniAL Acoustic Observatory in the Antarctic Ocean (PALAOA, Ekström Iceshelf, 70°31′S, 8°13′W) also suggest the acoustic presence of the aforementioned species in an Antarctic coastal area during both winter and summer. We also disregard the hourglass dolphin (*Lagenorhynchus cruciger*), sei whales (*B. borealis*), southern right whales (*Eubalaena australis*) and long-finned pilot whales (*Globicephala melas*). These are mostly pelagic and circumpolar in the Southern Ocean and occasionally come near the pack ice.

22.2.1 Antarctic minke whale

The Antarctic minke whale (*Balaenoptera bonaerensis*) is regularly associated with the Antarctic sea ice zone (Figure 22.2a). The species was named after a sailor called Meincke, who mistook a school of minke whales for blue whales, and thereafter small rorquals (all baleen whales with throat grooves) were called 'Minkie's whale'. The genus name means 'winged whale', and the specific name *bonaerensis* means 'Buenos Aires' where the first described specimen was found. They are also called 'little piked whales' and 'sharp-headed finner' from the pointed shape of the upper jaw, 'lesser rorqual' and 'lesser finback.' Antarctic minke whales grow to a length at physical maturity of 8.9–9.0 m and 8.5–8.6 m in females and males, respectively, and an estimated (from a known length:weight relationship) 9000–10 000 kg. Estimated longevity is >40 years (from growth layers in the ear plug) and age of sexual

Figure 22.2 Antarctic cetaceans. (a) Antarctic minke whale (photograph by R. Currey, UC); (b) blue whale (photograph by Jan Andries van Franeker, IMARES); (c) Arnoux's beaked whales (photograph by H. Bornemann, AWI); (d) southern bottlenose whale (photograph by A. Meijboom, IMARES); (e) sperm whale (photograph by Ken Findlay, MRI); (f) killer whale. Source: (photograph by Ken Findlay, MRI).

maturity is 5–9 years in both sexes. Since 1978/79 three circumpolar estimates of Antarctic minke whales south of 60°S produced figures of 608 000, 766 000 and 268 000, respectively. To what extent the significantly lower, most recent figure represents a real decline in the whale numbers is unclear, and a contributing factor to this might have been changes in survey methodologies.

The Antarctic minke whales have relatively small, sleek bodies with a sharply pointed head, a single longitudinal ridge on the rostrum, and a relatively tall, falcate dorsal fin, which may carry a grey flare on its trailing edge. There are grey streaks extending back from the double blowhole, a thin grey shoulder crescent and a grey shoulder region. A light grey flank patch extends high up the back in front of the dorsal fin which is set far back, the tip being 73–76% of the body back from the snout. The whole throat and belly are almost completely white, except for dark in the folds of the throat grooves, of which there are 44–76 that extend almost to the umbilicus. The outer surface of the flipper is either a plain steel-grey, or two tones of grey separated by a distinct line running across the flipper near its base, with the lighter tone covering most of the flipper. The left flipper is sometimes more brightly coloured than the right. There are 215–310 baleen plates on each side of the mouth; the baleen plates are black on the left beyond the first few plates, and on the right they are white in the first third of the row, and black in the remainder of the row.

The little information on the reproductive biology of Antarctic minke whales suggests that peak calving takes place in July and August in warmer waters at low latitudes after a pregnancy of ~10 months. The nursing

Figure 22.2 (*Continued*)

period appears to last 5–6 months, with female ovulation 4 months after parturition, with a peak in conceptions between August and November. The marked seasonality in male testes size and spermatogenic activity indicates that conception most probably takes place in winter in warmer waters. The calving interval of 14 months suggests that a female will conceive progressively later each year (equivalent to a calving rate of about four calves every 5 years), eventually getting out of phase with the breeding season, and pass into a resting stage once lactation has ended.

Antarctic minke whales have a circumpolar distribution and are abundant from 60°S to the ice edge during the austral summer, also occurring in the loose pack ice and polynyas, but also in more consolidated ice (Figure 22.2a). Some overwinter in the Antarctic, when most retreat to breeding grounds in mid-latitudes off northeastern and eastern Australia, off western South Africa and off the northeastern coast of Brazil. They feed predominantly on euphausiids, with no substantial fish component to their diet, both on their summer feeding grounds and during winter migrations. At the ice edge and offshore of the ice-edge zone, krill *Euphausia superba* predominates the diet. They feed by gulping, side- and lunge-feeding, near the surface. Their diving behaviour is unknown, but the intervals between visible blows average 1.5 minutes for single animals, and 0.11 minutes for groups of over 20 animals, the maximum interval of a group of three animals being 7.73 minutes. This suggests that dives are usually rather short (most are < 2 min) and shallow (perhaps < 60 m).

Most sightings of Antarctic minke whales have been of solitary whales, and groups are usually small, consisting of two to three animals, although larger aggregations of up to 400 may occasionally occur in high latitudes.

A smaller, more tropical, dwarf minke whale subspecies also exists, and can easily be confused with the larger Antarctic minke whale, despite some clearly distinguishing features. A taxonomic conundrum, the dwarf form is much closer in appearance, habits and genetics to the northern minke whale *B. acutorostrata*, and for want of better information is considered a subspecies of the northern minke whale. Within the Antarctic they may range as far south as 65°S, but generally are more common on the feeding grounds north of 60°S during the austral summer.

Threats

Major exploitation of minke whales began in the early 1970s with more than 100 000 minke whales *senso latu* killed in the Southern Ocean during the 20th century. It became fully protected in its range when the moratorium on commercial whaling came into effect in 1986/87. Currently there is no commercial harvest of the species and none are planned or likely to be considered. Japan continues to take Antarctic minke whales annually under a research permit issued under terms of the International Whaling Commission (IWC) for the purposes of scientific research. Although the non-aggregating nature and remote foraging habitat of Antarctic minke whales shelter them from other direct interactions with human activities, this is not the case during the winter breeding season in lower latitudes off continents. Perhaps the most important threat is loss of foraging habitat accompanying ocean climate warming and constriction of seasonal pack ice. Antarctic minke whales that use sea ice and ice-edge habitats are at risk, as they are preying on species linked with the pack ice ecosystem.

Conservation status

Antarctic minke whales are listed as 'Lower Risk, Conservation Dependent' by the IUCN. They are included on Appendix I of CITES, which lists species that are threatened with extinction or may become threatened by trade, and thus international trade in these species for commercial purposes is banned. They also appear in Appendix II of the Convention on Migratory Species (CMS) which recognizes it as a migratory species that would significantly benefit from international cooperation in its conservation.

22.2.2 (Antarctic) blue whale

The genus name of the blue whale (*Balaenoptera musculus*) means 'winged whale', and the specific name *musculus* (either 'muscular' or 'mouse') perhaps jokingly refers to its huge body. They are also called 'sulphur-bottoms' a reference to the yellowish sheen acquired from algae on the skin. The smaller form, 'pygmy' blue whale (*B. m. brevicauda*) does not have an Antarctic distribution and is not further considered. Antarctic blue whales (*B. m. intermedia*) grow to a length at physical maturity of 25.9 and 24.1 m in females and males, respectively, and are conservatively estimated to weigh (using known length:weight relationships) in excess of 96 000 kg. Body weights of adults generally range from 50 000 to 150 000 kg. The heaviest blue whale actually weighed (in pieces and adjusted for loss of blood and other body fluids) was 173 400 kg, the largest animal ever to have lived, although a 190 000 kg female was reportedly taken in 1947. Maximum life expectancy for both sexes might be about 65–70 years, and perhaps 80–90 years and even longer. The age of sexual maturity is probably about 8–10 years in both sexes. Brought to the brink of extinction by whaling, there is evidence for population increase in the higher latitudes of the Antarctic, with a still small population estimated at 1700 in 1996, though population estimates are rather uncertain.

The blue whales have huge bodies with a U-shaped head when seen from above, flat in profile, with a single central ridge extending forward from the large splashguard that rises before the two large blowholes (Figure 22.2b). A small dorsal fin, often sloped back with a rounded tip and concave trailing edge, is set far back on the body (74.7–75.5% of the way back from the tip of the snout). Mottled, blue-grey in colour, the large body appears aquamarine underwater. The mottling is most intense on the shoulder and flanks, where the marks may coalesce. The large mouth has 270–395 baleen plates per side that are blue-black with course black bristles, and 70–118 ventral (throat) pleats extend from just behind the tip of the lower jaw to at least as far as the umbilicus. The lower jaw extends some 1–1.7% of the body length beyond the tip of the snout. The flippers have a markedly convex leading edge and a somewhat S-shaped trailing edge. The outer surface of the flippers is dark with a white tip, and they are white on the inside. The broad fluke has a deep median notch, and the underside of the flukes is usually paler than the upper side.

The reproductive cycle is thought to last 2 or possibly 3 years, including a resting period. Calving takes place between late March and early July in warmer waters at low latitudes after a pregnancy of approximately 12 months, some calves being born en route to and from the Antarctic. The nursing period appears to last 7 months, followed by a resting stage, resulting in females being pregnant sometimes at intervals of more than 2 years.

Antarctic blue whales have a circumpolar distribution and are abundant south from 60°S, especially close to the edge of, and even within, the pack ice during the austral summer. While in high latitudes, they may range nearly halfway around Antarctica, although most animals are thought not to move more than 90° of longitude. Much of the population migrates to low latitudes in winter, although some overwinter in the Antarctic (Thomisch et al., 2016). Exactly where they go in winter is a mystery as no specific breeding ground has been discovered for blue whales in any ocean.

Being compulsive euphausiid feeders, krill *Euphausia superba* predominates the diet at the ice edge and in high latitudes, with copepods and amphipods also taken. They feed by a swallowing rather than a skimming technique, and side- and lunge-feeding has also been observed. Short shallow dives are most common, with animals breathing eight to 15 times between dives, lasting generally 8–15 minutes, although 20-minute dives are not uncommon. Shallow dives (10–20 consecutive dives) are often followed by a deep dive of 10–30 minutes. Foraging blue whales in the North Pacific dive to around 140 m for 7.8 minutes (with maxima of 204 m and 14.7 min), and to 68 m for 5 minutes when not foraging. Most sightings of Antarctic blue whales have been of solitary whales or small groups of two to five, although larger feeding aggregations of up to 50 or more may occasionally occur in high latitudes.

Threats

Major exploitation of Antarctic blue whales, as the largest of the whales and the most highly sought-after, began in 1904. From an estimated original population of 239 000 in 1905, stocks were reduced possibly to as few as 360 animals, a reduction of more than 99%. The International Convention for the Regulation of Whaling (1946) provided some relief with an introduction of quotas, but the blue whale only became fully protected from commercial whaling in its range when the IWC moratorium on commercial whaling came into effect in 1986/87. Perhaps the most important current threat is loss of foraging habitat accompanying ocean climate warming and constriction of seasonal pack ice. Antarctic blue whales that use ice-edge habitats are at risk, as they are preying on species linked with the pack ice ecosystem.

Conservation status

Antarctic blue whales are listed as 'Critically Endangered' by the IUCN. They are included in Appendix I of both CITES and CMS.

22.2.3 Arnoux's beaked whale

The origin of the genus and specific names of Arnoux's beaked whale (*Berardius arnuxii*) is a French corvette's Captain Bérard and the surgeon, Maurice Arnoux, who found the type specimen on a beach in New Zealand. The largest known Arnoux's beaked whale is 10.06 m long, and females are likely to be slightly longer than males. Body weights of adults are unknown, as are maximum life expectancy, the ages of sexual maturity and the reproductive cycle. It has never been exploited, and the population size is unknown.

A long, almost tubular body with a small, low dorsal fin positioned far behind (about two-thirds of the way down) on the body is characteristic for adults. Usually blunt or rounded at the peak, the low dorsal fin has a long leading edge and a somewhat falcate rear margin. The small flippers have rounded tips. The prominent melon is bulbous, sloping to the long beak, and the gape is long and curves upward at the back. The lower jaw protrudes noticeably past the upper jaw so that the first of two pairs of prominent mandibular teeth at the beak tip are exposed in front of the upper jaw also when the mouth is closed. The second pair of smaller teeth, which erupt behind the larger teeth, are concealed within the mouth. Both sexes have erupted teeth around sexual maturity. The blowhole is crescent-shaped but the arms of the crescent point backwards rather than forwards as in most other odontocetes. The V-shaped throat grooves may be accompanied by several smaller grooves behind them.

Most of the adult's body is brownish-black, but from the front of the melon to the level of the dorsal fin, it has slate-grey blotching, making the head slightly lighter in colour. Most of the beak remains dark, with the tip of the snout and lower jaw lighter. Increased scarring with age on the melon, back and sides give the skin a marbled appearance (Figure 22.2c). The tail fluke is slightly notched medially and has light grey scarring, especially near the leading edges and tips.

Arnoux's beaked whales have a circumpolar distribution in the southern hemisphere, ranging from the Antarctic continent and ice edge to temperate waters, with the northernmost record a stranding at 24°S. They are frequently seen in summer in the Antarctic, and in the pack ice in August. Individuals have sometimes been trapped by ice and forced to overwinter in the Antarctic. The simultaneous presence of these beaked whales off the coasts of southern continents and in the Antarctic in summer suggests different stocks, and it is unknown whether there is a definite migratory pattern. Both their wintering and breeding grounds are unknown.

Arnoux's beaked whales are presumably cephalopod feeders, based on unidentified beaks found in the stomach of only one individual (from New Zealand), and perhaps deep-sea fish. Accomplished divers, dive durations within leads in the Antarctic pack ice had a mode of 35–65 minutes, and a maximum of at least 70 minutes. Most sightings (off Australia) consisted of groups one to 16 animals, with a modal value of four. Larger groups of up to 80 animals were reported from the Antarctic, with an average 6.7 individuals per pod.

Threats
Never hunted directly, and with little direct involvement with fisheries, they are considered uncommon and of unknown number. Although underwater recordings at the ice edge suggested that these animals are highly vocal, there is no evidence that Arnoux's beaked whales have been involved in fatalities associated with the use of sonar. Loss of foraging habitat accompanying ocean climate warming and constriction of seasonal pack ice might play a role, especially for those (a separate stock perhaps?) that use ice-edge and pack ice habitats if preying on species linked with the pack ice ecosystem.

Conservation status
Arnoux's beaked whales are listed as 'Lower Risk, Conservation Dependent' by the IUCN. They are included in Appendix I of CITES, but not listed by the CMS.

22.2.4 Southern bottlenose whale
The genus name of the southern bottlenose whale (*Hyperoodon planifrons*) is from *Hyperoodon* (meaning 'teeth in the upper jaw'), clearly an error in observation, while the species is from *Plani* (= flat) and *frons* (= front). Based on a very small sample, adult males reach a length of 7.0 m and adult females 7.45 m. Body weights of adults are unknown but calculations from length:weight relationships suggest that a 6.0 m bottlenose whale would weigh at least 3400 kg. Maximum life expectancy, the ages of sexual maturity and the reproductive cycle are not known, apart from a female of 5.7 m which was reported lactating. It has never been exploited, apart from individuals being taken out of curiosity in commercial catches of other cetaceans. They are abundant and a Southern Ocean estimate (south of the Antarctic Convergence) of approximately 599 000 beaked whales is mostly attributed to the presence of southern bottlenose whales.

The body is comparatively long and robust, with the small (approximately 30 cm) dorsal fin falcate and located behind the midpoint of the back. The tail is broad without a median notch and the rear margin is straight. The flippers are short, elliptical in shape and situated well down on the side with a pair of grooves on the throat forming a forward-pointing V. It has a bulbous forehead that overhangs the prominent short robust beak that is well demarcated from the melon (Figure 22.2d) which may appear stubbier on males than on females or juveniles. A distinct dimple develops in the front of the melon of adult males. Only older males have a pair of conical teeth that erupt at the tip of the lower jaw and that are visible outside the closed mouth. A crescent-shaped blowhole is situated behind the forehead. The southern bottlenose whale is generally tan or greyish-brown in colour with a pale head, throat and belly. There are often white spots on the genital region and sides. The back and flanks of older males may be extensively scarred to the extent that they may appear white in very old animals.

Southern bottlenose whales have a circumpolar distribution in the southern hemisphere, ranging from the Antarctic ice edge to temperate waters generally south of about 30°S, although they range further north on occasion. They are frequently sighted in summer from October to March in the Antarctic, and can occur at the ice edge, although the highest densities of these whales are within 60 nm (111 km) of the ice edge. Sightings of bottlenose whales suggest a migratory pattern with a movement northwards from higher latitudes in late summer, and a movement back south again in spring, but their wintering grounds are unknown.

Southern bottlenose whales are cephalopod feeders and a total of 36 cephalopod species of 14 families were isolated from only two stomachs from South African (lower latitude) waters, including squids of Antarctic and subantarctic origin, although a stomach from Heard Island (15 species of nine families of squid) also had remains of Patagonian toothfish. Likely to be accomplished divers, dive durations of an hour or more are punctuated by blows every 30–40 seconds. Dive cycles recorded for 12 groups of beaked whales (presumably mostly bottlenose whales) in the Antarctic had mean submergences of 25.3 minutes, interspersed with surfacing bouts of, on average, 3.7 seconds. Their northern hemisphere counterpart, *H. ampullatus*, regularly dives to the seabed at depths of 800–1500 m.

Most Antarctic sightings ranged from groups of one to 10 animals, with means ranging from 1.7 to 3.6 in different areas, although groups of up to 25 have been seen.

Threats
Considered to be abundant, southern bottlenose whales have never been hunted directly, although 42 were caught in the Antarctic by Soviet whalers between 1970 and 1982. They also appear to have little direct involvement with fisheries. Loss of foraging habitat accompanying ocean climate warming and constriction of seasonal pack ice might play a role, especially for those that use ice-edge habitats if preying on species linked with the pack ice ecosystem.

Conservation status
Southern bottlenose whales are listed as 'Lower Risk, Conservation Dependent' by the IUCN. They are included in Appendix I of CITES, but not listed by the CMS.

22.2.5 Sperm whale
Widely distributed in both hemispheres, the sperm whale (*Physeter macrocephalus*) is the largest of the toothed whales, and the genus name *Physeter* comes from the Greek for 'blower', and the specific name *macrocephalus* refers to its 'big head.' Sperm whales grow to an adult length of up to 17.3 m, although males of 20 m or more were reported in earlier days. Females rarely exceed 11.7 m. Conservatively estimated (using known length:weight relationships) a 10 m sperm whale would weigh 10 800 kg, and a 15 m whale in excess of 32 700 kg. Maximum life expectancy for both sexes is at least 60–70 years. The age of sexual maturity is about 9 years in both sexes, although males only reach social maturity at the age of 25–27 years when they start to compete for access to schools of females. Heavily exploited from the early 18th century to the late 20th century, a worldwide population of 360 000 sperm whales is estimated, down from a pre-whaling population of about 1 110 000.

Sperm whales are sexually dimorphic, the heads of males are proportionally larger and in females the rear part of the body is longer. The head is box-like and the narrow, rod-like lower jaw is underslung and barely visible when viewed from the side (Figure 22.2e). The elongated S-shaped blowhole is set forward on the head and skewed strongly to the left. A small, triangular dorsal fin has a wide base, positioned well back along the body, with up to six smaller protuberances or knuckles set along the back between the dorsal fin and the triangular fluke, with a thick caudal peduncle on the tail stock. The fluke has a deep median notch and a roughened callus near the tip of the dorsal fin in most adult females (absent from adult males).

Blue-brown in colour with a steely dark grey sheen, the body surface behind the head appears loose and wrinkled. The lips are white and white patches may be found on the navel, posterior crest, back, flanks and tail. The narrow lower jaw carries 18–30 pairs of large conical teeth that erupt around puberty, commencing at the centre of the tooth row and progressively faster towards the back of the jaw. Some adult sperm whales have a few (up to nine to 10 pairs) erupted (maxillary) teeth in the upper jaw, although these are usually absent with empty sockets within the gums. The flippers are short and broad, with rounded tips, and are set low down on the sides of the body.

The reproductive cycle is thought to last 4–6 years. Calving takes place between January and April, after a pregnancy of 15–16 months. The nursing period appears to last on average 2 years, with the next calf being born after an interval of about 5 years. Younger females become pregnant more frequently, and older females lactate longer, perhaps nursing calves other than their own. Large, mature males in their late 20s and older return to tropical breeding grounds to mate, although the timing of such visits is largely unknown. They roam independently between groups of females, and are sometimes seen together within the same group of females. Males occasionally fight, which is presumably the source of heavy scarring in the head region.

In the southern hemisphere, sperm whales occur from the tropics to the ice edge. Only adult males move from tropical and temperate waters to the ice edge in summer, while the females and immatures are seldom found beyond 40°S, their distribution corresponding roughly to sea surface temperatures greater than 15°C. Sperm whales are migratory, at least off southern Africa, with medium-sized and large males being largely winter visitors from high latitudes. Movements and migrations are associated with food availability.

Sperm whales have a mostly catholic diet, but principally eat modestly sized squid, and also take demersal and mesopelagic fish, amongst others. Giant squid (*Architeuthis* spp.) and jumbo squid (*Dosidicus* spp.) are also taken, and males tend to eat larger individuals of the same taxa compared with females. In addition, males eat species that are largely restricted to higher latitudes, such as the colossal squid, *Mesonychoteuthis hamiltoni*, of Antarctic waters. In these areas, a smaller diversity of cephalopods are taken.

While foraging, sperm whales make repeated deep dives with a modal depth of 600 m and lasting about 45 minutes. Recovery periods between dives at or near the surface are about 9 minutes. Dives can be deeper (to over 1000 m), shallower (depending on the bathymetry) and longer (>1 hour). The longest recorded dive lasted at least 2 hours and 18 minutes, while the deepest was to 2035 m, with greater dive depths considered likely.

The basic social unit of sperm whales consists of 10–15 mature females and their associated offspring. Sperm whales are usually found singly (adult males) or in mixed schools of 15–50. Several types of social groups are formed, e.g. breeding groups and bachelor schools.

Threats
Sperm whales have been hunted extensively since the early 1800s. Decimated in certain areas, they remain the most abundant of the large whales. The sperm whale became fully protected from commercial whaling in its range after the IWC moratorium on commercial whaling from 1986/87. Interactions between sperm whales and longline fishing for Patagonian toothfish, *D. eleginoides*, in the South Atlantic, for example, is disconcerting, as resolution of such conflict without harming the whales is questionable. Being killed inadvertently by entrapment in fishing gear and collision with ships (all unlikely in the Antarctic) may have a negative effect. Ocean climate warming and the resultant constriction of seasonal pack ice may have a positive influence on adult males that use ice-edge habitats by opening up more benthic areas with the retreat of the ice edge, although this is speculative.

Conservation status
Sperm whales are listed as 'Vulnerable' by the IUCN. They are included in Appendix I of CITES, and in both Appendices I and II (presumably different populations) of the CMS.

22.2.6 Killer whale
The killer whale (*Orcinus orca*) is also known as Orca after 'Orcus', the other name for Pluto, the Roman god of the underworld, also called Hades, who tormented evil-doers in the afterlife. It is a cosmopolitan species with the highest densities in high latitudes and largest numbers occurring in Antarctica. The world population seems to comprise subpopulations, each adapted to live off resources within its particular home range. In the southern hemisphere, killer whales present a range of behavioural, social and morphological characteristics that are interpreted as evidence to categorize individuals or groups. Morphologically distinct forms (types A, B, C, and D) seem to occur in the Southern Ocean, evidently with specialized ecology and behaviours. However, it has been cautioned that ecotypic status classification for Southern Ocean killer whale morphotypes be reserved until more evidence-based ecological and taxonomic data are obtained (de Bruyn et al., 2012). We follow that sentiment here, and as a result we ignore the contentious naming in the 1980s of two new species of killer whale from the Antarctic, *O. nanus* and *O. glacialis*.

Killer whales grow to an adult length of up to 9.0 m, although females rarely exceed 7.9 m. Conservatively estimated (using known length:weight relationships), a 6.0 m killer whale would weigh 3079 kg, and an 8.0 m whale 7903 kg. Maximum longevity for males is 50–60 years, and for females about 80–90 years. The age of sexual maturity is about 12–16 years in females, and males reach sexual maturity at an estimated 10.5–17.5 years, mostly at about 15 years. Physical maturity is attained at a mean age of 21 years. Exploited to some extent regionally in the northern hemisphere, some 1287 were taken by USSR Antarctic whaling expeditions from 1947 to 1980, and the Antarctic summer population was estimated at about 80 400 in the period 1976–1988.

Killer whales are sexually dimorphic with a broad and robust body, black dorsally with white extending from the tip of lower jaw to beyond the anus. Anterior to the anus, the white extends upwards and backwards along the flanks. There is a white oval patch of varying size and shape above and posterior to the eye that runs back to about the anterior insertion of the flipper (Figure 22.2f). A saddle of grey is situated behind the dorsal fin. This saddle is sometimes connected to a grey cape marking. The characteristically large, roughly oval or paddle-shaped flippers are black, while the ventral surface of the tail flukes is usually white with a black edge. The tail has a deep median notch and a slightly concave rear margin.

The head in profile has traces of a snout or conical beak, weakly demarcated from the melon, straight mouthline and large gape. The upper jaw typically protrudes past the tip of the lower jaw. There are 10–14 pairs of large pointed conical teeth in both upper and lower jaws, which in exceptional cases may become flattened through wear virtually down to the gum. The blowhole is set slightly to the left of the centre of the forehead. The dorsal fin, which is situated around the middle of the back, is large, prominent and highly variable in shape. In adult males it is erect, with the tip pointing almost straight up to as much as 1.78 m in a large isosceles triangle. Dorsal fins are smaller and falcate in adult females and juveniles.

Calving may take place year-round, with peaks in autumn and spring depending upon locality, after gestation periods of 15–18 months. The nursing period lasts about 18 months, although calves may consume solid food from 6 to 7 months, perhaps even when a few weeks old. Average calving intervals for mature females in the Antarctic have been estimated at between 3.5 and 5.2 years. Captive females give birth for the first time between 8 and an estimated 17 years, and in the wild at 12–16 years.

In the southern hemisphere, killer whales occur from the tropics to the ice edge. An overall pattern of migration is proposed, with most killer whales being in warm waters in winter and in high latitudes in summer.

The southern migration apparently starts in September/October and is completed in January, followed by the return migration from the end of February, with most animals having left the higher latitudes by the end of April. Killer whales were recorded south of 60°S in midwinter within the pack ice, but this behaviour might be restricted to an ecotype(s) that might forego migration. Wind-blown or fast-forming ice may force killer whales to remain in small pools of open water for long periods.

Worldwide, killer whales feed on just about anything at sea, especially marine mammals (at least 20 cetacean species, 14 pinniped species, dugongs and sea otters), fish (at least 11 teleost species, 12 elasmobranch species), birds and turtles. The considerable degree of diet specialization amongst different communities of killer whales in the northern hemisphere, broadly classified as fish-eating (residents) and mammal-eating (transients), might have also developed amongst killer whales in the Southern Ocean and within the Antarctic. They hunt singly or cooperate to hunt large prey such as cetaceans and to maintain balls of baitfish. They also steal fish from long lines and scavenge on discarded fishery bycatch.

Foraging and travelling are the predominant activity states of killer whales, and they generally dive to depths below 150 m, reaching a maximum of 330+m, but remain mostly in the top 20 m of the water column. Group diving and surfacing can become synchronized and regular with a series of short, shallow dives (e.g. five to 12 dives of 10–30 seconds), followed by a dive, perhaps deeper, that can last 3–5 minutes. Dives of up to 30 minutes are not uncommon, but they do not have well-defined and characteristic blowing and surfacing behaviour.

The size of killer whale schools south of 60°S average 12.3 animals, with schools of 25 animals or larger not uncommon, including aggregations as large as 200. In the Antarctic, most schools contain one or more adult males, usually with one or more adult females and their calves, and animals are rarely seen alone. The extent to which the matrilineal grouping as the basic social unit of killer whales in the North Pacific, including the differences between 'residents' and 'transients', applies to killer whale groupings in the Antarctic is unknown.

Threats

As a species, the killer whale is not endangered, and threats such as pollution, hunting, heavy ship traffic, intense whale-watching operations and live-capture operations are of little consequence in the Antarctic. However, killer whales frequently scavenge fish from longline and other hook-and-line fisheries in the Southern Ocean, resulting in big losses of catch. Although few killer whales ever get entangled in fishing gear, unconfirmed reports suggest that fishermen retaliate by shooting offending killer whales. Ocean climate warming and the resultant constriction of seasonal pack ice may have an influence on a subpopulation(s) of killer whales that perhaps preferentially frequent pack ice habitats, although the nature of such potential influences is currently unknown.

Conservation status

Killer whales are listed as 'Lower Risk: Conservation Dependent' by the IUCN. They are included in Appendix II of CITES and the CMS.

22.3 Climate change

Simmonds and Isaac (2007), Nicol et al. (2008), Siniff et al. (2008), Van Franeker et al. (2008), Moore (2009), Evans et al. (2010), Massom and Stammerjohn (2010), Kaschner et al. (2011), Forcada et al. (2012), Kovacs et al. (2012), Parsons (2013), Bester (2014f), Turner et al. (2014) and others have all debated the influence of climate change on marine mammals. The dynamic relationship between sea ice and krill will mediate the impact of climate warming on top predators in the Southern Ocean (Nicol et al., 2008; Moore, 2009), as it is widely anticipated that impacts on marine mammals will be mediated primarily via changes in prey distribution and abundance, and that the more mobile (or otherwise adaptable) species may be able to respond to this to some extent (Simmonds & Isaac, 2007). Loss of Antarctic sea ice seems to have affected multiple levels of the marine food web in a complex fashion and has triggered cascading effects with impacts on primary production, Antarctic krill, fish, birds and marine mammals, both negative and positive (Massom & Stammerjohn, 2010). Siniff et al. (2008) concluded that the distribution and abundance of crabeater seals and Weddell seals will be negatively affected by changes in the extent, persistence and type of annual sea ice, while Ross seals and leopard seals will be the least negatively influenced by changes in pack ice characteristics. Southern elephant seals and (Antarctic) fur seals are expected to respond in ways opposite to the pack ice species. Kaschner et al. (2011) analysed patterns of marine mammal species richness based on predictions of global distributional ranges for 115 species, including all extant pinnipeds and cetaceans. Application of their model to explore potential changes in biodiversity under future perturbations of environmental conditions predicted that projected ocean warming and changes in sea ice cover until 2050 may have moderate effects on the spatial patterns of marine mammal richness. Increases in cetacean richness were predicted at $>40°$ latitude in both hemispheres, while decreases in both pinniped and cetacean richness were expected at lower latitudes. Here we present a case study of the more subtle likely impacts of climate change on southern elephant seals.

22.3.1 Case Study: Likely impacts of environmental shifts on southern elephant seals – climate change winners or losers?

Southern elephant seals typically spend more than 80% of their lives foraging at sea, hauling out twice a year as adults, once to breed (austral spring) and once to undergo their annual moult (austral summer). While they primarily breed on ice-free, subantarctic islands, breeding does take place on the Antarctic continent (e.g. McMahon & Campbell, 2000), and indeed occasionally on ice-covered areas (Laws, 1956). Although it is a comparatively well-studied marine mammal species, there is substantial uncertainty regarding how these seals are likely to respond behaviourally, and ultimately demographically, to changes in the Southern Ocean due to anthropogenic climate change (e.g. Bester, 2014f). A number of studies provided evidence for behavioural correlations in this species with ocean temperature, and although the species is not intimately bound to sea ice as compared with the other pinnipeds described earlier, it might well model the impact on these species as well.

The foraging ecology of southern elephant seals is known to correlate with a number of physical oceanographic characteristics, in particular sea ice cover and water temperature, although interpopulation differences seem to exist in the nature of such relationships.

For example, elephant seals tracked from a number of breeding populations appear to exploit, in particular, the marginal sea ice zone and areas of increased productivity associated with the ice edge, whilst evidently avoiding areas of persistent ice cover (Bailleul et al., 2007; Raymond et al., 2014). However, multiple tracking studies have illustrated the ability of elephant seals also to exploit areas of high sea ice concentrations. These seem mostly to be limited to seals tracked from high-latitude rookeries, such as those located around the Antarctic Peninsula (e.g. Bornemann et al., 2000; Tosh et al., 2009; Muelbert et al., 2013; McIntyre et al., 2014), and some of the southernmost islands (e.g. Biuw et al., 2010). Furthermore, seals from some populations (notably the small population from the Prince Edward Islands, southern Indian Ocean) rarely travel as far south as the ice edge, and undertake foraging trips covering a wide, ice-free area mostly south of the Subantarctic Front (McIntyre et al., 2011a, 2012).

Evidence for behavioural correlations in southern elephant seals with ocean temperature come from Marion Island (Prince Edward Islands) where they tend to perform deeper dives, characterized by shorter time at the bottom phases of dives, when foraging in warmer water masses (McIntyre et al., 2011b). Similar relationships between dive depth and water temperature were described for seals from the Kerguelen Islands (KI) and Macquarie Island, where seals were more likely to switch to a 'resident' (i.e. likely foraging vs. travelling) state in areas where the bottom temperature of dives were the coldest (Bestley et al., 2013). More recently, seals from KI were also shown to perform deeper dives in areas characterized by warmer deep-water (>200 m) temperatures, evidently also encountering fewer prey items while diving in such areas (Guinet et al., 2014). Importantly, seals that tended to forage in such (warmer) water masses showed similar mass gain to seals foraging elsewhere (and encountering more numerous prey items), suggesting that seals are capable of compensating for lower catch rates by consuming larger and/or richer prey.

While elephant seals display high levels of fidelity to their at-sea foraging grounds (Bradshaw et al., 2004), individual animals are also known to display substantial amounts of behavioural plasticity in their foraging strategies. For example, Figure 22.3 illustrates how a sub-adult male elephant seal switched his dive strategy when encountering changes in the temperature structure of the water column associated with the intrusion of a cold-core eddy. This seal, fitted with a conductivity–temperature–depth Satellite Relay Data logger (Sea Mammal Research Unit, University of St Andrews, UK), continuously foraged in a small area close to Marion Island during its post-moult foraging trip, where it mostly performed deep (~700–800 m) dives that showed little vertical diel variation in depth, and substantial overlap in depths between day and night (Figure 22.3a). However, when water temperatures cooled in October of that year (Figure 22.3b) due to the presence of the eddy, the seal switched to a dive pattern that displayed clear diel variation, performing much shallower dives at night than during the day (Figure 22.3c,d). Various zooplankton communities and their associated predators, such as myctophid fish, are known to display diel vertical migration patterns (e.g. Cisewski et al., 2010). The observed changes in dive behaviour of the tracked elephant seal are therefore most probably explained by a change in the available prey base when the eddy intruded, resulting in the seal changing its foraging strategy to exploit prey items that display vertical diel migration when the temperature structure of the water column changed.

The Southern Ocean is undergoing substantial changes associated with anthropogenically driven climate change. These include the warming of water masses, strengthening of westerly winds, poleward shifts in positions of frontal systems, as well as sea ice and shelf ice retreat in some regions (evidently increasing in others at present) (Chapter 10; Turner et al., 2014 and references therein). Furthermore, zooplankton community compositions are changing, with subtropical species evidently intruding polewards (Pakhomov & Chown, 2003). The individual behavioural plasticity and ability to switch between feeding strategies of elephant seals may permit elephant seals to adapt to climatic changes and associated changes in available prey resources quite easily. In fact, regional changes in climate are widely expected even to benefit southern elephant seals through the creation of additional ice-free beaches suitable for moulting and breeding, such as Anvers Island (Siniff et al., 2008) and Livingston Island (Gil-Delgado et al., 2013). However, there is evidence to suggest that climatic shifts have resulted in negative demographic consequences to some populations due to

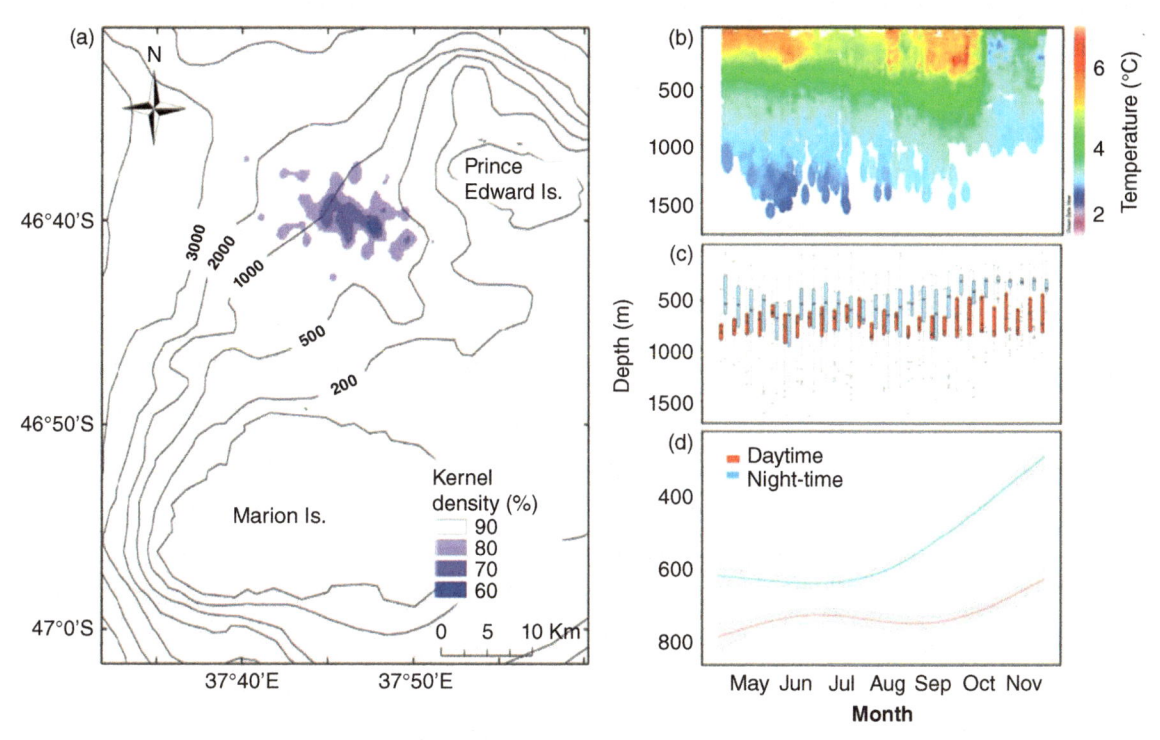

Figure 22.3 Example of behavioural plasticity of a southern elephant seal associated with hydrographic changes. (a) Dive locations of a tracked sub-adult male elephant seal, illustrating little spatial variation during the course of its 7-month post-moult foraging trip. (b) *In situ* temperature profiles collected by the device that the seal was instrumented with (conductivity–temperature–depth Satellite Relay Data logger, Sea Mammal Research Unit, University of St Andrews, UK), illustrating substantial cooling in October and November, associated with the intrusion of a cold-core eddy. (c) Boxplots (line = median, box = 25th and 75th percentiles; whiskers = 1.5 × interquartile range) of dive depths of the tracked seal during the study period. (d) Fit obtained (± 95% confidence intervals) from generalized additive models fitted to the dive depth data over time, illustrating the switch to a period of clear diel variation in dive depth at the same time as the cooling event (October–November).

decreased survival of pups in years of decreased ocean productivity (McMahon & Burton, 2005).

Furthermore, some of the northernmost populations (e.g. Marion Island) may be negatively affected by changes in prey distribution to ocean warming, due to increased physiological costs associated with deeper diving and/or more distant foraging migrations (McIntyre et al., 2011b). It is therefore difficult to predict what ongoing and further expected climatic shifts will mean for the global southern elephant seal population. However, we suggest that regionally specific influences can be expected, with some populations potentially benefiting through increased terrestrial habitat availability, while others may suffer declines in fitness due to required behavioural changes associated with changes in prey availability and distribution, ultimately incurring population losses.

22.4 Acknowledgements

We are indebted to Elke Burkhardt for access to the AWI photographic library of images taken from the bridge of the RV *Polarstern* during polar cruises. Jan van Franeker from the Institute for Marine Resources and Ecosystem Studies (IMARES) in the Netherlands generously allowed us to use his fur seal, Ross seal and southern blue whale images. André Meijboom (IMARES) kindly provided the southern bottlenose whale image. Through Regina Eisert, University of Canterbury, New Zealing, we located the minke whale image of Rohan Currey, who kindly allowed us to use it. All other images are from staff and students from the MRI and AWI: crabeater seal and leopard seal (Ryan Reisinger); Weddell seal (Chris Oosthuizen), Arnoux's beaked whales and elephant seals (Horst Bornemann); sperm whale and killer whale (Ken Findlay).

References

Bailleul, F., Charrassin, J.-B., Ezraty, R. et al. (2007) Southern elephant seals from Kerguelen Islands confronted by Antarctic Sea ice. Changes in movements and in diving behaviour. *Deep Sea Research Part II*, **54**, 343–355.

Berta, A., Sumich, J.L. & Kovacs, K.M. (Eds) (2006) *Marine Mammals: Evolutionary Biology*, 2nd edn. Academic Press, Burlington.

Best, P.B. (2007) *Whales and Dolphins of the Southern African Sub-region*. Cambridge University Press, Cape Town.

Bester, M.N. & Hofmeyr, G.J.G. (2007) Ross Seal. In: *Encyclopedia of the Antarctic* (Ed. B. Riffenburgh), pp. 815–816. Taylor & Francis Books Inc., New York.

Bester, M.N. (2014a) Crabeater seal. In: *Antarctica and the Arctic Circle: A Geographic Encyclopedia of the Earth's Polar Regions* (Ed. A. Hund), pp. 211–213. ABC-CLIO, Inc., Santa Barbara.

Bester, M.N. (2014b) Leopard seal. In: *Antarctica and the Arctic Circle: A Geographic Encyclopedia of the Earth's Polar Regions* (Ed. A. Hund), pp. 455–457. ABC-CLIO, Inc., Santa Barbara.

Bester, M.N. (2014c) Ross seal. In: *Antarctica and the Arctic Circle: A Geographic Encyclopedia of the Earth's Polar Regions* (Ed. A. Hund), pp. 618–620. ABC-CLIO, Inc., Santa Barbara.

Bester, M.N. (2014d) Southern elephant seal. In: *Antarctica and the Arctic Circle: A Geographic Encyclopedia of the Earth's Polar Regions* (Ed. A. Hund), pp. 670–672. ABC-CLIO, Inc., Santa Barbara.

Bester, M.N. (2014e) Weddell seal. In: *Antarctica and the Arctic Circle: A Geographic Encyclopedia of the Earth's Polar Regions* (Ed. A. Hund), pp. 747–749. ABC-CLIO, Inc., Santa Barbara.

Bester, M.N. (2014f) Marine mammals: natural and anthropogenic influences. In: *Handbook of Global Environmental Pollution* Vol. **1** (Ed. B. Freedman), pp. 167–174, Dordrecht, Springer-Verlag.

Bestley, S., Jonsen, I.D., Hindell, M.A., Guinet, C. & Charrassin, J.-B. (2013) Integrative modelling of animal movement: incorporating *in situ* habitat and behavioural information for a migratory marine predator. *Proceedings of the Royal Society B*, **280**, 2012–2262.

Biuw, M., Nøst, O.A., Stien, A., Zhou, Q., Lydersen, C. & Kovacs, K.M. (2010) Effects of hydrographic variability on the spatial, seasonal and diel diving patterns of southern elephant seals in the eastern Weddell Sea. *PLoS ONE*, **5**, e13816. doi:10.1371/journal.pone.0013816

Bornemann, H., Kreyscher, M., Ramdohr, S. et al. (2000) Southern elephant seal movements and Antarctic sea ice. *Antarctic Science*, **12**, 3–15.

Boyd, I.L. (2009) Antarctic marine mammals. In: *Encyclopedia of Marine Mammals*, 2nd edn. (Eds W.F. Perrin, B. Würsig & J.G.M. Thewissen), pp. 42–46. Academic Press, Burlington.

Bradshaw, C.J.A., Hindell, M.A., Sumner, M.D. & Michael, K.J. (2004) Loyalty pays: potential life history consequences of fidelity to marine foraging regions by southern elephant seals. *Animal Behaviour*, **68**, 1349–1360.

Burns, J.M., Costa, D.P., Fedak, M.A. et al. (2004) Winter habitat use and foraging behavior of crabeater seals along the Western Antarctic Peninsula. *Deep-Sea Research Part II*, **51**, 2279–2303.

Cisewski, B., Strass, V.H., Rhein, M. & Kragefsky, S. (2010) Seasonal variation of diel vertical migration of zooplankton from ADCP backscatter time series data in the Lazarev Sea, Antarctica. *Deep Sea Research Part I*, **57**, 78–94.

de Bruyn, M., Hall, B.L., Chauke, L.F., Baroni, C., Koch, P.L. & Hoelzel, A.R. (2009) Rapid response of a marine mammal species to Holocene climate and habitat change. *PLoS Genetics*, **5**, e1000554.

de Bruyn, M., Pinsky, M., Hall, B., Koch, P., Baroni, C. & Hoelzel AR (2014) Rapid increase in southern elephant seal genetic variation after a founder event. *Proceedings of the Royal Society B*, **281**, 1471–2954.

de Bruyn, P.J.N., Tosh, C.A. & Terauds, A. (2012) Killer whale ecotypes: is there a global model? *Biological Reviews*, **88**, 62–80.

Evans, P.G.H., Pierce, G.J. & Panigada, S. 2010. Climate change and marine mammals. *Journal of the Marine Biological Association of the United Kingdom*, **90**, 1483–1487.

Fontaine, P-H. (2007) *Whales and Seals*. Schiffer Publishing Ltd, Atglen, Pennsylvania.

Forcada, J., Trathan, P.N., Reid, K. & Murphy, E.J. (2005) The effects of global climate variability in pup production of Antarctic fur seals. *Ecology*, **86**, 2408–2417.

Forcada, J., Trathan, P.N., Boveng, P.L. et al. (2012) Responses of Antarctic pack-ice seals to environmental change and increasing krill fishing. *Biological Conservation*, **149**, 40–50.

Gentry, R.L. & Kooyman, R.L. (Eds) (1986) *Fur Seals: Maternal Strategies on Land and at Sea*. Princeton University Press, Princeton.

Gil-Delgado, J.A., Villaescusa, J.A., Diazmacip, M.E. et al. (2013) Minimum population size estimates demonstrate an increase in southern elephant seals (*Mirounga leonina*) on Livingston Island, maritime Antarctica. *Polar Biology*, **36**, 607–610.

Guinet, C., Vacquié-Garcia, J., Picard, B. et al. (2014) Southern elephant seal foraging success in relation to temperature and light conditions: insight into prey distribution. *Marine Ecology Progress Series*, **499**, 285–301.

Hund, A. (Ed) (2014) *Antarctica and the Arctic Circle: A Geographic Encyclopedia of the Earth's Polar Regions*. Vol. **1 & 2**, ABC-CLIO, Inc., Santa Barbara.

Kaschner, K., Tittensor, D.P., Ready, J., Gerrodette, T. & Worm, B. (2011) Current and future patterns of global marine mammal biodiversity. *PLoS ONE*, **6**, e19653, doi: 10.1371/journal.pone.0019653

Kovacs, K.M., Aguilar, A., Aurioles, D. et al. (2012) Global threats to pinnipeds. *Marine Mammal Science*, **28**, 414–436.

Laws, R.M. (1956) The elephant seal (*Mirounga leonina* Linn.). II. General, social and reproductive behaviour. *Falkland Islands Dependencies Survey, Scientific Reports*, **13**, 1–88.

Laws, R.M. (1993) *Antarctic Seals: Research Methods and Techniques*. Cambridge University Press, Cambridge.

Le Boeuf, B.J. & Laws, R.M. (Eds) (1994) *Elephant seals: Population Ecology, Behavior, and Physiology*. University of California Press, Berkeley.

Massom, R.A. & Stammerjohn, S.E. (2010) Antarctic sea ice change and variability – physical and ecological implications. *Polar Science*, **4**, 149–186.

McIntyre, T., Ansorge, I.J., Bornemann, H., Plötz, J., Tosh, C.A. & Bester, M.N. (2011b) Elephant seal dive behaviour is influenced by ocean temperature: implications for climate change impacts on an ocean predator. *Marine Ecology Progress Series*, **441**, 257–272.

McIntyre, T., Bornemann, H., de Bruyn, P.J.N. et al. (2014) Environmental influences on the at-sea behaviour of a major consumer, *Mirounga leonina*, in a rapidly changing environment. *Polar Research*, **33**, 23808. doi:https://dx.doi.org/10.3402/polar.v33.23808

McIntyre, T., Bornemann, H., Plötz, J., Tosh, C.A. & Bester, M.N. (2011a) Water column use and forage strategies of female southern elephant seals from Marion Island. *Marine Biology*, **158**, 2125–2139.

McIntyre, T., Bornemann, H., Plötz, J., Tosh, C.A. & Bester, M.N. (2012) Deep divers in even deeper seas: habitat use of male southern elephant seals from Marion Island. *Antarctic Science*, **24**, 561–570.

McMahon, C.R. & Burton, H.R. (2005) Climate change and seal survival: evidence for environmentally mediated changes in elephant seal, *Mirounga leonina*, pup survival. *Proceedings of the Royal Society B*, **272**, 923–928.

McMahon, C.R. & Campbell, D. (2000) Southern elephant seals breeding at Peterson Island, Antarctica. *Polar Record*, **36**, 51–52.

Moore, S.E. (2009) Climate change. In: *Encyclopedia of Marine Mammals*, 2nd edn (Eds W.F. Perrin, B. Würsig & J.G.M. Thewissen), pp 238–241. Academic Press, Burlington.

Muelbert, M.M. C., de Souza, R.B., Lewis, M.N. & Hindell, M.A. (2013) Foraging habitats of southern elephant seals, *Mirounga leonina*, from the Northern Antarctic Peninsula. *Deep Sea Research Part II: Topical Studies in Oceanography*, **88–89**, 47–60.

Nicol, S., Worby, A. & Leaper, R. (2008) Changes in the Antarctic sea ice ecosystem: potential effects on krill and baleen whales, *Marine and Freshwater Research*, **59**, 1323–1650.

Pakhomov, E.A. & Chown, S.L. (2003) The Prince Edward Islands: Southern Ocean oasis. In: *Ocean Yearbook*, Vol. 17, pp. 348–379. University of Chicago Press, Chicago, USA.

Parsons, E.C.M. (Ed.) (2013) *An Introduction to Marine Mammal Biology and Conservation*. Jones & Bartlett Learning, Burlington.

Perrin, W.F., Würsig, B. & Thewissen, J.G.M. (Eds) (2009) *Encyclopedia of Marine Mammals*, 2nd edn. Academic Press, Burlington.

Raymond, B., Lea, M.A., Patterson, T. et al. (2014) Important marine habitat off east Antarctica revealed by two decades of multi-species predator tracking. *Ecography* (May), 1–9. doi:10.1111/ecog.01021

Reeves, R.R., Stewart, B.S., Clapham, P.J. & Powell, J.A. (2002) *National Audubon Society Guide to Marine Mammals of the World*. Alfred A. Knopf, Inc. Chanticleer Press, New York.

Rice, D.W. (1998) *Marine Mammals of the World: Systematics and Distribution*. Special Publication No. 4, The Society for Marine Mammalogy, Allen Press, Inc., Lawrence.

Ropert-Coudert, Y., Hindell, M.A., Phillips, R., Charrassin, J.B., Trudelle, L. & Raymond, B. (2013) Biogeographic patterns of birds and mammals. In: *The Biogeographic Atlas of the Southern Ocean*. Scientific Committee on Antarctic Research (Eds. C. De Broyer & P. Koubbi P). Online: http://atlas.biodiversity.aq. (Accessed July 2016).

Simmonds, M.P. & Isaac, S.J. (2007) The impacts of climate change on marine mammals: early signs of significant problems. *Oryx*, **41**, 19–26.

Siniff, D.B., Garrott, R.A., Rotella, J.J., Fraser, W.R. & Ainley, D.G. (2008) Projecting the effects of environmental change on Antarctic seals. *Antarctic Science*, **20**, 425–35.

Southwell, C., Bengtson, J., Bester, M. et al. (2012) A review of data on abundance, trends in abundance, habitat use and diet of ice-breeding seals in the Southern Ocean. *CCAMLR Science*, **19**, 49–74.

Stewardson, C.L. (1997). *Mammals of the Ice*. Sedona Publishing, Braddon ACT, Australia.

Thomisch, K., Boebel, O., Clark, C.W., Hagen, W., Spiesecke, S., Zitterbart, D.P. & Van Opzeeland, I. (2016) Spatio-temporal patterns in acoustic presence and distribution of Antarctic blue whales *Balaenoptera musculus intermedia* in the Weddell Sea. *Endangered Species Research*, **30**, 239–253.

Tosh, C.A., Bornemann, H., Ramdohr, S. et al. (2009) Adult male southern elephant seals from King George Island utilize the Weddell Sea. *Antarctic Science*, **21**, 113–121.

Turner, J., Barrand, N., Bracegirdle, T. et al. (2014) Antarctic climate change and the environment: an update. *Polar Record*, **50**, 237–259.

Van Franeker, J., Fijn, R., Flores, H., Meijboom, A. & van Dorssen, M. (2008) Marine birds and mammals wintering in the Lazarev Sea: future evidence of a major role of sea ice in structuring the Antarctic. In: *The Expedition ANTARKTIS-XXIII/6 of the Research Vessel 'Polarstern' in 2006* (Ed. U. Bathmann). Berichte zur Polar- und Meeresforschung, Reports on Polar and Marine Research **580**, 79–86.

CHAPTER 23

A feathered perspective: the influence of sea ice on Arctic marine birds

Nina J. Karnovsky[1] and Maria V. Gavrilo[2]

[1] Pomona College, Department of Biology, Claremont, CA, USA
[2] National Park Russian Arctic, Archangelsk, Russia

23.1 Introduction

The lives of Arctic seabirds and the patterns of sea ice development and break-up are inextricably linked. Sea ice interacts with seabirds in a myriad important ways. On the one hand, sea ice allows access to a dependable food web which is associated with sea ice habitats. On the other hand, sea ice can be a physical barrier to being able to feed. Therefore, too much ice, or ice in the wrong place, or in the right place at the wrong time, can be devastating to Arctic seabirds which are wholly dependent on finding food in seasonally frozen seas. Sea ice influences everything from migration patterns, the timing of breeding, colony location, foraging behaviour, reproductive success and, ultimately, the population size of Arctic seabirds. The particular manner in which sea ice influences different seabird species depends primarily upon the type of prey and foraging method that the seabird uses to find and capture prey, as well as its migratory patterns and flight capabilities.

There are a number of papers reviewing the associations of seabirds and sea ice (Bradstreet & Cross, 1982; Hunt, 1991; Hunt et al., 1996; Gilg et al., 2012; Eamer et al., 2013; Mehlum, 1997) and so in this chapter we highlight some of the recent studies that elucidate the particular ways that sea ice influences seabird ecology in the Arctic, with a special emphasis on how Arctic marine birds are responding to changes in sea ice. The Arctic marine avifauna is a diverse assemblage of species. The highest breeding densities of seabirds in the northern hemisphere are found in the Arctic (Cairns et al., 2008). The Arctic supports several endemic seabird genera and species. We have decided to include several examples of how sea ice influences some endemic Arctic sea ducks – the eiders. Even though they are not technically seabirds, they are sea ice-dependent, marine-feeding avian species. The numbers of individuals and species diversity of marine birds increase greatly every spring and decline at the end of the breeding season; only a few species reside in the Arctic year-round. Some species, such as thick-billed murres (*Uria lomvia*) or dovekies (*Alle alle*), number in the many millions while others, such as ivory (*Pagophila eburnea*) and Ross's gulls (*Rhodostethia rosea*), are extremely rare (Ganter & Gaston, 2013). Of those that remain in the Arctic throughout the winter, some have populations that are migratory as well. Some species are circumpolar while others are confined to either Atlantic or Pacific Ocean basins. We focus our analysis on particular species that exemplify each of these situations (resident, migratory, rare, abundant, constrained distribution, circumpolar) but there are many other examples that could have been included.

23.2 The 'central' problem of breeding in the Arctic

Every spring, millions of seabirds migrate north to breed along Arctic coastlines and to feed in productive Arctic seas bordering the coastlines and islands on which they nest. A smaller subset of seabirds and sea ducks remain in the Arctic throughout the winter. Because the polar summer breeding season is short, seabirds

---

Sea Ice, Third Edition. Edited by David N. Thomas.
© 2017 John Wiley & Sons, Ltd. Published 2017 by John Wiley & Sons, Ltd.

A feathered perspective: the influence of sea ice on Arctic marine birds   557

Figure 23.1 Thick-billed murre (*Uria lomvia*) chicks leaving the nest and jumping into ice-choked waters near Franz Josef Land in 2014. Source: photograph by Yuri Krasnov.

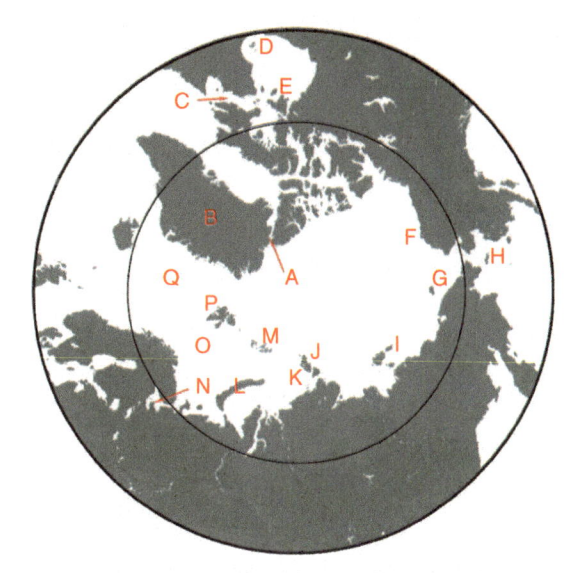

Figure 23.2 Map of some of the locations in the Arctic discussed in this chapter. The inner circle indicates the location of the Arctic Circle. A, Northwater Polynya; B, Greenland; C, Hudson Strait; D, Belcher Islands; E, Coates Island; F, Cooper Island; G, Chukchki Sea; H, St, Lawrence Island; I, East Siberian Sea; J, Severnaya Zemlya; K, Kara Sea with West Severnya Zemlya Polynya and Domashny Island, L, Novaya Zemlya; M, Franz Josef Land; N, White Sea; O, Barents Sea; P, Spitsbergen Island in the Svalbard archipelago; Q, Greenland Sea.

have to hatch and raise their chicks to independence in very short periods of time (often less than 2 months between arrival at nesting colonies and having chicks ready to leave the nest). Chicks of thick-billed murres even fledge before they are able to fly (they leap off the cliff ledge where they hatched, and bounce down rocks into the water) where they are cared for by their father as they swim south to their wintering areas. The chicks can't fly and so fledging into ice-covered waters is a hardship (Figure 23.1). Therefore the distribution of sea ice in August, when fledging occurs, delimits the northern distribution of nesting colonies of the thick-billed murres (Uspensky, 1969). This is probably the reason why the Russian Arctic areas of mid-Siberia and the Severnaya Zemlya archipelago (Figure 23.2) are devoid of thick-billed murres; there's too much ice during fledging. In a similar fashion, there is a gap in the breeding distribution of common eiders (*Somateri amollissima*) in mid-Siberia (Ganter & Gaston, 2013). Common eiders are shallow diving ducks that feed on bivalves and other benthic invertebrates in the intertidal zone. Vast areas of fast ice fringe the coasts of the central Siberian shelf seas for many months of the year. The fauna of the intertidal zone are repeatedly disturbed and ploughed over by the scouring action of ice floes and hummocks. The seafloor eventually becomes depauperate of common eider prey (Ganter & Gaston, 2013).

When seabirds are raising chicks, they become 'central place foragers' (Orians & Pearson, 1979); they commute from the nest where the chick is located to dependable feeding areas where abundant prey can be found. The Arctic seabird avifauna includes several species of seabirds from the family Alcidae that are well adapted for subsurface wing-propelled diving (much like penguins of the southern hemisphere). Their wings are designed for speed and agility underwater but they are unable to perform long foraging flights. Therefore these Alcids, such as thick-billed murres, black guillemots (*Cepphus grylle*) and dovekies, are especially dependent on finding large concentrations of energy-rich prey in waters close to their colonies due to their limited flight capabilities (Figure 23.3 and 23.4). Seabirds such as northern fulmars (*Fulmarus glacialis*) and black-legged kittiwakes (*Rissa tridactyla*) that are better adapted for flight are less constrained and can forage more widely and selectively (Figure 23.5; Sigler et al., 2012). Fledging success (the ability to raise a chick to the point it can leave the colony) is dependent on being able to provision chicks frequently with lipid-rich prey.

Raising chicks is very energetically costly to adults who have to work extremely hard in order to successfully raise their chicks. Even though feathers and

Figure 23.3 Thick-billed murres (*Uria lomvia*) commuting from the ice edge to their chicks at the colony. Note the fish that some birds are carrying in their bills. Source: photograph by Maria Gavrilo.

Figure 23.4 Black guillemots (*Cepphus grylle*) are distinctive seabirds with bright red feet and red inside their mouths. They have a circumpolar distribution and use their wings to 'fly' underwater to capture prey such as Arctic cod. They are constrained to finding food for their chicks close to their nest site. Source: photograph by Maria Gavrilo.

Figure 23.5 Black-legged kittiwakes (*Rissa tridactyla*) are able to search for prey at locations far from their nests. Source: photograph by Maria Gavrilo.

fat provide good insulation, they have high surface to volume ratio and therefore lose body heat. Arctic marine birds have elevated metabolic rates to meet the energetic demands of an active lifestyle in a cold environment in order to maintain thermal homeostasis (Gabrielsen & Mehlum, 1988). On top of meeting their own energetic needs, they must meet the demands of their growing chicks. Seabirds and sea ducks tend to be relatively long-lived and so they can still have high lifetime reproductive success even with several years of reproductive failures. However, with losses of sea ice and changes in the distribution, extent, timing of formation and break-up of sea ice, seabirds, whose ability to successfully raise chicks has been shaped by the dependable presence of sea ice, are facing new challenges.

23.3 Feeding at the edge

Sea ice influences the availability of food for marine birds in several important ways:
- It fosters high levels of primary production by allowing nutrients to rebuild in nutrient-depleted waters.
- Ice algae and moribund phytoplankton fall to the benthos and provide import inputs to the bivalve community which become import prey for sea ducks.
- A specialized community of fish and crustaceans develops in association with sea ice which provides critical prey for marine birds.
- Sea ice serves as a solid platform for roosting while digesting food (Figure 23.6).

Figure 23.6 Black-legged kittiwakes (*Rissa tridactyla*) resting on sea ice and feeding and bathing in the open water lead. Source: photograph by Maria Gavrilo.

Seasonally ice-covered waters in the Arctic get 'reset' in winter. The nutrients in the upper water column get replenished and then, when the ice retreats in spring, a large-scale phytoplankton bloom is triggered. The melting of the ice can also help to stratify the water column so that phytoplankton cells can be sustained in the euphotic zone. In an analysis of thick-billed murres breeding along the west coast of Greenland, Laidre et al. (2008) found that the size of the colony was directly related to the level of primary production (measured as mean density of chlorophyll *a*). The level of primary production was positively correlated with the sea ice levels directly adjacent to the colonies. Therefore, despite ample breeding habitat and the presence of open water early in the season, colonies further south were smaller (Laidre et al., 2008).

Both planktivorous and piscivorous seabirds are often seen feeding along ice edges (Bradstreet, 1979, 1980, 1982; Bradstreet & Cross, 1982; Matley et al., 2012). The ice edge is where the sympagic prey become accessible. Dovekies are known to frequently eat the sympagic amphipod *Apherusa glacialis* when they are available (Karnovsky & Hunt, 2002; Karnovsky et al., 2003, 2010; Boehnke et al., 2014) and Northern fulmars have been found with large numbers of Gammarids *Ischyrocerus anguipes* and *Gammarus setosus* in their stomachs (Mallory et al., 2010b). Predatory crustaceans such as the highly mobile amphipod *Themisto libellula* search for small crustaceans and fish to eat under the ice. This species is a common and important component of many seabird diets (Lønne & Gabrielsen, 1992; Karnovsky et al., 2008). Arctic cod (*Boreogadus saida*) is a staple for many seabirds (Lønne & Gabrielsen, 1992; Mehlum & Gabrielsen, 1993; Welch et al., 1993; Karnovsky et al., 2008, 2009; Mallory et al., 2010b) and is considered a keystone species in that all kinds of upper trophic predators (seabirds, seals, cetaceans) rely on this fish (Figure 23.7; Welch et al., 1992).

Black guillemots hunt by making wing-propelled dives for single items such as a fish or a large crustacean (Figure 23.4). Since 1975, the reproductive success of black guillemots nesting on Cooper Island has been monitored by George Divoky. Cooper Island is located in the Western Beaufort Sea, just East of Point Barrow, Alaska (Figure 23.2). During the breeding season these birds fly to the ice edge to find Arctic cod to bring back to their chicks. In the late 1980s, the ice edge began to retreat. In the following years, the reproductive success of the birds declined as the frequency of Arctic cod in their diets went down as a function of distance to the ice edge (Moline et al., 2008). The retreating ice edge brought other problems for the birds as well. Horned puffins (*Fratercula corniculata*), which are boreal seabirds, moved northwards to usurp the breeding sites of the black guillemots (sometimes even killing the guillemot chicks in the process). Horned puffins are adapted for feeding on fish that are not associated with the ice edge.

Figure 23.7 A black-legged kittiwake (*Rissa tridactyla*) captures the prey 'prize,' a lipid-rich Arctic cod, which is a fish associated with sea ice. Source: photograph by Maria Gavrilo.

Another indirect consequence of retreating ice is that polar bears (*Ursus maritimus*), who would normally be hunting for ringed seals (*Phoca hispida*) on the sea ice, come ashore to forage on seabirds. Polar bears hunting dovekies was first reported on Franz Josef Land in the early 1990s (Stempniewicz, 1993). Now, it is a common foraging habit of bears in colonies throughout the Arctic (M.V. Gavrilo & N.J. Karnovsky, personal observations). Polar bears are even capable of hunting cliff-breeding seabirds such as thick-billed murres, and ground nesting sea ducks, gulls and moulting geese are easy prey (Drent & Prop, 2008, Smith et al., 2010; Stempniewicz, 2006). Iverson et al. (2014) found that polar bear raids into common eider and thick-billed murre colonies have increased more than sevenfold since the 1980s. They also found an inverse correlation between ice season length and bear presence. In years of record low ice coverage (2010–2012), 34% of the eider colonies were visited and tested by polar bears, causing large-scale losses of eggs which exceeded the average number lost to the more customary nest predators (foxes and gulls).

Figure 23.8 Dovekies (*Alle alle*) flying along an ice edge during a foraging trip. Source: photograph by Maria Gavrilo.

Dovekies are abundant Alcids of the North Atlantic sector of the Arctic (Boertmann & Mosbech, 1998). They are so abundant that they may be considered the most numerous bird in the world (Stempniewicz, 2001). These seabirds 'fly' underwater in pursuit of crustaceans (primarily copepods such as *Calanus glacialis* and *Calanus hyperboreus*), amphipods and fish (Karnovsky et al., 2003, 2008, 2010). During chick-rearing they collect thousands of copepods during each foraging trip (Karnovsky et al., 2003). Jakubas et al. (2013) attached miniature global positioning system (GPS) loggers to dovekies nesting at two different colonies on the Island of Spitsbergen in Svalbard, Norway. They found that the chicks in the colony closest to sea ice grew faster and fledged earlier than those in the colony further from the ice edge (Figure 23.8). Karnovsky et al. (2010) found that dovekies foraging off the east coast of Greenland in the Greenland Sea (Figure 23.2) fed on large, energy-rich *Calanus* copepods associated with the East Greenland Current which flows south from the ice-covered Arctic Ocean. These birds also fed their chicks significantly more epontic zooplankton than their counterparts breeding on the west coast of Spitsbergen, probably because the marginal ice zone was close to their colony (Karnovsky et al., 2010). Karnovsky et al. (2011) compared the diving behavior of dovekies breeding in East Greenland with that of birds breeding on West Spitsbergen by attaching small time-depth recorders to adult birds. They found that the Spitsbergen dovekies that were further from sea ice worked harder to find food; they spent more time at sea, made more dives and made frequent long, deep dives compared with the birds nesting on Greenland where sea ice is more plentiful (Karnovsky et al., 2011).

With declines in sea ice and the retreat of ice edges away from breeding colonies, some seabirds have switched to find other sources of abundant food. For example, the ice has recently declined around Franz Josef Land (Figure 23.2) in summer, and in the low-ice seasons the dovekies changed their foraging habits from ice-edge foraging during the early 1990s, where they frequently ate the ice-associated amphipod *Apherusa glacialis* (Węsławski et al., 1994), to foraging at the face of a melting glacier on osmotically shocked copepods (Grémillet et al., 2015). Even though foraging trips were shorter than those measured at other dovekie colonies and chicks fed from the glacier fronts had high growth rates, the adult birds showed a 4% decline in body mass compared with when sea ice was present (Grémillet et al., 2015). Likewise, dovekies breeding along the west coast of Spitsbergen (Figure 23.2) far from ice edges continue to be successful at raising their chicks, but these adults spend more time replenishing their own energy reserves (Kongsfjorden colony; Brown et al., 2012) and even fly long distances to forage at an ice edge to recharge before resuming shorter foraging trips to collect food for their chicks (Magdalenefjorden colony; Jakubas et al., 2012). In a similar fashion, Provencher et al. (2012) found that the frequency of energy-rich Arctic cod in the diets of thick-billed murres breeding at lower latitudes in the Canadian Arctic has declined or disappeared entirely over a 20- to 30-year period. The change in diet from pagophilic, large Arctic cod has been accompanied by an increased reliance on boreal species which are smaller (capelin, *Mallotus villosus*) (Provencher et al., 2012). This change has led to increased energy expenditures during foraging and a lower energy rate of delivery to chicks, but has not yet been severe enough to lead to large-scale declines in reproductive success (Smith & Gaston, 2012).

23.4 Sometimes less (sea ice) is more: The importance of polynyas

Polynyas are areas that regularly become free of ice in high latitudes where sea ice persists. These regularly occurring ice-free areas allow seabirds guaranteed access to prey; it is along the shores next to polynyas that most of the large seabird colonies exist at high latitudes

(Brown & Nettleship, 1981; Stirling, 1997; Gilchrist & Robertson, 2000; Mallory & Gilchrist, 2005; Gavrilo & Popov, 2011; Gavrilo et al., 2011). For example, an estimated 60 million dovekies migrate to the North Water polynya located in northern Baffin Bay to breed along its shores (Figure 23.1; Karnovsky & Hunt, 2002). Once the ice opens, sunlight induces the food web to develop by triggering the onset of a phytoplankton bloom.

Seabirds that utilize regularly opening polynyas have evolved the life-history trait to time their migration so that they arrive when the food web is developed enough for them to feed. Karnovsky and Hunt (2002) found that dovekies arrived *en masse* at the North Water polynyas approximately 1 month after the polynya opened. Their arrival coincided with the vertical migration of their main prey at that time of year, the copepod *Calanus hyperboreus*. These copepods rise from deeper depths into the upper water column to feed on diatoms in the month of May. The dovekies not only time their arrival to when the copepods arrive but also land precisely where the copepods are found (Karnovsky & Hunt, 2002). The Greenland (east) side of the polynya experiences early stratification of the water column, which leads to high levels of primary production in the surface waters and high concentrations of grazers such as *C. hyperboreus*. It is there that the dovekies focus their feeding when they arrive (Karnovsky & Hunt, 2002, Karnovsky et al., 2007). An indication that the timing of the migration of dovekies to the North Water polynya is consistent from year to year is that the word for dovekie in Greenlandic is *Appaliarsuk*, which also refers to the month of May in the region.

In springtime, the system of flaw polynyas and shore leads serve as an ice-free water highway for migrating seabirds and sea ducks returning to their breeding grounds. The flyway along the coast of the East Siberian Sea (Figure 23.2) is used by hundreds of thousands of eiders, long-tailed ducks (*Clangula hyemalis*) and Alcids (Portenko, 1972; Dau & Kistchinski, 1977; Solovieva, 1999; Gavrilo & Popov, 2011). Sea ducks not only use polynyas as a pathway, but also stage there prior to dispersing to their tundra nesting grounds which remain snow-covered for a longer time (Solovieva, 1999).

23.5 Timing is everything

Given that seabirds have migratory patterns that have evolved to coincide with particular times of year when prey is available (McLaren, 1982), altering the timing of opening of polynyas, or break-up of ice in general, can lead to a mismatch between the development of the food web and the arrival of the seabirds (Visser et al., 1998; Stenseth & Mysterud, 2002; Michel et al., 2006). Gaston et al. (2009) found this type of mismatch to be occurring with thick-billed murres that breed on Coats Island and feed in the seasonally ice-free waters of Hudson Bay (Figure 23.2). Due to warmer temperatures, the timing of ice break-up in Hudson Bay is occurring earlier; from 1998 to 2007 the time when 50% of the ice has cleared has advanced by 17 days. The timing of hatching has not kept pace with the earlier ice break-up dates, and hatchlings have poor growth rates when they hatch after the peak in prey has passed (Gaston et al., 2009).

Loss of sea ice can benefit seabirds that nest in colonies that were once connected to the mainland by sea ice. Sea ice melt causes the islands to become isolated from mammalian predators that cannot make the journey across open water (Figure 23.9). In Svalbard, Mehlum (2012) found that in seasons with late ice break-up, fewer common eiders occupy their breeding islands than in seasons with an early ice break-up. Late sea ice break-up forces

Figure 23.9 An Arctic fox (*Vulpes lagopus*) with a dovekie it has captured. The fox is in its summer coat; in winter they are white. Arctic foxes are known to cache large numbers of marine birds caught in breeding colonies to eat during winter. Source: photograph by Witek Kaszkin.

the birds to nest in higher densities on a smaller number of islands that are less accessible to mammalian predators (Parker & Mehlum, 1991).

23.6 Sea ice in all seasons

Recurring polynyas are especially critical to seabirds and sea ducks that overwinter in the Arctic (Gilchrist & Robertson, 2000). A large population (approximately 100 000) of common eiders remain in Hudson Bay throughout the winter (Gilchrist et al., 2006). The wintering common eiders utilize the polynyas, leads in ice and ice edges to access mussel beds. They roost together in the larger polynyas near the Belcher islands throughout the night and use the edges of the polynyas to sit while digesting their prey. In winters of heavy ice, the common eiders are forced to compete for prey that they can access in the few polynyas that remain (Gilchrist et al., 2006). Prey rapidly becomes depleted and the common eiders may be forced to eat suboptimal prey. Subsequently large-scale die-offs of wintering adults have been documented during winters with heavy ice (as summarized in Mallory et al., 2010a). Due to regularly reoccurring polynyas in the White Sea, Russia (Figure 23.2), an endemic population of some 50 000 common eiders are year-round residents; they spend the winter among the sea ice in Onega and Dvina bays (Krasnov et al., 2010).

In a similar fashion, spectacled eiders (*Somateria fischeri*) spend their winters in the Bering Sea and access their preferred prey (*Nuculana radiata*) which are large energy-rich clams, by diving through leads and polynyas that form south of St Lawrence Island (Figure 23.10). When winds shifted during the winter of 2009, pack ice became consolidated over their foraging habitat and they had a harder time accessing profitable prey. The spectacled eiders were forced to feed on suboptimal prey (a smaller bivalve), which resulted in a significant decline in body fat. Body fat is critical to their overwinter survival and reproductive success the following year (Lovvorn et al., 2014).

Even though they are not tied to foraging around a nest site (they are no longer central place foragers) in winter, ice-associated marine birds are wholly dependent on winter sea ice forming in particular places with particular characteristics. Too much sea ice can block access to food, and losses of sea ice negatively impact wintering seabirds. Too much open water deprives sea ducks of critical resting places for digesting food and subjects them to high winds and waves.

Figure 23.10 Spectacled eiders (*Somateria fischeri*) gather in a small polynya south of St Lawrence Island in the Bering Sea (Figure 23.2). The polynya is critical for providing access to benthic prey for these rare marine ducks. Spectacled eiders rest on the ice while digesting. The activity of the birds helps to keep the water open but when even a thin layer of ice forms over the water, the birds can be prevented from feeding. Source: photograph by James Lovvorn.

Some black guillemots spend their entire winter in the Arctic. They utilize the small leads and polynyas created in the ice by tides to forage in the intertidal zone throughout the winter (Butyev, 1959; Renaud & Bradstreet, 1980). Not much is known about these birds in winter because very few studies on seabirds are carried out during the Arctic night of winter.

The Ross's gull is a small Arctic endemic gull which sometimes assumes a coral-pink color. In Russian, it is often referred to as a Firebird of the Arctic; many legends and romantic tales have been evoked around this little known seabird. It was encountered by Sir James Clark Ross during the expedition led by William Parry in the Canadian Arctic in 1823. More than 80 years later, in 1905, the first nests were found by Sergey Buturlin, in Yakutian swampy tundra in the lower Kolyma river reaches (Potapov, 1990). While breeding, Ross's gull is a typical tundra bird, feeding on small invertebrates and fish in freshwater ponds. As soon as breeding is completed, Ross's gulls, in contrast to most Arctic seabirds, perform a long northward migration to the ice of the Arctic Ocean, all the way to the North Pole (Gilg et al., 2015). Ross's gulls were found to be the most common seabird in pack ice in the Atlantic–Eurasian sector of the Arctic Basin (Hjort et al., 1997; Meltofte et al., 1981). Ross's gulls roam ice-covered waters during almost 10 months a year in search of sympagic crustaceans which give them their beautiful rosy color.

For Arctic seabirds that migrate south for winter, sea ice on their wintering grounds can provide important foraging opportunities. In a 20-year analysis of thick-billed murres breeding at Coates Island in Hudson Bay, Smith and Gaston (2012) found that adult survival and colony attendance were positively related to late winter and early spring ice conditions on their wintering grounds much further south. The more ice in the wintering grounds, the better. By tracking dovekies from their breeding colonies using miniaturized geolocators, Fort et al. (2013) found that they winter along ice edges of marginal ice zones in their southerly wintering grounds and may use the ice to rest on during winter moults.

An important function of sea ice is its role in creating a dampening effect on wind and waves. With losses of sea ice, storms with high wind speeds, wave heights and precipitation levels increase. Flying seabirds are susceptible to being blown off course and large-scale wrecks of seabirds often occur as a result of these storms (e.g. as described in relation to dovekies in Montevecchi & Stenhouse, 2002).

23.7 The sea ice-loving (pagophilic) seabird: the ivory gull

Of all the seabirds, the one considered most reliant on sea ice is the ivory gull (Figure 23.11). This seabird species is the epitome of a seabird whose appearance and life-history traits are closely associated with sea ice. The ivory gull species name in many different languages emphasizes their close relationship to sea ice: *Ismåke* (Norwegian), *Ismås* (Swedish), *Ismåge* (Danish) and *Eis Möve* (German) all mean the ice gull, while its scientific name (*Pagophila eburnea*) is translated as 'the one of ivory colour that loves ice'. It is the only bird species in the Northern Hemisphere found to breed occasionally on sea ice (like the Emperor penguin *Aptenodytes forsteri* in the Southern Hemisphere.) (Boertmann et al., 2010) or drifting icebergs (Nachtsheim et al., 2015).

Ivory gulls spend their entire life cycle within ice-filled Arctic waters. Life in this cold marine environment

Figure 23.11 Loss of sea ice has caused a decline in the numbers of the ice-obligate ivory gulls (*Pagophila eburnea*). Source: photograph by Maria Gavrilo.

has shaped the appearance of ivory gulls; they have relatively short legs (reduced surface area to decrease heat loss), mostly covered in feathers, with reduced webbing between their toes (Figure 23.11). These adaptations however, reduce their swimming abilities. Ivory gulls forage by hovering above ice floes, leads, polynyas or along the glacier fronts, as well as walking along the edge of ice floes or beaches, or wading in melt ponds where they pick up food items from water, ground or the ice surface. They avoid contact with the water when air temperatures are close to, or below, freezing; the ivory gull is peculiar among gulls in its avoidance of swimming and sitting in water. Their principal prey are sympagic fish and invertebrates, mostly Arctic cod and crustaceans (Divoky, 1976; Yudin & Firsova, 2002; Mallory et al., 2008; Karnovsky et al., 2009), but carrion and offal (mainly from seal carcasses) are also important diet components (Orr & Parsons, 1982; Renaud & McLaren, 1982; Mallory et al., 2008).

Because ivory gulls forage primarily in ice-covered waters, during both the breeding and non-breeding season they need sea ice. The ivory gulls have a semi-circumpolar breeding distribution in the Atlantic sector of the Arctic Ocean and nest on the remote islands of the Canadian Arctic Archipelago, Greenland, Svalbard, Franz Josef Land, Severnaya Zemlya archipelago and the Kara Sea offshore islands (Figure 23.2; Gilchrist et al., 2008). Spatial analysis revealed that the southern limit of the ivory gull nesting grounds is governed by the summer ice regime, the long-term mean position of ice edge in late August (Gavrilo, 2011). In other words, ivory gulls breed in areas where their ice-feeding habitats are likely guaranteed until chicks are fledged (this is the opposite of those required for thick-billed murres). During non-breeding periods of their life cycle, ivory gulls roam ice-covered Arctic oceans and have a circumpolar range. Using satellite telemetry, Gilg et al. (2010) found that after leaving their breeding colonies they move along the ice edges and then spend their winters at the northern ice edge of either the North Atlantic or North Pacific. Spencer et al. (2014), also using satellite telemetry, recently demonstrated a strong link between ivory gulls and ice habitats in all seasons, which can be attributed to their affinity for sea ice and their avoidance of open water.

Because ivory gulls are so ice-dependent, they are very sensitive to changes in ice conditions. Dalgety (1932) noticed that in the summer of 1930 there was very little ice around the island of Spitsbergen in the Svalbard archipelago (Figure 23.2) and ivory gulls had smaller clutch sizes (number of eggs in their nests) than in the following summer when average sea ice levels returned, a pattern that has been reported in other years and areas of the Arctic (Tomkovich, 1986; Bangjord et al., 1994; Volkov & de Korte, 2000). Long-term declines in summer sea ice observed during the past century in the northern Barents Sea (Figure 23.2; Zubakin et al., 2006) have affected ivory gull populations at the southern periphery of their nesting range, where the largest colonies have disappeared since the middle of the last century (Gavrilo, 2011). Existing colonies in Svalbard are small (<50 breeding pairs) (Gavrilo et al., 2007). In the past, these colonies had hundreds of pairs (Bakken & Tertitsky, 2000).

In 1897 the Jackson expedition discovered the largest ivory gull colony ever recorded of a couple of thousand pairs on Cape Mary Harmsworth in Franz Josef Land; currently no birds breed there (Gavrilo, 2009). In the Canadian Arctic, the ivory gull population has declined >80% over 30 years of surveys since the early 1980s (Gilchrist & Mallory, 2005; Gaston et al., 2012). Most of the remaining Canadian ivory gull colonies have retreated northwards, suggesting that the remaining birds are concentrated in areas of prolonged summer ice cover and colony size is significantly smaller than prior to 1991(Robertson et al., 2007).

However, even for the ice-loving ivory gull, there must be open water for them to access food. For example, the size of Western Severnaya Zemlya polynya just before the nesting season has a threshold relationship with the number of ivory gulls attempting to breed in the nearest colony on Domashny Island (Figure 23.2). A shrinkage in the area of open water below 10 000 km^2 in late May, when ivory gulls forage in the polynya to build up resources for further breeding, results in an abrupt drop in the number of nesting birds in following summer season (Gavrilo, 2011). The example of the ivory gull reveals some of the primary ways that sea ice is important for Arctic marine birds:

- providing energy rich prey associated with sea ice during the breeding season;
- providing foraging opportunities during the non-breeding season.

A decline in sea ice is the most probable cause of the endangered status of ivory gulls in Canada (Species at Risk Public Registry, 2011) and its threatened status in other places (Boertmann, 2008; Kålås et al., 2010) and globally (BirdLife International, 2012). Along with the polar bear (*Ursus maritimus*) the ivory gull is a top candidate for future extinction due to climate change (ACIA, 2005).

23.8 Conservation implications

It is important to note that many seabirds that are experiencing changes in their foraging habitat and prey due to loss of sea ice are exhibiting remarkable resilience (Harding et al., 2009; Karnovsky et al., 2010; Grémillet et al., 2012, 2015). The impacts can, however, have more subtle consequences, such as increased foraging effort (Karnovsky et al., 2011) and lower body condition (Grémillet et al., 2015). Sea ice-dependent birds may be able to take advantage of alternative prey or foraging locations in the short term; the tipping point will be different for different species. For example, in their analysis of seabirds' use of glaciers in the Svalbard archipelago, Lydersen et al. (2014) warn that further warming will result in fewer tidewater glaciers and lower calving fronts, and therefore seabird reliance on melting glaciers for foraging opportunities may be a temporary opportunity. There may be a point beyond which seabirds associated with sea ice cannot buffer the impacts of sea ice loss and they may show population declines or even extinctions.

Furthermore, changes in patterns of sea ice development and retreat are just one factor influencing the lives of Arctic seabirds. A combination of stressors could diminish their resilience. For example, the ivory gull is a top predator; it feeds at a high trophic level. Due to biomagnification of various toxins up through the Arctic food chain, ivory gulls carry a high load of toxins (Braune et al., 2006; Karnovsky et al., 2009; Miljeteig et al., 2009; Lucia et al., 2015). There may be multiple factors causing declines in the population of this rare Arctic seabird; the cumulative effect of carrying a high contaminant load and coping with losses of sea ice (Miljeteig et al., 2012).

Concomitant with losses of sea ice in the Arctic are increases in shipping, tourism and resource extraction, including oil, gas and fish. Seabirds are vulnerable to disturbances at their colonies by helicopter flights and boat traffic. Of particular concern is the fact that icebreakers and cargo vessels tend to use flaw polynyas – the areas of critical importance to seabirds in spring and early summer, i.e. during their pre-breeding and breeding seasons. Oil spills can be devastating, as only a small patch of oil can cause losses of waterproofing and lead quickly to hypothermia. Areas where seabirds and sea ducks gather to breed, feed or migrate need to be protected. Gaston et al. (2013) predicted the foraging area of thick-billed murres in the Canadian Arctic based on the assumption of central place foraging and direct measures of foraging range from murres instrumented with GPS loggers. The foraging area of the thick-billed murres had extensive overlap with documented shipping tracks and areas with oil and gas deposits that are likely to be developed. Of particular concern is Hudson Strait which has the highest level of shipping traffic and is where the largest thick-billed murre colony in eastern Canada is located.

Another critical area of concern is the Chukchi nearshore corridor where both the threatened (under the U.S. Endangered Species Act) spectacled eider and Steller's eider (*Polysticta stelleri*) migrate in spring (Lovvorn et al., 2015). This area is experiencing losses in sea ice and is also a proposed area for oil and gas development, including the placement of submerged pipes to transport oil that are likely to be breached by icebergs (Lovvorn et al., 2015).

There are large portions of the lives of Arctic seabirds and sea ducks that we know very little about. Much of the Arctic is still extremely undersampled. Greater efforts are needed to place seabird observers on oceanographic vessels and other ships of opportunity. Funding for studies that use miniaturized tracking technology needs to be increased. There also needs to be a greater focus on determining wintering locations and elucidating migration routes and winter foraging behaviour. Coordinated citizen science projects (such as those described in Gilchrist et al. (2006), which were conducted by local communities and students to study common eiders), need to be supported.

Some shipboard surveys (Durinck & Falk, 1996; Mckinnon et al., 2009; Wong et al., 2014) and results from tracking devices (Gilg et al., 2010; Fort et al., 2013; Spencer et al., 2014) have revealed particular 'hotspots' of seabird activity in winter. These areas should be given special conservation considerations, as large numbers of the populations are congregated in these small areas. For species with circumpolar distributions, it will be critical to forge coordinated conservation efforts amongst Arctic countries.

23.9 Acknowledgements

We would like to thank Pomona College undergraduate students M. Starr, A. Sartorius and B. Rodriguez for their help with the map, and M. Starr and E. Nassirinia for help with proofreading a draft of this paper.

References

ACIA. (2005) *Arctic Climate Impact Assessment*. Cambridge University Press, 1042p

Bakken, V. & Tertitsky, G.M. (2000) Ivory gull *Pagophila eburnea*. In: *The Status of Marine Birds Breeding in the Barents Sea Region, Norsk Polarinstitutt Rapportserie Nr. 1134* (Eds.T. Anker-Nilssen et al.), pp. 104–107. Norsk Polarinstittut, Tromso, Norway.

Bangjord, G., Korshavn, R. & Nikiforov, V. (1994) Fauna at Troynoy and influence of polar stations on nature reserve. Izvestija TsIK, Kara Sea, July 1994. *Working report. Norwegian Ornithological Society, Klaebu*, **3**, 1–55.

BirdLife International (2012). *Pagophila eburnea. The IUCN Red List of Threatened Species*. Version 2014.3. www.iucnredlist.org (Accessed July 2016).

Boehnke, R., Gluchowska M., Wojczulanis-Jakubas, K. et al. (2014) Supplementary diet components of little auk chicks in two contrasting regions on the West Spitsbergen coast. *Polar Biology*, **38**, 261–267.

Boertmann, D. (2008) *Grønlands Rødliste 2007* (The 2007 Greenland Red List). Greenland Home Rule and National Environmental Research Institute. http://www2.dmu.dk/Pub/Groenlands_Roedliste_2007_DK.pdf (Accessed July 2016).

Boertmann, D. & Mosbech, A. (1998) Distribution of little auk (*Alle alle*) breeding colonies in Thule District, northwest Greenland. *Polar Biology*, **19**, 206–210.

Boertmann D., Olsen K. & Gilg O. (2010) Ivory gulls breeding on ice. *Polar Record*, **46**, 86–88.

Bradstreet, M.S.W. (1979) Thick-billed murres and black guillemots in the Barrow Strait area, N.W.T., during spring: Distribution and habitat use. *Canadian Journal of Zoology*, **57**, 1789–1802.

Bradstreet, M.S.W. (1980) Thick-billed murres and black guillemots in the Barrow Strait area, N.W.T., during spring: Diets and food availability along ice edges. *Canadian Journal of Zoology*, **58**, 2120–2140.

Bradstreet, M.S.W. & Cross, W.E. (1982) Trophic relationships at high Arctic ice edges. *Arctic*, **35**, 1–12.

Bradstreet, M.S.W. (1982) Occurrence, habitat use, and behavior of seabirds, marine mammals, and Arctic cod at the Pond Inlet Ice Edge. *Arctic*, **35**, 28–40.

Braune, B.M., Mallory, M.L. & Gilchrist, H.G. (2006) Elevated mercury levels in a declining population of ivory gulls in the Canadian Arctic Baseline, *Marine Pollution Bulletin*. **52**, 969–987.

Brown, R.G.B. & Nettleship, D.N. (1981) The biological significance of polynyas to arctic colonial seabirds. In: *Polynyas in the Canadian Arctic* (Eds. I. Striling & H. Cleator). *Canadian Wildlife Service Occasional Paper*, **45**, 59–65.

Brown, Z.W., Welcker, J., Harding, A.M.A., Walkusz, W. & Karnovsky, N.J. (2012) Divergent diving behavior during short and long trips of a bimodal forager, the little auk (*Alle alle*). *Journal of Avian Biology*, **43**, 1–12.

Butyev, V.G. (1959) The wintering of birds at the north of Novaya Zemlya. Ornitologiya. Moscow State University Press, Moscow. **2** 99–101. (In Russian.)

Cairns, D.K., Gaston, A.J. & Heutemann, F. (2008). Endothermy, ectothermy and the global structure of marine vertebrate communities. *Marine Ecology Progress Series*, **356**, 239–250

Dalgety C.T. (1932) The ivory gull Spitsbergen. *British Birds*. **26**, 2–8.

Dau, C.P. & Kistchinski, A.A. (1977) Seasonal movements and distribution of the Spectacled Eider. *Wildfowl*, **28**, 65–75.

Divoky, G.J. (1976) Pelagic feeding-habits of ivory and Ross gulls. *Condor*, **78**, 85–90.

Drent, R.H. & Prop, J. (2008). Barnacle goose *Branta leucopsis* surveys on Nordenskiöldskysten, west Spitsbergen 1975–2007; breeding in relation to carrying capacity and predator impact. *Circumpolar Studies*, **4**, 59–83.

Durinck, J. & Falk, K. (1996) The distribution and abundance of seabirds off southwestern Greenland in autumn and winter 1988–1989. *Polar Research*, **15**, 23–42.

Eamer, J., Donaldson, G.M., Gaston, A.J. et al. (2013) *Life Linked to Ice: A Guide to Sea-Ice-Associated Biodiversity In This Time Of Rapid Change*. CAFF Assessment Series No. 10. Conservation of Arctic Flora and Fauna, Iceland.

Fort, J., Moe, B., Strøm, H., Grémillet, D. et al. (2013) Multicolony tracking reveals potential threats to little auks wintering in the North Atlantic from marine pollution and shrinking sea ice cover. *Diversity and Distributions*, **19**, 1322–1332.

Gabrielsen, G.W. & Mehlum, F. (1988) Thermoregulation and energetics of Arctic seabirds. In: *Physiology of Cold Adaptation in Birds* (Eds. C. Bech & R.E. Reinertsen, R.E.), pp. 137–152. NATO ASI Series. Series A, Life Sciences, Vol. 173. Plenum Press, New York.

Ganter, B. & Gaston, A.J. (2013) Birds. In: *Arctic Biodiversity Assessment 2013*. CAFF, pp. 142–180

Gaston, A.J., Elliott, K.H., Ropert-Coudert, Y. et al. (2013) Modeling foraging range for breeding colonies for thick-billed murres *Uria lomvia* in the Eastern Canadian Arctic and potential overlap with industrial development. *Biological Conservation*, **168**, 134–143.

Gaston, A.J., Gilchrist, H.G., Mallory, M.L. & Smith, P.A. (2009) Changes in seasonal events, peak food availability, and consequent breeding adjustment in a marine bird: a case of progressive mismatching. *Condor*, **111**, 111–119.

Gaston, A.J., Mallory, M.L. & Gilchrist, H.G. (2012) Populations and trends of Canadian Arctic seabirds. *Polar Biology*, **35**, 1221–1232.

Gavrilo, M.V. (2009) Breeding distribution of ivory gull in the Russian Arctic: difficulty when studying range of a rare and sporadically breeding high arctic species. *Problemy Arktiki and Antarctiki*, **3**, 127–151. [In Russian]

Gavrilo, M.V. (2011) Ivory gull *Pagophila eburnea* (Phipps, 1774) in the Russian Arctic: Breeding patterns of species within the current species range optimum. *PhD Thesis*. Saint Petersburg, Russia. [In Russian].

Gavrilo, M.V. & Popov A.V. (2011) *Sea Ice Biotopes and Biodiversity Hotspots of the Kara Sea and North-Eastern Barents Sea. Atlas Of Marine and Coastal Biological Diversities of the Russian Arctic Seas*. WWF, Moscow, pp. 34–35.

Gavrilo, M.V., Popov A.V. & Spiridonov, V.A. (2011) *Sea Ice Biotopes and Biodiversity Hotspots in the Laptev Sea. Atlas of Marine and Coastal Biological Diversities Of The Russian Arctic Seas*. WWF, Moscow, pp. 36–37.

Gavrilo, M.V., Strom, H. & Volkov, A.E. (2007) Population status of Ivory Gull populations in Svalbard and Western Russian Arctic: first results of joint Russian-Norwegian research project. Complex investigations of Spitsbergen nature. Kola Scientific Center RAS Publishers, Apatity, **7**, 220–234 [In Russian]

Gilchrist, H.G. & Robertson, G.J. (2000) Observations of marine birds and mammals wintering at polynyas and ice edges in the Belcher Islands, Nunavut, Canada. *Arctic*, **53**, 61–68.

Gilchrist, H.G. & Mallory, M.L. (2005) Declines in abundance and distribution of the ivory gull (*Pagophila eburnea*) in Arctic Canada. *Biological Conservation*, **121**, 303–309.

Gilchrist, H.G., Heath, J., Arragutainaq, L. et al. (2006) Combining science and local knowledge to study common eider ducks wintering in Hudson Bay. In: *Climate Change: Linking*

Traditional and Scientific Knowledge (Eds. R. Riewe & J. Oakes), pp. 284–303. Aboriginal Issues Press, Winnipeg, Manitoba.

Gilchrist, G., Strom, H., Gavrilo, M. & Mosbech, A. (2008) International Ivory Gull conservation strategy and action plan. CAFFs Circumpolar Seabird Group. *CAFF Technical report*. **18**: 1–20.

Gilg, O., Strøm, H., Aebischer, A., Gavrilo, M.V., Volkov, A.E., Miljeteig, C. & Sabard, B. (2010) Post-breeding movements of northeast Atlantic Ivory gull *Pagophila eburnea* populations. *Journal of Avian Biology*, **41**, 532–542.

Gilg, O., Kovacs, K.M., Aars, J. et al. (2012) Climate change and the ecology and evolution of Arctic vertebrates. *Annals of the New York Academy of Sciences*, **1249**, 166–190.

Gilg, O., Alexandreev, A., Aebischer, A, Kondratyev, A. Sokolov, A., & Dixon, A. (2015) Satellite tracking of Ross's Gull *Rhodostethia rosea* in the Arctic Ocean. *Journal of Ornithology*, 10.1007/s10336

Grémillet, D., Fort, J., Amélineau, F. et al. (2015) Arctic warming: nonlinear impacts of sea-ice and glacier melt on seabird foraging. *Global Change Biology*, **21**, 1116–1123

Grémillet, D., Welcker, J., Karnovsky, N.J. et al. (2012) Little auks buffer the impact of current Arctic climate change. *Marine Ecology Progress Series*, **454**, 197–206

Harding, A.M.A., Kitaysky, A.S., Hall, M.E. et al. (2009) Flexibility in the parental effort of an Arctic-breeding seabird. *Functional Ecology*, **23**, 348–358.

Hjort, C., Gudmundsson, G.A., and Elander M. (1997) Ross's gulls in the Central Arctic Ocean. *Arctic*, **50**, 289 – 292

Hunt, G.L., Jr., (1991) Marine birds and ice-influenced environments of polar oceans. *Journal of Marine Systems*, **2**, 233–240.

Hunt, G.L., Jr.,, Bakken, V. & Mehlum, F. (1996) Marine birds in the marginal ice zone of the Barents Sea in later winter and spring. *Arctic*, **49**, 53–61.

Iverson, S.A., Gilchrist, H.G., Smith, P.A., Gaston A.J., Forbes, M.R. (2014) Longer ice-free seasons increase the risk of nest depredation by polar bears for colonial breeding birds in the Canadian Arctic. *Proceedings of the Royal Society Series B*, **281**, 2013–3128.

Jakubas, D., Iliszko, L., Wojczulanis-Jakubas, K. & Stempniewicz, L. (2012) Foraging by little auks in the distant marginal sea ice zone during the chick-rearing period. *Polar Biology*, **35**, 73–81.

Jakubas, D., Trudnowska, E., Wojczulanis-Jakubas, K. et al. (2013) Foraging closer to the colony leads to faster growth in little auks. *Marine Ecology Progress Series*, **489**, 263–278.

Kålås, J., Gjershaug, J.O., Husby, M. et al. (Eds.) (2010) *The 2010 Norwegian Red List for Species*. Norwegian Biodiversity Information Centre, Norway

Karnovsky, N.J. & Hunt, G.L., Jr. (2002) Carbon flux through dovekies (*Alle alle*) in the North Water. *Deep Sea Research Part II*, **49**, 5117–5130.

Karnovsky, N.J., Kwaśniewski, S., Węsławski, J.M., Walkusz, W. & Beszczyńska-Möller. A. (2003) The foraging behavior of little auks in a heterogeneous environment. *Marine Ecology Progress Series*, **253**, 289–303.

Karnovsky N.J., Ainley, D. & Lee, P. (2007) The impact and importance of production in polynyas to top-trophic predators: three case histories. In: *Polynyas: Windows to the World's Oceans*, Elsevier Oceanography Series 74 (Eds.W.O. Smith Jr. & D.G. Barber), pp. 391–403. Elsevier, the Netherlands.

Karnovsky, N.J., Hobson, K.A., Iverson, S. & Hunt, G.L., Jr. (2008) Seasonal changes in diets of seabirds in the North Water polynya: a multi-indicator approach. *Marine Ecology Progress Series*, **357**, 291–299.

Karnovsky, N.J., Hobson, K.A., Brown, Z.W. & Hunt, G.L., Jr., (2009) Distribution and diet of ivory gulls (*Pagophila eburnea*) in the North Water Polynya. *Arctic*, **62**, 65–74.

Karnovsky, N.J., Harding, A., Walkusz, W. et al. (2010) Foraging distributions of little auks across the Greenland Sea: Implications of present and future Arctic climate change. *Marine Ecology Progress Series*, **415**, 283–293.

Karnovsky, N.J., Welcker, J., Harding, A.M.A. et al. (2011) Inter-colony comparison of diving behavior of an Arctic top predator: implications for warming in the Greenland Sea. *Marine Ecology Progress Series* **440**, 229–240

Krasnov, Y.V., Gavrilo, M.V., Shavykin, A.A. & Vashchenko, P.S. (2010) Sex and age structure of endemic White Sea population of common eider *Somateria mollissima*. *Reports of Russian Academy of Sciences*, **435**, 568 – 570. [In Russian].

Laidre, K.L., Heide-Jorgensen, M.P., Nyeland, J., Mosbech, A. & Boertmann, D. (2008) Latitudinal gradients in sea ice and primary production determine Arctic seabird colony size in Greenland. *Proceedings of the Royal Society B: Biological Sciences*, **275**, 2695–2702.

Lønne, O.J. & Gabrielsen, G.W. (1992) Summer diet of seabirds feeding in sea-ice-covered waters near Svalbard. *Polar Biology*, **12**, 685–692.

Lovvorn, J.R., Anderson, E.M., Rocha, A.R. et al. (2014) Variable wind, pack ice, and prey dispersion affect the long-term adequacy of protected areas for an Arctic sea duck. *Ecological Applications* **24**, 396–412.

Lovvorn, J.R., Rocha, A.R., Jewett, S.C., Dasher, D., Oppell, S. & Powell, A.N. (2015) Limits to viability of a critical Arctic migration corridor due to localized prey, changing sea ice, and impending industrial development. *Progress in Oceanography* **136**, 162–174

Lucia, M., Verboven, N., Strøm, H. et al. (2015) Circumpolar contamination in eggs of the high-Arctic ivory gull *Pagophila eburnean*. *Environmental Toxicology and Chemistry*, **34**, 1552–1561.

Lydersen, C., Assmey, P., Falk-Petersen, S. et al. (2014) The importance of tidewater glaciers for marine mammals and seabirds in Svalbard, Norway. *Journal of Marine Systems*, **129**, 452–471.

Mallory, M.L., Stenhouse, I.J., Gilchrist, G.J., Robertson, G.J., Haney, J.C. & MacDonald, S.D. (2008) Ivory Gull (*Pagophila eburnea*). In: *The Birds of North America Online* (Ed. A. Poole). Cornell Laboratory of Ornithology, Ithaca, New York. Available from http://bna.birds.cornell.edu/BNA/species/175/. (Accessed July 2016).

Mallory, M.L., Gaston, A.J., Gilchrist, H.G., Robertson, G.J. & Braune, B.M. (2010a) Effects of climate change, altered sea-ice distribution and seasonal phenology on marine birds. In: *A Little Less Arctic: Top Predators in the World's Largest Northern Inland Sea, Hudson Bay* (Eds. S.H. Ferguson, L.L. Loseto & M.L. Mallory), pp. 179–195. Springer, the Netherlands.

Mallory, M.L. & Gilchrist, H.G. (2005) Marine birds of the Hell Gate Polynya, Nunavut, Canada. *Polar Research*, **24**, 87–93.

Mallory, M.L., Karnovsky, N.J., Gaston, A.J. et al. (2010b) Temporal and spatial patterns in the diet of northern fulmars *Fulmarus glacialis* in the Canadian High Arctic. *Aquatic Biology*, **10**, 181-191.

Matley, J., Fisk, A. & Dick, T.A. (2012) Seabird predation on Arctic cod during summer in the Canadian Arctic. *Marine Ecology Progress Series*, **450**, 219– 228.

Mckinnon, L., Gilchrist, H.G. & Fifield, D. (2009) A pelagic seabird survey of Arctic and sub-Arctic Canadian waters during fall. *Marine Ecology Progress Series*, **37**, 77–84.

McLaren, P.L. (1982) Spring migration and habitat use by seabirds in eastern Lancaster Sound and western Baffin Bay. *Arctic*, **35**, 88– 111.

Mehlum, F. (1997) Seabird species associations and affinities to areas covered with sea ice in the northern Greenland and Barents Seas. *Polar Biology*, **18**, 116 – 127.

Mehlum, F. (2012) Effects of sea ice on breeding numbers and clutch size of a high arctic population of the common eider Somateria mollissima. *Polar Science*, **6**, 143–153.

Mehlum, F. & Gabrielsen, G.W. (1993). The diet of high arctic seabirds in coastal and ice-covered, pelagic areas near Svalbard archipelago. *Polar Research*, **11**, 1–20.

Meltofte, H., Edelstam, C., Granstrom, G., Hammar, J., & Hjort, C. (1981) Ross's Gulls in the Arctic pack-ice. *British Birds*, **74**, 316–320

Michel, C., Ingram, R.G. & Harris, L.R. (2006) Variability in oceanographic and ecological processes in the Canadian Arctic Archipelago. *Progress in Oceanography*, **71**, 379–401.

Miljeteig, C., Strom, H., Gavrilo, M., Volkov, A., Jenssen, B.M. & Gabrielsen, G.W. (2009) High Levels of contaminants in Ivory Gull *Pagophila eburnea* eggs from the Russian and Norwegian Arctic. *Environmental Science & Technology*, **43**, 5521–5528

Miljeteig, C., Gabrielsen, G.W., Strom, H., Gavrilo, M., Lie, E., Jenssen, B.M. (2012) Eggshell thinning and decreased concentrations of vitamin E are associated with contaminants in eggs of ivory gulls. *Science of the Total Environment*, **431**, 92–99.

Moline, M.A., Karnovsky, N.J., Brown, Z. et al. (2008) High latitude changes in ice dynamics and their impact on Polar marine ecosystems. In: *The Year in Ecology and Conservation Biology*. Annals of the New York Academy of Sciences (Eds. R.S. Ostfield & W.H. Schlesinger), **1134**, 267–319.

Montevecchi, W.A. & Stenhouse, I.J. (2002) Dovekie (*Alle alle*). In: *The Birds of North America Online*, No. 701 (Eds. A. Poole & F. Gill). The Birds of North America, Ithaca: Cornell Lab of Ornithology; Retrieved from the Birds of North America Online: http://bna.birds.cornell.edu/bna/species/701, doi:10.2173/bna.701

Nachtsheim, D.A., Joiris, C.R. & D'Hert, D. (2015) A gravel-covered iceberg provides an offshore breeding site for ivory gulls *Pagophila eburnea* off Northeast Greenland. *Polar Biology*, 1–4.

Orians, G.H. & Pearson, N.E. (1979) On the theory of central place foraging. In: *Analysis of Ecological Systems* (Eds. D.J. Horn, R.D. Mitchell & G.R. Stairs), pp. 154–177. Ohio State University Press, Columbus, Ohio.

Orr, C.D., Parsons, J.L. (1982) Ivory Gulls *Pagophila eburnea*, and ice edges in David Strait and the Labrador Sea. *Canadian Field Naturalist*, **96**, 323–328.

Parker, H. & Mehlum, F. (1991) Influence of sea-ice on nesting density in the Common Eider *Somateria mollissima* in Svalbard. *Norsk Polarinstitutt Skrifter*, **195**, 31–36.

Portenko, L.A. (1972) *The Birds of the Chukchi Peninsula and Wrangel Island*. Part 1. Nauka Publ., Leningrad: 1–423. (In Russian)

Potapov, E. (1990) Mysterious Ross's gull. *Birds International*, **2**, 72–83.

Provencher, J.F., Gaston, A.J., O'Hara, P.D. & Gilchrist, H.G. (2012) Seabird diet indicates changing Arctic marine communities in eastern Canada. *Marine Ecology Progress Series*, **454**, 171–182.

Renaud, W.E. & Bradstreet, M.S.W. (1980) Late winter distribution of black guillemots (*Cepphus grylle*) in northern Baffin Bay and the Canadian High Arctic. *Canadian Field Naturalist*, **94**, 421–425.

Renaud, W.E. & McLaren, P.L. (1982). Ivory Gull (*Pagophila eburnea*) distribution in late summer and autumn in eastern Lancaster Sound and western Baffin Bay. *Arctic*, **35**, 141–148.

Robertson, G.J., Gilchrist, H.G., & Mallory, M.L. (2007) Colony Dynamics and Persistence of Ivory Gull Breeding in Canada. *Avian Conservation and Ecology*, **2**, 8

Sigler, M.F., Kuletz, K.J., Ressler, P.H., Friday, N.A., Wilson, C.D. & Zerbini, A.N. (2012) Marine predators and persistent prey in the southeast Bering Sea. *Deep Sea Research Part II*, **65–70**, 292–303.

Smith, P.A., Elliott, K.H., Gaston, A.J. & Gilchrist, H.G. (2010). Has early ice clearance increased predation on breeding birds by polar bears? *Polar Biology*, **33**, 1149–1153.

Smith, P.A. & Gaston, A.J. (2012) Environmental variation and the demography and diet of thick-billed murres. *Marine Ecology Progress Series*, **454**, 237–249.

Solovieva, D.V. (1999) *Spring Stopover of Birds on the Laptev Sea Polynya* (Ed. H. Kasssens et al.), pp.189–195. Springer-Verlag, Berlin Heidelberg.

Species at Risk Public Registry. (2011) *Ivory gull*. http://www.registrelep-sararegistry.gc.ca/species/speciesDetails_e.cfm?sid=50

Spencer, N.C., Gilchrist, H.G. & Mallory, M.L. (2014) Annual Movement Patterns of Endangered Ivory Gulls: The Importance of Sea Ice. *PLoSONE* **9**, e115231. doi:10.1371/journal.pone.0115231

Stempniewicz, L. (2001) *Alle alle* little auk. BWP update. *Journal of Birds of the Western Palearctic*, **3**, 175–201.

Stempniewicz, L. (1993) The polar bear *Ursus maritimus* feeding in a seabird colony in Franz Josef Land. *Polar Research*, **12**, 33–36.

Stempniewicz, L. (2006) Polar bear predatory behaviour toward molting barnacle geese and nesting glaucous gulls on Spitsbergen. *Arctic*, **59**, 247–251

Stenseth, N.C. & Mysterud, A. (2002) Climate, changing phenology, and other life history traits: nonlinearity and match–mismatch to the environment. *Proceedings of the National Academy of Sciences, USA*, **99**, 13379–13381.

Stirling, I. (1997) The importance of polynyas, ice edges, and leads to marine mammals and birds. *Journal of Marine Systems*, **10**, 9–21.

Tomkovich, P.S. (1986) Data on biology of Ivory Gull at Graham Bell Island (Franz Josef Land). In: *Aktualnye Problemy Ornitologii (Actual Problems of Ornithology)* (Ed. V.D. Il'ichev), pp. 34 – 49.. Nauka, Moscow. [In Russian]

Uspensky, S.M. (1969) *Life in High Latitudes*. Demonstrated Mainly on Birds. Mysl', Moskva. [In Russian].

Visser, M.E., van Noordwijk, A.J., Tinbergen, J.M. & Lessells, C.M. (1998) Warmer springs lead to mistimed reproduction in Great Tits (*Parus major*). *Proceedings of the Royal Society Series B*, **265**, 1867–1870.

Volkov, A.E. & de Korte, J. (2000) Breeding ecology of the Ivory Gull (*Pagophila eburnea*) in Sedov Archipelago, Severnaya Zemlya. In: *Heritage of the Russian Arctic. Research, Conservation and International Cooperation* (Eds. B.S. Ebbinge et al.), pp. 483–500. Ecopros Publishers, Moscow.

Welch, H.E., Bergmann, M.A., Siferd, T.D., Martin, K.A., Curtis, M.F., Crawford, R.E., Conover, R.J. & Hop, H. (1992) Energy flow through the marine ecosystem of the Lancaster Sound region, Arctic Canada. *Arctic*, **45**, 343–357.

Welch, H.E., Crawford, R.E. & Hop, H. (1993) Occurrence of Arctic cod (*Boreogadus saida*) schools and their vulnerability to predation in the Canadian High Arctic. *Arctic*, **46**, 331–339.

Węsławski, J.M., Stempniewicz, L. & Galaktionov, K. (1994) Summer diet of seabirds from the Frans Josef Land archipelago, Russian Arctic. *Polar Research*, **13**, 173–181.

Wong, S.N.P., Gjerdrum, C., Morgan, K.H. & Mallory, M.L. (2014), Hotspots in cold seas: The composition, distribution, and abundance of marine birds in the North American Arctic, *Journal of Geophysical Research: Oceans*, **119**, 1691–1705.

Yudin, K.A. & Firsova, L.V. (2002) The ivory gull *Pagophila eburnea*. In: *Fauna of Russia and Adjacent Countries: Birds*. Vol. II (Ed.V.A. Paevsky), pp. 132–141. Nauka Publish., Moscow. [In Russian].

Zubakin, G.K., Buzin I.V., & Skutina E.A. (2006). *Seasonal Dynamics and Multi-Year Variability of Sea Ice Cover State of the Barents Sea. Ice formations in Western Arctic Seas*. AARI, Saint-Petersburg, pp. 10–25 [In Russian]

CHAPTER 24

Birds and Antarctic sea ice

David Ainley,[1] Eric J Woehler[2] and Amelie Lescroël[3]

[1] *H.T. Harvey & Associates, Los Gatos, CA, USA*
[2] *Institute for Marine and Antarctic Studies, University of Tasmania, Hobart, Tasmania, Australia*
[3] *Centre d'Écologie Fonctionnelle et Évolutive, Centre National de la Recherche Scientifique, Montpellier, France*

24.1 Introduction

The presence of sea ice has no analogue as a marine bird habitat feature in most of the world ocean outside of the high-latitude polar regions. With respect to the Southern Ocean, three species assemblages have been consistently identified among seabirds that frequent waters south of the Antarctic Polar Front (APF) as detailed from at-sea surveys around the Antarctic (Table 24.1). One assemblage is associated strictly with the presence of sea ice or with waters that within the year are covered by sea ice (ice-obligate species; see further detail later in the chapter), and the second is associated with open water and avoids sea ice (ice-avoiding species). The third assemblage, somewhat a subset of the second, includes species that mostly frequent open water but which can be found during summer within the outer portions of the sea ice field that rings the Antarctic continent (ice-tolerant species). In other words, especially for the latter group, membership can be affected by factors such as breeding phenology and distribution of nesting colonies, in which case, for example, a species may commute over sea ice to travel between a nesting colony and the open water but is not dependent on sea ice in the same sense as ice-obligate species.

These groups have been known for some time (Murphy, 1936; Bierman & Voous, 1950; Watson, 1975), but quantification at the community level is much more recent (Ribic & Ainley, 1988/89; Ainley et al., 1993, 1994; see also Ainley et al., 1984; Eakin et al., 1986; Ryan & Cooper, 1989; Hunt et al., 1990; Veit & Hunt, 1991; Woehler et al., 2003; Ainley et al., 2012). Depending on the timing of research cruises relative to the general summer seabird breeding season, and the proximity of breeding colonies, the species composition of these groups has varied among various authors. Eakin et al. (1986), for example, identified three seabird assemblages during a late autumn (May–June) voyage to the South Atlantic sector of the Southern Ocean, a period when there was no ice in the area visited. They identified an 'Antarctic assemblage' of Antarctic and Snow petrels (see Table 24.1 for Latin names of species), and three species of penguin (chinstrap, gentoo and macaroni); thus a mixture of ice-obligate and ice-tolerant species. A second assemblage identified by them was a 'wide-ranging' group of petrels, albatrosses, storm petrels and shearwaters, all ice-avoiding. The third assemblage was that of 'temperate species,' composed of other petrels, shearwaters and penguins – also ice-avoiding. Eakin et al. (1986) used sea-surface temperature (SST) ranges to separate the three assemblages, rather than characteristics of sea ice. Of course, SST $< -1.0°C$ is associated with sea ice.

Hunt et al. (1990) described five seabird assemblages in the southern Drake Passage and Bransfield Strait, based on observations of only 10 species or taxa. Their vessel also remained clear of the sea ice, just skirting it, and all five seabird assemblages described by them were associated with surface water properties and water mass characteristics, dominated by ice-avoiding and ice-tolerant species. They inferred a role of prey distributions related to water properties in influencing the seabird community compositions observed.

Fraser & Ainley (1986), on the other hand, working near and within the pack ice, described two seabird assemblages in the Scotia–Weddell Confluence area, one closely associated with the pack ice (emperor and Adélie penguins, Antarctic and snow petrels) and the other associated with adjacent open water (southern

Sea Ice, Third Edition. Edited by David N. Thomas.
© 2017 John Wiley & Sons, Ltd. Published 2017 by John Wiley & Sons, Ltd.

Table 24.1 The relative biomass (birds km^{-2} × body mass in kg) of seabird species encountered on eight cruises, covering all seasons, in the Weddell Sea, Scotia Sea and Antarctic Peninsula area (from Ainley et al., 1994). *, < 1.0 kg km^{-2}; **, 1–10 kg km^{-2}; ***, > 10 kg km^{-2}. Species shown generally occur circumpolar in the Southern Ocean; in the case of non-circumpolar species, counterpart species depending on sector are noted in brackets. Species are indicated by their four-letter code, common name and Latin name

	Open water			Pack ice		
	Spring	Autumn	Winter	Spring	Autumn	Winter
Species resident all year round (ice-obligate, ice-tolerant):						
PENE Emperor penguin, *Aptenodytes forsteri*		**		***	***	***
PENA Adélie penguin, *Pygoscelis adeliae*	**	***		***	***	***
FUSG Southern giant fulmar, *Macronectes giganteus*	**	*	**	**	*	*
FUAN Antarctic fulmar, *Fulmarus glacialoides*	**	***	**	*	**	*
PEAN Antarctic petrel, *Thalassoica antarctica*	*	**	*	***	**	***
PETS Snow petrel, *Pagodroma nivea*	**	**	***	***	**	***
GUDO Dominican (or Kelp) gull, *Larus dominicanus*	*	*	*	*	*	*
Species present year round (ice-tolerant, ice-avoiding)						
PENK King penguin, *Aptenodytes patagonica*	*	*	**			
PENC Chinstrap penguin, *Pygoscelis antarctica*	***	***	***	***		*
ALGH Grey-headed albatross, *Diomedea chrysostoma*		**	*	*		
ALLS Sooty [Light-mantled] albatross *Phoebetria spp.*		*	*	*		
PETC Cape petrel, *Daption capense*	**	**	*	*	*	*
PBL Blue petrel, *Halobaena caerulea*	**	*	***			*
PRAN Antarctic prion, *Pachyptila vitatta*	**	*	**			
PTKG Kerguelen [Mottled] petrel, *Lugensa spp.*	*	*	**			
DIPE Diving petrel *Pelecanoides urinatrix*	*	*	*			
Species resident seasonally (ice-avoiding, ice-tolerant)						
PENM Macaroni penguin, *Eudyptes chrysolophus*	*	*				
PENN Magellanic penguin, *Sphenicus magellanicus*		*				
PENR Rockhopper penguin, *Eudyptes chrysocome*		*				
ALWA Wandering albatross, *Diomedea exulans*		*	*			
ALBB Black-Browed albatross, *Diomedea melanophrys*		**	*			
PETH White-chinned petrel, *Procellaria aequinoctialis*		*	*			
SHSO Sooty [Short-tailed] shearwater, *Ardenna spp.*	*	*				
PRFA Fairy prion, *Pachyptila turtur*	*	*				
PTSP Soft-plumaged [White-headed] petrel, *Pterodroma spp.*		*				
STBB Black-bellied storm-petrel, *Fregetta tropica*		*	*			
STWI Wilson's storm-petrel, *Oceanites oceanicus*		*	*		*	*
SKBR Brown (or Subantarctic) skua, *Stercorarius skua*	*	*				
SKMA South Polar skua, *Stercorarius maccormicki*		*	*			
TEAN Antarctic tern, *Sterna vitatta*	*	*			*	*
TEAR Arctic tern, *Sterna paradisaea*	*	*			*	*
Total species	**28**	**31**	**14**	**12**	**11**	**10**

fulmar, cape petrel, blue petrel, Wilson's storm petrel and Kerguelen petrel). In the southwest Pacific sector, as a function of differences in circumpolar distribution of breeding, Ainley et al. (1984) found the same assemblages but with few blue petrels and mottled petrel substituted for Kerguelen petrel. The species' strong fidelities to their respective habitats were proposed as indicative of the important role that pack ice played in structuring Antarctic seabird communities.

Woehler et al. (2003) identified three assemblages in Prydz Bay (East Antarctica) and argued that their results supported earlier proposals (Ribic & Ainley, 1988/89; Ainley et al., 1993) that the pack ice seabird assemblage is consistent around the Antarctic, with observed differences in adjacent open water species composition indeed reflecting a lack of circumpolar distribution of many low-latitude species. To some extent, this consistency may be an artifact of the relatively low species diversity of high-latitude breeding seabirds present in continental Antarctica, where there are just nine breeding species.

The sea ice-obligate species (Figure 24.1) are found year-round in association with sea ice, mostly south of the APF and particularly south of the Southern Boundary of the Antarctic Circumpolar Current, and nest at snow-/ice-free sites on the continent (Laws, 1984). Proximity to sea ice is the overriding factor that explains their occurrence at sea (Ribic et al., 2011). The sea ice-tolerant and sea ice-avoiding species, on the other hand, are far more abundant south of the APF during the summer months than during the winter (when, in fact, many species are absent or large portions of their populations are absent) and, in general, nest on subantarctic or peri-Antarctic islands, such as South Georgia, South Sandwich Islands, Heard Island, and Iles Kerguelen and Crozet, all of which lie very close to the APF (Raymond & Woehler, 2003; Woehler et al., 2010). Proximity to open ocean fronts best explains their occurrence at sea (van Franeker, 1992; Woehler et al., 2006; Ribic et al., 2011). Those of the latter group that are absent from the Southern Ocean during winter can be found frequenting mostly cooler, productive waters to the north, especially eastern boundary currents such as Benguela and Humboldt currents (Murphy, 1936; Spear & Ainley, 2008).

Ribic & Ainley (1988/89) assessed species assemblages in several cruises that spanned the entire South Pacific Ocean and the South Pacific sector of the Southern Ocean. They used canonical correlation analysis to gauge the factors that affected at-sea species associations (using SST and salinity, depth, distance from land, presence of sea ice as variables) and concluded that:

> the species groups revealed … made biological sense. One species group represented a specialized community associated with pack ice or pack-ice-influenced ocean … [it] may well be … the most spatially and temporally coherent at the megascale (>1000 km) as well as at smaller scales [of all groups detected]; the same assemblage is also evident in the South Atlantic sector of the Antarctic, and at all seasons of the year' (as indicated in Ainley et al., 1994; Table 24.1, Figure 24.2).

Cimino et al. (2013) showed that sea ice concentration (especially 10–40% cover) was the overriding factor explaining ecological success of the ice-obligate Adélie penguin, while factors related to open water, i.e. temperature and ocean productivity, were important to explain success of the ice-tolerant gentoo and chinstrap penguins.

Woehler et al. (2003) showed that the three seabird assemblages identified in Prydz Bay (Figure 24.3) were stable in species composition at 5-, 10- and 20-year

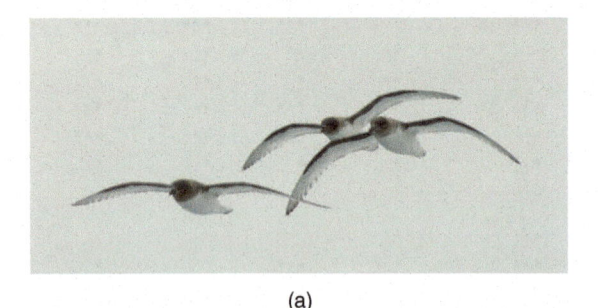

(a)

(b)

(c)

Figure 24.1 The four Antarctic species most closely associated with sea ice, year-round: (a) Antarctic petrels (almost always occurring in flocks; photography by E. Woehler); (b) snow petrel (normally a solitary species; photograph by E. Woehler); (c) Adélie and emperor penguins, which often coincide, normally occurring in flocks (photograph by D. Ainley).

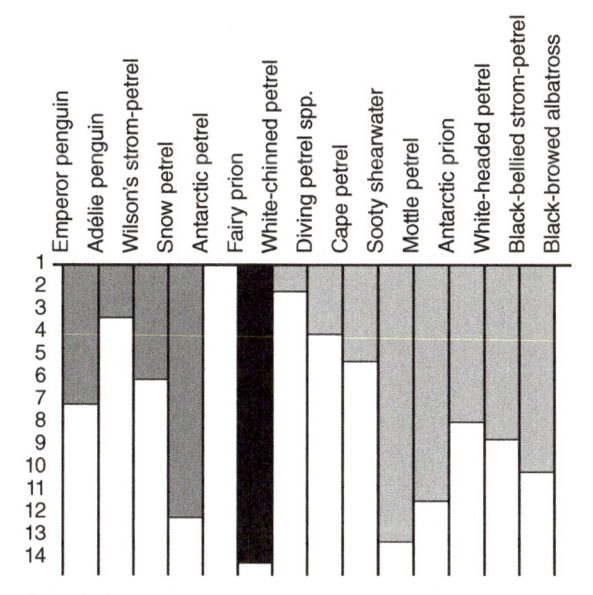

Figure 24.2 An 'icicle plot' showing correlation of species associations using Jaccard index for a cruise that crossed subantarctic and Antarctic waters in the western Pacific sector (New Zealand to Ross Sea, December–January). Numbers on the y-axis indicate the level of association; shading indicates the degree of connection among species. The pack ice (which was extensive in the Ross Sea) species (dark grey) and open-water species (light gray) assemblages are clearly evident.

periods. They noted that the variances increased at shorter timescales, but the assemblages were replicated at decadal and semi-decadal scales. In conjunction with the spatial consistency of seabird assemblages identified around the Antarctic continent by a number of studies (see earlier) and the temporal consistency identified by Woehler et al. (2003), it appears that the three Southern Ocean seabird assemblages identified (ice-obligate, ice-tolerant and ice-avoiding) are spatially and temporally persistent around the Antarctic.

24.2 Antarctic sea ice as a seabird habitat

What is it about sea ice that it leads to such a separate assemblage of seabird species, i.e. the ice-obligate assemblage: Adélie and emperor penguins, snow and Antarctic petrels (Figure 24.1)? Seemingly, these species must possess the anatomies, physiologies and behaviours needed to cope with sea ice, whereas other seabird species do not. Ainley et al. (1993), taking advantage of a 'natural experiment' in which storms led to a repeated sequence of mesoscale (100–1000 km) ice advances and retreats in the Weddell–Scotia Confluence area in summer, found that the seabird species assemblages alternated between ice-obligate and ice-tolerant/ice-avoiding as a function of ice presence or absence in the study area, but not the prey eaten by one assemblage or the other, i.e. diets were similar despite ice cover and species composition being different in the same area. Thus, it was the physical environment and not the prey type(s) or prey abundance that determined whether the ice-obligate or the ice-tolerant/ice-avoiding seabird assemblage was present in an area. Observations such as these of environmental variability influencing seabird assemblages in real time are rare [for results of a larger-scale 'natural experiment' (global climate change), see later]. Typically, seabird at-sea observations are confounded in real time by temporal considerations of research voyages, where spatial variability and relationships between seabirds and their environments cannot be distinguished from temporal determinants (Nicol et al., 2000; Woehler et al., 2010).

Although the temperature tolerances of most polar seabirds have not been investigated (but for early studies see Bech et al., 1991; Morgan et al., 1992; Weathers et al., 2000), this is one area of research that could be productive in answering this question: does temperature tolerance play a role in which species are ice-obligate or ice-avoiding? Much has been made of the huddling behaviour of penguins as a strategy against hypothermia (Gilbert et al., 2007, 2008 and references therein). Certainly, Adélie and emperor penguins begin to pant when air temperatures rise above about 2°C in still air (D. Ainley et al., personal observations). Therefore, it behooves these penguins to remain in association with ice even if just for thermoregulation. Earlier work by Chappell et al. (1990) showed that Adélie penguins showed little evidence of appreciable cold stress but their chicks were more likely to experience heat stress at ambient temperatures as low as 7–10°C. More recent work by Chapman et al. (2011), in fact, indicates that thermoregulation, and particularly hypothermia, plays an important role in the growth of Adélie penguin chicks; at that life stage, however, sea ice is not directly a part of their lives. Bech et al. (1991) suggested that higher standard metabolic rates of Antarctic petrel chicks could be a prerequisite for achieving a high

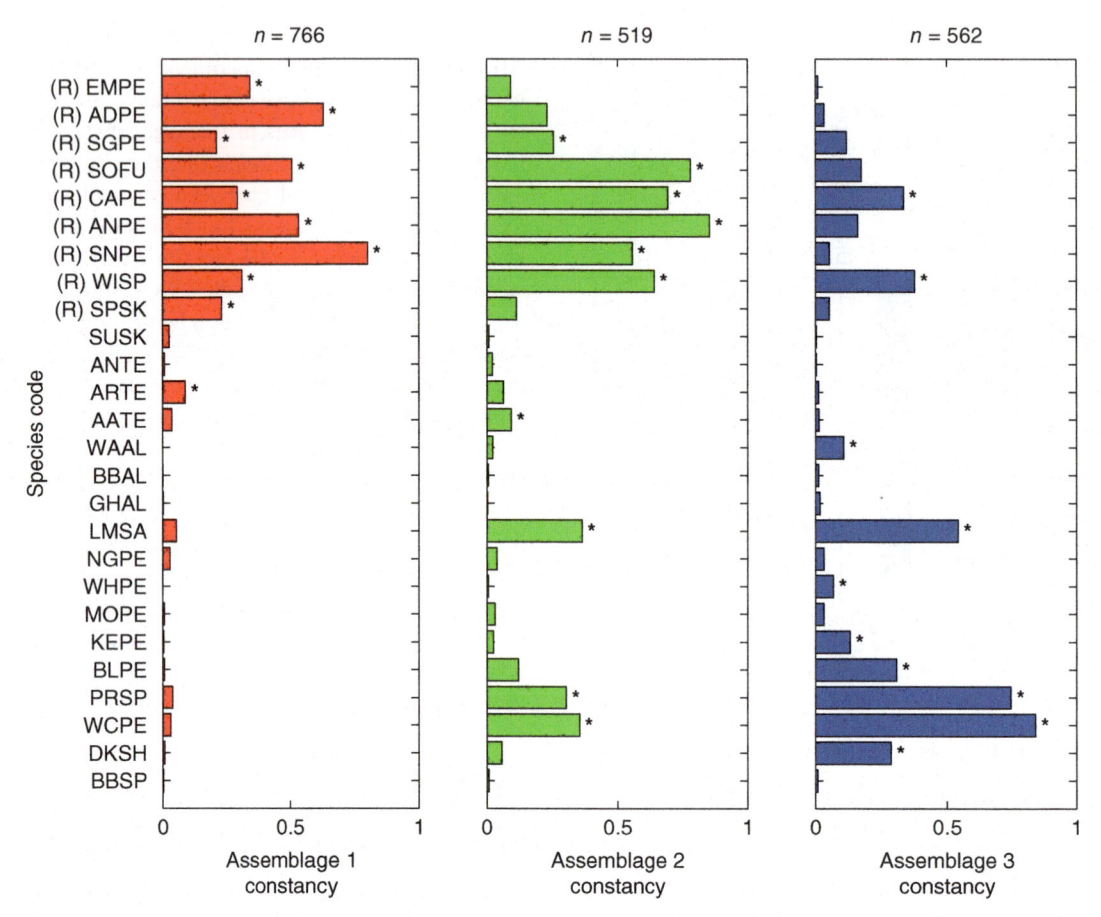

Figure 24.3 Species composition and constancy for the three seabird assemblages within Prydz Bay identified in Woehler et al. (2003). Species that breed within Prydz Bay are indicated by (R). Species/taxa with a fidelity-constancy product of 0.05 or greater are marked with an asterisk. The number of sites is shown above each assemblage. The top 13 species (EMPE to AATE) are ice-obligate and ice-tolerant; the remainder are ice-avoiding species. Resident species: EMPE, emperor penguin; ADPE, Adélie penguin; SGPE, southern giant petrel; SOFU, southern fulmar; CAPE, cape petrel; ANPE, Antarctic petrel; SNPE, snow petrel; WISP, Wilson's storm petrel; SPSK, south polar skua. Species/taxa that breed elsewhere: SUSK, subantarctic skua; ANTE, Antarctic tern; ARTE, Arctic tern; AATE, Arctic/Antarctic tern; WAAL, wandering albatross; BBAL, black-browed albatross; GHAL, grey-headed albatross; LMSA, light-mantled sooty albatross; NGPE, northern giant petrel; WHPE, white-headed petrel; MOPE, mottled petrel; KEPE, Kerguelen petrel; BLPE, blue petrel; PRSP, prion spp.; WCPE, white-chinned petrel; DKSH, dark shearwaters; BBSP, black-bellied storm petrel. Source: Woehler et al., 2003. Used under CC-BY-3.0 http://creativecommons.org/licenses/by/3.0/.

growth rate rather than for thermoregulation, thus allowing presence at high latitudes.

Another behavioural–physiological characteristic that Southern Ocean avian species appear to share, and especially ice-obligate species, is a very high capacity to quickly accumulate large amounts of subcutaneous fat, true not just among penguins (e.g. Ainley, 2002), but in petrels as well (Spear & Ainley 1998). This is beneficial for evolving natural history cycles that need to cope with long periods when the ocean is not accessible nearby owing to extensive ice cover, during which an individual needs to fast. In part, this ability is probably facilitated by the generally highly productive (albeit at times, highly patchy) nature of Antarctic coastal waters, leading to abundant food once a seabird is able to forage. The association of penguins, and other species, to highly productive peri-Antarctic, especially latent-heat, polynyas (Massom et al., 1998; Ainley, 2002; Arrigo & van Dijken, 2003) could be an expression of this pattern.

Thus, when an individual seabird is out foraging in the Southern Ocean, the chances are that it will find abundant food. The fact that sea ice, when extensive, can be a barrier to food acquisition (see later) has seemingly led, in an evolutionary sense, to the long periods during which individuals of ice-obligate species are able to remain with their eggs or chicks, while their mates necessarily forage at great distance away for days to weeks. Another manifestation of the necessity of an ease with fat accumulation is the rapidity by which these species can undergo their annual moult, in which old feathers are replaced by new ones and elevated prior nutrition is needed. In these pack ice species, the processes take only a few weeks, but during these they cannot forage, as they no longer possess the thermal insulation necessary to prevent major heat loss in icy-cold waters. In the case of the penguins, they stand still in the lee of rocks or ice hummocks and a pile of feathers accumulates around them as old feathers give way to new (Ainley 2002). In the case of the aerial species, they lose almost all of their flight feathers simultaneously during which they too spend most of their time roosting on icebergs or sea ice because flying is difficult. In more northern, less productive waters, closely related species take months to moult, losing one flight feather per wing at a time and thus preserving their flight and food-searching capabilities (Spear & Ainley 1998). The fact that ice is available on which to roost, either as icebergs or as sea ice, offers an unusual, predator-free opportunity to high-latitude polar seabirds.

Sea ice is especially critical moulting habitat for emperor penguins (Kooyman et al., 2000, 2004), with moult occurring in concentrated pack ice and on coastal ice floes, typically larger than $100\,m^2$. Emperor penguins moult late in the Antarctic summer when the sea ice is potentially at its minimum extent, and so larger floes and concentrated pack ice present greater safety to the moulting penguins who cannot enter the water for ~30 days (Kooyman et al., 2004; Massom et al., 2009). Large floes provide the greatest probability of persisting for the duration of the moult. Adélie penguins are somewhat less selective in their choice of moult site (Penney, 1967, 1968). Most Adélie penguins moult while sitting on ice floes (D. Ainley et al., personal observations) but in some cases will moult at their colonies (Ainley, 2002). Large flocks comprising several thousand migrating ice-tolerant Arctic terns have been observed roosting on pack ice and floes as small as $1\,m^2$; for long periods these terns are practically flightless, having moulted most of their flight feathers (D. Ainley et al., unpubl. data).

Regarding adaptations to their foraging habitat, the high-latitude Antarctic penguins are capable of longer breath-holding ability relative to body size than their open water counterparts. This allows them to find prey a long way underneath ice floes. Certainly, the emperor penguin, which by far is the deepest diving seabird of all, and can hold its breath for 20 minutes, is exemplary in this regard. The much smaller Adélie penguins can hold their breath for up to 6 minutes.

Antarctic petrels are also subsurface foragers, plunging on the fly into the waters of leads and elsewhere. Among the flighted, ice-obligate seabirds, seemingly by wing shape and loading, they are capable of easily exploiting the waters of leads in the sea ice, using the updrafts along the ice edges including ice floes and ice bergs for aerial lift. A Snow petrel, when foraging in an ice field, rarely has to flap, thus conserving energy, as it 'slope soars' along the edges of leads. The ice-tolerant southern giant petrel, on the other hand, has broader wings than similarly sized albatrosses and can therefore exercise flapping flight in a more energetically favourable manner; in pack ice-covered seas there are no ocean swells, which albatross need for their wave-crest soaring, and their long, narrow wings, suitable for that, are inefficient in flapping flight. Thus, southern giant petrels occupy the large scavenger niche within the pack ice, uncontested by other avian species that are prevalent in open water.

More than 99% of the global population of the sea ice-obligate emperor penguin breed on (fast) sea ice (Fretwell et al., 2012). Only a few small colonies nest on land, a few on glacial ice, and the remaining 45 colonies, contributing around 238 000 breeding pairs, are present on the fast ice occurring in coves and bays around the periphery of the Antarctic continent. Predictions of the loss of this species' breeding habitat arising from projected increases in SST indicate a concomitant decrease in the species' distribution and abundance (Ainley et al. 2010; Fretwell et al. 2012; Jenouvrier et al. 2014).

24.3 Sea ice as a hindrance or facilitator of seabird travel

While sea ice in most of the Southern Ocean is primarily seasonal in its extent, at minimum in February and at maximum in July–September, there are several areas around the Antarctic continent where sea ice never disappears at any time of the year, and for the winter and spring months this is exceedingly concentrated (Gloersen et al., 1992). These areas include the Amundsen Sea, southern Bellingshausen Sea, western Weddell Sea, southwestern Ross Sea and the ocean off George V Land. An examination of the distributions of seabird breeding colonies (Woehler & Croxall, 1997) quickly reveals that few colonies exist in these coastal areas of heavy sea ice (see also Ribic et al., 2011). The major problem for any seabirds that attempt to nest in these areas would be the lack of easy access to the ocean to forage during the breeding season when energy demands are highest. This was shown even for flighted birds by Jenouvrier et al. (2015), who reported that ice-tolerant southern fulmars had lower breeding success in years when coastal sea ice was extensive and they had to fly long distances to reach the outer marginal ice zone (MIZ) of the coastal ice pack.

Conversely, areas adjacent to polynyas, even in regions of concentrated, persistent sea ice, are where one finds penguin colonies and sometimes other seabird colonies, as noted earlier. In fact, there is a strong positive correlation between the presence and productivity of polynyas (in turn, a matter of polynya size and time of spring development) and the distribution of breeding colonies of both emperor (Massom et al., 1998) and Adélie penguins (Ainley, 2002; Arrigo & van Dijken, 2003). Polynyas, and especially their MIZs, are also important as foraging habitat for flighted seabirds (Ainley et al., 1998; Karnovsky et al., 2007).

The movements of pack ice can also facilitate or dictate seabird occurrence. This is well known with respect to the winter movements of penguins. During this season, the penguins and petrels spend most of their time residing on ice floes, spending only a few hours per day foraging in the water. Thus, as long as the ice floes move through productive waters, these birds do well, despite passively moving with the ice. This has been shown directly for Adélie penguins off East Antarctica (Clarke et al., 2003) and in the Ross Sea region (Ballard et al., 2010b), whereby the penguins take advantage of a regional oceanic gyre that returns them to about where they started several months previously. Emperor penguins in waters off East Antarctica may also exercise this strategy (Wienecke et al., 2004). In cases where the moving ice takes penguins to unproductive waters, e.g. north of the Southern Boundary of Antarctic Circumpolar Current, there can be negative demographic consequences (Wilson et al., 2001).

24.4 Sea ice and seabird foraging

Flighted species of seabirds are especially abundant in the MIZs of polynyas, owing to the increased prey availability, perhaps best demonstrated by the Ross Sea polynya (Ainley et al., 1984; Karnovsky et al., 2007; Figure 24.3). In the case of the high predator biomass in the eastern Ross Sea corresponding to the shelf-break and the MIZ, as shown in Figure 24.4 (see also Ainley & Jacobs, 1981), this is mostly contributed by millions of ice-obligate Antarctic and snow petrels that at the time are nesting several hundred kilometres to the southeast in the Rockefeller Mountains, Marie Byrd Land or in areas yet to be discovered. Thus petrels, by virtue of their great mobility, while constrained by the location of suitable breeding habitat (snow-free talus slopes in which to nest are scarce), can exploit even distant polynyas that they use for access to food. In Figure 24.4, the high biomass present along the western margin of the Ross Sea polynya is mostly contributed by Adélie and emperor penguins that nest along the Victoria Land coast (Woehler, 1993; Lyver et al., 2014; Lynch & La Rue, 2014). Similarly, Ribic et al. (2008) found that snow and Antarctic petrels were more associated with dispersed ice (a result of underwater, up-canyon warm water intrusion to the surface) than were Adélie penguins (which, as noted earlier, can access a greater volume of subsurface ocean than many other species) in Marguerite Bay, on the west coast of the Antarctic Peninsula, during winter.

In addition to the MIZs of polynyas within the pack ice field, seabirds are also attracted and associate their presence and travels to the MIZ at the large-scale (10s to 100s of kilometres) pack ice edge. A productivity pulse is associated with MIZs, both those bordering polynyas and those at the large-scale pack ice edge (Smith & Nelson, 1985; Smith et al., 1988), and accordingly seabird prey availability is also enhanced in MIZs. Interannual variability in the prevalence or location of the MIZs, especially during winter, can have profound effects on food availability for seabird predators (Ainley et al., 2006; Saba et al., 2014).

The increased availability of prey in these MIZs apparently has to do with the composition of the

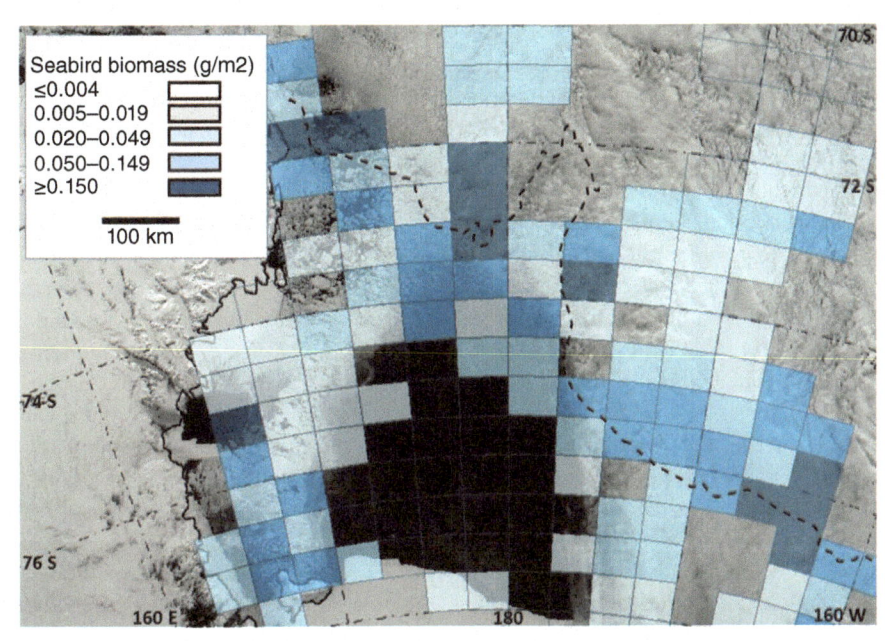

Figure 24.4 The Ross Sea on 15 November 1991 (AARC) showing the distribution of seabird biomass around the Ross Sea polynya (black cells) by 15′ latitude × longitude cell. If a cell has no coloration (clear), then biomass is ≤ 0.004. The dotted line is 1000 m isobaths or approximately the shelf break. The plain black line is the coastline.

phytoplankton, with single-celled diatoms frequenting the stratified waters of MIZs and colonial algae dominating the more well-mixed and turbulent open waters (Arrigo et al., 1998); grazers such as krill prefer the single-cell phytoplankton (a diagram of these two food webs, shelf vs slope, is presented in Ainley & DeMaster, 1990). Increased seabird density at large scale MIZs have been noted in the Ross (Ainley et al., 1984), Amundsen and Bellingshausen (Ainley et al. 1998), and Weddell Seas (Fraser & Ainley, 1986; Joiris, 1991; van Franeker, 1992), and several other sectors of the Southern Ocean (Hunt & Schneider, 1987). Away from the MIZs, where the sea ice is nonexistent or highly dispersed, they key on the increased prey availability associated with subsurface fronts, e.g. the shelf-break front, attracting ice-obligate and ice-tolerant species (Ainley & Jacobs, 1981; Ainley et al., 1998; Chapman et al., 2004; Ribic et al., 2011).

24.5 Antarctic penguins and sea ice: intensively researched perspective

Both Adélie and emperor penguins are sea ice-obligate species (Figure 24.1). Like many snow and Antarctic petrels, they nest only at continental margins or on continental islands not far off the Antarctic mainland coast (Fretwell et al., 2012; Lynch & LaRue, 2014). Owing to that characteristic alone, the penguins' breeding sites are surrounded by concentrated sea ice during at least the early part of their respective breeding seasons, and at no time during the year do they stray far from sea ice or at least from waters that sea ice had covered a few weeks earlier. The emperor penguin takes the ice association to the extreme – it forms its colonies on fast ice that has to stay in place from at least May to December with breeding birds not being territorial, making no nest and their chicks forming crèches while the parents are foraging. The only exception to these penguins' sea ice association is evident among the fledglings of emperor penguins who initially move north to frequent waters of the APF (Kooyman & Ponganis, 2008; Wienecke et al., 2010; Thiébot et al., 2013), there coinciding with their only congener, the largely ice-avoiding king penguin. This movement may be a response to cope with heavy predation pressure that exists in coastal waters at the time of fledging (Ainley & Ballard, 2011). Other alternative hypotheses include competitive exclusion by more efficient adult congeners, and exploitation of highly predictable and productive areas (i.e. a risk-adverse strategy) by fledglings which are still poorly efficient foragers (Thiébot et al., 2013). As yearlings, the emperor penguins move to the ice pack, from which they will never again stray in their lifetimes.

To cope with the highly concentrated seasonal sea ice near their breeding colonies, both penguin species reduce nest exchanges to a minimum, male Adélies taking about two-thirds of egg incubation duties (in two stints) and male emperors undertaking the entire incubation period (Prévost, 1961; Penney, 1968; Ainley, 2002). This is accomplished by males, in particular, accumulating body fat reserves before nesting that doubles their body mass or more. They then fast and exist without direct ingestion of water, living on metabolites for several weeks. Females, too, accumulate much fat prior to nesting, for egg production and in the event that a long walk over ice is required before and after laying the egg(s). As noted earlier, the colonies of both penguin species are closely associated with recurring polynyas, which normally reduce the distance of the required over-ice walking. If abnormally concentrated sea ice occurs, e.g. the polynya does not develop owing to lack of wind or other reasons, then breeding propensity and success both become significantly reduced (Ainley, 2002; Massom et al., 2006; Emmerson & Southwell, 2008; Dugger et al., 2014).

Where sea ice does not break out in the summer or when the distance to the nearest polynya is especially long, these penguins, particularly Adélies, feed by diving through cracks and holes in the ice, these features often being associated with grounded icebergs (Watanuki et al., 1993, 1997; Dugger et al., 2014). Emperor penguins, which breed during the winter and spring, almost always encounter extensive fast ice and heavily concentrated pack ice; they forage through the leads among the ice floes of the pack ice seaward of the fast ice (Kirkwood & Robertson, 1997; Wienecke & Robertson, 1997). In these situations, these penguins often trap prey against the undersurface of sea ice (Ponganis et al., 2000; Watanabe & Takahashi, 2013). During the winter, Adélie penguins associate with sea ice that ideally is ~15% cover, southward from the large-scale MIZ (Ballard et al. 2010b); during spring and summer, Adélie penguins forage most proficiently in the MIZ where ice concentration is 6–15% cover (Ballard et al., 2010a; Lescroël et al., 2014; see also Cinimo et al. 2013). Not only are prey predictably available in that habitat due to enhanced seasonal production (Ainley & DeMaster 1990; Smith & Sakshaug, 1990; Knox, 1994 and references therein), but so are the ice floes available for resting or avoiding predators if necessary.

24.6 Predictions for the future

Sea ice is currently disappearing along the northwestern Antarctic Peninsula, but it is growing in the Ross Sea sector (Zwally et al., 1998; Stammerjohn et al. 2008, 2012; Holland & Kwok, 2010; Chapter 10). Elsewhere in Antarctica, it is exhibiting little change except for the increasing persistence of polynyas in coastal waters (Parkinson, 2002; Stammerjohn et al., 2012). The changing polynyas may or may not be affecting coastal seabird populations, although in some cases this clearly appears to be the case. Growth of Adélie penguin colonies during the 1980s in the Ross Sea has been linked to expanding persistence of polynyas, associated with increasing winds (Ainley et al., 2005).

In response to the northward spread of persistent sea ice along the west coast of the Antarctic Peninsula during the Little Ice Age, Adélie penguin colonies also appeared farther north (Emslie, 2001). Nowadays, as the sea ice cover retreats in that area, Adélie colonies are disappearing in the north but are growing in the south as the normally highly concentrated sea ice there loosens (Scofield et al., 2010; Trivelpiece et al., 2011; Lynch et al. 2012). As these authors note, with the retreating sea ice, ice-tolerant and ice-avoiding species, such as gentoo penguins, are shifting southward themselves. The founding of penguin colonies as the pack ice loosens is a phenomenon documented elsewhere, e.g. in the Ross Sea during the Holocene (Emslie et al., 2003) and in East Antarctica (Emslie & Woehler, 2005). Variability in the extent of ice and particularly the MIZ has been documented to affect other ice-obligate species as well, e.g. snow petrels, over previous millennia (Ainley et al., 2006).

The physical and biotic patterns between Antarctic seabirds and their environment described above are expected to continue during the next few decades, though eventually, in response to global warming, sea ice should begin to decrease everywhere in the Southern Ocean (Turner et al., 2009; Ainley et al., 2010). Its last refuge may well be the Ross Sea, the southernmost extent of the Southern Ocean and the region in which sea ice cover is currently growing. With the retreat of sea ice – it will become largely nonexistent or the sea ice season will become very short, as in the Arctic – ice-obligate seabird species will disappear (Jenouvrier et al., 2009, 2014; Ainley et al., 2010), certainly replaced by the ice-tolerant and ice-avoiding species that will colonize sections of the coast that have become ice-free (see also Younger et al., 2015). In the near term, the ice-obligate species will emigrate to newly formed colonies where sea ice still persists (LaRue et al., 2014).

24.7 Acknowledgements

David Ainley's contribution was written with support from the National Science Foundation (grant ANT-0944411).

References

Ainley, D.G. (2002) *The Adélie Penguin: Bellwether of Climate Change*. Columbia University Press, NY.

Ainley, D.G. & Ballard, G. (2011) Non-consumptive factors affecting foraging patterns in Antarctic penguins: a review and synthesis. *Polar Biology*, **35**, 1–13.

Ainley, D.G, Clarke, E.D., Arrigo, K., Fraser, W.R., Kato, A., Barton, K.J. & Wilson, P.R. (2005) Decadal-scale changes in the climate and biota of the Pacific sector of the Southern Ocean, 1950s to the 1990s. *Antarctic Science*, **17**, 171–182.

Ainley, D.G. & DeMaster, D.P. (1990) The upper trophic levels in polar marine ecosystems. In: *Polar Oceanography* (Ed., W.O. Smith, Jr.,), pp. 599–630. Academic Press, San Diego.

Ainley, D.G. & Jacobs, S.S. (1981) Affinity of seabirds for ocean and ice boundaries in the Antarctic. *Deep-Sea Research, Part 1*, 28, 1173–1185.

Ainley, D.G., O'Connor, E.F. & Boekelheide, R.J. (1984) *The Marine Ecology of Birds in the Ross Sea, Antarctica*. Ornithological Monographs No. 32. American Ornithological Union, Washington, D.C.

Ainley, D.G., Ribic, C.A. & Fraser, W.R. (1994) Ecological structure among migrant and resident seabirds of the Scotia-Weddell Confluence region. *Journal of Animal Ecology*, **63**, 347–364.

Ainley, D.G., Ribic, C.A. & Spear, L.B. (1993) Species-habitat relationships among Antarctic seabirds: a function of physical and biological factors? *Condor*, **95**, 806–816.

Ainley, D.G., Ribic, C.A., Woehler, E.J. (2012) Adding the ocean to the study of seabirds: A brief history of at-sea seabird research. *Marine Ecology Progress Series*, **451**, 231–243.

Ainley, D.G., Jacobs, S.S., Ribic, C.A. & Gaffney, I. (1998) Seabird distribution and oceanic features of the Amundsen and southern Bellingshausen seas. *Antarctic Science*, **10**, 111–123.

Ainley, D.G., Hobson, K.A., Crosta, X., Rau, G.H., Wassenaar, L.I. & Augustinus, P.C. (2006) Holocene variation in the Antarctic coastal food web: linking δD and $\delta 13C$ in snow petrel diet and marine sediments. *Marine Ecology Progress Series*, **306**, 31–40.

Arrigo, K.R. & van Dijken, G.L. (2003) Phytoplankton dynamics within 37 Antarctic coastal polynya systems. *Journal of Geophysical Research*, **108**, 3271, doi:10.1029/2002JC001739.

Arrigo, K.R., Weiss, A.M. & Smith, W.O. Jr. (1998) Physical forcing of phytoplankton dynamics in the southwestern Ross Sea. *Journal of Geophysical Research*, **103**, No. C1, 1007–1021.

Ballard, G., Dugger, K.M., Nur, N. & Ainley, D.G. (2010a) Parental foraging and chick provisioning strategies of Adélie penguins in response to body condition and environmental variability. *Marine Ecology Progress Series*, **405**, 287–302.

Ballard G., Toniolo, V., Ainley, D.G., Parkinson, C.L., Arrigo, K.R. & Trathan, P.N. (2010b) Responding to climate change: Adélie penguins confront astronomical and ocean boundaries. *Ecology*, **91**, 2056–2069.

Bech, C., Mehlum, F. & Haftorn, S. (1991). Thermoregulatory abilities in chicks of the Antarctic Petrel (*Thalassoica antarctica*). *Polar Biology*, **11**, 233–238.

Bierman, W.H. & Voous, K.H. (1950) Birds observed and collected during the whaling expeditions of the 'Willem Barendsz' in the Antarctic, 1946–1947 and 1947–48. *Ardea*, **37** (Suppl.), 1–123.

Chapman, E., Hofmann, E., Patterson, D., Ribic, C. & Fraser, W. (2011) Marine and terrestrial factors affecting Adélie-penguin *Pygoscelis adeliae* chick growth and recruitment off the western Antarctic Peninsula. *Marine Ecology Progress Series*, **436**, 273–289.

Chapman, E.W., Ribic, C.A. & Fraser, W.R. (2004) The distribution of seabirds and pinnipeds in Marguerite Bay and their relationship to physical features during austral winter 2001. *Deep-Sea Research Part II*, **51**, 2261–2278.

Cimino, M.A., Fraser, W.R., Irwin, A.J. & Oliver, M.J. (2013) Satellite data identify decadal trends in the quality of *Pygoscelis* penguin chick-rearing habitat. *Global Change Biology*, **19**, 136–148.

Clarke, J., Kerry, K., Fowler, C., Lawless, R., Eberhard, S. & Murphy, R. (2003) Post-fledging and winter migration of Adélie penguins *Pygoscelis adeliae* in the Mawson region of East Antarctica. *Marine Ecology Progress Series*, **248**, 267–278.

Dugger, K.M., Ballard, G., Ainley, D.G., Lyver, P.O'B. & Schine, C. (2014) Adélie penguins coping with environmental change: results from a natural experiment at the edge of their breeding range. *Frontiers in Ecology and Evolution*, **2**, 68, doi:10.3389/fevo.2014.00068.

Eakin, R.R., Dearborn, J.H. & Townsend, W.C. (1986) Observations of marine birds in the South Atlantic Ocean in the late austral autumn. In: *Biology of the Antarctic Seas XVII, Antarctic*

Research Series No. 44 (Ed., L.S. Kornicker), pp. 69–86. American Geophysical Union, Washington, D.C.

Emmerson, L. & Southwell, C. (2008) Sea ice cover and its influence on Adélie penguin reproductive performance. *Ecology*, **89**, 2096–2102.

Emslie, S.D. (2001) Radiocarbon dates from abandoned penguin colonies in the Antarctic Peninsula region. *Antarctic Science*, **13**, 289–295.

Emslie, S.D. & Woehler, E.J. (2005) A 9000-year record of Adélie penguin occupation and diet in the Windmill Islands, East Antarctica. *Antarctic Science*, **17**, 57–66.

Emslie, S.D., Berkman, P.A., Ainley, D.G., Coats, L. & Polito, M. (2003) Late-Holocene initiation of ice-free ecosystems in the southern Ross Sea, Antarctica. *Marine Ecology Progress Series*, **262**, 19–25.

van Franeker, J.A. (1992) Top predators as indicators for ecosystem events in the confluence zone and marginal ice zone of the Weddell and Scotia seas, Antarctica, November 1988 to January 1989 (EPOS Leg 2). *Polar Biology*, **12**, 93–102.

Fraser, W.R. & Ainley, D.G. (1986) Ice edges and seabird occurrence in Antarctica. *BioScience*, **36**, 258–263.

Fretwell, P.T., LaRue, M.A., Morin, P. et al. (2012) An emperor penguin population estimate: the first global, synoptic survey of a species from space. *PloS ONE*, **7**, no. 4: e33751.

Gilbert, C., Blanc, S., Le Maho, Y. & Ancel, A. (2008) Energy saving processes in huddling emperor penguins: from experiments to theory. *Journal of Experimental Biology*, **211**, 1–8.

Gilbert, C., Le Maho, Y., Perret, M. & Ancel, A. (2007). Body temperature changes induced by huddling in breeding male emperor penguins. *American Journal of Physiology*, **292**, 176–185.

Gloersen P., Campbell W.J., Cavalieri, D.J., Comiso, J.C., Parkinson, C.L. & Zwally, H.J. (1992) *Arctic and Antarctic sea ice, 1978–1987: Satellite Passive-Microwave Observations and Analysis*. NASA SP-511. National Aeronautics and Space Administration, Washington, DC.

Holland, P.R. & Kwok, R. (2012) Wind-driven trends in Antarctic sea-ice drift. *Nature Geoscience*, http://www.nature.com/doifinder/10.1038/ngeo1627

Hunt, G.L., Jr., & Schneider, D.C. (1987). Scale-dependent processes in the physical and biological environment of marine birds. In: *Seabirds: Feeding Ecology and Role in Marine Ecosystems*, (Ed. J.P. Croxall), pp. 7–41. Academic Press, NY.

Hunt, G.L., Heinemann, D., Veit, R.R., Heywood, R.B. & Everson, I. (1990) The distribution, abundance and community structure of marine birds in southern Drake Passage and Bransfield Strait, Antarctica. *Continental Shelf Research*, **10**, 243–257

Jenouvrier, S., Caswell, H., Barbraud, C., Holland, M., Stoeve, J. & Weimerskirch, H. (2009) Demographic models and IPCC climate projections predict the decline of an emperor penguin population. *Proceedings of the National Academy of Sciences USA*, **106**, 1844–1847.

Jenouvrier, S., Holland, M., Stroeve, J. et al. (2014) Projected continent-wide declines of the emperor penguin under climate change. *Nature Climate Change*, doi:10.1038/NCLIMATE2280.

Jenouvrier, S., Péron, C. & Weimerskirch, H. (2015). Extreme climate events and individual heterogeneity shape lifehistory traits and population dynamics. *Ecological Monographs*, **85**, 605–624

Joiris C R. (1991) Spring distribution and ecological role of seabirds and marine mammals in the Weddell Sea, Antarctica. *Polar Biology*, **11**, 415–424.

Karnovsky, N., Ainley, D.G. & Lee, P. (2007) The impact and importance of production in polynyas to top-trophic predators: three case histories. In: *Polynyas* (Eds., W.O. Smith, Jr., & D.G. Barber), pp. 391–410. Elsevier Publishers, London.

Kirkwood, R. & Robertson, G. (1997) The foraging ecology of female emperor penguins in winter. *Ecological Monographs*, **67**, 155–176.

Knox, G.A. (1994) *The Biology of the Southern Ocean*. Cambridge University Press, Cambridge

Kooyman, G.L. & Ponganis, P. (2008) The initial journey of juvenile emperor penguins. *Aquatic Conservation: Marine and Freshwater Systems*, **17**, 37–43

Kooyman, G.L., Siniff, D.B., Stirling, I. & Bengtson, J.L. (2004) Moult habitat, pre-and post-moult diet and post-moult travel of Ross Sea emperor penguins. *Marine Ecology Progress Series*, **267**, 281–290.

LaRue, M., Kooyman, G., Lynch, H.J. & Fretwell, P. (2014) Emigration in emperor penguins: Implications for interpretation of long-term studies. *Ecography*, 10.1111/ecog.00990

Laws, R.M. (1984) *Antarctic Ecology*. Academic Press, London.

Lescroël, A., Ballard, G., Grémillet, D., Authier, M. & Ainley, D.G. (2014) Antarctic climate change: extreme events disrupt plastic phenotypic response in Adélie Penguins. *PLoS ONE*, **9**: e85291. doi:10.1371/journal.pone.0085291

Lynch, H.J. & LaRue, M.A. (2014). First global census of the Adélie Penguin. *Auk*, **131**, 457–466.

Lynch, H.J., Naveen, R., Trathan, P.N. & Fagan, W.F. (2012) Spatially integrated assessment reveals widespread changes in penguin populations on the Antarctic Peninsula. *Ecology*, **93**, 1367–1377.

Lyver, P.O'B., Barron, M., Barton, K.J. et al., (2014) Trends in the breeding population of Adélie penguins in the Ross Sea, 1981–2012: a coincidence of climate and resource extraction effects. *PloS ONE*, **9**, e91188.

Massom, R.A., Harris, P.T., Michael, K.J. & Potter, M.J. (1998) The distribution and formative processes of latent-heat polynyas in East Antarctica. *Annals of Glaciology*, **27**, 420–426.

Massom, R.A., Hill, K., Barbraud, C. et al. (2009) Fast ice distribution in Adélie Land, East Antarctica: interannual variability and implications for emperor penguins *Aptenodytes forsteri*. *Marine Ecology Progress Series*, **374**, 243–257.

Massom, R.A., Stammerjohn, S.E., Smith, R.C. et al. (2006) Extreme anomalous atmospheric circulation in the West Antarctic Peninsula region in Austral spring and summer 2001/02, and its profound impact on sea ice and biota. *Journal of Climate*, **19**, 3544–3571.

Murphy, R.C. (1936) *Oceanic Birds of South America*. Macmillan, New York.

Nicol, S., Pauly, T., Bindoff, N.L. et al. (2000) Sea ice, circulation and the East Antarctic ecosystem. *Nature*, **406**, 504–507.

Parkinson, C.L. (2002) Trends in the length of the Southern Ocean sea ice season, 1979–99. *Annals of Glaciology*, **34**, 435–440.

Penney, R.L. (1967) Molt in the Adélie Penguin. *Auk*, **84**, 61–71

Penney, R.L. (1968) Territorial and social behaviour in the Adélie Penguin. In: Antarctic Bird Studies, Antarctic Research Series No. 12 (Ed., O.L. Austin, Jr,), pp. 83–131. American Geophysical Union, Washington, DC.

Ponganis, P.J., Van Dam, R.P., Marshall, G., Knower, T. & Levenson, D.H. (2000) Sub-ice foraging behavior of emperor penguins. *Journal of Experimental Biology*, **203**, 3275–3278.

Prévost, J. (1961) *Ecologie du Manchot Empereur Aptenodytes forsteri Gray*. Hermann, Paris.

Raymond, B. & Woehler, E.J. (2003) Predicting seabirds at sea in the southern Indian Ocean. *Marine Ecology Progress Series*, **263**, 275–298.

Ribic, C.A. & Ainley, D.G. (1988/89) Constancy of seabird species assemblages: an exploratory look. *Biological Oceanography*, **6**, 175–202.

Ribic, C.A., Chapman, E., Fraser, W., Lawson, G.L. & Wiebe, P.H. (2008) Top predators in relation to bathymetry, ice and krill during austral winter in Marguerite Bay, Antarctica. *Deep-Sea Research II*, **55**, 485–499.

Ribic, C.A., Ainley, D.G., Ford, R.G., Fraser, W.R., Tynan, C.T. & Woehler, E.J. (2011) Water masses, ocean fronts, and the structure of Antarctic seabird communities: putting the eastern Bellingshausen Sea in perspective. *Deep-Sea Research Part II*, **58**, 1695–1709.

Ryan, P.G. & Cooper, J. (1989) The distribution and abundance of aerial seabirds in relation to Antarctic krill in the Prydz Bay region, Antarctica, during late summer. *Polar Biology*, **10**, 199–209.

Saba, G.K., Fraser, W.R., Saba, V.S. et al. (2014) Winter and spring controls on the summer food web of the coastal West Antarctic Peninsula. *Nature Communications*, doi:10.1038/ncomms5318.

Schofield, O., Ducklow, H.W., Martinson, D.G., Meredith, M.P., Moline, M.A. & Fraser, W.R. (2010) How do polar marine ecosystems respond to rapid climate change? *Science*, **328**, 1520–1523.

Smith, W.O. & Sakshaug, E. (1990) Polar phytoplankton. In: *Polar Oceanography* (Ed. W.O. Smith, Jr.,), pp 477–526. Academic Press, San Diego.

Smith, W.O., Jr., & Nelson, D.M. (1985) Phytoplankton bloom produced by a receding ice edge in the Ross Sea: spatial coherence with the density field. *Science*, **227**, 163–166.

Smith, W.O. Jr.,, Keene, N.K. & Comiso, J.C. (1988) Inter-annual variability in estimated primary productivity of the Antarctic marginal ice zone. In: *Antarctic Ocean and Resources Variability* (Ed. D. Sahrhage), pp. 131–139. Springer-Verlag, Berlin.

Spear, L.B. & Ainley, D.G. (1998) Morphological differences relative to ecological segregation in petrels (Family: Procellariidae) of the Southern Ocean and tropical Pacific. *Auk*, **115**, 1017–1033.

Spear, L.B. & Ainley, D.G. (2008) The seabird community of the Peru Current, 1980–1995, with comparisons to other eastern boundary currents. *Marine Ornithology*, **36**, 125–144.

Stammerjohn, S., Massom, R., Rind, D. & Martinson, D. (2012) Regions of rapid sea ice change: An inter-hemispheric seasonal comparison. *Geophysical Research Letters*, **39**, L06501, doi:10.1029/2012GL050874.

Stammerjohn, S.E., Martinson, D.G., Smith, R.C., Yuan, X. & Rind, D. (2008) Trends in Antarctic annual sea ice retreat and advance and their relation to ENSO and Southern Annular Mode variability. *Journal of Geophysical Research*, **113**, C03S90, doi:10.1029/2007JC004269.

Thiébot, J.B., Lescroël, A., Barbraud, C. & Bost, C.A. (2013) Three-dimensional use of marine habitats by juvenile Emperor penguins *Aptenodytes forsteri* during post-natal dispersal. *Antarctic Science*, **25**, 536–544.

Trivelpiece, W.Z., Hinke, J.T., Miller, A.K., Reiss, C.S., Trivelpiece, S.G. & Watters, G.M. (2011) Variability in krill biomass links harvesting and climate warming to penguin population changes in Antarctica. *Proceedings of the National Academy of Sciences USA*, **108**, 7625–7628.

Turner, J., Comiso, J.C., Marshall, G.J. et al. (2009) Non-annular atmospheric circulation change induced by stratospheric ozone depletion and its role in the recent increase of Antarctic sea ice extent. *Geophysical Research Letters*, **36**, L08502, doi:10.1029/2009GL037524.

Veit, R.R. & Hunt, G.L., Jr., (1991) Broadscale density and aggregation of pelagic birds from a circumnavigational survey of the Antarctic Ocean. *Auk*, **108**, 790–800.

Watanabe, Y.W. & Takahashi, A. (2013) Linking animal-borne video to accelerometer reveals prey capture variability. *Proceedings of the National Academy of Sciences USA*, **110**, 2199–204.

Watanuki, Y., Kato, A., Mori, Y. & Naito, Y. (1993) Diving performance of Adélie penguins in relation to food availability in fast sea-ice areas: comparison between years. *Journal of Animal Ecology*, **62**, 634–646.

Watanuki, Y., Kato, A., Naito, Y., Robertson, G. & Robinson, S. (1997) Diving and foraging behaviour of Adélie penguins in areas with and without fast sea-ice. *Polar Biology*, **17**, 296–304.

Watson, G.E. (1975) *Birds of the Antarctic and Subantarctic*. American Geophysical Union, Washington, DC.

Wienecke, B.C. & Robertson, G. (1997) Foraging space of emperor penguins *Aptenodytes forsteri* in Antarctic shelf waters in winter. *Marine Ecology Progress Series*, **159**, 249–263.

Wienecke, B., Kirkwood, R. & Robertson, G. (2004) Pre-moult foraging trips and moult locations of Emperor penguins at the Mawson Coast. *Polar Biology*, **27**, 83–91.

Wienecke, B., Raymond, B. & Robertson, G. (2010) Maiden journey of fledgling emperor penguins from the Mawson

Coast, East Antarctica. *Marine Ecology Progress Series*, **410**, 269–282.

Wilson, P.R., Ainley, D.G., Nur, N. et al. (2001) Adélie Penguin population change in the Pacific Sector of Antarctica: relation to sea-ice extent and the Antarctic Circumpolar Current. *Marine Ecology Progress Series*, **213**, 301–309.

Woehler, E.J. (1993) The Distribution and Abundance of Antarctic and Subantarctic Penguins. Scientific Committee on Antarctic Research (SCAR) Bird Biology Subcommittee, Cambridge, U.K.

Woehler, E.J. & Croxall, J.P. (1997) The status and trends of Antarctic and sub-Antarctic seabirds. *Marine Ornithology*, **25**, 43–66.

Woehler, E.J., Raymond, B. & Watts, D.J. (2003) Decadal-scale seabird assemblages in Prydz Bay, East Antarctica. *Marine Ecology Progress Series*, **251**, 299–310.

Woehler, E.J., Raymond, B. & Watts, D.J. (2006) Convergence or divergence: where do short-tailed shearwaters forage in the Southern Ocean? *Marine Ecology Progress Series*, **324**, 261–270.

Woehler, E.J., Raymond, B., Boyle, A. & Stafford, A. (2010) The role of environmental determinants on seabird assemblages observed during BROKE-west, January - March 2006. *Deep-Sea Research Part II*, **57**, 982–991.

Younger, J., Emmerson, L., Southwell, C., Lelliott, P. & Mille, K. (2015) Proliferation of East Antarctic Adélie penguins in response to historical deglaciation. *BMC Evolutionary Biology*, **15**, 236 doi 10.1186/s12862-015-0502-2

Zwally, H.J., Comiso, J.C., Parkinson, C.L., Cavalieri, D.J. & Gloersen, P. (2002) Variability of Antarctic sea ice 1979–1998. *Journal of Geophysical Research*, **107**, doi:10.1029/2000JC000733.

CHAPTER 25

Sea ice is our beautiful garden: indigenous perspectives on sea ice in the Arctic

Henry P. Huntington,[1] Shari Gearheard,[2] Lene Kielsen Holm,[3] George Noongwook,[4] Margaret Opie[5] and Joelie Sanguya[6]

[1] *Eagle River, AK, USA*
[2] *National Snow and Ice Data Center, University of Colorado Boulder, Boulder, CO, USA*
[3] *Lene Kielsen Holm, Greenland Climate Research Centre, Greenland Institute of Natural Resources, Nuuk, Greenland*
[4] *Native Village of Savoonga, Savoonga, AK, USA*
[5] *Barrow, AK, USA*
[6] *Clyde River, NU, Canada*

25.1 Introduction

'I refer to the sea ice as a beautiful garden. Much of our life depends on what our garden provides.'
(Wesley Aiken, Barrow, Alaska, 2008 – in Gearheard et al., 2013; see Figure 25.1).

Wesley Aiken's description of his connection to sea ice probably resonates strongly with those who live with, on and amid an ocean that is frozen much of the year. Sea ice is a means: a means of obtaining food, of travelling, of feeling at home. Above all, using sea ice as part of the daily and yearly cycle is part of living a complete, satisfying life as a member of a family and a community. Sea ice matters for what it makes possible, materially, spiritually and emotionally.

A great deal has been written about sea ice and Arctic peoples (see Section 25.3 for more details), especially Inuit (ICC, 2008; Krupnik et al., 2010a). Hunting practices have been described for many communities (e.g. Boas, 1888; Nelson, 1969; McDonald, 1997; Druckenmiller et al., 2010). Knowledge and terminology about sea ice have been documented in recent years from Chukotka to Greenland (Oozeva et al., 2004; Taverniers, 2010; Weyapuk & Krupnik, 2012; Bogoslovskaya & Krupnik, 2013). More recently, the impacts of climate change on sea ice and sea ice use have received a great deal of attention (ACIA, 2005; Huntington, 2013; Lovecraft, 2013). A few publications have looked beyond the material and physical aspects of sea ice use to the deeper relationships that at least some Arctic peoples have with sea ice (Gearheard et al., 2013).

This chapter, rather than restating at length what has been written before about indigenous peoples and Arctic sea ice, begins with three indigenous accounts of daily sea ice use, describing the mechanics of preparing for, travelling and hunting on or amid ice, and also what those acts mean to the people doing them. This aspect of indigenous perspectives on sea ice has not often been presented in the words of indigenous hunters and their families. We are not the first to recognize the value of indigenous perspectives on sea ice, yet their inclusion in a book about sea ice follows a relatively untrodden path and offers fresh insight into the ways indigenous peoples view, use and understand sea ice. Two of the three accounts begin by setting the context for sea ice-related activities, rather than by describing those activities and sea ice at the beginning. Sea ice is important, but as a means rather than as an end in itself, and is experienced directly and indirectly in various ways. These first-hand narratives illustrate the way that sea ice is perceived as part of the seasonal round of activities in Arctic communities and the relationships

---

Sea Ice, Third Edition. Edited by David N. Thomas.
© 2017 John Wiley & Sons, Ltd. Published 2017 by John Wiley & Sons, Ltd.

Figure 25.1 'Sea ice is a beautiful garden.' (Conceptualized by Wesley Aiken, illustrated by Dorothia Rohner, in Gearheard et al., 2013:xxxiii. Reproduced with permission of Dorothia Rohner.)

that connect people to people, and people to the ocean, their environment, their culture and their general well-being.

Sea ice is the defining characteristic of the Arctic marine environment (Loeng, 2005) and can be found well to the south of the Arctic Circle. While peoples throughout northern regions have used sea ice in one way or another, e.g. in northern Russia (e.g. Krupnik et al., 2010b) or among the Ainu in northern Japan (Tezuka, 2009), Inuit have the most extensive use of sea ice by area and through the seasons. (We use the term 'Inuit' to describe the peoples from the Bering Sea to Greenland, including Yup'ik, Cupik, St Lawrence Island and Siberian Yupik, Iñupiat, Inuvialuit, Inuit, Inughuit, Tunumiut and Kalallit, who are part of a cultural and linguistic continuum.) Our three accounts come from Clyde River, Baffin Island, Nunavut, amid Inuit territory; from Barrow, on the north coast of Alaska and home to the Iñupiat; and from Savoonga, Alaska, on St Lawrence Island in the northern Bering Sea, inhabited by St Lawrence Island Yupik (see Figure 25.2).

Following the three accounts, the chapter provides a short summary of research on sea ice and indigenous peoples, reviews the significance of harvests made from sea ice, and concludes with a discussion about how life amid sea ice is studied and described and how indigenous use of sea ice is being affected by climate change.

25.2 Indigenous accounts of sea ice use

25.2.1 *Siku* – surface of liquid water solidifies
Joelie Sanguya, Clyde River, Nunavut

It is the fall time, the 'waiting season', and Nature promises that the sea ice will form in about a month's time. *Piruqtut kaniqtut* (plants) are covered in the morning by frost and it is cool now when you walk out the door in the morning. You take a deep breath of the fresh air, enjoying the fact that the season's change is here. Small streams, puddles and the top layer of the ground slightly freeze and this fresh air quality energizes your body in a way that is different from the summer. A little bit of snow comes down now, but it doesn't stay. Later on comes the *qitiqquusiq*, snow that covers the tops of the mountains and partly down the slopes, but there is no snow from the sea up to the mountainsides. Then the snow does come and all of the land is covered with the white stuff.

The snow is on the ground now and the wind blows down the fjord toward the open sea because the land is colder than the water. It is now the ocean's turn to make some changes in the environment – without any question, it is still the beautiful season. Polar bears are now walking along the shoreline while most boaters are

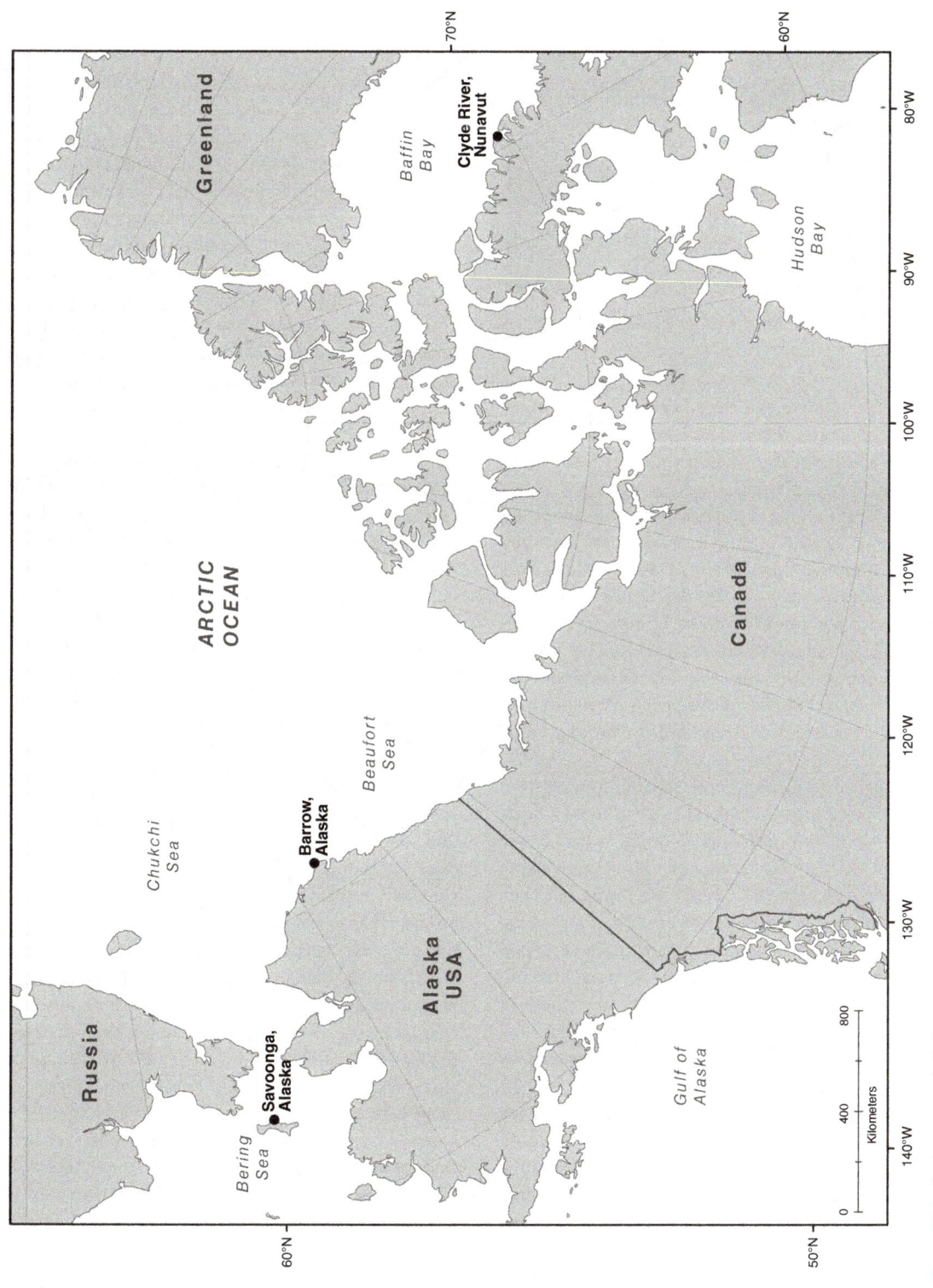

Figure 25.2 The locations of Clyde River (Nunavut, Canada), Barrow (Alaska) and Savoonga (Alaska), where the accounts in Section 25.2 are set.

Figure 25.3 Early freeze-up at Clyde River, Nunavut. Source: photograph by Shari Gearheard.

removing the snow out of their boats to go out hunting. As the temperature fluctuates, colder days are more noticeable and the season is destined to turn to winter while daylight is shortening.

On a day with wind, *quvviqquat* (tear-like streaks on the surface of the sea) are the mark that happens only in the late autumn. They are one of the first indicators of the ocean's change. Then, the ice begins to build up slightly along the shore while pieces of snow and ice drift off from the shoreline (Figure 25.3). This is the time of the year when we sometimes see dog teams, and more snowmobiles travel on the land as people who love fishing through the frozen lakes go out to fish and enjoy their time.

As the sea ice develops, people are still hunting from their boats, pushing through the thin ice. But polar bears are now walking out on the ice and that is part of my amazement – when very heavy polar bears are walking and running on ice that I find unsafe to walk on. They are going for the seals just as we are. There are some fresh holes through the sea ice which ringed seals pop their heads through. You now can be very stealthy and shoot at your target without causing any movement in your boat. Some people are now hunting seals, choosing exactly the kind they want, whether large or small seals, although others still shoot at any seal at all.

My father once said to me, 'Dogs feel what they walk on, but snowmobiles don't. This is the time to use your *unaaq* (harpoon) on the fresh new ice,' meaning, be careful of the ice you travel over by snowmobile because you might fall through. Use your *unaaq* to test the thickness. Travelling by dog team, dogs are working together pulling a *qamutiik* (sled), whether in daylight or in the dark without the moon. Dogs will spread out if they get on to very thin sea ice. My father worried about what kind of sea ice I was travelling on, whether it was thin or thick sea ice, as I was not supposed to be gone from his life. Once when I fell through thin ice about 30 miles away from home and had to walk for about 20 miles and wait for rescue, that's when I realized what my father had said to me was true. I became much more careful and used my *unaaq* on unknown thin sea ice.

The temperature and the snow play big roles on the condition and safety of new sea ice. When the sea ice is still thin, constant temperature will keep the texture of the ice the same, but when either warmer temperatures or snow comes, the thin sea ice becomes dangerous. *Ugjugajataktunaat* (bearded seal skin ropes) can be used as an indicator for this. The skin ropes will change in their texture before the temperature changes within the environment where you are. Before we started using thermometers, *ugjugajataktunaat* were traditional indicators of the temperature in winter. Using different indicators and our knowledge, as hunters we will talk about where the ice might be safer, or where it may not be so safe to travel while the sea ice is developing.

When the sea ice becomes safe to travel on, there are no leads or cracks although the ice goes up and down in high and low tide. Especially when clear cold days come, leads (cracks) form between points of land. Cold clear days cause the ice to shrink or contract and seals will find these leads and make breathing holes. As leads open, other new cracks will happen later on, some at any time and others not as often, depending on the location. Within the fjords leads will open fewer times, but closer to Baffin Bay leads will reopen throughout the year. When the spring comes, some of these leads will be too wide to cross over and we look for alternative spots to cross leads.

Tusaqtuut is the name of the moon in November when we are able to travel to other communities because of the new travel routes made possible by sea ice, even though parts of the sea ice may be carried off by strong ocean currents or the wind, both of Nature's strengths.

We explore the conditions of the sea ice in the vicinity where we are, although some people will travel farther than others. The smoothness or the roughness of certain areas of the sea ice becomes known to travellers and hunters. The snow becomes essential padding for the rough sea ice conditions and passable travel takes time to form, but it does occur.

On smooth sea ice and at the floe edge is where we hunt seals. And as we know, seals will claw through the ice to make their breathing holes. They also make dens in the sea ice and have their pups there – the early pups are born in February and most are born in March. Seals make their dens where the snow accumulates on top of the sea ice. They find the areas with more snow by seeing the difference in light and dark looking up at the sea ice. The ice is thinner where snow insulates the ice. Both male and female seals dig dens into the snowbanks on the ice and they maintain a hole to the ocean from inside the den. Seal meat is the one that keeps our bodies from getting cold so easily and we travel over the land and sea ice where we live to get this real traditional food. *Qujannamiik* (thank you).

25.2.2 Umiapiaq
Margaret Opie, Barrow, Alaska

The whaling skinboat or *umiapiaq*, still in use today for traditional whaling, is constructed much the same as when our ancestors used it. The process of maintaining the *umiaq* requires the bearded seal skins be changed every 3–5 years.

The hunt for *ugruk* (bearded seal) begins in early June, when the shorefast ice breaks up and hunters can go by boat through the pack ice. With 24-hour daylight, hunters are out as long as the weather permits. In addition to providing the much-needed skins for our traditional whaling boats, the *ugruk* is a prized catch in our community for the *misigaaq* (seal oil) that we can't do without during the long winter, and the dried meat (*kiniqtak*), which we love with our *misigaaq*.

An average boat will require six to seven *ugruk* skins, with an extra skin or two for any required patchwork. Soon as the catch is brought in, a team of women and men begin butchering the *ugruk*. The *ulu*, or women's knife, is already sharpened and ready to make the first cut. The younger women are immediately put to work under the guidance of the elders. Even the younger children help by doing whatever is asked of them, taking messages, emptying trash, or handing this and that to the women. The stronger young men do the heavy lifting and assist wherever they're needed. Each hunter in the boat gets a share of the catch, as well as the women who do most of the butchering.

Once butchering is done, the blubber is carefully removed with an *ulu* (Figures 25.4–25.7) . The blubber is then cut and placed generously on the skin. You can

Figure 25.4 Nancy Leavitt works to remove blubber from a bearded seal skin in preparation for covering an *umiapiaq*. Source: photograph by Margaret Opie.

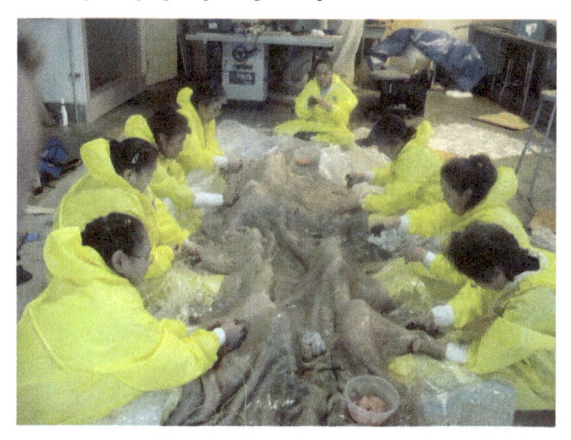

Figure 25.5 Women sew the skins together using caribou sinew. Source: photograph by Margaret Opie.

Figure 25.6 Crew members oil the frame with seal blubber in preparation for the new skin. Source: photograph by Margaret Opie.

Figure 25.7 The crew is almost done tying down the sides of the skin to the frame. Source: photograph by Margaret Opie.

now fold it into a good-sized bundle, tie it with a sturdy rope, and place inside a gunnysack. The more blubber you use, the better it is for the skin to properly cure. The gunnysack is placed in a container that will catch the seeping oil as it ferments.

Sinew from the *tuttu* (caribou) tendons and backstrap are used when sewing the *umiapiaq*. Collecting the tendons begins very early in the summer and even a couple years prior to braiding the sinew. The tendons are best dried outside during the months of October to January in the dark dry months of the Arctic, the time of year when sea ice is forming again. The backstrap is used to make headers that are thin enough to go through the sewing needle. The sinew from the tendons are spliced and then braided, overlapping a strand here and there to make a thread strong enough for the *umiapiaq* to carry up to 10 strong whalers. Sinew braiding can begin during the long winter months.

When the sinew is ready, the frozen bundles of *ugruk* skins, which have been stored outside all winter, are retrieved. If you have prepared the skins properly, they should be very pinkish and prime and the hair will easily peel off.

The women then select the suitable skins for the stern (*aqu*) and bow (*sivu*) of the boat. Large female skins are more suitable for their thickness and size for the *aqu* and *sivu*. The rest of the skins are placed in the middle. The flippers, including the *taliguq* (front flippers) and the tail (*pamiuq*), are left intact when preparing the skins. When stretching the skins over the frame, the thickness of the tail makes a stronghold to tie on both ends of the boat frame.

Sewing the *ugruk* skins for the boat usually occurs in February, when the men may start scouting trails to the edge of the shorefast ice. Trail-breaking will take place from February to May. The sinew is oiled with *natchiq* (seal) for waterproofing and makes it easier to pull through the thick hide. The waterproof stitches are made very cautiously, avoiding the needle from going through the whole skin.

When they are done sewing the top side, it is time for the women to have lunch. Most women opt to have the fermented flippers, which are a delicacy and provide the sewers with much-needed energy. Other delicious items on the menu include *tuttu* soup, duck soup, *nigliq* (goose) soup, *maktak* (whale skin and blubber), fish *quaq* (frozen fish), Eskimo donuts, breads, turkey with all the trimmings and desserts galore.

In the meantime, elder men and women drop by to eat and tell stories that keep them laughing. Occasionally, the elder women will inspect the sewing and provide assistance. After a delicious meal, we are ready to flip the skins and sew the inside part. This is done with a regular stitch, again being careful not to poke through the skin.

The men have brought the *umiapiaq* frame inside. They do minor repair and oil the frame with seal blubber. Once the women are done sewing, the skin is placed and tied onto the frame. The skins are very pliable and stretch with no problem. Women stand by to inspect the seams and do additional stitching as required.

The *umiapiaq* is ready and will bleach outside under the Arctic sun. The captain and crew have cleaned out their ice cellar, prepared their trails over jagged ice ridges and will await the gift of the whale, hunted from the edge of the shorefast ice in April, May and June, before the ice starts to break up and the cycle starts again.

Life on the edge of the sea ice provides peace and indescribable fulfilment to the Inupiat whalers of the Arctic. We love our ocean!

25.2.3 Spring hunting
George Noongwook, Savoonga, Alaska

There is a season for everything that happens in terms of hunting and gathering here in Savoonga. The bulk of our activities occur in the spring when the sea ice begins to recede up to the Arctic Ocean and along with it marine mammals commence their annual migration to their summering grounds in the Arctic. You can sense and feel the excitement of people here because it is the busiest time of the year when all able-bodied people start to prepare for their spring activities.

Preparation is the key to success in our outings. That means sharpening all of our equipment that needs to be honed, knives especially, because you rely on a sharp knife to butcher your catches in an efficient manner and sharpness reduces the chances of getting injured with a dull knife. Efficiency is a big factor in all areas of your activity and being able to butcher a walrus, *maklak* (bearded seal), or seal in very short order is a must, because the prevailing northerly water currents in the Bering Sea are unforgiving. If you are slow, the currents will begin to take you away from the island at a fast clip. All hooks, ice testers, motors, oars and navigation and communication gear are included, along with enough gasoline and lubricants to take you out and back.

Weather forecasts, scanning the horizon, monitoring communication between hunters and double-checking everything before you push off are required. Generally looking and glassing the horizon will help you decide where you want to go. You can see the blink in the sky where the ice systems are located and steer towards the far end of the ice (*sivulitangani*) because that is where the marine mammals are. You can also find a high point in the ice to look around for game occasionally (Figure 25.8). It is also prudent to look back and pinpoint where home is and mark it on your compass every time you stop. Monitor your communications radio for any news coming from other hunters for any concentration of herds of walrus. You must always check your gasoline usage if you have to go out further. We always make sure that we have more than half of what we tote to make sure we come home safely.

When we spot a herd of walrus, we look for a route that will take us close to the herd, downwind from the herd so that they will not smell us. We have to be very quiet when approaching them with our motor running almost idle as we get even closer. We look for the ideal walrus that can provide you with a hide if it is a good-sized female. It provides you food, ivory and all other parts, including walrus stomachs, whether they are full or not. You can also hear a herd of walrus if you are very quiet while glassing the environment. Females and pups make a lot of noise when congregated with others and are very much prized for food. You can pinpoint where the walrus calls are coming from and advance towards them. Hunters can often smell a herd directly upwind and head toward the smell.

Often the herds tend to be away from the ice edge. It is not prudent to waste valuable time in pursuit unless the ice is separating, indicating calm weather. It is better to look for a herd near the ice edge, saving time and energy: efficiency matters. The hunters usually steer clear of pack ice, as you can easily become trapped and the ice recedes north rapidly. The ice can also act as a wind barrier, allowing me to conduct my activities in windier conditions. But again, you just have to keep looking for an opportunity, taking advantage of the ideal ecological conditions. We stick close to the ice edge but keep away from pack ice unless we are alee of the ice, because the ice would be separating on the lee side. The herd can now be approached with ease and you can get back out just as easily.

When travelling north or northeast, depending on the ice pattern, I like to go point-to-point along the ice, trying to get to the front of the ice unless we spot a herd within bays or on the points. Going from point to point saves me gasoline, time and energy so I can concentrate on getting to the animals at or near the front of the ice. It all depends on the weather, ice and sea conditions and it takes constant observation, being vigilant, communicating with your crew on best practices and following through.

It is not all rosy out there in the vast open seas because weather happens whether it is on the rise or receding, but we rely on the ice to provide us safe passage and

Figure 25.8 Chester Noongwook atop an ice floe, scanning his environment for a hunting opportunity. If an animal is seen, he will determine if it can be retrieved, if it can be done safely, and how to be efficient and successful. Source: photograph by Collin Noongwook.

there is food there. This is where I am most comfortable and happy, when any success happens just like anywhere in the world. Having and wearing raingear provides safety for you and your crew. I always like to take the most direct route home, minding the conditions. If it means I need to stick close to the ice edge to protect us from weather, then I will take it. It provides comfort and saves time and energy.

Coming in from the offshore, we can see the Kukuklget Mountains and head to the western edge to compensate for the prevailing water currents. It is quite different near shore because the ice tends to pack with the current moving west to east constantly. We look for an opening to the west of the village and make better time working with the current. There are often lookouts on high points on shore directing traffic as they know the best route that will be easiest for all of us. Crews tend to form a group in tightly packed ice to manoeuvre over the ice as we look for the best path from a high point. Then, working cooperatively together, we can glide our boats over the ice. Helping each other uplifts each other's spirit.

The fog limits your ability to see ahead so you use your other senses to look for mammals. Sense of smell and hearing become very important. Communicating with others becomes very important and knowing where the others are. Manoeuvring through the ice, you see the reflection of open water mirrored above the water in the fog, and you can also see the bright reflection of ice. This will guide you through the ice, saving time and energy heading home. If it does not work out, head away from the ice and look for an opening to the west, remembering to work with the current.

To get to the marine mammals, you learn where to go, learn their environment, their behaviour, patterns and the most ideal conditions. Sometimes you encounter challenges that sap your energy, but it is best to do it together; after all, you are all family.

The camaraderie displayed by hunters forms bonds that are lifelong (Figure 25.9). The spirit of cooperation and sharing yields success and happiness to you, your family and your community. The desired nutrition, teamwork and benefits gained are quite an experience. There is a tremendous amount of respect for each other and for marine mammals, and enjoyment from the fruits of your work that is much desired in life.

Figure 25.9 The result of scanning for a hunting opportunity: a bowhead whale was successfully harvested and is being readied to tow to a butchering site. Source: photograph by Collin Noongwook.

25.3 Documentation of indigenous use of sea ice

25.3.1 Current research on indigenous use of sea ice

A great deal has been written about sea ice and Arctic peoples, especially Inuit. The use of sea ice for hunting has been documented going back to early anthropologists and other scientists (Boas, 1888; Rasmussen, 1927; Freuchen, 1936; Nelson, 1969). Ethnographies and accounts of Inuit and other Arctic peoples' life with sea ice can be found since these early times, but the last decade or so marks an increase in attention to Inuit knowledge and use of sea ice at a time when Arctic climate change has received tremendous interest. As the world began to understand the dramatic changes to sea ice in the Arctic, a new effort to understand these changes from Inuit and other residents who know the ice best was under way. One of the largest examples of this was the (SIKU) Sea Ice Knowledge and Use Project, which was organized during the International Polar Year 2007–2008 (project no. 166; Krupnik et al., 2010a). Dozens of researchers, institutions and communities collaborated under the SIKU umbrella and this resulted in over 50 scientific, popular and online publications in at least four languages, as well as a number of outreach and educational activities (Krupnik et. al, 2010a). Two recent special journal issues dedicated to Inuit knowledge and human interactions with sea ice (more broadly, but including local and indigenous case studies), respectively, also highlight the research trend: 'Geographies of Inuit Sea Ice Use' (Aporta et al., 2011) and 'The Human Geography of Arctic Sea Ice' (Lovecraft, 2013).

One of the most intensive areas of research on sea ice and indigenous peoples in recent times is documentation of observations and knowledge of sea ice. This includes documenting indigenous knowledge of sea ice types, processes and changes in various regions (e.g. Nichols et al., 2004; Gearheard et al., 2006; Laidler & Elee, 2008; Laidler & Ikummaq, 2008; Laidler et al., 2008, 2010; Kielsen Holm, 2010; Krupnik et al., 2010c; Taverniers, 2010; Bogoslovskaya & Krupnik, 2013), sea ice knowledge and use with specific regard to hunting success (e.g. George et al., 2004; Kapsch et al., 2010; Huntington et al., 2013), sea ice knowledge related to dangers and hazards using the ice (e.g. Ford et al., 2009; Laidler et al., 2009; Fienup-Riordan & Rearden, 2012) and reflections on how indigenous knowledge about sea ice is represented and understood (Wisniewski, 2010; Aporta, 2011).

Two other key areas of research in recent years include linking indigenous and local knowledge of sea ice with science and technology, and documenting indigenous language and terminology associated with sea ice. In the first case, a number of approaches have been taken, including bringing scientists and local residents together to design and conduct sea ice field research and monitoring studies (Norton, 2002; George et al., 2004; Norton & Gaylord, 2004; Tremblay et al., 2006; Carmack & MacDonald, 2008; Mahoney et al., 2009; Krupnik et al., 2010c), including indigenous and local knowledge in scientific sea ice field courses (e.g. Huntington et al., 2009), employing remote sensing techniques to study sea ice characteristics with connections to local knowledge and use (Druckenmiller et al., 2010; Kapsch et al., 2010), publishing multimedia products and online atlases to share community-based sea ice knowledge (Fox, 2003; Pulsifer et al., 2010), employing GPS and field computer technology to document sea ice travel and sea ice use (e.g. Aporta, 2003; Gearheard et al., 2010, 2011), and working with hunters to pull survey instruments on their sleds to measure sea ice thickness (Druckenmiller et al., 2010; Druckenmiller, 2011; Wilkinson, 2011; Christian Haas, personal communication).

Sea ice terms and definitions for describing many facets of sea ice are important for capturing sea ice knowledge and transmitting this knowledge to new generations. In recent years, a number of sea ice dictionaries and terminology studies have been conducted, the SIKU project noting that over 25 indigenous vocabulary lists of local terms have been documented during the IPY alone (Krupnik & Weyapuk, 2010, table 14.1). The work has resulted in several dictionaries (Weyapuk & Krupnik, 2012) and other lists, analyses of terms and illustrated terminologies (Johns, 2010; Krupnik & Müller-Wille, 2010; Tersis & Taverniers, 2010; Heyes, 2011; Gearheard et al., 2013).

Lastly, some recent work also highlights the value of indigenous writing and first-person accounts on sea ice by elders and hunters themselves. A published interview with Aipilik Inuksuk from Igloolik (Inuksuk, 2011) shows the richness of printing a full elder interview (as opposed to the soundbites edited or compiled by a researcher). *The Meaning of Ice* (Gearheard et al., 2013) provides a number of first-person accounts of life with sea ice written by some 40 Inuit contributors from Alaska, Canada and Greenland, and *Our Ice, Snow and Winds* (Bogoslovskaya & Krupnik, 2013) was a collaboration of mostly indigenous experts from seven communities on the Russian side of the Bering strait, also richly photographed and illustrated. This type of research and writing, with more direct community-to-community knowledge exchange, is a relatively new pathway for future research to follow.

25.3.2 Harvesting from sea ice

For indigenous hunters and fishers, sea ice serves many functions. Shorefast ice, in particular, is a platform for travel to and from harvesting sites, and a platform for harvesting itself (Figures 25.10 and 25.11). Pack ice is a place to hunt in and also provides refuge from strong winds and high seas or just as a place to rest. Sea ice is habitat for ice-associated marine mammals (e.g. Huntington & Moore, 2008), which provide a large proportion of the food harvested in many Arctic communities. Even during summer when the waters are open, the influence of sea ice on marine ecology persists in the form of the species available and, often,

Figure 25.10 Spring ice fishing through sea ice in Qaanaaq, Greenland. Source: photograph by Shari Gearheard.

Figure 25.11 Seal hunting on the floe edge near Clyde River, Nunavut. Source: photograph by Kelly Elder.

the productive conditions that attract marine mammals and fish.

If the importance of sea ice for marine mammals, seabirds and fish is qualitatively high, quantifying the harvests specifically associated with sea ice is harder. Data on hunting and fishing are sparse around the Arctic. Rigorous harvest surveys have been conducted in many communities in Arctic Alaska (Alaska Department of Fish and Game, no date), but most often only once, yielding a record that is often out of date and provides little insight into interannual variability. Burch (1985) provides a rare example of repeated subsistence studies in the same community, documenting harvest activity in each week of the years the studies were done, along with insights into the role of ice conditions in interannual variability of harvests. Harvest data are available for some communities in Arctic Canada (e.g. Wenzel, 1991), but more effort has been applied to documenting participation in hunting, fishing and trapping, rather than the outcomes of those activities (Northwest Territories Bureau of Statistics, 2009). In Greenland, more attention has been paid to the consumption of local foods, especially in relation to human health outcomes (Hansen et al., 2008). Furthermore, species can be harvested in open water as well as when ice is present, and harvest statistics often present annual totals rather than a seasonal breakdown, depending on ice and snow conditions. Even seasonal summaries may not reveal the differences between early summer hunting or fishing on ice and late summer hunting or fishing in open water.

With those considerations in mind, we can still note that in some communities, such as Savoonga, Alaska, ice-associated marine mammals account for 88% of the total subsistence harvest of over 400 kg per capita in 2009. Here, walrus may be half or more of the harvest by weight, with bearded seals and bowhead whales accounting for up to an additional third (Fall et al., 2013). Communities in northern Canada and Greenland may have similarly high dependence on marine mammals, more typically ringed seals, beluga and narwhal in those regions (e.g. Hansen et al., 2008). Polar bears are harvested throughout their range, accounting for a modest percentage of the total harvest by weight, but playing a much larger role in culture and imagination (e.g. D'Anglure, 1990; Kochneva, 2007; Zdor, 2007). In the Canadian Arctic, polar bears also play a large economic role, as many communities sell some of their polar bear hunting quota to visiting sport hunters (e.g. Dowsley, 2009).

Some harvest surveys and similar research have also examined seasonal patterns in hunting. Changes in sea ice timing and characteristics have caused alterations in hunting patterns throughout the Arctic. In some cases, these changes have produced opportunities, such as the development of an autumn bowhead whale hunt on St Lawrence Island, Alaska, thanks to delayed freeze-up of ice (Noongwook et al., 2007). In other cases, rapid break-up of ice reduces the period in which hunting in pack ice is possible, thus reducing harvest of species most commonly taken in that period, as has been the case for bearded seals in Barrow, Alaska (J. Leavitt, Barrow, Alaska, personal Communication, 2007). Determining the net effect of sea ice changes

Box 25.1 Travel routes

One of the characteristics of Inuit traditional culture is mobility; for generations families moved across the land and sea ice following the cycles of ice, snow, seasons and animals. Although Inuit live in settled communities today, travel is an important part of life and identity, especially sea ice travel for many. Although they may vary depending on the conditions year to year, sea ice trails have historical continuity and the same trails have been travelled for generations (Aporta, 2009). Sea ice trails take people along routes that avoid hazards and aim at particular camps or destinations on the coast (Aporta, 2009), and when it forms each year, sea ice connects people in different communities (Sanguya & Gearheard, 2010).

For many experienced Inuit in the Arctic, the ice trails are changing. One of the most dramatic examples of this may be in far northwest Greenland, where hunters already travel the 'ice foot', a narrow shelf of landfast sea ice that can be tens of metres wide to just a few metres wide (Figure 25.12). The ice foot is an important platform for travel and is increasingly the main option as more areas of sea ice are too thin or have open water (Gearheard et al., 2013). In some areas, the ice foot itself is disappearing as polynyas grow larger, cutting off travel routes and forcing dog teams over land. In some cases, land routes are impassable, require dangerous travel onto glaciers, or involve longer travel times (Gearheard et al., 2013).

Another kind of sea ice trail is found on the North Slope of Alaska, where Iñupiat whaling crews construct seasonal trails from the coast to the edge of the sea ice for the spring whaling season (Figure 25.13; Druckenmiller et al., 2010; Druckenmiller, 2011). These trails pass through ice ridges, rubbled ice and over cracks and flooded areas. The paramount concern for everyone is safety and the ice conditions and any changes are continually observed, with the whaling crews evacuating their camps if indicators point to a risk of shifting or breaking ice (George et al., 2004).

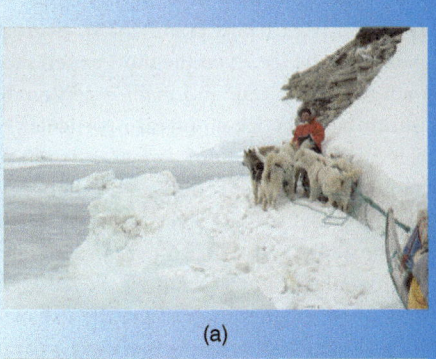

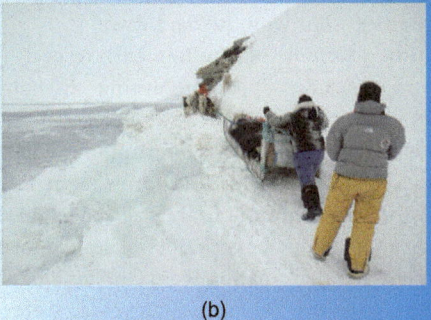

(a)

(b)

Figure 25.12 Travelling the narrow ice foot between Qaanaaq and Siorapaluk, Greenland. Source: photographs by Lars Poort.

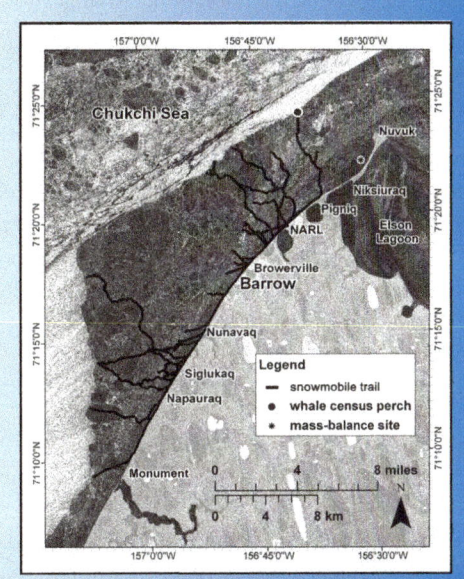

Figure 25.13 Map of Barrow's spring ice trails in 2010. A version of this map was provided to the hunting community as a navigational aid during spring whaling. Trails can be seen terminating at the edge of the shorefast ice, along open water and drifting pack ice. The satellite image from 1 May 2010 is a 12.5-m resolution synthetic aperture radar scene from the European Remote Sensing Satellite, ERS-2, provided by the Barrow Area Information Database. Also shown are the locations of a camp used by biologists to count migrating bowhead whales and the site of a seasonal research site, instrumented to continuously monitor ice and snow thicknesses. Source: Druckenmiller, 2011. Reproduced with permission of Matthew Druckenmiller.

on harvests is not straightforward, as indicated in a quote from Joelie Sanguya, Clyde River, Nunavut (the author of Section 25.2.1), which was used as the title of a paper on sea ice in the region of that community. In response to requests from visiting researchers to specify whether changing sea ice made things easier or harder, Joelie replied, 'It's not that simple' (Gearheard et al., 2006). Instead, he explained, the effect is more one of feeling out of sorts, hunting from a boat in open water at a time of year when one expects to be travelling by snowmobile over ice (see Box 25.1).

25.4 Discussion

Amid rapid social, economic, environmental, cultural and other changes in the Arctic, the use of sea ice continues to provide an emotional and spiritual foundation for many individuals, families and communities across the Arctic. Sea ice is a source of food and a platform for travel, serving a major role in physical access and nutritional well-being. The simple fact of the extensive and intensive use of ice by Inuit peoples from Greenland to Chukotka is enough to stimulate wonder and respect from those visitors fortunate enough to participate first-hand in a way of life reliant upon knowledge and skills about an ecosystem out of the range of experience of most people on Earth.

Thus, as outlined in Section 25.3, many papers and books have been written about the ways in which various Inuit groups use sea ice, describe sea ice, learn about sea ice and so on. Additional work has been done to connect Inuit understanding of sea ice with modern scientific understanding (Eicken, 2010), both to deepen our collective appreciation for the complexities and subtleties of ice, and also to understand better how sea ice is changing over time. While many such efforts demonstrate convergent understanding of sea ice in many ways (Druckenmiller et al., 2010), the significance of ice conditions may be seen in divergent ways. For example, the dramatic summer retreat of sea ice in 2007 made headlines and generated a great deal of scientific investigation, but it was largely a non-event for most coastal communities (Huntington, 2013). These topics are clearly enough to engage the curiosity of many researchers and travellers, not to mention those willing to fund studies of this kind.

All of these points are relevant and important in the communities where sea ice use is a way of life. And as the three accounts in this chapter show, there is much more to the story as well. Sea ice does indeed provide food, and, perhaps more importantly, it provides the right kind of food. Healthy food that has been relied upon for generations. Food to share with relatives and friends. Food to warm the heart as well as the body. Food that connects people with one another, with those here and now and with those in the past, whose traditions are continued. Sea ice does indeed provide a surface for travelling, for visiting others, for being part of the land and sea, for breathing fresh air, for feeling free and whole. For Arctic peoples, sea ice is more than an object, it is part of a way of life, part of living a full life, part of what it means to be human.

And of course sea ice is also dangerous. Many sea ice terms have to do with hazards, with features that require special attention and care. Knowledge of sea ice is never fully sufficient to eliminate danger. Warren Matumeak, a late elder from Barrow, recalled that his grandfather Momigana went out one day to check his seal net and never returned (Gearheard et al., 2013). Every hunter and traveller on the ice, even the most experienced, has a story about a close call, and many traditional stories about hunting on the ice emphasize particular hazards or ways to avoid or get out of danger. It is not obvious that people choosing to spend time in a changeable and unforgiving icescape are risk-averse, but whalers in Barrow will repeatedly, and at the cost of considerable effort, move their whaling camps back from the ice edge when conditions worsen rather than accept the risk of staying put (George et al., 2004). Safety is always a paramount concern, its lessons won the hard way.

Another example of caution and respect for the unknown became apparent during exchanges of hunters from Barrow, Clyde River and Qaanaaq (Gearheard et al., 2013). The visiting hunters were highly experienced and knew a great deal about ice. What they saw in the communities they visited made sense to them, in that the basic mechanics of sea ice formation, movement and deformation were familiar and comprehensible. None of the visitors, however, were willing to travel on the ice without a local guide, someone familiar with the particulars of sea ice conditions and characteristics for that location. The difference between knowing about sea ice in theory and knowing about it in practice can be the difference between life and death. Travelling alone in unfamiliar territory just to have a look around made no sense when local guides were available.

In this context, assessments of climate change and its impacts on sea ice use often emphasize increased hazard along with decreased hunting opportunities (Ford et al., 2008; Durkalec et al., 2014). The latter problem, that loss of sea ice means less hunting on, from and amid sea ice, captures only one side of the impacts of change (Gearheard et al., 2006). Instead, hunting can occur in open water, too, and peak hunting seasons often correspond with particular periods in the annual sea ice cycle. Whether those periods occur earlier or later on the calendar may be secondary, so long as the animals associated with those periods are available and so long as the ice conditions allow hunters to get out. On St Lawrence Island, poor hunting can be associated with too much ice as well as with too little ice and often is a result of poor weather rather than the ice itself (Huntington et al., 2013).

Current rates of change in sea ice are alarming in many respects, but not necessarily of direct and overriding concern to Arctic communities (Gearheard et al., 2006; Huntington, 2013). Annual variability in ice conditions has always been high, and anomalies have always occurred from time to time. Thinner ice poses a challenge for whalers seeking a place to haul a 50-ton bowhead whale out of the water, but the whale hunt continues and west winds are a bigger problem in Barrow than thin ice, as seen in the springs of 1992 and 2013 (C. George, personal communication). Hunters have had to be flexible, to deal with a wide variety of conditions and still provide food for their families (e.g. Huntington et al., 2013). The fact that Arctic communities have persisted through time is sufficient testament to the fortitude, ingenuity and adaptability of their residents to allay concerns about the immediate demise of traditional practices and the cultures sustained by those practices.

Of greater concern than environmental change in many communities, at least in the short term, is social change. Arctic languages are being lost at a rapid rate, as children no longer speak their ancestral tongues at home (Barry, 2013). In Barrow, English is the language of choice for most people under 50. On St Lawrence Island, the pattern may be delayed by a few decades but

Figure 25.14 Spring seal hunting near St Lawrence Island. Source: Courtesy of the Eskimo Heritage Program, Kawerak, Inc., Nome, Alaska. Reproduced with permission.

the trend is still apparent. In Clyde River and Qaanaaq, the situation is more promising for the present, though mass media and entertainment have a long reach and a strong influence. Sea ice terms may continue in use, appropriated into sentences in English, but nuances can easily be lost along with precision.

More important still is the practice of sea ice use. Here, as the three accounts show, there is room for optimism. People continue to use sea ice in many of the same ways as they always have (Figure 25.14). Not everyone does so in every community, but enough people continue traditional practices that they remain vital to community identity and well-being. Indigenous use of sea ice can only continue as long as indigenous people use sea ice. Rather than being a tautology, this statement reflects the idea that sea ice knowledge and sea ice practice are one and the same. Without practice, one cannot retain working knowledge of sea ice. Wisniewski (2010), in his work with Kigiqtaamiut (Shishmaref) hunters, highlights this, explaining that hunters' knowledge is better conceptualized as *knowing* rather than knowledge. Knowing emphasizes both the active application of knowledge and processual learning without delineating any differentiation between the two (Wisniewski, 2010). He takes it a step further by understanding it as *knowings*, as the hunters place great emphasis on individual personal experiences and perspectives for understanding the world. Without knowing the sea ice, one cannot continue to practise sea ice use safely and efficiently.

Ways of knowing the ice have been passed down to today's hunters going back to the original migrants to the Arctic and all the intervening generations up to the

Figure 25.15 Spring whaling near Pugughileq, St Lawrence Island, in a skin boat under sail, April 1976. Source: Courtesy of the Eskimo Heritage Program, Kawerak, Inc., Nome, Alaska. Reproduced with permission.

Figure 25.16 Qaanaaq hunter Toku Oshima leads her dogs team over rough sea ice. Source: photograph by Shari Gearheard.

present. Noongwook et al. (2007) recount how stories about whale hunting from Pugughileq on St Lawrence Island were passed down from the late 1800s to 1970, when Nathan Noongwook (uncle of George Noongwook, author of Section 25.2.3) returned to Pugughileq to start whaling there again. Nathan and others had continued whaling from Gambell and had continued hunting amid sea ice from Savoonga, but no one living had hunted from Pugughileq in spring. Nonetheless, the stories, combined with personal experience in other places around the island, were sufficient to allow Nathan and his family to hunt safely and, by 1972, successfully from the old hunting camp (Figure 25.15). It seems unlikely that the stories alone would have been sufficient, without extensive experience. And it is clear that extensive experience elsewhere around the island was aided by the stories specific to Pugughileq.

Changing conditions will continue to alter the ways people use sea ice. On St Lawrence Island, later arrival of ice has allowed an autumn whale hunt (Noongwook et al., 2007), as well as hunting by boat throughout the winter. In northern Greenland, hunters from Savissivik are no longer able to travel and hunt on the sea ice as they used to, and food caches are vulnerable to polar bears that remain on land during summer (Qaerngaaq Nielsen, in Gearheard et al., 2013). These changes are striking in one way, as traditional patterns are no longer possible, but they are also modest, in that hunters are still able to pursue the same animals using many of the same techniques. Going on the hunt has the same meaning, producing the same feelings of satisfaction and community. Hunters continue to learn as they always have, noting new conditions and sharing experiences with others.

Although the last decade or so has seen a great deal of writing and attention to Inuit knowledge and use of sea ice, it has for the most part focused on the knowledge of men. This is no surprise, as it is the men who have always travelled the ice and hunted for food, while the women primarily took care of preparing skins, making clothing and tending to children, in addition to travelling as part of the nomadic life or, in the modern era, to and from seasonal and other camps. However, women do have important contributions to make to many aspects of knowing the environment, including about sea ice (Dowsley et al., 2010; Gearheard et al., 2013), as well as about the activities that women have typically carried out in support of hunting and fishing. More research is needed into the roles of women and their participation in sea ice-related activities through time and today (Figure 25.16).

The sea ice remains a beautiful garden. The use of sea ice remains a beautiful form of art and practice, drawing together personal experience and knowledge from past generations, continuing to fulfil the relationship between people and their environment, which has long defined human existence in the Arctic. Scholarly and other writings about indigenous use of sea ice are valuable and interesting, and can help indigenous peoples recognize the value of what they might otherwise take for granted as just another part of their daily lives. But the future of indigenous use of sea ice lies in continuing to practise a way of life, and to experience the feelings of joy and worthiness that are apparent when indigenous people tell their own stories about sea ice.

25.5 Acknowledgements

We are grateful to David Thomas for the invitation to prepare this chapter. We thank the many sea ice users and sea ice researchers who have contributed so much to our collective understanding of sea ice, and to making connections across different ways of seeing and knowing ice. This chapter is a result of what we have learned from many people. Igor Krupnik reviewed a draft of the chapter and we are grateful for his constructive comments. We thank Jeremy Davies for preparing Figure 25.2. Any mistakes are ours, but any insights and value are the product of the generosity of all who have shared with us their lifetimes of experience, study and love of *siku* in all its forms and with all its meanings.

References

ACIA (2005) *Arctic Climate Impact Assessment*. Cambridge University Press, Cambridge.

Alaska Department of Fish and Game (n.d.) Community subsistence information system. http://www.adfg.alaska.gov/sb/CSIS/

Aporta, C. (2003) New ways of mapping: using GPS mapping software to plot placenames and trails in Igloolik (Nunavut). *Arctic*, **56**, 321–327.

Aporta, C. (2009) The trail as home: Inuit and their pan-Arctic network of routes. *Human Ecology*, **37**, 131–146.

Aporta, C. (2011). Shifting perspectives on shifting ice: documenting and representing Inuit use of sea ice. *The Canadian Geographer*, **55**, 6–19.

Aporta, C., Taylor, D.R.F. & Laidler, G.J. (Eds.) (2011) Geographies of Inuit sea ice use. *Canadian Geographer*, **55**, 1–142.

Barry, T. (2013) Linguistic diversity. In: *Arctic Biodiversity Assessment*, pp. 652–663. Conservation of Arctic Flora and Fauna, Akureyri, Iceland.

Boas, F. (1888) *The Central Eskimo*. University of Nebraska, Lincoln, Nebraska (reprint 1964).

Bogoslovskaya, L.S. & Krupnik, I. (Eds.) (2013) *Our Ice, Snow and Winds: indigenous and Academic Knowledge on Ice-Scapes and Climate of Eastern Chukotka* (in Russian). Russian Heritage Institute, Moscow and Washington.

Burch, E.S. Jr. (1985) *Subsistence Production in Kivalina, Alaska: A Twenty-Year Perspective*. Technical Paper No. 128. Alaska Department of Fish & Game, Juneau, Alaska.

Carmack, E. & MacDonald, R. (2008) Water and ice-related phenomena in the coastal region of the Beaufort Sea: Some parallels between Native experience and western science. *Arctic*, **61**, 265–280.

D'Anglure, B.S. (1990) Nanook, super-male: the polar bear in the imaginary space and social time of the Inuit of the Canadian Arctic. In: *Signifying Animals: Human Meaning in the Natural World* (Ed. R. Willis), pp. 178–195. Unwin Hyman, Boston, Massachusetts.

Dowsley, M. (2009) Inuit-organised polar bear sport hunting in Nunavut territory, Canada. *Journal of Ecotourism*, **8**, 161–175.

Dowsley, M., Gearheard, S., Johnson, N. & Inksetter, J. (2010) Should we turn the tent?: Inuit women and climate change. *Etudes Inuit/ Inuit Studies*, **34**, 151–165.

Druckenmiller, M.L. (2011) *Alaska shorefast ice: interfacing geophysics with local sea ice knowledge and use*. Unpublished Ph.D. dissertation. University of Alaska Fairbanks, Fairbanks, Alaska.

Druckenmiller, M., Eicken, H., George, J.C. & Brower, L. (2010) Assessing shorefast ice: Iñupiat whaling trails off Barrow, Alaska. In: *SIKU: Knowing Our Ice: Documenting Inuit Sea-Ice Knowledge and Use* (Eds. I. Krupnik et al.,), pp. 203–228. Springer, London.

Durkalec, A., Furgal, C., Skinner, M.W. & Sheldon, T. (2014) Investigating Environmental Determinants of Injury and Trauma in the Canadian North. *International Journal of Environmental Research and Public Health*, **11**, 1536–1548.

Eicken, H. (2010) indigenous knowledge and sea ice science: what can we learn from indigenous ice users? In: *SIKU: Knowing Our Ice: Documenting Inuit Sea-Ice Knowledge and Use* (Eds. I. Krupnik et al.,), pp. 357–376. Springer, London.

Fall, J.A., Braem, N.S., Brown, C.L., Hutchinson-Scarbrough, L.B., Koster, D.S. & Krieg, T.M. (2013). Continuity and change in subsistence harvests in five Bering Sea communities: Akutan, Emmonak, Savoonga, St Paul, and Togiak. *Deep-Sea Research Part II*, **94**, 274–291.

Fienup-Riordan, A. & Rearden, A. (2012) *Ellavut: Our Yup'ik World and Weather: Community and Change on the Bearing Sea Coast*. University of Washington, Seattle.

Ford, J.D., Pearce, T., Gilligan, J., Smit, B. & Oakes, J. (2008) Climate change and hazards associated with ice use in Northern Canada. *Arctic, Antarctic, and Alpine Research*, **40**, 647–659.

Ford, J.D., Gough, W.A., Laidler, G.J., MacDonald J., Irngaut, C. & Qrunnut, K. (2009) Sea ice, climate change, and community vulnerability in northern Foxe Basin, Canada. *Climate Research*, **38**, 137–154.

Fox, S. (2003) *When the Weather is Uggianaqtuq: Inuit Observations of Environmental Change*. CD-ROM. National Snow and Ice Data Center, Boulder, Colorado.

Freuchen, P. (1936) *Arctic Adventure: My Life in the Frozen North*. Echo Point, Brattleboro, Vermont (reprint 2013).

Gearheard, S., Matumeak, W., Angutikjuaq, I. et al. (2006) 'It's not that simple': A comparison of sea ice environments, uses of sea ice, and vulnerability to change in Barrow, Alaska, USA, and Clyde River, Nunavut, Canada. *Ambio*, **35**, 203–211.

Gearheard, S., Aipellee, G. & O'Keefe, K. (2010) The Igliniit Project: Combining Inuit knowledge and geomatics engineering to develop a new observation tool for hunters. In: *SIKU: Knowing Our Ice: Documenting Inuit Sea-Ice Knowledge and Use* (Eds. I. Krupnik et al.,), pp. 181–202. Springer, London.

Gearheard, S. Aporta, C., Aipellee, G. & O'Keefe, K. (2011) The Igliniit Project: Inuit hunters document life on the trail to map and monitor Arctic change. *The Canadian Geographer*, **55**, 42–56.

Gearheard, S., Kielsen Holm, L., Huntington, H.P. et al. (Eds.) (2013) *The Meaning of Ice: People and Sea ice in Three Arctic Communities*. International Polar Institute, Hanover, New Hampshire.

George, J.C., Huntington, H.P., Brewster, K., Eichen, H, Norton, D. & Glenn, R. (2004) Observations on shorefast ice dynamics in arctic Alaska and the responses of the Inupiat hunting community. *Arctic*, **57**, 363–374.

Hansen, J.C., Deutch, B. & Odland, J.Ø. (2008) Dietary transition and contaminants in the Arctic: emphasis on Greenland. *Circumpolar Health Supplements*, **2008**, 2.

Heyes, S.A. (2011) Cracks in the knowledge: sea ice terms in Kangiqsualujjuaq, Nunavik. *The Canadian Geographer*, **55**, 69–91.

Huntington, H.P. 2013. A question of scale: local versus pan-Arctic impacts from sea-ice change. In: *Media and the Politics of Arctic Climate Change* (Eds. M. Christensen, A.E. Nilsson & N. Wormbs), pp. 114–127. Palgrave MacMillan, Basingstoke, Hants., UK.

Huntington, H.P. & Moore. S.E. (Eds.) (2008) Assessing the impacts of climate change on arctic marine mammals. *Ecological Applications* **18** Supplement. http://www.esajournals.org/toc/ecap/18/sp2

Huntington, H.P., Gearheard, S., Druckenmiller, M. & Mahoney, A. (2009) Community-based observation programs and indigenous and local sea ice knowledge. In: *Field Techniques for Sea Ice Research* (Eds. H. Eicken et al.,), pp. 345–364. University of Alaska, Fairbanks, Alaska.

Huntington, H.P., Noongwook,G., Bond, N.A., Benter, B., Snyder, J.A. & Zhang, J. (2013) The influence of wind and ice on spring walrus hunting success on St Lawrence Island, Alaska. *Deep-Sea Research II*, **94**, 312–322.

ICC Canada. 2008. *The Sea Ice Is Our Highway: An Inuit Perspective on Transportation in the Arctic*. Inuit Circumpolar Council, Ottawa.

Inuksuk, A. 2011. On the nature of sea ice around Igloolik. *The Canadian Geographer*, **55**, 36–42.

Johns, A. (2010) Inuit sea ice terminology in Nuavut and Nunatsiavut. In: *SIKU: Knowing Our Ice: Documenting Inuit Sea-Ice Knowledge and Use* (Eds. I. et al.,), pp. 401–412. Springer, London.

Kapsch, M., Eicken, H. & Robards, M. (2010) Sea ice distribution and ice use by Indigenous walrus hunters on St Lawrence Island, Alaska. In: *SIKU: Knowing Our Ice: Documenting Inuit Sea-Ice Knowledge and Use* (Eds. I. Krupnik et al.,), pp. 115–144. Springer, London.

Kielsen Holm, L. (2010) Sila-Inuk: Study of the impacts of climate change in Greenland. In: *SIKU: Knowing Our Ice: Documenting Inuit Sea-Ice Knowledge and Use* (Eds. I. Krupnik et al.,), pp. 145–160. Springer, London.

Kochneva, S. (2007) *Polar Bear in Material and Spiritual Culture of the Native Peoples of Chukotka*. Chukotka Association of Traditional Marine Mammal Hunters, Anadyr, Chukotka.

Krupnik, I. & Müller-Wille, L. (2010) Franz Boas and Inuktitut terminology for ice and snow: From the emergence of the field to the 'Great Eskimo Vocabulary Hoax'. In: *SIKU: Knowing Our Ice: Documenting Inuit Sea-Ice Knowledge and Use* (Eds. I. Krupnik et al.,), pp. 377–400. Springer, London.

Krupnik, I. & Weyapuk Jr., W. (2010) Qanuqilitaavut: 'How we learned what we know' (Wales Inupiaq sea ice dictionary). In: *SIKU: Knowing Our Ice: Documenting Inuit Sea-Ice Knowledge and Use* (Eds. I. Krupnik et al.,), pp. 321–354. Springer, London.

Krupnik I., Aporta C., Gearheard S., Laidler G. & Kielsen Holm, L. (Eds.) (2010a) *SIKU: Knowing Our Ice: Documenting Inuit Sea-Ice Knowledge and Use*. Springer, London.

Krupnik, I., Aporta, C. & Laidler, G.J. (2010b) SIKU: International Polar Year Project #166 (An Overview). In: *SIKU: Knowing Our Ice: Documenting Inuit Sea-Ice Knowledge and Use* (Eds. I. Krupnik et al.,), pp. 1–28. Springer, London.

Krupnik, I. Apangalook Sr., L. & Apangalook, P. (2010c) 'It's cold, but not cold enough': Observing ice and climate change in Gambell, Alaska, *in IPY 2007–2008 and beyond. In:* SIKU: Knowing Our Ice: Documenting Inuit Sea-Ice Knowledge and Use (Eds. I. Krupnik et al.,), pp. 81–114. Springer, London.

Laidler, G.J. & Elee, P. (2008) Human geographies of sea ice: freeze/thaw processes around Cape Dorset, Nunavut, Canada. *Polar Record*, **44**, 51–76.

Laidler, G.J. & Ikummaq, T. (2008) Human geographies of sea ice: freeze/thaw processes around Igloolik, Nuanvut, Canada. *Polar Record*, **44**, 127–153.

Laidler, G. J., Dialla, A. & Joamie, E. (2008) Human geographies of sea ice: freeze/thaw processes around Pangnirtung, Nunavut, Canada. *Polar Record*, **44**, 335–361.

Laidler, G.J., Ford, J.D., Gough, W.A. et al. (2009) Travelling and hunting in a changing arctic: assessing Inuit vulnerability to sea ice change in Igloolik, Nunavut. *Climatic Change*, **94**, 363–397.

Laidler, G.J., Elee, P., Ikummaq, T., Joamie, E. & Aporta, C. (2010). Mapping Inuit sea ice knowledge, use, and change in Nunavut, Canada (Cape Dorset, Igloolik, Pangnirtung). In: *SIKU: Knowing Our Ice: Documenting Inuit Sea-Ice Knowledge and Use* (Eds. I. Krupnik et al.,), pp. 45–80. Springer, London.

Loeng, H. (2005) Marine systems. In: *Arctic Climate Impact Assessment*, pp. 453–538.

Lovecraft, A.H. (Ed.) (2013) The human geography of Arctic sea ice. *Polar Geography*, **36**, 1–162.

Mahoney, A., Gearheard, S., Oshima, T. & Qillaq, T. (2009) Ice thickness measurements from a community-based observing network. *Bulletin of the American Meteorological Society*, **90**, 370–377.

McDonald, M., Arragutainaq, L. & Novalinga, Z. (1997) *Voices from the Bay: Traditional Ecological Knowledge of Inuit and Cree in the James Bay Bioregion*. Canadian Arctic Resources Committee and Environmental Committee of the Municipality of Sanikiluaq, Ottawa.

Nelson, R.K. (1969). *Hunters of the Northern Ice*. University of Chicago, Chicago.

Nichols, T., Berkes, F., Jolly, D., Snow, N.B. & the community of Sachs Harbour (2004) Climate change and sea ice: local observations from the Canadian western Arctic. *Arctic*, **57**, 68–79.

Noongwook, G., the Native Village of Savoonga, the Native Village of Gambell, Huntington, H.P. & George, J.C. (2007) Traditional knowledge of the bowhead whale (*Balaena mysticetus*) around St. Lawrence Island, Alaska. *Arctic*, **60**, 47–54

Northwest Territories Bureau of Statistics. (2009) *State of the Environment Report*. NWT Bureau of Statistics, Yellowknife.

Norton, D.W. (2002) Coastal sea ice watch: Private confessions of a convert to indigenous knowledge. In: *The Earth Is Faster Now: Indigenous Observations of Arctic Environmental Change* (Eds. I. Krupnik & D. Jolly), pp. 126–155. Arctic Research Consortium of the United States, Fairbanks, Alaska.

Norton, D.W. & Graves, A.G. (2004) Drift velocities of ice floes in Alaska's northern Chukchi Sea flaw zone: Determinants of success by spring subsistence whalers in 2000 and 2001. *Arctic*, **57**, 347–362.

Oozeva, C., Noongwook, C., Noongwook, G., Alowa, C. & Krupnik, I. (2004) *Watching Ice and Weather Our Way*. Arctic Studies Center, Smithsonian Institution, Washington, DC.

Pulsifer, P., Laidler, G.J., Taylor, F.D.R. & Hayes, A. (2010). Creating an online cybercartographic atlas of Inuit sea ice knowledge and use. In: *SIKU: Knowing Our Ice: Documenting Inuit Sea-Ice Knowledge and Use* (Eds. I. Krupnik et al.,), pp. 229–254. Springer, London.

Rasmussen, K. (1927) *Across Arctic America: Narrative of the 5th Thule Expedition*. University of Alaska, Fairbanks, Alaska (reprint 1999).

Sanguya, J. & S. Gearheard. (2010) Preface. In: *SIKU: Knowing Our Ice: Documenting Inuit Sea-Ice Knowledge and Use* (Eds. I. Krupnik et al.,), pp. ix–x. Springer, London.

Taverniers, P. (2010) Weather variability and changing sea ice use in Qeqertaq, west Greenland 1987–2008. In: *SIKU: Knowing Our Ice: Documenting Inuit Sea-Ice Knowledge and Use* (Eds. I. Krupnik et al.,), pp. 31–44. Springer, London.

Tersis, N. & Taverniers, P. (2010) Two Greenlandic sea ice lists and some considerations regarding Inuit sea ice terms. In: *SIKU: Knowing Our Ice: Documenting Inuit Sea-Ice Knowledge and Use* (Eds. I. Krupnik et al.,), pp. 413–426.

Tezuka, K. (2009) Ainu sea otter hunting from the perspective of Sino-Japanese trade. In: *Human-Nature Relations and the Historical Backgrounds of Hunter-Gatherer Cultures in Northeast Asian Forests* (Ed. S. Sasaki), pp. 117–131. Senri Ethnological Studies 72.

Tremblay, M., Furgal, C., Lafortune, V. et al. (2006) Communities and ice: Linking traditional and scientific knowledge. In: *Climate Change: Linking Traditional and Scientific Knowledge* (Eds. R. Riewe & J. Oakes), pp. 185–201. Aboriginal Issues Press, University of Manitoba, Winnipeg.

Wenzel, G. (1991) *Animal Rights, Human Rights: Ecology, Economy and Ideology in the Canadian Arctic*. Belhaven Press, London.

Weyapuk Jr., W. & Krupnik, I. (2012) *Wales Inupiaq Sea Ice Dictionary*. Arctic Studies Center, Smithsonian Institution, Washington, DC.

Wilkinson, J.P., Hanson, S., Hughes, N.E. et al. (2011) Tradition and technology: Sea ice science on Inuit sleds. *EOS*, **92**, 1–4.

Wisniewski, J. (2010) Knowing about sigu: Kigiqtaamiut hunting as an experiential pedagogy. In: *SIKU: Knowing Our Ice: Documenting Inuit Sea-Ice Knowledge and Use* (Eds. I. Krupnik et al.,), pp. 275–294. Springer, London.

Zdor, E. (2007) Traditional knowledge about polar bear in Chukotka. *Études/Inuit/Studies*, **31**, 321–323.

CHAPTER 26
Advances in palaeo sea ice estimation

Leanne Armand,[1] Alexander Ferry[1] and Amy Leventer[2]

[1] *Department of Biological Sciences and MQ Marine Research Centre, Macquarie University, North Ryde, NSW, Australia*
[2] *Department of Geology, Colgate University, Hamilton, NY, USA*

26.1 Introduction

Sea ice is the most seasonal and expansive geophysical parameter of the Earth's surface. Annual sea ice growth and decay influence key oceanic and atmospheric processes, as well as global climate (Comiso, 2010; Brandon et al., 2010; Chapter 1). Sea ice exerts a significant influence on the radiative balance and ocean–atmosphere heat flux of the polar regions (Comiso, 2010; Chapters 5 and 6). The palaeo sea ice record for the world's oceans is therefore an important aspect of any effort focusing on the study of past (and future) climate change (Dieckmann & Hellmer, 2010; Vaughan et al., 2013; Chapter 12), mechanisms driving ice-age climates, and for testing palaeoclimate models (Gersonde et al., 2003, 2005). Since the 1970s, oceanographers have worked towards reconstructing past oceanographic conditions (CLIMAP Project Members 1976, 1981), including palaeo sea ice, with the field employing increasingly sophisticated methods, more highly resolved records and a wider suite of proxies through time (de Vernal et al., 2013a).

The objective of this updated review chapter is to cover both the growth and establishment of methods for reconstructing past sea ice distribution and its seasonal variability based on microfossil assemblage distribution and on geochemical and sedimentological tracers that have been established for polar regions, thus highlighting advances since our previous two reviews (Armand & Leventer, 2003, 2010). In addition, palaeo sea ice reconstructions for specific time-slices are reviewed. In general, in both the northern and southern high latitudes, the primary method through which sea ice cover is reconstructed is via the microscopic remains of marine organisms in marine sediments, with new techniques targeting biomarkers linked to microorganisms (Massé et al., 2011; Brown et al., 2014a). Although diatom assemblage data have been most useful in the Antarctic, in the Arctic, quantitative analysis of dinocyst assemblages has been a more powerful tool (de Vernal et al., 2013a, 2013b). Given the large differences in the utility of types of palaeo sea ice proxies that are useful in the northern versus southern polar regions, the Arctic and the Antarctic are reviewed separately.

26.2 Antarctic

The palaeoceanography of the Southern Ocean, in particular palaeo sea ice cover, is of considerable interest to a broad range of researchers, including climate modellers, palaeoclimatologists, oceanographers and palaeobiologists (Armand & Leventer, 2010). Efforts to find new, rapidly processed and analysed chemical proxies have embraced, to varying degrees, solutions involving the utility of ice core records, such as sea salt flux and methanesulphonic acid (MSA) (Abram et al., 2013), or biomarkers – specifically, highly branched isoprenoid (HBI) lipids, coined as 'the ice proxy with 25 carbons' by Brown et al. (2014a). Nonetheless, the

Sea Ice, Third Edition. Edited by David N. Thomas.
© 2017 John Wiley & Sons, Ltd. Published 2017 by John Wiley & Sons, Ltd.

traditional means of using sea ice-associated diatoms has continued to supply the most frequently referred records of sea ice cover or extent for the Holocene and the Late Quaternary (de Vernal et al., 2013a). The most recent Intergovernmental Panel on Climate Change report persists in emphasizing the lack of modern sea ice palaeo records from all sectors of the Southern Ocean, which hinders modelling efforts (Vaughan et al., 2013); however, major international initiatives have been targeting research addressing this issue (e.g. PAGES Sea Ice Proxy Working Group; Scientific Committee on Antarctic Research (SCAR) Six Priorities for Antarctic Science; Kennicutt et al., 2014).

This Antarctic section focuses initially on new advances within sea ice diatom ecology with respect to the estimation of past sea ice cover, and subsequently on the improvements to statistical approaches used to generate estimates of palaeo sea ice. Following this summary, we then discuss the evolving alternative sea ice proxies based on Antarctic ice core and marine sediment geochemical records. The alternative proxies we cover include the ice core-recovered sea salt flux, MSA and halogen (bromine and iodine)-based proxies, whilst HBIs represent the major geochemical sediment-based proxy. The robustness and utility of each sea ice proxy are briefly discussed and, where possible, compared with the traditional diatom proxy. We complete our summary of the Antarctic palaeo sea ice record by focusing on the past climatic reconstruction advances from the proximate pre-satellite record, the Early Holocene Climatic Optimum (11 000–9500 years before present (11–9.5 kyr BP)). Our summary would not be complete without considering the sea ice results covering the Last Glacial Maximum (LGM) in the Southern Ocean.

26.2.1 Microfossil diatom distributions with respect to sea ice

Diatoms, single-celled microscopic algae, are abundant throughout the Southern Ocean (Hart, 1942; Zernova, 1970). The physical, chemical and biological conditions of the Southern Ocean vary on seasonal and interannual timescales, with implications for marine phytoplankton biomass, abundance and composition (e.g. Iida & Odate, 2014; Chapter 14). The vertical stability of the water column and nutrient input, resulting from sea ice melt, are now considered major factors enhancing diatom productivity at the sea ice edge and throughout the marginal sea ice zone (Smith & Comiso, 2008; Taylor et al., 2013) and it is not uncommon to see reports of diatom blooms being associated with the retreating sea ice (e.g. Riaux-Gobin et al., 2011; Grigorov et al., 2014; Saba et al., 2014). It then follows that the distribution of fossil diatom assemblages throughout the Southern Ocean is largely representative of the environmental conditions prevailing in the surface water of each oceanographic zone and, in the case of the marginal sea ice zone, principally influenced by the seasonal sea ice cover (e.g. Armand et al., 2005; Esper et al., 2010; Olguín & Alder, 2011; Esper & Gersonde, 2014; Chapter 14). Importantly, diatom siliceous valves remain abundant within the sediments surrounding the Antarctic (Lisitzin, 1971; Burckle, 1984), and over the last 45 years the distribution of fossil diatom species has been incrementally linked to the sea ice environment as seafloor sediments are analysed and added to diatom seafloor distribution databases (Table 26.1). Therefore, diatom fossil assemblages have become a useful proxy for evaluating palaeo sea ice parameters (i.e. sea ice concentration), as well as past sea ice cover (e.g. Armand & Leventer, 2010; McKay et al., 2012; de Vernal et al., 2013a; Esper and Gersonde 2014; Ferry et al., 2015a).

The seafloor preservation of diatom valves has enabled sector and latitudinal understanding of the Southern Ocean diatom biogeography. Studies across the three major oceanic sectors (Atlantic, Indian and Pacific) have developed our understanding of the associations between diatom assemblage distribution and sea ice extent (Zielinski & Gersonde, 1997; Armand et al., 2005; Esper et al., 2010; Esper & Gersonde, 2014). The diatoms considered relevant to elucidating past sea ice cover are detailed in Figure 26.1 (and references within). Each species is considerably silicified and represents different regions of sea ice concentration, extent or influence. Each has been attributed a role within the estimation of past sea ice conditions with varying success, dependent on their proximity to the Antarctic coast and preservational qualities. Sediment distributions from all sectors of the Southern Ocean suggest that the winter sea ice extent boundary and near freezing sea surface temperatures (SSTs; −1.5–1°C) remain tightly bound to the distribution of *Actinocyclus actinochilus*, *Fragilariopsis curta*, the *Thalassiosira antarctica/scotia* group, and, in a negative exclusion relationship, to *Thalassiosira lentiginosa*. Species with clearly tied maximum abundances south of the sea

Table 26.1 List of published diatom seafloor surface studies in the Southern Ocean since 2008.

Reference	Region	Collection method	Analysis method
Pike et al. (2008)	Western Antarctic Peninsula	Core tops (sediment grabs, box, trigger and kasten cores)	Relative abundances, DCA, PCA
Armand et al. (2008)	Kerguelen Plateau, South Indian	Multicorer	Relative abundances, Diversity indicies
Cochran & Neil (2010)	South Western Pacific, New Zealand	Core tops (sediment grabs and cores)	Diatom concentrations, DCA, Correspondence and Cluster Analysis.
Esper et al. (2010)	Eastern and central South Pacific	Multicorer	Relative abundances, RDA
Mohan et al. (2011)	Coastal East Antarctic (off Kemp Land)	Peterson Grab	Relative abundances
Shukla & Sudhakar (2011)	Coastal East Antarctic (Kemp Land)	Peterson Grab	Relative abundances
Sañé et al. (2011)	Coastal western Weddell Sea (Larsen embayment)	Multicorer	Relative abundances

DCA, detrended correspondence analysis; PCA, principal components analysis; RDA, polynomial canonical redundancy analysis.

ice zone but with decreasing abundances north to the winter sea ice edge include *Fragilariopsis cylindrus* and *Fragilariopsis obliquecostata*. Additional species that have an association with the summer sea ice edge, but only observed from southeast Indian and southwest Pacific sectors sediments, include *Fragilariopsis sublinearis*, *Porosira glacialis* and *Stellarima microtrias*. *Chaetoceros* resting spores are predominant in the Antarctic coastal regions and within the sea ice zone where their abundances can attain up to 92% of the seafloor assemblage, yet due to their generic treatment their significance to sea ice conditions remains established, but unclear. Equally, the generic treatment of *Eucampia antarctica* varieties in sediment studies is likely to conceal a relationship between the sea ice variety *recta* to the maximum and minimum extents of annual sea ice. Spore formation in both *Chaetoceros* (Hyalochaete) spp. and the *Thalassiosira antarctica/scotia* group are likely to represent other environmental conditions related to sea ice zone dynamics (e.g. nutrient supply/exhaustion or sea ice cover closing a polynya).

For each species listed in Figure 26.1, the sediments provide one window to wider biogeographic distribution. Alternatively, regional snapshots by sediment traps affected by seasonal sea ice also detail the relationship and seafloor transfer history of these species. The best picture we have in this spatial context of biogeographic flux and relationship to sea ice comes from the Antarctic Environment and Southern Ocean Process Study longitudinal time-series (Grigorov et al., 2014). The results of this work extending from the Subantarctic Zone to south of the Southern Boundary of the Antarctic Circumpolar Current clearly observed the increase in the relative abundance (as a function of each species per year) of *A. actinochilus*, *F. curta*, *F. cylindrus*, *F. obliquecostata*, *S. microtrias* and *Eucampia antarctica* var. *recta*. Interestingly, no fluxes of *F. sublinearis*, *P. glacialis* or *Chaetoceros* resting spores were ever encountered across the AESOPS transect in the space of 1 year. In contrast, both *T. antarctica/scotia* group and *T. lentiginosa* showed increasing abundances when north of the winter sea ice edge; the latter species' absolute fluxes were noted to be closely tied to the retreating sea ice edge.

Seafloor studies and sediment-trap fluxes do not stand alone in their interpretations and utility to palaeoceanography and sea ice history. Modern oceanographic phytoplankton studies, satellite observations, geochemical biomarkers and under-way metagenomic surveys of the Southern Ocean (not covered here) are also playing a new role in the chapter of sea ice associations and spatiotemporal distributions of diatoms to sea ice. It is not impossible to imagine that within two decades, the seasonal cycle and flux of diatoms from the surface of the ocean through to the sediments will be modelled based upon the findings of both studies. The transfer by proxy between the surface conditions and palaeo records is progressively advancing to the point where careful study will deliver the key to unlocking ever more precise records of past climatic change.

Species	Atlantic/SW Indian	SE Indian/SW Pacific	SE Pacific	Combined Antarctic (not SE Indian/SW Pacific)
Actinocyclus actinochilus 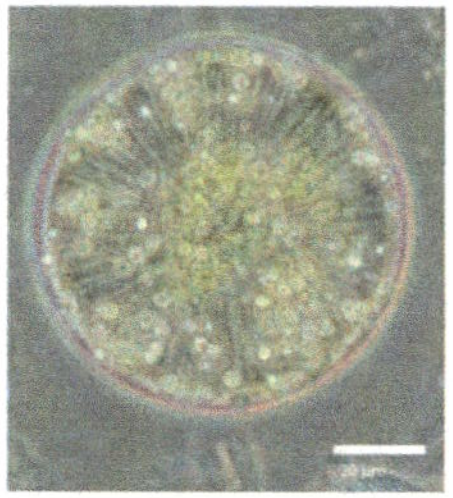	**Sediments:** Ref A *Max RA:* 4.7%, −1 to 0.5°C sSST *SI assoc.:* Year-round to WSI edge **Traps:** Ref E, F PB – low RA 0.86%. WS1-4 – no data available	**Sediments:** Ref B *Max RA:* 2.9%, 0 to 1°C fSST *SI assoc.:* between WSI and SSI edge *Max SSI, WSI conc @MRA:* 2%, 82% **Traps:** Ref G Low RA (≤ 0.15 RA year⁻¹) in all traps S of 60°S (PF) December–January	**Sediments:** Ref C *Max RA:* 11%, −1.5 to −0.5°C sSST *SI assoc.:* N of WSI *Max VAR:* N or S of WSI edge **Traps:** Ref E, F BST – ~8% RA (upper trap) to 1% RA (deep traps ≥ 1410 m) and seafloor sediment. DP – ~0.5% (upper trap) to 1.2% deep trap, 0.77% seafloor sediment. K1–K2 traps this species is considered associated with the occurrence of *Chaetoceros* spp	**Sediments:** Ref D *Max RA:* 4.4% *SI assoc.:* >1% RA in >60% WSI conc.

Figure 26.1 Key Southern Ocean diatom species used in modern proxy analysis to determine sea ice estimations. Summary observations from significant papers (sediments and sediment traps studies) are presented by oceanographic sectors, indicating the general circumpolar nature of the distributions observed against sea ice parameters and exported flux observations. Sediment trap coverage is poor in the Southeast (SE) Pacific sector and traps in the far Southwest (SW) Pacific/Antarctic Peninsula region have been used as the closest indicator. All photographs were taken by L. Armand from surface-water samples along the Sabrina Coast, East Antarctica, with the exception of *Thalassiosira antarctica* resting spores, which are from the Kerguelen Plateau. Scale bars are all 20μm. References: A, Zielinski and Gersonde (1997); B, Armand et al. (2005); C, Esper et al. (2010); D, Esper and Gersonde (2014); E, Abelmann and Gersonde (1991); F, Gersonde and Wefer (1987); G, Grigorov et al. (2014); H, Honjo et al. (2000; supplementary material); I, Crosta et al. (2005). Abbreviations: S, south; N, north; WSI, winter sea ice; SSI, summer sea ice; RA, relative abundance; SI, sea ice; MRA, maximum relative abundance; Max VAR, maximum valve accumulation rate; sSST, summer sea surface temperature; fSST, February sea surface temperature; PF, Polar Front; SAF, Subantarctic Front; sACCF, southern Antarctic Circumpolar Current front; SB-ACCZ, Southern Boundary-Antarctic Circumpolar Current zone; AESOPS, US JGOFS Antarctic Environment and Southern Ocean Process Study trap programme; BST, Bransfield Strait short-term trap 22.5 days; DP, Drake Passage short-term trap 20 day; PB, Powell Basin short-term trap; K1-K2, Bransfield Strait traps consecutive years; K3, Bransfield Strait 6-month trap; WS1-4, Weddell Sea trap consecutive ~1 year series.

Species	Atlantic/SW Indian	SE Indian/SW Pacific	SE Pacific	Combined Antarctic (not SE Indian/SW Pacific)
Fragilariopsis curta 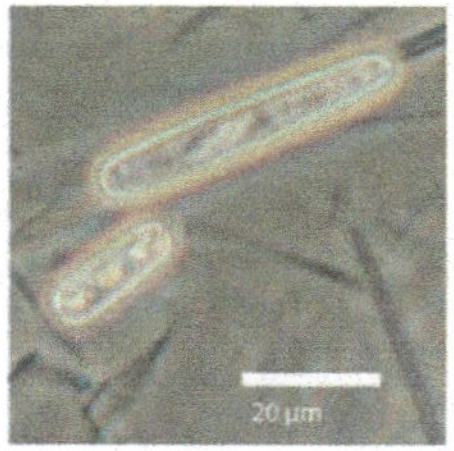	**Sediments:** Ref A *Max RA:* 57.4%, −2 to −1°C sSST *SI assoc.:* Year-round sea ice to WSI edge **Traps:** Ref E, F *WSI* – dominant in both open water (RA 25–40%) and SI covered (RA 20–30% upper trap, ~20% deep trap). *WS2-4* –dominant every second year in open water (av. RA 12%). *PB*- MRA 5%	**Sediments:** Ref B *Max RA:* 64.6%, 0.5 to 1°C fSST *SI assoc.:* S of WSI *Max SSI, WSI conc @ MRA:* 43%, 62% **Traps:** Ref G High RA (Max 31.7 RA year⁻¹) in all traps, increasing to S. Max flux in March but dominant all year-round in traps S of 65°S	**Sediments:** Ref C *Max RA:* 17.9%, −1.5 to 1°C sSST *SI assoc.:* S of WSI *Max VAR:* between WSI and SSI. **Traps:** Ref E, F *BST*- enrichment from surface trap RA (3.6% RA) to deepest trap (6.2% RA) and similar seafloor abundance (4.6% RA). *DP* – 25.4%RA (upper trap) decreasing to 19.3% deep trap, 1.9% seafloor sediment	**Sediments:** Ref D *Max RA:* 88.4% *SI assoc.:* 3–20% RA in 15–60% WSI conc.; > 20% RA > 75% WSI conc. Combined with *F. cylindrus* abundances.
Fragilariopsis cylindrus 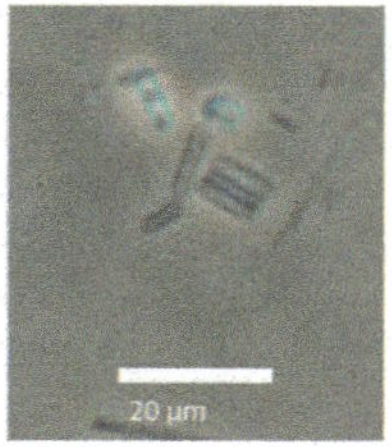	**Sediments:** Ref A *Max RA:* 29%, −1 to 1°C sSST *SI assoc.:* Year-round sea ice to WSI edge **Traps:** Ref E, F *WSI* - dominant in both open water (RA 10–25%) and SI covered (50–70%) *WS2-4* -dominant every 2nd year in open water (av. RA 25%). *PB* - dominant MRA 49%	**Sediments:** Ref B *Max RA:* 2.9%, 0.5 to 1°C fSST *SI assoc.:* Coastal Antarctica to SSI edge *Max SSI, WSI conc @ MRA:* 0%, 87% **Traps:** Ref G High av. fluxes (Max 23.3 RA year⁻¹) in all traps, increasing to S. Max flux in March but common all year round in traps S of 65°S.	**Sediments:** Ref C *Max RA:* 2.1%, −1.5 to 0°C sSST *SI assoc.:* Sea ice zone to POOZ, MRA S of SSI (SSI conc 31%, WSI conc = 84%). *Max VAR:* no data available **Traps:** Ref E, F *BST*– surface trap (1.4% RA) to constant RA (~5.2%) through the water column (539m–1835m). Seafloor sediments (1.6%) *DP*– 8.3%RA (upper trap) decreasing to 6.6% deep trap, 0% seafloor sediment.	**Sediments:** Ref D Data is combined with *F. curta* abundances

Figure 26.1 (*Continued*)

Species	Atlantic/SW Indian	SE Indian/SW Pacific	SE Pacific	Combined Antarctic (minus SE Indian/SW Pacific)
Fragilariopsis obliquecostata 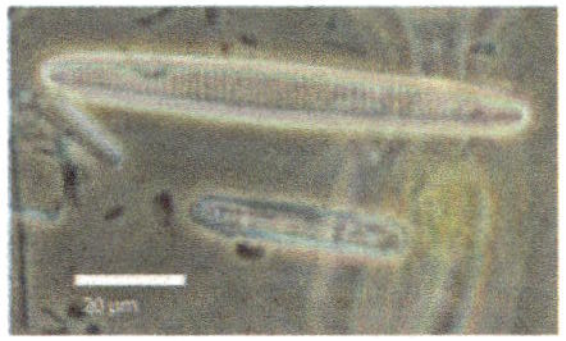	**Sediments:** Ref A *Max RA:* 16.9%, −2 to 1.5°C sSST *SI assoc.:* SIZ **Traps:** Ref F *PB* – low RA 0.22%. *WSI-4* – no data available	**Sediments:** Ref B *Max RA:* 10.2%, −1.3 to 2°C fSST *SI assoc.:* S of SSI edge *Max SSI, WSI conc @ MRA:* 0%, 90% **Traps:** Ref G Low av. fluxes (Max 0.36 RA year^{-1}) increasing flux S of 62°S traps (~ N of SB-ACC)	**Sediments:** Ref C *Max RA:* 0.3%, −0.44° sSST, SSI conc 0%, WSI conc 81% *SI assoc.:* Rarely observed, MRA in Seasonal Sea Ice Zone *Max VAR:* no data available **Traps:** Ref F *BST*- mid-water column presence at low levels (Av. ~0.5% RA) whilst seafloor sediments are enriched at ~1% RA. *DP*– not encountered	**Sediments:** Ref D *Max RA:* 13.7% *SI assoc.:* >1% RA in >75% WSI conc.
Fragilariopsis sublinearis 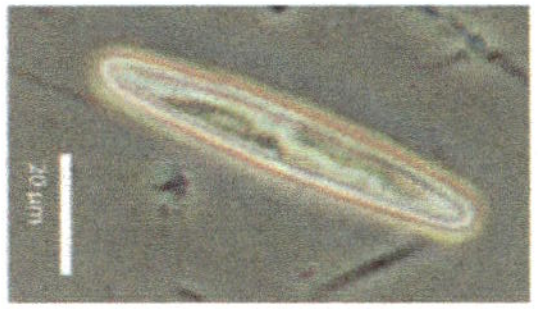	**Sediments:** Ref A *Max RA:* 13%, −2 to −1°C sSST *SI assoc.:* S of PF, (SI covered) **Traps:** Ref E, F *PB* – not encountered. *WSI-4* no data available	**Sediments:** Ref B *Max RA:* 6.1%, 0 to 1°C fSST *SI assoc.:* S of SSI edge. *Max SSI, WSI conc @ MRA:* 24%, 61% **Traps:** Ref G Not identified in any AESOPS traps	**Sediments:** Ref C *Max RA:* 1.6%, 0°C sSST *SI assoc.:* no data available *Max VAR:* no data available **Traps:** Ref F *BST*- mid-water column presence at low levels (Av. ~0.6% RA) and deepest trap RA of 0.3% RA. Seafloor sediments are similar at ~0.5% RA. *DP*- not encountered	**Sediments:** Ref D *Max RA:* 9.1% *SI assoc.:* >1% RA in >75% WSI conc.
Porosira glacialis 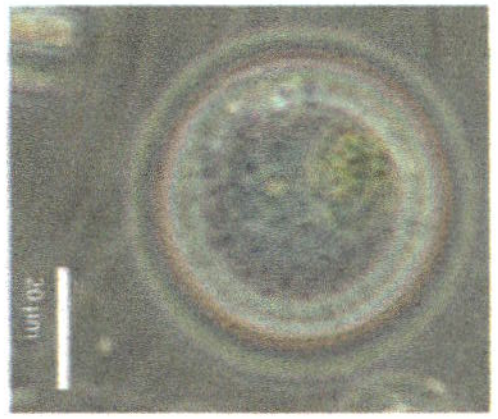	**Sediments:** Ref A *Max RA:* 2.0%, −1.5 to 0°C sSST *SI assoc.:* Year-round sea ice to WSI edge **Traps:** Ref E *PB* – not encountered. *WSI-4* no data available	**Sediments:** Ref B *Max RA:* 6.4%, 0 to 0.5°C fSST *SI assoc.:* S of SSI edge. *Max SSI, WSI conc @ MRA:* 0%, 90% **Traps:** Ref G Not identified in any AESOPS traps	**Sediments:** Ref C *Max RA:* 0.2%, no sSST data *SI assoc.:* none, only 1 specimen observed in Subantarctic Zone *Max VAR:* no data available **Traps:** Ref E, F *BST*– steady decrease in RA with depth from 0.5% at the top trap to 0.16% RA at the lowest, however, seafloor sediments show a slight increase (0.6% RA). *DP*- not encountered	**Sediments:** Ref D Species not considered in this study

Figure 26.1 (*Continued*)

Species	Atlantic/SW Indian	SE Indian/SW Pacific	SE Pacific	Combined Antarctic (minus SE Indian/SW Pacific)
Stellarima microtrias 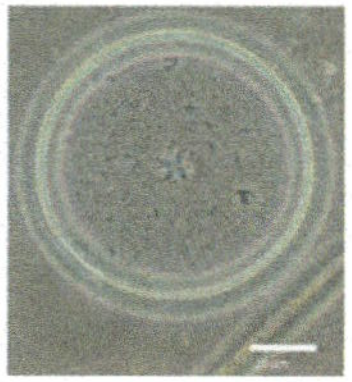	**Sediments:** Ref A *Max RA:* 2.6%, −2 to −1°C sSST *SI assoc.:* Year-round to WSI edge **Traps:** Ref E *PB* – not encountered. *WSI-4* – no data available	**Sediments:** Ref B *Max RA:* 3.2%, −0.5 to 0.5°C fSST *SI assoc.:* S of SSI edge *Max SSI, WSI conc @ MRA:* 0%, 88% **Traps:** Ref H *Stellarima* spp. only in traps S of SACCZ (MS 6 & 7), constant low levels in MS-6 (73°S) in Jan.	**Sediments:** Ref C *Max RA:* not encountered **Traps:** Ref F *BST* - a mid-water column appearance decreases from 1% to 0.5% RA within 500m, whilst sea floor sediments are 0.8% RA. *DP* – not encountered	**Sediments:** Ref D *Max RA:* 3.7% *SI assoc.:* >1% RA in >80% WSI conc.
Chaetoceros spp. (principally resting spores) 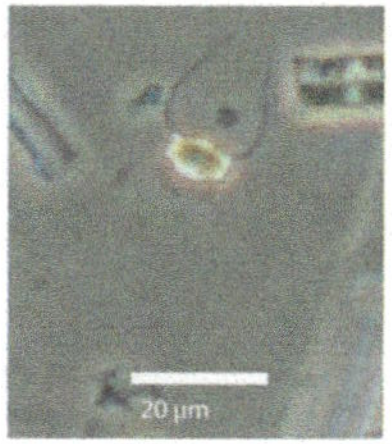	**Sediments:** Ref A *Max RA:* 92.6%, −1 to 1°C sSST *SI assoc.:* Unclear relationship to SI due to neritic preference and generic level treatment **Traps:** Ref E, F *PB* – MRA 6.4%.*WSI-4* – no data available, not dominant	**Sediments:** Ref B *Max RA:* 91.8%, −0.5 to 1.5°C fSST *SI assoc.:* Unclear relationship to SI due to neritic preference and generic level treatment *Max SSI, WSI conc @ MRA:* 43%, 62% **Traps:** Ref G, H *Chaetoceros* resting spores were not reported in any AESOPS traps	**Sediments:** Ref C *Max RA:* ~8% (in SIZ), −1.5 to 0°C sSST (in SIZ) *SI assoc.:* S of SSI (in SIZ) *Max VAR:* between sACCF and SSI (in SIZ) **Traps:** Ref E, F *BST*- surface trap (~32% RA), two deepest traps (av. 79% RA), seafloor sediments (73% RA), resuspension questioned. *DP* – 0.5% (upper trap) to 1% deep trap, 3% seafloor sediment. *K1* – surface trap (~50-75% RA) deep trap enriched 70-80% RA). *K2*- (50-60% RA)	**Sediments:** Ref D *Max RA:* 91.9% *SI assoc.:* up to 90% RA can occur in areas of 0% WSI conc. No discrimination of veg and spores No clear sea ice relationship due to lumping of resting spores
Eucampia antarctica (var. *recta* and var. *antarctica*) 	**Sediments:** Ref A *Max RA:* 12.1%, −2 to 5.5°C sSST *SI assoc.:* Highest abundances N of WSI, no direct relationship found as both varieties combined represent different environments N & S of WSI-PF **Traps:** Ref E, F *PB* – MRA 0.22%.*WSI-4* – no data available	**Sediments:** Ref B *Max RA:* 23.2%, −1 to 0.5°C fSST *SI assoc.:* Highest abundances N of WSI, no direct relationship found as both varieties combined in database *Max SSI, WSI conc @ MRA:* 16%, 92% **Traps:** Ref G, H Only observed in traps close to or S of SB-ACCZ (Traps MS4, 6 & 7), and only constantly at low levels in MS-6 (73°S) peaking in late Dec.	**Sediments:** Ref C *Max RA:* 12.5%, −1.5 to 0°C sSST *SI assoc.:* interpreted in this chapter as S of WSI (if var. *recta*) *Max VAR:* S of WSI at 120°W, N of WSI at 90°E and S of SSI at 90°E **Traps:** Ref F *BST*– surface traps (0.35–1.7% RA) and no mid or deep trap presence, whilst seafloor abundance was similar to top trap (0.32%RA). *DP* – 0% (upper trap) to 0.2% deep trap, ~6% seafloor sediment	**Sediments:** Ref D *Max RA:* 24.9% *SI assoc.:* found up to 11% RA in areas of no WSI cover; >11% RA in >90% WSI conc. Includes both varieties

Figure 26.1 (*Continued*)

Species	Atlantic/SW Indian	SE Indian/SW Pacific	SE Pacific	Combined Antarctic (minus SE Indian/SW Pacific)
***Thalassiosira antarctica-scotia* gp.** (*T. scotia* vegetative valves = *T. ant.* resting cells) 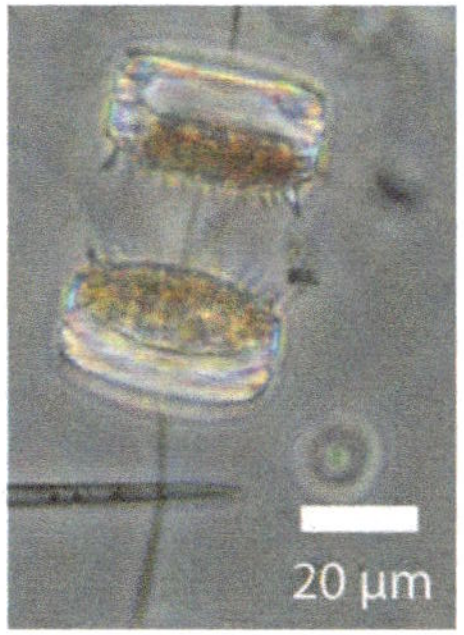	**Sediments:** Ref A *Max RA:* 32.7%, −2 to 1°C sSST *SI assoc.:* to WSI for *T. antarctica*, but difficult to separate from *T. scotia* – both species represent different environments S of PF (i.e. −2 to 9.5°C range) **Traps:** Ref E *PB* – Vegetative cells and resting spores both 0.11%RA. *WSI-4* – no data available	**Sediments:** Ref B *Max RA:* 31.8%, 0 to 0.5°C fSST *SI assoc.:* S of WSI *Max SSI, WSI conc @ MRA:* 3%, 92% **Traps:** Ref G Low abundances in all traps (Max 4.4 RA year^{-1}), increasing from traceoccurrences in SIZ to Max RA in PFZ.	**Sediments:** Ref C *Max RA:* 0.2%, no sSST data available *SI assoc.:* uncertain, observed at trace levels in two samples between Subantarctic to Seasonal Sea Ice Zone *Max VAR:* no data available **Traps:** Ref E, F BST– bloom levels of vegetative cells in surface trap (~23% RA, resting spore 1% RA), ratio reversed in deepest trap (resting spore 3.8%, vegetative cells 0.6% RA). Seafloor sediments similar (resting spore 8.7%, veg. cell 0.8%). DP– has no trapped cells but seafloor RA is 11.2%. K3– during flux peaks 15–20% RA vegetative cells, post flux period resting spores are ~50–80% RA	**Sediments:** Ref D *Max RA:* 32.7% *SI assoc.:* >1% RA in >40% WSI conc.
Thalassiosira lentiginosa (negative relationship to sea ice) 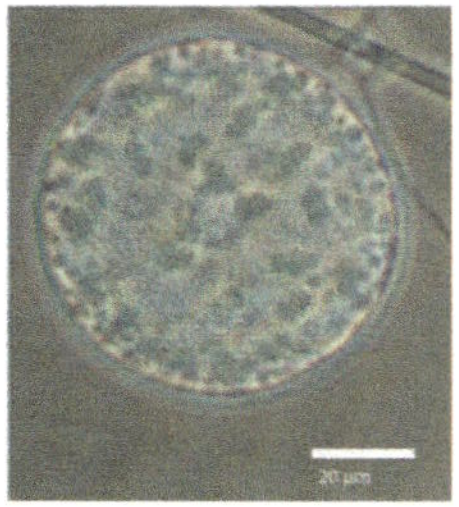	**Sediments:** Ref A *Max RA:* 26.0%; 0–7°C sSST *SI assoc.:* None, Open-Ocean species limited to S by WSI **Traps:** Ref E, F *PB* – MRA 0.43%. *WSI-4* – no data available. In text note in Ref E states: At WS-1 *T. lentiginosa, F. kerguelensis* and *T. oliverana* are low RA in upper trap but combined, contribute to ~15% in deeper trap	**Sediments:** Ref I *Max RA:* 36.7%, 1 to 1.5°C fSST (range–0.5 to 18°C) *SI assoc.:* Significant dampened abundances by WSI edge. *Max SSI, WSI conc @ MRA:* 0%, 55% **Traps:** Ref G, H Common and increases in all traps N of 66°S (Max 4.4 RA year^{-1}). Absolute flux of species peaks near the retreating sea ice edge in AAZ, but signal is diluted in sediments	**Sediments:** Ref C *Max RA:* ~15%, −1.5 to 0°C sSST in the SIZ; 29.7%, 5 to 7°C sSST N of SAF *SI assoc.:* muted S of sACCF but present in low abundances *Max VAR:* just north of the WSI and close (N/S) to the sACCF **Traps:** Ref F BST– indicate total removal of the species below 539m traps from >1% RA in surface traps. No sediment record of species was preserved. DP– 0% (upper trap) to 0.4% deep trap, 7.3% seafloor sediment	**Sediments:** Ref D *Max RA:* 29.9%, 4.5°C sSST *SI assoc.:* no sea ice association at MRA. MRA at SAF in western Indian Ocean (49.5°S)

Figure 26.1 (*Continued*)

26.2.2 Statistical proxy evolution

The estimation of Antarctic palaeo sea ice cover, based on diatom microfossil assemblages, has relied on the modern analogue technique (MAT) (Hutson, 1980; Crosta et al., 1998) and a form of principal component regression, known as the Imbrie and Kipp transfer function (IKTF) (Imbrie & Kipp, 1971). Since 2010, a small number of studies have utilized various statistical approaches to produce diatom-based sea ice estimates from cores covering the Holocene to the last 220 kyr BP (Table 26.2). There is a growing body of research suggesting that the application of IKTF and/or MAT methods for the estimation of palaeoceanographic variables is inappropriate (Telford & Birks, 2005; Ferry et al., 2015a). Therefore, MAT and IKTF may not be appropriate for estimating Antarctic palaeo sea ice. Generally, the computation of principal components (as is done when applying IKTF) for the purpose of making subsequent predictions is inefficient (Juggins & Birks, 2012). Computation of principal components may provide inappropriate ecological answers when species assemblage data contains many zero values (i.e. diatom species are absent at certain sample sites; Legendre & Birks, 2012). Ferry et al. (2015a), using the French sea ice diatom database (i.e. Crosta et al., 2004), illustrated that the assumptions required for a linear regression model, normality and constant error variance of model errors may not hold. More importantly, Ferry et al. (2015a) illustrate the poor fit of IKTF to existing training data set under tenfold, and a spatially independent, hold-out cross-validation. MAT is biased by the spatial structures within a training database and will provide overly optimistic estimates for the error of estimation if these spatial structures are ignored (Telford & Birks, 2005; Ferry et al., 2015a). Recent alternative approaches undertaken on the Alfred-Wegener-Institut (AWI) diatom database have recommended the application of MAT, but their approach has yet to account for spatial structures within their database to ensure this approach remains valid using their database (Esper & Gersonde, 2014).

26.2.3 Ice-core records

A significant effort has been made over the last decade to develop a series of sea ice proxies based on the measured fluxes of atmospheric aerosols within Antarctic ice cores (Wolff et al., 2010; Abram et al., 2011, 2013; Spolaor et al., 2013). We discuss each of these new ice core proxies, with a focus on their physical basis and limitations.

Sea salt aerosol flux

It has been hypothesized that sea salt flux is positively related to features of the sea ice surface (Röthlisburger et al., 2008; Russell & McGregor, 2010; Wolff et al., 2010), and hence marine sea salt flux may increase as the sea ice area increases (Röthlisburger et al., 2010). Interpretation of the sea salt flux recorded within Antarctic ice cores remains unclear. Currently, it is unknown if the sea salt aerosol flux originates from frost flowers or the flooding and wicking of brine on the sea ice surface (Wolff et al., 2010). It is also difficult to distinguish between sea spray and sea ice surface-derived sea salt fluxes (Spolaor et al., 2013). Furthermore, as atmospheric sea salt aerosol concentrations decline during aeolian transport, only a small percentage of the sea salt flux will remain after transportation over hundreds of kilometres. Therefore, the sea salt flux proxy does not appear to be registered during colder climates when sea ice areas are extensive (Fischer et al., 2007; Röthlisburger et al., 2008, 2010). Fischer et al. (2007) casted doubt on the utility of sea salt flux as a sea ice proxy after their sea salt flux record failed to reflect any expression of the Antarctic isotope maxima events during marine isotopic stage three.

Methanesulphonic acid

The MSA signature within Antarctic ice cores is derived from the oxidization of dimethylsulphide, which is produced almost exclusively by marine algae (Hezel et al., 2011; Abram et al., 2013). Therefore, as sea ice coverage influences biological activity, sea ice may also influence MSA production. To date, MSA has only provided a useful proxy for Antarctic sea ice between 80° and 140°E (Curran et al., 2003).

Ultimately, MSA is difficult to interpret as a proxy for Antarctic sea ice. Wind speed and direction are the primary factors controlling the concentration of MSA deposition over certain regions of the Antarctic (e.g. Rhodes et al., 2009; Hezel et al., 2011). Under modern conditions, areas with low snowfall accumulation rates experience a loss in methanesulphide concentration due to the slow evaporative loss of MSA (Spolaor et al., 2013). Similarly, MSA may be present as a non-volatile neutral salt rather than a volatile acid (Wolff et al., 2006). In light of the later complications surrounding the MSA sea ice proxy, recent ice core studies have chosen to focus on alternative ice core proxies (Wolff et al., 2010; Spolaor et al., 2013).

Table 26.2 List of key papers using diatoms to reconstruct Antarctic sea ice since 2010.

Reference/	Region	Method	Time period (BP)
Allen et al. (2010)	Neny Fjord, Antarctic Peninsula	PCA, to describe principal stratigraphic zones within the diatom data	Holocene (9–0.2 kyr)
Barbara et al. (2010)	Eastern Prydz Bay, East Antarctica	C25 ratio of HBI diene and triene and diatom assemblage analysis	11–9 kyr
Denis et al. (2010)	Adélie Land, East Antarctica	C25 ratio of HBI diene and triene and diatom assemblage analysis	7 kyr BP to present day
Röthlisburger et al. (2010)	Indian Ocean sector	MAT	Last 240 kyr
Allen et al. (2011)	Scotia Sea and Antarctic Peninsula	Calibration of Gersonde & Zielinski (2000)	LGM to Holocene
Collins et al. (2012)	Southwest Atlantic, Scotia Sea	Calibration of Gersonde & Zielinski (2000)	35–15 kyr
Maddison et al. (2012)	East Antarctic	Laminated diatom-rich sediments	1136–3122 calendar years
Barbara et al. (2013)	Northwestern and Northeastern Antarctic Peninsula	Q-mode PCA	1935–1950 AD
Collins et al. (2013)	Southwest Atlantic, Scotia Sea	Calibration of Gersonde & Zielinski (2000), HBI diene and triene	60–0 kyr
Etourneau et al. (2013)	West Antarctic Peninsula, Palmer Deep Basin	Calibration of Gersonde & Zielinski (2000)	Holocene (last 9 kyr)
Tolotti et al. (2013)	Western Ross Sea	Diatom assemblage analysis	Last 40 kyr
Esper & Gersonde (2014)	Atlantic and Pacific Oceans	MAT and IKTF	150 kyr
Ferry et al. (2015a)	Southwest Pacific sector	Performance testing of IKTF, MAT, GAM and WA PLS of Crosta et al. (1998). GAM estimates	Last 130 kyr
Ferry et al. (2015b)	Southwest Pacific sector	GAM	Last 220 kyr

GAM, generalized additive model; HBI, highly branched isoprenoids; IKTF, Imbrie and Kipp transfer function; LGM, Last Glacial Maximum; MAT, modern analogue technique; PCA, principal components analysis; WA PLS, weighted averaging partial least squares.

Halogens

Volatile halogenated organic compounds are produced by marine phytoplankton and ice algae during photosynthesis by the enzymatic removal of hydrogen peroxide by haloperoxides (Granfors et al., 2013). Recent work within the Amundsen and Ross Seas suggests that halocarbon production/concentration is mostly influenced by sea ice (Mattsson et al., 2012; Granfors et al., 2013). Recorded fluxes of the halogens bromine and iodine are of particular interest as a proxy for Antarctic sea ice, with enriched bromide concentrations used as an indicator for the presence of multi-year ice, whilst decreases in iodine indicate an extension of the sea ice extent (Spolaor et al., 2013, 2014). Sea ice algae produce organo-iodine, which in conjunction with bromine production at the sea ice margin, results in the sea ice margin becoming an area of active bromine and iodine chemistry. Recent *in situ* measurements confirmed the Weddell Sea sea ice environment as an iodine emission hotspot (Atkinson et al., 2012). Satellite measurements have shown a decrease in bromine oxide and iodine oxide concentration with increasing distance from the sea ice margin (Spolaor et al., 2013). Furthermore, Spolaor et al. (2014) have gone on to illustrate how halogen production captured by satellite observations are related to the preservation of a seasonal cycle of both halogens in century-old Antarctic ice cores. This ground-truthing allows application to the palaeo record in the future. Therefore, sea ice dynamics appear to act as the primary driver for bromine and iodine fluxes over glacial to interglacial cycles (Spolaor et al., 2013, 2014).

26.2.4 Marine sediment records
Highly branched isoprenoids

Highly branched isoprenoids diene and triene have received recent attention as a potential proxy for Antarctic sea ice (Denis et al., 2010; Massé et al., 2011; Etourneau et al., 2013; Barbara et al., 2013). The di-unsaturated isomer (diene) is believed to be synthesized by diatoms living within the sea ice matrix during the spring months (Collins et al., 2013), whilst the tri-unsaturated isomer (triene) may be biosynthesized within the open ocean areas of the marginal ice zone, near the sea ice edge (Denis et al., 2010; Collins et al., 2013; Etourneau et al., 2013). Therefore, the ratio of diene to triene has been used as a proxy for the ratio of sea ice to open ocean surrounding the west Antarctic Peninsula (Etourneau et al., 2013) and the Scotia Sea (Collins et al., 2013). Together, the HBIs diene and triene may provide a proxy for the duration of the sea ice season, rather than a strictly sea ice versus open ocean proxy. The coeval absence of both HBIs may provide an important Antarctic summer sea ice proxy (Collins et al., 2013). Similarly, the isotopically enriched ^{13}C HBI diene may provide a proxy for the contribution of organic matter, derived from sea ice diatoms, to the sediments of the Adélie Land continental shelf (Massé et al., 2011). The HBI proxy does, however, have current shortcomings. The specific diatoms responsible for the biosynthesis of diene are currently unknown, although the sea ice diatoms *F. curta* and *F. cylindrus* have been discounted as the source of diene (Collins et al., 2013). HBI preservation is altered by bacterial degradation and, whilst HBIs are resistant to degradation, a rapid sulphurization of HBIs under highly anoxic conditions does occur (Etourneau et al., 2013).

An effort to improve our understanding of the HBI proxy is required, with a focus on comparing HBI records with other down-core proxies between differing environments (Collins et al., 2013). The use of HBIs has also opened the door for tropic cascade determination related to the sea ice environment (Brown et al., 2014b), but this approach has yet to be applied to the Antarctic. New Antarctic HBI research by a collaborative UK, Australian and US team is currently under way in an effort to understand the diatom sources, transport and entrainment of HBIs in the sediments.

26.2.5 Holocene sea ice record

The climatic variability of the Holocene throughout the Southern Hemisphere remains poorly understood, as does our understanding of the various forcing mechanisms that have driven Holocene climatic variation (Divine et al., 2010). Developing our knowledge of Antarctic climatic variation during the Holocene is necessary for clarifying our understanding of recent global warming, ice sheet mass balance and sea level changes (Masson et al., 2000). Our discussion of the Holocene palaeo sea ice records for the Antarctic will focus on diatom proxy and climate model data that have been used to study the Holocene Climatic Optimum (Bentley et al., 2009; Waldmann et al., 2010; Renssen et al., 2012; Pike et al., 2013). We will also address ice core records, whale catch statistics and ship-based observations that help to unravel the nature of 20th-century, pre-satellite, sea ice cover.

The Early Holocene Climatic Optimum

A palaeo climatic signature of the Early Holocene Climatic Optimum (eHCO) has been recorded between 12 and 9 kyr BP over Antarctica (Masson et al., 2000), the west Antarctic Peninsula (Shevenell et al., 2011; Pike et al., 2013), Siple Dome (Das & Alley, 2008), East Antarctica (Ciais et al., 1992; Verleyen, 2012; Bentley et al., 2005) as well as the Atlantic (Divine et al., 2010), central Indian and Pacific sectors of the Southern Ocean (Katsuki et al., 2012).

The eHCO climate is considered to have been between 1 and 2°C warmer than the pre-industrial climate (Ljungqvist, 2011), and characterized by a concomitant sea ice decline throughout the Southern Ocean (Divine et al., 2010). Pike et al. (2013) suggested that the warming along the west Antarctic Peninsula was manifest with ice sheet retreat and thinning, and warmer sea-surface temperatures and noted a similar collapse of the George VI ice shelf in the East Antarctic at 9.6 kyr BP (Bentley et al., 2005). New Zealand palaeo records also suggest a shift to a warmer and a moister climate between 9.6 and 8.7 kyr BP (Turney et al., 2006). eHCO expression was greatest (weakest) in the higher (lower) latitudes, for both hemispheres. The results from palaeoclimate modelling of the Southern Ocean suggest that the greatest eHCO anomalies occurred during winter, with thinner sea ice permitting a stronger upward heat flux from the ocean. An enhancement of meridional latent heat transport, associated with a warmer atmosphere, may have supplemented the increased ocean heat flux associated with thinner sea ice. Additionally, the latitudinal contrast in the eHCO expression may have been attributable to orbital-forced insolation anomalies, which have a greater magnitude at higher latitudes (Renssen et al., 2012).

Pre-satellite (1971) sea ice concentration records

In September 2014, the largest extent of Antarctic sea ice cover ever observed since the inception of satellite monitoring occurred, hitting 20 million km², which is around three times the area of the Australian continent (Lieser personal communication; ACE CRC September 2014; Chapters 1 and 10). The increase in sea ice extent is considered a possible response to global warming and the increased melting of the Antarctic ice cap (Rignot et al., 2013; Marshall et al., 2014). Yet evidence from other proxy records indicate that the Southern Ocean was characterized by more expansive sea ice coverage prior to the modern satellite record. Two MSA-based records, one from the East Antarctic sector (Curran et al., 2003) and the other in the west Antarctic Peninsula (Abram et al., 2010), both suggest that sea ice extent has declined since the 1950s and generally over the 20th century.

On the East Antarctic ice cap, an ice core from Dome C contains a sea salt flux record supporting an increase in sea ice extent in the Indian Ocean sector (affecting the source of sea salt to Dome C) over the last 6 kyr BP (Wolff et al., 2003). Diatom records, specifically the ratio of *Porosira glacialis* resting spores to *Thalassiosira antarctica* resting spores, identified increases in Holocene winter sea ice concentration over Dumont d'Urville Trough (DUT) (Pike et al., 2009). Later, focusing on diatom-laminated sediments at DUT, Maddison et al. (2012) suggested that the sea ice extent had increased from ∼ 1.1 kyr BP.

Whaling and ship's log records are another alternative source for the estimation of recent pre-satellite sea ice extent. The southernmost catch statistics for blue and fin whales have been used as a proxy for sea ice extent (Cotté & Guinet, 2007). Whale catch statistics suggest that sea ice extent decreased between 1960 and the satellite records of 1970, particularly within the Weddell Sea sector (Cotté & Guinet, 2007), but also through the Indian Ocean sector and around to the Ross Sea (de la Mare, 2009). In the Atlantic sector, a comparison of early 20th century ship-based observations suggests sea ice extended to within 200 km of South Georgia whilst the satellite record indicated a sea ice extension to 325 km of the same polar island (Whitehouse et al., 2008).

26.2.6 Last Glacial Maximum: timing and extent

Timing for the LGM has previously been placed at 19–23 kyr BP (Mix et al., 2001) and ∼26.5–19 kyr BP (Clark et al., 2009). Evidence suggests that the LGM throughout eastern Australia, New Zealand and the South Pacific may have commenced by ∼30 kyr BP, with a duration of 10 kyr (Petherick et al., 2008; Stephens et al., 2012; Vandergoes et al., 2013). The climatic cooling between 29 and 19 kyr BP has been referred to as the extended LGM (eLGM; Newnham et al., 2007), with onset of the eLGM throughout the southern hemisphere possibly driven by atmospheric cooling, stratification of the Southern Ocean and the associated expansion of sea ice (Putnam et al., 2013). Adkins (2013) suggested that cooling of North Atlantic Deep Water (NADW), driven by the 65°N summer solar forcing, ultimately increased the salinity of Antarctic Bottom Water (AABW). The colder, more saline AABW increases in density relative to the NADW and therefore stratifies the deep ocean, producing a positive feedback. A stratified deep ocean acts an effective carbon trap, lowering atmospheric CO_2. Therefore, the deep ocean synchronizes both hemispheres, driving global climatic cooling (Adkins, 2013).

A considerable body of evidence suggests that Southern Ocean palaeo sea ice cover expanded during the LGM. It is believed that winter sea ice extent may have expanded to between 5° and 10° north of the modern winter sea ice extent (e.g. Gersonde et al., 2005; Allen et al., 2011). The study of southern Bull Kelp population genetics suggests that sea ice coverage during the LGM may have reached the Falkland Islands in the western Atlantic sector, and Macquarie Island in the east Pacific sector (Fraser et al., 2009). The recent testing of the semi-circumpolar MAT sea ice estimation model to cores in the Pacific Polar Frontal Zone (60°S) and the Antarctic Zone (52°S) of the eastern Atlantic (Esper & Gersonde, 2014) both indicate that maximum winter sea ice extended out over the core sites with concentrations reaching up to ∼60% (representative of consolidated sea ice cover) during the height of what appears to be the LGM. Although summer sea ice concentration estimates in this period were attempted, and hinted at occasional sea ice-influenced events, the authors cautioned their use and interpretation due to unacceptable statistical error margins in their application testing.

26.3 Arctic

Concern over the 'state of sea ice' in the Arctic has been highlighted by recent decreases in ice extent (Comiso,

2012; Chapter 11) and decreasing ice thickness (Kwok & Rothrock, 2009; Chapter 2) observed over the past several decades, with predictions of a summer sea ice-free Arctic by the late 2030s (Wang & Overland, 2009). These predictions and their potential regional to global consequences have heightened awareness of the impact of human-induced climate change.

One way to distinguish anthropogenically induced change versus natural variability is to study climate records that predate the industrial revolution. However, as with the Antarctic, historical data on sea ice extent in the Arctic are limited, with satellite-derived records of sea ice extent extending back only to the early 1970s. This relatively short timescale of the satellite record makes it difficult to understand the natural changes in sea ice extent and thickness versus anthropogenically induced warming. Thus researchers turn to proxy indicators of sea ice extent that provide a longer-term and valuable perspective on changes in palaeo sea ice distribution.

Similar to work in the Antarctic, estimates of palaeo sea ice in the high northern latitudes were first put forward by CLIMAP (1976, 1981) based on the absence of coccoliths, low accumulation rates of planktonic foraminifera, reduced bulk sedimentation rates and the occurrence of ice-rafted debris (IRT). The conclusions of CLIMAP have since been revised dramatically by more extensive work, and sea ice reconstructions have been extended farther back in time and at greater resolution, based on a suite of multi-proxy evidence. Here we present a brief review of northern high-latitude sea ice proxies with a focus on newer developments, which then allow us to examine the results of these methods under time-slice scenarios.

26.3.1 Methods for reconstructing past sea ice distribution and seasonal variability
Microfossil assemblage distribution: dinoflagellate cysts

In the high northern latitudes, dinoflagellate cysts remain the most powerful proxy recorder of palaeo sea ice, with reconstructions based primarily on modern analogue techniques (de Vernal et al., 2013a–c; see Table 26.3). The utility of these cysts, which form as a normal part of the dinoflagellate life cycle, results from several factors. First, dinoflagellate distribution is sensitive to changes within the upper water column, including fluctuations in sea ice cover, which restrict the light available for photosynthesis, thus disrupting metabolic activities. While dominantly photosynthetic, many dinoflagellates are also capable of heterotrophic behaviour. Second, dinoflagellate populations are relatively diverse and widespread in the Arctic, with changes in assemblage composition driven primarily by changes in surface water conditions, therefore making dinoflagellate populations ideal for transfer function analyses. For example, species commonly found in regions with extensive sea ice cover include *Brigantedinium* spp. and *Operculodinium centrocarpum*, both of which tolerate a broad range of environmental conditions, and others such as *Islandinium minutum*, more specifically adapted to colder and ice-covered conditions (de Vernal & Hillaire-Marcel, 2000; Head et al., 2001; de Vernal et al., 2013a). In contrast, the extremely low diversity exhibited by planktonic foraminiferal populations and coccolithophorids in the Arctic, and the often-poor preservation of diatoms make it difficult to track changes in sea ice cover by examining variability in their respective assemblage compositions. The combination of their heterotrophic behaviour, ability to encyst and wide tolerance to salinity levels results in their ability to survive even under very harsh environmental conditions (Mudie et al., 2001; de Vernal et al., 2001). Finally, many of the dinocysts are composed of dinosporin, a complex and refractory organic compound that is very resistant to dissolution (de Vernal et al., 2001; Zonneveld, 2001). Consequently, the records of dinocysts are much more continuous than those of siliceous and calcareous microfossils, in both glacial and interglacial sediments.

de Vernal et al. (2013b) summarize the current state of statistical assessment of dinocyst assemblages to reconstruct sea ice cover, reviewing the latest database of 1492 reference samples. Quantitative studies completed by Mudie and Short (1985) and Mudie (1992) were among the first to use Q-mode factor analysis to develop dinocyst-based transfer functions for SST and salinity, but not sea ice cover. These studies were based on the eastern Canadian continental margin, using fewer than 100 samples. Later work increased geographic coverage, producing a larger data set of almost 400 core tops, which were analysed with a best analogue technique. The primary advantage of this technique over the Imbrie and Kipp Q-mode factor analysis is in its logarithmic transformation of abundance data that allocates a greater significance to otherwise sparsely represented, ecologically significant

Table 26.3 List of dinocyst studies relevant to Arctic palaeo sea ice reconstruction.

Reference	Region	Time period
Mudie (1992)	Circum-Arctic	Neogene and Quaternary
de Vernal et al. (1993a)	Gulf of St Lawrence	Late glacial–Holocene
de Vernal et al. (1993b)	North Atlantic	Modern
Rochon & de Vernal (1994)	Labrador Sea	Modern and Holocene
de Vernal et al. (1996)	North Atlantic	Younger Dryas
Levac & de Vernal (1997)	Labrador Coast	Postglacial
de Vernal et al. (1997)	North Atlantic	Modern
Rochon et al. (1998)	North Sea	Late glacial–deglacial
Simard & de Vernal (1998)	North Atlantic	Holocene
Grøsfjeld et al. (1999)	Western Norway fjords	Late glacial–Holocene
Rochon et al. (1999)	North Atlantic	Modern
de Vernal & Hillaire-Marcel (2000)	North-west N. Atlantic	Last Glacial Maximum versus modern
de Vernal et al. (2000)	North Atlantic	Last Glacial Maximum
Boessenkool et al. (2001)	South-east Greenland Margin	Modern
Head et al. (2001)	Arctic/Kara and Laptev Seas	Modern
Kunz-Pirrung (2001)	Arctic/Laptev Sea	Modern
Kunz-Pirrung et al. (2001)	Arctic/Laptev Sea	Late Holocene
Levac et al. (2001)	Baffin Bay	Postglacial
Matthiessen et al. (2001)	Barents Sea	Oxygen isotope stage 5
Mudie & Rochon (2001)	Canadian Arctic	Modern
Mudie et al. (2001)	Arctic and high latitude circum-Arctic basins	Introduction to special volume
Peyron & de Vernal (2001)	Arctic and sub-Arctic seas	Modern and 25 000 year records
Radi et al. (2001)	Bering and Chukchi Seas	Modern
Voronina et al. (2001)	Barents Sea	Holocene
de Vernal et al. (2001)	North Atlantic, Arctic, sub-Arctic Seas compilation	Modern
Solignac et al. (2004)	North Atlantic	Holocene
de Vernal et al. (2005a)	Arctic Seas compilation	Modern, Last Glacial Maximum
de Vernal et al. (2005b)	Chukchi Sea	Holocene
Solignac et al. (2006)	Western North Atlantic	Holocene
de Vernal & Hillaire-Marcel (2006)	North Atlantic	Holocene
Ledu et al. (2008)	Eastern Arctic	Holocene
McKay et al. (2008)	Eastern Chukchi Sea	Holocene
Solignac et al. (2008)	Western North Atlantic	Holocene
de Vernal et al. (2008)	Arctic – subarctic	Holocene summary paper
Ledu et al. (2010a)	Northwest Passage, Canadian Arctic	Holocene
Ledu et al. (2010b)	Northwest Passage, Canadian Arctic	Late Holocene
Farmer et al. (2011)	Western Arctic	Holocene
Kinnard et al. (2011)	Arctic	Late Holocene
Bringué and Rochon (2012)	Beaufort Sea	Late Holocene
de Vernal et al. (2013a)	Summary paper	Summary paper
de Vernal et al. (2013b)	Summary paper	Summary paper

taxa. This technique is relatively conservative, requiring little manipulation and transformation of the data (de Vernal et al., 2005a). One of the most significant problems, however, with these analyses was the occurrence of non-analogue situations in the palaeo record. To address this issue, researchers expanded the database to include 1492 reference points (de Vernal et al., 2013b). In summary, over the past several decades, researchers have increased the number of sample sites and the total range of ecological variability, and have worked with these data using increasingly sophisticated statistical methods. Work must still be undertaken to deal with the problems of non-analogue situations and other known issues related to reconstruction of environmental variables, like sea ice cover, that exhibit a large amount of interannual and/or interdecadal variability.

Avenues for continued research, aimed at improving the accuracy and resolution of dinocyst assemblage-based reconstructions of sea ice cover, include continued work on: dinocyst taxonomy and better understanding of the ecological framework for each taxon; species distribution and population dynamics, which can be accomplished through a combination of water column and sediment trap work; and addressing the link between dinoflagellates and their cysts through incubation studies to confirm taxonomic relationships and the environmental controls on encystment and germination. Continuing challenges with the Arctic dinocyst proxy for sea ice estimation (de Vernal et al., 2013a, 2013b) centre on, first, difficulties in estimating perennial and multi-year ice. In this case, the absence of dinocysts, in part a situation also associated with preservation problems, hinders resolution of these sea ice expressions; it is inevitable that multi-proxy efforts will aid in addressing this issue. Second, the mismatch between the time period recorded in a surface sediment sample, compounded by questions concerning the reference 1953–2003 observational database, may result in both over- and underestimates of sea ice, based on the dinocyst proxy. This problem can be addressed by high-resolution time-series work (de Vernal et al., 2013b).

Microfossil assemblage distribution: diatoms

As described in Section 26.2.1, diatoms are excellent palaeo sea ice proxies in the Southern Ocean. However, their use in the northern high latitudes is more limited, primarily as a result of the more lightly silicified and easily dissolved frustules found in the nutrient-poor Arctic. Consequently, although diatoms are diverse and abundant in Arctic surface waters, their distribution in surface sediments is patchy enough that a circum-Arctic database is unlikely to be developed. Nevertheless, in specific regions – the Bering Sea and Sea of Okhotsk (Sancetta, 1979, 1981, 1982, 1983; Caissie et al., 2010), the Greenland, Iceland and Norwegian Seas (Koç Karpuz & Schrader, 1990; Koç et al., 1993; Jiang et al., 2001; Justwan & Koç, 2008), the Labrador Sea (De Sève, 1999), Baffin Bay, Frobisher Bay and the Davis Strait (Williams, 1990) and the Laptev Sea (Cremer, 1999; Bauch & Polyakova, 2000) – diatom abundances have provided important palaeoceanographic insights. With the exception of the Laptev Sea, these sites are outside the enclosed Arctic Ocean. Clearly, specific diatoms can be, and are, used to trace the distribution of sea ice over time. However, for large portions of the central and marginal Arctic Ocean, diatom abundances are so low, and preservation so poor, that other proxies must be utilized. Note that despite preservational problems, diatom data have proven to be crucial in tracking the initiation of sea ice development in the Arctic (Stickley et al., 2008). In addition, as described in the following section, advances in biomarker research show great promise in tracking sea ice-associated diatom species.

Geochemical tracers: biomarkers

The principal guiding use of geochemical work using photosynthetic biomarker proxies is the association of the chemistry with photosynthetic organisms. Consequently, the presence of photosynthetic biomarker proxies in marine sediment cores suggests an absence of permanent ice cover. Early work by Rosell-Melé and Koç (1997) measured the concentration of chlorins and porphyrins, diagenetic products of chlorophyll and C_{37} alkenones associated with the coccolithophorid *Emiliani huxleyi*, a photosynthetic organism common in waters that are ice-free for at least part of the year (Rosell-Melé et al., 1998).

Over the past decade, studies that utilize the biomarker IP_{25}, a monounsaturated hydrocarbon with a 25-carbon skeleton (Belt et al., 2007), have proliferated, with research on analytical methods, modern process and calibration studies, and palaeo sea ice reconstruction on timescales ranging from less than the past 100 years to over 2 million years (Table 26.4). A critical characteristic of IP_{25} is its identification as a biomarker specific to sea ice diatoms. Brown et al. (2014a) finally provided a conclusive link between several sea ice-endemic diatom species and the production of IP_{25}, data critical in the utilization of this biomarker as a sea ice proxy. Although their work documented that many of the more abundant diatom species found in sea ice do not produce IP_{25}, they observed a consistent contribution of sea ice diatoms that do produce IP_{25}, supporting the interpretation of down-core variability in IP_{25} concentration as indicative of changes in sea ice cover. IP_{25} is measurable in both surface and down-core sediment samples (Belt et al., 2007), indicating that it is stable enough and occurs in sufficient quantities to allow detection; consequently the method has a great deal of promise as a sea ice proxy. However, Belt et al. (2007) caution that, for now, IP_{25} can be used as a sea ice presence or absence indicator only and that more work needs to be done to derive quantitative estimates of sea ice cover.

The occurrence of IP_{25} reflects the presence of seasonal sea ice, so that identification of situations characterized by either perennial or non-seasonal sea ice conditions remains problematic (Belt & Müller, 2013, see their figure 3). Belt and Müller (2013) summarize studies that address IP_{25} production and downward flux, the use of the IP_{25} proxy in conjunction with other biomarkers, and palaeo sea ice reconstructions based on IP_{25} data. Study locations and studies that combine proxies are mapped out and denoted (see their figures 2 and 3, respectively). Given the potential limitations of the IP_{25} data, new research is directed toward combining the IP_{25} data with other sea ice proxies (Belt & Müller, 2013), including newly developing phytoplankton biomarkers, which can be used to identify ice-free surface conditions (Müller et al., 2009, 2011; Müller, 2011; Weckstrom et al., 2013; Xiao et al., 2013).

The use of PIP_{25}, a ratio of IP_{25} to combined IP_{25} and phytoplankton biomarkers was pioneered by Müller (2011), working with surface sediments from East Greenland and West Spitsbergen, and has been tested and applied more recently by several other researchers (Fahl & Stein, 2012; Müller et al., 2012; Stein et al., 2012; Navarro-Rodriguez et al., 2013; Cabedo-Sanz et al., 2013; Stoynova et al., 2013; Xiao et al., 2013). Another combined approach is based on the observation of an HBI diene 2 that co-occurs with IP_{25} (Belt et al., 2007, 2008), an approach that has been tested by Fahl and Stein (2012), Cabedo-Sanz et al. (2013) and Xiao et al. (2013). In general, their data sets suggest that more work needs to be completed in order to develop the diene/IP_{25} ratio into a robust proxy, but that it holds great potential for providing more detailed information about the history of sea ice than can be deduced from a single biomarker.

Microfossil data: calcareous microfossils – coccolithophorids, ostracodes and foraminifera

Calcareous microfossils, such as coccolithophorids, ostracodes and foraminifera, are used to track sea ice cover over time in the high northern latitudes, with the caveat that greater dissolution of $CaCO_3$ occurs as water temperature decreases, and species diversity decreases as latitude increases. The photosynthetic coccolithophorids in high latitudes are dominated by two species, *E. huxleyi* and *Coccolithus pelagicus* (Braarud, 1979; Samtleben & Schroder, 1992; Baumann et al., 2000) whose occurrence in marine sediments has been used to trace seasonally ice-free conditions (Rahman & de Vernal, 1994 – Labrador Sea; Hebbeln & Wefer, 1997 – Fram Strait; Hebblen et al., 1998 – North Atlantic).

Ostracodes, micro-arthropods with a low magnesium calcite carapace, are a new tool for reconstructing sea ice in the Arctic. Cronin et al. (2010) presented modern distribution data on *Acetabulastoma arcticum*, a sea ice-dwelling species associated exclusively with perennial sea ice, a consequence of its parasitic relationship with amphipods that live in perennial ice. The semi-quantitative *A. arcticum* sea ice index was developed based on a modern core-top database of 682 samples and comparison to mean September sea ice concentration between 1978 and 1991. The primary drawbacks to this new approach are the generally low abundance of ostracodes in Arctic sediments, and problems with dissolution below the lysocline (de Vernal et al., 2013a). However, application to down-core sediments is promising, with data from the Mendeleev Ridge extending back 600 kyr (Cronin et al., 2013). These data support the extreme warmth of MIS 11 and MIS 13, and demonstrate sea ice variability occurring at precessional (~22 kyr) and obliquity (~40 kyr) spectral frequencies. Further, Cronin et al. (2010) demonstrate the broad reach of ostracode-based sea ice studies with data from the Mendeleyev, Lomonosov and Gakkel Ridges, the Morris Jesup Rise, Yermak Plateau, and the Rockall and Iceland Plateaus. These data support both temporal and spatial variability in perennial Arctic sea ice, occurring over both glacial/interglacial and millennial time scales.

The use of planktonic foraminiferal assemblage data as a sea ice proxy in the Arctic has been limited due to the commonplace distribution of only a single species, *Neogloboquadrina pachyderma*. Also, as planktonic foraminifera can live deeper in the water column, their distribution is not directly linked to the presence or absence of sea ice. Sarnthein et al. (2003) used a similarity maximum modern-analog technique in the North Atlantic, first to reconstruct SST based on planktonic

Table 26.4 List of IP$_{25}$ studies relevant to Arctic palaeo sea ice reconstruction.

Reference	Region	Timescale
Belt et al. (2007)	Canadian Arctic	Modern sea ice and sediments Holocene–9000 years
Belt et al. (2008)	Arctic	Modern sediment traps and sediments
Massé et al. (2008)	Iceland	Late Holocene–past 1000 years
Andrews et al. (2009)	Iceland	Late Holocene–past 2000 years
Müller et al. (2009)	Fram Strait	30,000 years
Vare et al. (2009)	Canadian Arctic	Holocene–past 10,000 years
Belt et al. (2010)	Central Canadian archipelago	Holocene–7000 years
Gregory et al. 2010	Canadian Arctic, Barrow Strait	Holocene–past 10,000 years
Vare et al. (2010)	Barents Sea	Late Holocene–past 300 years
Axford et al. (2011)	Iceland	Late Holocene–past 2000 years
Brown (2011)	Arctic	Modern sea ice
Brown et al. (2011)	Canadian Beaufort Sea	Modern sea ice
Müller (2011)	Fram Strait	Modern calibration
Müller et al. (2011)	East Greenland and West Spitsbergen	Modern calibration
Belt et al. (2012a)	Arctic	Modern sediments
Belt et al. (2012b)	Arctic	Methods
Fahl & Stein (2012)	Central Arctic, Lomonosov Ridge	Modern 14 500 years
Müller et al. (2012)	Fram Strait	Holocene–8500 years
Stein & Fahl (2012)	Fram Strait, southern Yermak Plateau	2.2 million years ago
Stein et al. (2012)	Arctic	Review paper
Navarro-Rodriguez et al. (2013)	Barents Sea	Modern sediments
Belt and Müller (2013)	Arctic	Review paper
Cabedo-Sanz et al. (2013)	northern Norway	Younger *Dryas*, 13.8–7.2 calendar kyr bp
Stoynova et al. (2013)	Arctic	Modern sediments
de Vernal et al. (2013a)	Arctic	Review paper
Weckstrom et al. (2013)	SW Labrador Sea	Modern sediments, late Holocene–past 150 years
Xiao et al. (2013)	Kara and Laptev Seas	Modern sediments
Brown et al. (2014a)	Arctic	Modern sea ice

foraminiferal assemblage data and then to derive sea ice data, based on observed spatial relationships between SST data and sea ice. This strategy was applied toward an assessment of sea ice extent during the LGM in the northern North Atlantic (Sarnthein et al., 2003). Their results are in contrast to the interpretation provided by CLIMAP (1981), suggesting that sea ice covered a smaller area of the North Atlantic during the LGM summer than indicated by the CLIMAP reconstruction, while winter sea ice was quite extensive. One other approach has been to use phytodetritus-dependent benthic foraminiferal species as indirect indicators of overlying sea ice due to the controls of sea ice cover and/or extent on primary productivity (Wollenburg & Kuhnt, 2000; Wollenburg et al., 2001, 2004, 2007).

Use of multiple proxies, including foram data, expands the utility of individual proxies. For example, Scott et al. (2008) used a combination of planktonic and benthic foraminifera and planktonic ciliate (tintinnids) data from the Beaufort Sea, noting that the presence of both tintinnids and abundant agglutinated benthic foraminifera suggests freshwater input to the marine setting, a condition not possible with perennial sea ice cover. In addition, since freshwater input results in the decreased abundance of planktonic foraminifera, the planktonic to benthic foraminifera ratio can be used to evaluate if there was perennial sea ice, a strategy used by Schell et al. (2008) in the Amundsen Gulf. Gregory et al. (2010) combined foraminiferal data and the biomarker IP$_{25}$ to evaluate changes in oceanographic conditions in

Barrow Strait over the past 10 000 years, suggesting that an up-core increase in carbonate dissolution associated with higher IP$_{25}$ flux may be related to sea ice formation. Wollenburg et al. (2004) describe how brine rejection and increased seawater density during sea ice formation may increase oxygenation of bottom waters and lead to greater decay of organic matter and higher partial pressure of carbon dioxide (pCO$_2$) in pore waters, increasing carbonate dissolution. Finally, Polyak et al. (2013) present a 1.5-million-year multi-proxy record from the much under-studied western Arctic Basin; data include foraminiferal abundances, diversity and assemblages, with a focus on phytodetritus versus polar species. Their data track the progression of northern hemisphere cooling by estimating mostly seasonal sea ice cover in the Lower Pleistocene to mostly perennial sea ice cover by the Early–Middle Pleistocene boundary. Episodic sea ice expansion was marked by decreased foraminifera abundance and the decreased contribution of phytodetritus species (Polyak et al., 2013).

One advantage of working with CaCO$_3$-shelled organisms is the ability to conduct stable isotope analyses. Hillaire-Marcel and de Vernal (2008) provide a novel interpretation of planktonic foraminiferal stable isotopic data to evaluate the history of sea ice growth. They interpret 'off-equilibrium, low δ^{18}O values in mesopelagic planktic foraminiferal species' (p. 143) as the result of sea ice formation and the production of isotopically light brines. This rationale forms the framework for their reinterpretation of δ^{18}O data from cores from the northwestern North Atlantic, where heavy isotopic data from the LGM indicate little sea ice formation, while light isotopic data from the Heinrich layers suggest high rates of production of sea ice at that time (Hillaire-Marcel & de Vernal, 2008).

Sedimentological tracers: ice-rafted debris (IRD) including driftwood

Both sea ice and icebergs carry lithogenic material that is released as the ice melts (IRD). While icebergs transport material of all sizes, > 90% of the lithogenic material carried by sea ice consists of clay- to silt-sized particles (Pfirman et al., 1989; Wollenburg, 1993). Distinction of sea ice versus iceberg-rafted material is not always clear, particularly in the case of fine-grained lithogenic particles, and hence few studies adopt fine-grained IRD as a proxy for sea ice extent. Exceptions include work by Bond et al. (1997), Darby (2003) and Darby and Bischof (2004). Distinguishing transport modes can be done through observation of surface features of sand-sized quartz grains (St John, 2008). Glacially transported grains have features such as conchoidal fractures, angular edges and step fractures associated with mechanical breakage. Surface features associated with chemical weathering, including more rounded edges, were assumed to be associated with sea ice transport. Application of these criteria to sand-sized grains from IODP 302 cores from the central Arctic supports the major findings of this project regarding the timing of the initiation of both sea ice and glacial ice in the Arctic at 46 million years ago (Ma) and is used to track changes in ice extent since its initial appearance. Studies also utilize the composition of IRD to trace pathways of sea ice transport. Clay mineral studies provide information on both the source of the sediments and the transport path of the sea ice (Pfirman et al., 1997; Dethleff et al., 2000). Work by Hillaire-Marcel et al. (2013) utilizes the radiogenic isotopes ^{87}Sr/^{86}Sr and $\sum$Nd in tracing sea ice transport. Darby et al. (2012) reconstruct the Arctic Oscillation (AO) over the past 8000 years based on patterns of sea ice drift, noting that pulses of sediment from the Kara Sea, identified by IRD composition, can only reach the Alaska Coast during strongly positive AO.

Drifting ice also transports larger objects, such as wood. River bank erosion in North America and Eurasia results in delivery of wood to the Arctic that is then transported, with moving sea ice, via surface currents. After several years of drift as part of the moving pack ice, the wood can be stranded on distant coastlines. Radiocarbon dating of stranded wood has been used to identify changes in sea ice drift patterns over time (Dyke et al., 1997; Tremblay et al., 1997). Building on this approach, Funder et al. (2011) combined driftwood results with observations of the occurrence versus absence of beach ridges, which form only during seasonally open water periods, to develop a 10 000-year record of sea ice in the Arctic. Their findings support both millennial-scale variability, with a sea ice minimum between ~8500 and 6000 years ago and spatial variability probably linked to anomalies such as the AO.

Ice-core data

Physical and geochemical properties of glacial ice in the Arctic have been related to sea ice cover. Koerner (1977) found a positive relationship between melt layers in glacial ice of the Devon Island Ice Cap and maximum open water nearby, a relationship driven by warmer summers that produce both surface melting on the ice cap that subsequently refreezes as a melt layer and less summer sea ice. Fisher et al. (1995, 2006) similarly present melt layer data (thickness and number) from the Agassiz and Penny Ice Caps, and relate these data to summer warmth. The concentration of sea salts in glacial ice may also provide a record of sea ice cover (Grumet et al., 2001; Kinnard et al., 2006), through the transport and deposition of marine aerosols from open water to glacial ice. However, this relationship is also dependent on polar atmospheric circulation strength and salt transport (Mayewski et al., 1994; O'Brien et al., 1995), as also discussed for the Antarctic. In general, the full potential of Arctic ice core data in reconstructing sea ice remains to be fully utilized.

Other new approaches

The low sedimentation rates in the Arctic marine environment have necessitated an investigation into the utility of higher-resolution sea ice records (Polyak et al., 2010). Higher-resolution records are important, especially over the past thousand years, as they provide a window into the scale of climate change that has impacted human societies. Halfar et al. (2013) presented one such record based on data from crustose coralline algae with annual growth increments. Magnesium/calcium ratios and growth are both related to overlying sea ice cover, allowing the researchers to develop a nearly 650-year record. Their data document both extensive and highly variable sea ice cover during the Little Ice Age, a time of expansion of sea mammal hunters dependent on sea ice. In contrast, diminished sea ice cover and lower-frequency variability characterised warmer intervals. This kind of information is especially relevant to today's world of rapidly decreasing Arctic sea ice cover.

26.3.2 Time-slice palaeo sea ice reconstructions

Polyak et al. (2010) presented a comprehensive review of the development and history of Arctic sea ice, and de Vernal et al. (2013c) and Belt and Müller (2013) presented summaries of more recent Holocene sea ice history, based primarily on dinocysts and the biomarker IP$_{25}$, respectively. Therefore, here we present a more focused approach, addressing only two time periods: the initiation of Arctic sea ice ~46 Ma, and its recent history.

Initiation of sea ice and perennial ice in the Arctic Basin

A wealth of data concerning the history of sea ice and the initiation of its formation in the Arctic emerged from the Integrated Ocean Drilling Program (IODP) Arctic Coring Expedition (ACEX) on the Lomonosov Ridge during 2004. The results place the initiation of sea ice at ~46 Ma, much earlier than previously thought (Moran et al., 2006). Most importantly, the data indicate a synchroneity with both the Antarctic record of ice and with records of atmospheric CO_2, suggesting the key role of the atmosphere in driving both northern and southern hemisphere glaciation (Moran et al., 2006).

A variety of proxy data from the IODP 302 cores suggest the initiation of sea ice at ~46 Ma (Sangiorgi et al., 2008), including the IRD record of both quantitative information on sand accumulation rates and a qualitative evaluation of surface textures and grain composition (St John, 2008), which are in accordance with pebble-based data (Backman et al., 2006; Moran et al., 2006). Stickley et al. (2008) observed an increase in the abundance of a form of diatoms (needle-like *Synedropsis* spp.) which they interpret to be indicative of sea ice, based on morphological similarities to modern sea ice diatoms. Finally, Wadell and Moore (2008) used $\delta^{18}O$ data from fish bone carbonate to reconstruct middle Eocene salinities. An increase in salinity at ~46 Ma is interpreted to be the result of sea ice formation.

The second major finding of the ACEX project is that perennial ice in the central Arctic developed at ~14 Ma. Strong evidence comes from geochemical fingerprinting of detrital iron oxide grains (Darby, 2008). By identifying source regions for the iron oxide mineral grains <180 μm in size, following likely drift paths, and then estimating if the drift time was either greater than or less than 1 year, Darby (2008) distinguished between perennial and seasonal sea ice cover. Similarly, heavy and clay mineral analysis of the ACEX IODP 302 cores indicate a large-scale shift in provenance at ~13 Ma, which Krylov et al. (2008) interpreted as a shift to perennial sea ice at that time. Their rationale is similar to that of Darby (2008), which is that, in order for grains from the eastern Laptev-East Siberian seas to have been transported to the ACEX site, sea ice must have survived for more than a single season.

Holocene (0–12 kyr bp).

A focus on Holocene climate originates from both the well-documented relationship between human activities and climate over the past ~10 000 years (Grove, 2001; Jensen et al., 2004; Mudie et al., 2005) and the recognition of recent dramatic change in sea ice in the northern high latitudes (see Comiso, 2012). To date, the most extensive and highest-resolution records come from the Holocene and the most recent data can be compared with historical and instrumental records. Studies demonstrate both temporal and geographic heterogeneity in sea ice cover through the Holocene (Polyak et al., 2010; de Vernal et al., 2013c). These regional differences must be understood in order to evaluate the role of processes, such as the AO, that have influenced sea ice cover in the Arctic in the past and continue under warming conditions (Fisher et al., 2006).

Reduced early Holocene sea ice cover is a feature observed in many records (Dyke et al., 1996; Koç et al., 1993; Dyke & Savelle, 2001; Levac et al., 2001; de Vernal et al., 2001, 2005b; Vare et al., 2009) with ice-free summers as far north as 83°N (Funder et al., 2011), reflecting a greater extent of seasonally open water than observed today. As noted previously, patterns in changes following the early Holocene minimum vary geographically and are quite complex. Circum-Arctic synchrony is not observed, even within a single sub-basin. Data from the Chukchi Sea indicate that minimal sea ice cover before 12 000 calendar years BP was followed by extensive sea ice cover until 6000 calendar years BP and then generally reduced sea ice for the rest of the Holocene, though century-scale and millennial-scale oscillations, are also observed (de Vernal et al., 2005b; McKay et al., 2008; Farmer et al., 2011). Increased mid-Holocene sea ice has also been reported from Baffin Bay (Levac et al., 2001), the North Atlantic (Moros et al., 2006) and the Canadian Arctic (Vare et al., 2009). In the northwest North Atlantic, de Vernal and Hillaire-Marcel (2006) observed a sea ice minimum from 11 500 to 6000 calendar years BP, with slightly increased and variable sea ice for the remainder of the Holocene. A steady decrease in sea ice cover through the late Holocene is observed in the Beaufort Sea (Rochon et al., 2006) and on the Greenland and north Iceland shelves (Solignac et al., 2006).

Multi-proxy work in the North Atlantic (Bond et al., 1997) demonstrates ~1500-year cyclicity in North Atlantic circulation throughout the Holocene. Such cyclicity may have been responsible for changes in IRD distribution (i.e. via drifting sea ice) and can be linked to millennial-scale variability in the accumulation of *E. huxleyi* in phase with associated IRD peaks (Iceland – Giraudeau et al., 2000). Jennings et al. (2002) also found links to the millennial-scale cycles of Bond et al. (1997) in their multi-proxy data set from the East Greenland margin. Millennial- and century-scale variability in sea ice cover has been recorded in multi-proxy data sets from western Greenland (Möller et al., 2006; Moros et al., 2006; Seidenkrantz et al., 2007, 2008), where a combination of microfossil and X-ray fluorescence data demonstrates highly variable sea ice conditions through the Holocene. Despite the isolated fjord settings from which these cores were recovered, their records are similar to those from the Labrador Sea. Seidenkrantz et al. (2007, 2008) note that western Greenland data are in accordance with the North Atlantic climate seesaw that describes the air temperature differences between Greenland and northern Europe.

Increasingly evident is the need for increased temporal and spatial resolution of Holocene sea ice trends in the Arctic, a complex region characterized by dynamic processes. This is especially true for the central Arctic, where generally low accumulation rates have precluded development of adequate palaeo sea ice records (Polyak et al., 2010). These data will enhance our ability to model and predict the potentially heterogeneous response to modern atmospheric warming that has been observed and which is predicted to continue (Johannessen et al., 2004).

26.4 Conclusion

Our objective in this review chapter was to highlight those advances in palaeo sea ice estimation since our previous review (Armand & Leventer, 2010). Conclusions drawn 5 years ago regarding the need for improved coverage in both polar regions, or the combination of microfossil databases, has generally been achieved with perhaps better success in the Arctic than in the Antarctic. Extension of the sea ice record to longer-term variation over the Arctic region has also been addressed.

The effort required to collect samples of any type from the Antarctic remains one of the largest impediments to advancing coverage, although new international efforts through the International Ocean Discovery Program (IODP), the GEOTRACERS programme or indeed individual national efforts are conscious of the need to obtain wider sample coverage. The call for records from both polar regions and their various oceanographic sectors with data, such as sea ice concentrations, relevant to modelling needs, remains (Vaughan et al., 2013). However, with the recent advances in database consolidations and statistical treatments better adapted to estimating the sea ice, it is only a matter of a few years before a substantial revision of pre-existing core records can be treated with the same community-tested and agreed methodology. In terms of Antarctic records, significant advances have been made with the inclusion of the Pacific sector by AWI researchers to finally enable a circumpolar database to be constructed. A single step remains within this community; to join all available datasets together so that a full testing phase of current estimation methods can be assessed against latitudinal and sectorial representative cores. Summer sea ice estimation from diatoms remains problematic in the Antarctic, in part due to the low seafloor coverage of samples without preservational biases in the seasonal sea ice zone (between the maximum winter sea ice and the summer sea ice). It is highly likely that the recovery of a summer sea ice extent, particularly during glacial periods, will come from alternative biomarker or ice core proxy records in the future rather than a diatom proxy. Testing these alternative sea ice proxy records (e.g. derived from MSA, sea salt fluxes, HBIs, corals etc.) may also help with linking the early 20th century, pre-satellite, records from indigenous societies or whaling records, as is already under way in the Arctic.

Alternative proxy developments have significantly increased, with 'the ice proxy with 25 carbons' (HBIs) leading the charge. Although the ground-truthing of this proxy and its use as an ecosystem trophic passage indicator have been clearly established in the Arctic, it is only a short matter of time before the same ground-truthing is provided for the Antarctic. In contrast, ice core gas inclusions have matured and become refined in the Antarctic; however, no study has tried to compare diatom-based sea ice estimates with records derived from MSA or any of the promising halogens. So far, diatom-based sea ice estimates have only been compared with ice core sea salt Na flux, identifying where each record complements the limitations of the other. Inevitably, the ice core-derived records will become a useful independent comparison between atmospheric processes, sea ice conditions and extent, and the biological signature or productivity of the oceans preserved in the seafloor sediments. Following this thread, long-term sediment trap studies may help further our understanding of the transfer of the primary productivity signal under changing sea ice regimes, which in turn can be compared with modern satellite observations.

Eventually, our observations of the satellite-observed physical parameters, water column biota and chemical sampling, ice core gas records, new metagenomic analyses, phytoplankton sediment trap ecosystem succession and variability, and seafloor biological and geochemical signature characterization will meld to provide the most robust sea ice palaeo story from both polar regions.

26.5 Acknowledgements

The authors wish to acknowledge formal and informal discussions with many members of the palaeo sea ice community in the derivation of this work, which includes the participation in the PAGES-sponsored Sea Ice Proxy workshops by Alexander Ferry and Leanne Armand. We thank O. Esper and R. Gersonde for their personal communication of data included in Figure 26.1. An Australian Postgraduate Award provided by the Australian Government through Macquarie University supported Alexander Ferry's contribution.

References

Abelmann, A. & Gersonde, R. (1991) Biosiliceous particle flux in the Southern Ocean. *Marine Chemistry*, **35**, 503–536.

Abram, N.J., Thomas, E.R., McConnell, J.R. et al. (2010) Ice core evidence for a 20th century decline of sea ice in the Bellingshausen Sea Antarctica. *Journal of Geophysical Research*, **115**, doi:10.1029/2010JD014644.

Abram, N., Mulvaney, R. & Arrowsmith, C. (2011) Environmental signals in a highly resolved ice core from James Ross Island, Antarctica. *Journal of Geophysical Research: Atmospheres*, **116**, 1–15.

Abram, N.J., Wolff, E.W. & Curran, M.A.J. (2013) A review of sea-ice proxy information from polar ice cores. *Quaternary Science Reviews*. **79**(1), 168–183.

Adkins, J.F. (2013) The role of deep ocean circulation in setting glacial climates. *Palaeoceanography*, **28**, 539–561.

Allen, C.S., Oakes-Fretwell, L., Anderson, J.B. & Hodgson, D.A. (2010) A record of Holocene glacial and oceanographic variability in Neny Fjord, Antarctic Peninsula. *The Holocene*, **20**, 551–564.

Allen, C.S., Pike, J & Pudsey, C.J. (2011) Last glacial interglacial sea-ice cover in the SW Atlantic and its potential role in global deglaciation. *Quaternary Science Reviews*, **30**, 2446–2458.

Andrews, J.T., Belt, S.T., Ólafsdóttir, S., Massé, G. & Vare, L.L. (2009) Sea ice and climate variability for NW Iceland/Denmark Strait over the last 2000 cal. yr BP. *The Holocene*, **19**, 775–784.

Armand, L.K. & Leventer, A. (2003) Palaeo sea ice distribution; its reconstruction and significance. In: *Sea Ice*, (Eds. D. N. Thomas & G.S. Dieckmann, G.S.), pp. 333–372, Blackwell Publishing Ltd, Oxford.

Armand, L.K. & Leventer, A. (2010) Palaeo sea ice distribution and reconstruction derived from the geological record. In: *Sea Ice* (Eds. D. N. Thomas & G.S. Dieckmann, G.S.), pp. 469–530, Wiley-Blackwell Publishing Ltd, Oxford.

Armand, L.K., Crosta, X., Romero, O. & Pichon, J.J. (2005) The biogeography of major diatom taxa in Southern Ocean sediments: 1. Sea ice related species. *Palaeogeography, Palaeoclimatology, Palaeoecology*, **223**, 93–126.

Armand, L.K., Crosta, X., Quéguiner, B., Mosseri, J. & Garcia, N. (2008) Diatoms preserved in surface sediments of the northeastern Kerguelen Plateau. *Deep Sea Research Part II : Topical Studies in Oceanography*, **55**, 677–692.

Atkinson, H.M., Huang, R.-J., Chance, R. et al. (2012) Iodine emissions from the sea ice of the Weddell Sea. *Atmospheric Chemistry and Physics*, **12**, 11 229–11 244.

Axford, Y., Andresen, C.S., Andrews, J.T. et al.(2011) Do paleoclimate proxies agree? A test comparing 19 late Holocene climate and sea-ice reconstructions from Icelandic marine and lake sediments. *Journal of Quaternary Science*, **26**, 645–656.

Backman, J., Moran, K., McInroy, D.B., Mayer, L. & Expedition 302 Scientists (2006) *Proceedings of the Integrated Ocean Drilling Program*, **302**, doi.10.2204/iodp.proc.302.104.2006.

Barbara, L., Crosta, X., Massé, G. & Ther, O. (2010) Deglacial environments during the Holocene in East Antarctica. *Quaternary Science Reviews*, **29**, 2731–2740.

Barbara, L., Crosta, X., Schmidt, S. & Massé, G. (2013) Diatoms and biomarkers evidence for major changes in sea ice conditions prior the instrumental period in Antarctic Peninsula. *Quaternary Science Reviews*, **79**, 99–110.

Bauch, H.A. & Polyakova, Y.I. (2000) Late Holocene variations in Arctic shelf hydrology and sea ice regime: evidence from north of the Lena Delta. *International Journal of Earth Sciences*, **89**, 569–577.

Baumann, K.-H., Andruleit, H.A. & Samtleben, C. (2000) Coccolithophores in the Nordic Seas: comparison of living communities with surface sediment assemblages. *Deep-Sea Research*, **47**, 1743–1772.

Belt, S.T. & Müller, J. (2013) The Arctic sea ice biomarker IP$_{25}$: A review of current understanding, recommendations for future research and applications in palaeo sea ice reconstructions. *Quaternary Science Reviews*, **79**, 9–25.

Belt, S.T., Massé, G., Rowland, S.J., Poulin, M., Michel, C. & LeBlanc, B. (2007) A novel chemical fossil of palaeo sea ice: IP$_{25}$. *Organic Geochemistry*, **38**, 16–27.

Belt, S.T., Massé, G., Vare, L.L. et al. (2008) Distinctive ^{13}C isotopic signature distinguishes a novel sea ice biomarker in Arctic sediments and sediment traps. *Marine Chemistry*, **112**, 158–167.

Belt, S.T., Vare, L.L., Massé, G. et al. (2010) Striking similarities in temporal changes to seasonal sea ice conditions across the central Canadian Arctic Archipelago over the last 7,000 years. *Quaternary Science Reviews*, **29**, 3489–3504.

Belt, S.T., Brown, T.A., Cabedo Sanz, P. & Navarro Rodriguez, A. (2012a) Structural confirmation of the sea ice biomarker IP$_{25}$ found in Arctic marine sediments. *Environmental Chemistry Letters*, **10**, 189–192.

Belt, S.T., Brown, T.A., Navarro Rodriguez, A., Cabedo Sanz, P., Tonkin, A. & Ingle, R. (2012b) A reproducible method for the extraction, identification and quantification of the Arctic sea ice proxy IP$_{25}$ from marine sediments. *Analytical Methods*, **4**, 705–713.

Bentley, M.J., Hodgson, D.A., Sugden, D.E. et al. (2005) Early Holocene retreat of the George VI Ice Shelf, Antarctic Peninsula. *Geology*, **33**, 173–176.

Bentley, M.J., Hodgson, D.A., Smith, J.A. et al. (2009) Mechanisms of Holocene palaeoenvironmental change in the Antarctic Peninsula region. *The Holocene*, **19**, 51–69.

Boessenkool, K.P., Van Gelder, M.-J., Brinkhuis, H. & Troelstra, S.R. (2001) Distribution of organic-walled dinoflagellate cysts in surface sediments from transects across the Polar Front offshore southeast Greenland. *Journal of Quaternary Science*, **16**, 661–666.

Bond, G., Showers, W., Cheseby, M. et al. (1997) A pervasive millennial-scale cycle in North Atlantic Holocene and glacial climates. *Science*, **278**, 1257–1266.

Braarud, T. (1979) The temperature range of the non-motile stage of Coccolithus pelagicus in the North Atlantic region. *British Phycological Journal*, **14**, 349–352.

Brandon, M.A., Cottier, F.R. & Nilsen, F. (2010) Sea Ice and Oceanography. In: *Sea Ice* (Eds. D. N. Thomas & G.S. Dieckmann, G.S.), pp. 79–112. Wiley-Blackwell Publishing Ltd, Oxford.

Bringué, M. & Rochon, A. (2012) Late Holocene paleoceanography and climate variability over the Mackenzie Slope (Beaufort Sea, Canadian Arctic). *Marine Geology*, **291–294**, 83–96.

Brown, T.A. (2011) Production and preservation of the Arctic sea ice diatom biomarker IP$_{25}$. PhD Thesis. University of Plymouth, UK.

Brown, T.A., Belt, S.T., Mundy, C. et al. (2011) Temporal and vertical variations of lipid biomarkers during a bottom ice diatom bloom in the Canadian Beaufort Sea: further evidence for the use of the IP$_{25}$ biomarker as a proxy for spring Arctic sea ice. *Polar Biology*, **34**, 1857–1868.

Brown, T.A., Belt, S.T., Tatarekm, A. & Mundy, C.J. (2014a) Source identification of the Arctic sea ice proxy IP$_{25}$. *Nature Communications*, **5**, doi:10.1038/ncomms5197.

Brown, T.A., Yurkowski, D.J., Ferguson, S.H., Alexander, C. & Belt, S.T. (2014b) H-Print: a new chemical fingerprinting approach for distinguishing primary production sources in Arctic ecosystems. *Environmental Chemistry Letters*, **12**, 387–392.

Burckle, L.H. (1984) Diatom distribution and palaeoceanographic reconstruction in the Southern Ocean - present and last glacial maximum. *Marine Micropalaeontology*, **9**, 241–261.

Cabedo-Sanz, P., Belt, S.T., Knies, J.K. & Husum, K. (2013) Identification of contrasting seasonal sea ice conditions during the Younger Dryas. *Quaternary Science Reviews*, **79**, 74–86.

Caissie B. E., J. Brigham-Grette, Lawrence, K.T., Herbert, T.D. & Cook, M.S. (2010) Last Glacial Maximum to Holocene sea surface conditions at Umnak Plateau, Bering Sea, as inferred from diatom, alkenone, and stable isotope records. *Paleoceanography*, **25**, PA1206, doi:10.1029/2008PA001671.

Ciais, P., Petit, J.R., Jouzel, J. et al. (1992) Evidence for an early Holocene climatic optimum in the Antarctic deep ice-core record. *Climate Dynamics*, **6**, 169–177.

Clark, P. U., Dyke, A.S., Shakun, J.D. et al. (2009) The Last Glacial Maximum. *Science*, **325**, 710–714.

CLIMAP Project Members (1976) The surface of the Ice-Age Earth. *Science*, **191**, 1131–1137.

CLIMAP Project Members (1981) Seasonal reconstructions of the Earth's surface at the last glacial maximum. *Geological Society of America: Map and Chart Series*, **MC-36**.

Cochran, U.A. & Neil, H. (2010) Diatom (<63 micro-m) distribution offshore of eastern New Zealand : surface sediment record and temperature transfer function. *Marine Geology*, **270**(1–4), 257–271.

Collins, L. G., Pike, J., Allen, C.S. & Hodgson, D.A. (2012) High-resolution reconstruction of southwest Atlantic sea-ice and its role in the carbon cycle during marine isotope stages 3 and 2. *Palaeoceanography*, **27**, doi:10.1029/2011PA002264.

Collins, L.G., Allen, C.S., Pike, J., Hodgson, D.A., Weckström, K. & Massé, G. (2013) Evaluating highly branched isoprenoid (HBI) biomarkers as a novel Antarctic sea-ice proxy in deep ocean glacial age sediments. *Quaternary Science Reviews*, **79**, 87–98.

Comiso, J. (2010) Variability and trends of the global sea ice cover. In: *Sea Ice* (Eds. D.N. Thomas and G.S. Dieckmann), pp. 247–282. Wiley-Blackwell Publishing Ltd, Oxford.

Comiso, J. (2012) Large decadal decline of the Arctic multiyear ice cover, *Journal of Climate*, **25**, 1176–1193.

Cotté, C. & Guinet, C. (2007) Historical whaling records reveal major regional retreat of Antarctic sea ice. *Deep-Sea Research II*, **54**, 243–252.

Cremer, H. (1999) Distribution patterns of diatom surface sediment assemblages in the Laptev Sea (Arctic Ocean). *Marine Micropaleontology*, **38**, 39–67.

Cronin, T.M., Gemery, L., Briggs, W.M. Jr., Jakobsson, M., Polyak, L. & Brouwers, E.M. (2010) Quaternary sea-ice history in the Arctic Ocean based on a new sea-ice proxy. *Quaternary Science Reviews*, **29**, 3415–3429.

Cronin, T.M., Polyak, L., Reed, D., Kandiano, E.S., Marzen, R.E. & Council, E.A. (2013) A 600-ka Arctic sea-ice record from the Mendeleev Ridge based on ostracodes. *Quaternary Science Reviews*, **79**, 157–167.

Crosta, X., Pichon, J-J. & Burckle, L.H. (1998) Application of modern analog technique to marine Antarctic diatoms: reconstruction of maximum sea-ice extent at the Last Glacial Maximum. *Palaeoceanography*, **13**, 284–297.

Crosta, X., Sturm, A., Armand, L. & Pichon, J.-J. (2004) Late Quaternary sea ice history in the Indian sector of the Southern Ocean as recorded by diatom assemblages. *Marine Micropalaeontology*, **50**, 209–223.

Crosta, X., Romero, O., Armand, L.K. & Pichon, J.-J. (2005) The biogeography of major diatom taxa in Southern Ocean sediments: 2. Open-Ocean related species. Palaeogeography, Palaeoclimatology. *Palaeoecology*, **223**, 66–92.

Curran, M.A.J., van Ommen, T.D., Morgan, V., Phillips, K.L. & Palmer, A.S. (2003) Ice core evidence for Antarctic sea ice decline since the 1950s. *Science*, **302**, 1203–1206.

Darby, D.A. (2003) Sources of sediment found in sea ice from the western Arctic Ocean, new insights into processes of entrainment and drift patterns. *Journal of Geophysical Research*, **108**, doi:10.1029/2002JC001350.

Darby, D.A. (2008) Arctic perennial ice cover over the last 14 million years. *Paleoceanography*, **23**, doi:10.1029/2007PA001479.

Darby, D.A. & Bischof, J.F. (2004) A Holocene record of changing Arctic Ocean ice drift analogous to the effects of the Arctic Oscillation. *Paleoceanography*, **19**, PA1027, doi:10.1029/2003PA000961.

Darby, D.A., Ortiz, J. D., Grosch, C. & Lund, S. (2012) 1,500 year cycle in the Arctic Oscillation identified in Holocene Arctic sea-ice drift. *Nature Geoscience*, **5**, 897–900.

Das, S. B. & Alley, R.B. (2008) Rise in frequency of surface melting at Siple Dome through the Holocene: Evidence for increasing marine influence on the climate of West Antarctica. *Journal of Geophysical Research*, **113**, doi:10.1029/2007JD008790.

De Sève, M.A. (1999) Transfer function between surface diatom assemblages and sea-surface temperature and salinity of the Labrador Sea. *Marine Micropalaeontology*, **36**, 249–267.

Denis, D., Crosta, X., Barbara, L. et al. (2010) Sea ice and wind variability during the Holocene in East Antarctica: insight on middle-high latitude coupling. *Quaternary Science Reviews*, **29**, 3709–3719.

Dethleff, D., Rachold, V., Tintelnot, M. & Antonow, M. (2000) Sea ice transport of riverine particles from the Laptev Sea to Fram Strait based on clay mineral studies. *International Journal of Earth Sciences*, **89**, 496–502.

Dieckmann, G.S. & Hellmer, H.H. (2010) The importance of sea ice: an overview. In: *Sea Ice* (Eds. D. N. Thomas & G.S. Dieckmann), pp. 1–22. Wiley-Blackwell Publishing Ltd, Oxford.

Divine, D. V., Koç, N., Isaksson, E., Nielsen, S., Crosta, X. & Godtliebsen, F. (2010) Holocene Antarctic climate variability from ice and marine sediment cores: insights on ocean–atmosphere interaction. *Quaternary Science Reviews*, **29**, 303–312.

Dyke, A.S. & Savelle, J.M. (2001) Holocene history of the Bering Sea bowhead whale (*Balaena mysticetus*) in its Beaufort Sea summer grounds off southwestern Victoria Island, NWT, western Canadian Arctic. *Quaternary Research*, **55**, 371–379.

Dyke, A.S., Hooper, J. & Savelle, J.M. (1996) A history of sea ice in the Canadian Arctic Archipelago based on postglacial remains of the bowhead whale (*Balaena mysticetus*). *Arctic*, **49**, 235–255.

Dyke, A.S., England, J., Reimnitz, E. & Jette, H. (1997) Changes in driftwood delivery to the Canadian Arctic archipelago: the hypothesis of postglacial oscillations of the transpolar drift. *Arctic*, **50**, 1–16.

Esper, O. & Gersonde, R. (2014) New tools for the reconstruction of Pleistocene Antarctic sea ice. *Palaeogeography, Palaeoclimatology, Palaeoecology*, **399**, 260–283.

Esper, O., Gersonde, R. & Kadagies, N. (2010) Diatom distribution in southeastern Pacific surface sediments and their relationship to modern environmental variables. *Palaeogeography, Palaeoclimatology, Palaeoecology*, **27**, 1–27.

Etourneau, J., Collins, L.G., Willmott, V. et al. (2013) Holocene climate variations in the western Antarctic Peninsula: evidence for sea ice extent predominantly controlled by changes in insolation and ENSO variability. *Climate of the Past*, **9**, 1431–1446.

Fahl, K. & Stein, R. (2012) Modern seasonal variability and deglacial/Holocene change of central Arctic Ocean sea-ice cover: new insights from biomarker proxy records. *Earth and Planetary Science Letters*, **351–352**, 123–133.

Farmer, J.R., Cronin, T.M., de Vernal, A., Dwyer, G.S., Keigwin, L.D. & Thunell, R.C. (2011) Western Arctic Ocean temperature variability during the last 8000 years. *Geophysical Research Letters*, **38**, L24602, doi:10.1029/2011GL049714.

Ferry, A., Prvan, T., Jersky, B., Crosta, X. & Armand, L.K. (2015a) Statistical modeling of Southern Ocean marine diatom proxy and winter sea ice data: model comparison and developments, *Progress in Oceanography*, **131**, 100–112.

Ferry, A., Crosta, X., Quilty, P.G., Fink, D., Howard, W. & Armand L.K. (2015b) First records of winter sea-ice concentration in the southwest Pacific sector of the Southern Ocean. *Paleoceanography*, **30**, 1525–1539.

Fischer, H., Fundel, F., Ruth, U. et al. (2007) Reconstruction of millennial changes in dust emission, transport and regional sea ice coverage using the deep EPICA ice cores from the Atlantic and Indian Ocean sector of Antarctica. *Earth and Planetary Science Letters*, **260**, 340–354.

Fisher, D.A., Koerner, R.M. & Reeh, N. (1995) Holocene Climatic Records from Agassiz Ice Cap, Ellesmere Island, NWT, Canada. *The Holocene*, **5**, 19–24.

Fisher, D., Dyke, A., Koerner, R. et al. (2006) Natural variability of Arctic sea ice over the Holocene. *EOS, Transactions, American Geophysical Union*, **87**, doi:10.1029/2006EO280001.

Fraser, C.I., Nikula, R., Spencer, H.G. & Waters, J.M. (2009) Kelp genes reveal effects of subantarctic sea ice during the Last Glacial Maximum. *Proceedings National Academy of Sciences*, **106**, 3249–3253.

Funder, S., Goosse, H., Jepsen, H. et al. (2011) A 10,000-year record of Arctic Ocean sea-ice variability—view from the beach. *Science*, **333**, 747–750.

Gersonde, R. & Wefer, G. (1987) Sedimentation of biogenic siliceous particles in Antarctic waters from the Atlantic sector. *Marine Micropaleontology*, **11**, 311–332.

Gersonde, R. & Zielinski, U. (2000) The reconstruction of late Quaternary Antarctic sea-ice distribution—the use of diatoms as a proxy for sea-ice. *Palaeogeography, Palaeoclimatology, Palaeoecology*, **162**, 263–286.

Gersonde, R., Abelmann, A., Brathauer, U. et al. (2003) Last glacial sea surface temperatures and sea-ice extent in the Southern Ocean (Atlantic-Indian sector): A multiproxy approach. *Paleoceanography*, **18**, 1061, DOI:10.1029/2002PA000809.

Gersonde, R., Crosta, X., Abelmann, A. & Armand, L. (2005) Sea-surface temperature and sea ice distribution of the Southern Ocean at the EPILOG Last Glacial Maximum—a circum-Antarctic view based on siliceous microfossil records. *Quaternary Science Reviews*, **24**, 869–896.

Giraudeau, J., Cremer, M., Manthe, S., Labeyrie, L. & Bond, G. (2000) Coccolith evidence for instabilities in surface circulation south of Iceland during Holocene times. *Earth and Planetary Science Letters*, **179**, 257–268.

Granfors, A., Karlsson, A., Mattsson, E., Smith, W.O. & Abrahamsson, K. (2013) Contribution of sea ice in the Southern Ocean to the cycling of volatile halogenated organic compounds. *Geophysical Research Letters*, **40**, doi:10.1002/grl.50777.

Gregory T.R., Smart C.W., Hart M.B., Vare, L.L., Massé, G. & Belt S.T. (2010) Holocene palaeoceanographic changes in Barrow Strait, Canadian Arctic: foraminiferal evidence. *Journal of Quaternary Science*, **25**, 903–910.

Grigorov, I., Rigual-Hernandez, A.S., Honjo, S., Kemp, A.E.S. & Armand, L.K. (2014) Settling fluxes of diatoms to the interior of the Antarctic Circumpolar Current along 170°W. *Deep Sea Research Part I*, **93**, 1–13.

Grove, J.M. (2001) The initiation of the 'Little Ice Age' in regions round the North Atlantic. *Climatic Change*, **48**, 53–82.

Grøsfjeld K, Larsen E, Sejrup HP et al. (1999) Dinoflagellate cysts reflecting surface-water conditions in Voldafjorden, western Norway during the last 11 300 years. *Boreas*, **28**, 403–415.

Grumet, N.S., Wake, C.P., Mayewski, P.A. et al. (2001) Variability of sea-ice extent in Baffin Bay over the last millennium. *Climatic Change*, **49**, 129–145.

Halfar, J., Adey, W.H., Kronz, A., Hetzinger, S., Edinger, E. & Fitzhugh, W. (2013) Arctic sea-ice decline archived by multicentury annual-resolution record from crustose coralline algal proxy, *Proceedings National Academy of Sciences*, **110**, 19737–19741.

Hart, T.J. (1942) Phytoplankton periodicity in Antarctic surface waters. *Discovery Reports*, **21**, 261–356.

Head, M.J., Harland, R. & Matthiessen, J. (2001) Cold marine indicators of the late Quaternary: the new dinoflagellate cyst genus *Islandinium* and related morphotypes. *Journal of Quaternary Science*, **16**, 621–636.

Hebbeln, D., Henrich, R. & Baumann, K.-H. (1998) Palaeoceanography of the last interglacial/glacial cycle in the polar North Atlantic. *Quaternary Science Reviews*, **17**, 125–153.

Hebbeln, D. & Wefer, G. (1997) Late Quaternary palaeoceanography in the Fram Strait. *Palaeoceanography*, **12**, 65–78.

Hezel, P.J., Alexander, B., Bitz, C.M. et al. (2011) Modeled methanesulfonic acid (MSA) deposition in Antarctica and its relationship to sea ice. *Journal of Geophysical Research*, **116**, D23214, doi:10.1029/2011JD016383, 2011

Hillaire-Marcel, C. & de Vernal, A. (2008) Stable isotope clue to episodic sea ice formation in the glacial North Atlantic. *Earth and Planetary Science Letters*, **268**, 143–150.

Hillaire-Marcel, C., Maccali, J., Not, C. & Poirier, A. (2013) Geochemical and isotopic tracers of Arctic sea ice sources and export with special attention to the Younger Dryas interval. *Quaternary Science Reviews*, **79**, 184–190.

Honjo, S., Francois, R., Manganinia, S., Dymond, J., & Collier, R. (2000) Particle fluxes to the interior of the Southern Ocean in the Western Pacific sector along 170°W, *Deep Sea Res. Part II*, **47**, 3521–3548.

Hutson, W. H. (1980) The Agulhas Current during the Late Pleistocene: analysis of modern faunal analogs. *Science*, **207**, 64–66.

Iida, T. & Odate, T. (2014) Seasonal variability of phytoplankton biomass and composition in the major water masses of the Indian Ocean sector f the Southern Ocean. *Polar Science*, **8**, 283–297.

Imbrie, J. & Kipp, N.G. (1971) A new micropalaeontological method for quantitative palaeoclimatology: application to a late Pleistocene Caribbean core. In: *The Late Cenozoic Glacial ages* (Ed. K.K. Turekian), pp. 71–147. Yale University Press, London.

Jennings, A.E., Andrews, J.T., Knudsen, K.L., Hald, M. & Hansen, C.V. (2002) A mid-Holocene shift in Arctic sea-ice variability on the East Greenland Shelf. *Holocene*, **12**, 49–58.

Jensen, K.G., Kuijpers, A., Koc, N. & Heinemeier, J. (2004) Diatom evidence of hydrographic changes and ice conditions in Igaliku Fjord. *The Holocene*, **14**, 152–164.

Jiang, H., Seidenkrantz, M.-S., Knudsen, K.L. & Eiríksson, J. (2001) Diatom surface sediment assemblages around Iceland and their relationships to oceanic environmental variables. *Marine Micropaleontology*, **41**, 73–96.

Johannessen O.A., Bengtsson, L., Miles, M.W. et al. (2004) Arctic climate change: observed and modeled temperature and sea-ice variability. *Tellus A*, **56**, 328–341.

Juggins, S. & Birks, J.B. (2012) Quantitative Environmental Reconstructions from Biological Data. In: *Tracking Environmental Change Using Lake Sediments* (Eds. H.J.B. Birks et al.), pp. 431–494. Springer Science and Business Media, New York.

Justwan, A. & Koç, N. (2008) A diatom based transfer function for reconstructing sea ice concentrations in the North Atlantic. *Marine Micropaleontology*, **66**, 264–278.

Katsuki, K., Ikehara, M., Yokoyama, Y., Yamane, M. & Khim, B.E. (2012) Holocene migration of oceanic front systems over the Conrad Rise in the Indian Sector of the Southern Ocean. *Journal of Quaternary Science*, **27**, 203–210.

Kennicutt, M. C., Chown, S.L., Cassano, J.J. et al. (2014) Polar Research: Six priorities for Antarctic Science. *Nature*, **512**, 23–25.

Kinnard, C., Zdanowicz, C.M., Fisher, D.A., & Wake, C.P. (2006) Calibration of an ice-core glaciochemical (sea-salt) record with sea-ice variability in the Canadian Arctic. *Annals of Glaciology*, **44**, 383–390.

Kinnard, C., Zdanowicz, C.M., Fisher, D.A., Isaksson, E., de Vernal, A. & Thompson, L.G. (2011) Reconstructed changes in Arctic sea ice over the past 1,450 years. *Nature*, **479**, 509–513.

Koç, N., Jansen, E. & Haflidason, H. (1993) Palaeoceanographic reconstructions of surface ocean conditions in the Greenland, Iceland and Norwegian seas through the last 14 ka based on diatoms. *Quaternary Science Reviews*, **12**, 115–140.

Koç Karpuz, N. & Schrader, H. (1990) Surface sediment diatom distribution and Holocene palaeotemperature variations in the Greenland, Iceland and Norwegian Sea. *Palaeoceanography*, **5**, 557–580.

Koerner, R.M. (1977) Devon Island ice cap: core stratigraphy and palaeoclimate. *Science*, **196**, 15–18.

Krylov, A.A., Andreeva, I.A., Vogt, C. et al. (2008) A shift in heavy and clay mineral provenance indicates a middle Miocene onset of a perennial sea ice cover in the Arctic Ocean. *Paleoceanography*, **23**, PA1S06, doi: 10.1029/2007PA0014979.

Kunz-Pirrung, M. (2001) Dinoflagellate cyst assemblages in surface sediments of the Laptev Sea region (Arctic Ocean) and their relationship to hydrographic conditions. *Journal of Quaternary Science*, **16**, 637–649.

Kunz-Pirrung, M., Matthiessen, J. & de Vernal, A. (2001) Late Holocene dinoflagellate cysts as indicators for short-term climate variability in the Eastern Laptev Sea (Arctic Ocean). *Journal of Quaternary Science*, **16**, 711–716.

Kwok, R. & Rothrock, D. A. (2009) Decline in Arctic sea ice thickness from submarine and ICESat records: 1958–2008. *Geophysical Research Letters*, **36**, L15501, doi:10.1029/2009GL039035.

Ledu, D., Rochon, A., de Vernal, A. & St-Onge, G. (2008) Palynological evidence of Holocene climate oscillations in the Eastern Arctic: a possible shift in the Arctic Oscillation at the millennial time scale. *Canadian Journal of Earth Sciences*, **45**, 1363–1375.

Ledu, D., Rochon, A., de Vernal, A., Barletta, F. & St-Onge, G. (2010a) Holocene sea-ice history and climate variability along the main axis of the Northwest Passage, Canadian Arctic. *Paleoceanography*, **25**, PA2213, doi:10.1029/2009PA001817.

Ledu, D., Rochon, A., de Vernal, A. & St-Onge, G. (2010b) Holocene paleoceanography of the northwest passage, Canadian Arctic Archipelago. *Quaternary Science Reviews*, **29**, 3468–3488.

Legendre, P. & Birks, H.J.B. (2012) From classical to canonical ordination. In: *Tracking Environmental Change Using Lake Sediments*, (Eds. H.J.B. Birks et al.), pp. 201–248. Springer Science and Business Media, New York.

Levac, E. & de Vernal, A. (1997) Postglacial changes of terrestrial and marine environments along the Labrador coast: palynological evidence from cores 91-045-005 and 91-045-006, Cartwright Saddle. *Canadian Journal of Earth Sciences*, **34**, 1358–1365.

Levac, E., de Vernal, A. & Blake, W., Jr. (2001) Sea-surface conditions in northernmost Baffin Bay during the Holocene: palynological evidence. *Journal of Quaternary Science*, **16**, 353–363.

Lisitzin, A.P. (1971) Distribution of siliceous microfossils in suspension and in bottom sediments. In: *The Micropalaeontology of Oceans*, (Eds. B.M. Funnell & W.R. Riedel), pp. 173–195. Cambridge University Press, Cambridge.

Ljungqvist, F.C. (2011) The spatio-temporal pattern of the Mid-Holocene Thermal Maximum, *Geografie*, **116**, 91–110.

Maddison, E. J., Pike, J. & Dunbar, R. (2012) Seasonally laminated diatom-rich sediments from Dumont d'Urville Trough, East Antarctic margin: late-Holocene neoglacial sea-ice conditions. *The Holocene*, **22**, doi:10.1177/0959683611434223.

de la Mare, W.K. (2009) Changes in Antarctic sea-ice extent from direct historical observations and whaling records. *Climatic Change*, **92**, 461–493.

Marshall, J., Armour, K.C., Scott, J.R., Kostov, Y. et al. (2014) The ocean's role in polar climate change: asymmetric Arctic and Antarctic responses to greenhouse gas and ozone forcing. *Philosophical Transactions of the Royal Society A*, **372**: 20130040.

Massé, G., Rowland, S.J., Sicre, M.-A., Jacob, J., Jansen, E. & Belt, S.T. (2008) Abrupt climate changes for Iceland during the last millennium: Evidence from high resolution sea ice reconstructions. *Earth and Planetary Science Letters*, **269**, 565–569.

Massé, G., Belt, S., Crosta, X. et al. (2011) Highly branched isoprenoids as proxies for variable sea ice conditions in the Southern Ocean. *Antarctic Science*, **23**, 487–498.

Masson, V., Vimeux, F., Jouzel, J. et al. (2000) Holocene climate variability in Antarctica based on 11 ice-core isotopic records. *Quaternary Research*, **54**, 348–358.

Matthiessen, J., Knies, J., Nowaczyk, N.R. & Stein, R. (2001) Late Quaternary dinoflagellate cyst stratigraphy at the Eurasian continental margin, Arctic Ocean: indications for Atlantic water inflow in the past 150,000 years. *Global and Planetary Change*, **31**, 65–86.

Mattson, E., Karlsson, A. Smith, W.O. & Abrahamsson, K. (2012) The relationship between biophysical variables and halocarbon distribution in the waters of the Amundsen and Ross seas, Antarctica. *Marine Chemistry*, **141–142**, 1–9.

Mayewski, P.A., Meeker, L.D., Whitlow, S.I. et al. (1994) Changes in atmospheric circulation and ocean ice cover over the North Atlantic during the last 41,000 years. *Science*, **263**, 1747–1751.

McKay, J., de Vernal, A., Hillaire-Marcel, C., Not, C., Polyak, L. & Darby, D. (2008) Holocene fluctuations in Arctic sea-ice cover: dinocyst-based reconstructions for the eastern Chukchi Sea. *Canadian Journal of Earth Sciences*, **45**, 1399–1415.

McKay, R., Naish, T., Carter, L. et al. (2012). Antarctic and Southern Ocean influences on Late Pliocene global cooling. *Proceedings of the National Academy of Sciences USA*, **109**(17), 6423–6428.

Mix, A. C., Bard, E. & Schneider, R. (2001) Environmental processes of the ice age: land, oceans, glaciers (EPILOG). *Quaternary Science Reviews*, **20**, 627–657.

Mohan, R., Shukla, S.K., Patil, S.M., Shetye, S.S. & Kerkar, K.K. (2011) Diatoms from the surface sediments of the Enderby Basin of the Indian Sector of the Southern Ocean. *Journal Geological Society of India*, **78**, 36–44.

Möller, P., Lubinski, D.J., Ingólfsson, O. et al. (2006) Severnaya Zemlya, Arctic Russia: a nucleation area for Kara Sea ice sheets during the Middle to Late Quaternary. *Quaternary Science Reviews*, **25**, 2894–2936.

Moran, K., Backman, J., Brinkhuis, H. et al. (2006) The Cenozoic palaeoenvironment of the Arctic Ocean. *Nature*, **441**, 601–605.

Moros, M., Andrews, J.T., Eberl, D.E. & Jansen, E. (2006) Holocene history of drift ice in the northern North Atlantic: evidence for different spatial and temporal modes. *Paleoceanography*, **21**, doi:10.1029/2005PA001214.

Mudie, P.J. (1992) Circum-Arctic Quaternary and Neogene marine palynofloras: palaeoecology and statistical analysis. In: *Neogene and Quaternary Dinoflagellate Cysts and Acritarchs*, (Eds. M.J. Head & J.H. Wrenn), pp. 347–390. American Association of Stratigraphic Palynologists Foundation, Dallas, Texas.

Mudie, P.J. & Short, S.K. (1985) Marine palynology of Baffin Bay. In: *Quaternary Environments*, (Ed. J.T. Andrews), pp. 263–308. Allen and Unwin, Boston.

Mudie, P.J. & Rochon, A. (2001) Distribution of dinoflagellate cysts in the Canadian Arctic marine region. *Journal of Quaternary Science*, **16**, 603–620.

Mudie, P.J., Harland, R., Matthiessen, J. & de Vernal, A. (2001) Marine dinoflagellate cysts and high latitude Quaternary palaeoenvironmental reconstructions: an introduction. *Journal of Quaternary Science*, **16**, 595–602.

Mudie, P., Rochon, A. & Levac, E. (2005) Decadal-scale sea ice changes in the Canadian Arctic and their impacts on humans during the past 4,000 years. *Environmental Archaeology*, **10**, 113–126.

Müller, J. (2011) Last Glacial to Holocene variability in the sea ice distribution in Fram Strait/Arctic Gateway: A novel biomarker approach. *PhD Thesis*. University of Bremen. Germany.

Müller, J., Massé, G., Stein, R. & Belt, S.T. (2009) Variability of sea-ice conditions in the Fram Strait over the past 30,000 years. *Nature Geoscience*, **2**, 772–776.

Müller, J., Wagner, A., Fahl, K., Stein, R., Prange, M. & Lohmann, G. (2011) Towards quantitative sea ice reconstructions in the northern North Atlantic: a combined biomarker and numerical modelling approach. *Earth and Planetary Science Letters*, **306**, 137–148.

Müller, J.,Werner, K., Stein, R., Fahl, K., Moros, M. & Jansen, E. (2012) Holocene cooling culminates in sea ice oscillations in Fram Strait. *Quaternary Science Reviews*, **47**, 1–14.

Navarro-Rodriguez, A., Belt, S.T., Brown, T.A. & Knies, J. (2013) Mapping recent sea ice conditions in the Barents Sea using the proxy biomarker IP$_{25}$: implications for palaeo sea ice reconstructions. *Quaternary Science Reviews*, **79**, 26–39.

Newnham, R. M., Lowe, D.J., Giles, T. & Alloway, B.V. (2007) Vegetation and climate of Auckland, New Zealand, since ca. 32 000 cal. yr ago: support for an extended LGM. *Journal of Quaternary Science*, **22**, 517–534.

O'Brien, S., Mayewski, P.A., Meeker, L.D., Meese, D.A., Twickler, M.S. & Whitlow, S.I. (1995) Holocene climate as reconstructed from a Greenland ice core. *Science*, **270**, 1962–1964.

Olguín, H. F. & Alder, V.A. (2011) Species composition and biogeography of diatoms in Antarctic and Subantarctic (Argentine shelf) waters (37–76°S). *Deep-Sea Research II*, **58**, 139–152.

Petherick, L., McGowan, H. & Moss, P. (2008) Climate variability during the Last Glacial Maximum in eastern Australia: evidence of two stadials? *Journal of Quaternary Science*, **23**, 787–802.

Peyron, O. & de Vernal, A. (2001) Application of artificial neural networks (ANN) to high-latitude dinocyst assemblages for the reconstruction of past sea-surface conditions in Arctic and sub-Arctic seas. *Journal of Quaternary Science*, **16**, 699–709.

Pfirman, S.L., Colony, R., Nürnberg, D., Eicken, H. & Rigor, I. (1997) Reconstructing the origin and trajectory of drifting Arctic sea ice. *Journal of Geophysical Research*, **102**, 12 575–12 586.

Pfirman, S.L., Wollenburg, I., Thiede, J. & Lange, M.A. (1989) Lithogenic sediment on Arctic pack ice: potential aeolian flux and contributions to deep sea sediments. In: *Palaeoclimatology and Palaeometeorology: Modern and Past Pattern of Global Atmospheric Transport*, (Eds. M. Sarnthein & M. Leinen), pp. 463–493. Kluwer, Dordrecht.

Pike, J., Allen, C., Leventer, A., Stickley, C.E. & Pudsey, C. (2008) Comparison of contemporary and fossil diatom assemblages from the western Antarctic Peninsula shelf. *Marine Micropaleontology*, **67**, 274–287.

Pike, J., Crosta, X., Maddison, E.J. et al. (2009) Observations on the relationship between the Antarctic coastal diatoms *Thalassiosira antarctica* Comber and *Porosira glacialis* (Grunow) Jørgensen and sea ice concentrations during the late Quaternary. *Marine Micropaleontology*, **73**, 14–25.

Pike, J., Swann, G.E.A., Leng, M.J. & Snelling, A.M. (2013) Glacial discharge along the west Antarctic Peninsula during the Holocene. *Nature Geoscience*, **6**, doi:10.1038/NGEO1703.

Polyak, L., Alley, R.B., Andrews, J.T. et al. (2010) History of sea ice in the Arctic. *Quaternary Science Reviews*, **29**, 1757–1778.

Polyak, L., Best, K.M., Crawford, K.A., Council, E.A. & St-Onge, G. (2013) Quaternary history of sea ice in the western Arctic Ocean based on foraminifera. *Quaternary Science Reviews*, **79**, 145–156.

Putnam, A. E., Schaefer, J.M., Denton, G.H. et al. (2013) The Last Glacial Maximum at 44°S documented by a ^{10}Be moraine chronology at Lake Ohau, Southern Alps of New Zealand. *Quaternary Science Reviews*, **62**, 114–141.

Radi, T., de Vernal, A. & Peyron, O. (2001) Relationships between dinoflagellate cyst assemblages in surface sediment and hydrographic conditions in the Bering and Chukchi seas. *Journal of Quaternary Science*, **16**, 667–680.

Rahman, A. & de Vernal, A. (1994) Surface oceanographic changes in the eastern Labrador Sea: nannofossil record of the last 31,000 years. *Marine Geology*, **121**, 247–263.

Renssen, H., Seppä, H., Crosta, X., Goose, H. & Roche, D.M. (2012) Global characterization of the Holocene Thermal Maximum. *Quaternary Science Reviews*, **48**, 7–19.

Rhodes, R., Bertler, N.A.N., Baker, J.A., Sneed, S.B., Oerter, H. & Arrigo, K.R. (2009) Sea ice variability and primary productivity in Ross Sea, Antarctica, from methylsulphonate snow record. *Geophysical Research Letters*, **36**, L10704, doi:10.1029/2009GL037311, 2009.

Riaux-Gobin, C., Poulin, M., Dieckmann, G., Labrune, C. & Vétion, G. (2011) Spring phytoplankton onset after the ice break-up and sea-ice signature (Adélie Land, East Antarctica). *Polar Research*, **30**, doi:10.3402/polar.v30i0.5910

Rignot, E., Jacobs, S, Mouginot, J. & Scheuchl, B. (2013) Ice shelf melting around Antarctica. *Science*, **341**(6143), 266–270.

Rochon, A. & de Vernal, A. (1994) Palynomorph distribution in recent sediments from the Labrador Sea. *Canadian Journal of Earth Sciences*, **31**, 115–127.

Rochon, A., de Vernal, A., Sejrup, H.-P. & Haflidason, H. (1998) Palynological evidence of climatic and oceanographic changes in the North Sea during the last deglaciation. *Quaternary Research*, **49**, 197–207.

Rochon, A., de Vernal, A., Turon, J.L., Matthiessen, J. & Head, M.J. (1999) Distribution of recent dinoflagellate cysts in surface sediments from the North Atlantic Ocean and adjacent seas in relation to sea-surface parameters. *American Association of Stratigraphic Palynologists, Contribution Series*, **35**, 1–152.

Rochon, A., Scott, D.B., Schell, T.M., Blasco, S., Bennett, R. & Mudie, P.J. (2006) Evolution of sea surface conditions during the Holocene: comparison between Eastern (Baffin Bay and Hudson Strait) and Western (Beaufort Sea) Canadian Arctic. *American Geophysical Union Annual Meeting*, San Francisco, Calif. **U34B**, 867.

Rosell-Melé, A. & Koç N. (1997) Palaeoclimatic significance of the stratigraphic occurrence of photosynthetic biomarker pigments in the Nordic seas. *Geology*, **25**, 49–52.

Rosell-Melé, A., Weinelt, M., Sarnthein, M., Koç, N. & Jansen, E. (1998) Variability of the Arctic front during the last climatic cycle: application of a novel molecular proxy. *Terra Nova*, **10**, 86–89.

Röthlisberger, R., Mudelsee, M., Bigler, M. et al. (2008) The southern hemisphere at glacial terminations: insights from the Dome C ice core. *Climate of the Past*, **4**, 345–356.

Röthlisberger, R., Crosta, X., Abram, N.J., Armand, L. & Wolff, E.W. (2010) Potential and limitations of marine and ice core sea ice proxies: an example from the Indian Ocean sector. *Quaternary Science Reviews*, **29**, 296–302.

Russell, A. & McGregor, G.R. (2010), Southern hemisphere atmospheric circulation: impacts on Antarctic climate and reconstructions from Antarctic ice core data. *Climatic Change*, **99**, 155–192.

Saba, G. K., Fraser, W.R., Saba, V.S. et al. (2014) Winter and spring controls on the summer food web of the coastal west Antarctic Peninsula. *Nature Communications*, **5**, doi:10.1038/ncomms5318.

Samtleben, C. & Schroder, A. (1992) Living coccolithophore communities in the Norwegian–Greenland Sea and their record in sediment. *Marine Micropalaeontology*, **19**, 333–354.

Sancetta, C. (1979) Oceanography of the North Pacific during the last 18,000 years: evidence from fossil diatoms. *Marine Micropalaeontology*, **4**, 103–123.

Sancetta, C. (1981) Oceanographic and ecologic significance of diatoms in surface sediments of the Bering and Okhotsk seas. *Deep-Sea Research*, **28**, 789–817.

Sancetta, C. (1982) Distribution of diatom species in surface sediments of the Bering and Okhotsk Seas. *Marine Micropalaeontology*, **28**, 221–257.

Sancetta, C. (1983) Effect of Pleistocene glaciation upon oceanographic characteristics of the North Pacific Ocean and Bering Sea. *Deep-Sea Research*, **30**, 851–869.

Sañé, E., Isla, E., Pruski, A.M., Bárcena, M.A., Vétion, G. & DeMaster, D. (2011) Diatom valve distribution and sedimentary fatty acid composition in Larsen Bay, Eastern Antarctic Peninsula. *Continental Shelf Research*, **31**, 1161–1168.

Sangiorgi, F., van Soelen, E.E., Spofforth, D.J.A. et al. (2008) Cyclicity in the middle Eocene central Arctic Ocean sediment record: orbital forcing and environmental response. *Paleoceanography*, **23**, PA1S08, doi:10.1029/2007PA001487.

Sarnthein, M., Plaumann, U. & Weinelt, M. (2003) Past extent of sea ice in the northern North Atlantic inferred from foraminiferal palaeotemperature estimates. *Paleoceanography*, **18**, doi:10.1029/2002PA000771.

Schell, T.M., Moss, T.J., Scott, D.B. & Rochon, A. (2008) Paleo–sea ice conditions of the Amundsen Gulf, Canadian Arctic Archipelago: implications from the foraminiferal record of the last 200 years. *Journal of Geophysical Research*, **113**, doi:10.1029/2007JC004202.

Scott, D.B., Schell, T., Rochon, A. & Blasco, S. (2008) Modern benthic foraminifera in the surface sediments of the Beaufort shelf, slope and MacKenzie Trough, Beaufort Sea, Canada: taxonomy and summary of surficial distributions. *Journal of Foraminiferal Research*, **38**, 228–250.

Seidenkrantz, M.-S., Aagaard-Sørensen, S., Sulsbrück, H., Kuijpers, A., Jensen, K.G. & Kunzendorf, H. (2007) Hydrography and climate of the last 4400 years in a SW Greenland fjord: implications for Labrador Sea palaeoceanography. *The Holocene*, **17**, 387–401.

Seidenkrantz, M.-S., Roncaglia, L., Fischel, A., Heilmann-Clausen, C., Kuijpers, A. & Moros, M. (2008) Variable North Atlantic climate seesaw patterns documented by a late Holocene marine record from Disko Bugt, West Greenland. *Marine Micropaleontology*, **68**, 66–83.

Shevenell, A.E., Ingalls, A.E., Domack, E.W. & Kelly, C. (2011) Holocene Southern Ocean surface temperature variability west of the Antarctic Peninsula. *Nature*, **470**, 250–254.

Shukla, S.K. & Sudhakar, M. (2011) Comparison of modern and fossil diatom assemblages and their implication on sea-ice conditions in coastal Antarctic region. *MAUSAM: Quarterly Journal of Meteorology, Hydrology and Geophysics*, **62**(4), 659–664.

Simard, A. & de Vernal, A. (1998) Distribution des kystes de Alexandrium excavatum dans les sédiments récents et postglaciaires des marges est-canadiennes. *Géographie physique et Quaternaire*, **52**, 361–371.

Smith, W.O. & Comiso, J. (2008) Influence of sea ice on primary production in the Southern Ocean: A satellite perspective. *Journal of Geophysical Research Oceans*, **113**, doi:10.1029/2007JC004251.

Solignac, S., de Vernal, A. & Hillaire-Marcel, C. (2004) Holocene sea-surface conditions in the North Atlantic – contrasted trends and regimes in the western and eastern sectors (Labrador Sea vs. Iceland Basin). *Quaternary Science Reviews*, **23**, 319–334.

Solignac, S., Giraudeau, J. & de Vernal, A. (2006) Holocene sea surface conditions in the western North Atlantic: Spatial and temporal heterogeneities. *Paleoceanography*, **21**, PA2004, doi:10.1029/2005PA001175.

Solignac, S., Grelaud, M., de Vernal, A. et al. (2008) Reorganisation of the upper ocean circulation in the mid-Holocene in the northeastern Atlantic. *Canadian Journal of Earth Sciences*, **45**, 1417–1433.

Spolaor, A., Vallelonga, P., Plane, J.M.C. et al. (2013) Halogen species record Antarctic sea ice extent over glacial-interglacial periods. *Atmospheric Chemistry and Physics*, **13**, 6623–6635.

Spolaor, A., Vallelonga, P., Gabrieli, J. et al. (2014) Seasonality of halogen deposition in polar snow and ice. *Atmospheric Chemistry and Physics Discussions*, **14**, 81–85–8207.

St John, K. (2008) Cenozoic ice-rafting history of the central Arctic Ocean: terrigenous sands on the Lomonosov Ridge. *Paleoceanography*, **23**, PA1S05, doi:10.1029/2007PA001483.

Stein, R. & Fahl, K. (2012) Biomarker proxy IP$_{25}$ shows potential for studying entire Quaternary Arctic sea-ice history. *Organic Geochemistry*, **55**, 98–102.

Stein, R., Fahl, K. & Müller, J. (2012) Proxy reconstruction of Arctic Ocean sea ice history: from IRD to IP$_{25}$. *Polarforschung*, **82**, 37–71.

Stephens, T., Atkin, D., Augustinus, P. et al. (2012) A late glacial Antarctic climate teleconnection and variable Holocene seasonality at Lake Pupuke, Auckland, New Zealand. *Journal of Palaeolimnology*, **48**, 785–800.

Stickley, C. E., Koc, N., Brumsack, H.-J., Jordan, R.W. & Suto, I. (2008) A siliceous microfossil view of middle Eocene Arctic paleoenvironments: a window of biosilica production and preservation. *Paleoceanography*, **23**, PA1S14, doi:10.1029/2007PA001485.

Stoynova, V., Shanahan, T.M., Hughen, K.A. & de Vernal, A. (2013) Insights into circum-Arctic sea ice variability from molecular geochemistry. *Quaternary Science Reviews*, **79**, 63–73.

Taylor, M. H., Losch, M. & Bracher, A. (2013) On the drivers of phytoplankton blooms in the Antarctic marginal ice zone: A modeling approach. *Journal of Geophysical Research Oceans*, **118**, 63–75.

Telford, R.J. & Birks, H.J.B. (2005) The secret assumption of transfer functions: problems with spatial autocorrelation in evaluating model performance. *Quaternary Science Reviews*, **24**, 2173–2179.

Tolotti, R., Salvi, C., Salvi, G. & Bonci, M.C. (2013) Late Quaternary climate variability as recorded bymicropaleontological diatom data and geochemical data in the western Ross Sea. *Antarctic Science*, **25**, 804–820.

Tremblay, L.-B., Mysak, L.A. & Dyke, A.S. (1997) Evidence from driftwood records for century-to-millennial scale variations of the high latitude atmospheric circulation during the Holocene. *Geophysical Research Letters*, **24**, 2027–2030.

Turney, C. S. M., Haberle, S., Fink, D. et al. (2006) Integration of ice-core, marine and terrestrial records for the Australian Last Glacial Maximum and Termination: a contribution from the OZ INTIMATE group. *Journal of Quaternary Science*, **21**, 751–761.

Vandergoes, M. J., Newnham, R.M., Denton, G.H., Blaauw, M. & Barrell, D.J.A. (2013) The anatomy of Last Glacial Maximum climate variations in south Westland, New Zealand, derived from pollen records. *Quaternary Science Reviews*, **74**, 215–229.

Vare, L.L., Massé, G., Gregory, T.R., Smart, C.W. & Belt, S.T. (2009) Sea ice variations in the central Canadian Arctic Archipelago during the Holocene. *Quaternary Science Reviews*, **28**, 1354–1366.

Vare, L.L., Massé, G. & Belt, S.T. (2010) A biomarker-based reconstruction of sea ice conditions for the Barents Sea in recent centuries. *The Holocene*, **40**, 637–643.

Vaughan, D. G., Comiso, J.C., Allison, I. et al. (2013) Observations: cryosphere. In: *Climate Change 2013: The Physical Science Basis. Contribution of Working Group I to the Fifth Assessment Report of the Intergovernmental Panel on Climate Change*, (Eds. T.F. Stocker et al.), pp. 317–382. Cambridge University Press, Cambridge.

Verleyen, E. (2012) Post-glacial regional climate variability along the east Antarctic coastal margin – evidence from shallow marine and coastal terrestrial records. *Quaternary International*, **279–280**, 462–565.

de Vernal, A. & Hillaire-Marcel, C. (2000) Sea ice cover, sea-surface salinity and halothermocline structure of the northwest North Atlantic: modern versus full glacial conditions. *Quaternary Science Reviews*, **19**, 65–85.

de Vernal, A. & Hillaire-Marcel, C. (2006) Provincialism in trends and high frequency changes in the northwest North Atlantic during the Holocene. *Global and Planetary Change*, **54**, 263–290.

de Vernal, A., Guiot, J. & Turon, J.-L. (1993a) Late and postglacial palaeoenvironments of the Gulf of St. Lawrence: marine and terrestrial palynological evidence. *Géographie physique et Quaternaire*, **47**, 167–180.

de Vernal, A., Turon, J.-L. & Guiot, J. (1993b) Dinoflagellate cyst distribution in high-lati-tude marine environments and quantitative reconstruction of sea-surface salinity, temperature, and seasonality. *Canadian Journal of Earth Sciences*, **31**, 48–62.

de Vernal, A., Hillaire-Marcel, C. & Bilodeau, G. (1996) Reduced meltwater outflow from the Laurentide ice margin during the Younger Dryas. *Nature*, **381**, 774–777.

de Vernal, A., Rochon, A., Turon, J.-L. & Matthiessen, J. (1997) Organic-walled dinoflagellate cysts: palynological tracers of sea-surface conditions in middle to high latitude marine environments. *GEOBIOS*, **30**, 905–920.

de Vernal, A., Hillaire-Marcel, C., Turon, J.-L. & Matthiessen, J. (2000) Reconstruction of sea-surface temperature, salinity, and sea ice cover in the northern North Atlantic during the last glacial maximum based on dinocyst assemblages. *Canadian Journal of Earth Sciences*, **37**, 725–750.

de Vernal, A., Henry, M., Matthiessen, J. et al. (2001) Dinoflagellate cyst assemblages as tracers of sea-surface conditions in the northern North Atlantic, Arctic and sub-Arctic seas: the new 'n = 677' data base and its application for quantitative palaeoceanographic reconstruction. *Journal of Quaternary Science*, **16**, 681–698.

de Vernal, A., Eynaud, F., Henry, M. et al. (2005a) Reconstruction of sea-surface conditions at middle to high latitudes of the Northern Hemisphere during the Last Glacial Maximum (LGM) based on dinoflagellate cyst assemblages. *Quaternary Science Reviews*, **24**, 897–924.

de Vernal, A., Hillaire-Marcel, C. & Darby, D.A. (2005b) Variability of sea ice cover in the Chukchi Sea (western Arctic Ocean) during the Holocene. *Paleoceanography*, **20**, PA4018, doi: 10.1029/2005PA0011575.

de Vernal, A., Hillaire-Marcel, C., Solignac, S., Radi, T. & Rochon, A. (2008) Reconstructing sea ice conditions in the Arctic and Sub-Arctic prior to human observations, In: *Arctic Sea Ice Decline: Observations, Projections, Mechanisms, and Implications, Geophysical Monograph Series* **180** (Eds. E.T. DeWeaver, C.M. Bitz, & L-B. Tremblay), pp. 27–45. American Geophysical Union, Washington, D.C.

de Vernal, A., Gersonde, R., Goosse, H., Seidenkrantz, M.-S. & Wolff, E.W. (2013a) Sea ice in the palaeoclimate system: the challenge of reconstructing sea ice proxies – an introduction. *Quaternary Science Reviews*, **79**, 1–8.

de Vernal, A., Rochon, A., Fréchette, B., Henry, M., Radi, T. & Solignac, S. (2013b) Reconstructing past sea ice cover of the Northern Hemisphere from dinocyst assemblages: Status of the approach. *Quaternary Science Reviews*, **79**, 122–134.

de Vernal, A., Hillaire-Marcel, C., Rochon, A. et al. (2013c) Dinocyst-based reconstructions of sea ice cover concentration during the Holocene in the Arctic Ocean, the northern North Atlantic Ocean and its adjacent seas. *Quaternary Science Reviews*, **79**, 111–121.

Voronina, E., Polyak, L., de Vernal, A. & Peyron, O. (2001) Holocene variations of sea surface conditions in the southeastern Barents Sea, reconstructed from dinoflagellate cyst assemblages. *Journal of Quaternary Science*, **16**, 717–726.

Wadell, L.M. & Moore, T.C. (2008) Salinity of the Eocene Arctic Ocean from oxygen isotope analysis of fish bone carbonate. *Paleoceanography*, **23**, PA1S14, doi:10.1029/2007PA001451.

Waldmann, N., Ariztegui, D., Anselmetti, F.S. et al. (2010) Holocene climatic fluctuations and positioning of the Southern Hemisphere westerlies in Tierra del Fuego (54°S), Patagonia. *Journal of Quaternary Science*, **25**, 1063–1075.

Wang, M. & Overland, J.E. (2009) A sea ice free summer Arctic within 30 years? *Geophysical Research Letters*, **36**, L07502, doi:10.1029/2009GL037820.

Weckström, K., Massé, G., Collins, L.G. et al. (2013) Evaluation of the sea ice proxy IP$_{25}$ against observational and diatom proxy data in the SW Labrador Sea. *Quaternary Science Reviews*, **79**, 53–62.

Whitehouse, M. J., Meredith, M.P., Rothery, P., Atkinson, A., Ward, P. & Korb, R.E. (2008) Rapid warming of the ocean around South Georgia, Southern Ocean, during the 20th century: Forcings, characteristics and implications for lower trophic levels. *Deep-Sea Research I*, **55**, 1218–1228.

Williams, K.M. (1990) Late Quaternary palaeoceanography of the western Baffin Bay region: evidence from fossil diatoms. *Canadian Journal of Earth Sciences*, **27**, 1487–1494.

Wolff, E. W., Barbante, C., Becagli, S. et al. (2010) Changes in environment over the last 800,000 years from chemical analysis of the EPICA Dome C ice core. *Quaternary Science Reviews*, **29**, 285–295.

Wolff, E. W., Fischer, H., Fundel, F. et al. (2006) Southern Ocean sea-ice extent, productivity and iron flux over the past eight glacial cycles. *Nature*, **440**, doi:10.1038/nature04614.

Wolff, E. W., Rankin, M. & Röthlisberger, R. (2003) An ice core indicator of Antarctic sea ice production? *Geophysical Research Letters*, **30**, doi:10.1029/2003GL018454.

Wollenburg, I. (1993) Sediment transport durch das arktische Meereis: Die rezente lithogene und biogene Materialfracht. *Berichte zur Polarforschung*, **127**, 1–159.

Wollenburg, J.E. & Kuhnt, W. (2000) The response of benthic foraminifers to carbon flux and primary production in the Arctic Ocean. *Marine Micropaleontology*, **40**, 189–231.

Wollenburg, J.E., Kuhnt, W. & Mackensen, A. (2001) Changes in Arctic Ocean paleoproductivity and hydrography during the last 145 kyr: The benthic foraminiferal record. *Paleoceanography*, **16**, 65–77.

Wollenburg, J.E., Knies, J. & Mackensen, A. (2004) High-resolution palaeoproductivity fluctuations during the past 24 kyr as indicated by benthic foraminifera in the marginal Arctic Ocean. *Palaeogeography, Palaeoclimatology, Palaeoecology*, **204**, 209–238.

Wollenburg, J.E., Mackensen, A. & Kuhnt, W. (2007) Benthic foraminiferal biodiversity response to a changing Arctic palaeoclimate in the last 24.000 years. *Palaeogeography, Palaeoclimatology, Palaeoecology*, **255**, 195–222.

Xiao, X., Fahl, K. & Stein, R. (2013) Biomarker distributions in surface sediments from the Kara and Laptev seas (Arctic Ocean): indicators for organic-carbon sources and sea-ice coverage. *Quaternary Science Reviews*, **79**, 40–52.

Zernova, V.V. (1970) Phytoplankton of the Southern Ocean. In: *Antarctic Ecology*, (Ed. M.W. Holdgate), pp. 136–142. Academic Press, London.

Zielinski, U. & Gersonde, R. (1997) Diatom distribution in Southern Ocean surface sediments (Atlantic sector): Implications for paleoenvironmental reconstructions. *Palaeogeography, Palaeoclimatology, Palaeoecology*, **129**, 213–250.

Zonneveld, K.A.F., Versteegh, G.J.M. & de Lange, G.J. (2001) Palaeoproductivity and postdepositional aerobic organic matter decay reflected by dinoflagellate cyst assemblages of the Eastern Mediterranean S1 sapropel. *Marine Geology*, **172**, 181–195.

CHAPTER 27

Ice in subarctic seas

Hermanni Kaartokallio[1], Mats A. Granskog[2], Harri Kuosa[1] and Jouni Vainio[3]

[1] *Finnish Environment Institute (SYKE), Helsinki, Finland*
[2] *Norwegian Polar Institute, Fram Centre, Tromsø, Norway*
[3] *Finnish Meteorological Institute, Helsinki, Finland*

27.1 Introduction

The existence of sea ice is not limited to polar areas but also reaches to lower latitude sea areas with sufficiently cold climates. These seasonally ice-covered seas are all surrounded or bordered by continental land masses with below-freezing temperatures typical of continental climates of North America and Eurasia. As the Antarctic area is surrounded by the open Southern Ocean with no land masses experiencing below-freezing temperatures, the seasonal sea ice cover is restricted to the northern hemisphere. In this chapter we introduce some aspects of sea ice of these northern sea areas, termed 'subarctic' here to distinguish them from polar sea ice. Detailed aspects of sea ice formation, properties and ecology have been described thoroughly in the earlier chapters and we therefore only briefly introduce the characteristics and peculiarities of ice in subarctic seas, which often get much less attention than their polar counterparts.

What, then, can be considered as subarctic sea? There are many ways to define the polar regions, especially in the northern hemisphere (see the Arctic Monitoring and Assessment Programme, http://www.amap.no), and the definition is ambiguous. Several different criteria can be used, such as latitude (south of the Arctic Circle), temperature, vegetation, permafrost and marine boundaries, among others. Choosing any of these would result in a different set of seas being considered as subarctic. For practical reasons, we have simply opted here to describe those seasonally ice-covered seas that get less attention than sea ice at the poles and which are therefore less well known to the public, despite being close(st) to habited areas and having severe ice conditions at least in parts of the sea for part of the year (Figure 27.1, Table 27.1).

Polar and subarctic seas share common traits, such as pronounced seasonality, low water temperature for most of the year and at least seasonal presence of sea ice and snow cover. The major difference between polar and subarctic sea ice is in the higher amplitude of seasonal changes bringing about the seasonality of the ice cover. In subarctic seas, ice is always seasonal, whereas in polar seas, a continuous multi-year sea ice cap has existed for perhaps the last five million years (Polyak et al., 2010). In some subarctic seas, such as the Baltic Sea and the Caspian Sea, salinity of the parent water is also below that of oceanic seawater, which also affects the ice properties.

The sea ice environment itself is temperature-structured, as the annual temperature characteristics define the temporal and spatial occurrence of ice as well as many important ice characteristics (thickness, snow cover etc.). Subarctic ice-covered seas are located at lower latitudes than the polar region with generally milder climate conditions and thus more restricted sea ice formation and spatiotemporal existence than in polar areas. Subsequently they may also be very vulnerable to both human activities and climate change, as a minor shift in the climate regime may already result in a considerable reduction in annual ice cover extent and duration.

In polar areas, climate warming is expected to influence most the spatial and temporal occurrence of multi-year ice, but in subarctic areas, where sea ice occurrence is seasonal, the temperature increase is more likely to lead to diminution and disappearance of sea ice in general, as temperature and sea ice in the

---

Sea Ice, Third Edition. Edited by David N. Thomas.
© 2017 John Wiley & Sons, Ltd. Published 2017 by John Wiley & Sons, Ltd.

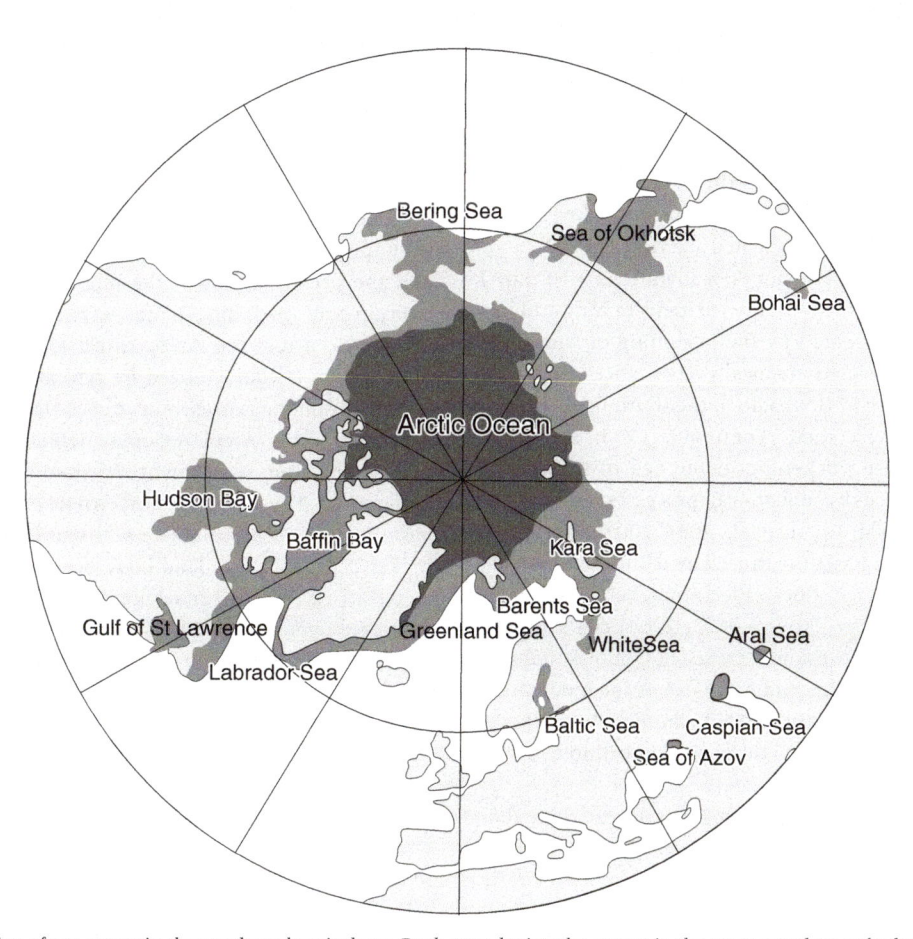

Figure 27.1 Map of sea extent in the northern hemisphere. Dark grey depicts the extent in the summer when only the perennial ice pack in the Arctic Ocean is frozen, and the lighter grey depicts the seasonal sea ice zone. Source: adapted from Leppäranta, 2004.

Table 27.1 Basic features of the sea ice cover in some major (best known) subarctic seas.

Sea	Total sea area ($\times 10^6$ km^2)	Maximum ice extent (% of area)	Highest latitude (°N)	Maximum ice thickness range (m)	Average ice season length (days with ice)
Sea of Okhotsk	1.530	50–90	62	0.5–1.5	180
Hudson Bay	0.830	95–100	64	1.0–2.0	275
Baltic Sea	0.377	10–100	66	0.1–1.2	180

middle latitudes may respond more sensitively in the context of global warming and the associated changes in atmospheric circulation. In this chapter, we analyse the differences and commonalities among subarctic seas and discuss the potential effects of climate warming on their ecosystems.

27.2 Sea of Okhotsk

By far the largest subarctic extent of sea ice is in Sea of Okhotsk (Table 27.1), a marginal sea in the western Pacific Ocean, located within latitudes of about 44–59°N and longitudes of 135–155°E (Figure 27.1). The Sea of

Okhotsk is bordered by the Siberian coast and Sakhalin Island to the west and north, the Kamchatka Peninsula and the Kuril Islands (all belonging to Russia) to the east and Hokkaido Island (Japan) to the south. Ice formation begins from the northwest shelf in November and the maximum ice extent is usually reached in late February or early March, when ice covers 50–90% of the surface area, with about 80% (about 1.2×10^6 km^2) being the average. Most of the ice melts in May and has completely disappeared by the beginning of June. The thickness of thermodynamically grown ice can range from 0.3 to 0.6 m in Hokkaido in the south up to 150 cm on the northwest shelf (Fukutomi, 1950; Shirasawa et al., 2005). Ice thickening in the Sea of Okhotsk is largely governed by dynamical processes (Kimura & Wakatsuchi, 2004; Toyota et al., 2004, 2007; Fukamachi et al., 2006) and is mainly caused by piling-up or rafting and ridging of ice floes. Electromagnetic induction measurements show (Toyota et al., 2004) characteristic ice thicknesses for the southern Sea of Okhotsk, namely 0.4–0.6 m for level ice and a 0.9–1.1 m for moderately deformed ice. In northern Sakhalin (level), pack ice floes can be 1.0–1.2 m thick (Polomoshnov, 1992; Kharitonov, 2008).

The salinity of Okhotsk Sea ice is lower than that reported for Arctic or Antarctic sea ice (about 2–3 units lower), which is thought to be a result of the moderate climate, with melting in the daytime (based on heat budget calculations), and the observed salinity is close to multi-year ice salinities from polar regions (Toyota et al., 2000). The total extent of sea ice in winter varies interannually by a factor of 2, from approximately 1.5×10^6 to 0.8×10^6 km^2. The maximum ice extent in 1996 was only about half of the maximum extent in 2001, which was the largest ice extent observed in the past 20+ years (Ohshima et al., 2006). Variability in atmospheric circulation patterns is known to cause large fluctuations in ice coverage. The variable ice conditions influence larger-scale atmospheric and oceanic processes in the region, and thus the Okhotsk sea ice extent plays an important role in the global climate system. In recent years, sea ice investigations have been increasing, largely due to the exploitation of oil and gas offshore from Sakhalin, where sea ice information is needed for platform construction, for transportation of oil and gas, and for environmental risk analysis for oil spills (Shirasawa et al., 2005).

27.3 Hudson Bay

Hudson Bay is a large inland sea that includes Hudson Bay and James Bay in Canada (Figure 27.1). Foxe Basin, Ungava Bay and Hudson Strait are also often considered to belong to the same system, as they provide the connection between Hudson Bay and the ocean. Although the meridional extension of Hudson Bay (including James Bay) spans roughly 14° of latitude across the Arctic Circle (50° to 64°N), the entire basin is characterized by typical Arctic (oceanic and atmospheric) conditions and is the southernmost extension of the Arctic region as defined by the Arctic Monitoring and Assessment Programme (http:/www.amap.no). The most typical Arctic feature of the basin is perhaps its complete seasonal ice cover, which makes it the largest inland body of water in the world ($\sim 0.83 \times 10^6$ km^2) with a complete cryogenic cycle, i.e. it freezes over completely is then ice-free in the summer, although for a relatively short period (Prinsenberg, 1988).

In the summer months, the ice-free bay creates a marine climate in the surrounding area, while in the winter it is insulated by ice and snow and permits cold polar air masses to extend south towards central Canada (Prinsenberg, 1984). This southern extension of Arctic conditions, well beyond the Arctic Circle, has a large impact on the climate of the surrounding areas, which is exemplified by the southern displacement of the treeline in eastern Canada. The extreme southerly presence of Arctic marine mammals is also characteristic of the Hudson Bay marine ecosystem.

Although belonging to the Arctic, from a climatological point of view, Hudson Bay has seldom received any detailed attention when the Arctic Ocean is considered. Hudson Bay is ice-covered for about 8–9 months every year, with nearly complete ice cover between November and June, and ice-free conditions prevailing for only 2–3 months in late summer. This reflects the cold continental climate in the region, in comparison to, for example, the Baltic Sea, which is located at roughly similar latitudes but has, on average, a considerably shorter ice season and less severe ice conditions (Table 27.1). Recent updates on the ice climatology of Hudson Bay for the satellite era have been produced by Hochheim and co-workers (Hochheim et al., 2011; Hochheim & Barber, 2014).

Ice formation starts in the northwestern corner of the bay in late October and the ice cover generally then spreads southwards along the west coast. By mid-November ice typically covers only the northwest corner of the bay around Southampton Island, and along a relatively narrow stretch along the western coast, almost all the way down to the mouth of James Bay. Thereafter, the ice cover spreads eastwards and within a month, by mid-December, the bay is ice-covered to 90–100% of its surface area. At this point, ice thickness for level landfast ice is from 0.8 m in the north to 0.3 m in James Bay in the south. The almost complete ice cover, with the exceptions of leads and openings in the mobile pack ice, is present until late May or early June, when ice starts to break up in James Bay. Fast ice fills bays along the western coast of Hudson Bay and forms around the Ottawa and Belcher Islands in southeastern parts of the bay. There is a steady growth in ice thickness from January until April, and April and May represent the maximum ice conditions, as ice thickness is close to its maximum and ice concentration is high.

The climatological maximum ice thickness in the region, i.e. thermodynamic growth of level ice, varies from 1.0 m in James Bay to 2.0 m in northwestern Hudson Bay (Markham, 1981), with 1.6 m as averaged over the whole bay (Prinsenberg, 1988). However, it has been shown that dynamic thickening of the ice cover in Hudson Bay is significant, and the volume of ice produced in the bay can be as much as 90% above values based on level ice (Prinsenberg, 1988). Melting usually starts in James Bay in late April, where the air temperature increases first above freezing but also because increased run-off caused by snowmelt from more southerly watersheds brings additional heat to melt ice. The rapid change in the length of day in northern spring results means that melting occurs over the whole bay by late June (Markham, 1986). After (southern) James Bay, the first areas to become free of ice are in the northwest, where the combined effect of winds moving pack ice offshore to the south/southeast and melting result in widening of the flaw lead.

The region along the eastern shore from James Bay to Mansel Island also becomes ice-free relatively early, which is attributed to the northward flow of river waters from the spring freshet in James Bay melting the ice. By mid-July the southwestern coast is free of ice and all remaining ice is generally located in the south-central parts of the bay. By the end of July this pack has shrunk considerably, with some remains floating in offshore waters between Nelson River and James Bay, and by mid-August all the ice has disappeared. Thereafter, the bay remains essentially ice-free until freezing starts again in late fall in the northwestern parts of the bay.

27.4 Baltic Sea

The Baltic Sea is one of the world's largest brackish water basins, with a surface area of about 422 000 km² and a volume of about 21 000 km³, and spans a relatively large latitudinal zone, from 53° to 65°N (roughly the same as Hudson Bay). The mean depth of the Baltic Sea is only 55 m, and in the eastern and northernmost basins, the Gulf of Finland and the Bothnian Bay, respectively, it is less than 40 m. The surface salinity varies from about 9 in the southern Baltic proper to < 1 in the innermost parts of the Gulf of Finland and the Bothnian Bay and in the proximity of larger rivers. The Baltic Sea is heavily influenced by river discharge, and the sea has a positive water balance, meaning that river run-off and precipitation exceeds evaporation. The brackish nature is maintained by intermittent inflows of saline North Sea water through the narrow and shallow Danish Straits, which is the only connection to the world ocean.

Seasonal sea ice is an important feature of the Baltic Sea system, although, as in other seasonally ice covered seas, the whole importance of the annual ice cover is not well known (for a review, see Granskog et al., 2006a). The ice regime in the Baltic can be divided into a landfast ice cover along the coasts and a mobile pack ice regime further offshore. The many islands and skerries bordering the coasts in the northern Baltic are decisive in the extent of landfast ice, which can usually be said to extend to the 5–15 m isobath, depending on local topography. As the ice becomes anchored to islands, skerries and shoals, it prevents wind break-up, even of thin ice. In the offshore ice pack, dynamic processes such as deformation, rafting and ridging are common. Ice formation usually begins at the northernmost Bothnian Bay and the easternmost Gulf of Finland in October–November. Ice forms first in the inner skerries and bays where the water is often fresher and shallower, and thus has a lower heat content, and where the ice cover can be anchored to islands and shoals. Next to freeze over are the Quark between the Bothnian Bay and Bothnian Sea, the entire Bothnian Bay and the coastal areas of the Bothnian Sea. In average winters, the ice also covers the Archipelago Sea, and the gulfs of Riga and Finland as well as the northern parts of the Baltic proper. In severe winters, the Danish Straits and the southern Baltic proper are also covered with ice (Granskog et al., 2006a). Annually, sea ice covers a mean of 40% of the Baltic Sea, and the median maximum ice extent for the period 1971–2000 was 157 000 km² (Granskog et al., 2006a). Typical ice-covered areas are presented in Figure 27.2.

The ice conditions in the Baltic Sea can be characterized by the large interannual variability in the severity of the winters (Figure 27.2), and depending on the severity of the winter, the ice covers 10–100% of the surface area at its maximum annual extent. There is also large variability in the date that freezing begins, in ice thickness and in break-up dates (Seinä & Palosuo, 1996; Haapala & Leppäranta, 1997; Jevrejeva et al., 2004; Granskog et al., 2006a). Evidently large-scale atmospheric circulation patterns are significantly correlated with the ice conditions. Average and mild winters occur when warm air masses associated with westerly moving cyclones from the Atlantic dominate, while in severe winters blocking anti-cyclonic patterns dominate. In very simple terms, the strength of the westerlies, bringing warm and moist air masses to Scandinavia, results in mild ice winters. Furthermore, it is the conditions during the winter months (December–February) that largely control the development of the ice cover, with the climate beforehand having little importance on the ice conditions.

While there is rather good information on the total ice thickness in the landfast sea ice domain – for example, Schmelzer et al. (2008) has compiled information on the ice conditions in the Baltic Sea for the period 1956–2005 – similar information is more limited for pack ice. Mean landfast ice thicknesses in the southern parts of the Baltic Sea (German and Polish coasts) and Archipelago Sea (northern Baltic proper) are generally less than 0.25 m, about 0.45 m in the Gulf of Finland and Bothnian Sea, and 0.55 m or more in the Quark and the Bothnian Bay (see Schmelzer et al., 2008). The maximum (level) ice thickness observed, in the northeastern part of the Bothnian Bay, is 1.1–1.2 m (e.g. Kalliosaari & Seinä, 1987; Schmelzer et al., 2008), whereas mean thicknesses in Bothnian Bay pack ice are much higher, from 0.4 to 1.8 m, due to strong ice deformation (Haas, 2004). Ridges can make up a large portion of ice volume in the central Bothnian Bay (Lewis et al., 1993; Kankaanpää 1997; Björk et al., 2008). In the Gulf of Finland and Bothnian Sea, the contribution of deformed ice appears to be much smaller (Kankaanpää, 1997).

Despite the brackish nature of the parent water, sea ice in the Baltic appears to be structurally similar and comparable to sea ice formed in Polar waters, although differences do exist because of the lower parent water salinity, and the ice becomes freshwater ice when the water salinity is lower than about 1 (Granskog et al., 2006a). Baltic Sea ice reflects the low water salinities: the bulk salinities in the northern Baltic Sea are generally less than 2, and even lower depending on the ambient water salinity, growth conditions and thermal history of the ice, and therefore much lower than in many other sea ice regions.

Another feature that distinguishes Baltic Sea ice from other regions is the significant contribution of meteoric ice, i.e. precipitation converted to ice, to the thickness growth. Two types of meteoric ice can be distinguished, snow ice and superimposed ice. Snow ice is formed when the weight of snow suppresses the ice surface below the water level, and seawater that floods the bottom of the snow pack freezes, while superimposed ice forms when snow meltwater drains to the ice surface and refreezes as a solid ice layer. Both mechanisms are important in the Baltic Sea, presumably because of the relatively thin ice cover that easily makes flooding possible (because relatively little snow is needed to submerge the ice surface) and because the mild climate conditions allow for periods of snow melt, and even rain, to occur throughout the winter. Strong diurnal cycling in the energy balance at the surface also reinforces melt–freeze cycling. Up to 50% of the total ice thickness and 35% of the total ice mass can be composed of meteoric ice in landfast sea ice (Granskog et al., 2006a); however, there is large interannual variability (Palosuo, 1963; Granskog et al., 2004). In spring the snow pack on the ice can be completely converted into a superimposed ice layer (Granskog et al., 2006b) before final ice melt. Due to the importance of the snow cover on the development of Baltic Sea ice, the response to future changes is somewhat uncertain, as it depends largely on the timing, amount and quality of precipitation during the ice season.

Ice in subarctic seas 635

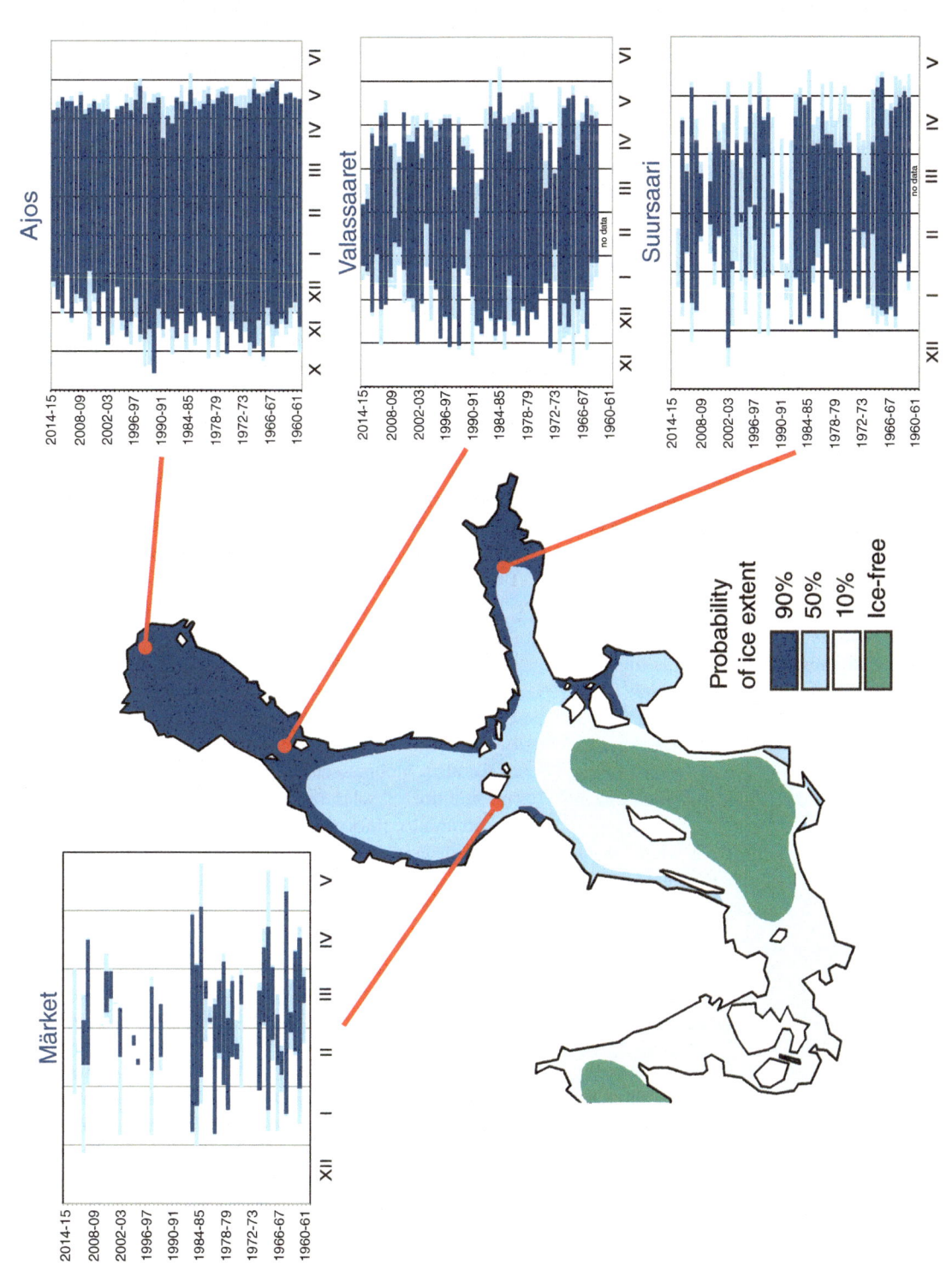

Figure 27.2 Map of the Baltic Sea, showing the probability of sea ice coverage and annual ice cover duration compiled from data collected by the Finnish Ice Service for the time period 1961–2015. The annual duration is plotted for four ice stations (from north to southeast: Ajos, Valassaaret, Märket, Suursaari). In each annual bar the ice-covered period is shown. The dark-blue regions of each bar represent the period of permanent ice cover, and the light-blue regions represent that of temporary ice cover. Absence of a coloured bar denotes complete absence of ice cover for the particular year.

27.5 Subarctic sea ice ecosystems

The gathering of information on sea ice biota from the Hudson Bay (mainly southeastern Hudson Bay), the Gulf of St Lawrence, Baltic Sea (the Gulfs of Finland and Bothnia), White Sea (Chupa inlet in the central White Sea) and the Sea of Okhotsk (Saroma-ko Lagoon in Hokkaido) has mainly been done in landfast level ice in recent decades (Legendre et al., 1996 and references therein; Granskog et al., 2006a; Krell et al., 2003; Ratkova & Wassmann, 2005; Sazhin et al., 2004; Melnikov et al., 2005). From many sea ice-covered areas outside the Arctic, such as the Caspian Sea, the Sea of Azov and the Bohai Sea, information on sea ice assemblages and ice ecosystems in general is, to our knowledge, unavailable.

As the subarctic sea ice is always seasonal, the length of the ice season is the most important factor governing the formation and succession of sea ice organism communities. Thus interannual variability is high in all subarctic areas in the ice extent, which creates variability specifically in the southern ice/open water conditions; the length of the ice season, which determines the lifetime of ice habitat; the ice mass, which gives rise to different ice layers; and snow cover, which is important factor in production dynamics and ice habitat formation (Sime-Ngando et al., 1999; Kuosa & Kaartokallio, 2006). Geographical location (latitude) and other factors contribute to the length of the ice season in subarctic ice-covered areas and are discussed earlier in this chapter.

The properties of parent water, from where the ice forms, are also important as they partly determine the properties of sea ice. The most critical parent water property is salinity, which is a key factor determining sea ice bulk salinity and subsequently ice main interior habitat, brine channel system characteristics. Salinity variability due to river inflows is a key factor in Baltic Sea sub-basin-scale variations in the sea ice properties (Meiners et al., 2002; Granskog et al, 2003) and regional-scale variation in coastal sea ice properties (Granskog et al., 2005). Bulk salinity of the ice bottom, typically reflecting the inshore–offshore salinity gradients in ice parent water, has been suggested to control the amount and distribution of sea ice algal biomass (as chlorophyll) as well as the composition of ice organism assemblages also in the southeastern Hudson Bay (Legendre et al., 1981, 1996), Saroma-ko Lagoon (Sea of Okhotsk; Robineau et al., 1997) and the White Sea (Krell et al., 2003; Melnikov et al., 2005).

Other chemical constituents dissolved in the parent water, such as inorganic nutrients and dissolved organic matter (DOM), contribute to the sea ice characteristics. Initial organism communities incorporated into growing ice form a basis for sea ice community succession. In the Baltic Sea, the typical succession sequence of organism communities in subarctic sea ice is analogous to Polar sea ice (Grossmann & Gleitz, 1993; Günther & Dieckmann, 1999), with the initial colonization during the sea ice formation followed by low-productive midwinter stage, the bloom of the sea ice algae and finally a heterotrophy-dominated stage late in the season (Figure 27.3). Biomass accumulation of the sea ice algae generally follows the seasonal increase in solar radiation, beginning at the transition of winter and spring and lasting until the onset of ice melt (Cota et al., 1991; Norrman & Andersson, 1994; Haecky & Andersson, 1999). Mass accumulation of ice algae at early spring is typical for all subarctic areas studied (Michel et al., 1993; Legendre et al., 1996; Sazhin, 2004).

A full succession sequence beginning with a low-productive midwinter stage followed by an algal bloom inside the ice and a heterotrophy-dominated post-bloom situation has been documented in the Baltic Sea and the Sea of Okhotsk (Sime-Ngando et al., 1997; Haecky & Andersson, 1999; Kaartokallio, 2004). Timing of ice algal bloom seems to be partly related to both the amount of solar radiation and the length of ice season and ice thickness. Algal bloom inside the sea ice occurs in February in the Saroma-ko Lagoon (located at 44°N, Sea of Okhotsk), and in the Baltic Sea either in March (the Gulf of Finland, 60°N; Kaartokallio, 2004; Kuosa & Kaartokallio, 2006) or in April (the Gulf of Bothnia, 64°N; Norrman & Andersson, 1994; Haecky et al., 1998; Haecky & Andersson, 1999) as is the case also in the southeastern Hudson Bay (55°N; Michel et al., 1993).

Under-ice algal blooms starting at the ice–water interface before ice break-up and facilitated by a stable melting water layer under the ice are a common phenomenon in subarctic sea ice areas (Legendre et al., 1981; Spilling, 2007). These blooms are assumed to contribute to the onset of the major phytoplankton spring bloom after ice break-up in all studied areas. In the Gulf of Finland, the Baltic Sea, these blooms are often dominated by dinoflagellates (Larsen et al., 1995;

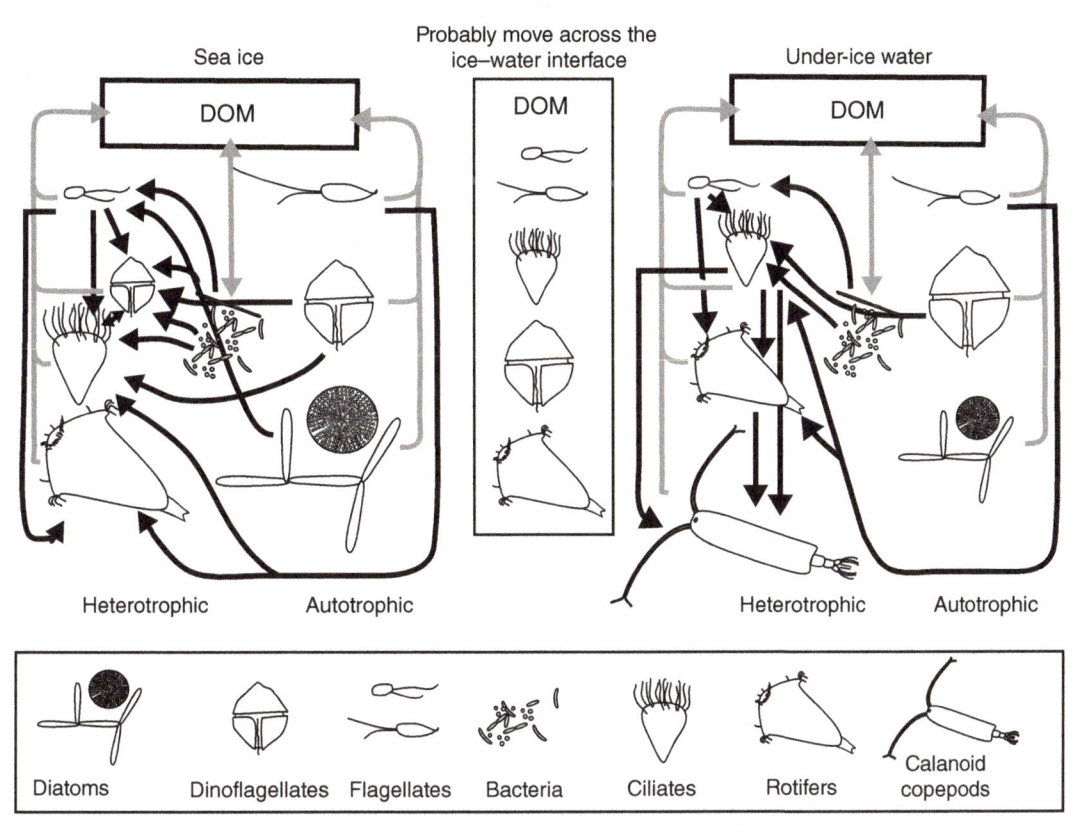

Figure 27.3 A schematic representation of the Baltic Sea sea ice and under-ice food webs. Black lines and arrows stand for trophic interaction, and grey lines indicate dissolved organic matter (DOM) flow in the food web. Source: Granskog et al., 2006a. Reproduced with permission of Elsevier.

Spilling, 2007), whereas in southeastern Hudson Bay (Michel et al., 1993), the White Sea (Ratkova & Wassmann, 2005) and the Sea of Okhotsk (Asami & Imada, 2001), blooms are dominated by the same diatom species also growing inside the ice sheet.

Owing to space limitation in the brine channels, internal sea ice food webs are considered to be truncated, meaning that organisms larger than the upper size limit of channels are lacking (Krembs et al., 2000). Clearly, this may be even more the case for subarctic ice, as all studied areas typically have a salinity below that of oceanic sea water. In the Baltic Sea ice, the largest metazoan animals are occasional copepods and copepod nauplii (Meiners et al., 2002; Kaartokallio, 2004; Werner & Auel, 2004; Kaartokallio et al., 2007). In the White Sea sea ice, which also has higher salinity, nematodes, copepods and polychaetes are encountered (Sazhin et al., 2004), although the abundance has been reported to be low (Melnikov et al., 2005). In contrast to the Baltic and White seas, nematodes and copepods are reported to occur in southeastern Hudson Bay lower ice layers with two to three times higher biomass than in under-ice water (Grainger, 1988). The complete absence, or low significance, of larger metazoans typical for polar sea ice simplifies the ecosystem by lowering the number of trophic interactions.

One of the key questions in sea ice food webs is the fate of organic matter, either autochthonous or allochthonous, in the ice environment. Different 'short circuits' are suggested to be typical of microbial food webs inside the sea ice. These include herbivory by ciliates and flagellates, ciliate bacterivory and direct utilization of DOM by heterotrophic flagellates (Marchant & Scott, 1993; Laurion et al., 1995; Sime-Ngando et al., 1997; Vezina et al., 1997; Haecky & Andersson, 1999; Kaartokallio 2004; Kaartokallio et al., 2007). Of these, at least direct utilization of DOM by flagellates (Haecky & Andersson, 1999) and ciliate grazing over several size classes have been suggested to be functional in Baltic Sea ice (Kaartokallio, 2004). In Baltic Sea sea ice, ciliates seem to be more important than in other areas, probably reflecting the paucity of higher animals in ice (Haecky & Andersson, 1999; Meiners et al., 2002; Kaartokallio, 2004), whereas in the White Sea (Sazhin, 2004) and the Sea of Okhotsk, heterotrophic flagellates seem to be more important in terms of biomass.

27.6 Changes in ice cover in subarctic seas in the coming decades

Recent studies have concluded that both the extent and the duration of the sea ice cover in Hudson Bay have been decreasing over the past few decades, probably as a result of climate change (e.g. increasing surface air temperature and changes in atmospheric circulation; Gough et al., 2004; Galbraith & Larouche, 2011; Hochheim and Barber, 2014). Hudson Bay is also considered to be one of the most vulnerable areas for climate change, and could be potentially undergoing a climate shift towards ice-free conditions, although there are still large uncertainties in future scenarios (Gagnon & Gough, 2005). The open water season was shown to have increased by about 3 weeks in 1996–2010 relative to 1980–1995 (Hochheim & Barber, 2014), with ice break-up (freeze-up) being 1.5 weeks earlier (later). A similar trend for earlier break-up (5.6 days per decade) was found by Galbraith and Larouche (2011) for the period 1971–2009.

Climate scenarios for 2041–2070 indicate that the ice cover in Hudson Bay is quite sensitive to climate warming, and the ice season would be significantly shorter than now (7–9 weeks), ice thickness would be reduced by 50% but the ice would cover all of the bay in winter (Joly et al., 2011). There are basically no observational data in the Hudson Bay to say whether the ongoing changes in the ice climatology have also affected the ice thickness.

Jevrejeva et al. (2004) examined the evolution of ice seasons in the Baltic Sea during the 20th century and indicated a general trend toward milder ice conditions, the largest change being the length of ice season, which has shortened by 14–44 days during the last century. There has also been a reduction of about 8–20 days per century in the earliest ice break-up, which the authors relate to a warming trend in winter air temperatures over Europe. It has been estimated that if the winter air temperature increased about 4°C, the Baltic Sea would become completely ice-free (Omstedt & Hansson, 2006). The recent review by Haapala et al. (2015) showed a considerable decrease in the length of the ice season along the Finnish coast. In the eastern Gulf of Finland, the length of the ice season has decreased by 41 days during the last 100 years. Also severe ice winters in the Baltic Sea have become rare and the maximum ice extent is decreasing at a rate of 340 km^2 year^{-1}, or 0.2% (Haapala et al., 2015).

27.7 Direct effect of sea ice loss

The sea ice biome is a central feature of polar and subarctic seas. Sea ice influences biochemical cycles through exchange of gases, liquids and radiative transfer at the sea ice–atmosphere interface and by subsistence of active biological and chemical processes within the ice (Vancoppennolle et al., 2013). Its presence modifies the productivity and trophic interactions (ultimately the food web structure) as well as the abundance, distribution, seasonality, and interactions of marine species (Post et al., 2013).

The degree of ice cover-induced changes in marine ecosystems can be expected to vary across the spectrum of ice cover presence from multi-year ice in high polar seas to short-term or sporadic seasonal ice occurrence in the southern fringes of subarctic ice occurrence. Sea ice loss can be categorized into direct effects such as loss of thickness and ice concentration (e.g. changes from multi-year to first-year ice regime in the Arctic) as well as into changes in ice cover duration and timing (Post et al., 2013). Melting of sea ice can deepen the halocline depth and increase stratification strength (McLaughlin & Carmack 2010). Primary production in the water column, regulated by light and nutrient availability, depends on a balance between stratification and mixing, which either keep the phytoplankton in the illuminated surface layer or fuel new nutrients to support production from below the mixed layer (Eilola et al., 2013). Changes in the duration of the ice cover affect biogeochemical cycles by shortening the productive period for ice algae, advancing the onset of open water phytoplankton blooms (Leu et al., 2011) and creating favourable circumstances for high biomass under ice algal blooms (Arrigo et al., 2012).

Changes in the timing of ice algal and open-water phytoplankton blooms may cause a mismatch between primary production and growth season of lipid-rich Arctic copepods which are key species in the transfer of energy to upper trophic levels (Leu et al., 2011). The magnitude of the ice algal bloom may also affect the amount of new production reaching the sea floor, which may change the trophic structure and biomass of benthic communities in shallower waters (Wassmann, 2006; Juhl et al., 2011; Boetius et al., 2013). Also later ice formation in the autumn may change the nutrient regime in the upper water column throughout the winter until the following spring (Meier et al., 2013). The anticipated biological impacts affecting the lower trophic levels are complex, with direct ice loss superimposed by changes in seasonality, stratification, lateral freshwater input and retreat of Arctic species into the north (Soreide et al., 2010; Meier et al., 2013). Arctic sea ice-associated food webs are speculated to become more similar to their Antarctic counterparts as first-year ice is dominant in the Southern Ocean. Along the same lines, Arctic wintertime food webs and productivity patterns may develop closer to those in the subarctic seas.

One example is the presence of high-biomass under-ice blooms in the Baltic Sea (Spilling et al., 2007). On the upper trophic levels, the sea ice loss is reverberated as changes in habitat extent or modified food availability (Gilg et al., 2012; Kovacs et al., 2012). Changes in sea ice cover in the Arctic are driving the system towards dominance in seasonal sea ice typical of subarctic seas. Climate warming is also affecting sea ice cover in the northern hemisphere subarctic seas, which can be seen as harbingers of the next stage of change transition from a seasonally ice-covered system to temperate systems with sporadic or nonexistent ice occurrence. Comparative research on the winter ecology of Arctic Ocean and subarctic seas can be seen as timely, enhancing the predictive capacity regarding consequences of further sea ice loss in the Arctic area.

Over the last few decades, ice concentration and ice cover duration in Hudson Bay have decreased and this trend is predicted to continue (Stroeve et al., 2007; Comiso et al., 2008). A recent modelling study on productivity of the Hudson Bay ecosystem implies there are different ice algal productivity regimes in the different parts of the Bay system (Sibert et al., 2011). Ice algal production is expected to be higher in western Hudson Bay and lower in Hudson Strait in the northeast.

Food web trophic structure is also possibly different, shifting from an algal production-driven food web in Hudson Strait and western Hudson Bay to a microbial-driven food web in nearshore areas of southeastern Hudson Bay and James Bay, where the influence of river water is greater (Sibert et al., 2011). High river water influence there is reminiscent of the brackish-water Baltic Sea. The onset of spring phytoplankton bloom is primarily driven by ice retreat and melting, with variable time lag after ice cover retreat (Sibert et al., 2010). Spring bloom is preceded by algal bloom under ice, where melting enables favourable light conditions for algal growth in a stable freshwater layer beneath thinning ice cover bloom development already under ice.

Although ice dynamics evidently exert important influence on the productivity, the effect of ice cover changes on Hudson Bay productivity and food webs remains to be studied (Sibert et al., 2011). Changing sea ice conditions can, however, be expected to lead to changes in productivity and trophic structures. For example, an abrupt change in ice cover conditions in the mid-1990s in Hudson Bay is suggested to have led to changes in seabird diet and nutritional status of the nestlings (Gaston et al., 2012). For seals, ice cover decline has been hypothesized to lead to replacement of ringed seals by harbour seals in the polar bear diet (Derocher et al., 2004; Ferguson et al., 2005). However, this speculation was not supported by Bajzak et al. (2012), who found that harbour seals stayed in shallow, nearshore waters irrespective to ice cover presence.

In the Sea of Okhotsk, there has been a declining trend in sea ice cover in recent decades and the period of maximum ice cover has shifted from March to February (McKinnell & Dagg, 2010; Radchenko et al., 2010; Figurkin 2011). Also in the Sea of Okhotsk, the main phytoplankton spring bloom commences after ice melt in April–May, although the primary production starts immediately after ice melt a month earlier (Saitoh et al., 1996; Matsumoto et al., 2004). The Sea of Okhotsk ecosystem supports extensive fisheries on diverse species (Kim, 2012) and the variability of climatic and oceanic patterns in general determines the dynamics of fish production. However, the natural variability is high and the current level of knowledge does not yet allow

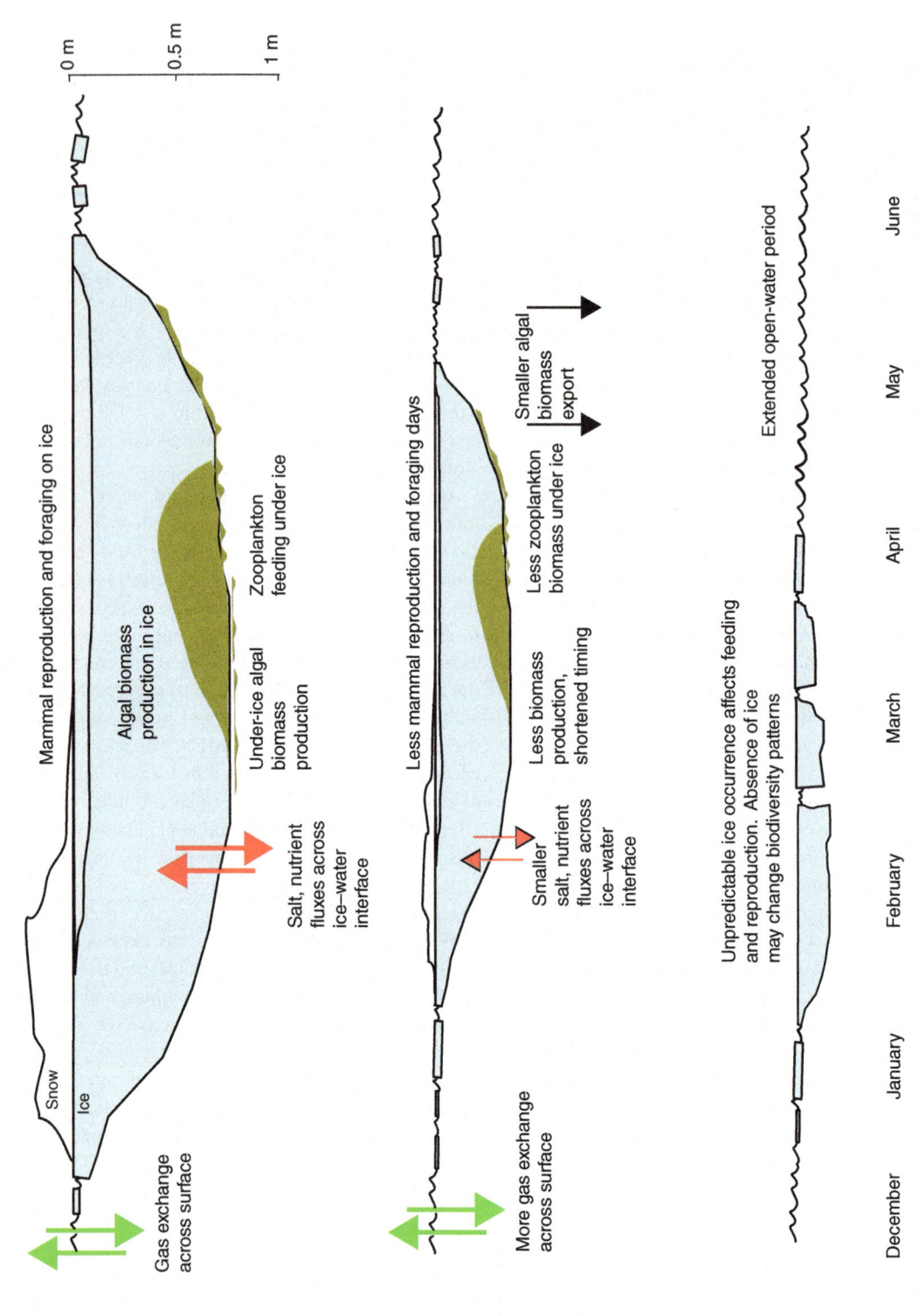

Figure 27.4 A schematic representation of selected possible ecosystem effects caused by a decrease in ice cover duration and thickness. Algal biomass build-up and release to water and benthos, exchange of gases and nutrients at interfaces, and marine mammal reproduction and foraging are prominent ecosystem features projected to change in the subarctic seas. The three scenarios can be thought to relate to the ice occurrence at the three Baltic Sea ice stations from north to south presented in Figure 27.2.

evaluation of the effect of the warming trend (Kim, 2012). The earlier retreat of sea ice has been suggested to favour Okhotsk herring (Zavernin, 1972) over the other major species important for fisheries.

Subarctic sea ice cover is, in analogy to polar sea ice, an important habitat for organisms under, above or inside it. In this chapter, we have mainly summarized the importance of organisms in the sea ice matrix. It is obvious that the disappearance of sea ice affects both species distribution and organic matter cycling (Figure 27.4). The effect on individual species depends on their capability to withstand a changing environment. In enclosed areas like the Baltic Sea, the dispersal of species from the Arctic area lacks suitable corridors, which may effectively prevent dispersal. Emergence from 'seed banks' or effective dispersal from areas that are still ice-covered may be a solution for a considerable period of time for many protists. However, our knowledge on these solutions as well as the vulnerability of different species is lacking in detail.

References

Arrigo, K.R., Perovich, D.K., Pickart, R.S. et al. (2012) Massive phytoplankton blooms under Arctic sea ice. *Science*, **336**, doi: *10.1126/science.1215065*

Asami, H. & Imada, K. (2001) Ice algae and phytoplankton in the late ice-covered season in Notoro Ko lagoon, Hokkaido. *Polar Bioscience*, **14**, 24–32.

Bajzak, C.E., Bernhardt, W., Mosnier, A. & Hammill, M.O. (2012) Habitat use by harbour seals (Phoca vitulina) in a seasonally ice-covered region, the western Hudson Bay. *Polar Biology*, **36**, 477–491.

Björk, G., Nohr, C., Gustafsson, B.G. & Lindberg, A.E.B (2008) Ice dynamics in the Bothnian Bay inferred from ADCP measurements. *Tellus A*, **60**, 178–188.

Boetius, A., Albrecht, S., Bakker, K. et al. (2013) Export of algal biomass from the melting Arctic sea ice. *Science*, **339**, 1430–1432.

Comiso, J.C., Parkinson, C.L., Gersten, R. & Stock, L. (2008). Accelerated decline in the Arctic sea ice cover. *Geophysical Research Letters*, **35**, L01703.

Cota, G.F., Legendre, L., Gosselin, M. & Ingram, R.G. (1991) Ecology of bottom ice algae. I. Environmental controls and variability. *Journal of Marine Systems*, **2**, 257–277.

Derocher, A.E., Lunn, N.J. & Stirling, I. (2004) Polar bears in a warming climate. *Integrative and Comparative Biology*, **44**, 163–176.

Eilola, K., Mårtensson, S. & Meier, H.E.M. (2013) Modeling the impact of reduced sea ice cover in future climate on the Baltic Sea biogeochemistry. *Geophysical Research Letters*, **40**, 149–154.

Ferguson, S.H., Stirling, I. & McLoughlin, P. (2005) Climate change and ringed seal (*Phoca hispida*) recruitment in Western Hudson Bay. *Marine Mammal Science*, **21**,121–135.

Figurkin, A.L. (2011) Variability of temperature and salinity for bottom waters in the northern Sea of Okhotsk. *Izvestiya TINRO*, **166**, 255–274.

Fukamachi, Y., Mizuta, G. Ohshima, K.I., Toyota, T., Kimura, N. & Wakarsuchi, M. (2006) Sea ice thickness in the southwestern Sea of Okhotsk revealed by a moored ice-profiling sonar. *Journal of Geophysical Research*, **111**, C09018, 10.1029/2005JC003327.

Fukutomi, T. (1950) Study of sea ice (The 4th Report). A theoretical study on the formation of sea ice in the central part of the Okhotsk Sea (in Japanese with English summary). *Low Temperature Science Series A*, **3**, 143–157.

Gagnon, A.S. & Gough, W.A. (2005) Climate change scenarios for the Hudson Bay region: an intermodel comparison. *Climatic Change*, **69**, 269–297.

Galbraith, P.S. & Larouche, P. (2011) Sea-surface temperature in Hudson Bay and Hudson Strait in relation to air temperature and ice cover breakup, 1985–2009. *Journal of Marine Systems*, **87**, 66–78.

Gaston A.J., Smith, P.A. & Provencher, J.F. (2012) Discontinuous change in ice cover in Hudson Bay in the 1990s and some consequences for marine birds and their prey. *ICES Journal of Marine Science*, **69**, 1218–1225.

Gilg, O., Kovacs, K.M., Aars, J. et al. (2012) Climate change and the ecology and evolution of Arctic vertebrates. *Annals of the New York Academy of Sciences*, **1249**, 166–90.

Gough, W.A. & Wolfe, E. (2001) Climate change scenarios for Hudson Bay, Canada from general circulation models. *Arctic*, **54**, 142–148.

Gough, W.A., Cornwell, A.R. & Tsuji, L.T.S. (2004) Trends in seasonal sea ice duration in southwestern Hudson Bay. *Arctic*, **57**, 299–305.

Grainger, E.H. (1988) Influence of a River Plume on the Sea ice Meiofauna in South-eastern Hudson Bay. *Estuarine Coastal and Shelf Science*, **27**, 131–141.

Granskog, M.A., Kaartokallio, H. & Shirasawa, K. (2003) Nutrient status of Baltic Sea ice: evidence for control by snow-ice formation, ice permeability, and ice algae. *Journal of Geophysical Research*, **108**, 3253, 10.1029/2002JC001386.

Granskog, M.A., Leppäranta, M., Kawamura, T., Ehn, J. & Shirasawa, S. (2004) Seasonal development of the properties and composition of landfast sea ice in the Gulf of Finland, the Baltic Sea. *Journal of Geophysical Research*, **109**, C02020, 10.1029/2003JC001874.

Granskog, M.A., Kaartokallio, H., Thomas, D.N. & Kuosa, H. (2005) The influence of freshwater inflow on the inorganic nutrient and dissolved organic matter within coastal sea ice and underlying waters in the Gulf of Finland (Baltic Sea). *Estuarine and Coastal Shelf Science*, **65**, 109–122.

Granskog, M., Kaartokallio, H., Kuosa, H., Thomas, D.N. & Vainio, J. (2006a) Sea ice in the Baltic Sea – a review. *Estuarine Coastal and Shelf Science*, **70**, 145–150.

Granskog, M.A., Vihma, T., Pirazzini, R. & Cheng, B. (2006b) Superimposed ice formation and surface energy fluxes on sea ice during the spring melt-freeze period in the Baltic Sea. *Journal of Glaciology*, **62**, 119–127.

Grossmann, S. & Gleitz, M. (1993) Microbial responses to experimental sea ice formation: implications for the establishment of Antarctic sea ice communities. *Journal of Experimental Marine Biology and Ecology*, **173**, 273–289.

Günther, S. & Dieckmann, G.S (1999) Seasonal development of algal biomass in snow-covered fast ice and the underlying platelet layer in the Weddell Sea, Antarctica. *Antarctic Science*, **11**, 305–315.

Haapala J. & Leppäranta M. (1997) The Baltic Sea ice season in changing climate. *Boreal Environment Research*, **2**, 93–108.

Haapala, J.J., Ronkainen, I., Schmelzer, N. & Sztobryn, M. (2015) Recent change – sea ice. In: *The BACC II Author Team: Second Assessment of Climate Change for the Baltic Sea Basin*. Regional Climate Studies. Springer International Publishing, pp. 145–153.

Haas, C. (2004) Airborne EM sea ice thickness profiling over brackish Baltic Sea water. *Proceedings of the 17th International IAHR Symposium on Ice, June 21–25, 2004, St. Petersburg, Russia*, All-Russian Research Institute of Hydraulic Engineering (VNIIG), Saint Petersburg, Russia, **2**, 12–17.

Haecky, P. & Andersson, A. (1999) Primary and bacterial production in sea ice in the northern Baltic Sea. *Aquatic Microbial Ecology*, **20**, 107–118.

Haecky, P., Jonsson, S. & Andersson, A. (1998) Influence of sea ice on the composition of the spring phytoplankton bloom in the northern Baltic Sea. *Polar Biology*, **20**, 1–8.

Hochheim, K.P., Barber, D.G. (2014) An update on the ice climatology of the Hudson Bay System. *Arctic, Antarctic, and Alpine Research*, **46**, 66–83.

Hochheim, K.P., Lukovich, J.V., and Barber, D.G. (2011) Atmospheric forcing of sea ice in Hudson Bay during the spring period, 1980–2005. *Journal of Marine Systems*, **88**, 476–487.

Jevrejeva, S., Drabkin, V.V., Kostjukov, J. et al. (2004) Baltic Sea ice seasons in the twentieth century. *Climate Research*, **25**, 217–227.

Joly, S., Senneville, S., Caya, D. & Saucier, F.J. (2011). Sensitivity of Hudson Bay Sea ice and ocean climate to atmospheric temperature forcing. *Journal of Climate Dynamics*, **36**, 1835–1849.

Juhl, A.R., Krembs, C. & Meiners, K.M. (2011) Seasonal development and differential retention of ice algae and other organic fractions in first-year Arctic sea ice. *Marine Ecology Progress Series*, **436**, 1–16.

Kaartokallio, H. (2004) Food web components, and physical and chemical properties of Baltic Sea ice. *Marine Ecology Progress Series*, **273**, 49–63.

Kaartokallio, H., Kuosa, H., Thomas, D.N., Granskog, M.A. & Kivi, K. (2007) Changes in biomass, composition and activity of organism assemblages along a salinity gradient in sea ice subjected to river discharge. *Polar Biology*, **30**, 186–197.

Kalliosaari, S. & Seinä, A. (1987) Ice winters 1981–1985 along the Finnish coast. *Finnish Marine Research*, **254**, 5–63.

Kankaanpää, P. (1997) Distribution, morphology and structure of sea ice pressure ridges in the Baltic Sea. *Fennia*, **175**, 139–240.

Kharitonov, V.V. (2008) Internal structure of ice ridges and stamukhas based on thermal drilling data. *Cold Regions Science and Technology*, **52**, 302–325.

Kim, S.T. (2012) A review of the Sea of Okhotsk ecosystem response to the climate with special emphasis on fish populations. *ICES Journal of Marine Sciences*, **69**, 1123–1133.

Kimura, N. & Wakatsuchi, M. (2004) Increase and decrease of sea ice area in the Sea of Okhotsk: ice production in coastal polynyas and dynamic thickening in convergence zones. *Journal of Geophysical Research*, **109**, C09S03, doi:10.1029/2003JC001901.

Kovacs, K.M., Aguilar, A., Aurioles, D. et al. (2012) Global threats to pinnipeds. *Marine Mammal Science*, **28**, 414–436.

Krell, A., Ummenhofer, C., Kattner, G. et al. (2003) The biology and chemistry of land fast-ice in the White Sea, Russia – a comparison of winter and spring conditions. *Polar Biology*, **26**, 707–719.

Krembs, C., Gradinger, R. & Spindler, M. (2000). Implications of brine channel geometry and surface area for the interaction of sympagic organisms in Arctic sea ice. *Journal of Experimental Marine Biology and Ecology*, **243**, 55–80.

Kuosa, H. & Kaartokallio, H. (2006) Experimental evidence on nutrient and substrate limitation of Baltic sea ice algae and bacteria. *Hydrobiologia*, **554**, 1–10.

Laurion, I., Demers, S. & Vézina, A.F. (1995) The microbial food web associated with the ice algal assemblage: biomass and bacterivory of nanoflagellate protozoans in Resolute Passage (High Canadian Arctic). *Marine Ecology Progress Series*, **120**, 77–87.

Legendre, L., Ingram, R.G. & Poulin, M. (1981) Physical control of phytoplankton production under sea ice (Manitounuk Sound, Hudson Bay). *Canadian Journal of Fisheries and Aquatic Science*, **38**, 1385–1392.

Legendre, L., Robineau, B., Gosselin, M. et al. (1996) Impact of freshwater on a subarctic coastal ecosystem under seasonal sea ice (southeastern Hudson Bay, Canada) II. Production and export of microalgae. *Journal of Marine Systems*, **7**, 233–250.

Leppäranta, M. (2004) *The Drift of Sea Ice*. Springer-Verlag, Berlin.

Leu, E., Søreide, J.E., Hessen, D.O., Falk-Petersen, S. & Berge, J. (2011) Consequences of changing sea-ice cover for primary and secondary producers in the European Arctic shelf seas: Timing, quantity, and quality. *Progress in Oceanography*, **90**, 18–32.

Lewis, J.E., Leppäranta, M. & Granberg, H.B. (1993) Statistical properties of sea ice surface topography in the Baltic Sea. *Tellus A*, **45**, 127–142.

Marchant, H.J. & Scott, F.J. (1993) Uptake of sub-micrometre particles and dissolved organic material by Antarctic choanoflagellates. *Marine Ecology Progress Series*, **92**, 59–64.

Markham, W.E. (1981) *Ice Atlas of Canadian Arctic Waterways*. Environment Canada Atmospheric Environment Service, Toronto.

Markham, W.E. (1986) The ice cover. In: *Canadian Inland Seas* (Ed. I.P. Martini), pp. 101–116. Elsevier Oceanography Series 44. Elsevier, New York.

Matsumoto, C., Saitoh, S., Takahashi, F. & Wakatsuchi, M. (2004) Use of multi sensor remote sensing to detect seasonal and interannual variability in chlorophyll a distribution in the Sea of Okhotsk. In: *Proceedings of the Third Workshop on the Sea of Okhotsk and Adjacent Areas, PICES Scientific Report* **26**, 151–154.

McKinnell, S.M. & Dagg, M.J. (Eds). 2010. *Marine Ecosystems of the North Pacific Ocean, 2003–2008*. PICES Special Publication, **4**.

McLaughlin, F.A. & Carmack, E.C. (2010) Deepening of the nutricline and chlorophyll maximum in the Canada Basin interior, 2003–2009. *Geophysical Research Letters*, **37**, L24602.

Meier, W.N., Hovelsrud, G.K., van Oort, B.E.H. et al. (2013) Arctic sea ice in transformation: A review of recent observed changes and impacts on biology and human activity. *Reviews of Geophysics*, 51, doi:10.1002/2013RG000431.

Meiners, K., Fehling, J., Granskog, M.A. & Spindler, M. (2002) Abundance, biomass and composition in Baltic Sea ice and underlying water (March 2000). *Polar Biology*, **25**, 761–770.

Melnikov, I.A., Dikarev, S.N., Egorov, V.G., Kolosova, E.G. & Zhitina, L.S. (2005) Structure of the coastal ice ecosystem in the zone of sea-river interactions. *Oceanology*, **45**, 511–519.

Michel, C., Legendre, L., Therriault, J.-C., Demers, S. & Vandevelde, T. (1993) Springtime coupling between ice algal and phytoplankton assemblages in southeastern Hudson Bay, Canadian Arctic. *Polar Biology*, **13**, 441–449.

Norrman, B. & Andersson, A. (1994) Development of ice biota in a temperate sea area (Gulf of Bothnia). *Polar Biology*, **14**, 531–537.

Ohshima, K.I., Nihashi, S., Hashiya, E. & Watanabe, T. (2006) Interannual variability of sea ice area in the Sea of Okhotsk: importance of surface heat flux in fall. *Journal of the Meteorological Society of Japan*, **84**, 907–919.

Omstedt, A. & Hansson, D. (2006) The Baltic Sea ocean climate system memory and response to changes in the water and heat balance components. *Continental Shelf Research*, **26**, 236–251.

Palosuo, E. (1963) The Gulf of Bothnia in winter. II. Freezing and ice forms. *Merentutkimuslaitoksen julkaisu/ Havsforskningsinstitutets skrift*, **209**, 64pp.

Polomoshnov, A. (1992) Seasonal variability of sea ice physicomechanical properties. In: *Proceedings of the 7th International Symposium on Okhotsk Sea and Sea Ice. The Okhotsk Sea and Cold Ocean Research Association, pp.* 336–339. Mombetsu, Japan.

Polyak L, Alley, R.B., Andrews, J.T. et al. (2010) History of sea ice in the Arctic. *Quaternary Science Reviews*, **29**, 1757–1778.

Post, E., Bhatt, U.M., Bitz, C.M. et al. (2013) Ecological Consequences of Sea-Ice Decline. *Science*, **34**, 519–524.

Prinsenberg, S.J. (1984) Freshwater contents and heat budgets of James Bay and Hudson Bay. *Continental Shelf Research*, **3**, 191–200.

Prinsenberg, S.J. (1988) Ice-cover and ice-ridge contributions to the freshwater contents of Hudson Bay and Foxe Basin. *Arctic*, **41**, 6–11.

Radchenko, V.I., Dulepova, E.P., Figurkin, A.L. et al. (2010) Status and trends of the Sea of Okhotsk region, 2003–2008. In: *Marine Ecosystems of the North Pacific Ocean* (Eds. McKinnell, S.M. & Dagg, M.J.) *2003–2008. PICES Special Publication*, **4**, 268–299

Ratkova, T. & Wassmann, P. (2005) Sea ice algae in the White and Barents seas: composition and origin. *Polar Research*, **24**, 95–110.

Robineau B., Legendre, L., Kishino, M. & Kudoh, S. (1997) Horizontal heterogeneity of microalgal biomass in the first-year sea ice of Saroma-ko Lagoon (Hokkaido, Japan). *Journal of Marine Systems*, **11**, 81–91.

Saitoh, S., Kishino, K. & Kiyofuji, S. (1996). Seasonal variability of phytoplankton pigment concentration in the Sea of Okhotsk. *Journal of the Remote Sensing Society of Japan*, **16**, 172–178.

Saucier, F.J., Senneville, S., Prinsenberg, S. et al. (2004) Modelling the sea ice-ocean seasonal cycle in Hudson Bay, Foxe Basin and Hudson Strait, Canada. *Climate Dynamics*, **23**, 303–326.

Sazhin A.F. (2004) Phototrophic and heterotrophic nano- and microorganisms of sea ice and sub-ice water in Guba Chupa (Chupa Inlet), White Sea, in April 2002. *Polar Research*, **23**, 11–18.

Schmelzer, N., Seinä, A., Lundqvist, J-E. & Sztobryn, M. (2008) Ice. In: *State and Evolution of the Baltic Sea, 1952–2005* (Eds. R. Feistel, G. Nausch & N. Wasmund), pp. 199–240. Wiley, New York.

Seinä, A. & Palosuo, E. (1996) The classification of the maximum annual extent of ice cover in the Baltic Sea 1720–1995. *Meri – Report Series of the Finnish Institute of Marine Research*, **27**, 79–91.

Shirasawa, K., Leppäranta M., Saloranta, T., Kawamura, T., Polomoshnov, A. & Surkov, G. (2005) The thickness of coastal fast ice in the Sea of Okhotsk. *Cold Regions Science and Technology*, **42**, 25–40.

Sibert, V., Zakardijan, B., Saucier, F., Gosselin, M., Starr, M. & Senneville, S. (2010) Spatial and temporal variability of ice algal production in a 3D ice ocean model of the Hudson Bay, Hudson Strait and Foxe Basin system. *Polar Research*, **29**, 353–378.

Sibert, V., Zakardjian, B., Le Clainche, Y., Gosselin, M., Starr, M. & Senneville, S. (2011) Spatial and temporal variability of primary production over the Hudson Bay, Foxe Basin and Hudson Strait marine system via coupled bio-physical models. *Journal of Marine Systems*, **88**, 401–422.

Sime-Ngando, T., Gosselin, M., Juniper, K.S. & Levasseur, M. (1997) Changes in phagotrophic microprotists (20–200 μm) during the spring algal bloom, Canadian Arctic Archipelago. *Journal of Marine Systems*, **11**, 163–172.

Sime-Ngando, T., Demers, S. & Juniper, S.K. (1999) Protozoan bacterivory in the ice and the water column of a cold temperate lagoon. *Microbial Ecology*, **37**, 95–106.

Soreide, J.E., Leu, E., Berge, J., Graeve, M. & Falk-Petersen, S. (2010) Timing of blooms, algal food quality and *Calanus glacialis* reproduction and growth in a changing Arctic. *Global Change Biology*, **16**, 3154–3163.

Spilling, K. (2007) Dense sub-ice bloom of dinoflagellates in the Baltic Sea, potentially limited by high pH. *Journal of Plankton Research*, **29**, 895–901.

Stroeve, J., Holland, M.M., Meier, W., Scambos, T. & Serreze M. (2007) Arctic sea ice decline: Faster than forecast. *Geophysical Research Letters*, **34**, L09501, doi: 10.1029/2007GL029703.

Toyota, T., Kawamura, T., Ohshima, K.I., Shimoda, H. & Wakatsuchi, M. (2004) Thickness distribution, texture and stratigraphy, and a simple probabilistic model for dynamical thickening of sea ice in the southern Sea of Okhotsk. *Journal of Geophysical Research*, **109**, C06001, doi: 10.1029/2003JC002090.

Toyota, T., Kawamura, T. & Wakatsuchi, M. (2000) Heat budget in the ice cover of the southern Okhotsk Sea derived from in-situ observations. *Journal of the Meteorological Society of Japan*, **78**, 585–596.

Toyota, T., Takatsuji, T., Tateyama, K., Naoki, K. & Ohshima, K.I. (2007) Properties of sea ice and overlying snow in the Southern Sea of Okhotsk. *Journal of Oceanography*, **63**, 393–411.

Vancoppennolle, M., Meiners, K.M., Michel, C. et al. (2013) Role of sea ice in global biogeochemical cycles: emerging views and challenges. *Quaternary Science Reviews*, **79**, 207–230

Vezina, A.F., Demers, S., Laurion, I., Sime-Ngando, T., Juniper, K.S. & Devine, L. (1997) Carbon flows through the microbial food web of the first-year ice in Resolute Passage (Canadian High Arctic). *Journal of Marine Systems*, **11**, 173–189.

Wassmann, P. (2006) Structure and function of contemporary food webs on Arctic shelves: An introduction. *Progress in Oceanography*, **71**, 123–128.

Werner, I. & Auel, H. (2004) Environmental conditions and overwintering strategies of planktonic metazoans in and below coastal fast ice in the Gulf of Finland (Baltic Sea). *Sarsia*, **89**, 102–116.

Zavernin, J.P. (1972) The effect of the hydrometeorological conditions on the time of shore approaching for spawning and on the generation quantity of the Okhotsk herring. *Izvestiya TINRO*, **81**, 44–51 (in Russian).

Index

Note: page numbers followed by f or t refer to Figures or Tables

absorption, 112f, 113, 114–15
advection, 46, 207
air-sea ice-ocean fluxes, 461–64
akaryotes. *See* bacteria
albatrosses, 570–71, 573–75
albedo, 117–24, 128–29
algae. *See also* modelling approaches
   abundance and biomass, 354–57
   biodiversity and, 353–54
   contaminants and, 481
   dissolved organic matter and, 426–27
   ice algae growth drivers, 506–10
   ice-algal bloom, 326, 330–32, 334–35, 338
   primary production and, 358–63
   sea ice as habitat, 352–53
   trophic connections, 405–6
altimetry, 252–54
amoeboid protozoans, 382–84
amphipods, 395, 398–99, 402–8
Amundsen Sea
   changes in, 407
   dipole trend patterns, 318
   primary producers, 357, 361
   seabirds, 576
   seasonal changes, 265–69, 271, 273–79, 281, 283
   snow and, 80f, 94–95, 96f, 97
Antarctic blue whales, 545–46
Antarctic Bottom Water (AABW), 222–26, 228, 611
Antarctic Circumpolar Current (ACC), 219–23, 229
Antarctic Dipole (ADP), 278, 281
Antarctic fur seals, 534–35
Antarctic Intermediate Water (AAIW), 222–23
Antarctic minke whales, 542–45
Antarctic sea ice. *See also* Southern Ocean sea ice
   albedo, 123–24
   atmospheric boundary layer and, 162f, 174, 178, 182, 184–86
   changes in, 315, 316–18, 407–9
   climate and, 276–81
   data sources, 264–65
   dissolved nutrients in, 417f, 419f
   drift patterns, 274–75
   formation of, 10
   frazil ice, 10
   historical perspectives, 281–84
   palaeo sea ice estimation, 600–611
   regional perspectives, 266–69
   seasonal cover, 265–66
   seasonal timing and feedbacks, 269–72
   snow and, 31, 65, 67, 70, 75–76, 81, 84–88, 93–101
   super-cooling, 6
   thickness, 48–51, 272–74
   variability of, 261–64
   winds and, 275–76
Antarctic shelf, 221, 223–25, 228–31
Antarctic Surface Water (AASW), 222–23
Antarctic Treaty, 53
antifreeze proteins, 343, 404
apparent optical properties (AOPs), 114, 117–25
Archaea, 326, 328, 331–33, 337–40, 345–46
Arctic Coring Expedition (ACEX), 618
Arctic inversion (AI) layer, 170–75, 179–81
Arctic oceanography
   geography, 199–201
   halcline, 205–10
   historical perspectives, 198–99
   oceanic heat, 203–5
   overview, 197–98
   polynyas, 210–11
   surface warming, 201–3
Arctic Oscillation, 57
Arctic sea ice
   albedo, 118–23, 127–29
   atmospheric boundary layer and, 161–62, 170–77, 180–88
   changes in, 315, 316–18, 407–9
   contaminants in, 473
   dissolved nutrients in, 417f, 419f
   energy budget, 163–67
   extent of, 290–95
   formation of, 10–11
   growth patterns, 8–9
   heat fluxes, 32–33
   ice-ocean boundary layer, 148–49
   indigenous people's use of, 590–97
   melt onset and freeze-up, 299, 300f
   motion of, 299–301
   multi-year ice, 129–31
   palaeo sea ice estimation, 611–19
   snow and, 31, 66f, 70, 75–76, 80f, 83–84, 93–100, 100–101, 298–99
   solar partitioning and, 131
   summer, 24f
   super-cooling and, 6
   thickness, 48–49, 55f, 56–57, 295–97
   type and age, 297–98
Argo floats, 232–33
argon, 439–42
Arnoux's beaked whales, 543f, 546–47
Atlantic Multidecadal Oscillation (AMO), 278

Atlantic water (AW), 199–200, 203–10
Atlantic white-sided dolphins, 522
atmospheric boundary layer (ABL)
    cloud-radiation-turbulence-wave interactions, 179–81
    energy budget, 161–68
    feedback processes, 201
    geographical differences, 178–79
    leads, polynyas, and thin ice, 175–77
    marginal ice zone, 177–78
    modelling, 182–86
    overview, 160–61
    surface-layer profiles, 168–70
    temperature and humidity profiles, 170–74
    wind and ice motion, 181–82
    wind profiles, 174–75

bacteria
    abundance and distribution, 330–32, 333f
    adaptations, 340–45
    diversity, 336–40
    frontiers, 345–46
    history and overview, 326–27
    nutrients and, 425–26
    sampling issues, 327–30
    seasonal dynamics, 332–36
bacteriophage. *See* viruses
bacterivory, 384–87
Baffin Bay, 42, 355, 358, 386, 520–22, 528, 561
Baird's beaked whales, 522
Baltic Sea, 97, 178–79, 631f, 633–41
Barents Sea, 199, 203–4, 207, 209, 211, 355, 358, 360, 517, 520, 564
Barrow, Alaska sea ice, 36f, 440–43, 446, 462, 592
beaked whales, 522, 548
bearded seals, 518, 519f, 522
Beaufort gyre, 49, 199–200, 407
Beaufort Sea ice
    albedo, 129
    contaminants, 475, 481–87
    decrease in, 619
    gas fluxes, 462
    ice thickness, 43f
    mammals and, 517, 519, 527
    primary production, 355, 358, 360
    snow depth, 92f
Beer-Lambert (BL) law, 499–501
Beer's law, 113
Bellingshausen Sea
    changes in, 407
    dipole trend patterns, 318
    gases, 462
    primary producers, 357
    seabirds, 576
    seasonal changes, 266–69, 271, 273–79, 281, 283
    snow and, 96f, 97
beluga whales, 517, 524
Bent, Silas, 198f
biogeochemical processes. *See also* modelling approaches
    macrograzers and, 427
    modelling approaches, 492, 496
    photosynthesis and, 502–3, 505–11
    protozoa and, 386–88, 427
biologically active layers (BALs), 497, 498f

biomarkers, 614–15
birds. *See* seabirds
black guillemots, 558f, 559, 562
black-legged kittiwakes, 557, 558f–559f
blue whales, 282, 522, 543f, 545–46
Bohai Sea, 290, 631f, 636
Bouguer-Lambert law, 113
Boundary layer *vs.* mushy layer theory, 459, 460
bowhead whales, 517
Brendan, 67
brine channels
    appearance, 3f
    brine movement and, 14–16, 18
    dissolved solutes and, 419–21, 424, 426–27
    macrograzers and, 394–95, 397f–398f, 401, 403–5
    micrograzers and, 371, 373–75, 379, 386
brine movement, 12–17, 498–99
brine skim, 332, 333f, 339
brine volume, 20
bromine explosion events (BEEs), 480
buoyancy flux, 145, 147

calcareous microfossils, 614–17
carbon cycles, 404–7
carbon dioxide, 450–53
carbon losses, 505
Caspian Sea, 630, 631f, 636
Central Arctic Index, 57
cetaceans, 516–17, 522, 523, 527–28, 542–51
chlorophyll *a* (Chl*a*), 354–55, 357–62
choanoflagellates, 377–78, 381–82, 385
Chukchi Sea, 78, 80f, 81f, 129, 355, 453–54, 519, 526, 619
ciliates, 377, 379–81, 384–87
Circumpolar Deep Water (CDW), 220–25, 229–30
circumpolar trough (CPT), 269, 271
climate change, 278–81, 551–53, 595, 630–31
cloud-mixed layer (CML), 170–75, 177, 179–80, 184
cloud-radiation-turbulence-wave interactions, 179–81
cnidarians, 395
coccolithophorids, 614–15
columnar ice, 5–8
contaminants
    in Arctic sea ice, 473
    freeze rejection and, 473–77
    hexachlorocyclohexane, 473–75, 479, 483–85, 488
    implications of, 488
    mercury, 472–83
    overview, 472
    particle entrapment and, 477
    petrogenic polycyclic aromatic hydrocarbons, 473–75, 481, 485–88
    transformation of, 480–81
    transport of, 479–80
    wet and dry deposition, 477–78
convection, 207–10
convergence, 2f, 42, 43f, 45f, 46–47, 49
Cook, James, 216
copepods, 395–96, 398–400, 402–7
crabeater seals, 535–37
cracks, 2
crude oil, 473–74, 485–88
crustaceans, 398–401, 403–5
crystal structure, 4–5

ctenophores, 395, 398f, 406
Cuvier's beaked whales, 522

Dall's porpoise, 522
deformation processes, 2f, 28, 35–36, 46–47, 57–60
degree-day models, 33–35
Delta-Eddington approach, 128, 499, 501
dense water production, 223–28
desalination
    brine movement, 12–17
    bulk salinity, 18
    permeability, 17–18
    phase fractions, 18–19
    pore microstructure, 15–16
    salinity profiles, 12
    temperature and, 18–21
diatoms, 353–54, 357, 601–12, 614, 618, 620
dielectric properties, 25–27
diffusivity, 460–61
dimethylsulfoniopropionate (DMSP), 453–59
dimethylsulphide (DMS), 454–59
dinoflagellate cysts, 612–14
Dipole Anomaly, 57, 318
Discovery Investigations, 216
discrete ordinates method (DOM), 127
dissolved organic matter (DOM), 415, 418–19, 422, 424–26, 636–38
dissolved organic nitrogen (DON), 418, 425
dissoslved inorganic carbon (DIC), 450–52, 461, 463–64
divergence, 2f, 42, 43f, 45f, 46, 49
double-diffusion, 153–55
dovekies, 556–57, 559–61, 563
drift patterns, 47–50, 58–59, 181–82, 274–76. *See also* motion/movement of sea ice
driftwood, 617
drill-hole measurement, 50

Earth system models (ESMs)
    Arctic vs Antarctic changes, 316–18
    feedbacks and, 314–16
    future evolution of sea ice and, 318–22
    general setup, 304–7
    movement of sea ice, 309–10
    numerical models and, 493–94
    thermodynamics, 307–9
    thickness distribution and, 310–11
    tuning, 311–12
    usefulness of, 312–14
eddy viscosity, 140, 142–45, 147–48
eiders, 556–57, 559, 561–62, 565
Ekman spiral, 139
electrically-scanning microwave radiometers (ESMR), 281–82
electromagnetic (EM) induction sounding, 53–56
electromagnetic spectrum (EM), 43, 111f, 239–43, 248, 252, 254–55
elephant seals, 539–41, 551–53
El Niño-Southern Oscillation (ENSO), 227, 231, 278–81
energy budgets, 32–33, 161–68
energy fluxes, 161, 163–65, 167, 183–85
enrichment, 331
enthalpy, 149–50
euglenoid flagellates, 377–78, 382
eukaryotes, 370–71, 376

euphausiids, 395, 398, 406
exopolymeric substances (EPS), 415, 419–20, 422–27
extinction coefficients, 125–26, 127f
extracellular polysaccharide substances (EPS), 329–32, 334–36, 343–45

false bottoms, 154–56
fauna. *See* metazoan fauna
feedbacks, 314–16
fin whales, 522
first-order closure models, 147
first-year white ice
    Arctic changes, 122–23, 167, 213
    heat flux and, 163
    ice-ocean boundary layer, 148
    photograph, 2f
    salinity, 12, 27
    seasonal changes, 332–36
    textures, growth, timescales, temperature and salinity, 7f
    thermal evolution, 9f
fish fauna, 400–401, 404
flooding, 2
food webs, 404–6, 639. *See also* trophic behaviors
foraminifera, 615–17
formation of sea ice, 10–11
Fram expedition, 139, 199
frazil ice, 7f, 10, 31, 124, 155, 211, 331, 353, 402, 416f
freeze rejection, 473–77
freezing
    Arctic sea ice, 299, 300f
    buoyancy flux and, 145
    columnar ice and, 5, 15
    frazil ice and, 9
    granular ice and, 8
    growth of sea ice and, 3
    ice growth, 2f
    latent heat and, 23–24, 32–35
    phase relations and, 11
    salinity and, 13–14, 150, 152–53, 155–56, 423–24
frequency-modulated, continuous-wave (FMCW) radars, 51
frost flowers, 328, 332, 333f, 339, 340f, 423

gases
    air-sea ice-ocean fluxes, 461–64
    argon, 439–42
    Boundary layer *vs.* mushy layer theory, 460
    carbon dioxide, 450–53
    composition measures, 438–39
    diffusivity and, 460–61
    dimethylsulphide, 454–59
    dissolved *vs.* gaseous state, 434
    environmental and climatic role, 433–34
    halogenated volatile compounds, 459
    methane, 453–54
    nitrogen, 442–43
    nitrous oxide, 454
    oxygen, 443–50
    solubility and, 435
    total content, 435–38
    Tsurikov's classification and, 464–65
general circulation models (GCMs), 304
glacial ice, 524–25
granular ice, 7f, 8–10

gray whales, 522
grazing impact, 406–7
grease ice, 2f, 10, 118
greenhouse gases, 317–18, 320, 322
grey seals, 522
ground-penetrating radar (GPR), 51
growth models
    degree-day models, 33–35
    energy budget and, 32–33
    ice deformation and, 35–36
    melting, 36–37
    snow cover and, 31–32
Gulf of Finland, 178, 633–34, 636, 638
Gulf of St. Lawrence, 520, 631f, 636
gyres
    Beaufort, 49, 199–200, 407
    in the Southern Ocean, 220–21, 224

halocline, 205–10
halogenated volatile compounds, 459, 609
harbour porpoises, 522
harbour seals, 522
harp seals, 520, 526–27
heat fluxes. *See also* thermodynamic processes
    in the Arctic, 48, 201–4, 207, 210
    Earth system models and, 307–9
    ice-ocean boundary layer and, 138, 151–57
    ice thickness and, 60
    melting and, 36, 58
    snow and, 31–34
herbivory, 384, 387
heterotrophic dinoflagellates, 377–78, 381, 384–86
heterotrophic flagellates, 377–78, 380–82, 385
heterotrophic protists. *See* protozoa
hexachlorocyclohexane (HCH), 473–75, 479, 483–85, 488
highly branched isoprenoids (HBI), 609–10
Holocene, 610–11, 619
hooded seals, 520
horned puffins, 559
Hudson Bay, 355, 631f, 632–33, 636–39
humpback whales, 522
hunting, 586, 588–91, 594–96
hydraulic roughness, 143, 146, 148

ice-albedo feedback, 3–4, 155
ice-algal bloom, 326, 330–32, 334–35, 338
ice-core records, 608, 617–18
ice evolution and melting, 2
ice floe measurement, 50–51
ice growth, 2
ice mass buoys (IMBs), 94–95
ice-ocean boundary layer (IOBL)
    double-diffusion during melting, 153–54
    enthalpy and salt balance, 149–50
    false bottoms, 154–55
    heat and salt exchange during freezing, 155–57
    interface approximations, 151–52
    measurements, 150–51
    overview, 138–39
    parameterization, 152–53
    surface warming and, 201–3
    turbulent exchange in, 139–49

ice profiling sonars (IPS), 52–53
ice-rafted debris (IRD), 617
ice shelf water (ISW), 224
Ice Station Weddell, 184, 217
ice strength, 27–30
Indian Ocean, 96f, 266–69, 271, 274, 283
indigenous peoples
    hunting, 586, 588–91, 594–96
    overview, 583–84
    Siku, 584–87
    umiapiaq use, 587–88
    use of sea ice, 590–97
inertial oscillation, 141–42
inherent optical properties (IOPs), 111, 114–17, 119, 128
Integrated Ocean Drilling Program (IODP), 618
Intergovernmental Panel on Climate Change (IPCC), 239
iodine emissions, 609
iodocarbons, 459
IP_{25}, 614–17
iron, 418, 423
ivory gulls, 563–64

killer whales, 522, 549–51
krill, 403–9

Labrador Sea, 290, 355, 357–58, 360, 619
lake ice, 1–5, 28
La Niña, 278–80
large-eddy simulation (LES) models, 147
large-scale sea ice models, 494
laser altimetry, 51f, 56
Last Glacial Maximum (LGM), 611
latent heat, 22–24, 211, 226
latitude, 49
leads, 2, 131, 175–77
leopard seals, 537–38
light limitation, 503
local turbulence closure, 147
long-finned pilot whales, 522
low-level jet (LLJ), 174–75, 177–81, 184
Lutzow-Holm Bay, 97, 355, 462

macrograzers
    biogeochemical role of, 427
    change and trends, 407–9
    diversity, distribution, abundance, 395–402
    environmental setting and forcing factors, 394–95
    food web and carbon cycle, 404–6
    grazing impact and carbon cycle, 406–7
    life cycles, 402–3
    physiology, 403–4
macronutrients, 416–18
mammals of the Antarctic
    cetaceans, 542–51
    climate change and, 551–53
    pinnipeds, 534–42
mammals of the Arctic
    cetaceans, 516–17, 522, 523, 527–28
    conservation and management, 526–28
    habitats, 522–26
    pinnipeds, 518–20, 522
    polar bears, 521–22

reproduction and foraging, 640f
species, 516
marginal ice zone (MIZ), 164, 175, 177–79, 576–77
Marginal Ice Zone Experiments (MIZEX), 150–51, 178
mat communities, 353
Maud Rise, 226
McMurdo Sound, 156, 340–41, 356, 359, 361, 376, 385
meiofauna, 394–95, 398f, 402–7
melting
   akaryotic sampling and, 327–29
   Arctic sea ice, 299, 300f
   brine movement, 16–17
   buoyancy and, 145
   double-diffusion during, 153–54
   equations, 24
   overview, 36–37
   satellite sensors and, 248, 255–56
   surface puddles, 2f
   thickness distribution and, 47
mercury (Hg), 472–83
mercury depletion events (MDEs), 477f–478f, 480–82
metazoan fauna. *See* macrograzers
meteoric ice, 634
methane, 453–54
methanesulphonic acid (MSA), 283, 608
methylmercury (MeHg), 474, 475f, 477f–478f, 481–83
microbially mediated redox reactions, 481
microstructure
   columnar ice, 5–8
   crystal structure, 4–5
   formation of sea ice, 10–11
   frazil ice, 10
   granular ice, 8–10
   phase fractions and, 18–19
   pore, 15–16
microwave radiometry and scatterometry, 243–48, 290–301, 313
minke whales, 522, 542–45
mixing length, 144f
mixotrophic flagellates, 381, 384, 386
mixotrophy, 386
modelling approaches. *See also* one-dimensional modelling
   biogeochemical studies and, 492, 496
   complexity of, 510
   earth system models, 493–94
   ice algae growth drivers, 509–10
   large-scale sea ice models, 494
   limitations of, 510–12
   numerical models and, 492–93
   sea ice process models, 494
   simulations, 506–9
Monte Carlo method, 127–28
motion/movement of sea ice, 181–82, 299–301, 309–10. *See also* drift patterns
multi-year Arctic ice
   albedos, 118–22
   atmospheric boundary layer and, 163, 167
   changes in, 129–31, 407–9, 488
   decline of, 297–98
   ice-ocean boundary layer, 148–49
   seasonal changes, 349
multi-year ice zone (MYZ), 58–59
mushy layer theory, 439, 460

Nansen, Fridtjof, 198–99
naphthalene, 473–75, 485–86
narwhals, 516–17, 524
near-surface temperature maximum (NSTM), 202–3
nematodes, 395–96, 398f, 403–6
nemerteans, 395
net community production (NCP), 444, 446–48
neutral stratification, 143–44
nilas, 2
nitrogen, 442–43
nitrous oxide, 454
North Atlantic Deep Water (NADW), 222–23, 611
North Atlantic right whales, 522
Northern bottlenose whales, 522
northern fulmars, 557, 559
North Pacific right whales, 522
N-P models, 495f, 502
N-P-Z/D models, 495f, 502
nudibranchs, 395, 398f, 403–4
numerical models, 492–93
nutrient limitation, 503–4
nutrient uptake, 505–6

ocean gliders, 235–36
ocean-sea ice-atmosphere (OSA) interface, 472, 479–82
old ice, 2, 48–49
omnivory, 384, 386
one-dimensional modelling
   biological processes, 501–6
   overview, 494–95
   physical processes, 497–501
   tracer framework, 495, 497
   vertical discretization, 497
Operation Ice Bridge, 56, 58f
optical and thermal infrared imaging, 254–56
optical properties, 24–25
organic matter synthesis, 502–3
ostracodes, 615
overturning circulation, 220–21, 231
oxygen, 443–50
ozone depletion events (ODEs), 459, 480

Pacific Ocean, 83t, 86f, 96f, 96t, 266–68, 271, 273–74, 277–78, 354
Pacific water (PW), 199, 204–5, 207
pack ice, 353, 355, 357–61, 576
palaeo sea ice estimation
   Antarctic, 600–611
   Arctic, 611–19
   overview, 600
pancake ice, 2f, 10
particle entrapment, 477
penguins, 570–78
permeability, 17–18, 24
persistent organic pollutants (POPs), 473, 478, 481, 483
petrels, 570–78
petrogenic polycyclic aromatic hydrocarbons (PAHs), 473–75, 481, 485–88
phase fractions, 18–19
phase functions, 115–17
phase relations, 11–12
photochemical degradation processes, 424–25, 480–81
photosynthesis, 358–63, 502–3, 505–11

phylogenic techniques, 336–40, 345
physical properties
  dielectric properties, 25–27
  ice strength, 27–30
  latent heat, 22–24
  optical properties, 24–25
  specific heat capacity, 22–24
  thermal conductivity, 21–22
phytoplankton blooms, 375, 406f
pinnipeds, 518–20, 522, 534–42. *See also* seals
planetary scale, 144, 146–47
platelet ice, 353–54, 356–61
polar bears, 28, 521–22, 524, 527, 559
polychaetes, 397–98, 400f, 403–5
polynyas
  Antarctic seabirds and, 576–78
  Arctic mammals and, 523–24, 525f
  Arctic oceanography and, 210–11
  atmospheric boundary layer and, 162
  divergence and, 46, 49
  heat fluxes and, 32
  photograph, 2f
  seabirds and, 560–63
  in the Southern Ocean, 225–26
  thin ice and, 58
pore structure, 7–8, 12, 15–16, 24
pressure ridges. *See* ridging
primary production, 358–63, 359–60, 511f
protist assemblages, 371–76
protozoa
  biogeochemical processes, 386–88
  biogeochemical role of, 427
  diversity and abundance of, 376–84
  ecology of, 384–86
  future studies, 388
  origin and fate of, 371–76
  overview, 370–71
pteropods, 395, 406
Pytheas, 67, 216

quota models, 502

radiative transfer modelling, 126–28, 499, 500f
radiative transfer theory, 112–13
rafting, 2
Redfield-Monod models, 502–3, 505–6
relative entropy, 321f
remote sensing, 93, 98–99
ribbon seals, 519–20
ridging, 2, 43–47, 44, 49, 56
ringed seals, 518, 522, 524, 526
Rossby similarity, 146–47
Ross Sea
  dense shelf waters and, 224–25
  dipole trend patterns, 318
  macrograzers, 407
  mammals and, 539, 541
  micrograzers, 372f, 374f, 375–76, 378, 380–82, 386
  primary production, 357, 361
  seabirds and, 573, 573f, 576–78
  sea ice production, 227f
  seasonal changes, 265–69, 271, 273–79, 281, 283
  snow and, 96f, 97

Ross seals, 538–39
Ross's gulls, 562–63
rotation effects, 140–41
Rothera Time Series (RaTS), 229–30
rotifers, 395, 398f, 399, 405, 408
rotten ice, 2f

salinity dependence, 504–5
salinity profiles, 12
salt/salinity
  adaptations to, 404
  bacterial and archaeal adaptations and, 340–43
  brine movement and, 13–15, 17
  crystal structure, 5–8
  flux, 608
  frazil ice, 10
  freezing and, 155–56, 423–24
  heat fluxes and, 151–53
  ice-ocean interface, 149–50
  measurements, 11–12
  parameterizations, 18
  phase relations and, 11
  release and enrichment of solutes, 421–22
  sea ice *vs.* lake ice, 1, 3–4
  snow and, 81–83
  temperature and, 14, 19–21, 24, 205–10, 229–31
satellite-passive-microwave (SPMW) remote sensing, 93, 98
satellite sensors
  altimetry, 252–54
  Antarctic sea ice and, 264–65, 281–84
  basic concepts and principles, 239–41
  EM spectrum and, 241–42
  melt onset and, 248
  microwave radiometry and scatterometry, 243–48
  milestones, 241
  optical and thermal infrared imaging, 254–56
  scattering mechanisms, 242–43
  synthetic aperture radar, 248–52
  uncertainties, 256
  validation, 256–57
scaling, 330
scattering, 112f, 113–14, 115–17
Scoresby, William, 67
Scotia Seas, 407, 571t, 610
seabirds of the Antarctic
  foraging patterns, 576–77
  future of, 578–79
  habitat, 573–75
  overview, 570–73, 574f
  penguins, 577–78
  travel patterns, 575–76
seabirds of the Arctic
  breeding, 556–58
  conservation of, 564–65
  feeding, 558–62
  ivory gulls, 563–64
  overview, 556
  polynyas and, 560–63
Sea Ice Knowledge and Use Project (SIKU), 590–91
sea ice process models, 494
sea level pressure trends, 319f
seals, 232–35, 518–20, 522, 524, 526–27, 534–42, 551–53, 586–87, 592

Sea of Azov, 631f, 636
Sea of Okhotsk, 97, 179, 355, 357, 631–32, 636–39, 641
sea otters, 522
sei whales, 522
semi-annual oscillation (SAO), 269, 271
shearwaters, 570–71, 573–74
shelf waters, 221, 223–25, 228–31, 525–26
Siku, 584–87
siphonophores, 395
skeletal layer, 5–8, 17
slab models, 147
snow
   Antarctic, 31, 65, 67, 70, 75–76, 81, 84–88, 93–100, 100–101
   Arctic, 31, 66f, 70, 75–76, 80f, 83–84, 93–100, 100–101, 298–99
   density and snow-water equivalent, 88, 89f
   deposition of, 2
   depth distribution, 91–98
   drifts, 82f
   grain types, 73–78
   layering, salinity and wetness, 78, 80–83
   metamorphism/metamorphic pathways, 71–73, 78, 79f
   northern cultures and, 67
   overview, 65–67
   remote sensing, 98–99
   satellite sensors and, 248
   sea ice and, 67–71
   sea ice growth and, 31–32
   significance of, 99–100
   studies of, 67
   thermal properties, 88–91
   thickness, 48
   uncertainties about, 100–101
snowflakes, 68f, 71, 72f
solar partitioning, 130f, 131
solutes. *See* dissolved organic matter
Southern Annular Mode (SAM), 227, 278–79, 281, 408
southern bottlenose whales, 543f, 547–48
southern elephant seals, 539–41, 551–53
southern fulmars, 570–71, 574, 576
Southern Ocean Observing System (SOOS), 235f, 236
Southern Ocean sea ice. *See also* Antarctic sea ice
   deep convection, 226–28
   future issues, 231–36
   geographic and oceanographic setting, 219–23
   heat fluxes, 32
   overview, 216–19
   polynyas, 225–26
   seasonal cycles, 228–31
   shelf waters, 223–25
   thickness, 48–49
species invasions, 408
specific heat capacity, 22–24
sperm whales, 522, 543f, 548–49
spotted seals, 519
stable boundary layer (SBL), 169–70, 175, 177–78, 181
Stejneger's beaked whales, 522
Steller's sea lions, 522
strand layers, 353–54, 360f
Subantarctic Mode Water (SAMW), 222–23
subarctic sea ice
   Baltic Sea, 633–35
   changes, 638
   ecosystems, 636–38

   Hudson Bay, 632–33
   loss of, 641
   overview, 630–31
   Sea of Okhotsk, 631–32
sulphur cycle, 454–59
sunlight
   absorption, 113, 114–15
   albedo, 117–24, 128–29
   extinction coefficients, 125–26
   multi-year ice vs first year ice, 129–31
   overview, 110–12
   radiative transfer modelling, 126–28
   radiative transfer theory, 112–13
   scattering, 113–14, 115–17
   solar partitioning, 131
   transmittance, 124–25
super-cooling, 5–6
Surface Heat Budget of the Arctic (SHEBA) project, 78, 80–81, 83–84, 89f, 98f
surface warming, 201–3
synthetic aperture radar (SAR), 248–51

temperature. *See also* freezing; melting
   adaptations to, 404
   Antarctic shelf and, 228–31
   atmospheric boundary layer and, 170–74
   bacterial and archaeal adaptations and, 340–43
   desalination and, 18–21
   first-year white ice and, 7f
   ocean heat and, 200–205
   protozoans and, 387
   salinity and, 14, 19–21, 24, 30, 205–10
temperature dependence, 505
terns, 571
thermal conductivity, 21–22, 150f
thermodynamic processes, 5, 13–14, 16, 22, 45–46, 307–9, 498. *See also* heat fluxes
thick-billed murres, 556–65
thickness distribution
   in the Arctic, 295–97
   comparisons and trends, 56–57
   deformation and, 46–47, 57–60
   divergence and advection, 46
   earth system models and, 310–11
   globally, 47–50
   laser altimetry, 51f, 56
   measurement techniques, 50–56
   melting and, 47
   overview, 42–44
   satellite sensors and, 247–48, 252–54, 255
   statistical description, 44–45
   thermodynamics, 45–46
thin ice, 175–77, 247–48
top-of-atmosphere (TOA) perspective, 316
total alkalinity (TA), 450–52, 461, 463
tracer framework, 495, 497
tracer transport, 499
transmittance, 124–25
Transpolar Drift, 49, 199–200
travel routes, 593
trophic behaviors, 384–88, 404–6
Tsurikov's classification, 435–36, 464–65
turbulence scales, 142–43, 145

turbulent exchange, 139–49
two-layer BL law, 500–501

umiapiaq use, 587–88
upward-looking sonars (ULS), 51–53

vertical discretization, 497
viruses
    abundance and distribution, 330–32
    adaptations, 344–45
    diversity, 336, 340
    frontiers, 345–46
    history and overview, 326–27
    sampling issues, 327–30
    seasonal dynamics, 333–36

walruses, 518–19, 522–23, 526, 589, 592
Warm Deep Water (WDW), 224–25
Weddell Sea ice
    atmospheric boundary layer and, 178, 184, 186
    dense shelf waters and, 224–25
    drift patterns, 46
    gases, 437–38, 444–45, 452, 457–58, 462, 464
    Ice Station Weddell, 217
    iodine emissions, 609
    macrograzers, 407
    mammals and, 539
    micrograzers, 386
    polynyas, 32, 226–28
    primary producers, 356–57, 359, 361
    production, 227f
    seabirds and, 571
    seasonal changes, 265–66, 265–69, 271, 273–78, 281–83
    snow and, 95f, 96f, 97
    surface roughness, 148
    thickness, 48–50
Weddell seals, 541–42
well-mixed layer (WML), 145, 147–48
western Antarctic Peninsula (wAP), 228–29
Western Pacific Ocean, 266–68, 271, 273–74, 277–78
whaling, 587–88, 590f, 592–96
white-beaked dolphins, 522
White Sea, 355, 520, 562, 636–38
wind, 181–82, 275–76
winter water (WW), 222–23
World Ocean Ccirculation Experiment (WOCE), 216–17

young ice, 2f, 118, 129–31